SOME USEFUL PHYSICAL PROPERTIES
(USCS UNITS AND SI UNITS)

Property[1]	USCS	SI
Aluminum		
specific weight	169 lb/ft^3	26.6 kN/m^3
mass density	5.26 slugs/ft^3	2710 kg/m^3
Concrete[2]		
specific weight	150 lb/ft^3	23.6 kN/m^3
mass density	4.66 slugs/ft^3	2400 kg/m^3
Steel		
specific weight	490 lb/ft^3	77.0 kN/m^3
mass density	15.2 slugs/ft^3	7850 kg/m^3
Water (fresh)		
specific weight	62.4 lb/ft^3	9.81 kN/m^3
mass density	1.94 slugs/ft^3	1000 kg/m^3
Acceleration of gravity (g)		
(at earth's surface)		
recommended value	32.2 ft/sec^2	9.81 m/sec^2
standard international value	32.1740 ft/sec^2	9.80665 m/sec^2
Atmospheric pressure		
(at sea level)		
recommended value	14.7 psi	101 kPa
standard international value	14.6959 psi	101.325 kPa

[1]Except as noted, the tabulated values are typical, or average, values.

[2]The specific weight of concrete made of cement, sand, and stone or gravel ranges from 140 to 150 lb/ft^3. The values tabulated here are average values for steel-reinforced concrete. The specific weight of concrete made with lightweight aggregate ranges from 90 to 115 lb/ft^3.

MECHANICS OF MATERIALS

Third Edition

ROY R. CRAIG, JR.

with
MDSolids software
by
TIMOTHY A. PHILPOT

JOHN WILEY & SONS

Named for the civil rights activist Lenny Zakim and the soldiers who fought the legendary Battle of Bunker Hill, the Leonard P. Zakim Bunker Hill Memorial Bridge serves as the northern entrance to and exit from Boston. One of the widest cable-stayed bridges in the world, the Leonard P. Zakim Bunker Hill Memorial Bridge is part of the "The Big Dig" project—a complex and technologically challenging highway project initiated in order to improve mobility and decrease traffic congestion in the city of Boston. http: www.leonardpzakimbunkerhillbridge.org

VICE PRESIDENT AND EXECUTIVE PUBLISHER Don Fowley
EXECUTIVE EDITOR Linda Ratts
EDITORIAL ASSISTANT Renata Marchione
PRODUCTION SERVICES MANAGER Dorothy Sinclair
SENIOR PRODUCTION EDITOR Janet Foxman
EXECUTIVE MARKETING MANAGER Christopher Ruel
EXECUTIVE MEDIA EDITOR Tom Kulesa
CREATIVE DIRECTOR Harry Nolan
DESIGNER Wendy Lai
PHOTO EDITOR Sheena Goldstein
SENIOR ILLUSTRATION EDITOR Anna Melhorn
PRODUCTION SERVICES Ingrao Associates
COVER IMAGE © Gordon Mills/Alamy

This book was set in Times Ten Roman by Aptara®, Inc. and printed and bound by RRD/Jefferson City. The cover was printed by RRD/Jefferson City.

This book is printed on acid-free paper. ∞

Founded in 1807, John Wiley & Sons, Inc. has been a valued source of knowledge and understanding for more than 200 years, helping people around the world meet their needs and fulfill their aspirations. Our company is built on a foundation of principles that include responsibility to the communities we serve and where we live and work. In 2008, we launched a Corporate Citizenship Initiative, a global effort to address the environmental, social, economic, and ethical challenges we face in our business. Among the issues we are addressing are carbon impact, paper specifications and procurement, ethical conduct within our business and among our vendors, and community and charitable support. For more information, please visit our website: *www.wiley.com/go/citizenship*.

Evaluation copies are provided to qualified academics and professionals for review purpose only, for use in their courses during the next academic year. These copies are licensed and may not be sold or transferred to a third party. Upon completion of the review period, please return the evaluation copy to Wiley. Return instructions and a free of charge return shipping label are available at *www.wiley.com/go/returnlabel*. Outside of the United States, please contact your local representative.

Library of Congress Cataloging-in-Publication Data:

Craig, Roy R., Jr. 1934-
  Mechanics of materials / Roy R. Craig, Jr. -- 3rd ed.
    p. cm.
  ISBN 978-0-470-48181-3 (hardback)
1. Strength of materials. I. Title.
  TA405.C89 2011
  620.1'1292--dc22

                          2010047234

Printed in the United States of America

10 9 8 7 6 5 4 3 2

PREFACE

TO THE STUDENT

This textbook is an introduction to the topic of **mechanics of materials,** an engineering subject that also goes by the names *mechanics of solids, mechanics of deformable bodies,* and *strength of materials.* You should already have a thorough background in **statics,** that is, you should know how to analyze the equilibrium of "rigid" bodies by constructing free-body diagrams and formulating and solving the appropriate equilibrium equations. In this course you will learn how to extend equilibrium analysis to **deformable bodies,** specifically to various members that make up structures and machines. This requires not only careful attention to equilibrium requirements, but also consideration of the behavior of the material (e.g., aluminum, steel, or wood) of which the member is made, and consideration of the geometry of deformation. Therefore, you will learn to apply the **three fundamental concepts of solid mechanics:** (1) *Equilibrium,* (2) *Force-Temperature—Deformation Behavior of Materials,* and (3) *Geometry of Deformation.*

You will learn a number of important new topics, such as: how the external forces that are applied to a body are distributed throughout the body, and whether the body will fail under the action of the applied forces (the topic of **stress**); how the body will deform under the action of the applied external forces and temperature changes (the topic of **strain**); what material properties affect the way that the body responds to the applied forces and temperature changes (the topic of **stress-strain-temperature behavior of materials**); and other important solid-mechanics topics.

With the aid of this textbook you will learn **systematic problem-solving methods,** including ways to assess the probable accuracy of your homework solutions. You will enjoy using the computer program **MDSolids** that is available for use with this textbook. Its intuitive graphical interface will help you develop problem-solving skills by showing you the important factors affecting various problem types, by helping you visualize the nature of internal stresses and member deformations, and by providing you an easy-to-use means of investigating a greater number of problems and variations. Nevertheless, the emphasis in this textbook remains on your developing an **understanding of the fundamentals of elementary solid mechanics,** not on writing computer programs or on using an existing computer program just to get immediate answers.

Please take time now to look at the color photographs that are included as a color-photo insert in Chapter 1. They illustrate how the **Finite Element Method,** a direct extension of this introduction to solid mechanics, is used to design and analyze everything from airplanes to cars, from skyscrapers and bridges to bicycle frames and tennis racquets, from offshore oil rigs to computer chips. The goal of this book is to prepare you to study further courses in solids and structures that will enable you to carry out such complex analyses, opening the door to the exciting world of **Computer-Aided Engineering,** whether your application is to aerospace engineering, architectural engineering, civil engineering, mechanical engineering, petroleum engineering, or even to electrical engineering.

SPECIAL FEATURES

The philosophy guiding the development of this introductory solid-mechanics textbook has been that students learn engineering topics best: (1) when they are made aware of the *fundamental concepts* involved in the subject, (2) when they are taught *systematic problem-solving procedures* and are provided many *example problems* that are solved in a systematic manner and are complete, (3) when they have ample opportunity for *drill and practice* in solving problems and obtaining feedback, and (4) when they are given *real engineering examples* and shown the relevance of what they are studying. To implement this philosophy, the following features have been incorporated in this textbook.

- **A Strong Emphasis on the Three Basic Concepts of Deformable-Body Mechanics.** Throughout this book students are reminded that solid mechanics problems involve three fundamental concepts: **Equilibrium, Material Behavior,** and **Geometry of Deformation.** In the Example Problems, the equations that correspond to each of these three concepts are highlighted and identified by name, so that the student should thoroughly understand the important role played by each one of these three fundamental concepts.

- **A Four-Step Problem-Solving Procedure.** The following four steps are included in the solution of most of the Example Problems in this book.
  - State the Problem
  - Plan the Solution
  - Solve the Problem
  - Review the Solution

  Once an engineering student leaves the university environment and becomes a practicing engineer, with powerful computer programs to carry out the detailed solution of complex problems, the importance of being able to **Plan the Solution** and **Review the Solution** for probable accuracy will become readily apparent.

- **Example Problems; Systematic Problem-Solving Procedures.** In this textbook, **over 140 Example Problems** provide the student with detailed illustrations of systematic procedures for solving solid-mechanics problems. In addition, as part of the accompanying **MDSolids** software, there are an **additional 90 Example Problems,** with complete solutions provided in the same notation and style as the solutions in the textbook itself.

  As noted above, the distinct contribution of each of the three fundamental concepts—Equilibrium, Material Behavior, and Geometry of Deformation—is highlighted and identified by name. Once the basic equations have been written down, solutions are completed by combining these equations to obtain the final answer(s). **Procedure lists** indicate convenient and systematic procedures for solving problems, and **flow charts** summarize these procedures graphically. These problem-solving procedures, labeled the *Basic Force Method,* the *Displacement Method,* and the *Force Method,* are first presented in Chapter 3.

Sign conventions for forces, displacements, etc. are established and are consistently followed, and wherever **equilibrium equations** are required, a complete **free-body diagram** is drawn. Because equilibrium analysis plays such a central role in mechanics of solids, a special section in Chapter 1 is devoted to reviewing statics of rigid bodies and to introducing the concept of internal resultants for deformable bodies.

- **Computer Exercises and Computer Software.** Thirty computer exercises are included as homework problems; these are identified by a C-superscripted problem number. Students are presented the opportunity to develop their own computer programs, and the award-winning **MDSolids** software is available via download for use on PCs running the Windows operating system.[1] Appendix G describes the software that accompanies this textbook. To download the software, you must register on the Wiley Student Companion Site for this book at www.wiley.com/college/craig. For details, see the registration card provided in the book.

- **Design.** There is a multifaceted treatment of the topic of **design.** For example, Section 2.8 discusses the philosophy of design, introduces the student to straightforward allowable-stress design, and closes with an example of optimum (minimum-weight) design of a simple statically determinate truss. The final chapter, Chapter 12, discusses three special design-related topics: stress concentrations, failure theories, and fatigue and fracture. Throughout the text and among the special MDSolids Example Problems there are design-related examples. Homework problems with design content are identified by a D-superscripted problem number.

- **Accuracy.** Special efforts have been made to provide as error-free a book as possible. At least two independent solutions have been obtained for every homework problem, and the *Solutions Manual* was prepared directly by the author in order to insure consistency and accuracy.

- **Communication with the Authors.** Comments regarding the book or the *Solutions Manual* may be addressed to the author at the following e-mail address: roy_craig@mail.utexas.edu. Please address your comments or inquiries regarding MDSolids by e-mail to Dr. Timothy Philpot: philpott@mst.edu.

NEW IN THIS EDITION

This third edition of *Mechanics of Materials* retains the hallmark features of the first two editions—a strong emphasis on the three fundamental types of equations of solid mechanics and on systematic procedures for solving problems. Listed below are the new features of the Third Edition, followed by the major modifications that have been incorporated into this edition

- From the student's standpoint, the most significant new feature is the addition of a Chapter Review table at the end of each chapter. These Chapter Reviews summarize the key points of the chapter, and they include the major equations and figures from the chapter. In addition, students will find a list of problems that will be useful to them in reviewing the chapter and preparing for examinations.

---

[1]The **MDSolids** software suite, developed by Dr. Timothy A. Philpot, was awarded the 1998 Premier Award for Excellence in Engineering Education Courseware.

- Two new topics have been added in this edition—**Mechanical Properties of Composite Materials** as a new Section 2.14, and **Wire Rope** as a topic in Section 3.3. This new material was written by the author's colleague, Dr. Eric Taleff, Professor of Mechanical Engineering at The University of Texas at Austin.

Other modifications and additions that have been incorporated into this Third Edition include:

- There are over 1300 homework problems in this edition–approximately one-third of them are in US Customary Units, one-third are in SI Units, and one-third are stated in symbolic form. Approximately 30% of the problems in this edition are either completely new or are significantly modified versions of homework problems from the two previous editions.

- The theory of axial deformation in Sections 3.2 through 3.4 has been reorganized, with examples of nonuniform axial deformation now in Section 3.3, preceding the examples of statically determinate structures with uniform axial-deformation members.

- The steps that are used in solving various problems are summarized in **Procedure Lists,** and these steps are often summarized graphically in **Flow Charts** (e.g., p. 146).

- As in the previous edition, the topic of statically indeterminate structures is introduced in the "classical" way.[2] There is still a strong emphasis on the three distinct equations of deformable-body mechanics: *equilibrium, force-temperature-deformation behavior of materials, and geometry of deformations.*

- There has been a substantial re-organization of Chapter 3–*Axial Deformation*–with the derivation of the force-deformation behavior of uniform linearly elastic axial-deformation elements, including the definitions of **flexibility coefficient** and **stiffness coefficient,** moved into Section 3.2–*Basic Theory of Axial Deformation.*

- As in the previous edition, sections that introduce the **Displacement Method** (Sections 3.8 and 4.7) are not considered to be "optional" sections, since this topic forms the basis of courses that quite likely will follow Mechanics of Materials in the student's curriculum (e.g., Matrix Structural Analysis and/or Finite Element Analysis).

- The topic of **Shear-Force and Bending-Moment Diagrams** has been divided into two sections, one (Section 5.4) treating the "Equilibrium Method" and the other (Section 5.5) treating the "Graphical Method." This change, together with the table in Section 5.5, will permit instructors to place special emphasis on graphical procedures for constructing and interpreting Shear-Force and Bending-Moment Diagrams.

SUPPLEMENTS

MDSolids Software with 90 Special Example Problems. The MDSolids computer program, winner of the 1998 Premier Award for Excellence in Engineering Education Courseware, is available to students and instructors. MDSolids has a

---

[2]The so-called "classical" solution procedure consists of simultaneously solving the equations of equilibrium and the compatibility equations written in terms of forces. This is labeled the "Basic Force Method" because, as with the Force Method of Section 3.9, the quantities that are obtained first in a solution are the unknown forces.

superb graphical user interface, is extremely user friendly, and covers a very broad range of mechanics of materials topics. Ninety special MDSolids example problems are closely linked to the examples in the book and to homework problems. (See Appendix G for further description of MDSolids.)

Solutions Manual. As was the case for previous editions, the *Solutions Manual* for the third edition includes original problem statements and text figures in addition to complete solutions. Instructors who have adopted this textbook for their course can visit www.wiley.com/college/craig and click on the Instructor Companion Website to download the *Solutions Manual*. Each solution consists of a separately-numbered (e.g., P6.3-18.pdf) PDF document, which can be downloaded from the Instructor Companion Website.

Website The publisher maintains a website where additional descriptions, feedback, and ordering information are located. The URL is: www.wiley.com/college/craig.

ACKNOWLEDGMENTS

Many colleagues in the engineering mechanics teaching profession reviewed parts or all of the manuscript of one or more editions of this book, and I am deeply indebted to them for their constructive criticism and words of encouragement.

I express my sincerest gratitude to Dr. A. L. Hale, formerly of Bell Laboratories, for his meticulous checking of practically every detail of the manuscript for the first edition, and for his careful editing of material that was revised for the second edition.

I am indebted to Dr. Timothy Philpot for allowing his award-winning **MDSolids** software to accompany the second edition and, again, the third edition of this book. He expanded the number of **MDSolids** modules from six to twelve, and created 90 special example problems. (See Appendix G.)

Professor Eric Taleff has been a constant source of encouragement to me during the preparation of this third edition. His expertise in materials science and engineering is reflected in his contribution of Section 2.14 on the **Mechanical Behavior of Composite Materials** and the **Wire Rope** text and figures in Section 3.3

Doug Bendele, Greg Cabe, Cathy Cantu, Michelle Caskey, Steve Golab, Aly Khawaja, Brett Pickin, Clint Slatton, and Ian Wagner assisted me with the solution of the homework problems of the first and second editions. Shawn Van DeWiele not only provided many excellent solutions of the homework problems of the first edition, but her sense of organization and her meticulous penmanship are reflected in the quality of the *Solutions Manual* write-ups. Jakub Jodkowski has been diligent in providing solutions for problems that are new or revised in the third edition.

Acquisitions Editors Charity Robey and Joe Hayton were very helpful throughout the editing and production stages of the first and second editions, respectively, and they provided excellent staff to assist in this work. For this third edition, I am grateful to Executive Editor Linda Ratts, who has been a ready source of guidance and encouragement; she had been ably assisted by Renata Marchione and Chris Teja. Sheena Goldstein has been most helpful in securing new photos and permissions to use both new photos and photos from the first two editions, and Wendy Lai has created an excellent cover design for this third edition. To Janet Foxman, Internal Project Editor for this third edition, I express my sincere gratitude. It has been a real pleasure to work with Susanne Ingrao, External Project Editor, who has provided copyediting services and managed the various stages of production of all three editions.

Throughout my 40-year career at The University of Texas at Austin, it was my privilege many times to teach the topics that are covered in this book. I am grateful to my colleagues in the Department of Aerospace Engineering and Engineering Mechanics and to all of the students who helped me learn and teach this course material.

Finally, for her constant love and support, I sincerely thank my wife, Dr. Jane S. Craig. I dedicate this third edition to our seven grandchildren: Talia, Kyle, and Hart Barron; and Alex, Brandon, Chase, and Dashyl Lemens.

ABOUT THE AUTHOR

Roy R. Craig, Jr. is the John J. McKetta Energy Professor Emeritus in Engineering in the Department of Aerospace Engineering and Engineering Mechanics at The University of Texas at Austin. He received his B.S. degree in Civil Engineering from the University of Oklahoma, and M.S. and Ph.D. degrees in Theoretical and Applied Mechanics from the University of Illinois at Urbana-Champaign. From 1961 until 2001 he was on the faculty of The University of Texas at Austin. His industrial experience was with the U.S. Naval Civil Engineering Laboratory, the Boeing Company, Lockheed Palo Alto Research Laboratory, Exxon Production Research Corporation, NASA, and IBM.

Dr. Craig's research and publications were principally in the areas of structural dynamics analysis and testing, structural optimization, control of flexible structures, and the use of computers in engineering education. He is the developer of the Craig-Bampton Method of component-mode synthesis, which is extensively used throughout the world for analyzing the dynamic response of complex structures. Dr. Craig has received several citations for his contributions to aerospace structures technology, including a NASA citation for contributions to the U.S. manned space flight program; and he has presented invited lectures in England, France, Taiwan, The Netherlands, and The People's Republic of China.

Dr. Craig has received numerous teaching awards and faculty leadership awards, including the General Dynamics Teaching Excellence Award in the College of Engineering, a University of Texas Students' Association Teaching Excellence Award, the Halliburton Award of Excellence, and the John Leland Atwood Award presented jointly by the Aerospace Division of the American Society for Engineering Education and by the American Institute of Aeronautics and Astronautics. Dr. Craig is the author of many technical papers and reports and of one other textbook, *Fundamentals of Structural Dynamics,* second edition. He is a licensed Professional Engineer, a Fellow of the American Institute of Aeronautics and Astronautics, and a Member of the American Society for Engineering Education, the American Society of Mechanical Engineers, and the Society for Experimental Mechanics.

Roy R. Craig, Jr.
Austin, Texas
November 2010

CONTENTS

INTRODUCTION TO MECHANICS OF MATERIALS

1

1.1 WHAT IS MECHANICS OF MATERIALS?

Mechanics is the physical science that deals with the conditions of rest or motion of bodies acted on by forces or by thermal disturbances. The study of bodies at rest is called *statics*, whereas *dynamics* is the study of bodies in motion. You have been introduced to the fundamental principles of statics and dynamics and have applied these principles to *particles* and to *rigid bodies*, which are both simplified idealizations of real physical systems. The principles of statics and dynamics are also fundamental to the *mechanics of solids* and to the *mechanics of fluids*, two major branches of applied mechanics that deal, respectively, with the behavior of solids and with the behavior of fluids. This book is an introduction to *mechanics of materials*, a topic that is also known by several other names, including "strength of materials," "mechanics of solids," and "mechanics of deformable bodies."

This book deals almost exclusively with the behavior of solids under static-loading conditions (the exception is Section 11.9). However, studies of the *dynamics of solids* (e.g., earthquake excitation of buildings and bridges) utilize most of the same concepts that are covered here. This topic is generally referred to as "structural dynamics" or "mechanical vibrations."[1]

Mechanics of Materials. We can begin to answer the question, "What is mechanics of materials?" by considering Fig. 1.1. First, a **deformable body** is a solid that changes size and/or shape as a result of loads that are applied to it or as a result of temperature changes. The diving board in Fig. 1.1 visibly changes shape due to the weight of the diver standing on it. Changes of size and/or shape are referred to as **deformation.** The deformation may even be so small that it is invisible to the naked eye, but it is still very important. To relate the deformation to the applied loading, it is necessary to understand how *materials* (i.e., solids) behave under loading.

Whereas it would be possible from rigid-body equilibrium alone, given the weight of the diver and the lengths L_1 and L_2, to determine the diving-board

---

[1]See, for example, *Fundamentals of Structural Dynamics*, Ref. [1-1] in the References section near the back of the book.

1

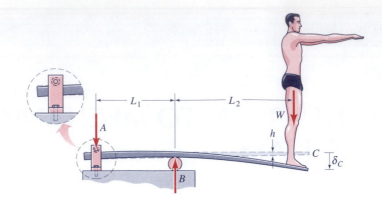

FIGURE 1.1 A diving board as an example of a deformable body.

support reactions at A and B in Fig. 1.1, questions of the following type can only be answered by employing the principles and procedures of mechanics of materials:

1. What weight W would cause the given diving board to break, and where would the break occur?
2. For a given diving board and given position of roller B, what is the relationship between the tip deflection at C, δ_C, and the weight, W, of the diver standing on the board at C?
3. Would a tapered diving board be "better" than one of constant thickness? If so, how should the thickness, h, vary from A to C?
4. Would a diving board made of fiberglass be "preferable" to one made of aluminum?

Stress and Strain. All of the preceding questions require consideration of the diving board as a *deformable body;* that is, they require consideration not only of the external forces applied to the diving board by the diver and by the diving-board supports at A and B, but they also involve the localized effects of these forces within the diving board (i.e., the **stress distribution** and the **strain distribution**) and the behavior of the material from which the diving board is constructed (i.e., the **stress-strain behavior** of the material). Stress and strain, the key concepts in the study of mechanics of materials, are formally defined in Chapter 2.

Finite Element Analysis; Color Photos. Throughout this course you will be learning how to analyze the distribution of stress and strain in various one-dimensional deformable bodies, and how they deform, but the principles and procedures you learn here are also the basis for computer programs that are used to analyze very complex deformable bodies. **Finite element analysis** is a very powerful procedure for analyzing complex structures and machines by treating them as assemblages of thousands of small parts, or elements.[2] The photos in the Color-Photo Insert illustrate analyses (e.g., stress and strain distributions) that have been obtained by use of various finite element computer programs. In the photos, different colors are used to indicate different levels of stress, etc.

Analysis and Design. All deformable-body mechanics statics problems fall into one of two categories—**strength** problems, or **stiffness** problems. A structure or machine must be "strong enough"; that is, it must satisfy prescribed strength criteria. It must also be "stiff enough"; that is, its deformation must be within acceptable limits. The first diving-board question is a strength question; the second one addresses stiffness.

The first two diving-board questions above fall under the category of **analysis.** That is, given the system (in this case the diving board) and the loads applied to it, your task

[2]See, for example, Refs. [1-2] and [1-3].

Finite Element Analyses

The solid mechanics principles and the computational procedures presented in this *Mechanics of Materials* textbook form the basis for the Finite Element Method, which was used to create these color photos.

CP-1 V-22 Osprey tilt-rotor aircraft structure. Shown are the complete aircraft and a wing-tip engine-mount section with finite element grids, and a wing-tip section that illustrates the use of color graphics to interpret strain contours. (*MSC/NASTRAN* finite element analysis by Bell Helicopter Textron. Reprinted with the permission of the American Institute of Aeronautics and Astronautics.)

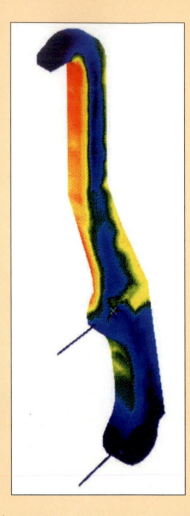

CP-2 An example of the use of finite element analysis in support of the design of a mechanical latch member from a Space Shuttle payload. (*MSC/NASTRAN* finite element analysis with *SDRC 1-DEAS* postprocessing. Courtesy NASA-Johnson Space Center.)

CP-3 Finite element analysis of the stress concentration due to a hole in an axially loaded flat bar, as described in Section 12.2. The adjacent color bar indicates that the stress depicted is the von Mises stress, which is discussed in Section 12.3. (*ABAQUS* finite element analysis by Courtesy Greg Swadener.)

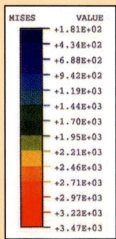

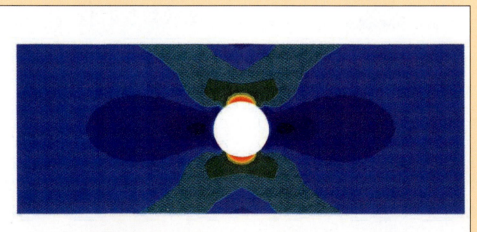

MISES	VALUE
	+1.81E+02
	+4.34E+02
	+6.88E+02
	+9.42E+02
	+1.19E+03
	+1.44E+03
	+1.70E+03
	+1.95E+03
	+2.21E+03
	+2.46E+03
	+2.71E+03
	+2.97E+03
	+3.22E+03
	+3.47E+03

is to analyze the behavior of <u>this particular system</u> subjected to <u>this particular loading condition</u>. Questions 3 and 4, on the other hand, are **design** questions. Given certain information about the loading and the performance criteria (e.g., the range of diver weight to be accommodated, and what constitutes a better diving board), your task is to select the configuration of the diving board and the material to be used in its construction.

The *design process* usually involves an iterative procedure whereby a *design* (a specific configuration made of specific materials) is proposed, the response of the designed system to given loads is analyzed, and the response is compared with the response of other designs and with the stated design criteria. The "best" of several candidate designs is then selected. Finally, there may be a requirement that a prototype based on the selected design be manufactured and tested to verify that the system meets all the design requirements in an acceptable manner. Figure 1.2 shows an airplane wing undergoing an ultimate-load test, that is, a test to determine the maximum wing loading that can be applied without causing the wing to break. The designers' expectations were exceeded when, during the test, the wings were pulled to approximately 24 ft above their normal position. (Of course, in normal service the wing will only undergo deflections that are much smaller than those experienced in an ultimate-load test like this.)

Applications of Mechanics of Materials. The applications of deformable-body mechanics are practically endless and can be found in every engineering discipline. The impressive Brooklyn Bridge, shown in Fig. 1.3, is truly an engineering marvel. It was designed by Johann Roebling in 1867–1869 and built under the supervision of his son Washington Roebling in 1870–1883 at a cost of $25 million. Its span of 1595 ft is suspended from towers that are 271 ft tall.

Since the 1960's, finite element analysis (FEA) computer programs (e.g., NASTRAN, ANSYS, ABAQUS, SAP 2000) have been employed extensively for carrying out the numerical computations required in applying deformable-body mechanics principles to the design of mechanical systems and structures. Very detailed

FIGURE 1.2 An airplane wing undergoing an ultimate-load test. (Copyright © Boeing)

FIGURE 1.3 The Brooklyn Bridge, New York, (Andreas Feininger/Time & Life Pictures/Getty Images, Inc.)

finite element models were used to create the color plots shown in the Color-Photo Insert. In designing modern airplanes, automobiles, and other mechanical and structural systems, Computer-Aided-Design (CAD) plays an essential role in defining the geometry of components, creating mathematical models of these components, and then performing the deformable-body analysis of these components.

Not only are the principles and procedures of deformable-body mechanics used to analyze and design large objects like bridges and airplanes (e.g., the tilt-rotor vehicle in color Fig. CP-1), but they also find application to very small objects as well. For example, FEA has been used to determine the stress distribution in a small computer chip at one stage in the heating-cooling cycle that occurs when the computer is turned on and off. Color Fig. CP-3 illustrates how FEA can be used to depict the effect of **stress concentration**, as discussed in Section 12.2.

Although many of the tasks involved in the design and analysis of the systems illustrated in the Color-Photo Insert require a knowledge of the mechanics of deformable bodies that is beyond the scope of this introductory textbook, the principles and procedures introduced in this book form a foundation on which more advanced topics build, and on which the design of complex applications, like those illustrated in the Color-Photo Insert, depends.

1.2 THE FUNDAMENTAL EQUATIONS OF DEFORMABLE-BODY MECHANICS

Throughout this textbook, the three *fundamental types of equations* that are used in solving strength and stiffness problems of deformable-body mechanics will be stressed repeatedly. They are:

1. The **equilibrium** conditions must be satisfied.
2. The **geometry of deformation** must be described.
3. The **material behavior** (i.e., the force-temperature-deformation relationships of the materials) must be characterized.

Here these fundamental equations are applied to fairly simple deformable bodies, but the same three basic types of equations apply, in more advanced mathematical form in many cases, to all studies of deformable solids.

Equilibrium. We have already noted that the principles of statics, that is, the *equations of equilibrium,* are fundamental to the study of deformable-body mechanics. Section 1.4 gives a brief review of static equilibrium and introduces the equilibrium concepts that are particularly important in the study of mechanics of solids. It also stresses the importance of drawing complete, accurate free-body diagrams. An entire chapter, Chapter 5, is devoted to the topic of equilibrium of beams.

Geometry of Deformation. There are several ways in which the *geometry of deformation* enters the solution of deformable-body mechanics problems, including:

1. Definitions of extensional strain and shear strain (Chapter 2).
2. Simplifications and idealizations (e.g., "rigid" member, "fixed" support, plane sections remain plane, displacements are small).
3. Connectivity of members, or geometric compatibility.
4. Boundary conditions and other constraints.

Several of these may be illustrated by a comparison of Fig. 1.1 with Fig. 1.4. In Fig. 1.1, the diving board itself was considered to be deformable, but the supports at A and B were assumed to be rigid. Therefore, the *idealized model* in Fig. 1.1 is a *deformable beam* with *rigid constraints* at A and B. By contrast, the beam BD in Fig. 1.4 is assumed to be "rigid" under the loading and support conditions shown. Although BD does actually deform, that is, change shape, its deformation is assumed to be small in comparison to the rotation, θ, that it undergoes if the rod AB stretches significantly when load W is applied to the beam at D. Hence, the *idealized model* depicted in Fig. 1.4 is a *rigid beam, BD,* connected by a frictionless pin at end B to a *deformable rod AB*. As rod AB stretches, beam BD rotates through a small angle about a fixed, frictionless pin at C.

Material Behavior. The third principal ingredient in deformable-body mechanics is *material behavior.* Unlike equilibrium and geometry of deformation, which are purely analytical in nature, the constitutive behavior of materials, that is, the *force-temperature-deformation relationships* that describe the materials, can only be established by conducting experiments. These are discussed in Chapter 2.

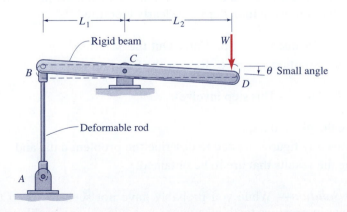

FIGURE 1.4 A system that illustrates several deformation assumptions.

It will be of great help to you in solving problems in the mechanics of deformable bodies if you will always keep in mind these three distinct ingredients: **equilibrium, geometry of deformation,** and **material behavior.**

1.3 PROBLEM-SOLVING PROCEDURES

A consistent, systematic procedure is required for solving most problems in engineering practice, and this certainly applies to solving problems involving the mechanics of deformable bodies. The five steps in such a problem-solving procedure are:

1. Select the **system** of interest. This may be based on an existing physical system, or it may be defined by a set of design drawings and specifications.
2. Make simplifying assumptions that reduce the real system to an **idealized model,** or idealization of the system. For example, Figs. 1.1 and 1.4 illustrate two different idealized models of diving boards.
3. Apply the principles of deformable-body mechanics to the idealized model to create a **mathematical model** of the system, and solve the resulting equations to predict the *response* of the system to the applied disturbances (applied forces and/or temperature changes). Interpret the results that you obtain, and seek to understand the behavior (stress and deformation) that the system exhibits in response to the applied disturbances.
4. Perform a *test* to compare the predicted responses to the behavior of the actual system. (A full-scale prototype or a scale model may have to be constructed if the physical system does not already exist.)
5. If the response predicted in step 3 does not agree with the response of the tested system, repeat steps 1–4, making changes as necessary until agreement is achieved.

In the future, as a practicing engineer you will find steps 1, 2, and 4 to be very important and very challenging. However, the main purpose of this textbook is to introduce you to the fundamental concepts of mechanics of solids and to enable you to solve deformable-body mechanics problems. Therefore, attention here will be devoted primarily to carrying out step 3, in which a mathematical model is formulated and its behavior analyzed.

So, how do you apply the principles of deformable-body mechanics to create mathematical equations, and how do you solve these equations to obtain the response of the system? A glance at the Example Problems in this textbook will indicate that the following four steps are clearly identified:

1. State the Problem   3. Carry Out the Solution
2. Plan the Solution   4. Review the Solution

1. State the Problem— This step involves:

- listing the given data,
- drawing any figures needed to describe the problem data, and
- listing the results that are to be obtained.

2. Plan the Solution— While you probably have not seen this step treated in a formal manner in previous textbooks, your success in solving problems quickly and accurately depends on how carefully you plan your solution strategy in advance. You should think about the given data and the results desired, identify the basic principles involved and recall the applicable equations, and plan the steps that will be needed to carry out the solution.

3. Carry Out the Solution— As noted in Section 1.2, your solution will involve three basic ingredients: equilibrium, geometry of deformation, and material behavior. Example problems and homework problems in this textbook appear in one of two forms—problems where numerical values are employed directly in the solution, and ones where algebraic symbols are used to represent the quantities involved and where the final answer is essentially a formula. One advantage of the symbolic form of solution is that a check of dimensional homogeneity may be easily made at each step of the solution. A second advantage is that the symbols serve to focus attention on the physical quantities involved. For example, the effect of a force designated by the symbol P can be traced through the various steps of the solution. Finally, since the end result is an equation in symbolic form, different numerical values can be substituted into the equation if desired.

The importance of checking for *dimensional homogeneity* at each step of a solution cannot be overemphasized! The principal quantities involved in static solid mechanics problems are force (dimension F) and length (dimension L). If, at some step in the solution of a problem, you check an equation for dimensional homogeneity and obtain the result that

$$F \cdot L = F/L^2$$

you should realize that some error has been made. It is a waste of time and effort to proceed further without rectifying the error and establishing a dimensionally homogeneous equation! Other checks for accuracy should also be made frequently as the solution progresses.

Appendix B provides a discussion of the *units* used in solving deformable-body mechanics problems, and Appendix A discusses the number of *significant digits* required.

4. Review the Solution— This step, like the Plan the Solution step, may be one that you have not previously encountered in a formal manner. However, it is important for you to get into the habit of always checking your results by asking yourself the following types of questions:

- Is the answer dimensionally correct?
- Do the quantities involved appear in the final solution in a reasonable manner?
- Is the sign of the final answer reasonable, and is the numerical magnitude reasonable?
- Is the final result consistent with the assumptions that were made in order to achieve the solution? (For example, an assumption that "the slope is small" is violated if the final slope turns out to be 45°.)

You may be tempted to think that Review the Solution means for you to compare your answer with one in the back of the book. However, in the "real world" there are no "answer books." Hence, you should begin now, if you have not already done so, to make a habit of testing by any means possible the reasonableness, dimensional homogeneity, and accuracy of your own answers. But, it is not enough just to solve problems and get correct answers. You should also learn to *interpret your answers,* so that you begin to develop that indispensable quality called *engineering judgment.*

Although the Carry Out the Solution step will undoubtedly occupy a major portion of your problem-solving time as you work on problems from this textbook, you will probably find that, as a practicing engineer, most of your time is involved in the other three steps, since a computer may quickly and obediently carry out the actual numerical solutions of your engineering problems. It then becomes your task to set up the problem correctly (State the Problem and Plan the Solution) and to evaluate carefully the results of the computer solution (Review the Solution).

Finally, you should heed the advice of the authors of a popular statics text:[3]

> It is also important that all work be neat and orderly. Careless solutions that cannot be easily read by others are of little or no value. The discipline involved in adherence to good form will in itself be an invaluable aid to the development of the abilities for formulation and analysis. Many problems that at first may seem difficult and complicated become clear and straightforward when begun with a logical and disciplined method of attack.

Communication is a vital part of any engineering project, and only work that is neat and orderly can serve to communicate technical information, like the solution of a deformable-body mechanics problem.

Finally, engineers who design machines and structures (e.g., commercial airplanes or high-rise office buildings) using the principles of deformable-body mechanics assume not only legal responsibility, but also moral and ethical responsibility that their designs will not fail in service in ways that could lead to serious loss of life or property.[4]

1.4 REVIEW OF STATIC EQUILIBRIUM; EQUILIBRIUM OF DEFORMABLE BODIES

In this textbook we consider deformable bodies at rest, that is, bodies whose acceleration and velocity are both zero. In your previous study of statics, you learned the equations of equilibrium and you learned how to apply these equations to particles and to rigid bodies through the use of free-body diagrams. In this section we will review the fundamental equations and problem-solving procedures of statics and will begin to indicate how they apply to the study of deformable bodies.

Equations of Equilibrium. Recall that the *necessary conditions* for equilibrium of a body (rigid or deformable) are:[5]

$$\sum F = 0, \qquad \left(\sum M \right)_o = 0 \qquad (1.1)$$

[3]See Ref. [1-4], p. 14.

[4]The National Transportation Safety Board "cited a design flaw as the likely cause of the collapse" of the I-35W Mississippi River Bridge in Minneapolis, Minnesota on August 1, 2007. Thirteen people were killed and 145 people were injured in the collapse of this bridge. See, for example, wikipedia.com.

[5]For a rigid body, Eqs. 1.1 also constitute *sufficient conditions* for equilibrium of the body. That is, if Eqs. 1.1 are satisfied, then the rigid body is in equilibrium. However, the *necessary and sufficient condition* for equilibrium of a deformable body is that the sets of external forces that act on the body and on every possible subsystem isolated out of the original body all be sets of forces that satisfy Eqs. 1.1. (See, for example, [Ref. 1-5] p. 16.)

- the sum of the external forces acting on the body is zero, and
- the sum of the moments, about any arbitrary point O, of all the external forces acting on the body is zero.

These equations are usually expressed in component form with the components referred to a set of rectangular Cartesian axes x, y, z. Then, the resulting scalar equations are:

$$\sum F_x = 0, \qquad \left(\sum M_x \right)_o = 0$$

$$\sum F_y = 0, \qquad \left(\sum M_y \right)_o = 0 \qquad (1.2)$$

$$\sum F_z = 0, \qquad \left(\sum M_z \right)_o = 0$$

When the number of *independent* equilibrium equations available is equal to the number of unknowns, the problem is said to be *statically determinate*. When there are more unknowns than available independent equations of equilibrium, the problem is said to be *statically indeterminate*. For example, if a body has more external supports or constraints than are required to maintain the body in a stable equilibrium state, the body is statically indeterminate. Supports that could be removed without destroying the equilibrium of the body are called *redundant supports,* and their reactions are called, simply, *redundants.*

In order to apply equilibrium equations to a body, it is <u>always</u> wise to draw a *free-body diagram* (FBD) of the body. However, before reviewing the procedure for drawing a free-body diagram, let us consider the types of external loads that may act on a body and several ways in which the body may be supported or connected to other bodies.

External Loads. The **external loads** acting on a deformable body are the known force and moments that are applied to the body. They may be classified in four categories, or types. These types, together with their appropriate dimensions, are:

- *Concentrated loads,* including *point forces* (F) and *couples* $(F \cdot L)$,
- *Line loads* (F/L),
- *Surface loads* (F/L^2), and
- *Body forces* (F/L^3).

The first three types of external loads are illustrated on the generic deformable body in Fig. 1.5*a*. Body forces are produced by action-at-a-distance. Like the force of gravity (weight), they are proportional to volume, and they act on particles throughout the body.

Although, in reality, all external loads that act on the surface of a deformable body must act on a finite area of that surface, line loads and concentrated forces are considered to act along a "line" or to act at a single "point," respectively, as indicated in Fig. 1.5*a*. Concentrated loads and line loads are, therefore, idealizations. Nevertheless, they permit accurate analysis of the behavior of the deformable body, except in the immediate vicinity of the loads.[6] In Fig. 1.5*b* a cross-beam at C (shown in end view as an I) exerts a downward concentrated force P_C on the horizontal

---

[6]See the discussion of St. Venant's Principle in Section 2.10.

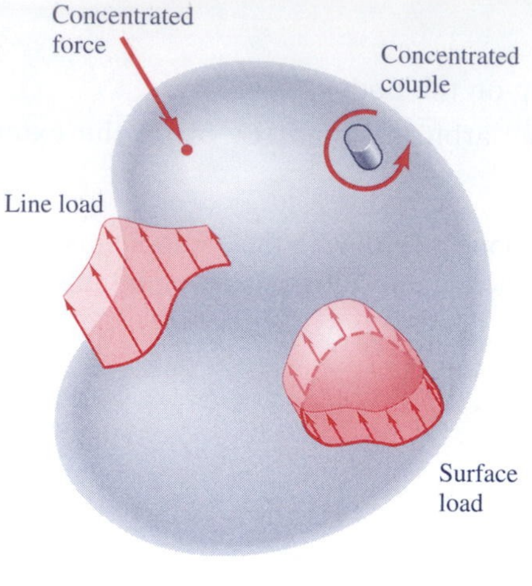

FIGURE 1.5 External loads acting on deformable bodies.

(a) A generic deformable body.

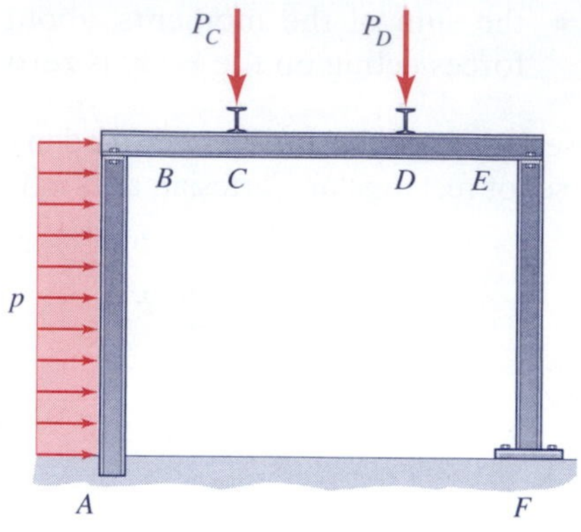

(b) A portal frame.

frame member BE, and a horizontal line load of uniform intensity p acts on the vertical frame member AB.

Concentrated forces have the units of force [e.g., newtons (N) or pounds (lb)], and line loads have the units of force per unit length (e.g., N/m or lb/ft). Other external loads are expressed in units appropriate to their dimensionality.

Support Reactions and Member Connections. The external loads that are applied to a member must generally be transmitted to adjacent members that are connected to the given member, or carried directly to some form of support.[7] For example, the vertical loads P_C and P_D that act on the horizontal beam BE in Fig. 1.5b are eventually "transmitted" to the ground at supports A and F. Where there is a *support,* as at points A and F in Fig. 1.5b, the displacement (i.e., the change of position) is specified to be zero, but the force is unknown. Therefore, forces (including couples) at supports are called **reactions,** since they react to the loads that are applied elsewhere. We say that the support enforces a *constraint,* that is, the support constrains (i.e., makes) the displacement to be zero.

Table 1.1 gives the symbols that are used to represent idealized supports and member connections. Also shown are the force components and the couples that correspond to these. In this textbook, all *reactions* are indicated by an arrow (straight for forces, curved for couples) that has a single slash through its shank, as illustrated in Table 1.1. For the most part, in this text we will consider loading and supports that lie in a single plane, that is, coplanar loading and support. Occasionally, however, we will consider a three-dimensional situation.

Internal Resultants. In the study of mechanics of deformable bodies, we must consider not only external forces and couples, that is, the applied loads and reactions, but we also must consider **internal resultants,** that is, forces and couples that are internal to the original body. For example, to analyze the L-shaped two-force linkage in Fig. 1.6a,[8] it is necessary to imagine a *cutting plane,* like the one indicated

---

[7]The exception is a self-equilibrated system, like an airplane, whose (upward) distributed lift force equals its (downward) weight.

[8]You may recall from statics that, if a body in equilibrium is subjected only to concentrated forces acting at two points in the body, the forces must be equal and opposite and must be directed along the line joining the points of application of the forces, as illustrated in Fig. 1.6a. Such a body is referred to as a *two-force member.*

Table 1.1 Support Reactions and Member Connections

Description	Symbol	Required Forces/Couples
REACTIONS – 2D		
1. Roller support	A A	A_y A_y y x
2. Cable or rod	θ A	F_A θ
3. Pin support	A A	A_x A_y A_x A_y y x
4. Cantilever support (fixed end)	A	M_A A_x A_y y x
REACTIONS – 3D		
5. Ball joint	A	A_z A_x A_y y x z
6. Cantilever support (fixed end)		F_y M_y F_x M_x M_z F_z y x z
CONNECTIONS – 2D		
7. Pinned connection	C	C_y C_x C_x C_y y x
8. Rigid connection (e.g., welded, bolted)	C	M_C C_y C_x C_x C_y M_C y x

in Fig. 1.6*a*, and to show the (unknown) internal resultants acting on this plane, as has been done in Fig. 1.6*b*. This procedure may be called the **method of sections.**

The engineering theories that are developed in this textbook apply to deformable bodies for which one dimension is significantly greater than the other two dimensions; that is, we will consider long, thin members.[9] The six internal resultants that result from general loading of such a member are indicated on the sketch in Fig. 1.7, where the x axis is taken to lie along the longitudinal direction of the member, and a cutting plane normal to the x axis, called a **cross section,** is passed through the member at coordinate $x(x < L)$.

On an arbitrary cutting plane through a body subject to general three-dimensional loading there will be three components of the *resultant force* and three

---

[9]The only exception is the thin-wall pressure vessels discussed in Section 9.2.

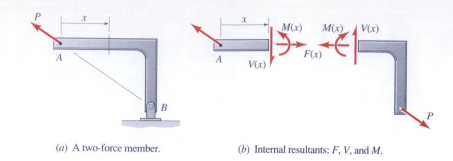

FIGURE 1.6 An illustration of internal resultants.

(*a*) A two-force member.

(*b*) Internal resultants: *F*, *V*, and *M*.

components of the *resultant moment.* When the body is slender, as in Fig. 1.7, these resultants are given special names. The force normal to the cross section, labeled $F(x)$, is called the *normal force,* or **axial force.** The two components of the resultant force that are tangent to the cutting plane, $V_y(x)$ and $V_z(x)$, are called transverse shear forces, or just **shear forces.** The component of moment about the axis of the member is called the **torque,** or *twisting moment,* and is labeled $T(x)$. Finally, the other two components of moment, $M_y(x)$ and $M_z(x)$, are called **bending moments.** Much of the remainder of this book is devoted to the determination of how these six resultants are distributed over the cross section.

Free-body Diagram (FBD).

Let us now review the steps that are involved in drawing a complete **free-body diagram.** They may be summarized as follows:

- Determine the extent of the body to be included in the FBD. Completely *isolate* this body from its supports and from any other bodies attached to it. When internal resultants are to be determined, pass a sectioning plane through the member at the desired location. Sketch the outline of the resulting *free body.*

- Indicate on the sketch all of the *applied loads,* that is, all known external forces and couples, acting <u>on</u> the body. These include distributed and concentrated forces applied to the body and also, when it is not negligible, the distributed weight of the body itself. The location, magnitude, and direction of each applied load should be clearly indicated on the sketch.

- Where the body is supported or is connected to other bodies, or where it has been sectioned, show the *unknown forces and couples* that are exerted <u>on</u> this body by the adjacent bodies. Assign a symbol to each such force (or force component or couple) and, where the direction of an unknown force or couple is

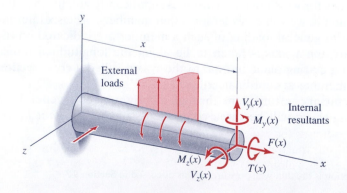

FIGURE 1.7 The six internal resultants on an arbitrary cross section of a slender member of length $L > X$.

known, use this information. Often there is a *sign convention* that establishes the proper sense to be assumed as positive. This is particularly true for the internal resultants. However, in some cases the sense of an unknown can be assumed arbitrarily.

- Label significant points and include significant *dimensions*. Also, if reference axes are needed, show these on the sketch.
- Finally, keep the FBD as simple as possible so that it conveys the essential equilibrium information quickly and clearly.

Except in Chapter 10, where we will examine *stability of equilibrium* and where it will be necessary to draw a free-body diagram of the deformed system, we will assume that deformations are small enough that the **free-body diagram can be drawn showing the body in its undeformed configuration,** even though the forces acting on it are those associated with the deformed configuration.

The following two example problems will serve as a review of the way that free-body diagrams are chosen, and will illustrate how equilibrium equations are used to determine internal resultants. Example Problem 1.1 reviews the *method of joints* and the *method of sections* for solving truss equilibrium problems. Example 1.2 applies the *method of sections* to solve a frame-type problem.

EXAMPLE 1.1

A sign weighing 600 N hangs from a pin-jointed planar truss, as shown in Fig. 1. (a) Use the *method of joints* to determine the internal axial forces F_1 and F_2 in members CD and DE, respectively. Neglect the weight of the truss members. (b) Use the *method of sections* to determine the axial force F_5 in member FG.

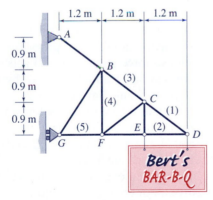

Fig. 1

Plan the Solution This problem asks for the internal axial force in each of three members, and specifies the types of free-body diagrams to use in obtaining the appropriate equilibrium equations. In your Statics course you were introduced to the two methods indicated, so this will give you a good opportunity to review those methods.

The weight of the sign will tend to pull the truss away from the wall at joint A, forcing the wall to push to the right at joint G. Therefore, members along the top of the truss will be in tension, while the axial forces in members along the bottom of the truss will be compressive.

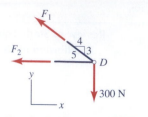

Fig. 2 Free-body diagram of joint D.

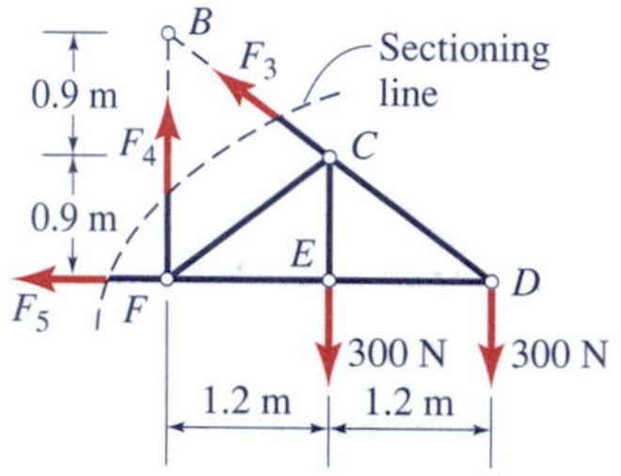

Fig. 3 Free-body diagram of section with cut through member *FG*.

Solution   (a) *Using the method of joints, determine F_1 and F_2.* We draw a free-body diagram of the joint at D (Fig. 2), taking F_1 and F_2 to be positive in tension. Also, we select reference axes x and y as shown. Equilibrium must be satisfied in both the x and y directions.

$$\overset{+}{\rightarrow} \sum F_x = 0: \qquad\qquad -F_2 - (4/5)F_1 = 0$$

$$+\uparrow \sum F_y = 0: \qquad\qquad (3/5)F_1 - 300 \text{ N} = 0$$

$$F_1 = (5/3)(300 \text{ N}) = 500 \text{ N}$$

$$F_2 = -(4/5)F_1 = -(4/5)(500 \text{ N}) = -400 \text{ N}$$

$$F_1 = 500 \text{ N (T)}, \quad F_2 = 400 \text{ N (C)} \qquad\qquad \textbf{Ans. (a)}$$

(b) *Using the method of sections, determine F_5, the force in truss member FG.* We make an imaginary sectioning "cut" as shown in Fig. 3, and draw a free-body diagram of the portion of the truss to the right of this cut, taking the axial forces in the three cut members to be <u>positive in tension</u>. With this free-body diagram, there will be three unknown member forces, but by taking moments about joint B, only one equation of equilibrium will be needed to obtain the required unknown force F_5.

$$+\circlearrowleft \left(\sum M \right)_B = 0: \quad (300 \text{ N})(1.2 \text{ m}) + (300 \text{ N})(2.4 \text{ m}) + F_5(1.8 \text{ m}) = 0$$

$$F_5 = -600 \text{ N, or } F_5 = 600 \text{ N (C)} \qquad\qquad \textbf{Ans. (b)}$$

Review the Solution   The force F_1 has turned out to be in tension, and forces F_2 and F_5 in compression, as we expected. Also, the magnitudes are reasonable, so our solution appears to be correct.[10]

Based on our experience in solving equilibrium problems, we could have assumed at the outset that the unknown force in member *FG* acts to the right on joint *F* (i.e., that member *FG* is in compression). Had we done so (by reversing the sense of the arrow representing force F_5), we would have gotten an answer of $F_5 = 600$ N, without the minus sign. Instead, we chose to show all unknown axial forces on the free-body diagram assuming tension to be positive. As a consequence, the answer for F_5 turned out to be $F_5 = -600$ N. That is, the minus sign indicates that the force F_5 is a compressive force rather than a tensile force.

As problems get more complex (e.g., several interconnected bodies) it will become impossible to mentally solve all of the resulting equilibrium equations to the extent that the "correct" sense of every force can be established at the outset when the free-body diagram is drawn. **The procedure of assuming internal axial forces to be positive in tension makes it both easy to draw the free-body diagram and easy to interpret the meaning of the answers (positive forces are tension; negative forces are compression).** Hence, this sign convention will be followed throughout this textbook.

---

[10]Note that the units (N) and (m) are stated when numerical values are used. It is good practice to show the proper force units (F) and length units (L) in the solution of numerical problems. Also, note that the answers are marked, and that tension (T) and compression (C) are identified in the answers.

EXAMPLE 1.2

An electrical worker stands in the bucket that hangs from a pin at end D of the boom of the cherry picker in Fig. 1. The worker and bucket together weigh a total of 300 lb. Between A and C the boom weighs 1.1 lb/in., and between C and D it weighs 0.8 lb/in. Assume that AC and CD are uniform beams.

Determine the normal force, the transverse shear force, and the bending moment that act at cross section E, midway between A and B.

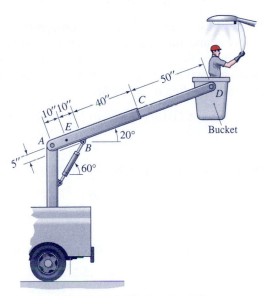

Fig. 1 (*a*) "Bucket truck."

(*b*) Telescopic aerial device. (Used with permission of Altec Industries)

Plan the Solution   Since there will be three unknowns on the cross section at E and, in addition, an unknown force at the pin B, we cannot solve for all four unknowns using a single free-body diagram. Hence, we will have to determine the pin force first using a separate free-body diagram; then we can determine the two components of the internal force at E and the moment at E.

Solution   *Pin Reaction at B:* First, we use the free-body diagram in Fig. 2 to determine the pin reaction at B.

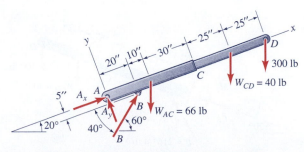

Fig. 2   A free-body diagram of boom AD.

15

$$W_{AC} = (1.1 \text{ lb/in.})(60 \text{ in.}) = 66 \text{ lb}$$

$$W_{CD} = (0.8 \text{ lb/in.})(50 \text{ in.}) = 40 \text{ lb}$$

$+\circlearrowleft \left(\sum M \right)_A = 0:$

$$-(66 \text{ lb})(30 \text{ in.})(\cos 20°) - (40 \text{ lb})(85 \text{ in.})(\cos 20°)$$

$$-(300 \text{ lb})(110 \text{ in.})(\cos 20°) + (B \cos 50°)(20 \text{ in.})$$

$$+ (B \cos 40°)(5 \text{ in.}) = 0$$

$$B = 2161.4 \text{ lb}$$

Internal Resultants: Next, we pass a section through the beam at E and determine the forces and moment on this cross section (Fig. 3). (Let point E be on the centerline of the beam AD.) On the section at E we show the unknown normal force F_E, the unknown transverse shear force V_E, and the unknown bending moment M_E.

$$W_{EC} = (1.1 \text{ lb/in.})(50 \text{ in.}) = 55 \text{ lb}$$

$+\nearrow \sum F_x = 0:$

$$-F_E + (2161.4 \text{ lb})(\sin 50°) - (55 \text{ lb} + 40 \text{ lb} + 300 \text{ lb}(\sin 20°) = 0$$

$$F_E = 1520.6 \text{ lb}$$

$+\nwarrow \sum F_y = 0:$

$$V_E + (2161.4 \text{ lb})(\cos 50°) - (55 \text{ lb} + 40 \text{ lb} + 300 \text{ lb})(\cos 20°) = 0$$

$$V_E = -1018.2 \text{ lb}$$

(i.e., 1018.2 lb acting opposite to the direction shown on the free-body diagram).

$+\circlearrowleft \left(\sum M \right)_E = 0:$

$$-M_E - (55 \text{ lb})(\cos 20°)(25 \text{ in.}) - (40 \text{ lb})(\cos 20°)(75 \text{ in.})$$

$$-(300 \text{ lb})(\cos 20°)(100 \text{ in.}) + (2161.4 \text{ lb})(\cos 50°)(10 \text{ in.})$$

$$+ (2161.4 \text{ lb})(\sin 50°)(5 \text{ in.}) = 0$$

$$M_E = -10{,}129.9 \text{ lb} \cdot \text{in.}$$

The answers, rounded to the proper number of significant digits are:

$$F_E = 1520 \text{ lb}, \quad V_E = -1020 \text{ lb}, \quad M_E = -10{,}100 \text{ lb} \cdot \text{in.} \qquad \textbf{Ans.}$$

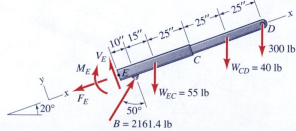

Fig. 3 A free-body diagram showing the resultants at section E.

Review the Solution Because of the long moment arm of the 300-lb load compared with the moment arm of the force at B, it is reasonable for the magnitude of B to be much larger than the magnitude of the total load. The magnitude and sense of F_E, V_E, and M_E also seem to be reasonable in view of the magnitude of B and the magnitude and location of the other loads.

1.5 PROBLEMS

▼ **REVIEW OF EQUILIBRIUM; DETERMINATION OF INTERNAL RESULTANTS**

Problems 1.4-1 through 1.4-6. *In these truss equilibrium problems, adopt the sign convention for axial force that a tensile force is positive. That is, on your free-body diagrams, show all unknown forces F_1 as tensile forces. Then, a force that is compressive will be negative.*

Prob. 1.4-1. The pin-jointed truss in Fig. P1.4-1 supports a load of $P = 2$ kips at joint C. Determine the axial forces F_1 through F_5 in the truss members.

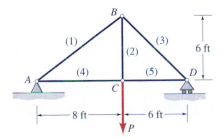

P1.4-1 and P2.2-12

Prob. 1.4-2. Each member in the pin-jointed planar truss in Fig. P1.4-2 is 6 ft long. The truss is attached to a firm base by a frictionless pin at A, and it rests on a roller support at B. For the loading shown, (a) determine the reactions at A and B, and (b) determine the force in each of the three members labeled (1) through (3).

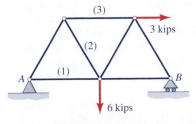

P1.4-2

Prob. 1.4-3. For the pin-jointed truss in Fig. P1.4-3, (a) determine the reactions at the supports at A and C, and (b)

determine the axial force in each of the following members: F_1 (in member AB), F_2 (in member BC), and F_3 (in member CD).

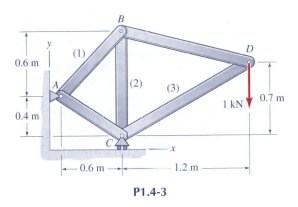

P1.4-3

Prob. 1.4-4. For the pin-jointed truss in Fig. P1.4-4, (a) determine the reactions at the supports at A and C; and (b) determine the axial force in each of the following members: F_1 (in member AB), F_2 (in member BC), and F_3 (in member CD). Express all of your answers in terms of the weight W. Note: CD and DE are two separate members that are pinned to member BD at D.

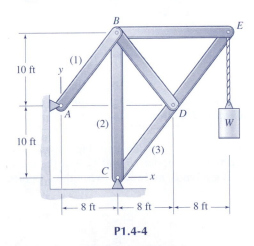

P1.4-4

Prob. 1.4-5. The pin-jointed truss in Fig. P1.4-5 supports equal vertical loads P at joints B, C, and D. (a) Determine the reactions at the supports A and E, and (b) use the *method of sections* to determine the forces in the following three truss members: F_1 (in member BC), F_2 (in member BG) and F_3 (in member HG). (Note: Joint E is supported by a frictionless roller on a 45° inclined plane.)

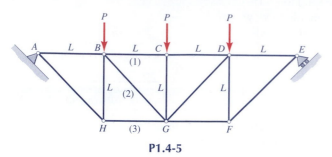

P1.4-5

*Prob. 1.4-6. A portion of the boom of a crane at a construction site is shown in Fig. P1.4-6. The cable from the lift motor to the cargo sling is parallel to BE and FI and passes over a pulley that is supported by a frictionless pin at A. The weight of the sling and pallet being lifted is $W = 6$ kN. Neglecting the weight of the truss members and the weight of the pulley and cable, (a) determine the axial force in each of the following members: F_1 (in member BC), F_2 (in member CF), F_3 (in member FG), and F_4 (in member DG). (b) If member 4 were to be attached between joints C and H, instead of between joints D and G, would the force F_4 be the same in either case? Show calculations to support your answer.

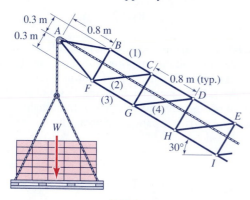

P1.4-6

> **In Probs. 1.4-7 through 1.4-8, adopt the sign convention for axial force F, shear force V, and bending moment M that is given in Fig. 1.6b.**

Prob. 1.4-7. For the beam shown in Fig. P1.4-7, determine: (a) the reactions at support B and C, and (b) the internal resultants at section E, midway between the supports.

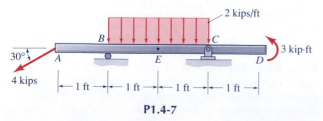

P1.4-7

Prob. 1.4-8. Weights W are attached by cables to beam BE at its midpoint D and at its end E, as shown in Fig. P1.4-8. The beam is supported by a fixed, frictionless pin at B and by an inclined cable from A to D. (a) Determine the tension in cable AD, and determine the reaction force at pin B. (b) Determine the internal resultants (axial force, shear force, and bending moment) on the cross section at C.

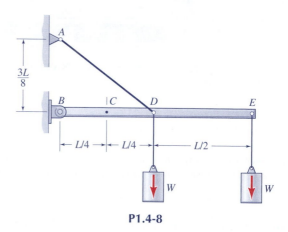

P1.4-8

Prob. 1.4-9. For the beam shown in Fig. P1.4-9, determine: (a) the reactions at supports B and C, and (b) the internal resultants—F_D, V_D, and M_D—at section D, midway between the supports.

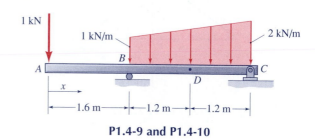

P1.4-9 and P1.4-10

*Prob. 1.4-10. For the beam shown in Fig. P1.4-10, determine: (a) the reactions at supports B and C, and (b) the internal resultants—$F(x)$, $V(x)$, and $M(x)$—for 1.6 m $< x <$ 4.0 m.

Prob. 1.4-11. Beam AD in Fig. P1.4-11 has a frictionless-pin support at A and a roller support at D. The beam has a linearly varying distributed load, of maximum intensity w_0 (force per unit length), over two-thirds of its length. (a) Determine the reactions at A and D, and (b) determine the internal resultants (axial force, shear force, and bending moment) on the cross section at B.

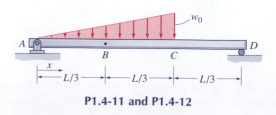

P1.4-11 and P1.4-12

Prob. 1.4-12. Beam AD in Fig. P1.4-12 has a frictionless-pin support at A and a roller support at D. It has a linearly varying distributed load, of maximum intensity w_0 (force per unit length), over two-thirds of its length. (a) Determine the reactions at A and D, and (b) determine expressions for the internal resultants—$F(x)$, $V(x)$, and $M(x)$—for $0 < x < 2L/3$.

Prob. 1.4-13. For the overhanging beam shown in Fig. P1.4-13, determine: (a) expressions for the internal resultants $V_1(x)$ and $M_1(x)$ in interval AB ($0 \leq x < 6$ in.), and (b) the internal resultants V_D and M_D at section D, midway between the two supports.

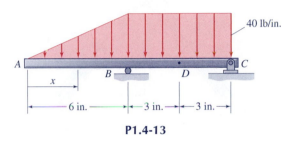

P1.4-13

*Prob. 1.4-14. Beam AD is supported by a rollers at B and C that are equidistant form the ends, as shown in Fig. P1.4-14. The beam supports a uniformly distributed downward load of intensity w_0 (force per unit length) along its entire length. (a) Determine an expression for the location a of the two supports that will minimize the maximum bending moment M_{max} in the beam. (b) Also determine an expression for the value of M_{max}.

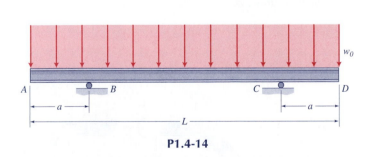

P1.4-14

Prob. 1.4-15. One of the lift arms of a fork-lift truck has the loading and support shown in Fig. P1.4-15. The load consists of two identical crates, each weighing $W = 400$ lb. (Each of the two lift arms supports half of the total load.) The "support" consists of a hoist cable attached to the lift arm at B and frictionless rollers that react against the truck frame at A and at C. Neglect the weight of the lift arm. (a) Determine the roller

reactions R_A and R_C at A and C, respectively. (b) Determine the internal resultants (axial force, shear force, and bending moment) on a horizontal cross section at D. (c) Determine the internal resultants on a vertical cross section at E.

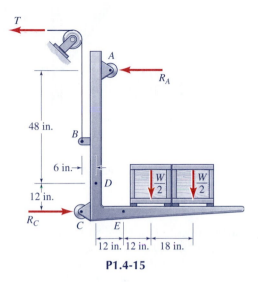

P1.4-15

Prob. 1.4-16. The right-angle frame in Fig. P1.4-16 has equal legs of length L and is supported by a pin and roller as shown. If the total weight of the frame is W, (a) determine the reactions at supports A and B, and (b) determine the internal resultants (axial force, shear force, and bending moment) on the cross section at point C.

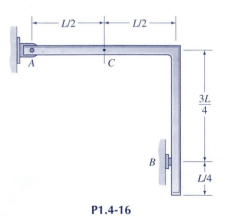

P1.4-16

Prob. 1.4-17. Determine the internal resultants F_G, V_G, and M_G on the cross section at G of the horizontal frame member in Fig. P1.4-17. The uniformly distributed load on member AC has a magnitude $w_0 = 220$ lb/ft. (See the inset for a definition of the resultants.)

19

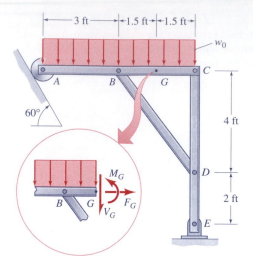

P1.4-17 and P2.2-16

Prob. 1.4-19. A vertical force $P = 10$ kips acts downward at the apex B of frame ABC, as shown in Fig. P1.4-19. The frame is supported by a frictionless pin at C and rests on a roller support at A. (a) Determine the reactions at A and C, and (b) determine the internal resultants (axial force, shear force, and bending moment) at arbitrary cross sections of legs AB and BC, using the respective notations shown in the two cutaways.

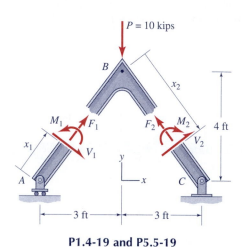

P1.4-19 and P5.5-19

*Prob. 1.4-18. A vertical force $P = 1$ kN acts on frame ABE, as shown in Fig. P1.4-18. The frame is supported, through frictionless pins, by a wheel at B and by a fixed-pin support block at E. (a) Determine the reaction at E, and (b) determine the internal resultants (axial force, shear force, and bending moment) on the cross section at point C. See the inset for a definition of the resultants. (Note: Use the given dimensions to eliminate the angle ϕ from your answers. The radius of the pulley at A is not negligible.)

Prob. 1.4-20. A uniformly distributed vertical load $w_0 = 3.2$ kN/m acts over the entire 2.5 m length of the horizontal member BC of frame ABC, as shown in Fig. P1.4-20. The frame is supported by a frictionless pin at C and rests on a roller support at A. (a) Determine the reactions at A and C, and (b) determine the internal resultants (axial force, shear force, and bending moment) at arbitrary cross sections of members AB and BC, using the respective notations shown in the two cutaways.

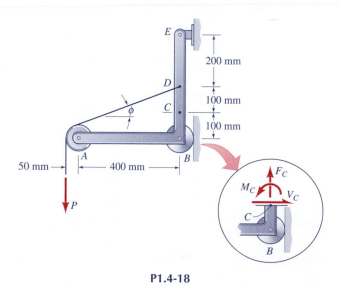

P1.4-18

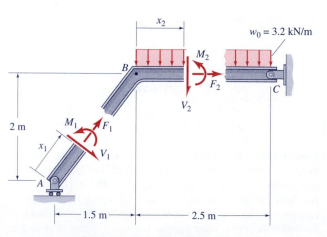

P1.4-20, P5.5-20 and P9.4-24

Problems 1.4-19 and 1.4-20. *For these frame equilibrium problems, use the sign convention that is indicated for each member of each frame.*

Section			Suggested Review Problems
1.1	Section 1.1 gives you definitions of the following: *mechanics of materials, deformable body, stress and strain, and analysis and design.* It also introduces you to the *finite element method*, a powerful compatational tool for analyzing complex structural and mechanical systems.	You should familiarize yourself with this material	
1.2	Section 1.2 Introduces the *three fundamental type of equation* that you will be studying throughout this course. They are: 1. **equilibrium** equations; 2. **geometry of deformation** equations; and 3. **material behavior** (i.e., force-termerature-deformation) equations.	You should always remember this **Big 3** of mechanics of materials!	
1.3	Section 1.3 discusses a four-step *problem-solving procedure* for you to use. The four steps are: 1. State the Problem; 2. Plan the Solution; 3. Carry Out the Solution; and 4. Review the Solution.	You should use this four-step problem-solving procedure.	
1.4	Section 1.4 reviews important topics that were covered in your *Statics* course. It emphasizes the importance of carefully drawn **free-body diagrams**. Section 1.4 also introduces the topic of **internal resultants — axial force, transverse shear force**, and **bending moment**. Table 1.1 describes symbols for **external reactions**.	The statics review problems involve: • pin-jointed planar trusses, • beams, and • planar frames. (a) A two-force member. (b) Internal resultants: *F*, *V*, and *M*. Internal resultants (Fig. 1.6)	1.4-5 1.4-11 1.4-17

2 STRESS AND STRAIN; INTRODUCTION TO DESIGN

The two most important concepts in mechanics of materials are the concepts of **stress** and **strain.** The photographs in the Color-Photo Insert give a visual indication of the complexity of the internal behavior of the members pictured in response to the given external loads. (See the Color-Photo Insert.) Figure 2.1a, repeated from Fig. 1.6, shows an L-shaped bracket loaded as a two-force member, and Fig. 2.1b shows the internal resultants, $F(x)$, $V(x)$, and $M(x)$, that are required to maintain the equilibrium of the two sectioned parts of the bracket. Although we could compute the internal resultants shown on Fig. 2.1b by using static equilibrium procedures (free-body diagrams and equations of equilibrium), those procedures are clearly insufficient for determining the complex *internal* force distribution making up those resultants. The concept of **stress** is introduced in this chapter to enable us to quantify internal force distributions.

The shape of the bracket also changes due to the applied loads; that is, the member deforms. The concept of **strain** is introduced to permit us to give a detailed analytical description of such deformation. In this chapter we will define the two key forms of stress and corresponding two forms of strain.

Finally, stress and strain are related to each other. This relationship, which depends on the material(s) used in the fabrication of the member, must be determined by performing certain **stress-strain tests,** which are described in this chapter. Also discussed are many other important **mechanical properties of materials** that must be determined by laboratory testing. Throughout the remainder of this book we will be

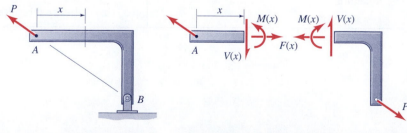

FIGURE 2.1 An illustration of internal resultants.

(a) A two-force member.

(b) Internal resultants: F, V, and M.

22

determining the stresses and strains produced in various structural members by the forces and temperature changes acting on them.

2.2 NORMAL STRESS

To introduce the concepts of stress and strain, we begin with the relatively simple case of a straight bar undergoing **axial loading,** as shown in Fig. 2.2.[1] In this section we consider the stress in the bar, and in Section 2.3 we treat the corresponding strain.

Equal and opposite forces of magnitude P acting on a straight bar cause it to *elongate,* and also to get narrower, as can be seen by comparing Figs. 2.2*a* and 2.2*b*. The bar is said to be in *tension.* If the external forces had been applied in the opposite sense, that is, pointing toward each other, the bar would have shortened and would then be said to be in *compression.*

Definition of Normal Stress. The thin red arrows in Figs. 2.2*c* and 2.2*d* represent the distribution of force on cross sections at A and B, respectively. (A *cross section* is a plane that is perpendicular to the axis of the bar.) Near the ends of the bar, for example at section A, the resultant normal force, F_A, is not uniformly distributed over the cross section; but at section B, farther from the point of application of force P, the force distribution is uniform. In mechanics, the term **stress** is used to describe the distribution of a force over the area on which it acts and is expressed as force intensity, that is, as force per unit area.

$$\text{Stress} = \frac{\text{Force}}{\text{Area}}$$

The units of stress are units of force divided by units of area. In the U.S. Customary System of units (USCS), stress is normally expressed in pounds per square inch (psi) or in kips per square inch, that is, kilopounds per square inch (ksi). In the International System of units (SI), stress is specified using the basic units of force (newton) and length (meter) as newtons per meter squared (N/m²). This unit, called the *pascal* (1 Pa = 1 N/m²), is quite small, so in engineering work stress is normally

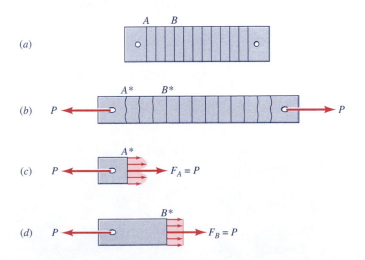

FIGURE 2.2 A straight bar undergoing axial loading. (*a*) The undeformed bar, with vertical lines indicating cross sections. (*b*) The deformed bar. (*c*) The distribution of internal force at section A. (*d*) The distribution of internal force at section B.

---

[1]*Axial loading* is discussed here in order to introduce the concepts of stress and strain and the relationship of stress to strain. Chapter 3 treats axial deformation in greater detail.

expressed in kilopascals (1 kPa = 10^3 N/m^2), megapascals (1 MPa = 10^6 N/m^2), or gigapascals (1 GPa = 10^9 N/m^2). For example, 1 psi = 6895 Pa = 6.895 kPa.

There are two types of stress, which are called **normal stress** and **shear stress.** In this section we will consider only normal stress; shear stress is introduced in Section 2.7. In words, normal stress is defined by

$$\textbf{Normal Stress} = \frac{\text{Force \textbf{normal} (i.e., perpendicular) to an area}}{\text{Area on which the force acts}}$$

The symbol used for normal stress is the lowercase Greek letter sigma (σ). The **normal stress at a point** is defined by the equation

$$\sigma(x, y, z) = \lim_{\Delta A \to 0} \left(\frac{\Delta F}{\Delta A} \right) \qquad \textbf{Normal Stress} \qquad (2.1)$$

where, as shown in Fig. 2.3a, ΔF is the *normal force* (assumed positive in tension) acting on an elemental area ΔA containing the point (x, y, z) where the stress is to be determined.

The **sign convention for normal stress** is as follows:

- A positive value for σ indicates **tensile stress,** that is, the stress due to a force ΔF that *pulls* on the area on which it acts.
- A negative value for σ indicates **compressive stress.**

Thus, the equation $\sigma = 6.50$ MPa signifies that σ is a *tensile stress* of magnitude 6.50 MPa, or 6.50 MN/m^2, and the equation $\sigma = -32.6$ ksi indicates a *compressive stress* of magnitude 32.6 kips/in^2.

Average Normal Stress. Even when the normal stress varies over a cross section, as it does in Fig. 2.2c, we can compute the **average normal stress** on the cross section by letting

$$\sigma_{\text{avg}} = \frac{F}{A} \qquad \textbf{Average Normal Stress} \qquad (2.2)$$

Thus, for Figs. 2.2c and 2.2d we get

$$(\sigma_{\text{avg}})_A = \frac{F_A}{A} = \frac{P}{A}, \qquad (\sigma_{\text{avg}})_B = \frac{F_B}{A} = \frac{P}{A}$$

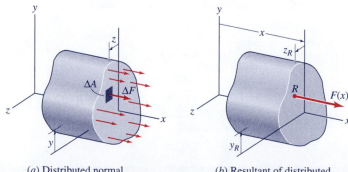

FIGURE 2.3 Normal force on a cross section.

(*a*) Distributed normal stress on a cross section.

(*b*) Resultant of distributed normal stress in (*a*).

Much of the rest of this textbook is devoted to determining how stress is distributed on cross sections of structural members under various loading conditions. However, in many situations the normal stress on a cross section is either constant or very nearly constant, as in the next two examples.

EXAMPLE 2.1

A shop crane consists of a boom AC that is supported by a pin at A and by a rectangular tension bar BD, as shown in Fig. 1. Details of the pin joints at A and B are shown in Views $a–a$ and $b–b$, respectively. The tension bar BD has a width $w = 1.5$ in. and a thickness $t = 0.5$ in. If the vertical load at C is $P = 5$ kips, what is the average tensile stress in the bar BD?

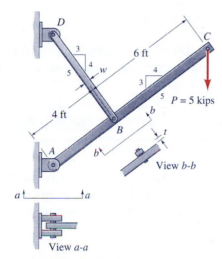

Fig. 1 A shop crane.

Plan the Solution   Equilibrium of boom AC can be used to determine the tensile force F_B in two-force member BD. Then, the average tensile stress can be calculated by using Eq. 2.2.

Solution

Equilibrium: Figure 2 shows a free-body diagram of the boom AC, with two components of the reaction force at pin A, and with the force F_B in two-force member BD shown as a <u>tensile force</u> acting at B along the direction BD. The 3–4–5 triangles in Fig. 2 can be used to establish that

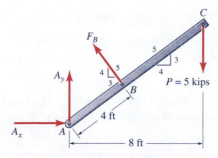

Fig. 2   Free-body diagram.

AC and BD are perpendicular to each other, and that the horizontal projection of boom AC is 8 ft.

$$\circlearrowleft + \sum M_A = 0: \qquad (5 \text{ kips})(8 \text{ ft}) - F_B(4 \text{ ft}) = 0$$

$$F_B = 10 \text{ kips}$$

and from Eq. 2.2,

$$\sigma_{\text{avg}} = \frac{F}{A} = \frac{F_B}{wt} = \frac{10 \text{ kips}}{(1.5 \text{ in.})(0.5 \text{ in.})} = 13.33 \text{ ksi}$$

Rounded to three significant figures, the average normal stress on a typical cross section of the bar BD is

$$\sigma_{\text{avg}} = 13.3 \text{ ksi} \ (T) \qquad \qquad \textbf{Ans.}$$

where the (T) stands for tension.

Review the Solution   The above calculations are very straightforward, but they should be double-checked. The answer seems reasonable.

EXAMPLE 2.2

The Washington Monument (Fig. 1a) stands 555 ft high and weighs 181,700 kips (i.e., approximately 182 million pounds). The monument was made from over 36,000 blocks of marble and granite. As shown in Fig. 1b, the base of the monument is a square that is 665.5 in. long on each side, and the stone walls at the base are 180 in. thick.

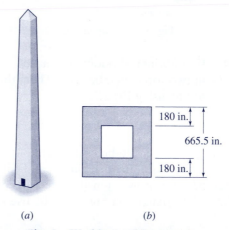

(a)   (b)

Fig. 1   Washington Monument.

Determine the compressive stress that the foundation exerts over the cross section at the base of the monument, assuming that this normal stress is uniform.

Solution   From the free-body diagram in Fig. 2, the total normal force on the base of the monument is equal to negative of the weight of the monument, so

$$\sum F = 0: \qquad \qquad F = -181,700 \text{ kips}$$

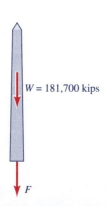

W = 181,700 kips

F

Fig. 2   Free-body diagram.

(Note: In accordance with the *sign convention* for normal stress, the normal force F is taken positive in tension. The negative value for F indicates that it is a compressive force, as is clearly evident in this case.) The cross-sectional area of the base is

$$A = (665.5 \text{ in.})^2 - (665.5 \text{ in.} - 360 \text{ in.})^2 = 349{,}600 \text{ in}^2$$

Therefore, from Eq. 2.2,

$$\sigma_{avg} = \frac{F}{A} = \frac{-181{,}700 \text{ kips}}{349{,}600 \text{ in}^2} = -519.8 \text{ psi}$$

Rounded to three significant figures, the average normal stress on the cross section of the monument at its base is

$$\sigma_{avg} = 520 \text{ psi } (C) \qquad \textbf{Ans.}$$

where the (C) stands for compression. This is a very low stress, even for stone. Figure 3 illustrates how this uniform compressive stress would be distributed over the foundation at the base of the monument.

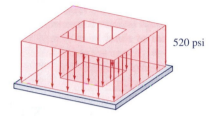

520 psi

Fig. 3   Compressive stress at the base.

The compressive stress that results when one object bears on another, like the stress that the monument exerts on the foundation in the above example problem, is frequently called **bearing stress.** Bearing stress is just a special case of compressive normal stress.

Stress Resultant. Internal resultants were introduced in Section 1.4, and Examples 1.1 and 1.2 show how equilibrium is used to relate these resultants on a cross section to the external loads. Equation 2.2 relates the average normal stress on a cross section to the normal force on the cross section. Let us now examine in greater detail the relationship between the distributed normal stress on a cross section and its resultant. Based on the definition of normal stress in Eq. 2.1, we can re-place the ΔF in Fig. 2.3a by an elemental force $dF = \sigma\, dA$. Referring again to Fig. 2.3a and following the right-hand rule for moments, we can see that this elemental force dF contributes a moment $z\,dF$ about the $+y$ axis and $y\,dF$ about the $-z$ axis. In Fig. 2.3b, the resultant normal force on the cross section at x is labeled $F(x)$, and it acts at point (y_R, z_R) in the cross section. Given the distribution of normal stress on a cross section, $\sigma \equiv \sigma(x, y, z)$, we can integrate over the cross section to determine the magnitude and point of application of the **resultant normal force:**[2]

$$\sum F_x: \qquad F(x) = \int_A \sigma\, dA$$

$$\sum M_y: \qquad z_R F(x) = \int_A z\sigma\, dA \qquad (2.3)$$

$$\sum M_z: \qquad -y_R F(x) = -\int_A y\sigma\, dA$$

---

[2]For generality, the normal force has been permitted to be a function of x in Fig. 2.3b and in Eqs. 2.3. Of course, $F(x) = P = \text{const}$ in the axial-loading case illustrated in Fig. 2.2.

The two moment equations are used to locate the line of action of the force $F(x)$. Note that the sign convention for σ implies that the force F in Eq. 2.3 is to be taken positive in tension. This is the reason that we will consistently, as we did in Chapter 1, take normal force resultants to be positive in tension.

Resultant of Constant Normal Stress on a Cross Section: Let us determine the resultant of normal stress on the cross section at x (Fig. 2.3a) if the normal stress is constant over the cross section. We will prove that **normal stress that is constant on a cross section corresponds to an axial force $F(x) = A\sigma(x)$ acting through the centroid of the cross section at x.** (In Section 3.2 you will learn the conditions under which $\sigma(x)$ is constant over the cross section.)

Let the resultant be assumed to be a force $F(x)$ acting parallel to the x axis and passing through point (y_R, z_R), as in Fig. 2.3b. We must show that

$$F(x) = A(x)\sigma(x), \qquad y_R = \bar{y}, \qquad z_R = \bar{z}$$

For this we can use Eqs. 2.3. Substituting the condition $\sigma(x, y, z) = \sigma(x)$ into Eqs. 2.3, we get

$$\sum F_x: \qquad F(x) = \sigma(x)\int_A dA = \sigma(x)A$$

$$\sum M_y: \qquad z_R F(x) = \sigma(x)\int_A z\, dA = \sigma(x)\bar{z}A$$

$$\sum M_z: \qquad -y_R F(x) = -\sigma(x)\int_A y\, dA = -\sigma(x)\bar{y}A$$

Therefore, if the normal stress is uniform over a cross section, the normal stress on the cross section, also called the **axial stress,** is given by

$$\boxed{\sigma(x) = \frac{F(x)}{A(x)}} \qquad \begin{array}{l}\textbf{Axial} \\ \textbf{Stress} \\ \textbf{Equation}\end{array} \qquad (2.4)$$

and corresponds to a force $F(x)$ (tension positive) acting at the centroid of the cross section, that is, at $z_R = \bar{z}. y_R = \bar{y}$.

We would certainly expect uniform stress on a circular rod to correspond to a force acting along the axis of the rod, and similarly for a square or rectangular bar. Hence, it is "reasonable" that a uniform normal stress distribution acting over a cross section of general shape produces a resultant force acting through the centroid of the cross section. In most cases, the cross-sectional area is constant throughout the length of the member, but Eq. 2.4 may also be used if the cross-sectional area varies slowly with x. (See Example 3.2.)

Uniform Normal Stress in an Axially Loaded Bar: Under certain assumptions, an axially loaded bar will have the same uniform normal stress on every cross section; that is, $\sigma(x, y, z) = \sigma = $ constant. These assumptions are:

- The bar is *prismatic;* that is, the bar is straight and it has the same cross section throughout its length.
- The bar is *homogeneous;* that is, the bar is made of the same material throughout.
- The load is applied as equal and opposite uniform stress distributions over the two end cross sections of the bar.

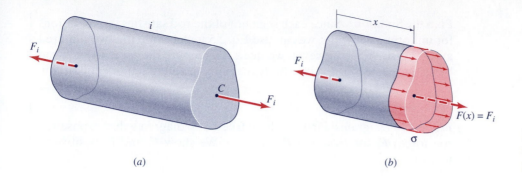

(a)   (b)

FIGURE 2.4 Uniform stress in an axially loaded prismatic bar.

So long as the resultant force at each end of the bar is applied at the centroid of the end cross section, the last assumption—that the loads are applied as uniform normal stress distributions on the end cross sections—can be relaxed. As illustrated in Fig. 2.2 *a–d*, the stress is uniform on every cross section, except on cross sections that are very near the points of application of load. This is an application of *Saint-Venant's Principle*, which is discussed further in Section 2.10.

The uniform, prismatic bar in Fig. 2.4a is labeled as member "*i*" and is subjected to equal and opposite axial forces F_i acting through the centroids at its ends. Its cross-sectional area is A_i.

The normal stress on cross sections of an axially loaded member, like the one in Fig. 2.4, is called the **axial stress.** Since, from the free-body diagram in Fig. 2.4b, the resultant force, $F(x)$, on every cross section of the bar is equal to the applied load F_i, and since the cross-sectional area is constant, from Eq. 2.4 we get the following formula for the **uniform axial stress:**

$$\boxed{\sigma_i = \frac{F_i}{A_i} = \text{const}}$$

Axial-Stress Equation (2.5)

Example 2.3 shows one application of the axial-stress equation. You will find several other examples in the computer program, **MDSolids,** which accompanies this book.

EXAMPLE 2.3

Two solid circular rods are welded to a plate at B to form a single rod, as shown in Fig. 1. Consider the 30-kN force at B to be uniformly distributed around the circumference of the collar at B and the 10 kN load at C to be applied at the centroid of the end cross section. Determine the axial stress in each portion of the rod.

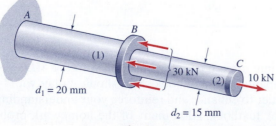

Fig. 1

Plan the Solution   Since each segment of the rod satisfies the conditions for uniform axial stress, we can use Eq. 2.5 to calculate the two required axial stresses. First, however, we need to compute the force in each rod by using an appropriate free-body diagram and equation of equilibrium.

Solution

Free-body Diagrams: First we draw free-body diagrams that expose the rod forces F_1 (or F_{AB}) and F_2 (or F_{BC}). We show F_1 and F_2 positive in tension.

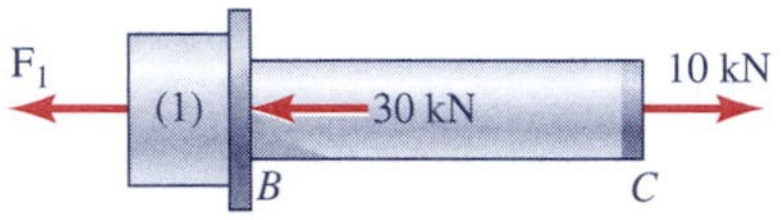

Equations of Equilibrium: From free-body diagram 1 (Fig. 2a),

$$\overset{+}{\rightarrow}\left(\sum F\right)_1 = 0: \qquad -F_1 - 30 \text{ kN} + 10 \text{ kN} = 0, \qquad F_1 = -20 \text{ kN}$$

and, from free-body diagram 2 (Fig. 2b),

$$\overset{+}{\rightarrow}\left(\sum F\right)_2 = 0: \qquad -F_2 + 10 \text{ kN} = 0, \qquad F_2 = 10 \text{ kN}$$

$$A_1 = \frac{\pi}{4}d_1^2 = \frac{\pi}{4}(20 \text{ mm})^2 = 314.2 \text{ mm}^2$$

$$A_2 = \frac{\pi}{4}d_2^2 = \frac{\pi}{4}(15 \text{ mm})^2 = 176.7 \text{ mm}^2$$

(a) Free-body diagram 1.

(b) Free-body diagram 2.

Fig. 2

Axial Stresses: Using Eq. 2.5, we obtain the axial stresses

$$\sigma_1 = \frac{F_1}{A_1} = \frac{-20 \text{ kN}}{314.2 \text{ mm}^2} = -63.7\frac{\text{MN}}{\text{m}^2}$$

$$\sigma_1 = \frac{F_2}{A_2} = \frac{10 \text{ kN}}{176.7 \text{ mm}^2} = 56.6\frac{\text{MN}}{\text{m}^2}$$

$$\left.\begin{array}{l} \sigma_1 = -63.7 \text{ MPa (63.7 MPa C)} \\ \sigma_2 = 56.6 \text{ MPa (56.6 MPa T)} \end{array}\right\} \qquad \textbf{Ans.}$$

Review the Solution   In this problem we could "mentally" solve the equilibrium problems and "see" that AB is in compression and that BC is in tension.

The free-body diagrams in Fig. 2 of Example 2.3 illustrate the application of the **method of sections** to structures that have axially loaded segments that are collinear. For example, the free-body diagram in Fig. 2a makes it possible with a single free-body diagram to relate the internal force F_1 to the external forces applied at the two nodes (joints) B and C; this is the most efficient way to determine F_1. In Chapter 3 there are many additional examples of this approach to equilibrium of axially loaded structures.

MDSolids

This is an excellent time for you to get acquainted with the **MDSolids** computer program that accompanies this textbook. You should "play with" each of the MDS examples in order to enhance and reinforce your understanding of the concepts presented in the textbook. Also, many of the homework problems can be checked by using MDSolids.

MDS2.1 **Segmented Axial** is a computer program for solving axial stress problems like Example 2.3.

MDS2.2 **Beam and Two Rods** is a computer program for solving statically determinate problems having two non-collinear axial-deformation members.

MDS2.3 **Two-Bar Assembly** is a program for analyzing the stresses in members of simple statically determinate truss-like structures.

MDS2.4 **Truss Analysis** is a program for analyzing the stresses in members of statically determinate trusses..

2.3 EXTENSIONAL STRAIN; THERMAL STRAIN

When a solid body is subjected to external loading and/or temperature changes, it deforms; that is, changes occur in the size and/or the shape of the body. The general term **deformation** includes both changes of lengths and changes of angles. For example, consider the axial deformation of a bar as shown in Fig. 2.5. To illustrate how the bar deforms locally when it is stretched, two squares are drawn on the surface of the undeformed bar (Fig. 2.5a); one square is aligned with the axis of the bar, and one square is oriented at 45° to the axis of the bar. Local changes in length, such as that of line segment BC when it elongates to become segment $B*C*$, are described by *extensional strain*. Asterisk superscripts denote "after deformation has occurred." Local angle changes, like the change of right angle DEF to become acute angle $D*E*F*$ are described by *shear strain*. Both types of strain are important in describing the *geometry of deformation* of deformable bodies. (Shear strain is discussed in Section 2.7.)

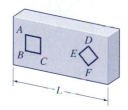

(a) The undeformed bar.

Definition of Extensional Strain. To define extensional strain, let us consider again the case of axial deformation, as illustrated by Fig. 2.5. The **total elongation** of the bar is designated by ΔL,[3] and the *extensional strain*, or *normal strain*, is designated by the lowercase Greek letter epsilon (ϵ).[4] The **average extensional strain** is defined as the ratio of the total elongation ΔL to the original length L, that is,

$$\boxed{\epsilon_{avg} = \frac{\Delta L}{L} = \frac{L* - L}{L}} \quad \begin{array}{l} \text{Average} \\ \text{Extensional} \\ \text{Strain} \end{array} \quad (2.6)$$

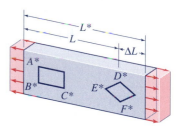

(b) The deformed bar.

FIGURE 2.5 The deformation of a bar under axial loading.

If the bar stretches (i.e., $L* > L$), the strain ϵ is positive and is called **tensile strain**. A shortening of the bar results in a negative value for ϵ and is referred to as **compressive strain.** Although strain is a dimensionless quantity, it is common practice to report strain values in units of in./in., or μin./in. (1 microinch per inch = 10^{-6} in./in.), or μm/m. The magnitude of extensional strain is generally quite small, say < 0.001, so the latter two microstrain units are appropriate. Frequently just the symbol μ is used, like 100 μ, which is read as 100 *microstrain*.

---

[3]The Greek capital letter delta (Δ) is frequently used to designate the change in a quantity, so ΔL is a change in the length L.
[4]In Section 2.4 it is shown that the extensional strain ϵ is related to the normal stress σ. Therefore, *extensional strain* is sometimes called *normal strain*.

For the remainder of this chapter we will consider only the case where the extensional strain is uniform along the length of a member. This type of uniform strain is also called **axial strain.** Thus, the axial strain is given by

$$\epsilon = \epsilon_{\text{avg}} = \frac{\Delta L}{L} \qquad \begin{array}{l} \textbf{Axial} \\ \textbf{Strain} \end{array} \qquad (2.7)$$

(Nonuniform extensional strain is discussed in Sections 2.12, 3.2 and 3.3.)

Strain-Displacement Analysis. Equations 2.6 and 2.7 are definitions that involve only the *geometry of deformation*. They enable us to relate strain quantities to displacement quantities, as will be illustrated in the following example of **strain-displacement analysis.** Later we will consider the causes of the deformation, the applied loads and/or temperature changes.

> ══════════ **EXAMPLE 2.4** ◄──────────

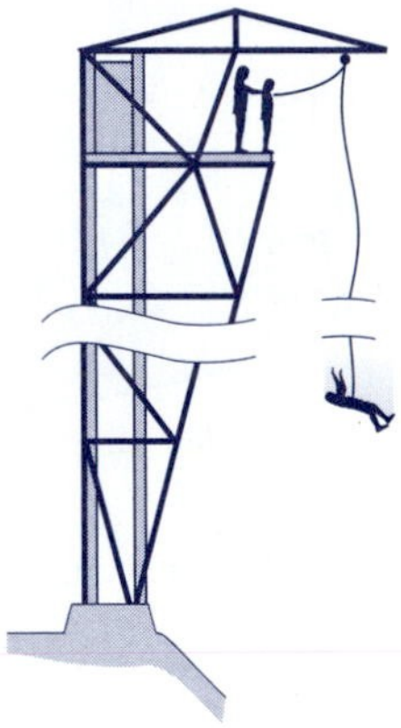

Fig. 1

When the bungee jumper in Fig. 1 stands on the jump platform, the unstretched length of the bungee cord is 12.0 ft. Assuming that the bungee cord stretches uniformly along its length, determine the extensional strain in the bungee cord when the jumper "hits bottom," where the extended length of the cord is 37.1 ft.

Plan the Solution The initial length of the bungee cord and its final length are given in the problem statement. Therefore, we can simply use Eq. 2.6 to calculate the average extensional strain in the bungee cord when the jumper hits bottom.

Solution From Eq. 2.6,

$$\epsilon = \frac{L^* - L}{L} = \frac{37.1 \text{ ft} - 12.0 \text{ ft}}{12.0 \text{ ft}} = 2.09 \frac{\text{ft}}{\text{ft}} \qquad \textbf{Ans.}$$

Review the Solution The final extensional strain of the bungee cord, 209%, is very large by comparison with the strain in a steel or aluminum rod when stretched until it fractures, which is about 10%. For information on the elongation of bungee cord, see www.bungeezone.com/equip.

> ══════════ **EXAMPLE 2.5** ◄──────────

When the "rigid" beam *AB* in Fig. 1 is horizontal, the rod *BC* is strain free. (a) Determine an expression for the average extensional strain in rod *BC* as a function of the angle θ of clockwise rotation of *AB* in the range $0 \leq \theta \leq \pi/2$. (b) Determine an approximation for $\epsilon(\theta)$ that gives acceptable accuracy for values of ϵ when $\theta \ll 1$ rad.

Plan the Solution The defining equation for extensional strain, Eq. 2.6, can be used to determine the required expression for the average extensional strain of rod *BC*. To determine the geometrical relationship between the extended length L^* and the angle θ, we can draw a sketch of the deformed rod-beam system (i.e., a *deformation diagram*).[5] Since beam AB is assumed to be rigid, end B moves in a circle about end A.

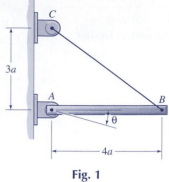

Fig. 1

Solution   (a) *Obtain an expression for* $\epsilon(\theta)$. From Eq. 2.6,

$$\epsilon = \frac{L^* - L}{L}$$

From the original figure,

$$L \equiv \overline{BC} = 5a$$

Deformation Diagram: Rigid beam AB rotates about A, while rod BC stretches and rotates about C, as indicated in Fig. 2.

Geometry of Deformation: Using the Pythagorean theorem and dimensions from the sketch of the deformed configuration in Fig. 2, we get

$$L^* = \sqrt{(3a + c^*)^2 + (b^*)^2}$$

where

$$b^* = 4a \cos \theta, \qquad c^* = 4a \sin \theta$$

Then, from Eq. 2.6,

$$\epsilon = \frac{L^* - L}{L} = \frac{a\sqrt{(3 + 4 \sin \theta)^2 + (4 \cos \theta)^2} - 5a}{5a}$$

This can be simplified to

$$\epsilon(\theta) = \sqrt{1 + \left(\frac{24}{25}\right)\sin \theta} - 1 \qquad \textbf{Ans. (a)} \quad (1)$$

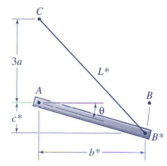

Fig. 2   Deformation diagram.

(b) *Approximate* $\epsilon(\theta)$ *for* $\theta \ll 1$ *rad.*[6] If $\theta \ll 1$, then $\sin \theta \approx \theta$. Furthermore, the term under the radical has the form $1 + \beta$, with $\beta \ll 1$. According to the binomial expansion theorem, for small values of β

$$\sqrt{1 + \beta} \approx 1 + \frac{\beta}{2}$$

Thus, for $\theta \ll 1$, $\epsilon(\theta)$ takes the form

$$\epsilon(\theta) = \left(\frac{12}{25}\right)\theta, \qquad \theta \ll 1 \text{ rad} \qquad \textbf{Ans. (b)} \quad (2)$$

Review the Solution   In both answers, $\epsilon(\theta)$ is dimensionless, as it should be. If Eq. (1) is evaluated at $\theta = \pi/2$, we get $\epsilon(\pi/2) = 2/5$. Since $L^* = 3a + 4a = 7a$ when B^* is directly below A, this value of $\epsilon(\pi/2)$ is correct.

---

[5]A *deformation diagram* is useful in the analysis of the geometry of deformation of a system, analogous to the way a free-body diagram is useful in an equilibrium analysis.
[6]See Appendix A.2 for information regarding approximations of this nature.

Note that nothing was said in the preceding example about what caused the deformation illustrated in Fig. 2. In fact, a downward force at B would cause beam AB to rotate clockwise; but heating of rod BC could also be the cause. Thus, Example 2.5 illustrates that we do not need to know what caused the deformation in order to relate the extensional strain ϵ to the rotation angle θ. That is <u>purely a geometry problem!</u>

Small-Displacement/Small-Strain Behavior. In Part (b) of Example 2.5, an expression for the strain $\epsilon(\theta)$ was obtained that is valid for very small values of the angular displacement θ, namely

$$\epsilon(\theta) = \left(\frac{12}{25}\right)\theta, \qquad \theta \ll 1 \text{ rad}$$

Two things may be noted about this approximation: (1) the strain is a linear function of the displacement variable θ, and (2) since the displacement angle θ is assumed to be very small, the strain ϵ will also be very small. This small-displacement/small-strain behavior is typical of the normal behavior of engineering structures (e.g., building structures and machines) for which deformations are usually too small to be seen with the naked eye, and for which strains are usually of the order of 0.001 in./in. (i.e., 1000 μ) or less.[7]

A small-displacement/small-strain situation that arises frequently is illustrated in Fig. 2.6. Consider an extensible rod BC that is pinned at B to a "rigid" beam AB, and let point B move downward by a small distance, δ_B, where $\delta_B \ll L$ and $\delta_B \ll a$. Since AB is assumed to be rigid, point B actually moves in a circular path around a center at A, as shown in Fig. 2.6a. Figure 2.6b illustrates a simplifying assumption: because $\dfrac{\delta_B}{a} \ll 1$ and $\dfrac{\delta_B}{L} \ll 1$, point B can be assumed to move vertically downward to point B', rather than along the circular path shown in Fig. 2.6a. By comparing the expressions that we obtain in each case for the strain in rod BC and the angle θ, we will show that it is acceptable to make the simplifying assumption that rod BC remains vertical, as shown in Fig. 2.6b.

From Fig. 2.6a, where point B follows a circular path about A,

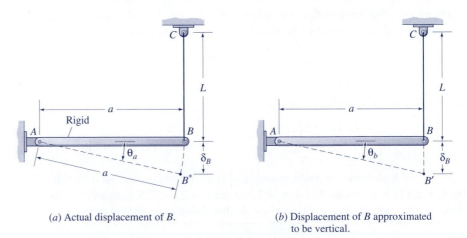

(a) Actual displacement of B.

(b) Displacement of B approximated to be vertical.

FIGURE 2.6   Small-displacement approximations.

[7]There are situations where large displacements occur, but the strain remains very small, like the deflection of the tip of a very flexible diving board under the weight of a heavy diver. However, throughout this textbook assume that displacements are small unless stated otherwise.

$$\epsilon_a = \frac{\overline{B^*C} - L}{L} = \frac{\sqrt{(L + \delta_B)^2 + [a(1 - \cos\theta_a)]^2} - L}{L}$$

$$= \frac{1}{L}\sqrt{L^2 + 2L\delta_B + \delta_B^2 + a^2\left(\frac{\theta_a^2}{2} - \frac{\theta_a^4}{4!} + \cdots\right)} - 1$$

By neglecting the squared terms and higher in δ_B and θ_a in the last line of the above equation (see Appendix A.2), we obtain the approximation

$$\epsilon_a = \frac{\delta_B}{L} \tag{1a}$$

Since

$$\theta_a = \sin^{-1}\left(\frac{\delta_B}{a}\right) \rightarrow \frac{\delta_B}{a} = \theta_a - \frac{\theta_a^3}{3!} + \cdots$$

therefore, angle θ_a can be approximated by

$$\theta_a = \frac{\delta_B}{a} \tag{2a}$$

From Fig. 2.6b, where rod BC is assumed to remain vertical,

$$\epsilon_b = \frac{\delta_B}{L} \tag{1b}$$

$$\theta_b = \tan^{-1}\left(\frac{\delta_B}{a}\right) \rightarrow \frac{\delta_B}{a} = \theta_b + \frac{\theta_b^3}{3!} + \cdots$$

Therefore, angle θ_b can be approximated by

$$\theta_b = \frac{\delta_B}{a} \tag{2b}$$

From the above, it is evident that the results of the two small-displacement analyses are the same; that is, Eq. (1a) is the same as Eq. (1b), and Eq. (2a) is the same as Eq. (2b). It is clear, therefore, that the analysis in situations like this is made much simpler if we just assume from the outset that points along rotating "rigid" bars move along straight lines perpendicular to the original bar, as in Fig. 2.6b, rather than in circular paths.

Thermal Strain. We turn now to the relationship between extensional strain and the change of temperature of a body, that is, to *thermal strain*.[8] Although there are some exceptions, most engineering materials respond to a uniform increase in temperature, ΔT, by *expanding* in all directions by a uniform amount

$$\boxed{\epsilon_T = \alpha\,\Delta T} \qquad \textbf{Thermal Strain} \tag{2.8}$$

---

[8]Although some texts refer to *thermal stress,* a uniform temperature change only results in stresses in a body if the body is restrained, that is, if it is not completely free to expand. Here we will assume that the solids that are heated or cooled are completely free to expand or contract. We will solve for thermally induced stresses in Section 3.6.

where ϵ_T is the **thermal strain,** α is the **coefficient of thermal expansion,** and ΔT is the change in temperature. (A positive ΔT corresponds to an *increase* in temperature above the reference temperature.) When the temperature of a body decreases, ΔT is negative, and the body shrinks a corresponding amount. The units of α and ΔT must be consistent. For example, the U.S. Customary units of α are 1/°F (the reciprocal of degrees Fahrenheit). The SI units are 1/K (the reciprocal of kelvins) or 1/°C (the reciprocal of degrees Celsius), depending on the units of ΔT.

Let the block of material in Fig. 2.7 be subjected to a uniform temperature increase ΔT. Since it is free to expand in all directions,

$$\epsilon_{xT} = \epsilon_{yT} = \epsilon_{zT} = \alpha\,\Delta T$$

Since this strain is uniform throughout the block, Eqs. 2.6 and 2.8 can be combined to give the elongations of the block:

$$\Delta L_{xT} = (\alpha\,\Delta T)L_x, \qquad \Delta L_{yT} = (\alpha\,\Delta T)L_y, \qquad \Delta L_{zT} = (\alpha\,\Delta T)L_z \qquad (2.9)$$

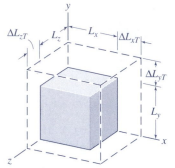

FIGURE 2.7 The free expansion of a uniform block caused by a uniform increase in the temperature of the block.

The coefficient of thermal expansion is a property of the material; it is determined experimentally by applying a change in temperature and measuring the change in dimensions of the specimen, as in Fig. 2.7. Values for α are given in tables of Mechanical Properties of Engineering Materials, such as those in Table F.3 in Appendix F.

EXAMPLE 2.6

If the rod BC in Example 2.5 is made of a high-strength steel for which $\alpha = 8.0 \times 10^{-6}$/°F, and if BC is uniformly heated by 100°F, through what angle θ will beam AB rotate? Assume that AB and BC are weightless, and that AB is horizontal before the heating of BC occurs.

Plan the Solution Equation 2.8 relates the temperature change ΔT to the thermal strain, and the answers in Example 2.5 relate strain to θ. Therefore, we can simply combine these.

Solution

$$\epsilon_T = \alpha\,\Delta T = (8.0 \times 10^{-6}/°F)(100°F) = 800 \times 10^{-6} \text{ in./in.}$$

Since the expression in Eq. (2) of Example 2.5 is simpler than the expression in Eq. (1), let us try Eq. (2) first.

$$\theta = \frac{25}{12}\epsilon = \frac{25}{12}(8.0 \times 10^{-4}) = 1.667 \times 10^{-3} \text{ rad}$$

or

$$\theta = 0.0955 \text{ deg} \qquad\qquad \textbf{Ans.}$$

Since this answer satisfies the requirement that $\theta \ll 1$ rad, we do not need to resort to Eq. (1).

Review the Solution Although a tenth of a degree rotation for a temperature increase of 100°F may seem to be very small, the formulas are so simple that all we can do to check our result is just to double-check the calculations and the conversion from radians to degrees.

2.4 STRESS-STRAIN DIAGRAMS; MECHANICAL PROPERTIES OF MATERIALS

In order to relate the loads on engineering structures to the deformation produced by the loads, experiments must be performed to determine the **load-deformation behavior** of the materials (e.g., aluminum, steel, and concrete) used in fabricating the structures. Many useful mechanical properties of materials are obtained from tension tests or from compression tests, and these properties are listed in tables like those in Appendix F. This section describes how a tension test is performed and discusses the material properties that are obtained from this type of test.

Tension Tests and Compression Tests. Figure 2.8 shows a computer-controlled, hydraulically actuated testing machine that may be used to apply a tensile load or a compressive load to a test specimen, like the steel tension specimen in Fig. 2.9a or the concrete compression specimen in Fig. 2.9c.[9] Figure 2.9b shows a close-up view of a ceramic tension specimen mounted in special testing-machine

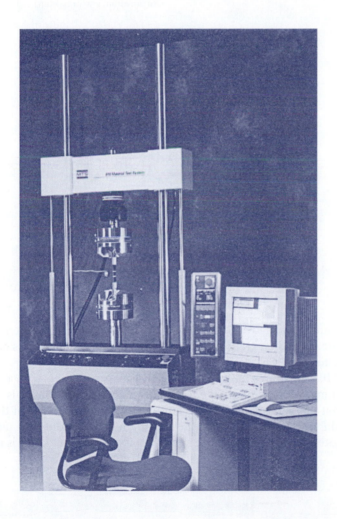

FIGURE 2.8 A computer-controlled hydraulically actuated testing machine. (Courtesy MTS Systems Corporation, www.mts.com)

---

[9]Specimen dimensions and procedures for preparing and testing specimens are prescribed by various standards organizations, like the *American Society for Testing Materials* (ASTM) and the *American Concrete Institute* (ACI).

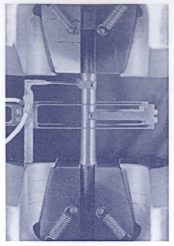

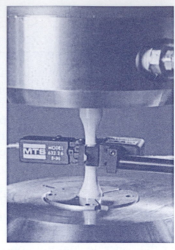

(*a*) A metal tension specimen with extensometer attached. (Courtesy Roy Craig)

(*b*) A ceramic tension specimen with extensometer attached. (Courtesy MTS Systems Corporation, www.mts.com)

FIGURE 2.9   Tension and compression test specimens.

(*c*) A concrete cylinder before and after compression testing. (Courtesy Roy Craig)

grips. Electromechanical *extensometers* are mounted on the specimens in Figs. 2.9*a* and 2.9*b* to measure the extension (i.e., the elongation) that occurs over the *gage length* of the test section.

Figure 2.10*a* illustrates an underformed tension specimen with two points on the specimen marking the *original gage length, L_0*. The notation L_0 is used here to emphasize that this is the original gage length, not the total length of the specimen. An axial load P causes the portion of the specimen between the gage marks to elongate, as indicated in Fig. 2.10*b*. As the specimen is pulled, the load P is measured by

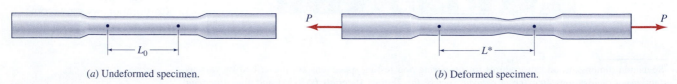

(*a*) Undeformed specimen.

(*b*) Deformed specimen.

FIGURE 2.10   A typical tension-test specimen.

the testing machine and recorded. The extensometer provides a simultaneous measurement of the corresponding length, $L^* \equiv L^*(P)$, of the test section, or else it directly measures the elongation

$$\Delta L = L^* - L_0 \tag{2.10}$$

In a *static tension test* the length of the specimen is increased very slowly, in which case the loading rate need not be measured. In some situations, however, a dynamic test must be performed. Then, the rate of loading must be measured and recorded, since material properties are affected by high rates of loading.

Stress-Strain Diagrams. A plot of stress versus strain is called a **stress-strain diagram,** and from such stress-strain diagrams we can deduce a number of significant **mechanical properties of materials.**[10] The values of normal stress and extensional strain that are used in plotting a *conventional stress-strain diagram* are the *engineering stress* (load divided by <u>original</u> cross-sectional area of the test section) and *engineering strain* (elongation divided by <u>original</u> gage length), that is,

$$\sigma = \frac{P}{A_0}, \qquad \epsilon = \frac{L^* - L_0}{L_0} \tag{2.11}$$

Mechanical Properties of Materials.   Figures 2.11*a* and 2.11*b* are stress-strain diagrams for *structural steel* (also called *mild steel,* or *low-carbon steel*), which is the metal commonly used in fabricating bridges, buildings, automotive and construction vehicles, and many other machines and structures. A number of important mechanical properties of materials that can be deduced from stress-strain diagrams are illustrated in Fig. 2.11. In Fig. 2.11*a* the stress is plotted accurately, but the strain is plotted to a variable scale so that all important features can be shown and discussed. In Fig. 2.11*b*, which gives typical numerical values of stress and strain for structural steel, one stress-strain curve, the lower one, is plotted against a strain scale that emphasizes the low-strain region; the upper curve is plotted against a strain scale that emphasizes the high-strain region and puts the entire stress-strain history into perspective.

Starting at the origin *A* in Fig. 2.11*a* and continuing to point *B*, there is a linear relationship between stress and strain. The stress at point *B* is called the **proportional limit,** σ_{PL}. The ratio of stress to strain in this linear region of the stress-strain diagram is called *Young's modulus,*[11] or the **modulus of elasticity,** and is given by

$$E = \frac{\Delta\sigma}{\Delta\epsilon}, \qquad \sigma < \sigma_{PL} \qquad \textbf{Young's Modulus} \tag{2.12}$$

Typical units for *E* are ksi or GPa.

At *B* the specimen begins *yielding,* that is, smaller and smaller increments of load are required to produce a given increment of elongation. The stress at *C* is called the *upper yield point,* $(\sigma_{YP})_u$, while the stress at *D* is called the *lower yield point,* $(\sigma_{YP})_l$. The upper yield point has little practical importance, so the lower yield point is usually referred to simply as the **yield point,** σ_{YP}. From *D* to *E* the specimen

---

[10]These material properties are also called *constitutive properties of materials.*

[11]In two volumes entitled *A Course of Lectures on Natural Philosophy and Mechanical Arts,* London, 1807, Thomas Young (1773–1829) introduced the modulus of elasticity and discussed many other interesting topics in mechanics of deformable bodies [Ref. 2-1].

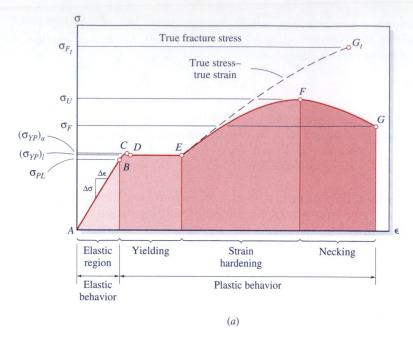

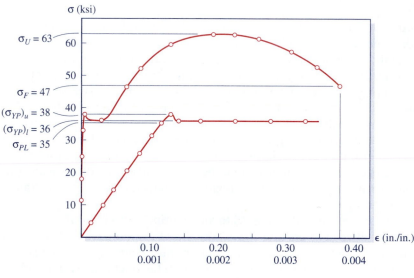

FIGURE 2.11 Stress-strain diagrams for structural steel in tension. (*a*) Strain not plotted to scale. (*b*) Strain plotted to two different scales.

continues to elongate without any increase in stress. The region DE is referred to as the *perfectly plastic* zone. The stress begins to increase at E, and the region from E to F is referred to as the zone of *strain hardening*. The stress at F is called the *ultimate stress*, or **ultimate strength,** σ_U. At F the load begins to drop, and the specimen begins to "neck down." This neck-down behavior continues until, at G, fracture occurs at the *fracture stress*, σ_F. Figure 2.12*a*, shows a hot-rolled steel specimen at three stages of tensile testing: (1) before testing, (2) as removed from the testing machine at a point between F and G with pronounced reduction in area referred to as *necking* or *neck-down*, and (3) after fracture. Figure 2.12*b* shows the typical *cup and cone fracture* of a hot-rolled steel tensile specimen. In Section 2.9 it is shown that ductile fracture, like the cup-and-cone fracture at 45° to the axis of the member in Fig. 2.12*b*, is due to shear stress on the fracture surface.

The *true stress*, σ_{true}, is the load at some instant during the test divided by the actual minimum cross-sectional area of the specimen at that instant. Thus, when a specimen starts to neck down, the true stress is taken as the load divided by the minimum cross-sectional area in the neck-down region. The *true strain*, ϵ_{true}, is the instantaneous change in length of a test section divided by the instantaneous length of that test section. True stress and true strain are given by the formulas

$$\sigma_{\text{true}} = \frac{P}{A_{\min}}, \qquad \epsilon_{\text{true}} = \ln(1 + \epsilon) \qquad (2.13)$$

True strain can also be expressed in terms of area change[12]

$$\epsilon_{\text{true}} = \ln\left(\frac{A_0}{A_{\min}}\right)$$

The solid-line curve in Fig. 2.11a is a *conventional stress-strain diagram* of engineering stress versus engineering strain, while the dashed curve is a sketch of true stress versus true strain. The curves differ only when strain is large and when the cross-sectional area is decreasing significantly.

Figures 2.11a and 2.11b both illustrate the stress-strain behavior of structural steel. Materials scientists and metallurgists have developed a number of processes for altering the mechanical properties of metals, including alloying, work-hardening, and tempering. Figure 2.13 contrasts the tensile stress-strain behavior of several ferrous metals. Stress-strain curves for several aluminum alloys are shown in Fig. 2.14.[13]

It is apparent from a comparison of Figs. 2.11 and 2.14 that several properties exhibited by structural steel, for example, a definite yield point followed by a significant zone of yielding at constant stress, are not characteristic of all other materials. For materials like aluminum that have no clearly defined yield point, a stress value called the **offset yield stress** σ_{YS}, is used in lieu of a yield-point stress.

As illustrated in Fig. 2.15, the offset yield stress is determined by first drawing a straight line that best fits the data in the initial (linear) portion of the stress-strain diagram. A second line is then drawn parallel to the original line but offset by a

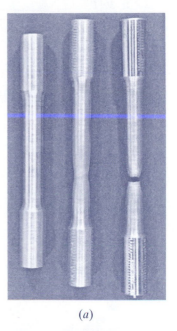

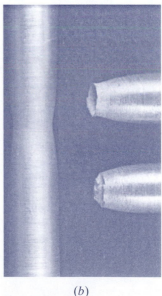

(a)

(b)

FIGURE 2.12   A hot-rolled steel tensile specimen. (Courtesy Roy Craig)

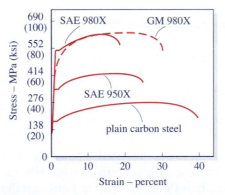

FIGURE 2.13   The stress-strain curves of plain carbon steel and three high-strength/low-alloy steels. (Used with the permission of ASM International.)

[12]This formula for true strain is given by D. C. Drucker in *Introduction to Mechanics of Deformable Solids*, [Ref. 2-2], p. 12.

[13]ASM International, Materials Park, OH, 44073-0002, is an excellent source for information on the mechanical properties of materials. ASM International publishes the *Metals Handbook*, the *Engineered Materials Handbook*, and many other reference works. The *ASM Handbook* is available on CD-ROM.

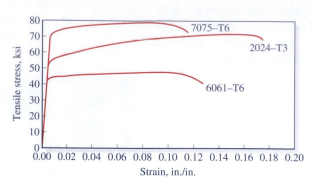

FIGURE 2.14 The stress-strain curves of three typical aluminum alloys. (Used with the permission of McGraw-Hill, Inc.)

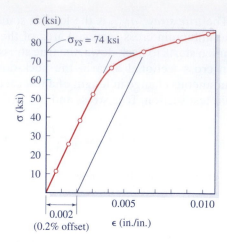

FIGURE 2.15 The procedure for determining the offset yield stress.

specified amount of strain. The intersection of this second line with the stress-strain curve determines the offset yield stress. In Fig. 2.15 the commonly used offset (strain) value of 0.002 (or 0.2%) is illustrated.

MDS2.5 & 2.6 **Stress-Strain Curves** is a computer program for plotting stress-strain curves from load-elongation data and for determining the yield stress from the plotted data.

Design Properties. Now that you have some idea of the stress-strain behavior of several common metallic materials, let us note the material properties that are of primary interest to an engineer designing some structure or machine. From the design standpoint the most significant stress-strain properties can be categorized under the three headings—*strength, stiffness, and ductility.*

- **Strength**—There are three strength values of interest.[14] (1) The *yield strength,* σ_Y, is the highest stress that the material can withstand without undergoing significant yielding. The yield-point stress or the offset yield stress, whichever is appropriate for the particular material, is taken as the yield strength (i.e., $\sigma_Y = \sigma_{YP}$ or $\sigma_Y = \sigma_{YS}$ as appropriate). (2) The *ultimate strength,* σ_U, is the maximum value of stress (i.e., the maximum value of engineering stress) that the material can withstand. Finally, (3) the *fracture stress,* σ_F, if different from the ultimate stress, may be of interest. It is the value of the stress at fracture.

- **Stiffness**—The *stiffness* of a material is basically the ratio of stress to strain. Stiffness is of interest primarily in the linearly elastic region; therefore, Young's modulus, E, is the value used to represent the stiffness of a material.

- **Ductility**—Materials that can undergo large strain before fracture are classified as *ductile materials;* those that fail at small values of strain are classified as *brittle materials.* Strictly speaking, the terms ductile and brittle refer to *modes of fracture,* and a material like structural steel, which behaves in a ductile manner at room temperature, may exhibit brittle behavior at very low

---

[14]The word *strength* is used to designate various critical stress quantities exhibited by materials.

temperatures. Therefore, when we speak of a "brittle material" or a "ductile material," we are referring to the normal (room temperature) behavior of the material.

Ductility. The two commonly used measures of ductility are the *percent elongation* (the final elongation expressed as a percentage of the original gage length) and the *percent reduction in area* at the section where fracture occurs (the area reduction expressed as a percentage of the original area). The *percent elongation* is given by the formula

$$\text{Percent elongation} = \left(\frac{L_F - L_0}{L_0}\right)(100\%)$$

where L_0 is the original gage length and L_F is the length of the gage section at fracture. When percent elongation is stated, the gage length should also be stated, since the elongation at fracture is not uniform over the entire gage length but is concentrated in the necked-down region. The *percent reduction in area* is given by the formula

$$\text{Percent reduction is area} = \left(\frac{A_0 - A_F}{A_0}\right)(100\%)$$

where A_0 is the original cross-sectional area of the test section, and A_F is the final cross-sectional area at the fracture location.

Structural steel is a highly ductile material with a percent elongation in 2 in. of about 30% and a percent reduction in area of about 50%. High ductility permits this type of steel to be pressed to form automobile sheet-metal parts and bent to form concrete-reinforcing bars. Ductile materials permit large local deformation to occur near cracks, rivet holes, and other stress concentrations, thereby preventing the occurrence of sudden, catastrophic failures. Other materials that may be classified as ductile include pure aluminum and some of its alloys, brass, copper, nickel, nylon, and teflon. Figure 2.16 shows the 1100% elongation of a *superplastically deformed* alloy-steel tensile specimen.

As can be seen in Fig. 2.13, high-strength steels generally tend to be far less ductile than mild steel; the dashed-line stress-strain curve labeled "GM 980X" is for a special high-strength, high-ductility steel. Although there is no sharp dividing line between ductile materials and brittle materials, glass, other ceramics, gray cast iron and concrete are among the materials that are classified as brittle materials. Figure 2.17a shows a typical stress-strain curve for a brittle material, with the curve for a ductile material shown for comparison. The fracture surface of a brittle tensile-test specimen is illustrated in Fig. 2.17b. Note that brittle fracture is a direct

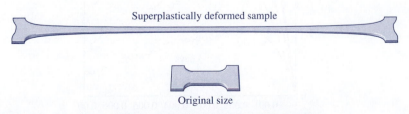

Superplastically deformed sample

Original size

A steel-alloy specimen tested to over 1100% elongation.

FIGURE 2.16 A superplastically deformed steel-alloy specimen tested to over 1100% elongation. (Courtesy of Prof. Oleg D. Sherby, Stanford University.)

result of the tensile stress on the cross section, and contrast this to the fracture of a ductile specimen (Fig. 2.12b) where, as is discussed in Section 2.9, the fracture occurs at an angle of 45°. Ordinary glass is a nearly ideal brittle material, exhibiting linear stress-strain behavior up to a fracture stress in the neighborhood of 10^4 psi. It has been found that the strength of glass is highly dependent on the type of glass, the size of the specimen, and surface defects. *Glass fibers* with diameters on the order of 10^{-4} in. may have an ultimate strength as high as 10^6 psi or more.

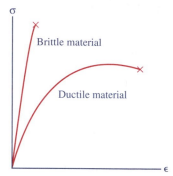

(a) Stress-strain diagrams.

Compression Tests. In *compression,* a ductile metal like steel or aluminum yields at a stress magnitude approximately equal to its tensile yield stress, and Young's modulus in compression is equal to the tensile modulus. Therefore, compression tests are seldom performed on these materials. By contrast, periodic casting and testing of concrete compression specimens, like the one pictured in Fig. 2.9c, is required for quality control on reinforced-concrete construction projects. The compressive strength of concrete increases with age, as illustrated by the compression stress-strain diagrams for Portland-cement concrete specimens, as shown in Figure 2.18. Concrete is a brittle material with very little tensile strength, so it is usually just assumed that the tensile strength of plain (i.e., unreinforced) concrete is zero.

Plastics and Composites. Since the introduction of Bakelite in 1906, hundreds of polymeric materials, called *plastics,* have been developed, and many of these, like nylon, have found structural application. Their advantages, which include their light weight, resistance to corrosion, ease of molding, and good electrical insulation properties, have made them increasingly popular. On the other hand, their low stiffness, tendency to creep (i.e., to continue to deform under constant load) and to absorb moisture, and the strong dependence of their strength and stiffness properties on temperature, are disadvantages that must be carefully weighed against their advantages.

Composite materials are materials that combine two constituent materials in a manner that leads to improved mechanical properties. Reinforced concrete can be

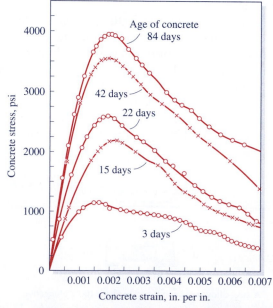

(b) Brittle-fracture surface.

FIGURE 2.17 Tensile-test behavior of a brittle material.

FIGURE 2.18 Stress-strain diagrams for concrete in compression. (From [Ref. 2-3], Bureau of Reclamation.)

considered to be a composite material, in which the steel reinforcement is used to overcome the inherent weakness of concrete in tension. It was previously noted that glass can be made extremely strong when drawn into thin fibers and properly protected against surface damage. It was also noted that some plastics have several undesirable properties, including low stiffness. A material that is stronger and stiffer than the plastic matrix material can be produced by embedding glass fibers or other reinforcing fibers in the matrix. Furthermore, by placing the fibers in specific orientations in the matrix, a material with direction-dependent properties can be fabricated. Figure 2.19 shows the ±45° lay-up of graphite-epoxy layers to form a tensile test specimen and the resulting tensile-test failure. In sports, composite materials are being used in many applications, including tennis racquets, bicycle frames, skis, and boat hulls.[15]

Section 2.14 discusses the *mechanical properties of composites.* Tables of *Mechanical Properties of Selected Engineering Materials* are provided in Appendix F. Because of the extremely large number of commercially available plastics, these tables include only a small sample of the properties of engineering plastics.

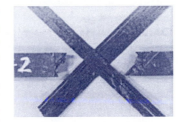

FIGURE 2.19 A graphite-epoxy tensile test specimen; ±45° graphite-fiber lamina. (Courtesy Roy Craig)

2.5 ELASTICITY AND PLASTICITY; TEMPERATURE EFFECTS

In the previous section we considered the stress-strain behavior of tension specimens under *loading.* Specifically, the tensile tests were assumed to be performed in a relatively short time with monotonically increasing tensile strain. We consider now what happens when the loading is reversed, that is, when the strain is allowed to decrease.

Elastic Behavior and Plastic Behavior. Consider the loading and unloading behavior of a material, as illustrated in Fig. 2.20. The upward-pointing arrows in Fig. 2.20*a* and Fig. 2.20*b* indicate the loading curves, that is, the curves that would be followed for monotonically increasing initial loading. The stress-strain behavior of a material is said to be **elastic** if the unloading path retraces the loading path. The

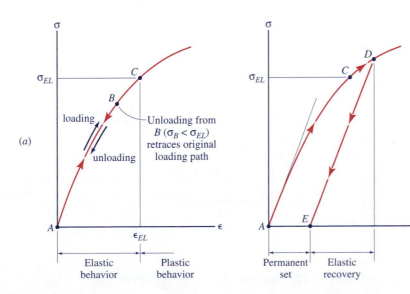

FIGURE 2.20 Illustrations of elastic and plastic stress-strain behavior.

---

[15]Excellent sources of information on engineering composites are the *International Encyclopedia of Composites,* [Ref. 2-4]; and *Composites* [Ref. 2-5].

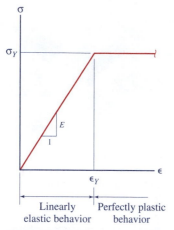

Linearly   Perfectly plastic
elastic behavior   behavior

FIGURE 2.21 Linearly
elastic, perfectly plastic
material model.

stress at C is called the *elastic limit*, σ_{EL}. Since the stress-strain curve from the origin A in Fig. 2.20a to the elastic limit at C is not a straight line, material behavior in this range is termed *nonlinear elastic behavior.* Unloading from a point below point C (e.g., point B) retraces the loading path indicated in Fig. 2.20a, but unloading from a point beyond C (e.g., point D) follows a path DE that is different than the loading path. Unloading from a point on the stress-strain curve beyond the elastic limit typically follows a straight-line path whose slope is parallel to the tangent to the stress-strain curve at the origin. The strain that remains at point E when the stress returns to zero is called the *permanent set*, or *residual strain*, as illustrated in Fig. 2.20b.

For structural steel the elastic limit is very close to the proportional limit and also to the yield point. Therefore, for structural steel the proportional limit stress and the elastic limit stress are usually assumed to be equal to the yield-point stress. It is sometimes convenient to approximate the stress-strain behavior of mild steel and similar materials by the *linearly elastic, perfectly plastic* representation of Fig. 2.21. In Sections 3.11, 4.11, and 6.7, we will use this model of stress-strain behavior in describing the inelastic behavior of members undergoing axial deformation, torsion, and bending, respectively.

Time-dependent Stress-Strain Behavior. Stress-dependent plastic deformation such as the yielding of structural steel described above, is referred to as *slip*. It is essentially an instantaneous process resulting from the slip that occurs within crystals and along the boundaries between crystals that make up the solid.[16] Depending on the material and its temperature, a time-dependent plastic deformation may occur. One form of time-dependent deformation is called **creep.** Creep behavior may be demonstrated by the *constant-stress experiment* illustrated in Fig. 2.22. A constant load is placed on a specimen, and the elongation (or strain) of the specimen is then plotted versus time, as illustrated in Fig. 2.22b. Creep elongation is negligible in many materials at room temperature. However, when heated sufficiently, most materials will exhibit creep behavior. Hence, creep is a significant design consideration in high-temperature applications like turbine engines, boilers, and so on. Since plastics exhibit creep behavior at much lower temperatures, they may be unsuitable for certain applications even though they possess adequate static strength and stiffness at room temperature.

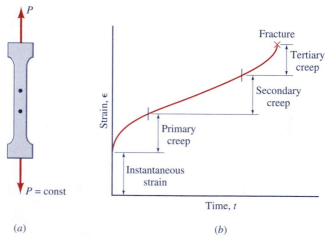

(a)   (b)

FIGURE 2.22   A constant-stress creep experiment.

---

[16]The slip that occurs during yielding will be discussed further in Section 2.9.

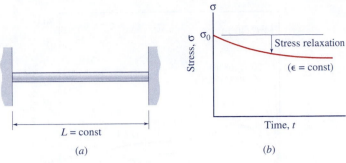

FIGURE 2.23   A constant-strain stress-relaxation experiment.

A related form of time-dependent stress-strain behavior, called *stress relaxation,* can be demonstrated by a *constant-strain experiment.* Figure 2.23*a* shows a specimen that has been stretched to give it an initial tensile stress σ_0 and then attached to immovable supports, and Fig. 2.23*b* is a graph of the stress in the specimen versus time. This *stress relaxation* at constant strain is a second manifestation of time-dependent stress-strain behavior.

Temperature Effects on Material Properties. The coefficient of thermal expansion, discussed in Section 2.3, is a measure of the direct effect that temperature change has on the deformation of a body. However, temperature also has a significant effect on material properties like yield strength and modulus of elasticity. Materials developed specifically to perform at very high temperatures are required for applications like turbine blades in aircraft engines or the structure of a hypersonic aircraft. Figure 2.24 illustrates the profound influence that temperature has on the strength and ductility properties of a particular stainless steel. Note that strength decreases with increasing temperature and vice versa. As indicated by the strain at fracture, ductility decreases with decreasing temperature and increases with increasing temperature. Temperature also affects the stiffness of a material (i.e., the slope of the initial portion of the stress-strain curve), as can be clearly seen in Fig. 2.25.

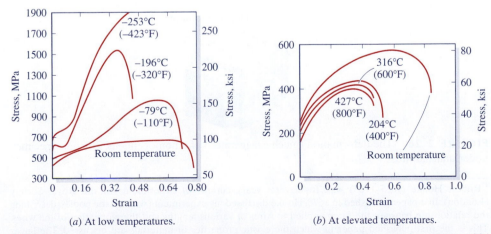

(*a*) At low temperatures.   (*b*) At elevated temperatures.

FIGURE 2.24   The influence of temperature on the stress-strain behavior of type 304 stainless steel. (Used with the permission of ASM International.)

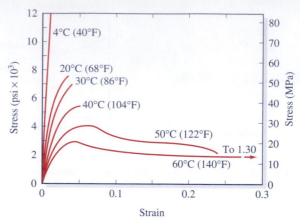

FIGURE 2.25 The influence of temperature on the stress-strain behavior of polymethyl methacrylate. (From [Ref. 2-6]. Reprinted with the permission of the American Society for Testing Materials.)

2.6 LINEAR ELASTICITY; HOOKE'S LAW AND POISSON'S RATIO

In order to keep deformations small and stresses at safe levels, most structures and machine parts are designed so that stresses remain well below the yield stress. Fortunately, most engineering materials exhibit a linear stress-strain behavior at these lower stress levels.

Hooke's Law. Let us, for the present, restrict our discussion to the case of uniaxial stress applied to a homogeneous (same properties throughout), isotropic (same properties in every direction) member oriented along the x axis (Fig. 2.26). The linear relationship between stress and strain, given by Eq. 2.12, applies for $0 \leq \sigma \leq \sigma_Y$. Therefore,

$$\sigma_x = E\epsilon_x \qquad \textbf{Hooke's Law} \qquad (2.14)$$

This equation is called **Hooke's Law.**[17] The subscripts on σ and ϵ identify the axis of the particular stress and strain.

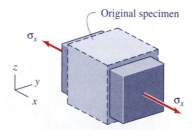

FIGURE 2.26 The deformation (much exaggerated) of a homogeneous, isotropic specimen under uniaxial stress.

---

[17]Robert Hooke (1635–1703) was, for several years, curator of experiments of the Royal Society (London). In a paper published in 1678, Hooke discussed his experiments with elastic bodies, describing the relationship between the force applied to wires of various lengths and the elongation of the wires. This is the first published paper in which the elastic properties of materials are discussed. The linear relationship between force and deformation, called *Hooke's Law*, became the foundation upon which further development of the mechanics of elastic bodies was built. [Ref. 2-1]

Hooke's Law is valid for uniaxial tension or compression within the linear portion of the stress-strain diagram. As noted earlier, E is called the *modulus of elasticity,* or *Young's modulus.* It has the units of stress, typically ksi (or psi) in U.S. Customary units, and GPa (or MPa) in SI units. Representative approximate values of E are: 30×10^3 ksi (200 GPa) for steel, 10×10^3 ksi (70 GPa) for aluminum, and 300 ksi (2 GPa) for nylon. Values of the modulus of elasticity for selected materials are listed in Table F.2 in Appendix F.

Poisson's Ratio. Associated with the elongation of a member in axial tension, there is a transverse contraction, which is illustrated in Fig. 2.26. The transverse contraction during a tensile test is related to the longitudinal elongation by

$$\boxed{\epsilon_{transv} = -\nu\epsilon_{longit}} \qquad \text{Poisson's Ratio} \qquad (2.15)$$

where ν (Greek symbol nu) is *Poisson's ratio.*[18] This expression also holds when the longitudinal strain is compressive; then, the lateral strain results in an expansion of the transverse dimensions. Poisson's ratio is dimensionless, with typical values in the 0.25–0.35 range. For the orientation of axes in Fig. 2.26 the transverse strains are related to the longitudinal strain by

$$\epsilon_y = \epsilon_z = -\nu\epsilon_x \qquad (2.16)$$

Equations 2.14 and 2.16 apply to the simple case of uniaxial stress. More general cases of linearly elastic behavior are treated in Section 2.13.

EXAMPLE 2.7

The cylindrical rod in Fig. 1 is made of steel with $E = 30 \times 10^3$ ksi, $\nu = 0.3$, and $\sigma_Y = 50$ ksi. If the initial length of the rod is $L = 4$ ft and its original diameter is $d = 1$ in., what is the change in length, ΔL, and what is the change in diameter, Δd, due to the application of an axial load $P = 10$ kips?

P = 10 kips

$L = 4$ ft $\qquad$ P $\updownarrow d^*$ $\quad d = 1$in.

ΔL

x

Fig. 1

Plan the Solution From the load P and the cross-sectional area A we can determine the stress σ. Then, if $\sigma \leq \sigma_Y$, we can use Hooke's Law to relate the stress σ to the strain ϵ. Then we can relate the uniform strain, ϵ, to the elongation ΔL. The change in diameter is due to the Poisson's-ratio effect.

---

[18]In 1829 and in 1831 S. D. Poisson (1781–1840), a French mathematician and physicist, wrote important memoirs on the mechanics of solids. He noted that, for simple tension of a prismatic bar, the axial elongation ϵ is accompanied by a lateral contraction of magnitude $\nu\epsilon$. [Ref. 2-1]

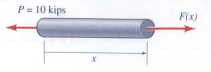

P = 10 kips

F(x)

x

Fig. 2 Free-body diagram.

Solution

Equilibrium: From the free-body diagram in Fig. 2,

$$\sum F_x = 0: \qquad F(x) = P = \text{const}$$

From Eq. 2.5,

$$\sigma_x = \frac{P}{A} = \frac{10 \text{ kips}}{\pi(0.5 \text{ in.})^2} = 12.73 \text{ ksi} < 50 \text{ ksi} \tag{1}$$

Therefore, linearly elastic behavior occurs when the load $P = 10$ kips is applied to the bar.

Material Behavior: From Eq. 2.14,

$$\epsilon_x = \frac{\sigma_x}{E} = \frac{P}{AE} \tag{2a}$$

and, from Eq. 2.15,

$$\epsilon_{\text{radial}} = -\nu\epsilon_x = -\frac{\nu P}{AE} \tag{2b}$$

Strain-Displacement: From Eq. 2.7,

$$\Delta L = L\epsilon_x = \frac{PL}{AE} \tag{3a}$$

and

$$\Delta d = d\epsilon_{\text{radial}} = -\frac{\nu P d}{AE} \tag{3b}$$

Substituting numerical values into Eqs. (3), we get

$$\Delta L = L\epsilon_x = \frac{(10 \text{ kips})(48 \text{ in.})}{\pi(0.5 \text{ in.})^2(30 \times 10^3 \text{ ksi})} = 20.4(10^{-3}) \text{ in.} \quad \textbf{Ans.} \tag{4a}$$

and

$$\Delta d = d\epsilon_{\text{radial}} = -\frac{0.3(10 \text{ kips})(1 \text{ in.})}{\pi(0.5 \text{ in.})^2(30 \times 10^3 \text{ ksi})} \quad \textbf{Ans.} \tag{4b}$$

$$= -127(10^{-6}) \text{ in.}$$

Review the Solution The changes in length and diameter are quite small in comparison with the original length and the original diameter, respectively, as they should be. Δd should definitely be smaller than ΔL, and the signs should be different, which is the case. The extensometer in Fig. *2.8a* is capable of measuring both ΔL and Δd.

MDS2.7 **Poisson's Ratio** is a program for determining the transverse deformation of an axial-deformation member due to the Poisson's ratio effect.

2.7 SHEAR STRESS AND SHEAR STRAIN; SHEAR MODULUS

In the preceding sections of Chapter 2, you were introduced to stress and strain through a discussion of normal stress and extensional strain. We turn now to a discussion of *shear stress* and *shear strain,* which are used, respectively, to quantify the distribution of force acting tangent to a surface and the angle change produced by tangential forces.

Definition of Shear Stress. Referring to Fig. 2.27*a*, we define the **shear stress at a point** by the equation[19]

$$\tau = \lim_{\Delta A \to 0} \left(\frac{\Delta V}{\Delta A} \right) \qquad \textbf{Shear Stress} \qquad (2.17)$$

where ΔV is the *tangential (shear) force* acting on an infinitesimal area ΔA at the point where the shear stress is to be determined. As in the case of normal stress σ, the units of shear stress are force/area; hence, usually psi or ksi in the USCS units, and kPa or MPa in the SI system.

The *resultant shear force,* shown in Fig. 2.27*b*, is obtained by summing the ΔV's over the cross section, giving

$$\sum F: \qquad V = \int_A \tau \, dA \qquad (2.18)$$

Equations 2.17 and 2.18 will be used in Chapters 4 and 6, where we will determine the shear stress distribution for torsion of circular rods and bending of beams, respectively.

Average Shear Stress. Even when the exact shear stress distribution on a surface cannot be readily determined, it is sometimes useful to calculate the **average shear stress** on the surface. This is given by

$$\tau_{\text{avg}} = \frac{V}{A_s} \qquad \begin{array}{l} \textbf{Average} \\ \textbf{Shear} \\ \textbf{Stress} \end{array} \qquad (2.19)$$

where V is the total shear force acting on area A_s. In order to determine τ_{avg} we must first determine what area has shear stress acting on it, and then, using a free-body diagram, determine the value of the shear force, V, acting on this area.

Direct Shear. The average shear stress can be readily calculated in the case of *direct shear,* examples of which are shear in bolts, pins, and rivets, and shear in welds and lap splices.[20] **Direct shear** (or *simple shear*) is caused by forces that act parallel to a particular surface of some part, with the direct result of shearing, or tending to shear (i.e., sever), the material at that surface. For example, in the case of the

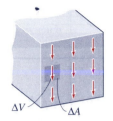

(*a*) The distribution of shear force on a sectioning plane.

(*b*) The resultant shear force on the sectioning plane.

FIGURE 2.27 Shear force on a sectioning plane.

---

[19]Here we assume that all ΔV's act in the same direction on the sectioning plane. More general cases are treated in Section 2.12.

[20]The actual shear stress distribution in these situations is quite nonuniform and would be very difficult to determine. Therefore, the average shear stress value is calculated, and an allowance is made, through factors of safety (see Section 2.8), for the approximate nature of the calculated average stress.

sheet-metal punch in Fig. 2.28*a*, the force *P* of the punch rod shears the sheet metal on a cylindrical surface, producing a coin-shaped metal slug (Fig. 2.28*b*). The shear stress for this example of direct shear is

$$\tau_{\text{avg}} = \frac{V}{A_s} = \frac{P}{(\pi d)t}$$

where $V = P$ is determined from the free-body diagram in Fig. 2.28*b*, A_s is the area of the cylindrical surface, t is the thickness of the sheet metal, and d is the diameter of the punch.

When the pliers in Fig. 2.28*c* are gripped, the pin that holds the two arms of the pliers together is subjected to direct shear on the pin cross section indicated in Fig. 2.28*d*.

Single Shear and Double Shear. Many of the circumstances that can be characterized as direct shear may be further classified as **single shear** or as **double shear.** This applies particularly to connections such as pinned, bolted, or welded joints. A **single-shear connection** is one where there is a single plane on which shear stress acts to transfer load from one member to the adjacent member. The pin of the pliers in Fig. 2.28*d* is one example of a single-shear connection.

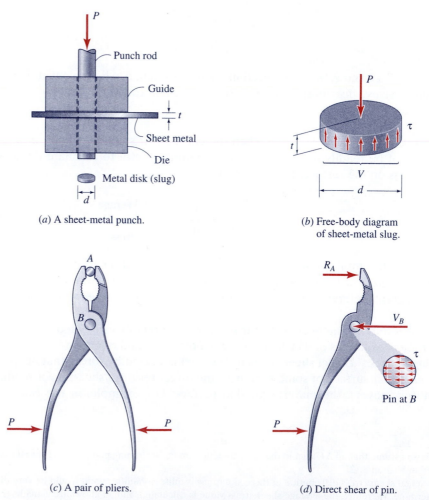

(*a*) A sheet-metal punch.

(*b*) Free-body diagram of sheet-metal slug.

(*c*) A pair of pliers.

(*d*) Direct shear of pin.

FIGURE 2.28   Examples of direct shear.

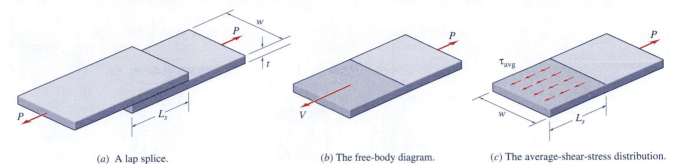

| (a) A lap splice. | (b) The free-body diagram. | (c) The average-shear-stress distribution. |

FIGURE 2.29   An illustration of direct shear—a lap splice.

As another example of single shear, consider the *lap joint,* or lap splice, in Fig. 2.29a, where two rectangular bars are glued together to form a tension member. (Assume that, because the bars are very thin, the moment Pt caused by misalignment of the tensile forces, P, may be neglected.) Applying $\Sigma F = 0$ to the free-body diagram in Fig. 2.29b, we get $V = P$. The area on which the shear V acts is $A_s = L_s w$. Therefore, from Eq. 2.19, the average shear stress on the splice area is

$$\tau_{\text{avg}} = \frac{V}{A_s} = \frac{P}{L_s w}$$

Although the transfer of load from one member to another through a riveted or bolted joint is not easily analyzed when there are several rows of bolts or rivets at the joint, the case of load transfer through a single bolt or a single pin can be treated as direct shear. The following example illustrates the calculation of direct shear stress in *single-shear* and *double-shear* joints.[21]

> **EXAMPLE 2.8**

In Fig. 1a, a bolted lap joint connects together two rectangular bars. In Fig. 1b, a pin passes through an eye at the end of a rod to connect the rod to a U-shaped support bracket. The diameter of the bolt is d_b and of the pin is d_p. If an axial load P is applied in each case, determine expressions for the average shear stress in the bolt on surface S_1 and the average shear stress in the pin on surfaces S_2 and S_3.

Plan the Solution   By drawing appropriate free-body diagrams, we can determine the shear force transmitted across S_1 from the upper rectangular bar to the bolt in Fig. 1a and the shear forces transmitted across surfaces S_2 and S_3 from the rod to the pin in Fig. 1b. Then we can apply Eq. 2.19, the formula for average shear stress.

---

[21]The book *Structures, or Why Things Don't Fall Down,* by J. E. Gordon [Ref. 2-7] contains an interesting discussion of joints. *Composites* [Ref. 2-5] contains a chapter on joints that discusses adhesively bonded lap joints.

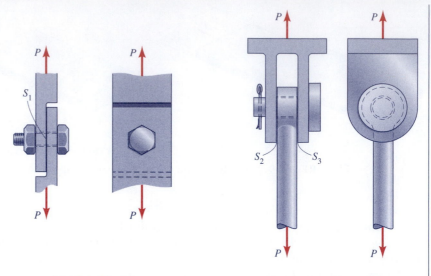

(a) A bolted lap joint.   (b) A bracket and hanger rod.

Fig. 1   Examples of single shear and double shear.

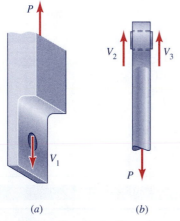

(a)   (b)

Fig. 2   Free-body diagrams.

Solution

Free-body Diagrams: Figures 2a,b show free-body diagrams for the single-shear joint and the double-shear joint, respectively.

Equilibrium Equations: We pass cutting planes forming S_1, S_2, and S_3, as shown on the free-body diagrams, and write the equation of equilibrium for each free body.

$$\left(\sum F\right)_{\text{bar}} = 0: \qquad\qquad V_1 = P$$

$$\left(\sum F\right)_{\text{rod}} = 0: \qquad\qquad V_2 + V_3 = p$$

Because of symmetry,

$$V_2 = V_3 = \frac{P}{2}$$

Therefore, for the bolt in Fig. 1a, the average shear stress is

$$(\tau_b)_{\text{avg}} = \frac{V_1}{A_b} = \frac{4P}{\pi d_b^2} \qquad\qquad \textbf{Ans.}$$

and, for the pin in Fig. 1b, the average shear stress is

$$(\tau_p)_{\text{avg}} = \frac{V_2}{A_p} = \frac{V_3}{A_p} = \frac{2P}{\pi d_p^2} \qquad\qquad \textbf{Ans.}$$

Review the Solution   The above answers account for the fact that a single-shear surface, S_1, transmits the force P across the lap splice, but that there are two shear surfaces to transmit the force from the rod to the bracket. The bolt in Fig. 1a is said to be in *single shear,* and the pin in Fig. 1b is said to be undergoing *double shear.*

In the preceding example of a bolted lap joint (Fig. 1*a* of Example 2.8), friction on the mated surfaces of the two rectangular bars was disregarded. If, in such bolted connections the nut is barely tightened, this assumption would certainly be valid. In any case, whatever friction might exist between two members that are joined by a bolt reduces the amount of force that must be transmitted across the joint by the bolt. Hence, it is wise to be conservative and just neglect such friction forces.

MDS2.8 **Hole Punch** is a program for determining the direct shear stress in a hole-punch situation like the one illustrated in Figs. 2.28*a* and 2.28*b*.

MDS2.9 & 2.10 **Bolted Connection** is a program for determining the direct shear stress in single shear and in double shear situations, like the ones analyzed in Example Problem 2.8.

MDS2.11 **Shear Key** is a program for determining the direct shear stress in a shear key.

Equilibrium Requirements for Shear Stresses; Pure Shear.

In our study of shear stress we have, so far, only considered the average shear stress on a particular surface. Before proceeding further to define shear strain and to discuss the distribution of shear stress under various loading conditions, we must examine the equilibrium requirements that must be satisfied by shear stresses.

As a typical shear-stress example, consider the thin plate-like member of deformable material shown in Fig. 2.30*a*, and let us examine the stresses that act on the six faces of the small (darker blue) elemental volume at point *A* when the upper part of the member is sheared to the right relative to the bottom. The lower part, after deformation, is depicted in Fig. 2.30*b*; and a free-body diagram of the elemental volume is shown in Fig. 2.30*c*. First, there is no stress, normal stress or shear stress, on the front ($+z$) face or the back ($-z$) face of the element, so we need only examine the stresses on the other four faces. In Fig. 2.30*b*, and on the free-body diagram, the shear force acting to the right on the top ($+y$) face of the element has been labeled ΔV_P. Let the average shear stress on the top face be τ_P. Then,

$$\Delta V_P = \tau_P \Delta A_y = \tau_P (t\,\Delta x)$$

To satisfy force equilibrium ($\Sigma F_x = 0$) for the free body in Fig. 2.30*c*, an equal force, ΔV_P, must act to the left on the bottom ($-y$) face. Since the top face and

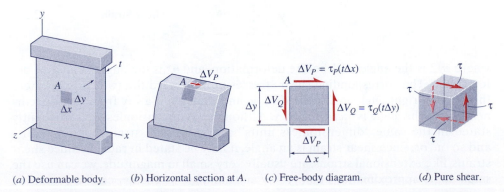

(*a*) Deformable body.   (*b*) Horizontal section at *A*.   (*c*) Free-body diagram.   (*d*) Pure shear.

FIGURE 2.30 Equilibrium requirements for shear stresses.

bottom face have the same area, the average shear stress on the bottom face is τ_P, the same as on the top face.

If the only forces on the free-body diagram in Fig. 2.30c were the equal and opposite ΔV_P forces on the top and bottom faces, moment equilibrium about an axis in the z direction, for example, at corner A, could not be satisfied. Therefore, in addition to the horizontal forces on the element there must also be equal and opposite vertical forces, which are labeled ΔV_Q on the free body. Let

$$\Delta V_Q = \tau_Q \Delta A_x = \tau_Q(t\,\Delta y)$$

So, to satisfy moment equilibrium, we get

$$\left(\sum M_z\right)_A = 0: \qquad (\tau_Q\, t\, \Delta y)\Delta x - (\tau_P\, t\, \Delta x)\Delta y = 0$$

Therefore,

$$\boxed{\tau_Q = \tau_P \equiv \tau} \tag{2.20}$$

The results of the above equilibrium analysis are summarized in Fig. 2.30d, which depicts an elemental volume in pure shear. We can conclude, therefore, that *the shear stresses on an element in pure shear satisfy the following statements:*

- The shear stresses on parallel faces are equal in magnitude and opposite in sense.
- On adjacent faces at right angles to each other, the shear stresses are equal in magnitude, and they must both point toward the intersection of the two faces or both point away from the intersection (i.e., the shear-stress arrows on an element must be "head-to-head" and "tail-to-tail").

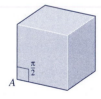

(a) Original (undeformed) element.

In Section 2.13 it is shown that these conclusions about shear stresses are valid even if normal stresses as well as shear stresses act on the faces of the element.

Shear Strain. Referring to Fig. 2.31 let us now consider the shear strain that is associated with shear stress. As a result of the shear stress τ, the original right angle at A becomes an acute angle θ^*. The **shear strain** γ (lowercase Greek letter gamma) at A is defined as the *change in angle* between two originally perpendicular line segments that intersect at A. Thus,

$$\boxed{\gamma = \frac{\pi}{2} - \theta^*} \qquad \textbf{Shear Strain} \tag{2.21}$$

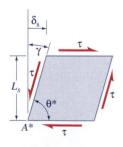

(b) Pure shear deformation.

FIGURE 2.31 Illustrations for a definition of shear strain.

where $\pi/2$ is the angle at A before deformation, and θ^* is the angle at A after deformation. The corresponding shear stresses point toward the two corners where the original right angle is *decreased* by γ, and they point away from the two corners where the angle is *increased* by γ. Although γ is dimensionless, it is frequently stated in the same "dimensionless units" as extensional strain, that is, in./in., and so on, or, since shear strain is an angle, it may be stated in radians. Since shear strains, like extensional strains, are usually very small in magnitude, we can use the small-angle approximations $\tan(\gamma) \approx \gamma$ and $\sin(\gamma) \approx \gamma$. Then, γ can be computed

$$\gamma = \frac{\pi}{2} - \theta^* \approx \tan\left(\frac{\pi}{2} - \theta^*\right) = \frac{\delta_s}{L_s} \tag{2.22}$$

where δ_s and L_s are defined in Fig. 2.31b.

Material Properties in Shear. Material properties relating to shear, like those for normal-stress-extensional-strain behavior, must be determined experimentally. The material properties in shear, such as yield stress in shear, shear modulus of elasticity, and so on, may be obtained from a torsion test, which will be discussed later in Section 4.4. For example, linearly elastic behavior in shear is described by **Hooke's Law for shear,** which

$$\tau = G\gamma \qquad \text{Hooke's Law} \qquad (2.23)$$
$$\text{for Shear}$$

The constant of proportionality, G, is called the *shear modulus of elasticity,* or, simply, the **shear modulus.** Like E, the shear modulus G is usually expressed in units of ksi or GPa. The shear properties are closely related to the extensional properties through equations of equilibrium and geometry of deformation. For example, it is shown in Section 2.11 that G, E, and ν are related through the equation

$$G = \frac{E}{2(1 + \nu)} \tag{2.24}$$

Values of the shear modulus for selected materials are given in Table F.2.

2.8 INTRODUCTION TO DESIGN—AXIAL LOADS AND DIRECT SHEAR

Design is the very heart of engineering. **Engineering design** *is the process of devising components, systems, or processes to meet the needs of society in the areas of housing, food production, transportation, communication, recreation, and many others.* To perform this important and fulfilling role, engineers must have an understanding of basic sciences, engineering sciences, manufacturing and construction methods, and economics; whether they are designing an automobile or an airplane, a bridge or a building, a microchip for a computer, or a chemical process.

Mechanics of materials topics are fundamental to the design of all objects that support or transmit loads, including bridges, buildings, storage tanks, machines, land vehicles, airplanes, etc., as well as the many individual components that make up each of these (beams and columns of a building, ribs and spars of an airplane wing, etc.). For simplicity, such objects will be referred to by the generic term *structures*.

Design of a structure *involves selecting a promising configuration for the structure and applying the principles and equations of deformable-body mechanics to select materials and dimensions of individual components so that no failure occurs under the prescribed loading conditions.*

In many cases, other factors such as weight, cost, or environmental impact must also be considered in the design of a structure or machine.

From the mechanics of materials standpoint, several possible **modes of failure** typically need to be considered. These can be grouped under three general headings:

design for strength, design for stiffness, and *design for ductility,* which are mentioned in Section 2.4. *Failure by yielding,* which is discussed in this section, and *fatigue failure,* which is discussed in Section 12.4, fall under the broad category of design for strength. Failure by *buckling,* which is the subject of Chapter 10, is related to the bending stiffness of compression members. In this chapter, we will restrict our attention primarily to strength design, with application to members that are subjected to axial loading or direct shear.

Factor of Safety. Suppose that your task as an engineer is to design the legs of the tension-leg offshore oil platform depicted in Fig. 2.32. The platform is basically a floating barge that is held in place by a number of large cables (tension legs) that are anchored to the ocean floor. In this design, as is typical of all designs, there are two factors to consider, which we will refer to as the **load** L, and the **resistance** R.[22] The structural member, in this case the tension leg, must have the capacity (i.e., *resistance*) R that exceeds the demand (i.e., *load*) L. That is, the design must satisfy the inequality

$$Resistance > Load \tag{2.25}$$

For the tension leg under consideration, the "load" L would be the maximum tensile force that could occur in the leg during the service life of the platform. Correspondingly, the "resistance" R would be the resisting force that a cable of a certain diameter made of a certain type of steel would be able to provide without exceeding its yield strength. There are many factors that enter into the determination of L and R, and all of these factors are subject to variability and uncertainty:

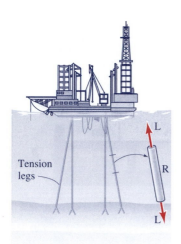

FIGURE 2.32 A tension-leg offshore oil platform.

- Some Uncertainties Affecting L:
  - Assumptions regarding external loads acting on the platform (location, magnitude, direction, etc.)
  - Assumptions and approximations made in computing estimates of L, the force in the cable (boundary conditions, linear behavior, etc.)
- Some Uncertainties Affecting R:
  - Assumptions made in calculating the strength of the cable
  - Manufacturing uncertainties (material properties, cable dimensions, tolerances, etc.)
  - Construction uncertainties (quality of workmanship, etc.)

In selecting member sizes and materials, the designer must account for uncertainties like those listed above and must ensure that the *failure load,* or resistance, R (the minimum value of the load required to cause failure of the member), is safely above the *allowable load,* L (the maximum load that the member is expected to see during its service lifetime). The inequality in Eq. 2.25 may be replaced by the following equation that can be used directly in design:

$$FS = \frac{\text{Failure load}}{\text{Allowable load}} \tag{2.26}$$

where FS is called the **factor of safety.** Of course, $FS > 1$.

---

[22]The term "load" here refers to some internal member force, not to an external load acting on the structure.

- Some Factors That Determine the Value of *FS:*
  - —Importance of failure of a single member to overall platform failure
  - —Accessibility of the member for inspection and repair
  - —Nature of possible failure(s) (e.g., "slow" ductile failure, or fast" brittle failure)
  - —"Cost" (economic and/or social) of failure.

The normal range of values for the factor of safety is from about 1.3 to 3.0, although a value as high as ten might occasionally be applied. The use of a small value of factor of safety (e.g., $FS = 1.1$) is justified only when it is possible, by analysis and testing, to sufficiently minimize uncertainties, and when there is no likelihood that failure will result in unacceptable circumstances such as serious personal injury or death. On the other hand, it is undesirable to use a factor of safety that is unnecessarily large (e.g., $FS > 3$), since that would lead to excess structural weight, which, in turn, entails excess initial costs and operating costs. Since the choice of a value of factor of safety has such important economic and legal implications, design specifications, including the relevant factor(s) of safety to be used, conform to design codes or other standards developed by groups of experienced engineers in engineering societies or in various government agencies. Examples of such codes and specifications are: (1) for steel: *Code of Standard Practice for Steel Buildings and Bridges,* by the American Institute of Steel Construction, and (2) for concrete: *Building Code Requirement for Structural Concrete (ACI 318-08),* by the American Concrete Institute.

Allowable-Stress Design. If there is a linear relationship between the loads on a structure and the stresses caused by the loads, it is permissible to define the factor of safety as the ratio of two stresses, the *failure stress* and the *allowable stress.* It is convenient to write this relationship in the form

$$\text{Allowable stress} = \frac{\text{Yield strength}}{FS} \qquad (2.27)$$

For axial deformation, the tensile (or compressive) **yield strength** σ_Y is taken as the stress corresponding to failure by yielding; in direct shear, the **shear yield strength** τ_Y is used.[23]

$$\sigma_{\text{allow}} = \frac{\sigma_Y}{FS}, \quad \text{or } \tau_{\text{allow}} = \frac{\tau_Y}{FS} \qquad (2.28)$$

For simple loading situations like axial loading and direct shear, Eq. 2.28 can also be used to define a factor of safety with respect to *ultimate failure* by using the **ultimate strength** σ_U (or τ_U) as the failure stress.

Design based on Eq. 2.28 is referred to as **allowable-stress design (ASD).** Equation 2.28 is the design equation that will be used most frequently in this text, but Eq. 2.26 is required for the design of columns (Section 10.7). Design based on either Eq. 2.26 or Eq. 2.28 may be referred to as **factor-of-safety design (FSD).** In either case, a single factor of safety is used to incorporate all of the uncertainty

---

[23]Section 12.3 discusses *Failure Theories* for more complex loadings.

related to the loads and all of the uncertainty associated with the structure's ability to resist the loads.[24]

There are two ways in which the above design information is used in practice:

1. *Evaluation of an Existing Structure or a Proposed Design:* The configuration of a structure is known, together with the sizes of all of its components, the materials used, the applicable code-specified *FS*, etc., and the **allowable load** is to be determined.

2. *Design of a New Structure:* The configuration of a structure is given, but the **sizes of components** (tension rods, pins, beams, etc.) are to be chosen so that maximum stresses in the components do not exceed the allowable stress (Eq. 2.28), or, if loads are to be considered, the maximum applied load does not exceed the allowable load (Eq. 2.26).

Example 2.9 illustrates the calculation of allowable load; Example 2.10 illustrates the sizing of components based on allowable-stress design.

EXAMPLE 2.9

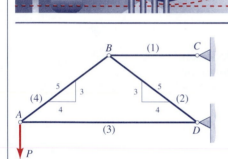

Fig. 1   A pin-jointed truss.

The pin-jointed planar truss in Fig. 1 is subjected to a single downward force P at joint A. All members have a cross-sectional area of 500 mm². The allowable stress in tension is $(\sigma_T)_{\text{allow}} = 300$ MPa, while the allowable stress (magnitude) in compression is $(\sigma_C)_{\text{allow}} = 200$ MPa.[25] Determine the *allowable load,* P_{allow}.

Plan the Solution   Equilibrium (e.g., the "method of sections") can be used to determine all member forces in terms of the load P. The stress in the member with the largest tensile axial force and the stress in the member with the largest compressive axial force should be set equal to $(\sigma_T)_{\text{allow}}$ and $(\sigma_C)_{\text{allow}}$, respectively, to determine values of P_{allow} based on each. The <u>smaller</u> of the two values governs.

Solution

Equilibrium: The results of an equilibrium analysis are shown in Fig. 2. (The reader should verify the values shown in Fig. 2.)

[24]In recent years, versions of *probability-based design* have been adopted by major standards organizations, for example, the *Load and Resistance Factor Design* (*LRFD*) method adopted in 1986 by the American Institute of Steel Construction [Ref. 2-8]. Such methods provide a more rigorous accounting for the uncertainties that affect L and R individually, resulting in separate *load factors* and *resistance factors*. Some typical references are: *Probabilistic Methods in Structural Engineering,* [Ref. 2-9]; *Methods of Structural Safety,* [Ref. 2-10]; and *Introduction to Reliability Engineering,* [Ref. 2-11]. Such methods are beyond the scope of this text.

[25]The magnitude of the compressive allowable stress is given. The allowable stress in compression is frequently governed by elastic buckling or inelastic buckling, topics that are discussed in Chapter 10.

Allowable Force Based on Member BC: Members AB and BC are in tension; member BC has the larger tensile force. Therefore,

$$\frac{8}{3}\frac{P_1}{A_1} = (\sigma_T)_{\text{allow}} = 300 \text{ MPa}$$

$$P_1 = \frac{3}{8}\left(300 \times 10^6 \frac{\text{N}}{\text{m}^2}\right)(500 \times 10^{-6} \text{ m}^2) = 56.25 \text{ kN}$$

Allowable Force Based on Member BD:

$$\frac{5}{3}\frac{P_2}{A_2} = (\sigma_C)_{\text{allow}} = 200 \text{ MPa}$$

$$P_2 = \frac{3}{5}\left(200 \times 10^6 \frac{\text{N}}{\text{m}^2}\right)(500 \times 10^{-6} \text{ m}^2) = 60.00 \text{ kN}$$

Since P_1, the force based on the tension allowable, is smaller than P_2, the allowable force is

$$P_{\text{allow}} = 56.2 \text{ kN} \qquad\qquad \textbf{Ans.}$$

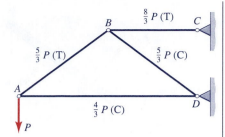

Fig. 2 Member forces in a pin-jointed truss.

EXAMPLE 2.10

The shop crane in Fig. 1 consists of a boom AC that is supported by a pin at A and a rectangular tension bar BD. Details of the pin joints at A and B are shown in Views $a - a$ and $b - b$, respectively. The tension bar is to be made of structural steel with $\sigma_Y = 36$ ksi, while the pins at A and B are to be of high-strength steel with $\tau_Y = 48$ ksi.

The design (i.e., allowable) load is $P = 5$ kips, and there is to be a factor of safety with respect to yielding of $FS = 3.0$. (a) If the width of the bar BD is $w = 2$ in., determine the required thickness, t, to the nearest 1/8 in. (b) Determine the required pin diameters at A and B to the nearest 1/8 in.

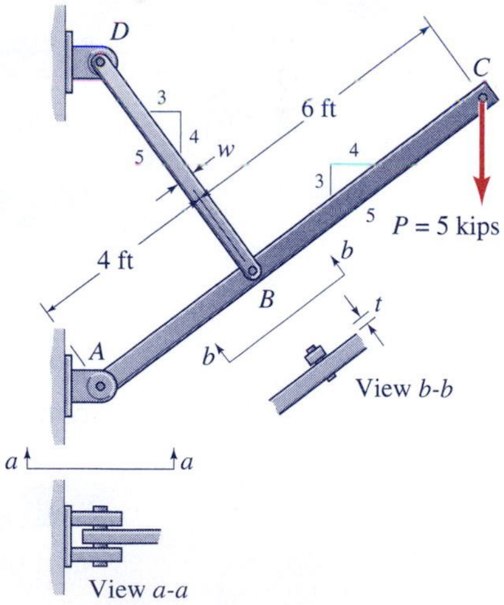

Plan the Solution Equilibrium of boom AC can be used to determine the tension F_B in two-force member BD and the resultant force F_A on the pin at A. Then we can use allowable-stress design (Eq. 2.28) to determine the cross-sectional area of bar BD and the required pin diameters, noting that the pin at A is in *double shear* and the pin at B is in *single shear*.

Fig. 1 A shop crane.

Solution

Equilibrium: Equilibrium equations were used to solve for the forces F_A and F_B that are shown on the free-body diagram in Fig. 2.

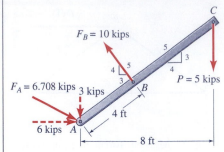

Fig. 2 Free-body diagram of boom AC.

(a) *Design of Bar BD:* From Eq. 2.28,

$$\sigma_{\text{allow}} = \frac{\sigma_\gamma}{FS} \rightarrow \sigma_{\text{allow}} = \frac{36\ \text{ksi}}{3.0} = 12\ \text{ksi}$$

$$F_{BD} = \sigma_{\text{allow}} A_{BD} \rightarrow A_{BD} = (2\ \text{in.}) t_{BD} = \frac{10\ \text{kips}}{12\ \text{ksi}}$$

$$t_{BD} = 0.417\ \text{in.} \qquad \text{Select } t_{BD} = 0.5\ \text{in.} \qquad \textbf{Ans. (a)}$$

(b) *Design of Pins at A and B:* From Eq. 2.28,

$$\tau_{\text{allow}} = \frac{\tau_Y}{FS} \rightarrow \tau_{\text{allow}} = \frac{48\ \text{ksi}}{3.0} = 16\ \text{ksi}$$

$$\frac{1}{2} F_A = \tau_{\text{allow}} A_A \rightarrow A_A = \frac{6.708\ \text{kips}}{2(16\ \text{ksi})} = 0.210\ \text{in}^2$$

$$A_A = \frac{\pi d_A^2}{4} \rightarrow d_A = 0.517\ \text{in.} \qquad \text{Select } d_A = 0.625\ \text{in.} \quad \textbf{Ans. (b)}$$

$$F_B = \tau_{\text{allow}} A_B \rightarrow A_B = \frac{10\ \text{kips}}{16\ \text{ksi}} = 0.625\ \text{in}^2$$

$$A_B = \frac{\pi d_B^2}{4} \rightarrow d_B = 0.892\ \text{in.} \qquad \text{Select } d_B = 1.0\ \text{in.} \qquad \textbf{Ans. (b)}$$

Review the Solution These calculations are very straightforward, but should be double-checked, especially to make sure that the *FS* has been properly applied. The answers seem to be "reasonable" numbers.

MDS2.12 **Beam and Strut**—a program for designing axial-load members and bolt/pin connections, like the ones analyzed in Example Problem 2.10.

MDS2.13 **Two-Bar Assembly**—a program for designing members of a two-bar truss or determining the allowable load on a given two-bar truss.

MDS2.14 **Bar and Pin**—a program for designing axial-load members and bolt/pin connections, like the ones analyzed in Example Problem 2.10.

MDS2.15 **Beam and Strut Design**—a program for designing axial-load members and bolt/pin connections, like the ones analyzed in Example Problem 2.10.

Optimal Design. In Example Problem 2.10 a straightforward application of allowable-stress design led to the proper sizes for tension bar *BD* and for the pins at *A* and *B*. In many design problems, however, there is not a unique design solution. There may be many *acceptable solutions,* so, from these acceptable solutions, the engineer would like to pick the "best" solution, that is, the **optimal solution.** Example Problem 2.11 illustrates such an optimal-design problem, where the *optimal solution is defined to be the minimum-weight solution among all solutions for which no member has a stress that exceeds the allowable stress.*

The pin-jointed planar truss shown in Fig. 1a is to be made of three steel two-force members and support a single vertical load $P = 1.2$ kips at joint B. The locations of joints A and B are fixed, but the vertical position of joint C can be changed by varying the lengths L_1 and L_3. For the steel truss members, the allowable stress in tension is $(\sigma_T)_{\text{allow}} = 20$ ksi, the allowable stress in compression is $(\sigma_C)_{\text{allow}} = 12$ ksi, and the weight density is 0.284 lb/in^3. You are to consider truss designs for which the vertical member AC has lengths varying from $L_1 = 18$ in. to $L_1 = 50$ in. (a) Show that, if each member has the minimum cross sectional area that meets the strength criteria stated above, the weight W of the truss can be expressed as a function of the length L_1 of member AC by the function that is plotted in Fig. 1b. (b) What value of L_1 gives the minimum-weight truss, and what is the weight of that truss?

Plan the Solution   The member forces F_1, F_2, and F_3 can be related to the applied force P by writing equilibrium equations for joints B and C. The length L_1, which determines the geometry of the triangle ABC, will enter into the expressions for these forces. Next we can use *allowable-stress design* to determine expressions for the cross-sectional areas of

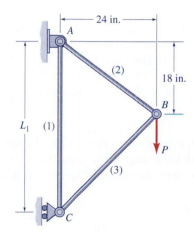

(a) A three-bar planar truss.

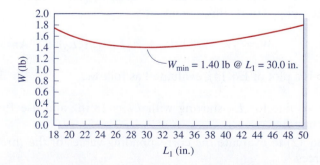

(b) Plot of truss weight vs length of member (1).

Fig. 1   Minimum-weight design for the three-bar planar truss.

the bars. Finally, we can write an expression for the total weight of the truss as the sum of the weights of its three members, an expression that will contain the *design variable* L_1; then evaluate this expression over the specified range 18 in. $\leq L_1 \leq$ 50 in.

Solution

Truss Geometry: The length L_2 is constant, $L_2 = 30$ in., and, from Fig. 1, the length L_3 is given by

$$L_3 = \sqrt{(24 \text{ in.})^2 + (L_1 - 18 \text{ in.})^2} \qquad (1)$$

Equilibrium:

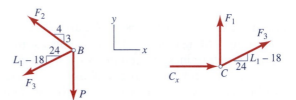

Fig. 2 Free-body diagrams.

By drawing free-body diagrams of joints B and C and writing equilibrium equations, we can show that the following formulas relate member forces F_1, F_2, and F_3 to the load P:

$$F_1 = \frac{(L_1 - 18 \text{ in.})P}{L_1}, \quad F_2 = \frac{(30 \text{ in.})P}{L_1}, \quad F_3 = -\frac{L_3 P}{L_1} \qquad (2)$$

Allowable-Stress Design Constraints: The allowable-stress criterion, Eq. 2.28, leads to the following three equations for the minimum areas that are required to carry the member forces given in Eqs. (2):

$$A_1 = \frac{F_1}{(\sigma_T)_{\text{allow}}}, \quad A_2 = \frac{F_2}{(\sigma_T)_{\text{allow}}}, \quad A_3 = \frac{|F_3|}{(\sigma_C)_{\text{allow}}} \qquad (3)$$

Minimum-Weight Optimal Design: The total weight of the truss is the sum of the weights of its three members:

$$W = \gamma V = \gamma(A_1 L_1 + A_2 L_2 + A_3 L_3) \qquad \textbf{Ans.} \qquad (4)$$

Figure 1b is a plot of Eq. (4), evaluated as follows:

1. Select a value for L_1, starting with $L_1 = 18$ in., and use Eq. (1) to determine L_3.
2. Use Eq. (2) to evaluate the corresponding values of the three member forces.
3. Use Eq. (3) to determine the resulting cross-sectional areas required to meet the stated allowable-stress criteria.

4. Finally, use Eq. (4) to evaluate the weight of the truss that corresponds to the given value of L_1.

5. Repeat Steps (1) through (4) for about fifty values of L_1 in the given range 18 in. $\leq L_1 \leq$ 50 in. These values of the function $W(L_1)$ are plotted in Fig. 1b.

A spreadsheet computer program was used to carry out the computations in the above optimal-design solution and to plot the curve of the *objective function* (weight W) versus the *design variable* (length L_1). From Fig. 1b, the *optimal design* (i.e., the minimum-weight design) is the design for which

$$L_1 = 30.0 \text{ in.} \qquad \textbf{Ans.}$$

and the corresponding minimum weight is

$$W_{\min} = 1.40 \text{ lb} \qquad \textbf{Ans.}$$

Review the Solution   From the plot of the weight function, $W(L_1)$, we can see how the weight depends on the configuration of the truss. The optimum truss configuration is close to that shown in Fig. 1. With $L_1 = 30$ in., the tension in member (2) and the compression in member (3) each has a significant vertical component that acts to support load P; yet, the lengths L_1 and L_3 are not so long as to make the truss excessively heavy.

2.9   STRESSES ON AN INCLINED PLANE IN AN AXIALLY LOADED MEMBER

Figure 2.5, illustrating the deformation of an axially loaded bar, clearly shows that there are shear strains in the bar. For example, the right angle DEF in Fig. 2.5a becomes the acute angle $D^*E^*F^*$ in Fig. 2.5b. Let us now consider how the normal stress and the shear stress on an oblique plane, such as the plane whose normal direction is labeled n in Fig. 2.33, are related to the axial stress, $\sigma_x = P/A$.

To satisfy equilibrium of the free body in Fig. 2.33b, we get

$$\nearrow \sum F_n = 0: \qquad\qquad N = P \cos\theta$$
$$\nwarrow \sum F_t = 0: \qquad\qquad V = -P \sin\theta$$

where θ is the angle measured *positive counterclockwise* from the x axis (normal to the cross section) to the n axis (normal to the oblique plane). The area of the

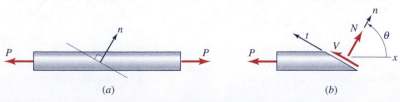

(a)   (b)

FIGURE 2.33   The force resultants on an oblique section through an axial tension member.

oblique plane is related to the cross-sectional area of the bar, A, by

$$A_n = \frac{A}{\cos \theta}$$

The average normal stress, σ_n, and the average shear stress, τ_{nt}, on the oblique plane are obtained by dividing N and V, respectively, by the area A_n on which they act:

$$\sigma_n = \frac{N}{A_n} = \sigma_x \cos^2 \theta$$

$$\tau_{nt} = \frac{V}{A_n} = -\sigma_x \cos \theta \sin \theta$$

(2.29)

where *the subscript n in σ_n and τ_{nt} designates the outward normal to the face on which these stresses act. The subscript t identifies τ_{nt} as the shear stress acting in the $+t$ direction on the $+n$ face.* (Shear stress τ_{nt} is negative in Eq. 2.29b, since the shear force actually acts in the $-t$ direction on the $+n$ face.)

Using trigonometric identities, we can write σ_n and τ_{nt} as functions of the double-angle 2θ. The resulting equations:

$$\sigma_n = (\sigma_x/2)(1 + \cos 2\theta)$$

$$\tau_{nt} = -(\sigma_x/2)\sin 2\theta$$

(2.30)

show that σ_n and τ_{nt} are periodic in θ with a period of 180°. These expressions are plotted in Fig. 2.34. For example, consider an element rotated at $\theta = 45°$ with respect to the x axis, as shown in Fig. 2.35. The subscripts 1 and 2 refer to the points designated 1 and 2 on Fig. 2.34 and to faces designated n_1 and n_2 in Fig. 2.35. The shear stresses on the 45°-rotated element in Fig. 2.35 account for the shear deformation of the 45°-rotated element at point E in Fig. 2.5.

Figure 2.35 illustrates the fact that shear stresses of magnitude $\sigma_x/2$ occur on the planes oriented at 45° to the axis of a member undergoing axial deformation with axial stress σ_x. One evidence of this shear can be obtained by performing a tensile test of a low-carbon steel bar with polished surfaces. When the bar is loaded to the yield point, *slip bands* can be observed to form at approximately 45° to the

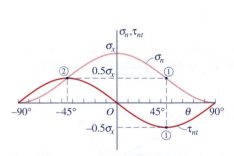

FIGURE 2.34 The normal stress and shear stress on arbitrary oblique planes.

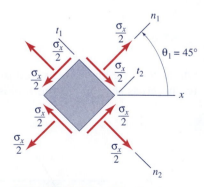

FIGURE 2.35 The stresses on a 45°-rotated element.

axis of the bar. These slip bands are called *Lüders' bands* or *Piobert's bands*.[26] Mild steel and other highly ductile materials exhibit this type of failure in shear, rather than a direct tensile failure.

Chapter 8 treats more general cases in which equilibrium is again used to relate the stresses on various planes to each other.

EXAMPLE 2.12

Several short pieces of timber are to be glued together end-to-end to form a single longer piece of timber. Figure 1 shows a simple diagonal splice joint. The glue that is to be used in the splice joint is 50% stronger in shear than in tension. Is it possible to take advantage of this higher shear strength by selecting a splice angle θ such that the magnitude of the average shear stress on the joint is 50% higher than the average normal stress? If so, what is the appropriate angle?

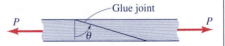

Fig. 1 A diagonal splice joint.

Plan the Solution The stresses on an oblique plane are given by Eqs. 2.29 and plotted in Fig. 2.34. From Fig. 2.34 it appears that the answer is, Yes, the cut should be somewhere between 50° and 60° (or between −50° and −60°).

Solution We want to determine a splice angle θ_s such that $|\tau_{nt}(\theta_s)| = 1.5\sigma_n(\theta_s)$ Therefore, from Eqs. 2.29 we have

$$1.5(\sigma_x \cos^2 \theta_s) = \pm \sigma_x \cos \theta_s \sin \theta_s$$

or

$$\tan \theta_s = \pm 1.5$$

Therefore, the shear stress exceeds the normal stress by 50% on planes oriented at

$$\theta_s = \pm 56.3° \qquad \textbf{Ans.}$$

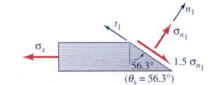

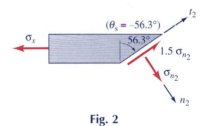

Review the Solution When $\theta = 0$ we get no shear stress on the splice joint. Therefore, a "long" splice, like one at ±56.3°, makes sense as a splice on which shear stress predominates over normal stress. The stresses for these two cases are illustrated in Fig. 2.

Fig. 2

2.10   SAINT-VENANT'S PRINCIPLE

Up to this point we have treated the distribution of normal stress σ_x on a cross section of a uniform member undergoing axial deformation as being uniform across the cross section. However, near points of application of load, plane sections do

---

[26]Such slip bands were first observed by G. Piobert in 1842 and then by W. Lüders in 1860.

not remain plane, and the normal stress is not uniform, as is the case depicted in Fig. 2.2c.

In Section 12.2 we will take up the topic of *stress concentration,* the increase in stress caused by abrupt changes in cross section, holes, and so on. Here, however, let us briefly examine the stress distribution near points of application of concentrated loads. Consider the short compression bar, with cross-sectional area $A = bt$, shown in Fig. 2.36.[27]

From Fig. 2.36 we can make the following three observation:

- The *average stress* is the same on all cross sections, namely $\sigma = -P/A$.
- Near the ends of the bar, where the concentrated load is applied (e.g., Fig. 2.36b), there is a *stress concentration,* with higher stress near the point of application of the load.
- At distances from the point of application of the load that are greater than the width of the compression bar (e.g., Fig. 2.36d), the stress distribution is essentially uniform.

The third observation above is referred to as **Saint-Venant's Principle**.[28] The significance of the principle can be stated as follows:

> The stresses and strains in a body at points that are sufficiently remote from points of application of load depend only on the static resultant of the loads and not on the distribution of the loads.

Thus, the stress distribution in Fig. 2.36d would not be greatly altered if the single load P at the bottom end of the bar in Fig. 2.36a were to be replaced by two symmetrically placed loads of magnitude $P/2$, shown as dashed arrows in the figure.

Throughout the remainder of this text on mechanics of deformable bodies we will obtain expressions for stress distributions in and deformations of various members under various types of loading. On the basis of Saint-Venant's Principle we can say that the expressions we derive are valid <u>except very near to points of loading or support,</u> or near to an abrupt change in cross section.

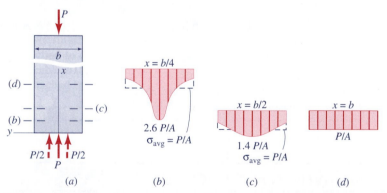

FIGURE 2.36   The effect of a concentrated load on the distribution of normal stress.

---

[27]Figure 2.36 is adapted from information in Section 24 of *Theory of Elasticity,* 3rd ed., [Ref. 2-12].
[28]Barré de Saint-Venant (1797–1886) is credited with many outstanding contributions to the theory of elasticity, especially his theories for torsion and bending of prismatic bars with various cross-sectional shapes. [Ref. 2-1].

2.11 HOOKE'S LAW FOR PLANE STRESS; THE RELATIONSHIP BETWEEN E AND G

For a *uniaxial stress state*, σ_x, the linear relationship between stress and strain was given in Section 2.6 by Hooke's Law, Eq. 2.14:

$$\epsilon_x = \frac{\sigma_x}{E} \qquad \text{(2.14)} \atop \text{repeated}$$

Poisson's ratio, Eq. 2.16, relates the transverse strain, ϵ_y, to ϵ_x:

$$\epsilon_y = \epsilon_z = -\nu\epsilon_x = -\nu\frac{\sigma_x}{E} \qquad \text{(2.16)} \atop \text{repeated}$$

In Section 2.7 it was noted that, for linearly elastic materials, the shear modulus G and Young's modulus E are related by the equation

$$G = \frac{E}{2(1+\nu)} \qquad \text{(2.24)} \atop \text{repeated}$$

For this equation to apply, the material must not only be linearly elastic, it must also be **isotropic**, that is, *its material properties like E and ν must be independent of orientation in the body.* We will first extend the stress-strain laws for linearly elastic materials to the two-dimensional loading case of plane stress. Then we will derive Eq. 2.24.

Plane Stress. *A body that is subjected to a two-dimensional state of stress with $\sigma_z = \tau_{xz} = \tau_{yz} = 0$, is said to be in a state of* **plane stress.** An element in plane stress is shown in Fig. 2.37.

If the material of which the body is composed is linearly elastic and isotropic, the effects of stresses σ_x, σ_y, and τ_{xy} can be superposed, giving *Hooke's Law for*

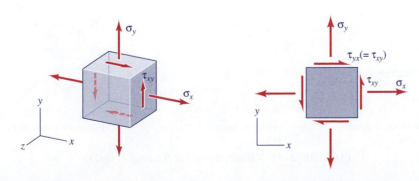

(a) Three-dimensional view.   (b) Two-dimensional view.

FIGURE 2.37 A state of plane stress depicted in 3-D and in 2-D.

$$\epsilon_x = \frac{1}{E}(\sigma_x - \nu\sigma_y)$$

$$\epsilon_y = \frac{1}{E}(\sigma_y - \nu\sigma_x)$$

$$\gamma_{xy} = \frac{1}{G}\tau_{xy}$$

**Hooke's Law
for
Plane Stress**　　　　(2.31)

It is because the material is isotropic, as well as linearly elastic, that E, G, and ν are independent of the orientation of the x and y axes and that the effects of σ_x and σ_y can be superposed in this manner.

The Relationship Between *E* and *G*.

By considering the case of *pure shear*, which is a special case of plane stress, and employing equations of equilibrium, geometry of deformation, and isotropic material behavior, we will now derive Eq. 2.24.[29] Consider a square plate of unit thickness subjected to pure shear relative to the x, y axes, as shown in Fig. 2.38*a*, and let the n and t axes be the diagonals of the square.

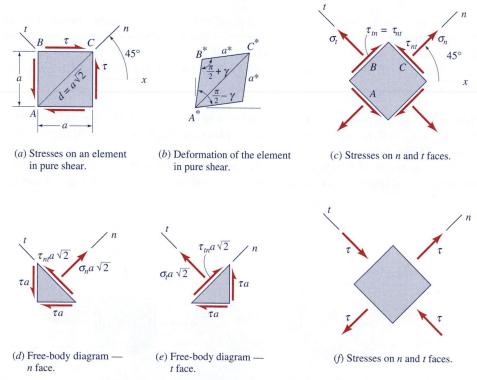

(*a*) Stresses on an element in pure shear.

(*b*) Deformation of the element in pure shear.

(*c*) Stresses on n and t faces.

(*d*) Free-body diagram — n face.

(*e*) Free-body diagram — t face.

(*f*) Stresses on n and t faces.

FIGURE 2.38　Illustrations for relating E and G.

---

[29]The relationship in Eq. 2.24 is a general one that applies to any stress state in an isotropic, linearly elastic body. It is most easily demonstrated by use of the case of *pure shear*, as is done here. Material properties of linearly elastic, isotropic bodies are discussed further in Section 2.13.

By considering the shear stresses on the faces of an n, t element and relating them to the elongation of the diagonal AC (i.e., the elongation in the n direction), we can derive Eq. 2.24.

Equilibrium: From the free-body diagram in Fig. 2.38d,

$$+\nearrow \sum F_n = 0: \qquad \sigma_n a\sqrt{2} - 2\tau a(\sqrt{2}/2) = 0, \quad \sigma_n = \tau$$
$$+\nwarrow \sum F_t = 0: \qquad \tau_{nt} a\sqrt{2} - \tau a(\sqrt{2}/2) + \tau a(\sqrt{2}/2) = 0, \quad \tau_{nt} = 0$$

and from the free-body diagram in Fig. 2.38e,

$$+\nearrow \sum F_n = 0: \qquad \tau_{tn} a\sqrt{2} - \tau a(\sqrt{2}/2) + \tau a(\sqrt{2}/2 = 0, \quad \tau_{tn} = 0$$
$$+\nwarrow \sum F_t = 0: \qquad \sigma_t a\sqrt{2} + 2\tau a(\sqrt{2}/2) = 0, \quad \sigma_t = -\tau$$

Material Behavior: From *Hooke's Law for Plane Stress* in Eq. 2.31 we have the following:

1. Relating the shear stresses shown in Fig. 2.38a to the shear strain shown in Fig. 2.38b is the equation

$$\tau_{xy} = G\gamma_{xy}, \quad \text{or} \quad \tau = G\gamma \tag{a}$$

2. Relating the normal stresses $\sigma_x = \sigma_y = 0$ in Fig. 2.38a to the extensional strains in Fig. 2.38b, we get

$$\epsilon_x = \epsilon_y = 0, \quad \text{so} \quad a^* = a$$

3. Relating the normal stresses, $\sigma_n = \tau$ and $\sigma_t = -\tau$, in Fig. 2.38f to the extensional strain ϵ_n along diagonal A^*C^* in Fig. 2.38b, we have

$$\epsilon_n = \frac{1}{E}(\sigma_n - \nu\sigma_t) = \frac{\tau}{E}(1 + \nu) \tag{b}$$

Geometry of Deformation: The law of cosines can be applied to the triangle $A^*B^*C^*$ in Fig. 2.38b to give

$$\overline{A^*C^*}^2 = \overline{A^*B^*}^2 + \overline{B^*C^*}^2 - 2\overline{A^*B^*}\,\overline{B^*C^*}\cos(\angle B^*)$$

or

$$[a\sqrt{2}(1 + \epsilon_n)]^2 = a^2 + a^2 - 2a^2 \cos\left(\frac{\pi}{2} + \gamma\right)$$

$$= 2a^2(1 + \sin\gamma)$$

$$1 + 2\epsilon_n + \epsilon_n^2 = 1 + \sin\gamma$$

For small ϵ_n and γ, the ϵ_n^2 term can be dropped, and $\sin\gamma$ can be approximated by γ. Then,

$$\epsilon_n = \frac{\gamma}{2} \tag{c}$$

Combining Eqs. (a), (b), and (c), we get the desired result, Eq. 2.24:

$$G = \frac{\tau}{\gamma} = \frac{E}{2(1 + \nu)}$$

(2.24)
repeated

2.12 GENERAL DEFINITIONS OF STRESS AND STRAIN

As has been illustrated in Section 2.9, the values of normal stress and shear stress depend on the orientation of the plane on which the stresses act; they may also depend on the point in the plane where the stresses are to be obtained. Therefore, we extend our previous definitions of normal stress (Eq. 2.1) and shear stress (Eq. 2.17) and give definitions for *normal stress and shear stress at an arbitrary point in an arbitrarily oriented plane* passing through an arbitrarily loaded three-dimensional body. The stresses on a plane are then related to the stress resultants on the plane.

Definitions of Normal Stress and Shear Stress.

To define the normal stress and shear stress at an arbitrary point P in a body, on the plane that has an outward normal vector $\boldsymbol{n}$ and that passes through point P, we start with the force vector $\Delta\boldsymbol{R}(P)$ at point P acting on an infinitesimal area ΔA of the n plane, as shown in Fig. 2.39. In Fig. 2.40*a* the force vector $\Delta\boldsymbol{R}(P)$, whose magnitude is denoted by $\Delta R(P)$,

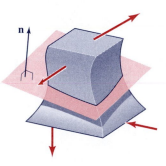

(*a*) A 3-D body with cutting plane.

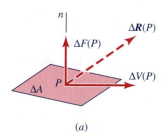

(*a*)

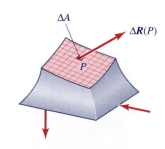

(*b*) The force on an infinitesimal area ΔA at point P in plane n.

FIGURE 2.39 The cutting plane whose normal vector is $\boldsymbol{n}$ and which passes through a given point P in a 3-D body.

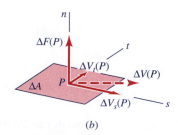

(*b*)

FIGURE 2.40 The components of force on infinitesimal area ΔA.

is resolved into normal and tangential components $\Delta F(P)$ and $\Delta V(P)$, respectively. In Fig. 2.40b the tangential component is further resolved into components along two orthogonal directions, s and t, in the n plane. Then, the **normal stress** σ_n and the **shear stresses** τ_{ns} and τ_{nt} at point P on the plane whose outward normal is n are defined by the expressions

$$\sigma_n(P) = \lim_{\Delta A \to 0} \left(\frac{\Delta F(P)}{\Delta A} \right), P \text{ always in } \Delta A$$

$$\tau_{ns}(P) = \lim_{\Delta A \to 0} \left(\frac{\Delta V_s(P)}{\Delta A} \right), P \text{ always in } \Delta A$$

$$\tau_{nt}(P) = \lim_{\Delta A \to 0} \left(\frac{\Delta V_t(P)}{\Delta A} \right), P \text{ always in } \Delta A$$

Stresses at a Point (2.32)

Sign Convention: The subscript n in σ_n, τ_{ns}, and τ_{nt} designates the direction of the outward normal to the face on which these stresses act. The shear stress τ_{ns} is the shear stress acting in the $+s$ direction on the $+n$ face and also the shear stress acting in the $-s$ direction on the $-n$ face. (If τ_{ns} is negative, it is a shear stress that acts in the $-s$ direction on the $+n$ face, etc.) The same shear stress sign convention applies to τ_{nt}, with, of course, t substituted for s.

Stress Resultants. Equations 2.32 are the basis for very important equations that relate the stresses on an area to the stress resultants on that area. Consider the stresses on a cross section of a slender member, as shown in Fig. 2.41. Here the subscripts xyz correspond to nst in Eqs. 2.32. The stresses σ_x, τ_{xy}, and τ_{xz} on a small element of area ΔA at point (x, y, z) give rise to forces $\Delta F = \sigma_x \Delta A$, $\Delta V_y = \tau_{xy} \Delta A$, and $\Delta V_z = \tau_{xz} \Delta A$. These forces, in turn, contribute to the force and moment resultants shown in Fig. 2.41b, which are related to the stresses by the following integrals over the cross section:

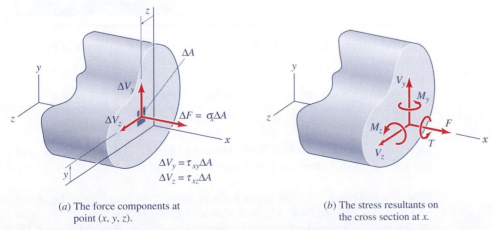

(a) The force components at point (x, y, z).

(b) The stress resultants on the cross section at x.

FIGURE 2.41   The stresses and stress resultants on a cross section.

Force Resultants:

$\sum F_x$:

$\sum F_y$

$\sum F_z$:

$$F(x) = \int_A \sigma_x \, dA$$

$$V_y(x) = \int_A \tau_{xy} \, dA$$

$$V_z(x) = \int_A \tau_{xz} \, dA$$

Force Resultants   (2.33)

Moment Resultants:

$\sum M_x$:

$\sum M_y$:

$\sum M_z$:

$$T(x) = \int_A y\tau_{xz} \, dA - \int_A z\tau_{xy} \, dA$$

$$M_y(x) = \int_A z\sigma_x \, dA$$

$$M_z(x) = -\int_A y\sigma_x \, dA$$

Moment Resultants   (2.34)

In Eqs. 2.33, F is the **normal force** on the x face, tension positive, while V_y and V_z are components of **shear force** that act on the x face in the y direction and the z direction, respectively. The resultant force components—F, V_y, and V_z—act through the point of intersection of the x axis with the cross section, as shown in Fig. 2.41b. In Eqs. 2.34, T is the **torque,** or twisting moment, while M_y and M_z are **bending moments** about y and z, respectively. The *right-hand rule* is used to establish the sign convention for the torque and the two bending moments. The dimensions of torque and bending moment are $F \cdot L$. Typical units used are $lb \cdot ft$, $kN \cdot m$, and so on.

The relationship of stresses to stress resultants is illustrated by the following example.

> ### EXAMPLE 2.13
>
> The stress distribution on the rectangular cross section shown in Fig. 1 is given by
>
> $$\sigma_x = (800y - 400z + 1200) \text{ psi}$$
>
> $$\tau_{xy} = 0, \quad \tau_{xz} = 300(9 - z^2) \text{ psi}$$
>
> Determine the net internal force system (i.e., the resultant forces and moments) on this cross section. Let $b = 6$ in. and $h = 8$ in.
>
> **Plan the Solution**   At the cross section there can, in general, be three components of force—F, V_y, and V_z—and three moments—T, M_y, and M_z. Using the given distribution of stresses and using Eqs. 2.33 and 2.34, we can calculate these stress resultants.

Solution

$$F = \int_A \sigma_x \, dA = 800 \int_A y \, dA - 400 \int_A z \, dA + 1200 \int_A dA$$

$$= 800 \, \bar{y} A - 400 \, \bar{z} A + 1200 A$$

$$= (1200 \text{ psi})(8 \text{ in.})(6 \text{ in.}) = 57{,}600 \text{ lb}$$

(Since the origin of y and z is at the centroid of the cross section, $\bar{y} = \bar{z} = 0$.) Continuing,

$$V_y = \int_A \tau_{xy} \, dA = 0$$

$$V_z = \int_A \tau_{xz} \, dA = 2700 \int_A dA - 300 \int_A z^2 \, dA$$

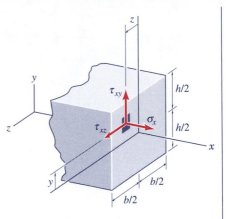

(a) The stresses at a point.

From Appendix C.2, for the rectangular cross section of "base" b and "height" h, $\int_A z^2 \, dA = \frac{1}{12} h b^3$ and $\int_A y^2 \, dA = \frac{1}{12} b h^3$. Therefore,

$$V_z = 2700 bh - \frac{300}{12}(h b^3)$$

$$= (2700 \text{ psi})(8 \text{ in.})(6 \text{ in.}) - (25 \text{ lb/in}^4)(8 \text{ in.})(6 \text{ in.})^3$$

$$= 86{,}400 \text{ lb}$$

$$T = \int_A y \tau_{xz} \, dA - \int_A z \tau_{xy} \, dA$$

$$= 2700 \int_A y \, dA - 300 \int_A y z^2 \, dA$$

$$= 2700 \bar{y} A - 300 \int_{-3}^{3} z^2 \int_{-4}^{4} y \, dy \, dz = 0$$

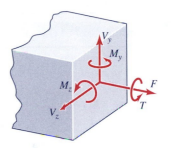

(b) The stress resultants.

Fig. 1

$$M_y = \int_A z \sigma_x \, dA$$

$$= 800 \int_A yz \, dA - 400 \int_A z^2 \, dA + 1200 \int_A z \, dA$$

$$= -\frac{400}{12}(h b^3) = -\left(\frac{400}{12} \text{ lb/in}^3\right)(8 \text{ in.})(6 \text{ in.})^3$$

$$= -57{,}600 \text{ lb} \cdot \text{in.}$$

$$M_z = -\int_A y \sigma_x \, dA = -800 \int_A y^2 \, dA$$

$$\qquad + 400 \int_A yz \, dA - 1200 \int_A y \, dA$$

$$= -\frac{800}{12}(b h^3) = -\left(\frac{800}{12} \text{ lb/in}^3\right)(6 \text{ in.})(8 \text{ in.})^3$$

$$= -204{,}800 \text{ lb} \cdot \text{in.}$$

In summary,

$$\left. \begin{array}{lll} F = 57.6 \text{ kips}, & V_y = 0, & V_z = 86.4 \text{ kips} \\ T = 0, & M_y = -57.6 \text{ kip} \cdot \text{in.}, & M_z = -205 \text{ kip} \cdot \text{in.} \end{array} \right\} \textbf{Ans.}$$

Review the Solution The only way to check the above answers is to go back over the calculations, using information from Appendix C to check all of the integrals. We can also spot-check some of the magnitudes. For example, the 1200-psi term in σ_x represents constant normal stress on the cross section, which would produce an axial force $F = \sigma_x A = (1200 \text{ psi})(8 \text{ in.})(6 \text{ in.}) = 57.6 \text{ kips}$. We can also see that $0 \leq \tau_{xz} \leq 2700 \text{ psi}$ everywhere on the cross section. Hence, V_z must be less than $(2700 \text{ psi})(8 \text{ in.})(6 \text{ in.}) = 129.6 \text{ kips}$. So the value of $V_z = 86.4 \text{ kips}$ seems reasonable.

Stress distributions like the one in this example will arise in Chapter 6 in the study of bending of beams.

► EXAMPLE 2.14 ◄

On a particular cross section of a rectangular bar (Fig. 1a) there is a constant tensile stress of magnitude σ_0 over the bottom half of the bar, and the stress tapers linearly to zero at the top edge of the cross section, as shown in Fig. 1b. (a) Determine expressions for the six stress resultants listed in Eqs. 2.33 and 2.34. (b) Replace the normal force F and moment M_z by a single resultant force, as depicted in Fig. 1c. Give the location, y_R, of the resultant force.

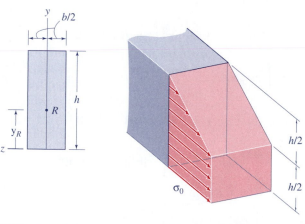

(a) The cross section.   (b) The stress distribution.

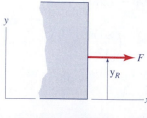

(c) The resultant force.

Fig. 1

Plan the Solution At the cross section there can, in general, be three components of force—F, V_y, and V_z—and three moments—T, M_y, and M_z. The only nonzero stress in this problem is σ_x. Using the given distribution of stress σ_x and using Eqs. 2.33 and 2.34, we can calculate these stress resultants.

Solution (a) *Determine expressions for the six stress resultants.*

Force Resultants: From Eq. 2.33,

$$F = \int_A \sigma_x \, dA$$

$$= \sigma_0 b \left(\frac{h}{2}\right) + b \int_{h/2}^{h} 2\sigma_0 \left(1 - \frac{y}{h}\right) dy \tag{1}$$

$$= \frac{1}{2} \sigma_0 bh + b\left[2\sigma_0\left(\frac{h}{2} - \frac{3h}{8}\right)\right]$$

$$= \frac{1}{2} \sigma_0 bh + \frac{1}{4} \sigma_0 bh = \frac{3}{4} \sigma_0 bh$$

Since $\tau_{xy} = \tau_{xz} = 0$, $V_y = V_z = 0$.

Moment Resultants: From Eq. 2.34a, since $\tau_{xy} = \tau_{xz} = 0$, $T = 0$. From Eq. 2.34b, since σ_x is symmetric in z, $M_y = 0$. Finally, from Eq. 2.34c,

$$M_z = -\int_A y\sigma_x \, dA$$

$$= -b\left[\int_0^{h/2} \sigma_0 y \, dy + \int_{h/2}^{h} 2\sigma_0\left(1 - \frac{y}{h}\right) y \, dy\right] \tag{2}$$

$$= -b\left[\sigma_0 \frac{h^2}{8} + 2\sigma_0\left(\frac{3h^2}{8} - \frac{7h^2}{24}\right)\right]$$

$$= -\frac{1}{8} \sigma_0 bh^2 - \frac{1}{6} \sigma_0 bh^2 = -\frac{7}{24} \sigma_0 bh^2$$

In summary, the six stress resultants are

$$\left.\begin{array}{ll} F = \dfrac{3}{4}\sigma_0 bh, & V_y = V_z = 0 \\[2ex] T = M_y = 0, & M_z = -\dfrac{7}{24}\sigma_0 bh^2 \end{array}\right\} \qquad \textbf{Ans. (a)}$$

(b) *Locate the single resultant force.* Since the z axis is at the bottom edge of the cross section, and since F is positive in tension, Fig. 1c indicates that

$$M_z = -y_R F \tag{3}$$

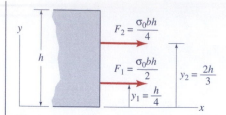

Fig. 2 Resultants of two stress blocks.

So,

$$y_R = \frac{7}{18} h \qquad \textbf{Ans. (b)}$$

Review the Solution Another way to determine the force resultant F and determine its location is to see that it is the sum of the two force resultants F_1 and F_2 shown in Fig. 2. Then,

$$y_1 F_1 + y_2 F_2 = y_R F$$

Let us turn our attention now to general definitions of extensional strain and shear strain to complement the definitions of normal stress and shear stress in Eqs. 2.32.

General Definition of Extensional Strain.

Recall that extensional strain is the change in length of a line segment divided by the original length of the line segment. To define the *extensional strain* in a direction n at a point P in a body, we take an infinitesimal line segment of length Δs, in direction n, starting at P as shown in Fig. 2.42a.[30] That is, we take the infinitesimal line segment PQ of length Δs as the original line segment. After deformation, the line segment PQ becomes the infinitesimal arc P^*Q^* with arclength Δs^*, as shown in Fig. 2.42b. To determine the extensional strain right at point P, we need, of course, to start with a very short length Δs; that is, we must pick Q very close to point P. By picking Q closer and closer to P, we get, in the limit as $\Delta s \to 0$, the extensional strain right at point P. Then, the **extensional strain** *at point P in direction n, denoted by $\epsilon_n (P)$, is defined by*

$$\boxed{\epsilon_n(P) = \lim_{Q \to P \text{ along } n} \left(\frac{\Delta s^* - \Delta s}{\Delta s} \right)} \qquad \begin{array}{l}\textbf{Extensional} \\ \textbf{Strain}\end{array} \qquad (2.35)$$

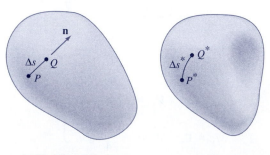

(a) Undeformed body.   (b) Deformed body.

FIGURE 2.42 The infinitesimal line segment used to define extensional strain.

---

[30]In the present discussion the term *line segment* refers to the collection of material particles that lie along a straight line connecting specified points, say P and Q, in the undeformed body. Such a line segment is sometimes referred to as a *fiber*.

EXAMPLE 2.15

The thin, square plate *ABCD* in Fig. 1a undergoes deformation in which no point in the plate displaces in the *y* direction. Every horizontal line in the plate, except line *AD*, is uniformly shortened as the edge *AB* remains straight and rotates clockwise. Determine an expression for $\epsilon_x(x, y)$.

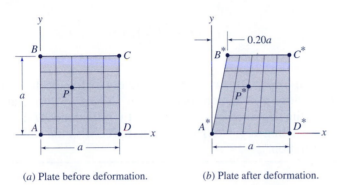

(*a*) Plate before deformation. (*b*) Plate after deformation.

Fig. 1

Plan the Solution We can use the definition of extensional strain $\epsilon_n(P)$, which, in this case, is $\epsilon_x(x, y)$. We will have to use the geometry of deformation to determine an expression for Δs^* in Eq. 2.35.

Solution To determine $\epsilon_x(x, y)$, let Eq. 2.35 be written as

$$\epsilon_x(x, y) = \lim_{\Delta x \to 0} \left(\frac{\Delta x^* - \Delta x}{\Delta x} \right)$$

where Δx and Δx^* are defined in Fig. 2.

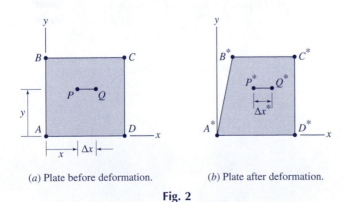

(*a*) Plate before deformation. (*b*) Plate after deformation.

Fig. 2

To get an expression for $\epsilon_x(x, y)$ we need an expression for Δx^*. We are told that every horizontal line is uniformly shortened, and we see that *AD* remains its original length while B^*C^* is shorter than *BC* by 20%. Furthermore, the shortening of horizontal lines is linearly related to y. Therefore,

$$\Delta x^* = \Delta x - \left(\frac{y}{a}\right)(0.2\ \Delta x)$$

So,

$$\epsilon_x(x, y) = \lim_{\Delta x \to 0} \left[\frac{-(y/a)(0.2\ \Delta x)}{\Delta x}\right]$$

or

$$\epsilon_x(x, y) = -\left(\frac{y}{a}\right)(0.2) \qquad \textbf{Ans.}$$

Review the Solution   The answer is negative, which indicates a shortening, and the answer is dimensionless, as it should be for strain. Also, according to the answer, there is no strain at $y = 0$ (along AD), and there is a 20% shortening at $y = a$. These results agree with the stated geometry of deformation.

Definition of Shear Strain.   When a body deforms, the change in angle that occurs between two line segments that were originally perpendicular to each other is called *shear strain*. To define the shear strain, let us consider the undeformed body in Fig. 2.43*a* and the deformed body in Fig. 2.43*b*. Let PQ and PR be infinitesimal line segments in the n direction and t direction, respectively, in the underformed body. After deformation, line segments PQ and PR become arcs P^*Q^* and P^*R^*. Secant lines P^*Q^* and P^*R^* define an angle θ^* in the deformed body. In the limit, as we pick Q and R closer and closer to P, the angle θ^* approaches the angle between tangents to the arcs at P^*, shown as dashed lines in Fig. 2.43*b*. The **shear strain** *between line segments extending from P in directions n and t is defined by the equation*

$$\gamma_{nt}(P) = \lim_{\substack{Q \to P \text{ along } n \\ R \to P \text{ along } t}} \left(\frac{\pi}{2} - \theta^*\right) \qquad \textbf{Shear Strain} \qquad (2.36)$$

This definition is illustrated in the following example.

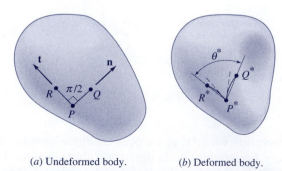

(*a*) Undeformed body.   (*b*) Deformed body.

FIGURE 2.43   The angles used to define shear strain.

For the square plate and the deformation described in Example 2.15, determine an expression for $\gamma_{xy}(x, y)$.

Plan the Solution By comparing the "before deformation" and "after deformation" figures, we can see that there is a definite change in the right angle between x lines and y lines. We need to determine this change in angle so that we can evaluate the shear strain from Eq. 2.36.

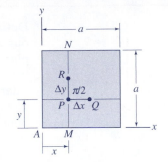

(*a*) Plate before deformation.

Solution Equation 2.36 can be written as

$$\gamma_{xy}(P) \equiv \gamma_{xy}(x, y) = \lim_{\substack{\Delta x \to 0 \\ \Delta y \to 0}} \left(\frac{\pi}{2} - \theta^*\right)$$

where Δx, Δy, and θ^* are indicated in Fig. 1.

$$\theta^* = \frac{\pi}{2} - \tan^{-1}\left(\frac{x_{N*} - x_{M*}}{a}\right)$$

$$x_{N*} = x + \left(1 - \frac{x}{a}\right)(0.2a)$$

$$x_{N*} - x_{M*} = \left(1 - \frac{x}{a}\right)(0.2a)$$

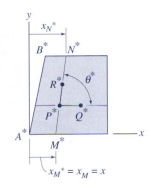

(*b*) Plate after deformation.

Fig. 1

Therefore,

$$\theta^* = \frac{\pi}{2} - \tan^{-1}\left[0.2\left(1 - \frac{x}{a}\right)\right]$$

Since the x lines and y lines remain straight, θ^* doesn't depend on the lengths of Δx and Δy, and we don't need the limit operation. So,

$$\gamma_{xy}(x, y) = \frac{\pi}{2} - \theta^* = \tan^{-1}\left[0.2\left(1 - \frac{x}{a}\right)\right] \qquad \textbf{Ans.}$$

Review the Solution From Fig. 1*b*, we see that γ_{xy} should be independent of y, as our final result indicates. Also, from Fig. 1*b* we see that γ_{xy} should be greatest when $x = 0$ and should be zero at $x = a$. These observations are consistent with our answer. Finally, the largest shear strain is at $x = 0$, where $\gamma_{xy}(x, y) = \tan^{-1}(0.2) = 0.1974 \approx 0.2$. Therefore, γ_{xy} could be approximated by $\gamma_{xy}(x, y) = 0.2\left(1 - \frac{x}{a}\right)$.

In Chapters 3, 4, and 6 we will make extensive use of the definitions of extensional strain and shear strain, Eqs. 2.35 and 2.36, respectively, to determine key strain-displacement equations for theories of axial deformation, torsion, and bending.

*2.13 CARTESIAN COMPONENTS OF STRESS; GENERALIZED HOOKE'S LAW FOR ISOTROPIC MATERIALS

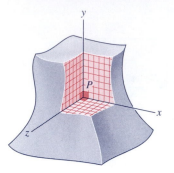

FIGURE 2.44 A set of three mutually orthogonal planes through an arbitrary point P.

Definitions of normal stress and shear stress were given in Eqs. 2.32 of the previous section. These equations define the components of stress on a particular plane, the n plane, at the given point. However, we can pass three mutually orthogonal planes through any point in a deformable body, so we need to consider the stresses on all three mutually orthogonal planes. Consider the x, y, and z planes passing through point P in Fig. 2.44. It is customary to sketch the components of stress, as defined by Eqs. 2.32, on a cube having x, y, and z faces, as shown in Fig. 2.45.

The sign convention for the stresses shown on Fig. 2.45 is the sign convention associated with Eqs. 2.32. The normal stresses are always taken positive in tension. Therefore, σ_x, σ_y, and σ_z are all shown in tension on the respective x, y, and z faces. The first subscript on a shear stress refers to the plane on which the shear stress acts, while the second subscript indicates the direction in which the shear stress acts. Hence, on the "$+x$ face," that is, the face whose outward normal is the $+x$ axis, the shear stress τ_{xy} acts in the $+y$ direction, and the stress τ_{xz} acts in the $+z$ direction. Conversely, on the "$-x$ face," τ_{xy} acts in the $-y$ direction, and τ_{xz} acts in the $-z$ direction. (To avoid "cluttering," the stresses acting on the hidden $(-)$ faces are not shown in Fig. 2.45.)

Shear Stress Equilibrium Requirements. In the discussion of pure shear in Section 2.7, you learned that, in order to satisfy moment equilibrium, the shear stresses on faces that intersect at right angles must be equal (e.g., see Fig. 2.30d). Let us now examine the relationship of shear stresses on perpendicular faces on which normal stresses also act. To simplify the free-body diagram, only those stresses that contribute to moment about the z axis are shown in Fig. 2.46. Stresses on exposed faces are indicated by a $+$ superscript; those on the three hidden faces

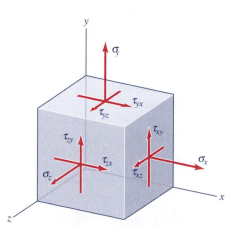

FIGURE 2.45 A three-dimensional state of stress referred to as rectangular Cartesian axes. (Stresses shown for visible faces only.)

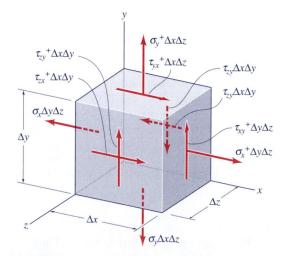

FIGURE 2.46 A three-dimensional free-body diagram. (Only stresses that contribute to M_z are shown.)

have no subscript.[31] Consider first the moment about the z axis due to σ_x and σ_x^+. We get

$$(\Delta M_z)_{\sigma_x} = (\sigma_x\,\Delta y\,\Delta z)\left(\frac{\Delta y}{2}\right) - (\sigma_x^+\,\Delta y\,\Delta z)\left(\frac{\Delta y}{2}\right)$$

Since Δx is small, we can write

$$\sigma_x^+ = \sigma_x + \Delta\sigma_x$$

where $\Delta\sigma_x$ is a small quantity of the same order as Δx. Then,

$$(\Delta M_z)_{\sigma_x} = -\frac{1}{2}\Delta\sigma_x\,\Delta y^2\,\Delta z$$

Note that the right-hand side is a product of four Δ-terms. The stresses σ_y, τ_{zx}, and τ_{zy} contribute similar "four-Δ" amounts to ΔM_z. On the other hand, consider the moment due to τ_{xy}^+ and τ_{yx}^+.

$$(\Delta M_z)_{\tau_{xy}} = (\tau_{xy}^+\,\Delta y\,\Delta z)\Delta x - (\tau_{yx}^+\,\Delta x\,\Delta z)\Delta y$$

$$= (\tau_{xy} - \tau_{yx})\Delta x\,\Delta y\,\Delta z + (\Delta\tau_{xy} - \Delta\tau_{yx})\Delta x\,\Delta y\,\Delta z$$

Collecting all contributions to moment about the z axis, we get

$$\sum M_z = (\tau_{xy} - \tau_{yx})\Delta x\,\Delta y\,\Delta z + \text{higher-order terms}$$

As Δx, Δy, and Δz approach zero, the higher-order terms become negligible, leaving only the first term. However, to satisfy moment equilibrium, we must have $\Sigma M_z = 0$, so it is necessary that

$$\tau_{xy} = \tau_{yx}$$

From the above analysis of moment equilibrium, we can conclude that, **even if there are normal stresses acting on an element, the shear stresses must satisfy the following equations:**

$$\boxed{\tau_{yx} = \tau_{xy}, \quad \tau_{yz} = \tau_{zy}, \quad \tau_{zx} = \tau_{xz}} \tag{2.37}$$

Generalized Hooke's Law for Isotropic Materials.

A solid whose material properties, for example E and v, are independent of orientation in the body, is said to be **isotropic.** We now consider the stress-strain-temperature relationships for a linearly elastic, isotropic body. Let the body be subjected to stresses σ_x, σ_y, σ_z, τ_{xy}, τ_{xz}, and τ_{yz}, and to a temperature change ΔT. By the *principle of linear superposition*[32] we can get the combined strain response by adding together the separate responses

---

[31]Since the $+x$ face and the $-x$ face are Δx distance apart, the stress on the $+x$ face may be slightly different than the stress on the $-x$ face; this distinction is indicated by the $+$ superscript notation.

[32]The strains may be added linearly if the deformation is small and the material remains linearly elastic.

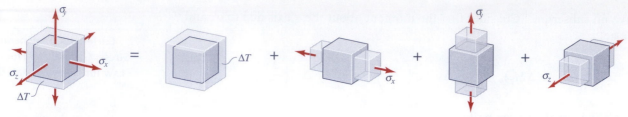

(a) The total extensional strain. (b) Uniform thermal strain. (c) Strains due to σ_x. (d) Strains due to σ_y. (e) Strains due to σ_z.

FIGURE 2.47 Superposition of extensional strains.

due to σ_x, σ_y, and so on. For example, the strains produced by σ_x are

$$\epsilon_x = \frac{\sigma_x}{E}, \quad \epsilon_y = \epsilon_z = \frac{-\nu\sigma_x}{E}$$

Figure 2.47a illustrates the combined effect of the σ's and ΔT. Figures 2.47b through 2.47e illustrate the strains produced separately by the three normal stresses, σ_x, σ_y, and σ_z, and the strain produced by a temperature increase ΔT. Let

$$\epsilon_{xT} = \epsilon_{yT} = \epsilon_{zT} = \alpha\Delta T \qquad \text{(Fig. 2.47b)}$$

$$\epsilon_x' = \frac{\sigma_x}{E}, \quad \epsilon_y' = \epsilon_z' = \frac{-\nu\sigma_x}{E} \qquad \text{(Fig. 2.47c)}$$

$$\epsilon_y'' = \frac{\sigma_y}{E}, \quad \epsilon_x'' = \epsilon_z'' = \frac{-\nu\sigma_y}{E} \qquad \text{(Fig. 2.47d)}$$

$$\epsilon_z''' = \frac{\sigma_z}{E}, \quad \epsilon_x''' = \epsilon_y''' = \frac{-\nu\sigma_z}{E} \qquad \text{(Fig. 2.47e)}$$

By the superposition principle, the total extensional strains are given by

$$\epsilon_x = \frac{1}{E}[\sigma_x - \nu(\sigma_y + \sigma_z)] + \alpha\Delta T$$

$$\epsilon_y = \frac{1}{E}[\sigma_y - \nu(\sigma_x + \sigma_z)] + \alpha\Delta T \qquad \begin{array}{l}\textbf{Generalized} \\ \textbf{Hooke's} \\ \textbf{Law}\end{array} \qquad (2.38)$$

$$\epsilon_z = \frac{1}{E}[\sigma_z - \nu(\sigma_x + \sigma_y)] + \alpha\Delta T$$

For an isotropic linearly elastic material, the shear stresses are related to the shear strains by the following equations:

$$\gamma_{xy} = \frac{1}{G}\tau_{xy}, \quad \gamma_{xz} = \frac{1}{G}\tau_{xz}, \quad \gamma_{yz} = \frac{1}{G}\tau_{yz} \qquad \begin{array}{l}\textbf{Gen. Hooke's} \\ \textbf{Law (cont.)}\end{array} \qquad (2.39)$$

These shear strains are illustrated in Fig. 2.48.

Equations 2.38 and 2.39 are referred to as the **generalized Hooke's Law** for isotropic materials. Note that, in an isotropic material, shear stresses do not enter

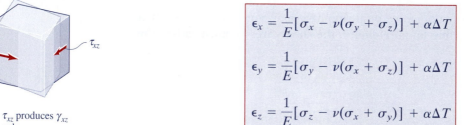

(a) τ_{xy} produces γ_{xy} only.

(b) τ_{xz} produces γ_{xz} only.

(c) τ_{yz} produces γ_{yz} only.

FIGURE 2.48 Illustration of shear strains.

into the expressions for extensional strains, and, likewise, normal stresses do not enter into the expressions for shear strains. In addition, the three components of shear are uncoupled.

Solving Eqs. 2.38 and 2.39 for the stresses in terms of the strains and ΔT, we get

$$\sigma_x = \frac{E}{(1 + \nu)(1 - 2\nu)}[(1 - \nu)\epsilon_x + \nu(\epsilon_y + \epsilon_z) - (1 + \nu)(\alpha \Delta T)]$$

$$\sigma_y = \frac{E}{(1 + \nu)(1 - 2\nu)}[(1 - \nu)\epsilon_y + \nu(\epsilon_z + \epsilon_x) - (1 + \nu)(\alpha \Delta T)] \quad \begin{matrix} \textbf{Gen.} \\ \textbf{Hooke's} \\ \textbf{Law} \end{matrix} \quad (2.40)$$

$$\sigma_z = \frac{E}{(1 + \nu)(1 - 2\nu)}[(1 - \nu)\epsilon_z + \nu(\epsilon_x + \epsilon_y) - (1 + \nu)(\alpha \Delta T)]$$

$$\tau_{xy} = G\gamma_{xy}, \quad \tau_{yz} = G\gamma_{yz}, \quad \tau_{zx} = G\gamma_{zx}$$

As noted in Section 2.7 and derived in Section 2.11, there is an equation that relates G to E and ν, namely

$$G = \frac{E}{2(1 + \nu)} \qquad \begin{matrix} (2.24) \\ \text{repeated} \end{matrix}$$

The following example illustrates the relationship of Poisson's ratio to the change in volume of a body subjected to triaxial stress. The change in volume per unit volume is called the **volumetric strain,** or **dilatation,** ϵ_V.

EXAMPLE 2.17

A rectangular parallelepiped of linearly elastic, isotropic material is subjected to general triaxial stress $\sigma_x, \sigma_y, \sigma_z$, as shown in Fig. 1. Determine the volumetric strain, ϵ_V. Assume that $\epsilon \ll 1$ for all three coordinate directions.

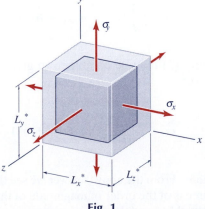

Fig. 1

Plan the Solution We can write the change in volume in terms of the change in length of each of the three sides of the body, which can, in turn, be expressed in terms of the extensional strains ϵ_x, ϵ_y, and ϵ_z. We can use Hooke's Law, Eqs. 2.38, to relate these three strains to the three stresses.

Solution The basic equation for volumetric strain (dilatation) is

$$\epsilon_V = \frac{\Delta V}{V} = \frac{L_x^* L_y^* L_z^* - L_x L_y L_z}{L_x L_y L_z}$$

where the L^* lengths are the after-deformation dimensions of the body, and the L's are the corresponding pre-deformation dimensions of the body. For uniform strain in each coordinate direction,

$$L^* = (1 + \epsilon)L$$

so

$$\epsilon_V = (1 + \epsilon_x)(1 + \epsilon_y)(1 + \epsilon_z) - 1$$
$$= \epsilon_x + \epsilon_y + \epsilon_z + \epsilon_x\epsilon_y + \epsilon_y\epsilon_z + \epsilon_z\epsilon_x + \epsilon_x\epsilon_y\epsilon_z$$

Since all three strains satisfy $\epsilon \ll 1$, the squared and cubed terms can be dropped in the preceding equation, leaving the approximation

$$\epsilon_V = \epsilon_x + \epsilon_y + \epsilon_z \tag{1}$$

From Eqs. 2.38,

$$\epsilon_x = \frac{1}{E}[\sigma_x - \nu(\sigma_y + \sigma_z)]$$

$$\epsilon_y = \frac{1}{E}[\sigma_y - \nu(\sigma_x + \sigma_z)] \tag{2}$$

$$\epsilon_z = \frac{1}{E}[\sigma_z - \nu(\sigma_x + \sigma_y)]$$

Finally, Eqs. (1) and (2) may be combined to give following expression for the dilatation ϵ_V:

$$\epsilon_V = \frac{1 - 2\nu}{E}(\sigma_x + \sigma_y + \sigma_z) \qquad \textbf{Ans.}$$

Review the Solution From the above answer we see that the change in volume per unit volume is of the order of magnitude of the strain, and is therefore small, as we would expect. It is interesting to note that, because of the $(1 - 2\nu)$ factor, a tensile stress causes an increase in volume, which is what we would expect to happen, but only if $0 \leq \nu \leq 0.5$. As noted earlier, most materials have a value of ν that falls within the range $\nu = 0.25$ to $\nu = 0.35$.

*2.14 MECHANICAL PROPERTIES OF COMPOSITE MATERIALS

Composite materials were mentioned briefly at the end of Section 2.4. In this section we further describe composites and work through the equations that describe the elastic modulus of the simplest composite cases, iso-stress and iso-strain.

In many engineering applications, the performance of a part in a structure or a machine can be improved if it is composed of more than one material. Parts that contain multiple materials working together are often called **composite materials.** Common examples of composite materials include steel-reinforced concrete and fiberglass-reinforced polymers. A composite material is generally considered to contain a *matrix* phase (distinct material) and one or more *reinforcement* phases. The primary purpose of the matrix phase is to distribute load to the reinforcement phase(s). By combining multiple materials into a composite, properties not possible in a single material can be achieved (e.g., greater strength or greater ductility).

Composite materials are often categorized by the shape of the reinforcement phase. General categories include continuous-fiber-reinforced, short-fiber-reinforced, particle-reinforced, and laminated composites. Examples are shown in Fig. 2.49.

When the behavior of composite materials is to be analyzed, two approaches are possible. One is to treat each phase in the composite separately, in which case the composite is viewed as a structure composed of more than one material. Such an analysis can be very difficult when the internal structure of the composite is complex. A second approach is to view the composite as a new material, with its own properties. This method of analysis treats the entire composite part as made of a single *composite material* with its own *effective* properties. It is typically valid to do this when the size of the reinforcement phase, e.g. fiber diameter or particle width, is much smaller than the size of the part made of the composite material.

We will now calculate the *effective Young's modulus* for two simple composite material cases, the iso-strain case and the iso-stress case, which are illustrated in Fig. 2.50. These cases are reasonable approximations for situations in which a reinforcement phase spans at least one dimension of the part. In order to calculate the effective Young's modulus in each case, we assume that the reinforcement and matrix phases are well bonded and have approximately the same Poisson's ratio. The effective properties depend strongly on how much reinforcement phase the composite contains. We consider a composite material containing a volume fraction V_f of reinforcement phase, such as stiff fibers. The volume fraction is the ratio of the reinforcement phase volume to the total volume of the composite, and this may vary from zero (no reinforcement) to one (all reinforcement). The reinforcement phase has a Young's modulus of E_f (subscript f for *fiber*), and the matrix phase has a Young's modulus of E_m. For the iso-strain case, load is applied parallel to the direction of reinforcement, as shown in Fig. 2.50(a). It can be seen in Fig. 2.50(a) that the strain along the loading direction is identical between the reinforcement and matrix phases, assuming that these are well bonded to each other. This is the origin of the **iso-strain** designation. The effective Young's modulus for

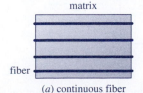
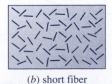

(a) continuous fiber   (b) short fiber   (c) particle reinforced   (d) laminated

FIGURE 2.49 Examples of different composite material types.

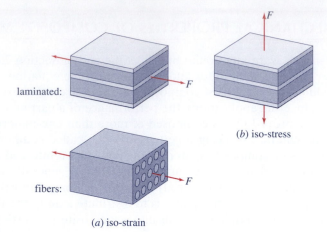

laminated:

F

F

(b) iso-stress

fibers:

F

(a) iso-strain

FIGURE 2.50   The two limiting cases of composite material construction.

the iso-strain case is

$$E_{\parallel} = V_f E_f + (1 - V_f)E_m$$

**Iso-strain
Modulus**   (2.41)

This is the case that has the greatest effective Young's modulus.

When load is applied perpendicular to the direction of reinforcement, as shown in Fig. 2.50(b), this configuration is called the **iso-stress** case. The stress in both phases is approximately the same, and the effective Young's modulus in this iso-stress case is

$$\frac{1}{E_{\perp}} = \frac{V_f}{E_f} + \frac{(1 - V_f)}{E_m}$$

**Iso-stress
Modulus**   (2.42)

The iso-stress case has the lowest effective Young's modulus. Composite cases between the extremes of iso-strain and iso-stress will have intermediate values of effective Young's modulus. A discontinuously reinforced composite, e.g. Fig. 2.49(b) or (c), with randomly oriented reinforcements might be approximated as an average of these two extremes, as shown in Fig. 2.51, which plots the effective Young's modulus as a function of the volume fraction of reinforcement phase. Although the effective modulus of a composite with randomly oriented reinforcement will typically be slightly less than this average, $E_c \leq \frac{1}{2}(E_{\perp} + E_{\parallel})$, the average value is a reasonable approximation from which to begin.

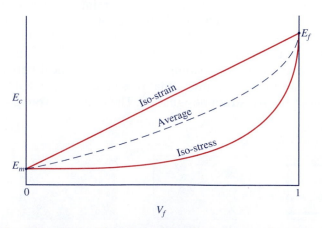

FIGURE 2.51   The effective Young's modulus, E_c, shown as a function of volume fraction of reinforcement, V_f, for different composite cases.

2.15 PROBLEMS

▼ **NORMAL STRESS; AXIAL STRESS**

> For all problems in Section 2.2, *assume unknown axial forces to be positive in tension. As in Examples 2.1 and 2.2, label tensile-stress answers with (T) and compressive-stress answers with (C).*

Prob. 2.2-1. A 1-in.-diameter solid bar (1), a square solid bar (2), and a circular tubular member with 0.2-in. wall thickness (3), each supports an axial tensile load of 5 kips. (a) Determine the axial stress in bar (1). (b) If the axial stress in each of the other bars is 6 ksi, what is the dimension, b, of the square bar, and what is the outer diameter, c, of the tubular member?

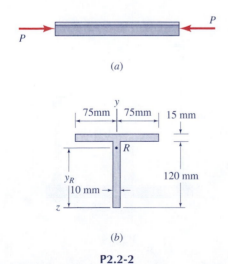

P2.2-1

Prob. 2.2-2. The structural tee shown in Fig. P2.2-2 supports a compressive load $P = 200$ kN. (a) Determine the coordinate y_R of the point R in the cross section where the load must act in order to produce uniform compressive axial stress in the member, and (b) determine the magnitude of that compressive stress.

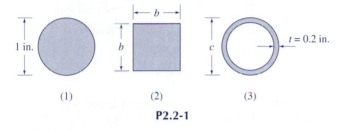

P2.2-2

Prob. 2.2-3. A steel plate is welded onto each end of the structural angle in Fig. P2.2-3 so that a load can be applied at point R, where it will produce uniform axial stress in the

member. (a) Determine the coordinates y_R and z_R of the point where the tensile load P must act in order to produce uniform tensile stress in the cross section of the structural angle, and (b) determine the magnitude of that tensile stress if $P = 18$ kips.

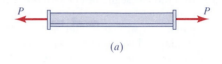

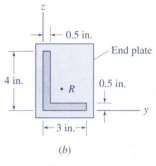

P2.2-3

*Prob. 2.2-4. Consider the free-hanging rod shown in Fig. P2.2-4. The rod has the shape of a conical frustum, with radius R_0 at its top and radius R_L at its bottom, and it is made of material with mass density ρ. The length of the rod is L. Determine an expression for the normal stress, $\sigma(x)$, at an arbitrary cross section x ($0 \leq x \leq L$), where x is measured downward from the top of the rod.

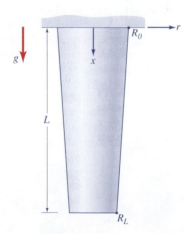

P2.2-4 and P2.3-6

Prob. 2.2-5. A solid brass rod AB and a solid aluminum rod BC are connected together by a coupler at B, as shown in Fig. P2.2-5. The diameters of the two segments are $d_1 = 60$ mm and $d_2 = 50$ mm, respectively. Determine the axial stresses σ_1 (in rod AB) and σ_2 (in rod BC).

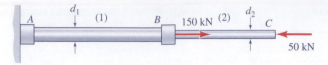

P2.2-5

Prob. 2.2-6. The three-part axially loaded member in Fig. P2.2-6 consists of a tubular segment (1) with outer diameter $(d_o)_1 = 1.00$ in. and inner diameter $(d_i)_1 = 0.75$ in., a solid circular rod segment (2) with diameter $d_2 = 1.00$ in., and another solid circular rod segment (3) with diameter $d_3 = 0.75$ in. The line of action of each of the three applied loads is along the centroidal axis of the member. Determine the axial stresses σ_1, σ_2, and σ_3 in each of the three respective segments.

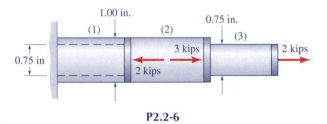

P2.2-6

Prob. 2.2-7. At a local marina the dock is supported on wood piling in the manner shown in Fig. P2.2-7a. The top part, AB, of one pile is above the normal waterline; the middle part, BC, is in direct contact with the water; and the part below C is underground. The original diameter of the pile is $d_0 = 12$ in., but action of the water and insects has reduced the diameter of the pile over the part BC. (a) If the axial force that the deck exerts on this pile is $P = 200$ kips, what is the axial stress in AB? Neglect the weight of the pile itself. (b) An inspector estimates that the diameter of the pile in segment BC has been eroded by 5%. What axial stress does the deck load of $P = 200$ kips produce in this damaged part of the pile? (c) If the maximum axial stress allowed in the wood piles is 7.5 ksi (in compression), what is the maximum deck load that this damaged pile can support?

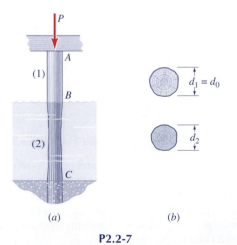

P2.2-7

Prob. 2.2-8. A column in a two-story building is fabricated from square structural tubing having the cross-sectional dimensions shown in Fig. P2.2-8b. Axial loads $P_A = 200$ kN and $P_B = 350$ kN are applied to the column at levels A and B, as shown in Fig. P2.2-8a. Determine the axial stress σ_1 in segment AB of the column and the axial stress σ_2 in segment BC of the column. Neglect the weight of the column itself.

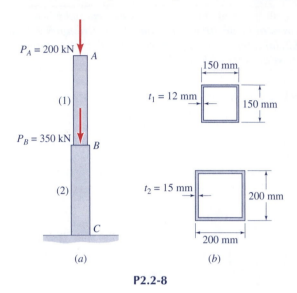

P2.2-8

Prob. 2.2-9. A rigid beam AB of total length 3 m is supported by vertical rods at its ends, and it supports a downward load at C of $P = 60$ kN, as shown in Fig. P2.2-9. The diameters of the steel hanger rods are $d_1 = 25$ mm and $d_2 = 20$ mm. Neglect the weight of beam AB and the rods. (a) If the load is located at $x = 1$ m, what are the stresses σ_{1a} and σ_{2a} in the respective hanger rods? (b) At what distance x from A must the load be placed such that $\sigma_2 = \sigma_1$, and what is the corresponding axial stress, $\sigma_{1b} = \sigma_{2b}$, in the rods?

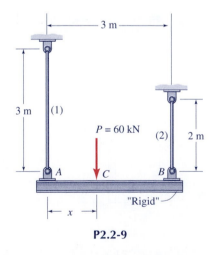

P2.2-9

Prob. 2.2-10. A 12-ft beam AB that weighs $W_b = 180$ lb supports an air conditioner that weighs $W_a = 1000$ lb. The beam, in turn, is supported by hanger rods (1) and (2), as shown in Fig. P2.2-10. (a) If the diameter of rod (1) is $\frac{3}{8}$ in., what is the stress, σ_1, in the rod? (b) If the stress in rod (2) is to be the same as the stress in rod (1), what should the diameter of rod (2) be (to the nearest $\frac{1}{32}$ in.)?

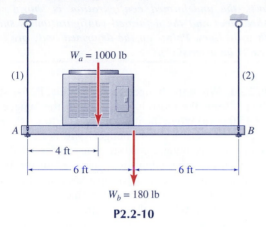

$W_a = 1000$ lb

(1)

(2)

A

B

|← 4 ft →|

|← 6 ft →|← 6 ft →|

$W_b = 180$ lb

P2.2-10

Prob. 2.2-11. A rigid, weightless beam BD supports a load P and is, in turn, supported by two hanger rods, (1) and (2), as shown in Fig. P2.2-11. The rods are initially the same length $L = 6$ ft and are made of the same material. Their rectangular cross sections have original dimensions ($w_1 = 1.5$ in., $t_1 = 0.75$ in.) and ($w_2 = 2.0$ in., $t_2 = 1.0$ in.), respectively. (a) At what location, b, must the load P act if the axial stress in the two bars is to be the same, i.e., $\sigma_1 = \sigma_2$? (b) What is the magnitude of this tensile stress if a load of $P = 40$ kips is applied at the location determined in Part (a)?

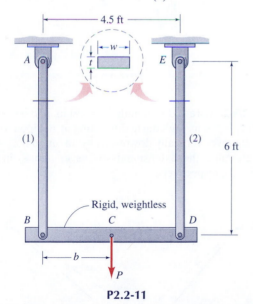

P2.2-11

Prob. 2.2-12. Each member of the truss in Fig. P1.4-1 is a solid circular rod with diameter $d = 0.50$ in. Determine the axial stresses σ_1, σ_2, and σ_3 in members (1), (2), and (3), respectively. (See Prob. 1.4-1.)

Prob. 2.2-13. Each member of the truss in Fig. P2.2-13 is a solid circular rod with diameter $d = 10$ mm. Determine the axial stress σ_1 in the truss member (1) and the axial stress σ_6 in the truss member (6).

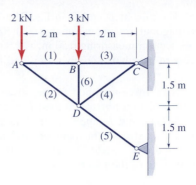

P2.2-13

Prob. 2.2-14. The three-member frame structure in Fig. P2.2-14 is subjected to a downward vertical load P at pin C. The pins at B, C, and D apply axial loads to members BD and CD, whose cross-sectional areas are $A_1 = 0.5$ in^2 and $A_2 = 1.0$ in^2, respectively. (a) If the axial stress in member BD is $\sigma_1 = 1200$ psi, what is the value of force P? (b) What is the corresponding axial stress, σ_2, in member CD?

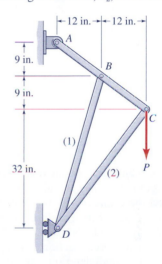

P2.2-14

Prob. 2.2-15. The three-member frame in Fig. P2.2-15 is subjected to a horizontal load P at pin E. The pins at C and D apply an axial load to cross-brace member CD, which has a rectangular cross section measuring 30 mm $\times$ 50 mm. If $P = 210$ kN, what is the axial stress in member CD?

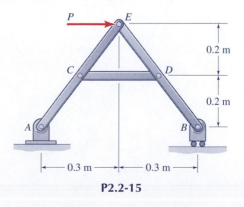

P2.2-15

91

Prob. 2.2-16. The pins at B and D in Fig. P1.4-17 apply an axial load to diagonal bracing member BD. If BD has a rectangular cross section measuring 0.50 in. $\times$ 2.00 in., what is the axial stress in member BD when the load is $w_0 = 220$ lb/ft?

Prob. 2.2-17. The three-member frame structure in Fig. P2.2-17 is subjected to a horizontal load $P = 500$ lb at pin C. The pins at B, C, and D apply axial loads to members BD and CD. If the stresses in members BD and CD are $\sigma_1 = 1200$ psi and $\sigma_2 = -820$ psi, what are the respective cross-sectional areas of the two members?

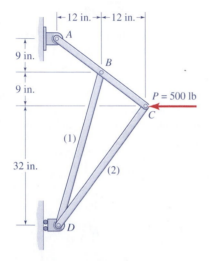

P2.2-17

***Prob. 2.2-18.** A cylinder of weight W and diameter d rests between thin, rigid members AE and BD, each of length L. Friction between the cylinder and its supports is negligible. The members are joined at their midpoint C by a frictionless pin, and they are prevented from collapsing by a restraining wire DE of length b and cross-sectional area A. Consider W, L, d, and A to be given, and determine an expression that relates the axial stress in wire DE to its length b.

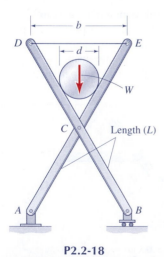

P2.2-18

▼ **EXTENSIONAL STRAIN**

> *Where both undeformed and deformed configurations are shown, the undeformed configuration is shown with dashed lines and the deformed configuration is shown with solid lines. Points on the deformed body are indicated by an asterisk(*).*

Prob. 2.3-1. When the bungee jumper in Fig. P2.3-1 stands on the platform, the unstretched length of the bungee cord is $L = 15.0$ ft. (a) When the jumper "hits bottom," the maximum extended length of the bungee cord is $L_m = 41.4$ ft. Assuming that the bungee cord stretches uniformly along its length, determine the extensional strain ϵ_m in the bungee cord at this point. (b) After bouncing a few times, the bungee jumper comes to rest with the final length of the bungee cord being $L_f = 32.4$ ft. What is the final strain, ϵ_f?

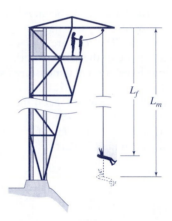

P2.3-1

Prob. 2.3-2. Wire AB of length $L_1 = 30$ in. and wire BC of length $L_2 = 36$ in. are attached to a ring at B. Upon loading, point B moves vertically downward by an amount $\delta_B = 0.25$ in. Determine the extensional strains ϵ_1 and ϵ_2 in wires (1) and (2), respectively.

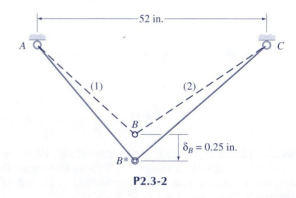

P2.3-2

Prob. 2.3-3. A wire is used to hang a lantern over a pool. Neglect the weight of the wire, and assume that it is taut, but strain free, before the lantern is hung. When the lantern is hung, it causes a 50-mm sag in the wire. Determine the extensional strain in the wire with the lantern hanging as shown.

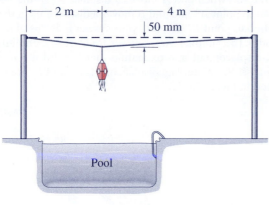

P2.3-3

^C**Prob. 2.3-4.** A "rigid" beam BC of length L is supported by a fixed pin at C and by an extensible rod AB, whose original length is also L. When $\theta = 45°$, rod AB is horizontal and strain free, that is, $\epsilon(\theta = 45°) = 0$. (a) Determine an expression for $\epsilon(\theta)$, the strain in rod AB, as a function of the angle θ shown in Fig. P2.3-4, valid for $45° \le \theta \le 90°$. (b) Write a computer program and use it to plot the expression for $\epsilon(\theta)$ for the range $45° \le \theta \le 90°$.

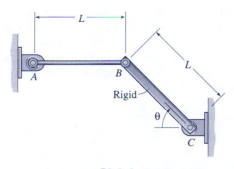

P2.3-4

Prob. 2.3-5. When a rubber band is uniformly stretched around the solid circular cylinder in Fig. P2.3-5b, its extensional strain is $\epsilon = 0.025 \frac{mm}{mm}$. If the diameter of the cylinder is $d = 100$ mm, what is the unstretched length of the rubber band (i.e., length L in Fig. P2.3-5a)? (Neglect the thickness of the rubber band.)

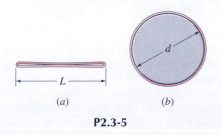

(a)   *(b)*

P2.3-5

***Prob. 2.3-6.** Determine an expression for the extensional strain $\epsilon(x)$ at cross section x ($0 \le x \le L$) of the hanging conical frustum in Prob. 2.2-4, and (b) determine an expression for the total elongation, e, of this hanging conical frustum in Prob. 2.2-4. The conical-frustum rod is made of material for which $\sigma = E\epsilon$, with $E = $ const.

Prob. 2.3-7. For small loads, P, the rotation of "rigid" beam AF in Fig. P2.3-7 is controlled by the stretching of rod AB. For larger loads, the beam comes into contact with the top of column DE, and further resistance to rotation is shared by the rod and the column. Assume (and later show that this is a valid assumption) that the angle θ through which beam AF rotates is small enough that points on the beam essentially move vertically, even though they actually move on circular paths about the fixed pin at C. (a) A load P is applied at end F that is just sufficient to close the 1.5-mm gap between the beam and the top of the column at D. What is the strain, ϵ_1, in rod AB for this value of load P? (b) If load P is increased further until $\epsilon_1 = 0.001 \frac{mm}{mm}$, what is the corresponding strain, ϵ_2, in column DE?

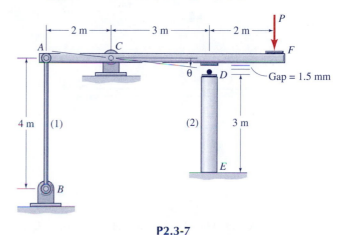

P2.3-7

Prob. 2.3-8. Vertical rods (1), (2), and (3) are all strain free when they are initially pinned to a straight, rigid,

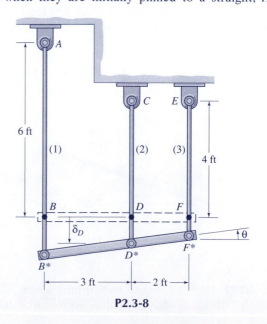

P2.3-8

horizontal beam *BF*. Subsequently, heating of the rods causes them to elongate and leaves the beam in the position denoted by *B*D*F**. Point *D* moves vertically downward by a distance $\delta_D = 0.20$ in., and the inclination angle of the beam is $\theta = 0.4°$ in the counterclockwise sense, as indicated on Fig. P2.3-8. Determine the strains ϵ_1, ϵ_2, and ϵ_3 in the three rods.

*Prob. 2.3-9.** A "rigid" beam *AD* is supported by a smooth pin at *D* and by vertical rods attached to the beam at points *A*, *B*, and *C*. The rods are all strain free when the beam is horizontal ($\theta = 0$). Subsequently, rod (2) is heated until its extensional strain reaches the value $\epsilon_2 = 80.0(10^{-6}) \frac{in.}{in.}$. (a) Determine the value of the (counterclockwise) beam angle θ that corresponds to the strain $\epsilon_2 = 80.0(10^{-6}) \frac{in.}{in.}$. (b) Determine the corresponding extensional strains ϵ_1 and ϵ_3.

$$L_1 = L_3 = 30 \text{ in.}, \quad L_2 = 40 \text{ in.}, \quad a = 20 \text{ in.}, \quad b = 60 \text{ in.}$$

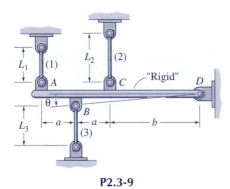

P2.3-9

Prob. 2.3-10. A rod *AB*, whose unstretched length is *L*, is originally oriented at angle θ counterclockwise from the $+x$ axis. (a) If the rod is free to rotate about a fixed pin at *A*, what is the extensional strain in the rod when end *B* moves a distance u in the $+x$ direction to point *B**? Express your answer in terms of the displacement u, the original length *L*, and the original angle θ. (b) Simplify the answer you obtained for Part (a), obtaining a small-displacement approximation that is valid if $u \ll L$. (Note: This result will be used in Section 3.10.)

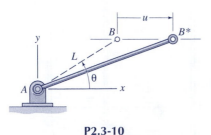

P2.3-10

Prob. 2.3-11. Rod *AB*, whose undeformed length is *L*, is originally oriented at angle θ counterclockwise from the $+x$

axis (the dashed line in Fig. P2.3-11). Both ends of the rod are free to move in the $x–y$ plane, while the rod remains straight. (a) Derive an expression for the extensional strain, ϵ, in the rod when end *A* moves a distance u_A in the $+x$ direction to point *A** and end *B* moves a distance v_B in the $+y$ direction to point *B**. Express your answer in terms of u_A, v_B, *L*, and θ. (b) Simplify your answer for Part (a), obtaining a small-displacement approximation that is valid if $u_A \ll L$ and $v_B \ll L$. (Note: The displacements u_A and v_B are exaggerated in Fig. P2.3-11.)

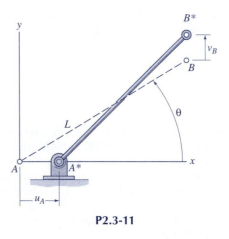

P2.3-11

Prob. 2.3-12. A thin sheet of rubber in the form of a square (Fig. P2.3-12a) is uniformly deformed into the parallelogram shape shown in Fig. P2.3-12b. All edges remain the same length, *b*, as the sheet deforms. (a) Compute the extensional strain ϵ_1 of diagonal *AC*. (b) Compute the extensional strain ϵ_2 of diagonal *BD*.

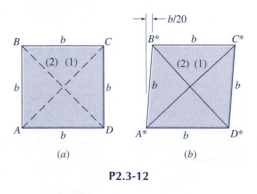

P2.3-12

Prob. 2.3-13. A thin sheet of rubber in the form of a square (Fig. P2.3-13a) is uniformly deformed into the parallelogram shape shown in Fig. P2.3-13b. All edges remains the same length, *b*, as the sheet deforms. (a) Compute the extensional strain ϵ_1 of diagonal *AC*. (b) Compute the extensional strain ϵ_2 of diagonal *BD*.

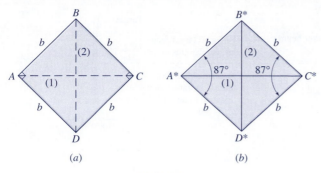

(a)

(b)

P2.3-13

▼ THERMAL STRAIN

Prob. 2.3-14. At the reference temperature, three identical rods of length L form a pin-jointed truss in the shape of an equilateral triangle, ABC, as shown in Fig. P2.3-14. Determine the angle θ as a function of ΔT, the temperature increase of tie rod AB. Rods AC and BC remain at the reference temperature. The properties of the rods are: area = A, modulus of elasticity = E, and coefficient of thermal expansion = α.

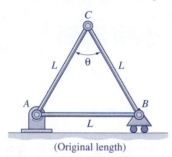

(Original length)

P2.3-14

Prob. 2.3-15. As shown in Fig. P2.3-15, a "rigid" beam AB of length L_{AB} = 30 in. is supported by a wire BC that is 40 in. long at the reference temperature of T_0 = 70°F. Determine an expression that relates the horizontal coordinate y_B of point B to the temperature T of the wire BC in °F if the coefficient of thermal expansion of the wire is $\alpha = 8 \times 10^{-6}/°F$. (Hint: Use the trigonometric *law of cosines*.)

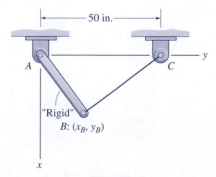

P2.3-15

Prob. 2.3-16. A steel pipe ($\alpha = 8 \times 10^{-6}/°F$) has a nominal inside diameter d = 4.06 in. at 70°F. (a) What is the inside diameter if the pipe carries steam that raises its temperature to 212°F? (Assume that the outside of the pipe is insulated so that the pipe reaches a uniform temperature of 212°F.) (b) How much would the steam-carrying pipe increase in length if it is originally 40 ft long, if its ends are not restrained against axial motion, and if the temperature increases from 70°F to 212°F?

Prob. 2.3-17. The angular orientation, θ, of a "rigid" mirror is controlled by the lengths of rods (1) and (2), as shown in Fig. P2.3-17. At the reference temperature, the rods are the same length: $L_1 = L_2 = 2.00$ m. The distance between the rods is a = 1.2 m. (a) If the thermal coefficient of the rods is $\alpha_1 = \alpha_2 = 14 \times 10^{-6}/°C$, determine an expression for the angle θ (in radians) as a function of ΔT_1 and ΔT_2. (b) If the maximum temperature difference that can be achieved between the temperature changes ΔT_1 and ΔT_2 of the rods is 30°C, what is the maximum mirror rotation angle that can be achieved? (Assume that the rods are uniformly heated or cooled along their lengths.)

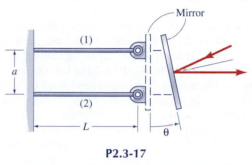

P2.3-17

▼ MEASUREMENT OF STRAIN

Prob. 2.4-1. A *mechanical extensometer* uses the lever principle to magnify the elongation of a test specimen enough to make the elongation (or contraction) readable. The extensometer shown in Fig. P2.4-1 is held against the test specimen by a spring that forces two sharp points against the specimen at A and B. The pointer AD pivots about a pin at C, so that the distance between the contact points at A and B is exactly L_0 = 6 in. (the *gage length*, or gauge length, of this extensometer) when the pointer points to the origin, O, on the scale. In a particular test, the extensometer arm points "precisely" at point O when the load P is zero. Later in the test, the 10-in.-long pointer points a distance d = 0.12 in. below point O. What is the current extensional strain in the test specimen at this reading?

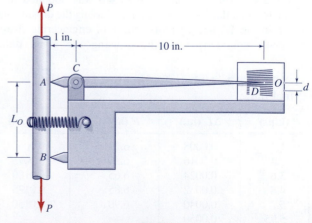

P2.4-1

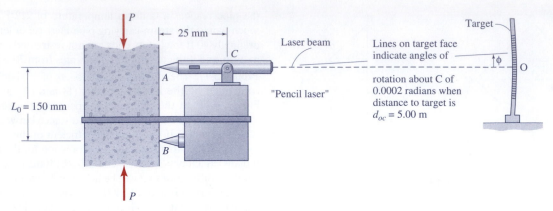

P2.4-2

Prob. 2.4-2. A "pencil" *laser extensometer*, like the mechanical lever extensometer in Prob. 2.4-1, measures elongation (from which extensional strain can be computed) by multiplying the elongation. In Fig. P2.4-2 the laser extensometer is being used to measure strain in a reinforced concrete column. The target is set up across the room from the test specimen so that the distance from the fulcrum, C, of the laser to the reference point O on the target is $d_{oc} = 5$ m. Also, the target is set so that the laser beam points directly at point O on the target when the extensometer points are exactly $L_0 = 150$ mm apart on the specimen, and the cross section at B does not move vertically. At a particular value of (compressive) load P, the laser points upward by an angle that is indicated on the target to be $\phi = 0.0030$ rad. Determine the extensional strain in the concrete column at this load value.

▼ **STRESS-STRAIN CURVES** | **MDS 2.5 & 2.6** |

Problems 2.4-3 through 2.4-6. *You are strongly urged to use a computer program (e.g., MDSolids or a spreadsheet program) to plot the stress-strain diagrams for these problems. In some cases it will be advantageous to make two plots, one covering the initial few points and one covering the entire dataset.*

[C]**Prob. 2.4-3.** The data in Table P2.4-3 was obtained in a tensile test of a flat-bar steel specimen having the dimensions shown in Fig. P2.4-3. (a) Plot a curve of engineering stress, σ, versus engineering strain, ϵ, using the given data.

TABLE P2.4-3. **Tension-test Data; Flat Steel Bar**

P (kips)	ΔL (in.)	P (kips)	ΔL (in.)
1.2	0.0008	6.25	0.0060
2.4	0.0016	6.50	0.0075
3.6	0.0024	6.65	0.0100
4.8	0.0032	6.85	0.0125
5.7	0.0040	6.90	0.0150
5.95	0.0050	—	—

(b) Determine the modulus of elasticity of this material. (c) Use the 0.2%-offset method to determine the yield strength of this material, σ_{YS}.

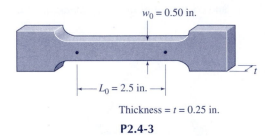

P2.4-3

[C]**Prob. 2.4-4.** A standard ASTM tension specimen (diameter = $d_0 = 0.505$ in., gage length = $L_0 = 2.0$ in.) was used to obtain the load-elongation data given in Table P2.4-4. (a) Plot a curve of engineering stress, σ, versus engineering strain, ϵ, using the given data. (b) Determine the modulus of elasticity of this material. (c) Use the 0.2%-offset method to determine the yield strength of this material, σ_{YS}.

TABLE P2.4-4. **Tension-test Data; ASTM Tension Specimen**

P (kips)	ΔL (in.)	P (kips)	ΔL (in.)
1.9	0.0020	10.0	0.0145
3.8	0.0040	10.4	0.0180
5.7	0.0060	10.65	0.0240
7.6	0.0080	11.00	0.0300
9.0	0.0100	11.05	0.0360
9.5	0.0120	—	—

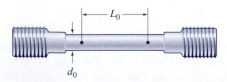

P2.4-4, P2.4-5, P2.4-6, and P2.4-8

[C]**Prob. 2.4-5.** A tension specimen (diameter = d_0 = 13 mm, gage length = L_0 = 50 mm) was used to obtain the load-elongation data given in Table P2.4-5. (a) Plot a curve of engineering stress, σ, versus engineering strain, ϵ, using the given data. (b) Determine the modulus of elasticity of this material. (c) Use the 0.2%-offset method to determine the yield strength of this material, σ_{YS}. (d) Determine the tensile ultimate stress, σ_{TU}.

TABLE P2.4-5. Tension-test Data

P (kN)	ΔL (mm)	P (kN)	ΔL (mm)
0.0	0.000	27.5	1.68
9.3	0.050	28.4	2.00
14.9	0.200	28.6	2.33
17.7	0.325	28.9	2.68
22.4	0.675	28.4	3.00
25.2	1.00	27.5	3.33
26.6	1.33	26.1	3.68

[C]**Prob. 2.4-6.** A tension specimen (diameter = d_0 = 0.500 in., gage length = L_0 = 2.0 in.) was used to obtain the load-elongation data given in Table P2.4-6. (a) Plot a curve of engineering stress, σ, versus engineering strain, ϵ, using the given data. (b) Determine the modulus of elasticity of this material. (c) Use the 0.2%-offset method to determine the yield strength of this material, σ_{YS}. (d) Determine the tensile ultimate stress, σ_{TU}.

TABLE P2.4-6. Tension-test Data

P (kips)	ΔL (in.)	P (kips)	ΔL (in.)
0.0	0.000	12.5	0.060
5.2	0.005	12.7	0.070
9.4	0.009	12.9	0.080
9.7	0.010	13.0	0.090
10.0	0.013	13.1	0.100
10.6	0.020	13.2	0.110
11.3	0.030	13.2	0.120
11.8	0.040	13.0	0.130
12.2	0.050	12.6	0.138

Prob. 2.4-7. Tension specimens (diameter = d_0 = 0.500 in., gage length = L_0 = 2.00 in.) made of structural materials A and B are tested to failure in tension. (a) At failure the distances between the gage marks are L_{Af} = 2.90 in. and L_{Bf} = 2.22 in.; the corresponding diameters at the failure cross sections are d_{Af} = 0.263 in. and d_{Bf} = 0.471 in., respectively. Determine the *percent elongation in 2 in.* and the *percent reduction in area* for these two materials, and classify each material as either *brittle* or *ductile*. (b) From these tensile

tests the following data are also obtained: E_A = 10.0 × 10³ ksi, $(\sigma_Y)_A$ = 5 ksi, $(\sigma_U)_A$ = 13 ksi; E_B = 10.4 × 10³ ksi, $(\sigma_Y)_B$ = 73 ksi, $(\sigma_U)_B$ = 83 ksi. From the data given here, make rough sketches (to scale) of the stress-strain diagrams of materials A and B.

Prob. 2.4-8. Numerous reference sources (e.g., see footnote 13 on p. 41) provide information on the mechanical properties of structural materials—aluminum alloys, copper, nylon, steel, titanium, etc.—including σ-ϵ curves, like those in Figs. 2.12 and 2.13. You will find such resources in your technical library or, perhaps, on the Internet. Obtain, for one particular material, a "room-temperature" stress-strain diagram and any other information on the material that you are able to find. (a) Make a copy of this information to hand in. Be sure to write down complete bibliographic information about your source (e.g., see References on p. R-1). (b) From the room-temperature stress-strain diagram, determine as many mechanical properties as you can (e.g., E, σ_Y). (c) Write a brief paragraph discussing appropriate uses for this material. For example, a particular aluminum alloy may be most useful in sheet form; another alloy is more widely used for extrusions.

▼ **MECHANICAL PROPERTIES OF MATERIALS**

MDS 2.7

In Problems 2.6-1 through 2.6-10, *dimensions that are shown on the figures, or dimensions that are labeled with subscript 0 (e.g., d_0, L_0), are dimensions of the specimen without any load applied.*

Prob. 2.6-1. A tensile test is performed on an aluminum specimen that is 0.505 in. in diameter using a gage length of 2 in., as shown in Fig. P2.6-1. (a) When the load is increased by an amount P = 2 kips, the distance between gage marks increases by an amount ΔL = 0.00196 in. Calculate the value of the modulus of elasticity, E, for this specimen. (b) If the proportional limit stress for this specimen is σ_{PL} = 45 ksi, what is the distance between gage marks at this value of stress?

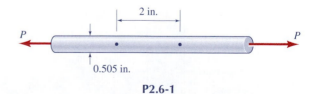

P2.6-1

Prob. 2.6-2. A short brass cylinder (d_0 = 15 mm, L_0 = 25 mm) is compressed between two perfectly smooth, rigid plates by an axial force P = 20 kN, as shown in Fig. P2.6-2. (a) If the measured shortening of the cylinder due to this force is 0.0283 mm, what is the modulus of elasticity, E, for this brass specimen? (b) If the increase in diameter due to the load P is 0.0058 mm, what is the value of Poisson's ratio, ν?

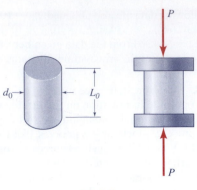

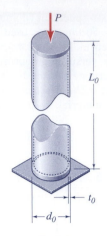

P2.6-2

P2.6-7

Prob. 2.6-3. A tensile specimen of a certain alloy has an initial diameter of 0.500 in. and a gage length of 8.00 in. Under a load $P = 4500$ lb, the specimen reaches its proportional limit and is elongated by 0.0118 in. At this load the diameter is reduced by $2.52(10^{-4})$ in. Determine the following material properties: (a) the modulus of elasticity, E, (b) Poisson's ratio, ν, and (c) the proportional limit, σ_{PL}.

Prob. 2.6-4. A tensile force of 400 kN is applied to a uniform segment of an ASTM-A36 structural steel bar. The cross section is a 50 mm × 50 mm square, and the length of the segment being tested is 200 mm. Using A36 steel data from Table F.2, (a) determine the change in the cross-sectional dimension of the bar, and (b) determne the change in volume of the 200 mm segment being tested.

Prob. 2.6-5. A cylindrical rod with an initial diameter of 8 mm is made of 6061-T6 aluminum alloy. When a tensile force P is applied to the rod, its diameter decreases by 0.0101 mm. Using the appropriate aluminum-alloy data from Table F.2, determine (a) the magnitude of the load P, and (b) the elongation over a 200 mm length of the rod.

Prob. 2.6-6. Under a compressive load of $P = 24$ kips, the length of the concrete cylinder in Fig. P2.6-6 is reduced from 12 in. to 11.9970 in., and the diameter is increased from 6 in. to 6.0003 in. Determine the value of the modulus of elasticity, E, and the value of Poisson's ratio, ν. Assume linearly elastic deformation.

kN, as shown in Fig. P2.6-7. If the steel has a modulus of elasticity $E = 200$ GPa and Poisson's ratio $\nu = 0.29$, determine: (a) the change, ΔL, in the length of the column, and (b) the change, Δt, in the wall thickness.

Prob. 2.6-8. The cylindrical rod in Fig. P2.6-8 is made of annealed (soft) copper with modulus of elasticity $E = 17 \times 10^3$ ksi and Poisson's ratio $\nu = 0.33$, and it has an initial diameter $d_0 = 1.9998$ in. For compressive loads less than a "critical load" P_{cr}, a ring with inside diameter $d_r = 2.0000$ in. is free to slide along the cylindrical rod. What is the value of the critical load P_{cr}?

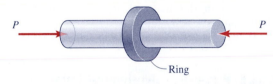

P2.6-8

Prob. 2.6-9. A rectangular aluminum bar ($w_0 = 2.0$ in., $t_0 = 0.5$ in.) is subjected to a tensile load P by pins at A and B (Fig. P2.6-9). Strain gages (which are described in Section 8.10) measure the following strains in the longitudinal (x) and transverse (y) directions: $\epsilon_x = 566\mu$, and $\epsilon_y = -187\mu$. (a) What is the value of Poisson's ratio for this specimen? (b) If the load P that produces these values of ϵ_x and ϵ_y is $P = 6$ kips, what is the modulus of elasticity, E, for this specimen? (c) What is the change in volume, ΔV, of a segment of bar that is initially 2 in. long? (Hint: $\epsilon_z = \epsilon_y$.)

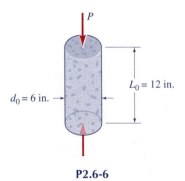

P2.6-6

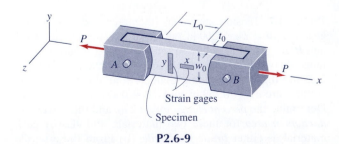

P2.6-9

Prob. 2.6-7. A steel pipe column of initial length $L_0 = 4$ m, initial outer diameter $d_0 = 100$ mm, and initial wall thickness $t_0 = 10$ mm is subjected to an axial compressive load $P = 200$

Prob. 2.6-10. A rigid, weightless beam BD supports a load P and is, in turn, supported by two hanger rods, (1) and (2), as shown in Fig. P2.6-10. The rods are initially the same length $L = 2$ m and are made of the same material. Their rectangular cross sections have original dimensions ($w_1 = 40$ mm, $t_1 = 20$ mm) and ($w_2 = 50$ mm, $t_2 = 25$ mm), respectively. $L_{BD} = 1.5$ m. (a) At what location, b, must the load P act if the rigid bar BD is to remain horizontal when the load is applied? (b) If the longitudinal strain in the hanger rods is $\epsilon_1 = \epsilon_2 = 500\mu$ when the load is $P = 205$ kN, what is the value of the modulus of elasticity, E? (c) If application of load P in the manner described in Parts (a) and (b) above causes the dimension w_2 of hanger rod (2) to be reduced from 50 mm to 49.9918 mm, what is the value of Poisson's ratio, ν, for this material?

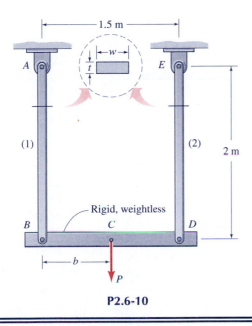

P2.6-10

Prob. 2.7-1. Two bolts are used to form a joint connecting rectangular bars in tension, as shown in Fig. P2.7-1. If the bolts have a diameter of 3/8 in., and the load is $P = 20$ kips, determine the average shear stress on the bolt surfaces that are subjected to direct shear.

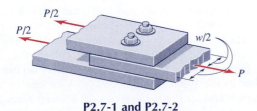

P2.7-1 and P2.7-2

ᴰProb. 2.7-2. Two bolts are used to form a joint connecting rectangular bars in tension, as shown in Fig. P2.7-2. Determine the required diameter of the bolts if the average shear stress for the bolts is not to exceed 140 MPa for the given loading of $P = 80$ kN.

Prob. 2.7-3. An angle bracket, whose thickness is $t = 12.7$ mm, is attached to the flange of a column by two 15-mm-diameter bolts, as shown in Fig. P2.7-3. A floor joist that frames into the column exerts a uniform downward pressure of $p = 2$ MPa on the top face of the angle bracket. The dimensions of the loaded face are $L = 152$ mm and $b = 76$ mm. Determine the average shear stress, τ_{avg}, in the bolts. (Neglect the friction between the angle bracket and the column.)

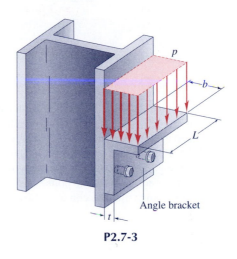

P2.7-3

▼ **SHEAR STRESS**   MDS 2.8–2.11

ᴰProb. 2.7-4. A 100-kip capacity hydraulic punch press is used to punch circular holes in a 3/8-in.-thick aluminum plate, as illustrated in Fig. P2.7-4. If the average punching shear resistance of this plate is 30 ksi, what is the maximum diameter of hole that can be punched?

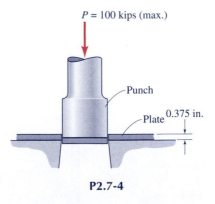

P2.7-4

Prob. 2.7-5. The hole in a hasp plate is punched out by a hydraulic punch press similar to the one in Prob. 2.7-4, with a punch in the shape of the "rectangular" hole as illustrated in Fig. P2.7-5. If the hasp plate is $\frac{1}{16}$-in.-thick steel with an average punching shear resistance of 38 ksi, what is the required punch force P?

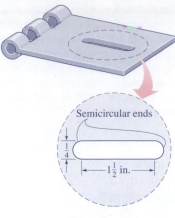

Semicircular ends

$\frac{1}{4}$

$1\frac{1}{2}$ in.

P2.7-5

Prob. 2.7-6. A pipe flange is attached by four bolts, whose effective diameter is 0.425 in., to a concrete base. The bolts are uniformly spaced around an 8-in.-diameter bolt circle, as shown in Fig. P2.7-6. If a twisting couple $T = 5000$ lb · in. is applied to the pipe flange, as shown in Fig. P2.7-6, what is the average shear stress in each of the four bolts? Neglect friction between the pipe flange and the concrete base.

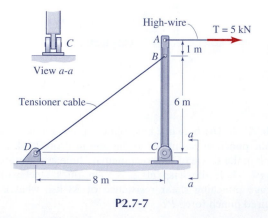

$T = 5000$ lb·in.

Pipe flange

Concrete base

8 in.

P2.7-6

Prob. 2.7-7. The high-wire for a circus act is attached to a vertical beam AC and is kept taut by a tensioner cable BD, as illustrated in Fig. P2.7-7. At C, the beam AC is attached by

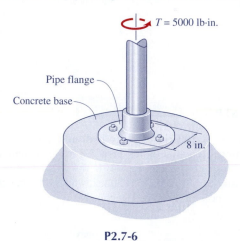

View a-a

High-wire

$T = 5$ kN

C

A

B

1 m

Tensioner cable

6 m

D

C

a

8 m

a

P2.7-7

a 10-mm-diameter bolt to the bracket shown in View a-a. Determine the average shear stress in the bolt at C if the tension in the high-wire is 5 kN. (Assume that the high-wire is horizontal, and neglect the weight of AC.)

Prob. 2.7-8. An angle bracket ABC is restrained by a high-strength steel wire CD, and it supports a load P at A, as shown in Fig. P2.7-8. The diameter of the wire CD is $d_w = \frac{1}{8}$ in., and the diameter of the pin at B is $d_p = \frac{1}{8}$ in. Determine the tensile stress in wire CD and the average direct shear stress in the pin at B.

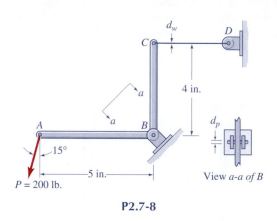

d_w

D

C

a

a

A

B

15°

4 in.

d_p

5 in.

$P = 200$ lb.

View a-a of B

P2.7-8

Prob. 2.7-9. Two $\frac{3}{4}$-in. nylon rods are spliced together by gluing a 2-in. section of plastic pipe over the rod ends, as shown in Fig. P2.7-9. If a tensile force of $P = 500$ lb is applied to the spliced nylon rod, what is the average shear stress in the glue joint between the pipe and the rods?

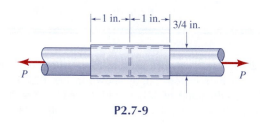

1 in. 1 in. 3/4 in.

P P

P2.7-9

Prob. 2.7-10. Loads P (pull) and V (vertical) in Fig. P2.7-10 are exerted on the ball of a trailer hitch by the trailer it is towing. The ball is bolted to a solid square steel bar which, in turn, fits into a square tubular steel receptacle that is attached to the tow vehicle. A steel pin transmits load from the solid bar to the tubular receptacle. (a) Visit one or more parking lots and see if you can spot such a trailer hitch. (You might need to visit a trailer rental agency or a boat dealership.) Briefly report on what you saw, including sketches that indicate (approximate) dimensions of the hitch parts. (b) Draw a free-body diagram of the solid steel hitch bar together with the ball, indicating how you think loads P and V are transmitted to the tubular receptacle. Indicate specifically what loading the steel pin transmits.

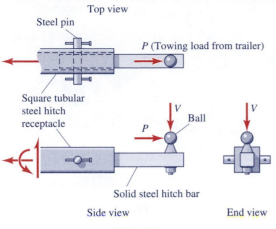

Top view

Steel pin

P (Towing load from trailer)

Square tubular
steel hitch
receptacle

Ball

P

V

V

Solid steel hitch bar

Side view

End view

P2.7-10

▼ SHEAR STRAIN

Prob. 2.7-11. A rectangular plate (dashed lines show original configuration) is uniformly deformed into the shape of a parallelogram (shaded figure) as shown in Fig. P2.7-11. (a) Determine the average shear strain, call it $\gamma_{xy}(A)$, between lines in the directions x and y shown in the figure. (b) Determine the average shear strain, call it $\gamma_{x'y'}$ (B), between lines in the directions of x' and y' shown in the figure. (Hint: Don't forget that shear strain is a signed quantity, that is, it can be either positive or negative.)

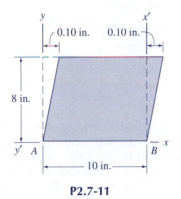

y

x'

0.10 in. 0.10 in.

8 in.

y' A

B

x

10 in.

P2.7-11

Prob. 2.7-12. Shear stress τ produces a shear strain γ_{xy} (between lines in the x direction and lines in the y direction) of $\gamma_{xy} = 1200\,\mu$ (i.e., $\gamma = 0.0012\,\frac{m}{m}$). (a) Determine the horizontal displacement δ_A of point A. (b) Determine the shear strain $\gamma_{x'y'}$ between the lines in the x' direction and the y' direction, as shown on Fig. P2.7-12.

Prob. 2.7-13. Two identical symmetrically placed rubber pads transmit load from a rectangular bar to a C-shaped bracket, as shown in Fig. P2.7-13. (a) Determine the average shear stress, τ, in the rubber pads on planes parallel to the top and bottom surfaces of the pads if $P = 250\,\text{N}$ and the dimensions of the rubber pads are: $b = 50$ mm, $w = 80$ mm, and $h = 25$ mm. (Although the load is transmitted predominately by shearing deformation, the pads are not undergoing pure shear. However, you can still calculate the *average*

shear stress and *average* shear strain.) (b) If the shear modulus of elasticity of the rubber is $G_r = 0.6$ MPa, what is the average shear strain, γ, related to the average shear stress τ computed in Part (a)? (c) Based on the average shear strain determined in Part (b), what is the relative displacement, δ, between the rectangular bar and the C-shaped bracket when the load $P = 250$ N is applied?

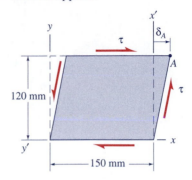

x'

y

δ_A

τ

A

120 mm

τ

x

y'

150 mm

P2.7-12

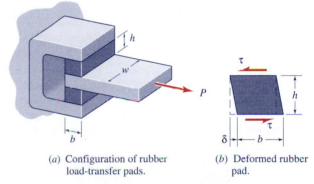

h

w

P

τ

h

b

δ

τ

b

(a) Configuration of rubber load-transfer pads.

(b) Deformed rubber pad.

P2.7-13 and P2.7-14

ᴰProb. 2.7-14. Two identical, symmetrically placed rubber pads transmit load from a rectangular bar to a C-shaped bracket, as shown in Fig. 2.7-14. The dimensions of the rubber pads are: $b = 3$ in., $w = 4$ in., and $h = 2$ in. The shear modulus of elasticity of the rubber is $G_r = 100$ psi. If the maximum relative displacement between the bar and the bracket is $\delta_{max} = 0.25$ in., what is the maximum value of load P that may be applied? (Use average shear strain and average shear stress in solving this problem.)

*Prob. 2.7-15. Vibration isolators like the one shown in Fig. P2.7-15 are used to support sensitive instruments. Each isolator consists of a hollow rubber cylinder of outer diameter D, inner diameter d, and height h. A steel center post of diameter d is bonded to the inner surface of the rubber cylinder, and the outer surface of the rubber cylinder is bonded to the inner surface of a steel-tube base. (a) Derive an expression for the average shear stress in the rubber as a function of the distance r from the center of the isolator. (b) Derive an expression relating the load P to the downward displacement of the center post, using G as the shear modulus of the rubber, and assuming that the steel post and steel tube are rigid (compared with the rubber). (Hint: Since the shear strain varies with the distance r from the center, an integral is required.)

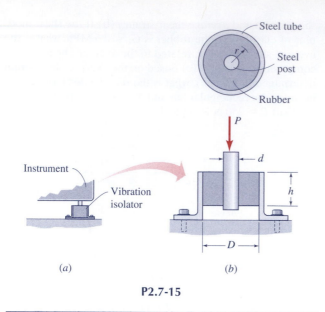

Steel tube

Steel post

Rubber

P

Instrument

Vibration isolator

d

h

D

(a)

(b)

P2.7-15

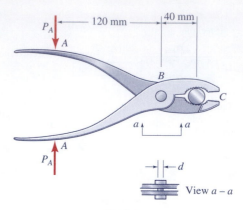

120 mm

40 mm

P_A

A

B

C

a a

P_A

A

d

View $a - a$

P2.8-3

▼ DESIGN FOR AXIAL LOADS AND DIRECT SHEAR

MDS 2.12–2.15

^D**Prob. 2.8-1.** A bolted lap joint is used to connect a rectangular bar to a hanger bracket, as shown in Fig. P2.8-1. If the allowable shear stress in the bolt is 15 ksi, and the allowable tensile load on the rectangular bar is to be $P_{allow.} = 2$ kips, what is the required minimum diameter of the bolt shank in inches?

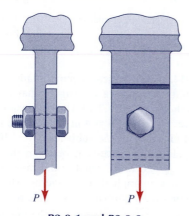

P P

P2.8-1 and P2.8-2

^D**Prob. 2.8-2.** A bolted lap joint is used to connect a rectangular bar to a hanger bracket, as shown in Fig. P2.8-2. If the allowable shear stress in the bolt is 80 MPa, and the diameter of the bolt shank 15 mm, what is the allowable tensile load on the rectangular bar, $P_{allow.}$, in kN?

^D**Prob. 2.8-3.** The pin that holds the two halves of a pair of pliers together at B has a diameter $d = 6.35$ mm and is made of steel for which $\tau_{allow.} = 75$ MPa. What is the allowable force $(P_C)_{allow.}$ (not shown) that can be exerted on the round rod at C by each jaw, assuming that the corresponding force P_A is applied to the handles at each of the two places marked A in Fig. P2.8-3?

^D**Prob. 2.8-4.** The brass eye-bar in Fig. P2.8-4a has a diameter $d_r = 0.500$ in. and is attached to a support bracket by a brass pin of diameter $d_p = 0.375$ in. If the allowable shear stress in the pin is 12 ksi and the allowable tensile stress in the bar is 18 ksi, what is the allowable tensile load $P_{allow.}$?

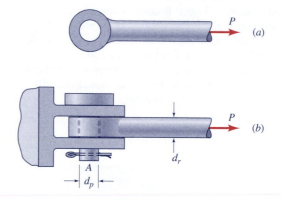

P

(a)

P

(b)

d_r

A

d_p

P2.8-4

^D**Prob. 2.8-5.** The forestay (cable) on a sailboat is attached to a tee-bracket on the deck of the boat by a (removable) stainless steel pin. If the allowable shear stress in the pin is $\tau_{allow.} = 11$ ksi, and the diameter of the pin is $d_p = 0.25$ in., what is the allowable tensile force, $T_{allow.}$, in the stay?

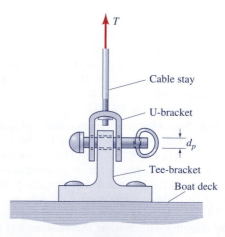

T

Cable stay

U-bracket

d_p

Tee-bracket

Boat deck

P2.8-5

ᴰProb. 2.8-6. A compressor of weight W is suspended from a sloping ceiling beam by long rods AB and CD of diameters d_1 and d_2, respectively, as shown in Fig. P2.8-6a. A typical bracket is shown in Fig. P2.8-6b. Using the data given below, determine the allowable compressor weight, $W_{allow.}$. (Neglect the weight of the platform between A and C, and neglect the weight of the two rods. Also, assume that rod AB and the pins at A and B are large enough that they do not need to be considered.)

$$\text{Rod } CD: d_2 = 10 \text{ mm}, \sigma_{allow.} = 85 \text{ MPa}$$

$$\text{Pins at } C \text{ and } D: d_p = 7 \text{ mm}, \tau_{allow.} = 100 \text{ MPa}$$

$$a = 0.75 \text{ m}, b = 0.50 \text{ m}$$

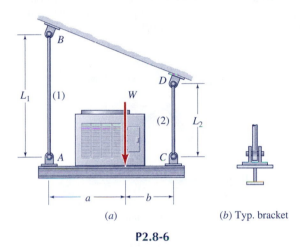

(a)   (b) Typ. bracket

P2.8-6

ᴰProb. 2.8-7. An angle bracket ABC is restrained by a high-strength steel wire CD, and it supports a load P at A, as shown in Fig. P2.8-7. The strength properties of the wire and the shear pin at B are $\sigma_y = 350$ MPa (wire), and $\tau_U = 300$ MPa (pin at B). If the wire and pin are to be sized to provide a factor of safety against yielding of the wire of $FS_\sigma = 3.3$ and a factor of safety against ultimate shear failure of the pin of $FS_\tau = 3.5$, what are the required diameters of the wire (to the nearest mm) and the pin (to the nearest mm)?

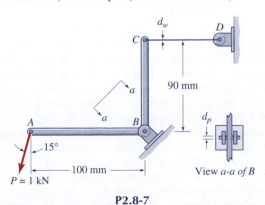

P2.8-7

ᴰProb. 2.8-8. Boom AC in Fig. P2.8-8 is supported by a rectangular steel bar BD, and it is attached to a bracket at C by

---

a high-strength steel pin. Assume that the pin at B is adequate to sustain the loading applied to it, and that the design-critical components are the bar BD and the pin at C. The factor of safety against failure of BD by yielding is $FS_\sigma = 3.0$, and the factor of safety against ultimate shear failure of the pin at C is $FS_\tau = 3.3$. (a) Determine the required thickness, t, of the rectangular bar BD, whose width is b. (b) Determine the required diameter, d, of the pin at C.

$$P = 2400 \text{ lb}, \quad L = 6 \text{ ft}, \quad h = 4 \text{ ft}$$

$$\text{Bar } BD: b = 1 \text{ in.}, \quad \sigma_Y = 36 \text{ ksi}, \quad \text{Pin } C: \tau_U = 60 \text{ ksi}$$

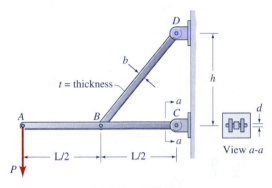

P2.8-8 and P2.8-9

ᴰProb. 2.8-9. Solve Prob. 2.8-8 using the following data:

$$P = 10 \text{ kN}, \quad L = 3 \text{ m}, \quad h = 2 \text{ m}$$

$$\text{Bar } BD: b = 25 \text{ mm}, \quad \sigma_Y = 250 \text{ MPa},$$

$$\text{Pin } C: \tau_U = 400 \text{ MPa}$$

ᴰProb. 2.8-10. A load W is to be suspended from a cable at end C of a rigid beam AC, whose length is $b = 3$ m. Beam AC, in turn, is supported by a steel rod of diameter $d = 25$ mm and length $L = 2.5$ m. Rod BD is made of steel with a yield point $\sigma_Y = 250$ MPa, and modulus of elasticity $E = 200$ GPa. If the maximum displacement at C is $(\delta_C)_{max} = 10$ mm, and there is to be a factor of safety with respect to yielding of BD of $FS_\sigma = 3.3$ and with respect to displacement of $FS_\delta = 3.0$, what is the allowable weight that can be suspended from the beam at C?

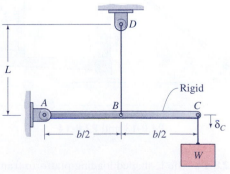

P2.8-10 and P2.8-11

ᴰProb. 2.8-11. A load of $W = 2$ kips is suspended from a cable at end C of a rigid beam AC, whose length is $b = 8$ ft. Beam AC, in turn, is supported by a steel rod BD ($E = 30 \times 10^3$ ksi) of diameter d and length $L = 12$ ft. The rod BD is to the sized so that there will be a factor of safety with respect to yielding of $FS_\sigma = 4.0$ and a factor of safety with respect to deflection of $FS_\delta = 3.0$. The yield strength of rod BD is $\sigma_Y = 50$ ksi, and the maximum displacement at C is limited to

$$(\delta_C)_{max} = 0.25 \text{ in.} \left(\text{i.e.,} (\delta_C)_{allow} = \frac{(\delta_C)_{max}}{FS_\delta} \right).$$ Determine the

required diameter, d, of rod BD to the nearest $\frac{1}{8}$ in.

ᴰProb. 2.8-12. A tension rod is spliced together by a pin-and-yoke type connector, as shown in Fig. P2.8-12. The tension rod is to be designed for an allowable load of $P_{allow.} = 3$ kips. If the allowable tensile stress in the rods is $\sigma_{allow.} = 15$ ksi, and the allowable shear stress in the pin is $\tau_{allow.} = 12$ ksi, determine (to the nearest $\frac{1}{16}$ in.) (a) the smallest diameter, d_r, of rod that can be used, and (b) the smallest diameter, d_p, of pin that can be used.

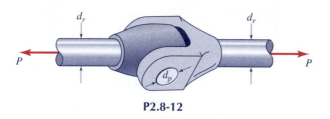

P2.8-12

ᴰProb. 2.8-13. The L-shaped loading frame in Fig. P2.8-13 is supported by a high-strength shear pin ($d_p = 0.5$ in, $\tau_U = 50$ ksi) and by a tie-rod AB ($d_r = 0.625$ in., $\sigma_Y = 50$ ksi). Both the tie-rod and the pin are to be sized with a factor of safety of $FS = 3.0$, the tie-rod with respect to tensile yielding, and the shear pin with respect to ultimate shear failure. Determine the allowable platform load, $W_{allow.}$. Let $L_1 = 3$ ft, $L_2 = L_3 = 4$ ft.

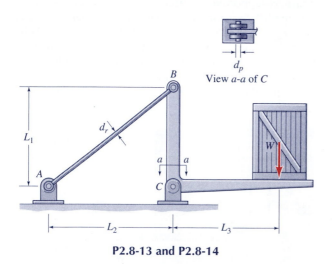

P2.8-13 and P2.8-14

ᴰProb. 2.8-14. The L-shaped loading-platform frame in Fig. P2.8-14 is supported by a high-strength steel shear pin at C

and by a tie-rod AB. Both the tie rod and the pin are to be sized with a factor of safety of $FS = 3.0$, the tie-rod with respect to tensile yielding and the pin with respect to shear failure. The strength properties of the rod and pin are: $\sigma_Y = 340$ MPa and $\tau_U = 340$ MPa; the respective lengths are: $L_1 = 1.5$ m and $L_2 = L_3 = 2.0$ m. (a) If the loading platform is to be able to handle loads W up to $W = 8$ kN, what is the required diameter, d_r, of the tie-rod (to the nearest millimeter)? (b) What is the required diameter, d_p, of the shear pin at C (to the nearest millimeter)?

ᴰProb. 2.8-15. A three-bar, pin-jointed, planar truss supports a single horizontal load P at joint B. Joint C is free to move horizontally. The allowable stress in tension is $(\sigma_T)_{allow.} = 140$ MPa, and the allowable stress in compression is $(\sigma_C)_{allow.} = -85$ MPa. If the truss is to support a maximum load $P_{allow.} = 50$ kN, what are the required cross-sectional areas, A_i, of the three truss members?

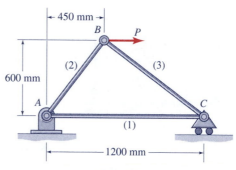

P2.8-15

▼ **COMPUTER-BASED DESIGN FOR AXIAL LOADS**

For **Problems 2.8-16 through 2.8-18** *you are to develop a computer program to generate the required graph(s) that will enable you to choose the "optimum design." You may use a spreadsheet program or other mathematical application program (e.g., TK Solver or Mathcad), or you may write a program in a computer language (e.g., BASIC or FORTRAN).*

ᶜProb. 2.8-16. The pin-jointed planar truss shown in Fig. P2.8-16a is to be made of two steel two-force members and support a single vertical load $P = 10$ kN at joint B. For the steel truss members, the allowable stress in tension is $(\sigma_T)_{allow.} = 150$ MPa, the allowable stress in compression is $(\sigma_C)_{allow.} = -100$ MPa, and the weight density is 77.0 kN/m³. You are to consider truss designs for which joint B can be located at any point along the vertical line that is 1 m to the right of AC, with y_B varying from $y_B = 0$ to $y_B = 2$m. (a) Show that, if each member has the minimum cross-sectional area that meets the strength criteria stated above, the weight W of the truss can be expressed as a function of y_B, the position of joint B, by the function that is plotted in Fig. P2.8-16b. (b) What value of y_B gives the minimum-weight truss, and what is the weight of that truss?

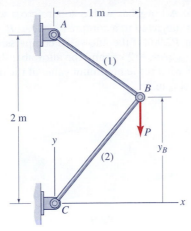

(a) A two-bar planar truss.

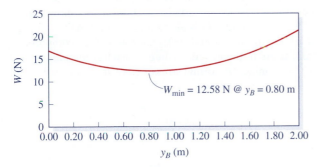

(b) Minimum-weight design for the two-bar planar truss.

$W_{min} = 12.58$ N @ $y_B = 0.80$ m

P2.8-16

is $(\sigma_C)_{allow} = -12$ ksi, and the weight density is 0.284 lb/in³. You are to consider truss designs for which the vertical member AC has lengths varying from $L_3 = 18$ in. to $L_3 = 50$ in. (a) Show that, if each member has the minimum cross-sectional area that meets the strength criteria stated above, the weight W of the truss can be expressed as a function of the length L_3 of member AC by a function that is similar to the one plotted in Fig. P2.8-16b. (Hint: Use the *law of cosines* to obtain expressions for the angle at joint A and the angle at joint C.) (b) What value of L_3 gives the minimum-weight truss, and what is the weight of that truss?

$^C*$**Prob. 2.8-18.** The pin-jointed planar truss shown in Fig. P2.8-18 is to be made of two aluminum two-force members and support a single horizontal load $P = 50$ kN at joint B. For the aluminum truss members, the allowable stress in tension is $(\sigma_T)_{allow} = 200$ MPa, the allowable stress in compression is $(\sigma_C)_{allow} = -130$ MPa, and the weight density is 28.0 kN/m³. You are to consider truss designs for which support C can be located at any point along the x axis, with x_C varying from $x_C = 1$ m to $x_C = 2.4$ m. (a) Show that, if each member has the minimum cross-sectional area that meets the strength criteria stated above, the weight W of the truss can be expressed as a function of x_C, the position of support C, by a function that is similar to the one plotted in Fig. 2.8-16b. (b) What value of x_C gives the minimum-weight truss, and what is the weight of that truss?

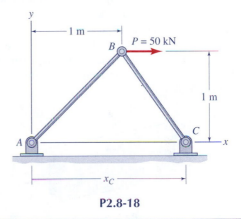

P2.8-18

$^C*$**Prob. 2.8-17.** The pin-jointed planar truss shown in Fig. P2.8-17 is to be made of three steel two-force members and is to support vertical loads $P_B = 2$ kips at joint B and $P_C = 3$ kips at joint C. The lengths of members AB and BC are $L_1 = 30$ in. and $L_2 = 24$ in., respectively. Joint C is free to move vertically. For the steel truss members, the allowable stress in tension is $(\sigma_T)_{allow} = 20$ ksi, the allowable stress in compression

▼ **STRESSES ON INCLINED PLANES**

In Problems 2.9-1 through 2.9-13, *use free-body diagrams and equilibrium equations to solve for the required stresses.*

Prob. 2.9-1. The plane NN' makes an angle $\theta = 30°$ with respect to the cross section of the prismatic bar shown in Fig. P2.9-1. The dimensions of the rectangular cross section

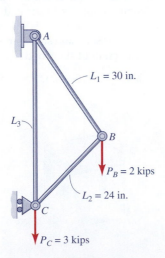

P2.8-17

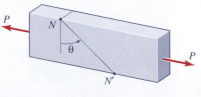

P2.9-1

of the bar are 1 in. × 2 in. Under the action of an axial tensile load P, the normal stress on the NN' plane is $\sigma_n =$ 8 ksi. (a) Determine the value of the axial load P; (b) determine the shear stress τ_{nt} on the NN' plane; and (c) determine the maximum shear stress in the bar. Use free-body diagrams and equilibrium equations to solve for the required stresses.

Prob. 2.9-2. The prismatic bar in Fig. P2.9-2 is subjected to an axial compressive load $P = -70$ kips. The cross-sectional area of the bar is 2.0 in². Determine the normal stress and the shear stress on the n face and on the t face of an element oriented at angle $\theta = 40°$. Use free-body diagrams and equilibrium equations to solve for the required stresses.

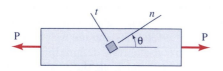

P2.9-2 and P2.9-3

Prob. 2.9-3. A prismatic bar in tension has a cross section that measures 20 mm × 50 mm and supports a tensile load $P = 200$ kN, as illustrated in Fig. P2.9-3. Determine the normal and shear stresses on the n and t faces of an element oriented at angle $\theta = 30°$. Use free-body diagrams and equilibrium equations to solve for the required stresses.

ᴰProb. 2.9-4. Determine the allowable tensile load P for the prismatic bar shown in Fig. P2.9-4 if the allowable tensile stress is $\sigma_{\text{allow}} = 135$ MPa and the allowable shear stress is $\tau_{\text{allow}} = 100$ MPa. The cross-sectional dimensions of the bar are 12.7 mm × 50.8 mm.

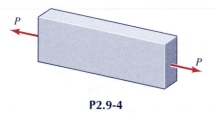

P2.9-4

Prob. 2.9-5. A bar with rectangular cross section is subjected to an axial tensile load P, as shown in Fig. P2.9-5. (a) Determine the angle, call it θ_{na}, of the plane NN' on which $\tau_{nt} = 2\sigma_n$, that is, the plane on which the magnitude of the shear stress is twice the magnitude of the normal stress. (b) Determine the angle, call it θ_{nb}, of the plane on which $\sigma_n = 2\tau_{nt}$. (Hint: You can get approximate answers from Fig. 2.34.)

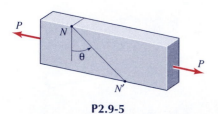

P2.9-5

ᴰProb. 2.9-6. A brass bar with a square cross section of dimension b is subjected to a compressive load $P = 10$ kips, as shown in Fig. P2.9-6. If the allowable compressive stress for the brass is $\sigma_{\text{allow.}} = -12$ ksi, and the allowable shear stress is $\tau_{\text{allow.}} = 7$ ksi, what is the minimum value of the dimension b, to the nearest $\frac{1}{16}$ in.?

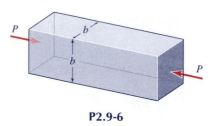

P2.9-6

Prob. 2.9-7. A 6-in.-diameter concrete test cylinder is subjected to a compressive load $P = 110$ kips, as shown in Fig. P2.9-7. The cylinder fails along a plane that makes an angle of 62° to the horizontal. (a) Determine the (compressive) axial stress in the cylinder when it reaches the failure load. (b) Determine the normal stress, σ, and shear stress, τ, on the failure plane at failure.

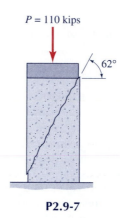

P2.9-7

Prob. 2.9-8. A wood cube that has dimension b on each edge is tested in compression, as illustrated in Fig. P2.9-8. The direction of the grain of the wood is shown in the figure. Determine the normal stress σ_n and shear stress τ_{nt} on planes that are parallel to the grain of the wood.

$$P = 5 \text{ kN}, \quad b = 150 \text{ mm}, \quad \alpha = 55°$$

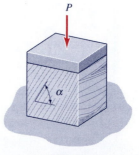

P2.9-8

D*Prob. 2.9-9. Either a finger-joint splice, Fig. P2.9-9a, or a diagonal lap-joint splice, Fig. P2.9-9b, may be used to glue two wood strips together to form a longer tension member. Determine the ratio of allowable loads, $(P_f)_{\text{allow.}}/(P_d)_{\text{allow.}}$ for the following two glue-strength cases: (a) the glue is twice as strong in tension as it is in shear, that is $\tau_{\text{allow.}} = 0.5\sigma_{\text{allow.}}$, and (b) the glue is twice as strong in shear as it is in tension, that is $\tau_{\text{allow.}} = 2\sigma_{\text{allow.}}$. (Hint: For each of the above cases, determine $P_{\text{allow.}}$ in terms of $\sigma_{\text{allow.}}$, using the given glue strength ratios.)

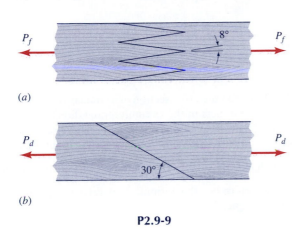

(a)

(b)

P2.9-9

Prob. 2.9-10. At room temperature (70°F) and with no axial load ($P = 0$) the extensional strain of the prismatic bar (Fig. P2.9-10) in the axial direction is zero, that is, $\epsilon_x = 0$. Subsequently, the bar is heated to 120°F and a tensile load P is applied. The material properties for the bar are: $E = 10 \times 10^3$ ksi and $\alpha = 13 \times 10^{-6}/°F$, and the cross-sectional area of the bar is 1.8 in². For the latter load-temperature condition, the extensional strain is found to be $\epsilon_x = 900 \times 10^{-6} \frac{\text{in.}}{\text{in.}}$. (a) Determine the value of the axial tensile load P. (b) Determine the normal stress and the shear stress on the oblique plane NN'. (Note: The total strain is the sum of strain associated with normal stress σ_x (Eq. 2.14) and the strain due to change of temperature ΔT (Eq. 2.8).)

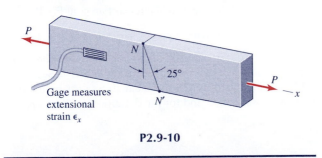

P2.9-10

▼ **STRESS RESULTANTS**

Prob. 2.12-1. The normal stress, σ_x, over the top half of the cross section of the rectangular bar in Fig. P2.12-1 is σ_0, while the normal stress acting on the bottom half of the cross section is $2\sigma_0$. (a) Determine the value of the resultant axial force, F_x. (b) Locate the point R in the cross section through which the resultant axial force, F_x, acts.

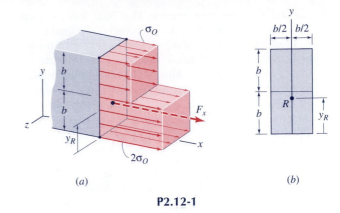

(a)

(b)

P2.12-1

Prob. 2.12-2 The normal stress on the rectangular cross section $ABCD$ in Fig. P2.12-2 varies linearly with respect to the y coordinate. That is, σ_x has the form $\sigma_x = a + by$, varying linearly from σ_{xb} at the bottom edge of the cross section to σ_{xt} at the top edge of the cross section. (a) Show that $M_y = 0$ for this symmetrical normal-stress distribution. (b) Determine an expression for the axial force F_x in terms of the stresses σ_{xb} and σ_{xt} and the dimensions of the cross section, width b and height h. (c) Determine an expression for the corresponding value of the bending moment M_z.

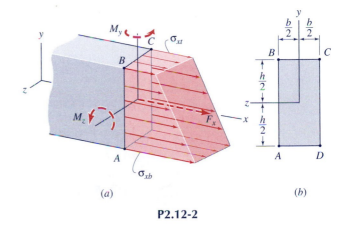

(a)

(b)

P2.12-2

Prob. 2.12-3. The normal stress on the rectangular cross section $ABCD$ in Fig. P2.12-3 varies linearly with respect to

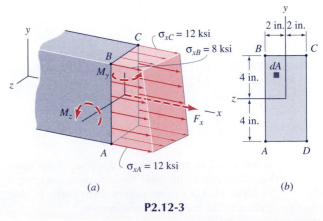

(a)

(b)

P2.12-3

position (y, z) in the cross section. That is, σ_x has the form $\sigma_x = a + by + cz$. The values of σ_x at corners A, B, and C are: $\sigma_{xA} = 12$ ksi, $\sigma_{xB} = 8$ ksi, and $\sigma_{xC} = 12$ ksi. (a) Determine the value of σ_{xD}, the normal stress at corner D. (b) Determine the axial force, F_x. (c) Determine the bending moment M_y.

Prob. 2.12-4. The stress distribution on the cross section shown in Fig. P2.12-4a is given by

$$\sigma_x = a + by \quad ; \quad \tau_{xy} = c\left[\left(\frac{h}{2}\right)^2 - y^2\right] \quad ; \quad \tau_{xz} = 0$$

Determine expressions for the resultant forces F_x and V_y and the bending moment M_z in terms of stress-related quantities a, b, and c and the dimensions d and h of the cross section. (See Example 2.13.)

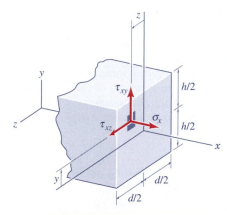

(a) The stresses on cross section x.

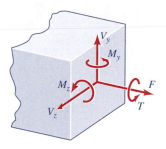

(b) The stress resultants at x.

P2.12-4

^D**Prob. 2.12-5.** On a particular cross section of the rectangular bar shown in Fig. 2.12-5 there is shear stress whose distribution has the form

$$\tau_{xy} = \tau_{max}\left[1 - \left(\frac{y}{50}\right)^2\right]$$

where y is measured in mm from the centroid of the cross section (Fig. P2.12-5b). If the shear stress τ_{xy} may not exceed $\tau_{allow.} = 50$ MPa, what is the maximum shear force V_y that may be applied to the bar at this cross section?

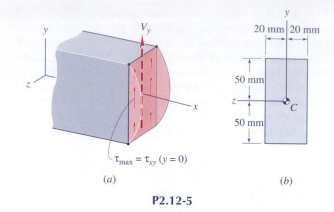

$\tau_{max} = \tau_{xy} \ (y = 0)$

(a)

(b)

P2.12-5

Prob. 2.12-6. On the cross section of a circular rod, the shear stress at a point acts in the circumferential direction at that point, as illustrated in Fig. P2.12-6. The shear stress magnitude varies linearly with distance from the center of the cross section, that is, $\tau = \dfrac{\tau_{max}\rho}{r}$. Using the ring-shaped area in Fig. P2.12-6b, determine the formula that relates τ_{max} and the resultant torque, T.

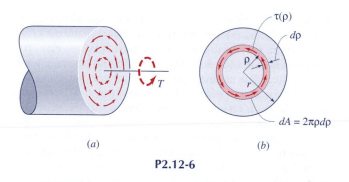

(a)

(b)

P2.12-6

Prob. 2.12-7. Determine the relationship between τ_{max} and T if, instead of acting on a solid circular bar, as in Fig. P2.12-6, the shear stress distribution $\tau = \dfrac{\tau_{max}\rho}{r_o}$ acts on a tubular cylinder with outer radius r_o and inner radius r_i. (The cross-sectional dimensions are shown in Fig. P2.12-7. See Prob. 2.12-6 for an illustration of the shear stress distribution on a circular cross section and for the definitions of T and ρ.)

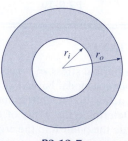

P2.12-7

*Prob. 2.12-8. If the magnitude of the shear stress on a solid, circular rod of radius r varies with radial position ρ as shown in Fig. P2.12-8, determine the formula that relates the resultant torque T to the maximum shear stress τ_Y. (See Prob. 2.12-6 for an illustration of the shear stress distribution on the cross section and for the definitions of T and ρ, and use the area shown in Fig. 2.12-6b.)

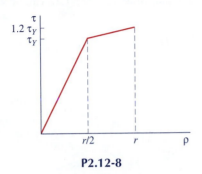

P2.12-8

▼ STRAIN-DEFORMATION EQUATIONS

Prob. 2.12-9. (a) Using Figs. P2.12-9 and the definition of extensional strain given in Eq. 2.35, show that the change in length, ΔL, of a thin wire whose original length is L is given by $\Delta L = \int_0^L \epsilon_x(x)\, dx$, where $\epsilon_x(x)$ is the extensional strain of the wire at x. (b) Determine the elongation of a 2-m-long wire if it has a coefficient of thermal expansion $\alpha = 20 \times 10^{-6}/°C$, and if the change in temperature along the wire is given by $\Delta T = 10x^2$ (°C).

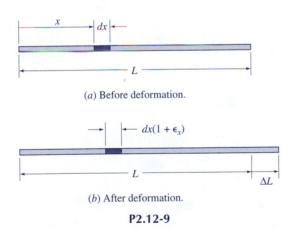

(a) Before deformation.

(b) After deformation.

P2.12-9

Prob. 2.12-10. The thin, rectangular plate $ABCD$ shown in Fig. P2.12-10a undergoes uniform stretching in the x direction,

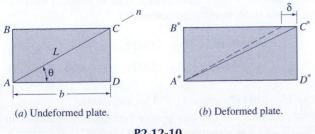

(a) Undeformed plate.　　　　(b) Deformed plate.

P2.12-10

so that it is elongated by an amount δ. Determine an expression for the (uniform) extensional strain ϵ_n of the diagonal AC. Express your answer in terms of δ, L, and the angle θ. Assume that $\delta \ll L$ and see Appendix A.2 for relevant approximations.

Prob. 2.12-11. A rectangular plate $ABCD$ with base b and height h is uniformly stretched an amount δ_x in the x direction and δ_y in the y direction to become the enlarged rectangle $AB^*C^*D^*$ shown in Fig. P2.12-11. Determine an expression for the (uniform) extensional strain ϵ_n of the diagonal AC. Express your answer in terms of δ_x, δ_y, L, and θ, where $L = \sqrt{b^2 + h^2}$ and $\tan \theta = h/b$. Base your calculations on the small-displacement assumptions, that is, assume that $\delta_x \ll L$ and $\delta_y \ll L$. (See Appendix A.2 for relevant approximations.)

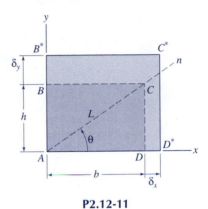

P2.12-11

Prob. 2.12-12. A thin, square plate $ABCD$ undergoes deformation in which no point in the plate moves in the y direction. Every horizontal line (except the bottom edge) is uniformly stretched as edge CD remains straight and rotates clockwise about D. Using the definition of extensional strain in Eq. 2.35, determine an expression for the extensional strain in the x direction, $\epsilon_x(x, y)$.

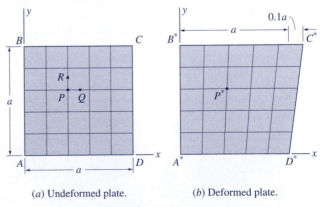

(a) Undeformed plate.　　　　(b) Deformed plate.

P2.12-12 and P2.12-13

*Prob. 2.12-13. Using the definition of shear strain, Eq. 2.36, and using the "undeformed plate" and "deformed plate" sketches in Fig. P2.12-13, determine an expression for the shear strain γ_{xy} as a function of position in the plate, that is, $\gamma_{xy}(x, y)$.

Prob. 2.12-14. A thin, square plate $ABCD$ undergoes deformation such that a typical point P with coordinates (x, y) moves horizontally an amount

$$u(x, y) \equiv \overline{PP^*} = \frac{1}{100}(a - x)\left(\frac{y}{a}\right)^2$$

The undeformed and deformed plates are shown in Figs. P2.12-14a and P2.12-14b, respectively. Using the definition extensional strain in Eq. 2.35, determine an expression for $\epsilon_x(x, y)$, the extensional strain in the x direction.

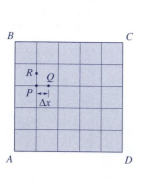 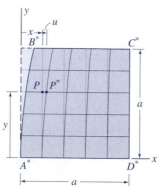

(a) Before deformation.   (b) After deformation.

P2.12-14 and P2.12-15

*__Prob. 2.12-15.__ Using the definition of shear strain, Eq. 2.36, and using the "before deformation" and "after deformation" sketches in Fig. P2.12-15, determine an expression for the shear strain γ_{xy} as a function of position in the plate, that is, determine $\gamma_{xy}(x, y)$.

*__Prob. 2.12-16.__ A typical point P at coordinates (x, y) in a flat plate moves through *small* displacements $u(x, y)$ and $v(x, y)$ in the x direction and the y direction, respectively. Using the definition of extensional strain, Eq. 2.35, and using the "before deformation" and "after deformation" sketches in Fig. P2.12-16, show that the formula for the extensional strain in the x direction, $\epsilon_x(x, y)$, is the partial differential equation

$$\epsilon_x = \frac{\partial u}{\partial x}$$

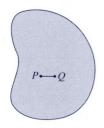 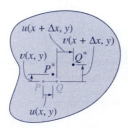

(a) Before deformation.   (b) After deformation.

P2.12-16

▼ HOOKE'S LAW FOR ISOTROPIC MATERIALS; DILATATION

Prob. 2.13-1. When thin sheets of material, like the top "skin" of the airplane wing in Fig. P2.13-1, are subjected to stress, they are said to be in a state of *plane stress*, with $\sigma_z = \tau_{xz} = \tau_{yz} = 0$. Starting with Eqs. 2.38, with $\Delta T = 0$, show that for the case of plane stress Hooke's Law can be written as

$$\sigma_x = \frac{E}{1 - \nu^2}(\epsilon_x + \nu\epsilon_y), \qquad \sigma_y = \frac{E}{1 - \nu^2}(\epsilon_y + \nu\epsilon_x)$$

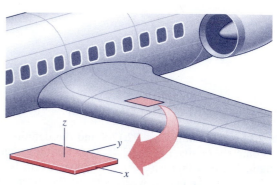

P2.13-1

Prob. 2.13-2. Figure P2.13-2 shows a small portion of a thin aluminum-alloy plate in plane stress ($\sigma_z = \tau_{xz} = \tau_{yz} = 0$). At a particular point in the plate $\epsilon_x = 600\mu$, $\epsilon_y = -200\mu$, and $\gamma_{xy} = 200\mu$. For the aluminum alloy, $E = 10 \times 10^3$ ksi and $\nu = 0.33$. Determine the stresses σ_x, σ_y, and τ_{xy} at this point in the plate. (Note: Start with Eqs. 2.38, not with Eqs. 2.40.)

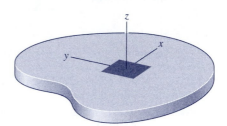

P2.13-2

Prob. 2.13-3. Determine the state of strain that corresponds to the following three-dimensional state of stress at a certain point in a steel machine component:

$$\sigma_x = 60 \text{ MPa}, \quad \sigma_y = 20 \text{ MPa}, \quad \sigma_z = 30 \text{ MPa}$$

$$\tau_{xy} = 20 \text{ MPa}, \quad \tau_{xz} = 15 \text{ MPa}, \quad \tau_{yz} = 10 \text{ MPa}$$

Use $E = 210$ GPa and $\nu = 0.30$ for the steel.

Prob. 2.13-4. The flat-bar plastic test specimen shown in Fig. P2.13-4 has a reduced-area "test section" that measures

0.5 in. × 1.0 in. Within the test section a strain gage oriented in the axial direction measures $\epsilon_x = 0.002 \frac{\text{in.}}{\text{in.}}$, while a strain gage mounted in the transverse direction measures $\epsilon_y = -0.0008 \frac{\text{in.}}{\text{in.}}$, when the load on the specimen is $P = 300$ lb. (a) Determine the values of the modulus of elasticity, E, and Poisson's ratio, ν. (b) Determine the value of the dilatation, ϵ_V, within the test section.

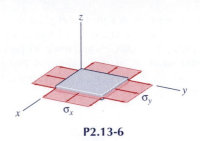

P2.13-6

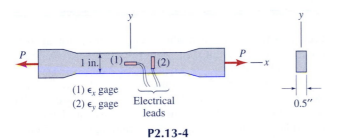

(1) ϵ_x gage
(2) ϵ_y gage — Electrical leads

P2.13-4

Prob. 2.13-5. A titanium-alloy bar has the following original dimensions: $a = 10$ in., $b = 4$ in., and $c = 2$ in. The bar is subjected to stresses $\sigma_x = 14$ ksi and $\sigma_y = -6$ ksi, as indicated in Fig. P2.13-5. The remaining stresses—σ_z, τ_{xy}, τ_{xz}, and τ_{yz}—are all zero. Let $E = 16 \times 10^3$ ksi and $\nu = 0.33$ for the titanium alloy. (a) Determine the changes in the lengths: Δa, Δb, and Δc, where $a^* = a + \Delta a$, etc. (b) Determine the dilatation, ϵ_V.

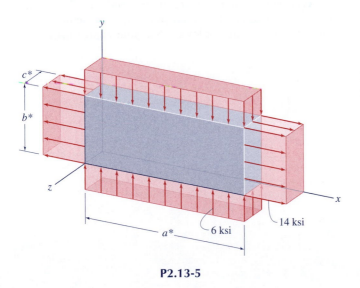

P2.13-5

Prob. 2.13-6. An aluminum-alloy plate is subjected to a biaxial state of stress, as illustrated in Fig. P2.13-6 ($\sigma_z = \tau_{xz} = \tau_{yz} = \tau_{xy} = 0$). For the aluminum alloy, $E = 72$ GPa and $\nu = 0.33$. Determine the stresses σ_x and σ_y if $\epsilon_x = 200\mu$, and $\epsilon_y = 140\mu$. (Note: Start with Eqs. 2.38, not with Eqs. 2.40.)

Prob. 2.13-7. A block of linearly elastic material (E, ν) is compressed between two rigid, perfectly smooth surfaces by an applied stress $\sigma_x = -\sigma_0$, as depicted in Fig. P2.13-7. The only other nonzero stress is the stress σ_y induced by the restraining surfaces at $y = 0$ and $y = b$. (a) Determine the value of the restraining stress σ_y. (b) Determine Δa, the change in the x dimension of the block. (c) Determine the change Δt in the thickness t in the z direction.

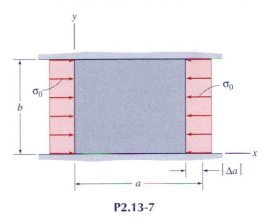

P2.13-7

Prob. 2.13-8. A thin, rectangular plate is subjected to a uniform biaxial state of stress (σ_x, σ_y). All other components of stress are zero. The initial dimensions of the plate are $L_x = 4$ in. and $L_y = 2$ in., but after the loading is applied, the dimensions are $L_x^* = 4.00176$ in., and $L_y^* = 2.00344$ in. If it is known that $\sigma_x = 10$ ksi and $E = 10 \times 10^3$ ksi, (a) what is the value of Poisson's ratio? (b) What is the value of σ_y?

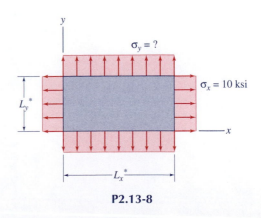

P2.13-8

111

Prob. 2.13-9. At a point in a thin steel plate in plane stress $(\sigma_z = \tau_{xz} = \tau_{yz} = 0)$, $\epsilon_x = 800\mu$, $\epsilon_y = -400\mu$, and $\gamma_{xy} = 200\mu$. For the steel plate, $E = 200$ GPa and $\nu = 0.30$. (a) Determine the extensional strain ϵ_z at this point. (b) Determine the stresses σ_x, σ_y and τ_{xy} at this point. (c) Determine the dilatation, ϵ_V, at this point.

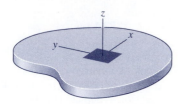

P2.13-9

*Prob. 2.13-10.** A block of linearly elastic material (E, ν) is placed under hydrostatic pressure: $\sigma_x = \sigma_y = \sigma_z = -p$; $\tau_{xy} = \tau_{xz} = \tau_{yz} = 0$, as shown in Fig. P2.13-10. (a) Determine an expression for the extensional strain $\epsilon_x (= \epsilon_y = \epsilon_z)$. (b) Determine an expression for the dilatation, ϵ_V. (c) The bulk

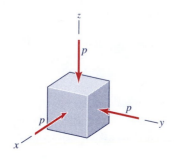

(Stresses on hidden faces not shown.)

P2.13-10

modulus, k_b, of a material is defined as the ratio of the hydrostatic pressure, p, to the magnitude of the volume change per unit volume, $|\epsilon_V|$, that is,

$$k_b \equiv \frac{p}{|\epsilon_V|}$$

Determine an expression for the bulk modulus of this block of linearly elastic material. Express your answer in terms of E and ν.

Prob. 2.14-1. You are to evaluate a new concept for an environmentally friendly building product, a laminated composite floor panel. This composite panel will use a new material consisting of a recycled polymer filled with recycled wood particles. This recycled material has an elastic modulus of 6 GPa and is produced in sheets 2 mm thick. These are laminated with thin, 0.5-mm-thick sheets of aluminum, $E_{Al} = 70$ GPa. The two different materials are firmly bonded by a strong adhesive to create the laminated composite panel. The final laminated composite panel contains 10 sheets of aluminum and 11 sheets of the recycled material in alternating layers.

Use the techniques discussed in Section 2.14 to calculate approximate values of elastic modulus in the plane of the laminated panel and through the thickness of the laminated panel. Before you begin your calculations, be sure to draw a simple schematic of the laminated composite structure, and use this to help you determine the volume fractions of each material.

Prob. 2.14-2. Consider a polymer matrix having $E_m = 2.8$ GPa, which is reinforced with $V_f = 0.2$ volume fraction of randomly oriented, short glass fibers having $E_f = 72$ GPa. (a) Calculate an approximate elastic modulus, E_c, for this composite material. (b) Would you expect the actual elastic modulus to be higher, or lower, than your approximation?

Section		Suggested Review Problems
2.1	Section 2.1 points out the need for definitions of **stress** and **strain** in order to explain how force is distributed throughout a deformable body under load and how the body deforms point by point.	You should familiarize yourself with this material.	
2.2	Section 2.2 defines **stress** as "Force divided by Area." **Normal stress** is normal i.e., perpendicular to the plane on which it acts, and it is denoted by the Greek symbol sigma (σ). Figure 2.2 and Eqs. 2.1 and 2.2 define: • **Normal stress at a point** (y, z) on cross section x. • **Average normal stress** on cross section x.	**Normal Stress** $$\sigma(x, y, z) = \lim_{\Delta A \to 0}\left(\frac{\Delta F}{\Delta A}\right) \quad (2.1)$$ $$\sigma_{\text{avg}} = \frac{F}{A} \quad (2.2)$$ (a) Distributed normal stress on a cross section. (b) Resultant of distributed normal stress in (a). Normal force on a cross section (Fig. 2.2).	
	The **sign convention** for normal stress is: • **Positive** normal stress is called **tensile stress.** • **Negative** normal stress is called **compressive stress.** The normal stress on cross sections of an axially loaded member is called **axial stress.** The resultant normal force on the cross section must act through the **centroid.**	**Axial Stress** $$\sigma_i = \frac{F_i}{A_i} \quad (2.5)$$ Normal force through centroid (Fig. 2.4b).	2.2-3 2.2-9 2.2-15
2.3	Section 2.3 defines **extensional strain,** the strain that goes with normal stress.	**Extensional Strain** $$\epsilon = \frac{\Delta L}{L} \quad (2.7)$$ (a) The undeformed bar. (b) The deformed bar. Extensional strain (Fig. 2.5).	2.3-7 2.3-11

Section			Suggested Review Problems
	Section 2.3 also defines **thermal strain,** the strain that is produced by a change in temperature.	$\epsilon = \alpha \Delta T$ where α is the *coefficient of thermal expansion* and ΔT is the change in temperature from the reference temperature.	2.3-15
2.4	Section 2.4 discusses **stress–strain diagrams** and the **mechanical properties of materials** that are obtained from testing tension specimens and compression specimens.	 (*a*) Undeformed specimen. (*b*) Deformed specimen. A typical tension-test specimen (Fig. 2.10).	
	A **stress-strain diagram** is a graph of the results of a tension test (or compression test): stress ($\sigma = P/A$ versus strain ($\epsilon = (\Delta L)/L$. From the stress-strain diagram you should be able to determine directly (or calculate) the following **mechanical properties:** • the **proportional limit** of the material, • the **modulus of elasticity** of the material, • the **yield point** of the material, and • the **ultimate strength** of the material.	**Stress-Strain Diagrams** (*a*) (*b*) Stress-strain diagrams for structural steel in tension (Fig. 2.11).	2.4-3

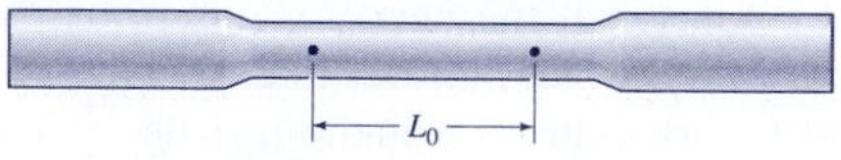

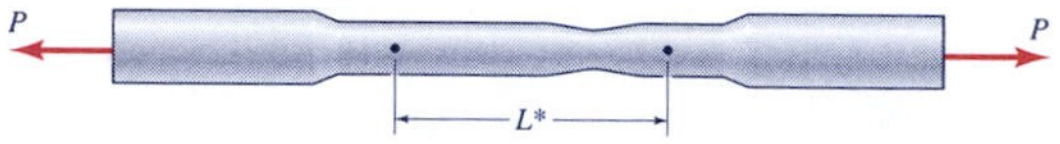

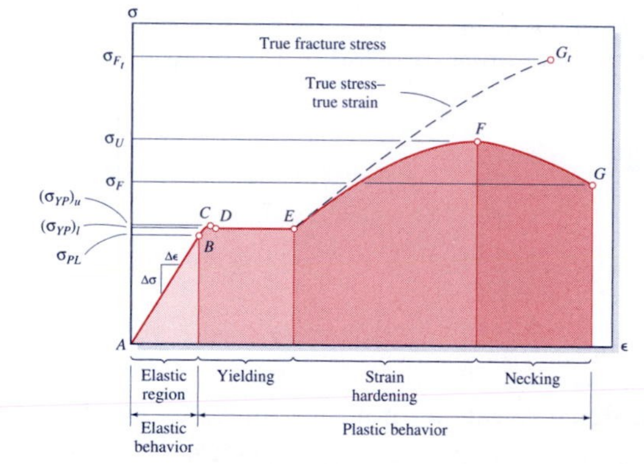

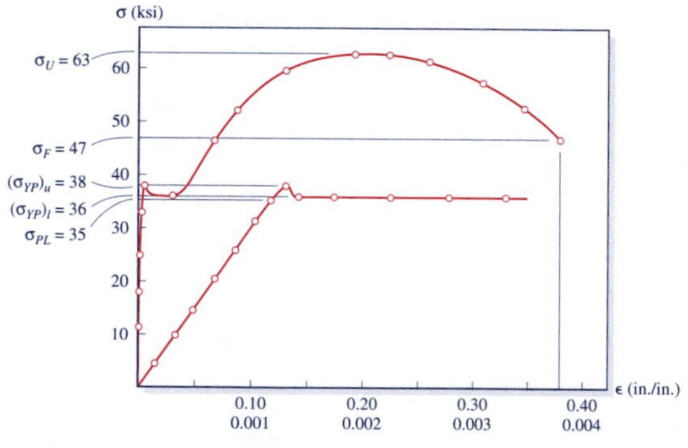

Section			Suggested Review Problems
	The principal **design properties** of materials are the following: • **Strength** • **Stiffness,** and • **Ductility.**	Define each of these design-related properties; discuss how each is determined from stress-strain diagrams; and discuss how each design property differs from the others.	
2.5	Section 2.5 discusses the differences between **elastic** behavior and **plastic** behavior of materials.	Define the following terms and indicate how each is determined: • **Elastic** behavior of a material, • **Plastic** behavior of a material, • **Linearly elastic** behavior of a material, and • **Permanent set.**	
2.6	Section 2.6 discusses **linearly elastic behavior.** The discussion is restricted to the case of uniaxial stress applied to **homogeneous, isotropic** materials. Define these two terms.	**Hooke's Law** $$\sigma_x = E\epsilon_x \qquad (2.14)$$ where E is the *modulus of elasticity,* also called *Young's modulus.* **Poisson's Ratio** $$\epsilon_y = \epsilon_z = -\nu\epsilon_x \qquad (2.15)$$ where ν is called *Poisson's ratio.* Linearly elastic behavior (Fig. 2.25).	2.6-1 2.6-7

Section		Suggested Review Problems
2.7	**Shear Stress** $$\tau = \lim_{\Delta A \to 0}\left(\frac{\Delta V}{\Delta A}\right) \qquad (2.17)$$ **Average Shear Stress** $$\tau_{\text{avg}} = \frac{V}{A_s} \qquad (2.19)$$ where A_s is the area on which the shear force V acts. 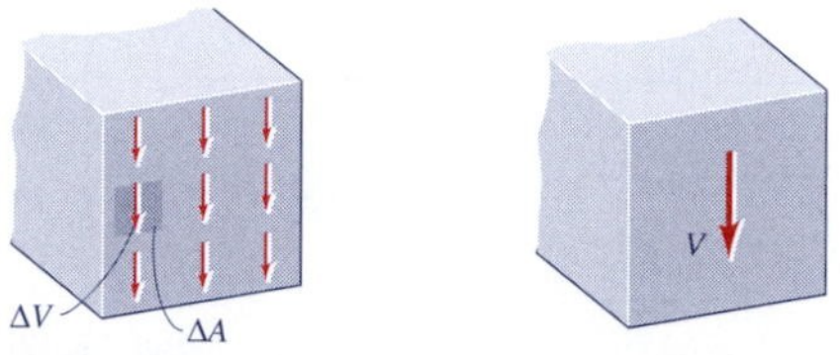 (a) The distribution of shear force on a sectioning plane. (b) The resultant shear force on the sectioning plane. Shear stress and its resultant shear force (Fig. 2.27).	2.7-1 2.7-7

Until now, Chapter 2 has discussed only *normal stress*, the stress that results in a force perpendicular to the surface on which the normal stress acts. Section 2.7 introduces the second form of stress, **shear stress,** whose resultant is <u>parallel</u> to the surface on which the shear stress acts.

| | **Shear Strain** $$\gamma = \frac{\pi}{2} - \theta^* \qquad (2.21)$$ where γ and θ^* are defined in the figure below. (b) Pure shear deformation. Definition of shear strain (Fig. 2.31b). | 2.7-11 |

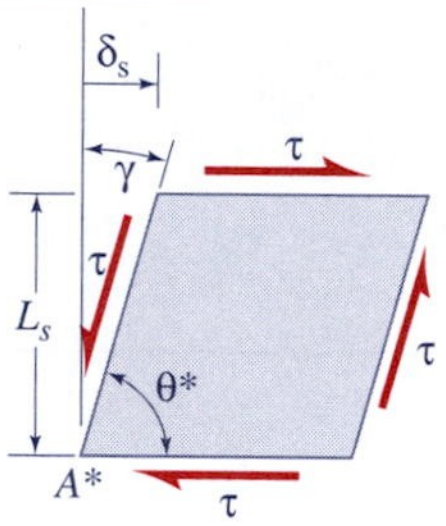

In Section 2.7 an equilibrium argument shows that the shear stress on perpendicular faces is required to be equal. Section 2.7 also gives the definition of **shear strain.**

You should be able to prove that the shear stresses on perpendicular faces must be equal to each other, as shown in Fig. 2.31*b*.

| | **Hooke's Law for Shear** $$\tau = G\gamma \qquad (2.23)$$ where G is the **shear modulus of elasticity,** which is discussed further in Section 2.11. | 2.7-13 |

Section 2.7 concludes with the **material properties in shear.**

Section		Suggested Review Problems
2.8	Section 2.8 is an **Introduction to Design.** There are two ways in which design information is used in engineering practice: • To evaluate an existing structure, or to evaluate a proposed design. • To design a new structure.	
2.8	Although **Allowable-Stress Design** is emphasized in this textbook, **Load and Resistance Design** is mentioned briefly. For allowable-stress design, the allowable stress in axial deformation is based on the tensile, or compressive, yield strength of the material; in direct shear, the allowable stress is based on the shear yield strength.	**Factor of Safety** $$FS = \frac{\text{Failure Load}}{\text{Allowable Load}} \quad (2.26)$$ **Allowable Stress** $$\sigma_{\text{allow.}} = \frac{\sigma_Y}{FS}, \text{ or } \tau_{\text{allow.}} = \frac{\tau_Y}{FS} \quad (2.28)$$ 2.8-1 2.8-7 2.8-13
	Review Example 2.11, which illustrates the process of **optimal design** of a minimum-weight structure.	
2.9	Section 2.9 introduces you to the fact that normal stress and shear stress depend on the orientation of the plane on which the stresses act. This topic is greatly expanded in Chapter 8.	**Transformation of Stresses** $$\left.\begin{array}{c} \sigma_n = (\sigma_x/2)(1 + \cos 2\theta) \\ \tau_{nt} = -(\sigma_x/2)\sin 2\theta \end{array}\right\} \quad (2.30)$$ (a) Inclined plane of axial-deformation member. (Fig. 2.33a) Derive Eqs. 2.30. 2.9-1 2.9-7
2.10	Section 2.10 introduces you to **Saint-Venant's Principle.**	State Saint-Venant's Principle, and discuss why it is an important principle in structural analysis (i.e., the analysis of components of machines and structures).
2.11	Section 2.11 shows how **Young's Modulus,** E, and the **Shear Modulus of Elasticity,** G, are related.	$$G = \frac{E}{2(1 + \nu)} \quad (2.24)$$ Derive Eq. 2.24.
2.12 2.13 2.14	Sections 2.12 through 2.14 discuss more advanced topics in stress and strain, and more advanced topics in the mechanical behavior of materials. These optional sections discuss: **General Definitions of Stress and Strain** Section 2.12, **Cartesian Components of Stress: Generalized Hooke's law for Isotropic Materials** Section 2.13, **Mechanical Properties of Composite Materials** Section 2.14.	There is no Chapter Review Problem for these optional sections.

3 AXIAL DEFORMATION

3.1 INTRODUCTION

In Chapter 2 the topic of uniform axial deformation was used to introduce the concepts of normal stress and extensional strain and to describe the experiments required to determine the stress-strain behavior of materials. In this chapter we will pursue the topic of axial deformation in greater detail. We begin with a **definition of axial deformation.**

*A structural member having a straight longitudinal axis is said to undergo **axial deformation** if, when loads are applied to the member or it is subjected to temperature change: (1) the axis of the member remains straight, and (2) cross sections of the member remain plane, remain perpendicular to the axis, and do not rotate about the axis as the member deforms.*

There are many examples of axial-deformation members: columns in buildings, hoist cables, and truss members in space structures, to name just a few. The picture in Fig. 3.1 illustrates several stages in the construction of columns (piers) for a highway interchange. On the left is an example of the steel reinforcement for a column, and on the right is a completed column with reinforcement protruding from the top of the column. The columns of bridges like the one in the background in Fig. 3.1 act primarily as axial-deformation members.

3.2 BASIC THEORY OF AXIAL DEFORMATION

Let us now develop the theory of axial deformation by applying the three types of equations that are fundamental to all of deformable-body mechanics: *equilibrium, geometry of deformation,* and *material behavior.* We begin by considering the geometry of deformation.

Geometry of Deformation; Strain-Displacement Analysis. The theory of axial deformation applies to a straight, slender member with cross section that is either constant or that changes slowly along the length of the member. Figure 3.2 shows such a member before and after it has undergone axial deformation caused by axial loading or temperature change.

FIGURE 3.1 Some reinforced concrete columns for highway interchange bridges. (Courtesy Roy Craig)

Axial deformation, as defined in Section 3.1, is characterized by two fundamental **kinematic assumptions:**

1. *The axis of the member remains straight.*
2. *Cross sections, which are plane and are perpendicular to the axis before deformation, remain plane and remain perpendicular to the axis after deformation. And, the cross sections do not rotate about the axis.*

These assumptions are illustrated in Fig. 3.2, where A and B designate cross sections at x and $(x + \Delta x)$ prior to deformation, and where A^* and B^* designate these same cross sections after deformation.

The distance that a cross section moves in the axial direction is called its **axial displacement.** The displacement of cross section A is labeled $u(x)$, while the neighboring section B displaces an amount $u(x + \Delta x)$. The displacement $u(x)$ is taken to be positive in the $+x$ direction. We can derive a **strain-displacement expression** that relates the axial strain ϵ to this axial displacement u by considering the fundamental definition of extensional strain:

$$\epsilon = \frac{\text{Final length} - \text{Initial length}}{\text{Initial length}}$$

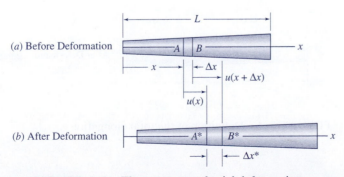

(a) Before Deformation

(b) After Deformation

FIGURE 3.2 The geometry of axial deformation.

The axial strain of any fiber[1] of infinitesimal length Δx that is parallel to the x axis and extends from section A to section B of the undeformed member may be determined from the fundamental definition of *extensional strain at a point*. By letting the initial length of a typical fiber be Δx, and then letting Δx approach zero, we can write the following expression for the axial strain (Eq. 2.35):

$$\epsilon_x(x) = \lim_{\Delta x \to 0} \left(\frac{\Delta x^* - \Delta x}{\Delta x} \right) = \lim_{\Delta x \to 0} \left[\frac{u(x + \Delta x) - u(x)}{\Delta x} \right] = \frac{du}{dx}$$

Therefore, the **axial strain** at section x is the derivative (with respect to x) of the axial displacement, or

$$\epsilon(x) = \frac{du(x)}{dx}$$

Strain-Displacement Equation (3.1)

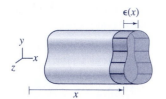

$\epsilon(x)$

FIGURE 3.3 Extensional strain distribution for a member undergoing axial deformation.

This equation relating axial strain to axial displacement is called the **strain-displacement equation** for axial deformation. The two fundamental kinematic assumptions stated above imply that the axial strain ϵ may be a function of x, but that it is not a function of position in the cross section, that is, of y or z. To emphasize this point, a plot of the **strain distribution** at an arbitrary cross section at x is shown in Fig. 3.3 superimposed on a sketch of a portion of the member. To reiterate, **axial deformation is characterized by extensional strain that is not a function of position in the cross section**.

As indicated in Fig. 3.4, the **total elongation** of the member is the difference between the displacements of its two ends, that is,

$$e = u(L) - u(0) \tag{3.2}$$

By summing up the changes in length of increments dx over the entire length of the member, we get the following equation for the **elongation** of an axial-deformation member of initial length L:

$$e = \int_0^L \epsilon(x) dx$$

Elongation Formula (3.3)

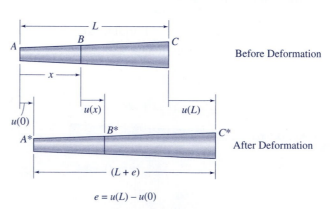

$e = u(L) - u(0)$

FIGURE 3.4 Definition of the total elongation e.

---

[1]The word *fiber* is used to signify a line of material particles.

How the strain ϵ varies with x depends on the loads that are applied to the member, whether is has a constant cross section or the cross section varies with x, and the behavior of the material (or materials) from which the member is constructed. Next, we will consider this material behavior.

Material Behavior. Having the strain distribution given by Eq. 3.1 and illustrated in Fig. 3.3. we can now employ the uniaxial stress-strain behavior of materials to determine the stress distribution in a member undergoing axial deformation. Let us first take the simplest case—**linearly elastic behavior**—and let us assume that the temperature remains constant (i.e., $\Delta T = 0$) and that $\sigma_y = \sigma_z = 0$. Then, from Eq. 2.38a (Section 2.13) we get the **uniaxial stress-strain equation**

$$\sigma \equiv \sigma_x = E\epsilon(x) \qquad \text{Hooke's Law} \qquad (3.4)$$

Nonlinear material behavior is treated in Section 3.11.

Equation 3.4 gives the distribution of the axial stress σ on the cross section at x. As indicated in Eq. 3.4, ϵ may only vary with x, but not with position in the cross section. Most axial-deformation members are homogeneous, so that Young's modulus, E, is constant throughout the member. Axial deformation of nonhomogeneous members is treated in Section 3.3.

Stress Resultant. Deformable-body mechanics problems are simplified by making assumptions that reduce a basically three-dimensional problem to a one-dimensional problem, like the axial-deformation kinematic assumptions discussed earlier in this section. Given the distribution of axial stress σ (Eq. 3.4), we can replace the distributed stress by a single **resultant axial force** and relate that force to the axial stress on the cross section.

Figure 3.5a shows the three stress resultants that are related to the axial stress σ, and defines the sign convention for these stress resultants. The resultants in Fig. 3.5a are shown for the $+x$ face; equal and opposite resultants act on the $-x$ face at the cross section. The **axial force,** F, on the cross section will always be taken to be <u>positive in tension</u>. The **bending moments** M_y and M_z are taken <u>positive according to the right-hand rule</u>. By summing up the contributions to these resultants of the forces dF on infinitesimal areas, dA, we get[2]

$$F(x) = \int_A \sigma \, dA$$

$$M_y(x) = \int_A z\sigma \, dA \qquad \text{Stress Resultants} \qquad (3.5)$$

$$M_z(x) = -\int_A y\sigma \, dA$$

Equations 3.5 define the three stress resultants associated with the axial stress σ. However, when the distribution of stress on the cross section is known, we can combine these to form a single resultant force on the cross section. Consider the most common case, a **homogeneous linearly elastic member.** Then, Young's modulus is constant throughout the entire member, so Eq. 3.4 takes the form

$$\sigma(x) = E\epsilon(x) \qquad \text{Stress Distribution } (E = \text{Const}) \qquad (3.6)$$

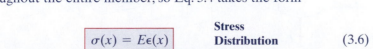

---

[2]Note that the three stress resultants, $F(x)$, $M_y(x)$ and $M_z(x)$, are defined as <u>scalar</u> quantities.

FIGURE 3.5 (a) Stress resultants on the cross section at x. (b) Axial stress at a point in the cross section at x.

When E = const, the stress distribution, like the strain distribution, is constant over the cross section.

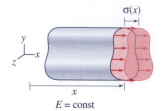

FIGURE 3.6 The stress distribution for a homogeneous member undergoing axial deformation.

This uniform axial stress distribution is illustrated in Fig. 3.6.

Whenever the normal stress is constant over a cross section, the resultant of that stress distribution is given by the following equations:

$$F(x) = \sigma(x) \int_A dA = \sigma(x)A(x)$$

$$M_y(x) = \sigma(x) \int_A z\, dA = \sigma(x)[\bar{z}A(x)] \tag{3.7}$$

$$M_z(x) = -\sigma(x) \int_A y\, dA = -\sigma(x)[\bar{y}A(x)]$$

since $\int_A y\, dA \equiv \bar{y}A$ and $\int_A z\, dA \equiv \bar{z}A$. We can simplify the axial-deformation problem by *choosing the x axis so that it passes through the centroid of the cross section.* Then, $\bar{y} = \bar{z} = 0$, and $M_y = M_z = 0$.

> **A member having E = const will undergo axial deformation if the applied load is an axial force that acts through the centroid of the cross section.**

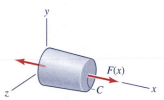

FIGURE 3.7 The resultant force on the cross section for axial deformation of a bar with E = const. passes through the centroid, C.

This resultant axial force is illustrated in Fig. 3.7.

Equilibrium. We have determined the resultant force on the cross section when a member undergoes axial deformation. Equilibrium is satisfied by drawing a free-body diagram and relating this internal axial force, $F(x)$, to the external forces applied to the member. Usually, **a free body of finite length** is used, as illustrated in Fig. 3.8a, where $F(x)$ is equal to the sum of the end force P and the weight of the member below section x.

Figure 3.8b shows a member with axisymmetric axial loading $p(x)$ per unit length. For such distributed-loading cases it is often convenient to use **a free body of infinitesimal length** Δx, as illustrated in Fig. 3.8c. Then, for equilibrium of the free body in Fig. 3.8c,

$$\xrightarrow{+} \sum F(x) = 0: \qquad \lim_{\Delta x = 0}\left[\frac{F(x + \Delta x) - F(x) + p(x)\Delta x}{\Delta x}\right] = 0$$

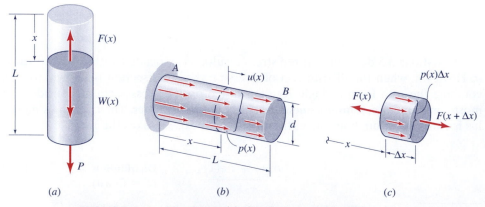

(a) (b) (c)

FIGURE 3.8 Members with distributed axial loading.

$$\frac{dF(x)}{dx} = -p(x)$$ Equilibrium Equation (3.8)

which can be integrated to give the axial force $F(x)$. Example 3.3 in Section 3.3 illustrates the use of Eq. 3.8.

Summary of Axial Deformation of a Homogeneous, Linearly Elastic Member.

To apply the theory of axial deformation to a member with **constant modulus of elasticity,** we need to relate the stress and the elongation of a member to the external loads acting on it. From the first of Eqs. 3.7 we get the formula for the **axial stress:**

$$\sigma \equiv \sigma_x(x) = \frac{F(x)}{A(x)}$$ Axial Stress Formula (3.9)

Then, from Eqs. 3.1, 3.6, and 3.9, the axial strain is given by

$$\epsilon = \frac{du}{dx} = \frac{F(x)}{EA(x)}$$ (3.10)

Finally, from Eqs. 3.3 and 3.10, the **total elongation** is

$$e = \int_0^L \frac{F(x)\,dx}{A(x)E}$$ Axial Force-Deformation Equation (3.11)

Equation 3.11 expresses the **force-deformation behavior** of a homogeneous member undergoing axial deformation.

If the displacement at an arbitrary section x is needed, we can employ a dummy variable of integration in Eq. 3.10 and write

$$u(x) = u(0) + \int_0^x \frac{F(\xi)\,d\xi}{A(\xi)E}$$ (3.12)

The **theory of axial deformation,** which has been summarized in Eqs. 3.8 through 3.12 above, will be illustrated by a number of examples. We begin in this section by discussing the force-deformation behavior of the uniform axial-deformation element, or *prismatic bar element.* This is followed by a brief discussion of another member whose structural function is largely axial-deformation in nature—the *linear spring.*

Uniform Axial-Deformation Element.

Many structures incorporate one or more **uniform axial-deformation elements,** that is, members that: (1) have constant cross-sectional area A, (2) have constant modulus of elasticity E, and (3) have axial forces applied only at the ends. The two hanger rods in Fig. 3.9a and the three numbered links in Fig. 3.9b are such uniform axial deformation members. The spring attached to the slider block in Fig. 3.9b also behaves as an axial-deformation member.

Because it forms such an important component of so many structures, we now derive the force-deformation equation for a single uniform linearly elastic element. In Section 3.3 we will consider the more general case of nonuniform axial deformation, but in Sections 3.4 through 3.10 systematic procedures for treating both *statically determinate structures* and *statically indeterminate structures* that have uniform axial-deformation members will be thoroughly examined.

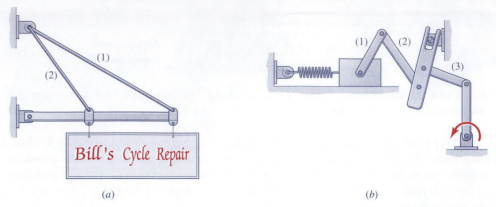

(a)   (b)

FIGURE 3.9   Two systems that employ uniform axial-deformation members.

To simplify the solution of problems involving uniform axial-deformation members, we will first apply the fundamental equations of equilibrium, material behavior, and geometry of deformation to determine the basic **force-deformation behavior of a typical uniform axial-deformation element,** like the element shown in Fig. 3.10a.

Using the free-body diagram in Fig. 3.10b, we get

$$\overset{+}{\rightarrow} \sum F_x = 0: \qquad\qquad F(x) = F = \text{const}$$

The derivation of Eq. 3.11 included both material behavior (linearly elastic) and geometry of deformation; hence, for the **elongation e** we can simply write

$$e = \int_0^L \frac{F(x)\,dx}{A(x)E} = \frac{F}{AE}\int_0^L dx$$

or

$$\boxed{e = \frac{FL}{AE}} \qquad\qquad (3.13)$$

It will often be convenient to write Eq. 3.13 in the simpler form

$$\boxed{e = fF, \text{ where } f \equiv \frac{L}{AE}} \qquad\qquad (3.14)$$

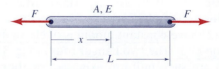

(a) A uniform axial-deformation member.

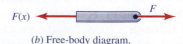

(b) Free-body diagram.

FIGURE 3.10   A typical uniform axial-deformation element.

or in the alternative form

$$F = ke, \text{ where } k = \frac{AE}{L} \tag{3.15}$$

Any of these last three equations characterizes the **force-deformation behavior of the uniform linearly elastic axial-deformation element**.

The parameter f in called the flexibility coefficient for the axial-deformation member, or element, and k is called the stiffness coefficient. From Eqs. 3.14 and 3.15, respectively, it can be seen that:

- The **flexibility coefficient, f,** is the *elongation produced when a unit force is applied to the member*. Its dimensions are L/F.
- The **stiffness coefficient, k,** is the *force required to produce a unit elongation of the member*. Its dimensions are F/L.

These physical interpretations of flexibility and stiffness can greatly assist us in interpreting how a structure responds to loads.

Linear Spring. One linear axial-deformation element is the prismatic bar shown in Fig. 3.10, whose behavior is given by Eqs. 3.13 through 3.15. A second common element that exhibits linear force-deformation behavior is the **linear spring,** shown in Fig. 3.11.

To elongate a linear spring by an amount e requires an axial force (tension positive)

$$F = ke \tag{3.16}$$

Alternatively, when an axial force F is applied to a linear spring, it elongates by an amount

$$e = fF, \text{ where } f = \frac{1}{k} \tag{3.17}$$

The parameter k is called the **spring constant;** it is the parameter that is usually stated to describe the force-deformation behavior of a spring. Some springs are designed to operate only in tension or only in compression. However, we will assume that Eqs. 3.16 and 3.17 hold for either tension or compression of the springs that are shown in this text. Figure 3.11c is a plot of F versus e for this linear behavior of a spring. The spring constant, k, is the slope of the line.

(a) A linear spring in its undeformed configuration.

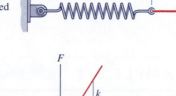

e = elongation
L_0 = free length

L_0

(b) The linear spring elongated by axial force F.

F

(c) The force-deformation plot for a linear spring.

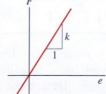

FIGURE 3.11 A linear spring.

The **theory of axial deformation,** which has been summarized in Eqs. 3.8 through 3.12 above, will now be illustrated by a number of examples in the following sections:

- Section 3.3-nonuniform axial deformation
- Section 3.4-statically determinate structures
- Section 3.5-statically indeterminate structures with external loads
- Section 3.6-statically indeterminate structures with temperature change
- Section 3.7-statically indeterminate structures with geometric "misfits"

3.3 EXAMPLES OF NONUNIFORM AXIAL DEFORMATION

In Section 3.4 we will concentrate on uniform axial deformation with application to statically determinate structures, and in Sections 3.5–3.9 we will consider uniform axial deformation of members in statically indeterminate structures. First, however, let us illustrate the fundamental axial-deformation equations that were derived in Section 3.2 by considering the following three examples of **nonuniform axial deformation:** Example 3.1 treats axial deformation of a member made of two linearly elastic materials; Example 3.2 discusses the deformation of a column with varying cross section; and, finally, Example 3.3 discusses equilibrium of an axial-deformation member with distributed external loading.

Concluding Section 3.3 is a brief introduction to the topic of **wire rope.** Although wire rope is uniform along its length, it is nonuniform in cross section, so it falls within the scope of this section on nonuniform axial deformation.

Axial Deformation of a Nonhomogeneous Member.

When an axial-deformation member is homogeneous, the modulus E is constant, and the axial stress σ is constant over the cross section and is given by Eq. 3.6, repeated here.

$$\sigma(x) = E\epsilon(x)$$

(3.6)
repeated

However, for the most general form of E, Eq. 3.4 may be stated in the form

$$\sigma(x, y, z) = E(x, y, z)\epsilon(x)$$

(3.18)

where it is noted that the axial strain is constant over the cross section but the axial stress is not.

Example 3.1 illustrates the special case of a single member made up of two side-by-side homogeneous, linearly elastic members such that $E = E(y)$. This type of problem can also be readily solved by treating the member as a statically indeterminate assemblage and using the method of solution discussed in Section 3.5.

EXAMPLE 3.1

A bimetallic bar is made of two linearly elastic materials, material 1 and material 2, that are bonded together at their interface, as shown in Fig. 1. Assume that $E_2 > E_1$. Determine the distribution of normal stress that must be applied at each end if the bar is to undergo axial deformation, and determine the location of the point in the cross section where the resultant force P must act. Express your answers in terms of P, E_1, E_2, and the dimensions of the bar.

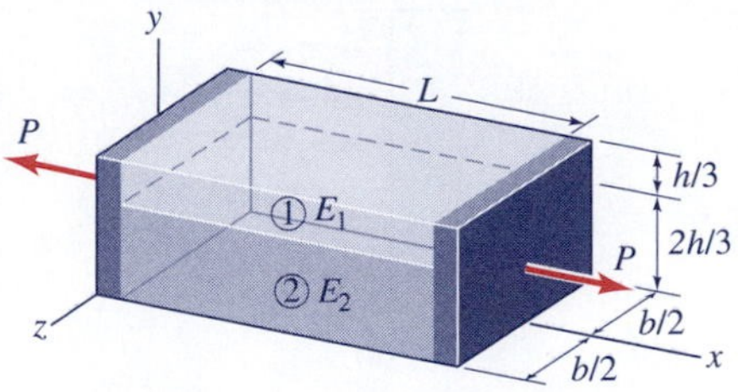

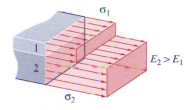

Fig. 1 A bimetallic bar.

Plan the Solution The bar is said to undergo axial deformation, so we know from Eq. 3.1 that the strain, $\epsilon(x)$, is constant on every cross section. And since the bar is prismatic (constant cross section) and is loaded only at its ends, we can assume that $\epsilon(x) = \epsilon =$ const for the entire bar. Since there are two values of E, Eq. 3.18 will lead to different values of stress in the two materials. We can use the stress-resultant integrals of Eqs. 3.5 to relate these two stresses to the resultant force.

Solution

Strain Distribution: For the reasons noted above,

$$\epsilon(x) = \epsilon = \text{const} \tag{1}$$

Stress Distribution: From Eq. (1) and Eq. 3.4, the stresses in the two parts of the bar will be

$$\sigma_1 = E_1\epsilon, \quad \sigma_2 = E_2\epsilon \tag{2}$$

as illustrated in Fig. 2.

Resultant Force: The resultant force and moments on the cross section are given by Eqs. 3.5.

$$F(x) = \int_A \sigma \, dA$$

$$M_y(x) = \int_A z\sigma \, dA \tag{3}$$

$$M_z(x) = -\int_A y\sigma \, dA$$

Fig. 2 The stress distribution in a bimetallic bar.

However, we know that the resultant of a constant normal stress distribution is a force acting through the centroid of the area on which the constant stress acts. Therefore, we can replace the stress distribution of Fig. 2 and Fig. 3a by two axial forces, P_1 and P_2, acting at the centroids of their respective areas of the cross section, as shown in Fig. 3b.

From Eqs. (2) and the $F(x)$ equation in Eqs. (3),

$$P_1 = E_1\epsilon A_1 = E_1\epsilon\left(\frac{bh}{3}\right), \quad P_2 = E_2\epsilon A_2 = E_2\epsilon\left(\frac{2bh}{3}\right) \tag{4}$$

Taking the summation of forces in the x direction, we get

$$\overset{+}{\rightarrow} \sum F_x: \qquad\qquad P_1 + P_2 = P \tag{5}$$

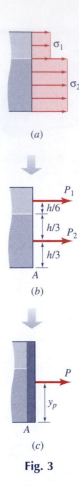

(a)

(b)

(c)

Fig. 3

Thus, the resultant force acting on the ends of the bar (and at every cross section, for that matter) is related to the extensional strain ε by the equation

$$P = \frac{\epsilon bh}{3}\left(E_1 + 2E_2\right) \tag{6}$$

Solving Eq. (6) for ε and inserting the result into the Eqs. (2), we get the stresses

$$\sigma_1 = \frac{3PE_1}{bh(E_1 + 2E_2)}, \quad \sigma_2 = \frac{3PE_2}{bh(E_1 + 2E_2)} \qquad \textbf{Ans.} \quad (7)$$

In Fig. 3b the forces P_1 and P_2 are shown acting at the centroids of the areas on which they act. The location of the single resultant force P in Fig. 3c can be determined by taking moments about a z axis passing through point A.

$$+\circlearrowleft\left(\sum M\right)_A: \qquad P_1\left(\frac{5h}{6}\right) + P_2\left(\frac{h}{3}\right) = Py_P \tag{8}$$

Thus, the resultant force acting on each end of the bar (and at every cross section) is a force P located at

$$y_P = h\left(\frac{5E_1 + 4E_2}{6E_1 + 12E_2}\right), \quad z_P = 0 \qquad \textbf{Ans.} \quad (9)$$

Review the Solution   A good check on the results above is to let $E_1 = E_2 = E$. Then, we get

$$\sigma_1 = \sigma_2 = \frac{P}{bh}, \quad y_P = \frac{h}{2}$$

which is consistent with Eq. 2.4.

Axial Deformation of a Column with Varying Cross Section.

Example 3.2 illustrates nonuniform axial deformation that results when the cross section of a member is not constant. The **total elongation** of a homogeneous, linearly elastic axial deformation member is given by Eq. 3.11, repeated here.

$$e = \int_0^L \frac{F(x)\,dx}{A(x)E} \tag{3.11 repeated}$$

EXAMPLE 3.2

The tapered column in Fig. 1 is subjected to a downward force P acting through the centroid of the top cross section at B. The column has a circular cross section, with a diameter that varies linearly from d_A at the bottom to d_B at the top. Determine an expression for the amount that the column shortens under the action of load P.

Plan the Solution The change in length of the column (in this case, the *shortening* of the column) can be computed by using Eq. 3.11.

Solution From Eq. 3.11, the shortening of the column, say δ, is

$$\delta = -\int_0^L \frac{F(x)\,dx}{A(x)E} \tag{1}$$

We need to use equilibrium to determine $F(x)$, and geometry to determine $A(x)$.

Equilibrium: From the finite free-body diagram in Fig. 2,

$$+\uparrow \sum F(x): \qquad F(x) = -P = \text{const} \tag{2}$$

Geometry: The cross-sectional area is

$$A(x) = (\pi/4)d^2(x)$$

By referring to Fig. 3 and employing similar triangles, we can determine $d(x)$ in terms of d_A and d_B.

$$\frac{d_A - d(x)}{x} = \frac{d_A - d_B}{L}$$

Thus,

$$d(x) = d_A - (x/L)(d_A - d_B)$$
$$A(x) = (\pi/4)[d_A - (x/L)(d_A - d_B)]^2 \tag{3}$$

Force-Deformation: Combining Eqs. (1) through (3), we get

$$\delta = \frac{4P}{\pi E}\int_0^L \frac{dx}{[d_A - (x/L)(d_A - d_B)]^2}$$

Finally, by evaluating this integral we obtain the answer

$$\delta = \frac{4PL}{\pi E d_A d_B} \qquad \textbf{Ans.} \quad (4)$$

Fig. 1

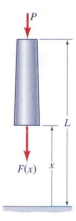

Fig. 2 Free-body diagram.

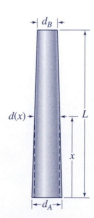

Fig. 3 Geometry.

The right side of Eq. (4) has the dimension of length, as it should. Also note that an increase in P or L causes an increase in δ, while a larger E or larger A (i.e., "fatter" column) decreases δ. These are reasonable effects. Finally, if $d_A = d_B \equiv d$, $\delta = PL/AE$, which is the expression we got in Example 2.7 (Section 2.6) for a uniform, linearly elastic member under axial loading.

Axial Deformation of a Member with Distributed Loading. Example 3.3 illustrates nonuniform axial deformation that results when an axial load is distributed along the member. This can be the result of the distributed weight of the member, or it can be the result of friction on the outer surface of the member, as is the case in Example 3.3. The distributed axial force enters into the **equilibrium equation,** which can be written for a free body of finite length or for a free body of infinitesimal length, as in Fig. 3.8c, for which the equilibrium equation is Eq. 3.8, repeated here.

$$\frac{dF(x)}{dx} = -p(x) \qquad \text{(3.8)}$$
$$\text{repeated}$$

EXAMPLE 3.3

Upon completion of a construction job, the contractor recovers some vertical piles by pulling them out of the ground, as illustrated in Fig. 1a. At one point in the pile-pulling operation, a length L of the pile is still in the ground, and friction between the pile and the ground exerts a distributed axial force on the pile of $p(x)$ per unit length, with the distribution of $p(x)$ assumed to be as shown in Fig. 1b. (a) Determine an expression for p_0 in terms of the axial force P. Assume that there is no force on the lower end of the pile, that is, that $F(L) = 0$. (b) Determine an expression for the <u>internal</u> axial force, $F(x)$, at an arbitrary depth x, expressing your answer in terms of P, L, and x.

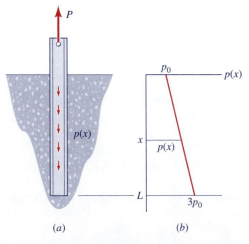

(a)   (b)

Fig. 1

Plan the Solution Figure 1b can be used to formulate an expression that relates $p(x)$ to x, and Eq. 3.8 can be used to relate $p(x)$ to the axial force $F(x)$. We will neglect the weight of the pile.

Solution (a) *Relate p_0 to P.* From Fig. 1a and the given information,

$$F(L) = 0, \quad F(0) = P \tag{1}$$

In Fig. 1b, $p(x)$ is a linear function that varies from p_0 at $x = 0$ to $3p_0$ at $x = L$. Therefore,

$$p(x) = p_0 + 2p_0(x/L) \tag{2}$$

Equation 3.8,

$$\boxed{\frac{dF}{dx} = -p(x)} \qquad \textbf{Equilibrium} \tag{3}$$

was based on the equilibrium of a free body of infinitesimal length (Fig. 3.8c). It can be integrated to give

$$F(L) = F(0) - \int_0^L [p_0 + 2p_0(x/L)]\,dx \tag{4}$$

Substituting Eqs. (1) into Eq. (4), we get

$$0 = P - p_0[x + (1/L)x^2]_0^L \tag{5}$$

which gives

$$p_0 = \frac{1}{2}\frac{P}{L} \qquad \textbf{Ans.(a)} \tag{6}$$

(b) *Determine an expression for F(x).*
Applying Eq. (3) now from $\xi = 0$ to $\xi = x$, we get

$$F(x) = P - \int_0^x [p_0 + 2p_0(\xi/L)]\,d\xi \tag{7}$$

Substituting Eq. (6) into Eq. (7) and carrying out the integration, we get

$$F(x) = P\left[1 - \frac{1}{2}\left(\frac{x}{L}\right) - \frac{1}{2}\left(\frac{x}{L}\right)^2\right] \qquad \textbf{Ans.(b)} \tag{8}$$

Review the Solution From Eq. 3.8 (Eq. (3) above) we know that, since $p(x)$ is linear in x, $F(x)$ will be quadratic in x. A good check on the result in Eq. (8) is to evaluate $F(x)$ at $x = 0$ and $x = L$ to make sure that we get the values given in Eq. (1). We do.

Instead of just using Eq. 3.8 for treating equilibrium in this problem, we could have used free bodies of finite length, similar to the one in Fig. 3.8a.

In this section we have applied the equations of the *basic theory of axial deformation,* from Section 3.2, to analyze three examples of *nonuniform axial deformation.* We conclude this section with an introduction to the topic of **wire rope.** Because of the complicated nature of wire rope, we cannot directly apply the equations of the basic theory of axial deformation in order to determine the stress distribution on the cross section or to determine an appropriate force-deformation equation. Instead, we will use characteristics of wire rope that are based on experimental data and are provided by the wire-rope manufacturer.

Wire Rope. Wire rope is a device that is useful for supporting tensile loads.[3] Static applications of wire rope include suspension bridges, like the Brooklyn Bridge in Fig. 1.3 and the *cable-stayed suspension bridge* shown on the front cover. Other important uses of wire rope include guy lines for towers and mooring lines for off-shore platforms. Although wire rope cannot bear compressive loads, its flexibility in bending gives it an important advantage for dynamic applications. This makes wire rope particularly useful for moving loads that are separated by great distance from a source of mechanical power. Example applications include cranes, hoists, elevators and drag lines for mining.

Wire rope is constructed of wire strands that are twisted about a core. Each strand is composed of several individual wires twisted together. The core of a wire rope can be composed of fibers, either natural or synthetic, a wire strand or an independent wire rope. Figure 3.12a shows the cross-section of a 6 × 7 wire rope, a common variety. Its fiber core conforms to the shape set by its strands, which complete a full helix in one *rope lay,* shown in Fig. 3.12b. The 6 × 7 designation indicates a construction of six strands, each containing seven wires. Wire rope applications that require repeated bending necessitate rope construction with a larger wire count, which can be achieved with a 6 × 19 construction. A large number of wire rope types are available to serve the wide range of applications for which wire rope is used. As a general rule, ropes with high wire counts are used in dynamic applications requiring repeated bending, and ropes with low wire count are used when repeated bending is not an issue. In static applications, even a single wire strand may be used to support a load (see the definition of *strand,* shown in Fig. 3.12a).

The two most important structural characteristics of a wire rope are its **force-elongation behavior** and its **strength.** The wires and strands of a wire rope interact and move relative to each other when a tensile load is applied, tightening the helix into which they are twisted, thus acting as a machine with moving parts. This action produces elongation under loading that can be separated into two categories, *permanent (constructional) stretch* and *elastic stretch.* Permanent stretch remains after the load is removed and is the result of wires and strands *seating* during service. In typical applications, permanent stretch reaches approximately 0.5% of the total wire rope length. In all applications except those with very small loads, wires and strands are fully seated within days to weeks of initial service, after which the permanent stretch is approximately constant. Pre-stretching a wire rope can greatly reduce the amount of

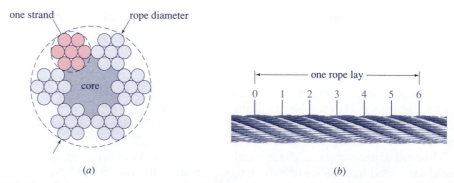

          (a)   (b)

FIGURE 3.12 Wire rope geometry: (*a*) The cross section of a 6 × 7 fiber-core wire rope, with properly measured rope diameter shown. (*b*) The helix formed by strands in a 6 × 7 wire rope completes one revolution within the rope *lay length.* (Eric M. Taleff)

---

[3]Although the names *cable* and *wire rope* are often used interchangeably, we use the term wire rope here because it is specific to load-bearing applications, whereas cable is a more general term, which can refer to non-structural applications, e.g., cable TV.

permanent stretch observed during service and is typically used for stationary applications, such as in suspension bridges. Elastic stretch increases as load increases and is removed as the load is removed. We are concerned here with calculating the elongation from elastic stretch only, i.e. elastic elongation, and we assume proper installation and break-in have already established a nearly constant permanent stretch.

Elastic elongation of a wire rope may be conveniently approximated as linearly elastic behavior. Elastic elongation is calculated using Eq. 3.14, so long as either the flexibility coefficient or stiffness coefficient is known. Figure 3.12a makes clear that the area of a wire rope calculated using its diameter, $\pi d^2/4$, is much greater than the area of material actually supporting the load applied to it. Thus, Eqs. 3.14 and 3.15 cannot be applied directly. Instead, the flexibility coefficient of a wire rope can be calculated as

$$f_r = \frac{L}{A_m E_r} \tag{3.19}$$

where A_m is the *metallic area,* also called the effective area, and E_r is the *rope modulus.* The stiffness coefficient of a wire rope, from Eq. 3.15, is then

$$k_r = \frac{A_m E_r}{L} \tag{3.20}$$

The metallic area may be calculated as follows

$$A_m = \text{CF} \times d^2 \tag{3.21}$$

where CF is the *compactness factor,* provided by the wire rope manufacturer, and d is the rope diameter, as illustrated in Fig. 3.12(a). Values of CF typically range from 0.380 to 0.405, but can reach 0.580 for guy strand. A conservative value of 0.380, typical of 6×7 wire rope, is recommended for working example problems. Some manufacturers simply provide tables of metallic areas for their wire rope products, which eliminates the need for calculation of A_m.

The **rope modulus,** E_r, depends on the wire type and the rope construction. Values of the rope modulus generally vary from 26% to 50% of the elastic modulus of an individual wire, a typical value being 40%. Using this typical value, which is valid for a standard 6×7 wire rope having a fiber core and steel wires (steel has an elastic modulus of $E = 200$ GPa), produces a rope modulus of $E_r = 80$ GPa. This simple conversion factor of 40% is sufficient for working example problems, but a rope modulus supplied by the wire rope manufacturer should always be used in engineering calculations.

The **allowable load** for a wire rope is of great importance, but is not easily calculated. Engineers rely upon tables of recommended maximum load values, which are compiled from experimental data. In applications involving very long wire ropes, such as mooring of offshore platforms, the weight of the wire rope itself can produce a large fraction of its allowable load. Thus, the weight of wire rope must be considered in such applications. Examples of recommended maximum load values and typical weights per unit length are shown, for several wire rope diameters, in Table 3.1.

TABLE 3.1 **Weight per Length and Recommended Allowable Loads (in tons of 2000 lb) Are Shown for 6 × 19 Seale IWRC (independent wire rope core) Construction Wire Ropes Using IPS (improved plough steel) Wires.**

Diameter (in.)	Weight, lb/ft.	Allowable load, tons
1/4	0.18	2.94
1/2	0.46	11.5
1	1.85	44.9
2	7.39	172

From: *Wire Rope Handbook,* Wire Rope Corporation of America, 1985, p. 13. (Now WireCo WorldGroup).

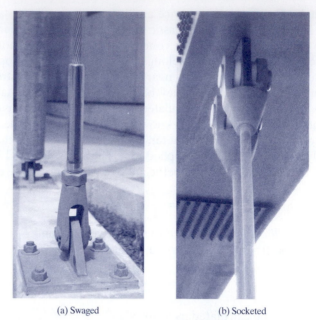

(a) Swaged   (b) Socketed

FIGURE 3.13   Examples of different end attachments used for wire ropes. (Eric M. Taleff)

To transfer load from the wire rope to some other part, many applications of wire rope require end attachments, such as those shown in Fig. 3.13(*a, b*). The *efficiency* of an end attachment determines how much the allowable load must be reduced because of the attachment, which may be weaker than the wire rope. For example, an end attachment that is 80% efficient will reduce the allowable load by 20%. Typical efficiencies of properly constructed end attachments range from 75% to 100%.

EXAMPLE 3.4

In the evening a contractor uses the wire rope on a large crane to lift a 1000-lb air compressor 10 ft above the ground, which will help prevent mischief occurring overnight. The wire rope has a diameter of $d = 1/4$ in. and is composed of IPS wires in a 6 × 19 Seale IWRC construction; see Table 3.1. The wire rope modulus is $E_r = 12{,}000$ ksi, and its compactness factor is CF = 0.40. The rope runs 10 ft between the crane winch at A and a pulley (sheave) at B, and it then runs $L = 40$ ft down to the attachment C at the compressor on the ground. The hook at the end of the wire rope, position C, has an attachment efficiency of 90%. (a) Calculate the elongation of the wire rope between the winch at A and the compressor at C, just as the compressor is lifted from the ground. This is the length of wire rope which must be wound onto the winch drum before the compressor will move. (b) Calculate the factor of safety for the wire rope in this operation, assuming that failure occurs at the recommended maximum load of Table 3.1.

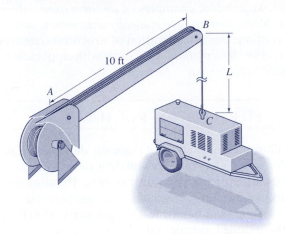

(a) It is necessary to first calculate the metallic area of the wire rope. From Eq. 3.21, the metallic area of the rope is

$$A_m = CF \times d^2 = 0.40 \times (0.25 \text{ in.})^2 = 0.025 \text{ in.}^2$$

The flexibility coefficient, from Eq. 3.19, is

$$f_r = \frac{L}{A_m E_r} = \frac{(10 \text{ ft} + 40 \text{ ft})}{(0.025 \text{ in}^2)(12{,}000 \text{ ksi})} = 0.167 \frac{\text{in.}}{\text{kip}}$$

Note that the entire length of the wire rope between the winch and the compressor (10 ft + 40 ft) elongates under the loading. You can verify this using free-body diagrams. The elongation of the wire rope is then

$$e = f_r P = \left(0.167 \frac{\text{in.}}{\text{kip}}\right)(1 \text{ kip}) = 0.167 \text{ in.} \qquad \textbf{Ans. (a)}$$

(b) Calculation of the expected failure load, P_F, must account for the efficiency of the hook end attachment at C. Using the maximum load suggested for this wire rope in Table 3.1 of 5.88 kips, the allowable load after adjusting for the attachment efficiency is

$$P_F = 0.90 \times 5.88 \text{ kips} = 5.29 \text{ kips}$$

The greatest load on the wire rope occurs at B and is the 1000 lbs. (1 kip) of the compressor plus the weight of the wire rope. The total weight of wire rope hanging from the sheave at B to the attachment at C is,

$$W = (0.18 \text{ lb/ft})(40 \text{ ft}) = 7.2 \text{ lb}$$

This is, of course, a negligibly small load compared to that of the compressor. Thus, we may reasonably neglect the weight of the wire rope for this relatively short length. The factor of safety is then

$$FS = \frac{5.29 \text{ kips}}{1 \text{ kip}} = 5.29 \qquad \textbf{Ans. (b)}$$

Next, in Section 3.4, we consider examples of *uniform axial deformation,* solving several example problems that involve *statically determinate structures.* Finally, in Sections 3.5–3.8, we continue our consideration of structural systems whose members undergo *uniform axial deformation,* this time with application to *statically indeterminate structures.*

3.4 STATICALLY DETERMINATE STRUCTURES

In this section, we will illustrate the analysis of **statically determinate structures** that have uniform axial-deformation elements. We will determine both the stresses in the elements and the deformation of the structural system. These example problems emphasize the importance of properly considering the fundamental equations: **equilibrium, element force-deformation behavior,** and **geometry of deformation.** Note carefully how each of these three fundamental types of equations enters into the solution of each problem. Also note how the *flexibility coefficient f,* which is defined in Eq. 3.14, is employed in these solutions.

EXAMPLE 3.5

Consider again the two-element axial-deformation structure of Example 2.3 (Section 2.2), shown here in Fig. 1. In Example 2.2 the axial stresses

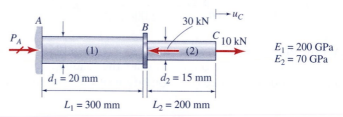

Fig. 1

σ_1 and σ_2 in elements (1) and (2), respectively, were determined. Since the two-element structure is statically determinate, these stresses are independent of the materials from which the elements are made. Now let us determine the displacement u_C of end C if element AB is steel ($E_1 = 200$ GPa) and element BC is aluminum ($E_2 = 70$ GPa).

Plan the Solution The nomenclature for a typical element is shown in Fig. 2. The axial forces were previously obtained in Example 2.3, and Eq. 3.14 can be used to express the elongation of each element in terms of its axial force. Finally, the displacement u_C of end C is simply the sum of the elongations of the two elements.

Solution

Equilibrium: From the free-body diagrams and equilibrium equations of Example 2.3 we have

$$F_1 = -20 \text{ kN}, \qquad F_2 = 10 \text{ kN} \qquad \textbf{Equilibrium} \qquad (1)$$

where F_i is the axial force in element i.

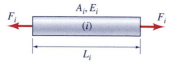

Fig. 2 A typical element.

Element Force-Deformation Behavior: For the two uniform elements, the force-deformation equation, Eq. 3.14, gives

$$e_i = f_i F_i, \quad \text{where} \quad f_i = \frac{L_i}{A_i E_i}, \quad i = 1, 2 \qquad \text{Element Force- Deformation} \qquad (2)$$

Inserting numerical values into Eq. (2) gives the two element flexibility coefficients

$$f_1 = \frac{L_1}{A_1 E_1} = \frac{(300 \text{ mm})}{(314.2 \text{ mm}^2)(200 \text{ kN/mm}^2)} = 4.77(10^{-3}) \text{ mm/kN}$$

$$f_2 = \frac{L_2}{A_2 E_2} = \frac{(200 \text{ mm})}{(176.7 \text{ mm}^2)(70 \text{ kN/mm}^2)} = 1.62(10^{-2}) \text{ mm/kN}$$

Note that element (2) is *over three times as flexible as* element (1), primarily because of its smaller cross-sectional area and smaller modulus of elasticity. Inserting these flexibility coefficients, together with the element axial forces from Eq. (1), into Eq. (2) gives the two element elongations

$$e_1 = f_1 F_1 = -0.0955 \text{ mm}$$
$$e_2 = f_2 F_2 = 0.1617 \text{ mm} \qquad (2')$$

Thus, element (1) is shortened due to its compressive force while element (2) elongates due to its tensile force.

Geometry of Deformation; Compatibility Equation: Since the two elements are attached end-to-end, the displacement of the right end of element (1) is compatible with (i.e., equal to) the displacement of the left end of element (2). Therefore, the displacement of end C is given by

$$u_C = e_1 + e_2 \qquad \text{Geometry of Deformation} \qquad (3)$$

This equation is called a **compatibility equation,** since it enforces the condition that the respective end displacements of the two elements joined together at B be compatible.

The final solution is

$$u_C = e_1 + e_2 = -0.0955 \text{ mm} + 0.1617 \text{ mm}$$
$$u_C = 6.62(10^{-2}) \text{ mm} \qquad \qquad \textbf{Ans.}$$

Review the Solution Note how each of the quantities L_i, A_i and E_i enters into the element flexibility coefficients f_i, and how these flexibilities and the element forces F_i together determine the elongations of these uniform elements.

For end-to-end assemblages of elements, such as the one in Example 3.5, the total elongation is given by the sum of the element elongations, that is,

$$e_{\text{total}} = \sum \frac{F_i L_i}{A_i E_i} \qquad (3.22)$$

MDS3.1   **Segmented Axial**—is a program for solving statically determinate problems like Example 3.5.

═══➤ **EXAMPLE 3.6** ◄───────

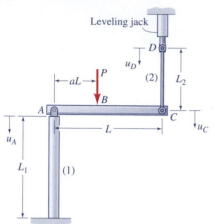

Leveling jack

Fig. 1

The rigid, weightless beam AC in Fig. 1 is supported at end A by a vertical column, and at end C it is supported by a vertical rod CD that is attached at D to a leveling jack, that is, a jack that can support the required load in rod CD but can also move the vertical position of D upward or downward. Point B, the point of application of load P, can be anywhere along the beam AC, that is, anywhere in the range $0 < a < 1$. The purpose of the leveling jack is to keep beam AC perfectly level, regardless of the location of B.

(a) Determine the axial stresses σ_1 in the column and σ_2 in the rod when the load P is on the beam. (b) Determine the downward displacement u_A of end A when the load is on the beam. (c) Finally, determine the displacement u_D required at the leveling jack for the beam AC to be level under the given loading, that is, to make $u_C = u_A$. The relevant dimensions and material properties are:

$$P = 2 \text{ kips}, L_1 = 10 \text{ ft}, L_2 = 5 \text{ ft}, L = 10 \text{ ft}, a = 0.4$$

$$A_1 = 2 \text{ in}^2, A_2 = 0.8 \text{ in}^2, E_1 = E_2 = 30 \times 10^3 \text{ ksi}$$

Plan the Solution   A free-body diagram of beam AC may be used in solving for the axial forces in the column and support rod. Equation 3.14 can be used to express the elongation of each element in terms of its axial force. Finally, the elongations of the two elements can be expressed in terms of the displacements $u_A = u_C$ and u_D.

Solution   (a) *Axial stresses σ_1 and σ_2.*

Equilibrium: The free-body diagram of beam AC in Fig. 2 will enable us to write two equilibrium equations that relate the axial force F_1 in the column and F_2 in the support rod to the external load P.

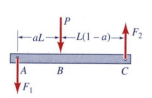

Fig. 2   Free-body diagram.

$$+\circlearrowleft \left(\sum M \right)_C = 0: \quad F_1 L + PL(1 - a) = 0$$

$$+\circlearrowleft \left(\sum M \right)_A = 0: \quad F_2 L - PaL = 0$$

or

$$F_1 = -P(1 - a) = -2 \text{ kips } (0.6)$$
$$= -1.2 \text{ kips} \qquad \textbf{Equilibrium} \qquad \text{(1a,b)}$$
$$F_2 = Pa = 2 \text{ kips} (0.4) = 0.8 \text{ kips}$$

Therefore, the axial stresses are

$$\sigma_1 = \frac{F_1}{A_1} = \frac{-1.2 \text{ kips}}{2 \text{ in}^2} = -0.6 \text{ ksi} = 0.6 \text{ ksi C}$$

Ans. (a)

$$\sigma_2 = \frac{F_2}{A_2} = \frac{0.8 \text{ kips}}{0.8 \text{ in}^2} = 1.0 \text{ ksi} = 1.0 \text{ ksi T}$$

(b) *The downward displacement u_A.* To determine the displacement u_A, we need to determine the change in the length of the column due to the compressive force acting on it. At the same time, we can determine the elongation of the rod due to the force F_2 acting on it.

Element Force-Deformation Behavior: For the two uniform elements, the force-deformation equation, Eq. 3.14. gives

$$e_i = f_i F_b \quad \text{where} \quad f_i = \frac{L_i}{A_i E_i}, \quad i = 1, 2 \qquad \text{Element Force-Deformation} \qquad (2)$$

Inserting numerical values into Eq. (2) gives the two element flexibility coefficients

$$f_1 = \frac{L_1}{A_1 E_1} = \frac{(120 \text{ in.})}{(2 \text{ in}^2)(30 \times 10^3 \text{ ksi})} = 2.00(10^{-3}) \text{ in./kip}$$

$$f_2 = \frac{L_2}{A_2 E_2} = \frac{(60 \text{ in.})}{(0.8 \text{ in}^2)(30 \times 10^3 \text{ ksi})} = 2.50(10^{-3}) \text{ in./kip}$$

Note that these two elements are almost *equally flexible,* even though one is twice as long as the other. Inserting these flexibility coefficients, together with the element axial forces from Eqs. (1a,b), into Eq. (2) gives the two elements elongations

$$e_1 = f_1 F_1 = -2.40(10^{-3}) \text{ in.}$$
$$e_2 = f_2 F_2 = 2.00(10^{-3}) \text{ in.} \qquad (2'\text{a,b})$$

Thus, the column is shortened due to its compressive force while the rod elongates due to its tensile force.

Geometry of Deformation: The elongation e_i of each element is the difference between the displacements of its two ends:

$$e_1 = -u_A$$
$$e_2 = u_C - u_D \qquad \text{Geometry of Deformation} \qquad (3\text{a,b})$$

Therefore, from Eqs. (2′a) and (3a), the vertical displacement at A is

$$u_A = 2.40(10^{-3}) \text{ in.} \downarrow \qquad \textbf{Ans. (b)} \quad (4)$$

(c) *The displacement u_D at the top of the support rod.* The beam BC is level whenever

$$u_C = u_A \qquad (5)$$

Therefore, from Eqs. (2′b), (3b), (4), and (5),

$$u_D = u_C - e_2 = 2.40(10^{-3}) \text{ in.} - 2.00(10^{-3}) \text{ in.}$$

so

$$u_D = 0.40(10^{-3}) \text{ in.} \downarrow \qquad \textbf{Ans. (c)}$$

Thus, to keep the beam AC level for the given position of the load, it is necessary for the leveling jack to support a load of 0.8 kips while allowing rod-attachment point D to move downward the small amount $0.40(10^{-3})$ in.

Review the Solution For statically determinate problems, like this one, the element forces can be determined from equilibrium alone, independent of the member sizes or the materials from which the members are made. However, wherever displacements are required, the additional equations come from the element force-deformation equations and the geometry-of-deformation equations.

MDS3.2 **Beam and Two Rods**—is a program for solving statically determinate problems similar to Example 3.6.

EXAMPLE 3.7

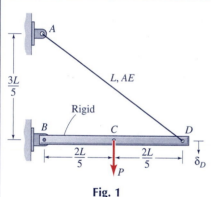

Fig. 1

A wire of length L, cross-sectional area A, and modulus of elasticity E supports a rigid beam of negligible weight, as shown in Fig. 1. When there is no load acting on the beam, the beam is horizontal. (a) Determine an expression for the axial stress in the wire AD when load P is applied at the midpoint of beam BD. (b) Determine an expression for the vertical displacement, δ_D, of end D, simplifying your solution by assuming that δ_D is small (i.e., $\delta_D \ll L$).

Plan the Solution Since this is a statically determinate structure. Part (a) is very straightforward. We simply use **equilibrium** to determine an expression for the force in the wire and divide that force by the cross-sectional area of the wire to get the axial stress in the wire. In Part (b), getting the correct geometrical relationship between the elongation of the wire and the tip displacement δ_D will be the challenge! A **deformation diagram** will help us visualize the geometry of deformation.

Solution (a) *Axial stress in wire AD.*

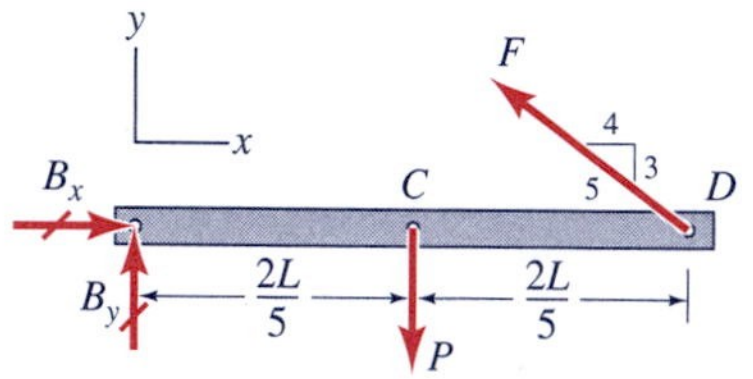

Equilibrium: Using the free-body diagram in Fig. 2, we can write an equilibrium equation that relates the tensile force F in the wire to the external load P.

Fig. 2 Free-body diagram.

$$+\curvearrowleft \left(\sum M \right)_B = 0: \qquad -P\left(\frac{2L}{5}\right) + \left(\frac{3}{5}F\right)\left(\frac{4L}{5}\right) = 0$$

or

$$F = \frac{5}{6}P \qquad \textbf{Equilibrium} \qquad (1)$$

Therefore, the axial stress in the wire is

$$\sigma = \frac{F}{A} = \frac{5}{6}\frac{P}{A} \qquad \textbf{Ans. (a)}$$

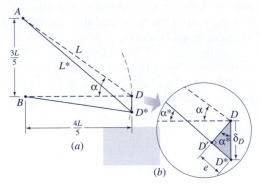

Fig. 3 Deformation diagram.

(b) *Tip displacement of the beam.* To determine the tip displacement of the beam, we need to determine the elongation of the support wire (force-deformation behavior) and then determine how far this elongation lets the beam rotate (geometry of deformation).

Element Force-Deformation Behavior: From the element force-deformation equation. Eq. 3.13, we have

$$e = \frac{FL}{AE} \qquad \text{\textbf{Element Force-Deformation}} \qquad (2)$$

Geometry of Deformation; Compatibility Equation: Figure 3 is a **deformation diagram** that enables us to relate the elongation e of the loaded support wire to the vertical displacement of D. The beam BD is rigid, so end D would actually follow a circular arc about end B when load P is applied to the beam, as indicated by the dashed circular arc in Fig. 3a. However, in Fig. 3b we have employed the simplifying assumption that point D moves vertically downward to D^*, which differs little from moving along the circular arc so long as δ_D is small (i.e., $\delta_D \ll L$).[4]

Figure 3b is an expanded view of the vicinity of D showing the deformation triangle $DD'D^*$, with side DD' drawn perpendicular to the wire in its final position AD^*. Side DD^* is the (approximated) vertical displacement of D, while $D'D^*$ is the elongation of the wire, $e = L^* - L$. Angle α in Figs. 3a,b is a vertex of the 3-4-5 triangle ADB. The two angles labeled α^* in Fig. 3b are equal to each other and clearly, if δ_D is very small, $\alpha^* \approx \alpha$. Therefore, we can identify triangle D'DD* as a 3-4-5 triangle, so

$$e = \frac{3}{5}\delta_D \qquad \text{\textbf{Geometry of Deformation}} \qquad (3)$$

---

[4]Example 2.5 in Sect. 2.3 shows how the Pythagorean Theorem can be used to calculate the exact elongation of the wire in terms of the rotation of the rigid beam.

This equation may be called a **compatibility equation,** since it enforces the condition that the end of the wire and the end of the beam undergo the same displacement from D to D^*.

Solution of the Equations: Since this is a statically determinate problem, we were able to determine the element force in Eq. (1) from equilibrium alone. This force can be substituted into the element force-deformation equation, Eq. (2), giving

$$e = \frac{5}{6}\left(\frac{PL}{AE}\right) \qquad (4)$$

Finally, this result can be substituted into the compatibility equation, Eq. (3), to give the tip displacement δ_D.

$$\delta_D = \frac{25}{18}\left(\frac{PL}{AE}\right) \qquad \textbf{Ans. (b)}$$

Review the Solution   The right-hand side of the answer has the correct dimension of length. A larger force should produce a larger tip displacement, which it does. Likewise, a larger area or a larger value of E should reduce the tip displacement, as is the case. Therefore, the answer we have obtained seems reasonable.

MDS3.3   **Pinned Beam and Strut**—is a program for solving statically determinate problems like Example 3.7.

All internal forces in the statically determinate structures in the previous three examples were determined through use of the equations of statics (i.e., equilibrium equations) alone. However, since the deformation of each structure was also to be determined, it was necessary to use force-deformation behavior and geometric compatibility. Whenever the geometry of deformation is not simple, a deformation diagram is very helpful, as is illustrated in Example 3.7.

In Sections 3.5 through 3.9 systematic procedures for treating **statically indeterminate systems** that have uniform axial-deformation members will be examined. Table 3.2 shows how various "inputs" (e.g., external forces) are related to the fundamental equations—*equilibrium, force-temperature-deformation behavior,* and *geometry of deformation.* The three cases that are listed in Table 3.2—external forces, temperature changes, and geometric misfits—are discussed in Sections 3.5, 3.6, and 3.7, respectively.

TABLE 3.2   A Summary of Basic Problem Types

Relevant Equation Set	"Input"	Sect.
(1) *Equilibrium*	External forces	3.5
(2) *Element force-temperature-deformation*	Temperature changes	3.6
(3) *Deformation geometry*	Geometric misfits	3.7

Figure 3.14 shows two structures, each consisting of two collinear elements. Acting on the structure in Fig. 3.14a are two known forces, P_B and P_C, and one reaction, P_A. The reaction at the left end and the axial force in each of the two elements of the structure in Fig. 3.14a can be determined from <u>statics alone</u>, that is, by drawing free-body diagrams and solving equilibrium equations. The values of these forces are independent of the materials involved and other member properties (e.g., cross-sectional area or length). Structures of this type are called **statically determinate structures.**

On the other hand, both ends of the structure in Fig. 3.14b are attached to rigid walls, so there are two unknown reactions P_A and P_C, but only one known load, P_B. Since there is only one useful equilibrium equation, summation of forces in the axial direction, it is not possible to determine both reactions from equilibrium alone. To determine the reactions and element forces for this case it is necessary to consider the deformation of the elements, and this involves member sizes and materials.[5] Such structures are classified as **statically indeterminate.** Analysis of statically indeterminate structures thus involves all of the three fundamental types of equations: **equilibrium, element force-deformation behavior,** and **geometry of deformation.** In this section we will illustrate the steps involved in the analysis of statically indeterminate structures.

Analysis of a Typical Statically Indeterminate Structure. Consider the simple two-element structure in Fig. 3.15a. Let us first determine the internal axial forces, F_1 and F_2, in the elements; then we will determine the displacement, u_B, of the node (joint) between the two elements.

Equilibrium: Since the forces F_1 and F_2 are constant along their respective elements, we can use the free-body diagrams shown in Fig. 3.15b. The free-body diagram of node B will enable us to write an equilibrium equation that relates the *external load* P_B at joint B to the *internal forces* F_1 and F_2 in the elements adjoining B.

$$\overset{+}{\rightarrow} \sum F = 0: \qquad\qquad F_1 - F_2 = P_B \qquad\qquad \textbf{Equilibrium} \qquad\qquad (1)$$

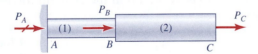

(*a*) A statically determinate structure.

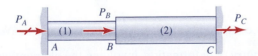

(*b*) A statically indeterminate structure.

FIGURE 3.14   Two two-element structures.

---

[5]If displacements are required, member sizes and materials must be considered for statically determinate structures as well as for statically indeterminate structures, as demonstrated in Examples 3.5–3.7.

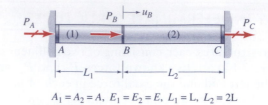

$$A_1 = A_2 = A, \ E_1 = E_2 = E, \ L_1 = L, \ L_2 = 2L$$

(a) A two-element structure with fixed ends.

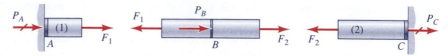

(b) Free-body diagrams of nodes A, B, and C.

FIGURE 3.15 A statically indeterminate structure.

The reactions at ends A and C are related to the element forces by equilibrium equations for node A and node C, respectively,

$$P_A = -F_1, \quad P_C = F_2 \tag{1'}$$

and they can easily be determined once we have solved for the element forces F_1 and F_2.

This structure is called **statically indeterminate,** because it is not possible to determine two unknown element forces from the single equilibrium equation, Eq. (1). The fact that bar AC is attached to rigid walls at <u>both</u> A and C prevents the length of the member from changing, and this constraint provides the needed additional equation, $e_{AC} = 0$. Such constraints on the deformation of a structure are called **compatibility** conditions, and they involve the geometry of deformation.

But the compatibility equation is expressed directly in terms of geometric quantities, not in terms of forces, like the equilibrium equation, Eq. (1). To address this, we first recall the force-deformation equations of the two elements, which relate forces to deformation quantities. Then we will be able to continue our discussion of the compatibility equation.

Element Force-Deformation Behavior: From the element force-deformation equation, Eq. 3.14, we have

$$e_1 = f_1 F_1, \quad e_2 = f_2 F_2 \qquad \begin{array}{l}\textbf{Element}\\ \textbf{Force-}\\ \textbf{Deformation}\\ \textbf{Behavior}\end{array} \tag{2a,b}$$

where the flexibility coefficients are $f_1 = \dfrac{L}{AE}$ and $f_2 = \dfrac{2L}{AE}$. We are now prepared to consider the geometry of deformation, involving the element elongations e_1 and e_2.

Geometry of Deformation; Compatibility Equation: Both ends of bar AC are fixed to rigid walls, so the two reactions not only support the bar AC, but they also force the total elongation of the bar to be zero. This constraint condition arises from the geometry of deformation and provides the following **compatibility equation:**

$$e_{AC} = e_1 + e_2 = 0 \qquad \textbf{Compatibility} \tag{3}$$

Solution of the Equations: In Eqs. (1) through (3) there are four equations and four unknowns. Therefore, it is a matter of solving these four equations simultaneously to determine the four unknowns. We are first interested in determining the **element forces**. Therefore, we can eliminate the e_i's by substituting Eqs. (2a,b) into Eq. (3), which gives the following **compatibility equation written in terms of element forces:**

$$f_1 F_1 + f_2 F_2 = 0$$

or

$$F_1\left(\frac{L}{AE}\right) + F_2\left(\frac{2L}{AE}\right) = 0 \qquad \begin{array}{l}\textbf{Compatibility}\\ \textbf{in Terms of}\\ \textbf{Element Forces}\end{array} \qquad (4)$$

Finally, the equilibrium equation (1) and the compatibility equation (4) can be **solved simultaneously** to give the following:

$$F_1 = \frac{2P_B}{3}, \quad F_2 = -\frac{P_B}{3} \qquad \textbf{Ans.} \quad (5a,b)$$

These element forces, then, are the primary solution for this statically indeterminate structure.

Note that, because element (2) is twice as long as element (1), it is twice as **flexible** (i.e., $f_2 = 2f_1$). Said another way, element (1) is twice as **stiff** as element (2) (i.e., $k_1 = 2k_2$). Therefore, a larger part of the load P_B is transferred by element (1) to the left wall than is transferred to the right wall by element (2). Had the two elements been of equal length, half of load P_B would have been transmitted to each wall. This illustrates the fact that, **for statically indeterminate structures, the reactions and the internal element forces depend on the element sizes and material properties—A_i, L_i, and E_i—and cannot be determined by statics alone.**

Displacement of Node B: In the problem statement we were asked not only to determine the element forces but also to determine the displacement of node B. From Fig. 3.15a we can see that u_B can be obtained from either of the following two geometry-of-deformation equations:

$$u_B = e_1, \quad u_B = -e_2 \qquad \begin{array}{l}\textbf{Geometry of}\\ \textbf{Deformation}\end{array} \qquad (6a,b)$$

Therefore, combining Eqs. (2a), (5a), and (6a) [or Eqs. (2b), (5b), and (6b)] we get

$$u_B = \frac{2P_B L}{3AE} \qquad \textbf{Ans.}$$

> **MDS3.4**   **End-to-End Bars**—is an MDS program for analyzing statically indeterminate structures like the one in Fig. 3.15a.

Solution Procedure; Basic Force Method.

The key steps that are required for analyzing statically indeterminate structures are: (1) to write down the *equilibrium equations* that will be needed for determining the unknown forces; and (2) to write down all of the *force-deformation equations* that relate the forces in the equilibrium equations to the displacement quantities in the compatibility equations. The third key step is (3) to write down the *compatibility equations* that characterize any *geometric constraints* on the deformation of the structure.

With the aid of the following Procedure, you should review the steps that were used to analyze the statically indeterminate structure shown in Fig. 3.15a. The flow chart that follows is just to provide a more graphic reminder of the role played by each of the three fundamental types of equations in the analysis of statically indeterminate structures, and to outline a straightforward solution procedure for you to use in solving (homework and exam) problems.

SOLUTION PROCEDURE: BASIC FORCE METHOD

▼ SET UP THE FUNDAMENTAL EQUATIONS:

(1) Let N_E, be the number of *independent* equilibrium equations. Using *free-body diagrams*, write down N_E **independent equilibrium equations.**

(2) Write an **element force-deformation equation** for each axial-deformation element. Equation 3.14 is the most convenient form to use.

$$e_i = f_i F_i, \quad \text{where } f_i = \frac{L_i}{A_i E_i} \qquad \begin{array}{c} (3.14) \\ \text{repeated} \end{array}$$

(3) Use *geometry of deformation* to write down the **compatibility equation(s)** in terms of the element elongations, e_i.

▼ SOLVE FOR THE UNKNOWN FORCES:

(4) Substitute the element force-deformation equations of Step 2 into the geometric-compatibility equations of Step 3. This gives the **geometric-compatibility equations in terms of the unknown element forces.**

(5) The final step in determining the unknown forces is to **solve simultaneously** the equilibrium equations (Step 1) and the compatibility equations written in terms of element forces (Step 4).

▼ SOLVE FOR THE DISPLACEMENTS:

(6) To obtain system displacements, if they are required, substitute the element forces into the force-deformation equations of Step 2. Finally, use geometry of deformation to relate element elongations to system displacements.

(7) Review the solution to make sure that all answers seem to be correct.

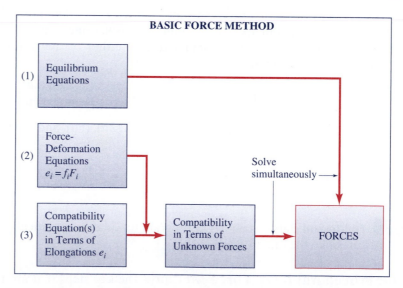

The method that is presented here is quite adequate for carrying out the analysis of the relatively simple statically indeterminate structures in this chapter. For reference purposes, this method has been given the label *Basic Force Method,* because the key (i.e., first) answers that are obtained are *forces*. Methods labeled the *Displacement Method* and the *Force Method,* more formalized methods that are suitable for computer-based analysis of very complex structures, are introduced in

Sections 3.8 and 3.9, respectively, and are treated in greater detail in textbooks on structural analysis and in textbooks on finite element analysis. Of course, the Displacement Method and the Force Method are also suited for solving simpler problems, such as the Example problems in the remainder of this chapter.

Each extra constraint that is added to a structure, beyond the support that is required for stable equilibrium, gives rise to an additional redundant force, so **there will always be as many constraint equations as there are unknown redundant forces.** This fact is illustrated in Examples 3.8 and 3.9.

EXAMPLE 3.8

A load P is hung from the end of a rigid beam AD, which is supported by a pin at end A and by two uniform hanger rods (see Fig. 1). How much load, to the nearest 100 pounds, can be applied without exceeding an allowable tensile stress of $\sigma_{allow} = 30$ ksi in either of the rods? Assume small-angle rotation of AD.

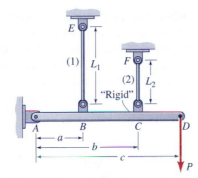

$a = 50$ in., $b = 100$ in., $c = 150$ in., $L_1 = 80$ in., $L_2 = 50$ in.
$E = 10 \times 10^3$ ksi, $A_1 = 1.0$ in^2, $A_2 = 0.5$ in^2

Fig. 1

Plan the Solution This problem is not quite as straightforward as the previous one that served to introduce statically indeterminate structures. Now the load is unknown, but the allowable axial stresses are known. Since the axial stress in an element is just the (internal) axial force in the element divided by its area, it is clear that we must eventually get equations that relate F_1 and F_2, the axial forces in the two rods, to the load P.

Solution Let us begin by writing down the three fundamental types of equations.

Equilibrium: We must ask ourselves the question: What free-body diagram(s) would lead to an equilibrium equation (or equations) that relate the external load P to the internal element forces F_1 and F_2? Clearly, a free-body diagram of the rigid beam AD should be used.

Since we are not specifically asked to determine the reactions A_x and A_y, we do not need to sum forces. A moment equation for the free-body diagram in Fig. 2 will directly relate F_1 and F_2 to P, so we write

$$+\zeta \left(\sum M \right)_A = 0: \qquad aF_1 + bF_2 - cP = 0 \qquad \textbf{Equilibrium} \quad (1)$$

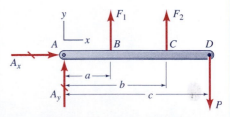

Fig. 2 Free-body diagram.

Since there are two unknown forces and only one equilibrium equation, the structure is statically indeterminate. Since the beam AD could be supported by just one hanger rod, one of the rods is redundant.

Element Force-Deformation Behavior: It will be convenient to use the form given in Eq. 3.14.

$$e_1 = f_1 F_1, \quad e_2 = f_2 F_2$$

Element Force-Deformation Behavior (2)

where the values of the flexibility coefficients for the hanger rods, labeled (1) and (2) in Fig. 1, are

$$f_1 = \left(\frac{L}{AE}\right)_1 = \frac{80 \text{ in.}}{(1.0 \text{ in}^2)(10 \times 10^3 \text{ ksi})} = 8.00(10^{-3}) \text{ in./kip}$$

$$f_2 = \left(\frac{L}{AE}\right)_2 = \frac{50 \text{ in.}}{(0.5 \text{ in}^2)(10 \times 10^3 \text{ ksi})} = 10.00(10^{-3}) \text{ in./kip}$$

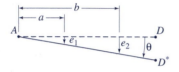

Fig. 3   Deformation diagram.

Geometry of Deformation: Because AD is assumed to be rigid, we can sketch a deformation diagram that relates the elongations of the hanger rods to the rotation of the beam AD about A (Fig. 3). Since the rotation angle θ is assumed to be small, we can assume that points B and C move vertically downward (instead of along arcs of circles with centers at A), and we can therefore use the properties of similar triangles to get the **compatibility equation**

$$e_2 = \left(\frac{b}{a}\right)e_1$$

Deformation Geometry (3)

Therefore, the rigid beam *constrains* the two hanger-rod elongations to satisfy this compatibility equation.

Solution of the Equations: To solve for the internal forces, we can eliminate the element elongations, e_1 and e_2, by substituting the force-deformation equations into the compatibility equation to get the following compatibility equation in terms of the unknown element forces.

$$f_2 F_2 = \left(\frac{b}{a}\right)f_1 F_1$$

Compatibility in Terms of Forces (4)

   In Eqs. (1) and (4) we have two equations in terms of the two unknown internal forces. These can be <u>solved simultaneously</u> to give expressions that relate the forces F_1 and F_2 to the external load P.

$$F_1 = \left(\frac{acf_2}{a^2 f_2 + b^2 f_1}\right)P$$

$$F_2 = \left(\frac{bcf_1}{a^2 f_2 + b^2 f_1}\right)P$$

(5a,b)

Allowable Load: Let us now substitute numerical values from Fig. 1 and numerical values for the flexibility coefficients f_1 and f_2 into Eqs. (5). From Eqs. (5),

$$F_1 = \left(\frac{15}{21}\right)P, \qquad F_2 = \left(\frac{24}{21}\right)P \qquad (6a,b)$$

Now we can relate the external load P to the allowable stress. With F_1 and F_2 given by Eqs. (6), we get the following expressions for the stresses in the rods:

$$\sigma_1 = \frac{F_1}{A_1} = \frac{(15/21)(P \text{ kips})}{(1.0 \text{ in}^2)}$$

$$\sigma_2 = \frac{F_2}{A_2} = \frac{(24/21)(P \text{ kips})}{(0.5 \text{ in}^2)} > \sigma_1 \qquad (7a,b)$$

Since rod (2) is more highly stressed than rod (1), we set σ_2 equal to the allowable stress and get

$$P_{\text{allow}} = \frac{21}{24}(0.5 \text{ in}^2)\sigma_{\text{allow}} = 13.1 \text{ kips} \qquad \textbf{Ans.}$$

Review the Solution   One way to verify the results is to check compatibility.

Is
$$f_2 F_2 = \left(\frac{b}{a}\right)f_1 F_1? \quad \text{Yes}$$

$$[10.00(10^{-3})\text{ in./kip}]\left(\frac{24}{21}\right)P\,(\text{kips}) =$$

$$\left(\frac{100 \text{ in.}}{50 \text{ in.}}\right)[8.00(10^{-3})\text{ in./kip}]\left(\frac{15}{21}\right)P\,(\text{kips})$$

MDS3.6   **Rigid Member and Two Bars**—is an MDS program for analyzing statically indeterminate structures like the one in Example 3.8.

In Example 3.8. there is only one equilibrium equation, but there are two unknown internal forces, F_1 and F_2, Therefore, there is **one redundant force.** How ever, to solve for both forces it was not necessary to identify either one of them specifically as the redundant force. Example 3.9 illustrates how to analyze statically indeterminate structural systems with more than one redundant force, that is, where the number of unknown forces in the equilibrium equation(s) exceeds the number of equilibrium equations by two or more.

EXAMPLE 3.9

The structural assemblage in Fig. 1 is made up of three uniform elements, or members. Element (1) is a solid rod. Element (2) is a pipe that surrounds element (3), which is a solid rod that is identical to element (1) and collinear with it. The three elements are all attached at B

to a rigid plate of negligible thickness. With no external force at B, the three-element assemblage exactly fits between the rigid walls at A and C; its ends are then attached to the two walls. Determine expressions for the axial forces in the three elements when an external force P_B is applied at B.

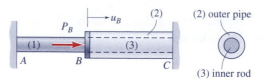

Fig. 1 The original structure.

Plan the Solution   We can think of this structure as being composed of three uniform elements and one connecting node (joint), and we can write an equilibrium equation for node B. Using Eq. 3.14, we can write a force-deformation equation for each of the three elements. We can relate the element elongations to each other through the displacement at B. Finally, we can use the Basic Force Method to combine these three sets of fundamental equations to get expressions for the forces in the individual elements.

If the external force P_B acts to the right (i.e., if it is positive), we should find that the left-hand element is in tension and the two right-hand elements are in compression.

Solution

Equilibrium: From the free-body diagram of node B in Fig. 2,

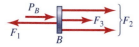

P_B and F_1 are collinear with F_3
F_2 is distributed around the circumference of the pipe.

Fig. 2 Free-body diagram of the plate at B.

$$\xrightarrow{+}\sum F_x = 0: \qquad -F_1 + F_2 + F_3 + P_B = 0 \qquad \textbf{Equilibrium} \qquad (1)$$

Equation (1) relates the three unknown internal element forces to the known external load. Since there are three unknown forces, but only one equilibrium equation, this system is *statically indeterminate* and there are **two redundant forces.**

Element Force-Deformation Behavior: We have three uniform, axial-deformation elements, and for each one we can write an element force-deformation equation like Eq. 3.14. We have called the element forces F_1, F_2, and F_3 (tension positive), so we have

$$
\begin{aligned}
e_1 &= f_1 F_1, & \text{where } f_1 &= (L_1/A_1 E_1) \\
e_2 &= f_2 F_2, & \text{where } f_2 &= (L_2/A_2 E_2) \\
e_3 &= f_3 F_3, & \text{where } f_3 &= (L_3/A_3 E_3)
\end{aligned}
$$

Element Force-Deformation Behavior (3a–c)

In Eqs. (2) the e's are the element elongations. A positive F_i (tension) produces a positive e_i (element gets longer), since the f_i's are, by definition, positive.

Geometry of Deformation: Referring to Fig. 1, we can easily relate the elongation of each of the three elements to the displacement u_B by using the definition of elongation of an element, that is, $e = u(L) - u(0)$. So,

$$e_1 = -e_2 = -e_3 = u_B$$

Here we have used the fact that the displacements at joints A and C are zero. Note that, since e is positive when an element gets longer, a displacement of joint B to the right by an amount u_B implies a shortening of elements (2) and (3) by that amount; hence the minus sign for e_2 and e_3. We can eliminate u_B and write the above as two **compatibility equations:**

$$e_2 = -e_1$$
$$e_3 = e_2$$

Geometry of Deformation (3a,b)

The fact that there are **two compatibility equations** is consistent with the fact that there are **two redundant forces** in the equilibrium equation. There will always be as many compatibility equations as there are redundant forces!

Solution of the Equations: If we count equations and unknowns, we find that we have six equations and six unknowns. Rather than just combine Eqs. (1) through (3) in some arbitrary order, we will follow the Basic Force Method.

   Substitute Eqs. (2) (element force-deformation) into Eqs. (3) (deformation compatibility) to obtain the compatibility equations written in terms of forces.

$$f_2 F_2 = -f_1 F_1$$
$$f_3 F_3 = f_2 F_2$$

Compatibility in Terms of Element Forces (4a,b)

We now have three equations, Eqs. (1) and (4a,b), in three unknowns, the three element forces. We solve these equations simultaneously to get the following expressions for the three unknown element forces:

$$F_1 = \left(\frac{f_2 f_3}{f_1 f_2 + f_2 f_3 + f_1 f_3} \right) P_B$$

$$F_2 = \left(\frac{-f_1 f_3}{f_1 f_2 + f_2 f_3 + f_1 f_3} \right) P_B \qquad \textbf{Ans.} \qquad (5a\text{--}c)$$

$$F_3 = \left(\frac{-f_1 f_2}{f_1 f_2 + f_2 f_3 + f_1 f_3} \right) P_B$$

Review the Solution   As one check of our work, we can substitute Eqs. (5) back into Eqs. (4a,b) to see if deformation compatibility is satisfied.

Are $f_2 F_2 = -f_1 F_1$ and $f_3 F_3 = f_2 F_2$?   Yes.

The fact that the compatibility equations, Eqs. (4a,b), are satisfied by our answers means that we have probably not made errors in our solution. Also, from Eqs. (5) we see that, when P_B is positive, element (1) is in tension and elements (2) and (3) are in compression. This is what we expected to find.

3.6 THERMAL EFFECTS ON AXIAL DEFORMATION

In Section 2.3 the formula given for **thermal strain** is

$$\boxed{\epsilon_T = \alpha \Delta T}$$ **Thermal Strain**   (3.23)

where α is the **coefficient of thermal expansion** and ΔT is the temperature <u>increase above the reference temperature</u> (e.g., room temperature). Generally a positive ΔT causes a tendency for the bar to expand, so α is positive. In a few cases there are materials for which α is zero or negative (i.e., materials that remain undeformed or actually contract when the temperature increases), but these are rare indeed.

In this section we examine thermal effects on slender members undergoing axial deformation. Consider the two axial-deformation members shown in Fig. 3.16. Member *AB* is supported along its length by a smooth surface and restrained only at end *A* so that, when heated or cooled, it is free to expand or contract. No axial force (or stress) is induced in this member, since it can freely expand or contract.[6] Member *CD* is assumed to be force-free when it is attached at both ends to a rigid base. If it is subsequently heated, it will tend to expand in accordance with Eq. 3.23. However, since it is prevented from expanding by the rigid walls, a compressive force will be induced in this member. Conversely, if it is cooled, it will tend to contract and pull away from the rigid walls, and in the process, tension will be induced in this member. Even if the base to which member *CD* is attached is not completely rigid, the presence of some restraining structure will cause compressive stress to be induced in *CD* as a result of heating the bar, and tension will be induced by cooling the bar.

If we consider only slender members, where it is reasonable to assume that the only significant normal stress is the axial stress $\sigma \equiv \sigma_x$ (i.e., we assume that σ_y and σ_z are negligible), than Eq. 2.38a (Section 2.13) gives the following **total strain equation:**

$$\boxed{\epsilon = \epsilon_\sigma + \epsilon_T = \frac{\sigma}{E} + \alpha \Delta T}$$ (3.24)

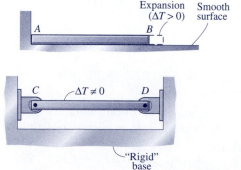

FIGURE 3.16   Restraint of thermal deformation.

---

[6]Although there would also he thermal strains in the lateral directions as well as in the axial direction, these will be ignored here. Since the members are assumed to be free to expand or contract laterally, these lateral thermal strains will not affect the axial stress in the member.

Equation 3.24 is the **stress-strain-temperature equation** for axial deformation of a slender, linearly elastic member.

From Eq. 3.24 it is clear that the total axial strain ϵ is the sum of a strain due to the axial stress σ and a thermal strain due to the temperature change ΔT. To get some feel for the magnitude of each of these terms, suppose that a fixed-end bar, like bar *CD* in Fig. 3.16, is made of structural steel ($E = 29 \times 10^3$ ksi, $\alpha = 6.5 \times 10^{-6}/°F$, and $\sigma_{YP} = 36$ ksi) and is uniformly heated by 100°F. If the walls to which the bar is attached are rigid, then $\epsilon = 0$, and, from Eq. 3.24, the following axial stress would be induced in the bar:

$$(\sigma)_{T=100°F} = -E\alpha\Delta T$$

$$= -(29 \times 10^3 \text{ ksi})(6.5 \times 10^{-6}/°F)(100°F)$$

$$= -18.85 \text{ ksi} = 18.85 \text{ ksi}(C)$$

Thus, a temperature change of only 100°F causes an induced compressive stress that is about half the yield stress. From these numbers we can conclude that, **in parts of a structure that are not free to expand or contract, significant stresses can be induced by reasonable changes in temperature.** For example, thermally induced stress is a very important consideration in the design of a spacecraft that is to spend part of each orbit directly exposed to the heat of the sun and part in pitch dark, where it is extremely cold. It is also a significant consideration in the design of bridges and many other structures and machines.

To solve axial-deformation problems that involve temperature change, we need to use the *stress-strain-temperature equation*, Eq. 3.24, but we also need to account for equilibrium and for the geometry of deformation. If we restrict our attention to axial-deformation situations where E, α, and ΔT may depend on x but are independent of position in the cross section, the axial stress is uniform over any cross section and is given by the familiar axial stress formula:

$$\boxed{\sigma \equiv \sigma(x) = \frac{F(x)}{A(x)}}$$

Axial Stress Formula (3.9) repeated

Combining Eqs. 3.9 and 3.24, we get the following expression for the total axial strain:

$$\epsilon(x) = \frac{F(x)}{A(x)E(x)} + \alpha(x)\Delta T(x) \tag{3.25}$$

The total elongation is related to the total axial strain by

$$e = \int_0^L \epsilon(x)\,dx \tag{3.3}$$
repeated

Finally, the **total elongation** of the member is

$$\boxed{e = \int_0^L \frac{F(x)\,dx}{A(x)E(x)} + \int_0^L \alpha(x)\Delta T(x)\,dx}$$

Force-Temperature-Deformation (3.26)

The following examples illustrate the application of Eqs. 3.9 and 3.26 Problems of this type are sometimes referred to as **thermal stress problems.**

EXAMPLE 3.10

The slender, uniform rod in Fig. 1 is attached to rigid supports at A and B, and it is stress-free when $\Delta T = 0$. It is surrounded by a heating element that is capable of producing the linearly varying temperature change $\Delta T(x)$ shown in Fig. 2. Determine the stress distribution $\sigma(x)$ that is induced by this nonuniform heating.

Plan the Solution   The only external forces applied to the rod AB are the reactions at A and B, as shown in Fig. 1. We can use a free-body diagram to relate $F(x)$ to P_A (or P_B). Since the supports at A and B are rigid, the total elongation of AB is zero. Since $\Delta T > 0$ all along the bar, we can expect a compressive stress to result.

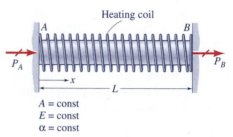

Heating coil

$A = \text{const}$
$E = \text{const}$
$\alpha = \text{const}$

Fig. 1   A heated, uniform bar.

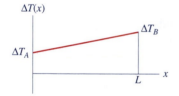

$\Delta T(x)$

ΔT_B

ΔT_A

L

x

Fig. 2   The temperature change.

P_A $F(x)$

x

Fig. 3   Free-body diagram.

Solution

Equilibrium: To determine how $F(x)$ varies with x, we draw the free-body diagram shown in Fig, 3, with a cut at an arbitrary section x,

$$\xrightarrow{+}\sum F_x = 0: \qquad \boxed{F(x) = -P_A = \text{const}} \qquad \textbf{Equilibrium} \qquad (1)$$

Force-Temperature-Deformation Behavior: From Eq. 3.26

$$e = \int_0^L \left(\frac{F}{AE} + \alpha \Delta T(x) \right) dx$$

From Fig. 2, the temperature distribution is given by

$$\Delta T(x) = \Delta T_A + (\Delta T_B - \Delta T_A)\left(\frac{X}{L}\right)$$

Then, since $F(x) = \text{const}$,

$$e = \frac{FL}{AE} + \alpha \int_0^L \left[\Delta T_A + (\Delta T_B - \Delta T_A)\left(\frac{x}{L}\right) \right] dx$$

or

$$\boxed{e = \frac{FL}{AE} + \alpha L \left(\frac{\Delta T_A + \Delta T_B}{2} \right)} \qquad \begin{array}{l}\textbf{Force-}\\\textbf{Temperature-}\\\textbf{Deformation}\\\textbf{Behavior}\end{array} \qquad (2)$$

Geometry of Deformation: Since the ends of the rod cannot move,

$$e = 0 \qquad \textbf{Geometry of Deformation} \qquad (3)$$

Solution: Combine Eqs. (2) and (3) to get

$$F = -AE\alpha\left(\frac{\Delta T_A + \Delta T_B}{2}\right) \qquad (4)$$

Then, since $\sigma = \dfrac{F}{A}$,

$$\sigma = -E\alpha\left(\frac{\Delta T_A + \Delta T_B}{2}\right) \qquad \textbf{Ans.} \quad (5)$$

Review the Solution   The right-hand side of Eq. (5) has the dimensions of stress (F/L^2), as it should. The stress is negative, which is what we would expect to happen when the bar is heated.

It may surprise you that the stress is independent of x, even though ΔT varies with x. This is the direct consequence of the fact that the equilibrium equation, Eq. (1), states that $F(x)$ is a constant, not a function of x, as ΔT is!

Uniform Axial-Deformation Element. Let us now consider the uniformly heated (or cooled), uniform axial-deformation element in Fig. 3.17. To determine the **force-temperature-deformation equation** for this uniform member subjected to an axial force and a uniform temperature change, we can essentially repeat the steps of Section 3.2.

Equilibrium: Draw a free-body diagram with the member cut at an arbitrary section x.

$$\overset{+}{\rightarrow}\sum F_x: \qquad\qquad F(x) = F = \text{const}$$

From Eq. 3.9,

$$\sigma = \frac{F(x)}{A(x)} = \frac{F}{A} = \text{const}$$

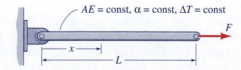

(*a*) A uniform element with axial force F and with $\Delta T = $ const.

(*b*) Free-body diagram.

FIGURE 3.17   A heated (cooled) axial-deformation element.

Force-Temperature-Deformation Equations: From Eq. 3.26, then,

$$e = \int_0^L \epsilon \, dx = \frac{FL}{AE} + \alpha L \Delta T$$

Thus, the **total elongation, e,** consists of the elongation due to the axial force F, plus the elongation due to the uniform temperature change ΔT.

$$e = fF + \alpha L \Delta T, \quad f = \frac{L}{AE} \tag{3.27}$$

In terms of the stiffness coefficient, we can write Eq. 3.27 as

$$F = k(e - \alpha L \Delta T), \quad k = \frac{AE}{L} \tag{3.28}$$

It is important to note that, in Eqs. 3.27 and 3.28, e is the total elongation. Therefore, when you are dealing with the geometry of deformation, this is the total change in length of the member, which is therefore directly related to the displacements of the two ends of the element.

The only fundamental equations that are affected by temperature change are the force-deformation equations, Eqs. 3.14 and 3.15, which are replaced by the **force-temperature-deformation equations,** Eqs. 3.27 and 3.28, respectively. **No changes are required in the equilibrium equations or deformation-geometry equations!**

Thermal-Stress Example Problems. Recall that a temperature change induces stresses in statically indeterminate structures, but not in statically determinate structures. We will now solve several example problems that involve temperature changes in statically indeterminate structures.

> ◄— ·—· ·—· ·—· —► **EXAMPLE 3.11** ◄— ·—· ·—· ·—·

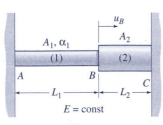

Fig. 1

Two elements are stress-free when they are welded together at B and welded to rigid walls at A and C (Fig. 1). Subsequently, element (1) is heated by an amount ΔT, while element (2) is held at the reference temperature. Determine an expression for the axial force induced in each element, and determine an expression for the displacement u_B of the joint at B.

Plan the Solution This problem can be solved by using the same Basic-Force-Method steps that were employed in Section 3.5. To incorporate the thermal strain, we just need to use Eq. 3.27 instead of Eq. 3.14. Since element (1) is heated, it will push on element (2), and both elements will therefore have an induced compressive stress.

Solution

Equilibrium: Although there are no external loads, that is, no external forces other than the reactions at A and C, we still need an equilibrium equation of the joint at B in order to relate the element forces to each other (Fig. 2). Note that tension is assumed positive.

$$\stackrel{+}{\rightarrow}\sum F_x = 0: \qquad -F_1 + F_2 = 0 \qquad \textbf{Equilibrium} \qquad (1)$$

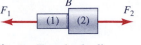

Fig. 2 Free-body diagram.

Since there are two unknown forces but only one equilibrium equation, this is a statically indeterminate system of bars. Therefore, we must turn to the geometry of deformation for an appropriate compatibility equation. To prepare for this, however, we first establish the correct form for relating the element forces to the total elongations of the elements, including thermal effects.

Element Force-Temperature-Deformation Behavior: These are the equations through which the temperature effect directly enters the solution, and they also relate the forces in the equilibrium equation to the elongations in the deformation-geometry equation. Equation 3.27 is the appropriate equation to use. We write this equation for each element, noting that $\Delta T_1 = \Delta T$ and $\Delta T_2 = 0$.

$$\begin{aligned} e_1 &= f_1 F_1 + \alpha_1 L_1 \Delta T \\ e_2 &= f_2 F_2 \end{aligned} \qquad \begin{array}{l} \textbf{Element Force-} \\ \textbf{Temperature-} \\ \textbf{Deformation} \\ \textbf{Behavior} \end{array} \qquad (2a,b)$$

where $f_i = (L/AE)_i$.

Geometry of Deformation: The elongations e_1 and e_2 in Eqs. (2) are the total elongations of the respective elements, due to both stress and temperature. Therefore, the **compatibility equation** is simply

$$e_{AC} = e_1 + e_2 = 0 \qquad \begin{array}{l} \textbf{Geometry of} \\ \textbf{Deformation} \end{array} \qquad (3)$$

Solution for Element Forces: Following the procedure of the Basic Force Method, we can eliminate the e's by substituting Eqs. (2) into Eq. (3), giving the following compatibility equation in terms of forces:

$$f_1 F_1 + f_2 F_2 = -\alpha_1 L_1 \Delta T \qquad \begin{array}{l} \textbf{Compatibility} \\ \textbf{in Terms of} \\ \textbf{Element Forces} \end{array} \qquad (4)$$

Inserting expressions for the f_i's into Eq. (4), and solving Eqs. (1) and (4) simultaneously, we get

$$F_1 = F_2 = \frac{-A_1 A_2 E \alpha_1 L_1 \Delta T}{A_1 L_2 + A_2 L_1} \qquad \textbf{Ans.} \quad (5)$$

Solution for the Displacement of B: To obtain an expression for the displacement u_B we can see from Fig. 1 that element (1) is elongated by an amount u_B while element (2) is shortened by an amount u_B, which leads to Eq. (3). Since Eq. (2b) is shorter than Eq. (2a), we will use

$$e_2 = -u_B \qquad \begin{array}{l} \textbf{Geometry of} \\ \textbf{Deformation} \end{array} \qquad (6)$$

so

$$u_B = -f_2 F_2 = \left(\frac{A_1 L_2}{A_1 L_2 + A_2 L_1}\right)(\alpha_1 L_1 \Delta T) \qquad \textbf{Ans.} \quad (7)$$

Review the Solution   As we would expect, heating element (1) causes a compression to be induced in both elements and, since there is no external force on the joint connecting the two elements, the same compressive force is induced in both elements. The right-hand side of Eq. (5) has the proper dimensions of force, and the right-hand side of Eq. (7) has the proper dimensions of displacement. Also, it can be seen that the displacement u_B is less than the free expansion of element (1) would be if element (2) were not there.

MDS3.7   **End-to-End Bars w/Temperature** is a program for analyzing temperature-induced forces in statically indeterminate structures, like the one in Example 3.11, by the Basic Force Method.

EXAMPLE 3.12

In Fig. 1 a cylindrical aluminum core (2) is surrounded by a titanium sleeve (1), and both are attached at each end to a rigid end-plate. Assume that the core and the outer sleeve are both stress-free at the reference temperature. Set up the fundamental equations and solve for the following: (a) the stress induced in each member if the entire composite bar is heated by 100°F, and (b) the resulting elongation of the composite bar.

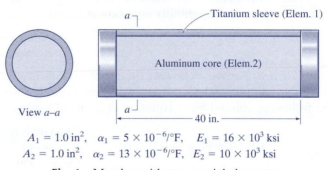

View *a–a*

$A_1 = 1.0 \text{ in}^2, \quad \alpha_1 = 5 \times 10^{-6}/°\text{F}, \quad E_1 = 16 \times 10^3 \text{ ksi}$
$A_2 = 1.0 \text{ in}^2, \quad \alpha_2 = 13 \times 10^{-6}/°\text{F}, \quad E_2 = 10 \times 10^3 \text{ ksi}$

Fig. 1   Member with two coaxial elements.

Plan the Solution   This two-element problem is simple enough so that we can reason that, because $\alpha_{\text{alum}} > \alpha_{\text{titan}}$, the aluminum will tend to expand more than the titanium and will therefore be in compression if both are heated the same amount. However, in a more complicated situation we would essentially have to solve the entire problem just to determine which elements are in compression and which are in tension. Fortunately, we do not have to do this! We just <u>assume tension in each element</u> and let the final solution tell us which element is in tension and which is in compression.

Solution   (a) *Solve for the stress in each component.*

Equilibrium: The basic question to ask in setting up the equilibrium equation is: What free-body diagram can I use that will relate the internal element forces to each other and to the external loads? (In this example, there are no external loads.) The answer is, of course, one of the end-plates to which both the core and the sleeve are attached, as shown in Fig. 2.

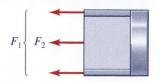

Fig. 2 Free-body diagram.

$$\stackrel{+}{\rightarrow} \sum F_x = 0: \qquad\qquad -F_1 - F_2 = 0 \qquad \textbf{Equilibrium} \qquad (1)$$

Element Force-Temperature-Deformation Behavior: We use the force-temperature-deformation (F-T-D) format of Eq. 3.27, which involves flexibility coefficients, f_i:

$$
\begin{aligned}
e_1 &= f_1 F_1 + \alpha_1 L_1 \Delta T_1 \\
e_2 &= f_2 F_2 + \alpha_2 L_2 \Delta T_2
\end{aligned}
$$

Element Force-Temperature-Deformation Behavior (2a,b)

where

$$f_1 = \left(\frac{L}{AE}\right)_1 = \frac{(40 \text{ in.})}{(1.0 \text{ in}^2)(16 \times 10^3 \text{ ksi})} = 2.50(10^{-3}) \text{ in./kip}$$

$$f_2 = \left(\frac{L}{AE}\right)_2 = \frac{(40 \text{ in.})}{(1.0 \text{ in}^2)(10 \times 10^3 \text{ ksi})} = 4.00(10^{-3}) \text{ in./kip}$$

$$\alpha_1 L_1 \Delta T_1 = (5 \times 10^{-6}/°\text{F})(40 \text{ in.})(100°\text{F}) = 0.020 \text{ in.}$$

$$\alpha_2 L_2 \Delta T_2 = (13 \times 10^{-6}/°\text{F})(40 \text{ in.})(100°\text{F}) = 0.052 \text{ in.}$$

Geometry of Deformation: Let δ be the displacement of the right-hand end-plate, as shown in the deformation diagram, Fig. 3, and recall that e_1 and e_2 are the <u>total elongations</u> of the respective elements. Then, the appropriate **compatibility equation** is

$$e_1 = e_2 \qquad \textbf{Geometry of Deformation} \qquad (3)$$

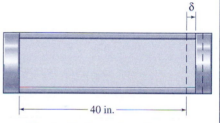

Fig. 3 Deformation diagram.

Solution for Element Forces: In Eqs. (1) through (3) we have four equations in four unknowns. Since we want to solve for the member stresses, we can eliminate the e's by substituting Eqs. (2) into Eq. (3), getting

$$f_1 F_1 + \alpha_1 L_1 \Delta T_1 = f_2 F_2 + \alpha_2 L_2 \Delta T_2 \qquad \textbf{Compatibility in Terms of Element Forces} \qquad (4)$$

This compatibility equation written in terms of forces can now be solved simultaneously with the equilibrium equation, Eq. (1), to give

$$F_1 = -F_2 = \frac{\alpha_2 L \Delta T - \alpha_1 L \Delta T}{f_1 + f_2}$$

$$= \frac{0.052 \text{ in.} - 0.020 \text{ in.}}{2.50(10^{-3}) \text{ in./kip} + 4.00(10^{-3}) \text{ in./kip}}$$

so, the forces in the titanium sleeve and aluminum core, respectively, are

$$\boxed{F_1 = 4.92 \text{ kips}, \quad F_2 = -4.92 \text{ kips}} \qquad (5)$$

The stresses are given by

$$\sigma_1 = \frac{F_1}{A_1}, \quad \sigma_2 = \frac{F_2}{A_2}$$

Therefore,

$$\sigma_1 = -\sigma_2 = 4.92 \text{ ksi} \qquad \textbf{Ans.} \quad (6)$$

(b) *Solve for the elongation of the member.* From the deformation diagram in Fig. 3, we see that

$$\delta = e_1 = e_2$$

and we can use either Eq. (2a) or Eq. (2b) to solve for an elongation. Then,

$$\delta = f_1 F_1 + \alpha_1 L \Delta T$$
$$= \left[2.50(10^{-3}) \text{ in/kip}\right](4.92 \text{ kips}) + 0.020 \text{ in}$$
$$= 0.0323 \text{ in.}$$

or

$$\delta = 0.0323 \text{ in.} \qquad \textbf{Ans.} \quad (b)$$

Review the Solution   The signs of member stresses σ_1 and σ_2 agree with our "Plan the Solution" discussion, and the value of the elongation, δ, is between the free-expansion values for aluminum and titanium, which is what we would expect. That is, the final elongation is a compromise between the amount that the aluminum "wants" to expand and the smaller amount that the titanium "wants" to expand.

The above examples clearly illustrate that **the solution of thermal-deformation problems involving axial deformation of uniform elements only requires a modification of the element force-displacement equations, but absolutely no change to either the equilibrium equations or the deformation-geometry equations.**

MDS3.8   **Coaxial Bars w/Temperature** is a program for analyzing temperature-induced forces in statically indeterminate structures, like the one in Example 3.12, by the Basic Force Method.

MDS3.9   **Rigid Member and Two Bars w/Temperature** is a program for analyzing statically indeterminate structures like the one in Example 3.8, with temperature change added.

Next, let us consider a problem that involves both an externally applied load and a temperature change of a member. Note that the external force enters through the equilibrium equation, while the temperature change enters through the force-temperature-deformation equation.

EXAMPLE 3.13

The rigid beam BD in Fig. 1 is supported by a wire AC of length L, cross-sectional area A, modulus of elasticity E, and coefficient of thermal expansion α; and by a rod DE of length $L/2$, cross-sectional area $2A$, and modulus E. Neglect the weight of the beam. When there is no load acting on the beam, the beam is horizontal and the wire and rod are stress-free. Assume that the tip displacement is very small, that is, $\delta_D \ll L$.

Determine expressions for the axial forces in wire AC and rod DE when a load of intensity w per unit length of beam is uniformly distributed over CD, and wire AC is <u>simultaneously cooled</u> by an amount ΔT (i.e., $\Delta T_1 = -\Delta T, \Delta T_2 = 0$).

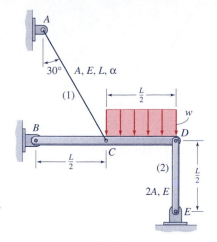

Fig. 1

Plan the Solution This is a statically indeterminate structure that is somewhat similar to the determinate structure in Example 3.3, but we will need to add rod DE and take into account the temperature change of wire AC. The Basic Force Method will provide a systematic way of combining the three sets of equations—equilibrium, element force-temperature-deformation behavior, and compatibility.

Cooling wire AC should cause tension in both the wire and the rod; applying a downward load over CD should cause tension in the wire, but it should cause compression of the rod. In the expressions for the force in the wire and the force in the rod, the temperature change should lead to terms having the form $AE\alpha\Delta T$.

Solution

Equilibrium: The free-body diagram in Fig. 2 will enable us to write an equilibrium equation that relates F_1 (the force in wire AC) and F_2 (the force in rod DE) to the external load. We do not need equations for the reaction components at B.

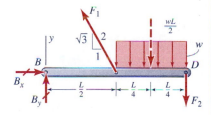

Fig. 2 Free-body diagram.

$$+\,\zeta\left(\sum M\right)_B = 0: \quad \left(\frac{\sqrt{3}}{2}F_1\right)\left(\frac{L}{2}\right) - \left(\frac{wL}{2}\right)\left(\frac{3L}{4}\right) - F_2L = 0$$

or

$$2\sqrt{3}F_1 - 8F_2 = 3wL \qquad \textbf{Equilibrium} \qquad (1)$$

Element Force-Temperature-Deformation Behavior. From Eq. 3.27, the total elongation of the wire and rod, respectively, are

$$e_1 = f_1F_1 + \alpha_1L_1\Delta T_1$$
$$e_2 = f_2F_2$$

Force-Temperature-Deformation Behavior (2a,b)

where it has been noted that $\Delta T_2 = 0$, and where

$$f_1 = \left(\frac{L}{AE}\right)_1 = \frac{L}{AE}, \quad f_2 = \left(\frac{L}{AE}\right)_2 = \frac{(L/2)}{(2AE)} = \frac{1}{4}\left(\frac{L}{AE}\right)$$

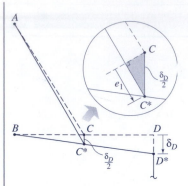

Fig. 3 Deformation diagram.

Geometry of Deformation; Compatibility Equation: Figure 3 is a **deformation diagram** that enables us to determine an equation that relates the total elongation, e_1, of wire AC to the vertical displacement of point D, and, hence, to the shortening of rod DE. From the deformation diagram in Fig. 3, we can write the following deformation equations:

$$e_1 = \frac{\sqrt{3}}{2}\left(\frac{\delta_D}{2}\right) = \frac{\sqrt{3}}{4}\delta_D \qquad \text{**Geometry of**}$$
$$\text{**Deformation**}$$
$$e_2 = -\delta_D$$

from which we can eliminate δ_D to obtain the **compatibility equation**

$$e_1 = -\frac{\sqrt{3}}{4}e_2 \qquad \text{**Compatibility**} \qquad (3)$$

Since it enforces the condition that the end of the wire AC and the top of the rod DE remain attached to a rigid beam that rotates about end A, Eq. (3) is called a compatibility equation.

Solution by the Basic Force Method: Following the procedure of the Basic Force Method, we can eliminate the e's by substituting Eqs. (2) into Eq. (3), giving the following compatibility equation in terms of forces:

$$f_1F_1 + \alpha_1L_1\Delta T_1 = -\left(\frac{\sqrt{3}}{4}\right)f_2F_2 \qquad \begin{array}{l}\text{**Compatibility**}\\ \text{**in Terms of**}\\ \text{**Element Forces**}\end{array} \qquad (4)$$

Noting that $\Delta T_1 = -\Delta T$, and substituting in the values of the flexibility coefficients f_1 and f_2, we can now solve Eqs. (1) and (4) simultaneously to get the forces

$$F_1 = \frac{64}{67}(AE\alpha\Delta T) + \frac{3\sqrt{3}}{134}(wL)$$

$$F_2 = \frac{16\sqrt{3}}{67}(AE\alpha\Delta T) - \frac{24}{67}(wL) \qquad \text{**Ans.**} \quad (5)$$

Review the Solution The expressions in Eqs. (5) for the unknown forces F_1 and F_2 have the proper forms that were anticipated in the "Plan the Solution" section. Therefore, the answers we have obtained seem reasonable.

Although this problem is made fairly difficult by the angle of the wire and the presence of both external distributed load and temperature change, its solution above by the Basic Force Method is very straightforward. The external force enters through the equilibrium equation; the temperature change enters through the force-temperature-deformation equations.

Section 3.5 treated problems involving structures that have uniform axial-deformation elements and are subjected to external loads. The external loads enter the solution through the *equilibrium equations,* which relate the element axial forces to each other and to the external loads. In Section 3.6 you learned that when axial-deformation members are heated or cooled, that information enters the problem solution through a modification of the *element force-deformation equations.* In this section we treat problems in which modification of the *deformation-geometry equation(s)* is required.

The word "misfit" is just a nickname for **geometric incompatibility.**[7] The misfit may be unintended, as in Fig. 3.18a, where rod (2) on the right has been fabricated an amount $\bar{\delta}$ too short. This misfit, or incompatibility, must be accounted for in relating the actual elongation of this member to the other geometric quantities (namely, the rotation of the beam and the elongations of the other two rods). Figure 3.18b represents a bolt surrounded by a sleeve (or pipe). In this case, the "misfit" is intentional. If the nut on the bolt is just snugged up against the washer, there will be no axial force in the bolt or the sleeve. However, if the nut is tightened further against the washer, the portion of the bolt between the nut and bolt head is, in effect, made "too short." This misfit must be accounted for in the deformation-geometry equation(s). It is emphasized that **no change is made directly either to the equilibrium equations or to the element force-deformation equations to account for misfits, which are solely a matter of geometric compatibility.**

It is not uncommon for parts of a structure to have small geometry errors that require that the components be forced to fit together, as would be required for the structure in Fig. 3.18a. The stresses that are induced when such force-fitting occurs can combine with the load-related stresses to cause failure of the structure to occur. Such stresses are called **initial stresses,** since they can be present before any external load is applied. Therefore, **it is very important to consider the effect of misfits on statically indeterminate structures.**

Analysis of a Typical Statically Indeterminate Structure with a Misfit.

Let us consider the two-element structure in Fig. 3.19, and let us assume that element (1) of the structure was fabricated $\bar{\delta}_1$ too short, so that the bar has to be

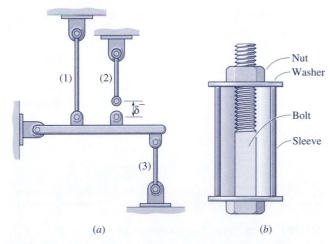

(a)   (b)

FIGURE 3.18   Two structures that exhibit geometric misfits.

---

[7]The topic of geometric misfits, or geometric incompatibility, is sometimes referred to by the name *prestrain effects* or by the name *initial stresses.*

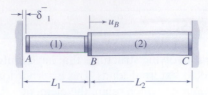

(a) Before elements are welded to the walls.

(b) After elements are welded to the walls.

$A_1 = A, \ A_2 = 2A, \ E_1 = E_2 = E, \ L_1 = L, \ L_2 = 2L$

FIGURE 3.19 A statically indeterminate structure.

stretched initially in order for the two ends to be welded to the rigid walls. Assume that the misfit is small, that is, $\bar{\delta}_1 \ll L_1$. Let us first determine the axial stresses, σ_1, and σ_2, in the elements without any applied external load P_B at node B. Then let us determine the displacement, u_B, of the node (joint) between the two elements.

Equilibrium: The three free-body diagrams in Fig. 3.20 are for the bar in Fig. 3.19b, that is, after the ends of the bar have been welded to the two walls. The free-body diagram of node B in Fig. 3.20 will enable us to write an equilibrium equation that relates the *internal forces* F_1 and F_2 in the elements adjoining B. (There is no external load P_B at joint B.)

$$\overset{+}{\rightarrow} \sum F = 0: \qquad\qquad -F_1 + F_2 = 0 \qquad \textbf{Equilibrium} \qquad (1)$$

This two-component system is statically indeterminate, since we are not able to determine the two element forces from the single equilibrium equation.

The reactions at ends A and C are related to the element forces by

$$P_A = -F_1, \qquad P_C = F_2 \qquad\qquad (1')$$

Element Force-Deformation Behavior: From the element force-deformation equation. Eq. 3.14, we have

$$e_1 = f_1 F_1, \quad e_2 = f_2 F_2 \qquad\qquad \begin{array}{l}\textbf{Element}\\\textbf{Force-}\\\textbf{Deformation}\\\textbf{Behavior}\end{array} \qquad (2a,b)$$

where $f_1 = \dfrac{L}{AE}$ and $f_2 = \dfrac{2L}{2AE} = \dfrac{L}{AE}$. [8]

FIGURE 3.20 Free-body diagrams of nodes A, B, and C.

---

[8] Actually, $f_1 = (L_1 - \bar{\delta}_1)/A_1 E_1 \approx L_1/A_1 E_1$. Since it has been assumed that $\bar{\delta}_1 \ll L_1$, the effect of $\bar{\delta}_1$, on the flexibility coefficient f_1 can be ignored in comparison with the effect that enters through the compatibility equation.

Geometry of Deformation; Compatibility Equation: Since the two elements are attached end-to-end at node B, and since the two-element bar has to be stretched by an amount $\bar{\delta}_1$ in order to attach its ends to the rigid walls at A and C, the **compatibility equation** for this problem is

$$e_1 + e_2 = \bar{\delta}_1 \qquad \text{Geometry of Deformation} \qquad (3)$$

Solution of the Equations: In Eqs. (1) through (3) there are <u>four equations and four unknowns</u>. Therefore, it is a matter of solving these four equations simultaneously to determine the four unknowns. We are first interested in determining the element forces. Therefore, we can eliminate the e_i's by substituting Eqs. (2) into Eq. (3), which gives the following **compatibility equation written in terms of element forces:**

$$f_1 F_1 + f_2 F_2 = \bar{\delta}_1$$

or

$$F_1 \left(\frac{L}{AE}\right) + F_2 \left(\frac{L}{AE}\right) = \bar{\delta}_1 \qquad \text{Compatibility in Terms of Element Forces} \qquad (4)$$

The equilibrium equation (1) and the compatibility equation (4) can be <u>solved simultaneously</u> to give the following element forces:

$$F_1 = F_2 = \frac{AE\bar{\delta}_1}{2L} \qquad (5a,b)$$

Finally, dividing the element forces by their respective cross-sectional areas gives the following expressions for the **initial stresses:**

$$\sigma_1 = \frac{E\bar{\delta}_1}{2L}, \qquad \sigma_2 = \frac{E\bar{\delta}_1}{4L} \qquad \textbf{Ans.} \quad (6)$$

In the problem statement we are asked not only to determine the element initial stresses, but also to determine the displacement of node B. From Fig. 3.19a we see that u_B can most easily be obtained from the **deformation-geometry equation**

$$u_B = -e_2 \qquad (7)$$

Therefore, combining Eqs. (2b), (4b), and (6) we get

$$u_B = -\frac{\bar{\delta}_1}{2} \qquad \textbf{Ans.} \quad (8)$$

The factor of one-half in Eq. (8) should not surprise us, since the two element flexibilities are equal to each other, that is, $f_1 = f_2$.

In summary, the above solution illustrates the fact that **geometric misfits enter into the compatibility equation(s),** that is, their <u>direct</u> effect is on the geometry of deformation. Equilibrium equations and force-temperature-deformation equations are not directly affected.

MDS3.10 & 3.11   **End-to-End Bars w/Gap and End-to-End Bars w/Misfit**—are programs for analyzing statically indeterminate structures with gaps, like the one in Fig. 3.20a, by the Basic Force Method.

An example will now be given to illustrate how "bolt problems" are solved as axial-deformation problems involving geometric "misfits."

EXAMPLE 3.14

A 14-mm-diameter steel bolt, and a steel pipe with 19-mm ID and 25-mm OD are arranged as shown in Fig. 1. For both the steel bolt and steel pipe, $E = 200$ GPa. The pitch of the (single) thread is 2 mm. What stresses will be produced in the steel bolt and sleeve if the nut is tightened by $\frac{1}{8}$ turn? (For a single thread the *pitch* is the distance the nut advances along the thread in one complete revolution of the nut.) Neglect the thickness of the washers.

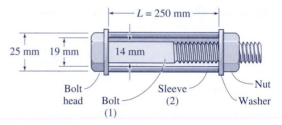

Bolt head | Bolt (1) | Sleeve (2) | Nut | Washer

Fig. 1

Plan the Solution   As we discussed earlier in this section, the misfit due to tightening of the nut will enter the solution through the deformation-geometry equations. Tightening the nut should put the bolt in tension and the sleeve in compression. We will use the Basic Force Method to solve this problem.

Solution

Equilibrium: We should ask the question: What free-body diagram can we draw that will enable us to relate the internal element forces to each other and to the external forces? (In this problem there are no external forces.) One answer is that a cut made just inside the washer at either end will expose the internal forces so that they can be included in the equilibrium equation. A free-body diagram of the left end is shown in Fig. 2. As always, for each of the elements we take tension to be positive. Then, we sum forces in the axial direction.

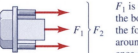

F_1 is the force in the bolt, and F_2 is the force distributed around the circumference of the sleeve.

Fig. 2   Free-body diagram.

$$\overset{+}{\rightarrow} \sum F_x = 0: \qquad\qquad F_1 + F_2 = 0 \qquad \textbf{Equilibrium} \qquad (1)$$

Element Force-Deformation Behavior: Since the Basic Force Method will be used, Eq. 3.14 is the most convenient form.

$$e_1 = f_1 F_1$$
$$e_2 = f_2 F_2$$

Element Force-Deformation Behavior (2)

where, as always, e is the <u>total elongation</u> of an element and is positive when the element gets longer.

$$f_1 = \left(\frac{L}{AE}\right)_1 = \frac{250 \text{ mm}}{\pi (7 \text{ mm})^2 (200 \text{ kN/mm}^2)} = 8.12(10^{-3}) \text{ mm/kN}$$

$$f_2 = \left(\frac{L}{AE}\right)_2 = \frac{250 \text{ mm}}{\pi [(12.5 \text{ mm})^2 - (9.5 \text{ mm})^2](200 \text{ kN/mm}^2)}$$
$$= 6.03(10^{-3}) \text{ mm/kN}$$

Deformation Geometry: We need to assess the deformation that results when the nut is tightened. To do this, let us suppose that the head of the bolt does not move, while the right-hand washer and nut move to the left an amount $\bar{\delta}$ when the nut is tightened. Therefore, we need to relate the elongations e_1 and e_2 in Eqs. (2) to the displacement $\bar{\delta}$. Figure 3 is a deformation diagram showing the shortening of the sleeve by an amount $\bar{\delta}$.

The sleeve is shortened by the amount $\bar{\delta}$ that the right-hand washer moves to the left, so

$$e_2 = -\bar{\delta}$$

If the sleeve were to be removed and the nut advanced $\frac{1}{8}$ turn, the working length of the bolt would be $[250 \text{ mm} - (\frac{1}{8})(2 \text{ mm})] = 250 \text{ mm} - 0.25 \text{ mm} = 249.75 \text{ mm}$. Thus, there is a "misfit" $\bar{\delta} = 0.25 \text{ mm}$, and we would have to stretch the bolt by 0.25 mm to restore it to the 250-mm length, from which the displacement $\bar{\delta}$ is measured. Therefore, the element elongations are related to the displacement δ by

$$e_1 = 0.25 \text{ mm} - \bar{\delta}$$
$$e_2 = -\bar{\delta}$$

Deformation Geometry (3)

Solution of the Equations by the Basic Force Method: We can eliminate the displacement $\bar{\delta}$ from Eqs. (3) to get the **compatibility equation**

$$e_1 - e_2 = 0.25 \text{ mm}$$

Substituting the element force-deformation equations into this equation gives the following compatibility equation in terms of element forces:

$$f_1 F_1 - f_2 F_2 = 0.25 \text{ mm}$$

Compatibility in Terms of Element Forces (4)

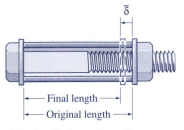

Fig. 3   Deformation diagram.

Solving this compatibility equation and the equilibrium equation simultaneously, we get

$$F_1 = 17.7 \text{ kN}$$
$$F_2 = -17.7 \text{ kN}$$

(5)

Finally, the element stresses are given by

$$\sigma_{\text{bolt}} \equiv \sigma_1 = \frac{F_1}{A_1} = \frac{17.7 \text{ kN}}{\pi (7.0 \text{ mm})^2} = 114.8 \text{ MPa}$$

$$\sigma_{\text{sleeve}} \equiv \sigma_2 = \frac{F_2}{A_2} = \frac{-17.7 \text{ kN}}{\pi [(12.5 \text{ mm})^2 - (9.5 \text{ mm})^2]}$$
$$= -85.2 \text{ MPa}$$

Ans.

Thus, the bolt has a tensile stress of 115 MPa, and the sleeve has a compressive stress of 85 MPa as a result of tightening the nut by $\frac{1}{8}$ turn.

Review the Solution   We expected the bolt to be in tension and the sleeve to be in compression, and this is the result that we obtained. The yield strength of the steel is not stated in the problem, but assuming that it is 250 MPa or greater (see Table F.3 in Appendix F), the values of σ_{bolt} and σ_{sleeve} are reasonable.

 **MDS3.12**   **Bolt and Sleeve**—is a program for analyzing statically indeterminate bolt/sleeve assemblages, similar to the one in Example 3.14, by the Basic Force Method.

3.8   DISPLACEMENT-METHOD SOLUTION OF AXIAL-DEFORMATION PROBLEMS

In Sections 3.5–3.7, a number of example axial-deformation problems were solved by a solution procedure that was labeled the *Basic Force Method*. That method is quite useful for solving fairly simple problems with only two or three unknown forces, such as those in Sections 3.5–3.7. In Sections 3.8 and 3.9, respectively, the **Displacement Method** and the **Force Method** are introduced. Table 3.3 lists the three fundamental sets of equations that enter into the solution of problems in the mechanics of

TABLE 3.3   **Comparison of Displacement Method and Force Method**

Fundamental Equation Sets	
(1) Equilibrium (2) Element force-temperature-deformation behavior (3) Deformation geometry	

Solution Method	Solution Procedure
Displacement Method:	(3) → (2) → (1) (Sect. 3.8)
Force Method:	(1) → (2) → (3) (Sect. 3.9)

deformable solids, and indicates the difference in the order that these two methods use to combine the three fundamental sets of equations.

This section presents a very important problem-solving approach called the **Displacement Method,** which is the solution procedure used in all commercial finite-element computer programs, like the ones used to produce the photos in the color photo section of Chapter 1. For such computer applications, matrix algebra operations are employed,[9] but the Displacement Method is also a <u>very systematic procedure</u> for solving problems "by hand." It can be used to solve statically determinate problems as well as statically indeterminate problems. As indicated in the flow chart on the next page, this method results in **solving simultaneously the equilibrium equations written in terms of the unknown system displacements.**

In Statics you were introduced to two procedures for solving equilibrium problems, especially truss equilibrium problems—the Method of Joints and the Method of Sections. These were reviewed in Example 1.1 in Section 1.4. As illustrated by the examples that follow in this section, the *Displacement Method* extends the *Method of Joints* to deformable-body analysis. In Section 3.9 we solve several statically indeterminate problems by the Force Method.[10] We show that the *Force Method* extends the *Method of Sections* to deformable-body analysis.

Solution Procedure: Displacement Method.

The steps in the Displacement Method are outlined in the solution-procedure list and flow chart that follow. Examples 3.14–3.16 are axial-deformation problems that are solved by the Displacement Method, and MD Solids has a module for solving axial-deformation problems by the Displacement Method. In Section 3.9 the Force Method is discussed and is compared with the Displacement Method.

SOLUTION PROCEDURE: DISPLACEMENT METHOD

▼ SET UP THE FUNDAMENTAL EQUATIONS:

(1) Draw a **free-body diagram** (or free-body diagrams) and write **equations of equilibrium** for these free bodies to relate the external loads to the (unknown) element forces. There should be one equation of equilibrium associated with each *independent system displacement variable*. Typical examples are: a ΣF-type equilibrium equation for each the nodal displacement of a nonuniform rod (see Example 3.14), or a ΣM-type equilibrium equation for each the angle of rotation of a "rigid" beam (see Example 3.15).

(2) Write an element **force-temperature-deformation equation** for each axial-deformation element. Equation 3.28 (Eq. 3.15 if $\Delta T_i = 0$),

$$F_i = k_i(e - \alpha L \, \Delta T)_i \quad \text{where } k_i = (AE/L)_i \qquad (3.28)$$
$$\text{repeated}$$

is the most convenient form to use for this step.

(3) Use **geometry of deformation** to relate the element elongations, e_i, to the system displacement variables that were identified in Step 1. Incorporate any "misfits," $\bar{\delta}_i$, here.

▼ SOLVE FOR THE UNKNOWN SYSTEM DISPLACEMENTS:

(4) Substitute the deformation-geometry equations (Step 3) into the force-deformation equations (Step 2). This gives **element forces in terms of system displacements.**

(5) Substitute the results of Step 4 into the equilibrium equations of Step 1. This gives **equilibrium equations written in terms of system displacements.**

(6) Solve the equations obtained in Step 5. The answer will be the **system displacements.**

▼ SOLVE FOR THE ELEMENT FORCES:

(7) Substitute the system displacements of Step 6 into the equations obtained in Step 4. The answer will be the **element forces.** If element axial stresses are required, the element forces can be divided by the respective element cross-sectional areas, that is $\sigma_i = F_i/A_i$.

(8) Review the solution to make sure that all answers are correct.

---

[9]See, for example, Refs. [3-1], [3-2].
[10]The so-called Basic Force Method employed in Sections 3.5 through 3.7 does not require one to identify specific forces as the "redundant" forces; the more formal Force Method, which is described in Section 3.9, does.

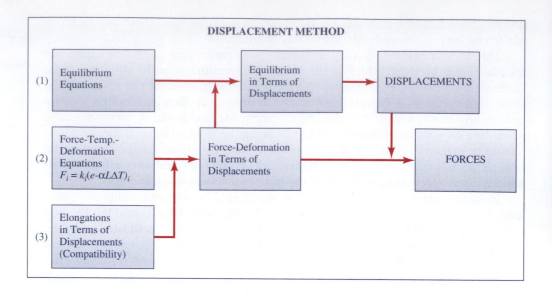

DISPLACEMENT METHOD

(1) Equilibrium Equations → Equilibrium in Terms of Displacements → DISPLACEMENTS

(2) Force-Temp.-Deformation Equations $F_i = k_i(e - \alpha L \Delta T)_i$ → Force-Deformation in Terms of Displacements → FORCES

(3) Elongations in Terms of Displacements (Compatibility)

Example 3.9 in Section 3.5 was solved by the Basic Force Method; in Example 3.15 the same problem is solved by the Displacement Method. If you compare these two solutions you will see that, for this problem, the Displacement-Method solution is more straightforward.

EXAMPLE 3.15

The structural assemblage in Fig. 1 is made up of three uniform elements, or members. Element (1) is a solid rod. Element (2) is a pipe that surrounds element (3), which is a solid rod that is identical to element (1) and collinear with it. The three elements are all attached at B to a rigid plate of negligible thickness. With no external force at B, the three-element assemblage exactly fits between the rigid walls at A and C; its ends are then attached to the two walls. (a) Use the Displacement Method to determine an expression that relates the displacement u_B at node B to the axial force P_B applied there. (b) Determine expressions for the axial forces in the three elements in terms of the external force P_B.

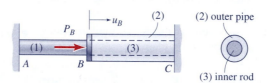

Fig. 1 The original structure.

Plan the Solution We can follow the steps outlined in the Displacement-Method Procedure. Step 1: We can think of this structure as being composed of three uniform elements and one connecting node (joint), and we can write an equilibrium equation for node B. Step 2: Using Eq. 3.15, we can write a force-deformation equation for each of the three elements. Step 3: We can relate the element elongations to the displacement at B. Steps 4–6: Finally, we can use the Displacement Method to combine these three sets of fundamental equations to get expressions for the displacement for node B. Step 7: The forces in the individual elements can then be determined by substituting this displacement into the element force-deformation equations.

Step 8: If the external force P_B acts to the right (i.e., if it is positive), we should find that node B moves to the right, and we should find that the left-hand element is in tension and that the two right-hand elements are in compression.

Solution   (a) *Determine the displacement u_B of node B.*

Equilibrium: For node B, shown in the free-body diagram in Fig. 2.

$$\overset{+}{\underset{\rightarrow}{}}\sum F_x = 0: \qquad -F_1 + F_2 + F_3 + P_B = 0 \qquad \textbf{Equilibrium} \qquad (1)$$

Equation (1) relates the three unknown internal element forces to the known external load. Since there are three unknown forces, but only one equilibrium equation, this problem is *statically indeterminate*. That is, the internal forces cannot be determined by statics alone. Thus, to find additional equations we must look to element force-deformation behavior and to the geometry of deformation.

Element Force-Deformation Behavior: We have three uniform, axial-deformation elements, and for each one we can write an element force-deformation equation like Eq. 3.15. We have called the element forces F_1, F_2, and F_3 (tension positive), so we have

$$
\begin{aligned}
F_1 &= k_1 e_1, \quad \text{where } k_1 = (A_1 E_1 / L_1) \\
F_2 &= k_2 e_2, \quad \text{where } k_2 = (A_2 E_2 / L_2) \\
F_3 &= k_3 e_3, \quad \text{where } k_3 = (A_3 E_3 / L_3)
\end{aligned}
\qquad
\begin{array}{l}
\textbf{Element} \\
\textbf{Force-} \\
\textbf{Deformation}
\end{array}
\qquad (2)
$$

In Eqs. (2) the e's are the element elongations. A positive F_i (tension) produces a positive e_i (element gets longer), since the k_i's are, by definition, positive.

So far, we have four equations in six unknowns—three F's and three e's. We must still enforce the deformation compatibility of the three elements with each other and with the walls at A and C.

Geometry of Deformation: Referring to Fig. 1, we can easily relate the elongation of each of the three elements to the displacement u_B by using the definition of elongation of an element, that is, $e = u(L) - u(0)$. So,

$$
\begin{aligned}
e_1 &= u_B \\
e_2 &= e_3 = -u_B
\end{aligned}
\qquad
\begin{array}{l}
\textbf{Geometry of} \\
\textbf{Deformation}
\end{array}
\qquad (3)
$$

Here we have used the fact that the displacements at nodes A and C are zero. Equations (3) can also be called **compatibility equations** since they express mathematically the fact that A and C are fixed ends and the fact that the three elements are joined together at B. (Note that, since e is positive when the element gets longer, a displacement of node B to the right by an amount u_B implies a shortening of elements (2) and (3) by that amount; hence the minus sign in the equations for e_2 and e_3.)

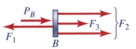

P_B and F_1 are collinear with F_3
F_2 is distributed around the circumference of the pipe.

Fig. 2   Free-body diagram of node B.

Displacement-Method Solution: Now, if we count equations and unknowns, we find that we have six equations and six unknowns. Rather than just combine Eqs. (1) through (3) in some arbitrary order, we can note that, since there is only one node, there is only <u>one equilibrium equation</u>; correspondingly, there is only <u>one nodal displacement</u>, u_B. By substituting Eqs. (3) into Eqs. (2), we are able to write the three F's in terms of the nodal displacement, u_B. If we then substitute these equations, call them Eqs. (4), into Eq. (1), we will have **one equilibrium equation expressed in terms of the one unknown nodal displacement.**

Substitute Eqs. (3) (deformation geometry) into Eqs. (2) (element force-deformation) to obtain

$$F_1 = k_1 u_B, \quad F_2 = -k_2 u_B, \quad F_3 = -k_3 u_B \tag{4}$$

Now substitute Eqs. (4) into Eq. (1).

$$(k_1 + k_2 + k_3) u_B = P_B \qquad \text{\textbf{Node Equilibrium in Terms of Node} } \tag{5}$$
Displacement

The solution of this equation is

$$u_B = \frac{P_B}{k_1 + k_2 + k_3} \qquad \textbf{Ans. (a)} \tag{6}$$

The strategy of substituting deformation geometry equations (3) into element force-deformation equations (2) into equilibrium equations (1) is called the **Displacement Method** because the major solution step, Eq. (6), gives an answer that is a displacement. It is also sometimes referred to as the *Stiffness Method* since stiffness coefficients appear in the final solution.

(b) *Determine the element forces.* This is simple to do, because we only need to substitute the nodal displacement u_B into Eqs. (4). Thus, we get

$$\left. \begin{aligned}
F_1 &= k_1 u_B = \frac{k_1 P_B}{k_1 + k_2 + k_3} \\[2mm]
F_2 &= -k_2 u_B = \frac{-k_2 P_B}{k_1 + k_2 + k_3} \\[2mm]
F_3 &= -k_3 u_B = \frac{-k_3 P_B}{k_1 + k_2 + k_3}
\end{aligned} \right\} \qquad \textbf{Ans. (b)} \tag{7}$$

Review the Solution   As one check of our work, we can substitute Eqs. (7) back into Eq. (1) to see if equilibrium is satisfied.

$$\text{Is } F_1 - F_2 - F_3 = P_B? \text{ Yes.}$$

The fact that equilibrium is satisfied by our answers means that we have probably not made errors in our solution. Also, from Eqs. (7) we see that, when P_B is positive, element (1) is in tension and elements (2) and (3) are in compression. This is what we expected to find.

Note how the stiffness coefficients enter into the answer for u_B in Eq. (6) of the above example and the answers for the element forces F_i in Eqs. (7) of the above example. From Eq. (6), an increase in any one of the stiffness coefficients will cause a decrease in the displacement of node B. Also, from Eqs. (7), an increase in any one of the k_i's causes the fraction of the load P_B that is transferred by that particular element to the supports to be increased.

After completing the analysis of any structure, whether it is statically determinate or statically indeterminate, your goal should be to **understand the behavior of the structure under load**—why are the stresses distributed the way they are, and why does the structure deform the way your analysis says that it does? You should always think first about how to obtain correct equations of all of the three fundamental types—equilibrium equations, force-deformation equations, and compatibility equations! Like the Basic Force Method, the Displacement Method, illustrated in Example Problem 3.15, is just one systematic procedure for combining the fundamental equations to get the final solution. However, since it is a very systematic procedure, it is the procedure that forms the basis of the finite-element programs that are used to solve structures problems involving thousands of equations with thousands of unknowns. It is also a very straightforward method to use for most "hand" calculations.

MDS3.13 **End-to-End Rods: Displacement Method** is a program for analyzing statically indeterminate structures, similar to the one in Example 3.15, by the Displacement Method.

The next example illustrates Displacement-Method solution of thermal-deformation problems like those in Section 3.6.

EXAMPLE 3.16

Three rods are attached to a rigid L-shaped bracket ABD, as shown in Fig. 1. Determine the force in each of the three rods if element (3) is cooled by 50°C. Assume that the rotation of the bracket is small and that all rods are force-free when the system is assembled.

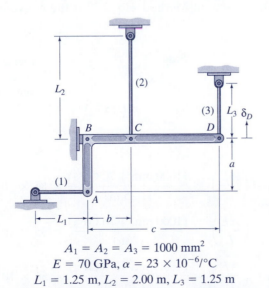

$A_1 = A_2 = A_3 = 1000 \text{ mm}^2$
$E = 70 \text{ GPa}, \alpha = 23 \times 10^{-6}/°C$
$L_1 = 1.25 \text{ m}, L_2 = 2.00 \text{ m}, L_3 = 1.25 \text{ m}$
$a = 1.25 \text{ m}, b = 1.00 \text{ m}, c = 2.50 \text{ m}$

Fig. 1

Plan the Solution Since it is the element forces that are required, it would seem best to try a Basic-Force-Method solution. There are three unknown rod forces and, by taking moments about point B, we will get one equilibrium equation. Therefore, the number of redundant internal forces is (# Redundants) = (# Unknown Forces) − (# Equilibrium Equations) = 3 − 1 = 2. A Basic-Force-Method solution would require us to solve an equilibrium equation and two compatibility equations simultaneously for the three rod forces. A Displacement-Method solution only requires that one equilibrium equation be solved for one displacement, say the angle of rotation of the bracket or the displacement at some point on it. Therefore, the Displacement Method will be used here. (See Homework Prob. 3.9-16 for a Force-Method version of this problem.)

Since rod (3) is cooled, it will tend to shorten, and the force induced in it will be tension. This will cause the bracket ABD to rotate counter-clockwise, putting rod (1) into tension and rod (2) into compression.

Solution

Equilibrium: Since we want a free-body diagram that will permit us to relate the internal element forces to each other and to the external loads (actually, there are no external loads in this problem), we select the rigid bracket ABD as shown in Fig. 2. As always, we select tension as positive for the force in each element.

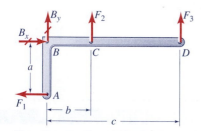

Fig. 2 Free-body diagram.

$$+\circlearrowleft\left(\sum M\right)_B = 0: \quad -F_1(a) + F_2(b) + F_3(c) = 0$$

The equilibrium equation is

$$-F_1 a + F_2 b + F_3 c = 0 \qquad \textbf{Equilibrium} \quad (1)$$

Element Force-Temperature-Deformation Equations: Since we will employ the Displacement Method, Eq. 3.28 is the convenient form to use.

$$
\begin{aligned}
F_1 &= k_1 e_1 \\
F_2 &= k_2 e_2 \\
F_3 &= k_3(e_3 - \alpha_3 L_3 \Delta T_3)
\end{aligned}
\qquad
\begin{matrix}
\textbf{Element} \\
\textbf{Force-} \\
\textbf{Temperature-} \\
\textbf{Deformation}
\end{matrix}
\quad (2)
$$

where

$$k_1 = \left(\frac{AE}{L}\right)_1 = \frac{(1000 \text{ mm}^2)(70 \text{ GPa})}{(1.25 \text{ m})} = 56 \times 10^3 \text{ kN/m}$$

$$k_2 = \left(\frac{AE}{L}\right)_2 = \frac{(1000 \text{ mm}^2)(70 \text{ GPa})}{(2 \text{ m})} = 35 \times 10^3 \text{ kN/m}$$

$$k_3 = \left(\frac{AE}{L}\right)_3 = \frac{(1000 \text{ mm}^2)(70 \text{ GPa})}{(1.25 \text{ m})} = 56 \times 10^3 \text{ kN/m}$$

$$\alpha_3 L_3 \Delta T_3 = (23 \times 10^{-6}/°\text{C})(1.25 \text{ m})(-50°\text{C}) = -1.4375 \text{ mm}$$

Geometry of Deformation: We need to sketch a **deformation diagram.** We will assume, as seems reasonable since element (3) is cooled, that bracket *ABD* rotates counterclockwise, and will use similar triangles in Fig. 3 to relate the elongations of the three elements to the displacement δ_D at the tip of the bracket.

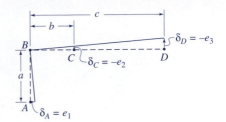

Fig. 3 Deformation diagram.

In Eqs. (2) a positive *e* corresponds to a lengthening of the element. Thus, on the deformation diagram in Fig. 3 a counterclockwise rotation of *ABD* corresponds to elongation of rod (1) but contraction of rods (2) and (3). The appropriate **deformation compatibility equations** are equations that relate the *e*'s to the rotation of the bracket. However, without introducing the beam's rotation angle, θ, we can make use of the geometry of similar triangles and write the following deformation-compatibility equations:

$$e_1 = \left(\frac{a}{c}\right)\delta_D, \; e_2 = -\left(\frac{b}{c}\right)\delta_D, \; e_3 = -\delta_D \qquad \begin{array}{l}\textbf{Deformation} \\ \textbf{Compatibility}\end{array} \qquad (3)$$

Solution of Equations by the Displacement Method: Now we use the Displacement-Method steps to combine Eqs. (1) through (3). First, we substitute Eqs. (3) into Eqs. (2) to get the element forces all written in terms of the one displacement δ_D.

$$\left.\begin{array}{l} F_1 = k_1\left(\dfrac{a}{c}\right)\delta_D \\[2ex] F_2 = k_2\left[-\left(\dfrac{b}{c}\right)\delta_D\right] \\[2ex] F_3 = k_3(-\delta_D - \alpha_3 L_3 \Delta T_3) \end{array}\right\} \qquad (4)$$

Finally, these equations are substituted into the equilibrium equation, Eq. (1), and rearranged to obtain the following **equilibrium equation in terms of the unknown displacement:**

$$(k_1 a^2 + k_2 b^2 + k_3 c^2)\delta_D = -k_3(\alpha_3 L_3 \Delta T_3)c^2 \qquad (5)$$

The solution of this equation gives the unknown displacement δ_D.

$$\boxed{\delta_D = \frac{-k_3(\alpha_3 L_3 \Delta T_3)c^2}{k_1 a^2 + k_2 b^2 + k_3 c^2}} \qquad (6)$$

Substituting numerical values into this equation, we get

$$\delta_D = 1.0648 \text{ mm}$$

Then, from Eqs. (4) we can directly compute the values of the three rod forces:

$$F_1 = (56 \times 10^3 \text{ kN/m})(0.50)(1.0648 \text{ mm}) = 29.81 \text{ kN}$$

$$F_2 = (35 \times 10^3 \text{ kN/m})(-0.4)(1.0648 \text{ mm}) = -14.907 \text{ kN}$$

$$F_3 = (56 \times 10^3 \text{ kN/m})(-1.0648 \text{ mm} + 1.4375 \text{ mm}) = 20.87 \text{ kN}$$

Rounding off these values, we have

$$F_1 = 29.8 \text{ kN}(29.8 \text{ kN T})$$
$$F_2 = -14.91 \text{ kN}(14.91 \text{ kN C}) \Bigg\} \qquad \textbf{Ans.} \quad (7)$$
$$F_3 = 20.9 \text{ kN}(20.9 \text{ kN T})$$

Review the Solution   As we expected, cooling element (3) puts elements (1) and (3) into tension, while element (2) is put into compression. Note that cooling of element (3) would have shortened it by 1.4375 mm if it were not attached to bracket ABD. However, since it must pull against the bracket as it tries to shorten, it actually shortens by only 1.0648 mm.

An equilibrium check can be used as one means of verifying the correctness of the solution.

$$\text{Is} - F_1(a) + F_2(b) + F_3(c) = 0? \text{ Yes.}$$

Finally, we will use the Displacement Method to solve an axial-deformation problem that involves an external load, a temperature change, and an element that was not manufactured the correct length.

EXAMPLE 3.17

A stepped rod is made up of three uniform elements, or members, as shown in Fig. 1. The rod was intended to exactly fit between rigid walls with the ends of the rod welded to the rigid walls at A and D. However, member (2) was manufactured δ_2 too short. First, the entire rod is stretched so that its ends can be welded to the rigid walls. Then, an axial load P_B is applied to the rod at B and member (3) is heated by an amount ΔT_3 (a) Determine expressions for the displacements u_B and u_C of the two nodes B and C measured, as indicated in Fig. 1, from the <u>intended</u> original locations of these nodes; and (b) determine expressions for the internal forces in the three elements.

Plan the Solution   We can think of this structure as being composed of three uniform elements and two connecting nodes, or joints, as shown in Fig. 2. We can write equilibrium equations for the joints and force-temperature-deformation equations for the separate elements. Finally,

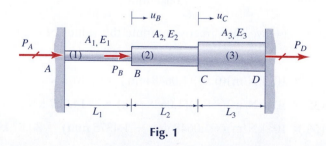

Fig. 1

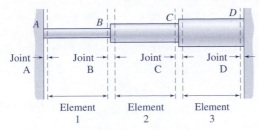

Fig. 2 The stepped rod divided into elements and joints (nodes).

we can relate element elongations to the displacements of the joints, taking into account the fact that member (2) was initially too short. We can then follow the Displacement-Method procedure to combine these three sets of equations to get the required answers.

Consider the effect of each of the three "loads" separately:

(1) *External Force* P_B: If P_B is positive, that is, if it acts to the right, we should find that the element (1) is in tension and that elements (2) and (3) are in compression.

(2) *Member* (2) *too short:* The three-element rod will have to be stretched in order to weld the ends to the rigid walls. Therefore, all three elements will exhibit tensile internal force due to this "load."

(3) *Member* (3) *heated:* Heating member (3) will cause it to push against the members to its left; they, in turn, will push back. Therefore, all three elements will exhibit a compressive internal force due to this "load."

Solution (a) *Determine the nodal displacements u_B and u_C.*

Equilibrium: In Fig. 3 we isolate each of the joints as a free body, and on each free body we show the external force (if there is one) acting on the joint and the internal forces exerted by the elements that connect to the joint. We label the element forces and, as usual, adopt the sign convention that element forces are taken to be positive in tension.

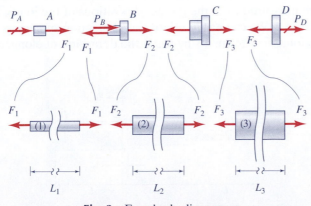

Fig. 3 Free-body diagrams.

For Node (Joint) B: $\xrightarrow{+} \Sigma F_x = 0$: $-F_1 + F_2 + P_B = 0$

For Node (Joint) C: $\xrightarrow{+} \Sigma F_x = 0$: $-F_2 + F_3 = 0$

$$
\begin{aligned}
F_1 - F_2 &= P_B \\
F_2 - F_3 &= 0
\end{aligned}
\qquad \textbf{Equilibrium} \qquad (1)
$$

Equations (1) relate the three unknown internal element forces to the known external load P_B and to each other. Since there are three unknowns, but only two equilibrium equations, this system is *statically indeterminate*. That is, the internal forces cannot be determined by statics alone. This situation arises, of course, because both ends of the stepped rod are attached to rigid walls. There is no way that we can find another equilibrium equation that will not be just a combination of these two joint-equilibrium equations. Therefore, to find additional equations we must look to the geometry of deformation and to element force-deformation behavior.

Element Force-Deformation Behavior: We have three uniform, axial-deformation elements, and for each one we can write an element force-temperature-deformation equation like Eq. 3.28. We have called the element forces $F_1, F_2,$ and F_3 (tension positive), so we have

$$
\begin{aligned}
F_1 &= k_1 e_1, \text{ where } k_1 = (A_1 E_1 / L_1) \\
F_2 &= k_2 e_2, \text{ where } k_2 = (A_2 E_2 / L_2) \\
F_3 &= k_3 (e_3 - \alpha_3 L_3 \Delta T_3). \\
&\quad \text{where } k_3 = (A_3 E_3 / L_3)
\end{aligned}
\qquad
\begin{aligned}
&\textbf{Element} \\
&\textbf{Force-} \\
&\textbf{Temp.-} \\
&\textbf{Deformation}
\end{aligned}
\qquad (2)
$$

In Eqs. (2) the e's are the *total* element elongations. A positive F_i (tension) produces a positive e_i (element gets longer), since the k_i's are, by definition, positive.

Although we have three new equations, we have introduced three new unknowns, so we now have five equations in six unknowns—three F's and three e's. Geometry of deformation provides the additional equations.

Geometry of Deformation: The joint displacements are labeled u_B and u_C and are measured from the <u>intended</u> positions of the joints, as shown in Figs. 1 and 4. We can easily relate the elongation of each element to these two joint displacements by using the definition of elongation of an

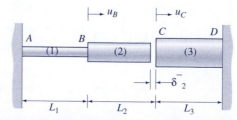

Fig. 4   Definition of displacement coordinates.

element, that is, $e = u(L) - u(0)$. But we also have to account for the fact that element (2) was initially too short by an amount $\bar{\delta}_2$ and would have to be stretched by this amount for its ends to occupy their intended positions. So,

$$
\begin{aligned}
e_1 &= u_B - u_A = u_B \\
e_2 &= \bar{\delta}_2 + u_C - u_B \\
e_3 &= u_D - u_C = -u_C
\end{aligned}
\qquad \text{\textbf{Geometry of Deformation}} \qquad (3)
$$

Here we have used the fact that the displacements at joints A and D are zero. (Note that, since e is positive when the element gets longer, a displacement of joint C to the right by an amount u_C implies a shortening of element (3) by that amount.)

Displacement-Method Solution: Now, if we count equations and unknowns, we find that we have eight equations and eight unknowns. Rather than just combine Eqs. (1) through (3) in some arbitrary order, we can note that, since there are two joints, there are two equilibrium equations, and also there are two joint displacements, u_B and u_C. By substituting Eqs. (3) into Eqs. (2), we will be able to write the three F's in terms of the two u's. If we then substitute these equations, call them Eqs. (4), into Eqs. (1), we will have *two equilibrium equations expressed in terms of two unknown joint displacements*.

Substitute Eqs. (3) (deformation geometry) into Eqs. (2) (element force-temperature-deformation) to obtain

$$
\begin{aligned}
F_1 &= k_1 u_B \\
F_2 &= k_2(\bar{\delta}_2 + u_C - u_B) \\
F_3 &= k_3(-u_C - \alpha_3 L_3 \Delta T_3)
\end{aligned}
\qquad (4)
$$

Now substitute Eqs. (4) into Eqs. (1).

$$
\begin{aligned}
k_1 u_B - k_2(\bar{\delta}_2 + u_C - u_B) &= P_B \\
k_2(\bar{\delta}_2 + u_C - u_B) - k_3(-u_C - \alpha_3 L_3 \Delta T_3) &= 0
\end{aligned}
$$

which can be written

$$
\begin{aligned}
(k_1 + k_2)u_B - k_2 u_C &= (P_B)_{\text{eff}} \\
-k_2 u_B + (k_2 + k_3)u_C &= (P_C)_{\text{eff}}
\end{aligned}
\qquad \text{\textbf{Joint Equil. in Terms of Joint Displs.}} \qquad (5)
$$

where

$$
\begin{aligned}
(P_B)_{\text{eff}} &\equiv P_B + k_2 \bar{\delta}_2 \\
(P_C)_{\text{eff}} &\equiv -k_2 \bar{\delta}_2 - k_3 \alpha_3 L_3 \Delta T_3
\end{aligned}
\qquad (5')
$$

Note how the $\bar{\delta}_2$ terms and the ΔT_3 term produce effective "loads" on the nodes at B and C.

The solution of Eqs. (5), a set of two equations in two unknowns, can easily be obtained by applying Cramer's Rule. The solution is

$$u_B = \frac{(k_2 + k_3)(P_B)_{\text{eff}} + k_2(P_C)_{\text{eff}}}{k_1k_2 + k_1k_3 + k_2k_3}$$

$$u_C = \frac{k_2(P_B)_{\text{eff}} + (k_1 + k_2)(P_C)_{\text{eff}}}{k_1k_2 + k_1k_3 + k_2k_3}$$

Ans. (a)   (6)

(b) *Determine the three element forces.* This is very straightforward, since we only need to substitute the above expressions for the nodal displacements, Eqs. (6), together with the definitions of $(P_B)_{\text{eff}}$ and $(P_C)_{\text{eff}}$ from Eqs. (5′), into Eqs. (4). Thus, we get

$$F_1 = \frac{k_1[(k_2 + k_3)P_B + k_2k_3(\bar{\delta}_2 - \alpha_3 L_3 \Delta T_3)]}{k_1k_2 + k_1k_3 + k_2k_3}$$

$$F_2 = \frac{k_2[-k_3 P_B + k_1k_3(\bar{\delta}_2 - \alpha_3 L_3 \Delta T_3)]}{k_1k_2 + k_1k_3 + k_2k_3}$$

$$F_3 = \frac{k_3[-k_2 P_B + k_1k_2(\bar{\delta}_2 - \alpha_3 L_3 \Delta T_3)]}{k_1k_2 + k_1k_2 + k_2k_3}$$

Ans. (b)   (7)

Review the Solution   As one check of our work, we can substitute Eqs. (7) back into Eqs. (1) to see if equilibrium is satisfied.

Is $F_1 - F_2 = P_B$? Yes.   Is $F_2 - F_3 = 0$? Yes.

The fact that equilibrium is satisfied by our answers means that we have probably not made errors in our solution. Also, by examining Eqs. (7) we can see that the tension/compression contributions of P_B, $\bar{\delta}_2$, and ΔT_3 to the various element axial forces are what we expected to find.

The Displacement Method was used as the basis for the development of a computer program to solve multisegment axial-deformation problems involving externally applied axial loads, element temperature changes, and misfits. This computer program is one of the *General Analysis* suite of the **MDSolids** computer program.

MDS3.14   **Multisegment Axial Rod: Displacement Method** is a program for analyzing multi-segment statically indeterminate structures, like the one in Example 3.17, by the Displacement Method. Temperature changes and misfits are included.

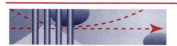

*3.9   FORCE-METHOD SOLUTION OF AXIAL-DEFORMATION PROBLEMS

In Sections 3.5–3.7 the analysis of relatively simple statically indeterminate systems followed a procedure called the Basic Force Method. As problem complexity increases, the number of simultaneous equations that arise—equilibrium equations, force-temperature-deformation equations, and compatibility equations—increases, and it becomes necessary to examine ways of solving these as systematically as possible. One of these ways is the Displacement Method, described in Section 3.8. Another is the **Force Method,** which is also referred to as the *Flexibility Method* or the *Method of Consistent Deformations*. The Basic Force Method does not require one to identify specific forces as the "redundant" forces; the more formal Force Method, described here, does. It is the natural extension of the *Method of Sections* that is used in statics to solve statically determinate problems.

Solution Procedure: Force Method.[11]

The Force Method, outlined in the following solution-procedure list and flow chart, is illustrated in Example Problems 3.18 and 3.19. and comparison is drawn between the Displacement Method and the Force Method.

SOLUTION PROCEDURE: FORCE METHOD

▼ **SET UP THE FUNDAMENTAL EQUATIONS:**

(1) Determine N_E the number of <u>independent</u> equilibrium equations that can be written. The number of **redundant internal forces** is equal to the total number of unknown internal forces minus the number of independent equilibrium equations that are available, that is,

$$N_R = N_U - N_E$$

Select N_R Internal forces to be the redundant internal forces; the other N_E forces will be the **determinate internal forces.** Using *free-body diagrams*, **write down N_E independent equilibrium equations by expressing each of the N_E determinate internal forces in terms of the N_R redundant internal forces and the external loads.** This is most conveniently done by imagining that the structure is made statically determinate by making a "cut" that releases each constraint that gives rise to a selected redundant force, thereby treating the redundant forces as though they were known loads.

(2) Write an **element force-temperature-deformation equation** for each axial-deformation element. Equation 3.27 (Eq. 3.14 if $\Delta T_i = 0$) is the most convenient form to use for this step:

$$e_i = f_i F_i + \alpha_i L_i \Delta T_i, \quad f_i = (L/AE)_i \qquad \begin{array}{r}(3.27)\\ \text{repeated}\end{array}$$

(3) Use *geometry of deformation* to write down the N_R **compatibility equations(s)** in terms of the element elongations, e_i.

▼ **SOLVE FOR THE UNKNOWN FORCES:**

(4) Substitute the N_E determinate-force equilibrium equations of Step 1 into the *element force-temperature-deformation equations* of Step 2. This produces N_U **force-temperature-deformation equations in terms of the N_R redundant forces** (and the external loads and ΔT's).

(5) Substitute the results of Step 4 into the N_R geometric-compatibility equations of Step 3. This gives N_R **geometric-compatibility equations in terms of the N_R redundant internal forces.**

(6) **Solve simultaneously** the equations obtained in Step 5. The answer will be the N_R **redundant internal forces.** The remaining N_R (determinate) internal forces are obtained by substituting the redundant forces into the equilibrium equations of Step 1.

▼ **SOLVE FOR THE DISPLACEMENTS:**

(7) To obtain system displacements, substitute the internal forces from Step 6 into the force-temperature-deformation equations of Step 2. Finally, use geometry of deformation to relate element elongations to system displacements.[12]

(8) Review the solution to make sure that all answers seem to be correct.

[11]Because the Force Method is far less widely used in computer-based structural analysis and finite element analysis than is the Displacement Method, this introduction to the Force Method is treated as an "optional" section.
[12]If displacements are required in addition to element forces and/or reactions at the supports, the *Displacement Method,* which is discussed in Section 3.8. is generally easier to use than the Force Method.

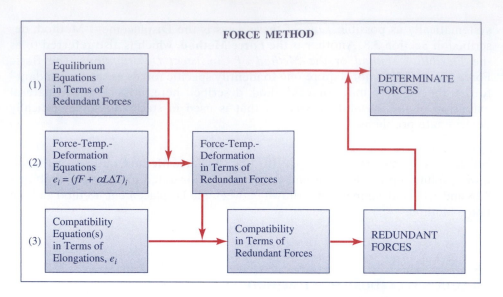

FORCE METHOD

In some cases, it is easier to solve a problem by the Force Method, rather than by the Displacement Method, as is demonstrated now in the Force-Method solution of Example Problem 3.18 as compared to the Displacement Method solution of Example Problem 3.17.

EXAMPLE 3.18

A stepped rod is made up of three uniform elements, or members, as shown in Fig. 1. The rod exactly fits between rigid walls when no external forces are applied, and the ends of the rod are welded to the rigid walls. (a) Determine the axial stress, σ_i, in each of the three elements. Let F_1, the internal force in member AB, be the redundant force. (b) Determine the joint displacements u_B and u_C.

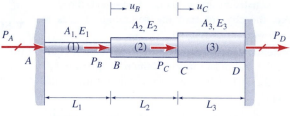

Fig. 1 The original structure.

Plan the Solution As always, we will first set up the three fundamental sets of equations—equilibrium equations, force-deformation equations, and compatibility equations. In solving these equations we are asked to let F_1 be taken as the redundant force. As illustrated in Fig. 2, by making a single (imaginary) "cut" anywhere along element (1) we make the resulting structure "statically determinate," with F_1 treated as a known load rather than as an unknown internal force. There is one compatibility (constraint) equation, namely, that the total elongation is zero. By writing this equation in terms of the redundant force F_1, we will be able to solve this one equation for the one unknown.

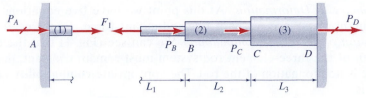

Fig. 2 The "cut" structure.

Solution (a) *Determine the axial stresses* σ_i.

Equilibrium: Having selected F_1 as the redundant internal force, we need to draw two free-body diagrams that can be used to express F_2 and F_3 in terms of the redundant force F_1. By cutting on both sides of the joint at B, we produce the free-body diagram of Fig. 3a, which relates F_2 to F_1. Similarly, by cutting through elements (1) and (3) we get the free-body diagram in Fig. 3b, which relates F_3 to the redundant force F_1. Note that we have applied the *method of sections* in drawing the free-body diagrams and writing the equilibrium equations. You will find that the method of sections is the best way to draw free-body diagrams for *Force-Method* solutions.

For Fig. 3a: $\overset{+}{\to} \Sigma F_x = 0: \to \qquad -F_1 + F_2 + P_B = 0$

For Fig. 3b: $\overset{+}{\to} \Sigma F_x = 0: \to \quad -F_1 + P_B + P_C + F_3 = 0$

As indicated in Step 1 of the Force-Method Procedure, we rearrange these equilibrium equations so that they express the determinate forces F_2 and F_3 in terms of the one redundant force F_1 and the external loads.

$$
\begin{aligned}
F_2 &= F_1 - P_B \\
F_3 &= F_1 - P_B - P_C
\end{aligned}
\qquad \textbf{Equilibrium} \qquad (1)
$$

Note that, once we have solved for the redundant force F_1, we will come back to Eqs. (1) to solve for the remaining unknown forces.

Element Force-Deformation Behavior: The force-deformation equations for the three elements can be written in the form given by Eq. 3.14, namely,

$$
\begin{aligned}
e_1 &= f_1 F_1, \quad f_1 = (L/AE)_1 \\
e_2 &= f_2 F_2, \quad f_2 = (L/AE)_2 \\
e_3 &= f_3 F_3, \quad f_3 = (L/AE)_3
\end{aligned}
\qquad
\begin{aligned}
&\textbf{Element} \\
&\textbf{Force-} \\
&\textbf{Deformation} \\
&\textbf{Behavior}
\end{aligned}
\qquad (2)
$$

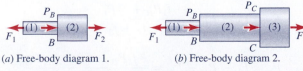

(*a*) Free-body diagram 1. (*b*) Free-body diagram 2.

Fig. 3 Free-body diagrams.

Geometry of Deformation: At this point we have five equations and six unknowns. Since we have one redundant force, F_1, we need one *equation of geometric compatibility*. We can see (Fig. 1) that the total length of the three-element rod system must remain constant, that is, there is no elongation of the rod. The appropriate **compatibility equation** is

$$e_{\text{total}} = e_1 + e_2 + e_3 = 0 \qquad \textbf{Compatibility} \qquad (3)$$

Force-Method Solution of the Fundamental Equations: Note that the equilibrium equations have been written in a form that expresses the determinate forces, F_2 and F_3, in terms of the redundant force, F_1, and the two nodal loads. Following Step 4 in the Force-Method Procedure, we first combine Eqs. (1) and (2) so that all of the force-deformation equations will be expressed in terms of the known loads and the one redundant force, F_1.

$$e_1 = f_1 F_1, \quad e_2 = f_2(F_1 - P_B), \quad e_3 = f_3(F_1 - P_B - P_C) \qquad (4)$$

Next (Step 5), we substitute Eqs. (4) into the compatibility equation, Eq. (3), to get

$$(f_1 + f_2 + f_3)F_1 = f_2 P_B + f_3(P_B + P_C) \qquad \begin{array}{l}\textbf{Compatibility} \\ \textbf{in Terms of} \\ \textbf{the Redundant} \\ \textbf{Force}\end{array} \qquad (5)$$

We have reduced six equations in six unknowns to <u>one equation in one unknown</u>, which we can easily solve for the **redundant force** (Step 6).

$$F_1 = \frac{(f_2 + f_3)P_B + f_3 P_C}{f_1 + f_2 + f_3} \qquad (6a)$$

This is the <u>key solution step</u>. We have determined the redundant internal force in terms of the given external loads and properties of the structure. The approach we have used is called the **Force Method** because the major solution step gives us a force quantity. It is sometimes referred to as the *Flexibility Method* because it is convenient to use flexibility coefficients, f_i, in the solution; and it is sometimes referred to as the *Method of Consistent Deformations* because the key equation(s) in the solution is (are) the deformation-compatibility equation(s).

Knowing the redundant force F_1, we can now use the equilibrium equations, Eqs. (1), to obtain the determinate forces F_2 and F_3, thus completing Step 6.

$$F_2 = F_1 - P_B = \frac{-f_1 P_B + f_3 P_C}{f_1 + f_2 + f_3}$$

$$F_3 = F_1 - P_B - P_C = \frac{-f_1 P_B - (f_1 + f_2)P_C}{f_1 + f_2 + f_3} \qquad (6b,c)$$

Since we were asked for the stresses in the three elements, we need to divide the forces in Eqs. (6) by their respective cross-sectional areas, that is,

$$\sigma_1 = \frac{F_1}{A_1}, \quad \sigma_2 = \frac{F_2}{A_2} \quad \sigma_3 = \frac{F_3}{A_3} \qquad \textbf{Ans. (a)} \quad (7)$$

(b) *Obtain the displacements u_B and u_C.* We need to use *geometry of deformation* equations. We can note (Fig. 1) that

$$u_B = e_1, \quad u_C = -e_3 \qquad \begin{array}{c}\textbf{Geometry of}\\ \textbf{Deformation}\end{array} \qquad (8)$$

The first and third of Eqs. (2) may be substituted into these to give

$$u_B = f_1 F_1, \quad u_C = -f_3 F_3 \qquad (9)$$

Then, forces F_1 and F_3 from Eqs. (6) can be substituted into Eqs. (9) giving

$$\left.\begin{array}{c} u_B = \dfrac{f_1[(f_2 + f_3)P_B + f_3 P_C]}{f_1 + f_2 + f_3} \\[2em] u_C = \dfrac{f_3[f_1 P_B + (f_1 + f_2)P_C]}{f_1 + f_2 + f_3} \end{array}\right\} \qquad \textbf{Ans. (b)} \quad (10)$$

Review the Solution   We can easily check the dimensions in each of the answers and see that the answers are dimensionally homogeneous. For example, dimensionally, the f's in the numerator and the f's in the denominator of Eqs. (6) cancel, leaving the dimensional equation $F = F$. As a second check, we can see (Fig. 1) that, if $P_B > 0$ and $P_C = 0$, element (1) will be in tension while elements (2) and (3) will be in compression. From Eqs. (6) we can see that this is the case.

You should always think first about how to obtain correct equations of all of the three fundamental types—equilibrium equations, force-deformation equations, and compatibility equations! The Force Method is just one systematic procedure for combining them to get the final solution. It is a direct extension of the procedure used for statically determinate structures: equilibrium first, then element force-temperature-deformation, and finally geometry of deformation.

Comparison of the Force Method and the Displacement Method.

Any method of solving the simultaneous equations—the Displacement Method or the Basic Force Method or the Force Method—may be used as long as you have all of the necessary equations at hand. However, it may be more convenient to use one procedure rather than another one. The Basic Force Method is best if there is only one equilibrium equation and one compatibility equation. For more complex situations, it is preferable to follow the Displacement Method or the Force Method. Table 3.3 in Section 3.8 compares the steps in these two procedures. The Displacement Method is the more straightforward of the two, because it is not necessary to select redundant forces and because the deformation-compatibility equations directly involve system displacements. However, there are circumstances, such as solving for the axial forces in the elements in Fig. 3.21, where the Force Method requires less

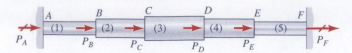

FIGURE 3.21 A multisegment statically indeterminate structure.

work. In this case we could just let F_1 be the redundant force and, using the Force Method, solve the deformation compatibility equation

$$e_1 + e_2 + e_3 + e_4 + e_5 = 0$$

once it has been written in terms of F_1.

In Example Problem 3.19 there are two independent equilibrium equations and one deformation-compatibility equation. Therefore, it is probably easier to solve it by using the Force Method, rather than the Displacement Method. It is "starred" as a more difficult problem because there are three unknown forces and because special care must be taken in order to obtain the correct compatibility equation. Nevertheless, the Force-Method solution is very straightforward.

EXAMPLE 3.19*

Two rods and a post support a rigid beam AD, as shown in Fig. 1. When assembled together as a system, the two rods and post would be stress-free and the beam AD would be horizontal, were it not for the fact that rod (1) was manufactured $\bar{\delta}_1$ too short. (a) Use the Force Method to determine the force in rod (1) after the whole system is assembled and the vertical load P is applied to the beam at C. Assume that the rotation of beam AD is small. (b) Determine the vertical displacement δ_A.

Plan the Solution Since the post at B and the rods at A and D are all deformable, we will have to allow for both vertical translation and rotation

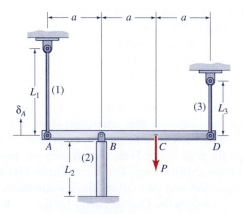

$A_1 = A_3 = 1000 \text{ mm}^2, A_2 = 1500 \text{ mm}^2$
$L_1 = 4.2 \text{ m}, L_2 = 2.1 \text{ m}, L_1 = 2.8 \text{ m}$
$a = 2 \text{ m}, \bar{\delta}_1 = 1.0 \text{ mm}, E = 70 \text{ GPa}, P = 25 \text{ kN}$

Fig. 1

of the rigid beam AD. We can write one equilibrium equation for the summation of forces in the vertical direction and, by taking moments about some point along the beam point (e.g., point B), we will get a second equilibrium equation. A more efficient approach, however, will be to use two moment equilibrium equations. Since there are three unknown internal forces, we will need to designate one of them as the redundant force, and one deformation-compatibility equation will have to be obtained and solved for this redundant force.

Because rod (1) is initially too short, it will tend to displace end A upward, and the force induced in rod (1) will be tension. This will cause post (2) and rod (3) to be in tension also. However, the effect of the load P is less clear, since both post (2) and rod (3) are deformable.

Solution (a) *Determine the force in rod (1).*

Equilibrium: Since we want a free-body diagram that will permit us to relate the internal element forces to each other and to the external load P, we select the rigid beam AD as shown in Fig. 2. As always, we select tension as positive for the force in each element. We need to designate one force as the **redundant** force. Since the problem statement only asks for F_1 and δ_A, the best approach is to designate F_1 as the redundant force and write equilibrium equations that express F_2 and F_3 in terms of F_1 and the external load P.

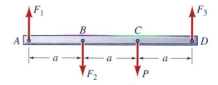

Fig. 2 Free-body diagram.

$$+\zeta \left(\sum M\right)_B = 0: \qquad -F_1(a) + F_3(2a) - P(a) = 0$$

$$+\zeta \left(\sum M\right)_D = 0: \qquad -F_1(3a) + F_2(2a) + P(a) = 0$$

With these, we can write equilibrium equations that give F_2 and F_3 in terms of the designated redundant force F_1 and the applied load. P.

$$\begin{array}{ll} F_2 = \tfrac{3}{2}F_1 - \tfrac{1}{2}P & \textbf{Equilibrium} \\ & \textbf{in Terms of the} \qquad (1) \\ F_3 = \tfrac{1}{2}F_1 + \tfrac{1}{2}P & \textbf{Redundant Force} \end{array}$$

Element Force-Temperature-Deformation Equations: Since we are to employ the Force Method, Eq. 3.14 is the convenient form to use.

$$\begin{array}{ll} e_1 = f_1 F_1 & \textbf{Element} \\ & \textbf{Force-} \\ e_2 = f_2 F_2 & \textbf{Deformation} \qquad \text{(2a–c)} \\ & \textbf{Behavior} \\ e_3 = f_3 F_3 & \end{array}$$

where

$$f_1 = \left(\frac{L}{AE}\right)_1 = \frac{(4.20\ \text{m})}{(1000\ \text{mm}^2)(70\ \text{GPa})} = 6.00 \times 10^{-2}\ \text{mm/kN}$$

$$f_2 = \left(\frac{L}{AE}\right)_2 = \frac{(2.10\ \text{m})}{(1500\ \text{mm}^2)(70\ \text{GPa})} = 2.00 \times 10^{-2}\ \text{mm/kN}$$

$$f_3 = \left(\frac{L}{AE}\right)_3 = \frac{(2.80\ \text{m})}{(1000\ \text{mm}^2)(70\ \text{GPa})} = 4.00 \times 10^{-2}\ \text{mm/kN}$$

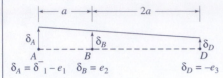

Fig. 3 Deformation diagram.

Geometry of Deformation: We need to sketch a **deformation diagram.** For simplicity, let us assume that beam AD displace upward throughout its entire length, and that it rotates clockwise. In Fig. 3 the three element elongations are related to the vertical displacements at their points of connection to the beam. (Recall that a positive e corresponds to a lengthening of the element.) We can now write a single compatibility equation that relates the elongations of the three elements to each other.

Without introducing the beam's rotation angle we can make use of the geometry of similar triangles in Fig. 3 and write

$$e_2 = \frac{1}{3}\delta_D + \frac{2}{3}\delta_A = \frac{1}{3}(-e_3) + \frac{2}{3}(\bar{\delta}_1 - e_1)$$

This deformation compatibility equation can be written in the form

$$2e_1 + 3e_2 + e_3 = 2\bar{\delta}_1 \qquad \textbf{Compatibility} \quad (3)$$

Solution of Equations by the Force Method: Now we use the Force-Method steps to combine Eqs. (1) through (3). First (Step 4), we substitute Eqs. (1) into Eqs. (2) to get the element force-displacement equations all written in terms of the one redundant force F_1.

$$e_1 = f_1 F_1$$
$$e_2 = f_2 F_2 = f_2(\tfrac{3}{2}F_1 - \tfrac{1}{2}P) \qquad (4)$$
$$e_3 = f_3 F_3 = f_3(\tfrac{1}{2}F_1 + \tfrac{1}{2}P)$$

Next (Step 5), these equations are substituted into the compatibility equation, Eq. (3), to obtain the following **compatibility equation in terms of the redundant force** F_1:

$$(4f_1 + 9f_2 + f_3)F_1 = (3f_2 - f_3)P + 4\bar{\delta}_1 \qquad \begin{array}{l}\textbf{Compatibility}\\ \textbf{in Terms of}\\ \textbf{the Redundant}\\ \textbf{Force}\end{array} \quad (5)$$

We have reduced six equations in six unknowns to <u>one equation in one unknown</u>, which we can easily solve for the **redundant force** (Step 6).

$$\boxed{F_1 = \frac{(3f_2 - f_3)P + 4\bar{\delta}_1}{4f_1 + 9f_2 + f_3}} \qquad (6)$$

Inserting numerical values into Eq. (6). we get

$$F_1 = \frac{[3(2 \times 10^{-2}\text{ mm/kN}) - (4 \times 10^{-2}\text{ mm/kN})](25\text{ kN}) + 4(1.0\text{ mm})}{4(6 \times 10^{-2}\text{ mm/kN}) + 9(2 \times 10^{-2}\text{ mm/kN}) + (4 \times 10^{-2}\text{ mm/kN})}$$

$$F_1 = 9.78\text{ kN} \qquad \textbf{Ans. (a)}$$

(b) *Determine the displacement at A.* From Fig. 3 and Eq. (2a),

$$\delta_A = \bar{\delta}_1 - e_1 = \bar{\delta}_1 - f_1 F_1 \qquad \begin{array}{l}\textbf{Geometry of}\\ \textbf{Deformation}\end{array} \quad (7)$$

Numerical values are inserted into Eq. (7) giving

$$\delta_A = 1.0 \text{ mm} - (6 \times 10^{-2} \text{ mm/kN})(9.78 \text{ kN})$$

$$\delta_A = 0.413 \text{ mm} \qquad \textbf{Ans. (b)}$$

Review the Solution From Eq. (6) it is clear that a tensile force is induced in rod (1) by the "misfit" δ_1. However, it is not so easy to see why the contribution of P is also tension. If the post at B were "rigid" rather than deformable (i.e., if $f_2 = 0$), the contribution of load P to F_1 should be compression, as it is in Eq. (6). If the rod at D were "rigid" rather than deformable (i.e., if $f_3 = 0$), the contribution of load P to F_1 should be tension, as it is in Eq. (6).

As an exercise, you can solve for F_2 and F_3 and show that the sum of the forces in the vertical direction is zero.

Notice that, compared with a solution by the Basic Force Method, where the number of equations to be solved simultaneously is equal to N_U, the total number of unknown forces, the Force Method described in this section makes that task easier by reducing the number of equations that must finally be solved simultaneously to N_R, the number of redundant forces.

*3.10 INTRODUCTION TO THE ANALYSIS OF PLANAR TRUSSES

Figure 3.22 shows a three-element planar truss. In this section we examine the fundamental equations for an inclined element that is pinned to a rigid base at one end and connected to another member or members at the other end, like the elements in the planar truss in Fig. 3.22.[13] We restrict our attention to uniform, linearly elastic elements.

Fundamental Equations. Figure 3.23 snows a typical uniform planar truss element that is assumed to be subjected to an axial force F and uniformly heated by an amount ΔT.[14] Let u be the horizontal displacement of the "free" end, positive in the $+x$ direction, and let v be the vertical displacement of the free end, positive in the $+y$ direction. Let F_x be the x component of force applied to the element at the free end, and F_y be the y component. Then, F is the total axial force (tension is assumed positive). The *angle θ is always taken positive counterclockwise and is measured from the x axis*, as shown in Fig. 3.23.

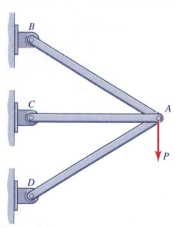

FIGURE 3.22 A Planar truss.

Equilibrium: When we show a free-body diagram of the joint to which the element in Fig. 3.23 is attached, we can show either the axial force F, or we can show the components F_x and F_y.

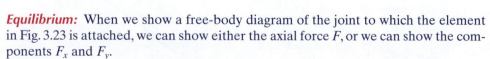

---

[13]This section exemplifies the systematic problem-solving procedure that is employed in *matrix structural analysis*. Since only elements that have one end pinned to is rigid base are considered here, the problems can be readily solved, even without the use of a computer.
[14]The temperature increment ΔT is measured from the *reference temperature,* the temperature at which the truss element is stress-free.

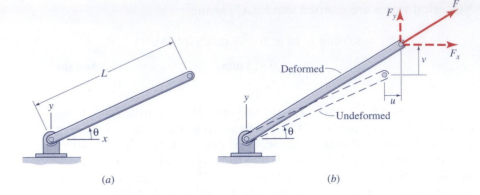

FIGURE 3.23 Planar truss element notation.

(a)

(b)

Element Force-Temperature-Deformation Behavior: Equations 3.27 and 3.28 relate the axial force F to the elongation of the element and to its (uniform) temperature change. That is,

$$e = fF + \alpha L \Delta T$$

(3.27) repeated

$$F = k(e - \alpha L \Delta T)$$

(3.28) repeated

Geometry of Deformation: Now we need to relate the elongation e to the joint's displacement components u and v. We will assume that u and v are very small in comparison with L. We need expression that relate e to u and v. Let us take u and v separately, as shown in Fig. 3.24.

Since u and v are assumed to be small, we can determine their contributions to e by projecting the displacements onto the original element direction, noting that the small triangles have a vertex angle θ.[15]

$$e = e_u + e_v = u \cos \theta + v \sin \theta$$

(3.29)

It is beyond the scope of this text to treat trusses with more than one node (joint), since this is, properly, a principal topic treated in courses in matrix structural analysis. We will, however, consider one-node trusses in order to show how easily the procedures used in Sections 3.4 and 3.5 through 3.9 to solve axial-deformation problems can be extended to the solution of planar truss problems.

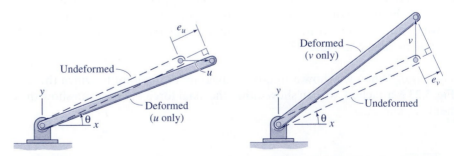

FIGURE 3.24 The contributions of displacements u and v to the elongation e.

---

[15]See Problem 3.10-26 for an exercise that illustrates the validity of this approximation.

Analysis of Simple Statically Determinate Trusses. If a one-node planar truss; has only two elements, it is statically determinate. The Force-Method Procedure outlined below is a very efficient way to solve this type of problem. If the one-node truss has more than two elements, like the three-element planar truss in Fig. 3.22, the Displacement-Method Procedure is preferable.

FORCE-METHOD PROCEDURE FOR STATICALLY DETERMINATE TRUSS PROBLEM

1. Draw, free-body diagram of the joint where the truss members are joined by a pin. Solve the two equilibrium equations for the unknown element forces. F_1 and F_2.

2. Use Eq. 3.27 to determine the elongations e_i due to forces F_i and temperature changes ΔT_i.

3. Write two geometry equations of the form given in Eq. 3.29 and solve these for the joint displacement components u and v.

In Example 3.20 we consider a statically determinate planar truss acted on by an external load and having one member cooled. In Example 3.21 we consider a statically indeterminate truss with one member that is not manufactured the correct length.

EXAMPLE 3.20

A (statically determinate) two-bar planar truss has the configuration shown in Fig. 1 when it is assembled. If a downward load $P = 20$ kN is applied to the pin at C, and, at the same time, element (1) is *cooled* by 20°C, (a) what are the stresses σ_1 and σ_2 in elements (1) and (2), respectively? (b) What are the horizontal and vertical displacements, u_C and v_C) respectively?

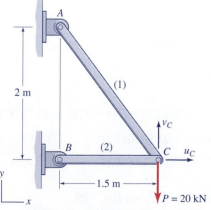

$A_1 = 1000$ mm^2, $E_1 = 200$ GPa, $\alpha_1 = 20(10^{-6})$/°C
$A_2 = 1000$ mm^2, $E_2 = 100$ GPa

Fig. 1

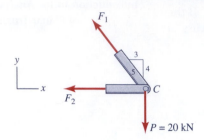

F_1

y

x

F_2

C

$P = 20$ kN

Fig. 2 Free-body diagram of joint C.

Solution (a) *Determine the element stresses σ_1 and σ_2.*

Equilibrium: We should use a free-body diagram of joint C shown in Fig. 2, summing forces in the x and y directions.

$\overset{+}{\rightarrow} \sum F_x = 0:$ $\qquad -\frac{3}{5}F_1 - F_2 = 0, \quad F_2 = -\frac{3}{5}F_1$

$+\uparrow \sum F_y = 0:$ $\qquad \frac{4}{5}F_1 - 20\,\text{kN} = 0, \quad F_1 = 25$ kN

$$F_1 = 25 \text{ kN}$$
$$F_2 = -15 \text{ kN}$$

Equilibrium (1)

Thus, the stresses are:

$$\sigma_1 = \frac{F_1}{A_1} = \frac{25\,000 \text{ N}}{0.001 \text{ m}^2} = 25 \text{ MPa}$$

$$\sigma_2 = \frac{F_2}{A_2} = \frac{-15\,000 \text{ N}}{0.001 \text{ m}^2} = -15 \text{ MPa}$$

$$\sigma_1 = 25 \text{ MPa (T)}, \quad \sigma_2 = 15 \text{ MPa (C)} \qquad \textbf{Ans. (a)}$$

(b) *Determine the joint displacement components u_C and v_C.*

Element Force-Temperature-Deformation Behavior: From Eq. 3.27,

$$e_i = f_i F_i + \alpha_i L_i \Delta T_i$$

Element Force-Temp.- Deformation Behavior (2)

where

$$f_1 = \left(\frac{L}{AE}\right)_1 = \frac{2.5 \text{ m}}{(0.001 \text{ m}^2)(200 \times 10^9\,\text{N/m}^2)} = 1.250(10^{-8})\,\text{m/N}$$

$$f_2 = \left(\frac{L}{AE}\right)_2 = \frac{1.5 \text{ m}}{(0.001 \text{ m}^2)(100 \times 10^9\,\text{N/m}^2)} = 1.500(10^{-8})\,\text{m/N}$$

Then, from Eq. (2),

$$e_1 = (1.250 \times 10^{-8}\,\text{m/N})(25\,000\,\text{N}) + (20 \times 10^{-6}/°\text{C})(2.5\,\text{m})(-20°\text{C})$$
$$= -0.000\,688 \text{ m} = -0.688 \text{ mm}$$

$$e_2 = (1.500 \times 10^{-8}\,\text{m/N})(-15\,000\,\text{N}) = -0.225 \text{ mm}$$

Geometry of Deformation: From Eq. 3.29,

$$e_i = u_C \cos\theta_i + v_C \sin\theta_i$$

Geometry of Deformation (3)

$$\cos \theta_1 = \frac{3}{5}, \sin \theta_1 = -\frac{4}{5}, \cos \theta_2 = 1, \sin \theta_2 = 0$$

$$e_1 = \frac{3}{5}u_C - \frac{4}{5}v_C = -0.688 \text{ mm}$$

$$e_2 = u_C = -0.225 \text{ mm}$$

Then,

$$u_C = -0.225 \text{ mm}, \quad v_C = 0.691 \text{ mm} \qquad \textbf{Ans. (b)}$$

Review the Solution   The displacement of joint C, upward and to the left, seems a bit strange, since the applied force P pulls downward. However, we must remember that since element (1) is cooled, it will tend to pull joint C upward and to the left, apparently overriding the downward displacement due to load P.

Note that Eq. (3), the geometry-of-deformation equation, is the only really "new" feature that distinguishes this truss problem from the axial-deformation problems treated in Section 3.5.

Analysis of Simple Statically Indeterminate Trusses.

If the truss is statically indeterminate, that is, if it has more than two members but only one joint, a Displacement-Method solution is the most straightforward. Of course, the Displacement Method could also be used to solve statically determinate truss problems.

DISPLACEMENT-METHOD PROCEDURE FOR TRUSS PROBLEMS

1. Draw a free-body diagram of the joint where the truss members are joined together by a pin. Write the two equilibrium equations in terms of the unknown element forces. F_i.

2. Write an element *force-temperature-deformation equation* for the each element Equation 3.28

$$F_i = k_i(e_i - \alpha L \Delta T_i), k_i = (AE/L)$$

   is the most convenient from to use for this step.

3. Use *geometry of deformation*, Eq. 3.29, to relate the element elongations to the joint displacement components u and v.

4. Substitute the deformation-geometry equations (Step 3) into the force-temperature-deformation equations

(Step 2). This gives *element forces in terms of system displacements*.

5. Substitute the results of Step 4 into the equilibrium equations of Step 1. This gives *equilibrium equations written in terms of system displacements*.

6. Solve the equations obtained in Step 5. The answer will be the *system displacements*.

7. Substitute the system displacements of Step 6 into the equations obtained in Step 4. The answer will be the *element forces*. If element normal stresses are required, the element forces can be divided by the respective element cross-sectional areas, that is $\sigma_i = F_i/A_i$.

The following example problem illustrates the solution of truss problems using the Displacement Method.

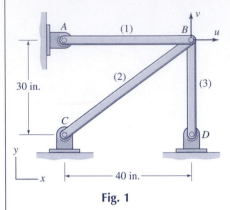

Fig. 1

A three-element truss has the configuration shown in Fig. 1. Each member has a cross-sectional area of 1.2 in², and all of them are made of aluminum, with $E = 10 \times 10^3$ ksi. When the truss members were fabricated, members (1) and (3) were manufactured correctly (i.e., correct lengths of $L_1 = 40$ in. and $L_3 = 30$ in.). However, member (2) has a distance between hole centers of $L_2 = 49.90$ in., rather than the correct value of 50.00 in. If member (2) is stretched so that a pin can be inserted to connect all three members together at B, and then the system is released, (a) what displacements, $u \equiv u_B$ and $v \equiv v_B$, will occur at node B? (b) What forces will be induced in the three members?

Plan the Solution   As always, we need to formulate equations for equilibrium, element force-deformation behavior, and geometry of deformation. We must incorporate the "misfit" condition in the deformation-compatibility equations.

Because member (2) will tend to return to its original length but will be restrained by members (1) and (3), we should expect member (2) to be in tension and the other two members to be in compression.

Solution   (a) *Use the Displacement Method to solve for the displacements u and v.*

Equilibrium: The node at B (Fig. 2) is acted on only by the element forces, since there is no external load. As usual, the axial forces in the elements are taken positive in tension.

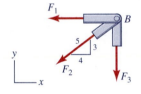

Fig. 2   Free-body diagram.

$$\overset{+}{\rightarrow} \sum F_x = 0: \qquad\qquad -F_1 - \frac{4}{5}F_2 = 0$$

$$+\uparrow \sum F_y = 0: \qquad\qquad -\frac{3}{5}F_2 - F_3 = 0$$

$$\boxed{\begin{aligned} F_1 + \frac{4}{5}F_2 &= 0 \\[2mm] \frac{3}{5}F_2 + F_3 &= 0 \end{aligned}} \qquad\qquad \textbf{Equilibrium} \qquad (1)$$

Element Force-Deformation Behavior: We have three elements, so we need to write Eq. 3.28 for each of these elements. It will be helpful to compile a table of element properties first.

$$k_1 = \left(\frac{AE}{L}\right)_1 = \frac{(1.2 \text{ in}^2)(10 \times 10^3 \text{ ksi})}{(40 \text{ in.})} = 300 \text{ kips/in.}$$

$$k_2 = \left(\frac{AE}{L}\right)_2 = \frac{(1.2 \text{ in}^2)(10 \times 10^3 \text{ ksi})}{(50 \text{ in.})} = 240 \text{ kips/in.}$$

$$k_3 = \left(\frac{AE}{L}\right)_3 = \frac{(1.2 \text{ in}^2)(10 \times 10^3 \text{ ksi})}{(30 \text{ in.})} = 400 \text{ kips/in.}$$

194

Table of Element Properties

Element, i	L_i (in.)	k_i (kips/in.)	$\cos \theta_i$	$\sin \theta_i$
1	40	300	1.0	0.0
2	50	240	0.8	0.6
3	30	400	0.0	1.0

Then, from Eq. 3.28,

$$F_1 = k_1 e_1 = 300 e_1$$
$$F_2 = k_2 e_2 = 240 e_2$$
$$F_3 = k_3 e_3 = 400 e_3$$

Element Force-Deformation Behavior (2)

Geometry of Deformation: We sketch a *deformation diagram* (Fig. 3) so that we can see how to relate the element elongations to the system displacements u and v. We greatly exaggerate the gap between the end of element (2) and node B, where it is supposed to connect to elements (1) and (3).

First, we can use Eq. 3.29 to express the elongation of each element in terms of the displacement at its "free" end, that is, end B.

$$e_i = u_i \cos \theta_i + v_i \sin \theta_i \quad i = 1, 2, 3$$

However, we will have to modify the equation for element (2) to account for the "misfit." Since all elements are pinned together at joint B,

$$u_1 = u_2 = u_3 \equiv u, \quad v_1 = v_2 = v_3 \equiv v$$

Combining these equations, and referring to the Table of Element Properties, we get

$$e_1 = u \cos \theta_1 + v \sin \theta_1 = u$$
$$e_2 = u \cos \theta_2 + v \sin \theta_2 + \bar{\delta}$$
$$\quad = 0.8u + 0.6v + \bar{\delta}$$
$$e_3 = u \cos \theta_3 + v \sin \theta_3 = v$$

Deformation Compatibility (3)

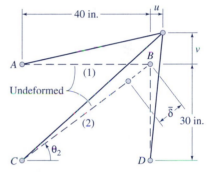

Fig. 3 Deformation diagram.

Displacement-Method Solution: All we need to do is combine Eqs. (1) through (3) in the Displacement-Method order: (3) → (2) → (1).

$$F_1 = 300u, \quad F_2 = 192u + 144v + 24.00, \quad F_3 = 400v \quad (4)$$

195

Then, from Eqs. (1) and (4), we have

$$300u + 0.8(192u + 144v + 24.00) = 0$$

$$0.6(192u + 144v + 24.00) + 400v = 0$$

or

$$453.6u + 115.2v = -19.20$$

$$115.2u + 486.4v = -14.40$$

Nodal Equilibrium in Terms of Nodal Displacement   (5)

Solving these simultaneous algebraic equations, we get

$$u = -3.70(10^{-2}) \text{ in.}$$

$$v = -2.08(10^{-2}) \text{ in.}$$

Ans. (a)   (6)

(b) *Solve for the element forces.* The displacements u and v can be substituted into Eqs. (4) to give

$$F_1 = 300u = -11.11 \text{ kips}$$

$$F_2 = 192u + 144v + 24.00 = 13.89 \text{ kips}$$

$$F_3 = 400v = -8.33 \text{ kips}$$

$$\left.\begin{array}{l} F_1 = 11.11 \text{ kips (C)} \\ F_2 = 13.89 \text{ kips (T)} \\ F_3 = 8.33 \text{ kips (C)} \end{array}\right\}$$ **Ans. (b)**   (7)

Review the Solution   One way to verify our results is to check equilibrium (using full calculator precision, not the reported rounded values).

$$\text{Is } F_1 + 0.8F_2 = 0? \quad \text{Yes.}$$

$$\text{Is } 0.6F_2 + F_3 = 0? \quad \text{Yes.}$$

Also, we note that all of the element forces have the signs that we expected them to have, that is, member (2) is in tension, while members (1) and (3) are in compression.

From the two truss examples of this section we can conclude that, by establishing a geometric relationship between the elongation of an element and the Cartesian components of displacement at its "free" end (Fig. 3.24), we can solve planar truss problems involving elements attached to "ground" at one end by exactly the same procedures used in Sections 3.4 through 3.9.

*3.11 INELASTIC AXIAL DEFORMATION

Up to this point in our analysis of axial deformation we have considered only linearly elastic behavior. It is important, however, that we also study the inelastic behavior of deformable bodies whose stress has exceeded the proportional limit. If properly designed, structures and machines do not completely fail when the stress at one or more points reaches the proportional limit, or yield point. This is true for statically indeterminate structures, as you will discover in the elastic-plastic analysis that follows. In many cases present design codes acknowledge the fact that "failure" does not necessarily correspond to the first yielding of a structural element or machine component.

Fundamental Equations. Of the three fundamental concepts of deformable-body mechanics—*equilibrium, geometry of deformation*, and *material behavior*—**only material behavior is treated in a manner different from that in the previous analysis of linearly elastic behavior.** Let as consider the axially loaded element in Fig. 3.25a. Assume that it is homogeneous (i.e., it has the same material properties throughout), and that it may be stressed beyond the proportional limit of the material.

(a) A slender member subjected to axial loading.

Geometry of Deformation: For inelastic axial deformation, just as for linearly elastic axial deformation, we assume that *the axis of the member remains straight and that plane sections remain plane.* Therefore, Eq. 3.1 holds, so the strain is independent of position in the cross section, and the *strain-displacement equation* for axial deformation is

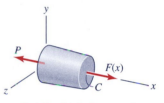

(b) A free-body diagram of a portion of the member.

$$\epsilon = \frac{du}{dx}$$

(3.1) repeated

FIGURE 3.25 An element undergoing axial deformation.

The *elongation* of the axial-deformation member is obtained by integrating Eq. 3.1, that is, by summing the elongations of infinitesimal fibers of length dx.

$$e = u(L) - u(0) = \int_0^L \epsilon(x)\, dx$$

(3.2, 3.3) repeated

Equilibrium: The extensional strain may vary with x, but it is constant at any cross section. Since the material is assumed to be homogeneous, the stress will also be a constant on the cross section. Then Eqs. 3.7 apply:

$$F = \int_A \sigma\, dA = \sigma A$$

$$M_y = \int_A z\sigma\, dA = \bar{z}\sigma A = \bar{z}F = 0$$

(3.7) repeated

$$M_z = -\int_A y\sigma\, dA = -\bar{y}\sigma A = -\bar{y}F = 0$$

Therefore, axial deformation occurs when the axial force $F(x)$ acts through the centroid of the cross section of the member, the loading situation shown in

Fig. 3.25. The axial stress is related to the (internal) axial force $F(x)$ by

$$\sigma = \frac{F(x)}{A(x)} \qquad (3.30)$$

just as in the case of a homogeneous, linearly elastic axial-deformation member.

Material Behavior: Figure 3.26 depicts three typical stress-strain curves that exhibit inelastic behavior beyond the proportional limit, σ_{PL}. Figure 3.26a is typical of metals like aluminum; Fig. 3.26b is an idealization of the behavior of a material with a definite yield point, like mild steel; and Fig. 3.26c is an idealization of a material that exhibits strain hardening after it yields. In each of the three cases illustrated in Fig. 3.26 the material is linearly elastic (and, therefore, it obeys Hooke's Law) up to the proportional limit. And, in each case, if unloading occurs from a point on the stress-strain diagram above the yield stress, the unloading will proceed along a path that is parallel to the original linearly elastic portion of the σ versus ϵ curve, as illustrated in Fig. 3.26.

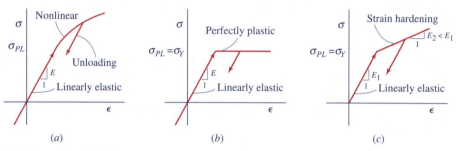

(a)   *(b)*   *(c)*

FIGURE 3.26 Three types of nonlinear stress-strain behavior: (a) elastic, nonlinearly plastic, (b) elastic, perfectly plastic, and (c) elastic, strain-hardening.

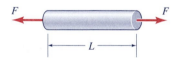

FIGURE 3.27 A prismatic bar with axial loading.

Uniform End-Loaded Element. If the element is prismatic and is loaded only by axial forces at its ends, as in Fig. 3.27 then

$$\sigma = \frac{F}{A} = \text{const}$$

$$e = \epsilon(\sigma)L \qquad (3.31)$$

where $\epsilon(\sigma)$ is the strain that corresponds to the stress σ.

Elastic-Plastic Analysis of Statically Indeterminate Structures. An important illustration of inelastic axial deformation is the behavior of it statically indeterminate system whose elements exhibit **elastic, perfectly plastic** behavior, as represented by the stress-strain diagram in Fig. 3.26b. Consider the statically indeterminate rod system in Fig. 3.28a. Let both members be made of a material whose stress-strain curve exhibits the elastic, perfectly plastic behavior shown in Fig. 3.28b. (It is assumed that the compressive yield stress has the same magnitude, σ_Y, as the tensile yield stress.) We wish to determine a load-displacement curve that relates the force P at B to the displacement u of point B in Fig. 3.28a.

We handle equilibrium and geometry of deformation just as in the previous analysis of linearly elastic axial deformation.

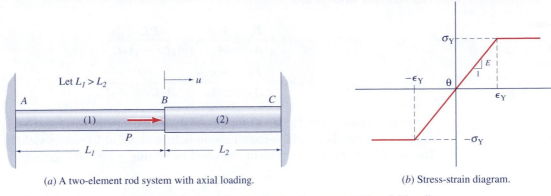

Let $L_1 > L_2$

(a) A two-element rod system with axial loading.

(b) Stress-strain diagram.

FIGURE 3.28   An elastic-plastic system with axial loading.

Equilibrium: From a free-body diagram of the node (joint) at B, we get

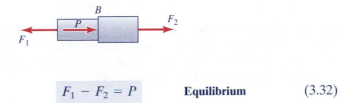

$$F_1 - F_2 = P \qquad \textbf{Equilibrium} \qquad (3.32)$$

Geometry of Deformation: By examining Fig. 3.28a, we see that element 1 elongates an amount u and element 2 shortens the same amount u when node B moves to the right by an amount u. Therefore, from deformation compatibility, we have

$$e_1 = u, \quad e_2 = -u \qquad \begin{array}{l}\textbf{Geometric}\\\textbf{Compatibility}\end{array} \qquad (3.33)$$

Material Behavior: Now let us consider the linear and nonlinear material behavior that occurs <u>as the load P is increased</u>. There are three cases: Case 1—both elements are linearly elastic; Case 2—one member has yielded, but the other is still linearly elastic; and Case 3—both elements have yielded.

Case 1—Both elements are linearly elastic. From Eq. 3.15 of Section 3.3,

$$F_i = k_i e_i = \left(\frac{A_i E_i}{L_i}\right) e_i \qquad \begin{array}{l}\textbf{Material}\\\textbf{Behavior—}\\\textbf{Case 1}\end{array} \qquad (3.34)$$

Combining Eqs. 3.32 through 3.34 in *Displacement-Method* fashion, we get

$$(k_1 + k_2)u = P \rightarrow \boxed{\; u = \frac{P}{\left(\dfrac{AE}{L}\right)_1 + \left(\dfrac{AE}{L}\right)_2} \;} \qquad (a)$$

The element stresses are

$$\sigma_1 = \frac{F_1}{A_1} = \frac{k_1 u}{A_1} = \frac{PL_2}{A_1 L_2 + A_2 L_1}$$

$$\sigma_2 = \frac{F_2}{A_2} = \frac{-k_2 u}{A_2} = \frac{-PL_1}{A_1 L_2 + A_2 L_1}$$

(b)

It was assumed in Fig. 3.28a that $L_1 > L_2$ Therefore, from Eqs. (b), $|\sigma_2| > |\sigma_1|$ so element 2 reaches yield in compression before element 1 reaches yield in tension The *yield load*, P_Y, is therefore determined by setting $\sigma_2 = -\sigma_Y$, giving

$$P_Y = \sigma_Y \left(\frac{A_1 L_2 + A_2 L_1}{L_1} \right)$$ **Yield Load**   (3.35)

We can also use the second of Eqs. (b) to obtain

$$u_Y = \frac{\sigma_Y L_2}{E}$$

(c)

Case 2—One element has yielded. Since element 2 yields first, for this case we must replace the linear force-elongation equation of element 2 to reflect a constant yield stress of $(-\sigma_Y)$. Element 1, however, is still linearly elastic, so the material behavior for Case 2 is represented by the equations

$$F_1 = k_1 e_1$$ **Material**
$$F_2 = -\sigma_Y A_2$$ **Behavior**   (3.36)
Case 2

Combining Eqs. 3.32 (equilibrium), 3.33 (geometric compatibility), and 3.36 (material behavior) in Displacement-Method fashion, we get

$$k_1 u = P + F_2 = P - \sigma_Y A_2$$

or

$$u = \frac{(P - \sigma_Y A_2) L_1}{A_1 E}$$

(d)

This phase of loading behavior extends from $P = P_Y$ to $P = P_P$, the *plastic load*, at which rod 1 also yields. Line segments (1) and (2) in Fig. 3.29 are defined by Eqs. (a) and (d), respectively.

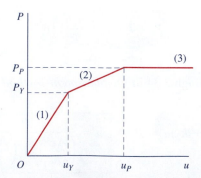

FIGURE 3.29 A load-displacement diagram for elastic- plastic axial deformation.

Case 3—Both elements have yielded. In this case both elements have yielded, and the appropriate material behavior equations (replacing Eqs. 3.34 and 3.36) are:

$$F_1 = \sigma_Y A_1$$
$$F_2 = -\sigma_Y A_2$$

Material Behavior— Case 3 (3.37)

Combining these with the equilibrium equation, Eq. 3.32, we get the value of the *plastic load* P_P.

$$P_P = \sigma_Y(A_1 + A_2)$$

Plastic Load (3.38)

This plastic load is reached when member 1 also yields, so equating the strain in member 1 to σ_Y/E, we get

$$\epsilon_{1P} = \frac{u_P}{L_1} = \frac{\sigma_Y}{E} \tag{e}$$

or

$$u_P = \frac{\sigma_Y L_1}{E} \tag{f}$$

Note that

$$\frac{P_P}{P_Y} = \frac{A_1 L_1 + A_2 L_1}{A_1 L_2 + A_2 L_1}, \quad \frac{u_P}{u_Y} = \frac{L_1}{L_2} \tag{g,h}$$

The above elastic-plastic analysis is summarized on the load-displacement diagram in Fig. 3.29, with the ranges of applicability of the three cases indicated on the plot.

EXAMPLE 3.22

The two support rods in Fig. 1 are made of structural steel that may be assumed to have a stress-strain diagram like Fig. 3.28b, with $E = 30(10^3)$

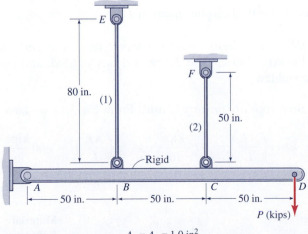

$A_1 = A_2 = 1.0$ in^2

Fig. 1

ksi and σ_Y = 36 ksi. (a) Construct a load-displacement diagram that relates the load P to the vertical displacement at D. Make the usual small-angle assumption for the rotation of the "rigid" beam AD. (b) Determine the allowable load if the factor of safety (see Eq. 2.26) for first yielding is FS_Y = 2.0, and the factor of safety for fully-plastic-behavior is FS_P = 2.5.

Plan the Solution The equilibrium analysis and the geometry-of-deformation analysis will be the same as in Example Problem 3.8. As in the above discussion of elastic-plastic behavior, the material behavior here falls under three cases—(1) both support rods are linearly elastic, (2) one rod has yielded, and (3) both rods have yielded. The load-displacement diagram should resemble Fig. 3.29. The allowable load is determined by applying the factors of safety as defined in Eq. 2.26, with the allowable load being the lower of the two loads $(P_{\text{allow}})_Y$ and $(P_{\text{allow}})_P$.

Solution (a) *Construct a load-displacement diagram relating load P to the vertical displacement at D.*

Equilibrium: For the free-body diagram in Fig. 2, the equation of equilibrium is

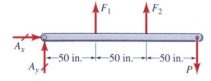
Fig. 2 Free-body diagram.

$$\left(\sum M\right)_A = 0: \qquad \boxed{F_1 + 2F_2 = 3P} \qquad \textbf{Equilibrium} \qquad (1)$$

Geometry of Deformation: Figure 3 is a deformation diagram for small rotation of the rigid beam AD. From it, the equations of geometric compatibility can be written as

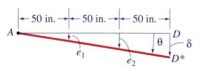
Fig. 3 Deformation diagram.

$$\boxed{\begin{array}{c} e_1 = \dfrac{1}{3}\delta \\[2mm] e_2 = \dfrac{2}{3}\delta \end{array}} \qquad \begin{array}{l}\textbf{Geometric}\\ \textbf{Compatibility}\end{array} \qquad (2)$$

where δ is the vertical displacement at D.

Material Behavior: There are three cases to be considered: Case 1, both rods are linearly elastic; Case 2, one rod has yielded; and Case 3, both rods have yielded.

Case 1—Both rods are linearly elastic. From Eq. 3.15 we have

$$k_1 = \left(\frac{AE}{L}\right)_1 = 375\,\frac{\text{kips}}{\text{in.}}, \quad k_2 = \left(\frac{AE}{L}\right)_2 = 600\,\frac{\text{kips}}{\text{in.}}$$

and

$$\boxed{\begin{array}{l} F_1 = k_1 e_1 = 375 e_1 \\[2mm] F_2 = k_2 e_2 = 600 e_2 \end{array}} \qquad \begin{array}{l}\textbf{Material}\\ \textbf{Behavior—}\\ \textbf{Case 1}\end{array} \qquad (3)$$

202

Equations (1) through (3) may be combined to give the following load-displacement equation for Case 1:

$$\delta = \frac{3}{925}P, \quad \text{or} \quad P = \frac{925}{3}\delta \qquad \text{Ans. (a)} \quad (4)$$

The corresponding stresses in the two rods are obtained by combining Eqs. (2) through (4). Thus,

$$\sigma_1 = \frac{F_1}{A_1} = \frac{15}{37}P, \quad \sigma_2 = \frac{F_2}{A_2} = \frac{48}{37}P \qquad (5)$$

Since $\sigma_2 > \sigma_1$, rod 2 will yield before rod 1 does. Therefore, the yield load P_Y is given by setting $\sigma_2 = \sigma_Y = 36$ ksi in Eq. (5b), the equation for σ_2. This gives

$$P_Y = 27.8 \text{ kips} \qquad \text{Ans. (a)} \quad (6)$$

The corresponding displacement at which first yield occurs is obtained by substituting Eq. (6) into Eq. (4a), the equation for δ, giving

$$\delta_Y = \frac{3}{925}(27.8) = 0.090 \text{ in.} \qquad \text{Ans. (a)} \quad (7)$$

The load-displacement formula for loads up to load P_Y is given by Eqs. (4).

Case 2—One rod has yielded; one is linearly elastic. After rod 2 yields and up until the load at which rod 1 yields, the material-behavior equations for rods 1 and 2 are:

$$F_1 = 375e_1 \text{ kips} \qquad \textbf{Material}$$
$$\textbf{Behavior—} \quad (8)$$
$$F_2 = \sigma_Y A_2 = 36 \text{ kips} \qquad \textbf{Case 2}$$

Combining Eqs. (1) (equilibrium equation), (2a) (deformation-geometry equation for e_1), and (8) (force-deformation equations), we get

$$375\left(\frac{1}{3}\delta\right) + 2\left(36\right) = 3P$$

or

$$P = \left(\frac{125}{3}\delta + 24\right)\text{kips} \qquad \text{Ans. (a)} \quad (9)$$

This formula characterizes the load-displacement behavior from the yield load P_Y up to the fully plastic load P_P, which corresponds to the yielding of rod 1.

Case 3—Both rods have yielded. Once rod 1 also yields, the material behavior equations become

$$F_1 = \sigma_Y A_1 = 36 \text{ kips} \qquad \textbf{Material} \qquad (10)$$
$$F_2 = \sigma_Y A_2 = 36 \text{ kips} \qquad \textbf{Behavior—Case 3}$$

These rod forces can be substituted into the equilibrium equation. Eq. (1), to give the plastic load P_P.

$$P_P = \frac{F_1 + 2F_2}{3} = \frac{36 + 2(36)}{3}$$

or

$$P_P = 36 \text{ kips} \qquad \textbf{Ans. (a)} \quad (11)$$

The corresponding displacement δ_P is obtained by substituting Eq. (11) into Eq. (9) to obtain

$$\delta_P = \frac{3}{125}(12) = 0.288 \text{ in.} \qquad \textbf{Ans. (a)} \quad (12)$$

We can now use Eqs. (4), (6), (7), (9), (11), and (12) to plot the load-displacement diagram, Fig. 4.

(b) *Determine the allowable load.* Based on first yield, the allowable load is given by

$$(P_{\text{allow}})_Y = \frac{P_Y}{FS_Y} \qquad (13a)$$

$$(P_{\text{allow}})_Y = \frac{27.75 \text{ kips}}{2.0} = 13.875 \text{ kips} \qquad (13b)$$

Based on the fully-plastic load, the allowable load is given by

$$(P_{\text{allow}})_P = \frac{P_P}{FS_P} \qquad (14a)$$

So,

$$(P_{\text{allow}})_P = \frac{36 \text{ kips}}{2.5} = 14.40 \text{ kips} \qquad (14b)$$

In this case, the *allowable load* is determined by the first-yield criterion. Therefore,

$$P_{\text{allow}} = 13.88 \text{ kips} \qquad \textbf{Ans. (b)} \quad (15)$$

Review the Solution The key results in the preceding analysis are the "break points," (P_Y, δ_Y) and (P_P, δ_P), in the load-displacement diagram.

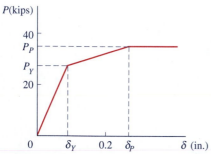

Fig. 4 A load-displacement diagram.

Let us check the first of these, starting with $\delta_Y = 0.090$ in. The compatibility equations, Eqs. (2) are easily checked by referring to Fig. 3, and these give

$$e_{1Y} = 0.030 \text{ in.,} \qquad e_{2Y} = 0.060 \text{ in.}$$

The strain $\epsilon_{2Y} = e_{2Y}/L_2 = 0.0012$ is in agreement with the yield strain σ_Y/E, so we have the correct elongations. We can get the rod forces by using Eqs. (3). Thus, $F_{1Y} = 11.25$ kips and $F_{2Y} = 36$ kips. When these forces are substituted into the equilibrium equation, Eq. (1), we do get $P_Y = 27.75$ kips, so our solution appears to be correct.

Residual Stress.

Let us now see what happens when the load is removed after yielding has occurred in one or more elements in an assemblage of axial deformation members made of an elastic-plastic material represented by the stress-strain curve in Fig. 3.30.

Statically Determinate Case: Consider a single axial-deformation member that has been stretched to a strain of ϵ_B. Upon removal of the load, the member will undergo *linearly elastic recovery* along the straight line BD to point C, where there is a residual strain, or *permanent set*, which is given by Fig. 3.30.

$$\epsilon_{PS} = \epsilon_B - \epsilon_Y$$

There is, however, no residual stress, because the load can be completely removed from the member, since the member is statically determinate. A similar situation would hold for a statically determinate assemblage of two or more members—there would be some permanent set, but there would not be any residual stress.

Statically Indeterminate Case: If the assemblage of axial-deformation members is statically indeterminate, it is possible to have self-equilibrating internal forces after all external loads have been removed. The corresponding member stresses are called **residual stresses,** because they are stresses that result from some prior loading, but they remain after removal of all external loads. The next example illustrates how residual stresses may result from inelastic material behavior.

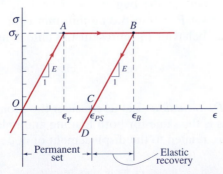

FIGURE 3.30   Loading and unloading paths for an elastic-plastic member.

Let us take a special case of the statically indeterminate system in Fig. 3.28. This special case is shown in Fig. 1a. If the load P is increased until $\epsilon_2 = -1.5\,\epsilon_Y$ and is then removed, what residual stresses will be left in the two elements? What is the permanent deformation? Let $A_1 = A_2 = A$, $L_1 = 2L_2 = 2L$, and let both elements be made of the elastic-plastic material depicted in Fig. 1b.

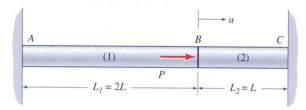

Fig. 1a A statically indeterminate assemblage.

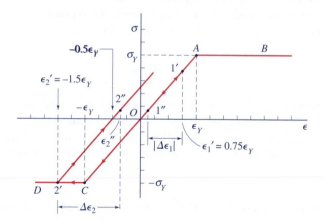

Fig. 1b Elastic-plastic material behavior.

Plan the Solution We can divide the analysis into a loading phase and an unloading phase. The analysis of the loading phase can be taken directly from the discussion of elastic-plastic analysis of statically indeterminate members at the beginning of Sect. 3.11. For the unloading phase we must set $P = 0$ in the equilibrium equation and use linearly elastic material behavior to represent the unloading $\sigma - \epsilon$ paths of the two members.

Solution

Geometry of Deformation: Let us consider the geometry of deformation first, since this is the same for both loading and unloading. The member elongations are related to the displacement of node B by

$$e_1 = u, \quad e_2 = -u \tag{1}$$

Since the strain is uniform along each member, the geometric compatibility equations can be written in terms of strains as

$$\epsilon_1 = \frac{e_1}{2L} = \frac{u}{2L}$$
$$\epsilon_2 = \frac{e_2}{L} = \frac{-u}{L}$$

Geometric Compatibility (2)

These strain equations will prove to be more convenient than Eqs. (1), since we have to carefully examine the loading and unloading paths on the stress-strain diagram.

Equilibrium: Equilibrium of node B (Fig. 2) gives

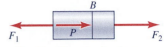

Fig. 2 Free-body diagram.

$$F_1 - F_2 = P$$

Equilibrium (3)

(At the end of the unloading phase, $P = 0$.)

Loading Phase: You should follow the loading and unloading strain paths on Fig. 1b. We are given that the load is increased until load P produces a strain $\epsilon_2' = -1.5\epsilon_Y$. For member 2 this corresponds to a path $O \rightarrow C \rightarrow 2'$ in Fig. 1b. Combining this information with Eq. (2b) we get

$$\epsilon_2' = -1.5\epsilon_Y = \frac{-u'}{L}$$

(4)

where the primes denote quantities when the load has been increased until it produces the given strain in member 2. Point 2' on Fig. 1b characterizes the state for member 2. From Eq. (4),

$$u' = 1.5\epsilon_Y L$$

(5)

Combining Eqs. (2a) and (5) gives

$$\epsilon_1' = \frac{1.5\epsilon_Y L}{2L} = 0.75\epsilon_Y$$

(6)

Therefore, member 1 is still elastic at the point designated 1' on Fig. 1b.

Material Behavior—Loading Phase: Since $|\epsilon_2'| > \epsilon_Y$ and $|\epsilon_1'| < \epsilon_Y$ the appropriate material-behavior equations are given by Case 2 of the earlier elastic-plastic analysis, namely

$$F_1 = A\sigma_1 = AE\epsilon_1$$
$$F_2 = A\sigma_2 = -\sigma_Y A$$

Material Behavior— Loading (7)

To determine the external load P' that corresponds to the given state (i.e., that makes $\epsilon_2' = -1.5\epsilon_Y$), we can combine Eqs. (7), (2a), and (5) to obtain the member forces

$$F_1' = \frac{AE}{2L}(1.5\epsilon_Y L) = 0.75\sigma_Y A$$

$$F_2' = -\sigma_Y A \tag{8}$$

These may be substituted into the equilibrium equation, Eq. (3), giving the external load

$$P' = 1.75\sigma_Y A \tag{9}$$

From Eq. (5) the corresponding displacement of node B is

$$u' = 1.5\frac{\sigma_Y L}{E} \tag{10}$$

Unloading Phase: When the load is removed, members 1 and 2 unload to points $1''$ and $2''$, respectively, on Fig. 1b. Since $P'' = 0$, the equilibrium equation, Eq. (1), becomes

$$F_1'' - F_2'' = 0 \tag{11}$$

The geometry of deformation is still described by Eqs. (1) and (2), but the material-behavior equations must now correspond to the unloading paths indicated in Fig. 1b. Point $1''$ lies on the original linear portion, AC, of the $\sigma - \epsilon$ diagram, but point $2''$ lies along the unloading path $2' - 2''$ that intersects the ϵ axis at $\epsilon_2' + \epsilon_Y = -0.5\epsilon_Y$. Hence, along these two respective paths,

$$
\begin{aligned}
F_1 &= A\sigma_1 = AE\epsilon_1 \\
F_2 &= A\sigma_2 = AE(\epsilon_2 + 0.5\epsilon_Y)
\end{aligned}
\qquad
\begin{aligned}
&\textbf{Material} \\
&\textbf{Behavior—} \\
&\textbf{Unloading}
\end{aligned}
\tag{12}
$$

Substituting Eqs. (2) into Eqs. (12), we get the following expressions for the internal forces:

$$F_1 = AE\left(\frac{u}{2L}\right)$$

$$F_2 = AE\left(-\frac{u}{L} + \frac{\sigma_Y}{2E}\right) \tag{13}$$

The *permanent displacement of point B, u''*, at which these forces satisfy the equilibrium equation, Eq. (11), is

$$u'' = \frac{1}{3}\frac{\sigma_Y L}{E} \qquad \textbf{Ans.} \tag{14}$$

208

with corresponding element forces

$$F_1'' = F_2'' = \frac{1}{6}\sigma_Y A \qquad (15)$$

Therefore, the two members are left with a *residual stress* of

$$\sigma_1'' = \sigma_2'' = \frac{1}{6}\sigma_Y \qquad \textbf{Ans.} \quad (16)$$

Review the Solution A key equation is the geometric-compatibility equation, Eq. (2). It tells us that for any increment of displacement Δu, $\Delta\epsilon_1 = \Delta u/2L$ and $\Delta\epsilon_2 = -\Delta u/L$. Therefore, we can easily check points $2'$ and $1'$ on Fig. 1b. From the corresponding stresses, $\sigma_1 = 3/4\sigma_Y$ and $\sigma_2 = -\sigma_Y$, and the nodal equilibrium equation we can verify that the force P' given in Eq. (9) is correct.

Next, to determine the residual stresses we know that $F_1'' = F_2''$, so $\sigma_1'' = \sigma_2''$. We also know that $\Delta\epsilon_1 = -\frac{1}{2}\Delta\epsilon_2$. We see in Fig. 1b that points $1''$ and $2''$ have equal stress and that the strain increments, $\Delta\epsilon_1 \equiv (\epsilon_{1''} - \epsilon_{1'})$ and $\Delta\epsilon_2 \equiv (\epsilon_{2''} - \epsilon_{2'})$, satisfy this strain-increment equation. Therefore, our residual stress answers in Eqs. (16) appear to be correct.

Following the elastic-plastic loading and unloading behavior of a statically indeterminate structure, as in the above analysis, is obviously not a very easy task. However, note that by considering separately the roles of *equilibrium*, of *geometry of deformation*, and of *material behavior*, we have been able to make the analysis both straightforward and tractable.

3.12 PROBLEMS

▼ AXIAL DEFORMATION—NONHOMOGENEOUS CROSS SECTION

Prob. 3.3-1. The bimetallic bar in Fig. 1 of Example 3.1 has length L. Using results from Example 3.1, develop an expression for the total elongation, e, of the bimetallic bar due to the axial force P.

Prob. 3.3-2. A bimetallic bar is made by bonding together two homogeneous rectangular bars, each having a width b and length L. The moduli of elasticity of the bars are E_1 and E_2, respectively. An axial force P is applied to the ends of the bimetallic bar at $(y = y_P, z = 0)$ such that the bar undergoes axial deformation. Let $L = 1.5$ m, $b = 50$ mm, $t_1 = 25$ mm, $t_2 = 15$ mm, $E_1 = 70$ GPa, $E_2 = 210$ GPa, and $P = 48$ kN. (a) Determine the normal stress in each material, that is, determine σ_1 and σ_2. (b) Determine the value of y_P. (c) Determine the total elongation, e, of the bar.

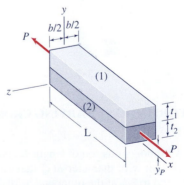

P3.3-2 and P3.3-3

Prob. 3.3-3. Solve Prob. 3.3-2 using the following dimensions, material constants, and load: $L = 20$ in., $b = 2$ in.,

$t_1 = 1.0$ in., $t_2 = 0.5$ in., $E_1 = 10 \times 10^3$ ksi, $E_2 = 30 \times 10^3$ ksi, and $P = 15$ kips.

Prob. 3.3-4. A steel pipe ($E_1 = 200$ GPa) surrounds a solid aluminum-alloy rod ($E_2 = 70$ GPa), and together they are subjected to a compressive force of 200 kN acting on rigid end caps. Determine the shortening of this bimetallic compression member, and determine the normal stresses in the pipe and in the rod.

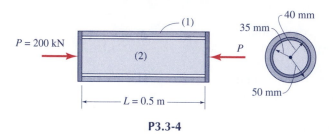

P3.3-4

Prob. 3.3-5. A steel pipe is filled with concrete, and the resulting column in subjected to a compressive load $P = 80$ kips. The pipe has an outer diameter of 12.75 in. and an inside diameter of 12.00 in. The elastic moduli of the steel and concrete are: $E_s = 30 \times 10^3$ ksi and $E_c = 3.6 \times 10^3$ ksi. (a) Determine the stress in the steel and the stress in the concrete due to this loading. (b) If the initial length of the column is $L = 12$ ft, how much does the column shorten when the load is applied? (Ignore radial expansion of the concrete and steel due to Poisson's ratio effect.)

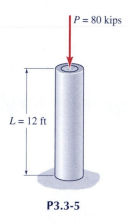

P3.3-5

▼ **AXIAL DEFORMATION—VARYING CROSS SECTION**

Prob. 3.3-6. A homogenous rod of length L and modulus E is a conical frustum with diameter $d(x)$ that varies linearly from d_0 at one end to $2d_0$ at the other end, with $d_0 \ll L$. An axial load P is applied to the rod, as shown in Fig. P3.3-6. (a) Determine an expression for the stress distribution, $\sigma(x)$, on an arbitrary cross section. (b) Determine an expression for the elongation of the rod.

210

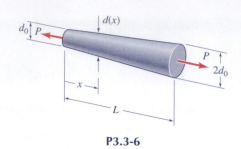

P3.3-6

Prob. 3.3-7. A flat bar with rectangular cross section has a constant thickness t and an unstretched length L. The width of the bar varies linearly from b_1 at one end to b_2 at the other end, and its modulus of elasticity is E. (a) Derive a formula for the elongation, e, of the bar when it is subjected to an axial tensile load P, as shown in Fig. P3.3-7. (b) Calculate the elongation for the following case: $b_1 = 50$ mm, $b_2 = 100$ mm, $t = 25$ mm, $L = 2$ m, $P = 250$ kN, $E = 70$ GPa.

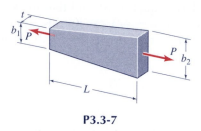

P3.3-7

Prob. 3.3-8. An axial load P is applied to a tapered rod, as shown in Fig. P3.3-8. The radius of the rod is given by

$$r(x) = \left[\frac{r_0}{1 + (x/L)} \right]$$

where $r_0 \equiv r(0)$. (a) Determine a symbolic expression for the elongation of this tapered bar in terms of the parameters P, L r_0, and E, and (b) solve for the elongation in inches if $P = 2$ kips, $L = 100$ in., $r_0 = 20$ in., and the rod is made of an aluminum alloy for which $E = 10 \times 10^3$ ksi.

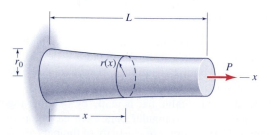

P3.3-8

Prob. 3.3-9. The tapered solid stone pier in Fig. P3.3-9 is 20 ft high, and it has a square cross section with side dimension that varies linearly from 36 in. at the top to 48 in. at the bottom. Assume that the stone is linearly elastic with modulus of elasticity $E = 4.0 \times 10^3$ ksi. Determine the shortening of the pier under a compressive load of $P = 150$ kips. (Neglect the weight of the stone.)

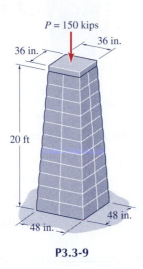

P3.3-9

▼ AXIAL DEFORMATION—DISTRIBUTED AXIAL LOAD

Prob. 3.3-10. A bar of unstretched length L and cross-sectional area A is made of material with modulus of elasticity E and specific weight γ (weight per unit volume). (a) Determine the elongation of the bar due to its own weight (i.e., with $P = 0$). (b) Determine the compressive axial force P that would be required to return the bar to its original length L. Express your answer in terms of $\gamma, A, E,$ and L.

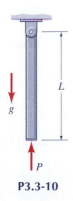

P3.3-10

Prob. 3.3-11. A uniform circular cylinder of diameter d and length L is made of material with modulus of elasticity E and specific weight γ. It hangs from a rigid "ceiling" at A, as shown in Fig. P3.3-11. Using Eq. 3.12, determine an expression for the (downward) vertical displacement, $u(x)$, of the cross section at distance x from end A.

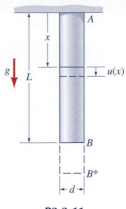

P3.3-11

Prob. 3.3-12. Nonuniform distributed axial loading of the circular rod AB in Fig. P3.3-12 causes an extensional strain that can be expressed in the form $\epsilon(x) = c_1\left[1 - \left(\dfrac{x}{L}\right)^2\right]$, where c_1 is a constant. (a) Determine an expression for the total elongation of the rod AB. (b) If the maximum axial stress in this rod is $\sigma_{max} = 15.0$ ksi, and its modulus of elasticity is $E = 10 \times 10^3$ ksi, what is the value of the constant c_1? (Give the proper units of c_1.) (c) Using your previous results, calculate the total elongation of rod AB if its length is $L = 10$ ft.

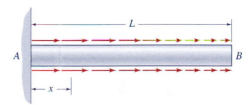

P3.3-12 and P3.3-13

Prob. 3.3-13. Nonuniform distributed axial loading of the circular rod AB in Fig. P3.3-12 causes an extensional strain that can be expressed in the form $\epsilon(x) = c_1\left[1 - \left(\dfrac{x}{L}\right)^2\right]$, where c_1 is a constant. (a) Determine an expression for the total elongation of the rod AB. (b) If the maximum axial force in this rod is $F_{max} = 10.0$ kN, its diameter is 40 mm, and its modulus of elasticity is $E = 70$ GPa, what is the value of the constant c_1? (Give the proper units of the constant c_1.) (c) Using your previous results, calculate the total elongation of rod AB if its length is $L = 4$ m.

Prob. 3.3-14. A uniform circular cylinder of diameter d and length L is made of material with modulus of elasticity E. It is fixed to a rigid wall at end A and subjected to distributed external axial loading of magnitude $p(x)$ per unit length, as shown in Fig. P3.3-14a. The axial stress, $\sigma_x(x)$, varies linearly with x as shown in Fig. P3.3-14b. (a) Determine an expression for the distributed loading, $p(x)$. (Hint: Draw a free-body diagram of the bar from section x to section $(x + \Delta x)$.)

(b) Using Eq. 3.12, determine an expression for the axial displacement, $u(x)$, of the cross section at distance x from end A.

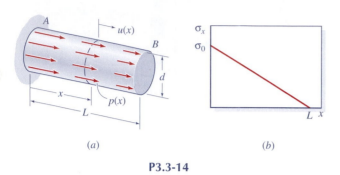

(a)   (b)

P3.3-14

Prob. 3.3-15. The stiffeners in airplane wings may be analyzed as uniform rods subjected to distributed loading. The stiffener shown below has a cross-sectional area of 0.60 in^2 and is made of aluminum ($E_a = 10 \times 10^6$ psi). Determine the elongation of section AB of the stiffener, whose original length is 20 in., if the stress in the stiffener at end A is $\sigma_A = 5,000$ psi, and a uniform distributed loading of 40 lb/in. is applied on either side of the stiffener over the 20-in. length from A to B.

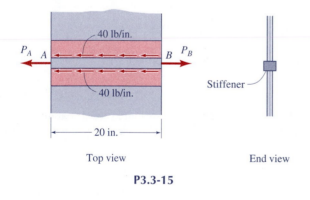

Top view   End view

P3.3-15

▼ **AXIAL DEFORMATION—DISTRIBUTED AXIAL LOAD & VARYING CROSS SECTION**

Prob. 3.3-16. The trapezoidal flat bar in Fig. P3.3-16 is suspended from above and loaded only by its own weight. The thickness of the bar is constant, and its width varies linearly from width b_0 at the bottom to $2b_0$ at the top. The weight density of the material is γ. For the following case, calculate the normal stress on the cross section midway between bottom and top of the bar: $b_0 = 10$ in., $L = 20$ ft, and $\gamma = 0.284$ lb/in^3.

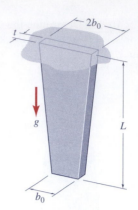

P3.3-16 and P3.3-17

*Prob. 3.3-17. (a) Derive a formula for the elongation of the bar in Fig. P3.3-16 if its unstretched length is L and its modulus of elasticity is E. (b) Calculate the elongation of the bar for the following case: $b_0 = 25$ mm, $L = 10$ m, $E = 200$ GPa, and $\gamma = 77$ kN/m^3.

▼ **SEGMENTED AXIAL— DETERMINATE**   MDS 3.1

Prob. 3.4-1. The two-segment steel rod in Fig. P3.4-1 has a circular cross section with diameter $d_1 = 15$ mm over one-half of its length, and diameter $d_2 = 10$ mm over the other half. The modulus of elasticity of the steel is $E = 200$ GPa. (a) How much will the rod elongate under the tensile load of $P = 20$ kN? (b) If the same volume of material were to be made into a rod of constant diameter and the same 2-m length, what would be the elongation of this rod under the same load P?

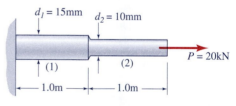

P3.4-1 and P11.7-5

Prob. 3.4-2. A column in a two-story building is fabricated from square structural steel tubing having a modulus of elasticity $E = 210$ GPa. The cross-sectional dimensions of the two segments are shown in Fig. P3.4-2b. Axial loads $P_A = 200$ kN and $P_B = 300$ kN are applied to the column at levels A and B, as shown in Fig. P3.4-2a. (a) Determine the axial stress σ_1 in segment AB of the column and the axial stress σ_2 in segment BC of the column. (b) Determine the amount δ by which the column is shortened.

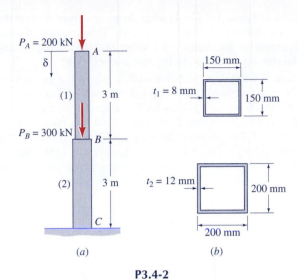

$P_A = 200$ kN

δ

(1) 3 m

$P_B = 300$ kN

(2) 3 m

$t_1 = 8$ mm

150 mm

150 mm

$t_2 = 12$ mm

200 mm

200 mm

(a)   (b)

P3.4-2

symbolic expressions for the downward displacements u_A and u_B at the respective column joints.

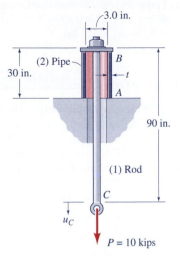

3.0 in.

(2) Pipe

30 in.

(1) Rod

90 in.

u_C

$P = 10$ kips

P3.4-4 and P3.4-7

Prob. 3.4-3. The three-part axially loaded member in Fig. P3.4-3 consists of a tubular segment (1) with outer diameter $(d_o)_1 = 1.25$ in. and inner diameter $(d_i)_1 = 0.875$ in., a solid circular rod segment (2) with diameter $d_2 = 1.25$ in., and another solid circular rod segment (3) with diameter $d_3 = 0.875$ in. The line of action of each of the three applied loads is along the centroidal axis of the member. (a) Determine the axial stresses σ_1, σ_2, and σ_3 in each of the three respective segments. (b) If $L_1 = L_2 = L_3 = 20$ in. and $E = 30 \times 10^3$ ksi, what are the nodal displacements u_B, u_C, and u_D?

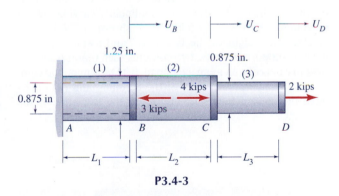

U_B U_C U_D

1.25 in.   0.875 in.

(1)   (2)   (3)

4 kips   2 kips

0.875 in

3 kips

A   B   C   D

L_1 L_2 L_3

P3.4-3

Prob. 3.4-4. A 1.2 in.-diameter aluminum-alloy hanger rod is supported by a steel pipe with an inside diameter of 3.0 in. and a wall thickness of $t = 0.125$ in. Determine the displacement u_C at the lower end of the hanger rod when the hanger rod is supporting a load $P = 10$ kips. Let $E_1 = E_{al} = 10 \times 10^3$ ksi and $E_2 = E_{st} = 30 \times 10^3$ ksi.

Prob. 3.4-5. A two-story parking garage has steel columns labeled (1) and (2), as shown in Fig. P3.4-5. The roof load P_A and the load P_B from the second floor can both be assumed to act downward directly through the centroids of the cross sections of the two columns, as shown. Each column has an effective length L, and the cross-sectional areas of the two columns are A_1 and A_2, respectively. Let the modulus of elasticity of the steel be designated as $E_1 = E_2 = E_{st}$. Determine

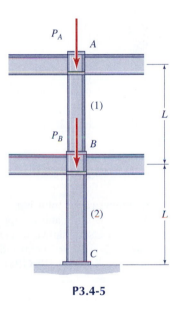

P_A

A

(1)   L

P_B

B

(2)   L

C

P3.4-5

DProb. 3.4-6. A uniform rod of diameter d is subjected to axial loads at the three cross sections, as illustrated in Fig. P3.4-6. If the displacement at the right end, u_D, cannot exceed 5 mm, and the maximum axial stress in the rod cannot exceed 80 MPa, what is the minimum allowable diameter of the (circular) cylindrical rod? Use $E = 70$ GPa.

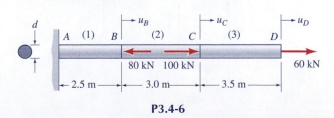

d

A (1) B u_B u_C u_D

(2) C   (3) D

80 kN 100 kN   60 kN

2.5 m   3.0 m   3.5 m

P3.4-6

213

DProb. 3.4-7. A 1.2-in.-diameter aluminum-alloy hanger rod is supported by a steel pipe with an inside diameter of 3.0 in., as shown in Fig. P3.4-4. Determine the minimum wall thickness of the steel pipe if the maximum displacement, u_C at the lower end of the hanger rod is 0.10 in. Let $E_1 = E_{al} = 10 \times 10^3$ ksi and $E_2 = E_{st} = 30 \times 10^3$ ksi.

▼ BEAM AND TWO RODS   MDS 3.2

Prob. 3.4-8. A 12-ft beam AB that weighs $W_b = 180$ lb supports an air conditioner unit that weighs $W_a = 900$ lb. The beam, in turn, is supported by hanger rods (1) and (2), as shown in Fig. 3.4-8. If the diameters of the rods are $d_1 = d_2 = \frac{3}{8}$ in., their lengths are $L_1 = L_2 = 6$ ft, and they are both made of steel with $E_1 = E_2 = 30 \times 10^3$ ksi, what will be the downward displacements u_A and u_B of the two ends of the beam AB?

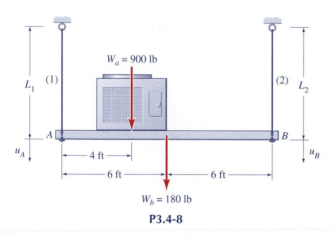

P3.4-8

Prob. 3.4-9. A rigid beam AB of total length 3 m is supported by vertical rods at its ends, and it supports a downward load at C of $P = 60$ kN. as shown in Fig. P3.4-9. The diameters of the hanger rods are $d_1 = 25$ mm and $d_2 = 20$ mm, and both are made of steel ($E = 210$ GPa). Neglect the

weight of beam AB. (a) At what distance x from A must the load be placed such that $u_B = u_A$? (b) How much is this displacement of the beam? (c) What are the corresponding axial stresses σ_1 and σ_2 in the two hanger rods?

DProb. 3.4-10. A rigid beam AB is supported by vertical rods at its ends, and it supports a downward load at C of $P = 60$ kN as shown in Fig. P3.4-10. The diameter of the support rod at A is $d_1 = 25$ mm. Both hanger rods are made of steel ($E = 210$ GPa). Neglect the weight of beam AB. (a) If it is found that $u_B = 2u_A$, what is the diameter, d_2, of the hanger rod at B? (b) What is the corresponding displacement at the load point, C?

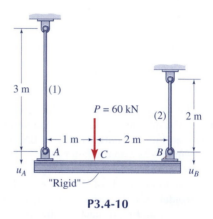

P3.4-10

Prob. 3.4-11. Two steel pipe columns ($E = 30 \times 10^3$ ksi, $A_1 = A_2 = 3.0$ in^2) support a rigid beam AC that has a 12-ft overhang AB. The beam supports a linearly varying load of maximum intensity $w_0 = 5$ kips/ft, as shown in Fig. P3.4-11. Neglect the weight of the beam AC, and assume that the beam is supported on the two columns such that it transmits load to each column as an axial load. (a) Determine the compressive axial stress in each of the columns, and (b) determine the displacements u_A and u_C at the two ends of the beam AC.

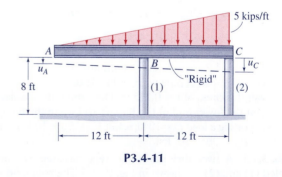

P3.4-11

Prob. 3.4-12. A hanger rod CD is supported by a rigid beam AB that is, in turn, supported by identical vertical rods at A and B (see Fig. P3.4-12). If the three rods all have the same cross-sectional area, A, and modulus of elasticity, E, determine a symbolic expression that relates the load P to the

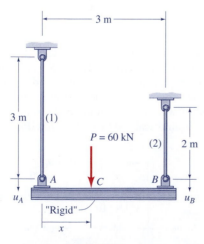

P3.4-9

vertical displacement u_D. Write your answer in the form $u_D = f_D P$.

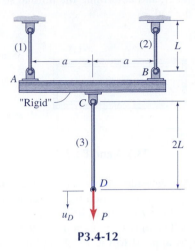

P3.4-12

allowable stress in rod CD is $\sigma_{\text{allow}} = 100$ MPa and the maximum allowable deflection at A is $\delta_{\text{allow}} = 10$ mm.

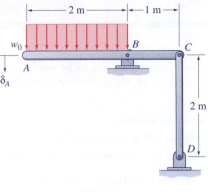

P3.4-14

▼ **SPECIAL TRUSS**

Prob. 3.4-15. Rod AB in Fig. P3.4-15a is pinned to a fixed support at A and is pinned at B to a block that is forced to move horizontally when load P is applied, as shown in Fig. P3.4-15b. Assume that the displacement of point B satisfies $u_B \ll L$. (a) Determine an expression for the axial stress in rod AB as a function of P, L, E, A, and the angle θ. (b) Similarly, determine an expression for the horizontal displacement u_B.

▼ **PINNED BEAM AND STRUT** | MDS 3.3 |

Prob. 3.4-13. As shown in Fig. P3.4-13, wire CD of length L, cross-sectional area A, and modulus of elasticity E supports a rigid beam of negligible weight, which, in turn, supports a sign whose weight is W. Attachment pin D is directly above pin A, and when there is no load acting on the beam, the beam is horizontal. (a) Determine an expression for the axial stress in the wire CD when the sign is hanging from the beam at B and C. (b) Determine an expression for the vertical displacement δ_C of the end of the beam. Simplify your solution by assuming that $\delta_C \ll L$.

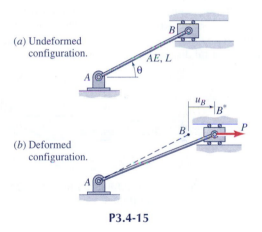

P3.4-15

Prob. 3.4-16. The pin-jointed truss ABC in Fig. P3.4-16 consists of three aluminum-alloy members, each having a

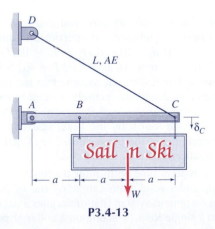

P3.4-13

ᴰProb. 3.4-14. As shown in Fig. P3.4-14, vertical rod CD of length $L = 2$ m, cross-sectional area $A = 150$ mm^2. and modulus of elasticity $E = 70$ GPA supports a rigid beam AC of negligible weight, which, in turn, supports a uniformly distributed load over segment AB. When there is no load, the beam is horizontal. Determine the maximum load intensity w_0 (kN/m) that can be placed on the beam if the maximum

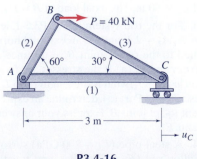

P3.4-16

cross-sectional area of $A = 1000$ mm^2 and a modulus of elasticity $E = 70$ GPa. If a horizontal load $P = 40$ kN acts to the right on the pin at B, calculate the displacement u_C of the roller support at C.

▼ AXIAL DEFORMATION— INDETERMINATE ‖ MDS 3.4 & 3.5

Prob. 3.5-1. Two uniform, linearly elastic members are joined together at B, and the resulting two-segment rod is attached to rigid supports at ends A and C. A single external force, P_B, is applied symmetrically around the joint at B. Member (1) has modulus E_1, cross-sectional area A_1, and length L_1; member (2) is made of material with modulus E_2, and has a cross-sectional area A_2 and length L_2. (a) Develop symbolic expressions for the axial stresses σ_1 and σ_2 in the respective segments, and (b) develop a symbolic expression for the horizontal displacement, u_B, of joint B.

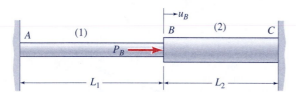

P3.5-1, P3.5-2, P3.5-3, P3.8-1, P3.8-2, P3.8-24 and P3.9-1

Prob. 3.5-2. Solve Prob. 3.5-1 for the following two-segment rod: member (1) is steel with modulus $E_1 = 30 \times 10^3$ ksi, cross-sectional area $A_1 = 2.0$ in^2, and length $L_1 = 80$ in.; member (2) is made of aluminum alloy, with $E_2 = 10 \times 10^3$ ksi, $A_2 = 3.6$ in^2, and $L_2 = 60$ in. The axial load at B is $P_B = 10$ kips.

Prob. 3.5-3. For the rod system in Fig. P3.5-1, let $E_1 = E_2 = E$ and $A_1 = A$, $A_2 = 2A$; and let an external horizontal force P_B be applied to the rod at joint B, at distance L_1 from end A and L_2 from end C. (a) If the stresses are found to satisfy the equation $|\sigma_2| = 3\sigma_1$, what is the length ratio L_1/L_2? (b) For the case described in Part (a), determine an expression for the displacement u_B, of joint B. Express your answer in terms of P_B, L_1, A, and E.

Prob. 3.5-4. A steel pipe ($E_1 = 200$ GPa) surrounds a solid aluminum-alloy rod ($E_2 = 70$ GPa), and together they are

---

subjected to a compressive force of 200 kN acting on rigid end caps. Determine the shortening of this bimetallic compression member, and determine the normal stresses in the pipe and in the rod.

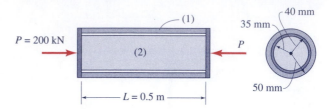

P3.5-4 and P3.8-13

Prob. 3.5-5. A magnesium-alloy rod ($E_1 = E_m = 6.5 \times 10^3$ ksi) of diameter $d_1 = 1$ in. is encased by, and securely bonded to, a brass sleeve ($E_2 = E_b = 15 \times 10^3$ ksi) with outer diameter $d_2 = 1.5$ in; the length is $L_1 = L_2 = L = 20$ in. An axial load $P = 10$ kips is applied to the resulting bimetallic rod. (a) Treating the magnesium-alloy rod as Element (1) and the brass sleeve as Element (2) of a statically indeterminate system, determine the normal stresses σ_1 and σ_2 in the two materials, and (b) determine the elongation of the bimetallic rod.

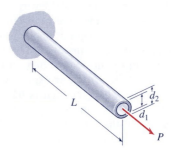

P3.5-5 and P3.5-6

Prob. 3.5-6. The magnesium-alloy rod and brass sleeve in Fig. P3.5-5 have the following properties and dimensions: $E_1 = E_m = 45$ GPa), $d_1 = 30$ mm, ($E_2 = E_b = 100$ GPa); both have length $L = 500$ mm. An axial load $P = 40$ kN is applied to the resulting bimetallic rod. Assume that the magnesium rod and the brass sleeve are securely bonded to each other. (a) If three-fourths of the load P is carried by the magnesium rod, and one-fourth by the brass sleeve, what is the outer diameter, d_{bo}, of the brass sleeve? (b) What is the resulting elongation of the bimetallic rod?

Prob. 3.5-7. A steel pipe with outer diameter d_o and inner diameter d_i, and a solid aluminum-alloy rod of diameter d form a three-segment system that undergoes axial deformation due to a single load P_C acting on a collar at point C, as shown in Fig. P3.5-7. (a) Calculate the axial stresses σ_1, σ_2, and σ_3, in the three elements, and (b) determine the displacements u_B and u_C.

$$d_o = 2 \text{ in.}, \quad d_i = 1.5 \text{ in.}, \quad d = 0.75 \text{ in.}$$

$$L_1 = L_2 = 30 \text{ in.}, \quad L_3 = 50 \text{ in.}$$

$$P_C = 12 \text{ kips}, \quad E_1 = 30 \times 10^3 \text{ ksi}, \quad E_2 = E_3 = 10 \times 10^3 \text{ ksi}$$

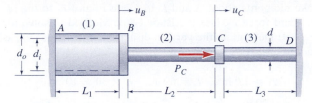

P3.5-7, P3.5-8, P3.8-3, P3.8-4, P3.8-26, P3.9-4,
and P11.7-11

Prob. 3.5-8. Solve Prob. 3.5-7 using the dimensions, material properties, and load given below.

$$d_o = 50 \text{ mm}, \quad d_i = 36 \text{ mm}, \quad d = 20 \text{ mm}$$

$$L_1 = L_2 = 1 \text{ m}, \quad L_3 = 2 \text{ m}, \quad P_C = 50 \text{ kN}$$

$$E_1 = 210 \text{ GPa}, \quad E_2 = E_3 = 70 \text{ GPa}$$

Prob. 3.5-9. A three-segment rod is initially stress free after it is attached to rigid supports at ends A and D. It is subjected to equal and opposite external axial loads P at nodes (joints) B and C, as shown in Fig. P3.5-9. The rod is homogeneous and linearly elastic, with modulus of elasticity E. Let $A_1 = A_3 = A$, and $A_2 = 2A$; let $L_1 = 2L$, and $L_2 = L_3 = L$. (a) Determine symbolic expressions (i.e., formulas) for the axial stresses σ_1, σ_2, and σ_3 in the three segments; and (b) determine formulas for the horizontal displacements u_B and u_C at nodes B and C, respectively.

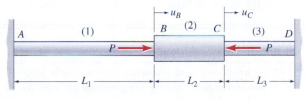

P3.5-9, P3.5-10, and P3.9-2

Prob. 3.5-10. For the three-segment rod in Fig. 3.5-9, the respective dimensions and material properties are: $A = 500 \text{ mm}^2$, $L = 800 \text{ mm}$, and $E = 200 \text{ GPa}$. If the magnitude of the force is $P = 35 \text{ kN}$, (a) determine the reactions at ends A and D; and (b) determine the amount by which segment (2) (i.e., BC) is shortened.

▼ **PINNED RIGID BAR AND TWO RODS** | MDS 3.6 |

Prob. 3.5-11. The rigid beam AD in Fig. P3.5-11 is supported by a smooth pin at D and by vertical steel rods attached to the beam at points A and C. Neglect the weight of the beam, and assume that the support rods are stress-free when $P = 0$. (a) Solve for the forces F_1 and F_2 in the support rods after

load P is applied, and (b) determine the corresponding elongation of support rod (1). Assume that the rotation of the beam is very small, so $e_1 \ll L_1$.

$$A_1 = A_2 = 500 \text{ mm}^2, \quad L_1 = 1 \text{ m}, \quad L_2 = 2 \text{ m}$$

$$E_1 = E_2 = 210 \text{ GPa}, \quad P = 50 \text{ kN}$$

$$a = 0.50 \text{ m}, \quad b = 1.5 \text{ m}$$

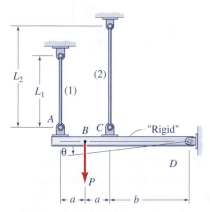

**P3.5-11, P3.5-12, P3.5-13, P3.8-15,
P3.9-5, and P11.7-14**

Prob. 3.5-12. With Fig. P3.5-12 and the data below, (a) determine the axial stresses in support rods (1) and (2) after load P is applied, and (b) solve for the corresponding rotation, θ, of the beam AD. Assume that angle θ is very small.

$$A_1 = A_2 = 1.0 \text{ in}^2. \quad L_1 = 40 \text{ in.}, \quad L_2 = 60 \text{ in.}$$

$$E_1 = E_2 = 30 \times 10^3 \text{ ksi}, \quad P = 10 \text{ kips}$$

$$a = 20 \text{ in.}, \quad b = 60 \text{ in.}$$

D Prob. 3.5-13. (a) With Fig. P3.5-11 and the data below, determine the <u>minimum</u> cross-sectional area of support rod (1) such that the maximum normal stress in that rod does not exceed $\sigma_{max} = 6 \text{ ksi}$ when the load is $P = 10 \text{ kips}$. (b) Derive formulas to show that the normal stresses in rods (1) and (2) are independent of the material used in the rods, as long as they have the same modulus of elasticity, that is, $E_1 = E_2$.

$$A_1 = ?, \quad A_2 = 1.0 \text{ in}^2, \quad L_1 = 40 \text{ in.}, \quad L_2 = 60 \text{ in.}$$

$$E_1 = E_2 = 30 \times 10^3 \text{ ksi}, \quad P = 10 \text{ kips}$$

$$a = 20 \text{ in.}, \quad b = 60 \text{ in.}$$

D*Prob. 3.5-14. Wires (1) and (2) support a rigid, weightless beam AD that is hinged to a wall bracket at A and supports a load P at end D. as shown in Fig. P3.5-14. The wires are taut, but stress-free, when $P = 0$. The wires are made of the same material (i.e., same E). but have areas A_1 and A_2 and lengths L_1 and L_2. What must be the relationship among the parameters a, b, c, A_1, A_2, L_1 and L_2 if the wires are designed to be *fully stressed*, that is, designed such that the stresses in the two wires are equal? Discuss your answer; that is, try to justify the way various parameters enter into your answer. (Assume that the rotation of the beam is very small.)

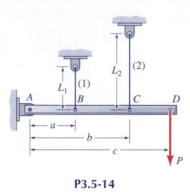

P3.5-14

*Prob. 3.5-15.** A sign whose weight is W hangs from bolts at points C and D on the rigid, weightless beam in Fig. 3.5-15. The beam is supported at C and D by wires having cross-sectional area A and made of material with modulus E, and it is hinged to a wall bracket at end B. Before the sign is attached to it, beam BD is horizontal and the two support wires are taut but stress-free. Let $h = b$. (a) Determine expressions for the tensile stresses σ_1 and σ_2 in the two wires, and (b) determine an expression for the vertical displacement, δ_D, at D. Assume that the angle of rotation of the beam BD is very small. Express your answers in terms of P, b, A, and E.

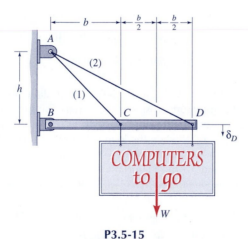

P3.5-15

▼ **RIGID BEAM WITH THREE SUPPORT RODS**

*Prob. 3.5-16.** A rigid beam AD is supported by three identical vertical rods that are attached to the beam at points A, C, and D. When the rods are initially attached to the beam and $P = 0$, the rods are stress-free. With Fig. P3.5-16 and the data below, (a) solve for the axial forces F_1, F_2, and F_3, in the support rods. (Hint: Select F_2 as the redundant force, and

take moments about A and D to relate F_3 and F_1 to the redundant force F_2. Use the Basic Force Method to solve for F_2.) (b) Determine the vertical displacements u_A and u_D, as identified in Fig. P3.5-16.

$$A = 1.0 \text{ in}^2, \quad L = 60 \text{ in.}, \quad E = 30 \times 10^3 \text{ ksi}$$

$$a = 20 \text{ in.}, \quad b = 40 \text{ in.}, \quad c = 60 \text{ in.}, \quad P = 10 \text{ kips}$$

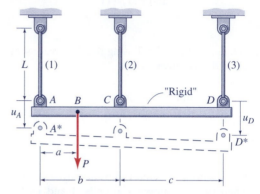

P3.5-16, P3.5-17, P3.8-16 and P3.9-7

*Prob. 3.5-17.** With Fig. P3.5-17 and the data below, (a) determine the axial forces in the three support rods after load P is applied, and (b) solve for the corresponding displacement u_B at the point of application of the load. Assume that the angle of rotation of the beam AD is very small.

$$A = 500 \text{ mm}^2, \quad L = 2 \text{ m}, \quad E = 100 \text{ GPa}$$

$$a = 0.5 \text{ m}, \quad b = 1.0 \text{ m}, \quad c = 1.5 \text{ m}, \quad P = 20 \text{ kN}$$

All of the problems in this section are statically indeterminate problems that involve temperature change. Solve these problems by writing appropriate equations of each of the following three fundamental types:

1. *Equilibrium equations.*
2. *Force-temperature-deformation relationships.*
3. *Geometry of deformation; Compatibility equation.*

Use the Basic Force Method to solve these simultaneous equations.

Prob. 3.6-1. (a) Solve the thermal-stress problem stated in Example 3.10, replacing the linear temperature distribution in Fig. 2 with the following quadratic temperature distribution:

$$\Delta T(x) = \Delta T_A \left[1 - \left(\frac{x}{L} \right)^2 \right]$$

(b) Determine an expression for the axial displacement $u(L/2)$ of the center cross section of the rod AB.

Prob. 3.6-2. A square aluminum-alloy bar is attached to rigid walls at ends A and B when the temperature of the bar is 90°F. If the bar is cooled to 40°F, what is the maximum normal stress in the bar and what is the maximum <u>shear</u> stress? (Recall Section 2.9.) Let $\alpha = 13 \times 10^{-6}/°F$ and $E = 10 \times 10^3$ ksi. and give your answers in kips/in^2.

P3.6-2 and P3.6-3

Prob. 3.6-3. If the square steel bar in Fig. P3.6-3 is stress free when it is attached to the rigid walls at A and B, how much must the temperature of the bar be raised, in °C; to cause the maximum <u>shear</u> stress in the bar to be 35 MPa? (Recall Section 2.9). Let $\alpha = 12 \times 10^{-6}/°C$ and $E = 200$ GPa.

▼ **END-TO-END BARS WITH ΔT**   **MDS 3.7**

Prob. 3.6-4. The two rod elements in Fig. P3.6-4 are stress free when they are assembled together and attached to the rigid walls at A and C. Determine expressions for the stresses $\sigma_1(\Delta T)$ and $\sigma_2(\Delta T)$ that would result from a uniform temperature increase $\Delta T_1 = \Delta T_2 = \Delta T$. Express your answers in terms of $E_1, E_2, A_1, A_2, L_1, L_2, \alpha_1, \alpha_2,$ and ΔT.

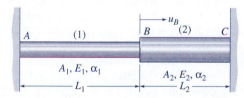

P3.6-4, P3.6-5, P3.6-6, and P3.8-5

Prob. 3.6-5. Two uniform, linearly elastic rods are joined together at B, and the resulting two-segment rod is attached to rigid supports at A and C. Element (1) is steel with modulus $E_1 = 30 \times 10^3$ ksi, cross-sectional area $A_1 = 2.0$ in^2, length $L_1 = 80$ in., and coefficient of thermal expansion $\alpha_1 = 7 \times 10^{-6}/°F$; the corresponding values for the aluminum element (2) are: $E_2 = 10 \times 10^3$ ksi, $A_2 = 3.6$ in^2, $L_2 = 60$ in., and $\alpha_2 = 13 \times 10^{-6}/°F$. (a) Determine the axial stresses σ_1 and σ_2 in the rods if the temperature of both is raised by $\Delta T_1 = \Delta T_2 = 60°F$. (b) Does node B move to the right, or does it move to the left? How much does it move?

Prob. 3.6-6. For the two-element rod system in Fig. P3.6-6, element (1) is steel with modulus $E_1 = 210$ GPa, area $A_1 = 1000$ mm^2, length $L_1 = 2$ m, and coefficient of thermal expansion $\alpha_1 = 12 \times 10^{-6}/°C$; element (2) is of titanium alloy with $E_2 = 120$ GPa, $A_2 = 1500$ mm^2, $L_2 = 1.5$ m, and $\alpha_2 = 8 \times 10^{-6}/°C$. Determine the axial stresses σ_1 and σ_2 in the rods if the temperature of both is raised by $\Delta T_1 = \Delta T_2 = 30°C$.

Prob. 3.6-7. When a linear spring (1), with spring constant k, and rod (2), with cross-sectional area A, length L, modulus E, and coefficient of thermal expansion α, are fixed between rigid walls, as shown in Fig. P3.6-7, there is no stress in the spring-rod assembly. If the temperature of the rod is subsequently raised uniformly by an amount ΔT, determine an expression for the axial force induced in the rod. Assume that the temperature of the spring remains unchanged.

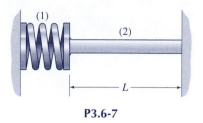

P3.6-7

Prob. 3.6-8. A steel pipe with outer diameter d_o and inner diameter d_i, and a solid aluminum-alloy rod of diameter d form a three-segment system. When the system is welded to rigid supports at A and D, it is stress free. Subsequently, the aluminum rod is cooled by 100°F (i.e., $\Delta T_2 = \Delta T_3 = -100°F$), while $\Delta T_1 = 0$. Determine the axial stresses $\sigma_1, \sigma_2,$ and σ_3 in the elements.

$$d_o = 2 \text{ in.}, \quad d_i = 1.5 \text{ in.}, \quad d = 0.75 \text{ in.}$$

$$L_1 = L_2 = 30 \text{ in.}, \quad L_3 = 50 \text{ in.}$$

$$E_1 = 30 \times 10^3 \text{ ksi}, \quad E_2 = E_3 = 10 \times 10^3 \text{ ksi}$$

$$\alpha_2 = \alpha_3 = 6.5 \times 10^{-6}/°F$$

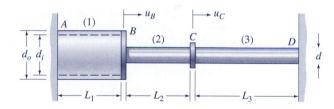

P3.6-8, P3.8-6, and P3.9-9

Prob. 3.6-9. Two flat steel bars are supposed to be connected together by a pin at B, and the system is supposed to be stress free after the pin is inserted (see Fig. 3.6-9a). However, the distance between the rigid walls to which the two bars are attached was measured incorrectly, so that a misalignment of $\bar{\delta}$ (exaggerated in Fig. 3.6-9b) results. Determine the amount $\Delta T(°C)$ by which the two bars would need to be uniformly heated to bring the holes in the two bars into alignment. The mechanical properties and dimensions of the two bars are

$$E_1 = E_2 = 200 \text{ GPa}, \quad \alpha_1 = \alpha_2 = 12 \times 10^{-6}/°C$$

$$A_1 = 25 \text{ mm}^2, \quad A_2 = 50 \text{ mm}^2, \quad L_1 = 3 \text{ m},$$
$$L_2 = 5 \text{ m}, \quad \bar{\delta} = 1 \text{ mm}$$

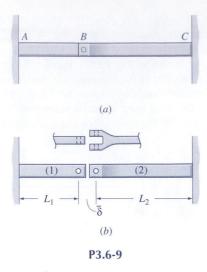

(a)

(b)

P3.6-9

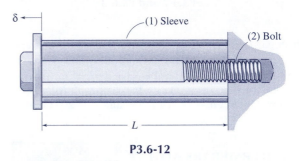

nated as A_1, E_1, and α_1, and those of the bolt A_2, E_2, and α_2. (Neglect the reduction in bolt area A_2 due to the threads, and neglect the contribution of the washer's thickness to the effective length of the bolt.) Assume that the coefficient of thermal expansion of the sleeve is greater than that of the bolt, that is, $\alpha_1 > \alpha_2$. (a) Determine expressions for the stresses σ_1 and σ_2 that would be induced in the sleeve and the bolt, respectively, when the temperature of the bolt-sleeve assembly is uniformly increased by an amount ΔT. (b) Determine an expression for the resulting elongation, δ, of the sleeve and bolt.

δ — (1) Sleeve

(2) Bolt

P3.6-12

▼ **COAXIAL BARS WITH ΔT**   MDS 3.8

Prob. 3.6-10. The sleeve of the titanium-sleeve/aluminum-core system described in the problem statement of Example 3.12 is heated by $\Delta T_1 = 100°F$, but the aluminum core is maintained at the reference temperature. (a) Determine the axial stress induced in each material, and (b) determine the total elongation, e, of the sleeve/core system. (Assume that the core and the sleeve are both stress-free at the reference temperature.)

Prob. 3.6-11. The pipe-sleeve element (1) and cylindrical-rod core element (2) in Fig. P3.6-11 are both stress-free after they are attached to rigid end-plates. (a) Determine expressions for the stresses σ_1 and σ_2 in the pipe and rod, respectively, if the temperature of the system is uniformly increased by ΔT. Express your answer in terms of E_1, E_2, A_1, A_2, $L_1 = L_2 \equiv L$, α_1, α_2, and ΔT, and (b) determine an expression for the total elongation, e, of the pipe/rod system due to the uniform temperature increase ΔT.

▼ **RIGID MEMBER AND 2 OR 3 BARS WITH ΔT**

MDS 3.9

Prob. 3.6-13. The mechanical system of Fig. P3.6-13 consists of a "rigid" beam ABC and two identical ASTM A-36 steel bars with cross-sectional area $A = 1.2$ in^2. (For E and α, see Tables F.2 and F.3.) The beam is supported by a smooth pin at B and is connected to "ground" by the bars at A and C. After the pins have been inserted at A and C, the two bars are stress free. (a) Using the dimensions listed below, determine the axial stresses induced in bars (1) and (2) if the temperature of bar (1) is decreased by 50°F. (b) Determine the (small) angle θ through which the beam AC would rotate due to this temperature change. (c) Re-solve Part (a) if the cross-sectional area of bar (1) is doubled to $A_1 = 2.4$ in^2, while the cross-sectional area of bar (2) remains unchanged.

$$a = 20 \text{ in.}, \quad L_1 = L_2 = 20 \text{ in.}$$

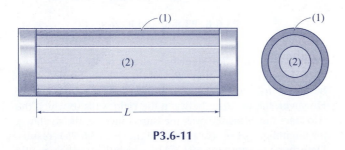

P3.6-11

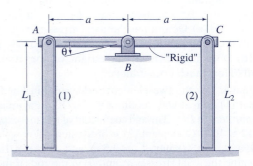

Prob. 3.6-12. A pipe sleeve (1) of length L is slipped over a bolt (2), with a heavy washer between the bolt head and the sleeve. The bolt is initially advanced until the sleeve is just brought into firm contact with the heavy machine part, as shown in Fig. 3.6-12 (i.e., the sleeve is held in contact with the heavy machine part, but no stress is induced in either the sleeve or the bolt). Let the properties of the sleeve be desig-

P3.6-13

220

Prob. 3.6-14. The mechanical system of Fig. P3.6-14 consists of two identical steel rods ($A = 40$ mm^2, $E = 200$ GPa, $\alpha = 12 \times 10^{-6}$/°C) and a "rigid" beam AC. The beam is supported by a smooth pin at B. The nuts at A and C are initially tightened just enough to remove the slack, leaving the two rods stress-free. (a) Determine the axial stresses induced in rods (1) and (2) if the temperature of <u>both</u> rods is <u>decreased</u> by 50°C. (b) Determine the (small) angle θ through which the beam AC would rotate due to this temperature change.

three support rods. To solve this problem, let F_1 be the redundant force. (Hint: Take $(\Sigma M)_B$ and $(\Sigma M)_C$ for the equilibrium equations.)

$$A_1 = A, \quad A_2 = A_3 = 2A, \quad E_1 = E_2 = E_3 = E$$

$$\alpha_2 = \alpha_3 = \alpha$$

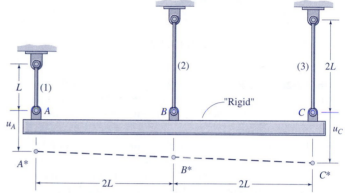

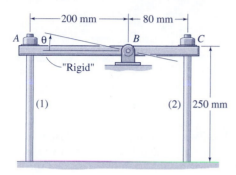

P3.6-14

P3.6-17, P3.8-19, P3.9-11, P11.7-15, and P11.7-16

Prob. 3.6-15. Steel rods ($A = 0.050$ in^2, $E = 30 \times 10^3$ ksi, $\alpha = 6.5 \times 10^{-6}$/°F) are attached at points A and C to the "rigid" right-angle bracket in Fig. P3.6-15. The bracket is supported by a smooth pin at B, and the rods at A and C are initially stress-free. (a) Determine the axial stresses induced in rods (1) and (2) if the temperature of rod (1) is <u>decreased</u> by 50°F. (b) Determine the transverse shear force V_D and bending moment M_D on the cross section at D that result from this temperature change.

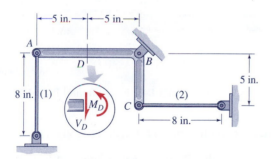

P3.6-15 and P3.9-10

Prob. 3.6-16. Solve the problem as stated in Example 3.13 if the structure in Fig. 1 of that example is modified by having support A directly above point C on the beam, rather than directly above point B. The length of the vertical wire AC is still L.

*Prob. 3.6-17. A "rigid" beam AC is supported by three vertical rods, as shown in Fig. P3.6-17. When originally assembled, the rods are all stress free and beam AC is horizontal. If rod (1) is maintained at the assembly temperature (i.e., $\Delta T_1 = 0$), and rods (2) and (3) are heated the same amount ($\Delta T_2 = \Delta T_3 = \Delta T$), determine the stresses induced in the

The problems in this section are statically indeterminate problems that involve misfits and, in a few cases, temperature change. Solve these problems by writing appropriate equations of each of the following three fundamental types:

1. *Equilibrium equations.*
2. *Force-temperature-deformation relationships.*
3. *Geometry of deformation; Compatibility equations.*

Use the **Basic Force Method** *to solve these simultaneous equations.*

▼ **END-TO-END BARS WITH "MISFIT"**

MDS 3.10 & 3.11

Prob. 3.7-1. Elements (1) and (2) were supposed to fit exactly between the rigid walls at A and C, as indicated in Fig. P3.7-1a. Unfortunately, element (1) was manufactured a small amount, $\bar{\delta} \ll L$, too long (exaggerated in Fig. 3.7-1b). However, an ingenious technician was able to cool element (1) just enough to be able to insert it into the space between A and B (indicated by dashed lines). (a) Determine expressions for the stress σ_1 induced in element (1) and the stress σ_2 induced in element (2) when element (1) returns to its original temperature; and (b) determine the amount u_B by which element (2) is shortened when element (1) is thus forced into the dashed-line position.

$$L_1^* = L_2 = L, A_1 = A, A_2 = 1.5A, E_1 = E_2 = E$$

*Assume $\bar{\delta} \ll L$.

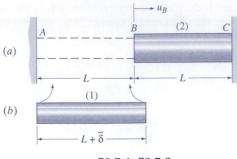

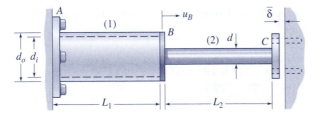

(a)

(b)

$L + \bar{\delta}$

P3.7-1, P3.7-2

Prob. 3.7-2. Solve Parts (a) and (b) of Prob. 3.7-1 for the following rod systems: System S1: $E_1 = E$, $E_2 = 2E$, and System S2: $E_1 = 2E$, $E_2 = E$. Discuss the difference in the stresses induced in the rods in these two cases.

Prob. 3.7-3. A pipe-rod system with flanges at ends A and C was supposed to fit exactly between two rigid walls, as shown in Fig. P3.7-3. Element (1) is a steel pipe with outside diameter d_o and inside diameter d_i. Element (2) is a solid steel rod with diameter $d_2 = d$. Bolts hold the flange at A against a rigid wall. Other bolts are installed in the flange at C and are tightened until the gap $\bar{\delta}$ is closed. Dimensions and material properties are given below. (a) Determine the axial stress induced in each element of the system as a result of closing the gap at C; and (b) determine the displacement u_B of joint B when the gap at C has been closed.

$$d_o = 2 \text{ in.}, \ d_i = 1.5 \text{ in.}, \ d = 0.75 \text{ in.}$$

$$L_1 = L_2 = 50 \text{ in.}, \bar{\delta} = 0.1 \text{ in.}$$

$$E_1 = E_2 = 30 \times 10^3 \text{ ksi}$$

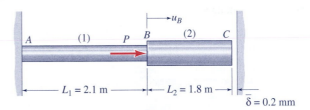

P3.7-3, P3.7-4, P3.7-7, P3.8-8, P3.8-11
P3.8-27, P3.9-12, and P3.9-14

Prob. 3.7-4. Solve Problem 3.7-3 for a pipe-rod system with the following dimensions and material property:

$$d_o = 50 \text{ mm}, \ d_i = 40 \text{ mm}, \ d = 20 \text{ mm}$$

$$L_1 = L_2 = 1 \text{ m}, \bar{\delta} = 2 \text{ mm}$$

$$E_1 = E_2 = 200 \text{ GPa}$$

Prob. 3.7-5. A two-segment linearly elastic rod has a circular cylindrical cross section of area $A_2 = 1.5A$. Half of the rod, from A to B, is turned down so that $A_1 = A$. (Note:

$E_1 = E_2 = E$, and $L_1 = L_2 = L$.) At end A the rod is connected to a rigid wall, but at end C there is a small gap $\bar{\delta}$ between the original position of end C and the rigid wall, as shown in Fig. P3.7-5. (a) Derive an expression for the width of the gap if, when an axial force P is applied at B as shown in the figure, the magnitude of the reaction at C is equal to the reaction at A. (b) Derive an expression for the amount, u_B, that joint B moves under the conditions stated in Part (a). (Note: The gap $\bar{\delta}$ shown in the figure below is the gap **before** load P is applied at B.)

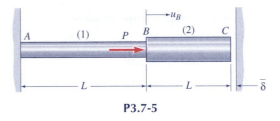

P3.7-5

Prob. 3.7-6. Two uniform, linearly elastic members are joined together at B, and the resulting two-segment rod is attached to a rigid support at end A. When there is no load on the 2-element bar (i.e., when $P = 0$) there is a gap of $\bar{\delta} = 0.2$ mm between the end of element (2) and the rigid wall at C. Element (1) is steel with modulus $E = 210$ GPa, cross-sectional area $A_1 = 1000$ mm^2, and length $L_1 = 2.1$ m: element (2) is titanium alloy with $E_2 = 120$ GPa, $A_2 = 1000$ mm^2, and $L_2 = 1.8$ m. A single external force $P = 50$ kN is applied at node B. (a) Determine the axial stresses σ_1 and σ_2 induced in the respective rod elements when the load P is applied, and (b) determine the corresponding displacement u_B of the joint B.

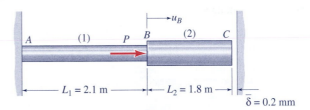

P3.7-6, P3.7-8, P3.8-9, P3.8-10, and P3.9-13

▼ **END-TO-END BARS WITH MISFIT AND ΔT**

Prob. 3.7-7. Solve Prob. 3.7-3 if, <u>in addition to</u> having the flange tightened against the wall at C, the pipe-rod system ABC is heated uniformly by 50°F. Use $\alpha = 6.5 \times 10^{-6}/°F$.

Prob. 3.7-8. Solve Prob. 3.7-6 if <u>instead of</u> having a force P applied at B, the entire 2-segment rod is heated by 20°C. Let $\alpha_1 = 12 \times 10^{-6}/°C$ and $\alpha_2 = 10 \times 10^{-6}/°C$.

▼ **RIGID MEMBER AND RODS WITH "MISFIT"**

Prob. 3.7-9. A rigid beam is supported by three aluminum-alloy cylindrical posts, which are identical except that the center post is slightly shorter than the two outer posts by an amount $\bar{\delta}$, as shown in Fig. P3.7-9. The dimensions and material properties of the three posts are given below. Since the

222

two outer posts are identical, and since the system and loading are symmetrical, the two outer posts are both labeled (1). (a) A load P acts downward directly over the center of the middle post. What is the width of the gap if, when an axial force $P = 14$ kN is applied to the beam, as shown in the figure, the gap $\bar{\delta}$ is just closed. (b) Determine the total amount, u_B, that the beam moves downward when the load is increased to $P = 28$ kN. (c) Sketch a plot of P vs. u_B for $0 \leq P \leq 28$ kN, including appropriate scales for P and for u_B.

$$A_1 = A_2 = 2000 \text{ mm}^2, L_1 = L_2 = 2 \text{ m},$$

$$E_1 = E_2 = 70 \text{ GPa}$$

(Note: The gap $\bar{\delta}$ shown in the figure below is the gap <u>before</u> load P is applied to the beam.)

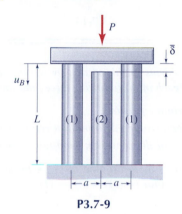

P3.7-9

Prob. 3.7-10. A "rigid" beam AD is supported by a smooth pin at D and by vertical steel rods attached to the beam at points A and C. Neglect the weight of the beam. Element (1) was fabricated the correct length, and, when only rod (1) is attached to beam AD, the beam is horizontal (i.e., $\theta = 0$). Element (2), on the other hand, was fabricated $\bar{\delta}$ too short, and it has to be stretched manually in order to connect it to the beam by a pin at C. (a) Determine expressions for the axial stresses, σ_1 and σ_2, in the two support rods when load P is applied as shown in Fig. P3.7-10, and (b) determine an expression for the corresponding rotation of the beam AD. Express your answers in terms of the following parameters: $A, L, a, E, P,$ and $\bar{\delta}$.

$$A_1 = A_2 = A, L_1 = L_2 = L$$

$$E_1 = E_2 = E, b = 2a$$

P3.7-10

Prob. 3.7-11. The mechanical system of Fig. P3.7-11 consists of a brass rod ($E_1 = 15 \times 10^3$ ksi) with cross-sectional area $A_1 = 1.5$ in^2, a structural steel rod ($E_2 = 30 \times 10^3$ ksi) with cross-sectional area $A_2 = 0.75$ in^2, and a "rigid" beam AC. The beam is supported by a smooth pin at B. The nuts at A and C are initially tightened just enough to remove the slack, leaving the two rods stress free and $\theta = 0$. (a) Determine the axial stress σ_1 induced in rod (1) when the nut at C is advanced (i.e., tightened) by one turn (0.1 in. of thread length), and (b) determine the corresponding rotation angle θ of the beam AC. Assume that θ is small.

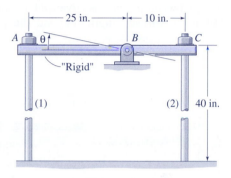

P3.7-11 and P3.8-22

Prob. 3.7-12. A "rigid" beam AD is supported by a smooth pin B and by vertical rods that are attached to the beam at points A and C, as shown in Fig. P3.7-12. Neglect the weight of the beam. The beam is horizontal when only rod (1) is attached and $P = 0$, but rod (2) was manufactured $\bar{\delta} = 0.5$ mm too short. Determine the axial stresses σ_1 and σ_2 induced in the respective support rods after rod (2) is attached to the upper bracket and the load $P = 20$ kN is applied. Assume that θ is small.

$$A_1 = A_2 = 200 \text{ mm}^2, \quad L_1 = L_2 = 2 \text{ m}$$

$$E_1 = 70 \text{ GPa}, \quad E_2 = 100 \text{ GPa}$$

$$P = 20 \text{ kN}, \quad a = 1 \text{ m}, \quad b = 1.5 \text{ m}$$

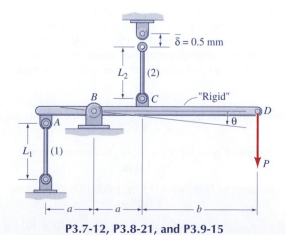

P3.7-12, P3.8-21, and P3.9-15

▼ BOLT AND SLEEVE ### MDS 3.12

Prob. 3.7-13. A brass pipe sleeve ($E_1 = 15 \times 10^3$ ksi) with outer diameter $d_o = 3$ in., and inner diameter $d_i = 2.5$ in. is held in compression against a rigid machine wall by a high-strength steel bolt ($E_2 = 30 \times 10^3$ ksi) with diameter $d = 1$ in. The head of the bolt bears on a 0.25-in.-thick (rigid) washer, which, in turn, bears on the brass sleeve. The bolt is initially advanced until there is no slack (i.e., the sleeve is held in contact with the rigid machine wall, but there is no stress induced in the bolt or the sleeve). Determine the axial stresses induced in the sleeve (σ_1) and the bolt (σ_2) when the bolt is advanced an additional $\frac{1}{4}$ turn (0.01 in. of thread length) beyond the just-light position.

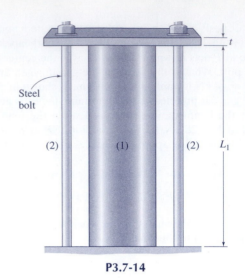

P3.7-14

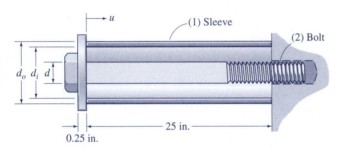

P3.7-13 and P3.8-23

Prob. 3.7-14. A "rigid" circular flat plate and two steel bolts hold a brass cylinder firmly in place against a "rigid" base, as shown in Fig. 3.7-14. Data for the bolt-cylinder assembly are given below. The two holes in the plate are fitted over the two bolts until the plate rests on the top of the cylinder. The two nuts are then "snugged" against the plate until they are just "finger tight," that is, until the plate makes contact around the entire circumference of the cylinder without inducing any stresses in the bolts or the cylinder.

Determine the normal stresses induced in the cylinder (σ_1) and in the two bolts (σ_2) if the two nuts are slowly advanced alternately until each has been advanced an additional 1/2 turn. Assume that the cylinder behaves as a linearly elastic axial-deformation member; that is, it does not buckle under the axial load imposed by the plate and bolts.

outer diameter of cylinder: $d_o = 4$ in., inner diameter: $d_i = 3.5$ in.

length of cylinder: $L_1 = 12$ in., thickness of plate: $t = 1.0$ in.

diameter of bolts: $d_2 = 0.5$ in., pitch of bolt threads: $p = 0.04$ in.

$E_1 = E_{\text{brass}} = 15 \times 10^3$ ksi, $E_2 = E_{\text{steel}} = 30 \times 10^3$ ksi

---

Solve the problems in this section by writing appropriate equations of each of the following three fundamental types:

1. *Equilibrium equations.*

2. *Force-Temperature-deformation relationships.*

3. *Geometry of deformation; Compatibility equations.*

Use the **Displacement Method** *to solve these simultaneous equations, that is, (3) → (2) → (1).*

▼ END-TO-END BARS—DISPLACEMENT METHOD

MDS 3.13 & 3.14

Prob. 3.8-1. Using the data and figure of Prob. 3.5-2, (a) solve for the horizontal displacement u_B, and (b) solve for the axial stresses σ_1 and σ_2 in the two elements that comprise the stepped rod.

Prob. 3.8-2. Using the data below and the figure of Prob. 3.5-1, (a) solve for the horizontal displacement u_B, and (b) solve for the axial stresses σ_1 and σ_2 in the two elements that comprise the stepped rod. The axial load at B is $P_B = 40$ kN. The dimensions and material properties of the system are

Member (1) is steel: $E_1 = 210$ GPa, $A_1 = 1000$ mm^2, $L_1 = 2$ m;

Member (2) is titanium alloy: $E_2 = 120$ GPa, $A_2 = 1200$ mm^2, $L_2 = 1.8$ m

Prob. 3.8-3. Using the data and figure of Prob. 3.5-7 (a) solve for the horizontal displacements u_B and u_C, and (b) solve for the axial stresses $\sigma_1, \sigma_2,$ and σ_3 in the three elements that comprise the stepped rod.

Prob. 3.8-4. Using the data and figure of Prob. 3.5-8, (a) solve for the horizontal displacements u_B and u_C, and (b) solve for the axial stresses $\sigma_1, \sigma_1,$ and σ_3 in the three elements that comprise the stepped rod.

Prob. 3.8-5. Using the data and figure of Prob. 3.6-5, (a) solve for the horizontal displacement u_B, and (b) solve for the axial stresses σ_1 and σ_2 in the two elements that comprise the stepped rod.

Prob. 3.8-6. The three-element pipe-rod system in Fig. P3.6-8 is stress free when it is welded to rigid supports at A and D. Subsequently, segment (2) is cooled by an amount $\Delta T_2 = -20°C$, while $\Delta T_1 = \Delta T_3 = 0$. The dimensions and material properties of the system are:

$$d_o = 50 \text{ mm}, \quad d_i = 36 \text{ mm}, \quad d = 20 \text{ mm},$$

$$L_1 = L_2 = 1 \text{ m}, \quad L_3 = 2 \text{ m}$$

$$E_1 = 210 \text{ GPa}, \quad E_2 = E_3 = 70 \text{ GPa}.$$

$$\alpha_2 = 23 \times 10^{-6}/°C$$

(a) Use the *Displacement Method* to solve for the horizontal displacements u_B and u_C at joints B and C, respectively, and (b) solve for the stresses σ_1, σ_2, and σ_3 that are induced in the three segments.

Prob. 3.8-7. The three-element stepped-rod system in Fig. P3.8-7 is stress free when ends A and D are attached to rigid walls. Subsequently, the middle segment is heated to an amount $\Delta T_2 = \Delta T$, while $\Delta T_1 = \Delta T_3 = 0$. Use the *Displacement Method*: (a) to obtain analytical expressions for the horizontal displacements u_B and u_C at joints B and C, respectively; and (b) to obtain expressions for the axial stresses σ_1, σ_2, and σ_3 induced in the three segments. Let $A_1 = A_3 = A$, $A_2 = 2A$, $L_2 = L_3 = L$, $L_1 = 2L$, $E = $ const., $\alpha = $ const.

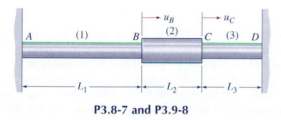

P3.8-7 and P3.9-8

Prob. 3.8-8. Using the data and figure for the pipe-rod system in Prob. 3.7-3, (a) solve for the horizontal displacement u_B when the gap at C has been closed, and (b) solve for the corresponding axial stresses σ_1 and σ_2 in the two elements.

Prob. 3.8-9. Using the data and figure for the multisegment rod system in Prob. 3.7-6, (a) solve for the horizontal displacement u_B when the load P is applied at B, and (b) solve for the corresponding axial stresses σ_1 and σ_2 in the two elements.

Prob. 3.8-10. Using the data and figure of Prob. 3.7-6. (a) solve for the horizontal displacement u_B when the load $P = 50$ kN is applied at B, and, in addition the entire rod is heated by 20°C. Let $\alpha_1 = 12 \times 10^{-6}/°C$ and $\alpha_2 = 10 \times 10^{-6}/°C$. (b) Solve for the corresponding axial stresses σ_1, and σ_2 in the two elements that comprise the stepped-rod system.

Prob. 3.8-11. Using the data and figure of Prob. 3.7-3, (a) solve for the horizontal displacement u_B if, in addition to having the flange tightened against the wall at C, the entire pipe-rod system is heated by 50°F. Let $\alpha = 6.5 \times 10^{-6}/°F$. (b) Solve for the corresponding axial stresses σ_1 and σ_2, respectively, in the pipe element AB and rod element BC.

▼ VARIOUS CONFIGURATIONS— DISPLACEMENT METHOD

Prob. 3.8-12. Three members that are attached together and supported by two rigid walls undergo axial deformation due to a single applied load P, as shown in Fig. P3.8-12. Member AB is a solid brass rod with diameter $d_1 = 1.0$ in. and modulus $E_1 = 15 \times 10^3$ ksi; member BC is a steel pipe with outer diameter $(d_o)_2 = 1.0$ in., inner diameter $(d_i)_2 = 0.75$ in., and modulus $E_2 = 30 \times 10^3$ ksi; and member BD is a solid aluminum-alloy rod with diameter $d_3 = 0.50$ in. and modulus $E_3 = 10 \times 10^3$ ksi. $L_1 = L_2 = 20$ in., $L_3 = 50$ in. (a) Determine the displacement u_D of the point D where the load $P = 2$ kips is applied, and (b) calculate the axial stress, σ_2, in the pipe section, BC.

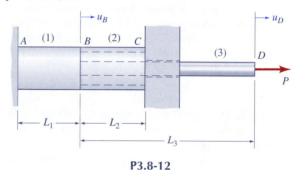

P3.8-12

Prob. 3.8-13. Using the *Displacement Method*, solve Prob. 3.5-4.

Prob. 3.8-14. The rigid beam AD in Fig. P3.8-14 is supported by a smooth pin at B and by vertical rods attached to the beam at points A and C. Neglect the weight of the beam, and assume that the rods are stress-free when $P = 0$. (a) Use the *Displacement Method* to solve for the rotation, θ, of the beam AD when load P is applied, (b) Determine the axial stresses in support rods (1) and (2) when load P is applied. Assume that θ is very small.

$$A_1 = A_2 = 200 \text{ mm}^2, \quad L_1 = L_2 = 2 \text{ m}$$

$$E_1 = 70 \text{ GPa}, \quad E_2 = 100 \text{ GPa}$$

$$a = 1 \text{ m}, \quad b = 2 \text{ m}, \quad P = 20 \text{ kN}$$

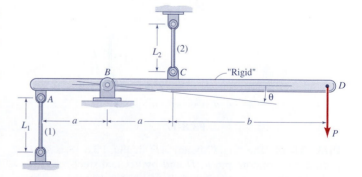

P3.8-14, P3.9-6, and P11.7-13

Prob. 3.8-15. The rigid beam AD in Fig. P3.5-11 is supported by a smooth pin at D and by vertical steel rods attached to the beam at points A and C. Neglect the weight of the beam, and

225

assume that the support rods are stress-free when $P = 0$. (a) Use the *Displacement Method* to solve for the rotation, θ, of the beam AD when load P is applied. (b) Determine the axial stresses in support rods (1) and (2) when load P is applied.

$$A_1 = A_2 = 1.0 \text{ in}^2, \quad L_1 = 40 \text{ in.}, \quad L_2 = 60 \text{ in.}$$

$$E_1 = E_2 = 30 \times 10^3 \text{ ksi}, \quad P = 10 \text{ kips}$$

$$a = 20 \text{ in.}, \quad b = 60 \text{ in.}$$

*Prob. 3.8-16. The rigid beam AD in Fig. P3.5-16 is supported by three identical vertical rods that are attached to the beam at points A, C, and D. When the rods are initially attached to the beam and $P = 0$, the rods are stress-free. (a) Use the *Displacement Method* to solve for the vertical displacements u_A and u_D of points A and D when the load P is applied to the beam at point B. (b) Calculate the axial stress in each of the three support rods when the load is applied.

$$A = 1.0 \text{ in}^2, \quad L = 60 \text{ in.}, \quad E = 30 \times 10^3 \text{ ksi}$$

$$a = 20 \text{ in.}, \quad b = 40 \text{ in.}, \quad c = 60 \text{ in.}, \quad P = 10 \text{ kips}$$

Prob. 3.8-17. The "rigid" beam AD in Fig. P3.8-17 is supported by a smooth pin at B and by vertical rods at A, C, and D. Neglect the weight of the beam, and assume that the rods are stress free when the system is assembled. Rods (1) and (2) are kept at the assembly temperature (i.e., $\Delta T_1 = \Delta T_2 = 0$), and rod (3) is cooled by an amount $\Delta T_3 = -50°C$.

$$A_1 = A_2 = A_3 = 200 \text{ mm}^2, L_1 = L_2 = L_3 = 2 \text{ m}$$

$$a = 1 \text{ m}, b = 1.5 \text{ m}, E_1 = E_3 = 70 \text{ GPa}$$

$$E_2 = 100 \text{ GPa}, \alpha_3 = 20 \times 10^{-6}/°C$$

(a) Use the *Displacement Method* to solve for the angle of rotation, θ, of the beam caused by cooling rod (3). (b) Determine the stresses σ_1, σ_2, and σ_3 induced in the three rods by the cooling of rod (3).

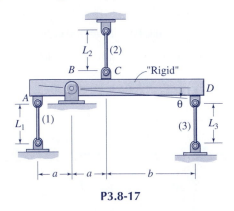

P3.8-17

Prob. 3.8-18. The "rigid" beam AD in Fig. P3.8-18 is supported by a smooth pin at D and by vertical steel rods attached to the beam at points A, B, and C. Neglect the weight of the beam, and assume that the support rods are stress-free when the system is assembled. Use the *Displacement Method* to solve this problem. (a) Solve for the rotation, θ, of the beam AD when element (3) is cooled by 80°F (i.e., $\Delta T_3 =$

$-80°F$). (b) Solve for the axial stresses σ_1, σ_2, and σ_3 induced in the three rods by the cooling of rod (3).

$$A_1 = A_2 = A_3 = 1.0 \text{ in}^2, \quad L_1 = L_3 = 40 \text{ in.}, \quad L_2 = 60 \text{ in.}$$

$$a = 20 \text{ in.}, \quad b = 60 \text{ in.}$$

$$E_1 = E_2 = E_3 = 30 \times 10^3 \text{ ksi}, \quad \alpha_3 = 6.5 \times 10^{-6}/°F$$

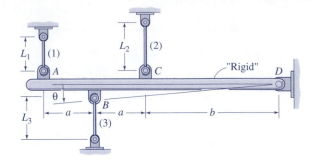

P3.8-18

Prob. 3.8-19. The "rigid" beam AC in Fig. P3.6-17 is supported by three vertical rods. When originally assembled, all of the rods are stress-free and beam AC is horizontal. Subsequently, rod (1) is maintained at the assembly temperature (i.e., $\Delta T_1 = 0$), and rods (2) and (3) are heated the same amount ($\Delta T_2 = \Delta T_3 = \Delta T$). (a) Use the *Displacement Method* to solve for the displacements u_A and u_C of the ends of the beam, and (b) solve for the stresses induced in the three support rods.

$$A_1 = A, \quad A_2 = A_3 = 2A, \quad E_1 = E_2 = E_3 = E$$

$$\alpha_2 = \alpha_3 = \alpha$$

Prob. 3.8-20. A "rigid beam AD is supported by a smooth pin at D and by vertical steel rods attached to the beam at points A and C; neglect the weight of the beam. Element (1) was fabricated the correct length, and, when only rod (1) is attached to beam AD, the beam is horizontal (i.e., $\theta = 0$). Element (2), on the other hand, was fabricated $\bar{\delta} = 0.02$ in. too short, and it has to be stretched manually in order to connect it to the beam by a pin at C. (a) Use the *Displacement Method* to solve for the rotation θ of the beam AD after rod (2) has been connected and, in addition, the load $P = 2$ kips has been applied at B. (b) Solve for the corresponding axial stresses σ_1 and σ_2 in the two support rods.

$$A_1 = A_2 = 1.0 \text{ in}^2, \quad L_1 = 40 \text{ in.}, \quad L_2 = 60 \text{ in.}$$

$$E_1 = E_2 = 30 \times 10^3 \text{ ksi}, \quad a = 20 \text{ in.}, b = 60 \text{ in.}, \quad P = 2 \text{ kips}$$

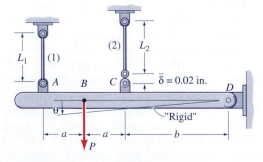

P3.8-20

Prob. 3.8-21. With the data and figure of Prob. 3.7-12, (a) use the *Displacement Method* to solve for the rotation θ of the beam AD after rod (2) has been attached to the bracket at the rod's upper end and, in addition, the load $P = 20$ kN has been applied at D. (b) Solve for the corresponding axial stresses σ_1 and σ_2 in the two support rods.

Prob. 3.8-22. With the data and figure of Prob. 3.7-11 (a) use the *Displacement Method* to solve for the (small) rotation angle θ through which the rigid beam AC would rotate if the nut at C were to be advanced (tightened) by one turn (i.e., 0.1 in. of thread length). (b) Determine the corresponding axial stress σ_1 in rod (1).

Prob. 3.8-23. With the data and figure of Prob. 3.7-13, (a) use the *Displacement Method* to solve for the amount u by which the sleeve is shortened when the bolt is advanced $\frac{1}{4}$ turn (i.e., 0.01 in. of thread length). (b) Solve for the corresponding stresses σ_1 in the sleeve and σ_2 in the bolt.

For **Problems 3.8-24 through 3.8-27,** *you are to write a computer program that carries out a* **Displacement-Method solution** *of the stated problems. Your computer program must incorporate equilibrium equations, element force-temperature-deformation equations, and deformation compatibility equations, and should include the following steps:*

1. *Input given data.*
2. *Solve equilibrium equation(s) for nodal displacement(s).*
3. *Solve force-temperature-displacement equations to obtain element stresses.*
4. *Output computed results.*

See the **Displacement-Method** *flowchart on p. 170.*

---

^C**Prob. 3.8-24.** Write a computer program for the two-element statically indeterminate system in Fig. P3.5-1, and illustrate its use by solving Probs. 3.8-1 and 3.8-2. See Example 3.15.

^C**Prob. 3.8-25.** Write a computer program for the three-element statically indeterminate system in Example 3.17. Illustrate the use of your computer program with $P_B = 10$ kips and $\Delta T_3 = -100°F$, and with the following physical data:

Elem.	Area A_i	(Intended) Length L_i	Modulus of Elasticity E_i	Coefficient of Thermal Expansion α_i	Gap $\bar{\delta}_i$
1	2.00 in²	30.00 in.	15(10³) ksi	—	—
2	1.50 in²	20.00 in.	30(10³) ksi	—	0.01 in.
3	1.00 in²	20.00 in.	30(10³) ksi	7(10⁻⁶)/°F	—

^C**Prob. 3.8-26.** Write a computer program for the three-element statically indeterminate system in Fig. P3.5-7, and illustrate its use by solving Probs. 3.8-3 and 3.8-4. See Example 3.17.

^C**Prob. 3.8-27.** Write a computer program for the two-element statically indeterminate system in Fig. P3.7-3, and illustrate its use by solving Probs. 3.8-8 and 3.8-11. See Example 3.15, but note that you must include the "gap" in the geometric compatibility equations, as discussed in Example 3.17.

Solve the problems in this section by writing appropriate equations of each of the following three fundamental types:

1. *Equilibrium equations.*
2. *Force-Temperature-deformation relationships.*
3. *Geometry of deformation; Compatibility equations.*

Use the **Force Method** *to solve these simultaneous equations, that is,* (1) → (2) → (3).

▼ **FORCE-METHOD PROBLEMS**

Prob. 3.9-1. For Fig. P3.9-1, see Prob. 3.5-1. Two uniform, linearly elastic members are joined together at B, and the resulting two-segment rod is attached to rigid supports at ends A and C. A single external force, P_B, is applied to the joint at B. Member (1) is steel with modulus $E_1 = 30 \times 10^3$ ksi, cross-sectional area $A_1 = 2.0$ in² and length $L_1 = 80$ in.; member (2) is made of aluminum alloy with $E_2 = 10 \times 10^3$ ksi, $A_2 = 3.6$ in², and $L_2 = 60$ in. Let $P_B = 16$ kips, and use the *Force Method:* (a) to solve for the axial stresses σ_1 and σ_2 in the two elements, and (b) to solve for the horizontal displacement, u_B, of joint B.

Prob. 3.9-2. For Fig. P3.9-2, see Prob. 3.5-9. A three-segment rod is attached to rigid supports at ends A and D and is subjected to equal and opposite external loads P at nodes (joints) B and C, as shown in Fig. P3.9-2. The rod is homogeneous and linearly elastic, with modulus of elasticity E. Let $A_1 = A_3 = A$. and $A_2 = 2A$; let $L_1 = 2L$ and $L_2 = L_3 = L$. (a) Let the force in member (2) be the *redundant force,* and use the *Force Method* to determine the axial stresses σ_1, σ_2. and σ_3 in the elements. (b) Determine expressions for the horizontal displacements u_B and u_C, at nodes B and C, respectively.

Prob. 3.9-3. The five-segment stepped rod shown in Fig. P3.9-3 is stress free when it is attached to rigid supports at ends A and F. Subsequently, it is subjected to two external loads of magnitude P at nodes B and D, as shown in Fig. P3.9-3. The bar is homogeneous with modulus of elasticity E, and the cross-sectional areas are $A_1 = A_5 = A$, and $A_2 = A_4 = 2A$, and $A_3 = 3A$. Use the *Force Method* to solve for the reaction P_F at end F. (Hint: Let $P_F = F_5$ be the *redundant force,* and use the Force Method to solve directly for this force.)

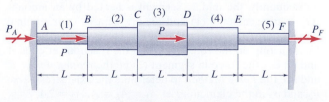

P3.9-3

Prob. 3.9-4. For Fig. P3.9-4, see Prob. 3.5-7. Consider the three-segment system shown in Fig. P3.9-4. Using the dimensions, material properties, and load given below, and using the *Force Method*, (a) determine the element axial forces F_1, F_2, and F_3, and (b) calculate the two nodal displacements u_B and u_C.

$$d_o = 50 \text{ mm}, \quad d_i = 36 \text{ mm}, \quad d = 20 \text{ mm}$$

$$L_1 = L_2 = 1 \text{ m}, \quad L_3 = 2 \text{ m}, \quad P_C = 50 \text{ kN}$$

$$E_1 = 210 \text{ GPa}, \quad E_2 = E_3 = 70 \text{ GPa}$$

Prob. 3.9-5. With Fig. P3.9-5 (See Prob. 3.5-11) and the data below, (a) use the *Force Method* to solve for the forces F_1 and F_2 in the support rods. Use force F_1 as the *redundant force*. (b) Determine the elongation of support rod (1).

$$A_1 = A_2 = 500 \text{ mm}^2, \quad L_1 = 1 \text{ m}, \quad L_2 = 2 \text{ m}$$

$$E_1 = E_2 = 210 \text{ GPa}, \quad P = 50 \text{ kN}$$

$$a = 0.50 \text{ m}, \quad b = 1.5 \text{ m}$$

Prob. 3.9-6. With Fig. P3.9-6 (See Prob. 3.8-14) and the data below, (a) use the *Force Method* to solve for the forces F_1 and F_2 in the support rods. Use force F_1 as the *redundant force*. (b) Determine the axial stress in each support rod. (c) Calculate the elongation of support rod (1).

$$A_1 = 1.0 \text{ in}^2, \quad A_2 = 0.50 \text{ in}^2, \quad L_1 = L_2 = 5 \text{ ft}$$

$$E_1 = E_2 = 10 \times 10^3 \text{ ksi}$$

$$a = 2 \text{ ft}, \quad b = 4 \text{ ft}, \quad P = 5 \text{ kips}$$

*Prob. 3.9-7. With Fig. P3.9-7 (See Prob. 3.5-16) and the data below. (a) use the *Force Method* to solve for the axial force F_1, F_2, and F_3 in the support bars. Use F_2 as the *redundant force*. (Hint: Take moments about A and D to relate F_3 and F_1 to F_2.) (b) Determine the vertical displacement of point B, the point of application of the load.

$$A = 500 \text{ mm}^2, \quad L = 2 \text{ m}, \quad E = 100 \text{ GPa}$$

$$a = 0.5 \text{ m}, \quad b = 1.0 \text{ m}, \quad c = 1.5 \text{ m}, \quad P = 20 \text{ kN}$$

Prob. 3.9-8. For Fig. P3.9-8, see Prob. 3.8-7. A three-segment rod is attached to rigid supports at ends A and D. When the rod is initially attached to the supports, it is stress free. Subsequently, the middle segment is heated by an amount ΔT_2, while segments (1) and (3) are kept at their original temperature (i.e., $\Delta T_1 = \Delta T_3 = 0$). There are no external forces acting on the rod, other than the reactions at ends A and D. Let the force in segment (2) be the *redundant force*, and use the *Force Method* to determine the axial stresses σ_1, σ_2, and σ_3 in the elements. Let $A_1 = A_3 = A, A_2 = 2A, L_2 = L_3 = L, L_1 = 2L, E = \text{const}, \alpha = \text{const}$.

Prob. 3.9-9. For Fig. P3.9-9, see Prob. 3.6-8. A steel pipe with outer diameter d_o and inner diameter d_i, and a solid aluminum-alloy rod of diameter d form a three-segment system. When the system is welded to rigid supports at A and D, it is stress-free. Subsequently, the steel pipe is cooled by $100°\text{F}$ (i.e., $\Delta T_1 = -100°\text{F}$). while $\Delta T_2 = \Delta T_3 = 0$. Let the force in segment (2) be the *redundant force*, and use the *Force Method* to determine the axial stresses σ_1, σ_2, and σ_3 in the elements.

$$d_o = 2 \text{ in.}, \quad d_i = 1.5 \text{ in.}, \quad d = 0.75 \text{ in.}$$

$$L_1 = L_2 = 30 \text{ in.}, \quad L_3 = 50 \text{ in.}$$

$$E_1 = 30 \times 10^3 \text{ ksi}, \quad E_2 = E_3 = 10 \times 10^3 \text{ ksi}$$

$$\alpha_1 = 6.5 \times 10^{-6}/°\text{F}$$

Prob. 3.9-10. With the figure and data of Prob. 3.6-15, (a) use the *Force Method* to solve for the axial forces F_1 and F_2 in the two rods. Use F_2 as the *redundant force*. (b) Determine the (counterclockwise) rotation θ of the bracket ABC.

Prob. 3.9-11. For Fig. P3.9-11, see Prob. 3.6-17. A "rigid" beam AC is supported by three vertical rods, as shown in Fig. P3.9-11. When originally assembled, the rods are all stress-free and beam AC is horizontal. If rod (1) is maintained at the assembly temperature (i.e., $\Delta T_1 = 0$), and rods (2) and (3) are heated the same amount ($\Delta T_2 = \Delta T_3 = \Delta T$), determine the stresses induced in the three support rods. Use the *Force Method* to solve this problem, letting F_1 be the *redundant force*. (Hint: Take $(\Sigma M)_B$ and $(\Sigma M)_C$ for the equilibrium equations.)

$$A_1 = A, \quad A_2 = A_3 = 2A, \quad E_1 = E_2 = E_3 = E$$

$$\alpha_2 = \alpha_3 = \alpha$$

Prob. 3.9-12. For Fig. P3.9-12, see Prob. 3.7-3. For the piperod system of Prob. 3.7-3, use the *Force Method* to determine the stresses σ_1 and σ_2 induced in elements (1) and (2), respectively, when the gap at C is closed.

Prob. 3.9-13. For Fig. P3.9-13, see Prob. 3.7-6. With the data of Prob. 3.7-6, use the *Force Method* to solve for the axial stresses, σ_1 and σ_2, in the two rod segments when the load $P = 50 \text{ kN}$ is applied. Let F_1 be the *redundant force*.

Prob. 3.9-14. For Fig. P3.9-14, see Prob. 3.7-3. Solve Prob. 3.7-3. if, in addition to having the flange tightened against the wall at C, the pipe-rod system ABC is heated uniformly by $50°\text{F}$. Use $\alpha = 6.5 \times 10^{-6}/°\text{F}$.

Prob. 3.9-15. For Fig. P3.9-15, see Prob. 3.7-12. With the data of Prob. 3.7-12, use the *Force Method* to solve for the axial stresses, σ_1 and σ_2, in the two support rods after rod (2) is attached to the upper bracket and the load $P = 20 \text{ kN}$ is applied. Let F_1 be the *redundant force*.

Prob. 3.9-16. With the figure and data of Example 3.16, (a) use the *Force Method* to solve for the axial forces F_1, F_2, and F_3 in the three rods. Use F_1 and F_2 as the two *redundant forces*. (b) Determine the (counterclockwise) rotation θ of the bracket ABD.

▼ **STATICALLY DETERMINATE TRUSSES**

Prob. 3.10-1. Bar AC is $L_1 = 8$ ft long, has a cross sectional area $A_1 = 1.0$ in^2, and is made of steel with a modulus of elasticity $E_1 = 30 \times 10^3$ ksi. Member BC has the following properties: $L_2 = 10$ ft, $A_2 = 1.4$ in^2, and $E_2 = 10 \times 10^3$ ksi. A load $P = 10$ kips acts downward on the pin at joint C. (a) Determine the axial stresses in rods (1) and (2). (b) Determine the horizontal and vertical displacements of joint C, u_C and v_C, respectively.

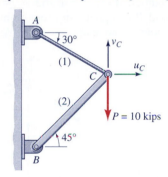

P3.10-1 and P3.10-2

Prob. 3.10-2. Repeat Prob. 3.10-1 if member (1) is heated by 80°F and, at the same time, the load $P = 10$ kips is applied. Member (2) is held at the reference temperature. Use $\alpha_1 = 8 \times 10^{-6}$/°F.

Prob. 3.10-3. A two-bar planar truss has the geometry shown in Fig. P3.10-3. A force $P = 200$ kN is applied to the truss at joint C. (a) What are the resulting stresses in elements (1) and (2)? (b) What are the horizontal and vertical displacements of joint C, u_C and v_C, respectively?

$$A_1 = A_2 = 1500 \text{ mm}^2, \quad E_1 = E_2 = 70 \text{ GPa}$$

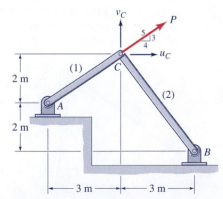

P3.10-3, P3.10-4, and P3.10-5

Prob. 3.10-4. Repeat Prob. 3.10-3 if member (2) is heated by 25°C and, at the same time, the load $P = 200$ kN is applied to the truss at C. Member (1) is held at the reference temperature, that is, $\Delta T_1 = 0$. Use $\alpha_2 = 23 \times 10^{-6}$/°C.

ᴰProb. 3.10-5. A force $P = 400$ kN is applied at joint C of the planar truss in Fig. P3.10-5. (a) Determine the required cross-sectional areas of members (1) and (2) if the allowable stresses are 100 MPa in tension and 75 MPa in compression. (b) For the truss with areas determined in Part (a), determine the displacements u_C and v_C of joint C. $E_1 = E_2 = 70$ GPa.

Prob. 3.10-6. Each of the two bars of the planar truss in Fig. P3.10-6 has length L, and they are made of a material with modulus of elasticity E. If $A_1 = A$, and $A_2 = 2A$, and a horizontal load P is applied at joint C, determine: (a) the stresses σ_1 and σ_2 in the two bars, and (b) the horizontal displacement, u_C, and vertical displacement, v_C, of the pin joint at C.

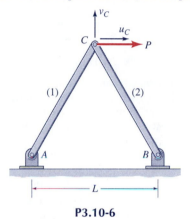

P3.10-6

Prob. 3.10-7. A planar truss ACB is part of an apparatus used to lift a weight W, as shown in Fig. P3.10-7. If $W = 4$ kips, determine: (a) the stresses σ_1 and σ_2 in the two truss members, and (b) the horizontal and vertical displacements of joint C, u_C and v_C, respectively.

$$A_1 = 0.8 \text{ in}^2, \quad A_2 = 1.0 \text{ in}^2$$
$$E_1 = 10 \times 10^3 \text{ ksi}, \quad E_2 = 30 \times 10^3 \text{ ksi}$$

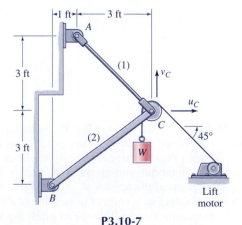

P3.10-7

ᴰProb. 3.10-8. The two tie rods in Fig. P3.10-8 support a maximum horizontal load $P = 10$ kips at joint B. Rod (1) is made of structural steel with a modulus of elasticity $E_2 = 29 \times 10^3$ ksi

229

and a yield point of $\sigma_{Y1} = 36$ ksi. The corresponding properties for the aluminum rod (2) are $E_2 = 10 \times 10^3$ ksi and $\sigma_{Y2} = 60$ ksi. (a) If the two-bar truss is to have a factor of safety of 2.5 with respect to failure by yielding, what are the required areas A_1 and A_2? (b) If the truss is sized according to the requirements stated in Part (a), and loaded by a force $P = 10$ kips, what will be the values of u_B and v_B, the horizontal and vertical displacements at joint B, at the maximum-load condition?

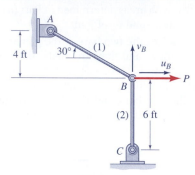

P3.10-8

*Prob. 3.10-9. The rigid beam BD in Fig. P3.10-9 is supported by a steel rod that is connected to the beam by a smooth pin at C. The beam is horizontal when there is no weight W hanging at D. The properties of rod AC are: $A = 500$ mm^2 and $E = 200$ GPa. (a) If a weight $W = 10$ kN is suspended from the beam at D, what is the stress in rod AC? Neglect the weight of the beam. (b) What is the elongation of rod AC? (c) What is the vertical displacement of D? (Make the assumption that beam BD rotates through an angle that is small, so points C and D can be assumed to move only vertically.)

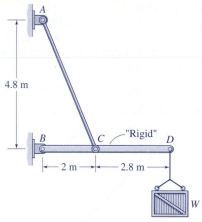

P3.10-9

^{D}**Prob. 3.10-10.** The crane hoist in Fig. P3.10-10 consists of a "rigid," "weightless" boom BC supported by a tie-rod AB. The rod has a cross-sectional area $A = 0.75$ in^2 and is made of steel with modulus of elasticity $E = 29 \times 10^3$ ksi. (a) If the yield point of the steel is $\sigma_Y = 50$ ksi and there is to be a factor of safety with respect to yielding of $FS = 2.5$, what is the maximum weight W that can safely be hoisted by the cable suspended from the pin at B? (b) If a load $W = 2$ kips is suspended from the pin at B, what will be the vertical displacement v_B? (Note: Assume that the angle through which boom BC rotates is small, so point B moves

along the dashed line perpendicular to the original line BC. That is $v_B = -\frac{4}{3}u_B$.)

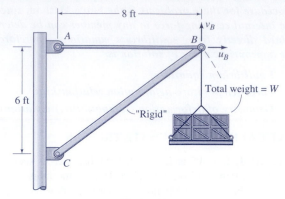

P3.10-10

Prob. 3.10-11. Truss members (1) and (2) in Fig. P3.10-11 have the same modulus of elasticity E. The cross-sectional areas of the members are $A_i = A$ and $A_2 = 2A$. At what angle θ must load P be applied in order to make $v_B = u_B$, that is, to make joint B move upward to the right at 45°?

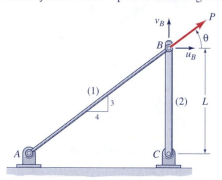

P3.10-11

*Prob. 3.10-12. The cross-sectional area of tie rod AB is $A_1 = A$, and the cross-sectional area of post BC is $A_2 = 8A$. Both are made of the same material with modulus of elasticity E. A load P is applied horizontally at B. If the length, $L_2 = L$, of the post is fixed, but the length L_1 and position of the bracket at A are both variable, determine the angle θ of the tie rod that will minimize the horizontal displacement u_B at B. (Hint: Obtain an expression for u_B as a function of θ; differentiate this expression with respect to θ; and set $du_B/d\theta$ equal to 0. You will obtain an equation involving $\sin \theta$ and $\cos \theta$ that can be solved by trial and error or by computer.)

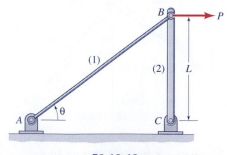

P3.10-12

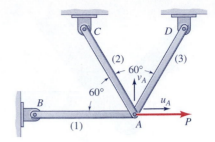

P3.10-16, P3.10-17, P3.10-18, and P11.5-54

▼ **STATICALLY INDETERMINATE TRUSSES**

Prob. 3.10-13. The three truss members in Fig. P3.10-13 all have cross-sectional area A and modulus of elasticity E. A horizontal load P is applied to the truss at joint A. (a) Determine expressions for the horizontal and vertical displacements of joint A, that is, u_A and v_A. (b) Determine expressions for the member axial forces F_1, F_2, and F_3.

Prob. 3.10-17. For the truss in Fig. P3.10-17: $L_1 = L_2 = L_3 = 2.1$ m, $E_1 = E_2 = E_3 = 70$ GPa, $A_1 = A_2 = 600$ mm^2, and $A_3 = 900$ mm^2. (a) If the truss is subjected to a horizontal force $P = 60$ kN at joint A, what are the resulting horizontal displacement, u_A, and vertical displacement, v_A, at that joint? (b) What are the resulting axial forces in the members—F_1, F_2, and F_3?

Prob. 3.10-18. Re-solve Problem 3.10-17, but this time let $P = 0$. $\Delta T_1 = \Delta T_3 = 0$, and $\Delta T_2 = -20°C$. Let $\alpha_2 = 23 \times 10^{-6}/°C$.

Prob. 3.10-19. For the truss in Fig. P3.10-19: $A_1 = A_2 = A_3 = 1.0$ in^2, $E_1 = E_2 = E_3 = 30 \times 10^3$ ksi, and $P = 15$ kips. (a) Determine the horizontal displacement, u_A, and vertical displacement, v_A, at joint A. (b) Determine the member axial forces—F_1, F_2, and F_3.

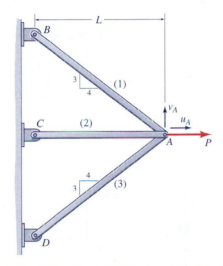

P3.10-13, P3.10-14, P3.10-15, and P11.5-53

Prob. 3.10-14. Repeat Prob. 3.10-13 with $A_1 = 2A$, $A_2 = A_3 = A$.

Prob. 3.10-15. Repeat Prob. 3.10-13 with $A_1 = 2A$, $A_2 = A_3 = A$. Let $P = 0$, but let member (2), whose coefficient of thermal expansion is α_2, be heated by an amount $\Delta T_2 = \Delta T$, with $\Delta T_1 = \Delta T_3 = 0$.

Prob. 3.10-16. Each of the three truss members in Fig. P3.10-16 has a length L and modulus of elasticity E. The cross-sectional areas of the members are $A_1 = A_2 = A$ and $A_3 = 2A$. (a) Determine expressions for the horizontal and vertical displacements, u_A and v_A, at joint A when a horizontal load P is applied to the right to the pin at joint A. (b) Determine expressions for the axial forces—F_1, F_2, and F_3—in the three truss members when load P is applied.

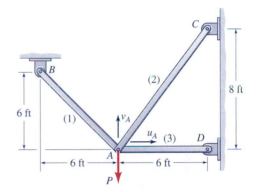

P3.10-19, P3.10-20, P3.10-21, P3.10-22(a), and P11.5-55

Prob. 3.10-20. If $P = 0$, how much must member (2) in Prob. 3.10-19 be cooled to cause joint A to move upward by 0.05 in. (i.e., $v_A = 0.05$ in.)? Let $\alpha_2 = 6.5 \times 10^{-6}/°F$?

Prob. 3.10-21. Consider the truss in Prob. 3.10-19. Suppose that member (3) was originally fabricated 0.05 in. too short (i.e., $\bar{\delta}_3 = 0.05$ in.) so that it had to be temporarily stretched enough to insert the pin at A. (a) If $P = 0$, what displacements u_A and v_A will result when the truss is forcefully assembled as described above? (b) What "initial stresses"—σ_1, σ_2, and σ_3—will be induced in the members when the truss is forcefully assembled?

*Prob. 3.10-22. Re-solve Prob. 3.10-19, this time assuming that there is a clearance of 0.02 in. in the hole at A in member (2) so that the pin does not act on member (2) at A until this clearance (gap) has been closed. See Prob. 3.10-19 for Fig. P3.10-22a.

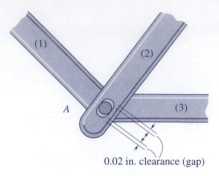

0.02 in. clearance (gap)

P3.10-22(b)

Prob. 3.10-23. Three members form the pin-jointed truss in Fig. P3.10-23. The joint at A is constrained by a slider block to move only in the horizontal direction (i.e., $v_A = 0$). All members have the same cross-sectional area $A = 400$ mm^2, and the modulus of elasticity is $E = 70$ GPa for all three members. (a) Determine the horizontal load P that would be required to move joint A to the right by 5 mm (i.e., $u_A = 5$ mm). (b) Determine the member forces—F_1, F_2, and F_3—corresponding to the load P determined in Part (a), (c) Determine the vertical reaction between the slider block at A and the track.

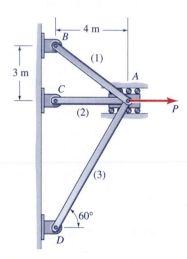

P3.10-23, P3.10-24, and P3.10-25

Prob. 3.10-24. The coefficient of thermal expansion for all members of the truss in Prob. 3.10-23 is $\alpha = 23 \times 10^{-6}/°$C. (a) If $P = 0$, but all three members of the truss are uniformly heated by $\Delta T = 20°$C, how much will joint A move? That is, what horizontal displacement, u_A, will occur as a result of the heating of all members? (b) What stresses—σ_1, σ_2, and σ_3—will be induced in the truss members due to the uniform heating?

Prob. 3.10-25. Member (2) of the truss in Prob. 3.10-23 was manufactured 2 mm too short (i.e., $\delta_2 = -2$ mm), so that members of the truss have to be forcefully assembled. (a) What will be the "initial displacement" of joint A, that is, what is the value of u_A prior to the application of any external load P? (b) What "initial stresses"—σ_1, σ_2, and σ_3—will be induced in the truss members due to the error in the original length of member (2)?

232

Prob. 3.10-26. Using the Pythagorean Theorem, show that if u/L is "small" (i.e., $u/L \ll 1$), the change in length, e_u, of member AB, of initial length L, is given by

$$e_u = u \cos \theta$$

if end B moves in the x-direction from B to B^* by an amount u, while end A does not move. (Note: This problem relates to Fig. 3.22 and Eq. 3.26.)

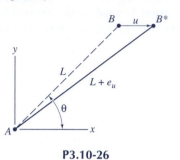

P3.10-26

▼ INELASTIC AXIAL DEFORMATION

*Prob. 3.11-1. The two-segment bar in Fig. P3.11-1a has a constant cross section and is homogeneous, with bilinear stress-strain behavior given by the diagram in Fig. P3.11-1b. External loads of magnitude P are applied at B and C as shown. (a) If $A = 1.0$ in^2, determine the value of P that causes first yielding in the bar AC. (Call this load P_{Y1}.) (b) Determine the total elongation, u_C, of the bar AC if $P = 20$ kips and $L = 20$ in.

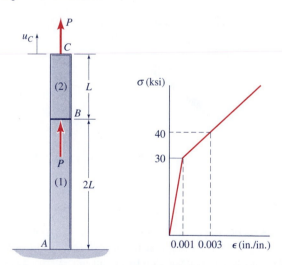

(a) A two-segment bar. (b) A bilinear stress-strain curve.

P3.11-1 and P3.11-2

*Prob. 3.11-2. For the two-segment bar in Prob. 3.11-1, (a) determine the value, P_{Y1}, of load P that causes first yielding in the two-segment bar AC. (b) Determine the value, P_{Y2}, of load P that causes the other segment to yield also. (c) Sketch a load-elongation diagram (i.e., P vs. u_C) for loads up to $2P_{Y2}$.

Prob. 3.11-3. The two-segment statically indeterminate bar in Fig. P3.11-3a has a constant cross-sectional area $A = 0.8$ in². It is made entirely of material that has an elastic, perfectly-plastic stress-strain behavior illustrated in Fig. P3.11-3b, with $\sigma_Y = 36$ ksi and $E = 30 \times 10^3$ ksi. (a) Determine the load P_Y at which first yielding occurs, and determine the corresponding displacement u_Y of section B where the load P is applied. (b) Determine the load P_U at which yielding occurs in the remaining segment of the bar, and determine the corresponding displacement u_U of section B. (c) Sketch a load-displacement diagram, that is, sketch a diagram of P versus u up to P_U.

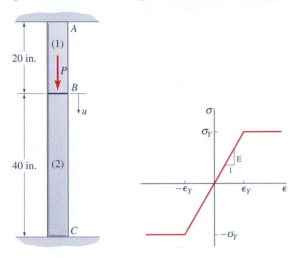

(a) A two-segment bar.   (b) A stress-strain diagram for an elastic, perfectly-plastic material.

P3.11-3 and P3.11-7

Prob. 3.11-4. Repeat Prob. 3.11-3 for the two-segment bar in Fig. 3.11-4. Let the bar have a stress-strain diagram of the form illustrated in Fig. P3.11-3b, with $\sigma_Y = 250$ MPa and $E = 200$ GPa. The cross-sectional areas of the respective segments of the bar are $A_1 = 500$ mm² and $A_2 = 800$ mm².

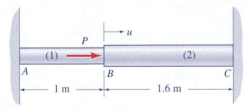

P3.11-4 and P3.11-8

*Prob. 3.11-5. A symmetric three-bar planar truss is loaded by a single horizontal force P at joint A. The members of the truss are all made of the same linearly elastic, perfectly-plastic material (see Fig. P3.11-3b) with $\sigma_Y = 36$ ksi and $E = 30 \times 10^3$ ksi, and all have a cross-sectional area $A = 2.0$ in². (a) Determine the load P_Y at which first yielding occurs, and determine the corresponding displacement u_Y of joint A, where the load P is applied. (b) Determine the load P_U at which yielding occurs in the remaining members of the truss, and determine the corresponding displacement u_U of joint A. (c) Sketch a load-displacement diagram, that is, sketch a diagram of P versus u up to P_U.

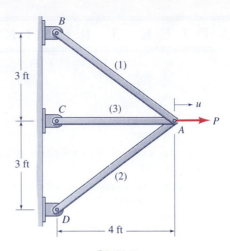

P3.11-5

*Prob. 3.11-6. When there is no load on the two-segment bar in Fig. P3.11-6, there is a 1-mm gap between the bar and the rigid wall at C. The two-segment rod is made of linearly elastic, perfectly plastic material with a stress-strain diagram like the one in Fig. P3.11-3b, with $\sigma_Y = 250$ MPa and $E = 200$ GPa. The bar has a constant cross-sectional area $A = 500$ mm². (a) Determine the value of P, say P_R, where the right end of the bar initially makes contact with the rigid wall at C. Determine the displacement u_R of node B that corresponds to the load P_R. (b) Determine the load P_Y at which first yielding occurs, and determine the corresponding displacement u_Y of section B where the load P is applied. (c) Determine the load P_U at which yielding occurs in the remaining segment of the bar, and determine the corresponding displacement, u_U, of section B. (d) Sketch a load-displacement diagram, that is, sketch a diagram of P versus u up to P_U.

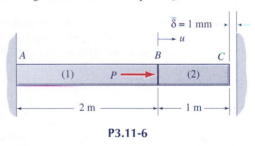

P3.11-6

> Problems 3.11-7 through 3.11-8 are statically indeterminate, and they involve unloading as well as loading. You are strongly urged to follow the procedure used in Example Problem 3.22 of tracking the loading and unloading stress-strain behavior of each element on a plot like the one in Fig. 1b of that example problem.

▼ RESIDUAL STRESSES

*Prob. 3.11-7. Determine the residual stresses in segments (1) and (2) of the bar AC in Prob. 3.11-3 if a force $P = 50$ kips is applied and then completely removed.

*Prob. 3.11-8. Determine the residual stresses in segments (1) and (2) of the bar AC in Prob. 3.11-4 in a force $P = 300$ kN is applied and then completely removed.

Section		Suggested Review Problems
3.2	**Axial deformation** of a member has two characteristics: (1) the axis of the member remains straight, and (2) plane cross sections remain plane and remain perpendicular to the axis after deformation. Geometry of axial deformation (Fig. 3.2).	
	The **extensional strain** ϵ is related to the **axial displacement** u by the equation $$\epsilon(x) = \frac{du(x)}{dx} \qquad (3.1)$$ The **total elongation** of a member undergoing axial deformation is given by $$e = \int_0^L \epsilon(x)\,dx$$	Derive Eq. 3.1.
	The **stress-strain equation** for linearly elastic behavior is called **Hooke's law.** $$\sigma(x) = E\epsilon(x)$$	
	The **normal stress** on the cross section at x of a linearly elastic member with $E = $ constant (or $E = E(x)$) that is undergoing axial deformation is given by the **axial stress formula** $$\sigma(x) = \frac{F(x)}{A(x)}$$ A member having $E = $ const. will undergo axial deformation if the applied loading results in a resultant axial force $F(x)$ that acts through the centroid at cross section x. Resultant force for axial deformation of a bar with $E = $ constant (or $E = E(x)$) (Fig. 3.7).	Prove this
	The **elongation** of a linearly elastic member undergoing axial deformation is given by $$e = \int_0^L \frac{F(x)\,dx}{A(x)E} \qquad (3.11)$$	State Eq. 3.11 restrictions.

Section		Suggested Review Problems
	The **force-deformation equation** for linearly elastic behavior of a uniform* member undergoing axial deformation can be stated in the form or, in the form (*uniform: E = const., A = const.)	$$e = fF \quad \text{where } f = \frac{L}{AE} \qquad (3.14)$$ f is the *flexibility coefficient* $$F = ke \quad \text{where } k = \frac{AE}{L} \qquad (3.15)$$ k is the *stiffness coefficient* A uniform* axial-deformation member (Fig. 3.10a).	See Probs. for Sects. 3.4 & 3.5.
3.3	Section 3.3 treats three examples of **nonuniform linearly elastic axial deformation**: (1) Nonhomogeneous members, $E = E(y, z)$ (Example 3.1), (2) Members with varying cross section, $A = A(x)$ (Example 3.2), and (3) Members with distributed axial loading, $F = F(x)$ (Example 3.3). For all of these cases of axial deformation, the **extensional strain is constant at any cross section x;** that is, $\epsilon = \epsilon(x)$. So the **axial stress** has the form	 $$\sigma(x, y, z) = E(x, y, z)\epsilon(x)$$	3.3-3 3.3-5 3.3-7 3.3-13
3.4	A **statically determinate structure** is one for which all reactions and all internal forces can be determined by the use of *equilibrium equations* (i.e., statics) alone.	 Typical statically determinate structures with uniform axial-deformation elements	3.4-3 3.4-7 3.4-9 3.4-16

Section		Suggested Review Problems
3.5	Analysis of a **statically indeterminate structure** requires the use of (1) *equilibrium equations*, (2) member *force-deformation equations*, and (3) *geometry of deformation equations (compatibility equations)*. Review the *Solution Procedure* and the *Flow Chart* for the **Basic Force Method** before solving the Suggested Review Problems. Typical statically indeterminate structures with uniform axial-deformation elements	3.5-1 3.5-9 3.5-11 3.5-17
3.6	Section 3.6 treats the effect of **temperature change** on axial deformation. The **stress-strain-temperature equation** for axial deformation is $$\epsilon_T = \alpha \Delta T \qquad (3.23)$$ $$\epsilon = \epsilon_\sigma + \epsilon_T = \frac{\sigma}{E} + \alpha \Delta T \qquad (3.24)$$	3.6-1 3.6-5 3.6-11 3.6-13
3.7	Section 3.7 treats the effect of **geometric "misfits"** on axial deformation. The *equilibrium equation(s)* and *force-deformation equation(s)* are not directly affected.	3.7-3 3.7-5 3.7-9 3.7-13
3.8	Section 3.8 introduces the **Displacement Method** and illustrates its use for analyzing simple structures that consist of uniform axial-deformation members. Review the *Solution Procedure* and the *Flow Chart* for the **Displacement Method** before solving the Suggested Review Problems. $$F_i = k_i(e - \alpha \Delta T)_i \qquad (3.28)$$ where e is the <u>total</u> elongation and $$k_i = \left(\frac{AE}{L}\right)_i$$	3.8-3 3.8-7 3.8-15 3.8-17

Sections 3.9–3.11 are all "optional" sections.

TORSION

4.1   INTRODUCTION

In Chapter 3 we considered the behavior of slender members subjected to axial loading, that is, to forces applied along the longitudinal axis of the member. In this chapter we will concentrate on the behavior of slender members subjected to torsional loading, that is, loading by couples that produce twisting of the member about its axis.

Figure 4.1 shows a common example of torsional loading and indicates the shear stresses and the stress resultant associated with torsion. A torque (couple) of magnitude $2Pb$ is applied to the lug-wrench shaft AB by the application of equal and opposite forces of magnitude P at the ends of arm CD (Fig. 4.1a,b). We say that shaft AB is a *torsion member*. As indicated in Fig. 4.1c, the shaft AB is subjected to equal and opposite *torques* of magnitude T that twist one end relative to the other, and, as shown in Fig. 4.1d, the torque T acting on a cross section between A and B is the resultant of distributed shear stresses.

Another common example of a torsion member is a power transmission shaft, like the drive shaft of an automobile or a truck. Several names are applied to torsion members, depending on the application: *shaft, torque tube, torsion rod, torsion bar*, or simply *torsion member*. This chapter deals primarily with torsion of slender members with circular cross sections, such as solid or tubular circular cylinders. However, torsion of noncircular thin-wall tubular members and torsion of noncircular prismatic bars are treated in Sections 4.9 and 4.10, respectively.

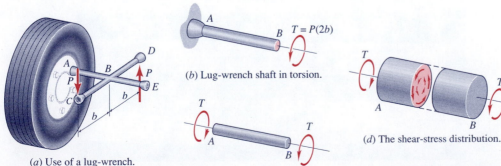

(*a*) Use of a lug-wrench.

(*b*) Lug-wrench shaft in torsion.

(*c*) A torsion rod.

(*d*) The shear-stress distribution.

FIGURE 4.1   An example of torsion.

TABLE 4.1 Analogy Between Axial Deformation and Torsion	
Axial Deformation	**Torsion**
Axial Force (F)	Torque (T)
Elongation (e)	Twist angle (ϕ)
Normal stress (σ)	Shear stress (τ)
Extensional strain (ϵ)	Shear strain (γ)
Modulus of elasticity (E)	Shear modulus (G)

You will find that the two key concepts—*stress* and *strain*—and the three fundamental types of equations—*equilibrium, material behavior*, and *geometry of deformation*—which were stressed in Chapter 3, have their counterparts in the analysis of torsion. There is a direct analogy between axial deformation and torsion, as indicated by the entries in Table 4.1. Thus, although there is new theory to be learned in Chapter 4, particularly in Sections 4.2, and 4.3 you should quickly feel at home solving problems in the same systematic manner used to solve the axial-deformation problems of Chapter 3.

4.2 TORSIONAL DEFORMATION OF CIRCULAR BARS

Geometry of Deformation of Circular Bars. As noted in the previous section, most of this chapter concerns torsion of members with circular cross sections. Figure 4.2 shows the deformation patterns of a circular steel rod (Fig. 4.2a) and a square steel bar (Fig. 4.2b) that have been subjected to torsion. In each case, the upper figure is of the undeformed member, on which lines have been scribed parallel to the axis and perpendicular to the axis; the latter represent cross sections. **When a circular shaft, whether solid or tubular, is subjected to torsion, each cross section remains plane and simply rotates about the axis of the member,** as you can see by comparing the upper figure and lower figure of Fig. 4.2a. On the other hand, as can be seen in Fig. 4.2b, cross sections of the square bar become warped when the bar is twisted about its axis. Because of the mathematical simplicity of the theory of torsion

(a) A steel torsion rod with circular cross section.

(b) A steel torsion bar with square cross section.

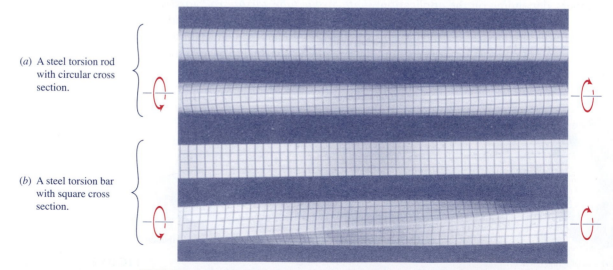

FIGURE 4.2 Examples of torsional deformation. (Roy Craig)

of members with circular cross section, and because of the widespread application of such members, we will now develop the *theory of torsion for circular members.* Torsion of solid noncircular bars is treated later in Section 4.10.

We begin with an analysis of the geometry of deformation and develop a **strain-displacement equation for torsion of circular members.** Figure 4.3 shows a circular cylinder before and after the application of equal and opposite torques to the ends of the member. Note that the circumferential lines, which represent plane cross sections before deformation, remain in a plane after deformation, but that longitudinal lines, which are parallel to the axis of the member before deformation, become helical as a result of torsional deformation. Furthermore, right angles, such as the before-deformation angle ABC, are no longer right angles after deformation. This angle change is evidence of shear deformation due to torsion. We will now develop a mathematical expression for the *torsional strain-displacement relationship*.

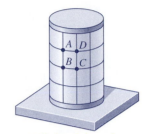

(a) Before deformation.

Strain-Displacement Analysis. On the basis of observation of how a circular bar deforms when subjected to twisting about its axis, as illustrated in Figs. 4.2 and 4.3, torsion of circular members can be characterized by three fundamental **torsional-deformation assumptions:**

1. *The axis remains straight and remains inextensible.*
2. *Every cross section remains plane and remains perpendicular to the axis.*
3. *Radial lines remain straight and radial as the cross section rotates about the axis.*

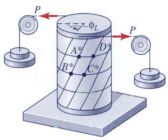

The torsional-deformation assumptions are illustrated in Fig. 4.4a where, before deformation, $ABFE$ is a radial plane. When torque T_L is applied to the end $x = L$, the radial lines CD and EF remain straight and rotate to positions CD^* and EF^*, respectively, as the cross sections rotate through angles $\phi(x)$ and $\phi_L \equiv \phi(L)$, respectively.

(b) After deformation.

FIGURE 4.3 The deformation of a circular cylinder.

A **sign convention for torsion** is defined as follows:

- The *longitudinal axis* of the bar is labeled the x axis, with one end of the member being taken as the origin. (In Fig. 4.4a the left end is taken as the origin, with the x axis going from left to right.)

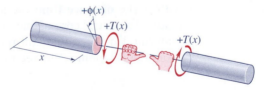

(b) Sign convention for internal (resisting) torque $T(x)$.

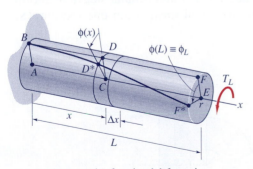

(a) An example of torsional deformation.

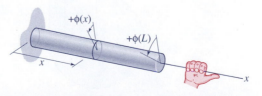

(c) Sign convention for angle of rotation $\phi(x)$.

FIGURE 4.4 Torsional deformation; Sign convention for torsion.

- A *positive torque, T(x)*, is a moment that acts on the cross section at x in a right-hand-rule sense about the outer normal to the cross section. On a cross-sectional cut at x there will be equal and opposite torques $T(x)$, as indicated in Fig. 4.4b.

- A *positive angle of rotation, $\phi(x)$*, is a rotation of the cross section at x in a right-hand-rule sense about the x axis, as illustrated in Fig. 4.4c.

Our task now is to establish a mathematical relationship between torsional deformation (rotation angle ϕ) and the resulting shear strain γ. To determine the shear strain associated with twisting of a circular cylinder, like the one in Fig. 4.4a, we use Fig. 4.5a, where we have redrawn the portion of the shaft between the crass sections at x and $(x + \Delta x)$. In Fig. 4.5b we concentrate on the central core of radius p. The angle QRS in Fig. 4.5b is a right angle. However, as a result of torsional deformation the angle QRS becomes angle $Q*R*S*$, which is no longer a right angle, but is smaller by the shear-strain angle

$$\gamma \equiv \gamma(x, \rho) = \frac{\pi}{2} - \angle Q*R*S* = \angle S'R*S*$$

as seen in Fig. 4.5b. Since γ is small, we can approximate the angle by its tangent, at the same time taking the limit as $\Delta x \to 0$, to get

$$\gamma = \lim_{\Delta x \to 0} \frac{S*S'}{R*S'} = \lim_{\Delta x \to 0} \frac{\rho \Delta \phi}{\Delta x} = \rho \frac{d\phi}{dx}$$

Therefore, the **strain-displacement equation** for torsional deformation of a circular member is

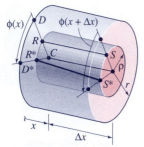

(a) An element of length Δx.

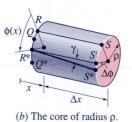

(b) The core of radius ρ.

FIGURE 4.5 Torsional deformation details.

$$\boxed{\gamma \equiv \gamma(x, \rho) = \rho \frac{d\phi}{dx}} \qquad \begin{array}{l} \textbf{Strain-} \\ \textbf{Displacement} \\ \textbf{Equation} \end{array} \qquad (4.1)$$

Here γ is the shear strain at cross section x at a distance ρ from the axis. The derivative $d\phi/dx$ is called the **twist rate.** Figure 4.6 shows plots of this *shear-strain distribution* at a typical cross section of a solid circular cylinder (Fig. 4.6a) and of a tubular circular cylinder (Fig. 4.6b). Note that, in each case, **the shear strain, γ, varies linearly with ρ, the distance from the axis.** Even if a circular torsion member were to be made of a central core of one material bonded to an outer tubular sleeve of another material. Eq. 4.1 would still represent the torsional shear-strain distribution. (See Example Problem 4.1 in Section 4.3.)

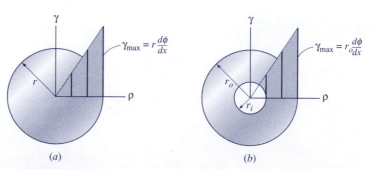

FIGURE 4.6 Examples of torsional shear-strain distribution.

Examples that involve the torsional strain-displacement behavior given by Eq. 4.1 and illustrated in Fig. 4.6 follow in Section 4.3.

4.3   TORSION OF LINEARLY ELASTIC CIRCULAR BARS

Shear Stresses Due to Torsion. The twisting of a circular shaft produces shear strains throughout the shaft, as was described in Section 4.2. This torsional deformation, which is illustrated again in Fig. 4.7a, results in shear stresses in the shaft. The shear stress on a typical cross section is illustrated in Fig. 4.7b; and Fig. 4.7c shows the shear stress, τ, and the corresponding shear strain, γ, at typical points. As indicated by Fig. 4.7d, the shear stress has the same distribution, $\tau(x, \rho)$, along every radial line in the cross section at x. Also note, in Figs. 4.7c and 4.7d, that **there is shear stress not only on cross sections, but there is always an accompanying shear stress that acts on radial planes** (from Eq. 2.37).

Stress Resultant and Equilibrium. Figure 4.8 shows the increment of shear force, dF_s, contributed by the shear stress τ acting on an incremental area dA at distance ρ from the center. The resultant of the incremental torques $dT = \rho \, dF_s$ is the torque $T(x)$ at the cross section, which is given by

$$T = \int_A \rho \, dF_s = \int_A \rho \tau \, dA \qquad \begin{array}{l} \textbf{Resultant} \\ \textbf{Torque} \end{array} \qquad (4.2)$$

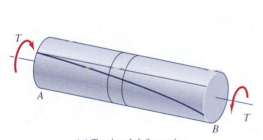

(a) Torsional deformation.

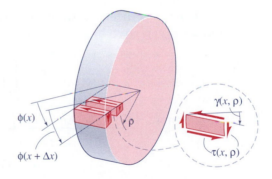

(c) Shear stress and shear strain at typical points.

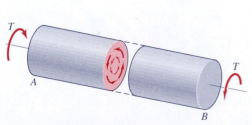

(b) Shear stress due to torsion.

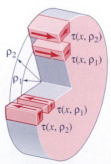

(d) Shear stresses along two typical radial lines in a cross section, and shear stress on radial planes.

FIGURE 4.7   Relationship of shear strain to shear stress in torsion of a circular shaft.

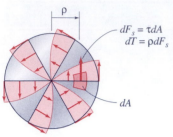

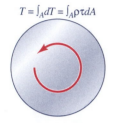

(a) The shear stress distribution.

$$T = \int_A dT = \int_A \rho\tau\, dA$$

(b) The resultant torque.

FIGURE 4.8 The relationship of torque to shear stress.

Equation 4.2 is general, since the distribution of τ is not specified in this equation. We will now determine the shear-stress distribution $\tau(x, \rho)$ for a linearly elastic torsion bar, and will relate the shear stress to the resultant torque, $T(x)$, at the cross section.

Linearly Elastic Stress-Strain Behavior.

To continue our development of the theory of torsion of circular bars we need a relationship between the shear strain. $\gamma(x, \rho)$, and the shear stress, $\tau(x, \rho)$. Chapter 2 describes how a tension test may be used to determine the uniaxial stress-strain behavior of materials, that is, σ versus ϵ. We will defer discussion of torsion testing until later in this chapter (Section 4.4) and will here confine our attention to the simplest case — **linearly elastic material behavior**.[1] For this case, Eq. 2.23 gives Hooke's Law for shear as

$$\boxed{\tau = G\gamma} \qquad \text{Hooke's Law for Shear} \qquad (4.3)$$

where G is called the shear modulus of elasticity, or simply the *shear modulus*.

By combining Eqs. 4.1 and 4.3 we obtain

$$\boxed{\tau = G\rho\,\frac{d\phi}{dx}} \qquad (4.4)$$

This equation gives the *shear-stress distribution* at a typical cross section of a linearly elastic bar. **If the torsion bar is homogeneous** (i.e., G = const), **then the shear stress varies linearly with the distance ρ from the center of the shaft,** with the maximum shear stress acting at the outer edge of the cross section, as illustrated in Fig. 4.9. As indicated in Fig. 4.7d, this same shear-stress distribution acts along each radial line in the cross section.

For the particular case of linearly elastic behavior, where Eq. 4.4 holds, the torque $T(x)$ is given by

$$T = \int_A \rho\left(G\rho\,\frac{d\phi}{dx}\right) dA \qquad (4.5)$$

Furthermore, <u>if G is independent of ρ we get</u>

$$T = G\,\frac{d\phi}{dx}\int_A \rho^2\, dA \qquad (4.6)$$

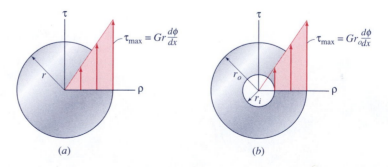

FIGURE 4.9 Torsional shear-stress distribution along a typical radial line — homogeneous, linearly elastic case.

(a) (b)

---

[1]In Section 4.11 we consider nonlinear stress-strain behavior.

The integral in Eq. 4.6 is called the *polar moment of inertia*, for which we will use the symbol I_p.[2] That is,

$$I_p = \int_A \rho^2 \, dA \tag{4.7}$$

It is shown in Appendix C.2 that, for a solid cross section of radius r

$$I_p = \frac{\pi r^4}{2} = \frac{\pi d^4}{32} \tag{4.8}$$

and for a tubular shaft of outer radius r_o and inner radius r_i

$$I_p = \frac{\pi}{2} (r_o^4 - r_i^4) = \frac{\pi}{32} (d_o^4 - d_i^4) \tag{4.9}$$

The combined symbol GI_p is referred to as the **torsional rigidity.**

Finally, Eq. 4.6 can be written in the form

$$\boxed{\frac{d\phi}{dx} = \frac{T}{GI_p}} \quad \begin{array}{l}\textbf{Torque-Twist}\\\textbf{Equation}\end{array} \tag{4.10}$$

This is the derivative form of the **torque-twist equation.** Note its similarity to the axial-deformation analogue, Eq. 3.10. When applying Eq. 4.10, it is important to remember that a sign convention is associated with both T and ϕ—**a positive torque T produces a positive twist rate $d\phi/dx$.** We have chosen to let the right-hand rule establish the positive sense of each (Figs. 4.4b and 4.4c).

By combining Eqs. 4.4 and 4.10 we get the **torsion formula**

$$\boxed{\tau = \frac{T\rho}{I_p}} \quad \begin{array}{l}\textbf{Torsion}\\\textbf{Formula}\end{array} \tag{4.11}$$

Note that this formula is valid only if the shear modulus is constant over the cross section. In Example 4.1 we will analyze the stress distribution in a composite bar for which $G = G(\rho)$; then we will consider torsion of homogeneous linearly elastic members.

EXAMPLE 4.1

A bimetallic torsion bar consists of an aluminum shell ($G_a = 4 \times 10^3$ ksi) bonded to the outside of a steel core ($G_s = 11 \times 10^3$ ksi). The shaft has the dimensions shown in Fig. 1 and is loaded by end torques of magnitude $T = 10$ kip · in. (a) Determine the maximum shear stress in the steel core and the maximum shear stress in the aluminum shell. (b) Determine the total twist angle of the composite torsion bar.

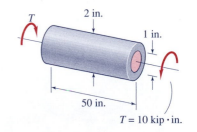

Fig. 1

Plan the Solution   Since the shear modulus varies with radial position, we must use Eq. 4.4 to express the shear-stress distribution in the composite shaft; then we can determine the corresponding torque from Eq. 4.5.

---

[2]Some texts use the symbol J. This sometimes leads to confusion, because J is also used in the discussion of torsion of noncircular members (e.g., Sections 4.9 and 4.10), where it does <u>not</u> refer to the polar moment of inertia.

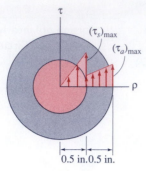

$(\tau_s)_{max}$

$(\tau_a)_{max}$

ρ

0.5 in. 0.5 in.

Fig. 2

Solution (a) *Determine* $(\tau_s)_{max}$ *and* $(\tau_a)_{max}$, *the two maximum stresses indicated in Fig. 2.* Since both parts of the shaft twist together, they have the same twist rate $d\phi/dx$. Then, from Eq. 4.4

$$\tau_s = G_s\rho\frac{d\phi}{dx}, \quad 0 \le \rho \le 0.5 \text{ in.} \tag{1a}$$

$$\tau_a = G_a\rho\frac{d\phi}{dx}, \quad 0.5 \text{ in.} < \rho \le 1.0 \text{ in.} \tag{1b}$$

These shear-stress distributions are plotted on the sketch in Fig. 2. Substituting these expressions for τ into Eq. 4.5, we get

$$T = \int_A \rho\tau\, dA = G_s\frac{d\phi}{dx}\int_{As} \rho^2\, dA + G_a\frac{d\phi}{dx}\int_{Aa} \rho^2\, dA$$

$$T = \frac{d\phi}{dx}[G_sI_{ps} + G_aI_{pa}]$$

Then, the *torque-twist equation* for this composite shaft can be written as

$$\frac{d\phi}{dx} = \frac{T}{[G_sI_{ps} + G_aI_{pa}]}$$

So,

$$\frac{d\phi}{dx} = \frac{10 \text{ kip} \cdot \text{in.}}{[(11\times10^3 \text{ ksi})(\pi/2)(0.5 \text{ in.})^4 + (4\times10^4 \text{ kis})(\pi/2)[(1.0 \text{ in.})^4 - (0.5 \text{ in.})^4]]}$$

$$= 1.4346(10^{-3}) \text{ rad/in.}$$

Evaluating Eq. (1a) at $\rho = 0.5$ in., we get

$$(\tau_s)_{max} = (11 \times 10^3 \text{ ksi})(0.5 \text{ in.})[1.4346(10^{-3}) \text{ rad/in.}]$$

$$(\tau_s)_{max} = 7.89 \text{ ksi} \hspace{3cm} \textbf{Ans. (a)} \quad (5a)$$

and evaluating Eq. (1b) at $\rho = 1.0$ in., we get

$$(\tau_a)_{max} = (4 \times 10^3 \text{ ksi})(1.0 \text{ in.})[1.4346(10^{-3}) \text{ rad/in.}]$$

$$(\tau_a)_{max} = 5.74 \text{ ksi} \hspace{3cm} \textbf{Ans. (a)} \quad (5b)$$

As is indicated in Fig. 2, the maximum shear stress in the steel core is larger than the maximum shear stress in the aluminum shell for the given dimensions of this composite torsion bar.

(b) *Determine the total angle of twist of the torsion bar.* We cannot use the torque-twist equation, Eq. 4.10, since it is for a homogeneous shaft. For this composite shaft we obtained a different equation, Eq. (3), for the twist rate. Since the twist rate $d\phi/dx$ is constant along the shaft, we get

$$\phi = \int_0^L \frac{d\phi}{dx}\, dx = \frac{d\phi}{dx}L = (1.4346 \times 10^{-3} \text{ rad/in.})(50 \text{ in.})$$

$$\phi = 7.17(10^{-2}) \text{ rad} \hspace{3cm} \textbf{Ans. (b)} \quad (6)$$

Review the Solution If the torsion bar was homogeneous (either aluminum or steel), its maximum shear stress would be at the outer radius, $\rho = 1.0$ in., and would be given by the elastic torsion formula, Eq. 4.11.

$$\tau_{max} = \frac{T_r}{I_p} = \frac{(10 \text{ kip} \cdot \text{in.})(1.0 \text{ in.})}{\pi/2(1.0 \text{ in.})^4} = 6.37 \text{ ksi}$$

Because the steel core is stiffer than the aluminum shell, it is reasonable that $(\tau_a)_{max}$ is somewhat less than this value.

Summary of Torsion Theory for Homogeneous Linearly Elastic Bars.

Let us summarize the theory of torsion as applied to the special case of a homogeneous linearly elastic member with circular cross section. While the derivation of the torsion theory summarized in Eqs. 4.10 and 4.11 was based on deformation of a circular cylinder, it can also be applied to approximate the behavior of a torsion member whose radius varies slowly with x. Figure 4.10 shows such a member with distributed and concentrated external torques applied to the member. Thus, for either cylindrical torsion members or ones whose radius varies slowly, we can write the **torsion formula**, Eq. 4.11, in the form

$$\tau \equiv \tau(x, \rho) = \frac{T(x)\rho}{I_p(x)} \qquad \text{Torsion Formula} \qquad (4.12)$$

indicating that both T and I_p may vary with x. We are usually most interested in the maximum shear stress in a torsion member, in which case the torsion formula becomes

$$\tau_{max} = \frac{T_{max}r}{I_p} \qquad \text{Maximum Shear-Stress Formula} \qquad (4.13)$$

where τ_{max} means the __magnitude__ of the maximum shear stress, T_{max} stands for the maximum absolute value of $T(x)$, and r is the radius of the shaft if it is a solid shaft or the outer radius if the shaft is tubular.

In Eq. 4.10, the torque-twist equation, T, G, and I_p may each vary with x. This derivative form of the torque-twist equation may be integrated over the length of

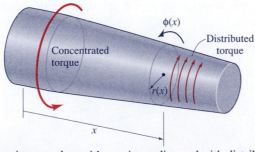

FIGURE 4.10 A torsion member with varying radius and with distributed torsional loading and concentrated torque.

the member to give

$$\phi = \int_0^L \frac{T(x)\,dx}{G(x)I_p(x)} \qquad \begin{array}{l}\text{Torque-}\\ \text{Twist}\\ \text{Equation}\end{array} \qquad (4.14)$$

where $\phi \equiv \phi(L) - \phi(0)$ is the **total angle of twist.**

The **angle of rotation**, $\phi(x)$, at cross section x is given by

$$\phi(x) = \phi(0) + \int_0^x \frac{T(\xi)\,d\xi}{G(\xi)I_p(\xi)} \qquad (4.15)$$

where ξ is a dummy variable of integration locating cross sections between 0 and x. Note the similarity of Eqs. 4.12 and 4.14 to the axial-deformation analogues, Eqs. 3.9 and 3.11, respectively.

Uniform Torsion Member. Figure 4.11 shows a typical **uniform torsion member** and defines the sign convention. Let the element have constant G and constant I_p along its length, and let the member be subjected to equilibrating end torques, T, as shown. Finally, let

$$\phi \equiv \phi(L) - \phi(0)$$

be the **twist angle** (i.e., the rotation of the end $x = L$ relative to the end $x = 0$), with positive sense as shown in Fig. 4.11.

From Eq. 4.14,

$$\phi = \int_0^L \frac{T\,dx}{GI_p}$$

Since the torque T is constant and the torsion member is uniform.

$$\phi = \frac{TL}{GI_p} \qquad \begin{array}{l}\text{Torque-}\\ \text{Twist}\\ \text{Equation}\end{array} \qquad (4.16)$$

The following example illustrates two key equations of the theory of torsion, the *maximum shear-stress formula*. Eq. 4.13, and the *torque-twist equation*, Eq. 4.16. It also illustrates the **strength-to-weight comparison** of two torsion members; one is a solid shaft, the other is a tubular shaft. The strength-to-weight advantage of tubular shafts relative to solid shafts is the reason that they are frequently used as torsion members in many applications, such as the drive shafts of sport utility vehicles and trucks (see Prob. 4.8-9).

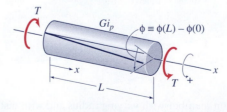

FIGURE 4.11 A typical end-loaded uniform torsion member.

EXAMPLE 4.2

A tubular shaft of length L has an outer radius r_1 and an inner radius $r_1/2$, and it is made of material with shear modulus G. It is subjected to end torques of magnitude T_0, as shown in Fig. 1a. (a) Determine an expression for τ_{max} for the shaft in Fig. 1a. (b) Determine an expression for the twist angle, ϕ, for this shaft. (c) If the same torque T_0 is applied to the solid circular shaft in Fig. 1b producing the same maximum shear stress τ_{max} as in the tubular shaft, what is the radius, r_2, of the solid shaft? What is the ratio of the weight W_2 of the solid shaft to the weight W_1 of the tubular shaft?

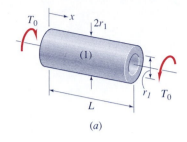

(a)

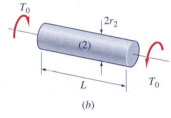

(b)

Fig. 1

Plan the Solution We can use the torsion formula, Eq. 4.13, to determine an expression for τ_{max}, and the torque-twist equation, Eq. 4.16, to determine the total twist angle, ϕ. Since there are only end torques, from equilibrium we get $T_{max1} = T_{max2} = T_0$.

Solution (a) *Determine an expression for τ_{max} for shaft (1).* From the torsion formula, Eq. 4.13,

$$\tau_{max} = \frac{T_{max}r}{I_p} \tag{1}$$

From Eq. 4.9, the polar moment of inertia of the tubular shaft is

$$I_{p1} = \frac{\pi}{2}\left[r_1^4 - \left(\frac{r_1}{2}\right)^4\right] = \frac{15}{32}\pi r_1^4$$

Therefore,

$$\tau_{max1} = \frac{T_0 r_1}{\frac{15}{32}(\pi r_1^4)} = \frac{32T_0}{15\pi r_1^3} \qquad \textbf{Ans. (a)} \tag{2}$$

(b) *Determine an expression for the twist angle ϕ for shaft (1).* From the torque-twist equation, Eq. 4.16,

$$\phi_1 = \frac{T_0 L}{GI_{p1}} = \frac{32T_0 L}{15\pi Gr_1^4} \qquad \textbf{Ans. (b)} \tag{3}$$

(c) *Determine the radius r_2 such that $\tau_{max2} = \tau_{max1}$ when $T_2 = T_1 = T_0$.* From the torsion formula, Eq. 4.13,

$$\tau_{max2} = \frac{T_{max2}r_2}{I_{p2}} = \frac{T_0 r_2}{\frac{\pi}{2}(r_2)^4} = \frac{2T_0}{\pi r_2^3} \tag{4}$$

Equating τ_{max1} from Eq. (2) with τ_{max2} from Eq. (4) we get

$$\frac{32T_0}{15\pi r_1^3} = \frac{2T_0}{\pi r_2^3}$$

So

$$r_2 = \left(\frac{15}{16}\right)^{1/3} r_1 = 0.979 r_1 \qquad \textbf{Ans. (c)} \quad (5)$$

Therefore, the ratio of the weights of the two shafts is

$$\frac{W_2}{W_1} = \frac{A_2 L_2}{A_1 L_1} = \frac{\pi r_2^2 L}{\pi [r_1^2 - (r_1/2)^2] L} = \frac{4}{3}\left(\frac{r_2}{r_1}\right)^2 = 1.277 \quad \textbf{Ans. (c)} \quad (6)$$

Review the Solution The answer to Part (c) seems questionable, since it says that the radius of the solid shaft must be almost equal to the radius of the tubular shaft even though both have the same "strength." However, a recheck of Eqs. (2) and (4) shows that both are correct. From answers (5) and (6) we can conclude that the tubular shaft has a definite strength-to-weight advantage over the solid shaft, since the solid shaft weighs 28% more than the tubular shaft, even though both have the same strength.

MDS4.1 – 4.4 **Simple Torsion** is a computer program module for calculating the maximum shear stress in, and the angle of twist of, a single uniform statically determinate torsion member (shaft or tube). There are four MDS examples.

EXAMPLE 4.3

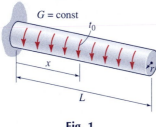

Fig. 1

A uniform shaft of radius r and length L is subjected to a uniform distributed external torque t_0 (moment per unit length). (See Fig. 1.) (a) Determine an expression for the maximum shear stress τ_{max}. (b) Determine an expression for the total twist angle $\phi \equiv \phi_L$.

Plan the Solution We need to determine $T(x)$ from equilibrium so that we can apply Eqs. 4.13 to determine the maximum shear stress and Eq. 4.14 to determine the twist angle. The maximum internal torque occurs at the wall at the left end, so the maximum shear stress occurs there also.

Solution (a) *Determine τ_{max}.*

Equilibrium: We need to draw a free-body diagram and write the appropriate moment equilibrium equation. An appropriate free-body diagram is shown in Fig. 2. On the section at x we show the internal torque $T(x)$ in the positive sense according to the right-hand rule.

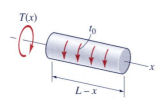

Fig. 2. Free-body diagram.

$$\sum M_x = 0: \qquad T(x) = t_0(L - x) \qquad (1)$$

Shear Stress: The maximum shear stress occurs at $x = 0$, where $T(0) = T_{max} = t_0 L$. Then, from the torsion formula, Eq. 4.13,

$$\tau_{max} = \frac{T_{max} r}{I_p} = \frac{t_0 L r}{I_p} = \frac{t_0 L r}{\pi/2 (r^4)}$$

or

$$\tau_{max} = \frac{2t_0 L}{\pi r^3} \qquad \textbf{Ans. (a)} \quad (2)$$

(b) *Solve for the twist angle* $\phi \equiv \phi_L$.

Torque-Twist: From the torque-twist relationship, Eq. 4.14,

$$\phi_L = \int_0^L \frac{T(x)\,dx}{GI_p} = \int_0^L \frac{t_0(L-x)\,dx}{GI_p}$$

$$= \frac{t_0}{GI_p}\left[\left(Lx - \frac{x^2}{2}\right)\right]_0^L$$

or

$$\phi_L = \frac{t_0 L^2}{2GI_p} = \frac{t_0 L^2}{\pi r^4 G} \qquad \textbf{Ans. (b)} \quad (3)$$

Review the Solution In this problem we can check to see that the answers are dimensionally correct, that is, F/L^2 in Eq. (2) and dimensionless in Eq. (3). We can also observe that t_0, L, r, and G have the proper effect on the answers (e.g., a longer bar will have a larger twist angle).

4.4 STRESS DISTRIBUTION IN CIRCULAR TORSION BARS; TORSION TESTING

Stress Distribution. In Section 2.9 we found that associated with pure normal stresses on the cross section of an axial-deformation member there are shear stresses and normal stresses on inclined cuts. Figure 4.7*d* shows the shear-stress distribution on a cross section due to torque and the corresponding shear stress on radial planes, and Fig. 4.12 shows that this leads to pure shear on an element between cross sections. We will now examine the stress distribution on faces that are inclined to the axis of a torsion bar.

Figure 4.12*a* shows a torsion bar and indicates the orientation of a "cut" inclined at angle θ with respect to the cross section. To determine the stresses on the inclined cut, we can take a triangular "wedge," as shown in Fig. 4.13. The stresses are shown on Fig. 4.13*b*, while Fig. 4.13*c* is a free-body diagram of the element; that is, it shows the forces that act on the respective faces of the element.

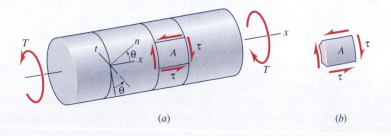

(a)   (b)

FIGURE 4.12 (a) A bar in torsion; (b) a pure-shear element.

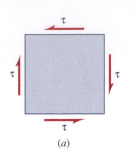

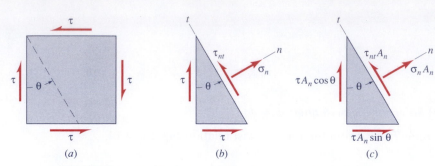

FIGURE 4.13   The state of stress for a pure-shear element.

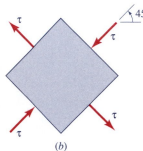

FIGURE 4.14   A pure-shear element, and the associated maximum-normal-stress element.

To determine expressions for σ_n and τ_{nt}, the normal stress and the shear stress on an inclined cut, we can write equilibrium equations for the free body in Fig. 4.13c. The n and t axes are taken normal to the inclined cut and tangential to the inclined cut, respectively.

$$\nearrow \sum F_n = 0: \qquad \sigma_n A_n + \tau A_n \cos\theta(\sin\theta) + \tau A_n \sin\theta(\cos\theta) = 0$$

$$\sigma_n = -2\tau \sin\theta\cos\theta$$

$$\nwarrow \sum F_r = 0: \qquad \tau_{nt} A_n + \tau A_n \cos^2\theta - \tau A_n \sin^2\theta = 0$$

$$\tau_{nt} = -\tau(\cos^2\theta - \sin^2\theta)$$

The expressions for σ_n and τ_{nt} can be simplified by introducing the trigonometric identities

$$\sin(2\theta) = 2\sin\theta\cos\theta, \qquad \cos(2\theta) = \cos^2\theta - \sin^2\theta$$

Therefore,

$$\boxed{\begin{aligned} \sigma_n &= -\tau\sin(2\theta) \\ \tau_{nt} &= -\tau\cos(2\theta) \end{aligned}} \qquad (4.17)$$

From Eqs. 4.17 it is clear that τ_{nt} has a maximum magnitude of τ for $\theta = 0°$ or $\theta = 90°$. On the other hand, σ_n has a maximum magnitude for $\theta = \pm45°$. From Eqs. 4.17, $\sigma_{45°} = -\tau$, $\sigma_{-45°} = \tau$, and $\tau_{nt} = 0$ for $\theta = \pm45°$. Figure 4.14 shows the original pure-shear element and an element rotated at 45° to this element.

Strain Distribution.   Hooke's Law relating shear stress to shear strain, Eq. 2.23, gives

$$\gamma = \frac{\tau}{G} \qquad (4.18)$$

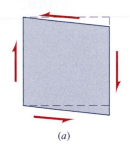

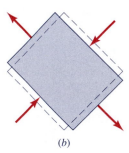

FIGURE 4.15   Deformation of a pure-shear element, and the associated maximum-normal-stress element.

Thus, the shear stress due to torsion, depicted in Fig. 4.14a, produces the shear deformation illustrated in Fig. 4.15a.

Along the $\pm45°$ directions, which are the directions of maximum compression and maximum tension, respectively, as depicted in Fig. 4.14b, the extensional strains may be obtained by applying Hooke's Law in the form of Eqs. 2.38, namely

$$\epsilon_{45°} = \frac{1}{E}[-\tau - \nu(\tau)] = \frac{-\tau}{E}(1 + \nu)$$

$$\epsilon_{-45°} = \frac{1}{E}[\tau - \nu(-\tau)] = \frac{\tau}{E}(1 + \nu)$$

Since E is related to G by Eq. 2.24, the above equations can be written in the form

$$\boxed{\begin{aligned}
\epsilon_{\text{max.comp.}} &= \epsilon_{45°} = -\frac{\tau}{2G} \\
\epsilon_{\text{max.tens.}} &= \epsilon_{-45°} = \frac{\tau}{2G}
\end{aligned}}$$

(4.19)

The deformation of a $\pm 45°$ element is illustrated in Fig. 4.15b.

Torsion Testing. To determine the shear modulus of a material, to determine its shear strength properties, or to determine other properties associated with torsion, bars may be tested in a torsion testing machine like the one shown in Fig. 4.16. The torque and twist angle are recorded as the torsion bar is twisted.

Figure 4.17 shows specimens tested to failure in a torsion testing machine. This figure illustrates the importance of Eq. 4.17 and the fact that σ_n and τ_{nt} vary with orientation. Both shafts in Fig. 4.17 were tested to failure in pure torsion. The failure of the mild-steel bar in Fig. 4.17a is quite different from that of the cast-iron bar in Fig. 4.17b. By comparing the failure planes in Figs. 4.17 with the stress elements in Fig. 4.14, we can conclude that the mild steel bar failed due to shear, while the cast iron fracture occurred on a maximum-tensile-stress surface. This is consistent with the types of tensile-test failures discussed in Section 2.4, leading to the conclusion that **brittle members are weaker in tension than in shear, while ductile members are weaker in shear.**

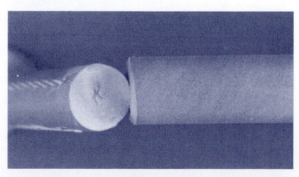

(a) Mild steel bar.

(b) Cast iron bar.

FIGURE 4.16 A torsion testing machine. (Photo courtesy of Tinius Olsen)

FIGURE 4.17 The failure surfaces of mild-steel and cast-iron torsion bars. (Courtesy Roy Craig)

EXAMPLE 4.4

The torque tube in Fig. 1, with outer diameter $d_o = 1.25$ in. and inner diameter 1.00 in., is subjected to a torque $T = 1000$ lb · in. A strain gage oriented at an angle $\theta = -45°$ with respect to the axis, which measures the extensional strain along this direction, gives a reading of $\epsilon_{-45°} = 190$ μin./in. (a) Determine the value of the maximum shear stress, τ_{max}. (b) Determine the shear modulus of elasticity, G. (c) Determine the angle of twist in a section of the tube of length $L = 30$ in.

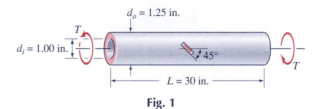

Fig. 1

Plan the Solution   The maximum shear stress, which occurs at the outer surface, can be determined directly from Eq. 4.13. Since the extensional strain along the direction $\theta = -45°$ is given, Eq. 4.19b can be used to compute G, the shear modulus. Finally, the angle of twist can be calculated from Eq. 4.16.

Solution   (a) *Determine* τ_{max}. The maximum shear stress occurs at the outer surface and is given by

$$\tau_{max} = \frac{Tr_o}{I_p}$$

where

$$I_p = \frac{\pi(r_o^4 - r_i^4)}{2} = \frac{\pi[(0.625 \text{ in.})^4 - (0.5 \text{ in.})^4]}{2} = 0.1415 \text{ in}^4$$

Then,

$$\tau_{max} = \frac{(1000 \text{ lb} \cdot \text{in.})(0.625 \text{ in.})}{0.1415 \text{ in}^4} = 4417 \text{ psi}$$

or, rounding to three significant figures,

$$\tau_{max} = 4420 \text{ psi} \qquad \text{**Ans. (a)**}$$

(b) *Determine* G. From Eq. 4.19b,

$$G = \frac{\tau_{max}}{2\epsilon_{-45°}} = \frac{4417 \text{ psi}}{2(0.000190 \text{ in./in.})} = 11.6(10^6) \text{ psi} \qquad \text{**Ans. (b)**}$$

(c) *Determine* ϕ. From Eq. 4.16,

$$\phi = \frac{TL}{GI_p} = \frac{(1000 \text{ lb} \cdot \text{in.})(30 \text{ in.})}{(11.6 \times 10^6 \text{ psi})(0.1415 \text{ in}^4)} = 0.0182 \text{ rad} \qquad \text{**Ans. (c)**}$$

MDS4.5 **Simple Torsion: Stress Distribution and Strain Distribution** illustrates
the determination of σ_{max} and ϵ_{max} in a circular bar in simple torsion.

253

**Stress Distribution in Circular
Torsion Bars; Torsion Testing**

4.5 STATICALLY DETERMINATE ASSEMBLAGES OF UNIFORM TORSION MEMBERS

In Section 3.2 Eqs. 3.14 and 3.15 were derived to describe the elastic-deformation behavior of a uniform, linearly elastic, axially loaded element. This led to the definition of stiffness and flexibility coefficients for relating the axial force to the elongation of a uniform member. In the present section we examine an analogous theory for analyzing the **linearly elastic behavior of a uniform torsion bar,** or *torsion element,* and we analyze **statically determinate assemblages** of such elements, like the pulley and shaft assemblage in Fig. 4.18a. In Section 4.6 we will analyze statically indeterminate assemblages, like the one in Fig. 4.18b.

Torque-Twist Equations. Figure 4.11 (Section 4.3) shows a typical uniform torsion member and defines the sign convention. The element has constant G and constant I_p along its length, and is subjected to equilibrating end torques, T, as shown. From Eq. 4.16 the torque-twist equation for the uniform element is

$$\phi_i = \left(\frac{TL}{GI_p}\right)_i \qquad \textbf{Torque-Twist} \qquad (4.16)$$
$$\textbf{Equation} \qquad \text{repeated}$$

As in Section 3.2, we can express this equation two ways. For a torsion member, or *torsion element,* designated as element i, two useful forms of the **element torque-twist equation** are

$$\phi_i = f_{ti}T_i, \quad \text{where } f_{ti} = \left(\frac{L}{GI_p}\right)_i \qquad (4.20)$$

and

$$T_i = k_{ti}\phi_i, \quad \text{where } k_{ti} = \left(\frac{GI_p}{L}\right)_i \qquad (4.21)$$

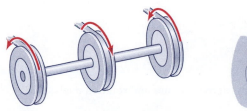
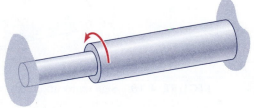

(a) A statically determinate assemblage.

(b) A statically indeterminate assemblage.

FIGURE 4.18 Examples of assemblages of torsion bars.

where the subscript t stands for torsion and subscript i identifies the particular torsion element. We call f_t the **torsional flexibility coefficient** and k_t the **torsional stiffness coefficient.** Note the similarity between Eqs. 3.14 and 4.20 and between Eqs. 3.15 and 4.21.

Once the torque T_i in an individual member has been determined, the maximum shear stress for that member can be computed by using Eq. 4.13.

Notation Convention and Sign Convention for Torsion.

Figure 4.18a illustrates a statically determinate torsion-bar assemblage. To solve torsion problems, we can employ the same problem-solving strategies used for axial deformation in Chapter 3. To establish a systematic problem-solving procedure, we first define a **notation convention** and **a sign convention.** Then we set up the three fundamental types of equations: *equilibrium, element torque-twist* (i.e., *force-deformation*) *behavior,* and *geometry of deformation.* Finally, we combine these equations to determine *twist angles,* and use Eq. 4.13 to determine *maximum shear-stress values.*

Notation Conventions:[3]

(a) Member-identification subscripts will always be either numbers or the generic subscript i, as in Eqs. 4.20 and 4.21. For example, T_i is the torque at every cross section of element i, and specifically at each end of element i; ϕ_i is the relative twist angle between the two ends of element i.

(b) Externally applied torques and nodal (joint) rotation angles will always be denoted by uppercase letter subscripts or will be given numerical values.

Sign Conventions:

(a) Figure 4.11 establishes the right-hand-rule sign convention for the torque T_i acting on an element, and the sign convention for the corresponding twist angle ϕ_i.

(b) When several elements form an assemblage, a sign convention is needed for external torques and nodal rotation angles. This sign convention is illustrated in Fig. 4.19. First, a direction for $+x$ is selected: then positive external torques and positive nodal rotations follow the right-hand rule with respect to the $+x$ axis.

Statically Determinate Assemblages of Elements.

Several statically determinate torsion problems will now be solved.

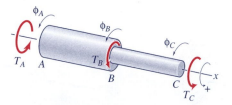

FIGURE 4.19 Sign convention for external torques and rotation angles.

---

[3]These "notation conventions" are necessitated by the fact that the symbol T is conventionally used for both externally applied load and reaction torques and also for internal resisting torques. Similarly for the symbol ϕ.

EXAMPLE 4.5

A statically determinate two-element torsion bar is shown in Fig. 1. T_A and T_B are external torques applied at nodes A and B, respectively, and ϕ_A and ϕ_B are the rotation angles at these two nodes. An x axis has been established, and these T's and ϕ's have been taken to be positive in the right-hand-rule sense about this x axis. (a) Determine the internal (resisting) torques in members (1) and (2). (b) Determine the rotation angles at A and B.

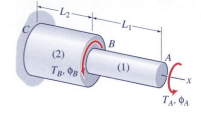

Fig. 1

Plan the Solution To determine the internal torques T_1 and T_2 in elements (1) and (2). respectively, we can draw free-body diagrams and write the corresponding equilibrium equations. Then we can use the element torque-twist equation, Eq. 4.20, and deformation compatibility to determine the two rotation angles.

Solution (a) *Determine the internal torques in members (1) and (2).*

Equilibrium: Free-body diagrams that expose the unknown element torques T_1 and T_2 are shown in Fig. 2. We can write a moment-equilibrium equation for each of these. (Note that the sense of T_1 and of T_2 is established by the right-hand-rule sign convention in Fig. 4.11.)

For FBD 1: $\qquad \sum M_x = 0: \qquad T_A - T_1 = 0$

For FBD 2: $\qquad \sum M_x = 0: \qquad T_A + T_B - T_2 = 0$

Hence,

$$T_1 = T_A$$
$$T_2 = T_A + T_B \qquad \qquad \textbf{Equilibrium} \quad \textbf{Ans. (a)} \quad (1)$$

(b) *Determine the rotation angles at A and B.*

Element Torque-Twist Behavior: Since we now know T_1 and T_2, Eq. 4.20 is the appropriate torque-twist equation to use.

$$\phi_i = f_{ti} T_i, \qquad i = 1, 2 \qquad \begin{array}{l} \textbf{Torque-Twist} \\ \textbf{Behavior} \end{array} \quad (2)$$

where

$$f_{ti} = \left(\frac{L}{GI_p} \right)_i$$

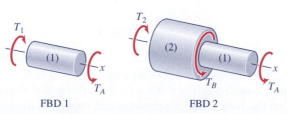

FBD 1 $\qquad\qquad\qquad\qquad$ FBD 2

Fig. 2 Free-body diagrams.

Geometry of Deformation: We can relate the nodal rotation angles ϕ_A and ϕ_B to the element twist angles ϕ_i as follows:

$$\phi_1 = \phi_A - \phi_B$$
$$\phi_2 = \phi_B - \phi_C = \phi_B$$

or

$$\boxed{\begin{aligned} \phi_A &= \phi_1 + \phi_2 \\ \phi_B &= \phi_2 \end{aligned}} \qquad \begin{array}{l}\textbf{Geometry} \\ \textbf{of} \\ \textbf{Deformation}\end{array} \qquad (3)$$

Solution: To determine ϕ_A and ϕ_B we can substitute (1) $\rightarrow$ (2) $\rightarrow$ (3). Finally,

$$\phi_A = f_{t1}T_A + f_{t2}(T_A + T_B)$$
$$\phi_B = f_{t2}(T_A + T_B)$$

or

$$\left. \begin{aligned} \phi_A &= \frac{T_A L_1}{G_1 I_{p1}} + \frac{(T_A + T_B)L_2}{G_2 I_{p2}} \\ \phi_B &= \frac{(T_A + T_B)L_2}{G_2 I_{p2}} \end{aligned} \right\} \qquad \textbf{Ans. (b)} \quad (4)$$

EXAMPLE 4.6

A stepped steel shaft AC is subjected to external torques at sections B and C, as shown in Fig. 1. The shear modulus of the steel is $G = 11.5 \times 10^3$ ksi, and the diameter of element (1) is $d_1 = 2$ in. Diameter d_2 is to be determined such that the maximum shear stress in the stepped shaft does not exceed the allowable shear stress $\tau_{\text{allow}} = 8$ ksi, and the total angle of twist does not exceed $\phi_{\text{allow}} = 0.06$ rad. Express your answer to the nearest $\frac{1}{8}$ in. that satisfies both of these criteria.

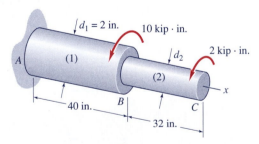

Fig. 1

Plan the Solution   There are two criteria: one on maximum shear stress and one on maximum total twist angle. We will first write down the fundamental equations. Then we will determine the required diameter d_2 based on the maximum-shear-stress criterion, and then on the maximum-twist-angle criterion.

Solution

Equilibrium: Free-body diagrams that expose the unknown element torques T_1 and T_2 are shown in Fig. 2. We can write a moment-equilibrium equation for each of these. (Note that the sense of T_1 and of T_2 is established by the right-hand-rule sign convention in Fig. 4.11.)

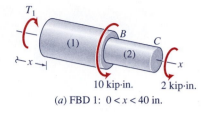

(a) FBD 1: $0 < x < 40$ in.

For FBD1: $\sum M_x = 0$: $T_1 = 10$ kip $\cdot$ in. $+ 2$ kip $\cdot$ in.

$$= 12 \text{ kip} \cdot \text{in.}$$

For FBD2: $\sum M_x = 0$: $T_2 = 2$ kip $\cdot$ in.

$$
\begin{aligned}
T_1 &= 12 \text{ kip} \cdot \text{in.} \\
T_2 &= 2 \text{ kip} \cdot \text{in.}
\end{aligned}
\qquad \textbf{Equilibrium} \qquad (1)
$$

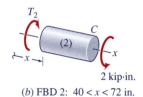

(b) FBD 2: $40 < x < 72$ in.

Fig. 2   Free-body diagrams.

Element Torque-Twist Behavior: Since we know T_1 and T_2, Eq. 4.20 is the appropriate torque-twist equation to use.

$$
\begin{aligned}
\phi_1 &= f_{t1} T_1 \\
\phi_2 &= f_{t2} T_2
\end{aligned}
\qquad
\begin{array}{l}
\textbf{Torque-} \\
\textbf{Twist} \\
\textbf{Equations}
\end{array}
\qquad (2)
$$

where

$$
f_{t1} = \left(\frac{L}{GI_p}\right)_1 = \frac{32L_1}{\pi G d_1^4}, \qquad f_{t2} = \left(\frac{L}{GI_p}\right)_2 = \frac{32L_2}{\pi G (d_2)^4}
$$

Geometry of Deformation: The total twist angle is simply the sum of the element twist angles, so

$$
(\phi_C)_{\text{allow}} \equiv \phi_{\text{allow}} = \phi_1 + \phi_2 = 0.06 \text{ rad} \qquad
\begin{array}{l}
\textbf{Geometry of} \\
\textbf{Deformation}
\end{array}
\qquad (3)
$$

Maximum-Shear-Stress Criterion: The maximum torsional shear stress, must be checked in each element by using Eq. 4.13.

$$
(\tau_{\max})_1 = \frac{|T_1|(d_1/2)}{I_{p1}} = \frac{16|T_1|}{\pi d_1^3} = \frac{16(12 \text{ kip} \cdot \text{in.})}{\pi (2.0 \text{ in.})^3} = 7.6394 \text{ ksi} \, (< 8 \text{ ksi})
$$

$$
(\tau_{\max})_2 = \frac{|T_2|(d_2/2)}{I_{p2}} = \frac{16|T_2|}{\pi (d_{2\tau})^3} = \tau_{\text{allow}}
$$

Therefore, based on the maximum-shear-stress criterion,

$$
d_{2\tau} = \left(\frac{16|T_2|}{\pi \tau_{\text{allow}}}\right)^{1/3} = \left[\frac{16(2 \text{ kip} \cdot \text{in.})}{\pi (8 \text{ ksi})}\right]^{1/3} = 1.0839 \text{ in.}
$$

Maximum-Twist-Angle Criterion: Now we need to determine diameter d_2 based on the maximum-twist-angle criterion. Combining Eqs. (1) through (3), we get

$$
\phi_{\text{allow}} = \left(\frac{32L_1}{\pi G d_1^4}\right) T_1 + \left(\frac{32L_2}{\pi G (d_{2\phi})^4}\right) T_2
$$

so

$$d_{2\phi} = \left[\left(\frac{\pi G \phi_{\text{allow}}}{32 L_2 T_2} - \frac{L_1 T_1}{L_2 T_2 d_1^4} \right)^{-1} \right]^{1/4}$$

$$= \left\{ \left[\frac{\pi (11.5 \times 10^3 \text{ ksi})(0.06 \text{ rad})}{32(32 \text{ in.})(2 \text{ kip} \cdot \text{in.})} - \frac{(40 \text{ in.})(12 \text{ kip} \cdot \text{in.})}{(32 \text{ in.})(2 \text{ kip} \cdot \text{in.})(2 \text{ in.})^4} \right]^{-1} \right\}^{1/4}$$

$$= 1.1412 \text{ in.}$$

We are to determine, to the nearest $\frac{1}{8}$ in., the diameter d_2 that is greater than both $d_{2\tau}$ and $d_{2\phi}$. Therefore,

$$(d_2)_{\min} = 1.250 \text{ in.} \qquad \textbf{Ans.}$$

MDS4.6 & 4.7 **Multiple Torques** combines an equilibrium analysis to determine segment internal torques T_i with routines for determining the shear stress in a shaft segment due to simple torsion and for determining the node rotation angles.

4.6   STATICALLY INDETERMINATE ASSEMBLAGES OF UNIFORM TORSION MEMBERS

If the number of unknown torques exceeds the number of applicable equilibrium equations, an assemblage of torsion members is said to be **statically indeterminate.** An example of a statically indeterminate shaft is shown in Fig. 4.20. The analysis of statically indeterminate torsion assemblages is virtually identical to the analysis of statically indeterminate assemblages of axial-deformation members covered in Sections 3.5, 3.7, and 3.8. (Thermal strain, the topic of Section 3.6, does not directly enter into torsion problems.)

Analysis of a Typical Statically Indeterminate Torsion Bar.   Consider the simple two-element torsion bar in Fig. 4.20. Let us first determine the internal torques, T_1 and T_2, in the elements; then we will determine the rotation, ϕ_B, of the node (joint) between the two elements.

Equilibrium: The free-body diagram of node B in Fig. 4.21 will enable us to write a moment equilibrium equation that relates the *external torque,* T_B, at joint B to the *internal torques,* T_1 and T_2, in the elements adjoining B.

$$\sum M_x = 0: \qquad \boxed{T_1 - T_2 = T_B} \qquad \textbf{Equilibrium} \qquad (1)$$

This problem is statically indeterminate, since we are not able to determine the two element torques from the single moment equilibrium equation. There is a **redundant torque,** that is, a torque that is not absolutely essential to maintain stable equilibrium of the shaft.

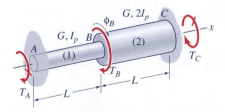

$$I_{p1} = I_p, \quad I_{p2} = 2I_p, \quad G_1 = G_2 = G, \quad L_1 = L_2 = L$$

FIGURE 4.20   A statically indeterminate shaft in torsion.

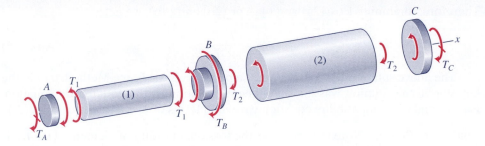

FIGURE 4.21 Free-body diagrams of nodes and elements.

The reaction torques at ends A and C are related to the element torques by

$$T_A = -T_1, \qquad T_C = T_2$$

Element Torque-Twist Behavior: From the element torque-twist equation, Eq. 4.20, we have

$$\phi_1 = f_{t1}T_1, \quad \phi_2 = f_{t2}T_2 \qquad \text{\begin{array}{l}\textbf{Element} \\ \textbf{Torque-Twist} \\ \textbf{Behavior}\end{array}} \qquad \text{(2a,b)}$$

where, using data from Fig. 4.20, we have $f_{t1} = \dfrac{L}{GI_p}$ and $f_{t2} = \dfrac{L}{2GI_p}$.

Geometry of Deformation; Compatibility Equation: Since the two elements are attached end-to-end, and since the ends of the elements are not free to rotate, the total twist angle is zero. Therefore, the **compatibility equation** for this problem is

$$\phi_{\text{total}} = \phi_1 + \phi_2 = 0 \qquad \text{\begin{array}{l}\textbf{Geometry of} \\ \textbf{Deformation}\end{array}} \qquad \text{(3)}$$

Solution of the Equations: In Eqs. (1) through (3) there are <u>four equations and four unknowns</u>. Therefore, it is a matter of solving these four equations simultaneously to determine the four unknowns. We are first interested in determining the **element torques.** Therefore, we can eliminate the ϕ_i's by substituting Eqs. (2) into Eq. (3), which gives the following **compatibility equation written in terms of element torques:**

$$f_{t1}T_1 + f_{t2}T_2 = 0$$

or

$$T_1\left(\frac{L}{GI_p}\right) + T_2\left(\frac{L}{2GI_p}\right) = 0 \qquad \text{\begin{array}{l}\textbf{Compatibility} \\ \textbf{in Terms of} \\ \textbf{Element} \\ \textbf{Torques}\end{array}} \qquad \text{(3′)}$$

Finally, the equilibrium equation (1) and the compatibility equation (3′) can be **solved simultaneously** to give the following:

$$\boxed{T_1 = \frac{T_B}{3}, \quad T_2 = -\frac{2T_B}{3}} \qquad \textbf{Ans.} \quad \text{(4a,b)}$$

These element torques, then, are the primary solution of this statically indeterminate problem.

Rotation Angle: In the problem statement we were asked not only to determine the element torques but also to determine the rotation of node B. From Fig. 4.20 we can see that ϕ_B can be obtained from either of the following two *geometry of deformation* equations:

$$\phi_B = \phi_1, \quad \phi_B = -\phi_2$$

Therefore, combining Eqs. (2a) and (4a), or (2b) and (4b), we get

$$\phi_B = \frac{T_B L}{3 G I_p} \qquad \textbf{Ans.} \quad (5)$$

The above solution has followed the steps of the Basic Force Method, that is, the key step in the solution was the simultaneous solution of the equilibrium equation, Eq. (1), and the compatibility equation written in terms of element torques, Eq. (3′).

Analysis of Results: Note that, because the torsional rigidity of element (1), GI_{p1}, is half that of element (2), it is twice as **flexible** (i.e., $f_{t1} = 2f_{t2}$). Said another way, element (2) is twice as **stiff** in torsion as is element (1) (i.e., $k_{t2} = 2k_{t1}$). Therefore, a larger part of the external torque T_B is transferred by element (2) to the right "wall" than is transferred to the left "wall" by element (1). Had the two elements been identical, half of the torque T_B would have been transmitted to each "wall." This illustrates the fact that, for statically indeterminate structures, the reactions and the internal element torques depend on the factors I_{pi}, G_i, and L_i.

Solution Procedure: Basic Force Method—Torsion.

The preceding introductory statically indeterminate torsion problem was solved by the *Basic Force Method,* the method used in Section 3.5 to solve statically indeterminate axial-deformation problems.[4] The key steps that are required for analyzing statically indeterminate structures are to: (1) write down the *equilibrium equations* that will be needed for determining the unknown torques, and (2) write down all of the *torque-twist equations* that relate the torques in the equilibrium equations to the rotations that appear in the compatibility equations. The third key step is (3) to use *geometry of deformation* in order to write down the *compatibility equations* that characterize any *geometric constraints* on the deformation of the structure (i.e., on the rotation at various sections of the torsion member).

With the aid of the following Procedure, you should review the steps that were used to analyze the statically indeterminate torsion assemblage shown in Fig. 4.20. The flow chart that follows presents a more graphic representation of this Procedure.

SOLUTION PROCEDURE: BASIC FORCE METHOD—TORSION

▼ **SET UP THE FUNDAMENTAL EQUATIONS:**

(1) Let N_E be the number of <u>independent</u> equilibrium equations. Using *free-body diagrams,* write down N_E **independent equilibrium equations** of the form $\Sigma M_x = 0$.

(2) Write an **element torque-twist equation** for each torsion element. Equation 4.20 is the most convenient form to use.

$$\phi_i = f_{ti} T_i, \quad \text{where } f_{ti} = \frac{L_i}{I_{pi} G_i} \qquad \begin{array}{c}(4.20)\\ \text{repeated}\end{array}$$

is the *torsional flexibility* of the ith element.

(3) Use *geometry of deformation* to write down the **compatibility equation(s)** in terms of the element twist angles, ϕ_i.

▼ **SOLVE FOR THE UNKNOWN TORQUES:**

(4) Substitute the element torque-twist equations of Step 2 into the geometric-compatibility equations of Step 3.

This gives the **geometric-compatibility equations in terms of the unknown element torques**.

(5) The final step in determining the unknown torques is to **solve simultaneously** the equilibrium equations (Step 1) and the compatibility equations written in terms of element torques (Step 4).

▼ **SOLVE FOR THE ROTATION ANGLES:**

(6) To obtain rotation angles at the nodes (joints), if they are required, substitute the element torques into the torque-twist equations of Step 2. Finally, use geometry of deformation to relate element twist angles to the required rotation angles.

(7) Review the solution to make sure that all answers seem to be correct.

---

[4]The word "Force" in Basic Force Method is used here to denote the "force-like" quantity in torsion problems, namely *torque*.

The next flow chart gives the corresponding steps in the solution of statically indeterminate torsion problems, emphasizing the importance of the three fundamental types of equations: *equilibrium, torque-twist behavior of elements*, and *geometry of deformation (compatibility)*.

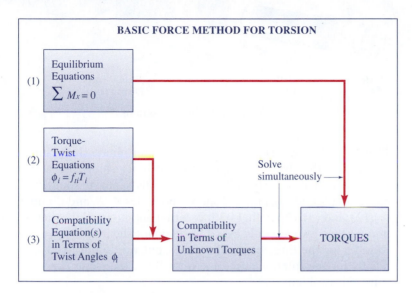

BASIC FORCE METHOD FOR TORSION

(1) Equilibrium Equations
$$\sum M_x = 0$$

(2) Torque-Twist Equations
$$\phi_i = f_{ti} T_i$$

(3) Compatibility Equation(s) in Terms of Twist Angles ϕ_i

Compatibility in Terms of Unknown Torques

Solve simultaneously

TORQUES

MDS4.8 **Statically Indeterminate Torsional Assemblages: Basic Force Method** combines *equilibrium, torque-twist*, and *compatibility equations* to solve for the unknown internal torques T_i. In the compatibility equation(s), Eq. (4.20) is used to express the element twist angles ϕ_i in terms of the element torques.

Example Problem 4.7 will now be solved by the Basic Force Method. Note that there is only one equilibrium equation, but three unknown internal torques. Therefore, there are two *redundant torques,* so we must formulate two compatibility equations.

EXAMPLE 4.7

A high-strength steel shaft ($G = 11.5 \times 10^6$ psi) of radius $r = 1.0$ in. and length $L = 30$ in. is sheathed over 20 in. of its length by an aluminum alloy tube ($G = 3.9 \times 10^6$ psi) of outer radius $r_o = 1.5$ in., as shown in Fig. 1. Ends (nodes) A and C are fixed. An external torque $T_B = 5000$ lb · in. is applied at node B as shown. Determine the maximum shear stress in the steel and the maximum shear stress in the aluminum.

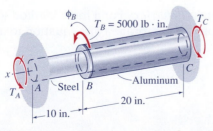

ϕ_B
$T_B = 5000$ lb · in.
T_C
x
T_A
A Steel B Aluminum C
10 in.
20 in.

Fig. 1

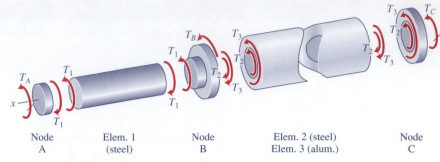

| Node
A | Elem. 1
(steel) | Node
B | Elem. 2 (steel)
Elem. 3 (alum.) | Node
C |

Fig. 2   Free-body diagrams.

Plan the Solution   This is a statically indeterminate shaft. Since the section *BC* of the shaft is composed of an inner steel core and an outer aluminum shell, we can call these portions separate elements, namely elements (2) and (3), Figure 2 shows the nodes *A*, *B*, and *C* and elements (1), (2), and (3). (Note that both external and internal torques act on the nodes.) Since we are only asked for element shear stresses, we will need to solve for the three element torques, but we do not need to solve explicitly for the rotation angle ϕ_B. We will use the *Basic Force Method* to solve this problem.

Solution

Equilibrium: The torsion-bar assemblage is separated into free-body diagrams of nodes and elements in Fig. 2. The nodal equilibrium equations are:

For Node A: $\sum M_x = 0$: $T_A - T_1 = 0$

For Node B: $\sum M_x = 0$: $T_B + T_1 - T_2 - T_3 = 0$

For Node C: $\sum M_x = 0$: $T_C + T_2 + T_3 = 0$

The equation for node *B* is the "active" equilibrium equation, that is, the one for a node that has nonzero rotation. The equilibrium equations for nodes *A* and *C* are not really needed, since they just provide the obvious equations for the reaction torques T_A and T_C.

$$-T_1 + T_2 + T_3 = T_B \qquad \textbf{Equilibrium} \quad (1)$$

Note that there is only one independent equilibrium equation, but there are three unknown torques. Therefore, there are <u>two</u> redundant torques.

Element Torque-Twist Behavior: For a force-method solution, Eq. 4.20 is the appropriate form of the torque-twist equation to use.

$$\phi_i = f_{ti}T_i, \quad i = 1, 2, 3 \qquad \begin{array}{l}\textbf{Torque-Twist}\\ \textbf{Equations}\end{array} \quad (2)$$

where $f_{ti} = \left(\dfrac{L}{GI_p}\right)_i$.

$$I_{p1} = I_{p2} = \frac{\pi(1.0 \text{ in.})^4}{2} = 1.571 \text{ in}^4$$

$$I_{p3} = \frac{\pi(1.5 \text{ in.})^4}{2} - \frac{\pi(1.0 \text{ in.})^4}{2} = 6.381 \text{ in}^4$$

$$f_{t1} = \left(\frac{L}{GI_p}\right)_1 = \frac{(10 \text{ in.})}{(11.5 \times 10^6 \text{ psi})(1.571 \text{ in}^4)} = 5.536(10^{-7})\frac{\text{rad}}{\text{lb} \cdot \text{in.}}$$

$$f_{t2} = \left(\frac{L}{GI_p}\right)_2 = \frac{(20 \text{ in.})}{(11.5 \times 10^6 \text{ psi})(1.571 \text{ in}^4)} = 1.107(10^{-6})\frac{\text{rad}}{\text{lb} \cdot \text{in.}}$$

$$f_{t3} = \left(\frac{L}{GI_p}\right)_3 = \frac{(20 \text{ in.})}{(3.9 \times 10^6 \text{ psi})(6.381 \text{ in}^4)} = 8.036(10^{-7})\frac{\text{rad}}{\text{lb} \cdot \text{in.}}$$

Geometry of Deformation: The element relative twist angles are

$$\phi_1 = \phi_A - \phi_B = -\phi_B$$

$$\phi_2 = \phi_B - \phi_C = \phi_B$$

$$\phi_3 = \phi_B - \phi_C = \phi_B$$

We can eliminate ϕ_B and write the following <u>two</u> **compatibility equations in terms of twist angles:**

$$\phi_1 = -\phi_2$$
$$\phi_3 = \phi_2$$

Compatibility (3)

Solution for the Element Torques: Using the solution steps of the Basic Force Method, we substitute the torque-twist equations, Eqs. (2), into the compatibility equations, Eqs. (3), and we get the following two compatibility equations in terms of the unknown torques:

$$f_{t1}T_1 = -f_{t2}T_2$$
$$f_{t3}T_3 = f_{t2}T_2$$

Compatibility in Terms of Torques (3′)

which we **solve simultaneously** with the equilibrium equation, Eq. 1, to obtain the three element torques:

$$T_1 = \frac{-(f_{t2}f_{t3})T_B}{f_{t1}f_{t2} + f_{t1}f_{t3} + f_{t2}f_{t3}}$$

$$T_2 = \frac{(f_{t1}f_{t3})T_B}{f_{t1}f_{t2} + f_{t1}f_{t3} + f_{t2}f_{t3}}$$ (4)

$$T_3 = \frac{(f_{t1}f_{t2})T_B}{f_{t1}f_{t2} + f_{t1}f_{t3} + f_{t2}f_{t3}}$$

When numerical values of T_B and the torsional flexibility coefficients are substituted into these equations, the resulting torques are

$$
\begin{aligned}
T_1 &= -2284 \text{ lb} \cdot \text{in.} \\
T_2 &= 1142 \text{ lb} \cdot \text{in.} \\
T_3 &= 1574 \text{ lb} \cdot \text{in.}
\end{aligned}
\tag{5}
$$

Maximum Shear Stresses: Now we can determine the maximum shear stress in each part. Since elements (1) and (2) are both steel, and both have the same radius, and since $|T_2| < |T_1|$, the maximum shear stress in the steel will occur in the outer fibers of element (1), that is, in shaft AB.

$$
(\tau_{\max})_{\text{steel}} = \frac{|T_1| r_1}{I_{p1}} = \frac{(2284 \text{ lb} \cdot \text{in.})(1.0 \text{ in.})}{1.571 \text{ in}^4} = 1454 \text{ psi}
$$

The maximum shear stress in the aluminum sleeve is given by

$$
(\tau_{\max})_{\text{alum}} = \frac{T_3 (r_3)_o}{I_{p3}} = \frac{(1574 \text{ lb} \cdot \text{in.})(1.5 \text{ in.})}{6.381 \text{ in}^4} = 369.9 \text{ psi}
$$

Rounding these answers to the appropriate number of significant figures, we get the following:

$$
\left.
\begin{aligned}
(\tau_{\max})_{\text{steel}} &= 1454 \text{ psi} \\
(\tau_{\max})_{\text{alum}} &= 370 \text{ psi}
\end{aligned}
\right\} \quad \textbf{Ans.} \quad (6)
$$

Review the Solution One way to check results is to check the overall equilibrium of the shaft. From the equilibrium equations for nodes A and C we get the following values for the torque reactions at A and C:

$$
T_A = -2280 \text{ lb} \cdot \text{in.}, \quad T_C = -2720 \text{ lb} \cdot \text{in.}
$$

These reactions are shown, along with couple-arrows indicating the proper sense, in Fig. 3.

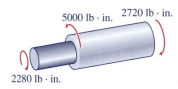

5000 lb · in. 2720 lb · in.

2280 lb · in.

Fig. 3 A free-body diagram.

$$
\sum M_x = 0: \qquad \text{Is } (-2280 + 5000 - 2720) = 0? \qquad \text{Yes.}
$$

It does seem reasonable that more torque would be transmitted to C than to A.

Homework Problem 4.7-6 is a Displacement-Method version of Example 4.7.

Initial Shear Stresses Due to Angular Misfits.

In Section 3.7 it was shown that axial misfits (e.g., members initially made too short or too long) lead to initial stresses in statically indeterminate assemblages. Likewise, angular misalignments in statically indeterminate torsional assemblages induce initial shear stresses. Table 4.2 summarizes the basic three equation sets and indicates that applied torques enter through the equilibrium equation(s) and that angular misfits enter

TABLE 4.2 A Summary of Torsion Equations

Fundamental Equation Sets	Corresponding "Inputs"
(1) Moment equilibrium	External torques
(2) Element torque-twist behavior	—
(3) Angular-deformation geometry	Angular misfits

through the geometry of deformation, that is, through the deformation-compatibility equation(s).

For simplicity, we will consider an angular misfit of the two-shaft system in Fig. 4.22, and will consider only the formulation of the deformation-geometry equations. Before the flanges are bolted together at B, the two segments of shaft were welded to the rigid supports at A and C, respectively. Assume that there is no axial gap between the two flanges (or only an insignificant axial gap; just enough so that the flanges can be rotated relative to each other), but that the holes in the two flanges are initially out of alignment by an angle $\bar{\phi}$. To complete the assembly, a temporary external torque can be applied to the flange B_1 to make its holes line up with the holes in flange B_2 while bolts are inserted into the four bolt holes and firmly tightened. When the temporary external torque is removed, there will be self-equilibrating torques in the two segments, resulting in so-called **initial shear stresses.**

Geometry of Deformation: Let the rotation angle at B be measured from the initial angular orientation of flange B_2. The twist angles of the two shaft segments are, therefore,

$$\phi_1 = \bar{\phi} + \phi_B - \phi_A$$

$$\phi_2 = \phi_C - \phi_B$$

Since $\phi_C = \phi_A = 0$,

$$\boxed{\begin{aligned}\phi_1 &= \bar{\phi} + \phi_B \\ \phi_2 &= -\phi_B\end{aligned}}$$
  **Geometry of
  Deformation**   (4.22)

As indicated in Table 4.2, angular alignment "misfits" are accounted for in the *deformation-geometry equation(s)*, as in Eq. 4.22.

See Homework Problems 4.6-16 and 4.6-17 for problems involving misaligned shafts. Similar homework problems to be solved by the Displacement Method may be found in Problems 4.7-7 and 4.7-8.

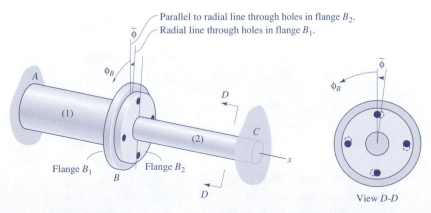

FIGURE 4.22 Angular misalignment of shaft segments.

*4.7 DISPLACEMENT-METHOD SOLUTION OF TORSION PROBLEMS

In Section 4.6 statically indeterminate torsion problems were solved by the Basic Force Method. In this section two example problems are solved by the Displacement Method, whose steps are outlined in the following flowchart. It will be apparent that the procedure used here for solving torsion problems is exactly analogous to the procedure used in Section 3.8 for solving axial-deformation problems.

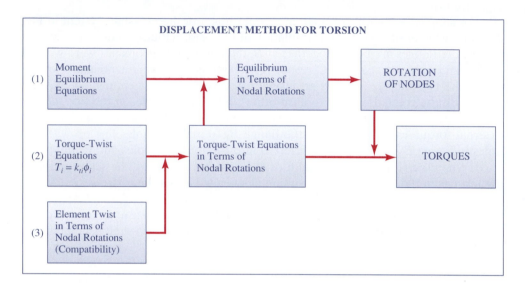

DISPLACEMENT METHOD FOR TORSION

(1) Moment Equilibrium Equations → Equilibrium in Terms of Nodal Rotations → ROTATION OF NODES

(2) Torque-Twist Equations $T_i = k_{ti}\phi_i$ → Torque-Twist Equations in Terms of Nodal Rotations → TORQUES

(3) Element Twist in Terms of Nodal Rotations (Compatibility)

EXAMPLE 4.8

A steel shaft ($G = 78$ GPa) of radius $r = 25$ mm and length $L = 0.5$ m is sheathed by an aluminum alloy tube ($G = 28$ GPa) of outer radius $r_o = 37.5$ mm, as shown in Fig. 1. End B is fixed, and an external torque $T_A = 350$ N · m is applied at node A as shown. Using the *Displacement Method,* (a) determine the angle of rotation at end A, and (b) determine the maximum shear stress in the steel core and the maximum shear stress in the aluminum shell.

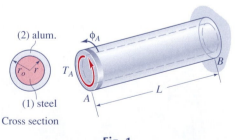

(2) alum.

r_o r T_A

(1) steel

Cross section

ϕ_A

B

L

A

Fig. 1

Plan the Solution   This is a statically indeterminate shaft assemblage. There is only one unknown nodal displacement, ϕ_A. Since the shaft is composed of an inner steel core and an outer aluminum shell, we can call these portions separate elements, namely elements (1) and (2), respectively. For a Displacement-Method solution we will solve an equilibrium equation that is eventually written in terms of the unknown nodal rotation angle ϕ_A. Then we can determine the two element torques and use them to solve for the maximum shear stresses.

Solution   (a) *Determine ϕ_B.*

Equilibrium: The free-body diagram of node A in Fig. 2 will enable us to relate the two element torques to the applied torque T_A.

Fig. 2   Free-body diagram of node A.

$$T_1 + T_2 = T_A \qquad \textbf{Equilibrium} \qquad (1)$$

Element Torque-Twist Behavior: For a Displacement-Method solution, Eq. 4.21 is the appropriate form of the torque-twist equation to use.

$$T_i = k_{ti}\phi_i, \quad i = 1, 2 \qquad \begin{array}{l}\textbf{Torque-Twist}\\\textbf{Behavior}\end{array} \qquad (2)$$

where $k_{ti} = \left(\dfrac{GI_p}{L}\right)_i$.

$$I_{p1} = \frac{\pi(25\ \text{mm})^4}{2} = 613.6(10^3)\ \text{mm}^4$$

$$I_{p2} = \frac{\pi(37.5\ \text{mm})^4}{2} - \frac{\pi(25\ \text{mm})^4}{2} = 2493(10^3)\ \text{mm}^4$$

$$k_{t1} = \left(\frac{GI_p}{L}\right)_1 = \frac{\left(78 \times 10^9\dfrac{\text{N}}{\text{m}^2}\right)(613.6 \times 10^3\ \text{mm}^4)}{(0.5\ \text{m})} = 95.72(10^3)\ \frac{\text{N} \cdot \text{m}}{\text{rad}}$$

$$k_{t2} = \left(\frac{GI_p}{L}\right)_2 = \frac{\left(28 \times 10^9\dfrac{\text{N}}{\text{m}^2}\right)(2493 \times 10^3\ \text{mm}^4)}{(0.5\ \text{m})} = 139.59(10^3)\ \frac{\text{N} \cdot \text{m}}{\text{rad}}$$

Geometry of Deformation: The element relative twist angles are

$$\phi_1 = \phi_2 = \phi_A - \phi_B$$

But, since $\phi_B = 0$, we get the following **compatibility equations:**

$$\phi_1 = \phi_A, \quad \phi_2 = \phi_A \qquad \begin{array}{l}\textbf{Geometry of}\\\textbf{Deformation}\end{array} \qquad (3a,b)$$

Displacement-Method Solution: The Displacement-Method steps for determining ϕ_A consist of substituting Eqs. (3) into (2) into (1), and then solving the resulting equilibrium equation for the unknown nodal displacement.

$$T_1 = k_{t1}\phi_A, \quad T_2 = k_{t2}\phi_A \qquad (2')$$

$$(k_{t1} + k_{t2})\phi_A = T_A \qquad (1')$$

Therefore, the unknown nodal displacement ϕ_A is

$$\boxed{\phi_A = \frac{T_A}{k_{t1} + k_{t2}} = 1.487(10^{-3})\,\text{rad}} \qquad \textbf{Ans. (a)} \quad (4)$$

(b) *Determine the maximum shear stress in each part.* Combining Eqs. (2′) and (4), we get

$$T_1 = 142.37 \text{ N} \cdot \text{m}, \quad T_2 = 207.63 \text{ N} \cdot \text{m}$$

In each case, steel and aluminum, the maximum shear stress occurs at the outer radius and is given by Eq. 4.13.

$$(\tau_{\max})_{\text{steel}} = \frac{T_1 r_1}{I_{p1}} = \frac{(142.37 \text{ N} \cdot \text{m})(0.25 \text{ m})}{613.6(10^{-9})\,\text{m}^4} = 5.801 \text{ MPa}$$

$$(\tau_{\max})_{\text{alum}} = \frac{T_2 (r_2)_o}{I_{p2}} = \frac{(207.63 \text{ N} \cdot \text{m})(0.0375 \text{ m})}{2493(10^{-9})\,\text{m}^4} = 3.123 \text{ MPa}$$

Rounding to three significant figures, we get

$$(\tau_{\max})_{\text{steel}} = 5.80 \text{ MPa}$$

$$(\tau_{\max})_{\text{alum}} = 3.12 \text{ MPa} \qquad \textbf{Ans. (b)} \quad (5)$$

MDS4.9 **Statically Indeterminate Torsional Assemblages: Displacement Method** combines *equilibrium, torque-twist,* and *compatibility equations* to solve for the unknown nodal rotation angles. Equation 4.21 is used to express the element torques T_i in terms of the element twist angles ϕ_i.

MDS4.10 **Statically Indeterminate Torsional Assemblages: Initial Stress** combines *equilibrium, torque-twist,* and *compatibility equations* to solve for the unknown nodal rotation angles. Equation 4.21 is used to express the element torques T_i in terms of the element twist angles ϕ_i. Similar to Eqs. 4.22, the angular misfit is incorporated in the *deformation-geometry equations*.

In the previous Example Problems the assembled torsion members have been collinear. However, torsion members may also be coupled together through gears or through belts and pulleys. Example Problem 4.9 shows how such problems can be solved using the fundamental equations of *equilibrium, torque-twist behavior,* and *deformation compatibility;* it will be solved by the Displacement Method.

EXAMPLE 4.9

Shafts AB and CE in Fig. 1 have the same diameter and are made of the same material. A torque T_E is applied to the shaft-gear system at E. Assume that torque is transmitted from shaft CE to shaft AB by a single gear-tooth contact force, and neglect the thickness of the gears.

Using the *Displacement Method:* (a) determine an expression for the rotation of gear B; (b) determine an expression for the shaft rotation at E: and (c) determine the torque transmitted to the base at C.

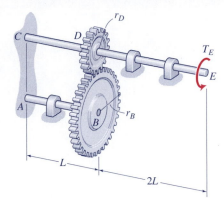

Fig. 1

Plan the Solution We have three uniform elements, which we can number as shown in Fig. 2. There is a relationship between torques applied to the two gears (equilibrium), and also a relationship between their angles and directions of rotation (deformation compatibility). Let x and x' axes be designated for the shafts so that we can adopt a consistent sign convention for the various torques and rotation angles that enter into the solution.

The torque in element (3) can be determined from statics (i.e., equilibrium) alone, but the problem is statically indeterminate, since there are restraints at both A and C.

Solution   (a) *Determine ϕ_B, the rotation angle of gear B.*

Equilibrium: We can separate the system into shaft elements and nodes, and we can then write a moment-equilibrium equation for each node. The gear-tooth force F is assumed to be acting normal to the radius at the point of contact. Note that all torques in Fig. 3 are shown in the positive sense according to the sign convention stated in Section 4.4. The sense assumed for the gear-tooth contact force is arbitrary, so long as Newton's third law of "action and reaction" is observed. (Only torques and gear-tooth forces are shown on the "free-body diagrams." Other forces, such as reaction forces at the bearings and at the wall, are omitted to reduce the complexity of the diagrams.) We now write the moment-equilibrium equation for each node.

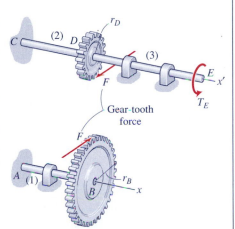

Gear-tooth force

Fig. 2

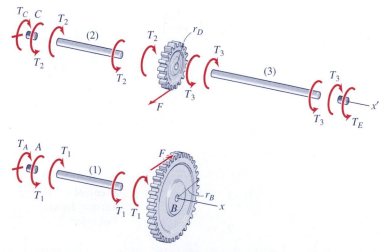

Note: All torques are shown, but force reactions
(except for the gear-tooth force) are omitted for clarity.

Fig. 3   Free-body diagrams.

269

For Node A: $\sum M_x = 0$: $T_1 - T_A = 0$

For Node B: $\sum M_x = 0$: $-T_1 - Fr_B = 0$

For Node C: $\sum M_{x'} = 0$: $T_2 - T_C = 0$

For Node D: $\sum M_{x'} = 0$: $-T_2 + T_3 - Fr_D = 0$

For Node E: $\sum M_x = 0$: $-T_3 + T_E = 0$

Summarizing the nodal equilibrium equations, we have

$$
\begin{aligned}
T_1 &= T_A \\
-T_1 &= Fr_B = 0 \\
T_2 &= T_C \\
-T_2 + T_3 - Fr_D &= 0 \\
T_3 &= T_E
\end{aligned}
$$ **Equilibrium** (1)

Equation (1e) enables us to determine T_3 from the given torque T_E. Equations (1a) and (1c) allow us to determine the reaction torques T_A and T_C once T_1 and T_2 have been determined. This leaves Eqs. (1b) and (1d) as the primary equilibrium equations to be used in determining T_1, T_2, and F. That is, we have two equations in three unknowns, so this is a statically indeterminate system.

Element Torque-Twist Behavior: Since we are to use the Displacement Method, Eq. 4.21 is the appropriate form for the torque-twist behavior. The sign convention for T_i and ϕ_i is shown in Fig. 4.

$$ T_i = k_{ti}\phi_i, \quad i = 1, 2, 3 $$

Fig. 4 A torsion element.

where

$$ k_{ti} = \left(\frac{GI_p}{L}\right)_i $$

All members have the same GI_P, and the length of element (3) is twice that of the other two elements. Let $k_t \equiv GI_P/L$. Then,

$$ k_{t1} = k_{t2} \equiv k_t, \quad k_{t3} = \frac{GI_p}{2L} \equiv \frac{k_t}{2} $$

$$
\begin{aligned}
T_1 &= k_t\phi_1 \\
T_2 &= k_t\phi_2 \\
T_3 &= \left(\frac{k_t}{2}\right)\phi_3
\end{aligned}
$$ **Element Torque-Twist Behavior** (2)

Deformation Geometry: We need two types of displacement information: (1) the relationships of the element twist angles ϕ_i to the nodal rotations,

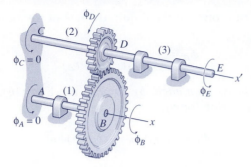

Fig. 5   Deformation diagram.

and (2) the relationship of the rotation of gear B to the rotation of gear D. The sign convention for nodal rotation angles is shown on the deformation sketch in Fig. 5.

From Fig. 5,

$$
\begin{aligned}
\phi_1 &= \phi_B - \phi_A = \phi_B \\
\phi_2 &= \phi_D - \phi_C = \phi_D \\
\phi_3 &= \phi_E - \phi_D
\end{aligned}
$$

Element Twist-Angle Definitions   (3)

In Fig. 6 the gear rotation angles ϕ_B and ϕ_D are shown in the positive sense. When gear B rotates through a positive angle ϕ_B, gear D will rotate in a negative sense by an amount such that the circumferential contact lengths will be equal. Thus,

$$
r_B\phi_B = -r_D\phi_D
$$

Rotational-Displacement Compatibility   (4)

Displacement-Method Solution: Since we are to use the Displacement Method to determine ϕ_B, we can use Eq. (4) to eliminate ϕ_D and then substitute (3) → (2) → (1).

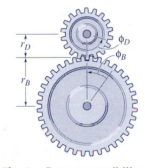

Fig. 6   Gear compatibility.

$$
\begin{aligned}
\phi_1 &= \phi_B \\
\phi_2 &= -(r_B/r_D)\phi_B \\
\phi_3 &= \phi_E + (r_B/r_D)\phi_B
\end{aligned}
$$

(3′)

$$
\begin{aligned}
T_1 &= k_t\phi_B \\
T_2 &= -k_t(r_B/r_D)\phi_B \\
T_3 &= (k_t/2)[\phi_E + (r_B/r_D)\phi_B]
\end{aligned}
$$

(2′)

We can eliminate T_3 and F from Eqs. (1b), (1d), and (1e) to get

$$
-T_2 + (r_D/r_B)T_1 = -T_E
$$

and then we can combine this equation with Eqs. (2′a) and (2′b) to get

$$
k_t(r_B/r_D)\phi_B + (r_D/r_B)(k_t\phi_B) = -T_E \qquad (1')
$$

from which we can solve for the required angle of rotation ϕ_B

$$\phi_B = \frac{-T_B/k_t}{(r_B/r_D) + (r_D/r_B)} \qquad \textbf{Ans. (a)} \quad (5)$$

(b) *Determine the rotation angle ϕ_E.* We can use Eqs. (1e), (2′c) and (5) to obtain an expression for ϕ_E.

$$\phi_E = (2/k_t)T_E - (r_B/r_D)\phi_B$$

or

$$\phi_E = (T_E/k_t)\left[2 + \frac{(r_B/r_D)}{(r_B/r_D) + (r_D/r_B)}\right] \qquad \textbf{Ans. (b)} \quad (6)$$

(c) *Determine the reaction torque T_C.* T_C may be obtained by combining Eqs. (1c), (2′b), and (5) to get

$$T_C = -T_2 = \frac{-(r_B/r_D)T_E}{(r_B/r_D) + (r_D/r_B)} \qquad \textbf{Ans. (c)} \quad (7)$$

Review the Solution Because this is a fairly complex problem, we should first check each answer to see if it has the proper sign and proper dimensions. The answers (5) through (7) satisfy these two requirements.

Next, we can observe that the magnitude of the torque transmitted from T_E to the base at C should be less than T_E, since part of the reaction to T_E is via the gear at B to the base at A. From Eq. (7) we can verify that $|T_C| < T_E$ as expected.

4.8 POWER-TRANSMISSION SHAFTS

Solid circular shafts and tubular circular shafts are frequently employed to transmit power from one device to another, for example from a turbine to a generator in a power plant or from the motor to the wheels of a car, truck, or other vehicle. Figure 4.23 shows a motor, a pulley, and the circular shaft connecting them. The motor at A

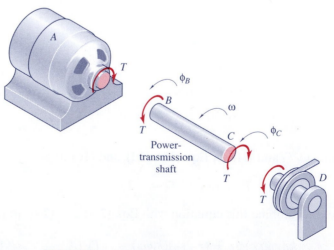

FIGURE 4.23 A power-transmission shaft.

supplies power to the pulley at D through the shaft BC, which is rotating at constant speed ω and which exerts a torque T at C, where the shaft is attached to the pulley D.

The **work** that the shaft does on the pulley at D is the torque times the angle (in radians) through which the shaft rotates, that is,

$$\mathcal{W}_{\text{on}D} = T\phi_C$$

The **power** delivered at C is given by

$$\mathcal{P} = \frac{dW}{dt} = T\frac{d\phi_C}{dt}$$

or

$$\boxed{\mathcal{P} = T\omega} \tag{4.23}$$

where ω is the rotational speed of the shaft in radians per second. The rotational speed is frequently expressed in revolutions per second or in revolutions per minute; the conversions are:

$$\omega(\text{rad/s}) = 2\pi f(\text{rev/s}) = \frac{2\pi n\,(\text{rpm})}{60}$$

where f is the rotational speed in revolutions per second (rev/s) and n is the speed in revolutions per minute (rpm).

When U.S. Customary units are used, the power is usually expressed in *horsepower* (hp), where

$$1\ \text{hp} = 550\ \text{lb} \cdot \text{ft/s} = 6600\ \text{lb} \cdot \text{in./s}$$

When SI units are used and T is expressed in $\text{N} \cdot \text{m}$, the power will be in $\text{N} \cdot \text{m/s} = J/s$, or watts (W). The conversion factor is

$$1\ \text{hp} = 745.7\ \text{W}$$

Note, from Eq. 4.23, that the torque is related to the power and shaft rotation speed by $T = \mathcal{P}/\omega$, so that, for a given power, the <u>slower</u> the speed, the higher the torque. Once the torque has been determined, the angle of twist of a shaft and the complete stress distribution in the shaft can be determined. Two examples will now be given to illustrate how Eq. 4.23 is employed in designing power-transmission shafts.

EXAMPLE 4.10

An electric motor delivers 10 hp to a pump through a solid circular shaft that is rotating at 875 rpm. If the shaft has an allowable shear stress of $\tau_{\text{allow}} = 20$ ksi, what is the minimum required diameter of the shaft as a multiple of 1/16 in.?

Solution

Torque: Equation 4.23 relates power to torque.

$$\mathcal{P} = T\omega \tag{1}$$

To get the torque T in units of lb · in., we will have to introduce consistent units for $\mathcal{P}$ and ω. Thus,

$$\mathcal{P} = (10 \text{ hp})\left(\frac{6600 \text{ lb} \cdot \text{in./s}}{1 \text{ hp}}\right) = 66{,}000 \text{ lb} \cdot \text{in./s}$$

$$\omega = (875 \text{ rpm})\left(\frac{2\pi \text{ rad}}{1 \text{ rev}}\right)\left(\frac{1 \text{ min}}{60 \text{ sec}}\right) = 91.63 \text{ rad/s}$$

Therefore,

$$T = \frac{\mathcal{P}}{\omega} = \frac{66{,}000 \text{ lb} \cdot \text{in./s}}{91.63 \text{ rad/s}} = 720.3 \text{ lb} \cdot \text{in.} \tag{2}$$

Allowable-Stress Design: Using the torsion formula for a solid shaft, Eq. 4.13, we can now determine the required diameter of the shaft.

$$\tau_{\text{allow}} \geq \tau_{\text{max}} = \frac{Tr}{I_p}: \qquad d_{\text{min}}^3 = \frac{16T}{\pi\tau_{\text{allow}}} = \frac{16(720.3 \text{ lb} \cdot \text{in.})}{\pi(20 \times 10^3)\text{psi}} \tag{4}$$

So, the required diameter is $d_{\text{min}} = 0.568$ in., or, to the next greater 1/16 in.,

$$d_{\text{min}} = \frac{5}{8} \text{ in.} \qquad\qquad \textbf{Ans.} \tag{5}$$

EXAMPLE 4.11

A truck driveshaft is to be a tube with an outer diameter of 46 mm. It is to be made of steel ($\tau_{\text{allow}} = 80$ MPa), and it must transmit 120 kW of power at an angular speed of 40 rev/s. Determine, to the nearest even millimeter (that satisfies the design allowable), the maximum inner diameter that the shaft may have.

Solution The key equations are the torque-power equation, Eq. 4.23, and the torsion formula, Eq. 4.13.

Torque:

$$\mathcal{P} = T\omega = T(2\pi f) \tag{1}$$

so

$$T = \frac{\mathcal{P}}{2\pi f} = \frac{120 \text{ kW}}{2\pi(40 \text{ rev/s})} = 477.5 \text{ N} \cdot \text{m} \tag{2}$$

Allowable-Stress Design:

$$\tau_{\text{allow}} \geq \tau_{\max} = \frac{T r_o}{I_p} \qquad (3)$$

For a circular tube,

$$I_p = \frac{\pi}{32}(d_o^4 - d_i^4) \qquad (4)$$

where d_o is the outer diameter of the shaft, and d_i is the inner diameter. Combining Eqs. (3) and (4) we get

$$d_i^4 \leq d_o^4 - \frac{16 d_o T}{\pi \tau_{\text{allow}}} \qquad (5)$$

Finally, from Eqs. (2) and (5),

$$d_i^4 \leq (46 \times 10^{-3}\,\text{m})^4 - \frac{16(46 \times 10^{-3}\,\text{m})}{\pi}\left(\frac{477.5\,\text{N} \cdot \text{m}}{80 \times 10^6\,\text{N/m}^2}\right) \qquad (6)$$

from which, $(d_i)_{\max} = 41.89$ mm. Although this value for $(d_i)_{\max}$ is very close to 42 mm, we must pick an inner diameter of

$$d_i = 40\,\text{mm} \qquad \textbf{Ans.} \quad (7)$$

to keep $\tau_{\max} \leq \tau_{\text{allow}}$.

MDS4.11–4.13 **Power-Transmission Shafts** is a computer program module for solving torsion problems when the torque in a shaft is related to the power $\mathcal{P}$ transmitted by the shaft and to its rotation speed ω through Eq. 4.23.

*4.9 THIN-WALL TORSION MEMBERS

Although power-transmission shafts and many other torsion members are solid circular cylinders or tubular members with circular cross section, not every important torsion member has a circular cross section. For example, airplane fuselages, wings, and tails are all thin-wall members that are subjected to loading that includes torsion. Figure 4.24 depicts the structure of the wing of a light aircraft. The part

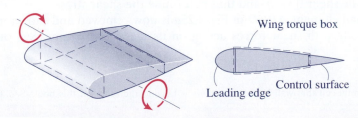

FIGURE 4.24 An airplane wing with torque box outlined.

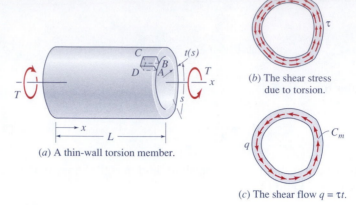

(b) The shear stress
due to torsion.

(a) A thin-wall torsion member.

(c) The shear flow $q = \tau t$.

FIGURE 4.25   A closed, thin-wall, single-cell torsion member.

of the wing structure that resists torsion is referred to as the *torque box*. The wing in Fig. 4.24 has a *single-cell torque box,* which is outlined with dashed lines.[4]

Under certain conditions, the shear stress distribution on the cross section of a thin-wall torsion member can be determined easily. These simplifying conditions are:

- The member is cylindrical, that is, the cross section does not vary along the length of the member.
- The cross section is "closed," that is, there is no longitudinal slit in the member.
- The wall thickness is small compared with the cross-sectional dimensions of the member.
- The member is subjected to end torques only.
- The ends are not restrained from warping.

Shear Flow. Figure 4.25 shows a portion of a typical closed, single-cell thin-wall torsion member. The thickness, t, may vary with circumferential position. The key assumption that is made in order to simplify the analysis of the stress distribution in closed, thin-wall members is—*the shear stress is constant through the thickness and is parallel to the median curve defining the cross section.* Figures 4.25b and 4.25c illustrate the significance of this assumption, which permits us to define a quantity called **shear flow,** q, by the equation

$$q = \tau t \qquad\qquad \textbf{Shear Flow} \qquad (4.24)$$

where the wall thickness t can be a function of circumferential location s. The name *shear flow* comes from the analogy of shear in thin-wall members and the flow of a fluid in a closed pipe or channel. Here, the shear "flows" in a closed path around the cross section, as you will see from the following derivation. Now we can proceed to determine the shear flow, q, and thus determine the shear stress τ.

The small element *ABCD* in Fig. 4.25a is now removed and shown in Fig. 4.26 as a free body with shear forces acting on its faces. The shear stress on the face

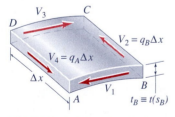

FIGURE 4.26 Free-body diagram of a thin-tube element.

---

[4]Analysis of multi-cell thin-wall torsion members is beyond the scope of this book. See, for example, [Ref. 4-1].

AB at A is τ_A. The same shear stress must also be acting on the face AD. Hence, the shear force on face AD is

$$V_4 = \tau_A t_A \Delta x = q_A \Delta x$$

Similarly, the shear force on face BC is

$$V_2 = \tau_B t_B \Delta x = q_B \Delta x$$

Since it was assumed that the member is subjected to torque only (thus, there is no axial stress σ_x), the equation of axial equilibrium is

$$\searrow \sum F_x = 0: \qquad V_4 - V_2 = 0, \qquad (q_A - q_B)\Delta x = 0$$

or

$$q_A = q_B \qquad (4.25)$$

Since A and B are arbitrary points in the cross section, Eq. 4.25 implies that q is independent of location in the cross section: that is, q is independent of s. Hence.

$$\boxed{q = \tau t = \text{const}} \qquad (4.26)$$

Shear Stress Resultant; Torque. Equilibrium of forces in the x-direction leads to a shear flow that is constant (Eq. 4.26). Now we need to determine the relationship between the shear flow, q, and the torque, T. The incremental force, dF_s, due to the shear flow, q, on a differential element of area of the cross section is illustrated in Fig. 4.27a. The force on this element of cross section,

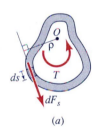
(a)

$$dF_s = q\, ds$$

acts tangent to the *median curve*, C_m, as indicated in Fig. 4.27a, and the moment of this force dF_s about an arbitrary point O in the cross section is

$$dT = \rho\, dF_s = q\rho\, ds$$

where ρ is the perpendicular distance from point O to the line of action of dF_s. To get the torque on the cross section, we sum the contributions around the curve C_m, that is,

(b)

$$T = \oint_{C_m} q\rho\, ds \qquad (4.27)$$

Since q is constant, we are left with evaluating $\oint \rho\, ds$ around C_m. This integral is a purely geometrical quantity. Figure 4.27b illustrates the fact that the origin O and base ds form a triangle whose area is given by

(c)

$$d\mathcal{A}_m = \frac{1}{2}(\rho\, ds) \qquad (4.28)$$

FIGURE 4.27 The torque due to shear flow.

Hence,

$$\oint_{C_m} \rho \, ds = 2\mathcal{A}_m \qquad (4.29)$$

where $\mathcal{A}_m$ is the <u>shaded area enclosed by the median curve</u> C_m, as illustrated in Fig. 4.27c. (Note: The area $\mathcal{A}_m$ is not the area of the material cross section!) Finally, combining Eqs. 4.27 and 4.29, we get

$$q = \frac{T}{2\mathcal{A}_m} \qquad (4.30)$$

When combined with Eq. 4.24, Eq. 4.30 gives the following expression for the shear stress acting on the cross section of a closed, single-cell, thin-wall torsion member:

$$\tau = \frac{T}{2t\mathcal{A}_m} \qquad \text{**Shear-Stress**} \qquad (4.31)$$
$$\text{**Formula**}$$

where $\tau \equiv \tau(s)$ is the average shear stress at location s in the cross section, and where $t \equiv t(s)$ is the local thickness at the point where the shear stress is evaluated.

Angle of Twist. To determine the angle of twist of a closed, thin-wall torsion member, like the one in Fig. 4.25, we can employ basic strain-displacement and stress-strain relationships, or we can employ energy methods. The latter approach is discussed in Section 11.4. The result is

$$\phi = \frac{TL}{4\mathcal{A}_m G} \oint_{C_m} \frac{ds}{t(s)} \qquad (4.32)$$

where G is the shear modulus, and where the integral is evaluated around the median curve C_m shown in Fig. 4.27c.

In the discussion of Eqs. 4.7 and 4.10 it was indicated that the relationship between torque and twist rate for a circular shaft is

$$\frac{d\phi}{dx} = \frac{T}{GI_p} \qquad \qquad (4.10)$$
$$\text{repeated}$$

where I_p is the polar moment of inertia. Furthermore, it was indicated in the footnote that the relationship between the torque T and the twist rate $d\phi/dx$ is sometimes written as

$$\frac{d\phi}{dx} = \frac{T}{GJ} \qquad \text{**Torque-**} \qquad (4.33)$$
$$\text{**Twist**}$$
$$\text{**Equation**}$$

where GJ is called the **torsional rigidity** and where, for noncircular cross sections, J is <u>not the polar moment of inertia</u>. From Eqs. 4.33 and 4.32 we can see that, for a closed, single-cell, thin-wall torsion member, the torsional rigidity, GJ, is given by

$$GJ = \frac{4\mathcal{A}_m^2 G}{\oint_{C_m} \dfrac{ds}{t(s)}} \qquad (4.34)$$

(a) Determine the maximum torque that may be applied to an aluminum-alloy tube if the allowable shear stress is $\tau_{allow} = 100$ MPa. The tube has the cross section shown in Fig. 1a. (b) When the aluminum tube was extruded, the hole was not perfectly centered and the tube was found to have the actual dimensions shown in Fig. 1b. If the torque determined in Part (a) is applied to the imperfect tube, what will the maximum shear stress be? Where will this maximum shear stress occur?

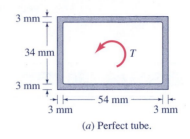

(a) Perfect tube.

Plan the Solution Equation 4.31 can be used to determine the allowable torque, given the allowable shear stress and the cross-sectional dimensions.

Solution (a) *Determine the allowable torque for the perfect tube.* Equation 4.31 can be written in the form

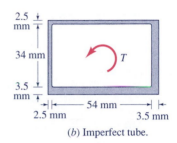

(b) Imperfect tube.

Fig. 1

$$T_{allow} = 2\mathcal{A}_m t_{min} \tau_{allow} \qquad (1)$$

Figure 2 shows the median curve used in calculating $\mathcal{A}_a = \mathcal{A}_b$.

$$\mathcal{A}_a = \mathcal{A}_b = (37 \text{ mm})(57 \text{ mm}) = 2.109(10^{-3}) \text{ m}^2 \qquad (2)$$

The minimum wall thickness is 3 mm. Therefore, from Eqs. (1) and (2) we get

$$(T_a)_{allow} = 2\mathcal{A}_a (t_a)_{min} \tau_{allow}$$

$$= 2[2.109(10^{-3}) \text{ m}^2](0.003 \text{ m})[100(10^6) \text{ N/m}^2]$$

$$(T_a)_{allow} = 1265 \text{ N} \cdot \text{m} \qquad \textbf{Ans. (a)} \quad (3)$$

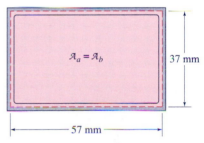

Fig. 2 Area enclosed by the median curve for tubes (a) and (b).

(b) *Determine the maximum shear stress in the imperfect tube.* In this case, Eq. 4.31 can be written in the form

$$T_b = 2\mathcal{A}_b (t_b)_{min} (\tau_b)_{max} \qquad (4)$$

Since $\mathcal{A}_a = \mathcal{A}_b$ and the two torque tubes are to have the same torque, Eqs. (3) and (4) can be combined to give

$$(\tau_b)_{max} = \tau_{allow} \left[\frac{(t_a)_{min}}{(t_b)_{min}} \right]$$

$$= 100 \text{ MPa} \left(\frac{0.003 \text{ m}}{0.0025 \text{ m}} \right)$$

Therefore, the maximum shear stress in the imperfect tube is in the minimum-thickness section of the tube wall and has the value

$$(\tau_b)_{max} = 120 \text{ MPa} \qquad \textbf{Ans. (b)}$$

Note that the shear stresses in the two tubes are inversely proportional to their minimum wall thicknesses.

Figure 4.25*b* indicates how shear stresses "flow" in the thin wall of a torque tube. Note that the rectangular tubes in Fig. 1 of the preceding example problem have filleted inner corners to minimize the stress concentration that occurs due to the 90° change in direction of the shear stress at each corner of the rectangular cross section. (See Section 12.2 for a discussion of stress concentrations.)

EXAMPLE 4.13

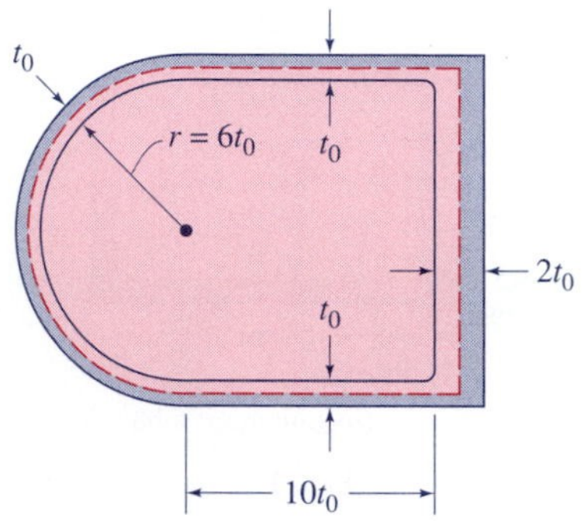

Determine the torsional rigidity, GJ, for the thin-wall tubular member whose cross section is shown in Fig. 1. The shear modulus is G.

Solution   This is a straightforward application of Eq. 4.34, but, since the thickness of the wall of the torque tube is piecewise-constant, we can write Eq. 4.34 in the form

$$GJ = \frac{4\mathcal{A}_m^2 G}{\sum_i \left(\dfrac{\Delta s_i}{t_i}\right)} \tag{1}$$

The area $\mathcal{A}_m$ enclosed by the dashed median curve in Fig. 1 is

$$\mathcal{A}_m = \frac{\pi}{2}(6.5t_0)^2 + (11t_0)(13t_0) = 209.4t_0^2 \tag{2}$$

Referring to Fig. 1, we can evaluate the sum in the denominator of Eq. (1) as

$$\sum_i \left(\frac{\Delta s_i}{t_i}\right) = \frac{\pi(6.5t_0)}{t_0} + \frac{2(11t_0)}{t_0} + \frac{13t_0}{2t_0} = 48.92 \tag{3}$$

Combining Eqs. (1) through (3), we get

$$GJ = \frac{4(209.4t_0^2)^2 G}{48.92} = 3584\,Gt_0^4$$

or, rounded to three significant figures,

$$GJ = 3580\,Gt_0^4 \qquad \textbf{Ans.}$$

Fig. 1

Cross-section figure with dimensions: t_0, $r = 6t_0$, t_0, $2t_0$, t_0, $10t_0$.

*4.10   TORSION OF NONCIRCULAR PRISMATIC BARS

In Section 4.2 the deformation of a circular cylinder twisted by equal and opposite torques applied at its ends was described. That discussion, supported by the photos in Fig. 4.2, pointed out that, for torsion members with circular cross sections, plane sections remain plane and simply rotate around the axis of the member. It is clear from the photo of the deformed square torsion bar in Fig. 4.2*b* that plane sections do not remain plane when a member with noncircular cross section is subjected to

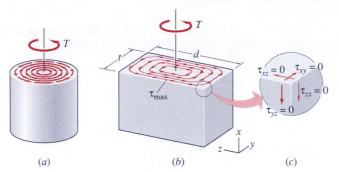

FIGURE 4.28   Torsion of circular and rectangular members.

torsional loading. An important feature of the torsional deformation of noncircular prismatic bars is the *warping* of the cross sections.

The theory of elasticity may be used to relate the torque applied to such non-circular prismatic members to the resulting stress distribution and angle of twist.[5]

Stress Distribution and Angle of Twist.

The shear-stress distribution in non-circular torsion bars is quite different than the shear stress distribution in circular torsion members. Figure 4.28 compares the stress distribution in a circular bar with that in a rectangular bar. The shear stress on the circular cross section varies linearly with distance from the center and reaches its maximum at the outer surface (Eq. 4.11). In contrast, the shear stress at the corners of the rectangular torsion member in Fig. 4.28b must be zero. (Recall that $\tau_{xy} = \tau_{yx}$.) In fact, the maximum shear stress on a rectangular cross section occurs at the middle of the longer edge, which is the point on the periphery of the cross section that is nearest the center!

The maximum shear stress in a rectangular prismatic bar subjected to torsion may be expressed in the form

$$\tau_{max} = \frac{T}{\alpha dt^2} \tag{4.35}$$

where α is a dimensionless constant obtained by a theory of elasticity solution and listed in Table 4.3. and where the dimensions d and t satisfy $d/t \geq 1$. The angle of twist of a bar of length L can be expressed by

$$\phi = \frac{TL}{GJ}, \quad \text{where } J = \beta dt^3 \tag{4.36}$$

where β is a dimensionless constant with value as listed in Table 4.3.

TABLE 4.3   Torsion Constants for Rectangular Bars

d/t	1.00	1.50	1.75	2.00	2.50	3.00	4	6	8	10	∞
α	0.208	0.231	0.239	0.246	0.258	0.267	0.282	0.298	0.307	0.312	0.333
β	0.141	0.196	0.214	0.229	0.249	0.263	0.281	0.298	0.307	0.312	0.333

---

[5]Saint-Venant (see Section 2.10) developed the theory of torsion for noncircular bars. He presented his famous memoir on torsion to the French Academy of Sciences in 1853. See, for example. Chapter 10 of [Ref. 4-2] for a discussion of Saint-Venant's theory of torsion.

The shear-stress distribution on the cross section of a shaft with elliptical cross section is illustrated in Fig. 4.29. The maximum shear stress occurs at the boundary at the two ends of the minor axis of the ellipse and is given by

$$\tau_{max} = \frac{2T}{\pi a b^2} \tag{4.37}$$

The angle of twist for an elliptical shaft of length L is given by

$$\phi = \frac{TL}{GJ}, \quad \text{where } J = \frac{\pi a^3 b^2}{a^2 + b^2} \tag{4.38}$$

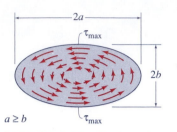

FIGURE 4.29 Torsion of an elliptical shaft.

Finally, the area of an ellipse is

$$A = \pi a b$$

It is important to note that the torsional behavior of circular bars is very special. If a channel section or a wide-flange section or any other noncircular cross section is subjected to torsional loading, its behavior must be analyzed by analytical methods like the Saint-Venant solution that produced Eqs. 4.35 through 4.38. Finite-element analysis may also be used to solve specific torsion problems.

EXAMPLE 4.14

(a)

(b)

(c)

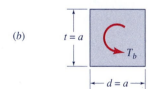

Fig. 1

If torsion members having the cross sections shown in Fig. 1 have the same cross-sectional area and are subjected to torques that produce the same maximum shear stress, τ_{max}, in each, what is the torque carried by each?

Solution Since the areas are to be the same, that is $A_a = A_b = A_c = a^2$, the radius of the circular bar is given by

$$\pi c^2 = a^2, \quad c = 0.5642a$$

For the circular bar $J = I_p = \frac{1}{2}\pi c^4$, and, from Eq. 4.13,

$$\tau_{max} = \frac{T_a c}{J} \tag{1}$$

Therefore,

$$T_a = \frac{\tau_{max} J}{c} = \frac{\pi \tau_{max} c^3}{2} = \frac{\pi \tau_{max}(0.5642a)^2}{2} \tag{2}$$

or

$$T_{circle} \equiv T_a = 0.282\tau_{max}a^3 \tag{3}$$

For rectangular bars, Eq. 4.35 gives

$$\tau_{max} = \frac{T}{\alpha d t^2} \qquad (4)$$

For the square bar in Fig. 1b, $d/t = 1$, and Table 4.3 gives $\alpha = 0.208$. Therefore,

$$T_{square} \equiv T_b = 0.208\tau_{max}a^3 \qquad (5)$$

Finally, for the rectangle in Fig. 1c, $d/t = 4$, so Table 4.3 gives $\alpha = 0.282$. Therefore, Eq. (2) gives

$$T_{4:1rect} \equiv T_c = 0.282\tau_{max}(2a)(a/2)^2 - 0.141\tau_{max}a^3 \qquad (6)$$

Summarizing the above results, we get

$$\frac{T_{circle}}{T_{square}} = 1.36, \qquad \frac{T_{circle}}{T_{4:1rect}} = 2.00 \qquad \textbf{Ans.} \quad (7a,b)$$

That is, the circular bar can support 36% more torque than a square bar of equal area; the circular bar can support 100% higher torque than can a rectangle with a 4:1 ratio of sides.

EXAMPLE 4.15

If the two torsion members in Fig. 1 have the same length L and the same cross-sectional area a^2, and if they are subjected to torques T_a and T_b that produce the same angle of twist, $\phi_a \equiv \phi_b = \phi$, what is the ratio of the two torques, T_a/T_b?

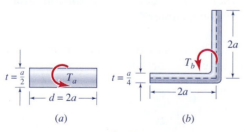

$t = \frac{a}{2}$ T_a $\leftarrow d = 2a \rightarrow$

$t = \frac{a}{4}$ T_b $2a$ $\leftarrow 2a \rightarrow$

(a)   (b)

Fig. 1

Solution   For rectangular bars, Eq. 4.36 gives

$$T = \left(\frac{GJ}{L}\right)\phi, \quad \text{where } J = \beta d t^3 \qquad (1)$$

For the rectangular cross section in Fig. 1a, $d/t = 4$, for which Table 4.3 gives $\beta = 0.281$.

For thin, open cross sections, like the equal-leg angle cross section in Fig. 1b, the dimension d can be taken to be the length of the centerline, as indicated by the dashed line in Fig. 1b. For this particular cross section, then, $\dfrac{d}{t} = \dfrac{4a}{a/4} = 16$. For this d/t ratio, we must extrapolate from the values given in Table 4.3. A reasonable estimate is $\beta = 0.323$.

$$\frac{T_{4:1rect}}{T_{16:1angle}} = \frac{J_a}{J_b} = \frac{0.281(2a)(a/2)^3}{0.323(4a)(a/4)^3} = 3.48 \qquad \textbf{Ans.} \quad (2)$$

That is, for the same angle of twist the 4:1 rectangular bar can support 248% more torque than an angle cross section with an equivalent 16:1 ratio of sides. Thin-wall, open cross sections do not make good torsion members. Compared to more compact sections, they will have much higher maximum shear stress and much larger angle of twist for a given torque, cross-sectional area, and length.

*4.11   INELASTIC TORSION OF CIRCULAR RODS

In the preceding sections of Chapter 4, we have considered torsion of linearly elastic members, the simplest case being the torsion of rods with circular cross section. Now we will examine the behavior of circular rods that are subjected to torques that produce shear stresses beyond the proportional limit. Inelastic torsion is similar in many respects to the inelastic axial deformation discussed in Section 3.11, with one very important difference. In the case of axial deformation, the strain and stress are uniform over the entire cross section of the axial-deformation member, but in the case of torsion, both shear strain and shear stress vary with distance from the center of the torsion rod.

Fundamental Equations. Of the three fundamentals of deformable-body mechanics—*equilibrium, geometry of deformation,* and *material behavior*—only the material behavior differs when we consider inelastic torsion rather than the linearly elastic behavior treated so far in Chapter 4.

Geometry of Deformation: The *strain-displacement equation,* Eq. 4.1, holds for inelastic as well as for linearly elastic torsion.

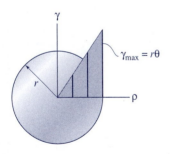

$$\boxed{\gamma(x, \rho) = \rho \frac{d\phi}{dx} \equiv \rho\theta} \qquad (4.39)$$

where

γ = the shear strain due to torsion.

ρ = the distance from the center of the rod to the point in the cross section where the strain is to be determined.

ϕ = the angle of twist at section x; $\theta \equiv \dfrac{d\phi}{dx}$ is the *twist rate.*

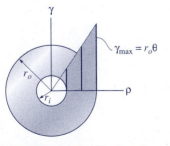

FIGURE 4.30 Torsional shear-strain distribution.

This linear strain distribution is sketched in Fig. 4.30 (repeat of Fig. 4.6).

Equilibrium: The shear stress $\tau(x, \rho)$ is related to the resultant torque $T(x)$ by the definition of the *resultant torque,* Eq. 4.2.

$$T(x) = \int_A \rho\tau \, dA \qquad (4.2)$$
$$\text{repeated}$$

Figure 4.31 illustrates how this integral can be evaluated for a solid shaft by using

$$\boxed{T(x) = 2\pi \int_0^r \tau\rho^2 \, d\rho} \qquad (4.40)$$

This internal torque $T(x)$ is related to the external load and reaction torques through free-body diagrams and equations of equilibrium.

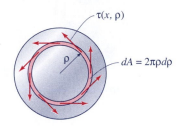

(*a*) Shear stress τ on dA ring at radius ρ.

Material Behavior: So far in Chapter 4 we have only considered linearly elastic behavior characterized by Hooke's Law for shear, $\tau = G\gamma$. Now, however, we will consider stresses beyond the proportional limit τ_{PL} for materials that have stress-strain curves in shear like those in Fig. 4.32. For the general case, therefore, the stress-strain diagram must be employed to establish the appropriate *material behavior* in one of the following two forms:

$$\boxed{\tau = \tau(\gamma), \quad \text{or} \quad \gamma = \gamma(\tau)} \qquad (4.41)$$

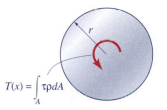

$T(x) = \int_A \tau\rho \, dA$

(Later, in the discussion of residual stresses, we will discuss how shear stress and shear strain are related during unloading.)

(*b*) Resultant internal torque at section *x*.

FIGURE 4.31 The relationship of torque to shear stress.

Elastic-Plastic Torque-Twist Analysis.

Equations 4.39 (strain-displacement), 4.40 (equilibrium), and 4.41 (material behavior) permit a complete solution of any problem of torsion of circular rods. The procedure will be illustrated here with the analysis of a rod made of elastic-plastic material having a stress-strain curve like that in Fig. 4.32*b*. As in the case of inelastic axial deformation (Section 3.11), there are also three cases to be considered for torsion: Case 1—linearly elastic behavior, Case 2—partially plastic behavior, and Case 3—fully plastic behavior.

Case 1—Linearly elastic behavior. In the linearly elastic range, that is, up to $\tau = \tau_Y$, the material obeys Hooke's Law,

$$\tau = G\gamma \qquad (4.42)$$

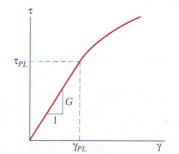

(*a*) Linearly elastic, nonlinearly plastic material.

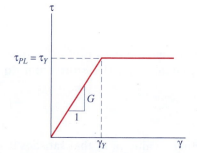

(*b*) Linearly elastic, perfectly plastic material.

FIGURE 4.32 Typical diagrams of shear stress versus shear strain.

In Section 4.3 the following torque-stress (Eq. 4.11) and torque-twist (Eq. 4.16) equations were derived for a homogeneous, linearly elastic rod.

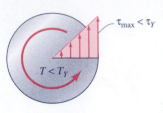

(a) Linearly elastic case.

$$\tau = \frac{T\rho}{I_p} \tag{4.43}$$

$$\theta = \frac{T}{GI_p} \tag{4.44}$$

where I_p is the polar moment of inertia of the cross section, and $\theta = \phi/L$. Linearly elastic behavior is depicted in Fig. 4.33a.

The maximum elastic torque, or **yield torque** (Fig. 4.33b), is obtained by setting $\tau = \tau_Y$ and $\rho = r$ in Eq. 4.43. Thus,

$$T_Y = \frac{\tau_Y I_p}{r} \tag{4.45}$$

Then, from Eq. 4.44

$$\theta_Y = \frac{T_Y}{GI_p} = \frac{\tau_Y}{Gr} \tag{4.46}$$

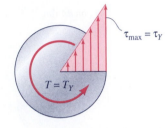

(b) Maximum elastic torque case.

FIGURE 4.33 Shear-stress distribution–elastic behavior.

The shear-stress distribution for this maximum elastic torque case is depicted in Fig. 4.33b.

Case 2—Partially plastic behavior. When the torque exceeds T_Y, the shear strain over a portion of the cross section exceeds the yield shear strain γ_Y. The shear-stress distribution then exhibits an *elastic core* of radius r_Y and a *plastic annulus*, as depicted in Fig. 4.34a. This stress-strain behavior for the partially plastic condition is given by the equations

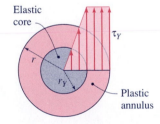

Elastic core

Plastic annulus

(a) Partially-plastic-torsion stress distribution.

$$\tau = \tau_Y\left(\frac{\rho}{r_Y}\right) \quad 0 \le \rho \le r_Y$$
$$\tau = \tau_Y \quad\quad r_Y \le \rho \le r \tag{4.47}$$

The corresponding torque is obtained by substituting Eqs. 4.47 into Eq. 4.40 giving

$$T = 2\pi \int_0^{r_Y}\left(\frac{\tau_Y\rho}{r_Y}\right)\rho^2\,d\rho + 2\pi \int_{r_Y}^r \tau_Y\rho^2\,d\rho$$

$$T = \frac{\pi\tau_Y}{6}(4r^3 - r_Y^3) \tag{4.48}$$

To determine the twist rate, θ, for this value of torque we can set the shear strain at $\rho = r_Y$ equal to γ_Y. Therefore, from Eq. 4.39,

$$\theta = \frac{\tau_Y}{Gr_Y} \tag{4.49}$$

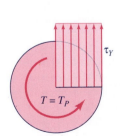

(b) Fully-plastic-torsion stress distribution.

FIGURE 4.34 Partially plastic torsion and fully plastic torsion.

Selecting a value of r_Y that satisfies $0 < r_Y < r$, we can eliminate r_Y from Eqs. 4.48 and 4.49 and thereby get a torque-twist relationship.

Case 3—Fully plastic torque. An elastic-plastic material, like mild steel, can experience very large shear strain while the shear stress remains constant at τ_Y. Therefore, it is possible to have $\gamma_{\max} \equiv \gamma(x, r) \gg \gamma_Y$. Then, $r_Y \rightarrow 0$ and we get the *fully plastic stress distribution* depicted in Fig. 4.34*b*. The material equation for the fully plastic case is

$$\tau = \tau_Y \quad 0 < \rho \leq r \tag{4.50}$$

The **plastic torque**, T_P, can be obtained by setting $r_Y = 0$ in Eq. 4.48, giving

$$T_P = \frac{2\pi\tau_Y r^3}{3} \tag{4.51}$$

This value can be approached, but not actually reached, since the strain-displacement equation. Eq. 4.39, does not permit the shear strain to actually reach γ_Y at $\rho = 0$ for a finite twist angle. Comparing Eqs. 4.45 and 4.51 we see that, for a solid circular shaft,

$$T_P = \frac{4}{3}T_Y \tag{4.52}$$

The above elastic-plastic torque-twist analysis is summarized in the **torque-twist curve** in Fig. 4.35.

Unloading; Residual Stress.

If the torque is allowed to exceed T_Y, say to T_B in Fig. 4.35, and then the torque is removed, the unloading curve will parallel the initial linearly elastic portion, OY, of the torque-twist curve. When the torque is completely removed, there will be *residual stresses* and a residual angle of twist, or *permanent set,* left in the rod (See Fig. 4.2).

Suppose that the maximum shear strain (at $\rho = r$) at torque T_B (Fig. 4.35) is γ_B, as indicated in Fig. 4.36. A subsequent decrease in torque would cause unloading at $\rho = r$ along the path BC indicated in Fig. 4.36. At all radii, whether in the elastic core or in the plastic annulus, unloading will occur in a linearly elastic manner along lines satisfying Hooke's Law in the form

$$\Delta\tau = G\Delta\gamma \tag{4.53}$$

where the Δ's indicate increments of stress and strain.

Since unloading takes place elastically, we can solve for the residual stresses in a torsion rod by superposing the plastic stress distribution due to torque T_B and an elastic stress distribution due to an equal and opposite torque $(-T_B)$. For the fully plastic case, we get the stress distributions shown in Fig. 4.37. Then, the maximum

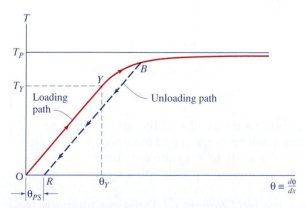

FIGURE 4.35 The torque-twist curve for an elastic-plastic circular torsion bar.

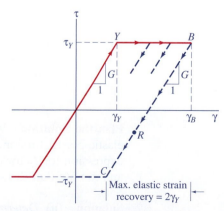

FIGURE 4.36 A stress-strain diagram illustrating elastic recovery.

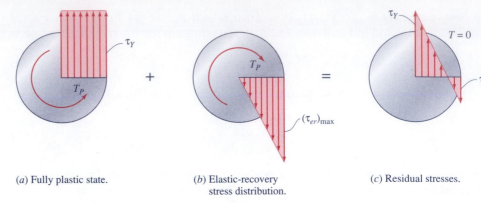

(*a*) Fully plastic state. (*b*) Elastic-recovery (*c*) Residual stresses.
                          stress distribution.

FIGURE 4.37 Determination of residual shear stresses in torsion by superposition.

elastic-recovery stress $(\tau_{er})_{max}$ in Fig. 4.37*b* is given by setting $T = T_P$ in the elastic-shear-stress formula, Eq. 4.43. Thus,

$$(\tau_{er})_{max} = \frac{T_P r}{I_P} = \frac{4}{3}\tau_Y \tag{4.54}$$

The clockwise-acting shear stress at $\rho = r$ in Fig. 4.37*c* is the shear stress at point R on the elastic recovery curve BC in Fig. 4.36. Its magnitude is $\tau_Y/3$.

➤ EXAMPLE 4.16 ◄

A tubular shaft with an outer diameter of 120 mm and an inner diameter of 100 mm is made of elastic-plastic material whose τ vs. γ curve is shown in Fig. 1*b*. The shear modulus is $G = 200$ GPa, and the yield stress in shear is $\tau_Y = 100$ MPa. (a) Determine the yield torque, T_Y, for this tubular shaft. (b) Determine the fully plastic torque, T_P. (c) Determine the distribution of residual shear stress if the torque is completely removed following loading to T_P.

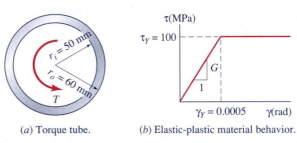

(*a*) Torque tube. (*b*) Elastic-plastic material behavior.

Fig. 1

Plan the Solution We can use the results of the above discussion of elastic-plastic torsion, but it will be necessary to incorporate the correct expression for I_P for a tubular shaft, to integrate only over $r_i \leq \rho \leq r_o$, and to note that the shaft becomes fully plastic when $\tau(r_i) = \tau_Y$.

Solution (a) *Determine the yield torque.* T_Y. The yield torque is the torque that makes $\tau(r_o) = \tau_Y$. From Eq. 4.45,

$$T_Y = \frac{T_Y I_p}{r_o} = \frac{T_Y(\pi/2)(r_o^4 - r_i^4)}{r_o}$$

$$= \frac{[100(10^6)\text{N/m}^2](\pi/2)[(0.06\text{ m})^4 - (0.05\text{ m})^4]}{(0.06\text{ m})} \qquad (1)$$

$$T_y = 17.57\text{ kN}\cdot\text{m} \qquad \textbf{Ans. (a)} \quad (2)$$

The stress distribution produced by the yield torque is shown in Fig. 2a.
(b) *Determine the fully plastic torque, T_P.* The shear stress distribution produced by the fully plastic torque, T_P, is shown in Fig. 2b. The fully plastic torque may be obtained by setting $\tau = \tau_Y$ in Eq. 4.40 but integrating from r_i to r_o.

$$T_P = 2\pi\tau_Y \int_{r_i}^{r_o} \rho^2\, d\rho$$

$$= \frac{2\pi[100(10^6)\text{ N/m}^2]}{3}[(0.06\text{ m})^3 - (0.05\text{ m})^3] \qquad (3)$$

$$T_P = 19.06\text{ kN}\cdot\text{m} \qquad \textbf{Ans. (b)} \quad (4)$$

(Note that $T_P = 1.08T_Y$, whereas $T_P = 1.33T_Y$ for a solid shaft.)
(c) *Determine the residual shear-stress distribution.* We can use the superposition of the linearly elastic recovery stress distribution of Fig. 3a and the fully plastic stress distribution of Fig. 2b to obtain the residual stress distribution in the torque tube when T_P is completely removed. Because the recovery is linearly elastic, we can set $T = T_P$ and use Eq. 4.43 to evaluate $(\tau_{er})_i$ and $(\tau_{er})_o$, the elastic recovery shear stresses at the r_i and r_o, respectively.

$$(\tau_{er}) = \frac{T_P\rho}{I_P} = \frac{(19.06\text{ kN}\cdot\text{m})\rho}{(\pi/2)[(0.06\text{ m})^4 - (0.05\text{ m})^4]} \qquad (5)$$

Therefore,

$$(\tau_{er})_i = (\tau_{er})_{\rho=0.05\text{ m}} = 90.4\text{ MPa}$$
$$(\tau_{er})_o = (\tau_{er})_{\rho=0.06\text{ m}} = 108.5\text{ MPa} \qquad (6)$$

Finally, the maximum residual stresses are:

$$\left.\begin{array}{l}(\tau_r)_o = \tau_Y - (\tau_{er})_o = -8.5\text{ MPa} \\[4pt] (\tau_r)_i = \tau_Y - (\tau_{er})_i = 9.6\text{ MPa}\end{array}\right\} \qquad \textbf{Ans. (c)}$$

The residual stress distribution is shown in Fig. 3b.

Review the Solution Figures 2 and 3 are very helpful in checking our solution. The calculations in Eq. (1) should be rechecked to make sure that T_Y is correct. Because the tube is relatively thin-walled, the stresses in the fully plastic state (Fig. 2b) are only a small amount larger than the stresses produced by the yield torque (Fig. 2a). Therefore, the ratio of $T_P/T_Y = 1.08$ is reasonable.

Finally, for Part (c), we know that the resultant torque is zero. Therefore, some shear stresses must be acting counterclockwise and some acting clockwise. The stresses nearer to the center of the shaft must be larger than those near the outer surface, because the latter stresses have a longer moment arm ρ. Therefore, the values of $(\tau_r)_i = 9.6$ MPa (ccw) and $(\tau_r)_o = 8.5$ MPa (cw) seem reasonable.

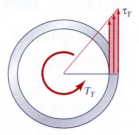

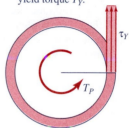

(a) Stress distribution for yield torque T_Y.

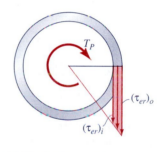

(b) Stress distribution for fully plastic torque T_P.

Fig. 2

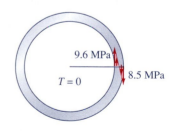

(a) Elastic recovery stresses.

(b) Residual stress distribution.

Fig. 3

▼ TORSIONAL DEFORMATION

Prob. 4.2-1. The bimetallic shaft in Fig. P4.2-1 consists of a solid inner core of diameter d_i surrounded by a tubular sleeve of inner diameter d_i and outer diameter d_o. Let subscripts C and S designate the core material and the sleeve material, respectively. Assume that this bimetallic shaft satisfies the *torsional-deformation assumptions*. (a) The maximum shear strain (due to torque T) in the core material is $(\gamma_c)_{max} = 180 \times 10^{-6}$ radians. Sketch the shear strain $\gamma(\rho)$ in the core and in the sleeve. Indicate on your sketch the value of $(\gamma_s)_{max}$, the maximum shear strain in the sleeve, (b) Determine the angle of twist of a bimetallic shaft of length L.

$$d_o = 2.0 \text{ in.}, \quad d_i = 1.50 \text{ in.}, \quad L = 48 \text{ in.}$$

P4.2-1 and P4.2-2

Prob. 4.2-2. Repeat Prob. 4.2-1 for a bimetallic shaft with $d_o = 80$ mm, $d_i = 60$ mm, and $L = 1$ m.

▼ SIMPLE TORSION | MDS 4.1 & 4.4 |

Prob. 4.3-1. A bimetallic torsion bar consists of a steel core $(G_s = 75 \text{ GPa})$ of diameter $d_i = 25$ mm around which is bonded a titanium sleeve $(G_t = 45 \text{ GPa})$ of inner diameter $d_i = 25$ mm and outer diameter $d_o = 40$ mm. (a) If the maximum shear stress in the steel is 50 MPa, what is the total torque, T, applied to the bimetallic bar? (b) What is the total twist angle of the composite bar if it is 2 m long?

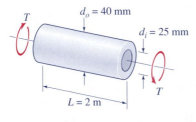

P4.3-1

Prob. 4.3-2. In Fig. P4.3-2 a mechanic (not shown) is using an X-style lug wrench to remove an automobile wheel by exerting equal forces P downward at C and upward at D. This creates a pure torque on arm AB, which is a rod 0.5 in. in diameter and 6.0 in. in length. (See also Fig. 4.1 at the beginning of this chapter.) The total length between the forces at C and D is $2b = 15$ in. (a) Because the lug nut at A is very balky, the mechanic must exert equal forces $P = 40$ lbs downward at C and upward at D. What is the maximum shear stress that this creates in arm AB? (b) If the lug wrench is made of steel with a shear modulus $G = 11.2 \times 10^3$ ksi, what is the relative angle of twist (in degrees) in the 6.0-in.-long arm AB?

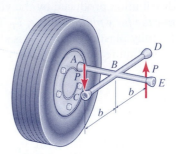

P4.3-2

Prob. 4.3-3. A solid, homogeneous, linearly elastic shaft of diameter $d = 4.0$ in. is subjected to a torque $T = 120$ kip · in. (Fig. P4.3-3). (a) Determine the maximum shear stress in the shaft. (b) Determine the percentage of the total torque that is carried by an inner core of diameter $d_c = 2.0$ in. (shown darkly shaded in the figure). (c) Determine the percentage of the total weight of the shaft that lies within this inner core.

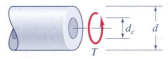

P4.3-3 and P4.3-4

Prob. 4.3-4. The solid, homogeneous, linearly elastic shaft in Fig. P4.3-4 has a diameter of $d = 80$ mm and a maximum shear stress of $\tau_{max} = 40$ MPa. (a) Determine the torque T acting on the shaft. (b) Determine the percentage of the total torque that is carried by an inner core of diameter $d_c = 40$ mm (shown darkly shaded in the figure).

ᴰProb. 4.3-5. The solid shaft in Fig. P4.3-5 is made of brass that has an allowable shear stress $\tau_{allow} = 100$ MPa and a shear modulus of elasticity $G = 39$ GPa. The length of the shaft is $L = 2$ m, and over this length the allowable angle of twist is $\phi_{allow} = 0.10$ rad. If the shaft is to be subjected to a maximum torque of $T = 25$ kN · m, what is the required diameter of the shaft?

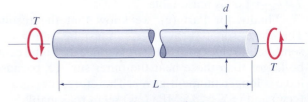

P4.3-5, P4.3-6, and P4.3-7

DProb. 4.3-6. The solid shaft in Fig. P4.3-6 is made of an aluminum alloy that has an allowable shear stress $\tau_{\text{allow}} = 10$ ksi and a shear modulus of elasticity $G = 3800$ ksi. The diameter of the shaft is $d = 1.5$ in., and its length is $L = 32$ in. If the allowable angle of twist over the 32-in length of the shaft is $\phi_{\text{allow}} = 0.10$ rad, what is the value of the allowable torque, T_{allow}?

DProb. 4.3-7. The shaft in Fig. P4.3-7 has a diameter d and is made of steel having a shear modulus G. The allowable shear stress is τ_{allow}, and the allowable twist rate is $(d\phi/dx)_{\text{allow}}$. (a) Determine an expression for the maximum torque that can be applied to the shaft without exceeding the allowable-shear-stress criterion; call this torque T_τ. (b) Similarly, determine an expression for the maximum torque that can be applied to the shaft without exceeding the allowable-twist-rate criterion; call this torque T_ϕ.

DProb.4.3-8. The shaft in Fig. P4.3-8 has an outside diameter $d_o = 60$ mm and is made of a steel alloy that has an allowable shear stress of $\tau_{\text{allow}} = 80$ MPa. The shaft is to be subjected to a torque $T = 2$ kN · m. (a) Relative to the weight of a solid shaft, by what percentage could the weight of the shaft be reduced by drilling out a core of diameter d_i? (b) Repeat Part (a) if the allowable shear stress is $\tau_{\text{allow}} = 100$ MPa.

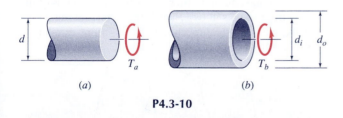

P4.3-8 and P4.3-9

DProb. 4.3-9. Solve Prob. 4.3-8 with $d_o = 4.0$ in. and $T = 60$ kip · in. for the following values of τ_{allow}: (a) $\tau_{\text{allow}} = 12$ ksi, and (b) $\tau_{\text{allow}} = 20$ ksi.

Prob. 4.3-10. The solid circular shaft of diameter d in Fig. P4.3-10a has a maximum shear stress $\tau_{\text{max}} = \tau_a$ under the action of a torque T_a. If the solid shaft is replaced by a tubular shaft with a ratio of outside diameter to inside diameter of $d_o/d_i = 1.2$ but weighing the same as the solid shaft (Fig. P4.3-10b), by what percentage would the torque have to be increased in order to produce the same maximum shear stress?

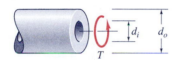

(a)   (b)

P4.3-10

▼ TORSION OF TAPERED SHAFTS

Prob. 4.3-11. The solid shaft in Fig. P4.3-11 has a diameter that varies linearly from d_o at $x = 0$ to $2d_o$ at $x = L$. It is subjected to a torque T_0 at $x = 0$, and is attached to a rigid wall at $x = L$. (a) Determine an expression for the maximum (cross-sectional) shear stress in the tapered shaft as a function of the distance x from the left end. (b) Determine an

expression for the total angle of twist of the shaft, ϕ_0. The shear modulus of elasticity is G.

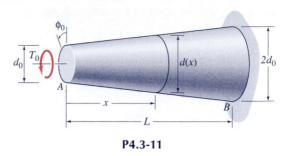

P4.3-11

▼ SHAFTS WITH DISTRIBUTED EXTERNAL TORQUE

Prob. 4.3-12. The solid circular shaft in Fig. P4.3-12 is subjected to a distributed external torque that varies linearly from intensity of t_0 per unit of length at $x = 0$ to zero at $x = L$. The shaft has a diameter d and shear modulus G and is fixed to a rigid wall at $x = 0$. (a) Determine an expression for the maximum (cross-sectional) shear stress in the shaft as a function of the distance x from the left end. (b) Determine an expression for the total angle of twist, ϕ_B, at the free end. The shear modulus of elasticity is G.

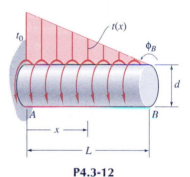

P4.3-12

Prob. 4.3-13. Solve Prob. 4.3-12 if the stated externally applied torque distribution is replaced by the torque distribution $t(x) = t_0 \cos\left(\dfrac{\pi x}{2L}\right)$, shown in Fig. P4.3-13.

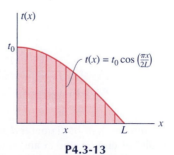

P4.3-13

Prob. 4.3-14. A uniform shaft of diameter d, length L, and shear modulus G, is subjected to a uniformly distributed torque t_0 over half of its length, as shown in Fig. P4.3-14. (a) Determine an expression for τ_{max}, and indicate the location(s) where τ_{max} occurs. (b) Determine the angle of rotation at A, ϕ_A, and the angle of rotation at B, ϕ_B.

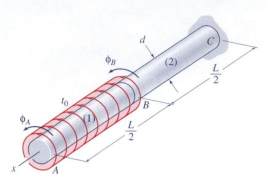

P4.3-14

*Prob. 4.3-15. A lunar soil sampler consists of a tubular shaft that has an inside diameter of 45 mm and an outside diameter of 60 mm. The shaft is made of stainless steel, with a shear modulus $G = 80$ GPa. In Fig. P4.3-15, the sampler is being removed from the sampling hole by two 150-N forces that the astronaut exerts at right angles to arm DE and parallel to the lunar surface. Assume that the external torque that is exerted by the soil from A to B is uniformly distributed, with magnitude t_0. (a) Determine the maximum shear stress in the shaft of the sampler tube AC, and indicate the location(s) where it acts. (b) Determine the angle of twist of end C with respect to end A, that is, determine $\phi_{C/A} \equiv \phi_C - \phi_A$.

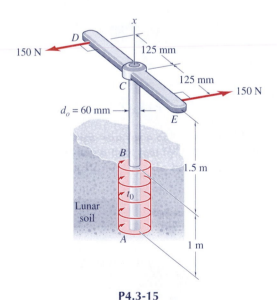

P4.3-15

*Prob. 4.3-16. A uniform shaft of diameter d and length L is fixed to rigid "walls" at ends A and B and is subjected to a quadratically varying distributed external torsional loading

$$t(x) = t_0[1 - (x/L)^2]$$

(Note that Fig. P4.3-16 shows reaction torques T_A and T_B and this external torsional loading $t(x)$ all acting in the positive right-hand-rule sense with respect to the x axis.)

By using a free-body diagram of the entire shaft, together with the two fixed-end constraint condition, $\phi(0) = \phi(L) = 0$, determine expressions for the reaction torques T_A and T_B at the fixed ends. Assume that linearly elastic behavior results from the torsional loading $t(x)$. (Note that this is a statically indeterminate torsion member since there are two unknown reaction torques but only one equilibrium equation.)

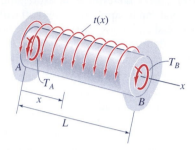

P4.3-16 and P4.3-17

*Prob. 4.3-17. Repeat Prob. 4.3-16 for a distributed torsional loading

$$t(x) = t_0 \cos\left(\frac{\pi x}{2L}\right)$$

▼ **SIMPLE TORSION: STRESS DISTRIBUTION AND STRAIN DISTRIBUTION**

MDS 4.5

Prob. 4.4-1. (a) Using Eq. 4.17 and the information in Fig. 4.14, discuss the type of stress that actually caused the failure of the cast iron bar in Fig. 4.17b. Was it shear stress, or tensile stress, or compressive stress; or was it possibly some combination? (b) As illustrated in Fig. P4.4-1, grip the two ends of a piece of blackboard chalk that is approximately 2.5–3 in. long; then, with your fingers, twist the piece of chalk (i.e., apply equal and opposite torques to the two ends of the chalk) until it breaks in two. Sketch the two resulting pieces of chalk, noting whether the break is more similar to Fig. 4.17a or to Fig. 4.17b.

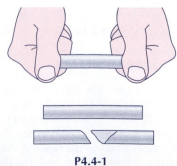

P4.4-1

ᴰProb. 4.4-2. A round wood dowel is made of oak having an allowable stress in shear, parallel to the grain, of $\tau_{allow} = 1.5$ MPa. (a) If the diameter of the dowel is 50 mm, what is the maximum permissible torque T that can be applied? (b) Why would the wood dowel fail in shear, rather than in tension or compression? Describe the shear failure surface(s).

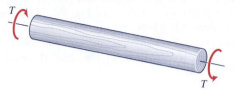

P4.4-2 and P4.4-3

ᴰProb. 4.4-3. Solve Prob. 4.4-2 if $\tau_{allow} = 200$ psi and the diameter of the dowel is 2 in.

This figure of a generic torsion member applies to all of the remaining problems for Section 4.4.

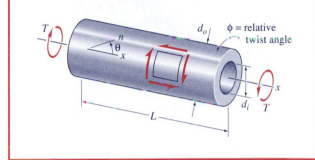

Problems 4.4-4 through 4.4-7. *For these four problems, use the generic torsion member shown above. For each problem: (a) determine the maximum tensile stress, the maximum compressive stress, and the maximum shear stress; and (b) show the stresses determined in Part (a) on properly oriented stress elements (e.g., see Fig. 4.14).*

In solving these problems, make the sense of the stresses consistent with the sense of the torque T shown on the generic torsion member figure.

Prob. 4.4-4. (See the "generic torsion member" figure.) $T = 15$ kip $\cdot$ in., $d_o = 1.5$ in., and $d_i = 0$ in.

Prob. 4.4-5. (See the "generic torsion member" figure.) $T = 200$ N $\cdot$ m, $d_o = 30$ mm, and $d_i = 0$ mm.

Prob. 4.4-6. (See the "generic torsion member" figure.) $T = 500$ N $\cdot$ m, $d_o = 40$ mm, and $d_i = 30$ mm.

Prob. 4.4-7. (See the "generic torsion member" figure.) $T = 3.0$ kip $\cdot$ in., $d_o = 2.0$ in., and $d_i = 1.75$ in.

Problems 4.4-8 through 4.4-11. *For these four problems, use the generic torsion member and determine (a) the value of the applied torque, T, and (b) the value of the maximum tensile stress, σ_{maxT}.*

Prob. 4.4-8. $L = 30$ in., $d_o = 0.75$ in., $d_i = 0$ in., $\phi = 0.09$ rad, and $G = 11 \times 10^3$ ksi.

Prob. 4.4-9. $L = 1.0$ m, $d_o = 20$ mm, $d_i = 0$ mm, $\phi = 0.10$ rad, and $G = 80$ GPa.

Prob. 4.4-10. $L = 2.0$ m, $d_o = 40$ mm, $d_i = 30$ mm, $\phi = 0.20$ rad, and $G = 37$ GPa.

Prob. 4.4-11. $L = 8.5$ ft, $d_o = 3.0$ in., $d_i = 2.5$ in., $\phi = 0.15$ rad, and $G = 15 \times 10^3$ ksi.

Prob. 4.4-12. For the generic torsion member, $T = 500$ lb $\cdot$ in., $d_o = 0.625$ in., and $d_i = 0$ in. The extensional strain for $\theta = 45°$ is $\epsilon_{45°} = -1300$ μ. (a) Determine the value of the maximum shear stress, τ_{max}. (b) Determine the shear modulus of elasticity, G.

Prob. 4.4-13. For the generic torsion member, $T = 600$ N $\cdot$ m, $G = 28$ GPa, $d_o = 50$ mm, and $d_i = 40$ mm. Determine (a) the maximum shear strain, γ_{max}; and (b) the maximum tensile strain, ϵ_{maxT}.

ᴰProb. 4.4-14. A steel shaft ($G = 11.8 \times 10^6$ psi) has the form of the generic torsion member pictured on this page. The outside diameter of the shaft is $d_o = 1.0$ in., and it is required to carry torques up to $T_{max} = 1200$ lb $\cdot$ in. If the maximum allowable extensional strain for this shaft is 400 μ, what is the minimum wall thickness of the shaft to the nearest $\frac{1}{32}$ in.?

▼ **MULTIPLE TORQUES** | MDS 4.6 |

Prob. 4.5-1. A stepped steel shaft AC ($G = 12 \times 10^6$ psi) is subjected to torsional loads at sections B and C as shown in Fig. P4.5-1. The diameters are: $d_1 = 1.75$ in. and $d_2 = 1.0$ in. (a) Determine the maximum shear stress in the shaft. Identify the location(s) where this maximum shear stress occurs. (b) Determine the angle of rotation at C, ϕ_C.

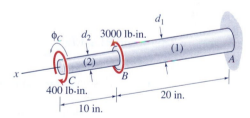

P4.5-1 and P4.5-12

Prob. 4.5-2. A brass shaft AC ($G = 5.6 \times 10^3$ ksi) is subjected to torsional loads at sections B and C, as shown in Fig. P4.5-2. The outside diameter of the shaft is $d_1 = d_{o2} = d = 1.25$ in. A central hole of diameter $d_{i2} = 0.75$ in. has been drilled from C to B, creating a two-segment shaft with lengths $L_1 = L_2 = 10$ in. If the two applied torques are $T_B = 400$ lb $\cdot$ in. and $T_C = 200$ lb $\cdot$ in., respectively, (a) determine the maximum shear stress in the two-segment shaft, and (b) determine the angle of rotation ϕ_C (in radians) at the end of the shaft. (Note that both torque T_B and T_C are positive, meaning that both torques act in the sense shown in Fig. P4.5-2.)

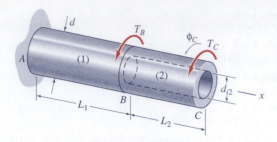

P4.5-2 and P4.5-3

Prob. 4.5-3. Solve Prob. 4.5-2 using the following values: $d =$ 40 mm, $d_{i2} = 30$ mm, $L_1 = 2.0$ m and $L_2 = 1.5$ m, $G = 37$ GPa, $T_B = 600$ N $\cdot$ m, and $T_C = -200$ N $\cdot$ m. (Note that torque T_C is negative, meaning that it acts in the opposite sense to that shown in Fig. P4.5-3.)

Prob. 4.5-4. The aluminum-alloy shaft AC in Fig. P4.5-4 ($G =$ 26 GPa) has an 800-mm-long solid segment AB and a 400-mm-long tubular segment BC. The shaft is subjected to the torsional loading shown in the figure. The diameters are $d_1 = (d_o)_2 = 50$ mm and $(d_i)_2 = 20$ mm. (a) Determine the maximum shear stress in the shaft, and identify the location(s) where this maximum shear stress occurs. (b) Determine the angle of rotation at B. ϕ_B. and the angle of rotation at C. ϕ_C.

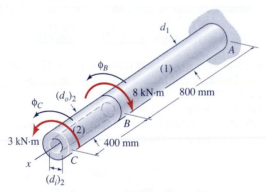

P4.5-4 and P4.5-13

Prob. 4.5-5. A uniform 1-in.-diameter steel shaft ($G = 12 \times 10^6$ psi) is supported by frictionless bearings and is used to transmit torques from gear B to gears at A and C as shown in Fig. P4.5-5 (a) Determine the maximum shear stress in

element (1), (i.e., in segment AB), and the maximum shear stress in element (2). (b) Determine the relative rotation between the two ends: that is, determine $\phi_{C/A} \equiv \phi_C - \phi_A$.

Prob. 4.5-6. A torque is applied to gear A of a two-shaft system and is transmitted through gears at B and C to a fixed end at D. The shafts are made of steel ($G = 80$ GPa). Each shaft has a diameter $d = 32$ mm, and they are supported by frictionless bearings as shown in Fig. P4.5-6. If the torque applied to gear A is 400 N $\cdot$ m, and D is restrained, (a) determine the maximum shear stress in each shaft, and (b) determine the angle of rotation of the gear A relative to its no-load position.

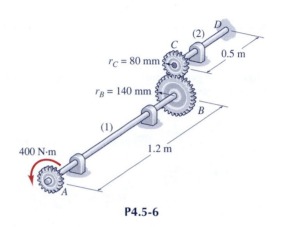

P4.5-6

Prob. 4.5-7. The gears and splined ends apply torques T_A through T_D to a steel shaft, as shown in Fig. P4.5-7. (a) Determine the maximum shear stress in the shaft, and (b) determine the relative twist angle between ends A and D, that is, determine $\phi_{D/A} = \phi_D - \phi_A$.

$$G = 75 \text{ GPa}, d = 25 \text{ mm}, L_1 = 200 \text{ mm}, L_2 = 300 \text{ mm},$$

$$L_3 = 400 \text{ mm}, T_A = 250 \text{ N} \cdot \text{m}, T_B = -400 \text{ N} \cdot \text{m},$$

$$T_C = -350 \text{ N} \cdot \text{m}, T_D = 500 \text{ N} \cdot \text{m}$$

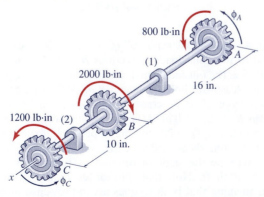

P4.5-5

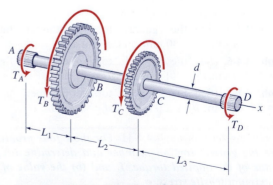

P4.5-7, P4.5-8, and P4.5-9

ᴰProb. 4.5-8. Use the following data with Fig. P4.5-8:

$G = 11 \times 10^6$ psi, $L_1 = 15$ in., $L_2 = 20$ in., $L_3 = 25$ in.,

$T_A = 200$ lb $\cdot$ ft, $T_B = -350$ lb $\cdot$ ft, $T_C = -150$ lb $\cdot$ ft,

$T_D = 300$ lb·ft

(a) To the nearest 1/16 in., what is the required diameter d of a solid shaft AD if the maximum shear stress in the shaft is 5000 psi? (b) To the nearest 1/16 in., what is the required outside diameter d_o of a tubular shaft AD if its inner diameter is $d_i = 1$ in. and the maximum shear stress in the shaft is 5000 psi?

Prob. 4.5-9. Use the following data with Fig. P4.5-9:

$G = 11 \times 10^6$ psi, $L_1 = 10$ in., $L_2 - 15$ in., $L_3 = 20$ in.,

$T_A = 200$ lb $\cdot$ ft, $T_B = -350$ lb $\cdot$ ft, $T_C = -150$ lb $\cdot$ ft,

$T_D = 300$ lb $\cdot$ ft

If AD is a solid shaft with diameter $d = 1.5$ in., determine the relative twist angle (in radians) between ends A and D, that is, determine $\phi_{D/A} \equiv \phi_D - \phi_A$.

Prob. 4.5-10. The steel shaft AD in Fig. P4.5-10 ($G = 80$ GPa) is subjected to torsional loads at sections B and D, as shown in the figure. The diameters are: $d_1 = d_2 = 40$ mm., and $d_3 = 30$ mm. (a) Determine the value of the torque T_D added at D that would make the rotation at C equal to zero, that is, make $\phi_C = 0$. (b) For the loading as determined in Part (a), determine the maximum shear stress in each of the three rod segments.

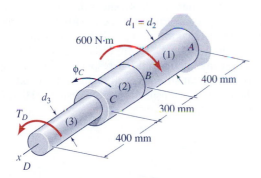

P4.5-10 and P4.5-14

Prob. 4.5-11. A plumber is cutting threads on the end of a 2-ft-long section of 1-in.-diameter* pipe. This section, AB, has already been attached to a 2-in.-diameter* section, BC, by a reducing coupler and, in turn, to a 3-in.-diameter* section CD. The shear modulus of the steel pipe is $G = 11.5 \times 10^3$ ksi. Neglecting the dimensions and flexibility of the couplers, (a) determine the maximum shear stress in the assembly, (b) determine the angle of rotation at end A, ϕ_A, and (c) determine the torsional stiffness, k_t, of the assembly. That is, determine the torque that must be applied at A to produce a unit rotation (i.e., one radian rotation) at A. (*The stated diameters are nominal pipe diameters. See Table D.7 for section properties of the pipe.)

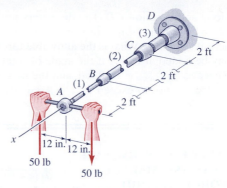

P4.5-11

Prob. 4.5-12. Determine the torsional stiffness. k_t, of the stepped shaft AC in Fig. P4.5-1.

Prob. 4.5-13. Determine the torsional stiffness, k_t, of the shaft AC in Fig. P4.5-4.

Prob. 4.5-14. Determine the torsional stiffness, k_t, of the shaft AD in Fig. P4.5-10.

ᴰProb. 4.5-15. A two-element aluminum shaft AC ($G = 27$ GPa) is subjected to external torques at sections B and C as shown in Fig. P4.5-15. The outer diameter of the shaft is $d_1 = d_2 = 50$ mm. The inner diameter of element (2), d_{i2}, is to be determined such that the maximum shear stress in the shaft does not exceed the allowable shear stress $\tau_{\text{allow}} = 35$ MPa and the total angle of twist does not exceed $\phi_{\text{allow}} = 0.08$ rad. Express your answer, d_{i2}, to the nearest even mm that satisfies these two criteria.

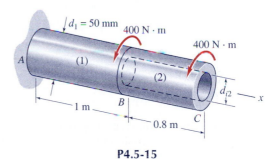

P4.5-15

ᴰProb. 4.5-16. The aluminum-alloy shaft AD in Fig. P4.5-16 ($G = 3.8 \times 10^3$ ksi) has a 40-in. long tubular section AC and

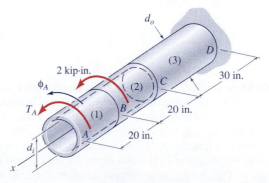

P4.5-16

a 30-in.-long solid section CD. The diameters are $(d_o)_1 = (d_o)_2 = d_3 = 2.0$ in. and $(d_i)_1 = (d_i)_2 = 1.75$ in. A 2 kip $\cdot$ in. torque acts at section B. What is the allowable torque $T_A \geq 0$ that may be added at end A if the angle of rotation at A is not to exceed $(\phi_A)_{\text{allow}} = 0.10$ rad, and the magnitude of the shear stress is not to exceed $\tau_{\text{allow}} = 4$ ksi anywhere in the shaft?

▼ STATICALLY INDETERMINATE
TORSIONAL ASSEMBLAGES:
BASIC FORCE METHOD | MDS 4.8

Prob. 4.6-1. A stepped steel shaft AC ($G = 12 \times 10^6$ psi) is subjected to an external torque T_B at B and is fixed to rigid supports at ends A and C, as shown in Fig. P4.6-1. (a) Determine. T_1 and T_2, the internal torques in segments AB and BC, respectively. (b) Determine the maximum shear stress in each segment. (c) Finally, determine ϕ_B, the angle of rotation of the shaft at joint B.

$T_B = 4000$ lb $\cdot$ in., $d_1 = 1.0$ in., $L_1 = 8.0$ in., $d_2 = 1.5$ in.,

$L_2 = 16.0$ in.

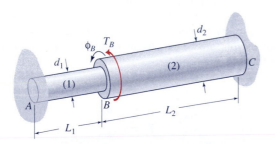

P4.6-1, P4.6-2, P4.6-3, and P4.7-1

Prob. 4.6-2. A stepped aluminum-alloy shaft AC ($G = 26$ GPa) is subjected to an external torque T_B at B and is fixed to rigid supports at ends A and C, as shown in Fig. P4.6-2. (a) Determine T_1 and T_2, the internal torques in segments AB and BC, respectively. (b) Determine the maximum shear stress in each segment. (c) Finally, determine ϕ_B, the angle of rotation of the shaft at joint B.

$T_B = 500$ N $\cdot$ m, $d_1 = 25$ mm, $L_1 = 150$ mm, $d_2 = 40$ mm,

$L_2 = 200$ mm

Prob. 4.6-3. A solid circular shaft AC of total length $L = L_1 + L_2$ is fixed against rotation at ends A and C and is loaded by a torque T_B at joint B, as shown in Fig. P4.6-3. If the diameters of the two segments of the shaft are in the ratio d_2/d_1, what ratio of lengths, L_2/L_1, will cause the maximum shear stress to be the same in the two segments, that is, will make $(\tau_{\text{max}})_1 = (\tau_{\text{max}})_2$? In your answer, express L_2/L_1 in terms of d_2/d_1.

Prob. 4.6-4. As illustrated in Fig. P4.6-4, a composite shaft of length $L = L_1 + L_2$ is made by connecting together at joint B two shafts of the same diameter d. Shaft segment AB has a shear modulus G_1, and segment BC has a shear modulus G_2. The ends A and C are fixed against rotation, and then an external torque T_B is applied to the shaft at joint B. (a) Determine expressions for T_1 and T_2, the internal torques in segments AB and BC, respectively. (b) Determine the maximum shear stress in each segment. (c) Finally, determine ϕ_B, the angle of rotation of the shaft at joint B.

Express all of your answers in terms of the given torque T_B and the physical parameters of the composite shaft: d, G_1, G_2, L_1, and L_2.

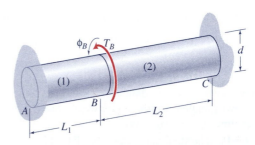

P4.6-4 and P4.7-3

Prob. 4.6-5. As illustrated in Fig. P4.6-5, a composite shaft is made by connecting together at joint B two tubular segments of the same outer diameter $d_o = 75$ mm and the same inner diameter $d_i = 50$ mm. Aluminum-alloy segment AB has a shear modulus $G_1 = 27$ GPa and length $L_1 = 0.8$ m, and titanium segment BC has a shear modulus $G_2 = 43$ GPa and length $L_2 = 1.0$ m. Ends A and C are fixed against rotation, and then an external torque $T_B = 10$ kN $\cdot$ m is applied to the shaft at joint B. (a) Determine T_1 and T_2, the internal torques in segments AB and BC, respectively. (b) Determine the maximum shear stress in each segment. (c) Finally, determine the maximum tensile stress in the shaft. (Hint: Recall Section 4.4.)

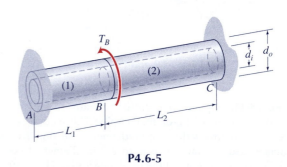

P4.6-5

Prob. 4.6-6. A composite shaft of length $L = 6$ ft is made by shrink-fitting a titanium-alloy sleeve (element 1) over an aluminum-alloy core (element 2), as shown in Fig. P4.6-6. The two-component shaft is fixed against rotation at end B, and a torque $T_A = 20$ kip $\cdot$ in. is applied at end A. (a) Determine the maximum shear stresses $(\tau_{\text{max}})_1$ in the titanium sleeve and $(\tau_{max})_2$ in the aluminum core. (b) Determine the angle of rotation at A, ϕ_A.

$$G_1 = 6 \times 10^3 \text{ ksi}, \quad G_2 = 4 \times 10^3 \text{ ksi}, \quad (d_o)_1 = 2.25 \text{ in.},$$

$$(d_i)_1 = d_2 = 2.0 \text{ in.}$$

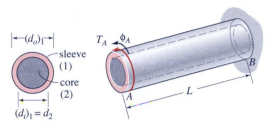

P4.6-6, P4.6-7, P4.6-8, P4.6-9, P4.7-4, and P4.7-5

Prob. 4.6-7. Solve Prob. 4.6-6 if, instead of being a solid core with outer diameter $d_2 = 2.0$ in., the inner element is tubular with an outer diameter $d_2 = 2.0$ in. and an inner diameter $(d_i)_2 = 1.0$ in.

Prob. 4.6-8. As illustrated in Fig. P4.6-8, a composite shaft of length L is made by shrink-fitting a sleeve (element 1) over a solid core (element 2). A torsional load T_A is applied to the shaft at end A, and end B is fixed against rotation. Let $G_2 > G_1$, and express your answers in terms of the shear-modulus ratio G_2/G_1. Determine expressions for the ratio of diameters $(d_o/d_i)_1$: (a) if the torque T_A is evenly divided between the sleeve and the core, and (b) if the maximum shear stress in the sleeve, $(\tau_{max})_1$ is the same as the maximum shear stress in the core, $(\tau_{max})_2$.

Prob. 4.6-9. A composite shaft is made by shrink-fitting a brass sleeve ($G_1 = 39$ GPa) over an aluminum-alloy core ($G_2 = 26$ GPa). and the shaft is subjected to an end torque T_A, as shown in Fig. P4.6-9. The diameter of the aluminum core is 50 mm. What thickness would be required for the brass sleeve in order to reduce the angle of twist of the composite shaft to one-half the angle of twist of the aluminum core alone, if the composite shaft and the all-aluminum shaft are subjected to the same magnitude of end torque, T_A?

Prob. 4.6-10. A solid aluminum-alloy rod of diameter $d_1 = 1.0$ in. is enclosed by a concentric brass tube of inner diameter $d_{2i} = 1.5$ in. and outer diameter $d_{2o} = 1.75$ in., and both are attached to a rigid support at end A and to a rigid flat plate at end B. The rod and tube form a composite torsion member of length $L = 20$ in. The shear moduli are $G_1 = 4.0 \times 10^3$ ksi and $G_2 = 5.6 \times 10^3$ ksi, respectively. A torque $T_B = 2500$ lb $\cdot$ in. is applied to the end plate at B. (a) Determine the torques T_1 and T_2 in the rod and tube, respectively. (b) Determine the maximum shear stresses τ_1 and τ_2 in the rod and tube, respectively. (c) Determine the angle of rotation, ϕ_B, (in radians) of end B. (d) Finally, determine the torsional stiffness k_t of the composite torsion member.

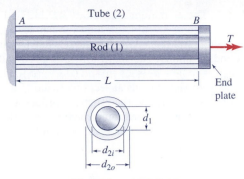

P4.6-10 and P4.6-11

Prob. 4.6-11. Repeat all four parts of Prob. 4.6-10 if both the rod and the tube are made of steel with shear modulus $G_1 = G_2 = 11.0 \times 10^3$ ksi.

Prob. 4.6-12. The aluminum-alloy shaft AC in Fig. P4.6-12 ($G = 26$ GPa) has a 400-mm-long tubular section AB and an 800-mm-long solid segment BC. The shaft is subjected to a 10 kN $\cdot$ m external torque at section B, and it is fixed to rigid supports at ends A and C. The diameters are $(d_o)_1 = d_2 = 60$ mm and $(d_i)_1 = 30$ mm. (a) Determine T_1 and T_2, the internal torques in the two segments. (b) Determine the maximum shear stress in each segment. (c) Determine ϕ_B, the angle of rotation at joint B.

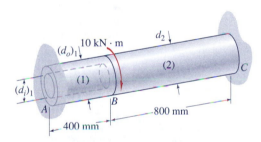

P4.6-12 and P4.7-2

Prob. 4.6-13. The uniform 1-in.-diameter steel shaft in Fig. P4.6-13 ($G = 12 \times 10^6$ psi) is supported by frictionless bearings and is used to transmit torques from the gear at B to the gears at A and C, as shown. Determine the maximum shear stress in each of the two segments of the shaft: (a) if the relative rotation between gears A and C is $\phi_{A/C} \equiv \phi_A - \phi_C = 0$, and (b) if the relative rotation between gears A and C is $\phi_{A/C} \equiv \phi_A - \phi_C = 0.01$ rad.

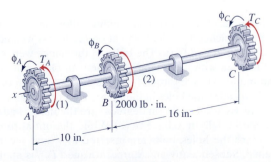

P4.6-13

Prob. 4.6-14. The aluminum-alloy shaft AD in Fig. P4.6-14 ($G = 3.8 \times 10^3$ ksi) has a 40-in.-long tubular segment AC and a 30-in.-long solid segment CD, and it is fixed against rotation at ends A and D. The shaft, whose respective diameters are $(d_o)_1 = (d_o)_2 = d_3 = 2.5$ in. and $(d_i)_1 = (d_i)_2 = 2.0$ in., is subjected to torsional loads at sections B and C, as shown in the figure, (a) Determine the reaction torques at ends A and D. (b) Determine the maximum shear stress in the shaft. (c) Determine the angle of rotation at section C, ϕ_0.

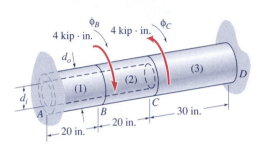

P4.6-14 and P4.7-11

Prob. 4.6-15. A uniform shaft with fixed ends at A and D is subjected to external torques of magnitude T_0 and $2T_0$, as shown in Fig. P4.6-15. The diameter of the shaft is d, and its shear modulus is G. (a) Determine expressions for the maximum shear stress in each of the three segments of the shaft: $(\tau_{max})_1$, $(\tau_{max})_2$, and $(\tau_{max})_3$, (b) Determine an expression for the angle of rotation at B, ϕ_B.

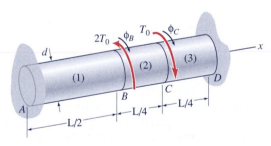

P4.6-15 and P4.7-10

▼ STATICALLY INDETERMINATE TORSIONAL ASSEMBLAGES: INITIAL STRESS

***Prob. 4.6-16.** The uniform aluminum-alloy shaft in Fig. P4.6-16 is attached to a rigid wall at end A and is welded to a rigid flange at end C. (a) The holes in the flange were supposed to align with holes tapped in the wall plate, but actually an initial external torque $(T_B)_i$ must be applied in order to perfectly align the holes in the flange with the holes in the wall plate. By what angle (in degrees) are the flange and the wall plate initially misaligned? (b) While the initial torque $(T_B)_i$ aligns the holes, bolts are inserted at C and securely tightened. Subsequently an external torque $(T_B)_f$ is applied at B. Determine the resulting maximum shear stress $(\tau_{max})_1$

in segment AB and the maximum shear stress $(\tau_{max})_2$ in segment BC.

$$G = 3.8 \times 10^3 \text{ ksi}, \ d = 1.5 \text{ in.}, \ L_1 = 60 \text{ in.}, \ L_2 = 40 \text{ in.},$$

$$(T_B)_i = 820 \text{ lb} \cdot \text{in.}, \ (T_B)_f = 7500 \text{ lb} \cdot \text{in.}$$

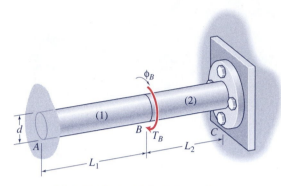

P4.6-16, P4.6-17, P4.7-7, and P4.7-8

ᴰProb. 4.6-17. The uniform aluminum-alloy shaft in Fig. P4.6-17 is attached to a rigid wall at end A and is welded to a rigid flange at end C. (a) The holes in the flange were supposed to align with holes tapped in the wall plate, but actually an initial external torque $(T_B)_i$ must be applied in order to rotate end C through an angle $\phi_C = 5$ degrees to perfectly align the holes in the flange with the holes in the wall plate. What initial torque $(T_B)_i$ (in kN $\cdot$ m) is required? (b) While the initial torque $(T_B)_i$ aligns the holes, bolts are inserted at C and securely tightened. Subsequently an external torque $(T_B)_f$ is applied at B. Determine the maximum torque $(T_B)_f$ that may be applied if the resulting maximum allowable shear stress in the shaft is $\tau_{allow} = 120$ MPa.

$$G = 26 \text{ GPa}, \ d = 50 \text{ mm}, \ L_1 = 2 \text{ m}, \ L_2 = 1 \text{ m}$$

▼ STATICALLY INDETERMINATE TORSIONAL ASSEMBLAGES: DISPLACEMENT METHOD   MDS 4.9

Prob. 4.7-1. A stepped steel shaft AC ($G = 12 \times 10^6$ psi) is subjected to an external torque T_B at B and is fixed to rigid supports at ends A and C, as shown in Fig. P4.7-1 (see Prob. 4.6-1). (a) Using the *Displacement Method*, determine ϕ_B, the angle of rotation of the shaft at joint B. (b) Determine T_1 and T_2, the internal torques in segments AB and BC, respectively. (c) Finally, determine the maximum shear stress in each segment.

$$T_B = 4000 \text{ lb} \cdot \text{in.}, \ d_1 = 1.0 \text{ in.}, \ L_1 = 10.0 \text{ in.}, \ d_2 = 2.0 \text{ in.},$$

$$L_2 = 20.0 \text{ in.}$$

Prob. 4.7-2. The aluminum-alloy shaft AC ($G = 26$ GPa) in Fig. P4.7-2 (see Prob. 4.6-12) has a 400-mm-long tubular section AB and an 800-mm-long solid segment BC. The shaft is subjected to a 10 kN · m external torque at section B, and it is fixed to rigid supports at ends A and C. The diameters are $(d_o)_1 = d_2 = 60$ mm and $(d_i)_1 = 30$ mm. (a) Using the *Displacement Method*, determine ϕ_B, the angle of rotation of the shaft at joint B. (b) Determine T_1 and T_2, the internal torques in segments AB and BC, respectively. (c) Finally, determine the maximum shear stress in each segment.

Prob. 4.7-3. As illustrated in Fig. P4.7-3 (see Prob. 4.6-4), a composite shaft of length $L = L_1 + L_2$ is made by connecting together at joint B two shafts of the same diameter d. Shaft segment AB has a shear modulus G_1, and segment BC has a shear modulus G_2. An external torque T_B is applied to the shaft at joint B, and ends A and C are fixed against rotation. (a) Using the *Displacement Method*, determine an expression for ϕ_B, the angle of rotation of the shaft at joint B. (b) Determine expressions for T_1 and T_2, the internal torques in segments AB and BC, respectively. (c) Finally, determine the maximum shear stress in each segment.

Express all of your answers in terms of the given torque T_B and the physical parameters of the composite shaft: d, G_1, G_2, L_1, and L_2.

Prob. 4.7-4. As illustrated in Fig. P4.7-4 (see Prob. 4.6-6), a composite shaft of length $L = 6$ ft is made by shrink-fitting a titanium-alloy sleeve (element 1) over an aluminum-alloy core (element 2). The two-component shaft is fixed against rotation at end B. The shear moduli are $G_1 = 6 \times 10^3$ ksi and $G_2 = 4 \times 10^3$ ksi, respectively, and the diameters are $(d_o)_1 = 2.25$ in., and $(d_i)_1 = d_2 = 2.0$ in. (a) Using the *Displacement Method*, determine the angle of rotation at A, ϕ_A, produced by an applied torque $T_A = 20$ kip · in. (b) Determine the maximum shear stresses $(\tau_{max})_1$ in the titanium sleeve and $(\tau_{max})_2$ in the aluminum core.

^DProb. 4.7-5. A composite shaft is made by shrink-fitting a brass sleeve ($G_1 = 39$ GPa) over an aluminum-alloy core ($G_2 = 26$ GPa), and the shaft is subjected to an end-torque, T_A, as shown in Fig. P4.7-5 (see Prob. 4.6-6). The diameter of the aluminum core is 50 mm. Using the *Displacement Method*, determine the thickness that would be required for the brass sleeve in order to reduce the angle of twist of the composite shaft to one-half the angle of twist of the aluminum core alone, if the composite shaft and the all-aluminum shaft are subjected to the same magnitude of end torque, T_A.

*Prob. 4.7-6. As illustrated in Fig. P4.7-6, a composite shaft AC is made by shrink-fitting a titanium-alloy sleeve of length $L_3 = 4$ ft over segment BC of a 6-ft-long aluminum-alloy rod. This makes a three-component shaft, which is fixed

against rotation at ends A and C. The shear moduli are $G_1 = G_2 = 6 \times 10^3$ ksi and $G_3 = 4 \times 10^3$ ksi, respectively, and the diameters and $d_1 = d_2 = (d_i)_3 = 2$ in., and $(d_o)_3 = 2.5$ in. (a) Using the *Displacement Method*, determine the angle of rotation at B, ϕ_B, produced by an applied torque $T_B = 50$ kip · in. (b) Determine the maximum shear stress $(\tau_{max})_1$ in the aluminum segment AB, $(\tau_{max})_2$ in the aluminum core segment BC, and $(\tau_{max})_3$ in the titanium sleeve BC.

▼ STATICALLY INDETERMINATE TORSIONAL ASSEMBLAGES: INITIAL STRESS

> MDS 4.10

*Prob. 4.7-7. The uniform aluminum-alloy shaft in Fig. P4.7-7 (see Prob. 4.6-16) is attached to a rigid wall at end A and is welded to a rigid flange at end C. (a) The holes in the flange were supposed to align with holes tapped in the wall plate, but actually an initial external torque $(T_B)_i$ must be applied in order to perfectly align the holes in the flange with the holes in the wall plate. By what angle (in degrees) are the flange and the wall plate initially misaligned? (b) While the initial torque $(T_B)_i$ aligns the holes, bolts are inserted at C and securely tightened. Subsequently an external torque $(T_B)_f$ is applied at B. Using the *Displacement Method*, determine the resulting maximum shear stress $(\tau_{max})_1$ in segment AB and be maximum shear stress $(\tau_{max})_2$ in segment BC.

$$G = 3.8 \times 10^3 \text{ ksi}, d = 1.5 \text{ in.}, L_1 = 60 \text{ in.}, L_2 = 40 \text{ in.}$$

$$(T_B)_i = 820 \text{ lb} \cdot \text{in.}, (T_B)_f = 7500 \text{ lb} \cdot \text{in.}$$

*Prob. 4.7-8. The uniform aluminum-alloy shaft in Fig. P4.7-8 (see Prob. 4.6-16) is attached to a rigid wall at end A and is welded to a rigid flange at end C. (a) The holes in the flange were supposed to align with holes tapped in the wall plate, but actually an initial external torque $(T_B)_i$ must be applied in order to rotate end C through an angle $\phi_C = 5$ degrees to perfectly align the holes in the flange with the holes in the wall plate. What initial torque $(T_B)_i$ (in kN · m) is required? (b) While the initial torque $(T_B)_i$ aligns the holes, bolts are inserted at C and securely tightened. Subsequently an external torque $(T_B)_f$ is applied at B. Using the *Displacement Method*, determine the maximum torque $(T_B)_f$ that may be applied if the resulting maximum allowable shear stress in the shaft is $\tau_{allow} = 120$ MPa.

$$G = 26 \text{ GPa}, d = 50 \text{ mm}, L = 2 \text{ m}, L_2 = 1 \text{ m}$$

*Prob. 4.7-9. A torque T_A is applied to gear A of the two-shaft system in Fig. P4.7-9, producing a rotation $\phi_A = 0.05$ rad. The shafts are made of steel ($G = 80$ GPa), and each has a diameter of $d = 32$ mm. The shafts are supported by frictionless bearings, and end D of shaft CD is restrained. (a) Using the *Displacement Method*, determine the angle of rotation of gear C and the angle of rotation at gear B. (b) Determine the internal torques in shafts (1) and (2). (c) Determine the maximum shear stress in the two-shaft system.

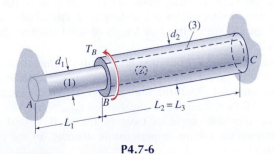

P4.7-6

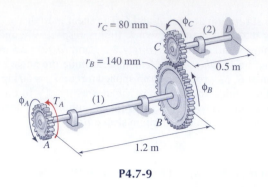

$r_C = 80$ mm

ϕ_C

(2) D

C

$r_B = 140$ mm

0.5 m

ϕ_A T_A

(1)

ϕ_B

B

A

1.2 m

P4.7-9

an outer diameter of $d_o = 120$ mm and an inside diameter of $d_i = 100$ mm, and it is made of steel with a shear modulus of elasticity $G = 80$ GPa. (a) Determine the maximum shear stress in the shaft under the above operating conditions. (b) If the shaft is 3 m long, what is the angle of twist of the shaft?

Prob. 4.8-5. The rotary mower in Fig. P4.8-5 attaches at the rear of a tractor and receives its power through a drive shaft that connects the power take off (PTO) of the tractor to the transmission at the center of the deck of the mower. The drive shaft rotates at 540 rpm inside a tubular safety shield and delivers 50 hp to the mower. If the drive shaft is a tubular steel shaft with outer diameter $d_o = 2.00$ in. and inner diameter $d_i = 1.675$ in., what is the maximum shear stress in the shaft at these operating conditions?

Prob. 4.7-10. A uniform shaft with fixed ends at A and D is subjected to external torques of magnitude T_0 and $2T_0$, as shown in Fig. P4.7-10 (see Prob. 4.6-15). The diameter of the shaft is d, and its shear modulus is G. (a) Using the *Displacement Method*, determine expressions for the rotation angles ϕ_B at section B and ϕ_C at section C. (b) Determine expressions for the maximum shear stress in each of the three segments of the shaft: $(\tau_{max})_1$, $(\tau_{max})_2$, and $(\tau_{max})_3$.

Express all of your answers in terms of T_0, L, d, and G.

Prob. 4.7-11. The aluminum-alloy shaft AD in Fig. P4.7-11 (see Prob. 4.6-14) ($G = 3.8 \times 10^3$ ksi) has a 40-in.-long tubular segment AC and a 30-in.-long solid segment CD, and it is fixed against rotation at ends A and D. The shaft, whose respective diameters are $(d_o)_1 = (d_o)_2 = d_3 = 2.5$ in. and $(d_i)_1 = (d_i)_2 = 2.0$ in., is subjected to torsional loads at sections B and C, as shown in the figure. (a) Using the *Displacement Method*, determine the rotation angles ϕ_B at section B and ϕ_C at section C. (b) Determine the maximum shear stress in the shaft.

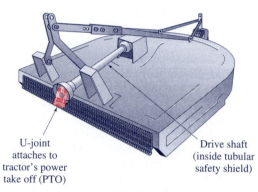

U-joint attaches to tractor's power take off (PTO)

Drive shaft (inside tubular safety shield)

P4.8-5

▼ **POWER-TRANSMISSION SHAFTS**

MDS 4.11–4.13

Prob. 4.8-1. A solid turbine shaft delivers 1000 hp at 60 rpm. The diameter of the shaft is 8 in., its length is 4 ft, and it is made of steel with a shear modulus of elasticity $G = 11.2 \times 10^3$ ksi. (a) What is the value of the maximum shear stress under the above operating conditions? (b) What is the angle of twist between the two ends of the shaft?

Prob. 4.8-2. A solid circular shaft having a diameter $d = 1.5$ in. rotates at an angular speed of 100 rpm. If the shaft is made of steel with an allowable shear stress of $\tau_{allow} = 8$ ksi, what is the maximum power, $\mathcal{P}$ (hp), that can be delivered by the shaft?

Prob. 4.8-3. A hollow bronze shaft ($\tau_{allow} = 100$ MPa) has an outside diameter of $d_o = 60$ mm and an inside diameter of $d_i = 50$ mm. How much power, $\mathcal{P}$ (kW), can be delivered by this shaft rotating at a speed of 5 rev/s?

Prob. 4.8-4. A hollow drive shaft delivers 150 kW of power to a rock crusher at a rotating speed of 5 rev/s. The shaft has

DProb. 4.8-6. Power is delivered through a tubular steel shaft of outer diameter $d_o = 2.50$ in. and inner diameter $d_i = 2.25$ in. What is the <u>slowest</u> angular speed (in rpm) at which the shaft can be allowed to rotate if the allowable shear stress in the shaft is 24 ksi and it is to deliver 200 kW of power?

DProb. 4.8-7. A shaft is required to deliver 500 kW of power at a rotational speed of 30 $\frac{rev}{sec}$. (a) If the shaft is to be solid, and made of bronze with an allowable shear stress of $\tau_{allow} = 50$ MPa, what is the required diameter of the shaft (to the nearest mm)? (b) What is the weight of a 1-m length of the solid bronze shaft from Part (a)? The specific weight of the bronze is 80 kN/m³. (c) Determine (to the nearest mm) the required outer diameter, d_o, of a tubular shaft made of the same bronze material if the tubular shaft is to have an inner diameter to outer diameter ratio of $d_i/d_o = 0.6$. (d) Compare the weight of a 1-m length of this tubular shaft with the weight of the 1-m length of solid shaft as determined in Part (b).

DProb. 4.8-8. A solid circular drive shaft 2 m long transmits 560 kW of power from the turbine engine of a helicopter to its rotor at 15 $\frac{rev}{s}$. If the shaft is steel, with shear modulus $G = 80$ GPa and allowable shear stress $\tau_{allow} = 60$ MPa, and if the allowable angle of twist of the shaft is 0.03 rad, determine the minimum permissible diameter of the drive shaft.

ᴰProb. 4.8-9. The electric motor A in Fig. P4.8-9 provides power to a pump through a belt drive from the pulley at B. The shaft between the motor and the pulley delivers 2 hp to the pulley and rotates at 1750 rpm. If the allowable shear stress in the solid steel shaft is $\tau_{allow} = 10$ ksi, what is the required diameter of the shaft (to the nearest 1/16 in.)?

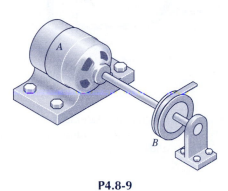

P4.8-9

ᴰProb. 4.8-10. The drive shaft of an inboard motor boat (Fig. P4.8-10) is required to deliver 100 hp at 300 rpm. A shaft is to be selected from the following group of available shafts: A($d_{oA} = 1.900$ in., $d_{iA} = 1.610$ in.), B($d_{oB} = 1.900$ in., $d_{iB} = 1.500$ in.), C($d_{oC} = 1.900$ in., $d_{iC} = 1.202$ in.). Select the lightest shaft that meets an allowable shear stress requirement of $\tau_{allow} = 20$ ksi.

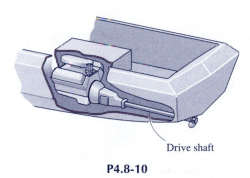

Drive shaft

P4.8-10

ᴰProb. 4.8-11. A tubular shaft of outer diameter $d_o = 50$ mm and inner diameter $d_i = 40$ mm drives a wind turbine that is producing 7 kW of power. (Assume 100% efficiency of the turbine.) If the allowable shear stress in the steel shaft is $\tau_{allow} = 50$ MPa, what is the <u>slowest</u> speed, ω_{min}, at which the blades may be allowed to rotate? (Neglect stresses in the shaft other than ones that are directly due to torsion.)

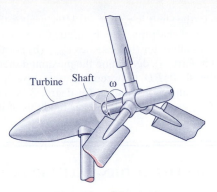

P4.8-11 and P9.4-12

ᴰProb. 4.8-12. A hollow circular shaft is to be designed to deliver 120 kW of power at a rotational speed of 60 $\frac{rev}{s}$. The allowable shear stress is 50 MPa, and the shaft is to be designed so that the outside diameter and the inside diameter have the ratio $d_o/d_i = 1.25$. Calculate the minimum allowable outside diameter, and express your answer to the nearest millimeter.

ᴰProb. 4.8-13. A solid circular drive shaft 20 in. long transmits 60 hp from a motorcycle's transmission to its rear wheel at 4600 rpm. If the shaft is made of steel with shear modulus $G = 11 \times 10^3$ ksi and allowable shear stress $\tau_{allow} = 10$ ksi, and if the allowable angle of twist of the drive shaft is 0.04 rad, determine the minimum permissible diameter of the drive shaft. Express your answer to the nearest $\frac{1}{16}$ in.

Prob. 4.8-14. The solid steel shaft AD in Fig. P4.8-14 delivers 5.5 kW of power to the gear at B and 4.0 kW of power to the gear at D. The shaft has a diameter of 7/8 in., is supported by a frictionless bearing at C, and rotates at an angular speed of 1725 rpm. Determine the maximum shear stress $(\tau_{max})_1$ in the shaft between the motor at A and the gear at B, and the maximum shear stress $(\tau_{max})_2$ in the shaft BD between the two drive gears.

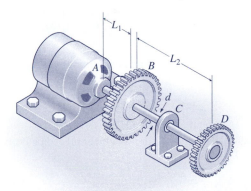

P4.8-14 and P4.8-15

Prob. 4.8-15. The solid steel shaft AD in Fig. P4.8-15 delivers 6 hp to the gear at B and 4 hp to the gear at D. The shaft is supported by a frictionless bearing at C, and it rotates at an angular speed of 1725 rpm. (a) If the maximum allowable shear stress in the shaft is $\tau_{allow} = 40$ MPa, what is the minimum diameter of shaft that may be used? Give your answer to the nearest millimeter that satisfies τ_{allow}. (b) For the shaft determined in Part (a), determine the maximum shear stress $(\tau_{max})_2$ in the shaft BD between the two drive gears. (c) Determine the angle of rotation (in degrees) between the two gears if $L_2 = 0.5$ m and the shear modulus of the steel shaft is $G = 76$ GPa.

▼ TORSION OF CLOSED, THIN-WALL MEMBERS

> **Problems 4.9-1 through 4.9-7.** *For the thin-wall tubular sections shown in Figs. P4.9-1 through P4.9-7; (a) determine the maximum shear stress in the cross section, and (b) determine the value of the torsion constant, J.*

Prob. 4.9-1. See the problem statement above, and use $T = 500$ kip · in.

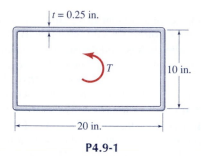

P4.9-1

Prob. 4.9-2. See the problem statement preceding Prob. 4.9-1, and use $T = 5$ kN · m.

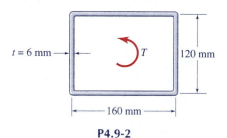

P4.9-2

Prob. 4.9-3. See the problem statement preceding Prob. 4.9-1, and use $T = 5$ kN · m.

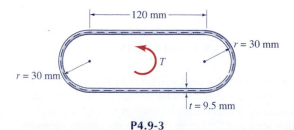

P4.9-3

Prob. 4.9-4. See the problem statement preceding Prob. 4.9-1, and use $T = 800$ kip · in.

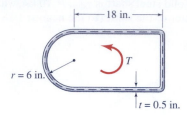

P4.9-4

Prob. 4.9-5. See the problem statement preceding Prob. 4.9-1 and use $T = 4$ kN · m, $a = 90$ mm, $b = 120$ mm, $c = 150$ mm, $t_1 = 3$ mm, and $t_2 = 5$ mm.

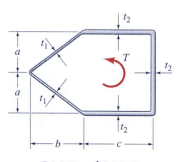

P4.9-5 and P4.9-6

Prob. 4.9-6. Repeat Prob. 4.9-5 using $T = 500$ kip · in., $a = 9$ in., $b = 12$ in., $c = 15$ in., $t_1 = 0.375$ in., and $t_2 = 0.625$ in.

Prob. 4.9-7. The "nose" of the cross section in Fig. P4.9-7 is semielliptical. See the problem statement preceding Prob. 4.9-1 and use $T = 800$ N · m.

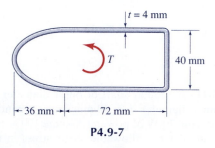

P4.9-7

ᴰProb. 4.9-8. A diamond-shaped torsion member has the midline dimensions shown in Fig. P4.9-8 and is subjected to a torque $T = 75$ kip $\cdot$ in. If the maximum allowable shear stress in the cross section is $\tau_{allow} = 8$ ksi, what is the minimum wall thickness? Express your answer to the nearest $\frac{1}{16}$ in.

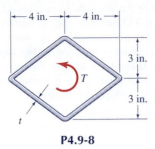

P4.9-8

Prob. 4.9-9. The torque tube in Fig. P4.9-9 has an elliptical cross section. Express the shear stress on the cross section in terms of the following parameters: T, a, b, and t. (See the Table of Geometric Properties of Plane Areas inside the back cover.)

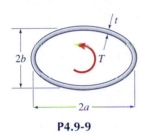

P4.9-9

Prob. 4.9-10. A tubular shaft having an inside diameter of $d_i = 2.0$ in. and a wall thickness of $t = 0.20$ in., is subjected to a torque $T = 5$ kip $\cdot$ in. (Note: The wall thickness is one-tenth of the inner diameter so that you can examine the range of validity of thin-wall torsion theory.) Determine the maximum shear stress in the tube: (a) using the exact theory for torsion of circular shafts, and (b) using the (approximate) thin-wall torsion theory of this section.

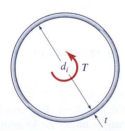

P4.9-10 and P4.9-11

Prob. 4.9-11. For the same tubular shaft of Prob. 4.9-10, determine the angle of twist per unit length: (a) using the exact theory for torsion of circular shafts, and (b) using the (approximate) thin-wall torsion theory of this section. Let $T = 5$ kip $\cdot$ in., as before, and let $G = 4 \times 10^3$ ksi.

Prob. 4.9-12. A 4-in.-square steel tube has a wall thickness $t = 0.25$ in. You are to compare the torsion behavior of the square tube to that of a tube with circular cross section having the same median-curve length ($L_m = 16$ in.). Use thin-wall torsion theory to (a) determine the shear-stress ratio τ_c/τ_s, where τ_c and τ_s are the maximum shear stresses in the circular tube and the square tube, respectively, when both members are subjected to a torque $T = 80$ kip $\cdot$ in. (b) Determine the ratio J_c/J_s, where J_c and J_s are the respective circle and square area properties defined by the torsional rigidity equation, Eq. 4.34.

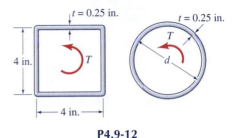

P4.9-12

Prob. 4.9-13. A thin-wall tube has uniform thickness t and cross-sectional dimensions b and h, measured to the median line of the cross section, Let the length of the median curve, $L_m = 2b + 2h$, and the thickness t be constant (hence, the weight will be constant), but let the ratio $\alpha \equiv \frac{b}{h}(b \geq h)$ vary. (a) Determine an expression that relates the maximum shear stress in a rectangular tube to the ratio α. (b) From your result in (a), show that the maximum shear stress will be smallest when the tube is square (i.e., when $\alpha = 1$).

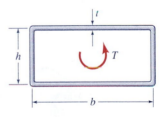

P4.9-13 and P4.9-14

Prob. 4.9-14. Repeat Prob. 4.9-13, but consider the angle of twist, rather than maximum shear stress. That is, (a) determine an expression that relates the angle of twist in a rectangular tube to the ratio α, and (b) show that the angle of twist per unit length is smallest for a square tube.

▼ TORSION OF NONCIRCULAR PRISMATIC BARS

Prob. 4.10-1. A torsion member with square cross section is subjected to end torques, as shown in Fig. P4.10-1. Let $d = 1.0$ in., $L = 20$ in., $G = 11 \times 10^3$ ksi, and $T = 50$ lb $\cdot$ in. (a) Determine the maximum shear stress in the shaft. (b)

Determine the angle of twist of the shaft. (c) Determine the cross-sectional area of a circular shaft that would carry the same torque as this square shaft without any increase in maximum shear stress.

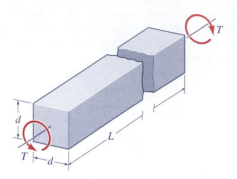

P4.10-1 and P4.10-2

Prob. 4.10-2. Repeat Prob. 4.10-1 using $d = 50$ mm, $L = 2$ m, $G = 80$ GPa, and $T = 500$ N $\cdot$ m.

Prob. 4.10-3. A torsion member with solid rectangular cross section is subjected to end torques T, as depicted in Fig. P4.10-3. Let the cross-sectional area of the torsion member be 5000 mm^2, and let $T = 2.0$ kN $\cdot$ m, $L = 2.0$ m, and $G = 25$ GPa. (a) Determine the maximum shear stress in torsion members having the two width/thickness ratios $d/t = 2.00$ and $d/t = 4.00$, respectively, where d and t are defined in Fig. P4.10-3. (b) Determine the angle of twist for torsion members having these two d/t ratios.

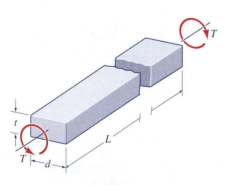

P4.10-3 and P4.10-4

^{D}Prob. 4.10-4. A torsion member with solid rectangular cross section is subjected to end torques T, as depicted in Fig. P4.10-4. Let the cross-sectional area of the torsion member be 4 in^2, let $L = 8$ ft, and let $G = 11 \times 10^3$ ksi. If the allowable shear stress is $\tau_{\text{allow}} = 6$ ksi, what is the maximum allowable torque that can be applied to torsion members

304

with width/thickness ratios $d/t = 2.00$ and $d/t = 4.00$, respectively, where d and t are defined in Fig. P4.10-4.

^{D}Prob. 4.10-5. A torsion member with elliptical cross section is subjected to end torques, as illustrated in Fig. P4.10-5. Let $a = 1.0$ in., $b = 0.5$ in., $L = 24$ in., $G = 11 \times 10^3$ ksi, and $T = 3000$ lb $\cdot$ in. (a) Determine the maximum shear stress in the shaft. (b) Determine the angle of twist of the shaft. (c) Determine the cross-sectional area of the elliptical cross section of this shaft. If the same torque T were to be applied to a shaft with circular cross section having the same cross-sectional area as this elliptical shaft, by what percent would the maximum shear stress decrease?

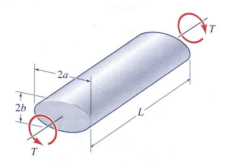

P4.10-5 and P4.10-6

^{D}Prob. 4.10-6. Repeat Prob. 4.10-5 using $a = 40$ mm, $b = 30$ mm, $L = 2$ m, $G = 80$ GPa, and $T = 2.5$ kN $\cdot$ m.

Prob. 4.10-7. (a) Determine the ratio of the maximum shear stress in a rectangular cross section to the maximum shear stress in an elliptical cross section,

$$\frac{(\tau_{\text{max}})_{\text{rect}}}{(\tau_{\text{max}})_{\text{ell}}}$$

if

$$\left(\frac{d}{t}\right)_{\text{rect}} = \left(\frac{a}{b}\right)_{\text{ell}} = 2$$

and if both torsion bars have the same cross-sectional area and are subjected to the same torque. (b) Determine the ratio of twist rates, $(\phi/L)_{\text{r}}/(\phi/L)_{\text{e}}$ for the same two torsion bars.

▼ **TORSION OF OPEN, THIN-WALL MEMBERS**

> **Problems 4.10-8 through 4.10-10.** *The torsion of open, thin-wall prismatic members, like angle sections and channel sections, can be analyzed by using Eqs. 4.35 and 4.36, taking the dashed centerline length, as in Example 4.15, as the dimension d in the formulas.*

Prob. 4.10-8. For the open, thin-wall angle section in Fig. P4.10-8, (a) determine the maximum shear stress due to torsion, and (b) determine the angle of twist of a torsion bar of length L. Use $G = 11 \times 10^3$ ksi. (You will have to use the values given in Table 4.3 to estimate α and β for the d/t ratio of this cross section.)

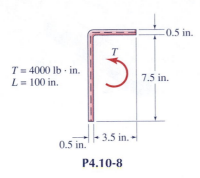

$T = 4000$ lb · in.
$L = 100$ in.

0.5 in.

7.5 in.

0.5 in.   3.5 in.

P4.10-8

Prob. 4.10-9. Repeat Prob. 4.10-8 for a channel-shaped torsion member whose cross section is shown in Fig. P4.10-9. Use $G = 80$ GPa.

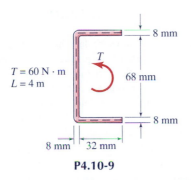

$T = 60$ N · m
$L = 4$ m

8 mm

68 mm

8 mm

8 mm   32 mm

P4.10-9

Prob. 4.10-10. (a) A closed tubular shaft having an inside diameter of $d_i = 2.0$ in. and a wall thickness of $t = 0.125$ in. is subjected to a torque $T = 5$ kip · in. Using the (approximate) thin-wall torsion theory of Section 4.9, determine the maximum shear stress in this closed, thin-wall tube. (b) If the tubular shaft in Part (a) is slit longitudinally, it becomes an open, thin-wall torsion member. Determine the maximum shear stress in this torsion member.

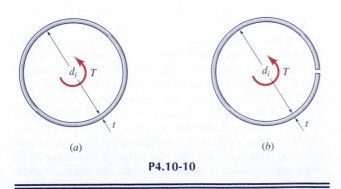

(a)   (b)

P4.10-10

▼ INELASTIC TORSION OF CIRCULAR RODS

Prob. 4.11-1. A 3-in.-diameter solid circular shaft made of elastic, perfectly plastic material is subjected to a torque T, that produces partially plastic deformation with an elastic core of radius $r_Y = 0.5$ in. (a) Sketch the shear-stress distribution in the shaft. (b) Determine the torque required to produce this partially plastic stress distribution. (c) Determine the angle of twist of the shaft, which is 4 ft. long.

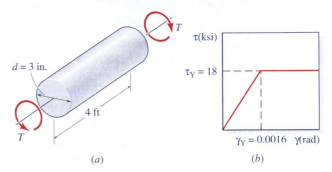

$d = 3$ in.

4 ft

τ(ksi)

$\tau_Y = 18$

$\gamma_Y = 0.0016$ γ(rad)

(a)   (b)

P4.11-1 and P4.11-2

Prob. 4.11-2. For the shaft in Fig. P4.11-2. (a) determine the value of the yield torque T_Y. (b) Determine the angle of twist, ϕ, for the following values of torque, T: T_Y, 1.1 T_Y, 1.2 T_Y, and 1.3 T_Y. Sketch the T versus ϕ curve for $0 \leq T \leq 1.3T_Y$. (c) Determine the fully plastic torque, T_P, for this shaft.

Prob. 4.11-3. A 50-mm-diameter shaft is made of elastic, perfectly plastic material and is subjected to a torque $T = 4$ kN · m. (a) Verify that this torque produces a partially-plastic stress distribution, that is, show that $T_Y \leq 4$ kN · m $\leq T_P$. (b) Determine the radius, r_Y, of the elastic core produced by the 4 kN · m torque. (c) Determine the angle of twist of this 2-m-long shaft.

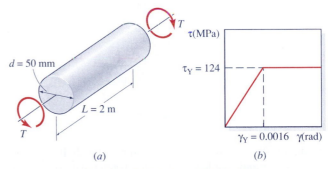

$d = 50$ mm

$L = 2$ m

τ(MPa)

$\tau_Y = 124$

$\gamma_Y = 0.0016$ γ(rad)

(a)   (b)

P4.11-3, P4.11-4, and P4.11-10

Prob. 4.11-4. For the shaft in Fig. P4.11-4, (a) determine the value of the yield torque, T_Y. (b) Determine the angle of twist, ϕ, for the following values of torque, T: T_Y, 1.1 T_Y, 1.2 T_Y, and 1.3 T_Y. Sketch the T versus ϕ curve for $0 \leq T \leq 1.3T_Y$. (c) Determine the fully-plastic torque, T_P, for this shaft.

*Prob. 4.11-5. A tubular shaft is made of elastic, perfectly plastic material with shear modulus $G = 6 \times 10^3$ ksi and yield stress $T_Y = 4$ ksi. The dimensions of the shaft are: $d_o = 1$ in., $d_i = 0.6$ in., and $L = 30$ in. (a) Determine the yield torque, T_Y, and the fully plastic torque. T_P, for this shaft. (b)

What percent of the cross-sectional area of the shaft has yielded when $T = 1.1\ T_Y$? (c) Determine the angle of twist of the shaft when the torque is just enough to cause yielding at the inner surface of the tubular shaft.

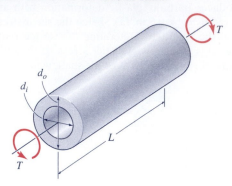

P4.11-5, P4.11-6, P4.11-11, and P4.11-12

*Prob. 4.11-6. A tubular shaft is made of elastic, perfectly-plastic material with shear modulus $G = 40$ GPa and yield stress $\tau_Y = 30$ MPa. The dimensions of the shaft are: $d_o = 40$ mm, $d_i = 25$ mm, and $L = 1$ m. (a) Determine the yield torque, T_Y, and the fully plastic torque, T_P, for this shaft. (b) What percent of the cross-sectional area of the shaft has yielded when $T = 1.1\ T_Y$? (c) Determine the angle of the twist of the shaft when the torque is just enough to cause yielding at the inner surface of the tubular shaft.

Prob. 4.11-7. Derive the formula, similar to Eq. 4.48, that relates the partially plastic torque. T, to the outer radius, r_Y, of the elastic core of an elastic, perfectly plastic tubular shaft (Fig. P4.11-7) having outer radius r_o, inner radius r_i, and yield stress in shear, r_Y.

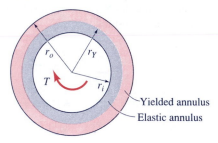

Yielded annulus
Elastic annulus

P4.11-7

Prob. 4.11-8. The two ends of a solid circular shaft of diameter $d = \frac{7}{8}$ in and length $L = 18$ in. are rotated with respect to each other by exactly one revolution (i.e., $\phi = 2\pi$ rad). The shaft is made of elastic, perfectly-plastic material with τ versus γ curve as shown in Fig. P4.11-8b, (a) Determine the maximum shear strain for this loading condition. (b) Determine the radius of the elastic core for this loading condition.

*Prob. 4.11-9. The two ends of the solid circular shaft in Fig. P4.11-9a are rotated relative to each other by an angle of $\phi = 6°$. The shaft is made of elastic, perfectly-plastic material with τ versus γ curve as shown in Fig. P4.11-9b. (a) Determine the torque required to produce this twist angle. (b) If the torque determined in Part (a) is completely removed, what is the permanent angle of twist left in the

shaft? (c) What is the residual stress distribution in the shaft after the torque is removed?

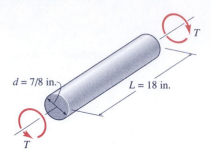

$d = 7/8$ in.
$L = 18$ in.

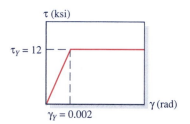

τ (ksi)

$\tau_Y = 12$

$\gamma_Y = 0.002$

γ (rad)

P4.11-8 and P4.11-9

*Prob. 4.11-10. The solid circular shaft in Fig. P4.11-10 (see Prob. 4.11-3) is subjected to a torque T that produces a twist angle $\phi = 10°$ between the two ends of the shaft. (a) Determine the torque required to produce this twist angle. (b) If the torque determined in Part (a) is completely removed, what is the permanent angle of twist left in the shaft? (c) What is the residual stress distribution in the shaft after the torque is removed?

*Prob. 4.11-11. A tubular shaft with dimensions $d_o = 1$ in., $d_i = 0.6$ in., and $L = 30$ in. (see Prob. 4.11-5) is made of elastic, perfectly plastic material with shear modulus $G = 6 \times 10^3$ ksi and yield stress $\tau_Y = 4$ ksi, The shaft is subjected to a torque T that is sufficient to cause yielding from $r_Y = 0.4$ in. to the outer surface. (a) Determine the value of the torque T required to produce this state of stress in the shaft. (b) What is the shear strain at the outer surface for this loading? (c) If the torque determined in Part (a) completely removed, what is the permanent angle of twist of the shaft? (d) Sketch the residual stress distribution after removal of the original torque. Indicate the value of the shear stress at the outer surface of the tubular shaft and the shear stress at the inner surface.

*Prob. 4.11-12. A tubular shaft with dimensions $d_o = 40$ mm, $d_i = 20$ mm, and $L = 1$ m (see Prob. 4.11-5) is made of elastic, perfectly plastic material with shear modulus $G = 40$ GPa and yield stress $\tau_Y = 30$ MPa. The shaft is subjected to a torque T that is sufficient to cause yielding from $r_Y = 15$ mm to the outer surface. (a) Determine the value of the torque. T, required to produce this state of stress in the shaft. (b) What is the shear strain at the outer surface for this loading? (c) If the torque determined in Part (a) is completely removed, what is the permanent angle of twist of the shaft? (d) Sketch the residual stress distribution after removal of the original torque. Indicate the value of the shear stress at the outer surface of the tubular shaft and the shear stress at the inner surface.

Section		Suggested Review Problems
4.2	**Torsion** of a circular member (rod, bar) has three characteristics: (1) the axis of the member remains straight, (2) plane cross sections remain plane and remain perpendicular to the axis, and (3) radial lines remain straight and radial as cross sections rotate about the axis.	
	 (a) An example of torsional deformation. (b) Sign convention for internal (resisting) torque $T(x)$. (c) Sign convention for angle of rotation $\phi(x)$. Geometry of torsional deformation and Sign Conventions for torsion (Fig. 4.4)	
	In torsion, the *shear strain* γ is related to the *rotation* $\phi(x)$ of cross section x by the **strain-displacement equation**	Derive Eq. 4.1
	$$\gamma(x, \rho) = \rho \frac{d\phi(x)}{dx} \qquad (4.1)$$ $\gamma_{max} = r\frac{d\phi}{dx}$ $\gamma_{max} = r_o\frac{d\phi}{dx}$ (a) (b) Torsional shear-strain distributions (Fig. 4.6)	
	The **total angle of twist** ϕ of a member undergoing torsional deformation is given by	
	$$\phi = \int_0^L \frac{d\phi(x)}{dx}\, dx$$	
4.3	The *shear stress—shear strain equation* for linearly elastic behavior is **Hooke's Law for Shear.**	4.3-3 4.3-7 4.3-17
		$$\tau = G\gamma$$	
	The *shear stress* on the cross section at x of a linearly elastic member undergoing torsional deformation is given by the **torsion formula**	
	$$\tau(x, \rho) = \frac{T(x)\rho}{I_p(x)}$$ where $I_p = \displaystyle\int_A \rho^2\, dA$ is the *polar moment of inertia.*	
	The **torque-twist equation** for a linearly elastic member undergoing torsional deformation is given by	
	$$\frac{d\phi}{dx} = \frac{T(x)}{GI_p(x)}$$	
4.4	The *maximum normal stresses* (tension and compression) occur on planes at $\pm 45°$ to the cross section, on which there is pure shear due to torsion (Section 4.4):	4.4-5
	 Pure shear due to torsion (Fig. 4.12a) Maximum normal stresses due to torsion (Fig. 4.14b)	

Section		Suggested Review Problems
4.5	The **torque-twist equation** for linearly elastic behavior of a uniform member undergoing torsional deformation due to end torques T can be stated in the form or in the form $$T = k_t\phi \quad \text{where} \quad k_t = \frac{GI_p}{L}$$ k_t is the *torsional stiffness coefficient* $$\phi = f_t T \quad \text{where} \quad f_t = \frac{L}{GI_p}$$ f_t is the *torsional flexibility coefficient* An end-loaded uniform torsion member (Fig. 4.11)	
	A **statically determinate system** is one for which all reactions and all internal torques can be determined by the use of *equilibrium equations* (i.e., statics) alone. A statically determinate assemblage of torsion members (Fig. 4.18a)	4.5-1 4.5-15
4.6	Analysis of **a statically in determinate system** requires the use of (1) *equilibrium equations,* (2) member *torque-twist equations,* and (3) *geometry of deformation equations.* Methods that may be used to analyze statically indeterminate torsion assemblages are: the **Basic Force Method** (Section 4.6) and the **Displacement Method** (Optional Section 4.7). A statically indeterminate assemblage of torsion members (Fig. 4.18b)	4.6-1 4.6-9 4.6-17
4.8	The **power** transmitted by a shaft that is rotating at an angular speed ω (rad/sec) is given by $$\mathcal{P} = \mathcal{T}\omega$$ A shaft delivering power (Fig. 4.23)	4.8-3 4.8-7 4.8-15
Sections 4.7 and 4.9–4.11 are all "optional" sections.		

EQUILIBRIUM OF BEAMS

5

5.1 INTRODUCTION

The behavior of slender members subjected to axial loading and to torsional loading was discussed in Chapter 3 and Chapter 4, respectively. Now we turn our attention to the problem of determining the stress distribution in, and the deflection of, beams.

> *A **beam** is a structural member that is designed to support transverse loads, that is, loads that act perpendicular to the longitudinal axis of the beam. A beam resists applied loads by a combination of internal transverse shear force and bending moment.*

Figure 5.1 shows steel beams (lower right) and prestressed concrete beams (center) that will support the bridge decks (roadways) and the vehicles that pass over the bridges.

FIGURE 5.1 Several bridge beams during bridge construction. (Courtesy Roy Craig)

309

Types of Beams; Loads and Reactions. Figure 5.2 illustrates beams with several types of support and several types of loading. The force and moment components at a support are called **reactions** since they react to the applied loads. As in Fig. 5.2, beam reactions will be indicated by slashes across their arrow symbols to distinguish reactions from applied loads.

Supports: Three types of support are shown in Fig. 5.2:

- *Roller Support*—prevents displacement in the transverse (i.e., y) direction, but permits z-rotation and displacement in the axial direction; the reaction is a force in the $+y$ or $-y$ direction. The support at end A in Fig. 5.2a is a roller support.
- *Pin Support*—prevents displacement in the axial direction and in the transverse direction, but permits z-rotation; the reaction is a force with both axial and transverse components. The support at end B in Fig. 5.2a is a pin support.
- *Cantilever Support* (or *Fixed End*)—prevents displacement in the axial direction and in the transverse direction, and also prevents z-rotation; the reaction consists of a force with both axial and transverse components, plus a couple. The support at end C in Fig. 5.2b is a cantilever support.

Beams are normally classified by the manner in which they are supported.

- *Simply Supported Beam*—a beam with a pin support at one end and a roller support at the other end. The beam in Fig. 5.2a is a simply supported beam. Simply supported beams are statically determinate.
- *Cantilever Beam*—a beam with a cantilever support (i.e., fixed end) at one end and free at the other. The beam in Fig. 5.2b is a cantilever beam. Cantilever beams are statically determinate.

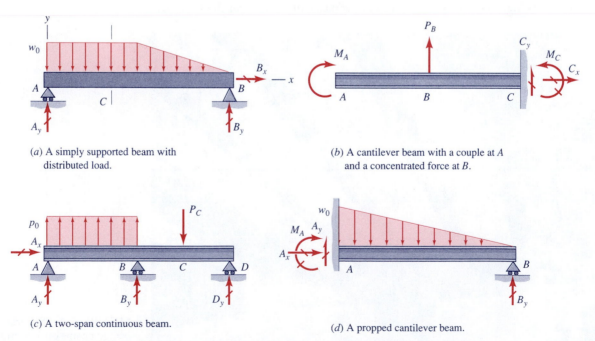

(*a*) A simply supported beam with distributed load.

(*b*) A cantilever beam with a couple at A and a concentrated force at B.

(*c*) A two-span continuous beam.

(*d*) A propped cantilever beam.

FIGURE 5.2 Examples of several types of beams with various types of loads.

- *Continuous Beam*—a beam with a pin support at one end, a roller support at the other end, and one or more intermediate roller supports. The beam in Fig. 5.2c is a continuous beam. Continuous beams are statically indeterminate.
- *Propped Cantilever Beam*—a beam with a cantilever support (i.e., fixed end) at one end and a roller at the other end. The beam in Fig. 5.2d is a propped cantilever beam. Propped cantilever beams are statically indeterminate.
- *Overhanging Beam*—a beam that extends beyond the support at one end (or at both ends). The beam in Fig. 5.2a would be an overhanging beam if the roller support at end *A* were to be moved to the right, leaving a part of the beam to the left of the roller as an overhang.

External Loads: The *loads* that are applied to beams may be classified as: *distributed transverse loads, concentrated transverse forces,* or *couples.* Figure 5.2a illustrates a simply supported beam with distributed loading; a cantilever beam with a concentrated force at *B* and a concentrated couple at *A* is shown in Fig. 5.2b. Since axial forces produce axial displacement (Chapter 3), not bending, axial forces will not be included among the types of loads applied to beams in the present chapter. Chapters 9 and 10 consider cases of combined axial loading and bending.

Stress Resultants—Bending Moment and Transverse Shear. How does

a beam respond to these external loads, and how does it transmit these loads to its supports? Let us consider a simple example, the handle of the shovel in Fig. 5.3a, and let us perform three equilibrium "experiments." The two hands of a worker are holding the shovel in a horizontal position. To simplify the discussion, let us assume that the weight of the shovel handle is negligible in comparison with the weight *W* of the shovel blade and its contents, and that the forces R_A and R_B of the hands on the shovel handle are vertical concentrated forces, as shown in the free-body diagram in Fig. 5.3b. From this free-body diagram, $\Sigma F = 0$ and $(\Sigma M)_A = 0$, we get

$$R_A = W\left(\frac{L}{a} - 1\right), \quad R_B = W\left(\frac{L}{a}\right)$$

The three experiments are labeled E1, E2, and E3.

E1. For the first experiment move the hand at *B* (i.e., reaction force R_B) <u>toward end *C*</u> of the shovel handle, that is, let $a \rightarrow L$. Then,

$$R_A \rightarrow 0, \quad R_B \rightarrow W, \quad \text{with } R_B a = WL = \text{const}$$

In this experiment, we would eventually need no force at *A*, and the hand at *B* would be directly under the load *W*.

E2. For the second experiment move the hand at *B* (i.e., reaction force R_B) <u>toward end *A*</u> of the shovel handle, that is, let $a \rightarrow 0$. Then,

$$R_A \rightarrow \infty, \quad R_B \rightarrow \infty, \quad \text{with } R_B a = WL = \text{const}$$

In this experiment, both R_A and R_B would get quite large, much larger than *W*, but the moment $R_B a$ would still remain the same.

(*a*) A shovel supported by a worker's hands.

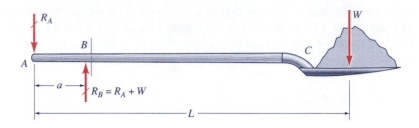

(*b*) A free-body diagram of the entire shovel.

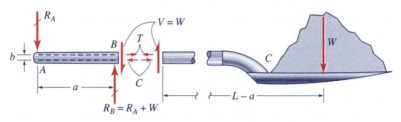

(*c*) Free-body diagrams showing stress resultants
at a cross section of the handle.

FIGURE 5.3 A shovel handle used to illustrate a beam.

E3. Finally, for the third experiment consider the free-body diagrams in Fig. 5.3*c*, so we can determine what internal stress resultants on a cross section are required for equilibrium. (Recall that this procedure is called the **method of sections.**) Imagine that the shovel handle is cut just to the right of *B*. From the two free-body diagrams in Fig. 5.3*c* we see that:

- A **transverse shear force V,** of magnitude W, is required to satisfy equilibrium in the vertical direction.

- There can be no <u>net</u> horizontal force on the cross section, since there are no horizontal external loads on either free body in Fig. 5.3*c*. Therefore, $C = T$.

- Finally, the cross section itself must supply a couple of magnitude $W(L - a)$. This couple is called the **bending moment M.** To form this couple there must be equal-and-opposite horizontal forces $C = T$ that <u>act on the cross section</u>, with a moment arm $b \ll L$, such that

$$C = T = W\left(\frac{L - a}{b}\right)$$

Therefore, within the depth of the cross section there must be both compressive normal stresses and tensile normal stresses, with the magnitude of the resultant compressive force C and the resultant tensile force T being much greater than the external load W.

Let's estimate the magnitude of R_A, R_B, and $C = T$ for the shovel in Fig. 5.3a. Let $W = 12$ lb, $L = 50$ in., $a = 10$ in., and $b = 1$ in. Then, $R_A = 48$ lb, $R_B = 60$ lb, and $C = T = 480$ lb. If a were to be reduced to 5 in., then R_B would be 120 lb and $C = T$ would be 540 lb. As you can imagine, few workers would hold the shovel with $a < 10$ in.!

In Chapter 6 we will determine the distribution of normal stress, whose resultant is the bending moment M, and the distribution of shear stress, whose resultant is the transverse shear force V.

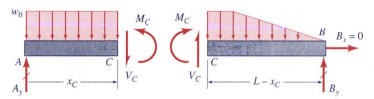

FIGURE 5.4   The transverse shear force V_C and bending moment M_C at cross section C.

Consider now the simply supported beam in Fig. 5.2a. The downward distributed load gives rise to upward reactions at the supports at A and B. The roller symbol at A implies that the reaction force can have no horizontal component. If we pass an imaginary cutting plane at C, as indicated in Fig. 5.2a, and we draw separate free-body diagrams of AC and CB (Fig. 5.4), we see that a **transverse shear force $\mathbf{V}_C$** and a **bending moment $\mathbf{M}_C$** must act on the cross section at C to maintain the force equilibrium and moment equilibrium of these two adjoining free bodies. *Newton's Law of action and reaction* determines the relationship of the directions of V_C and M_C on the two free-body diagrams.

The **internal stress resultants** that are associated with bending of beams are shown in Fig. 5.5 and are defined by the following equations:

$$
\begin{aligned}
V(x) &= -\int_A \tau_{xy}(x, y)\,dA \\[2mm]
M(x) &= -\int_A y\sigma_x(x, y)\,dA
\end{aligned}
\qquad \text{\textbf{Stress Resultants}} \qquad (5.1)
$$

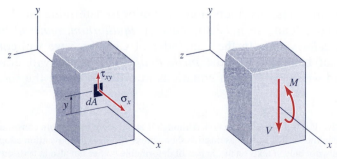

FIGURE 5.5   Definition of stress resultants—transverse shear force $V(x)$ and bending moment $M(x)$.

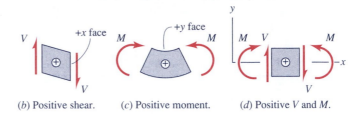

(*a*) Positive *V* and *M* on section "*x*."

FIGURE 5.6 The sign convention for internal stress resultants $V(x)$ and $M(x)$.

(*b*) Positive shear.   (*c*) Positive moment.   (*d*) Positive *V* and *M*.

The **sign conventions** for the internal stress resultants in beams are illustrated in Fig. 5.6. The sign conventions may be stated in words as follows:

- A *positive shear force, V*, acts in the $-y$ direction on a $+x$ face.[1]
- A *positive bending moment, M*, makes the $+y$ face of the beam concave.

Figures 5.6*b* and 5.6*c* illustrate the physical meaning of positive shear force and positive bending moment, while Fig. 5.6*d* summarizes the sign conventions for the internal stress resultants in beams. **It is very important to observe these sign conventions for *V* and *M*,** because equations will be developed to relate the stress distribution in beams and the deflection of beams to these two stress resultants.

The discussion of beams is divided into three chapters. In the present chapter we concentrate on *equilibrium of beams*, and we solve for the shear force and bending moment in various types of beams subjected to various loading conditions. In Chapter 6 we introduce displacement assumptions that permit us to determine the *normal-stress distribution* associated with the bending moment *M*; then we determine the *shear-stress distribution* associated with the transverse shear force *V*. Finally, in Chapter 7 we solve for the *deflection of beams*, including statically indeterminate beams.

5.2   EQUILIBRIUM OF BEAMS USING FINITE FREE-BODY DIAGRAMS

To determine the stress distribution in a beam or to determine the deflected shape of a beam under load, we need to consider *equilibrium, material behavior,* and *geometry of deformation*. In the remainder of Chapter 5 we will concentrate on equilibrium of beams. Using the *method of sections,* we will draw free-body diagrams and write equilibrium equations in order to relate the shear-force and

---

[1]Note that positive *V* acts in the $-y$ direction. Although it might seem to be more consistent to define positive shear force in the same direction as positive shear stress, τ_{xy}, the sign convention adopted here is used almost universally, not only in texts on mechanics of deformable bodies but also in texts on statics and texts on structures. The justification for this shear-force sign convention is not only its widespread adoption, but also the simplification that it introduces in the drawing of shear and moment diagrams (Section 5.4).

bending-moment stress resultants on beam cross sections to the external loads acting on the beam. Several examples that illustrate the use of finite-length free-body diagrams are given in this section. In Section 5.3 we will employ infinitesimal free-body diagrams, and in Sections 5.4 and 5.5 shear and moment diagrams are discussed. Finally, in Section 5.6 discontinuity functions are used to represent loads, shear, and moment.

EXAMPLE 5.1

The cantilever beam AD in Fig. 1 is subjected to a concentrated force of 5 kN at C and a couple of 4 kN $\cdot$ m at D. Determine the shear V_B and bending moment M_B at a section 2 m to the right of the support A.

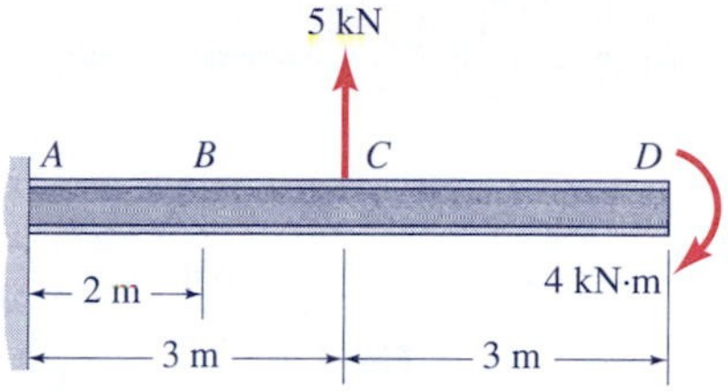

Plan the Solution We can use either a free-body diagram of AB or a free-body diagram of BD. Since the former would require us to compute the support reactions at A, we will, instead, use a free-body diagram of BD.

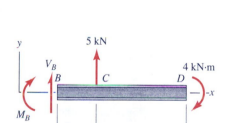

Fig. 1

Solution Figure 2 is a free-body diagram of BD. The shear force V_B and bending moment M_B are shown in the positive sense according to the sign convention in Fig. 5.6.

$+\uparrow \sum F_y = 0:$ $\qquad V_B + 5 \text{ kN} = 0$

$\qquad\qquad\qquad V_B = -5 \text{ kN}$ $\qquad$ **Ans.**

$+\circlearrowleft \left(\sum M\right)_B = 0:$ $\quad M_B - (5 \text{ kN})(1 \text{ m}) + 4 \text{ kN} \cdot \text{m} = 0$

$\qquad\qquad\qquad M_B = 1 \text{ kN} \cdot \text{m}$ $\qquad$ **Ans.**

Fig. 2 A free-body diagram.

Review the Solution To satisfy force and moment equilibrium of the cantilever beam, there are internal stress resultants at B as shown in Fig. 3. As a check, we should be able to satisfy $\sum M = 0$ about any point, for example point C.

$$\text{Is} \left(\sum M\right)_C = 1 \text{ kN} \cdot \text{m} - (5 \text{ kN})(1 \text{ m}) + 4 \text{ kN} \cdot \text{m} = 0? \quad \text{Yes}$$

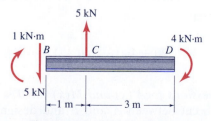

Fig. 3 A free-body diagram.

In the preceding example, the shear force and bending moment were required at a specific cross section. Using a finite free body permits these values to be determined directly from the corresponding equilibrium equations. The finite-free-body approach is also useful when expressions for $V(x)$ and $M(x)$ are required over some portion of the beam. Example 5.2 illustrates this type of problem and also illustrates a way to handle relatively simple distributed loads.

EXAMPLE 5.2

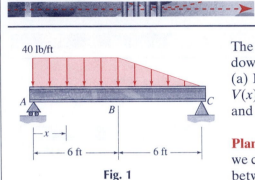

40 lb/ft

A

B

C

x

6 ft

6 ft

Fig. 1

The simply supported beam AC in Fig. 1 is subjected to a distributed downward loading as shown. The load varies linearly between B and C. (a) Determine the reactions at A and C, (b) determine expressions for $V(x)$ and $M(x)$ for $0 < x \le 6$ ft, and (c) determine expressions for $V(x)$ and $M(x)$ for 6 ft $\le x <$ 12 ft.

Plan the Solution   By using a free-body diagram of the entire beam AC, we can determine the reactions at A and C. We will need to make a "cut" between A and B to determine the shear force and bending moment required in Part (b), and we will need to make a "cut" between B and C to answer Part (c). The distributed loads can be replaced, on each free-body diagram, by their resultants.

Solution   (a) *Determine the reactions.* On the free-body diagram in Fig. 2, the resultants of the uniform distributed load on AB and the linearly varying distributed load on BC are shown with dashed-arrow symbols. The reactions at A and C can now be determined from three equilibrium equations.

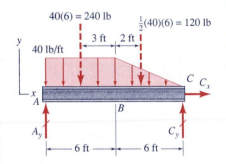

40(6) = 240 lb

$\frac{1}{2}(40)(6) = 120$ lb

3 ft

2 ft

y

40 lb/ft

x

A

B

C C_x

A_y

C_y

6 ft

6 ft

Fig. 2   A free-body diagram of beam AC.

$$+\zeta \left(\sum M \right)_A = 0: \quad (240 \text{ lb})(3 \text{ ft}) + (120 \text{ lb})(8 \text{ ft}) - C_y(12 \text{ ft}) = 0$$

$$C_y = 140 \text{ lb} \qquad \textbf{Ans. (a)}$$

$$+\zeta \left(\sum M \right)_C = 0: \quad A_y(12 \text{ ft}) - (240 \text{ lb})(9 \text{ ft}) - (120 \text{ lb})(4 \text{ ft}) = 0$$

$$A_y = 220 \text{ lb} \qquad \textbf{Ans. (a)}$$

$$\xrightarrow{+} \sum F_x = 0: \qquad C_x = 0 \qquad \textbf{Ans. (a)}$$

(Note: From now on we will ignore axial reactions and internal axial forces on beams that have no axial applied loads.)
   Before going on to Parts (b) and (c), we should check the above answers.

$$+\uparrow \sum F_y = 0: \qquad \text{Is} \quad 220 + 140 - 240 - 120 = 0? \qquad \text{Yes.}$$

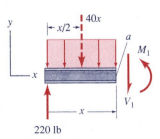

y

x/2

40x

a

M_1

x

V_1

x

220 lb

Fig. 3   A free-body diagram for Interval 1.

(b) *Determine expressions for $V(x)$ and $M(x)$ for $0 < x \le 6$ ft.* To do this, we can make a cut between A and B and designate this portion "Interval 1," giving the free-body diagram in Fig. 3. Let us use the symbols V_1 and M_1 for expressions that are valid in this interval. The resultant of the uniform distributed load is shown as a dashed arrow. The shear and moment are shown in the positive sense according to the sign conventions of Fig. 5.6.

$+\uparrow \sum F_y = 0$: $220\text{ lb} - (40x)\text{lb} - V_1 = 0$

$$V_1 = (220 - 40x)\text{lb}, \quad 0 < x \le 6\text{ ft} \qquad \textbf{Ans. (b)}$$

$+\circlearrowleft \left(\sum M \right)_a = 0$: $(220\text{ lb})(x\text{ ft}) - (40x\text{ lb})(\tfrac{x}{2}\text{ ft}) - M_1 = 0$

$$M_1 = (220x - 20x^2)\text{lb} \cdot \text{ft}, \quad 0 < x \le 6\text{ ft} \qquad \textbf{Ans. (b)}$$

(c) *Determine expressions for $V(x)$ and $M(x)$ for $6\text{ ft} \le x < 12\text{ ft}$.* To do this, we can take either a free-body diagram from A to a cut between B and C, or a free-body diagram from the cut to the right end, C. The latter, which is shown in Fig. 4, will be easier because we will then only have to deal with a single triangular load.

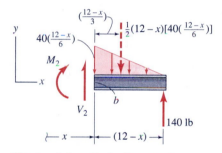

$+\uparrow \sum F_y = 0$: $V_2 - \dfrac{40}{12}(12 - x)^2 \text{ lb} + 140\text{ lb} = 0$

$$V_2 = \left[-140 + \frac{10}{3}(12 - x)^2 \right]\text{lb}, \quad 6\text{ ft} \le x < 12\text{ ft} \qquad \textbf{Ans. (c)}$$

$+\circlearrowleft \left(\sum M \right)_b = 0$:

Fig. 4 A free-body diagram for Interval 2.

$$M_2 + \left[\frac{40}{12}(12 - x)^2 \text{ lb} \right]\left[\left(\frac{12 - x}{3} \right)\text{ft} \right] - (140\text{ lb})[(12 - x)\text{ft}] = 0$$

$$M_2 = \left[140(12 - x) - \frac{10}{9}(12 - x)^3 \right]\text{lb} \cdot \text{ft}, \quad 6\text{ ft} \le x < 12\text{ ft} \qquad \textbf{Ans. (c)}$$

Review the Solution   We have already, in Part (a), performed an equilibrium check on the reactions. Since there is no concentrated transverse load or couple at B, we should have $V_1(x = 6\text{ ft}) = V_2(6\text{ ft})$ and $M_1(6\text{ ft}) = M_2(6\text{ ft})$.

$$V_1(6\text{ ft}) = 220 - 40(6) = -20\text{ lb}$$

$$V_2(6\text{ ft}) = -140 + \frac{10}{3}(6)^2 = -20\text{ lb} = V_1(6\text{ ft})$$

$$M_1(6\text{ ft}) = 220(6) - 20(6)^2 = 600\text{ lb} \cdot \text{ft}$$

$$M_2(6\text{ ft}) = 140(6) - \frac{10}{9}(6)^3 = 600\text{ lb} \cdot \text{ft} = M_1(6\text{ ft})$$

In Sections 5.4 and 5.5 shear and moment diagrams will be used to graphically represent $V(x)$ and $M(x)$.

In both Example 5.1 and Example 5.2 it was possible for us to solve the equilibrium equations and to determine values of (or expressions for) shear and moment, since in each case the beam is statically determinate. The next example illustrates the type of equilibrium results that are obtained for statically indeterminate beams.

EXAMPLE 5.3

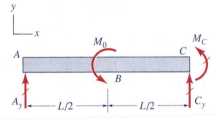

Fig. 1

The propped cantilever beam AC in Fig. 1 has a couple applied at its center B. Determine expressions for the reactions (i.e., the shear force and bending moment) at C in terms of the applied couple M_0 and the reaction at A.

Plan the Solution   A free-body diagram of the whole beam will permit us to relate the reactions at C to the reaction at A.

Solution

$$+\uparrow \sum F_y = 0: \qquad A_y + C_y = 0$$

$$C_y = -A_y \qquad \qquad \textbf{Ans.}$$

$$+\curvearrowleft \left(\sum M \right)_C = 0: \qquad A_y L - M_0 - M_C = 0$$

$$M_C = A_y L - M_0 \qquad \qquad \textbf{Ans.}$$

Fig. 2   A free-body diagram of beam AC.

Review the Solution   The results above are typical of *statically indeterminate problems,* that is, problems where the equations of statics (i.e., equilibrium) are not sufficient to determine all of the unknowns. Here we have three unknowns, but only two equilibrium equations, other than the trivial one for horizontal equilibrium. Therefore, two reactions can be written in terms of the third reaction (here A_y). This one is called a *redundant.* That is, the support at A is not essential to prevent collapse of the beam.

Just as for statically indeterminate axial deformation and torsion problems, in order to solve statically indeterminate beam problems we must consider the deformation of the beam. Statically indeterminate beam problems are examined in Chapter 7, *Deflection of Beams.*

5.3   EQUILIBRIUM RELATIONSHIPS AMONG LOADS, SHEAR FORCE, AND BENDING MOMENT

In the previous section we used finite free-body diagrams to determine values of shear force and bending moment at specific cross sections, and to determine expressions for $V(x)$ and $M(x)$ over specified ranges of x. Here we use *infinitesimal free-body diagrams* to obtain equations that relate the external loading to the internal shear force and bending moment. These expressions will be especially helpful in

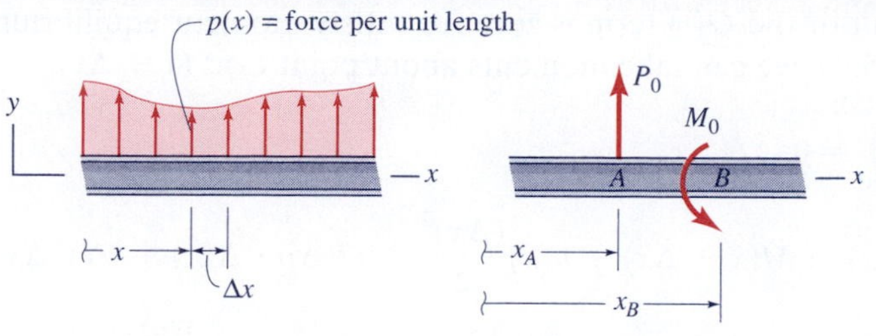

$p(x)$ = force per unit length

(a) Distributed load.

(b) Concentrated force and couple.

FIGURE 5.7 The sign convention for external loads on a beam.

Section 5.5, where we discuss shear and moment diagrams. In addition to the sign conventions for shear force and bending moment, given in Fig. 5.6, we need to adopt a **sign convention for external loads** (Fig. 5.7).

- *Positive distributed loads* and *positive concentrated loads* act in the $+y$ direction (e.g., loads $p(x)$ and P_0 in Fig. 5.7).
- A *positive external couple* acts in a right-hand-rule sense with respect to the z axis, that is, counterclockwise as viewed in the xy plane (e.g., the external couple M_0 in Fig. 5.7b).

First, let us consider a portion of the beam where there are no concentrated external loads, and let us establish equilibrium equations for an infinitesimal free-body diagram. Take the segment of beam from x to $(x + \Delta x)$ in Fig. 5.7a, as redrawn in Fig. 5.8a. For equilibrium of the free body in Fig. 5.8a.

$$+\uparrow \sum F_y = 0: \qquad V(x) + p(x)\,\Delta x + O\,(\Delta x^2) - V(x + \Delta x) = 0$$

where $O(\cdots)$ means "of the order of," and $\Delta p \sim \Delta x$. Collecting terms and dividing by Δx we get

$$\frac{V(x + \Delta x) - V(x)}{\Delta x} = p(x) + O(\Delta x)$$

By taking the limit as $\Delta x \to 0$, we get

$$\boxed{\frac{dV}{dx} = p(x)} \qquad (5.2)$$

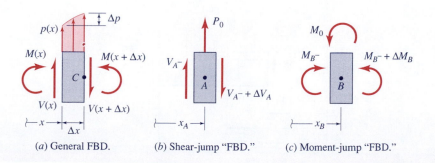

(a) General FBD.

(b) Shear-jump "FBD."

(c) Moment-jump "FBD."

FIGURE 5.8 Infinitesimal free-body diagrams.

since the limit of the $O(\cdot)$ term is zero. To satisfy moment equilibrium for the free body in Fig. 5.8a, we can take moments about point C at $(x + \Delta x)$.

$$+\circlearrowleft\left(\sum M\right)_C = 0:$$

$$M(x) - M(x + \Delta x) + p(x)\frac{(\Delta x)^2}{2} + O(\Delta p \cdot \Delta x^2) + V(x)\Delta x = 0$$

Dividing through by Δx and taking the limit as $\Delta x \to 0$, we obtain

$$\boxed{\frac{dM}{dx} = V(x)} \tag{5.3}$$

Wherever there is an external concentrated force, such as P_0 in Fig. 5.7b, or a concentrated couple, such as M_0 in Fig. 5.7b, there will be a step change in shear or moment, respectively. From the partial free-body diagram in Fig. 5.8b (moments have been omitted for clarity),

$$+\uparrow\sum F_y = 0: \qquad V_{A^-} - (V_{A^-} + \Delta V_A) + P_0 = 0$$

$$\boxed{\Delta V_A = P_0} \tag{5.4}$$

where V_{A^-} is the (internal) shear force just to the left of the point x_A where P_0 is applied. That is, a concentrated force P_0 at coordinate x_A will cause a step change ΔV_A in shear <u>having the same sign as P_0</u>.

An external couple M_0 at coordinate x_B causes a step change in the moment at x_B. From Fig. 5.8c (shear forces have been omitted for clarity),

$$+\circlearrowleft\left(\sum M\right)_B = 0: \qquad M_{B^-} - (M_{B^-} + \Delta M_B) - M_0 = 0$$

or

$$\boxed{\Delta M_B = -M_B} \tag{5.5}$$

Equations 5.2 and 5.3 are differential equations relating the distributed load $p(x)$ to the shear force $V(x)$, and the shear force $V(x)$ to the bending moment $M(x)$. Let $x_1 \le x \le x_2$ be a portion of the beam that is free of concentrated forces or couples (Fig. 5.9). We can integrate Eq. 5.2 over this portion of the beam to get

$$\int_{x_1}^{x_2}\frac{dV}{dx}\,dx = V_2 - V_1 = \int_{x_1}^{x_2} p(x)\,dx$$

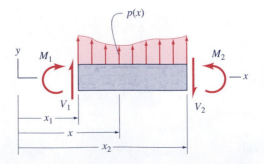

FIGURE 5.9 A free-body diagram of a finite portion of a beam.

or

$$V_2 - V_1 = \int_{x_1}^{x_2} p(x)\, dx \tag{5.6}$$

Similarly, from Eq. 5.3,

$$M_2 - M_1 = \int_{x_1}^{x_2} V(x)\, dx \tag{5.7}$$

Equations 5.6 and 5.7 can be stated in words as follows:

- *The **change in shear** from Section 1 to Section 2 is equal to the **area under the load curve** from 1 to 2.* (The "area" that results from negative $p(x)$ is negative.)
- *The **change in moment** from Section 1 to Section 2 is equal to the **area under the shear curve** from 1 to 2.* (The "area" that results from negative $V(x)$ is negative.)

Equations 5.2 through 5.7 will be very useful to us in Sections 5.4 and 5.5, where we draw shear and moment diagrams. And we can employ modifications of Eqs. 5.6 and 5.7 to determine expressions for $V(x)$ and $M(x)$. Thus,

$$V(x) = V_1 + \int_{x_1}^{x} p(\xi)\, d\xi \tag{5.8}$$

and

$$M(x) = M_1 + \int_{x_1}^{x} V(\xi)\, d\xi \tag{5.9}$$

Equations 5.2 through 5.9 will be used extensively in Section 5.5 in constructing shear and moment diagrams (Examples 5.7 through 5.9). Also, whenever shear and moment expressions must be obtained for a beam with a distributed load other than a simple uniform load or a triangular distributed load, it is much easier to use Eqs. 5.2 through 5.9 than to use a finite free-body diagram, as is illustrated in Example 5.6.

5.4 SHEAR-FORCE AND BENDING-MOMENT DIAGRAMS: EQUILIBRIUM METHOD

In Example 5.2 we obtained expressions for $V(x)$ and $M(x)$ for a simply supported beam with distributed loading. However, to *design* a beam (i.e., to select a beam of appropriate material and cross section) we need to ask questions like "What is the maximum value of the shear force, and where does it occur?" and "What is the maximum value of the bending moment, and where does it occur?" These questions are much more readily answered if we have a plot of $V(x)$ and a plot of $M(x)$. These plots are called the **shear diagram** and the **moment diagram,** respectively.

In Sections 5.4 and 5.5, two methods for constructing shear and moment diagrams are described:

- Method 1—**Equilibrium Method - (Section 5.4):** Use finite free-body diagrams or Eqs. 5.8 and 5.9 to obtain shear and moment functions, $V(x)$ and $M(x)$; then plot these expressions.
- Method 2—**Graphical Method - (Section 5.5):** Make use of Eqs. 5.2 through 5.7 to sketch $V(x)$ and $M(x)$ diagrams (see Table 5.1).

As you study the example problems in these two sections, observe that **maximum positive and negative bending moments** *can occur at any of the following cross sections of a beam*: (1) a cross section where the shear force is zero (Examples 5.4 and 5.9); (2) a cross section where a concentrated couple is applied (Examples 5.5, 5.6, and 5.8); (3) a cross section where a concentrated load is applied and where the shear force changes sign (Example 5.7); and (4) a point of support where there is a reaction force and where the shear force changes sign (Example 5.9).

The following examples illustrate the two procedures. Examples 5.4 through 5.6 illustrate the *Equilibrium Method;* Examples 5.7 through 5.9 illustrate the *Graphical Method.* A third method, the *Discontinuity-Function Method,* is presented in Section 5.6.

Shear-Force and Bending-Moment Diagrams: Equilibrium Method.

Examples 5.4 through 5.6 illustrate the **Equilibrium Method** for constructing shear and moment diagrams.

EXAMPLE 5.4

Figure 1 shows the simply supported beam of Example 5.2, including the reactions. The expressions for $V(x)$ and $M(x)$ obtained in Example 5.2 are

$$V_1 = (220 - 40x)\,\text{lb}, \quad 0 < x \le 6\,\text{ft}$$

$$V_2 = \left[-140 + \tfrac{10}{3}(12 - x)^2\right]\text{lb}, \quad 6 \le x < 12\,\text{ft}$$

$$M_1 = (220x - 20x^2)\,\text{lb}\cdot\text{ft}, \quad 0 < x \le 6\,\text{ft}$$

$$M_2 = \left[140(12 - x) - \tfrac{10}{9}(12 - x)^3\right]\text{lb}\cdot\text{ft}, \quad 6 \le x < 12\,\text{ft}$$

(a) Using the above expressions for $V(x)$ and $M(x)$, plot shear and moment diagrams for this simply supported beam. (b) Determine the location of the section of maximum bending moment, and calculate the value of the maximum moment.

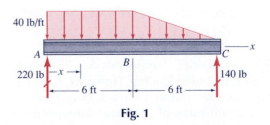

Fig. 1

Plan the Solution It is straightforward to plot the shear and moment diagrams from the given expressions (e.g., using a computer). From the shear diagram, the location of the section where $V(x) = 0$ can be determined. Then the appropriate moment equation can be used to determine the value of the moment at this critical section. Since there is more load over the left half of the beam than over the right half, the maximum moment should occur to the left of $x = 6$ ft.

Solution   (a) *Plot the shear diagram and the moment diagram.* Figure 2 shows the plots of shear and bending moment.

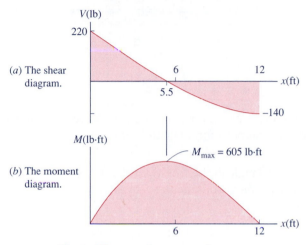

(a) The shear diagram.

(b) The moment diagram.

Fig. 2   Shear and moment diagrams.

(b) *Determine the maximum bending moment.* The maximum moment occurs where the shear vanishes. From the shear diagram and the equation for $V_1(x)$, it can be observed that $V(x) = 0$ in the interval $0 < x < 6$ ft. Therefore,

$$V_1(x_m) = 220 - 40x_m = 0 \rightarrow x_m = 5.50 \text{ ft}$$

Then, the maximum moment is

$$M_{max} = M_1(5.50 \text{ ft}) = 605 \text{ lb} \cdot \text{ft} \qquad \textbf{Ans. (b)}$$

Review the Solution   The downward load on this simply supported beam bends it downward, so it is concave upward everywhere. This is consistent with the fact that the bending moment is positive for the entire length of the beam. As expected, since there is more load over the left half of the beam than over the right half, the maximum moment does occur to the left of $x = 6$ ft. This is an example of a maximum moment that occurs where $V = dM/dx = 0$.

Note how much easier it is to get a "feel" for the distribution of shear force and bending moment in a beam from shear-force and bending-moment diagrams than it is from the shear and moment functions alone.

EXAMPLE 5.5

Derive expressions for $V(x)$ and $M(x)$ for the cantilever beam with linearly varying load shown in Fig. 1. Use these expressions to plot shear and moment diagrams for this beam.

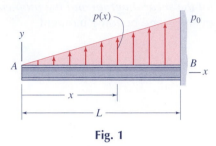

Fig. 1

Plan the Solution We can use a finite free-body diagram to determine the required expressions for $V(x)$ and $M(x)$.

Solution The triangle in the free-body diagram in Fig. 2 is similar to the triangle in the problem statement; so, by similar triangles,

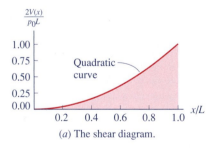

Fig. 2 A free-body diagram.

$$\frac{p(x)}{x} = \frac{p_0}{L} \rightarrow p(x) = \left(\frac{x}{L}\right)p_0$$

$$+\uparrow\sum F_y = 0: \qquad \left(\frac{x}{2}\right)p(x) - V(x) = 0$$

$$V(x) = \frac{p_0 x^2}{2L} \qquad \textbf{Ans.}$$

$$+\,\circlearrowleft\left(\sum M\right)_a = 0: \qquad [p(x)]\left(\frac{x}{2}\right)\left(\frac{x}{3}\right) - M(x) = 0$$

$$M(x) = \frac{p_0 x^3}{6L} \qquad \textbf{Ans.}$$

To plot these expressions for the shear force $V(x)$ and bending moment $M(x)$, we first calculate $V(L)$ and $M(L)$. Figure 3 shows the plots of $V(x)$ and $M(x)$.

$$V(L) = \frac{p_0 L}{2}, \quad M(L) = \frac{p_0 L^2}{6}$$

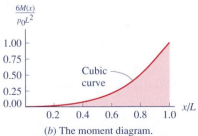

Fig. 3 The shear diagram and the moment diagram.

Review the Solution The shear is positive everywhere, as we expect from the free-body diagram. The maximum shear occurs at the cantilever support at B and is equal to the <u>total area under the load curve</u> in Fig. 1. The upward load will bend the beam upward, so it will be concave upward everywhere. This is consistent with the fact that the bending moment is positive for the entire length of the beam. The maximum bending moment occurs at B. Finally, the shear and moment have the proper dimensions, (F) and $(F \cdot L)$, respectively.

324

EXAMPLE 5.6

A cantilever airplane wing (represented as a beam in Fig. 1) has a distributed load given by

$$p(x) = p_0\left[1 - \left(\frac{x}{L}\right)^3\right]$$

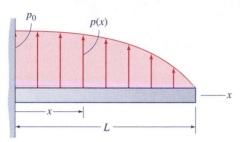

Fig. 1 An airplane wing.

Derive expressions for $V(x)$ and $M(x)$ for the cantilever wing. Use these expressions to plot shear and moment diagrams for this wing.

Plan the Solution It would be difficult to use a finite free-body diagram to determine the required expressions for $V(x)$ and $M(x)$, as was done in Example 5.4, since the load is not represented by a simple expression. Therefore, we will make use of Eqs. 5.8 and 5.9 to obtain expressions for $V(x)$ and $M(x)$. First, however, it is a good idea to draw a free-body diagram to define terms.

Solution Figure 2 shows a free-body diagram of the wing from cross-section x to the tip of the wing. All important terms are labeled on this free-body diagram. At the wingtip, $x = L$, there is no shear or bending moment, so $V(L) = M(L) = 0$. From Eq. 5.8, we can write the following expression for $V(x)$:

$$V(L) = V(x) + \int_x^L p(\xi)d\xi$$

where $p(\xi)$ is just the given load function with dummy variable ξ substituted for x. Since $V(L) = 0$,

$$V(x) = -\int_x^L p(\xi)d\xi$$

$$= -p_0\int_x^L\left[1 - \left(\frac{\xi}{L}\right)^3\right]d\xi$$

$$= p_0L\left[\frac{1}{4}\left(\frac{\xi}{L}\right)^4 - \frac{\xi}{L}\right]_x^L$$

So,

$$V(x) = p_0L\left[-\frac{3}{4} + \frac{x}{L} - \frac{1}{4}\left(\frac{x}{L}\right)^4\right] \qquad \textbf{Ans.}$$

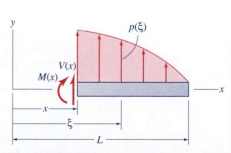

Fig. 2 A free-body diagram.

325

Similarly, from Eq. 5.9,

$$M(L) = M(x) + \int_x^L V(\xi)d\xi$$

Since $M(L) = 0$,

$$M(x) = -\int_x^L V(\xi)d\xi$$

$$= -p_0 L \int_x^L \left[-\frac{3}{4} + \frac{\xi}{L} - \frac{1}{4}\left(\frac{\xi}{L}\right)^4 \right] d\xi$$

$$= p_0 L^2 \left[\frac{3}{4}\left(\frac{\xi}{L}\right) - \frac{1}{2}\left(\frac{\xi}{L}\right)^2 + \frac{1}{20}\left(\frac{\xi}{L}\right)^5 \right]_x^L$$

So,

$$M(x) = p_0 L^2 \left[\frac{3}{10} - \frac{3}{4}\left(\frac{x}{L}\right) + \frac{1}{2}\left(\frac{x}{L}\right)^2 - \frac{1}{20}\left(\frac{x}{L}\right)^5 \right] \qquad \textbf{Ans.}$$

The shear and moment at the wing root, $V(0)$ and $M(0)$, are:

$$V(0) = -\frac{3}{4}p_0 L, \quad M(0) = \frac{3}{10}p_0 L^2$$

Finally, a computer was used to plot the above expressions for $V(x)$ and $M(x)$ that are shown in Fig. 3.

Review the Solution The shear is negative everywhere, as we expect from the free-body diagram. The maximum shear (magnitude) occurs at the wing root A and is equal to the total area under the load curve in the problem statement. The magnitude, $(3/4)p_0 L$, is reasonable compared with the value $p_0 L$ for a uniform load. The upward load will bend the wing upward, so it will be concave upward everywhere. This is consistent with the fact that the bending moment is positive for the entire length of the wing. The maximum bending moment occurs at the wing root A. Finally, the shear and moment have the proper dimensions, (F) and $(F \cdot L)$, respectively.

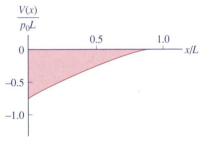

(a) The shear diagram

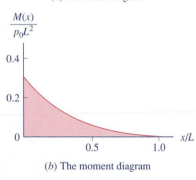

(b) The moment diagram

Fig. 3 Shear and moment diagrams.

5.5 SHEAR-FORCE AND BENDING-MOMENT DIAGRAMS: GRAPHICAL METHOD

In the previous section, Examples 5.4 through 5.6 illustrated how the *equilibrium method* can be used to construct shear-force and bending-moment diagrams. In this section, you will see how the *graphical method* facilitates the drawing of shear and moment diagrams.

Shear-Force and Bending-Moment Diagrams: Graphical Method. Examples 5.7 through 5.9 illustrate the **graphical method** for constructing shear and moment diagrams. Table 5.1 illustrates how Eqs. 5.2 through 5.7 are used in constructing (i.e., sketching) the diagrams, <u>proceeding from the left end to the right end of the beam</u>. If you study carefully the numbered steps that are given in the following examples and relate them to the entries in Table 5.1, you should be able to

TABLE 5.1 Shear and Moment Diagram Features

Equation	Load Diagram p	Shear Diagram V	Moment Diagram M
1. Slope of shear diagram equals value of load			
$\dfrac{dV}{dx} = p(x)$ (Eq. 5.2)	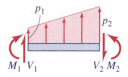	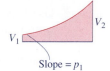	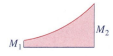
2. Jump in shear equals value of concentrated load			
$\Delta V = P_0$ (Eq. 5.4)	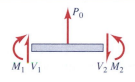	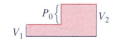	
3. Change in shear equals area under distributed-load diagram			
$V_2 - V_1 = \displaystyle\int_{x_1}^{x_2} p(x)\,dx$ (Eq. 5.6)	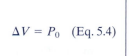	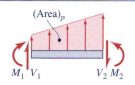	
4. Slope of moment diagram equals value of shear			
$\dfrac{dM}{dx} = V(x)$ (Eq. 5.3)	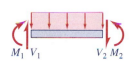	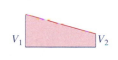	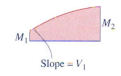
5. Jump in moment equals – (value of concentrated couple)			
$\Delta M = -M_0$ (Eq. 5.5)	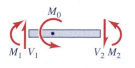		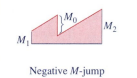
6. Change in moment equals area under shear diagram			
$M_2 - M_1 = \displaystyle\int_{x_1}^{x_2} V(x)\,dx$ (Eq. 5.7)	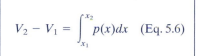	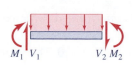	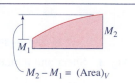

327

construct shear and moment diagrams for any beam with simple loading, once you have determined the loads and reactions acting on the beam.[2] There are also several MDSolids examples that employ the graphical method.

EXAMPLE 5.7

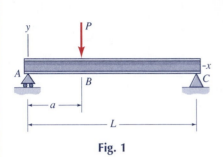

Fig. 1

Use Eqs. 5.2 through 5.7 to sketch shear and moment diagrams for the simply-supported beam shown in Fig. 1.

Plan the Solution   We can use a free-body diagram of the beam AC to determine the reactions at A and C. Since there is no distributed load on the beam, $p(x) = 0$ everywhere. Because of the concentrated load at B. we need to consider two spans, $0 < x < a$ and $a < x < L$.

Solution

Equilibrium—Reactions: To determine the reactions A_y and C_y, we first draw the free-body diagram of the entire beam AC (Fig. 2).

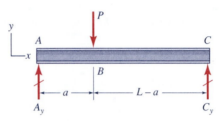

Fig. 2   A free-body diagram.

$$+\circlearrowleft\left(\sum M\right)_A = 0: \quad Pa - C_yL = 0, \quad C_y = P\left(\frac{a}{L}\right)$$

$$+\circlearrowleft\left(\sum M\right)_C = 0: \quad A_yL - P(L - a) = 0, \quad A_y = \frac{P(L - a)}{L}$$

Shear Diagram: Equations 5.2, 5.4, and 5.6 involve the shear. Using these equations, we can sketch $V(x)$ progressively from $x = 0$ to $x = L$. It is convenient to sketch the shear and moment diagrams directly below the load diagram (Fig. 3a). Each step involved in sketching $V(x)$ is numbered in Fig. 3b.

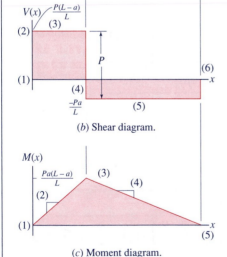

(a) Load diagram.

(b) Shear diagram.

(c) Moment diagram.

Fig. 3   Shear and moment diagrams.

1. The shear at $x = 0^-$ is zero.
2. The shear at $x = 0^+$ is determined from Eq. 5.4, that is, $\Delta V_A = A_y = \dfrac{P(L - a)}{L}$. Note that, because of the sign convention for shear, *an upward concentrated force causes an upward jump in the shear diagram.*
3. For $0 < x < a, p(x) = 0$. Therefore, from Eq. 5.2, $dV/dx = 0$.

---

[2]The graphical method is most useful when the "areas" in Eqs. 5.6 and 5.7 are simple rectangles or triangles, that is, when the loads on the beam are either concentrated loads or uniform distributed loads. The graphical method is also useful in interpreting the results of an equilibrium-method solution.

4. At $x = a$ there is a downward force P, so $\Delta V_B = -P$.

5. For $a < x < L$, $p(x) = 0$, so $\dfrac{dV}{dx} = 0$.

6. The reaction at C causes $\Delta V_C = \dfrac{Pa}{L}$, which closes the shear diagram back to zero at $x = L^+$.

Moment Diagram: Equations 5.3, 5.5, and 5.7 relate to $M(x)$ and can be used to sketch the moment diagram in Fig. 3c. Steps in the construction of the moment diagram are explained and numbered.

1. The moment at $x = 0$ is zero [simply supported beam].

2. For $0 < x < a$, Eq. 5.3 gives $\dfrac{dM}{dx} = V(x) = \dfrac{P(L - a)}{L} = $ constant.

3. At $x = a$, $M(a)$ can be determined from Eq. 5.7 as the area of the rectangle under the shear curve from $x = 0$ to $x = a$. Therefore
$$M(a) - 0 = \int_0^a V(x)\,dx = \dfrac{P(L - a)}{L}(a). \quad M(a) = \dfrac{Pa(L - a)}{L} \text{ is the}$$
maximum bending moment.

4. For $a < x < L$, Eq. 5.3 gives $\dfrac{dM}{dx} = V(x) = \dfrac{-Pa}{L} = $ constant.

5. Equation 5.7 gives $M(L) - M(a) = \displaystyle\int_a^L V(x)\,dx = -\dfrac{Pa}{L}(L - a)$,

   which closes the moment diagram back to zero at $x = L$, as it should [simple support at C].

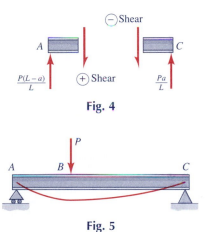

Fig. 4

Review the Solution   The dimensions on the shear diagram (F) and the moment diagram ($F \cdot L$) are correct. If we draw finite free-body diagrams of the ends of the beam, we get Fig. 4. Therefore, the shear diagram in Fig. 3b has the correct signs according to the free-body sketches in Fig. 4 and the sign convention in Fig. 5.6. The downward force will bend the beam as shown in Fig. 5, which is consistent with the fact that the bending moment is positive everywhere. The maximum bending moment occurs at the cross section where the force P is applied and where the shear force changes sign.

Fig. 5

MDS5.1 – 5.6   **V & M Diagrams**—*Determinate Beams* is a computer program module that may be used to plot shear-force and bending-moment diagrams for statically determinate beams. The solutions of MDS Examples 5.1–5.6 illustrate the *graphical method* for constructing V and M diagrams.

EXAMPLE 5.8

Use Eqs. 5.2 through 5.7 to sketch shear and moment diagrams for simply supported beam shown in Fig. 1.

Plan the Solution   We can use a free-body diagram of the beam AC in Fig. 2 to determine the reactions at A and C. Since there is no distributed

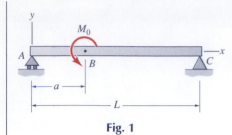

Fig. 1

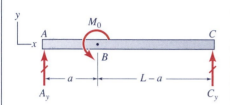

Fig. 2   A free-body diagram.

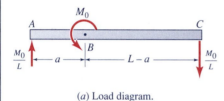

(a) Load diagram.

$V(x)$

(2) (3)

$\dfrac{M_0}{L}$

(4)

(1)

(b) Shear diagram.

$M(x)$

(2) (3) $\dfrac{M_0 a}{L}$

(1)

(6)

$-\dfrac{M_0(L-a)}{L}$ (5)

(4)

(c) Moment diagram.

Fig. 3   Load, shear, and moment diagrams.

load on the beam, $p(x) = 0$ everywhere. Because of the concentrated couple at B, we need to consider two spans, $0 < x < a$ and $a < x < L$.

Solution

Equilibrium—Reactions: We first determine the reactions A_y and C_y.

$$+\circlearrowleft \left(\sum M \right)_A = 0: \qquad -M_0 - C_y L = 0 \to C_y = \frac{-M_0}{L}$$

$$+\circlearrowleft \left(\sum M \right)_C = 0: \qquad A_y L - M_0 = 0 \to A_y = \frac{M_0}{L}$$

Shear Diagram: Equations 5.2, 5.4, and 5.6 involve the shear. Using these equations and the load diagram in Fig. 3a, we can sketch $V(x)$ progressively from $x = 0$ to $x = L$. It is convenient to sketch the shear and moment diagrams directly below the load diagram. Each step involved in sketching $V(x)$ is numbered in Fig. 3b.

1. The shear at $x = 0^-$ is zero.
2. The shear at $x = 0^+$ is determined from Eq. 5.4, that is, $\Delta V_A = A_y = M_0/L$. Note that, because of the sign convention for shear, *an upward concentrated force causes an upward jump in the shear diagram.*
3. For $0 < x < L$, $p(x) = 0$. Therefore, from Eq. 5.2, $dV/dx = 0$, $V(x) = M_0/L = $ constant.
4. The reaction at C causes $\Delta V_C = -M_0/L$, which closes the shear diagram back to zero at $x = L^+$.

Moment Diagram: Equations 5.3, 5.5, and 5.7 relate to $M(x)$ and can be used to sketch the moment diagram in Fig. 3c. Steps in the construction of the moment diagram are explained and numbered.

1. The moment at $x = 0$ is zero (simply supported beam).
2. For $0 < x < a$, Eq. 5.3 gives $dM/dx = V(x) = M_0/L = $ constant.
3. At $x = a^-$, $M(a^-)$ can be determined from Eq. 5.7 as the area of the rectangle under the shear curve from $x = 0$ to $x = a$. Therefore $M(a) - 0 = \int_0^a V(x)\,dx = M_0 a/L$.
4. At $x = a$ there is a negative jump in moment given by Eq. 5.5. So

$$M(a^+) = \frac{M_0 a}{L} - M_0 = -\frac{M_0(L-a)}{L}.$$

5. For $a < x < L$, Eq. 5.3 gives $dM/dx = V(x) = M_0/L = $ constant.
6. Equation 5.7 gives $M(L) - M(a^-) = \int_a^L V(x)\,dx = (M_0/L)(L - a)$, which closes the moment diagram back to zero at $x = L$, as it should [simple support at C].

Review the Solution   The dimensions on the shear diagram (F) and the moment diagram $(F \cdot L)$ are correct. Note that both the maximum moment, $M_0 a/L$, and the minimum bending moment, $-M_0(L - a)/L$, occur at the cross section where the concentrated couple acts. Compare this example with the previous one, where there was a concentrated force.

Determine the reactions and sketch the shear and moment diagrams for the beam shown in Fig. 1. (This beam is said to have an *overhang BC.*) Show all significant values (that is, maxima, minima, positions of maxima and minima, etc.) on the diagrams.

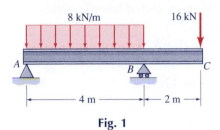

Fig. 1

Plan the Solution   We can use a free-body diagram of the whole beam to compute the reactions. Then we can use Eqs. 5.2 through 5.7 to sketch the $V(x)$ and $M(x)$ diagrams, as we did in Examples 5.7 and 5.8.

Solution

Equilibrium—Reactions: The reactions must be determined first. Figure 2 shows the appropriate free-body diagram.

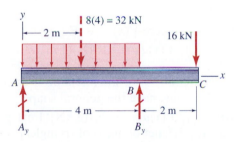

Fig. 2   A free-body diagram.

$$+\,\circlearrowleft\left(\sum M\right)_A = 0:$$

$$(8 \text{ kN/m})(4 \text{ m})(2 \text{ m}) + (16 \text{ kN})(6 \text{ m}) - B_y(4 \text{ m}) = 0$$

$$B_y = 40 \text{ kN} \qquad\qquad \textbf{Ans.}$$

$$+\,\circlearrowleft\left(\sum M\right)_B = 0:$$

$$A_y(4 \text{ m}) - (8 \text{ kN/m})(4 \text{ m})(2 \text{ m}) + (16 \text{ kN})(2 \text{ m}) = 0$$

$$A_y = 8 \text{ kN} \qquad\qquad \textbf{Ans.}$$

Check: Is $\sum F_y = 0$? $8 - 32 + 40 - 16 = 0$?   Yes

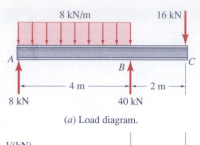

(a) Load diagram.

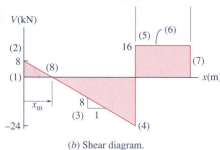

(b) Shear diagram.

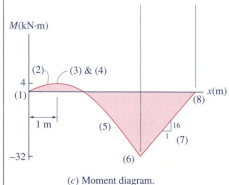

(c) Moment diagram.

Fig. 3 Shear and moment diagrams.

It is convenient to sketch the V (Fig. 3b) and M (Fig. 3c) diagrams directly below a sketch of the beam that has all of the loads and reactions shown (Fig. 3a).

Shear Diagram: The following steps are used in sketching the shear diagram (Fig. 3b).

1. $V(0^-) = 0$ [no shear at end of beam].
2. $V(0^+) = 8$ kN [Eq. 5.4].
3. $dV/dx = -8$ kN/m [Eq. 5.2].
4. $V(4^-) = V(0^+) + (-8 \text{ kN/m})(4 \text{ m}) = 8 - 32 = -24$ kN [Eq. 5.6].
5. $V(4^+) - V(4^-) = 40$ kN [Eq. 5.4].
6. $dV/dx = 0$ [Eq. 5.2].
7. $V(6^+) = V(6^-) - 16 = 0$.
8. Since $dV/dx = -8$ for $0 < x < 4$ m, and since $V(0^+) = 8$ kN, $V = 0$ at $x_m = 1$ m [Eq. 5.6].

Moment Diagram: The steps employed in constructing the moment diagram (Fig. 3c) using Eqs. 5.2 through 5.7 will now be described:

1. $M(0) = 0$ [no moment at end of beam].
2. From $dM/dx = V(x)$ we have the slope of $M(x)$ going from $+8$ kN · m/m at $x = 0^+$ to zero at $x = x_m = 1$ m. Therefore, the moment diagram for $0 < x < 1$ m must have the general shape [Eq. 5.3].
3. $M(x)$ is maximum where $V(x) = 0$ [Eq. 5.3].
4. $M(1) = M(0) + \int_0^1 V(x)dx = \frac{1}{2}(8 \text{ kN})(1 \text{ m}) = 4$ kN · m [Eq. 5.7; area of triangle].
5. From $x = 1$ m to $x = 4$ m, $V(x)$ gets progressively more negative. Therefore, $M(x)$ must have the general shape [Eq. 5.3].
6. $M(4) = M(0) + \int_0^4 V(x)dx = 0 + \frac{1}{2}(8 \text{ kN})(1 \text{ m}) + \frac{1}{2}(-24 \text{ kN})(3 \text{ m}) = -32$ kN · m [Eq. 5.7; net of areas of triangles]
7. $dM/dx = V(x) = 16$ kN [Eq. 5.3].
8. $M(6) = M(4) + \int_4^6 V(x)dx = -32 \text{ kN} \cdot \text{m} + (16 \text{ kN})(2 \text{ m}) = 0$ [Eq. 5.7; no moment at end of beam].

The maximum shear occurs just to the left of the support at B and has a magnitude of 24 kN. The maximum positive moment occurs where $V = 0$ at $x = 1$ m and has a magnitude of 4 kN · m; and the maximum negative moment occurs at the support B, and it has a magnitude of 32 kN · m.

Review the Solution By imagining cuts just to the right of A (Fig. 4a), just to the left of B (Fig. 4b), and just to the right of B (Fig. 4c), we can check the sign of the shear at these points.

The moment diagram is best checked by seeing if the sign of the moment diagram corresponds to a reasonable deflected shape, that is, concave upward where $M(x)$ is positive and concave downward where

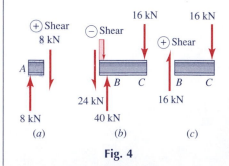

Fig. 4

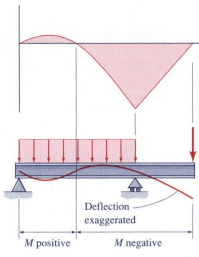

Fig. 5 A sketch showing the deflection of beam *AC*.

$M(x)$ is negative, according to the sign convention that is given Fig. 5.6c. Where $M(x) = 0$, the beam is locally straight, that is, it is neither concave upward nor concave downward. We are able to sketch (Fig. 5) a plausible deflection curve that passes over the supports at A and B and that is concave upward where $M(x)$ is positive and concave downward where M is negative. The distributed load between A and B and the concentrated load at C could, indeed, cause the beam to deflect as sketched.

The idea that a positive bending moment makes a beam concave toward the $+y$ side, whereas a negative bending moment causes the beam to be concave toward the $-y$ side, is in accord with the definition of positive bending moment in Fig. 5.6c. This fact was used in Examples 5.7 and 5.9 above to check the bending moment diagrams. In the next chapter we derive a mathematical relationship between bending moment and curvature, and in Chapter 7 we use this relationship to obtain expressions for the deflection of beams.

*5.6 DISCONTINUITY FUNCTIONS TO REPRESENT LOADS, SHEAR, AND MOMENT

Wherever there is a discontinuity in the loading on a beam or where there is a support, there will be a discontinuity in integrals that involve the loads and reactions. Between these discontinuities the integrals will be continuous. For example, the beam in Fig. 5.10 has four intervals—AB, BC, CD, and DE. Consequently, four separate expressions $V_i(x)$ and four expressions $M_i(x)$ would be required to specify $V(x)$ and $M(x)$ for this beam. The introduction of discontinuity functions simplifies the process of determining expressions for shear and bending moment (Section 5.6), and it greatly simplifies the process of solving for the slope and deflection of a beam (Section 7.5). For example, for the beam in Fig. 5.10, the shear $V(x)$ can be written as a single compact expression, valid for $0 \leq x \leq L$, rather than as four separate expressions. Similarly, the load $p(x)$ and the moment $M(x)$ can each be written as a single expression that is valid for $0 \leq x \leq L$.

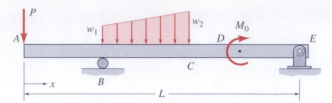

FIGURE 5.10   A beam with several applied loads.

Macaulay functions will be used to represent distributed loads on beams, and *singularity functions* will be used to represent concentrated external forces and concentrated couples. Together, they are referred to as **discontinuity functions,** and are symbolized by angle brackets, called *Macaulay brackets,*[3] that have the form: $\langle x - a \rangle^n$.

Macaulay Functions. Macaulay functions are useful in expressing functions that are zero up to some particular value of the independent variable and nonzero for larger values of the independent variable. For integer values of $n \geq 0$, the Macaulay functions $\langle x - a \rangle^n$ are defined by the following expressions:

$$\langle x - a \rangle^n = \begin{cases} 0 & \text{for } x < a \\ (x - a)^n & \text{for } x \geq a \end{cases} \quad n = 0, 1, 2, \ldots \tag{5.10}$$

For example, Fig. 5.11 shows the *unit step function* $\langle x - a \rangle^0$ and the *unit ramp function* $\langle x - a \rangle^1$, where a is the value of the independent variable x at which the discontinuity occurs. As illustrated by the unit ramp function in Fig. 5.11*b*, these functions have the value zero for $x < a$ and the value $(x - a)^n$ for $x \geq a$. The units of $\langle x - a \rangle^n$ are the units of x^n (e.g., ft^n, m^n, etc.).

Singularity Functions. The two singularity functions of interest here are the *unit doublet function,* $\langle x - a \rangle^{-2}$, which can be used to represent a concentrated couple, and the *unit impulse function,* $\langle x - a \rangle^{-1}$, which can be used to represent a

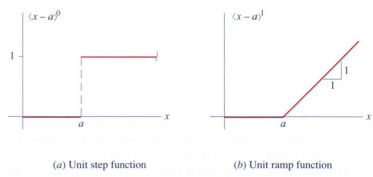

(*a*) Unit step function   (*b*) Unit ramp function

FIGURE 5.11   Examples of Macaulay functions.

[3]The English mathematician W. H. Macaulay (1857–1936) introduced the use of special brackets to represent these discontinuity functions. It has been common practice to use angle brackets for this purpose and to refer to them as *Macaulay brackets.*

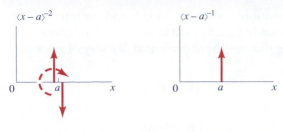

(a) Unit doublet function.   (b) Unit impulse function.

FIGURE 5.12   Examples of singularity functions.

concentrated force. These are illustrated in Fig. 5.12. Singularity functions are similar in some respects to the Macaulay functions, but they are defined for negative values of n. They become singular (i.e., infinite) at $x = a$, and they are zero for $x \neq a$. The units of singularity functions are the same as the units of x^n, even though n is negative.

Integrals of Discontinuity Functions. The rules of integration for discontinuity functions are given in Eq. 5.11.

$$\int \langle x - a \rangle^n \, dx = \begin{cases} \langle x - a \rangle^{n+1} & \text{for } n \leq 0 \\ \dfrac{1}{n+1} \langle x - a \rangle^{n+1} & \text{for } n > 0 \end{cases} \qquad (5.11)$$

Since we will use discontinuity functions in conjunction with beams of finite length with their origin at $x = 0$, as illustrated in Figs. 5.10 and 5.13, we need not be concerned with values of x less than $x = 0$. The origin, $x = 0$, will therefore be assumed to be located such that, for $x < 0, \langle x - a \rangle^n \equiv 0$ for all n.

Use of Discontinuity Functions to Represent Loads, Shear, and Moment.
Figure 5.13 illustrates how Macaulay functions can be used to represent distributed loads on beams. It must be remembered that Macaulay functions continue indefinitely for $x > a$. Therefore, when a particular load pattern terminates at some value of x, a new Macaulay function must be introduced to cancel out the effect of that previous Macaulay function. Note how the dimensions of each term in $p(x)$ are preserved even though the term $\langle x - a \rangle^0$ is dimensionless whereas the term $\langle x - a \rangle^1$ has the dimensions of length.

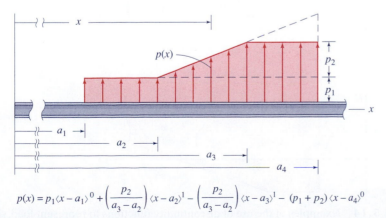

$$p(x) = p_1 \langle x - a_1 \rangle^0 + \left(\frac{p_2}{a_3 - a_2} \right) \langle x - a_2 \rangle^1 - \left(\frac{p_2}{a_3 - a_2} \right) \langle x - a_3 \rangle^1 - (p_1 + p_2) \langle x - a_4 \rangle^0$$

FIGURE 5.13   Distributed loads represented by Macaulay functions.

The sign conventions for external loads and for internal shear force and bending moment are given in Figs. 5.6, and in Section 5.3 relationships among loads, shear force, and bending moment were presented. By way of review,

$$\Delta V = P_0 \tag{5.4} repeated$$

$$\Delta M = -M_0 \tag{5.5} repeated$$

$$V(x) = V_1 + \int_{x_1}^{x} p(\xi)d\xi \tag{5.8} repeated$$

$$M(x) = M_1 + \int_{x_1}^{x} V(\xi)d\xi \tag{5.9} repeated$$

To apply the latter two when $p(x)$ and $V(x)$ are represented by discontinuity functions, we let $x_1 = 0^-$ and let $V_1(0^-) = M_1(0^-) = 0$. Therefore, Eqs. 5.8 and 5.9 can be written as

$$V(x) = \int_0^x p(\xi)\,d\xi, \qquad M(x) = \int_0^x V(\xi)\,d\xi \tag{5.12}$$

The load function for the beam in Fig. 5.14 can be written in terms of discontinuity functions by referring to Table 5.2; and we can use Eqs. 5.11 to integrate these

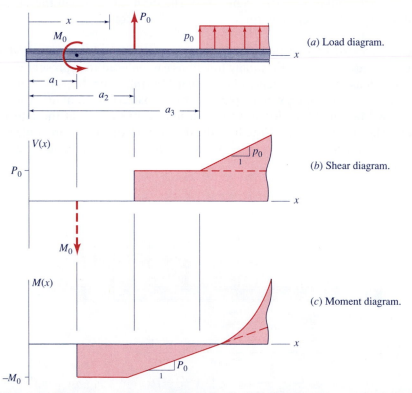

FIGURE 5.14 Examples of the use of discontinuity functions to represent load, shear, and moment.

TABLE 5.2 A Summary of Loads, Shear, and Moment Represented by Discontinuity Functions

Case	Load	Shear	Moment
1	$-M_0 \langle x - a \rangle^{-2}$	$-M_0 \langle x - a \rangle^{-1}$	$-M_0 \langle x - a \rangle^{0}$
2	$P_0 \langle x - a \rangle^{-1}$	$P_0 \langle x - a \rangle^{0}$	$P_0 \langle x - a \rangle^{1}$
3	$p_0 \langle x - a \rangle^{0}$	$p_0 \langle x - a \rangle^{1}$	$\dfrac{p_0}{2} \langle x - a \rangle^{2}$
4	$\left(\dfrac{p_1}{b}\right)\langle x - a \rangle^{1}$	$\left(\dfrac{p_1}{2b}\right)\langle x - a \rangle^{2}$	$\left(\dfrac{p_1}{6b}\right)\langle x - a \rangle^{3}$
5	$\left(\dfrac{p_2}{b^2}\right)\langle x - a \rangle^{2}$	$\left(\dfrac{p_2}{3b^2}\right)\langle x - a \rangle^{3}$	$\left(\dfrac{p_2}{12b^2}\right)\langle x - a \rangle^{4}$
6	$\left(\dfrac{p_n}{b^n}\right)\langle x - a \rangle^{n}$	$\left(\dfrac{p_n}{b^n\,(n+1)}\right)\langle x - a \rangle^{n+1}$	$\left(\dfrac{p_n}{b^n\,(n+1)(n+2)}\right)\langle x - a \rangle^{n+2}$

discontinuity functions to get

$$p(x) = -M_0\langle x - a_1\rangle^{-2} + \underline{P_0\langle x - a_2\rangle^{-1}} + p_0\langle x - a_3\rangle^0$$

$$V(x) = \int_{0^-}^{x} p(\xi)\,d\xi = \underline{-M_0\langle x - a_1\rangle^{-1}} + P_0\langle x - a_2\rangle^0 + p_0\langle x - a_3\rangle^1 \quad (5.13)$$

$$M(x) = \int_{0^-}^{x} V(\xi)\,d\xi = -M_0\langle x - a_1\rangle^0 + P_0\langle a - a_2\rangle^1 + \frac{p_0}{2}\langle x - a_3\rangle^2$$

Note that the beam in Fig. 5.14 is shown to extend on the left to $x = 0^-$. Singularity functions are underscored in the above equations for $p(x)$ and $V(x)$.

Table 5.2 summarizes load, shear, and moment relationships represented by discontinuity functions. The terms in this table that represent singularity functions are underscored to emphasize the fact that, strictly speaking, singularities have infinite values at the point $x = a_i$ and zero values everywhere else. Whereas it is common practice to represent concentrated couples and forces in the manner indicated in Case 1 and Case 2 of the "Load" column of Table 5.2, the moment M_0 in the "Shear" column of Case 1 is not a true concentrated shear force. Its effect is to cause the jump in moment at $x = a$ in the "Moment" column of Case 1, but otherwise it can be ignored. Therefore, it is shown dashed in Table 5.2, Case 1. Note how the amplitude of the load function is defined in Cases 4 through 6 to assure proper dimensionality.

In this section, only statically determinate problems are considered. Therefore, the reactions are treated as known quantities that have been obtained by the use of equilibrium equations. In Section 7.5, both statically determinate problems and statically indeterminate problems are examined.

EXAMPLE 5.10

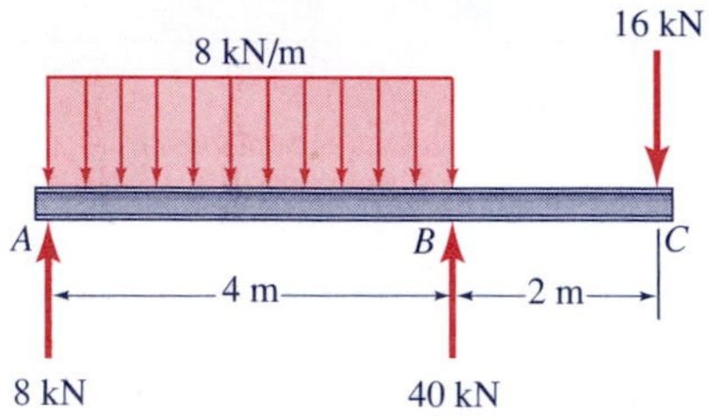

Fig. 1

For the beam in Fig. 1 of Example Problem 5.9, (a) use discontinuity functions to obtain expressions for $p(x)$, $V(x)$, and $M(x)$, and (b) use the discontinuity functions from Part (a) to construct shear and moment diagrams for the beam, indicating the contribution of each term in the discontinuity-function expressions.

The loads and reactions from Example Problem 5.9 are given in Fig. 1.

Plan the Solution We can refer to Cases 2 and 3 of Table 5.2 to construct the load function $p(x)$ and then to perform the required integrations to get $V(x)$ and $M(x)$.

Solution (a) By referring to the Load column for Cases 2 and 3 in Table 5.2, we can write

$$p(x) = (8\text{ kN})\langle x - 0\text{ m}\rangle^{-1} - (8\text{ kN/m})[\langle x - 0\text{ m}\rangle^0 - \langle x - 4\text{ m}\rangle^0]$$
$$+ (40\text{ kN})\langle x - 4\text{ m}\rangle^{-1} - (16\text{ kN})\langle x - 6\text{ m}\rangle^{-1} \quad \textbf{Ans.} \quad (1)$$

Integrating Eq. 1 by referring to Eqs. 5.11 or to the Shear column of Table 5.2, we get

$$V(x) = \int_{0-}^{x} P(\xi)\,d\xi:$$

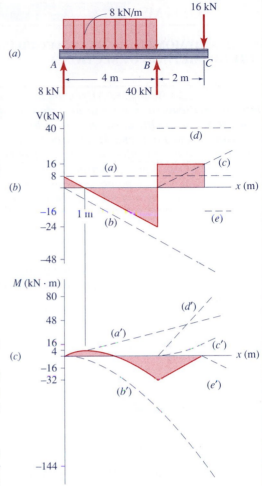

$$V(x) = \overset{(a)}{(8\text{ kN})\langle x\rangle^{0}} - \overset{(b)}{(8\text{ kN/m})}[\overset{(c)}{\langle x\rangle^{1} - \langle x - 4\text{ m}\rangle^{1}}]$$

$$+ \overset{(d)}{(40\text{ kN})\langle x - 4\text{ m}\rangle^{0}} - \overset{(e)}{(16\text{ kN})\langle x - 6\text{ m}\rangle^{0}} \quad \textbf{Ans.} \quad (2)$$

and, from the Moment column of Table 5.2, we get

$$M(x) = \int_{0-}^{x} V(\xi)\,d\xi:$$

$$M(x) = \overset{(a')}{(8\text{ kN})\langle x\rangle^{1}} - \overset{(b')}{(8\text{ kN/m})}[\overset{(c')}{\tfrac{1}{2}\langle x\rangle^{2} - \tfrac{1}{2}\langle x - 4\text{ m}\rangle^{2}}]$$

$$+ \overset{(d')}{(40\text{ kN})\langle x - 4\text{ m}\rangle^{1}} - \overset{(e')}{(16\text{ kN})\langle x - 6\text{ m}\rangle^{1}} \quad \textbf{Ans.} \quad (3)$$

(b) In Fig. 2, Eqs. (2) and (3) are plotted term-by-term, starting at the left end of the beam; the separate terms are then summed so the results can be compared with the shear diagram and the moment diagram obtained in Example Problem 5.9.

Review the Solution Since the shear diagram in Fig. 2b and the moment diagram in Fig. 2c both close to zero at the right end, our results are probably correct. For this problem we could use Eqs. 5.2 through 5.7 to check the shear and moment diagrams above. That is, the procedure used to construct the shear diagram and the moment diagram in Example Problem 5.9 can be used to check the results obtained by the discontinuity-function method.

Fig. 2 Load, shear, and moment diagrams.

Observe that it is a very straightforward procedure to obtain the discontinuity-function expressions for $V(x)$ and $M(x)$. Each term in $p(x)$, $V(x)$ and $M(x)$ can be evaluated separately, and the results summed to get the final discontinuity-function expressions. Likewise, graphs of $p(x)$, $V(x)$, and $M(x)$ can be easily constructed from the discontinuity-function expressions. This makes the discontinuity-function method an ideal one to serve as a basis for a computer program to evaluate and plot shear diagrams and moment diagrams. In Section 7.5 discontinuity functions will be used to solve beam deflection problems, including analysis of statically indeterminate beams.

MDS5.7–5.9 **V & M Diagrams**—*Determinate Beams* is a computer program module that may be used to plot shear-force and bending-moment diagrams for statically determinate beams. The *discontinuity-function method*, described in Section 5.6, was used in the development of the computer program module that solves the key equations and plots the shear and moment diagrams. The solutions of MDS Examples 5.7–5.9 illustrate this method.

▼ **SHEAR AND MOMENT IN BEAMS: EQUILIBRIUM METHOD (METHOD OF SECTIONS)**

> In Problems 5.2-1 through 5.2.11 *you are to determine the internal resultants (transverse shear force V and bending moment M) at specific cross sections of beams. Use the sign conventions for V and M given in Fig. 5.6, p. 314, and always draw complete and correct free-body diagrams. The minus-sign superscript signifies "just to the left of" the referenced point; the plus-sign superscript means "just to the right of" the referenced point.*

Prob. 5.2-1. For the simply supported beam AE in Fig. P5.2-1, (a) determine V_{C^-} and M_{C^-}, the internal resultants just to the left of the 20-kN load at C, and (b) determine V_{D^-} and M_{D^-}, the internal resultants just to the left of the 20-kN load at D.

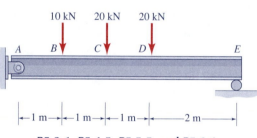

P5.2-1, P5.4-5, P5.5-5, and P5.6-1

Prob. 5.2-2. For the beam AD in Fig. P5.2-2, (a) determine the transverse shear force V_{B^-} and the bending moment M_{B^-} at a section just to the left of the support at B, and (b) determine V_E and M_E at section E.

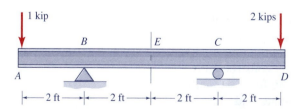

P5.2-2, P5.4-3, P5.5-3, and P5.6-2

Prob. 5.2-3. Transverse loads are applied to the beam in Fig. P5.2-3 at A, C, and E, and a concentrated couple $3Pa$ is applied to the beam at E. Determine expressions for (a) the transverse shear force V_{C^-} and bending moment M_{C^-} at a section just to the left of the load at C, and (b) shear V_{D^-} and moment M_{D^-} just to the left of the support at D. Express your answers in terms of P and a.

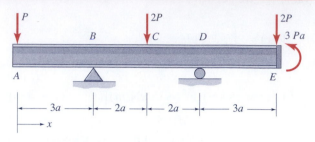

P5.2-3, P5.4-2, P5.5-2, and P5.6-3

Prob. 5.2-4. The shaft in Fig. P5.2-4 is supported by bearings at B and D that can only exert forces normal to the shaft. Belts that pass over pulleys at A and E exert parallel forces of 150 N and 300 N, respectively, as shown. Determine the transverse shear force V_C and the bending moment M_C at section C, midway between the two supports.

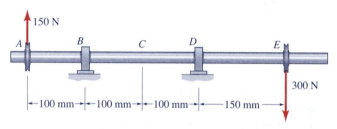

P5.2-4, P5.4-4, and P5.5-4

Prob. 5.2-5. Two transverse forces and a couple are applied as external loads to the cantilever beam AC in Fig. P5.2-5. Determine the transverse shear force V_C and the bending moment M_C at the fixed end C.

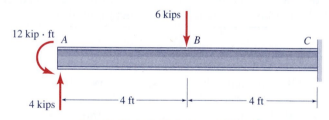

P5.2-5, P5.4-8, P5.5-8, and P5.6-4

Prob. 5.2-6. For the cantilever beam AD in Fig. P5.2-6, determine the reactions at D; that is determine V_D and M_D. Express your answers in terms of P and a.

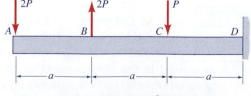

P5.2-6, P5.4-7, and P5.5-7

Prob. 5.2-7. A drilling engineer wishes to support pipes of length L on blocks so that the magnitude of the bending moment in the pipes directly over the supports is equal to the magnitude of the bending moment at the center of the pipes (see Fig. P5.2-7). Let w be the weight of the pipes per unit length, and neglect the width of the support blocks. (a) Determine the distance from each end, aL, at which the engineer should place the supports. (b) Determine the corresponding magnitude of the bending moments at the supports and at the center.

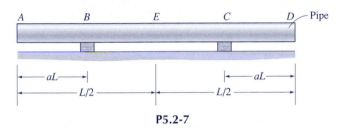

P5.2-7

Prob. 5.2-8. A uniformly distributed load of 1 kip/ft and a concentrated transverse load of 8 kips are applied to the simply supported beam in Fig. P5.2-8. Determine the transverse shear force V_D and the bending moment M_D at section D, midway between the supports.

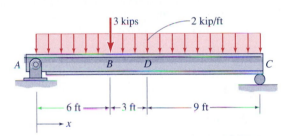

P5.2-8, P5.2-15, P5.4-9, P5.5-9, and P5.6-5

Prob. 5.2-9. A concentrated couple of 4 kN · m und a uniformly distributed load of 1 kN/m are applied to beam AE, as shown in Fig. P5.2-9. Determine the transverse shear force V_C and the bending moment M_C at section C, the middle of the beam.

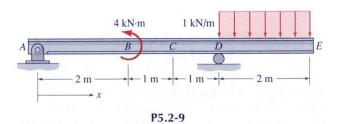

P5.2-9

Prob. 5.2-10. An L-shaped frame BD is welded to the beam AC in Fig. P5.2-10. A downward vertical load P is applied at D as shown. Neglecting the width of the connection at B, determine expressions for the following: (a) the transverse shear force V_{B^-} and bending moment M_{B^-} just to the left of

B, and (b) the transverse shear force V_{B^+} and bending moment M_{B^+} just to the right of B.

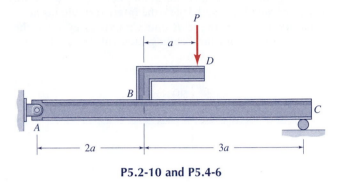

P5.2-10 and P5.4-6

Prob. 5.2-11. Two beam segments, AC and CD, are connected together at C by a frictionless pin as shown in Fig. P5.2-11. Segment CD is cantilevered from a rigid support at D, and segment AC has a roller support at A. (a) Determine the reactions at A and D. (b) Determine V_E and M_E, the internal resultants at section E, which is 5 ft to the left of the support D.

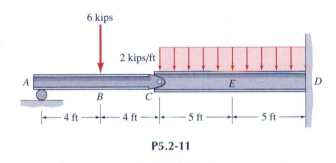

P5.2-11

> In Problems 5.2-12 through 5.2-25, you are to determine expressions for the internal resultants as functions of position x. That is, expressions for shear force V(x), bending moment M(x), and, occasionally, axial force F(x) are to be determined. *Always draw correct free-body diagrams.* Use the sign conventions for V and M given in Fig. 5.6 on p. 303, and let F be assumed positive in tension.

Prob. 5.2-12. For the beam in Fig. P5.2-12, determine the following: (a) V_A and M_A, the reactions at the cantilever end A: (b) $V_1(x)$ and $M_1(x)$, the internal resultants at an arbitrary section between A and B, that is, for $(0 < x \le 3L/4)$; and (c) $V_2(x)$ and $M_2(x)$, the internal resultants at an arbitrary section between B and C, that is, for $(3L/4 \le x \le L)$.

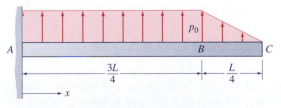

P5.2-12, P5.4-13, and P5.6-9

Prob. 5.2-13. For the cantilever beam AC in Fig. P5.2-13, (a) determine $V_1(x)$ and $M_1(x)$, the internal resultants at an arbitrary section between A and B, that is, for $(0 < x < 2$ m); (b) determine $V_2(x)$ and $M_2(x)$, the internal resultants at an arbitrary section between B and C; and (c) evaluate the shear force V_C and the bending moment M_C at the cantilever end C.

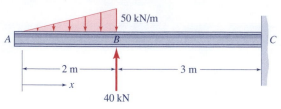

P5.2-13 and P5.4-18

Prob. 5.2-14. A linearly varying load of maximum intensity w_0 is applied to the simply supported beam AB in Fig. P5.2-14. Determine expressions for the transverse shear force $V(x)$ and the bending moment $M(x)$ at an arbitrary section x.

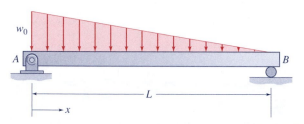

P5.2-14, P5.5-14, and P5.6-10

Prob. 5.2-15. For the beam in Fig. P5.2-8, determine the following: (a) $V_1(x)$ and $M_1(x)$, the internal resultants at an arbitrary section in "Interval 1" between A and B, that is, for $(0 < x < 6$ ft), and (b) $V_2(x)$ and $M_2(x)$, the internal resultants at an arbitrary section in "Interval 2" between B and C.

Prob. 5.2-16. The beam AB in Fig. P5.2-16 has a distributed load that varies linearly from w_0 at $x = 0$ to $2w_0$ at $x = L$. (a) Determine the reactions at A and B, and (b) determine $V(x)$ and $M(x)$, the internal resultants at an arbitrary section between A and B.

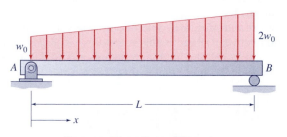

P5.2-16, P5.5-15, and P5.6-11

Prob. 5.2-17. The simply supported beam AD supports a concentrated load of 200N and a linearly varying load of maximum intensity 2 N/mm, as shown in Fig. P5.2-17.

Determine the following: (a) $V_1(x)$ and $M_1(x)$, the internal resultants at an arbitrary section between A and B, that is, for $(0 < x < 150$ mm); and (b) $V_2(x)$ and $M_2(x)$, the internal resultants at an arbitrary section between B and C.

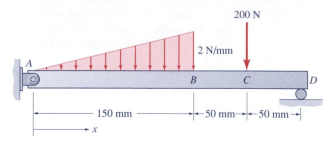

P5.2-17 and P5.4-15

Prob. 5.2-18. A hanger bar supports a load of 450 N and is, in turn, supported by a pin AF that passes through the hanger bar and a support bracket, as shown in Fig. P5.2-18a. Assume that the pin is a beam subjected to distributed loading from the bracket and the hanger rod as indicated in Fig. P5.2-18b. (a) Determine the maximum distributed-load intensities, p_t and p_b, on the top and bottom of the pin, respectively, (b) Determine expressions for $V_3(x)$ and $M_3(x)$, the shear and moment in the pin in the span between C and D, that is, for $(25$ mm $\leq x \leq 45$ mm).

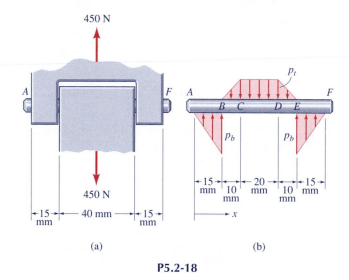

P5.2-18

Prob. 5.2-19. A boy and girl position themselves as shown in Fig. P5.2-19, so that their canoe remains level. Assume that the canoe behaves like a beam, and assume that the buoyant force of the water is uniformly distributed over the 16-ft waterline length, (a) If the weight of the boy is $W_b = 200$ lb, determine the weight, W_g, of the girl, (b) Determine the buoyant force per unit length, p_0. (c) Determine expressions for $V(x)$ and $M(x)$ at an arbitrary cross section of the canoe between B and D, that is, for $(-3$ ft $< x < 5$ ft).

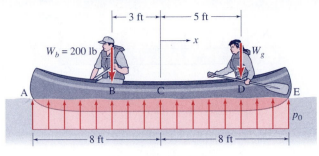

$W_b = 200$ lb

P_0

3 ft — 5 ft

8 ft — 8 ft

P5.2-19 and P5.5-13

Prob. 5.2-20. The boom of the shop crane shown in Fig. P5.2-20 is supported by a frictionless pin at A, and its elevation angle is controlled by the hydraulic cylinder between pins at B and D. (a) Determine the reactions at A and B when the boom is horizontal and supports a load of 900 N at C. (b) For the same boom angle and loading, determine expressions for internal resultants $F_1(x)$, $V_1(x)$, and $M_1(x)$ at an arbitrary cross section of the boom between A and B, that is, for $(0 < x < 0.75$ m).

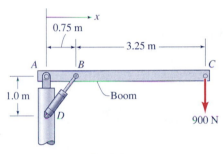

0.75 m
3.25 m
1.0 m
Boom
900 N

P5.2-20

Prob. 5.2-21. An engine that weighs 3.5 kN is suspended from point A of the engine hoist shown in Fig. P5.2-21. The hydraulic lift is positioned so that the hoist boom AC is horizontal. (a) Determine expressions for the transverse shear force $V_1(x)$ and the bending moment $M_1(x)$ at an arbitrary section that is x distance to the right of the support point A, that is, for $0 < x < 1.6$ m. (b) What is the axial force F_2 in the section BC of the hoist boom? Neglect the weight and bending of the members of the hoist frame.

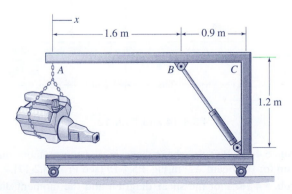

1.6 m — 0.9 m

1.2 m

P5.2-21

*__**Prob. 5.2-22.**__ A frame in Fig. P5.2-22 is supported by a frictionless pin at A and a smooth roller at D. (a) Determine the reactions at A and D. (b) Determine expressions for internal resultants $F_2(x)$, $V_2(x)$, and $M_2(x)$, at an arbitrary cross section of the frame between B and C, that is, for (24 in. $< x <$ 64 in.). Note carefully the dimensions in this problem. (The diameter of the pulley at E is not negligible.)

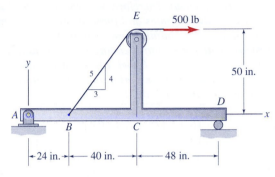

E — 500 lb

50 in.

24 in. — 40 in. — 48 in.

P5.2-22 and P5.5-6

Problems 5.2-23 through 5.2-25 are statically indeterminate problems. In each case, draw a correct free-body diagram and write the equilibrium equations for the internal resultants $V(x)$ and $M(x)$ in terms of the specified redundant reaction.

Prob. 5.2-23. For the propped-cantilever beam AB in Fig. P5.2-23, determine expressions for the internal resultants $V(x)$ and $M(x)$ in terms of the reaction force at A.

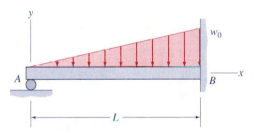

w_0

L

P5.2-23 and P5.2-24

Prob. 5.2-24. For the propped-cantilever beam AB in Fig. P5.2-24, determine expressions for the internal resultants $V(x)$ and $M(x)$ in terms of the moment at B.

Prob. 5.2-25. For the continuous beam ABC shown in Fig. P5.2-25, determine expressions for the internal resultants in each span in terms of the reaction force at A.

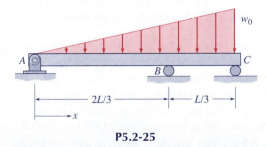

w_0

$2L/3$ — $L/3$

P5.2-25

343

▼ SHEAR-FORCE AND BENDING-MOMENT DIAGRAMS: EQUILIBRIUM METHOD

Problems 5.4-1 through 5.4-20. *Use the Equilibrium Equation Method to obtain and plot expressions for $V(x)$ and $M(x)$, as illustrated in Examples 5.4 and 5.5. Most of the problems require several expressions $V_i(x)$ for shear and $M_i(x)$ for moment. Therefore, you will need to draw several free-body diagrams and form the necessary equilibrium equations to obtain these expressions. Draw the requested shear-force and bending-moment diagrams approximately to scale. Label all critical ordinates, including the maximum and minimum values, and indicate the sections at which these occur.*

Prob. 5.4-1. The shaft AD in Fig. P5.4-1 is supported by bearings at A and C. Assume that these bearings produce concentrated reaction forces that are normal to the shaft. Plot the shear diagram $V(x)$ and the moment diagram $M(x)$ for this shaft.

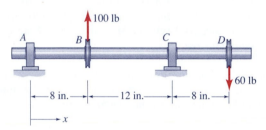

P5.4-1 and P5.5-1

Prob. 5.4-2. Plot the shear diagram and the moment diagram for the beam AE as shown in Fig. P5.4-2 (see Prob. 5.2-3).

Prob. 5.4-3. Plot the shear diagram and the moment diagram for the beam AD as shown in Fig. P5.4-3 (see Prob. 5.2-2).

Prob. 5.4-4. Plot the shear diagram and the moment diagram for the beam AE as shown in Fig. P5.4-4 (see Prob. 5.2-4).

Prob. 5.4-5. Plot the shear diagram and the moment diagram for the beam AE as shown in Fig. P5.4-5 (see Prob. 5.2-1).

Prob. 5.4-6. An L-shaped frame BD is welded to the beam ABC in Fig. P5.4-6 (see Prob. 5.2-10). A downward vertical load of $P = 5$ kips is applied to the frame at D as shown. Let dimension $a = 2$ ft, and neglect the width of the connection at B. Plot the shear diagram and the moment diagram for beam ABC.

Prob. 5.4-7. Plot the shear diagram and the moment diagram for the beam AD as shown in Fig. P5.4-7 (see Prob. 5.2-6).

Prob. 5.4-8. Plot the shear diagram and the moment diagram for the beam AC as shown in Fig. P5.4-8 (see Prob. 5.2-5).

Prob. 5.4-9. Plot the shear diagram and the moment diagram for the beam AC as shown in Fig. P5.4-9 (see Prob. 5.2-8).

Prob. 5.4-10. Plot the shear diagram and the moment diagram for the beam AC as shown in Fig. P5.4-10.

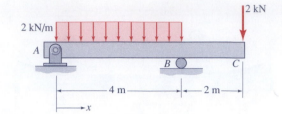

P5.4-10, P5.5-10, and P5.6-6

Prob. 5.4-11. Plot the shear diagram and the moment diagram for the beam AE in Fig. P5.4-11.

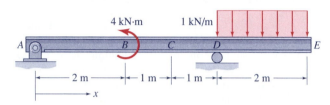

P5.4-11, P5.5-11, and P5.6-7

Prob. 5.4-12. Plot the shear diagram and the moment diagram for the beam AC in Fig. P5.4-12.

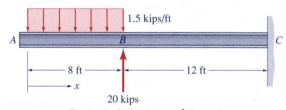

P5.4-12, P5.5-12, and P5.6-8

Prob. 5.4-13. Plot the shear diagram and the moment diagram for the beam AC in Fig. P5.4-13 (see Prob. 5.2-12).

Prob. 5.4-14. Plot the shear diagram and the moment diagram for the beam AD in Fig. P5.4-14.

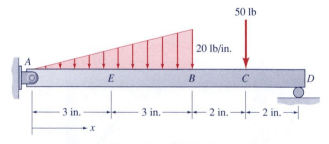

P5.4-14 and P5.6-12

Prob. 5.4-15. Plot the shear diagram and the moment diagram for the beam AD in Fig. P5.4-15 (see Prob. 5.2-17).

Prob. 5.4-16. Plot the shear diagram and the moment diagram for the beam AC in Fig. P5.4-16.

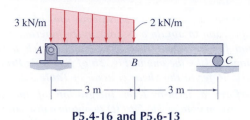

P5.4-16 and P5.6-13

Prob. 5.4-17. Plot the shear diagram and the moment diagram for the beam *AB* in Fig. P5.4-17.

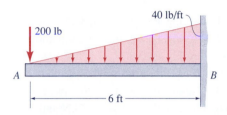

P5.4-17 and P5.6-14

Prob. 5.4-18. Plot the shear diagram and the moment diagram for the beam *AC* in Fig. P5.4-18 (see Prob. 5.2-13).

Prob. 5.4-19. Solve Example Problem 5.6 replacing the given load $p(x)$ with

$$p(x) = p_0\left[1 - \left(\frac{x}{L}\right)^2\right]$$

*Prob. 5.4-20. Solve Example Problem 5.6 replacing the given load $p(x)$ with

$$p(x) = p_0\cos\left(\frac{\pi x}{2L}\right)$$

▼ **SHEAR-FORCE AND BENDING-MOMENT DIAGRAMS: GRAPHICAL METHOD**

> **MDS 5.1–5.6**

> **Problems 5.5-1 through 5.5-20.** *Use the Graphical Method to sketch V(x) and M(x), as illustrated in Examples 5.7 through 5.9. That is, use the information in Table 5.1 in sketching the shear diagram and the moment diagram. Draw the requested shear-force and bending-moment diagrams approximately to scale. Label all critical ordinates, including the maximum and minimum values, and indicate the sections at which these occur.*

Prob. 5.5-1. Sketch the shear diagram and the moment diagram for the shaft *AD* in Prob. 5.4-1.

Prob. 5.5-2. Sketch the shear diagram and the moment diagram for the beam *AE* in Prob. 5.2-3.

Prob. 5.5-3. Sketch the shear diagram and the moment diagram for the beam *AD* in Prob. 5.2-2.

Prob. 5.5-4. Sketch the shear diagram and the moment diagram for the shaft *AE* in Prob. 5.2-4.

Prob. 5.5-5. Sketch the shear diagram and the moment diagram for the beam *AE* in Prob. 5.2-1.

*Prob. 5.5-6. Sketch the shear diagram and the moment diagram for the frame member *AD* in Prob. 5.2-22.

Prob. 5.5-7. Sketch the shear diagram and the moment diagram for the beam *AD* in Prob. 5.2-6.

Prob. 5.5-8. Sketch the shear diagram and the moment diagram for the beam *AC* in Prob. 5.2-5.

Prob. 5.5-9. Sketch the shear diagram and the moment diagram for the beam *AC* in Prob. 5.2-8.

Prob. 5.5-10. Sketch the shear diagram and the moment diagram for the beam *AC* in Prob. 5.4-10.

Prob. 5.5-11. Sketch the shear diagram and the moment diagram for the beam *AE* in Prob. 5.4-11.

Prob. 5.5-12. Sketch the shear diagram and the moment diagram for the beam *AC* in Prob. 5.4-12.

Prob. 5.5-13. Sketch the shear diagram and the moment diagram for the canoe (beam *AE*) in Prob. 5.2-19.

Prob. 5.5-14. Sketch the shear diagram and the moment diagram for the simply supported beam *AB* in Prob. 5.2-14.

Prob. 5.5-15. Sketch the shear diagram and the moment diagram for the simply supported beam *AB* in Prob. 5.2-16.

*Prob. 5.5-16. The distribution of the force between the bottom of a ski and the snow is approximated by the piecewise-linear distribution (force per unit length) depicted in Fig. P5.5-16. The skier's boot exerts a downward force R_C and a couple M_C that result in the load distribution shown. (Neglect the fact that R_C and M_C are actually distributed over the length of the boot and the bindings that attach the boot to the ski.) Sketch the shear diagram and the moment diagram for the ski, treating it as a straight beam *AE*.

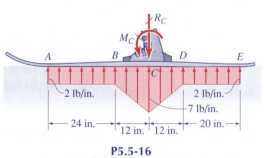

P5.5-16

Prob. 5.5-17. The railroad tracks in Fig. P5.5-17*a* rest on tie plates that are assumed to uniformly distribute the load from each wheel to the cross tie, which in turn distributes the load uniformly to the ballast below, as indicated in Fig. P5.5-17*b*. Sketch the shear and moment diagrams for the cross tie subjected to the indicated wheel loading of 32 kips per wheel.

Prob. 5.5-18. The frame in Fig. P5.5-18 has a roller-type support at *A* and a frictionless pin support at *E*. The roller at *A* can provide either an upward reaction or a downward reaction. Using the sign conventions shown in Fig. P5.5-18. sketch the shear diagrams and the moment diagrams for the frame members *AB*, *BD*, and *DE*.

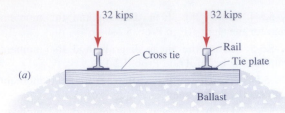

(a)

(b)

P5.5-17

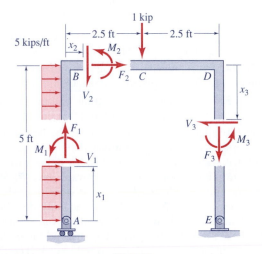

P5.5-18

Prob. 5.5-19. Sketch the shear diagrams and the moment diagrams for the two frame members AB and BC in Fig. P5.5-19 (see Prob. 1.4-19).

Prob. 5.5-20. Sketch the shear diagrams and the moment diagrams for the two frame members AB and BC in Fig. P5.5-20 (see Prob. 1.4-20).

▼ **SHEAR-FORCE AND BENDING-MOMENT DIAGRAMS: DISCONTINUITY-FUNCTION METHOD**

MDS 5.7–5.9

Problems 5.6-1 through 5.6-8. *A beam and its loading are shown in the referenced figure. For each of these problems:*

(a) *Use equilibrium to verify the reactions.*

(b) *Using discontinuity functions from the Load column of Table 5.2, write an expression for the intensity $p(x)$ of the equivalent distributed load. Include the reactions in your expression for the equivalent load.*

(c) *Perform a term-by-term integration of the load expression to obtain a discontinuity-function expression for the shear force $V(x)$, and sketch a shear diagram like the one in Fig. 2b of Example Problem 5.10. Refer to the Shear column of Table 5.2.*

(d) *Perform a term-by-term integration of the shear expression obtained in Part (c) to obtain a discontinuity-function expression for the moment $M(x)$. Sketch a moment diagram like the one in Fig. 2c of Example Problem 5.10. Refer to the Moment column of Table 5.2.*

Prob. 5.6-1. Use Fig. P5.2-1. $A_y = 28$ kN and $E_y = 22$ kN.
Prob. 5.6-2. Use Fig. P5.2-2. $B_y = 1/2$ kip and $C_y = 5/2$ kips.
Prob. 5.6-3. Use Fig. P5.2-3. $B_y = 2P$ and $D_y = 3P$.
Prob. 5.6-4. Use Fig. P5.2-5. $V_C = -2$ kips and $M_C = -4$ kip · ft.
Prob. 5.6-5. Use Fig. P5.2-8. $A_y = 20$ kips and $C_y = 19$ kips.
Prob. 5.6-6. Use Fig. P5.4-10. $A_y = 3$ kN and $B_y = 7$ kN.
Prob. 5.6-7. Use Fig. P5.4-11. $A_y = 0.5$ kN and $D_y = 1.5$ kN.
Prob. 5.6-8. Use Fig. P5.4-12. $V_C = 8$ kips and $M_C = 48$ kip · ft.

Problems 5.6-9 through 5.6-18. *Carry out the same steps outlined above for Problems 5.6-1 through 5.6-8, with the exception of omitting Part (d).*

Prob. 5.6-9. Use Fig. P5.2-12. $V_A = -7p_0L/8$ and $M_A = 37p_0L^2/96$.
Prob. 5.6-10. Use Fig. P5.2-14. $A_y = w_0L/3$ and $B_y = 5w_0L/6$.
Prob. 5.6-11. Use Fig. P5.2-16. $A_y = 2w_0L/3$ and $B_y = 5w_0L/6$.
Prob. 5.6-12. Use Fig. P5.4-14. $A_y = 46$ lb and $D_y = 64$ lb.
Prob. 5.6-13. Use Fig. P5.4-16. $A_y = 5.75$ kN and $C_y = 1.75$ kN.
Prob. 5.6-14. Use Fig. P5.4-17. $V_B = -320$ lb and $M_B = -1440$ lb · ft.
Prob. 5.6-15. Use Fig. P5.6-15. $A_y = B_y = 6.75$ kips.

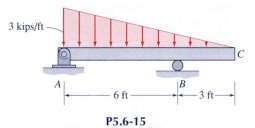

P5.6-15

Prob. 5.6-16. Use Fig. P5.6-16. $A_y = 46$ kN and $D_y = 38$ kN.

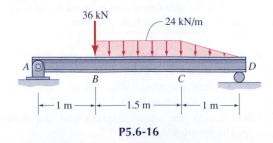

P5.6-16

Prob. 5.6-17. Use Fig. P5.6-17. $V_C = -7$ kips and $M_C = -17$ kip $\cdot$ ft.

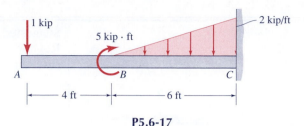

P5.6-17

Prob. 5.6-18. Use Fig. P5.6-18. $A_y = 4$ kN and $C_y = 8$ kN.

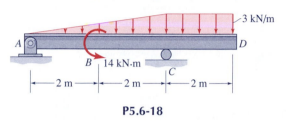

P5.6-18

▼ **SHEAR-FORCE AND BENDING-MOMENT DIAGRAMS: DISCONTINUITY-FUNCTION METHOD**   **MDS 5.7–5.9**

Problems 5.6-19 through 5.6-22. *A beam and its loading are shown in the referenced figure. For each of these problems, you are to write a computer program that carries out the following steps. Use the programming language or math application software of your own choice, unless your instructor indicates otherwise.*

(a) Input the given problem data (e.g., L, x_B).

(b) Use equilibrium to determine the reactions.

(c) Use discontinuity functions from the Load column of Table 5.2 to form an expression for the intensity p(x) of the equivalent distributed load. Include the calculated reactions in your expression for the equivalent load Plot a load diagram like the one in Fig. 2a of Example Problem 5.10.

(d) Perform a term-by-term integration of the load expression obtained in Part (c) to form a discontinuity-function expression for the shear force V(x). Refer to the Shear column of Table 5.2. Plot a shear diagram like the one in Fig. 2b of Example Problem 5.10.

(e) Perform a term-by-term integration of the shear expression obtained in Part (d) to obtain a discontinuity-function expression for the moment M(x). Refer to the Moment column of Table 5.2. Plot a moment diagram like the one in Fig. 2c of Example Problem 5.10.

[C]**Prob. 5.6-19.** (a) Write your computer program for the simply supported beam in Fig. P5.6-19. (b) Illustrate the use of your computer program for the following data: $L = 10$ m, $x_B = 2$ m, and $P_B = 25$ kN.

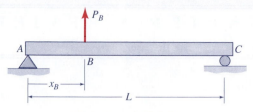

P5.6-19 and P7.5-21

[C]**Prob. 5.6-20.** (a) Write your computer program for the simply supported beam in Fig. P5.6-20. (b) Illustrate the use of your computer program for the following data: $L = 10$ m, $x_B = 2$ m, $x_C = 4$ m, and $p_0 = 12$ kN/m.

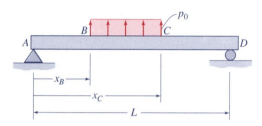

P5.6-20 and P7.5-22

[C]**Prob. 5.6-21.** (a) Write your computer program for the simply supported beam in Fig. P5.6-21. (b) Illustrate the use of your computer program for the following data: $L = 12$ ft, $x_B = 2$ ft, $x_C = 5$ ft, and $p_0 = 2$ kips/ft.

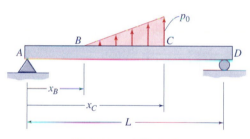

P5.6-21 and P7.5-23

[C]**Prob. 5.6-22.** (a) Write your computer program for the cantilever beam in Fig. P5.6-22. (b) Illustrate the use of your computer program for the following data: $L = 15$ ft. $x_B = 3$ ft, $x_C = 9$ ft. $P_A = -4$ kips (i.e., 4 kips downward), and $p_0 = 2$ kips/ft.

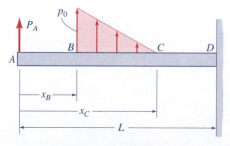

P5.6-22 and P7.5-24

Section		Suggested Review Problems
5.1	**Types of Supports** • *Roller Support* • *Pin Support* • *Fixed End* **Types of Beams & Loads** • *Simply supported* • *Cantilever* • *Multi-span continuous* • *Propped cantilever*	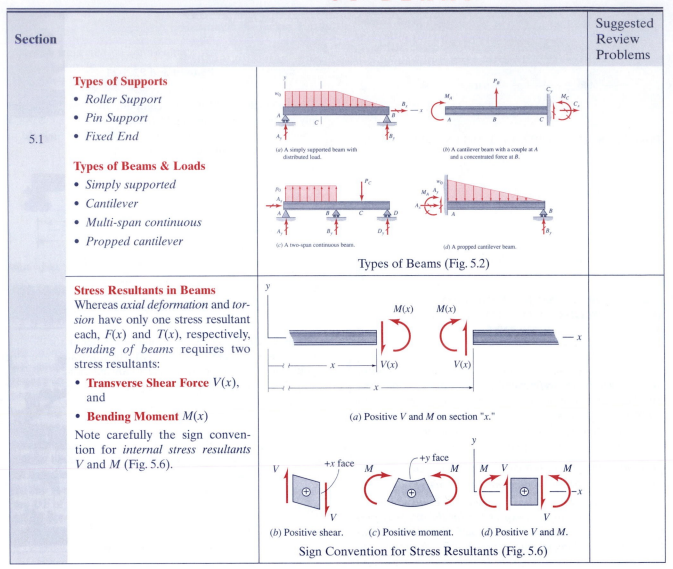 (a) A simply supported beam with distributed load. (b) A cantilever beam with a couple at A and a concentrated force at B. (c) A two-span continuous beam. (d) A propped cantilever beam. Types of Beams (Fig. 5.2)	
	Stress Resultants in Beams Whereas *axial deformation* and *torsion* have only one stress resultant each, $F(x)$ and $T(x)$, respectively, *bending of beams* requires two stress resultants: • **Transverse Shear Force** $V(x)$, and • **Bending Moment** $M(x)$ Note carefully the sign convention for *internal stress resultants* V and M (Fig. 5.6).	(a) Positive V and M on section "x." (b) Positive shear. (c) Positive moment. (d) Positive V and M. Sign Convention for Stress Resultants (Fig. 5.6)	

Section		Suggested Review Problems
5.2	**Equilibrium of beams** is a topic that you studied in your *Statics* course. The two basic procedures for solving beam equilibrium problems are: • **Finite Free-body Diagram Method (Sect. 5.2)** • **(See below)** Note carefully the sign conventions for *external loads* (Fig. 5.7). (*a*) Distributed load.　　(*b*) Concentrated force and couple. Sign Convention for External Loads (Fig. 5.7)	5.2–3 5.2–11 5.2–19
5.3	• Infinitesimal FBD Method (Sect. 5.3)	 (*a*) General FBD.　(*b*) Shear-jump "FBD."　(*c*) Moment-jump "FBD." Infinitesimal Free-body Diagrams (Fig. 5.8)	Derive Eqs. 5.2–5.5. and state Eqs. 5.2–5.5 in words.
		 A Finite Free-body Diagram (Fig. 5.9)	Derive Eqs. 5.6–5.7. and state Eqs. 5.6–5.7 in words.
5.4	In Sections 5.4 and 5.5, two methods are given for constructing **Shear-Force Diagrams** and **Bending-Moment Diagrams:** • Equilibrium Method (Sect. 5.4) • Continued below • Graphical Method (Sect. 5.5)		5.4–11 5.4–15

Section			Suggested Review Problems
5.5	Study carefully the *Shear and Moment Diagram Features* illustrated graphically in Table 5.1. The equations that are illustrated graphically in Table 5.1 are given in the next column.	$\dfrac{dV}{dx} = p(x)$ (Eq. 5.2) $\Delta V = P_0$ (Eq. 5.4) $V_2 - V_1 = \displaystyle\int_{x_1}^{x_2} p(x)\,dx$ (Eq. 5.6) $\dfrac{dM}{dx} = V(x)$ (Eq. 5.3) $\Delta M = -M_0$ (Eq. 5.5) $M_2 - M_1 = \displaystyle\int_{x_1}^{x_2} V(x)\,dx$ (Eq. 5.7)	5.5–1 5.5–11 5.5–17
Section 5.6 is an "optional" section.			

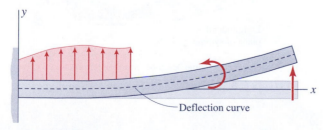

STRESSES IN BEAMS

6

6.1 INTRODUCTION

In this chapter we continue our study of beams by determining how the stress re-sultants, the bending moment $M(x)$ and the transverse shear force $V(x)$, are related to the normal stress and the shear stress at section x. Loads (transverse forces or couples) applied to a beam cause it to deflect laterally, as illustrated in Fig. 6.1. This lateral deflection, or **bending,** changes the initially straight longitudinal axis of the beam into a curve that is called the **deflection curve,** shown dashed in Fig. 6.1. By relating the curvature of the deflection curve to the bending moment M, we can determine the distribution of the normal stress σ_x. You will discover that this derivation includes all three of the fundamental types of equations: *geometry of deformation* (in the strain-displacement analysis), *material behavior* (in the stress-strain relations), and *equilibrium* (in the definition of stress resultants and in relating stress resultants to the external loads and reactions).

Beam-Deformation Terminology. To simplify this study of beams, we initially consider only straight beams that have a **longitudinal plane of symmetry** (LPS), and for which the loading and support are symmetric with respect to this plane, as illustrated in Fig. 6.2. Under these conditions, this longitudinal plane of symmetry is the **plane of bending.** (In Section 6.6 the more general case of unsymmetric bending is considered.) As indicated in Fig. 6.2a, **coordinate axes** will be assigned as follows: the plane of bending is labeled the xy plane; the **longitudinal axis** of the beam is labeled the x axis, with the positive x axis directed to the right; the positive

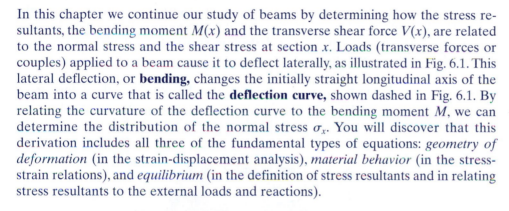

FIGURE 6.1 Transverse deflection of a beam.

351

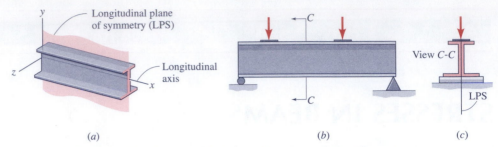

(a)   (b)   (c)

FIGURE 6.2   A symmetric beam with symmetric loading and support.

y axis points upward; and, finally, the z axis forms a right-handed coordinate system with the other two axes.

To investigate the distribution of stresses in a beam, like the one in Fig. 6.1, it is convenient to imagine the beam to be a bundle of many *longitudinal fibers* parallel to the x axis. Figure 6.3*a* depicts a few of these imaginary "fibers." (A wood beam, of course, has real fibers.) Under the action of an applied bending moment M, there is a shortening of the upper fibers and a stretching of the lower fibers, causing the beam segment to be curved upward. But some longitudinal fibers retain their original length. These are said to form the **neutral surface** (NS), which is identified in Fig. 6.3*b*. For convenience, we position the undeformed beam in the xyz coordinate frame with the xz plane corresponding to the neutral surface.[1] Then, the longitudinal axis of the beam, which is the intersection of the plane of bending (xy plane) and the neutral surface (xz plane), is the **deflection curve** of the deformed beam, as indicated in Figs. 6.1 and 6.3*b*.

Pure Bending.   Let us begin our analysis of beams by examining the deformation of a uniform beam segment subjected to pure bending, that is, a segment whose material properties are constant along its length, and for which $M(x)$ is constant.

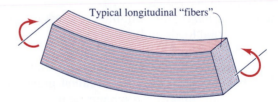

(*a*) A beam represented as a "bundle" of longitudinal "fibers."

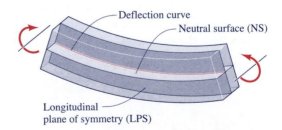

(*b*) Illustration of some beam-deformation terminology.

FIGURE 6.3   Beam-deformation terminology.

---

[1]In Section 6.3 the position of the *neutral surface* (i.e., the xz plane) with respect to the beam is determined.

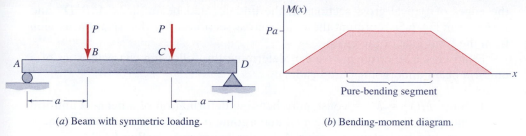

(a) Beam with symmetric loading.

(b) Bending-moment diagram.

FIGURE 6.4 A beam whose loading produces pure bending.

One type of beam loading that produces a segment subjected to pure bending is shown in Fig. 6.4a, where the segment between B and C is in pure bending, as shown in the corresponding bending-moment diagram in Fig. 6.4b.

If equal couples M_0 are applied to the ends of an otherwise unloaded segment of beam, as in Figs. 6.5c through e, the moment is constant along the segment and the segment is said to be in **pure bending.** Lines ABD and EFG in Fig. 6.5b represent

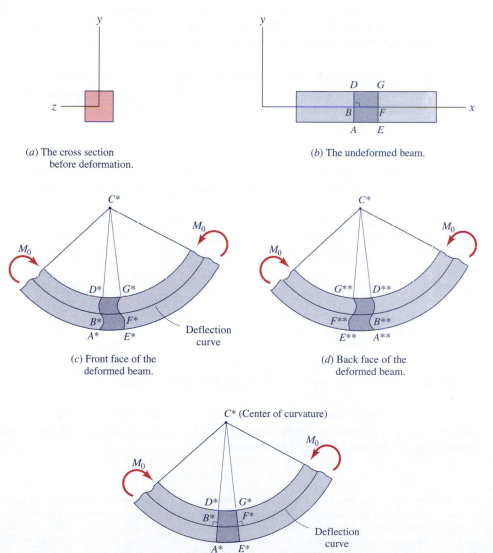

(a) The cross section before deformation.

(b) The undeformed beam.

(c) Front face of the deformed beam.

(d) Back face of the deformed beam.

(e) Front face showing that plane sections remain plane.

FIGURE 6.5 A uniform beam segment undergoing pure bending.

the edges of typical cross sections in the undeformed beam; lines $A*B*D*$ and $E*F*G*$ in Fig. 6.5e represent these same cross sections after deformation, as seen from the front face of the beam.

From Figs. 6.5c through e we can determine the following characteristics of a uniform beam undergoing pure bending:

1. Since $M(x) = M_0 = $ const, pure-bending deformation of a beam is uniform along the length of the segment undergoing pure bending; so whatever happens at a typical cross-section ABD also happens at section EFG.

2. The curvature of the deflection curve at $F*$ is the same as the curvature at $B*$ (Fig. 6.5c, e). Therefore, **the deflection curve forms a circular arc, with center of curvature at $C*$.**

3. Pure bending has front-to-back symmetry. Assume that straight lines ABD and EFG corresponding to two typical cross sections of the undeformed beam become the s-shaped curves $A*B*D*$ and $E*F*G*$ in Fig. 6.5c when viewed from the front side, and, correspondingly, the curves $A**B**D**$ and $E**F**G**$ in Fig. 6.5d when viewed from the back side. But, if the beam were to be viewed from the back side, as in Fig. 6.5d, curve $E**F**G**$ should look exactly like curve $A*B*D*$ does in Fig. 6.5c. The only way that this can be possible is for **all cross sections,** like ABD and EFG, to **remain plane and remain perpendicular to the deflection curve,** as illustrated in Fig. 6.5e.

4. In summary, **when a beam undergoes pure bending, its deflection curve forms a circular arc, and its cross sections remain plane and remain perpendicular to the deflection curve.** Experiments show that this is, indeed, the way that beams deform when subjected to pure bending.

The above description of beam deformation applies rigorously only to the case of pure bending, that is, when $dM/dx = V(x) = 0$, as in segment BC of the beam in Fig. 6.4. However, even in the case of **nonuniform bending,** where, as in segments AB and CD of the beam in Fig. 6.4, $dM/dx = V(x) \neq 0$, the assumption that cross sections remain plane and remain perpendicular to the deformed axis leads to expressions for extensional strain ϵ_x and for normal stress σ_x that are quite accurate if the beam is long compared with its cross-sectional dimensions; that is, if the beam is "slender."

6.2 STRAIN-DISPLACEMENT ANALYSIS

Let us continue our analysis of the deformation of a uniform beam segment subjected to pure bending, that is, a segment for which $M(x)$ is constant.

Kinematic Assumptions of Bernoulli-Euler Beam Theory.
The previous discussion of pure bending can be summarized in the following *four deformation assumptions* of **Bernoulli-Euler beam theory:**[2]

---

[2]The development of the beam theory presented here is attributed principally to the work of Jacob Bernoulli (1654–1705) and Leonard Euler (1707–1783). Jacob Bernoulli, a prominent member of the famous Bernoulli family of mathematicians and physicists, studied the deflection of beams, finding that the curvature of an elastic beam at any point is proportional to the bending moment at that point. The work of Jacob Bernoulli and his nephew Daniel Bernoulli (1700–1782) led Euler to his discovery of the differential equation of the elastic curve of a beam [Ref. 6-1].

1. The beam possesses a longitudinal plane of symmetry, and is loaded and supported symmetrically with respect to this plane. This plane is called the **plane of bending.**

2. There is a longitudinal plane perpendicular to the plane of bending that remains free of strain (i.e., $\epsilon_x = 0$) as the beam deforms. This plane is called the **neutral surface** (NS). The intersection of the neutral surface with a cross section is called the **neutral axis** (NA) of the cross section. The intersection of the neutral surface with the plane of bending is called the **axis** of the beam; it forms the **deflection curve** of the deformed beam.

3. **Cross sections, which are plane and are perpendicular to the axis of the undeformed beam, remain plane and remain perpendicular to the deflection curve of the deformed beam.**

4. Deformation in the plane of a cross section (i.e., transverse strains ϵ_y and ϵ_z) may be neglected in deriving an expression for the longitudinal strain ϵ_x.

The third of the preceding assumptions is crucial to the development of the Bernoulli-Euler beam theory; it leads to a **practical theory of bending of beams** that is comparable to the theories of axial deformation and torsion covered previously.

Strain-Displacement Analysis; Longitudinal Strain. Because of Assumptions 1 and 4, the fibers in any plane parallel to the xy plane behave identically to the corresponding fibers that lie in the xy plane (i.e., the plane of bending). Therefore, **bending deformation is independent of the coordinate z,** so the drawing in Fig. 6.6 represents the deformation of any plane in the beam parallel to the xy plane. Using Fig. 6.6 and the preceding four deformation assumptions, we can develop an expression for the extensional strain ϵ_x in a longitudinal fiber at coordinates (x, y, z) in the beam. In Fig. 6.6a, points A and P lie in the cross-sectional plane at coordinate x in the undeformed beam; similarly, points B and Q lie in the cross-sectional plane at $(x + \Delta x)$ in the undeformed beam. Line segment PQ is parallel to the x axis and lies at distance $+y$ above the NS (xz plane). Therefore, in the undeformed beam the infinitesimal fibers AB and PQ are both of length Δx. From Assumption 3, points A^* and P^* lie in a plane that is perpendicular to the neutral surface of the deformed beam, and points B^* and Q^* lie in a plane that also is perpendicular to the deformed neutral surface. According to Assumption 2, however, the length of A^*B^*, a fiber lying in the neutral surface, is unchanged; that

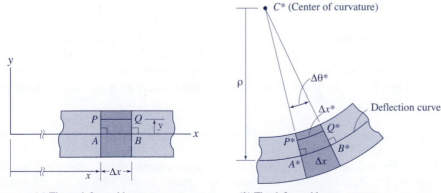

(a) The undeformed beam segment.　　　(b) The deformed beam segment.

FIGURE 6.6 The geometry of deformation of a beam segment, showing the plane of bending.

is, the length of A^*B^* is still Δx, as indicated in Fig. 6.6b. Finally, by virtue of Assumption 4, $\overline{A^*P^*} = \overline{AP} = y$, and $\overline{B^*Q^*} = \overline{BQ} = y$.

Figure 6.6, therefore, embodies all four deformation assumptions of Bernoulli-Euler beam theory, so we can use it in deriving an expression for the extensional strain of a longitudinal fiber.

From the general definition of extensional strain in Eq. 2.35, we can express the extensional strain in the longitudinal fiber PQ as

$$\epsilon_x \equiv \epsilon_x(x, y, z) = \lim_{Q \to P} \left[\frac{(\overline{P^*Q^*} - \overline{PQ})}{\overline{PQ}} \right] = \lim_{\Delta x \to 0} \left[\frac{(\Delta x^* - \Delta x)}{\Delta x} \right] \tag{6.1}$$

Considering A^*B^* to be the arc of a circle of radius $\rho(x) \equiv \rho$ subtending an angle $\Delta \theta^*$, we get

$$\overline{A^*B^*} = \Delta x = \rho \Delta \theta^*$$

Similarly,

$$\overline{P^*Q^*} = \Delta x^* = (\rho - y)\Delta \theta^*$$

Combining these two equations with Eq. 6.1, we get

$$\epsilon_x = \lim_{\Delta x \to 0} \left[\frac{[(\rho - y)\Delta \theta^* - \rho \Delta \theta^*]}{\rho \Delta \theta^*} \right] = -\frac{y}{\rho} \tag{6.2}$$

That is, the extensional strain ϵ_x at point (x, y, z) in the beam is independent of z and is related to the local **radius of curvature**, $\rho(x)$, by the following **strain-displacement equation:**

$$\epsilon_x = -\frac{y}{\rho(x)} \qquad \text{**Strain-Displacement Equation**} \tag{6.3}$$

The reciprocal of the radius of curvature, $\kappa \equiv 1/\rho$, is called the *longitudinal curvature,* or just the *curvature.*

As noted earlier, the assumptions that lead to this strain-displacement equation are strictly valid for the case of pure bending, that is, when $V \equiv 0$. However, Eq. 6.3 can also be used for analyzing the bending of beams with $V \neq 0$ [i.e., $M = M(x)$] if the beam is slender.

Transverse Strains. The longitudinal strain at a point in the beam is given by Eq. 6.3. In Section 6.3 we will show that the transverse stresses σ_y and σ_z are negligible in comparison with the normal stress σ_x. Therefore, from the generalized Hooke's Law, Eq. 2.38, there is a *Poisson's-ratio effect* that produces transverse strains

$$\epsilon_y = \epsilon_z = -\nu \epsilon_x \tag{6.4}$$

The deformed shape of a rectangular beam segment undergoing pure bending is shown in Fig. 6.7. Consider the transverse strain ϵ_z for this beam. Above the neutral axis (i.e., where y is positive), the beam is compressed axially, so ϵ_x is negative there. Then, from Eq. 6.4, ϵ_z will be positive. Therefore, since ϵ_z is positive,

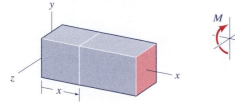

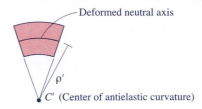

─ Deformed neutral axis

ρ'

C' (Center of antielastic curvature)

(a) The undeformed beam segment
and its cross section.

(b) The deformed beam segment and its cross section.

FIGURE 6.7 Transverse deformation of a beam segment in pure bending.

fibers oriented in the z direction and lying above the NA elongate. Correspondingly, fibers oriented in the z direction and lying below the NA contract, since ϵ_z is negative there. This produces the *antielastic curvature* of the beam that is illustrated in Fig. 6.7. In Homework Problem 6.2-5 you will be asked to prove that the *transverse curvature*, $\kappa' \equiv 1/\rho'$, is related to the longitudinal curvature $\kappa \equiv 1/\rho$ by the equation

$$\kappa' = \nu\kappa \tag{6.5}$$

You can easily demonstrate this antielastic curvature if you take a soft eraser with rectangular cross section (e.g., a "pink pearl" eraser) and bend it between your thumb and index finger.

As noted previously in Assumption 4, deformation in the plane of a cross section may be neglected in the derivation of the strain-displacement equation, Eq. 6.3.

Summary of Strain-Displacement Analysis.

Before we go on to discuss the normal stress σ_x and its relationship to $\rho(x)$ and $M(x)$, let us examine thoroughly the expression for extensional strain, Eq. 6.3, which we obtained from the four Bernoulli-Euler deformation assumptions and the fundamental definition of extensional strain. We observe the following:

1. **The extensional strain ϵ_x is independent of z, the thickness coordinate of the beam,** that is, $\epsilon_x = \epsilon_x(x, y)$.

2. The extensional strain ϵ_x is inversely proportional to the radius of curvature at cross section x. Figure 6.8 illustrates the fact that as ρ decreases, the curvature κ increases (i.e., the beam becomes more curved), and the strain at any given fiber increases. For example, the length $\overline{A^*B^*}$ (which is equal to $\overline{AB}$ since this is identified as lying in the neutral surface) is the same for all three figures in Fig. 6.8. But, since $\overline{D_1^*E_1^*} < \overline{D_2^*E_2^*} < \overline{D_3^*E_3^*}$, the length change in the bottom fiber as a result of bending increases from left to right in Fig. 6.8, and therefore the strain increases from the least-curved beam in Fig. 6.8a to the progressively more-curved beams in Figs. 6.8b and 6.8c.

3. The signs of $\rho(x)$ and y govern the sign of ϵ_x. If $\rho(x)$ is positive, the *center of curvature* lies above the beam, that is, on the $+y$ side of the beam. Therefore, when $\rho(x)$ is positive, the deformed beam is concave upward. Because of the minus sign in Eq. 6.3, the fibers above the neutral surface (i.e., fibers having positive y) are in compression, while the fibers below the neutral surface are in tension. This is the case for the beam segment in Fig. 6.9a. If $\rho(x)$ is negative, the center of curvature lies below the beam, and the beam is curved concave downward, as shown as Fig. 6.9b.

4. Finally, **the strain ϵ_x is proportional to the distance y from the neutral surface.** This *linear strain distribution* is illustrated in Fig. 6.10.

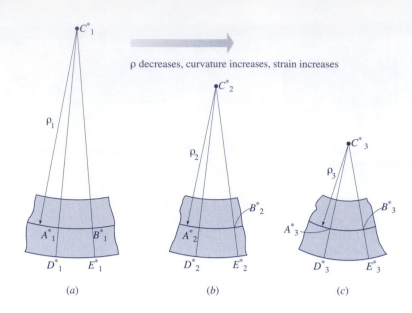

ρ decreases, curvature increases, strain increases

FIGURE 6.8 The relationship of the curvature of a beam to the extensional strain, ϵ_x.

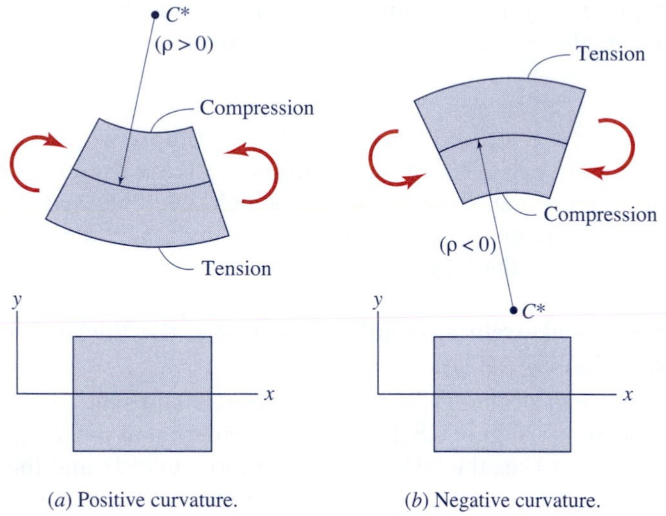

(a)   (b)   (c)

FIGURE 6.9 Illustrations of positive and negative curvature.

(a) Positive curvature.   (b) Negative curvature.

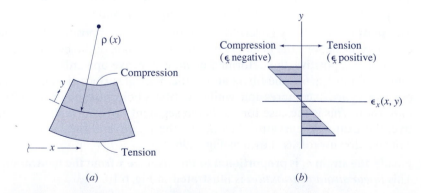

FIGURE 6.10 The strain distribution at a cross section where $\rho(x)$ is positive.

(a)   (b)

The fact that ϵ_x varies linearly with y is completely independent of the material properties of the beam. For example, the beam could consist of two or more different materials, like concrete and steel (Section 6.5), or the beam could be partly elastic and partly plastic (Section 6.7). As long as the deformation can be characterized by the four assumptions listed in this section, the *extensional strain* ϵ_x *is proportional to* y, as given by Eq. 6.3 and illustrated in Fig. 6.10b.

EXAMPLE 6.1

A couple M_0 acts on the end of a slender cantilever beam as shown in Fig. 1. Take $\dfrac{L}{c} = 30$, where L is the original length of the beam and $2c$ is the depth of the beam. (a) Determine an expression for the normalized radius of curvature (ρ/c) if the bottom fiber (at $y = -c$) is at the tensile yield strain of the material, ϵ_Y. (b) Determine the ratio ($\delta_{\max}/c$) for this loading condition. (c) Determine ρ and $\delta_{\max}$ if $L = 15$ ft and $\epsilon_Y = 0.001$.

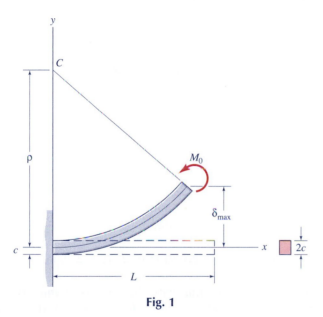

Fig. 1

Plan the Solution   Part (a) is a straightforward application of Eq. 6.3, which relates extensional strain ϵ_x to the radius of curvature of the deflection curve. To determine $\delta_{\max}$ in Part (b), we need, in addition, to use the geometric properties of a circle.

Solution   (a) From the strain-displacement equation, Eq. 6.3,

$$\epsilon_x = -\frac{y}{\rho}$$

The bottom fiber is in tension. Therefore,

$$\rho = -\frac{(-c)}{\epsilon_Y}$$

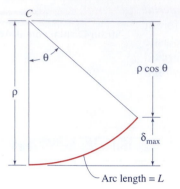

Fig. 2 The geometry of a circular arc.

so

$$\frac{\rho}{c} = \frac{1}{\epsilon_Y}$$ **Ans. (a)**

(b) From the sketch, in Fig. 2, of the deflection curve,

$$L = \rho\theta$$

and

$$\delta_{max} = \rho(1 - \cos\theta)$$

Therefore, for this particular beam,

$$\theta = \frac{L}{\rho} = \frac{L}{c}\frac{c}{\rho} = 30\,\epsilon_Y$$

or,

$$\frac{\delta_{max}}{c} = \frac{1}{\epsilon_Y}[1 - \cos(30\,\epsilon_Y)]$$ **Ans. (b)**

(c) For $L = 15$ ft and $\epsilon_Y = 0.001$, we get

$$\theta = 30\,\epsilon_Y = 0.03 \text{ rad}$$

Finally,

$$\rho = \frac{L}{30\,\epsilon_Y} = 500 \text{ ft} = 6000 \text{ in.}$$ **Ans. (c)**

$$\delta_{max} = \rho(1 - \cos\theta) = 0.225 \text{ ft} = 2.70 \text{ in.}$$ **Ans. (c)**

Review the Solution   Note that with the numerical value of $L = 15$ ft and the given ratio $L/c = 30$, $c = L/30 = 6$ in. Hence, for this 12-in.-depth beam, a maximum deflection that is less than one-fourth of the depth of the beam causes the extreme fiber of the beam to reach the yield strain $\epsilon_Y = 0.001$. (Figures 1 and 2 obviously exaggerate the deflection.)

6.3   FLEXURAL STRESS IN LINEARLY ELASTIC BEAMS

In the previous section, assumptions were made about the geometry of deformation of slender beams, and an expression for the resulting extensional strain ϵ_x was derived, Eq. 6.3. The corresponding normal stress in beams, σ_x, is often called the **flexural stress.** To obtain an expression for the flexural stress, we need to consider the material behavior, that is, the stress-strain-temperature behavior of the material. To simplify our initial study of stresses in beams, let us assume that the material is **linearly elastic and isotropic,** and that the temperature remains constant. Then, the following two assumptions permit us to determine the flexural stress σ_x:

1. The material obeys Hooke's law, Eq. 2.38a, with $\Delta T = 0$.
2. The transverse normal stresses, σ_y and σ_z, may be neglected in comparison with the flexural stress, σ_x.

By combining these two assumptions, we find that the uniaxial stress-strain equation

$$\sigma_x = E\epsilon_x \tag{6.6}$$

applies to bending of linearly elastic beams. When Eqs. 6.3 and 6.6 are combined, we obtain the following expression for the flexural stress:

$$\sigma_x = \frac{-Ey}{\rho} \tag{6.7}$$

If $E = \text{const}$, or if $E = E(x)$, the normal stress on a cross section is linear in y, as given by Eq. 6.8 and indicated in Fig. 6.11.[3]

$$\boxed{\sigma_x(x, y) = \frac{-E(x)y}{\rho(x)}} \tag{6.8}$$

As indicated in Fig. 6.11, the stress resultants that are related to the normal stress σ_x acting on the cross section are:

$$F(x) = \int_A \sigma_x \, dA, \quad M(x) = -\int_A y\sigma_x \, dA \tag{6.9}$$

A positive moment produces compression in the $+y$ fibers of the beam.

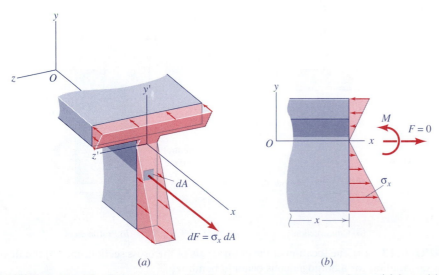

(a)   (b)

FIGURE 6.11   The flexural stress distribution at a cross section where $\rho(x)$ is positive.

---

[3]In Section 6.5 we will consider stresses in nonhomogeneous beams, that is, stresses in beams that are made of more than one material.

In Section 9.4 we will consider axial deformation combined with bending, but, for the present discussion of bending alone, let $F \equiv 0$. Therefore, substituting Eq. 6.8 into Eqs. 6.9, we get

$$F = -\frac{E}{\rho} \int_A y \, dA = 0, \quad M = \frac{E}{\rho} \int_A y^2 \, dA \tag{6.10}$$

The integrals appearing in Eqs. 6.10 are section properties that are defined in Appendix C:

$$\int_A dA = A, \quad \int_A y \, dA = \bar{y} A, \quad \int_A y^2 \, dA = I_z \tag{6.11}$$

where A is the cross-sectional *area,* $\bar{y}$ is the y coordinate of the *centroid* of the cross section, and I_z is the **area moment of inertia** about the z axis of the cross section.

The most useful formula for moment of inertia, that of a rectangular area, may be easily derived as follows. For the rectangle in Fig. 6.12, Eq. 6.11c gives

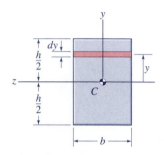

$$I_z = \int_A y^2 \, dA = \int_{-h/2}^{h/2} y^2 (b \, dy) = b \left. \frac{y^3}{3} \right|_{-h/2}^{h/2} = \frac{bh^3}{12}$$

FIGURE 6.12 Notation for calculating moment of inertia of a rectangle.

where the z axis passes through the centroid of the cross section and is parallel to the two sides of length b. See Appendix C.2 for further discussion of moments of inertia. A table listing formulas for coordinates of the centroid and for moments of inertia of a variety of shapes may be found inside the back cover of this book.

In order to satisfy the condition $F = 0$, we must make $\bar{y} = 0$. That is, the z axis of the cross section (labeled the z' axis in Figs. 6.11a and 6.13a) must pass through the centroid of the cross section. Thus, **the x axis passes through the centroid of each cross section of the undeformed beam.** The z' axis is called the *neutral axis of the cross section,* or simply the **neutral axis** (NA), because it is the boundary between the portion of the cross section that is in compression and the portion that is in tension, as indicated in Fig. 6.13.[4]

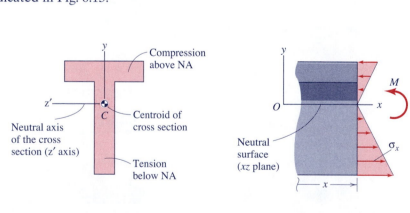

(a) Cross section.   (b) Profile view.

FIGURE 6.13 (a) The location of the neutral axis of the cross section, and (b) the flexural stress distribution for a homogeneous beam in bending.

---

[4]In the future, the "z axis of the cross section" will just be labeled z, not z', even though the true z axis does not lie in the particular cross section under consideration.

Combining Eqs. 6.10b and 6.11c, we obtain the **moment-curvature equation** of Bernoulli-Euler beam theory, namely

$$M = \frac{EI}{\rho} = EI\kappa \qquad \text{Moment-Curvature Equation} \qquad (6.12)$$

The *curvature* $\kappa(x)$ is related to the *radius of curvature* $\rho(x)$ by $\kappa(x) = \dfrac{1}{\rho(x)}$. The product EI is called the **flexural rigidity** of the beam. (In Eq. 6.12 the subscript has been dropped from I_z to simplify the remainder of the discussion of bending of symmetric beams. Subscripts will be needed again in the Section 6.6 on Unsymmetric Bending.)

We can relate the moment-curvature equation, Eq. 6.12, to the deformed-beam segments in Fig. 6.9 by noting that a positive bending moment, $M(x)$, leads to a positive value of $\rho(x)$, which means that the beam is concave upward, as shown in Fig. 6.9a. Conversely, a negative moment produces a negative curvature, which means that the center of curvature lies in the $-y$ direction, as shown in Fig. 6.9b.

Finally, Eqs. 6.8 and 6.12 may be combined to give the important **flexure formula** of Bernoulli-Euler beam theory.[5]

$$\sigma_x = \frac{-My}{I} \qquad \text{Flexure Formula} \qquad (6.13)$$

By making the assumptions that plane sections remain plane and that the material is linearly elastic with $E = E(x)$, we have obtained an expression for the stress distribution on a cross section subjected to bending moment $M(x)$. This is the linear stress distribution illustrated in Fig. 6.14.[6]

An assumption made in the derivation of the flexure formula, Eq. 6.13, is that σ_x is much greater than either σ_y or σ_z. It is left as an exercise for the reader to show that this is a reasonable assumption if the beam is long in comparison with its cross-sectional dimensions. (Homework Problem 6.3-37)

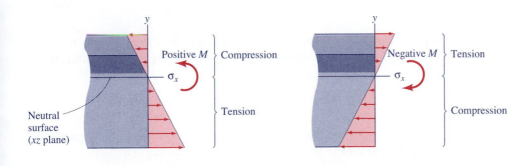

(a) Positive moment.   (b) Negative moment.

FIGURE 6.14   The flexural stress distribution in a linearly elastic beam.

---

[5]According to the sign convention adopted in this text and illustrated in Figs. 5.6 and 6.9, a positive moment produces compression in the $+y$ fibers of the beam. This results in a minus sign in Eq. 6.13. Some textbooks adopt a different sign convention that leads to a plus sign in the flexure formula.

[6]Compressive stresses as well as tensile stresses may be shown acting *on* the cross section, as in Fig. 6.13b. However, to emphasize here that σ_x is linear in y, compressive stresses are shown in Fig. 6.14 as a continuation of the straight-line plot that depicts tensile stresses.

EXAMPLE 6.2

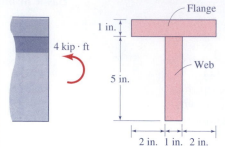

4 kip · ft

Flange

1 in.

Web

5 in.

2 in. 1 in. 2 in.

Fig. 1

The cross section of a beam is a T with the dimensions shown in Fig. 1. The moment at the section is $M = 4$ kip · ft. Determine (a) the location of the neutral axis of the cross section, (b) the moment of inertia with respect to the neutral axis, and (c) the maximum tensile stress and the maximum compressive stress on the cross section.

Plan the Solution To use the flexure formula, Eq. 6.13, we need first to locate the centroid of the cross section and then compute the moment of inertia about an axis through the centroid (Appendix C). Since M is positive, the maximum compressive stress occurs in the top fibers, and the maximum tensile stress occurs in the bottom fibers.

Solution (a) *Locate the neutral axis.* As indicated in Fig. 2, we can pick an arbitrary origin at the bottom and let a coordinate in the y-direction be called η. Then, by summing area contributions to the first moment, we have

$$\int_A \eta \, dA = \int_{A_1} \eta \, dA + \int_{A_2} \eta \, dA$$

or

$$\bar{\eta}A = \bar{\eta}_1 A_1 + \bar{\eta}_2 A_2 = (5.5 \text{ in.})(5 \text{ in}^2) + (2.5 \text{ in.})(5 \text{ in}^2) = 40 \text{ in}^3$$

where

$$A = A_1 + A_2 = (5 \text{ in}^2) + (5 \text{ in}^2) = 10 \text{ in}^2$$

Then,

$$\bar{\eta} = \frac{40 \text{ in}^3}{10 \text{ in}^2} = 4.0 \text{ in.} \qquad \textbf{Ans. (a)}$$

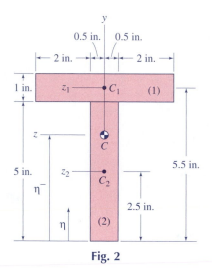

0.5 in. 0.5 in.

2 in. 2 in.

1 in.

z_1 C_1 (1)

z

C

5 in.

z_2 C_2

5.5 in.

2.5 in.

$\bar{\eta}$

η (2)

Fig. 2

(b) *Determine the moment of inertia with respect to the neutral axis.* The moment of inertia of a rectangle about an axis through its own centroid is

$$I_C = \frac{1}{12} bh^3$$

and, from the parallel-axis theorem, the moment of inertia about an axis through C' parallel to the axis through the centroid C is

$$I_{C'} = I_C + Ad_{CC'}^2$$

Therefore,

$$I = (I_{C_1} + A_1 d_{C_1 C}^2) + (I_{C_2} + A_2 d_{C_2 C}^2)$$

$$= \left[\frac{1}{12}(5 \text{ in.})(1 \text{ in.})^3 + (5 \text{ in.})(1 \text{ in.})(1.5 \text{ in.})^2 \right]$$

$$+ \left[\frac{1}{12}(1 \text{ in.})(5 \text{ in.})^3 + (1 \text{ in.})(5 \text{ in.})(1.5 \text{ in.})^2 \right]$$

so,

$$I = 33.3 \text{ in}^4 \qquad\qquad \textbf{Ans. (b)}$$

(c) *Compute $\sigma_{\max T}$ and $\sigma_{\max C}$.* The maximum compression occurs at the top of the beam, and the maximum tension occurs at the bottom of the beam. From the flexure formula, Eq. 6.13,

$$\sigma_x = \frac{-My}{I}:$$

$$\sigma_{\max C} = \sigma(x, 2 \text{ in.}) = \frac{-(4 \text{ kip} \cdot \text{ft})(12 \text{ in./ft})(2 \text{ in.})}{33.3 \text{ in}^4}$$

$$= -2.88 \text{ ksi (2.88 ksi C)}$$

$$\sigma_{\max T} = \sigma(x, -4 \text{ in.}) = \frac{-(4 \text{ kip} \cdot \text{ft})(12 \text{ in./ft})(-4 \text{ in.})}{33.3 \text{ in}^4}$$

$$= 5.76 \text{ ksi (5.76 ksi T)}$$

$$\sigma_{\max T} = 5.76 \text{ ksi}, \qquad \sigma_{\max C} = -2.88 \text{ ksi} \qquad\qquad \textbf{Ans. (c)}$$

Review the Solution   The centroid must lie between the centroids of the two areas A_1 and A_2. Furthermore, since the areas are equal, the combined centroid C lies midway between the individual centroids, as we obtained above.

To see if the order of magnitude of I is reasonable, we can compare our answer with the moment of inertia of a 1 in. × 6 in. web about its own centroid ($\frac{1}{12}(1)(6)^3 = 18 \text{ in}^4$). Thus, the value of $I = 33.3 \text{ in}^4$ appears reasonable.

Finally, since the T-section is not symmetric about the neutral axis, the maximum tension and maximum compression are not equal in magnitude, but, since σ is linear in y, their magnitudes are in the ratio of the distances to the top and bottom fibers.

(See Homework Problem 6.6-6, where the effect of misalignment of the load is examined.)

MDS6.1   **Beam Cross-Sectional Properties**—*Section Properties* is an MDS computer program module for calculating section properties of plane areas: area, location of centroid, moments of inertia, product of inertia, orientation of principal axes, etc. Use it to determine the section properties of the T section in Example Problem 6.2, and similar problems.

EXAMPLE 6.3

For the beam of Example 6.2, (a) determine the resultant compressive force F_C, the resultant tensile force F_T, and the distance that separates them, as indicated in Fig. 1; and (b) show that F_C and F_T form a couple of magnitude $M = 4 \text{ kip} \cdot \text{ft}$.

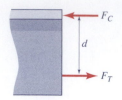

Fig. 1

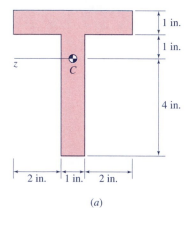

(a)

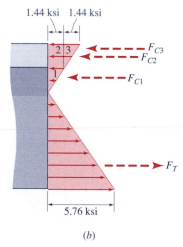

1.44 ksi   1.44 ksi

5.76 ksi

(b)

Fig. 2

Plan the Solution We can sketch the stress distribution on a figure like Fig. 6.13b. Then we can apply the standard procedures for locating the resultant of distributed forces to get the magnitude and the location of F_C and F_T.

Solution (a) *Determine the resultant compressive force, the resultant tensile force, and the distance between them.* We first represent the area under various portions of the stress distribution by forces F_{C1}, F_{C2}, and so forth, as indicated in Fig. 2b. For example, from the fact that F_T is the resultant of the tensile stress, we have

$$F_T = \frac{1}{2}(5.76 \text{ kips/in}^2)(1 \text{ in.})(4 \text{ in.}) = 11.52 \text{ kips}$$

Similarly,

$$F_{C1} = \frac{1}{2}(1.44 \text{ kips/in}^2)(1 \text{ in.})(1 \text{ in.}) = 0.72 \text{ kips}$$

$$F_{C2} = (1.44 \text{ kips/in}^2)(5 \text{ in.})(1 \text{ in.}) = 7.20 \text{ kips}$$

$$F_{C3} = \frac{1}{2}(1.44 \text{ kips/in}^2)(5 \text{ in.})(1 \text{ in.}) = 3.60 \text{ kips}$$

$$F_C = F_{C1} + F_{C2} + F_{C3} = 11.52 \text{ kips}$$

Since F_C is the resultant of F_{C1}, F_{C2}, and F_{C3}, we determine its location by computing first moments using the dimensions shown in Fig. 3a:

$$F_{C1}(d_{C1}) + F_{C2}(d_{C2}) + F_{C3}(d_{C3}) = F_C d_C$$

So,

$$d_C = \frac{(0.72 \text{ kips})(0.67 \text{ in.}) + (7.20 \text{ kips})(1.50 \text{ in.}) + (3.60 \text{ kips})(1.67 \text{ in.})}{11.52 \text{ kips}}$$

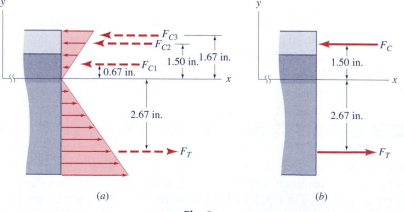

(a)   (b)

Fig. 3

or

$$d_C = 1.50 \text{ in.}$$

as indicated in Fig. 3b. Therefore,

$$d = d_C + d_T = 1.50 \text{ in.} + 2.67 \text{ in.} = 4.17 \text{ in.}$$

$$F_C = F_T = 11.52 \text{ kips}, \quad d = 4.17 \text{ in.} \qquad \textbf{Ans. (a)}$$

(b) *Determine the magnitude of the couple formed by F_C and F_T.* Since $F_C = F_T$, the resultant force on the cross section is zero. The couple formed by F_C and F_T is given by

$$M = F_C d = F_T d = (11.52 \text{ kips})(4.17 \text{ in.}) = 48.0 \text{ kip} \cdot \text{in.} \qquad \textbf{Ans. (b)}$$

which is the value of the applied moment.

Review the Solution The results of this example agree with our solution in Example 6.2. We can conclude that the stresses on a cross section due to a bending moment acting on the cross section are equivalent to a couple consisting of equal tensile and compressive forces acting on their respective portions of the cross section.

While the previous two examples illustrate the calculation of flexural stresses on a particular cross section with prescribed moment, it is also important for the absolute maximum tensile stress and absolute maximum compressive stress in a beam under given support and loading conditions to be determined. This is where a bending-moment diagram is very useful, as you will see in the following example.

EXAMPLE 6.4

A beam whose cross section is the T section of Example Problem 6.2 is subjected to the loading shown in Fig. 1a. The shear diagram that corresponds to this loading is given in Fig. 1b. (a) Using the procedure illustrated in Examples 5.7 through 5.9 (Section 5.5), sketch the moment diagram for this beam. (b) Compute the maximum compressive flexural stress and the maximum tensile flexural stress in this beam.

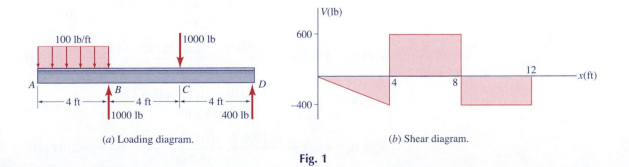

(a) Loading diagram.

(b) Shear diagram.

Fig. 1

Plan the Solution Once we have constructed the moment diagram, using Eqs. 5.3 and 5.7, we can determine the cross sections at which we must calculate maximum and minimum flexural stresses. The flexure formula, Eq. 6.13, may then be used to calculate the maximum compressive stress and the maximum tensile stress.

Solution (a) *Sketch the moment diagram.* Use

$$V = \frac{dM}{dx}$$

and

$$M_2 - M_1 = \int_{x_1}^{x_2} V(x)\,dx$$

to construct the moment diagram (Fig. 2a). From the diagram, the maximum negative moment is $M_B \equiv M\,(4\text{ ft}) = -800\text{ lb} \cdot \text{ft}$, and the maximum positive moment is $M_C \equiv M\,(8\text{ ft}) = 1600\text{ lb} \cdot \text{ft}$.

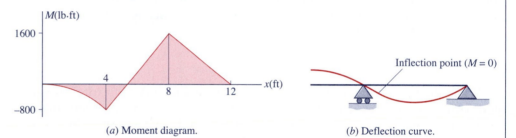

(a) Moment diagram.　　　　　(b) Deflection curve.

Fig. 2

(b) *Compute the maximum compressive flexural stress and the maximum tensile flexural stress.* In Example Problem 6.2, the location of the centroid and the value of $I \equiv I_z$ were calculated. These are shown in Fig. 3. At section B, the top of the beam is in tension and the bottom is in compression; at section C the top of the beam is in compression and the bottom is in tension. Just to be safe, we can compute σ_x at these four points and then select the maxima. We use the flexure formula, Eq. 6.13.

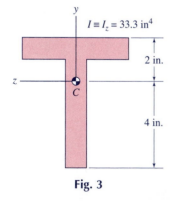

$I \equiv I_z = 33.3\text{ in}^4$

2 in.

4 in.

Fig. 3

$$\sigma_x = \frac{-My}{I}:$$

$$\sigma_x(4\text{ ft}, 2\text{ in.}) = \frac{-(-800\text{ lb} \cdot \text{ft})(12\text{ in./ft})(2\text{ in.})}{33.3\text{ in}^4} = 576\text{ psi}$$

$$\sigma_x(4\text{ ft}, -4\text{ in.}) = \frac{-(-800\text{ lb} \cdot \text{ft})(12\text{ in./ft})(-4\text{ in.})}{33.3\text{ in}^4} = -1152\text{ psi}$$

$$\sigma_x(8\text{ ft}, 2\text{ in.}) = \frac{-(1600\text{ lb} \cdot \text{ft})(12\text{ in./ft})(2\text{ in.})}{33.3\text{ in}^4} = -1152\text{ psi}$$

$$\sigma_x(8\text{ ft}, -4\text{ in.}) = \frac{-(1600\text{ lb} \cdot \text{ft})(12\text{ in./ft})(-4\text{ in.})}{33.3\text{ in}^4} = 2304\text{ psi}$$

Therefore, the maximum tensile flexural stress occurs at the bottom of the beam at section C. By coincidence, the particular loading and cross section of this beam produce equal maxima of compressive flexural stress at the bottom of the beam at B and at the top of the beam at C. The flexural stresses at sections B and C, rounded to the nearest 10 psi, are indicated on Fig. 4.

Fig. 4

$$\sigma_{\max T} = 2300 \text{ psi}, \qquad \sigma_{\max C} = -1150 \text{ psi} \qquad \textbf{Ans. (c)}$$

Review the Solution   First, as a check on the moment diagram, we can see if the moment distribution corresponds to a reasonable shape of the deflected beam, according to Fig. 6.9 and Eq. 6.12. A sketch of the deflection curve (Fig. 2b) is drawn adjacent to the moment diagram. The shape of the deflection curve does seem reasonable. The values of bending moments M_B and M_C can be checked easily by using free-body diagrams based on Fig. 1a. Each does have the correct magnitude and sign. The magnitudes and signs of the four calculated stresses are easily checked.

MDS6.2 & 6.3   **Beam Normal Stresses**—*Flexure* is an MDS computer program module that combines a program for plotting shear and moment diagrams for statically determinate beams, a program for computing section properties, and a program for computing *flexural stress in beams*.

In Example Problems 6.2 and 6.4, both the loading and the dimensions, including the shape of the cross section, were given. In Section 6.4 we will consider a method for selecting an appropriate cross section when the loading on the beam is given. This, then, is a design exercise.

6.4   DESIGN OF BEAMS FOR STRENGTH

The design of beams for specific applications is usually governed by detailed specifications and codes involving design requirements and procedures that are beyond the scope of this text.[7] In this section, however, we will consider the design of beams based on **allowable flexural stress.** That is, given the bending-moment distribution in a beam, and given the allowable tensile stress and allowable compressive stress of the material to be used, an appropriate beam cross section is to be selected. We will return to the topic of design of beams in Chapter 9, where the combined effect of flexural stress (σ_x) and transverse shear stress (τ_{xy}) will be considered.

Before we consider the actual process of selecting the cross section of a beam, let us first look at the types of beams available for selection. Standard sizes are available for beams made of wood and for beams made of steel, aluminum, and other metals. Appendices D.1 through D.10 give the properties of selected steel, aluminum, and wood structural shapes. More extensive tables may be found, for example, in publications of the American Institute of Steel Construction (AISC) [Ref. 6-2] and the Aluminum Association [Ref. 6-3]. Figure 6.15 illustrates five structural steel shapes.

---

[7]For example, the *Manual of Steel Construction* [Ref. 6-2], published by the American Institute of Steel Construction, describes how to design steel beams for structural applications.

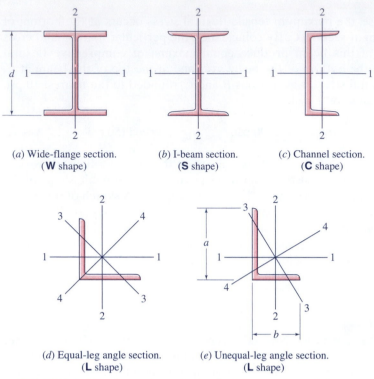

(a) Wide-flange section.
(**W** shape)

(b) I-beam section.
(**S** shape)

(c) Channel section.
(**C** shape)

(d) Equal-leg angle section.
(**L** shape)

(e) Unequal-leg angle section.
(**L** shape)

FIGURE 6.15   Several standard rolled structural steel shapes.

Shapes like the ones illustrated in Fig. 6.15 are produced by passing a hot billet of metal between sets of rollers that, after several passes, produce the desired shape. The most commonly used shape is the wide-flange section illustrated in Fig. 6.15a. The American Standard beam, commonly called the I-beam (Fig. 6-15b), is less frequently used because it tends to have excessive material in the web, and its flanges are generally too narrow to provide adequate lateral stiffness. A **W**-shape, **S**-shape, or **C**-shape is designated by its symbol, followed by its *nominal depth* in inches and its *weight per foot* in pounds (e.g., **W**12×96, **S**24×100, **C**10×25). In SI units, the *nominal depth* is given in mm and the *mass* is given in kg/m. Angles are designated by the symbol **L**, followed by the leg lengths (longer leg first), followed by the thickness (e.g., **L**8×8×3/4, **L**5×3×1/2).

In the case of wood beams, it is important to note that the quoted dimensions of lumber are nominal, rough-cut, dimensions. The actual net dimensions, or *finish dimensions*, which are given in Appendix D.8, are smaller. Thus, the finish dimensions should be used in all structural calculations.

The flexure formula, Eq. 6.13, forms the basis for beam design based on flexural stress. Consider a general situation like Example Problem 6.4, where the beam is not doubly symmetric and where the maximum positive bending moment and the maximum negative bending moment have different magnitudes. From Eq. 6.13 the fiber stresses at the extreme (i.e., top and bottom) fibers may be written, respectively, as

$$\sigma_{\text{top}} \equiv \sigma_1 = \frac{-M(c_1)}{I} = \frac{-Mc_1}{I} \equiv -\frac{M}{S_1}$$

$$\sigma_{\text{bot}} \equiv \sigma_2 = \frac{-M(-c_2)}{I} = \frac{Mc_2}{I} \equiv \frac{M}{S_2}$$

(6.14)

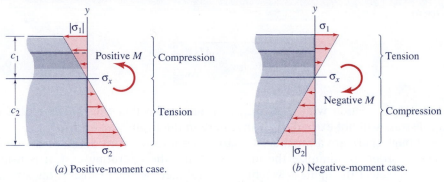

(a) Positive-moment case. (b) Negative-moment case.

FIGURE 6.16 Examples of the stress distribution in an unsymmetric beam.

where

$$S = \frac{I}{c} \tag{6.15}$$

In this expression, I is the moment of inertia about the neutral axis, and c is the distance to an extreme fiber (see Fig. 6.16a). The quantity S, called the **elastic section modulus,** is a property of the cross-sectional dimensions. For most structural shapes, the values of S are tabulated, along with the location of the centroid of the cross section and the moment of inertia values (e.g., see Appendices D.1 through D.10).

Allowable-Stress Design.[8] Consider first the simple case of selecting a beam with doubly symmetric cross section made of material whose *allowable stress* (with the same magnitude in both tension and compression) is σ_{allow}. The value of σ_{allow} is determined by applying a *factor of safety* to the value of the yield strength of the material, as given in a table of material properties (see Appendix F). Let M_{max} be the maximum absolute value of the bending moment $M(x)$, that is, let

$$M_{\text{max}} \equiv \max_x |M(x)| \tag{6.16}$$

as illustrated in Fig. 6.17. For design purposes, it is convenient to combine Eqs. 6.14, 6.15, and 6.16 and write the resulting **allowable-stress design equation** in

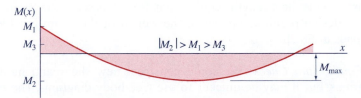

FIGURE 6.17 An example bending-moment diagram.

---

[8]Although the *Load and Resistance Factor Design Method*, mentioned briefly in Section 2.8, is now widely used in designing beams (e.g., the *AISC Manual of Steel Construction*, [Ref. 6-2]), application of that method is beyond the scope of this textbook.

the form

$$S_{\text{design}} = \frac{M_{\text{max}}}{\sigma_{\text{allow}}} \tag{6.17}$$

By selecting a section with $S \geq S_{\text{design}}$, we guarantee that the magnitude of the flexural stress will not exceed σ_{allow} anywhere in the beam.

Although there are a number of factors that must be considered in any design process in order to minimize the initial cost and the operating cost, it is usually desirable to select the lightest-weight structural member that satisfies the strength requirement (and all other requirements that are applicable). The following example problem illustrates this design process.

EXAMPLE 6.5

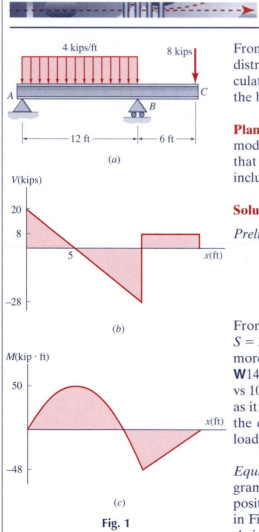

(a)

(b)

(c)

Fig. 1

From Appendix D.1, select a wide-flange steel beam to support the load distribution shown in Fig. 1a. Include the weight of the beam in your calculations. The moment diagram for the beam, neglecting the weight of the beam, is shown in Fig. 1c. Let $\sigma_{\text{allow}} = 19$ ksi.

Plan the Solution We can use Eq. 6.17 to compute the required section modulus. We need to pick the lightest section, with some margin in S so that we can accommodate the maximum moment with the beam weight included.

Solution

Preliminary Shape Selection: From Eq. 6.17,

$$S_{\text{design}} = \frac{M_{\text{max}}}{\sigma_{\text{allow}}} = \frac{(50 \text{ kip} \cdot \text{ft})(12 \text{ in./ft})}{(19 \text{ kips/in}^2)} = 31.6 \text{ in}^3$$

From Appendix D.1, there are two candidate sections, a **W**10×30 with $S = 32.4$ in^3 and a **W**14×26 with $S = 35.3$ in^3.[9] The **W**10×30 weighs 4 lb/ft more than the **W**14×26, so it seems that the best choice would be the **W**14×26. However, since this beam is deeper than the **W**10×30 (13.91 in. vs 10.47 in.), there may be justification for choosing the **W**10×30 as long as it meets the strength requirement. So, let us see if the **W**10×30 meets the design requirements when the weight of the beam is added to the loads in Fig. 1a.

Equilibrium Check: We could construct new shear and moment diagrams, but it may be quicker to use free-body diagrams. The maximum positive moment will occur at the point x_m, where $V(x_m) = 0$ ($x_m = 5$ ft in Fig. 1b). We first need to compute the reaction at A, which is labeled A_y in Fig. 2.

---

[9]There are additional candidate sections in the complete AISC tables [Ref. 6-2].

$$+\zeta\left(\sum M\right)_B = 0:$$

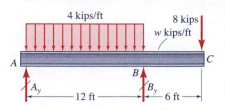

$$(4 \text{ kips/ft})(12 \text{ ft})(6 \text{ ft}) - (8 \text{ kips})(6 \text{ ft})$$
$$+ (w \text{ kips/ft})(18 \text{ ft})(3 \text{ ft}) - A_y(12 \text{ ft}) = 0$$
$$A_y = (20 + 4.5w) \text{ kips}$$

Fig. 2 Free-body diagram.

We can now determine the maximum positive moment using Fig. 3.

$$V(x_m) = (20 + 4.5w) - (4 + w)x_m = 0$$

Therefore, the maximum positive moment occurs at

$$x_m = \frac{20 + 4.5w}{4 + w}$$

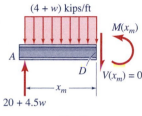

Fig. 3

$$\left(\sum M\right)_D = 0:$$

$$M(x_m) = (20 + 4.5w)\left(\frac{20 + 4.5w}{4 + w}\right) - (4 + w)\left(\frac{20 + 4.5w}{4 + w}\right)^2\left(\frac{1}{2}\right)$$

$$M(x_m) = \frac{(20 + 4.5w)^2}{2(4 + w)} \text{ kip} \cdot \text{ft}$$

The maximum negative moment will still occur at B, and its value can be computed using Fig. 4.

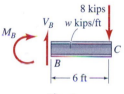

$$\left(\sum M\right)_B = 0: \qquad M_B = (-48 - 18w) \text{ kip} \cdot \text{ft}$$

Fig. 4

For the **W**10×30 beam, $w = 30 \text{ lb/ft} = 0.030 \text{ kips/ft}$. Therefore,

$$M(x_m) = 50.3 \text{ kip} \cdot \text{ft}, \qquad M_B = -48.5 \text{ kip} \cdot \text{ft}$$

So $M_{\text{max}} = 50.3 \text{ kip} \cdot \text{ft}$, which would require

$$S_{\text{design}} = \frac{(50.3 \text{ kip} \cdot \text{ft})(12 \text{ in./ft})}{19 \text{ ksi}} = 31.8 \text{ in}^3$$

Final Shape Selection: Since the **W**14×26 beam is lighter than the **W**10×30, and since its section modulus is greater (35.3 in^3 vs 32.4 in^3), the **W**14×26 would be the best design, unless its added depth is, for some reason, undesirable. In that case, the **W**10×30 would be a perfectly satisfactory design choice.

Review the Solution It is interesting to note that an 18-ft **W**14×26 beam, weighing 468 lb is able, in this case, to support a total load of 56,000 lb. (We will again consider the design of the beam in Fig. 1 in Section 9.3, after we have taken up the topics of shear stress in beams and the state of stress at a point.)

The previous discussion and Example Problem 6.5 assumed the same allowable stress in tension and compression and assumed that the choice of beam cross section was to be made among candidate doubly symmetric sections. If the allowable stresses in tension and compression are different, this must be taken into account, and stresses must be computed where $M(x)$ is a maximum and also where it is a minimum. Also, if the cross section is not symmetric about the neutral axis, Eqs. 6.14 must be applied at the sections of maximum positive moment and maximum negative moment.

MDS6.4 & 6.5 Beam Design—*Flexure* is an MDS computer program module that combines a program for plotting shear and moment diagrams for statically determinate beams, a program for computing section properties, and a program for computing normal stresses in beams.

In Example Problem 6.5 you learned that it is desirable to select a beam that maximizes S, the elastic section modulus, and minimizes the weight. The next example problem indicates how the shape of the cross section affects these two quantities and, more importantly, the **strength-to-weight ratio.**

EXAMPLE 6.6

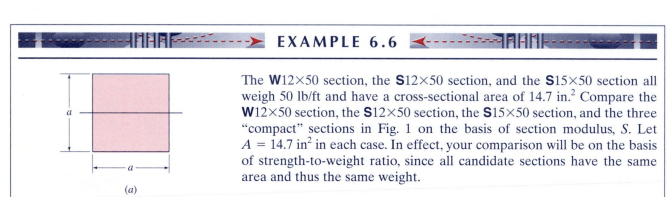

The **W**12×50 section, the **S**12×50 section, and the **S**15×50 section all weigh 50 lb/ft and have a cross-sectional area of 14.7 in.² Compare the **W**12×50 section, the **S**12×50 section, the **S**15×50 section, and the three "compact" sections in Fig. 1 on the basis of section modulus, S. Let $A = 14.7$ in² in each case. In effect, your comparison will be on the basis of strength-to-weight ratio, since all candidate sections have the same area and thus the same weight.

Plan the Solution Equation 6.15 gives the formula for the elastic section modulus. S, and Appendix C.2 and inside back covers give formulas for the area moments of inertia, I, for the compact sections. Appendices D.1 and D.3 give S values for the structural steel shapes.

Solution From Eq. 6.15,

$$S = \frac{I}{c}$$

and, from Appendix C.2,

Square: $\qquad S = \dfrac{I}{c} = \left(\dfrac{a^4}{12}\right)\left(\dfrac{2}{a}\right) = \dfrac{a^3}{6}$.

Rectangle: $\qquad S = \dfrac{I}{c} = \left[\dfrac{b(2b)^3}{12}\right]\left(\dfrac{1}{b}\right) = \dfrac{2b^3}{3}$

Circle: $\qquad S = \dfrac{I}{c} = \left(\dfrac{\pi r^4}{4}\right)\left(\dfrac{1}{r}\right) = \dfrac{\pi r^3}{4}$

Fig. 1 Three "compact" shapes.

It should be carefully noted that I_1 and I_2 are the moments of inertia of areas 1 and 2 <u>about the neutral axis</u>, defined by Eq. 6.20, not about their respective centroidal axes. Equation 6.22 is the **moment-curvature equation** for a beam that is nonhomogeneous within the cross section. Equations 6.20 and 6.23 can readily be extended to accommodate additional materials, but such situations are rare.

Flexure Formulas: Finally, we substitute Eq. 6.22 into Eqs. 6.18 to get the following expressions for the flexural stresses in materials 1 and 2.

$$\sigma_{x_1} = \frac{-ME_1 y}{\overline{EI}}, \qquad \sigma_{x_2} = \frac{-ME_2 y}{\overline{EI}} \qquad \text{Flexure Formulas} \qquad (6.24)$$

►►► EXAMPLE 6.7 ◄◄◄

A nonhomogeneous beam having the dimensions shown in Fig. 1*a* is constructed by gluing a thin aluminum plate to the top side of a square wood beam. Take $E_1 \equiv E_{alum.} = 70$ GPa and $E_2 \equiv E_{wood} = 12$ GPa.

Determine the maximum flexural stress in the aluminum and the maximum flexural stress in the wood when a moment $M = 3$ kN · m is applied in the manner indicated in Fig. 1*b*.

Plan the Solution We must use Eq. 6.20 to locate the neutral axis, and we can then use Eqs. 6.24 to compute the flexural stresses in the two materials.

Solution

Neutral-Axis Location: The neutral axis of the beam is located at distance c_1, from the top of the beam, as indicated in Fig. 2. From Eq. 6.20.

$$E_1 \bar{y}_1 A_1 + E_2 \bar{y}_2 A_2 = 0$$

$$(70 \text{ GPa})(1000 \text{ mm}^2)(c_1 - 5 \text{ mm})$$

$$+ (12 \text{ GPa})(10\,000 \text{ mm}^2)(c_1 - 60 \text{ mm}) = 0$$

Then,

$$c_1 = 39.74 \text{ mm}, \qquad c_2 = 70.26 \text{ mm}$$

and

$$\bar{y}_1 = 34.74 \text{ mm}, \qquad \bar{y}_2 = -20.26 \text{ mm}$$

Moment of Inertia: Since the moments of inertia I_1 and I_2 are to be taken about the z axis (neutral axis), we must use the parallel-axis theorem.

$$I_1 = \frac{b_1 h_1^3}{12} + A_1 \bar{y}_1^2$$

$$= \frac{(100 \text{ mm})(10 \text{ mm})^3}{12} + (100 \text{ mm})(10 \text{ mm})(34.74 \text{ mm})^2$$

$$= 1.215(10^6) \text{ mm}^4 = 1.215(10^6) \text{ mm}^4 (10^3 \text{ mm/m})^{-4}$$

$$I_1 = 1.215(10^{-6}) \text{ m}^4$$

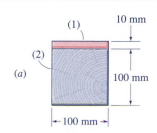

(a)

(b)

Fig. 1 A beam made of two materials.

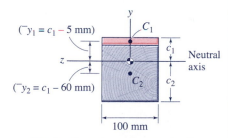

Fig. 2 Location of neutral axis.

Similarly,

$$I_2 = 12.439(10^{-6}) \text{ m}^4$$

Before calculating the flexural stresses we can check these moments of inertia by using the equation

$$I_z = I_1 + I_2 = 13.65(10^6) \text{ mm}^4$$

But,

$$I_z = (I_z)_{\substack{\text{area} \\ \text{above}}} + (I_z)_{\substack{\text{area} \\ \text{below}}}$$

$$= \frac{1}{3}(100 \text{ mm})(39.74 \text{ mm})^3 + \frac{1}{3}(100 \text{ mm})(70.26 \text{ mm})^3$$

$$= 13.65(10^6) \text{ mm}^4$$

Flexural Stresses: The flexural stresses in materials 1 and 2 are given by Eqs. 6.24, with $\overline{EI}$ defined by Eq. 6.23.

$$\overline{EI} = E_1 I_1 + E_2 I_2 = [70(10^9) \text{ N/m}^2][1.215(10^{-6}) \text{ m}^4]$$

$$+ [12(10^9) \text{ N/m}^2][12.439(10^{-6}) \text{ m}^4]$$

$$= 234.3(10^3) \text{ N} \cdot \text{m}^2 = 234.3 \text{ kN} \cdot \text{m}^2$$

Finally, the maximum stresses in materials 1 and 2, which occur at $y = c_1$ and $y = -c_2$, respectively, are

$$(\sigma_{x_1})_{\text{max}} = \frac{-M E_1(c_1)}{\overline{EI}}$$

$$= \frac{-(3 \text{ kN} \cdot \text{m})(70 \times 10^6 \text{ kN/m}^2)(39.74 \times 10^{-3} \text{ m})}{234.3 \text{ kN} \cdot \text{m}^2}$$

$$= -35.6(10^3) \text{ kN/m}^2 = -35.6 \text{ MPa}$$

$$(\sigma_{x_2})_{\text{max}} = \frac{-M E_2(-c_2)}{\overline{EI}}$$

$$= \frac{-(3 \text{ kN} \cdot \text{m})(12 \times 10^6 \text{ kN/m}^2)(-70.26 \times 10^{-3} \text{ m})}{234.3 \text{ kN} \cdot \text{m}^2}$$

$$= 10.79(10^3) \text{ kN/m}^2 = 10.79 \text{ MPa}$$

$$(\sigma_{x_1})_{\text{max}} = -35.6 \text{ MPa}, \qquad (\sigma_{x_2})_{\text{max}} = 10.79 \text{ MPa} \qquad \textbf{Ans.}$$

Review the Solution   This is the type of problem where we should check our result at each major step, as we did for the moment of inertia. We should expect the stiffer aluminum to experience higher stresses than those in the wood, and this is the case here.

This same problem is solved, in Example Problem 6.8, by the *trans-formed-section method.*

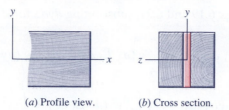

(a) Profile view.   (b) Cross section.

FIGURE 6.20 A form of nonhomogeneous beam.

A nonhomogeneous beam may also be constructed by forming a sandwich beam that is symmetric about the plane of loading, as illustrated in Fig. 6.20. For example, beams of this type may be constructed by bonding wood beams to a steel-plate core. In this case Eqs. 6.23 and 6.24 are used, but it is unnecessary to use Eq. 6.20 since the location of the neutral axis is obvious.

Transformed-Section Method. Equations 6.18, 6.20, and 6.22 are the key equations in the analysis of nonhomogeneous beams since they completely relate the flexural stress distribution to the stress resultants $F(x) = 0$ and $M(x)$. By creating a *transformed section* it is possible to treat a nonhomogeneous beam essentially the same way as a homogeneous beam. That is, *the neutral axis passes through the centroid of the transformed section, and the flexural stress is determined by a simple flexure formula* of the form

$$\sigma = \frac{-My}{I} \qquad (6.13)$$

repeated

Figure 6.21 shows the original two-material cross section (Figs. 6.21*a*, *b*) and the transformed section with *material 2 as the reference material* (Fig. 6.21*c*). (Material 1 is taken to be the stiffer material, as indicated in Fig. 6.19.) Note that the transformed material (1) is "widened" (since $E_1 > E_2$) only in the z direction; it is unchanged in the y direction. We want y-distances to be the same in the original and the transformed section so that the distance y in the flexure formula will be unaltered.

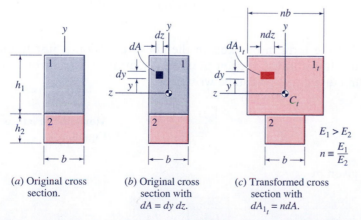

(a) Original cross section.   (b) Original cross section with $dA = dy\,dz$.   (c) Transformed cross section with $dA_{1_t} = n\,dA$.

FIGURE 6.21 Basic geometry of the transformed section representing a nonhomogeneous beam.

Also, if we rewrite Eqs. 6.19 and 6.21, substituting $dy\,dz$ for dA,

$$F(x) = \int_A \sigma_x\, dA = -\frac{1}{\rho}\int_{A_1} y\,(dy\,E_1\,dz) - \frac{1}{\rho}\int_{A_2} y\,(dy\,E_2\,dz) = 0$$

$$M = -\int_A y\sigma_x\, dA = \frac{1}{\rho}\left[\int_{A_2} y^2\,(dy\,E_1\,dz) + \int_{A_2} y^2\,(dy\,E_2\,dz)\right]$$

it is clear that, in order to leave y terms unchanged, we should group moduli E_1 and E_2 with dz, as illustrated in Fig. 6.21c.

Neutral Axis of the Transformed Section: Let the ratio of moduli, with E_2 as the reference, be

$$n \equiv \frac{E_1}{E_2} \tag{6.25}$$

Then, Eq. 6.19, which locates the neutral axis, can be written as

$$\int_{A_1} y\,(n\,dA) + \int_{A_2} y\,dA = 0$$

or, for the transformed cross section A_t simply

$$\boxed{\int_{A_t} y\,dA_t = 0} \tag{6.26}$$

Therefore, the *neutral axis passes through the centroid of the transformed section*, just as it would pass through the centroid of a homogeneous beam.

Moment-Curvature Equation: Equation 6.21 is the moment-curvature equation for a two-material, nonhomogeneous beam, and we want this equation to hold for the transformed section. Introducing the *modulus ratio n*, we can write Eq. 6.21 as

$$M = \frac{E_2}{\rho}\left[\int_{A_1} y^2(n\,dA) + \int_{A_2} y^2\,dA\right]$$

It is clear that the term in brackets in the above equation is just the moment of inertia of the transformed section about its neutral axis (centroid). Hence, the moment-curvature equation can be written as

$$\boxed{M = \frac{E_2 I_t}{\rho}} \qquad \text{Moment-Curvature Equation} \tag{6.27}$$

where I_t is given by

$$\boxed{I_t = \int_{A_t} y^2\,dA_t} \tag{6.28}$$

Flexure Formulas: Finally, we can substitute Eq. 6.27 into Eqs. 6.18 to determine the stress distribution in each material. We get

$$\sigma_{x_1} = -\frac{E_1 y}{\rho} = n\left(\frac{-My}{I_t}\right)$$

$$\sigma_{x_2} = -\frac{E_2 y}{\rho} = -\frac{My}{I_t}$$

Flexure Formulas (6.29)

Thus, the stress in the reference material is computed using the standard flexure formula, but the stress in the transformed material must be multiplied by the modulus ratio, n.

Material 2 was taken as the reference material for the transformed-section analysis in Eqs. 6.25 through 6.29. Usually the less-stiff material is taken as the reference material, so that $n > 1$, but this is not essential. However, since the labeling of materials as "material 1" and "material 2" is completely arbitrary, the less-stiff material can always be labeled as "material 2."

EXAMPLE 6.8

Solve the problem stated in Example Problem 6.7 using the transformed-section method, with material 2 as the reference material.

Plan the Solution The following four steps are required to solve this problem:

1. Sketch the transformed section using the modulus ratio n as the "z-stretch factor."
2. Locate the neutral axis (the centroid of the transformed section).
3. Compute I_t for the transformed section.
4. Use Eqs. 6.29 to compute the stresses.

Solution

Transformed Section: The modulus ratio (Eq. 6.25) is

$$n = \frac{E_1}{E_2} = \frac{70 \text{ GPa}}{12 \text{ GPa}} = 5.833$$

Therefore, the aluminum part of the cross section must be stretched by a factor $n = 5.833$. The resulting transformed section is shown in Fig. 1.

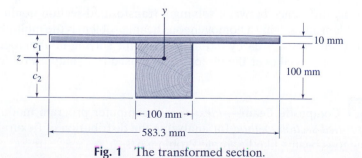

Fig. 1 The transformed section.

Neutral Axis: Since the transformed beam is "homogeneous" with $E_1 = E_2$, the neutral axis passes through the centroid of the transformed section. Therefore,

$$(583.3 \text{ mm})(10 \text{ mm})(c_1 - 5 \text{ mm}) + (100 \text{ mm})(100 \text{ mm})(c_1 - 60 \text{ mm}) = 0$$

$$c_1 = 39.74 \text{ mm}, \qquad c_2 = 70.26 \text{ mm}$$

and

$$\bar{y}_1 = 34.74 \text{ mm} \qquad \bar{y}_2 = -20.26 \text{ mm}$$

Moment of Inertia:

$$I_t = I_{1t} + I_2 = \sum \left(\frac{1}{12} b_t h^3 + A_t d^2 \right)$$

$$= \frac{1}{12}(583.3 \text{ mm})(10 \text{ mm})^3 + (583.3 \text{ mm})(10 \text{ mm})(34.74 \text{ mm})^2$$

$$+ \frac{1}{12}(100 \text{ mm})(100 \text{ mm})^3 + (100 \text{ mm})(100 \text{ mm})(-20.26 \text{ mm})^2$$

$$= 19.53(10^6) \text{ mm}^4 = 19.53(10^{-6}) \text{ m}^4$$

Flexural Stresses: Using Eqs. 6.29, we get

$$\sigma_{x_1} = n\left(\frac{-Mc_1}{I_t} \right) = \frac{5.833 \, (-3 \text{ kN} \cdot \text{m})(39.74 \times 10^{-3} \text{ m})}{(19.53 \times 10^{-6} \text{ m}^4)}$$

$$= -35.6 \text{ MPa}$$

$$\sigma_{x_2} = \frac{-M(-c_2)}{I_t} = \frac{(-3 \text{ kN} \cdot \text{m})(-70.26 \times 10^{-3} \text{ m})}{(19.53 \times 10^{-6} \text{ m}^4)} = 10.79 \text{ MPa}$$

$$\sigma_{x_1} = -35.6 \text{ MPa}, \qquad \sigma_{x_2} = 10.79 \text{ MPa} \qquad\qquad \textbf{Ans.}$$

Review the Solution   We have obtained the same answers as in Example Problem 6.7, so we can assume that they are probably correct.

Since the only difference between solving a transformed-section nonhomogeneous beam problem and solving a homogeneous beam problem is the multiplication of stress in the transformed area by n, this method for solving nonhomogeneous beam problems is preferred over the direct method used in Example Problem 6.7.

MDS6.6 & 6.7   **Composite Beams**—*Flexure* is a computer program module that uses the *transformed-section method* for analyzing the distribution of flexural stress in nonhomogeneous beams, like Example Problem 6.8.

*6.6 UNSYMMETRIC BENDING

Thus far in Chapter 6, we have been considering flexural stress and strain in beams whose cross-sectional shape and whose loading and support conditions produce bending that is confined to a longitudinal plane of symmetry (LPS) of the beam. This simplifies the analysis in two important respects. First, the deflection of the beam can be characterized by a deflection curve in the LPS (e.g., Fig. 6.1); second, there is no tendency of the beam to twist. However, we also need to be able to analyze the behavior of beams that are not loaded and supported in this simple manner.

In Chapter 9 we will consider *combined bending and torsion,* as would be experienced, for example, by a beam that is loaded parallel to, but not along, an axis of symmetry, like the doubly symmetric box beam in Fig. 6.22. Here, however, we will consider loading that does not produce bending in a single longitudinal plane of symmetry. Two examples of this are a channel beam loaded parallel to its web (Fig. 6.23a), and a Z section (Fig. 6.23b). The former beam has a longitudinal plane of symmetry, but the loading is perpendicular to, not in or parallel to, the plane of symmetry. The latter beam has no longitudinal plane of symmetry.

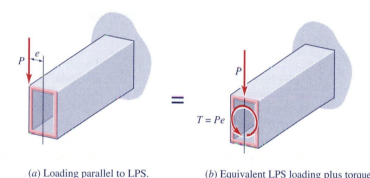

(a) Loading parallel to LPS.   (b) Equivalent LPS loading plus torque.

FIGURE 6.22   A beam with transverse loading parallel to an axis of symmetry.

Doubly Symmetric Beams with Inclined Loads.   Before we study the general case of unsymmetric bending, let us generalize the results of Section 6.3 to the case of a **doubly symmetric beam** whose loading does not lie in either longitudinal plane of symmetry. Figure 6.24 shows a doubly symmetric beam with such an **inclined load,** that is, a load that simultaneously produces bending about both axes of symmetry in the cross section.

Flexural Stress: For a doubly symmetric, linearly elastic beam with $E = E(x)$, the **flexure formula,** Eq. 6.13, can be applied separately for M_y and M_z, and the two

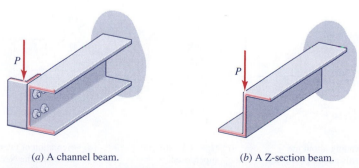

(a) A channel beam.   (b) A Z-section beam.

FIGURE 6.23   Two beams that are not loaded in a plane of symmetry.

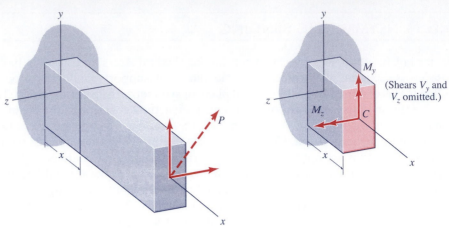

FIGURE 6.24 A doubly symmetric beam with inclined loading.

(a) Components of load in two planes of symmetry.

(b) Bending moments due to an inclined load (positive M_y and positive M_z shown).

expressions can then, by the principle of linear superposition, be added to give

$$\sigma_x = \frac{M_y z}{I_y} - \frac{M_z y}{I_z}$$

Flexure Formula (6.30)

The second term on the right corresponds to bending about the z axis of symmetry and comes directly from Eq. 6.13. The first term is a modification of Eq. 6.13 for bending about the y axis of symmetry. Note that this M_y term in Eq. 6.30 is positive. As illustrated in Fig. 6.25a, a positive M_y produces tension (i.e., positive σ_x) where z is positive. Since the y axis and z axis are both axes of symmetry of the cross section, their origin is the centroid of the cross section, as indicated in Fig. 6.24b. Figure 6.25 illustrates the superposition of the stresses due to moment components M_y and M_z acting on a rectangular cross section.

Orientation of the Neutral Axis: Due to the combined stresses, the beam will bend about the inclined neutral axis indicated by NA in Figs. 6.25c and 6.25d. The orientation of the neutral axis in a cross section may be determined by setting $\sigma_x = 0$ in Eq. 6.30. Then, if (y^*, z^*) are the coordinates of points that lie on

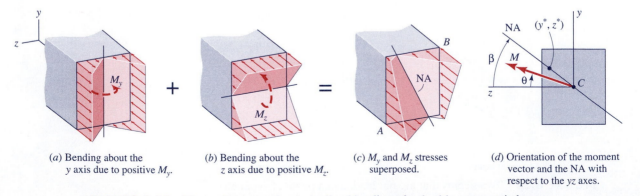

(a) Bending about the y axis due to positive M_y.

(b) Bending about the z axis due to positive M_z.

(c) M_y and M_z stresses superposed.

(d) Orientation of the moment vector and the NA with respect to the yz axes.

FIGURE 6.25 Flexural stresses due to inclined loading of a doubly symmetric beam.

$$\left(\frac{M_y}{I_y}\right)z^* - \left(\frac{M_z}{I_z}\right)y^* = 0 \tag{6.31}$$

is the equation of the **neutral axis** in the yz plane. Let the angles θ and β be defined by

$$\tan\theta \equiv \frac{M_y}{M_z}, \qquad \tan\beta \equiv \frac{y^*}{z^*} \tag{6.32}$$

as illustrated in Fig. 6.25d. That is, the moment vector is oriented at angle θ measured clockwise from the positive z axis, and the neutral axis is oriented at angle β measured clockwise from the positive z axis. Then Eq. 6.31 may be conveniently expressed as

$$\tan\beta = \left(\frac{I_z}{I_y}\right)\tan\theta \tag{6.33}$$

From Eq. 6.33 it is clear that β will lie in the same quadrant as θ, and that the relative orientation of the *NA* and *M* will depend on whether $I_z > I_y$ or $I_z < I_y$.

Figure 6.25c illustrates how flexural stress changes from tension to compression at the neutral axis, where $\sigma_x = 0$. Unlike the case of loading in a plane of symmetry, inclined loading produces a neutral-axis orientation that depends, at each cross section, on the orientation of the bending-moment vector at the particular cross section, that is, on the ratio of $M_y(x)$ to $M_z(x)$.

Maximum Tensile and Compressive Stresses: Figure 6.25c illustrates the fact that **the points of maximum tension and maximum compression on the cross section are the two points that are farthest from the neutral axis.** These points are labeled A and B on Fig. 6.25c. We use the yz coordinates of these points in order to find the maximum stresses from Eq. 6.30.

Example Problem 6.9 illustrates the effect of load inclination on the flexural stress in a beam with doubly symmetric cross section.

EXAMPLE 6.9

The beam in Fig. 1, an **S**12×50 I-beam, is subjected to a moment $M = 150$ kip · in. that is supposed to lie along the z axis (e.g., due to loading in the xy plane). (a) Determine the effect of a "load misalignment" of $\theta = 2°$ on the orientation of the neutral axis. Show the neutral axis on a sketch of the cross section; and (b) determine the maximum tensile stress and the maximum compressive stress on the section. Note the large difference in I_y and I_z, and note how this affects the solution.

Plan the Solution We can use Eq. 6.33 (or Eq. 6.31) to determine the orientation of the neutral axis. Then we can identify the two points farthest from the NA and use Eq. 6.30 to compute the flexural stress at these points.

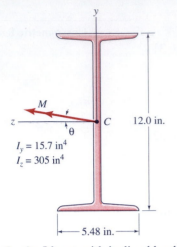

Fig. 1 An I-beam with inclined loading.

Solution   (a) *Locate the neutral axis.* From Eq. 6.33,

$$\tan \beta = \left(\frac{I_z}{I_y}\right) \tan \theta \tag{1}$$

where, from Fig. 1,

$$\tan \theta = \frac{M_y}{M_z} = \frac{M \sin \theta}{M \cos \theta} \tag{2}$$

Therefore,

$$\tan \beta = \left(\frac{305 \text{ in}^4}{15.7 \text{ in}^4}\right) \tan \theta \tag{3}$$

When $\theta = 0°, \beta = 0°$, and the z axis is the neutral axis, but when $\theta = 2°$ we get

$$\tan \beta = \left(\frac{305 \text{ in}^4}{15.7 \text{ in}^4}\right)(0.0349) = 0.678$$

So, for $\theta = 2°$,

$$\beta = 34.2° \qquad\qquad \textbf{Ans. (a)} \quad (4)$$

The orientation of the NA at this cross section is shown in Fig. 2. (b) *Determine the maximum tensile stress and maximum compressive stress.* The maximum stresses will occur at the extreme points A and B. From Eq. 6.30,

$$\sigma_x = \frac{M_y z}{I_y} - \frac{M_z y}{I_z} \tag{5}$$

$$\sigma_{x_A} \equiv \sigma_x(6.0, -2.74) = \frac{(150 \text{ kip} \cdot \text{in.})(\sin 2°)(-2.74 \text{ in.})}{15.7 \text{ in}^4}$$

$$- \frac{(150 \text{ kip} \cdot \text{in.})(\cos 2°)(6.0 \text{ in.})}{305 \text{ in}^4}$$

$$= -3.86 \text{ ksi}$$

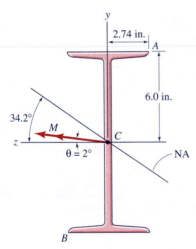

Fig. 2 Neutral axis orientation and points of maximum and minimum flexural stress.

From symmetry,

$$\sigma_{x_B} \equiv \sigma_x(-6.0, 2.74) = 3.86 \text{ ksi}$$

Had there been no misalignment, that is, for $\theta = 0$

$$\sigma_x(y, z) = \frac{-M_z y}{I_z} \tag{6}$$

so, the maximum compression is at $y = 6.0$ in., or

$$(\sigma_x)_{\mathrm{max}\,C} = \frac{-(150\ \mathrm{kip} \cdot \mathrm{in.})(6.0\ \mathrm{in.})}{305\ \mathrm{in}^4} = -2.95\ \mathrm{ksi}$$

and

$$(\sigma_x)_{\mathrm{max}\,T} = 2.95\ \mathrm{ksi}$$

In summary,

θ	β	$(\sigma_x)_{\mathrm{max}\,C}$	$(\sigma_x)_{\mathrm{max}\,T}$
$0°$	$0°$	-2.95 ksi	2.95 ksi
$2°$	$34.2°$	-3.86 ksi	3.86 ksi

Thus, a small *misalignment of 2° increases the maximum stresses by 31%.* This is due to the large ratio of I_z/I_y.

Review the Solution   The answers look reasonable, since a nonzero value of M_y, together with a large ratio of I_z/I_y, will cause a significant rotation of the neutral axis, that is, a large value of β. The values of the maximum stresses are not unreasonable.

   (See Homework Problem 6.6-5 where the same loading is applied to a very similar sized beam, a **W**12×50.)

MDS6.8   **Unsymmetric Bending**—The Unsymmetric option in the MDS computer program module *Flexure* combines a program for determining section properties with a program for computing the flexural stresses in a beam with arbitrarily oriented moment vector. The yz axes must be principal axes (i.e., $I_{yz} = 0$) with their origin at the centroid of the cross section.

Product of Inertia; Principal Axes of Inertia.   In studying unsymmetric bending we will need to make use of several geometric properties of plane areas. (These are discussed in greater detail in Appendix C.) The **product of inertia** of an area A (Fig. 6.26a) with respect to a yz reference frame in the plane of the area

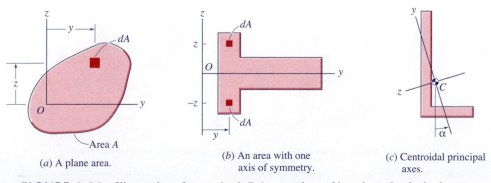

(a) A plane area.

(b) An area with one axis of symmetry.

(c) Centroidal principal axes.

FIGURE 6.26   Illustrations for use in defining product of inertia and principal axes.

is defined by

$$I_{yz} \equiv \int_A yz \, dA \tag{6.34}$$

If $I_{yz} = 0$, the y and z axes are said to be **principal axes of inertia** of the area. If either of the axes is an axis of symmetry, like the y axis in Fig. 6.26b, then $I_{yz} = 0$, since the contributions of symmetrically located dA's cancel in the integral for I_{yz}. Therefore, the axis of symmetry is one principal axis and <u>any</u> axis perpendicular to it is also a principal axis.

Finally, as discussed in Appendix C.3, for any planar area and for any origin in the plane, it is possible to orient a pair of orthogonal axes such that they are principal axes. For example, Fig. 6.26c shows the principal axes of inertia of an unequal-leg angle section with respect to an origin at the centroid. These are called **centroidal principal axes.** Appendix D.6 gives the angle of inclination of the centroidal principal axes of several unequal-leg angle sections.

Unsymmetric Bending: Arbitrary-Axis Method.
So far, we have shown that Eq. 6.30 applies to bending of beams with doubly symmetric cross section. Let us now determine an expression for the flexural stress σ_x in a beam with arbitrary cross section, including unsymmetric cross sections like those in Figs. 6.23b and 6.26c. In the remainder of this chapter axis labels y and z always refer to centroidal principal axes in the cross section, as in Fig. 6.26c, while y' and z' refer to arbitrarily oriented axes in the cross section, such as the axes in Fig. 6.27.

Flexure Formula: Consider the beam in Fig. 6.27a, which has no net axial force (i.e., $F = 0$), but which has both $M_{y'}$ and $M_{z'}$; that is, consider an arbitrarily oriented bending moment acting on an arbitrarily shaped cross section. The contribution of the flexural stress σ_x on an elemental area dA is shown in Fig. 6.27b. The stress resultants on the cross section are obtained by summing the contributions of elemental forces $dF = \sigma_x \, dA$ over the area of the cross section. This gives

$$F(x) = \int_A \sigma_x \, dA$$

$$M_{y'}(x) = \int z' \sigma_x \, dA \tag{6.35}$$

$$M_{z'}(x) = -\int_A y' \sigma_x \, dA$$

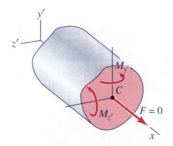

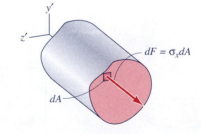

(a) The stress resultants related to σ_x.   (b) The normal force on an elemental area.

FIGURE 6.27 A beam with unsymmetric cross section and with arbitrarily oriented bending moment.

Let us assume that σ_x has the bilinear form

$$\sigma_x = a_0 + a_1 y' + a_2 z' \tag{6.36}$$

Then, combining Eqs. 6.35 with Eq. 6.36, we get

$$F(x) = a_0 \int_A dA + a_1 \int_A y' \, dA + a_2 \int_A z' \, dA$$

$$M_{y'}(x) = a_0 \int_A z' \, dA + a_1 \int_A z'y' \, dA + a_2 \int_A (z')^2 \, dA \tag{6.37}$$

$$M_{z'}(x) = -a_0 \int_A y' \, dA - a_1 \int_A (y')^2 \, dA - a_2 \int_A y'z' \, dA$$

Equations 6.37 can be simplified if we select the centroid of the cross section as the origin of the $y'z'$ reference frame, that is, let $\bar{y}' = \bar{z}' = 0$. Noting that $F = 0$, and that, for arbitrarily oriented centroidal axes $I_{y'z'}$ is not necessarily zero, we get

$$a_0 = 0, \quad a_1 = \frac{-(M_{z'}I_{y'} + M_{y'}I_{y'z'})}{I_{y'}I_{z'} - I_{y'z'}^2}, \quad a_2 = \frac{M_{y'}I_{z'} + M_{z'}I_{y'z'}}{I_{y'}I_{z'} - I_{y'z'}^2} \tag{6.38}$$

So, combining Eqs. 6.36 and 6.38, we get the following **flexure formula:**

$$\sigma_x = \frac{-y'(M_{z'}I_{y'} + M_{y'}I_{y'z'}) + z'(M_{y'}I_{z'} + M_{z'}I_{y'z'})}{I_{y'}I_{z'} - I_{y'z'}^2} \qquad \text{Flexure Formula} \tag{6.39}$$

where the "prime" axes may be any convenient set of rectangular coordinate axes with <u>origin at the centroid</u> of the cross section.

Orientation of the Neutral Axis: The orientation of the **neutral axis** in the cross section can be determined by setting $\sigma_x = 0$ in Eq. 6.39. Let (y'^*, z'^*) be coordinates of points that lie on the neutral axis. Then,

$$-y'^*(M_{z'}I_{y'} + M_{y'}I_{y'z'}) + z'^*(M_{y'}I_{z'} + M_{z'}I_{y'z'}) = 0 \tag{6.40}$$

or

$$\frac{y'^*}{z'^*} = \frac{M_{y'}I_{z'} + M_{z'}I_{y'z'}}{M_{z'}I_{y'} + M_{y'}I_{y'z'}} \tag{6.41}$$

is the equation of the neutral axis in $y'z'$ coordinates.

Maximum Tensile and Compressive Stresses: Figure 6.25c illustrates the fact that the points of maximum tension and maximum compression on a cross section are the two points that are farthest from the neutral axis. These points are labeled A and B on Fig. 6.25c and on the unsymmetric cross section in Fig. 6.28. Therefore, in order to compute the maximum tensile and compressive flexural stresses acting on a cross section it is necessary first to determine the orientation of the neutral axis and then to locate the two points in the cross section that are the farthest from the NA.

Example Problem 6.10 illustrates the calculation of flexural stresses on an unsymmetric cross section.

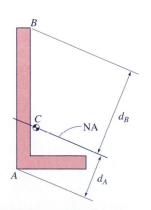

FIGURE 6.28 Location of points of maximum tension and maximum compression in a typical unsymmetric cross section.

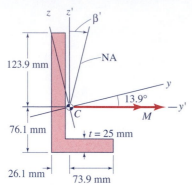

$I_{y'} = 27.20(10^6) \text{ mm}^4$
$I_{z'} = 4.55(10^6) \text{ mm}^4$
$I_{y'z'} = -5.97(10^6) \text{ mm}^4$

Fig. 1 An unequal-leg angle.

An unequal-leg angle section has the dimensions shown in Fig. 1.[11] At this cross section the moment is $M = 10$ kN · m and is oriented parallel to the short leg of the angle, as shown. (a) Determine the orientation of the neutral axis of the cross section, and show this orientation on a sketch; and (b) determine the maximum tensile stress and the maximum compressive stress on the cross section.

Plan the Solution We can use Eq. 6.41 to determine the orientation of the neutral axis. Then, we can locate the points in the cross section that are farthest from the NA and use Eq. 6.39 to compute the flexural stress at these points.

Solution (a) *Locate the neutral axis.* Since $M_{z'} = 0$, from Eq. 6.41,[12]

$$\tan \beta' \equiv \frac{y'^*}{z'^*} = \frac{I_{z'}}{I_{y'z'}} = \frac{4.55(10^6) \text{ mm}^4}{-5.97(10^6) \text{ mm}^4} \tag{1}$$

so

$$\beta' = -37.3° \qquad \textbf{Ans. (a)} \quad (2)$$

The orientation of the NA is shown in Fig. 2.
(b) *Calculate the maximum and minimum flexural stresses.* Points A and B are the two points that are farthest from the neutral axis. To compute the flexural stress σ_x at these points we need their coordinates in $y'z'$ reference frame. From Fig. 1 we get

$$(y'_A, z'_A) = (-26.1 \text{ mm}, -76.1 \text{ mm})$$
$$(y'_B, z'_B) = (-1.1 \text{ mm}, 123.9 \text{ mm})$$

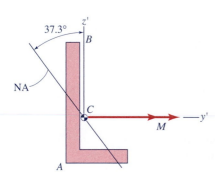

Fig. 2 The orientation of the neutral axis (NA).

Since $M_{z'} = 0$, from Eq. 6.39, we get

$$\sigma_x = \left[\frac{-y'I_{y'z'} + z'I_z}{I_{y'}I_{z'} - I_{y'z'}^2} \right] M_{y'} \tag{3}$$

so,

$$\sigma_{xA} \equiv \sigma_x(-26.1 \text{ mm}, -76.1 \text{ mm})$$

$$= \left[\frac{-(-26.1)(-5.97) - 76.1(4.55)}{27.20(4.55) - (5.97)^2} \right] 10$$

$$= -57.0 \text{ MPa}$$

---

[11]See Example Problems C1 through C3 for calculation of the section properties of a similar unequal-leg angle cross section. The MDS module *Section Properties* may be used to compute section properties for a wide variety of cross-sectional shapes.
[12]Note that β' is measured *from the z' axis in the direction toward the y' axis,* as shown in Fig 1.

$$\sigma_{xB} \equiv \sigma_x(-1.1 \text{ mm}, 123.9 \text{ mm})$$

$$= \left[\frac{-(-1.1)(-5.97) + 123.9(4.55)}{27.20(4.55) - (5.97)^2} \right] 10$$

$$= 63.2 \text{ MPa}$$

In summary,

$$\left. \begin{array}{l} \sigma_{\max T} = \sigma_{xB} = 63.2 \text{ MP} \\ \sigma_{\max C} = \sigma_{xA} = -57.0 \text{ MPa} \end{array} \right\} \qquad \textbf{Ans. (b)} \quad (4)$$

Review the Solution   The best way to check the results of Part (a) and Part (b) is to draw the cross section to scale in order to estimate the perpendicular distances from points A and B, respectively, to the neutral axis. Since the calculated flexural stresses at A and B have opposite signs, as they are supposed to, and since the ratio of their magnitudes is in agreement with their perpendicular distances from the NA, then our results are probably correct.

MDS6.9   **Unsymmetric Bending**—The Unsymmetric option in the MDS computer program module *Flexure* combines a program for determining section properties with a program for computing the flexural stresses in an unsymmetric beam with arbitrarily oriented moment vector. The origin of the $y'z'$ axes must be at the centroid of the cross section, but their orientation is arbitrary.

Unsymmetric Bending: Principal-Axis Method.   So far, we have shown that Eq. 6.30 applies to bending of beams with doubly symmetric cross section, and that Eq. 6.39 is the flexure formula for beams with arbitrary shape of cross section and with arbitrary orientation of the resultant moment vector. However, for centroidal principal axes, as illustrated in Fig. 6.29a, $\bar{y} = \bar{z} = I_{yz} = 0$, so Eq. 6.39 reduces to Eq. 6.30.

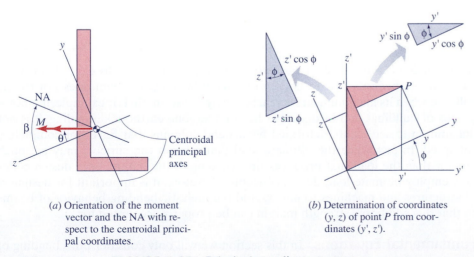

(a) Orientation of the moment vector and the NA with respect to the centroidal principal coordinates.

(b) Determination of coordinates (y, z) of point P from coordinates (y', z').

FIGURE 6.29   Principal-coordinate geometry.

$$\boxed{\sigma_x = \frac{M_y z}{I_y} - \frac{M_z y}{I_z}}$$

Flexure Formula (6.30) repeated

where M_y and M_z are the components of the moment vector M for bending about the y axis and z, axis, respectively, and I_y and I_z are the centroidal principal moments of inertia. (It is arbitrary which of the two is I_{max} and which is I_{min}.)

Therefore, we can conclude that **the flexural stress σ_x in a beam with arbitrary cross section and with an arbitrarily oriented bending moment is given by Eq. 6.30, provided that the y and z axes are <u>centroidal principal axes.</u>** Beams with doubly symmetric cross section (e.g., Fig. 6.24), and beams with a single longitudinal plane of symmetry (e.g., Fig. 6.26b) are two special cases. For the latter, loading in the longitudinal plane of symmetry (e.g., Sections 6.2 and 6.3) and loading perpendicular to the plane of symmetry (e.g., Section 6.12) are special cases.

As in the case of beams with doubly symmetric cross section, Eq. 6.31 or Eq. 6.33 can be used to determine the orientation of the neutral axis relative to the yz principal axes.

$$\tan \beta = \left(\frac{I_z}{I_y}\right) \tan \theta$$

(6.33) repeated

where the angle β is defined in Fig. 6.29a.

The difficult aspects of using the principal-axis method to solve for flexural stresses are (1) determining the orientation of the principal axes, and (2) determining the coordinates (y, z) of a point relative to the principal-axis reference frame. Figure 6.29b shows how these y, z coordinates can be related to arbitrarily oriented coordinates (y', z'). Using the shaded triangles, we get

$$y = y' \cos \phi + z' \sin \phi$$
$$z = -y' \sin \phi + z' \cos \phi$$

(6.42)

where the angle ϕ is measured counterclockwise from the y' axis to the y axis.

Homework Problem 6.6-11 calls for the principal-axis method to be used to solve Example Problem 6.10.

*6.7   INELASTIC BENDING OF BEAMS

If the loads on a beam are large enough to cause the stress to exceed the yield strength, the beam is said to undergo **inelastic bending.** A beam does not totally collapse when its maximum stress reaches the yield strength. For example, in photographs of buildings and bridges that have undergone earthquake loading, it is not uncommon to see highly deformed beams and columns that have not completely collapsed. In fact, since the ultimate load that a beam can support may be much greater than the load that produces first yielding of the outer fibers, design codes now employ ultimate-load design concepts. Therefore, it is important for designers of structures and machines to understand the inelastic-bending behavior of beams so that this additional strength margin can be properly accounted for.

Fundamental Equations. In this section we will only consider pure bending of beams that are symmetrical about a longitudinal plane of symmetry (Fig. 6.30a,b).

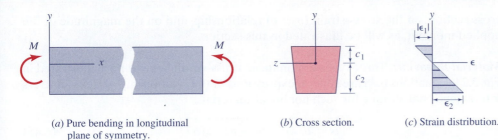

(a) Pure bending in longitudinal plane of symmetry.

(b) Cross section.

(c) Strain distribution.

FIGURE 6.30 Pure bending of a beam.

As in Sections 6.2 and 6.3, our analysis of pure bending will involve the three essentials of deformable-body mechanics—*geometry of deformation, equilibrium,* and *material behavior.*

Geometry of Deformation: The assumption that *plane sections remain plane* is valid for inelastic pure bending of beams as well as for linearly elastic bending (Section 6.2). This assumption leads to the **strain-displacement equation** (Eq. 6.3 repeated).

$$\epsilon = -\frac{y}{\rho} = -y\kappa \tag{6.43}$$

where ϵ is the extensional strain of fibers at distance y from the neutral surface, ρ is the radius of curvature of the deformed axis of the beam, and $\kappa = 1/\rho$ is the curvature. This linear strain distribution is shown in Fig. 6.30c

Equilibrium: The stress resultants that are related to the bending of beams are the axial force F and the bending moment M, shown in Fig. 6.31. From Eqs. 6.9,

$$F = \int_A \sigma \, dA = 0$$

$$M \equiv M_z = -\int_A y\sigma \, dA \tag{6.44}$$

As in the case of linearly elastic behavior, Eq. 6.44a is used to locate the neutral axis in the cross section. As long as the flexural stress does not exceed the proportional limit, the neutral axis passes through the centroid of the cross section (Section 6.3), For inelastic bending, the location of the neutral axis depends on the shape of the

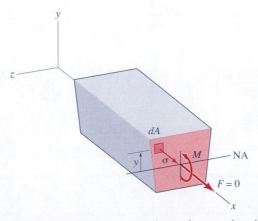

FIGURE 6.31 The stress resultants for pure bending.

cross section, on the stress-strain ($\sigma - \epsilon$) relationship, and on the magnitude of the applied moment, as will be illustrated in this section.

Material Behavior: Three types of nonlinear stress-strain behavior are illustrated in Fig. 3.24. Equations 6.45 are general expressions of the relationship of flexural stress σ to extensional strain ϵ for such nonlinear materials.

$$\sigma = \sigma(\epsilon), \quad \text{or} \quad \epsilon = \epsilon(\sigma) \tag{6.45}$$

To illustrate inelastic bending of beams we will examine only the case of elastic-plastic bending.

Elastic-Plastic Bending.

Consider pure bending of a beam made of elastic-plastic material whose stress-strain curve is given in Fig. 6.32. The material follows Hooke's Law up to the proportional limit, which is assumed to be the same stress as the yield point (σ_Y). It is assumed that the material has a distinct yield point in compression at ($-\sigma_Y$), as indicated in Fig. 6.32. The corresponding yield strains in tension and compression are $\epsilon_Y = \sigma_Y/E$ and ($-\epsilon_Y$), respectively.

Figure 6.33 illustrates the strain distribution and the stress distribution for three levels of loading of a beam with a singly symmetric cross section (e.g., Fig. 6.30b).[13]

Location of the Neutral Axis: Since, for pure bending, the axial force F is zero, the equation that determines the location of the neutral axis is Eq. 6.44a, which can be written as

$$\int_A \sigma \, dA = 0 \rightarrow T = C \tag{6.46}$$

where T is the tension on the cross section and C is the compression, as illustrated in Fig. 6.33. As the moment changes, the stress blocks that represent T and C

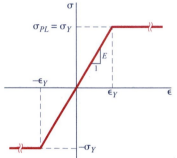

FIGURE 6.32 The stress-strain curve for an elastic-plastic material.

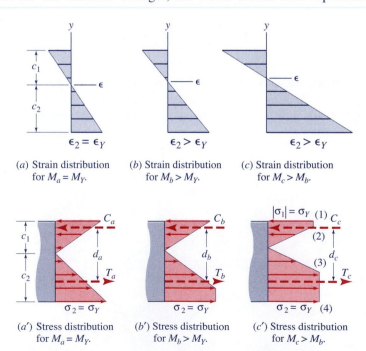

(a) Strain distribution for $M_a = M_Y$.

(b) Strain distribution for $M_b > M_Y$.

(c) Strain distribution for $M_c > M_b$.

(a′) Stress distribution for $M_a = M_Y$.

(b′) Stress distribution for $M_b > M_Y$.

(c′) Stress distribution for $M_c > M_b$.

FIGURE 6.33 Strain distribution and stress distribution for an elastic-plastic beam.

[13]Unloading behavior is discussed later under the topic of *residual stresses*.

change shape, and this change of shape leads to a shift in the location of the neutral axis, unless the cross section has two axes of symmetry. (Note in Figs. 6.33a through 6.33c and Figs. 6.33a' through 6.33c' that the neutral axis moves upward slightly with each increase in the applied moment.)

Maximum Elastic Moment: The maximum elastic moment, or **yield moment,** M_Y, is the moment that causes the fiber that is farthest from the elastic neutral axis to yield. Since the behavior is linearly elastic up the yield moment, we can use the flexure formula, Eq. 6.13, in the form

$$M_Y = \frac{\sigma_Y I}{c} \tag{6.47}$$

where c, the larger of the values c_1 and c_2, is measured from the *elastic neutral axis* (which passes through the centroid of the cross section).

Fully Plastic Moment: Materials like mild steel, that exhibit elastic-plastic behavior similar to that depicted in Fig. 6.32. are able to undergo extensional strains that are an order of magnitude or more greater than the yield strain ϵ_Y. In Fig. 6.33c' there is an *elastic core* (2)–(3), a *compressive plastic zone* (1)–(2), and a *tensile plastic zone* (3)–(4). With increasing bending moment, the magnitude of the strain in the outer fibers increases, and the elastic zone shrinks in depth until, in the limit, there is a fully plastic compression zone above the (fully plastic) neutral axis and a fully plastic tensile zone below the neutral axis, as illustrated in Fig. 6.34a, Since the neutral axis is located by setting $T = C$ (Eq. 6.46), and since both tensile and compressive stress blocks have constant stress of magnitude σ_Y, the **plastic neutral axis** is determined by the purely geometric condition that

$$A_C = A_T = \frac{1}{2} A \tag{6.48}$$

The resultant forces C and T act at the centroids of the compression zone and tensile zone, respectively, so the **plastic moment,** M_P, is given by

$$M_P = \frac{\sigma_Y A}{2} (d_C + d_T) \tag{6.49}$$

where d_C and d_T are the distances from the plastic neutral axis to the centroids of the compression zone and the tensile zone, respectively.

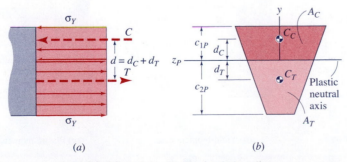

(a)   (b)

FIGURE 6.34 Fully plastic bending.

The value of the plastic moment can be expressed in the compact form

$$M_P = \sigma_Y Z$$

where Z is the **plastic section modulus** for the cross section. Values of Z are tabulated in Appendix D for selected structural shapes. Recall from Section 6.4 that the yield moment can be expressed in terms of the *elastic section modulus S* by the formula

$$M_Y = \sigma_Y S$$

The ratio of the plastic moment M_P to the yield moment M_Y is called the **shape factor,** f.

$$f = \frac{M_P}{M_Y} = \frac{Z}{S} \tag{6.50}$$

This factor indicates the additional moment capacity of the beam beyond the moment that causes first yielding. The value of f for wide flange beams is typically in the range 1.1 to 1.2.

EXAMPLE 6.11

Determine expressions for the yield moment, the plastic moment, and the shape factor for a beam with rectangular cross section (Fig. 1a).

Solution   Because the rectangular cross section has two axes of symmetry, the neutral axis passes through the centroid for all values of applied moment. The stress state corresponding to the yield moment is shown in Fig. 1b, and the stress state for the plastic-moment case is shown in Fig. 1c.

Yield Moment: From Eq. 6.47,

$$M_Y = \frac{\sigma_Y I}{c} = \frac{\sigma_Y \left(\dfrac{bh^3}{12} \right)}{\left(\dfrac{h}{2} \right)} \tag{1}$$

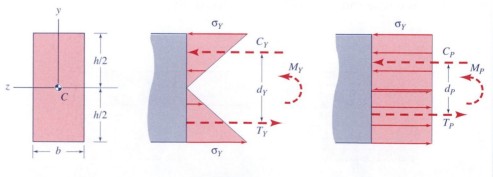

(a) Cross section.   (b) Yield-moment case.   (c) Plastic-moment case.

Fig. 1

or

$$M_Y = \frac{1}{6}\sigma_Y bh^2 \qquad\qquad \textbf{Ans.} \quad (2)$$

Plastic Moment: From Eq. 6.49,

$$M_P = \frac{\sigma_Y A}{2}(d_C + d_T) = \frac{\sigma_Y(bh)}{2}\left(\frac{h}{4} + \frac{h}{4}\right) \qquad\qquad (3)$$

or

$$M_P = \frac{1}{4}\sigma_Y bh^2 \qquad\qquad \textbf{Ans.} \quad (4)$$

Shape Factor: From Eq. 6.50,

$$f = \frac{M_P}{M_Y} = \frac{6}{4} = 1.5 \qquad\qquad \textbf{Ans.} \quad (5)$$

Thus, for a rectangular beam, the plastic moment is 50% higher than the yield moment.

EXAMPLE 6.12

The tee beam of Example Problem 6.2 is made of elastic-plastic material having a stress-strain diagram like Fig. 6.32 with $E = 29(10^3)$ ksi and $\sigma_Y = 40$ ksi. Determine the following: (a) the yield moment M_Y; (b) the location of the plastic neutral axis, and the value of the plastic moment M_P; and (c) the shape factor f.

The section properties calculated in Example Problem 6.2 are shown in Fig. 1.

Solution (a) *Determine the yield moment, M_Y.* From Eq. 6.47, the yield moment is given by

$$M_Y = \frac{\sigma_Y I}{c} \qquad\qquad (1)$$

The bottom fibers are farthest from the neutral axis, so

$$M_Y = \frac{(40 \text{ ksi})(33.3 \text{ in}^4)}{(4 \text{ in.})} = 333 \text{ kip} \cdot \text{in}$$

$$M_Y = 333 \text{ kip} \cdot \text{in} \qquad \textbf{Ans.(a)} \quad (2)$$

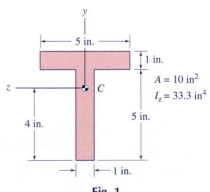

Fig. 1

(b) *Locate the plastic neutral axis.* The plastic neutral axis divides the cross section into two equal areas (Eq. 6.48). Therefore, for this problem, the plastic neutral axis falls at the flange-web interface, as illustrated in

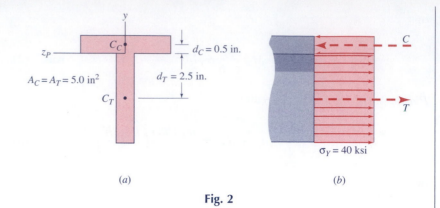

(a)   (b)

Fig. 2

Fig. 2. From Eq. 6.49,

$$M_P = \frac{\sigma_Y A}{2}(d_C + d_T) = \frac{(40 \text{ ksi})(10 \text{ in}^2)}{2}(0.5 \text{ in.} + 2.5 \text{ in.})$$

or

$$M_P = 600 \text{ kip} \cdot \text{in} \qquad\qquad \textbf{Ans. (b)} \quad (3)$$

(c) *Calculate the shape factor.* The shape factor is given by Eq. 6.50.

$$f = \frac{M_P}{M_Y} = \frac{600 \text{ kip} \cdot \text{in}}{333 \text{ kip} \cdot \text{in}} = 1.80$$

$$f = 1.8 \qquad\qquad \textbf{Ans. (c)} \quad (4)$$

MDS6.10   **Inelastic Bending**—An MDS Example Problem that uses the *Section Properties* module to determine cross-sectional properties of a beam and uses Eqs. 6.47 and 6.49 to solve for the yield moment M_Y and the fully plastic moment M_P.

Moment-Curvature Formulas. The moment-curvature equation for linearly elastic bending was derived in Section 6.3 and stated in Eq. 6.12, which we write here in terms of the curvature κ rather than the radius of curvature ρ.

$$M = EI\kappa \qquad\qquad (6.51)$$

The yield moment M_Y is, therefore, related to the yield curvature κ_Y by the equation

$$M_Y = EI\kappa_Y \qquad\qquad (6.52)$$

so the **moment-curvature equation** for linearly elastic behavior is

$$\boxed{\frac{M}{M_Y} = \frac{\kappa}{\kappa_Y}}, \qquad 0 \le M \le M_Y \qquad\qquad (6.53)$$

Let us now determine an expression that relates M/M_Y to κ/κ_Y for inelastic bending, and plot a curve of M/M_Y versus κ/κ_Y. The simplest case to consider is the rectangular beam, since the location of its neutral axis does not vary with the value of the applied moment and since it has a constant width. Figure 6.35 shows the stress

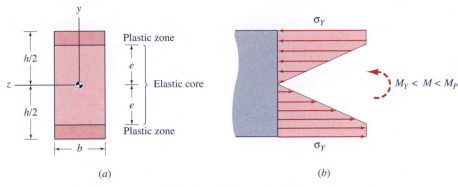

FIGURE 6.35 Partially plastic bending of a rectangular beam.

distribution for the partially plastic case, with $2e$ being the depth of the elastic core. Summing the moment contributions of the elastic core and the two plastic zones we get

$$M = \sigma_Y b\left[\left(\frac{h}{2}\right)^2 - \frac{1}{3}e^2\right] \tag{6.54}$$

When $e = h/2$, we get the value of the **yield moment** M_Y as

$$M_Y = \frac{1}{6}\sigma_Y bh^2 \tag{6.55}$$

just as we previously obtained in Example Problem 6.11. When $e = 0$, Eq, 6.54 reduces to the formula for the **plastic moment, M_P.**

$$M_P = \frac{1}{4}\sigma_Y bh^2 \tag{6.56}$$

In order to determine a moment-curvature equation corresponding to the moment expression in Eq. 6.54, we need to relate the curvature κ to the value e. From the strain-displacement equation, Eq. 6.43 (Eq. 6.3). we get

$$\epsilon_Y = \frac{h}{2}\kappa_Y = e\kappa \tag{6.57}$$

since the yielding occurs at the outer fiber for M_Y and at $y = e$ for $M > M_Y$. (Note that $\kappa \geq \kappa_Y$.) Therefore,

$$e = \frac{h}{2}\left(\frac{\kappa_Y}{\kappa}\right) \tag{6.58}$$

Finally, we can combine Eqs. 6.54, 6.55, and 6.58 to get the following **moment-curvature equation** for a partially plastic rectangular beam:

$$\frac{M}{M_Y} = \frac{3}{2} - \frac{1}{2}\left(\frac{\kappa_Y}{\kappa}\right)^2, \quad M_Y \leq M \leq M_P \tag{6.59}$$

Moment-curvature formulas for the rectangular beam, Eq. 6.53 and 6.59, are plotted as the solid-line curve in Fig. 6.36; the dashed-line curve is the moment-curvature curve for the T-beam of Example 6.12.

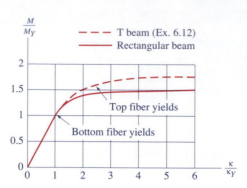

FIGURE 6.36 Moment-curvature plot for elastic-plastic beams.

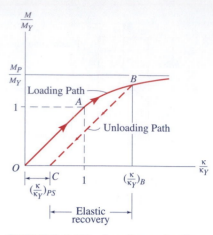

FIGURE 6.37 Loading-unloading curve of an elastic-plastic beam.

Residual Stresses. The elastic-plastic analysis summarized in the moment-curvature plots of Fig. 6.36 applies only when the load continues to increase. If the applied moment is reduced after exceeding the yield moment M_Y, unloading takes place along a moment-curvature path that is parallel to the original linear portion of the moment-curvature diagram, as illustrated in Fig. 6.37. This unloading behavior is very similar to the unloading behavior of circular torsion rods that was described in Section 4.10. Therefore, we can use the same procedure of adding the elastic-recovery stresses to the stresses due to plastic loading.

Figure 6.38 illustrates the superposition of stresses for a rectangular beam that has been loaded to the fully plastic state and then completely unloaded. The maximum elastic recovery stress for the rectangular beam in Fig. 6.38 is obtained by setting $M = -M_P$ and $y = h/2$ in the flexure formula, Eq. 6.13, giving[14]

$$(\sigma_{\text{er}})_{\text{max}} = \frac{-(-M_P)\left(\dfrac{h}{2}\right)}{\left(\dfrac{bh^3}{12}\right)} = \frac{\left(\dfrac{1}{4}\sigma_Y bh^2\right)\left(\dfrac{h}{2}\right)}{\left(\dfrac{bh^3}{12}\right)} = \frac{3}{2}\sigma_Y$$

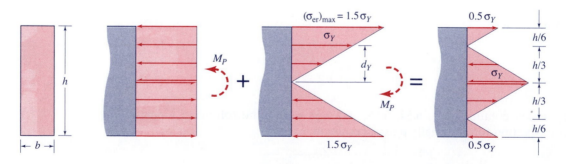

(a) Cross section.   (b) Fully plastic stress state.   (c) Elastic recovery stress state.   (d) Residual stress state.

FIGURE 6.38 Residual stresses determined by superposition.

---

[14]The maximum elastic recovery stress is also called the *modulus of rupture*.

The point of zero stress in Fig. 638d may be determined by examining similar triangles in Fig. 6.38c. Thus,

$$\frac{\sigma_Y}{d_Y} = \frac{(\sigma_{er})_{max}}{(h/2)} \rightarrow d_y = \left(\frac{\sigma_Y}{1.5\sigma_Y}\right)\left(\frac{h}{2}\right) = \frac{h}{3}$$

In spite of the fact that all externally applied moment has been removed from the beam in Fig. 6.38d, it is obvious that the beam has self-equilibrating **residual stresses.** Any reloading would begin from this state, rather than from the stress-free state. After being completely unloaded, the beam would be left with a residual curvature called the **permanent set.** Figure 6.37 illustrates the permanent set, at point C, that would result from unloading from point B on the moment-curvature diagram, a point well short of the (infinite) curvature required to provide the fully plastic state.

The next example problem illustrates the calculation of residual stresses.

EXAMPLE 6.13

A wide-flange beam is subjected to a fully plastic moment, and then the moment is completely removed. Determine the distribution of residual stresses. The beam is made of (elastic-plastic) steel with a yield stress of $\sigma_Y = 200$ MPa.

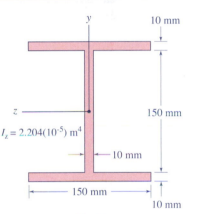

Fig. 1

$I_z = 2.204(10^{-5})$ m^4

Solution We can follow the superposition-of-stresses procedure that was illustrated in Fig. 6.38 for a rectangular beam. We first calculate M_P using Figs. 2a and 2d and the beam dimensions given in Fig. 1.

$$C_1 = T_1 = (200 \text{ MPa})(0.150 \text{ m})(0.010 \text{ m}) = 300 \text{ kN} \tag{1}$$

$$C_2 = T_2 = (200 \text{ MPa})(0.010 \text{ m})(0.075 \text{ m}) = 150 \text{ kN}$$

$$M_P = (300 \text{ kN})(0.160 \text{ m}) + (150 \text{ kN})(0.075 \text{ m}) = 59.25 \text{ kN} \cdot \text{m} \tag{2}$$

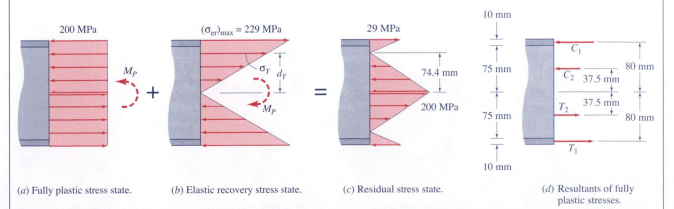

(a) Fully plastic stress state. (b) Elastic recovery stress state. (c) Residual stress state. (d) Resultants of fully plastic stresses.

Fig. 2 Stress distributions and stress resultants.

The maximum elastic recovery stress, or modulus of rupture, is obtained by setting $M = M_P$ in the flexure formula, Eq. 6.13, giving

$$(\sigma_{er})_{max} = \frac{M_P\left(\dfrac{h}{2}\right)}{I} = \frac{(59.25 \text{ kN} \cdot \text{m})(0.085 \text{ m})}{2.204(10^{-5}) \text{ m}^4} = 229 \text{ MPa} \qquad (3)$$

To sketch the residual stress state in Fig. 2c we add the stress blocks in Figs. 2a and 2b. The point of zero stress in Fig. 2c may be determined by using similar triangles in Fig. 2b.

$$\frac{\sigma_Y}{d_Y} = \frac{(\sigma_{er})_{max}}{(h/2)} \rightarrow d_Y = \left(\frac{200}{229}\right)(85 \text{ mm}) = 74.4 \text{ mm} \qquad (4)$$

Therefore, the residual stress distribution has the form shown in Fig. 2c.

6.8 SHEAR STRESS AND SHEAR FLOW IN BEAMS

Recall from Chapter 5 that transverse loads on a beam give rise not only to a bending moment $M(x)$ but also to a *transverse shear force* $V(x)$, as indicated in Fig. 6.39a. The bending moment on a cross section is related to the flexural stress σ (Fig. 6.39b); the transverse shear force $V(x)$ is the resultant of distributed transverse shear stresses (Fig. 6.39c). We need to determine how the *transverse shear stress* τ varies with position in the cross section, that is, with respect to y and z. We will also find that surfaces other than cross sections (e.g., the neutral surface) also experience shear stress.

Basic Assumption. In developing the theories of axial deformation (Section 3.2), torsion (Sections 4.2 and 4.3), and flexure (Sections 6.2 and 6.3), we began by making kinematic assumptions and performing a strain-displacement analysis. Then we introduced the constitutive (stress-strain-temperature) equations to determine the respective stress distributions. However, when we seek to determine the distribution of shear stress in a beam and to relate the shear stress to the transverse shear

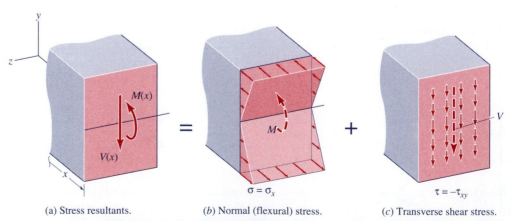

(a) Stress resultants. (b) Normal (flexural) stress. (c) Transverse shear stress.

FIGURE 6.39 The stress resultants and stresses on a cross section.

force $V(x)$, we are immediately faced with a dilemma. Shear stress produces a change in angle, but in Section 6.2 we assumed that **plane sections remain plane and remain perpendicular to the deformed axis of the beam** (i.e., without change of angle). Therefore, in order to derive an expression for the distribution of transverse shear stress, we make the following assumption:[15]

- *The distribution of flexural stress on a given cross section is not affected by the deformation due to shear.*

Thus, for example, we can analyze the shear stress in homogeneous, linearly elastic beams, using the flexural stress given by the flexure formula. Eq. 6.13.

$$\sigma = \frac{-My}{I}$$

(6.13)
repeated

Because we have no convenient way to characterize the displacement due to shear (like the "plane sections" assumption that was used to characterize displacement due to flexure) we must follow a different approach. Fortunately, we can employ an equilibrium analysis to develop a shear-stress theory for beams.[16]

Shear-Stress Distribution. Before developing formulas for the distribution of shear stress in a beam, let us examine physical descriptions of shear stress and shear strain in beams. Figure 6.40 shows a cantilever beam subjected to a transverse load P at $x = 0$. (To simplify the drawings, a rectangular cross section is used in Fig. 6.40.) On an arbitrary cross section at x, the load P produces a positive bending moment and a positive shear force (as in Fig. 6.39a).

With linearly elastic behavior, the bending moment $M(x)$ produces the flexural stress distribution at section x as shown in Fig. 6.40b. Imagine now that a horizontal sectioning plane separates the portion of the beam up to section x into two parts; one from $y = y_1$ to the top of the beam, the other from $y = y_1$ to the bottom of the

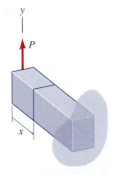

(a) A cantilever beam.

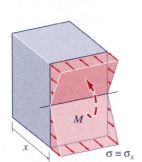

(b) Flexural-stress distribution (linearly elastic).

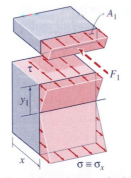

(c) Unbalanced flexural stresses lead to shear stresses on a longitudinal section.

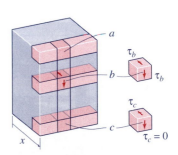

(d) Transverse and longitudinal shear stresses.

FIGURE 6.40 Shear-stress distribution in a beam.

---

[15]There are some limitations on the theory that is based on this assumption. These are discussed in Section 6.9.

[16]A Russian engineer, D. J. Jourawski (1829–1891) was the first person to develop the elementary shear stress theory presented here. He developed this theory, which is based on equilibrium, while designing timber railroad bridges [Ref. 6-1].

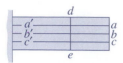

(*a*) A beam made of separate "planks."

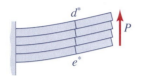

(*b*) Slip between non-bonded "planks."

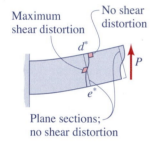

Maximum shear distortion

No shear distortion

Plane sections; no shear distortion

(*c*) Shear deformation of a uniform beam (or bonded-layer beam).

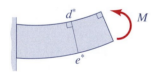

(*d*) Pure bending of a cantilever beam.

FIGURE 6.41 Some illustrations of shear deformation in beams.

beam, as illustrated in Fig. 6.40*c*. The resultant of the flexural stress on A_1, the cross-sectional area above $y = y_1$, is designated as F_1. It is clear from Fig. 6.40*c* that the only way to satisfy $\Sigma F_x = 0$ for either the portion of beam above $y = y_1$ or the portion below $y = y_1$ is for there to be **longitudinal shear stress** on the horizontal sectioning plane, as indicated in Fig. 6.40*c*. The resultant of this longitudinal shear stress balances the normal force F_1. But this says that the unbalanced flexural stresses on this beam free body lead to *longitudinal* shear stresses on the free body. What about *transverse* shear stress on the cross section of the beam? Consider Fig. 6.40*d*, and recall from Section 2.7 that $\tau_{yx} = \tau_{xy}$; that is, the shear stress on a vertical plane must be equal to the shear stress on a horizontal plane at the intersection of the two planes. Thus, at point *b* in Fig. 6.40*d* the longitudinal shear stress and the vertical shear stress can both be labeled τ_b, because they are equal. This leads to two very interesting conclusions:

1. The **transverse shear stress** at an arbitrary level *y* in the cross section can be calculated by determining the *longitudinal shear stress* at this level.
2. The transverse shear stress must vanish at points *a* and *c*, since there is no horizontal shear stress on the top surface or on the bottom surface of the beam.

Shear-Strain Distribution. From the preceding discussion of beams subjected to transverse loads, and especially from Fig. 6.40*c*, it should be clear that planes parallel to the neutral surface (i.e., horizontal planes) must be able to transmit shear. Consider a cantilever beam like the one in Fig. 6.41*a*. In Fig. 6.41*b* the cantilever beam is made up of four "planks" that are not bonded together at levels $a - a'$, $b - b'$, and $c - c'$, but are free to slip along the surfaces of contact at these levels. Clearly, there is slip along the plank interfaces, and plane sections, like *de*, do not remain plane through the entire thickness of the beam in Fig. 6.41*b*. If the planks are bonded together to form a single beam (or if the beam is originally homogeneous throughout its depth), the beam will undergo shear deformation as illustrated in Fig. 6.41*c*. Because of shear deformation, plane sections do not remain plane, as they do in the case of pure bending (Fig. 6.41*d*). However, as noted earlier, the **shear deformation has little effect on the distribution of flexural stress as long as the beam is slender** (length greater than ten times depth).

Shear Flow in Beams. To derive an expression relating the shear stress τ to the resultant shear force V we will first define and derive an expression for *shear flow*. Consider the segment from x to $(x + \Delta x)$ of a beam subjected to transverse loads (Fig. 6.42*a*). The segment from x to $(x + \Delta x)$ is enlarged in Fig. 6.43*a*, and

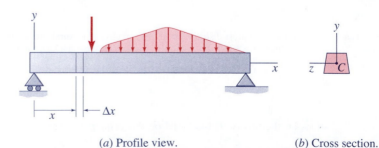

(*a*) Profile view.　　　　　　　　(*b*) Cross section.

FIGURE 6.42 A beam subjected to transverse loading.

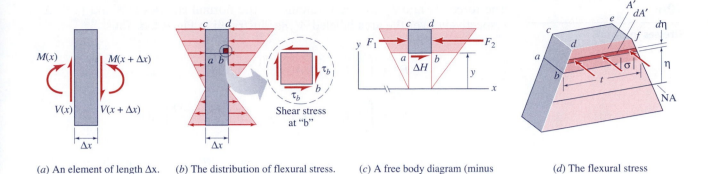

(a) An element of length Δx.   (b) The distribution of flexural stress.   (c) A free body diagram (minus vertical shear on ac and bd).   (d) The flexural stress contributing to F_2.

FIGURE 6.43   The stress resultants and stresses on segments of a beam.

the normal stresses acting on this segment are shown in Fig. 6.43b. (Note that $V(x)$ is taken in the positive sense according to the sign convention given in Chapter 5.) From Eq. 5.3, repeated here,

$$V = \frac{dM}{dx}$$

(5.3) repeated

Thus, when $V > 0$, $M(x + \Delta x) > M(x)$. Consequently, the flexural stresses at $(x + \Delta x)$ in Fig. 6.43b are shown larger in magnitude than the corresponding stresses at x.

Our goal is to determine an expression $\tau(x, y)$ that relates the shear stress at level y on the cross section at x to the transverse shear force $V(x)$. In deriving the expression for shear stress $\tau(x, y)$ it is useful to first define a shear quantity called shear flow. As was suggested earlier, the key to the derivation of the distribution of transverse shear stress on a cross section is the fact that **the transverse shear stress is equal to the longitudinal shear stress at the same point** (Figs. 6.40d and 6.43b). Figures 6.43c and 6.43d focus on the portion above the level y, where the shear flow and the shear stress are to be determined. In Fig. 6.43c the resultant of the normal stresses on the area A' above level y at cross section x is labeled F_1; the corresponding resultant at $(x + \Delta x)$ is labeled F_2. The total shear force on the horizontal section at level y is labeled ΔH. Equilibrium requires that $\Sigma F_x = 0$, and since there is no x-component of force on the sides or top of the beam,

$$\Delta H = F_2 - F_1$$

(6.60)

The **shear flow,** q, which is the *shear force per unit length,* can be defined by taking the limit

$$q \equiv \lim_{\Delta x \to 0} \frac{\Delta H}{\Delta x}$$

(6.61)

which, from Eq. 6.60, can be calculated by using the expression

$$q \equiv q(x, y) = \lim_{\Delta x \to 0} \frac{(F_2 - F_1)}{\Delta x}$$

(6.62)

405

The forces F_1 and F_2 are the resultants of the normal stresses at x and $(x + \Delta x)$, respectively, over the area labeled A', as illustrated in Fig, 6.43d. Thus,

$$F_1 = \int_{A'} |\sigma(x, \eta)|\, dA = \frac{M(x)}{I} \int_{A'} \eta\, dA$$

$$(6.63)$$

$$F_2 = \int_{A'} |\sigma(x + \Delta x, \eta)|\, dA = \frac{M(x + \Delta x)}{I} \int_{A'} \eta\, dA$$

where the variable of integration, η, locates a differential strip within A' (Fig. 6.43d). The area of integration, A', is from level y, where the shear flow is to be calculated, to the top (free surface) of the beam. The integral in Eq. 6.63 is just the *first moment of the area A' with respect to the neutral axis*. It is given the symbol Q, that is

$$Q(y) \equiv \int_{A'} \eta\, dA = A'\bar{y}' \qquad (6.64)$$

where $\bar{y}'$ is the y coordinate of the centroid of area A'. (Note that $A' \equiv A'(y)$ and $\bar{y}' \equiv \bar{y}'(y)$, that is, both depend on the level y at which the shear flow and shear stress are being evaluated.) Combining Eqs. 6.62 through 6.64, and recalling from Eq. 5.3 that $V = dM/dx$, we get

$$q = \lim_{\Delta x \to 0} \left[\frac{[M(x + \Delta x) - M(x)]}{\Delta x} \right] \frac{Q}{I} = \frac{V(x)Q(y)}{I}$$

or, simply,

$$q = \frac{VQ}{I} \qquad \textbf{Shear-Flow Formula} \qquad (6.65)$$

In the derivation of Eq. 6.65 it has been assumed that the beam is prismatic, that is, that the moment of inertia, I, is constant, and Q is a function of y only.

We will now use this expression for shear flow to obtain an expression for the distribution of transverse shear stress. Later, we will also make use of Eq. 6.65 in Sections 6.10 through 6.12, where we examine shear stresses in thin-wall beams and built-up beams.

Shear-Stress Formula. Note that the shear flow, q, in Eq. 6.65 was derived by dividing the shear force, ΔH, by the length Δx over which it acts. If, instead, we divide ΔH by the area over which it acts, we get an average shear stress on the longitudinal plane at level y. From the equality of longitudinal and transverse shear, we get[17]

$$\tau_{\text{avg}}(x, y) = \lim_{\Delta x \to 0} \frac{\Delta H}{t \Delta x} = \frac{V(x)Q(x, y)}{I(x)t(x, y)} \qquad (6.66)$$

---

[17]Here we allow the beam to be nonprismatic, so I, t, and Q may be functions of x.

where $t(x, y)$ is the thickness, or width, of the beam at level y in section x, as indicated in Fig. 6.43d. The formula for **transverse shear stress** is generally written without its xy qualifiers as, simply

$$\tau = \frac{VQ}{It} \qquad \text{Shear-Stress Formula} \qquad (6.67)$$

where

τ = the *average* transverse shear stress at level y in section x,

$Q = A'\overline{y}'$ = the first moment, with respect to the neutral axis, of the cross sectional area *above* level y,[18]

I = the moment of inertia *of the entire cross section*, taken with respect to the neutral axis, and

t = the width of the cross section at level y.

The *sign convention* implied in Eq. 6.67 is that the shear stress τ acts in the same direction as the resultant shear force V. (Since a positive V acts in the $-y$ direction, $\tau_{xy} = -\tau$). Sometimes the absolute values of V and Q are used in the shear-stress formula to compute the *magnitude* of the shear stress, and the direction is simply assigned as that of the resultant shear force.

After illustrating how to use the transverse-shear-stress formula (in Example Problem 6.14), we will discuss the limitations of the formula.

> ### EXAMPLE 6.14

The rectangular beam of width b and height h (Fig. 1) is subjected to a transverse shear force V. (a) Determine the average shear stress as a function of y, (b) sketch the shear-stress distribution, and (c) determine the maximum shear stress on the cross section.

Plan the Solution   This is a straightforward application of the shear-stress formula, Eq. 6.67. Since V, I, and b are constant, τ_{max} will occur at the neutral axis, where Q has its maximum value.

Solution   (a) *Determine an expression for the transverse shear stress.* The shear-stress formula, Eq. 6.67, is

$$\tau = \frac{VQ}{It}$$

As indicated in Fig. 2, the neutral axis of the rectangular cross section is at its mid-depth, and the area to be used in calculating Q is the area above level y.

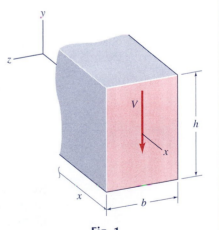

Fig. 1

---

[18]This derivation has been based on calculating Q using the area A' *above* level y. Since the first moment of the total area of the cross section, taken about the neutral axis, is zero (by definition of the neutral axis) the first moment of the area *below* level y is just the negative of the value Q in Eq. 6.64.

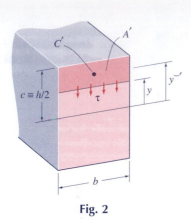

Fig. 2

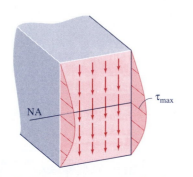

Fig. 3 The shear-stress distribution on a rectangular cross section.

Since $Q = A'\bar{y}'$, $t = b$, and $I = bh^3/12$,

$$\tau = \frac{VA'\bar{y}'}{Ib} = \frac{V\left[b\left(\dfrac{h}{2} - y\right)\right]\left[\dfrac{1}{2}\left(\dfrac{h}{2} + y\right)\right]}{\left(\dfrac{bh^3}{12}\right)b}$$

or

$$\tau = \frac{6V}{bh^3}\left(\frac{h^2}{4} - y^2\right) = \frac{3}{2}\frac{V}{A}\left(\frac{c^2 - y^2}{c^2}\right) \qquad \textbf{Ans. (a)} \quad (1)$$

(b) *Sketch the shear-stress distribution.* From Eq. 1 we see that the distribution of τ is parabolic. As we noted earlier when discussing Fig. 6.38d. τ vanishes at the top edge ($y = +c$) and at the bottom edge ($y = -c$). The parabolic distribution of τ is illustrated in Fig. 3.

(c) *Determine the maximum shear-stress.* From Eq. 1 and Fig. 3 it is obvious that τ_{max} **occurs at the neutral axis,** as expected. Therefore.

$$\tau_{max} = \frac{3}{2}\frac{V}{A} \qquad \textbf{Ans. (c)}$$

Review the Solution Unlike the case of direct shear (Section 2.7), where the shear stress is just the shear force divided by the area on which the shear stress acts, we have found a parabolic distribution of shear stress on the cross section of a rectangular beam. Since we know (Fig. 6.40d) that the shear stress must vanish at the top and bottom surfaces ($y = \pm c$), it is reasonable for the maximum shear stress to be 50% greater than the overall average shear stress, V/A.

MDS6.11 **Shear Stress in Beams**—The Shear Stress option in the MDS computer program module *Flexure* combines a program for determining section properties with equations for computing the transverse shear stress in beams.

6.9 LIMITATIONS ON THE SHEAR-STRESS FORMULA

The shear-flow formula and the shear-stress formula, Eqs. 6.65 and 6.67, respectively, may be applied to calculate the distribution of shear in a wide variety of beam shapes under a wide variety of loading conditions. These formulas were based on the flexure formula. Therefore, the limitations on the applicability of the flexure formula (slender beam, linearly elastic behavior, etc.) apply to these shear formulas also. However, there are some additional limitations on the shape of the beam and on the load distribution.

Effect of Cross-Sectional Shape. Consider first a beam with rectangular cross section. Figure 6.44 contrasts a "wide beam" ($b = 4h$ in Fig. 6.44a) and a

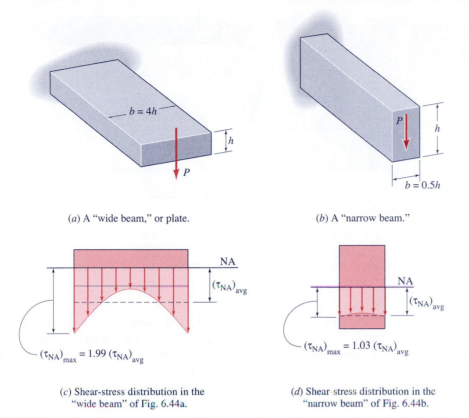

(*a*) A "wide beam," or plate.

(*b*) A "narrow beam."

(*c*) Shear-stress distribution in the
"wide beam" of Fig. 6.44a.

(*d*) Shear-stress distribution in the
"narrow beam" of Fig. 6.44b.

FIGURE 6.44 The shear-stress distribution in "beams" with rectangular cross section.

"narrow beam" ($b = 0.5h$ in Fig. 6.44*b*). The corresponding distributions of shear stress at the neutral axis of the two rectangular shapes come from a *theory of elasticity* solution [Ref. 6-4, Section 124]. It is clear that the elementary shear-stress theory represented by Eq. 6.67 is applicable only to narrow beams (e.g., $h \geq 2b$).

In mechanical applications, circular shafts often act as beams (Fig. 6.45). Although Fig. 6.43*d* and Eq. 6.66 deal with beams whose width may vary with *y*, the shear stress formula does not apply where the width, $t(x, y)$, varies rapidly. For example, the shear stresses at a line across a solid circular cross section, like line

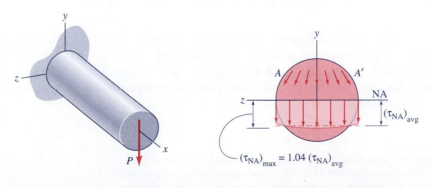

(*a*) A beam with circular cross section.

(*b*) The shear-stress distribution
on a circular cross section.

FIGURE 6.45 The shear-stress distribution in a solid circular beam.

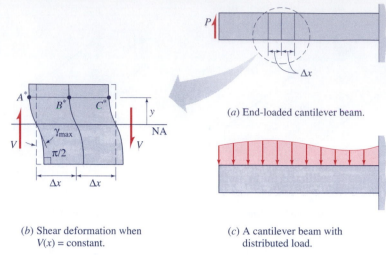

(a) End-loaded cantilever beam.

(b) Shear deformation when $V(x)$ = constant.

(c) A cantilever beam with distributed load.

FIGURE 6.46 An illustration of the effect of shear deformation.

AA' in Fig. 6.45b, are not parallel to the y axis and cannot be determined by the shear stress formula.[19] However, along the neutral axis (diameter) of the circular cross section, the theory of elasticity [Ref. 6-4, Section 122] shows that the elementary solution is quite accurate, as indicated in Fig, 6.45b.[20]

Here we have considered only "compact" sections, rectangular and circular. Limitations on the applicability of shear-stress formulas will also be discussed in Section 6.10 on thin-wall beams.

Effect of the Load Distribution. Consider first a cantilever beam with end load P, as illustrated in Fig. 6.46a. For this loading, $V(x) = P$ = constant. The shear deformation of a central portion of the beam, of length $2\Delta x$, is shown in Fig. 6.46b. Since V = constant, the shear stresses on cross sections away from the loaded end and away from the supported end produce the same warped shape at neighboring cross sections. Hence, considering shear deformation alone, $A*B* = B*C* = \Delta x$. Therefore, although the cross sections do not remain plane, the extensional strain based on the "plane-sections" assumption is not altered by the presence of shear. For loading conditions other than pure bending or constant shear, like the distributed loading in Fig. 6.46c. error in the flexural stress calculated by the flexure formula is introduced by the presence of shear deformation, but this error is small if the length of the beam is large in comparison with its depth.[21]

Effect of Length of Beam. The elementary expressions for flexural stress and transverse shear stress are accurate to within about 3% for beams whose length-to-depth ratio, L/h, is greater than 4.[22] A more elaborate analytical theory should be used to calculate the stresses in shorter spans, or the stresses should be computed by using a detailed finite-element solution.[23]

---

[19]Beams having a circular tubular cross section are considered in Section 6.10.

[20]See Homework Problem 6.8-10 for the elementary solution based on Eq. 6.67.

[21]See Ref. 6-4, Section 22 for a solution based on the theory of elasticity.

[22]This error estimate is based on an elasticity solution for bending of a rectangular beam by a uniform distributed load. (See Section 22 of Ref. 6-4.)

[23]See Ref. 6-5. Section 7.5 for illustrations based on finite-element solutions.

Many of the structural shapes used as beams can be classified as having thin-wall sections. That is, the wall thickness, t, is significantly smaller than the overall dimensions of the cross section, like the outer diameter, d_o, of a tubular beam (Fig. 6.47a) or the depth, h, of a wide-flange beam (Fig. 6.47b). The dashed curve in Fig. 6.47c is called the *centerline,* or *midline,* of the cross section. *A centerline coordinate* s, measured from an origin on the centerline, is used to locate points in the cross section at x. The wall thickness at location (x, s). identified as $t(x, s)$, is usually, but not always. independent of both x and s.

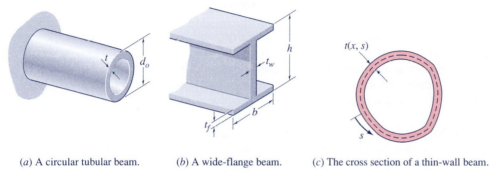

(a) A circular tubular beam.   (b) A wide-flange beam.   (c) The cross section of a thin-wall beam.

FIGURE 6.47   Dimensions of thin-wall beams.

Some examples of thin-wall beams are ones with cross sections that are symmetric about two axes, like **W**-shapes (Fig. 6.48a), **S**-shapes (Fig. 6.48b), box beams (Fig. 6.48c), and circular pipe beams (Fig. 6.48d); channel sections (Fig. 6.48e), T-sections (Fig. 6.48f), and equal-leg angles (Fig. 6.48g), which have only one axis of symmetry in the cross section: and unequal-leg angle sections (Fig. 6.48h) and Z-sections (Fig. 6.48i), which have no axis of symmetry in the cross section.[24]

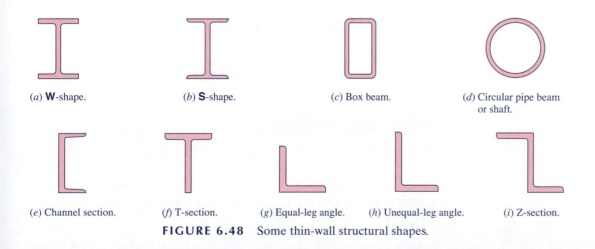

(a) **W**-shape.   (b) **S**-shape.   (c) Box beam.   (d) Circular pipe beam or shaft.

(e) Channel section.   (f) T-section.   (g) Equal-leg angle.   (h) Unequal-leg angle.   (i) Z-section.

FIGURE 6.48   Some thin-wall structural shapes.

---

[24]See Appendix D, *Section Properties of Selected Structural Shapes.*

The principal feature that distinguishes the theory of thin-wall beams is the fact that the *shear flow* $q(x, s)$ can be computed at the location (x, s), and, because the section is thin, it can be assumed that:

- *the shear flow, $q(x, s)$, is locally tangent to the centerline of the section,*
- *the average shear stress is given by*

$$\tau(x, s) = \frac{q(x, s)}{t(x, s)} \tag{6.68}$$

and

- *this average shear stress accurately reflects the shear stress distribution through the thickness at location s in cross section x.*

Figure 6.49*a* illustrates shear flow along a thin-wall section, and the corresponding average shear stress is depicted in Fig. 6.49*b*.

In this section we will illustrate the analysis of thin-wall beams by considering the stress distributions in wide-flange beams, in box beams, and in thin-wall beams with circular cross section.

Shear Stress in Web-Flange Beams.

Let us begin our analysis of stresses in thin-wall beams by considering the important case of doubly symmetric, open beams[25] that consist of a *web* lying in the longitudinal plane of symmetry and two equal *flanges*, as illustrated in Fig. 6.50. Wide-flange beams and I-beams fall in this category, as do steel plate girders used, for example, in buildings and bridges.

In Example Problem 6.6 it was shown that thin-wall beams, like **W**-sections and **S**-sections, are much more efficient than compact sections, such as rectangular sections and circular sections, in their resistance to bending, as expressed by the ratio S of bending moment to maximum flexural stress. Let us now consider the distribution of shear stress in beams of this type. To determine the shear-stress distribution in the flanges and in the web, we can follow the steps of the equilibrium analysis given in Section 6.8. In each case, that is, for flange shear and for web shear, we need to select the appropriate free-body diagram. The segment of wide-flange beam in Fig. 6.50*a* from x to $(x + \Delta x)$ is enlarged in Fig. 6.51*a*, and the flexural stresses are shown on the cross sections at x and $(x + \Delta x)$.

Shear Flow in Flanges:

Let us first determine the shear flow, q_f, and associated shear stress, τ_f, in a flange.[26] Figure 6.51*b* shows that location of the cutting plane

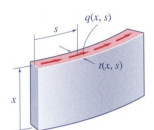

(*a*) Shear flow.

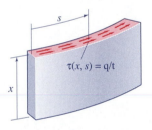

(*b*) Average shear stress.

FIGURE 6.49 An illustration of shear in a thin-wall beam.

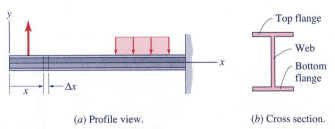

(*a*) Profile view.   (*b*) Cross section.

FIGURE 6.50 A doubly symmetric, thin-wall, web-flange beam.

---

[25]An *open cross section* is distinguished from *closed*, thin-wall sections like the box beam in Fig. 6.48*c* and the circular pipe beam in Fig. 6.48*d*.
[26]Here we are considering shear flow that is parallel to the flange (i.e., parallel to the z axis). This is much more important than the flange shear that is parallel to the web.

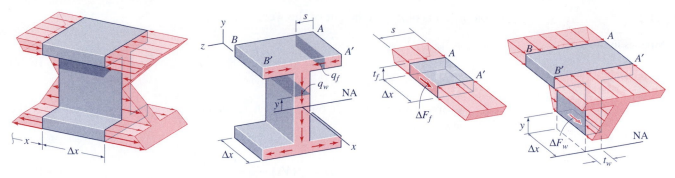

(a) A beam segment showing flexural stresses due to $M(x)$ and due to $M(x + \Delta x)$.

(b) Cutting planes used to determine the flange shear flow q_f and the web shear flow q_w.

(c) A free-body diagram for determining flange shear flow.

(d) A free-body diagram for determining web shear flow.

FIGURE 6.51   Illustrations of stresses in a web-flange beam.

that isolates the free body to use in determining q_f. Figure 6.51c shows the appropriate free-body diagram.[27] Note carefully how the **shear force ΔF_f on the cutting plane is required to balance the flexural stresses** acting on the flange outboard of the cutting plane. Note also that the direction of q_f in Fig. 6.51b is dictated by the direction of ΔF_f on the free-body diagram in Fig. 6.51c.

Having used 3-D sketches to clearly identify the free body that serves as a basis for this shear-flow analysis, we can now resort to a profile view (Fig. 6.52a) and a cross-sectional view (Fig. 6.52b) as we use Eq. 6.65 to calculate the flange shear flow q_f. From Eq. 6.65, the **flange shear flow** is given by

$$q_f = \frac{VQ_f}{I} \qquad (6.69)$$

The moment of inertia of the entire cross section, taken about the neutral axis, is

$$I = \frac{bh^3}{12} - \frac{(b - t_w)h_w^3}{12}$$

or

$$I = \frac{1}{12}(bh^3 - bh_w^3 + t_wh_w^3) \qquad (6.70)$$

where $h_w = h - 2t_f$. From Eq. 6.64,

$$Q_f = A_f'\bar{y}_f' \qquad (6.71)$$

where A_f' and $\bar{y}_f'$ are defined in Fig. 6.52b. Thus,

$$Q_f = st_f\left(\frac{h}{2} - \frac{t_f}{2}\right)$$

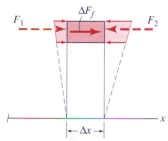

(a) Profile view of flange free-body diagram.

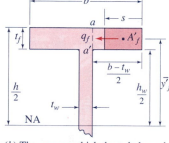

(b) The area on which the unbalanced flexural stresses act.

FIGURE 6.52   Information for use in calculating flange shear flow.

---

[27]Drawing 3-D free-body diagrams, like the ones in Figs. 6.51c and 6.51d, can be tedious, but the secret to developing a thorough understanding of shear stress in beams lies in the success with which you are able to visualize the manner in which **unbalanced flexural stresses lead to shear stresses.**

or, in terms of h_w, the height of the web,

$$Q_f = \frac{s}{8}(h^2 - h_w^2) \tag{6.72}$$

Therefore, combining Eqs. 6.69 through 6.72, we get the following expression for the *flange shear flow:*

$$q_f = \frac{3}{2}(Vs)\frac{h^2 - h_w^2}{bh^3 - bh_w^3 + t_w h_w^3} \tag{6.73}$$

If the flange is thin, say $t_f \leq b/5$, the **flange shear stress** can be computed by using Eq. 6.68, that is

$$\tau_f = \frac{q_f}{t_f} = \frac{3}{2}\left(\frac{Vs}{t_f}\right)\frac{h^2 - h_w^2}{bh^3 - bh_w^3 + t_w h_w^3} \tag{6.74}$$

After deriving an expression for shear flow and shear stress in the web, we will plot and discuss both flange shear and web shear.

Shear Flow in the Web: Starting with the free-body diagram in Fig. 6.51*d*, also shown in profile view in Fig. 6.53*a*, and following the same steps just used to derive expressions for flange shear flow and flange shear stress, we will now derive expressions for web shear flow and web shear stress. From Eq. 6.65, the **web shear flow** is given by

$$\boxed{q_w = \frac{VQ_w}{I}} \tag{6.75}$$

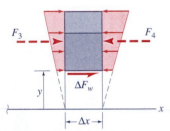

(*a*) Profile view of web-flange free-body diagram.

where, from Eq. 6.64 and Fig. 6.53*b*,

$$Q_w = A_w'\bar{y}_w' = A_1'\bar{y}_1' + A_2'\bar{y}_2'$$

So, in terms of h_w,

$$Q_w = b\left(\frac{h}{2} - \frac{h_w}{2}\right)\left(\frac{1}{2}\right)\left(\frac{h}{2} + \frac{h_w}{2}\right) + t_w\left(\frac{h_w}{2} - y\right)\left(\frac{1}{2}\right)\left(\frac{h_w}{2} + y\right)$$

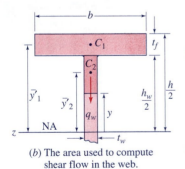

(*b*) The area used to compute shear flow in the web.

FIGURE 6.53 Information for use in calculating web shear flow.

or,

$$Q_w = \frac{b}{2}\left[\left(\frac{h}{2}\right)^2 - \left(\frac{h_w}{2}\right)^2\right] + \frac{t_w}{2}\left[\left(\frac{h_w}{2}\right)^2 - y^2\right] \tag{6.76}$$

Therefore, combining Eqs. 6.70, 6.75, and 6.76, we get the following expression for the *web shear flow:*

$$q_w = \frac{3V}{2}\left(\frac{bh^2 - bh_w^2 + t_w h_w^2 - 4t_w y^2}{bh^3 - bh_w^3 + t_w h_w^3}\right) \tag{6.77}$$

If the web is thin, then the **web shear stress** may be computed by using Eq. 6.68. So,

$$\tau_w = \frac{3V}{2t_w}\left(\frac{bh^2 - bh_w^2 - t_w h_w^2 - 4t_w y^2}{bh^3 - bh_w^3 + t_w h_w^3}\right) \tag{6.78}$$

Summary of Shear in Web-Flange Beams: From Eqs. 6.73 and 6.74 we see that:

1. The flange shear flow q_f and shear stress τ_f vary linearly with distance from the outer edge of the flange.
2. The maximum flange shear flow (and shear stress) occurs at the web-flange intersection (section $a - a'$ in Fig. 6.52b).
3. For the flange free body analyzed, shear flow is directed from the outer edge of the flange toward the web (Fig. 6.52b).
4. The maximum flange shear stress is given by[28]

$$(\tau_f)_{max} = \frac{3V(b - t_w)(h^2 - h_w^2)}{4t_f(bh^3 - bh_w^3 + t_w h_w^3)} \tag{6.79}$$

From similar flange shear analyses for the other three flange areas we can conclude that:

5. Equations 6.73 and 6.74 are valid for any of the four flange areas.
6. The shear flow in each of the four flange areas "flows" (in the sense of fluid flowing in a pipe) in the directions indicated on Fig. 6.51b and Fig. 6.54. and the shear flow in the web is in the direction of the resultant transverse shear force, V.

From Eqs. 6.77 and 6.78 we can conclude that:

7. The web shear flow and shear stress are parabolic, that is, they are quadratic functions of y.
8. The maxima of q_w and τ_w occur at the neutral axis ($y = 0$), and the minima occur at the web-flange junction ($y = \pm h_w/2$). (Because actual beam cross sections have fillets at the web-flange junctions to reduce stress concentration there, the minima are only theoretical minimum values.)
9. Expressions for the maximum and minimum web shear stresses are[29]

$$(\tau_w)_{max} = \frac{3V}{2t_w}\left(\frac{bh^2 - bh_w^2 + t_w h_w^2}{bh^3 - bh_w^3 + t_w h_w^3}\right) \tag{6.80}$$

and

$$(\tau_w)_{min} = \frac{3Vb}{2t_w}\left(\frac{h^2 - h_w^2}{bh^3 - bh_w^3 + t_w h_w^3}\right) \tag{6.81}$$

The magnitudes and directions of the web and flange shear stresses are indicated on Fig. 6.54. The actual stress distribution in the vicinity of the reentrant

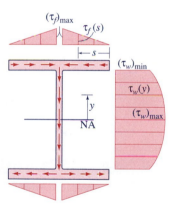

FIGURE 6.54 The shear stresses in a wide-flange beam.

---

[28]If V is negative, $(\tau_f)_{max}$ will be the magnitude of this quantity.
[29]Absolute values are taken if V is negative.

corner at a flange-web junction is complex and depends on the fillet radius and other dimensions. For a discussion of this corner stress distribution, see Ref. 6-4, Section 128.

From Eqs. 6.80 and 6.81 we note that, as $t_w/b \rightarrow 0$, $(\tau_w)_{min} \rightarrow (\tau_w)_{max}$, that is, the web shear stress approaches a constant shear stress. For beams of practical dimensions, the maximum web shear stress typically exceeds the minimum web shear stress by 10%–50%.

It is of interest to determine what percent of the vertical shear force V is carried by the web. We can determine this value by integrating the shear flow over the depth of the web, that is,

$$V_w = \int_{-h_w/2}^{h_w/2} q_w(y)\, dy \tag{6.82}$$

but we can also determine this value by multiplying the area under the τ_w curve in Fig. 6.54 by t_w, noting that the area consists of a rectangle and a parabola. Thus,

$$V_w = h_w t_w \left\{ (\tau_w)_{min} + \frac{2}{3} \left[(\tau_w)_{max} - (\tau_w)_{min} \right] \right\}$$

or

$$V_w = \frac{h_w t_w}{3} \left[2(\tau_w)_{max} + (\tau_w)_{min} \right] \tag{6.83}$$

For practical beams, V_w typically accounts for more than 90% of the total vertical shear force. The remainder is carried by vertical shear in the flanges. This vertical shear in the flanges is quite small and cannot be computed by the shear flow method above.

The following Example Problem illustrates the calculation of shear stresses in wide-flange beams.

EXAMPLE 6.15

(a) Determine the shear stress distribution in the flanges and the web of a **W**14 × 26 beam[30] subjected to the shear force $V = -28$ kips, as shown in Fig. 1.[31] (b) Determine the percent of the vertical shear that is carried by the web of this cross section.

Plan the Solution Although it would be possible to solve Part (a) by using the formulas in Eqs. 6.74 and 6.78, it will be more instructive to use the basic shear-flow formulas, Eqs. 6.69 and 6.75, and the average shear-stress formula, Eq. 6.68. We can determine the total web shear force by integrating the shear flow, as indicated in Eq. 6.82. Since the shear on the cross section is negative (i.e., it points in $+y$ direction in Fig. 1) the shear flows and shear stresses will act in the opposite sense to those shown in Fig. 6.54.

---

[30]The properties of the **W**14 × 26 section are given in Appendix D.1.
[31]This is the shear force just to the left of the support at B in Example Problem 6.5.

Solution (a) *Determine the shear flow and the shear stress in the flanges and in the web.*

Flange Shear Flow and Shear Stress: The flange shear flow is given by Eq. 6.69:

$$q_f = \frac{VQ_f}{I} = \frac{VA_f'\bar{y}_f'}{I} = \frac{VA_1\bar{y}_1}{I} \tag{1}$$

$$q_f = \frac{(-28 \text{ kips})(0.420 s \text{ in}^2)(6.745 \text{ in.})}{245 \text{ in}^4} = -0.324 s \text{ kips/in.} \tag{2}$$

$$\tau_f = \frac{q_f}{t_f} = \frac{-0.324 s \text{ kips/in.}}{0.420 \text{ in.}} = -0.771 s \text{ ksi} \tag{3}$$

$$(\tau_f)_{max} = |(\tau_f)_{s=2.385 \text{ in.}}| = 1.84 \text{ ksi} \tag{4}$$

$$\tau_f = -0.771 s \text{ ksi}, \quad (\tau_f)_{max} = 1.84 \text{ ksi} \qquad \textbf{Ans. (a)}$$

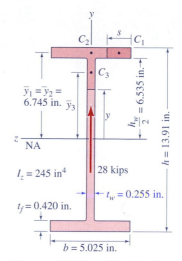

Fig. 1 A **W**14 × 26 wide-flange beam section.

Web Shear Flow and Web Shear Stress: The web shear flow is given by Eq. 6.75:

$$q_w = \frac{VQ_w}{I} = \frac{VA_w'\bar{y}_w'}{I} = \frac{V}{I}(A_2\bar{y}_2 + A_3\bar{y}_3) \tag{5}$$

where A_2 is the area of the entire top flange. Therefore,

$$q_w = \left(\frac{-28 \text{ kips}}{245 \text{ in}^4}\right)[(5.025 \text{ in.})(0.420 \text{ in.})(6.745 \text{ in.})$$
$$+ (6.535 \text{ in.} - y)(0.255 \text{ in.})(0.5)(6.535 \text{ in.} + y)]$$

or

$$q_w = \frac{(-28)[14.235 + 0.1275(42.71 - y^2)]}{245}$$

$$= -(2.249 - 0.0146y^2) \text{ kips/in.}$$

$$\tau_w = \frac{q_w}{t_w} = -(8.82 - 0.0571y^2) \text{ ksi} \tag{6}$$

$$\left.\begin{array}{l}(\tau_w)_{max} = |(\tau_w)_{y=0}| = 8.82 \text{ ksi} \\[6pt] (\tau_w)_{min} = |(\tau_w)_{y=6.535 \text{ in.}}| = 6.38 \text{ ksi}\end{array}\right\} \quad \textbf{Ans. (a)} \qquad \begin{array}{l}(7a)\\[6pt](7b)\end{array}$$

The results of the flange shear analysis and web shear analysis are summarized on Fig. 2. The negative shear ($V = -28$ kips) gives shear stresses in the opposite direction to those in Fig. 6.54.

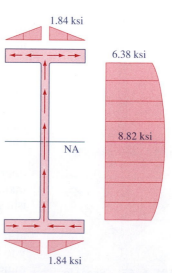

Fig. 2 The shear stresses in a **W**14 × 26 beam.

(b) *Determine the percent of vertical shear that is carried by the web.* Let us use Eq. 6.82.

$$V_w = \int_{-h_w/2}^{h_w/2} q_w(y)\,dy = -\int_{-6.535}^{6.535} (2.249 - 0.0146y^2)\,dy$$

$$= -2.249(6.535)(2) + \frac{2}{3}(0.0146)(6.535)^3$$

$$= -29.40 + 2.71 = -26.69 \text{ kips}$$

Comparing V_w with V, we see that 95% *of the vertical shear is carried by the web of this beam.*

Review the Solution We can spot-check the above results by using Eq. 6.80 to compute $(\tau_w)_{max}$. When we do so, we get $(\tau_w)_{max} = 9.02$ ksi, which does not agree very well with the value of 8.82 ksi that we got in Eq. (7a). So let us compare the value of $I = 245$ in^4 from Appendix D.1 with the value calculated from Eq. 6.70. For the latter we get $I = 240$ in^4. The difference in I values accounts for the difference in $(\tau_w)_{max}$ values. This difference stems from the fact that the actual cross section has web-to-flange fillets that provide a slight increase in the value of the moment of inertia. This was not accounted for in Eq. 6.70.

MDS6.12 & 6.13 **Shear Stress in Beams**—The Shear Stress option in the MDS computer program module *Flexure* combines a program for determining section properties with equations for computing the transverse shear stress in beams.

The analysis procedure employed above to determine the shear stress in the web and the flanges of a doubly symmetric, thin-wall, open beam can also be applied to thin-wall open beams with one axis of symmetry and with loading in the plane of symmetry. For example, Fig. 6.55 shows a T-section and an equal-leg V-section. Since these sections have only one axis of symmetry (the longitudinal plane of symmetry) it is necessary to locate the neutral axis first before proceeding with a shear flow/shear stress analysis based on Eq. 6.65. Here, as before, **the maximum shear flow occurs at the neutral axis,** since the imbalance of normal forces is a maximum at the neutral axis.

Stresses in Closed, Thin-Wall Beams. Box beams and circular pipe beams, as illustrated in Figs. 6.48c and 6.48d, may be classified as **closed, thin-wall sections.** Let us determine the shear-stress distribution in this type of beam, assuming that the beam has at least one longitudinal plane of symmetry and that the loads lie in this plane, as illustrated by the generic closed, thin-wall section in Fig. 6.56a.

The analysis of shear stress in closed, thin-wall beams differs from the analysis of open thin-wall beams, like the wide-flange beam illustrated in Fig. 6.51, in only one major respect, namely, the location of a point in the cross section where the shear vanishes. For the flange free-body diagram in Fig. 6.51c, the outboard edge AA' is free of shear, and the plot of $\tau_f(s)$ in Fig. 6.54 reflects this fact. Likewise, the free-body diagram in Fig. 6.51d includes the portion of beam from the web sectioning-plane at level y to the free surfaces of the flange. For a closed section we need to be able to determine the shear flow (and shear stress) at any arbitrary location in the cross section, say the locations identified by coordinates z, y, and s in Fig. 6.56a. But, how do we choose an appropriate free body? One answer is that we can make use of symmetry

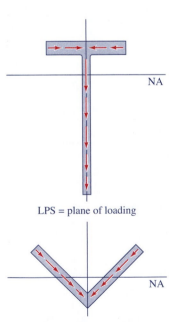

FIGURE 6.55 Shear in open beams with one axis of symmetry in the cross section.

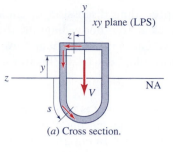

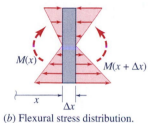

(a) Cross section.

(b) Flexural stress distribution.

FIGURE 6.56 A closed,
thin-wall section.

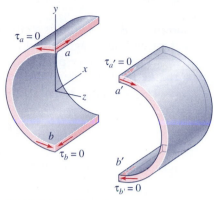

FIGURE 6.57 Shear stress on a
sectioning plane in the xy-plane.

about the xy-plane and always *select a free body that is symmetric in z*. Because of symmetry, there can be no longitudinal shear on surfaces exposed by a cutting plane in the LPS, and thus there can be no shear on the cross section at $z = 0$, as indicated in Fig. 6.57. (Can you explain why symmetry about the LPS makes $\tau = 0$ at $z = 0$?) The shear flow will be directed so that the resultant shear force on the cross section is V, as indicated in Fig. 6.56a. Finally, the sectioning cut should be made normal to the midline of the thin-wall section, like the cuts at z, y, and s in Fig. 6.56a. In the next example problem we will illustrate how to select the appropriate free-body diagram and how to calculate shear flow and shear stress on a closed, thin-wall section.

EXAMPLE 6.16

A pipe conveying fluid over a narrow stream crossing must act as a beam as well as a conduit, as indicated in Fig. 1.

Assuming that the ratio of mean diameter to wall thickness satisfies the requirement $d/t \gg 1$. determine the shear flow and shear stress distribution at a section due to the transverse shear force, V, at that section, Also, determine the maximum shear stress on the cross section, and express $\tau_{\max}$ in terms of V/A.

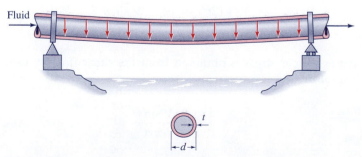

Fig. 1 A pipe transporting fluid and acting as a beam.

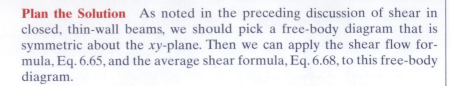

Plan the Solution As noted in the preceding discussion of shear in closed, thin-wall beams, we should pick a free-body diagram that is symmetric about the xy-plane. Then we can apply the shear flow formula, Eq. 6.65, and the average shear formula, Eq. 6.68, to this free-body diagram.

Solution Because this is a circular cross section, it is convenient to select a free-body diagram that is defined by radial cutting planes at angle θ either side of the xy-plane, as indicated in Fig. 2. In Fig. 2b there are two surfaces that have equal shear forces ΔH which serve to balance the net force $(F_2 - F_1)$ due to the flexural stresses. Therefore, we get

$$2\Delta H = F_2 - F_1 \tag{1}$$

where, from the definition of shear flow

$$\Delta H = q\Delta x \tag{2}$$

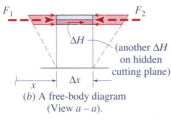

(a) Cross-sectional view.

Comparing Eq. (1) to Eq. 6.60, and following the same steps that were used to arrive at Eq. 6.65, we get

$$2q = \frac{VQ}{I} \tag{3}$$

for the symmetric free-body diagram in Fig. 2. Here, as in Eq. 6.65,

$$Q = \int_{A'} y \, dA = A'\bar{y}' \tag{4}$$

F_1 F_2

ΔH (another ΔH on hidden cutting plane)

Δx

x

(b) A free-body diagram
(View a – a).

Fig. 2 Selection of a symmetric free body.

is the first moment, about the neutral axis, of the area acted on by the unbalanced flexural stresses, and I is the moment of inertia of the entire cross section about its neutral axis. The geometric properties for a thin ring and for a sector of a thin ring are shown in Fig. 3. Using the formulas in Fig. 3b, we have, for $t \ll r$,

$$Q = A'\bar{y}' \approx 2r^2 t \sin \theta \tag{5}$$

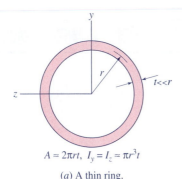

$A \approx 2\pi rt, \quad I_y = I_z \approx \pi r^3 t$

(a) A thin ring.

Combining the moment of inertia from Fig. 3a with Eqs. (3) and (5), we get

$$q = \frac{V(2r^2 t \sin \theta)}{(2\pi r^3 t)} = \frac{V \sin \theta}{\pi r} \qquad \textbf{Ans.} \tag{6}$$

The average shear stress is obtained from the shear flow by using Eq. 6.68. Therefore,

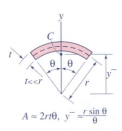

$A \approx 2rt\theta, \quad \bar{y}^- \approx \frac{r \sin \theta}{\theta}$

(b) A sector of a thin ring.

Fig. 3 Some geometric properties of plane areas.

$$\tau = \frac{q}{t} = \frac{V \sin \theta}{\pi rt} \qquad \textbf{Ans.} \tag{7}$$

Figure 4 illustrates the distribution of shear flow (and shear stress).

The maximum shear stress is given by

$$\tau_{max} = \tau_{(\theta = \pi/2)} = \frac{V}{\pi r t}$$

Since the area of the cross section can be approximated by

$$A \approx 2\pi r t$$

for a *thin-wall pipe beam with t* $\ll$ *r*, the maximum shear stress is[32]

$$\tau_{max} = \frac{2V}{A} \qquad \textbf{Ans.} \quad (8)$$

Review the Solution   As a check on the expression that we obtained for q, Eq. (6). we can see if the resultant of this distribution is V. as it is supposed to be. The force on the elemental area highlighted in Fig. 4 is

$$dF = q\,ds = qr\,d\theta$$

The vertical component, as shown on the insert in Fig. 4, is, therefore,

$$dF_y = qr\sin\theta\,d\theta$$

So, due to symmetry, we have

$$F_y = 2\int_0^\pi qr\sin\theta\,d\theta$$

$$= \frac{2V}{\pi}\int_0^\pi \sin^2\theta\,d\theta$$

$$= \frac{2V}{\pi}\left[\frac{\theta}{2} - \frac{1}{4}\sin 2\theta\right]_0^\pi = V$$

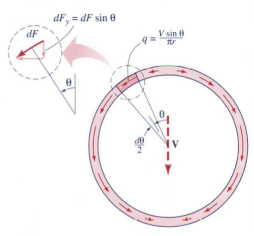

Fig. 4   The distribution of shear flow corresponding to a resultant shear force V.

Since the distribution looks "reasonable" (the shear flow has its maximum at the neutral axis and is zero on the plane of symmetry), and since it gives the correct resultant, we can assume that our answer is correct.

6.11   SHEAR IN BUILT-UP BEAMS

In Sections 6.8 and 6.10 we assumed the beams to be homogeneous, so it was appropriate for us to derive formulas for the *average shear stress* on a given surface. However, there are a number of important applications of beams where welds, glue, rivets, bolts, or nails are used to join two or more structural components to form a single beam. Several examples of such **built-up beams** are shown in Fig. 6.58. They are: (a) a glued-laminated timber beam, (b) a welded plate girder, (c) a wooden box

---

[32]See Homework Problem 6.8-12 for shear in a "thick-wall" pipe beam.

(a) A glued, laminated wood beam.

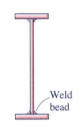

Weld bead

(b) A welded steel plate girder.

Nail

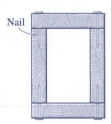

(c) A wood box beam.

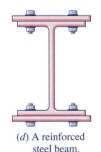

(d) A reinforced steel beam.

FIGURE 6.58 Several examples of built-up beams.

beam constructed of several planks nailed together, and (d) a beam with reinforcing flange plates bolted to the basic beam. In such cases it is assumed that the beam acts as a unit (e.g., plane sections remain plane) and that the joining medium (nails, welds, bolts, etc.) is capable of transferring shear across the longitudinal junctions between component parts of the beam.

Shear may be transferred between adjoining parts of a built-up beam in three ways:[33]

1. by *shear distributed over the interface areas,* as when surfaces are glued together (Fig. 6.58a).
2. by *shear distributed along a line,* as when two metal parts are jointed by weld beads (Fig. 6.58b), and
3. by *discrete shear connectors,* as when two parts are nailed, riveted, or bolted together (Figs. 6.58c,d).

As was pointed out earlier in the discussion of Fig. 6.41, if a beam is composed of several pieces, it will only behave as a single beam (i.e., satisfy the "plane sections" hypothesis) if some provision is made to transfer shear between adjacent parts. When shear is distributed over a bond area, like the glued joints in Fig. 6.58a, the analysis is based on the *shear stress* in the bond layer, as though the beam were homogeneous (Section 6.8). In the case of a linear shear connector (e.g., the weld beads in Fig. 6.58b), and the case of discrete shear connectors (e.g., the nails in Fig. 6.58c and the bolts in Fig. 6.58d), an analysis based on shear flow is appropriate. We will now consider the latter two cases.

The shear flow equation, Eq. 6.65, is used to compute the *required* shear flow, that is, the shear force per unit length that must be transferred from one part of the beam to the adjacent part under the given loading condition.

$$q_r = \frac{VQ}{I} = \frac{VA'\bar{y}'}{I} \tag{6.84}$$

where, as before, the area A' is the area on which the unbalanced flexural stresses responsible for the shear flow q_r are exerted. Knowing q_r, it is a design exercise to select the size weld or the number of discrete connectors (nails, rivets, or bolts) that will provide the required shear strength.

Linear Shear Connectors. In the case of a linear shear connector, Eq. 6.84 determines the actual shear force per unit length that must be transferred through the linear shear connector(s). The shear connector (e.g., a weld bead) can exert an <u>allowable</u> shear force per unit length, which we will designate as q_a. Therefore, the joint will fail locally unless

$$q_a \geq q_r \tag{6.85}$$

Example Problem 6.17 illustrates this type of problem.

Discrete Shear Connectors. If discrete connectors, each having a shear-force capacity V_a, are spaced along a joint at a spacing Δs, as illustrated in Fig. 6.59a, then the joint will fail unless

$$V_a \geq q_r \Delta s \tag{6.86}$$

---

[33]The analysis of the transfer of shear from steel reinforcing bars to the surrounding concrete is similar to the analysis of shear presented in this section, but is beyond the scope of this text.

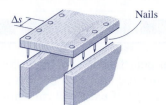

(a) Nails on a uniform spacing of Δs.

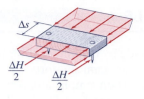

(b) A free-body diagram for determining shear force in discrete connectors.

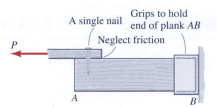

(c) A direct-shear test to determine the shear capacity of a single nail.

FIGURE 6.59 A nailed shear joint.

Figure 6.59b illustrates how the two nails in a beam segment of length Δs provide the longitudinal shear force ΔH equal to the force due to the unbalanced flexural stresses. The <u>allowable</u> shear-force capacity of a single discrete shear connector, like the nail in Fig. 6.59c, is designated as V_a, whether this available strength represents shearing the nail in two (which is highly unlikely) or whether the joint fails by pull-out of the nail or crushing of the adjacent wood. Direct-shear tests, as illustrated in Fig. 6.59c, may be performed to determine the ultimate shear capacity of a certain size and type of nail used to connect two pieces of wood of a certain type.

Example Problem 6.18 illustrates this discrete-shear-connector type of problem.

EXAMPLE 6.17

A built-up plate girder has the dimensions shown in Fig. 1. If the shear force at this cross section is $V = 300$ kips, what is the shear flow in each fillet weld bead? $I_z = 54,800$ in^4.

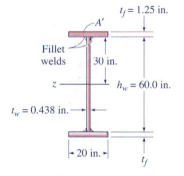

Fig. 1 (Drawing not to scale.)

Plan the Solution Since there are two weld beads that attach each flange to the web, we can use Eq. 6.84 modified by a factor of two.

Solution From Eq. 6.84 (modified).

$$2q_r = \frac{VQ}{I} = \frac{VA'\overline{y}'}{I}$$

where, as indicated, Q is the first moment of the entire flange area, taken about the NA.

$$q_r = \frac{(300 \text{ kips})(20 \text{ in.})(1.25 \text{ in.})(30.625 \text{ in.})}{2(54,800 \text{ in}^4)}$$

$$q_r = 2.10 \text{ kips/in.} \qquad\qquad \textbf{Ans.}$$

Review the Solution A simple recheck of the calculations is all that is necessary here.

Interesting design information on steel girders and weld beads can be found in Ref. 6-2, but we will not pursue the design of welds further in this text.

EXAMPLE 6.18

A box beam is made by nailing together four boards in the configuration shown in Fig. 1a. The beam supports a concentrated load of 1000 lb at its midspan, and it rests on simple supports (Fig. 1b). If each nail can withstand an allowable shear force of 150 lb, what is the maximum spacing that can be used?

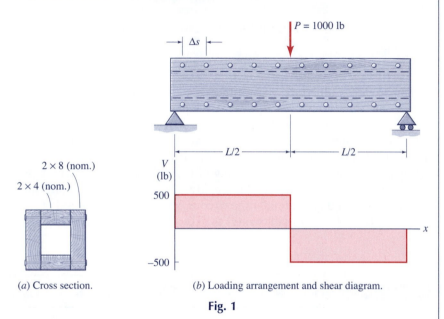

2 × 8 (nom.)

2 × 4 (nom.)

(a) Cross section.

(b) Loading arrangement and shear diagram.

Fig. 1

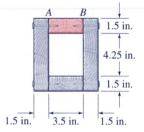

Fig. 2 Surfaced lumber design dimensions.

Plan the Solution Since the shear force V is constant over each half of the beam, we can use a constant nail spacing Δs. The shear flow along each row of nails can be computed using the shaded area between A and B in Fig. 2. Equations 6.84 and 6.86 can be combined to compute the maximum spacing. The finish dimensions of the boards are given in Appendix D.8.

Solution Since there are two rows of nails transferring shear between the shaded area in Fig. 2 and the vertical boards, we need to modify Eq. 6.84 to read

$$2q_r = \frac{VQ}{I} = \frac{VA'\bar{y}'}{I} \tag{1}$$

in order to calculate the shear flow along a single line of nails. The spacing along each row of nails is then given by Eq. 6.86.

$$\Delta s_{\max} = \frac{V_a}{q_r} \tag{2}$$

The moment of inertia can be computed by taking a 6.5 in. × 7.25 in. rectangle and subtracting a 3.5 in. × 4.25 in. rectangle.

$$I = \frac{6.5(7.25)^3}{12} - \frac{3.5(4.25)^3}{12} = 184.0 \text{ in}^4$$

The shear flow is given by Eq. (1),

$$q_r = \frac{VA'\bar{y}'}{2I} = \frac{(500 \text{ lb})[(3.5 \text{ in.})(1.5 \text{ in.})](2.875 \text{ in.})}{2(184.0 \text{ in}^4)}$$

$$q_r = 20.5 \text{ lb/in.}$$

Then, from Eq. (2),

$$\Delta s_{\max} = \frac{V_a}{q_r} = \frac{150 \text{ lb}}{20.5 \text{ lb/in.}} = 7.32 \text{ in.}$$

$$\text{Use } \Delta s = 7 \text{ in.} \qquad \textbf{Ans.}$$

Review the Solution   The answer seems reasonable, and the calculations are easily rechecked and found to be correct.

Of course, factors other than shear strength should be considered in the design of joints. For example, nails cannot be driven too close to the edge of a piece of timber, and nails cannot be spaced too close together, or the wood is likely to split.

*6.12   SHEAR CENTER

In Section 6.6 "Unsymmetric Bending," you discovered that *loading of a beam, even an unsymmetric beam, parallel to a principal axis of inertia of the cross section produces bending that is confined to the corresponding longitudinal principal plane.* We considered the *direction of loading,* but we did not establish the *line of action of loading* that would produce bending without simultaneous twisting. Consider bending of a beam that has a thin-wall channel cross section (Fig. 6.60a) and is loaded

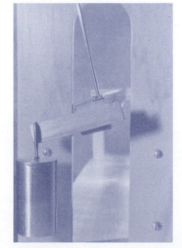

(a) Cross section of cantilever beam in Figs. 6.60b, c, d.

(b) Bending plus counter-clockwise twisting.   (c) Bending plus clockwise twisting.   (d) Bending without twisting.

FIGURE 6.60   Loading of a thin-wall, open beam normal to its plane of symmetry. (Courtesy Roy Craig)

FIGURE 6.61 The shear center of a wing cross section.

normal to the axis of symmetry of the cross section. Figure 6.60 shows this situation, with the load placed in three different positions. In Fig. 6.60*b* the load is too far to the left, so it produces both bending (which is not really visible) and a counterclockwise twisting, as indicated by the position of the pointer. In Fig. 6.60*c* a load that acts to the right of the centroid of the cross section clearly produces twisting in a clockwise sense. Finally, by trial-and-error, a location is found where the load produces bending (again, too small to see) without any twisting (Fig. 6.60*d*).

> *The **shear center** of a cross section is defined as that point in the cross section through which the resultant shear force V must pass if the beam is to bend without twisting.*

It is particularly important that the shear center of open, thin-wall beams be located, since, as Fig. 6.60 shows, such members are very flexible in torsion (See Section 4.9). There are other situations, as well, where locating the shear center of the cross section of a beam is very important. For example, the shear center of an airplane wing, which may in some cases be analyzed as a thin-wall closed beam, must be properly located with respect to the imposed aerodynamic loads (Fig. 6.61).

Shear-center location is important for all beams, but especially for thin-wall open beams. Therefore, we will begin our analysis of shear-center location by considering *singly symmetric, thin-wall beams,* similar to the channel section in Fig. 6.60. If a cross section has an axis of symmetry, the shear center will be located somewhere along the axis of symmetry of the section. An inclined load may be resolved into one component lying in the plane of symmetry, which contains one principal axis, and one component in the perpendicular (principal) direction, as illustrated in Fig. 6.62. Therefore, to locate the shear center, S, we need only consider shear that is perpendicular to the axis of symmetry of the cross section.

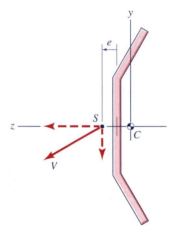

FIGURE 6.62 The shear center is on the axis of symmetry.

Consider the cantilever beam with open, thin-wall, singly symmetric section shown in Fig. 6.63. At an arbitrary cross section there is a resultant shear force V, as shown in the pictorial view of Fig. 6.63*a* and in the cross-sectional view in Fig. 6.63*b*. The single shear force V is the resultant of a shear force V_2 in the web and the shear forces V_1 and V_3 in the two inclined flanges (Fig. 6.63*c*). Finally, these shear forces are the resultants of shear flows (Fig. 6.63*d*) that we can

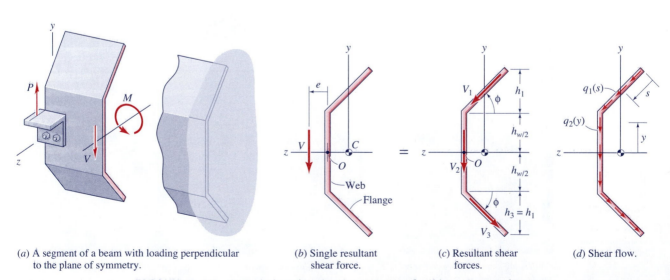

(*a*) A segment of a beam with loading perpendicular to the plane of symmetry.

(*b*) Single resultant shear force.

(*c*) Resultant shear forces.

(*d*) Shear flow.

FIGURE 6.63 Steps in locating the shear center of a thin-wall, open beam.

- *Determine the location of the shear center, S, so that the shear force acting through S is the resultant of shear stresses due to bending only (i.e., bending without twisting).*

First, we can relate the *eccentricity, e,* of the shear center to the shear resultants in Fig. 6.63c by setting the moment about O in Fig. 6.63b equal to the moment about O in Fig. 6.63c.

$$+\circlearrowleft\left(\sum M\right)_o = Ve = \underset{(b)}{(V_1 \cos \phi)}\left(\frac{h_w}{2}\right) + \underset{(c)}{(V_3 \cos \phi)}\left(\frac{h_w}{2}\right) \qquad (6.87)$$

We also must satisfy the force-equilibrium equation

$$\xleftarrow{+}\sum F_z = 0 = \underset{(b)}{V_1 \cos \phi} - \underset{(c)}{V_3 \cos \phi} \qquad (6.88)$$

Combining these two equations, we get

$$e = \frac{V_1 h_w \cos \phi}{V} \qquad (6.89)$$

Therefore, to locate the shear center for this cross section we need to determine the shear force V_1 in the flange. From Eq. 6.65, the shear flow in the inclined flange is given by

$$q_1(s) = \frac{VQ_1(s)}{I} \qquad (6.90)$$

Figure 6.64 can be used in the determination of the cross-sectional properties that are required in the calculation of the shear flow $q_1(s)$. We assume that the thicknesses of the flanges and of the web are small in comparison with other dimensions.

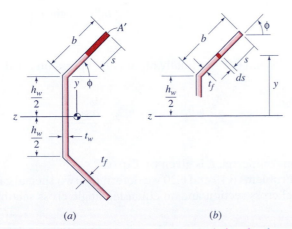

(a)   *(b)*

FIGURE 6.64 Cross-sectional dimensions used in calculating the shear flow in the flange.

First, we need the moment of inertia of the entire cross section, taken with respect to the z axis.

$$I \equiv I_z = (I_z)_{\text{web}} + 2(I_z)_{\text{flange}} \tag{6.91}$$

We can use the definition of moment of inertia, together with Fig. 6.64b, to determine an expression for $(I_z)_{\text{flange}}$.

$$(I_z)_{\text{flange}} = \int_{A_{\text{flange}}} y^2\, dA = \int_0^b \left[\frac{h_w}{2} + (b-s)\sin\phi \right]^2 (t_f\, ds)$$

Carrying out the integration, we get[34]

$$(I_z)_{\text{flange}} = bt_f\left[\left(\frac{h_w}{2}\right)^2 + \left(\frac{bh_w}{2}\right)\sin\phi + \left(\frac{b^2}{3}\right)\sin^2\phi \right]$$

Therefore, for the complete cross section,

$$I = \frac{t_w h_w^3}{12} + 2bt_f\left[\left(\frac{h_w}{2}\right)^2 + \left(\frac{bh_w}{2}\right)\sin\phi + \left(\frac{b^2}{3}\right)\sin^2\phi \right] \tag{6.92}$$

We can use Fig. 6.64a to determine $Q_1(s)$.

$$Q_1(s) = \int_{A'} y\, dA = A'\overline{y}'$$

or

$$Q_1(s) = (t_f s)\left[\frac{h_w}{2} + \left(b - \frac{s}{2}\right)\sin\phi \right] \tag{6.93}$$

Therefore, the flange shear flow is

$$q_1(s) = \frac{VQ_1(s)}{I} = \frac{Vt_f s}{I}\left[\frac{h_w}{2} + \left(b - \frac{s}{2}\right)\sin\phi \right] \tag{6.94}$$

This distribution of shear flow in the flange is illustrated in Fig. 6.65. The web shear flow distribution is also illustrated.

The total shear force in the flange is given by the integral

$$V_1 = \int_0^b q_1(s)\, ds = \frac{Vt_f b^2}{I}\left(\frac{h_w}{4} + \frac{b\sin\phi}{3} \right) \tag{6.95}$$

Combining Eqs. 6.89 and 6.95 we finally obtain an expression for the location of the shear center.

$$e = \frac{b^2 t_f h_w \cos\phi}{I}\left(\frac{h_w}{4} + \frac{b\sin\phi}{3} \right) \tag{6.96}$$

FIGURE 6.65 The shear-flow distribution in an open, thin-wall beam.

where the moment of inertia, I, is given by Eq. 6.92.

In Example Problems 6.19 and 6.20 we determine two special cases of the above analysis, a channel cross section and an equal-leg angle cross section, respectively.

---

[34]Check this expression by evaluating it for $\phi = 0$ and $\phi = 90°$.

EXAMPLE 6.19

Use the fundamental equation of shear flow analysis for beams. Eq. 6.65, to determine the eccentricity of the shear center for the channel section in Fig. 1. Neglect the thickness in comparison with the other cross-sectional dimensions. For dimensional purposes. let V be in newtons.

Plan the Solution We can follow the basic procedure that was followed in deriving Eq. 6.96, that is, we can derive an expression for e, like Eq. 6.89, based on resultant moments about O. This will involve the flange shear force. Next we can determine an expression for the flange shear flow, and then use the flange shear flow to determine the flange shear force needed in the expression for e. The shear center should lie "out-side" the cross section.

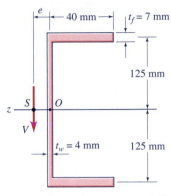

Fig. 1 (Drawing not to scale.)

Solution Neglecting a small flange contribution ($\frac{1}{12}bt_f^3$), we get the following value for the moment of inertia:

$$I = \frac{t_w h_w^3}{12} + 2A_f d_f^2$$

$$I = \frac{(4\ \text{mm})(250\ \text{mm})^3}{12} + 2(40\ \text{mm})(7\ \text{mm})(125\ \text{mm})^2$$

$$I = 13.96(10^6)\ \text{mm}^4 = 13.96(10^{-6})\ \text{m}^4$$

From resultant moments about point O, through which V_w passes (Fig. 2),

$$+\zeta\left(\sum M\right)_o = Ve = V_f h_w$$

so

$$e = \frac{V_f h_w}{V} \tag{1}$$

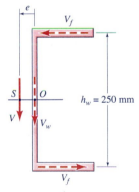

Fig. 2

From Eq. 6.65,

$$q_f = \frac{VQ_f}{I} \tag{2}$$

with Q_f based on the flange area A' in Fig. 3, that is,

$$Q_f = A'\bar{y}' = (7\ \text{mm})(s)(125\ \text{mm})$$

$$Q_f = 875s\ \text{mm}^3$$

So,

$$q_f = \frac{(VN)(875s\ \text{mm}^3)}{13.96(10^6)\ \text{mm}^4}$$

$$(q_f)_{\text{max}} = (q_f)_{s=40\ \text{mm}}$$

$$= 2.51(10^{-3})\ V\ \text{N/m} \tag{3}$$

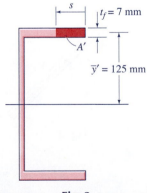

Fig. 3

Since the shear flow varies linearly, as illustrated in Fig. 4, the flange shear force V_f is

$$V_f = \frac{1}{2}[2.507(10^{-3})\,V\,\text{N/mm}](40\text{ mm})$$

$$V_f = 0.0501\,V\,\text{N} \tag{4}$$

Combining Eqs. (1) and (4) we get

$$e = 0.0501 h_w = 12.5\text{ mm} \qquad\qquad \textbf{Ans.}$$

Review the Solution   The eccentricity e should be positive so that the shear center lies outside the cross section. Our answer has the correct sign, and the magnitude seems reasonable in comparison with the dimensions of the beam. Many commercial channel sections have an e/h_w ratio in the 0.05–0.15 range.

Fig. 4

> **EXAMPLE 6.20** <

(a) Prove that the shear center of an equal-leg angle (Fig. 1) is at the corner of the angle, and prove that the resultant of the shear flow in the legs of the angle is equal to the total shear force on the section. The shear force acts normal to the axis of symmetry of the cross section. Assume that $t \ll b$. (b) Determine the maximum shear stress in the cross section in Fig. 1.

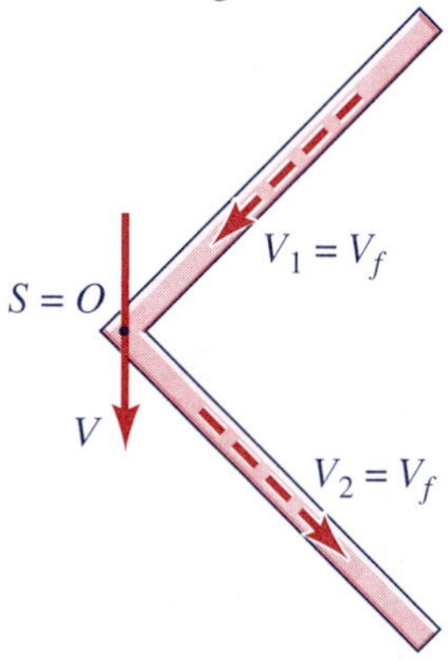

Fig. 1

Plan the Solution   There is a resultant shear force along each leg of the angle. Since the lines of action of these two forces pass through the corner O, this point must be the shear center. We could develop an expression for this using basic shear-flow concepts (i.e., Eq. 6.65), or we can make use of Eq. 6.95, setting $h_w = 0$ and $\phi = 45°$. The maximum shear stress should occur at the neutral axis.

Solution   (a) *Determine shear-center location and prove that the resultant of flange shears is V.* Since V_1 and V_2 both pass through point O, as indicated in Fig. 2, the resultant, V, must pass through this point. Therefore, the shear center lies at the intersection of the legs of the angle.[35] The flange shear is given by Eq. 6.95, with $h_w = 0$.

$$V_f = \frac{Vtb^3 \sin 45°}{3I} \tag{1}$$

where the moment of inertia is obtained by specializing Eq. 6.92 to give

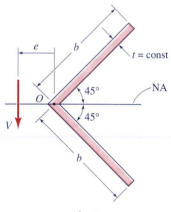

Fig. 2

$$I = \frac{2}{3}b^3 t \sin^2 45° = \frac{b^3 t}{3} \tag{2}$$

---

[35]This argument also holds for unequal-leg angles, but the shear-flow distribution in an unequal-leg angle must be obtained using the unsymmetric bending theory of Section 6.6, so this problem, which asks for the shear flow, only treats equal-leg angles.

Combining Eqs. (1) and (2) we get

$$V_f = \frac{V}{\sqrt{2}} \tag{3}$$

Since

$$+\downarrow \sum F_y = 2V_f \sin 45° = V$$

we have shown that combining Eqs. 6.92 and 6.95 leads to the correct resultant shear force.

(b) *Determine the maximum shear stress.* The shear flow distribution is given by Eq. 6.94, which specializes to

$$q_f(s) = \frac{Vts}{I}\left(b - \frac{s}{2}\right)\sin 45° \tag{4}$$

where I is given by Eq. (2). This shear-flow distribution is depicted in Fig. 3. Since $\tau_f = q_f/t$,

$$(\tau_f)_{\max} = \frac{(q_f)_{s=b}}{t}$$

or

$$(\tau_f)_{\max} = \left(\frac{3\sqrt{2}}{4}\right)\frac{V}{bt} \qquad \textbf{Ans.} \tag{5}$$

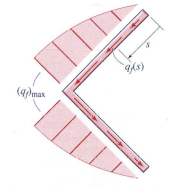

Fig. 3 Shear-flow distribution.

Review the Solution As a rough check on our answer for $\tau_{\max}$, we can compare the preceding result, Eq. (5), with $\tau_{\max}$ for a rectangle. From Example Problem 6.14,

$$(\tau_{\max})_{\text{rectangle}} = \frac{3}{2}\frac{V}{A}$$

Since the area of the equal-leg angle is $2bt$, Eq. (5) can be expressed as

$$(\tau_{f\max})_{\text{angle}} = \frac{3\sqrt{2}}{2}\frac{V}{A}$$

Since the shear flows in the angle are not parallel with the resultant V, it is reasonable that the shear stress in the angle section is greater than that in a rectangle by a factor of $\sqrt{2}$.

See Homework Problem 6.12-16 for a chance to use the basic shear flow formula, Eq. 6.65, to obtain the above results.

▼ STRAIN-DISPLACEMENT ANALYSIS OF BEAMS

Prob. 6.2-1. A steel band-saw blade with rectangular cross section passes over a pulley of radius $r = 600$ mm, as shown in Fig. P6.2-1. Neglect the teeth of the saw blade, and assume that the neutral surface passes through the center of the saw blade's cross section. Determine the maximum strain, ϵ_{max}, in the saw blade (a) if the thickness of the blade is $h = 1.25$ mm, and (b) if the thickness of the blade is $h = 2.0$ mm.

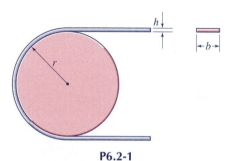

P6.2-1

Prob. 6.2-2. Couples M_0 are applied to a steel strap of length $L = 30$ in. and thickness $h = \frac{1}{8}$ in. to bend it into the form of a circular are, as shown in Fig. P6.2-2. The deflection δ at the midpoint of the are, measured from a line joining the tips, is 3.5 in. (a) Determine the radius of curvature, ρ, of the strap, measured from the center of curvature C to the neutral axis of the strap (which passes through the center of the cross section of the strap). (b) Calculate the maximum extensional strain, ϵ_{max}, in the curved strap.

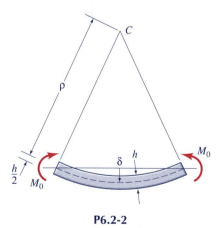

P6.2-2

Prob. 6.2-3. A steel strap of length L (meters) is bent to form a circle, and its ends are then butt-welded together, as shown in Fig. P6.2-3. (a) Determine the shortest length of strap that can be used if the maximum permissible strain in the strap is $\epsilon_{max} = 2.0 \times 10^{-3}$ mm/mm and if the thickness of

the strap is $h = 1$ mm. (b) If the length of the steel strap is $L = 3$ m, and if its thickness is $h = 1.5$ mm, what is the maximum strain in the strap?

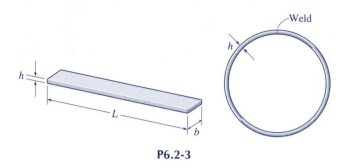

P6.2-3

Prob. 6.2-4. A couple M_0 acts on the end of a slender cantilever beam to bend it into the arc of a circle of radius ρ, as shown in Fig. P6.2-4. If the depth of the beam is $h = 2c$, the length of the beam is L, and the extensional strain in the bottom fibers ($y = -c$) is ϵ_{max}, (a) determine an expression for the radius of curvature ρ, and (b) determine an expression for the tip deflection δ_{max}. Your answers should be in terms of ϵ_{max}, c, and L.

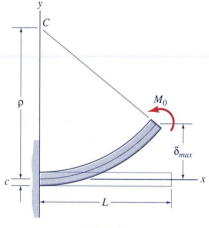

P6.2-4

Prob. 6.2-5. Prove that the transverse curvature, $\kappa' \equiv 1/\rho'$, is related to the longitudinal curvature, $\kappa \equiv 1/\rho$, by the equation $\kappa' = \nu\kappa$ that is, derive Eq. 6.5.

▼ FLEXURAL STRESS IN LINEARLY ELASTIC BEAMS

Prob. 6.3-1. A metal strap (Young's modulus = E) with rectangular cross section is bent around a solid circular cylinder of radius r, as shown in Fig. P6.3-1. Determine an expression for the maximum tensile stress in the curved portion of the strap. Express your answer in terms of the dimensions of the strap, the radius r, and the modulus E.

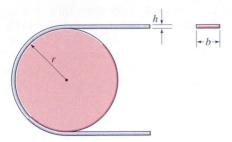

P6.3-1, P6.3-2, and P6.3-3

^D**Prob. 6.3-2.** A steel strap ($E = 210$ GPa) with rectangular cross section of dimensions $b = 6$ mm and $h = 1$ mm is bent around a solid cylinder of radius r as shown in Fig. P6.3-2. If the maximum permissible tensile stress in the steel strap is $\sigma_{max} = 200$ MPa, what is the minimum radius, r, of cylinder that can be used?

^D**Prob. 6.3-3.** A steel strap ($E = 29 \times 10^3$ ksi) with rectangular cross section is bent around a solid circular cylinder of radius $r = 20$ in., as shown in Fig. P6.3-3. If the maximum allowable flexural stress is not to exceed the yield strength ($\sigma_Y = 36$ ksi), what is the maximum thickness h that the strap can have?

Prob. 6.3-4. A high-strength steel strap ($E = 210$ GPa) with thickness $h = 4$ mm and length $L = 1.5$ m is bent into the arc of a circle by end couples M_0, as shown in Fig. P6.3-4. (a) If the resulting circular arc subtends an angle $\alpha = 30°$, what is the maximum flexural stress $(\sigma_{max})_a$ in the strap? (b) If the moment M_0 is decreased so that the resulting circular arc subtends an angle $\alpha = 24°$, what is the maximum flexural stress $(\sigma_{max})_b$ in the strap?

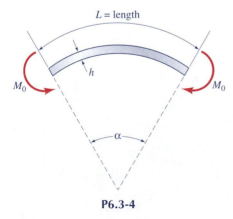

P6.3-4

Prob. 6.3-5. A couple $M_0 = 20$ kN · m acts on the end of a slender steel cantilever beam ($E = 210$ GPa) to bend it into the arc of a circle of radius ρ, as shown in Fig. P6.3-5. The cross section of the beam is a square of dimension $d = 120$ mm. (a) Calculate the maximum tensile stress in the beam. (b) Calculate the radius of curvature of the beam. (c)

Calculate the maximum deflection, δ_{max}, of the beam if its length is $L = 4$ m.

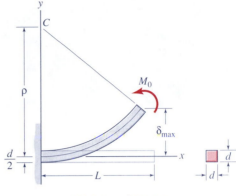

P6.3-5 and P6.3-6

Prob. 6.3-6. Solve Prob. 6.3-5 if $E = 30 \times 10^3$ ksi, $d = 4$ in., $L = 10$ ft, and $M_0 = 20$ kip · ft.

▼ **FLEXURAL STRESS IN**
LINEARLY ELASTIC BEAMS   MDS 6.1–6.3

Prob. 6.3-7. A timber beam consists of four planks fastened together with screws to form a box section 5.5 in. wide and 8.5 in. deep, as shown in Fig. P6.3-7. If the flexural stress at point B in the cross section is 900 psi (T), (a) determine the flexural stress at point A in the cross section; (b) determine the stress at point C in the cross section; and (c) determine the total force on the top plank.

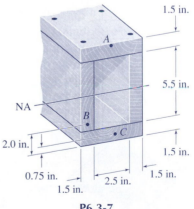

P6.3-7

^D**Prob. 6.3-8.** The structural steel wide-flange beam shown in Fig. P6.3-8 is a **W**10 × 60 section (See Table D.1). The allowable flexural stress in the beam is $\sigma_{allow} = 28$ ksi. (a) Calculate the maximum moment M_z, that can be applied (producing deflection in the xy-plane); and (b) calculate the

433

maximum moment M_y that can be applied (producing deflection in the xz-plane).

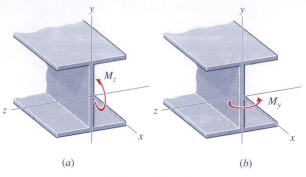

(a)   (b)

P6.3-8 and P6.3-9

DProb. 6.3-9. Repeat Prob. 6.3-8 if the wide-flange beam is a **W250 × 89** section (see Table D.2) and the allowable flexural stress is $\sigma_{\text{allow}} = 200$ MPa.

DProb. 6.3-10. An extruded aluminum machine part has the cross section in Fig. P6.3-10. Determine the maximum moment M that can be applied to the member if the allowable flexural stress in tension is $(\sigma_{\text{allow}})_T = 200$ MPa and the allowable flexural stress in compression is $(\sigma_{\text{allow}})_C = 100$ MPa.

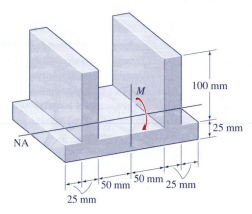

P6.3-10

Prob. 6.3-11. Determine the flexural stresses at points A and B in the cross section in Fig. P6.3-11 if the bending moment at this section is $M = 10$ kip · ft. The dimensions of the cross section are $b_f = 8$ in., $t_f = 2$ in., $h_w = 6$ in., and $t_w = 2$ in.

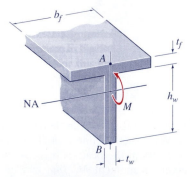

P6.3-11 and P6.3-12

Prob. 6.3-12. Repeat Prob. 6.3-11 if $M = 20$ kN · m and the dimensions of the beam are $b_f = 150$ mm, $t_f = 50$ mm, $h_w = 200$ mm, and $t_w = 50$ mm.

Prob. 6.3-13. A beam is made from three boards that are glued together to form a single beam, as shown in Fig. P6.3-13. The moment acting at this cross section is M. (a) Determine an expression for the distance from the bottom of the beam to the neutral axis (NA); (b) determine an expression for the maximum tensile stress in the beam; and (c) determine an expression for the total force (compressive) acting on the top board.

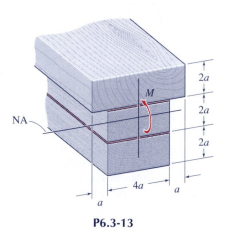

P6.3-13

Prob. 6.3-14. An extruded aluminum machine part has the cross section shown in Fig. P6.3-14. The bending moment M acts in the sense shown in the figure. If the magnitude of the maximum tension at the top and the magnitude of the maximum compression at the bottom are in the ratio of 3:1, what is the thickness t of the two vertical webs?

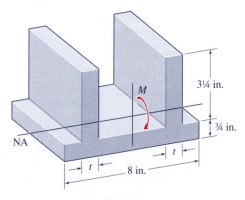

P6.3-14

Prob. 6.3-15. The solid shaft AD in Fig. P6.3-15 is supported by bearings at A and C. The belt forces on the pulleys at B and D are both perpendicular to the shaft. Assume that the bearings produce concentrated reaction forces that are perpendicular to the shaft. Determine the maximum flexural stress in the shaft at cross section B.

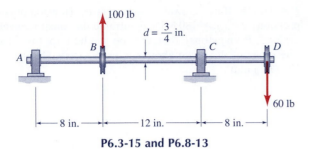

P6.3-15 and P6.8-13

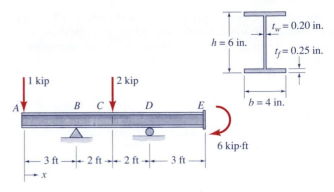

P6.3-18, P6.3-29, and P6.3-39

Prob. 6.3-16. A rectangular timber beam AE has the cross-sectional dimensions and loading shown in Fig. P6.3-16. (a) Determine the flexural stress distribution on the cross section at C; (b) make a two-dimensional sketch of this stress distribution; and (c) determine the total compressive force at this cross section.

Prob. 6.3-19. A $W12 \times 65$ wide-flange beam supports the distributed load and concentrated load shown in Fig. P6.3-19. See Table D.1 for the cross-sectional properties of the beam. (a) Determine the maximum tensile stress on the cross section just to the left of the 5-kip load at B, and (b) determine the maximum compressive stress on the cross section at C.

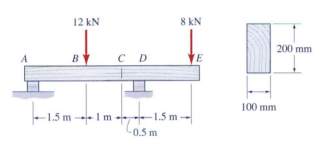

P6.3-16 and P6.3-27

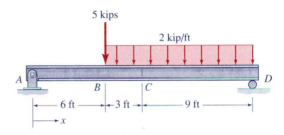

P6.3-19, P6.3-30 and P6.3-40

Prob. 6.3-17. A simply supported wide-flange beam has the loading and cross-sectional dimensions shown in Fig. P6.3-17. Determine the maximum tensile stress on the cross section at C. Neglect the weight of the beam.

Prob. 6.3-20. A $W150 \times 24$ wide-flange beam supports the distributed load and concentrated load shown in Fig. P6.3-20. See Table D.2 for the cross-sectional properties of the beam, (a) Determine the maximum tensile stress on a cross section just to the left of B, where the 8 kN · m couple acts, and (b) determine the maximum tensile stress on a cross section just to right of B.

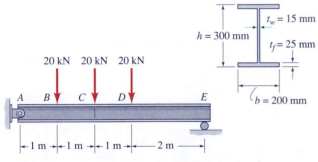

P6.3-17, P6.3-28, and P6.3-38

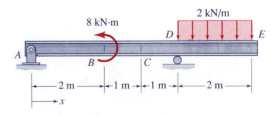

P6.3-20, P6.3-31, and P9.3-5

Prob. 6.3-18. A wide-flange beam with overhangs is shown in Fig. P6.3-18. Determine the maximum tensile stress on the cross section at C. Neglect the weight of the beam.

Prob. 6.3-21. The solid steel shaft AE in Fig. P6.3-21 is supported by bearings at B and D. Assume that the pulley loads at A and E are parallel to each other and that the bearings produce concentrated reaction forces that are normal to the shaft. (a) Sketch shear-force and bending-moment diagrams for the shaft, and (b) determine the maximum flexural stress in the shaft if its diameter is $d = 0.625$ in.

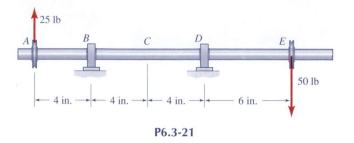

P6.3-21

Prob. 6.3-22. A structural tee section is used as a cantilever beam to support a triangular distributed load of maximum intensity $w = 600$ N/m and a concentrated load $P = 1$ kN, as shown in Fig. P6.3-22. Determine the maximum tensile flexural stress and the maximum compressive flexural stress at end B. The relevant dimensions of the cross section are shown in the figure, and the moment of inertia about the neutral axis (NA) is $I = 13.8 \times 10^6$ mm⁴.

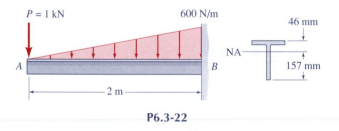

P6.3-22

Prob. 6.3-23. A channel section is used as a cantilever beam to support a uniformly distributed load of intensity $w = 25$ lb/ft and a concentrated load $P = 50$ lb, as shown in Fig. P6.3-23. Determine the maximum tensile flexural stress and the maximum compressive flexural stress at end A. The relevant dimensions of the cross section are shown in the figure, and the moment of inertia about the neutral axis (NA) is $I = 5.14$ in⁴.

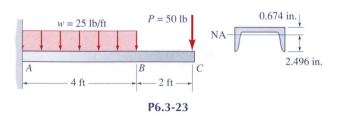

P6.3-23

ᴰProb. 6.3-24. A cantilever beam is subjected to two concentrated loads, and its cross section has the dimensions shown in Fig. P6.3-24. (a) Sketch shear and moment diagrams for the beam, expressing the shear values in terms of P and the bending moment values in terms of PL, where

436

$L = 4$ m. (b) If the allowable flexural stress (tension or compression) is $\sigma_{allow} = 180$ MPa, determine the maximum value of load P (newtons) that can be applied to the beam. The moment of inertia about the neutral axis (NA) is $I = 122 \, (10^6)$ mm⁴.

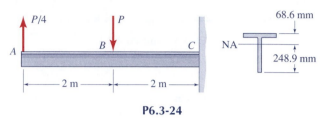

P6.3-24

Prob. 6.3-25. A **W**24 × 94 wide-flange member is used as a cantilever beam to support a uniformly distributed load of intensity $w = 1.5$ kips/ft and a concentrated load $P = 20$ kips as shown in Fig. P6.3-25. (a) Sketch shear-force and bending-moment diagrams for the beam AC, and (b) determine the maximum flexural stress in the beam. Include the weight of the beam in your calculation of the flexural stress.

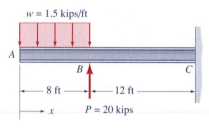

P6.3-25 and P9.3-4

Prob. 6.3-26. To limit the downward motion of D, a micro-switch is installed as shown in Fig. P6.3-26a. When fully depressed, the switch arm AC can be modeled as a cantilever beam with a concentrated force $P = 0.50$ oz applied at end C and a reaction of $2P = 1.00$ oz at the contact point B, as shown in Fig. P6.3-26b. (a) Sketch shear and moment diagrams for the beam AC, and (b) determine the maximum flexural stress in the beam when the switch is fully activated. Express the stress in psi. The cross section of the beam is shown in Fig. P6.3-26c.

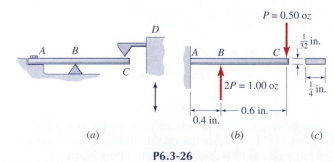

P6.3-26

Prob. 6.3-27. For beam *AE* in Fig. P6.3-16, (a) sketch shear and moment diagrams, and (b) determine the maximum flexural stress in the beam.

Prob. 6.3-28. For beam *AE* in Fig. P6.3-17, (a) sketch shear and moment diagrams, and (b) determine the maximum flexural stress in the beam.

Prob. 6.3-29. For beam *AE* in Fig. P6.3-18, (a) sketch shear and moment diagrams, and (b) determine the maximum flexural stress in the beam.

Prob. 6.3-30. For beam *AD* in Fig. P6.3-19, (a) sketch shear and moment diagrams, and (b) determine the maximum flexural stress in the beam.

Prob. 6.3-31. For beam *AE* in Fig. P6.3-20, (a) sketch shear and moment diagrams, and (b) determine the maximum flexural stress in the beam.

Prob. 6.3-32. A structural tee section is used as a cantilever beam to support a triangularly distributed load of maximum intensity $w_0 = 3$ kN/m and a concentrated load $P = 500$ N as shown in Fig. P6.3-32. (a) Sketch shear and moment diagrams for the beam, and (b) determine the maximum tensile flexural stress and the maximum compressive flexural stress in the beam. The relevant dimensions of the cross section are shown below, and the moment of inertia about the beam's neutral axis (NA) is $I = 895(10^3)$ mm^4.

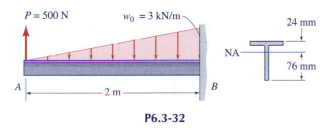

P6.3-32

Prob. 6.3-33. A channel section is used as a cantilever beam to support a uniformly distributed load of intensity $w = 100$ lb/ft and a concentrated load of $P = 100$ lb, as shown in Fig. P6.3-33. (a) Sketch shear and moment diagrams for the beam, and (b) determine the maximum tensile flexural stress and the maximum compressive flexural stress in the beam. The relevant dimensions of the cross section are shown below, and the moment of inertia about the neutral axis (NA) is $I = 5.14$ in^4.

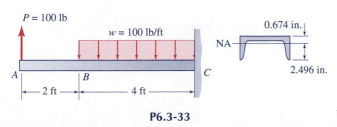

P6.3-33

DProb. 6.3-34. Determine the maximum uniform distributed load, *w*, that can be applied to a cantilever beam with tee section as shown in Fig. P6.3-34. The allowable tensile stress is $(\sigma_{allow})_T = 20$ ksi; and the allowable compressive stress is $(\sigma_{allow})_C = 16$ ksi.

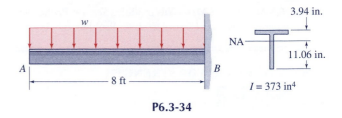

P6.3-34

DProb. 6.3-35. Determine the maximum uniform distributed load, *w*, that can be applied to the beam with overhang shown in Fig. P6.3-35. The allowable stress (magnitude) in tension or compression is 150 MPa, and the beam is a **W**310 × 97 (see Table D.2 of Appendix D).

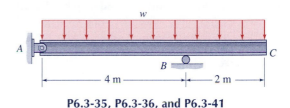

P6.3-35, P6.3-36, and P6.3-41

DProb. 6.3-36. Repeat Prob. 6.3-35 for a uniformly loaded **W**360 × 79 beam with length *AB* of 5 m and length *BC* of 2 m (see Table D.2 of Appendix D).

Prob. 6.3-37. One of the assumptions made in deriving the flexure formula, Eq. 6-13, was that σ_y and σ_z are much smaller than σ_x. (a) Using the cantilever beam with uniformly distributed load, shown in Fig. P6.3-37, derive an expression for the ratio of the magnitude of the maximum flexural stress at the top of the beam, $\sigma_{xm} \equiv |\sigma_x(x, y = h/2)|$, at an arbitrary cross section *x* to the maximum transverse normal stress in the beam, $\sigma_{ym} = p/b$. (b) Use your results from Part (a) to show that the stated assumption is satisfied practically everywhere in this cantilever beam with uniform distributed load.

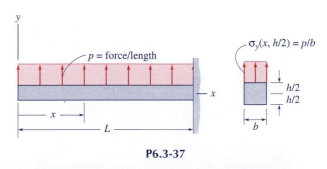

P6.3-37

▼ **FLEXURAL STRESS IN LINEARLY ELASTIC BEAMS** ┃ MDS 6.1–6.3 ┃

CProb. 6.3-38. Use the **MDSolids** modules *Section Properties, Determinate Beams,* and *Flexure* to solve Problem 6.3-28.

CProb. 6.3-39. Use the **MDSolids** modules *Section Properties, Determinate Beams,* and *Flexure* to solve Problem 6.3-29.
CProb. 6.3-40. Use the **MDSolids** modules *Section Properties, Determinate Beams,* and *Flexure* to solve Problem 6.3-30.
C*Prob. 6.3-41. Use the **MDSolids** modules *Section Properties, Determinate Beams,* and *Flexure* to solve Problem 6.3-35.

▼ DESIGN OF BEAMS FOR STRENGTH

MDS 6.4 & 6.5

> *In* **Problems 6.4-1 through 6.4-16** *ignore the weight of the beam in comparison with the applied loading. It is suggested that you use shear and moment diagrams to determine the locations of critical sections.*

DProb. 6.4-1. The simply-supported beam in Fig. P6.4-1 is subjected to a uniform downward load of $w = 8$ kips/ft on a span of $L = 10$ ft. The allowable stress (magnitude) in tension or compression is $\sigma_{allow} = 20$ ksi. From Table D.1 of Appendix D, select the lightest wide-flange steel beam that may be used for this application.

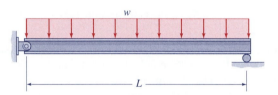

P6.4-1 and P6.4-2

DProb. 6.4-2. The simply supported beam in Fig. P6.4-2 is subjected to a uniform downward load of $w = 90$ kN/m on a span of $L = 4$ m. The allowable stress (magnitude) in tension or compression is $\sigma_{allow} = 150$ MPa. From Table D.2 of Appendix D, select the lightest wide-flange steel beam that may be used for this application.

DProb. 6.4-3. The simply supported beam in Fig. P6.4-3 is subjected to a concentrated load of $P = 200$ kN at the center of its span of $L = 4$ m and a uniformly distributed downward load of intensity $w = 50$ kN/m over the half-span AB. The allowable stress (magnitude) in tension or compression is $\sigma_{allow} = 150$ MPa. From Table D.2 of Appendix D select the lightest wide-flange steel beam that may be used for this application.

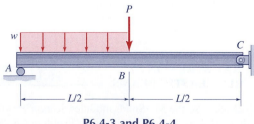

P6.4-3 and P6.4-4

DProb. 6.4-4. The simply supported beam in Fig. P6.4-4 is subjected to a concentrated load of $P = 40$ kips at the center of its span of $L = 8$ ft and a uniformly distributed downward load of intensity $w = 5$ kips/ft over the half-span AB. The allowable stress (magnitude) in tension or compression is $\sigma_{allow} = 20$ ksi. From Table D.1 of Appendix D select the lightest wide-flange steel beam that may be used for this application.

DProb. 6.4-5. (a) A glued, laminated (glulam) beam is made by gluing together n planks of width b and thickness t, as illustrated in Fig. P6.4-5a. Derive a formula that relates the value of the section modulus S to the number of planks used to make the beam. Assume that the thickness of each glue layer is negligible and that the laminated beam behaves like an ordinary wood beam. (b) A simply supported glulam beam of width $b = 200$ mm and length $L = 6$ m supports a concentrated load $P = 8$ kN at its center, as shown in Fig. P6.4-5b. The magnitude of the allowable stress of the wood in tension or compression is $\sigma_{allow} = 6$ MPa. Determine the number of planks of thickness $t = 50$ mm required for the lightest glulam timber beam that can be used for this application.

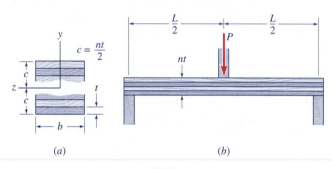

(a)   (b)

P6.4-5

DProb. 6.4-6. A simply supported timber beam supports a triangularly distributed load, as shown in Fig. P6.4-6. The magnitude of the allowable stress in tension or compression is $\sigma_{allow} = 800$ psi. From Table D.8 of Appendix D select the lightest timber beam that may be used for this application if the nominal depth of the beam may not exceed twice its nominal width.

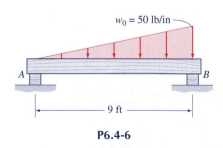

P6.4-6

DProb. 6.4-7. A simply supported timber beam supports a linearly varying load, as shown in Fig. P6.4-7. The magnitude of the allowable stress in tension or compression is

σ_{allow} = 6 MPa. Determine, to the nearest 10 mm, the dimension b of the lightest timber beam with square cross section that can be used for this application.

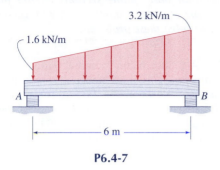

P6.4-7

allowable stress magnitude in compression is $(\sigma_{allow})_C$ = 15 ksi, what is the lightest angle that could be used to make the beam shown in Fig. P6.4-10a? (Note: Assume that $t/b \ll 1$ in obtaining expressions for the moment of inertia, etc.)

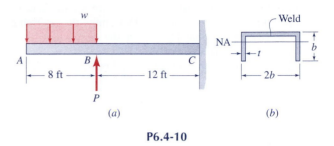

(a)   (b)

P6.4-10

DProb. 6.4-8. From Table D.2 of Appendix D, select the lightest wide-flange steel beam that may be used for the application shown in Fig. P6.4-8. The allowable stress (magnitude) in tension or compression is σ_{allow} = 150 MPa.

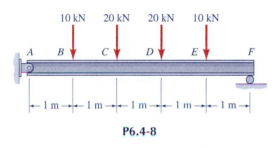

P6.4-8

DProb. 6.4-9. From Table D.1 of Appendix D, select the lightest wide-flange steel beam that may be used for the application shown in Fig. P6.4-9. The allowable stress (magnitude) in tension or compression is σ_{allow} = 20 ksi.

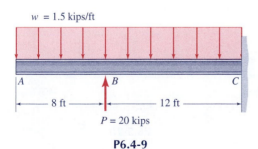

P6.4-9

D*Prob. 6.4-10. Two angle sections with equal legs (from Table D.5 of Appendix D) are to be welded together to form a beam with a channel cross section as illustrated in Fig. P6.4-10b. The loads are w = 1.0 kip/ft and P = 12 kips. If the allowable stress in tension is $(\sigma_{allow})_T$ = 20 ksi and the

DProb. 6.4-11. A sawmill cuts rectangular timber beams from circular logs. If two beams of width b and depth h are to be cut from a single log of diameter d, what b and h would give the strongest beams, that is, the beams with maximum value of S? (Neglect the width of saw cuts.)

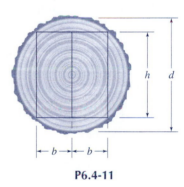

P6.4-11

DProb. 6.4-12. An 18-ft-long wood beam is subjected to two concentrated loads, as shown in Fig P6.4-12. The allowable tensile (and compressive) stress for the wood is σ_{allow} = 960 psi. Select the best beam for this application from the lists of 2× and 4× (i.e., nominal 2-in.-wide or 4-in.-wide) wood beams in Table D.8.

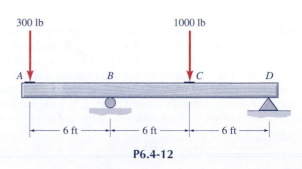

P6.4-12

ᴰProb. 6.4-13. A structural designer wishes to locate the roller support B such that the beam AD in Fig. P6.4-13 will have the same maximum magnitude of positive moment and negative moment when subjected to concentrated loads P_A and P_C acting on the beam as shown. Once the overhang length b has been determined, the designer must choose an appropriate beam to satisfy a flexural stress allowable of σ_{allow} (tension or compression). (a) Determine the overhang length b. (b) From Table D.2, select the best wide-flange section for this application.

Let $P_A = 30$ kN, $P_C = 40$ kN, $L = 1.5$ m, $\sigma_{allow} = 80$ MPa.

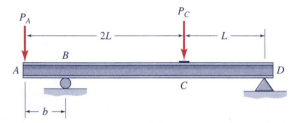

P6.4-13 and P6.4-14

ᴰProb. 6.4-14. Solve Prob. 6.4-13 for the following parameters: $P_A = 4$ kips, $P_C = 12$ kips, $L = 4$ ft, $\sigma_{allow} = 16$ ksi. For Part (b), use Table D1.

ᴰProb. 6.4-15. An inverted structural steel tee beam is supported by a pin at A and rests on a roller at B, as shown in Fig. P6.4-15a. What is the maximum load P that can be hung from the beam at C if the cross section of the tee has the dimensions shown in Fig. P6.4-15b, with $t_w = 0.420$ in., and the allowable bending stress (in tension and compression) is $\sigma_{allow} = 20$ ksi?

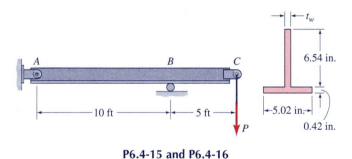

P6.4-15 and P6.4-16

ᴰProb. 6.4-16. An inverted structural steel tee beam is supported by a pin at A and rests on a roller at B, as shown in Fig. P6.4-16a. What is the maximum load P that can be hung from the beam at C if the cross section of the tee has the dimensions shown in Fig. P6.4-16b, with $t_w = 0.500$ in., and the allowable bending stress in tension is $(\sigma_{allow})_T = 20$ ksi and the allowable bending stress (magnitude) in compression is $(\sigma_{allow})_C = 9$ ksi?

▼ **DIRECT COMPOSITE-BEAM METHOD**

In Probs. 6.5-1 through 6.5-12 assume that the component parts of the beams are securely bonded together so that the assumption that "plane sections remain plane" is valid. Use the Direct Composite-Beam Method of Eqs. 6.18-6.24 to solve these problems.

Prob. 6.5-1. A timber beam (nominal 6 in. × 8 in.) is reinforced by attaching $\frac{3}{8}$-in.-thick steel plates to its top and bottom surfaces, as shown in Fig. P6.5-1. The ratio of elastic moduli of the steel and wood is $E_s/E_w = 20$. If the beam is subjected to a bending moment $M = 200$ kip · in, what are the maximum stresses in the steel and wood, $(\sigma_{max})_s$ and $(\sigma_{max})_w$?

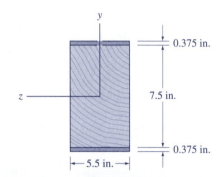

P6.5-1 and P6.5-2a

Prob. 6.5-2. A steel-reinforced timber beam has the cross section shown in Fig. P6.5-2a (see Prob. 6.5-1) and is used as an 18-ft-long simply supported beam to carry a uniformly distributed load $w = 500$ lb/ft, as shown in Fig. P6.5-2b. Determine the maximum normal stresses in the steel and the wood, $(\sigma_{max})_s$ and $(\sigma_{max})_w$, respectively, under this loading condition. The ratio of elastic moduli of the steel and wood is $E_s/E_w = 20$. (In your calculations, you may neglect the weight of the beam.)

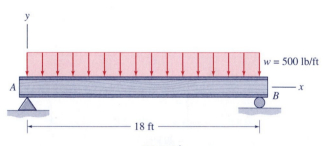

P6.5-2b

ᴰProb. 6.5-3. The reinforced timber beam shown in Fig. P6.5-3a is simply supported and is subjected to quarter-point loading, as shown in Fig. P6.5-3b. If the allowable stresses in the steel and wood are $(\sigma_{allow})_s = 124$ MPa and $(\sigma_{allow})_w = 8$ MPa, respectively, and the ratio of elastic moduli is $E_s/E_w = 20$, determine the (maximum) allowable load P. (In your calculations, you may neglect the weight of the beam.)

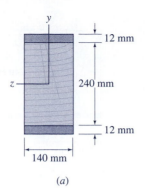

(a)

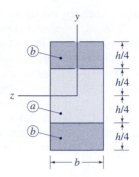

P6.5-5 and P6.5-15

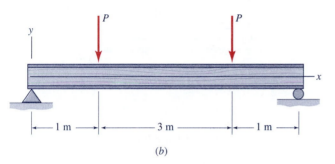

(b)

P6.5-3 and P6.5-4a

Prob. 6.5-4. A steel-reinforced timber beam has the cross section shown in Fig. P6.5-4a (see Fig. P6.5-3a) and is used as a simply supported beam to carry a uniformly distributed loading of intensity $w = 2.0$ kN/m, as shown in Fig. P6.5-4b. Determine the maximum normal stresses in the steel and the wood, $(\sigma_{max})_s$ and $(\sigma_{max})_w$, under this loading condition. The ratio of elastic moduli of the steel and wood is $E_s/E_w = 20$. (You may neglect the weight of the beam.)

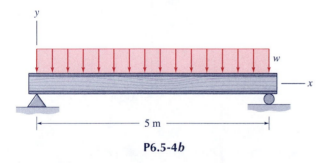

P6.5-4b

Prob. 6.5-5. A three-layer composite beam consists of an interior component of material "a" and outer components of material "b" bonded together to form the cross section shown in Fig. P6.5-5. The elastic moduli of the two materials are E_a and E_b, respectively, with $E_b = n\,E_a$. Determine formulas relating the maximum stresses in the two materials to the value of the bending moment at the given cross section and the dimensions shown in Fig. P6.5-5.

Prob. 6.5-6. An off-shore oil pipeline consists of a steel pipe of inside diameter $d_i = 10.00$ in. and wall thickness $t_p = 0.375$ in., as shown in Fig. P6.5-6a. For protection from abrasion and the salt water environment, the pipe is coated by a plastic sleeve of thickness $t_s = 0.25$ in. The ratio of elastic moduli of the pipe and sleeve is $E_p/E_s = 70$. While it is being transferred from a ship to the floor of the ocean (Fig. P6.5-4b), the pipeline bends, and may be subjected to a maximum bending moment of $M = 300$ kip $\cdot$ in. What are the corresponding maximum stresses in the steel pipe and the plastic sleeve, $(\sigma_{max})_p$ and $(\sigma_{max})_s$, respectively?

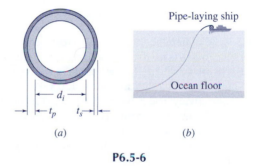

(a)   (b)

P6.5-6

Prob. 6.5-7. A timber beam (actual dimensions $b \times h$) is reinforced by bonding a steel splice plate of width h and thickness t_s on either side, as shown in Fig. P6.5-7. If the bending

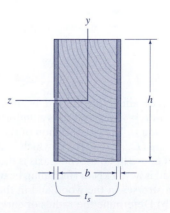

P6.5-7 and P6.5-13

moment that is to be resisted by the beam is M, and the ratio of elastic moduli is E_s/E_w, determine expressions for the flexural stresses σ_s in the steel and σ_w in the timber.

^DProb. 6.5-8. A steel/timber sandwich beam is fabricated from two 80 mm × 160 mm timber beams and a 10 mm × 160 mm steel plate, as illustrated in Fig. P6.5-8. The ratio of elastic moduli is $E_s/E_w = 20$. If the allowable stresses in the steel and wood are $(\sigma_{\text{allow}})_s = 120$ MPa and $(\sigma_{\text{allow}})_w = 8$ MPa, respectively, what is the maximum moment M_z that can be safely applied to this sandwich beam?

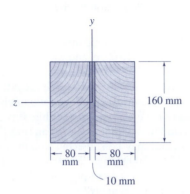

P6.5-8 and P6.5-14

Prob. 6.5-9. A bimetallic strip, whose cross section is depicted in Fig. P6.5-9, is used as the sensing element in a temperature-activated switch. The two metals are copper and nickel, whose elastic moduli are $E_c = 120$ GPa and $E_n = 210$ GPa, respectively. The z axis is the NA. (a) If the strip is subjected to a bending moment $M_z = 2$ N · m, what are the maximum stresses $(\sigma_c)_{\text{max}}$ and $(\sigma_n)_{\text{max}}$ in the copper and nickel, respectively? (b) What radius of curvature, ρ, is produced by this bending moment?

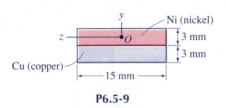

P6.5-9

Prob. 6.5-10. The simply supported beam in Fig. P6.5-10 spans a length of $L = 4$ m and carries a uniformly distributed load of magnitude w. The cross section of the beam consists of a timber beam ($E_w = 10$ GPa) to which is attached a steel bottom-plate ($E_s = 210$ GPa). The z axis is the NA. (a) If the distributed load is $w = 4$ kN/m (including beam weight), find the maximum stresses in the wood and in the steel. $(\sigma_{\text{max}})_w$ and $(\sigma_{\text{max}})_s$. (b) Determine the radius of curvature, ρ, at the midspan.

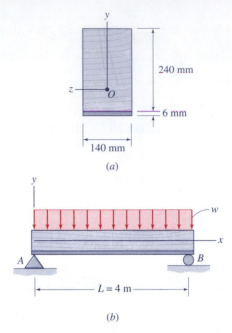

P6.5-10 and P6.5-19

*Prob. 6.5-11. A simply supported beam spans a length of 20 ft and supports equal downward loads, P, as shown in Fig. P6.5-11. To strengthen the 6 × 12 timber beam (finished dimensions are 5.5 in. × 11.5 in., as shown) a 4 in. × $\frac{3}{8}$ in. steel plate is bonded to the bottom of the wood beam. The elastic moduli are $E_w = 1.2 \times 10^6$ psi and $E_s = 30 \times 10^6$ psi for the wood and steel, respectively. The specific weights of the wood and steel are $\gamma_w = 40$ lb/ft^3 and $\gamma_s = 490$ lb/ft^3. (a) If

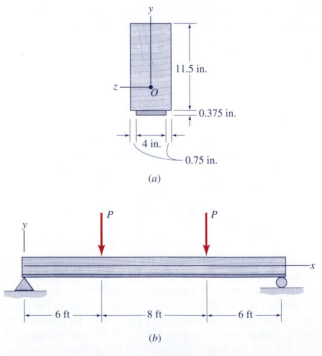

P6.5-11 and P6.5-12

the loads on the beam are $P = 2$ kips each, find the maximum stresses in the wood and in the steel, $(\sigma_{max})_w$ and $(\sigma_{max})_s$. Include the weight of the beam in your calculations. (b) Determine the radius of curvature in the section of the beam between the two loads.

D*Prob. 6.5-12. If the allowable stresses for the beam in Prob. 6.5-11 are $(\sigma_{allow})_w = 1{,}200$ psi and $(\sigma_{allow})_s = 18{,}000$ psi for the wood and steel, respectively, what is the allowable value, P_{allow}, of each of the two loads? (You may neglect the weight of the beam in your calculations.)

▼ TRANSFORMED-SECTION METHOD

| MDS 6.6 & 6.7 |

> **In Probs. 6.5-13 through 6.5-21** *assume that the component parts of the beams are securely bonded together so that the assumption that "plane sections remain plane" is valid. Use the* **Transformed-Section Method** *of Fig. 6.21 and Eqs. 6.25–6.29 to solve these problems.*

Prob. 6.5-13. (a) Use the *transformed-section method* to solve Prob. 6.5-7. (b) Letting $b = 5.5$ in., $h = 11.5$ in., $t_s = 0.375$ in., and $M = 20$ kip · ft, determine numerical values for the maximum flexural stresses in the steel and the timber.

DProb. 6.5-14. Use the *transformed-section method* to solve Prob. 6.5-8.

Prob. 6.5-15. Use the *transformed-section method* to solve Prob. 6.5-5.

Prob. 6.5-16. Aluminum-alloy cover sheets are bonded to a plastic core to form the sandwich beam whose cross section is shown in Fig. P6.5-16, with $t = 0.25$ in. The elastic moduli of the aluminum and plastic are $E_a = 10 \times 10^3$ ksi and $E_p = 0.40 \times 10^3$ ksi. respectively. Determine the maximum normal stresses in the aluminum skin and plastic core, $(\sigma_{max})_a$ and $(\sigma_{max})_p$, respectively, if the cross section is subjected to a bending moment $M_z = 10$ kip · in.

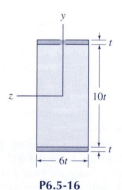

P6.5-16

DProb. 6.5-17. The simply supported timber beam in Fig. P6.5-17a is reinforced by aluminum-alloy cover plates and is subjected to a midspan load P, as shown in Fig. P6.5-17b. If the allowable stresses in the aluminum and wood are $(\sigma_{allow})_a = 120$ MPa and $(\sigma_{allow})_w = 8$ MPa, respectively, and the ratio of

elastic moduli is $E_a/E_w = 6$, determine the (maximum) allowable load P.

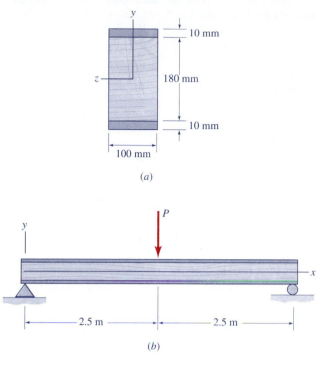

(a)

(b)

P6.5-17

Prob. 6.5-18. The cross section of a bimetallic beam is shown in Fig. P6.5-18. The elastic moduli of the two metallic components are $E_1 = 15{,}000$ ksi and $E_2 = 10{,}000$ ksi, and the z axis is the NA. (a) In which material does the maximum normal stress occur when the beam is subjected to a bending moment M_z? (b) Determine the value of the minimum elastic section modulus, where $S_{min} \equiv M_z/\sigma_{max}$.

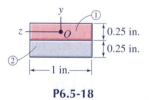

P6.5-18

DProb. 6.5-19. If the allowable stresses for the beam in Prob. 6.5-10 are $(\sigma_{allow})_w = 6$ MPa and $(\sigma_{allow})_s = 120$ MPa in the wood and steel, respectively, what is the allowable value of the distributed load w (including the weight of the beam)?

D*Prob. 6.5-20. The cross section of a steel-reinforced concrete beam of width b is shown in Fig. P6.5-20a. The allowable compressive stress in the concrete is σ_c, and the allowable tensile stress in the steel is σ_s. The diameter of each of the three steel bars is d, and the ratio of the moduli of elasticity is E_s/E_c. Assume that the area of the steel bars is

443

concentrated along a horizontal line at a distance of h from the top of the beam, and assume that the concrete is only effective in compression. That is, tension is carried solely by the steel bars. Calculate the maximum allowable bending moment for the beam. Let $b = 12$ in., $h = 16$ in., $d = 0.875$ in., $(\sigma_{allow})_c = 1.5$ ksi, $(\sigma_{allow})_s = 18$ ksi, and $E_s/E_c = 10$.

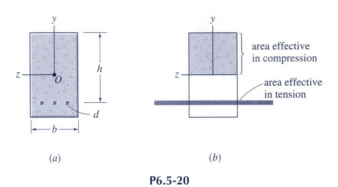

(a)   (b)

P6.5-20

DProb. 6.5-21. A wood beam ($E_w = 1,500$ ksi) is to be reinforced by the addition of a steel plate ($E_s = 30,000$ ksi), as shown in Fig. P6.5-21. The allowable stresses in the wood and steel are $(\sigma_{allow})_w = 1$ ksi and $(\sigma_{allow})_s = 12$ ksi, respectively. The dimensions of the wood beam are given (5.5 in. × 7.5 in.), as is the thickness of the steel plate ($t = 0.25$ in.). You are to determine the width of the steel plate such that the allowable stresses in the steel and wood are reached simultaneously (i.e., for the same value bending moment). Let the z axis be the NA.

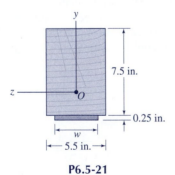

P6.5-21

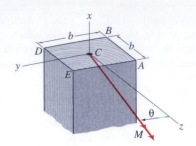

P6.6-1 and P6.6-2

Prob. 6.6-2. A square timber beam whose cross-sectional dimension is $b = 150$ mm is subjected to a resultant bending moment of magnitude $M = 2$ kN · m at an angle $\theta = 30°$, where θ is the angle indicated in Fig. P6.6-2. Determine the flexural stress at each corner of the cross section, and sketch the stress distribution (as in Fig. 6.25c).

DProb. 6.6-3. A rectangular beam with cross-sectional dimensions $b = 55$ mm and $h = 40$ mm is subjected to a resultant bending moment M oriented at an angle $\theta = 45°$, where θ is the angle indicated in fig. P6.6-3. (a) Locate the neutral axis for the given cross section and the given orientation of moment. (b) Determine the maximum moment that can be applied if the allowable flexural stress (tension or compression) is $\sigma_{allow} = 120$ MPa.

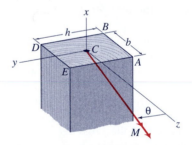

P6.6-3 and P6.6-4

Prob. 6.6-4. A beam with rectangular cross section of dimensions $b = 4$ in. and $h = 2$ in, is subjected to a resultant bending moment of magnitude $M = 2$ kip · ft oriented at an angle $\theta = 20°$, where θ is the angle shown in Fig. P6.6-4. (a) Determine the orientation of the neutral axis for this situation and sketch its location with respect to the cross section. (b) Determine the maximum tensile flexural stress and indicate where it occurs.

Prob. 6.6-5. Solve Example Problem 6.9, substituting a **W**12 × 50 wide-flange beam for the **S**12 × 50 I-beam in the present text example. Compare your answers with those of Example 6.9, and briefly discuss the similarities and differences in the stresses in the **S**-shape and the **W**-shape.

Prob. 6.6-6. The **W**250 × 45 wide-flange beam shown in Fig. P6.6-6 is subjected to a moment M that is intended to lie along the z axis (due to loading in the xy plane). (See Table D.2 for section properties.) (a) Determine the effect a "load misalignment" of $\theta = 2°$ has on the orientation of the neutral

▼ **INCLINED LOADING OF DOUBLY-SYMMETRIC BEAMS**   MDS 6.8

Prob. 6.6-1. A 4 × 4 timber beam (see Table D.8 for finish dimension b) in subjected to a resultant bending moment of magnitude $M = 5$ kip · in. oriented at angle $\theta = 20°$, where θ is the angle shown in Fig. P6.6-1. Determine the flexural stress at each corner of the cross section, and sketch the stress distribution (as in Fig. 6.25c).

axis. Show the neutral axis on a sketch of the cross section. (b) Determine the maximum tensile flexural stress acting on the cross section and compare this stress with the maximum stress in the beam if the load were properly aligned (i.e., $\theta = 0°$). Let $M = 30$ kN · m.

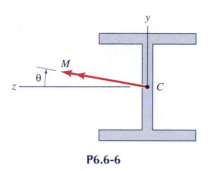

P6.6-6

C-22b, determine the moments of inertia $I_{y'}$ and $I_{z'}$ and the product of inertia $I_{y'z'}$. (b) Determine the orientation of the neutral axis at this section, and indicate the orientation of the neutral axis on a sketch of the cross section. (c) If the magnitude of the maximum allowable flexural stress (tensile or compressive) is $\sigma_{allow} = 16$ ksi, what is the maximum allowable value of the moment, $(M_{y'})_{max}$?

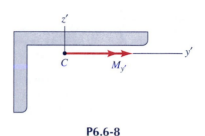

P6.6-8

▼ **UNSYMMETRIC BENDING** [MDS 6.9]

> *Use the **Arbitrary-Axis Method** of Eq. 6.39 to solve Probs. 6.6-7 through 6.6-10. Use the **Principal-Axis Method** of Eq. 6.30 to solve Prob. 6.6-11. Note that Probs. 6.6-8 through 6.6-11 involve bending of beams that have no cross-sectional axis of symmetry.*

*Prob. 6.6-7. A $4 \times 4 \times \frac{1}{2}$ steel equal-leg angle (See Table D.5) is subjected to loading in the $x'z'$ plane that produces a moment $M_{y'} = 12$ kip · in. at the section shown in Fig. P6.6-7. (a) Determine the product of inertia $I_{y'z'}$. (Hint: Use Eqs. C-28 and C-22b.) (b) Determine the orientation of the neutral axis, and show the neutral axis on a sketch of the cross section. (c) Determine the maximum tensile flexural stress acting on the cross section, and indicate the location of this maximum tensile stress.

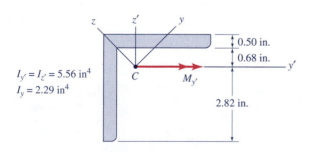

$I_{y'} = I_{z'} = 5.56$ in^4
$I_y = 2.29$ in^4

P6.6-7 and P6.6-17

D*Prob. 6.6-8. As shown in Fig. P6.6-8, the vector representing the bending moment on the $5 \times 3 \times \frac{1}{2}$ unequal-leg structural steel angle cross section is along the y' direction. (a) Using information from Table D.6, and using Eqs. C-28 and

*Prob. 6.6-9. As shown in Fig. P6.6-9, the bending moment M acts at an angle of 20° counterclockwise from the y' axis of an unequal-leg angle section. (a) Locate the centroid of the cross section, and determine the moments of inertia $I_{y'}$ and $I_{z'}$ and the product of inertia $I_{y'z'}$. You may use the results of Examples C-1 through C-3 in Appendix C. (b) Determine the orientation of the neutral axis at this section, and indicate the orientation of the neutral axis on a sketch of the cross section. (c) Calculate the maximum tensile flexural stress and the maximum compressive flexural stress at this cross section.

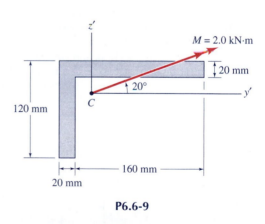

P6.6-9

D*Prob. 6.6-10. An aluminum **Z** section has the dimensions shown in Fig. P6.6-10. The yield strength (tension or compression) of the aluminum is $\sigma_Y = 40$ ksi, and a factor of safety of F.S. = 2.0 is to be maintained. (a) Determine $I_{y'}, I_{z'}$, and $I_{y'z'}$. (See Appendices C-2 and C-3.) (b) Determine the orientation of the neutral axis for the given orientation of bending moment. (c) Determine the maximum allowable magnitude of bending moment $M_{y'}$.

445

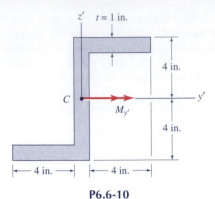

P6.6-10

Prob. 6.6-11. An unequal-leg angle section has the dimensions shown in Fig. 1 of Example 6.10. At this cross section the moment is $M_{y'} = 10$ kN · m and is oriented parallel to the short leg of the angle. (a) Using the *principal-axis method,* determine the orientation of the neutral axis of the cross section, and show this orientation on a sketch. (b) Determine the maximum tensile stress and the maximum compressive stress on the cross section, and indicate their respective locations on the cross section.

▼ INCLINED LOADING OF SINGLY SYMMETRIC BEAMS

Problems 6.6-12 through 6.6-17 *involve bending of beams that have one cross-sectional axis of symmetry. Use the* **Principal-Axis Method** *of Eq. 6.30 to solve these problems.*

Prob. 6.6-12. A moment M acts through the centroid of the T-section in Fig. P6.6-12 at an inclination angle $\theta = 10°$. The depth of the web is $h_w = 9a$. (a) Determine the orientation of the neutral axis, and show the neutral axis on a sketch of the cross section. (b) Determine the maximum tensile flexural stress acting on the cross section, and compare this with the maximum tensile flexural stress the beam would experience if $\theta = 0°$. Express your answers in terms of M and the dimension a.

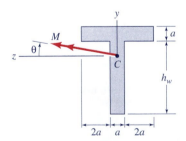

P6.6-12 and P6.6-13

Prob. 6.6-13. Repeat Prob. 6.6-12 using the following parameters: $h_w = 6a$ and $\theta = 4°$.

Prob. 6.6-14. The structural tee whose cross section is shown in Fig. P6.6-14b is used as a cantilever beam to support a concentrated load P that passes through the centroid of the end cross section, as shown in Fig. P6.6-14a. The load acts in the

446

$x'z'$-plane, but the beam is rotated through a 2° angle about the $x(x')$ axis as indicated. Calculate the increase (or decrease, if that is the case) in the maximum compressive flexural stress that is caused by a 2° rotation of the beam. Let $I_y = 7.33 \times 10^5$ mm^4, and $I_z = 2.77 \times 10^6$ mm^4.

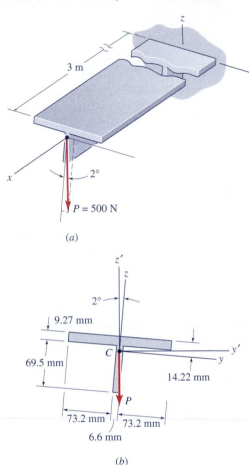

(a)

(b)

P6.6-14

[D]**Prob. 6.6-15.** Due to load misalignment, the bending moment acting on the channel sections in Fig. P6.6-15 is inclined at an angle of 4° with respect to the y axis. If the allowable flexural stress for this beam is $\sigma_{allow} = 16$ ksi, what is the maximum moment, M_{max}, that may be applied?

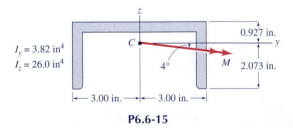

P6.6-15

[D]**Prob. 6.6-16.** The bending moment in Fig. P6.6-16 is inclined at an angle of $\theta = 6°$ with respect to the z axis of the 6.00 × 2.50 aluminum **C** section. (See Table D.10 of Appendix D for the properties of the cross section.) If the allowable flexural stress for this beam is $\sigma_{allow} = 16.0$ ksi, what is the maximum moment M_{max} that may be applied to the beam?

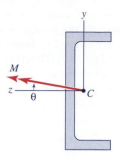

P6.6-16

Prob. 6.6-17. A $4 \times 4 \times \frac{1}{2}$ equal-leg angle (See Table D.5) is subjected to loading in the $x'z'$ plane that produces a moment $M_{y'} = 12$ kip · in. at the section shown in Fig. P6.6-17 (see Prob. 6.6-7.). (a) Determine the orientation of the neutral axis, and show the neutral axis on a sketch of the cross section. (b) Determine the maximum tensile flexural stress acting on the cross section, and indicate the location of this maximum stress. (Note: From Eq. C-28, $I_y + I_z = I_{y'} + I_{z'}$.)

▼ **INELASTIC BENDING**   `MDS 6.10`

For Problems 6.7-1 through 6.7-20, *assume that the material is elastic-plastic (i.e., linearly elastic, perfectly plastic), and has the same stress-strain properties for both tension and compression.*

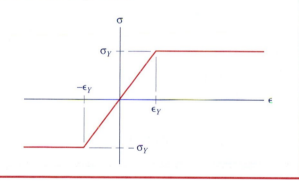

Prob. 6.7-1. A beam with rectangular cross section ($b = 1$ in., $h = 2$ in.) is made of structural steel with $\sigma_Y = 36$ ksi and

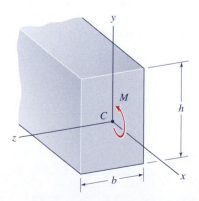

P6.7-1 and P6.7-2

$E = 30 \times 10^3$ ksi. (a) Calculate numerical values for the yield moment M_Y, the plastic moment M_P, and the shape factor f. (b) Plot the moment-curvature diagram for this particular beam for values of curvature up to $\kappa = 3\kappa_Y$.

Prob. 6.7-2. Repeat Problem 6.7-1 for a rectangular beam with $b = 50$ mm, $h = 150$ mm, $E = 210$ GPa, and $\sigma_Y = 250$ MPa.

Prob. 6.7-3. A rectangular box beam with height $h = 12$ in. and width $b = 8$ in. has a constant wall thickness $t = 0.50$ in. It is made of structural steel with $\sigma_Y = 36$ ksi and $E = 30 \times 10^3$ ksi. Calculate numerical values for the yield moment M_Y, the plastic moment M_P, and the shape factor f.

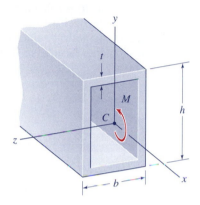

P6.7-3, P6.7-4, and P6.7-18

Prob. 6.7-4. Repeat Prob. 6.7-3 for a rectangular box beam with $b = 300$ mm, $h = 400$ mm and $t = 12.7$ mm, made of steel with $\sigma_Y = 250$ MPa and $E = 210$ GPa.

Prob. 6.7-5. A wide-flange beam with dimensions $b_f = 8.060$ in., $t_f = 0.660$ in., $h = 13.92$ in., and $t_w = 0.370$ in. is made of structural steel with $\sigma_Y = 36$ ksi and $E = 30 \times 10^3$ ksi. Calculate numerical values for the yield moment M_Y, the plastic moment M_P, and the shape factor f.

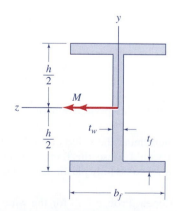

P6.7-5, P6.7-6, P6.7-13, and P6.7-14

Prob. 6.7-6. Repeat Prob. 6.7-5 for a wide-flange beam with dimensions $b_f = 400$ mm, $t_f = 30$ mm, $h = 850$ mm, and $t_w = 18$ mm. The beam is made of structural steel with $\sigma_Y = 250$ MPa and $E = 210$ GPa. Calculate numerical values for the yield moment M_Y, the plastic moment M_P, and the shape factor f.

Prob. 6.7-7. A **W**10 × 45 wide-flange beam is made of structural steel with $\sigma_Y = 36$ ksi and $E = 30 \times 10^3$ ksi. Calculate numerical values for the yield moment M_Y, the plastic moment M_P, and the shape factor f. (See Table D.1 for the section properties.)

Prob. 6.7-8. A **W**200 × 71 wide-flange beam is made of structural steel with $\sigma_Y = 250$ MPa and $E = 210$ GPa. Calculate numerical values for the yield moment M_Y, the plastic moment M_P, and the shape factor f. (See Table D.2 for the section properties.)

Prob. 6.7-9. The structural tee section shown in Fig. P6.7-9 has the dimensions $d = 5.05$ in., $b_f = 8.02$ in., $t_w = 0.350$ in., $t_f = 0.620$ in., and it is made of steel with $\sigma_Y = 36$ ksi and $E = 30 \times 10^3$ ksi. Calculate numerical values for the yield moment M_Y, the plastic moment M_P, and the shape factor f.

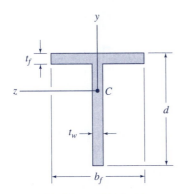

P6.7-9 and P6.7-10

Prob. 6.7-10. Repeat Prob. 6.7-9 for a structural tee beam with dimensions $d = 430$ mm, $b_f = 400$ mm, $t_w = 18$ mm, $t_f = 30$ mm that is made of structural steel with $\sigma_Y = 250$ MPa and $E = 210$ GPa.

Prob. 6.7-11. A beam with solid circular cross section of diameter d is made of elastic-plastic steel with yield point σ_Y modulus of elasticity E. Determine expressions for the yield moment M_Y, the plastic moment M_P, and the shape factor f.

Prob. 6.7-12. A circular tube of outer diameter d_o and inner diameter d_i is made of elastic-plastic steel with yield point σ_Y and modulus of elasticity E. Determine expressions for the yield moment M_Y, the plastic moment M_P, and the shape factor f.

Prob. 6.7-13. Determine the value of the bending moment, call it M_F, that would cause the flange of the wide-flange beam in Prob. 6.7-5 to be fully plastic while the web remains linearly elastic.

Prob. 6.7-14. Repeat Prob. 6.7-13 for the wide flange beam in Prob. 6.7-6.

Prob. 6.7-15. Repeat Prob. 6.7-13 for the wide flange beam in Prob. 6.7-7.

Prob. 6.7-16. Repeat Prob. 6.7-13 for the wide flange beam in Prob. 6.7-8.

Prob. 6.7-17. A beam with solid circular cross section of diameter d is subjected to the plastic moment M_P; then the load is completely removed. The beam is made of elastic-plastic steel with yield point σ_Y, Determine an expression for the maximum residual stress, and sketch the residual-stress state (as is done for a rectangular beam in Fig. 6.38d).

Prob. 6.7-18. Repeat Prob. 6.7-17 for the rectangular box beam in Prob. 6.7-3.

Prob. 6.7-19. Repeat Prob. 6.7-17 for the **W**10 × 45 wide-flange beam in Prob. 6.7-7.

Prob. 6.7-20. Repeat Prob. 6.7-17 for the **W**200 × 71 wide-flange beam in Prob. 6.7-8.

*Prob. 6.7-21. A beam with rectangular cross section of width b and height h, as shown in Fig. P6.7-21a, is made of material whose stress-strain curve (in tension or compression) is approximately the form shown in Fig. P6.7-21b. Let $\sigma_{Y2} = \sigma_Y$, $\sigma_{Y1} = \frac{3}{4}\sigma_Y$, $\epsilon_{Y2} = \epsilon_Y$, and $\epsilon_{Y1} = \frac{1}{2}\epsilon_Y$. (a) Determine an expression for the moment that causes first yielding, M_{Y1}. (b) Determine an expression for the moment M_{Y2} that causes the outer fibers to reach the second yield stress, $\sigma_{Y2} = \sigma_Y$. (c) Determine an expression for the fully plastic moment, M_P, the moment that causes all fibers (except an infinitesimally small core near the neutral axis) to reach the second yield stress.

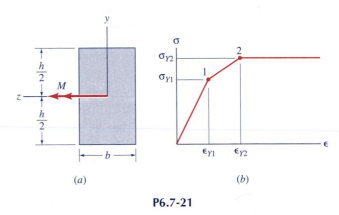

P6.7-21

MDS 6.11

▼ **SHEAR STRESS IN BEAMS**

[D]**Prob. 6.8-1.** If the allowable shear stress (for shear parallel to the grain of the wood) for a 4 in. × 6 in. timber beam (see Appendix D.8 for actual finish dimensions b and h) is

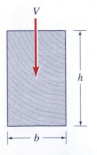

P6.8-1 and P6.8-2

$(\tau_{allow})_w = 400$ psi, what is the maximum value of transverse shear force, V_{max}, that the beam can sustain, based on this shear-stress criterion?

DProb. 6.8-2. A timber beam is to be selected to sustain a maximum transverse shear of $V_{max} = 50$ kN without exceeding an allowable shear stress of 2 MPa (for shear parallel to the grain of the wood). If wood beams are available with cross sections having $h = 2b$, what is the minimum value of b (to the nearest even mm) that satisfies the shear stress criterion?

Prob. 6.8-3. A simply supported beam AC of length $L = 4$ m supports a concentrated load $W = 1.0$ kN that hangs from the beam at B, as shown in Fig. P6.8-3. The cross section of the beam is 40 mm × 60 mm. Determine the flexural stress $\sigma(x, y)$ and shear stress $\tau(x, y)$ at three levels—$y = 0$ mm, $y = 10$ mm, and $y = 20$ mm—on the cross section just to the left of the load point B.

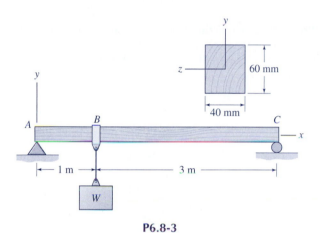

P6.8-3

Prob. 6.8-4. A simply supported beam AC of length $L = 15$ ft. supports a concentrated load $P = 15$ kips at B, as shown in Fig. P6.8-4. The cross section of the timber beam has nominal dimensions 8 in. × 8 in. (See Appendix D.8 for the actual cross-sectional dimensions.) Determine the flexural stress $\sigma(x, y)$ and shear stress $\tau(x, y)$ at three levels—$y = 0$ in., $y = 1.25$ in., and $y = 2.50$ in.—on the cross section just to the left of the load P.

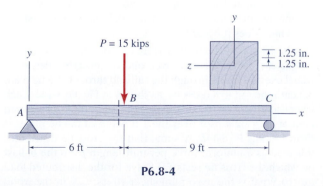

P6.8-4

DProb. 6.8-5. A simply supported beam of length L with rectangular cross section of width b and depth h is required

to support a midspan concentrated load P, as shown in Fig. P6.8-5. (a) Determine an expression for P_σ, the maximum allowable load based on an allowable flexural stress σ_{allow}. (b) Determine an expression for P_τ, the maximum allowable load based on an allowable shear stress τ_{allow}. (c) Form the ratio P_σ/P_τ; then discuss the implications of the beam dimensions that enter into the resulting ratio. [In Parts (a) through (c), neglect the weight of the beam.]

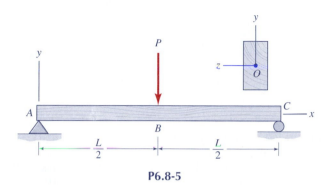

P6.8-5

DProb. 6.8-6. A simply supported timber beam, whose depth is $h = 9.5$ in., supports a concentrated load P at its midspan. The allowable flexural stress for the wood is $\sigma_{allow} = 1.2$ ksi, and the allowable shear stress is $\tau_{allow} = 150$ psi. Determine the length (call it L_{cr}) below which the shear-stress criterion governs the allowable load and above which the flexural stress criterion governs.

Prob. 6.8-7. A beam with 200 mm × 300 mm rectangular cross section supports loads $P_B = 15$ kN and $P_C = 25$ kN, as shown in Fig. P6.8-7. (a) Determine the maximum flexural stress, σ_{max}, in the beam, and (b) determine the maximum shear stress, τ_{max}, in the beam. (Neglect the weight of the beam.)

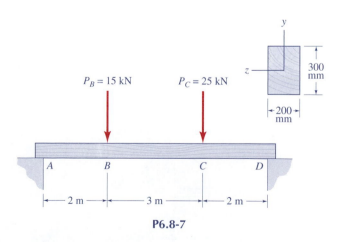

P6.8-7

DProb. 6.8-8. A wood beam with nominal dimensions 4 in. × 8 in. is supported and loaded, as shown in Fig. P6.8-8. If the allowable shear stress for the wood is $(\tau_{allow})_w = 120$ psi, and if the load at C is always three times the load at A, that is, $3P$

and P, respectively, calculate the maximum load value P that may be applied to this beam. Include the weight of the beam in your calculations, using $\gamma = 36$ lb/ft³ for the specific weight of the wood beam. (See Appendix D.8 for the finish dimensions of structural lumber.)

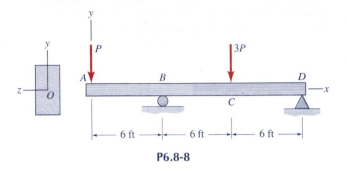

P6.8-8

Prob. 6.8-9. A wood beam with 4 in. × 6 in. (nominal) rectangular cross section supports loads $P_B = 6$ kips and $P_D = 3$ kips, as shown in Fig. P6.8-9. (a) Determine the maximum flexural stress, σ_{max}, in the beam, and (b) determine the maximum shear stress, τ_{max}, in the beam. (Assume that the beam is on simple supports at A and C, and neglect the weight of the beam.)

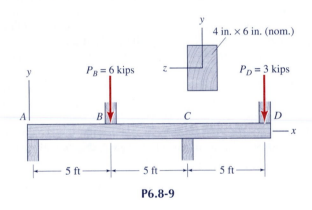

P6.8-9

Prob. 6.8-10. Using the shear-stress formula, Eq. 6.67. determine an expression for the shear stress at the neutral-axis level on the circular cross section shown in Fig. P6.8-10. Compare your results with the information given in Fig. 6.45b.

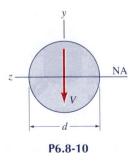

P6.8-10

ᴰProb. 6.8-11. A timber company wishes to increase its sales of timber by supplying log-cabin kits that can be assembled

450

and used as vacation cabins. As the structural engineer for the timber company, you must (a) determine an expression that relates the maximum shear stress in a log (circular cross section) to the transverse shear force V and the diameter d, and (b) determine the maximum span, L_{max}, of a simply supported log beam that must support a uniformly distributed load w (including beam weight) without exceeding the allowable shear stress τ_{allow}.

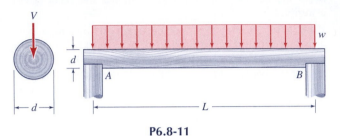

P6.8-11

Prob. 6.8-12. Using the shear stress formula, Eq. 6.67, determine an expression for the shear stress at the neutral axis level on the thick-wall circular tubular cross section shown in Fig. P6.8-12.

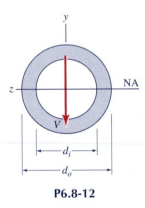

P6.8-12

Prob. 6.8-13. (a) Draw a shear diagram and a moment diagram for the shaft in Fig. P6.8-13 (see Fig. P6.3-15). (b) If the shaft has a diameter of 0.75 in., what is the maximum shear stress in the shaft, and where does it occur (i.e., at which cross section, and where in that cross section)? (c) What is the maximum flexural stress in this 0.75-in.-diameter shaft, and where does it occur?

*Prob. 6.8-14. As a set of wheels on one axle of a railway freight car passes directly over one of the cross ties, each wheel exerts a force through the rail and through the tie plate beneath the rail to a cross tie, as shown in Fig. P6.8-14a. Each wheel load can be represented by a uniformly distributed load of $p_w = 32$ kips/ft over the 1 ft width of the tie plate, as shown in Fig. P6.8-14b. Assume that the cross tie distributes this load as a uniform force-per-unit-length p_b to the ballast on which the cross tie rests. (a) Solve for the distributed load p_b. (b) What is the maximum shear stress, τ_{max}, in the wood cross tie and where does it occur? (c) Determine the maximum flexural stress, σ_{max}, and indicate where it occurs.

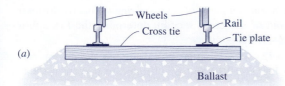

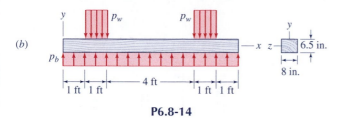

(a)

(b)

P6.8-14

▼ SHEAR STRESS IN THIN-WALL BEAMS

MDS 6.12 & 6.13

Problems 6.10-1 through 6.10-5. *A vertical shear force V is applied to a web-flange beam with cross-sectional dimensions as labeled in the figure below. Following Example Problem 6.15, for each problem below: (a) Calculate (or look up in Appendix D for commercial shapes) the moment of inertia of the cross section. Neglect the fillets at the junctions between the flanges and the web. (b) Calculate the maximum shear stress, τ_{max}, in the web. (c) Calculate the minimum shear stress, τ_{min}, in the web. (d) Calculate the vertical shear force, V_w, carried by the web, and determine the ratio of V_w to the total shear force V.*

Prob. 6.10-1. A beam with cross-sectional dimensions $h = 12.5$ in., $h_w = 11.75$ in., $b = 6.5$ in., and $t_w = 0.25$ in., and with shear force $V = 40$ kips.

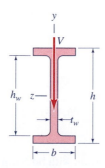

P6.10-1 through P6.10-5

Prob. 6.10-2. A beam with cross-sectional dimensions $h = 410$ mm, $h_w = 390$ mm, $b = 260$ mm, and $t_w = 15$ mm, and with shear force $V = 280$ kN.

Prob. 6.10-3. A W24 × 94 steel wide-flange beam, with shear force $V = 160$ kips. (See Table D.1.)

Prob. 6.10-4. An aluminum-alloy beam with cross-sectional dimensions $h = 8.00$ in., $h_w = 7.18$ in., $b = 5.00$ in., and $t_w = 0.25$ in., and with shear force $V = 12$ kips. (See Table D.9.)

Prob. 6.10-5. A W360 × 79 steel wide-flange beam, with shear force $V = 220$ kN. (See Table D.2.)

Prob. 6.10-6. A W12 × 50 wide-flange beam and a W16 × 50 wide-flange beam are each subjected to the same magnitude of vertical shear force, V. Determine the ratio of the maximum shear stresses $(\tau_{max})_a/(\tau_{max})_b$. (See Table D.1 for the section properties of these beams.)

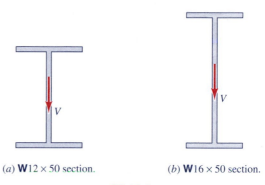

(a) W12 × 50 section.   (b) W16 × 50 section.

P6.10-6

Prob. 6.10-7. A wide-flange beam has the relative dimensions shown in Figs. P6.10-7. If the beam is to be subjected to a transverse shear force V, what is the ratio $(\tau_{max})_a/(\tau_{max})_b$ for the beam in orientations (a) and (b)? In each case, indicate where τ_{max} occurs.

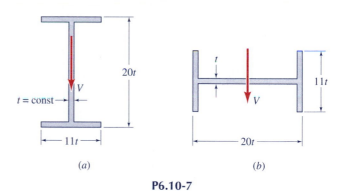

(a)   (b)

P6.10-7

Prob. 6.10-8. A tee beam has the relative cross-sectional dimensions shown in Fig. P6.10-8, and it is subjected to a

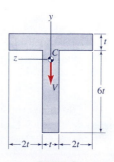

P6.10-8

451

vertical shear force V. (a) Determine an expression for $\tau_w(y)$, the (vertical) shear stress in the web, and (b) sketch the function $\tau_w(y)$ obtained in Part (a).

Prob. 6.10-9. A structural tee beam has the dimensions shown in Fig. P6.10-9. (a) Determine an expression for $\tau_w(y)$, the (vertical) shear stress in the web, and (b) determine the percent of the shear force that is carried by the web (i.e., $V_w/V \times 100\%$).

along the beam is a force of 8 kips acting vertically through the centroid of the cross section, determine the maximum shear stress in the beam.

Prob. 6.10-12. A rectangular box beam has the dimensions shown in Fig. P6.10-12 and is subjected to a vertical shear force $V = 60$ kN. Determine the web shear stresses τ_A and τ_B at the locations A and B indicated in the figure.

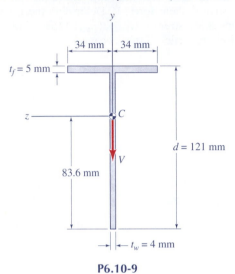

P6.10-9

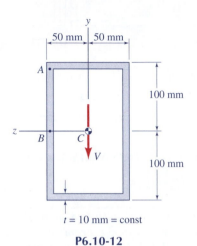

P6.10-12

Prob. 6.10-10. A channel section is subjected to a vertical shear force $V = 10$ kN at the section shown in Fig. P.6.10-10. Determine the values of the horizontal shear stress τ_A at the point designated A, and the vertical shear stress τ_B at the point labeled B. Assume the thickness t_f to be constant.

Prob. 6.10-13. A rectangular box beam has the relative dimensions shown in Fig. P6.10-13. If the beam is to be subjected to a transverse shear force V, what is the ratio $(\tau_{max})_a/(\tau_{max})_b$ for the beam in orientations (a) and (b)? In each case, indicate where τ_{max} occurs.

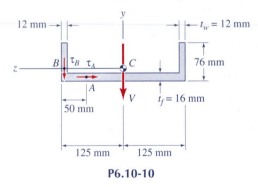

P6.10-10

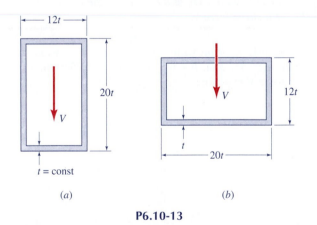

(a)   **(b)**

P6.10-13

Prob. 6.10-11. A beam has a channel-shaped cross section, as shown in Fig. P6.10-11. If the maximum shearing force

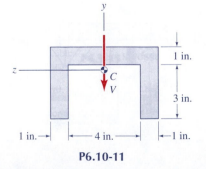

P6.10-11

Prob. 6.10-14. A tee beam with both top and bottom flanges has the dimensions shown in Fig. P6.10-14. The thickness of the web is $t_w = 40$ mm. If the beam is subjected to a vertical shear force $V = 10$ kN, determine the following: (a) the location of the centroid C, and the moment of inertia, I_C, about the neutral axis; and (b) the shear stress in the web at C, and the shear stresses in the web at points A and B where the web connects to the top flange and to the bottom flange, respectively.

452

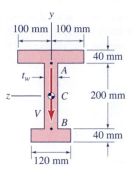

P6.10-14 and P6.10-15

Prob. 6.10-15. Repeat Parts (a) and (b) of Prob. 6.10-14 if the thickness of the web is $t_w = 50$ mm, instead of the thickness of the web in Prob. 6.10-14.

Prob. 6.10-16. The beam shown in Fig. P6.10-16a is a **W**12 × 35 wide-flange beam. (a) Determine the web shear stresses τ_C and τ_D at locations C and D in the cross section just to the left of the support at B; and (b) determine the maximum flexural stress, σ_{max}, and state where this occurs.

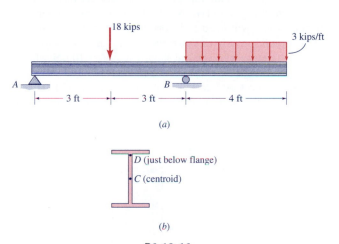

(a)

(b)

P6.10-16

Prob. 6.10-17. The beam in Fig. P6.10-17 supports a concentrated downward load $P = 20$ kN at its left end and a uniformly distributed downward load $w = 40$ kN/m over the span BC. If the beam is a **W**310 × 52 wide-flange beam, (a) determine the maximum flexural stress, σ_{max}, and (b) determine the maximum transverse shear stress, τ_{max}.

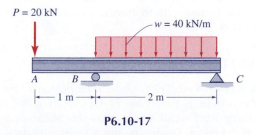

P6.10-17

Prob. 6.10-18. For the **W**12 × 35 wide-flange beam loaded as shown in Fig. P6.10-18, let $w = 2$ kips/ft. (a) Determine the

maximum flexural stress, σ_{max}, and (b) determine the maximum transverse shear stress, τ_{max}.

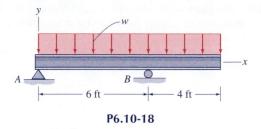

P6.10-18

Prob. 6.10-19. A tubular beam with circular cross section has an outer diameter $d_o = 10.75$ in. and wall thickness $t = 0.365$ in., as indicated in Fig. P6.10-19. If the beam is subjected to a vertical shear force $V = 40$ kips at this section, what is the maximum shear stress on the cross section?

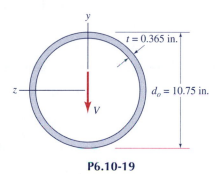

P6.10-19

DProb. 6.10-20. The tongue, AB, of a small utility trailer is the square box section shown in Fig. P6.10-20b. Assume that the total weight of the trailer and its load is W, and that its line of action is 1 ft forward of the axle of the trailer, as shown in Fig. P6.10-20a. Furthermore, assume that the tongue is effectively cantilevered from the trailer body at B. Neglect any axial towing loads on the tongue, and consider only the static loads when the trailer hitch at A is resting on a stationary trailer ball. Determine the maximum weight W (trailer plus load) if the allowable flexural stress in the tongue is $\sigma_{allow} = 12$ ksi and the allowable transverse shear stress in the tongue is $\tau_{allow} = 8$ ksi.

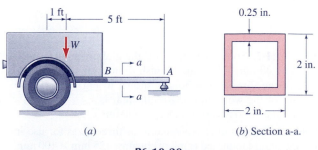

(a)   (b) Section a-a.

P6.10-20

DProb. 6.10-21. Two 2 in. × 6 in. (nominal dimensions) boards are nailed and glued together to form a tee beam, as shown in Fig. P6.10-21b. Assume that the nails and glue are sufficient to cause the two planks to function together as a single beam. The allowable stress in horizontal shear (i.e., shear parallel to the grain of the wood) is $\tau_{\text{allow}} = 80$ psi. If this beam is to be used as a cantilever to support a triangularly distributed load, as shown in Fig. P6.10-21a, what is the maximum load intensity w_0 that can be supported by this beam? In your calculations neglect the weight of the beam.

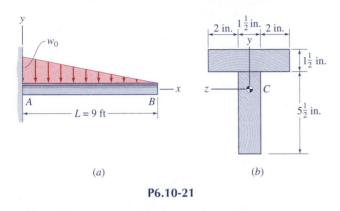

(a)   (b)

P6.10-21

▼ SHEAR STRESS IN BUILT-UP BEAMS

DProb. 6.11-1. A glued-laminated (glulam) timber beam is made from three 2×4 ($1\frac{1}{2} \times 3\frac{1}{2}$ in. finish dimensions) boards, as shown in Fig. P6.11.1. The strength of the wood in horizontal shear is $\tau_{\text{allow}} = 80$ psi (which takes into account a factor of safety). What is the required shear strength of the glue joints (psi). If there is to be a factor of safety against failure of the glue joints of $FS = 3.0$? That is, how strong must the glue joint be so that the joints do not fail before the wood itself does? Neglect the thickness of the glue joint.

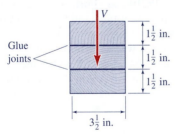

P6.11-1 and P6.11-2

DProb. 6.11-2. Solve Prob. 6.11-1 if the glulam beam is made from five $40 \text{ mm} \times 140 \text{ mm}$ (finish dimensions) boards, instead of the three 2×4 boards shown in Fig. P6.11-1. The strength of the wood in horizontal shear is $\tau_{\text{allow}} = 550$ kPa (which takes into account a factor of safety).

DProb. 6.11-3. Lumber is available in three sizes for use in making glued-laminated (glulam) beams—$25 \text{ mm} \times 100 \text{ mm}$, $50 \text{ mm} \times 100 \text{ mm}$, and 100 mm square (all are finished-lumber dimensions). "Company A" fabricates the $100 \text{ mm} \times 150 \text{ mm}$ beam shown in Fig. P6.11-3a, and "Company B" fabricates beams like the one in Fig. P6.11-3b. Both beams sell

for the same price and are made of the same type and grade of lumber. On the basis of shear strength of the beams (i.e., the allowable transverse shear force V), which one would you choose? How much stronger (in shear) is it than its competitor?

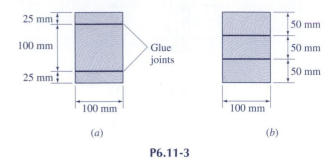

(a)   (b)

P6.11-3

*Prob. 6.11-4. A wood beam ($b = 90$ mm, $d = 250$ mm, $E_w = 14$ GPa) is strengthened and stiffened by the addition of steel cover plates ($b = 90$ mm, $t = 6$ mm, $E_s = 210$ GPa), which are glued to the top and bottom surfaces of the beam, as shown in Fig. P6.11-4. (Note: This is a nonhomogeneous beam, so you will need to incorporate the bending analysis of Section 6.5 in solving this problem.) (a) If this "sandwich" beam is to support a vertical shear force $V = 10$ kN, what is the shear stress that must be transferred between the cover plates and the wood core by the layer of glue? (b) What is the maximum shear stress in the wood for the given shear force V?

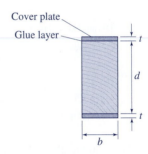

P6.11-4 and P6.11-5

*Prob. 6.11-5. Solve Prob. 6.11-4 for a beam that has a 3×8 wood-beam core ($b = 2.5$ in., $d = 7.5$ in., $E_w = 2 \times 10^3$ ksi), to which are glued two aluminum cover plates with dimensions $b = 2.5$ in., $t = 0.25$ in. and modulus $E_a = 10 \times 10^3$ ksi. The beam is required to handle a vertical shear force $V = 1400$ lb.

▼ SHEAR FLOW IN BUILT-UP BEAMS

DProb. 6.11-6. A steel plate girder is fabricated by the welding of two $16 \text{ in.} \times 1 \text{ in.}$ flange plates to a $70 \text{ in.} \times \frac{3}{8} \text{ in.}$ web plate using fillet welds whose allowable shear strength is $q_{\text{allow}} = 2$ kips/in. Determine the allowable vertical shear force, V, for this plate girder if the four fillet welds run continuously along the length of the girder.

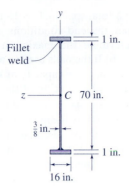

P6.11-6

Prob. 6.11-7. A steel plate girder is fabricated by the welding of two 300 mm × 25 mm flange plates to a 1 m × 12 mm web plate using fillet welds whose allowable shear strength is $q_{allow} = 500$ kN/m. If the four fillet welds run continuously along the length of the girder, what is the allowable vertical shear force, V, for this plate girder?

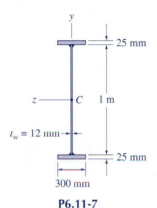

P6.11-7

Prob. 6.11-8. A **W**200 × 71 steel wide-flange beam is strengthened and stiffened by the addition of 175 mm × 10 mm steel plates that are welded to the flanges of the wide-flange beam by continuous fillet welds, as shown in Fig. P6.11-8. (a) Determine the moment of inertia I and the section modulus S for the modified beam. (Neglect the contribution of the weld area to these section properties.) (b) If the allowable shear flow for the weld heads is $q_{allow} = 400$ kN/m, determine the maximum shear force, V, allowed for this modified section.

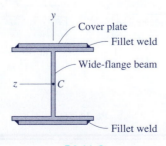

P6.11-8

Prob. 6.11-9. A **W**18 × 60 steel wide-flange beam is strengthened and stiffened by capping the flanges with **C**10 × 20 steel channels that are bolted to the flanges as shown in Fig. P6.11-9. If each bolt has an allowable (direct) shear strength of 2.0 kips and the bolts are spaced at 12 in. longitudinally, what is the maximum allowable vertical shear force V for this modified-section beam? (Hint: Use Table D.4 to obtain the area and centroidal location for the channel for use in calculating its Q.)

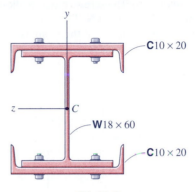

P6.11-9

Prob. 6.11-10. Four 2 × 6 boards (nominal dimensions) are nailed together to form the box beam shown in Fig. P6.11-10. If the nails are spaced at regular intervals of $\Delta x = 6$ in. along the beam, and if each nail has an allowable force in shear of $V_{nail} = 300$ lb, what is the maximum vertical shear force. V, for this built-up beam? ($d = 5\frac{1}{2}$ in., and $t = 1\frac{1}{2}$ in.)

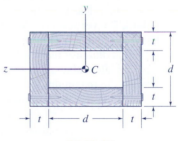

P6.11-10

Prob. 6.11-11. Four 30 mm × 180 mm (actual dimensions) boards are attached together by wood screws to form a box beam (Fig. P6.11-11). If each screw has an allowable shear-force

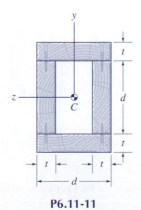

P6.11-11

capacity of $V_s = 1$ kN and the beam is to be subjected to a vertical shear force $V = 5$ kN, what is the maximum permissible longitudinal spacing, Δx, of the screws? ($t = 30$ mm and $d = 180$ mm.)

DProb. 6.11-12. A wood box beam is constructed of four boards ($t_1 = 30$ mm, $d = 250$ mm; $t_2 = 40$ mm, $b = 150$ mm) in the configuration shown in Fig. P6.11-12. The boards are joined by placing wood screws at regular intervals, Δx, in the four locations indicated. The screws have an allowable shear-force capacity of $V_s = 1.2$ kN. Determine the maximum longitudinal screw spacing, Δx, if the beam is to be designed for a maximum vertical shear force $V = 10$ kN.

equally spaced on a total span of $L = 16$ ft, as shown in Fig. P6.11-12a. Consider the end conditions at A and E to be "pinned." (a) Determine the maximum horizontal shear stress in the beam. (b) If the nails used to assemble the beam have an allowable shear-force capacity of $V_{nail} = 500$ lb. what is the maximum allowable spacing of the nails in each segment of the beam, that is, Δx_{AB}, Δx_{BC}, and so on?

DProb. 6.11-15. A wood box beam is fabricated by attaching four boards together in the configuration shown in Fig. P6.11-15a (see Prob. 6.11-12). The dimensions of the boards are: $t_1 = 30$ mm, $d = 250$ mm; $t_2 = 40$ mm, $b = 150$ mm. Each screw has an allowable shear capacity of 1.5 kN. If the box beam is to be loaded and supported as indicated in Fig. P6.11-15b, calculate the maximum allowable longitudinal spacing of the screws for each interval along the beam; that is, determine Δx_{AB}, Δx_{BC}, and Δx_{CD}.

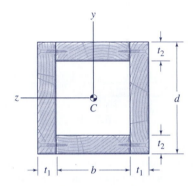

P6.11-12, P6.11-13, and P6.11-15a

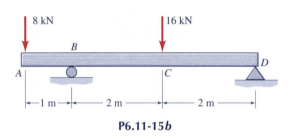

P6.11-15b

DProb. 6.11-13. A wood box beam is fabricated by placing wood screws at regular intervals, Δx, in the four locations indicated in Fig. P6.11-13. The vertical boards are 2×10's (actual dimensions: $t_2 = 1\frac{1}{2}$ in., $d = 9\frac{1}{2}$ in.) and the horizontal boards are 2×6's (actual dimensions: $t_2 = 1\frac{1}{2}$ in., $b = 5\frac{1}{2}$ in.), and the screws have an allowable shear-force capacity of $V_s = 360$ lb. Determine the maximum longitudinal spacing, Δx, if the beam is to be designed for a maximum vertical shear force $V = 4$ kips.

DProb. 6.11-14. A box beam is constructed of four 2×6 planks that are nailed together in the configuration shown in Fig. P6.11-14b. The beam supports three equal loads $P = 500$ lb

DProb. 6.11-16. Two 2×8 boards are attached together by screws that are spaced at regular intervals $\Delta x = 12$ in. along the length of the beam to form the inverted tee beam shown in Fig. P6.11-16b. Each screw has an allowable shear-force capacity of $V_s = 480$ lb. If the beam supports a single concentrated load P at the center and is simply supported, what is the maximum allowable load P?

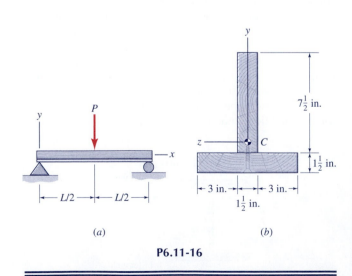

P6.11-16

(a)

P6.11-14

Probs. 6.12-1 through 6.12-3. Determine the eccentricity, e, of the shear center, S, of each channel section. The eccentricity is measured, in each problem, from the centerline of the web, which is taken as the y axis for these problems.

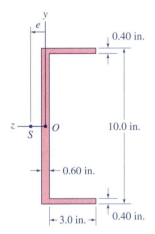

P6.12-1

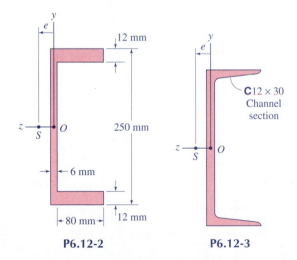

P6.12-2   **P6.12-3**

Prob. 6.12-4. The "hat" section in Fig. P6.12-4 has a constant wall thickness t. (a) Determine expressions for the shear flows $q_1(y)$ and $q_2(z)$. Express your answers in terms of the total shear force V and the dimensions of the cross section.

(b) Determine expressions for V_1 and H_2, the resultants of $q_1(y)$ and $q_2(z)$, respectively. (c) Determine an expression for the eccentricity, e, of the shear center, S.

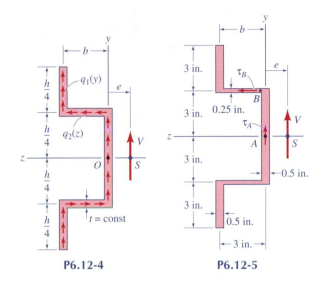

P6.12-4   **P6.12-5**

Prob. 6.12-5. (a) Locate the shear center S of the "hat" section in Fig. P6.12-5 by determining the eccentricity, e. (b) If a vertical shear force $V = 10$ kips acts through the shear center of this hat section, what are the values of the shear stresses τ_A and τ_B at the locations and in the directions indicated in Fig. P6.12-5?

Prob. 6.12-6. Consider the cross section shown in Fig. P6.12-6. (a) Determine expressions for the flange shear flows $q_1(z)$ and $q_2(z)$, as indicated on the figure. (b) Determine expressions for the resultant shear forces H_1 and H_2 corresponding to the shear flows q_1 and q_2. (c) Determine an expression for the eccentricity, e, of the shear center, S.

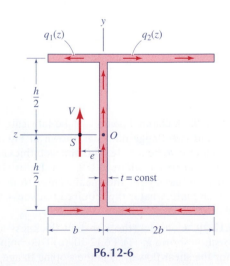

P6.12-6

Prob. 6.12-7. Determine the shear-center location (i.e., the eccentricity e) for the open-box section shown in Fig. P6.12-7.

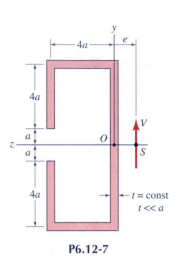

P6.12-7

Prob. 6.12-8. Determine the eccentricity, e, of the shear center, S, for the thin-wall box beam having a very thin longitudinal slit along one side, as shown in Fig. P6.12-8.

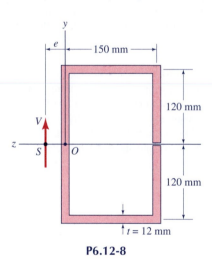

P6.12-8

^D***Prob. 6.12-9.** A channel beam is to be fabricated from a web plate and two flange plates as shown in Fig. P6.12-9. The total depth is to be $d = 18.0$ in., the web thickness $t_w = 0.50$ in., and the flange width $b_f = 6.0$ in. If, for structural reasons, it is necessary for the shear center to lie 0.90 in. from the web plate, what is the required thickness, t_f, of the flange plates?

Prob. 6.12-10. Consider the constant-thickness ($t \ll a$) channel section shown in Fig. P6.12-10. (a) Determine an expresion for the shear flow $q_f(s)$ in the sloping flange, and (b)

locate the shear center by determining an expression for the eccentricity, e. Express your answer in terms of the dimension a.

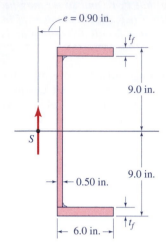

P6.12-9

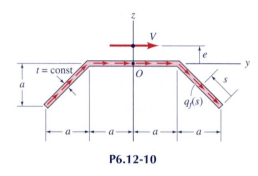

P6.12-10

*Prob. 6.12-11. Consider the modified hat section shown in Fig. P6.12-11. (a) Determine an expression for the shear flows $q_1(y)$ and $q_2(y)$, and (b) determine an expression relating the shear-center eccentricity, e, to the dimension a.

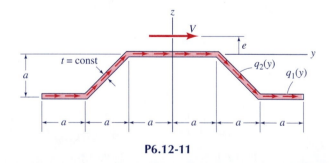

P6.12-11

*Prob. 6.12-12. Vertical extension plates with cross-sectional dimensions 0.400 in. $\times$ 2.00 in. are welded to the flanges of a **W**8 $\times$ 21 wide-flange beam to form the cross section illustrated in Fig. P6.12-12a. (a) Determine an expression for the shear flows $q_1(z)$ and $q_2(z)$ in the two sections of flange, and determine an expression $q_3(y)$ for the shear flow in the vertical flange extensions, as indicated in Fig. P6.12-12b. (b) Determine the eccentricity, e, of the shear center of this modified wide-flange beam.

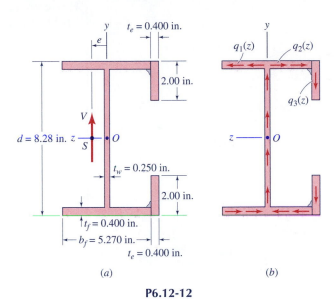

$t_e = 0.400$ in.

2.00 in.

V

$d = 8.28$ in. z S O

$t_w = 0.250$ in.

2.00 in.

$t_f = 0.400$ in.

$b_f = 5.270$ in.

$t_e = 0.400$ in.

(a)

$q_1(z)$ $q_2(z)$

$q_3(z)$

z O

(b)

P6.12-12

*__Prob. 6.12-13.__ A pipe-beam has a very thin longitudinal slit along one side that makes it an open, thin-wall beam. The thickness t is much less than the mean radius r. (a) Determine an expression for the shear flow $q(\theta)$. (b) Determine the location of the shear center, S, by determining its eccentricity, e, from the center of the circular cross section.

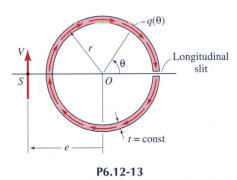

$q(\theta)$

r

θ

V

S O

Longitudinal slit

e

$t = $ const

P6.12-13

*__Prob. 6.12-14.__ A segment of a thin-wall pipe, having an excluded angle 2α, acts as a beam, with loading parallel to the xy plane and passing through the shear center, S. (a) Determine an expression for the shear flow $q(\theta)$. (b) Determine the location of the shear center, S, by determining its eccentricity, e, from the center of the circular cross section.

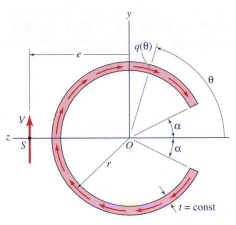

y

e

$q(\theta)$

θ

V

z S O α α

r

$t = $ const

P6.12-14

__Prob. 6.12-15.__ The special thin-wall channel section shown in Fig. P6.12-15 is formed by bending a 6-mm-thick steel plate. (a) Determine an expression for $q_f(s)$, the shear flow in the flanges. (b) Determine the location of the shear center, S.

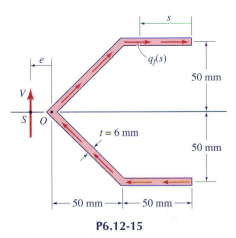

s

e

$q_f(s)$

50 mm

V

S O

$t = 6$ mm

50 mm

$\leftarrow$ 50 mm $\rightarrow$ $\leftarrow$ 50 mm $\rightarrow$

P6.12-15

__Prob. 6.12-16.__ An equal-leg, thin-wall angle section is subjected to a transverse shear force V at an angle of $45°$ to the legs of the angle, as shown in Fig. P6.12-16. (a) Determine an expression for $q(s)$, the shear flow in the legs of the angle. (b) By integration, show that the resultant of this shear flow is the shear force V passing through the point S shown in the figure. (c) Determine an expression for the maximum shear stress.

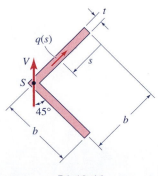

t

$q(s)$

V

S s

$45°$

b

b

P6.12-16

Section		Suggested Review Problems
6.1	The **flexural stress,** σ_x, can be visualized, as acting on individual **fibers** that are parallel to the axis of the beam. The simplest form of beam bending is called **pure bending.** The *longitudinal plane of symmetry, LPS,* the *neutral surface, NS*, and the *deflection curve, v(x)*, are identified in Fig. 6.3b.	 (a) A beam represented as a "bundle" of longitudinal "fibers." (b) Illustration of some beam-deformation terminology. Pure Bending (Fig. 6.3)	
6.2	The <u>four</u> *kinematic assumptions* of **Bernoulli-Euler Beam Theory** are: 1. The beam has a *longitudinal plane of symmetry—LPS,* call it the *x-y* plane, and it is loaded and supported symmetrically with respect to this plane of symmetry. 2. There is a plane, call it the *x-z* plane, that is free of extensional strain, i.e., where $\epsilon_x = 0$. This is called the *neutral surface—NS*. 3. **Plane cross sections remain plane and remain perpendicular to the deformed axis of the beam.** 4. Deformation in cross sectional planes may be neglected.	Review the **strain-displacement analysis** for beams as presented in Section 6.2. (a) The undeformed beam segment. (b) The deformed beam segment. Strain-Displacement Derivation (Fig. 6.6) $$\epsilon_x = -\frac{y}{\rho(x)} \qquad (6.3)$$	Derive Eq. 6.3.

Section		Suggested Review Problems
6.2	The **strain-displacement equation,** Eq. 6.3, gives an extensional strain due to flexure that is a *linear function of distance y from the neutral surface.* Linear strain distribution (Fig. 6.10)	6.2–1
6.3	Sections 6.3 and 6.4 deal with **flexural stresses in homogeneous, linearly elastic beams.** The **neutral axis** of a homogeneous, linearly elastic beam passes through the centroid of the cross section. (*a*) Location of the neutral axis, (*b*) distribution of flexural stress for homogeneous, linearly elastic beam (Fig. 6.13)	
	Two of the most important equations of Bernoulli-Euler Beam theory are: • the **moment-curvature equation,** and • the **flexure formula** Memorize these two equations! $$M = \frac{EI}{\rho} \qquad (6.12)$$ $$\sigma_x = -\frac{My}{I} \qquad (6.13)$$ where I is the *moment of inertia* of the cross section about its neutral axis.	Derive Eqs. 6.12 and 6.13 6.3–5 6.3–13 6.3–17 6.3–33 6.3–35
6.4	The **design of beams for strength** is based on selecting a beam cross section that will limit the flexural stress in the beam to some value of **allowable stress.** $$S_{design} = \frac{M_{max}}{\sigma_{allow}} \qquad (6.17)$$ where S is the *elastic section modulus*, defined by $$S = \frac{I}{c} \qquad (6.15)$$ Tables D.1–D.10 list S values for various shapes of beams.	6.4–1 6.4–7 6.4–13
6.5	Section 6.5 discusses **flexural stress in non-homogeneous, linearly elastic beams.** Two analysis methods are presented: • the **direct method,** and • the **transformed-section method.** The key assumption is that **plane sections remain plane.**	You should review the two methods as they are applied to a beam that is made of two linearly elastic materials. See Eqs. 6.18–6.24 and Eqs. 6.25–6.29.	6.5–1 6.5–5 6.5–7 6.5–15 6.5–19

Section		Suggested Review Problems
6.8	Sections 6.6 and 6.7 are optional sections. Sections 6.8–6.12 deal with **shear stress in beams.** The key idea is that shear stress τ_{xy} at location y on the cross section at x is equal to the shear stress τ_{yx} at that (x, y) location, and equilibrium can be used to determine the latter (Fig. 6.13c). (a) An element of length Δx. (b) The distribution of flexural stress. (c) A free body diagram (minus vertical shear on ac and bd). (d) The flexural stress contributing to F_2. Use of equilibrium to calculate shear stress (Fig. 6.43)	
	Shear flow q is defined as the shear force per unit length along the beam. $$q(x, y) = \frac{V(x)Q(y)}{I} \quad (6.65)$$ where $Q = A'\overline{y}'$	Derive Eq. 6.65 State the meaning of Q
	Shear stress τ is shear flow q divided by the local thickness $t(y)$. $$\tau(x, y) = \frac{V(x)Q(y)}{It} \quad (6.67)$$	6.8–1 6.8–5 6.8–13
6.9	Review the **Limitations on the Shear Stress Formula** in Section 6.9.	
6.10	Section 6.10, **Shear Stress in Thin-Wall Beams,** and Section 6.11, **Shear in Built-Up Beams,** treat shear stress in two very important classes of beams. These sections illustrate how	6.10–1 6.10–9 6.10–13 6.10–19
6.11	$$Q = A'\overline{y}'$$ is calculated in these special situations.	6.11–1 6.11–7 6.11–15
	Sections 6.6, 6.7, and 6.12 are optional sections.	

DEFLECTION OF BEAMS

7.1 INTRODUCTION

Couples and transverse forces applied to beams cause them to deflect (i.e., become curved), as illustrated in Fig. 6.1 and in Fig. 7.1. In Chapter 6 we determined a relationship between the curvature of the deflection curve of a beam and the bending moment at a cross section. In this chapter we will relate the deflection and slope of beams to their loading and support. As indicated in Fig. 7.1a, the **deflection curve** is characterized by a function $v(x)$ that gives the **transverse displacement** (i.e., displacement in the y direction) of the points that lie along the axis of the beam. The **slope** of the deflection curve is labeled $\theta(x)$.

There are several reasons for considering the deflection of beams. For example, we may need to know the maximum deflection of a given beam under a given set of loads. As illustrated in Fig. 7.2, the maximum deflection might occur at an

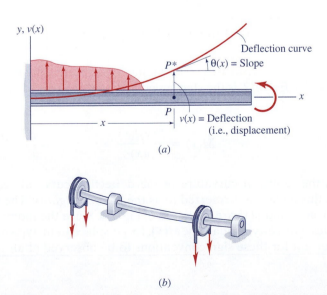

(a)

(b)

FIGURE 7.1   Examples of the deflection of beams and shafts (deflection is exaggerated).

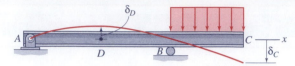

FIGURE 7.2   Points of maximum deflection.

Heavy load

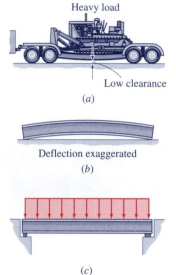

Low clearance

(a)

Deflection exaggerated

(b)

(c)

FIGURE 7.3   Illustrations of beam deflection.

unsupported end of the beam, (δ_C), or it might occur at an interior section where the slope vanishes, (δ_D). For example, the beams of the equipment trailer in Fig. 7.3a must not deflect so much under load that the clearance between the trailer and the ground becomes unacceptably small. Also, the beams that support a bridge deck should be designed to have just the right initial upward camber (deflection) (Fig. 7.3b) so that they become straight when they are loaded (Fig. 7.3c).

Statically indeterminate beams provide another important reason for studying beam deflection. If a beam is statically indeterminate, its reactions and its internal stresses cannot be determined without considering deflections. Analysis of statically indeterminate beams is discussed in Sects. 7.4 through 7.7. Finally, we will need to relate the *transverse* deflection $v(x)$ to *axial* loads when we study the buckling of columns (Chapter 10).

Starting with the moment-curvature equation that was derived in Chapter 6, Eq. 6.12, we will develop *differential equations* that relate the deflection $v(x)$ to the bending moment $M(x)$, the transverse shear force $V(x)$, and the transverse load $p(x)$. Then we will discuss the *boundary conditions* and *continuity conditions* that must accompany these differential equations. Next, we will solve for the deflection and slope of statically determinate and statically indeterminate beams by *integration of the differential equations,* incorporating the given loads and the given boundary and continuity conditions (Section 7.3 and 7.4). *Discontinuity functions* are introduced in Section 7.5 to facilitate the integration of the differential equations for beams having several loads and/or supports. Finally, we will use *superposition of known solutions* to solve beam deflection problems, first using a *Force-Method* approach (Section 7.6) and then a *Displacement-Method* approach (Section 7.7).

7.2   DIFFERENTIAL EQUATIONS OF THE DEFLECTION CURVE

Moment-Curvature Equation. The moment-curvature equation that we derived in Section 6.3, Eq. 6.12, is the starting point for an analysis of the deflection of linearly elastic beams.

$$M(x) = \frac{E(x)I(x)}{\rho(x)} \tag{7.1}$$

where $\rho(x)$ is the radius of curvature of the deflection curve at section x.[1] The product EI in this equation is referred to as the *flexural rigidity*. The sign conventions for $M(x)$, $v(x)$, and $\rho(x)$ are reviewed in Fig. 7.4. Since the moment-curvature equation relates a force-type quantity, $M(x)$, to a displacement-type quantity, $\rho(x)$, it is very important for these sign conventions to be observed at all times.

---

[1]For nonhomogeneous, linearly elastic beams, Eq. 6.22 or Eq. 6.27 of Section 6.5 must be used.

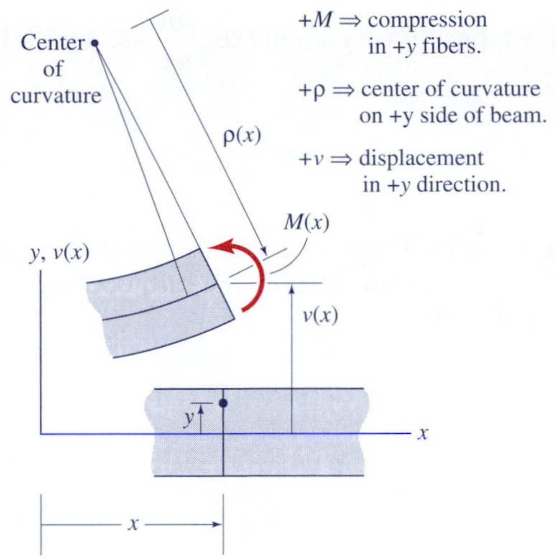

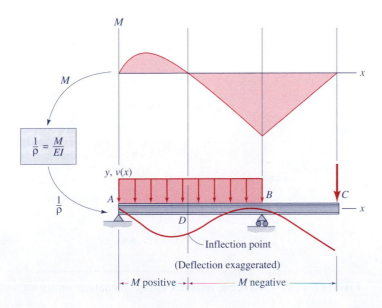

+M ⇒ compression
in +y fibers.

+ρ ⇒ center of curvature
on +y side of beam.

+v ⇒ displacement
in +y direction.

FIGURE 7.4   The sign convention for beam deflection analysis.

Before actually calculating the slope and deflection of a beam, it is helpful to sketch the *anticipated deflection curve* of the beam. This can be done with the help of Eq. 7.1 and a moment diagram. In fact, we did this in Chapter 5 as a means of checking moment diagrams. Figure 7.5 is taken from Example 5.9. First, there can be no deflection of the beam at supports A and B, but the slope of the beam is unrestricted at the supports and everywhere else along the beam. At any section where the moment is negative, that is, to the right of section D, the center of the curvature is on the $-y$ side of the beam, that is, the beam is concave downward. There is an *inflection point* at D, where the moment changes from positive to negative.

The inclination angle, $\theta(x)$, is related to the derivative of the deflection by

$$\theta = \arctan\left(\frac{dv}{dx}\right) \tag{7.2}$$

FIGURE 7.5   Use of the moment-curvature equation in sketching the deflection curve.

The slope of a beam is typically very small (i.e., $\dfrac{dv}{dx} \ll 1$), so the approximation

$$\theta \approx \frac{dv}{dx} \tag{7.3}$$

may be treated as an equality. <u>Deflection $v(x)$</u>, <u>slope $\theta(x)$</u>, and <u>radius of curvature $\rho(x)$</u> are illustrated in Fig. 7.6. From calculus, the radius of curvature is related to the derivatives of the deflection by

$$\frac{1}{\rho} = \frac{\dfrac{d^2v}{dx^2}}{\left[1 + \left(\dfrac{dv}{dx}\right)^2\right]^{3/2}} \tag{7.4}$$

Again, where the slope is small compared with unity, the approximation

$$\frac{1}{\rho} \approx \frac{d^2v}{dx^2} \tag{7.5}$$

can be treated as an equality. Then the moment-curvature equation (Eq. 7.1) becomes

$$\frac{d^2v(x)}{dx^2} = \frac{M(x)}{E(x)I(x)} \tag{7.6}$$

To simplify the equations in this chapter, we will adopt the notation of using primes to denote differentiation with respect to x. Thus,

$$v' \equiv \frac{dv}{dx}, \, v'' \equiv \frac{d^2v}{dx^2}, \, M' \equiv \frac{dM}{dx}, \, \text{etc.} \tag{7.7}$$

With the above prime notation, Eq. 7.6 may be written as

$$\boxed{EIv'' = M} \qquad \textbf{Moment-Curvature Equation} \tag{7.8}$$

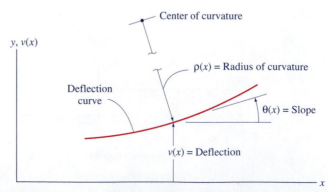

FIGURE 7.6 Geometric relationships of the deflection curve.

Although the curvature, $\kappa \equiv 1/\rho$, does not appear explicitly in it, we will refer to this equation as the **moment-curvature equation.** It is a *second-order, ordinary differential equation* that we can integrate to determine the slope and deflection, given an expression for $M(x)$ and appropriate boundary conditions and continuity conditions.

Load-Deflection Equation. Equations relating load, transverse shear force, and bending moment were derived in Chapter 5. Equations 5.3 and 5.2, repeated here, are

$$M'(x) = V(x) \tag{7.9}$$

$$V'(x) = p(x) \tag{7.10}$$

The sign conventions for p, V, and M are reviewed in Fig. 7.7.

Combining Eqs. 7.8, 7.9, and 7.10, we get the *shear-deflection equation*

$$\boxed{(EIv'')' = V} \qquad \text{**Shear-Deflection Equation**} \tag{7.11}$$

and the *load-deflection equation*

$$\boxed{(EIv'')'' = p} \qquad \text{**Load-Deflection Equation**} \tag{7.12}$$

The **load-deflection equation,** which is a *fourth-order, ordinary differential equation*, is especially useful. Although it must be integrated four times, this is a very straight forward process, one that can be readily programmed for computer solution (see Section 7.5).

If the flexural rigidity, EI, is constant, then Eqs. 7.11 and 7.12 become the following third-order and fourth-order differential equations, respectively.

$$EIv''' = V \tag{7.13}$$

$$EIv'''' = p \tag{7.14}$$

We will refer to Eq. 7.11 (or 7.13) as the *shear-deflection equation,* and to Eq. 7.12 (or 7.14) as the *load-deflection equation.*

A beam that requires more than one expression for bending moment because of the presence of concentrated loads, discontinuities in distributed loads, or interior supports will be referred to as a *multi-interval beam.* Consider the two-interval beam in Fig. 7.8 (repeated from Example Problem 5.2). Each interval has its own differential equations, and each interval will have its own expression for the deflection

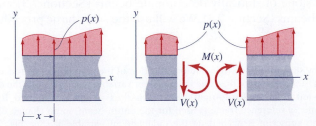

FIGURE 7.7   The sign conventions for positive load, shear, and moment.

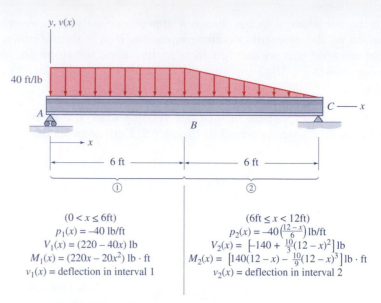

$(0 < x \le 6\text{ft})$
$p_1(x) = -40 \text{ lb/ft}$
$V_1(x) = (220 - 40x) \text{ lb}$
$M_1(x) = (220x - 20x^2) \text{ lb} \cdot \text{ft}$
$v_1(x) = $ deflection in interval 1

$(6\text{ft} \le x < 12\text{ft})$
$p_2(x) = -40\left(\frac{12-x}{6}\right) \text{lb/ft}$
$V_2(x) = \left[-140 + \frac{10}{3}(12-x)^2\right] \text{lb}$
$M_2(x) = \left[140(12-x) - \frac{10}{9}(12-x)^3\right] \text{lb} \cdot \text{ft}$
$v_2(x) = $ deflection in interval 2

FIGURE 7.8 A two-interval beam.

curve. As indicated in Fig. 7.8, we will use numerical subscripts to indicate the range of validity of each variable.[2]

Boundary Conditions and Continuity Conditions.

When the above differential equations are integrated, there will be *constants of integration* that must be evaluated. For each (second-order) moment-curvature equation, there will be two constants of integration; for each (fourth-order) load-deflection equation, there will be four constants of integration. These constants of integration are evaluated by making use of **boundary conditions** (BC), which are listed in Table 7.1. In the case of multi-interval beams (e.g., Fig. 7.8), there will also be **continuity conditions** (CC) that apply where two intervals meet. Table 7.2 lists the possible continuity conditions.

We will use only the second-order equation and the fourth-order equation, so let us consider the specific boundary and continuity conditions that apply to each. Recall from your study of differential equations that the constants of integration of a certain order ODE can only be evaluated using known values, or relationships of values, of the lower derivatives, including the function itself. Hence, only values of v and v' can be used in boundary conditions and continuity conditions for evaluating constants of integration for a second-order equation. Values of v, v', M, and V are used to evaluate constants of integration when the load-deflection equation (a fourth-order ODE) is used. Tables 7.1 and 7.2 list the various boundary conditions and continuity conditions that may arise.

In the next two sections we integrate differential equations to solve for the deflection and slope of statically determinate beams (Section 7.3) and of statically indeterminate beams (Section 7.4). We will use the same basic procedure to solve all

---

[2]Some texts employ shifting of the origin and/or reflection of the coordinate reference frame. A shift of origin necessitates the use of more than one independent variable (e.g., x_1, x_2). In the case of a reflected x axis (i.e., right-to-left instead of left-to-right), "tricky" changes of signs are required. To simplify matters, in this text the same xy reference frame is used for the entire beam; so no change of sign convention is required, and no subscript is needed on the independent variable x. (See Fig. 7.8 and Example Problem 7.3.)

TABLE 7.1   **Boundary Conditions**

	Type	Symbol*	2nd Order	4th Order
BC	Fixed end		$v = 0$ $v' = 0$	$v = 0$ $v' = 0$
	Simple support		$v = 0$	$v = 0$ $M = 0$
	Free end		No BC	$V = 0$ $M = 0$
	Concentrated force		No BC	$V = P_0$ $M = 0$
	Concentrated couple		No BC	$V = 0$ $M = -M_0$

*These boundary conditions also apply if the boundary under consideration is the other end of the beam (i.e., $x = L$).

TABLE 7.2   **Continuity Conditions***

	Type	Symbol	2nd Order	4th Order
CC	Roller		$v_1 = v_2 = 0$ $v_1' = v_2'$	$v_1 = v_2 = 0$ $v_1' = v_2'$ $M_1 = M_2$
	Discontinuity in load function		$v_1 = v_2$ $v_1' = v_2'$	$v_1 = v_2$ $v_1' = v_2'$ $V_1 = V_2$ $M_1 = M_2$
	Concentrated force		$v_1 = v_2$ $v_1' = v_2'$	$v_1 = v_2, \quad v_1' = v_2'$ $V_2 - V_1 = P_0$ $M_1 = M_2$
	Concentrated couple		$v_1 = v_2$ $v_1' = v_2'$	$v_1 = v_2, \quad v_1' - v_2'$ $V_1 = V_2$ $M_2 - M_1 = -M_0$
	Pin, with force		$v_1 = v_2$	$v_1 = v_2$ $V_2 - V_1 = P_0$ $M_1 = M_2 = 0$

*The displacement (v) and slope (v') continuity conditions that are listed in Table 7.2 are obtained by inspection, that is, by simply looking at the figures in the "Symbol" column. The continuity conditions on shear force (V) and bending moment (M) are obtained by taking a local free-body diagram of the "joint" that is common to beam segments (1) and (2).

beam-deflection problems, with only a minor modification required in the treatment of statically indeterminate problems.

7.3 SLOPE AND DEFLECTION BY INTEGRATION—STATICALLY DETERMINATE BEAMS

Let us now illustrate the solution of slope and deflection problems by integration of either the second-order differential equation, Eq. 7.8, or the fourth-order differential equation, Eq. 7.12. Since all solutions follow a straightforward procedure, we will outline that procedure here and will dispense with the *Plan the Solution* discussion of each problem.

PROCEDURES FOR DETERMINING SLOPE AND DEFLECTION OF STATICALLY DETERMINATE BEAMS BY INTEGRATION

1. Sketch the anticipated **deflection curve.** In some cases this can be done by reference to the original load diagram. In other situations, a moment diagram is very helpful (e.g., Fig. 7.5).

2. *Second-order method*—Sketch free-body diagrams and develop a **moment function** $M_i(x)$ for each interval of the beam. *Fourth-order method*—From the load diagram, develop a **load function** $p_i(x)$ for each interval of the beam. (This includes $p_i(x) = 0$ for intervals that have no distributed loading.)

   In Step 2 the *Sign conventions* in Fig. 7.7 must be rigorously adhered to.

3. *Second-order method*—Using Eq. 7.8 and the expressions for $M_i(x)$ developed in Step 2, write down a **moment-curvature equation** for each interval.

$$(EIv'')_i = M_i(x) \qquad (7.8)$$
$$\text{repeated}$$

   Fourth-order method—Using Eq. 7.12 and the expressions for $p_i(x)$ developed in Step 2, write down a **load-deflection equation** for each interval.

$$(EIv'')_i'' = p_i(x) \qquad (7.12)$$
$$\text{repeated}$$

4. **Integrate the differential equations** developed in Step 3. Include appropriate constants of integration.

5. Write down the appropriate **boundary conditions** and **continuity conditions** using Tables 7.1 and 7.2 as a guide. *Second-order method*—Use the 2nd Order" columns of

Tables 7.1 and 7.2 as a guide in setting up boundary conditions and continuity conditions. Note that all boundary conditions involve only the slope, v', or the deflection, v, since Eq. 7.8 is a second-order differential equation. *Fourth-order method*—Use the "4th Order" columns of Tables 7.1 and 7.2 as a guide. In addition to conditions on v' and v, boundary and continuity conditions on the shear. V, and the bending moment, M, may also apply.

6. Use the boundary and continuity conditions of Step 5 to **evaluate the constants of integration** introduced in Step 4.

   In the fourth-order method, the first integral of Eq. 7.12 leads to an equation of the form

$$(EIv'')_i' = \int p_i(x) \, dx + C_1$$

   Since $(EIv'')' = V$ (Eq. 7.11), boundary and continuity conditions on the shear, V, are appropriate. A second integration leads to an equation of the form

$$(EIv'')_i = \int \left[\int p_i(x)dx + C_1 \right] dx + C_2$$

   From Eq. 7.8, EIv'' is just the bending moment, M, so boundary and continuity conditions on M apply. Tables 7.1 and 7.2 list these conditions on V and M.

7. Complete the solution by evaluating the slope and deflection at points where they are required, by evaluating deflection maxima, etc.

Example Problem 7.1 is solved by both the second-order method and the fourth-order method. Others are solved by only one of the two methods.

EXAMPLE 7.1

A uniform cantilever beam is subjected to a transverse load P_B and a couple M_B at its "free" end, as shown in Fig. 1. Using the second-order method and the fourth-order method: (a) Determine expressions for the slope, $v'(x)$, and the deflection, $v(x)$. (b) Determine expressions for the deflection and slope at B.

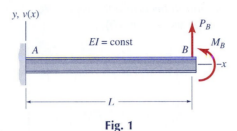

Fig. 1

Solution A (2nd-Order Method)

(A.a) Determine formulas for slope and deflection.

Sketch the Deflection Curve: Both loads in Fig. 1 will bend the beam upward. Therefore, the deflection should resemble the deflection curve in Fig. 2.

Determine $M(x)$: The free-body diagram in Fig. 3 can be used to determine an expression for $M(x)$.

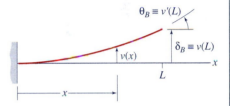

Fig. 2 The anticipated deflection curve.

$$+\circlearrowleft \left(\sum M \right)_a = 0: \quad M(x) - P_B(L - x) - M_B = 0 \tag{1}$$

Write the Moment-Curvature Equation: From Eqs. 7.8 and (1),

$$EIv'' = M(x) = M_B + P_B(L - x) \tag{2}$$

Integrate the Differential Equation: We need to integrate Eq. (2) twice.

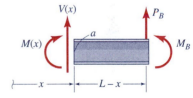

Fig. 3 A free-body diagram.

$$EIv' = M_Bx + P_BLx - P_B\left(\frac{x^2}{2}\right) + C_1 \tag{3a}$$

$$EIv = M_B\left(\frac{x^2}{2}\right) + P_BL\left(\frac{x^2}{2}\right) - P_B\left(\frac{x^3}{6}\right) + C_1x + C_2 \tag{3b}$$

Identify the Boundary Conditions: End A is "fixed." That is, the slope and the deflection must be zero at $x = 0$. So, from Table 7.1,

$$v'(0) = v(0) = 0 \tag{4a,b}$$

Evaluate the Constants of Integration: From Eqs. (3a) and (4a),

$$EIv'|_{x=0} = C_1 = 0 \tag{5a}$$

and, from Eqs. (3b), (4b), and (5a),

$$EIv|_{x=0} = C_2 = 0 \tag{5b}$$

Therefore,

$$v'(x) = \frac{1}{EI}\left[M_B x + P_B L x - P_B\left(\frac{x^2}{2}\right) \right]$$

$$v(x) = \frac{1}{EI}\left[M_B\left(\frac{x^2}{2}\right) + P_B L\left(\frac{x^2}{2}\right) - P_B\left(\frac{x^3}{6}\right) \right]$$

Ans. (A.a) (6)

(A.b) *Determine the tip deflection and slope.* We can now evaluate the tip deflection and slope using Eqs. (6).

$$\delta_B \equiv v(L) = \frac{1}{EI}\left[M_B\left(\frac{L^2}{2}\right) + P_B\left(\frac{L^3}{3}\right) \right]$$

$$\theta_B \equiv v'(L) = \frac{1}{EI}\left[M_B(L) + P_B\left(\frac{L^2}{2}\right) \right]$$

Ans. (A.b) (7)

For example, due to P_B alone, we get

$$\delta_B = \frac{P_B L^3}{3EI}, \quad \theta_B = \frac{P_B L^2}{2EI} \tag{8a,b}$$

These results, together with the results that are obtained when M_B is applied alone, are listed in Table E.1 of Appendix E.

Solution B (4th-Order Method)

Sketch the Deflection Curve: See Fig. 2.

Determine $p(x)$: From the load diagram, Fig. 1,

$$p(x) = 0 \tag{9}$$

Write the Load-Deflection Equation: From Eqs. 7.12 and (9),

$$(EIv'')'' = 0 \tag{10}$$

Integrate the Differential Equation: We need to integrate Eq. (10) four times. We can also make use of Eqs. 7.11 and 7.8.

$$(EIv'')' = V(x) = D_1 \tag{11a}$$

$$EIv'' = M(x) = D_1 x + D_2 \tag{11b}$$

$$EIv' = D_1\left(\frac{x^2}{2}\right) + D_2 x + D_3 \tag{11c}$$

$$EIv = D_1\left(\frac{x^3}{6}\right) + D_2\left(\frac{x^2}{2}\right) + D_3 x + D_4 \tag{11d}$$

Identify the Boundary Conditions: We can refer to Table 7.1. We know the displacement and slope at $x = 0$ and the moment and shear at $x = L$.

Therefore, observing all sign conventions, we get

$$V(L) = -P_B, \quad M(L) = M_B \tag{12a,b}$$

$$v'(0) = 0, \quad v(0) = 0 \tag{12c,d}$$

(Note: The minus sign in Eq. (12a) results from the sign convention for shear, Fig. 7.7.)

Evaluate the Constants of Integration: Combining Eqs. (11) and (12), we get

$$D_1 = -P_B \tag{13a}$$

$$D_1 L + D_2 = M_B \rightarrow D_2 = M_B + P_B L \tag{13b}$$

$$D_3 = 0, \quad D_4 = 0 \tag{13c,d}$$

Finally, combining Eqs. (11) and (13), we get

$$\left.\begin{array}{l} v'(x) = \dfrac{1}{EI}\left[M_B x + P_B L x - P_B\left(\dfrac{x^2}{2}\right)\right] \\[4mm] v'(x) = \dfrac{1}{EI}\left[M_B\left(\dfrac{x^2}{2}\right) + P_B L\left(\dfrac{x^2}{2}\right) - P_B\left(\dfrac{x^3}{6}\right)\right] \end{array}\right\} \qquad \textbf{Ans. (B.a)}$$

as we did in the second-order solution, Eqs. (6).

Review the Solution   A check of the dimensionality of each term in the slope and deflection expressions shows that each term has the proper dimensions.

To get some idea of the magnitude of the slope and the deflection that might occur in an actual beam in service, consider a cantilever beam of length $L = 10$ ft that supports a load $P_B = 4.5$ kips. Let the beam be an A-36 steel wide-flange beam that has been sized with a factor of safety $FS = 2$. For A-36 steel $\sigma_{YP} = 36$ ksi and $E = 29(10^3)$ ksi. Based on Eq. 6.17, the required section modulus is

$$S_{\text{design}} = \frac{M_{\text{max}}}{\sigma_{\text{allow}}} = \frac{P_B L}{\dfrac{\sigma_{YP}}{FS}} = \frac{(4.5 \text{ kips})\left[(10\,\text{ft})\left(12\,\dfrac{\text{in.}}{\text{ft}}\right)\right]}{\dfrac{36 \text{ ksi}}{2}} = 30.0 \text{ in}^3$$

From Table D.1, a suitable wide-flange beam would be a **W**10×30, with moment of inertia $I = 170$ in^4. Then, from Eqs. (8), the deflection and slope at the end of the cantilever beam are given by

$$\delta_B = \frac{P_B L^3}{3EI} = \frac{(4.5 \text{ kips})(120 \text{ in.})^3}{3(29 \times 10^3 \text{ ksi})(170 \text{ in}^4)} = 0.526 \text{ in.}$$

$$\theta_B = \frac{P_B L^2}{2EI} = \frac{(4.5 \text{ kips})(120 \text{ in.})^2}{2(29 \times 10^3 \text{ ksi})(170 \text{ in}^4)} = 0.00657 \text{ rad}$$

Clearly $|\theta_B| \ll 1$ and, even more so, $|\theta_B|^2 \ll 1$, so it is valid to treat the approximations in Eqs. 7.3 and 7.5 as equalities and use Eq. 7.8 as the basis for solving for the deflection of beams.

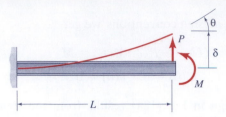

FIGURE 7.9 Load-deflection relationships for a uniform cantilever beam.

The results that were obtained in Example Problem 7.1 can be quite useful in solving beam deflection problems by superposition (Sections 7.6 and 7.7). For the cantilever beam of Fig. 7.9, Eqs. (7) of Example Problem 7.1 may be written in the form:

$$\delta = \left(\frac{L^3}{3EI}\right)P + \left(\frac{L^2}{2EI}\right)M$$

$$\theta = \left(\frac{L^2}{2EI}\right)P + \left(\frac{L}{EI}\right)M$$

(7.15)

On the other hand, we can solve these equations for P and M in terms of δ and θ and get

$$P = \left(\frac{12EI}{L^3}\right)\delta - \left(\frac{6EI}{L^2}\right)\theta$$

$$M = -\left(\frac{6EI}{L^2}\right)\delta + \left(\frac{4EI}{L}\right)\theta$$

(7.16)

The next example problem illustrates the usefulness of the fourth-order method when a beam is subjected to a distributed loading that is not simply constant or linear. Observe that by using the fourth-order method we avoid having to determine an expression for $M(x)$ at the outset.

EXAMPLE 7.2

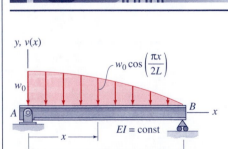

Fig. 1 A simply supported beam with cosine load.

Use the fourth-order method to analyze the uniform, simply supported beam in Fig. 1. (a) Determine expressions for the slope and deflection of the beam. (b) Determine the maximum deflection of the beam.

Solution (a) *Determine formulas for the slope and deflection of the beam.*

Sketch the Deflection Curve: The beam will obviously deflect downward over its entire length, as sketched in Fig. 2. The maximum deflection should occur to the left of the center of the beam.

Determine $p(x)$: The load distribution, $p(x)$, is given in Fig. 1. There is a negative sign because the load acts in the $-y$ direction.

$$p(x) = -w_0 \cos\left(\frac{\pi x}{2L}\right)$$

(1)

Write the Load-Deflection Equation: From Eqs. 7.12 and (1).

$$(EIv'')'' = p(x) = -w_0 \cos\left(\frac{\pi x}{2L}\right) \qquad (2)$$

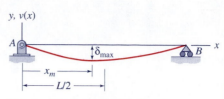

Fig. 2 The expected deflection curve.

Integrate the Differential Equation: We need to integrate Eq. (2) four times, making use of Eqs. 7.11 and 7.8 and introducing constants of integration.

$$V(x) = (EIv'')' = -w_0\left(\frac{2L}{\pi}\right)\sin\left(\frac{\pi x}{2L}\right) + C_1 \qquad (3a)$$

$$M(x) = EIv'' = w_0\left(\frac{2L}{\pi}\right)^2 \cos\left(\frac{\pi x}{2L}\right) + C_1 x + C_2 \qquad (3b)$$

$$EIv' = w_0\left(\frac{2L}{\pi}\right)^3 \sin\left(\frac{\pi x}{2L}\right) + C_1\left(\frac{x^2}{2}\right) + C_2 x + C_3 \qquad (3c)$$

$$EIv = -w_0\left(\frac{2L}{\pi}\right)^4 \cos\left(\frac{\pi x}{2L}\right) + C_1\left(\frac{x^3}{6}\right) + C_2\left(\frac{x^2}{2}\right) + C_3 x + C_4 \quad (3d)$$

Identify the Boundary Conditions: The moment vanishes at both ends, and the displacement is also zero at both ends (see Table 7.1). Therefore,

$$M(0) = M(L) = 0 \qquad (4a,b)$$

$$v(0) = v(L) = 0 \qquad (4c,d)$$

Evaluate the Constants of Integration: Combining Eqs. (3) and (4) we get (the source equations are cited in square brackets at the left margin):

[4a, 3b] $\qquad M(0) = w_0\left(\frac{2L}{\pi}\right)^2 + C_2 = 0$

so

$$C_2 = -w_0\left(\frac{2L}{\pi}\right)^2 \qquad (5a)$$

[4b, 3b] $\qquad M(L) = C_1 L + C_2 = 0$

so

$$C_1 = \frac{w_0}{L}\left(\frac{2L}{\pi}\right)^2 \qquad (5b)$$

[4c, 3d] $\qquad EIv(0) = -w_0\left(\frac{2L}{\pi}\right)^4 + C_4 = 0$

so

$$C_4 = w_0\left(\frac{2L}{\pi}\right)^4 \tag{5c}$$

[4d, 3d]
$$EIv(L) = C_1\left(\frac{L^3}{6}\right) + C_2\left(\frac{L^2}{2}\right) + C_3L + C_4 = 0$$

$$C_3 = \frac{w_0L^3}{3\pi^4}(4\pi^2 - 48) \tag{5d}$$

Finally, from Eqs. (3) and (5),

$$v' = \frac{w_0L^3}{\pi^4EI}\left[8\pi\,\sin\left(\frac{\pi x}{2L}\right) + 2\pi^2\left(\frac{x}{L}\right)^2 - 4\pi^2\left(\frac{x}{L}\right) + \frac{4}{3}\pi^2 - 16\right] \tag{6a}$$

$$v = \frac{w_0L^4}{\pi^4EI}\left[-16\cos\left(\frac{\pi x}{2L}\right) + \frac{2\pi^2}{3}\left(\frac{x}{L}\right)^3 - 2\pi^2\left(\frac{x}{L}\right)^2\right.$$
$$\left. + \left(\frac{4\pi^2}{3} - 16\right)\left(\frac{x}{L}\right) + 16\right] \qquad \textbf{Ans. (a)} \tag{6b}$$

(b) *Determine the maximum deflection of the beam.* The maximum deflection occurs where the slope is zero. Figure 2 suggests that this should occur in the interval $0 < x_m < L/2$. This means that we must set the expression in square brackets in Eq. (6a) equal to zero and find a root x_m for which $v'(x_m) = 0$. We can do this by trial and error or by using some root-finder, and we get

$$x_m = 0.485L$$

Using Eq. (6b) to evaluate $v(x_m)$, we get

$$\delta_{\max} = |v(0.485L)| = 8.70(10^{-3})\frac{w_0L^4}{EI} \qquad \textbf{Ans. (b)} \tag{7}$$

Review the Solution The dimensionality of all terms is correct. We get the correct signs by spot-checking $v'(0)$, $v'(L)$ and $v(L/2)$. We can use Table E.2 to check the magnitude of $\delta_{\max}$ in Eq. (7), since the cosine load is between a triangular load (Fig. 3a) and a uniform load (Fig. 3b). Therefore, the answer seems to be correct.

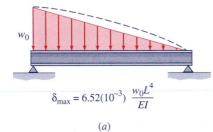

w_0

$\delta_{\max} = 6.52(10^{-3})\,\dfrac{w_0L^4}{EI}$

(a)

w_0

$\delta_{\max} = 13.02(10^{-3})\,\dfrac{w_0L^4}{EI}$

(b)

Fig. 3 Comparison deflections.

Example Problems 7.1 and 7.2 are single-span problems involving only one loading interval; the first is a cantilever-beam problem, and the second is a simply supported beam problem. To solve single-interval problems by the second-order method or by the fourth-order method, we need to incorporate the appropriate *boundary conditions,* which are catalogued in Table 7.1.

Let us now solve a two-interval problem by integrating the second-order moment-curvature equation and applying the appropriate **boundary conditions** from Table 7.1 *and* the appropriate **continuity conditions** as catalogued in Table 7.2. From this example you will observe that multi-interval problems can become very tedious to solve because of the large number of constants of integration that must be evaluated. For this reason, multi-interval problems are most often solved by one of the methods discussed in Sections 7.5–7.7.

> **EXAMPLE 7.3**

A uniform, simply supported beam has a uniform load over half of its length, as shown in Fig. 1. (a) Using the second-order integration method, determine expressions for the slope and deflection in load intervals AB and BC. (b) Determine the maximum deflection.

Solution

(a) *Determine formulas for the slope and deflection of the beam.*

Sketch the Deflection Curve: The deflection will be downward over the entire length, with the maximum deflection occurring between A and B, as illustrated in the deflection diagram, Fig. 2.

Equilibrium—Determine $M_1(x)$ and $M_2(x)$: Since this beam has two load intervals, we need two moment equations. First, however, we need an overall free-body diagram in order to determine the reactions at A and C. From the free-body diagram in Fig. 3a, we get the equilibrium equations

$$\left(\sum M\right)_A = 0: \qquad C_y = \frac{w_0 L}{8}$$

$$\left(\sum M\right)_C = 0: \qquad A_y = \frac{3 w_0 L}{8}$$

From the free-body diagrams in Figs. 3b and 3c, respectively, we get[3]

$$\left(\sum M\right)_a = 0: \qquad M_1(x) = \frac{3 w_0 L x}{8} - \frac{w_0 x^2}{2} \tag{1a}$$

$$\left(\sum M\right)_b = 0: \qquad M_2(x) = \frac{w_0 L}{8}(L - x) \tag{1b}$$

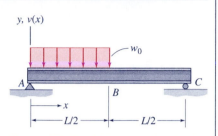

Fig. 1 The load diagram of a simply supported beam.

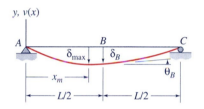

Fig. 2 The expected deflection curve.

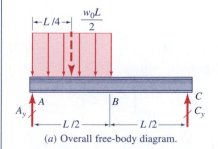

(a) Overall free-body diagram.

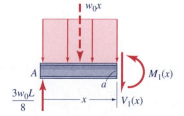

(b) Free-body diagram for load interval 1 – $(0 < x < L/2)$.

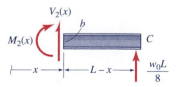

(c) Free-body diagram for load interval 2 – $(L/2 < x < L)$.

Fig. 3 Free-body diagrams.

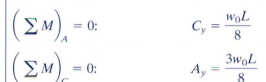

---

[3]See Footnote 2.

Write the Moment-Curvature Equations: There is a moment-curvature equation (Eq. 7.8) for each interval.

$$EIv_1'' = M_1(x) = \frac{3w_0Lx}{8} - \frac{w_0x^2}{2} \tag{2a}$$

$$EIv_2'' = M_2(x) = \frac{w_0L}{8}(L - x) \tag{2b}$$

Integrate the Differential Equations

$$EIv_1' = \frac{3w_0Lx^2}{16} - \frac{w_0x^3}{6} + C_1 \tag{3a}$$

$$EIv_2' = -\frac{w_0L}{16}(L - x)^2 + C_2 \tag{3b}$$

$$EIv_1 = \frac{w_0Lx^3}{16} - \frac{w_0x^4}{24} + C_1x + D_1 \tag{4a}$$

$$EIv_2 = \frac{w_0L}{48}(L - x)^3 + C_2x + D_2 \tag{4b}$$

Identify the Boundary Conditions and Continuity Conditions: For the simply supported beam in Fig. 1, the deflection is zero at the two ends, *A* and *C*. From Table 7.1,

$$v_1(0) = 0, \quad v_2(L) = 0 \tag{5a,b}$$

The displacement and slope are continuous at *B*, where the loading is discontinuous, since the beam is not broken there. Therefore, from Table 7.2,

$$v_1'\left(\frac{L}{2}\right) = v_2'\left(\frac{L}{2}\right), \quad v_1\left(\frac{L}{2}\right) = v_2\left(\frac{L}{2}\right) \tag{5c,d}$$

Evaluate the Constants of Integration: Using the source equations indicated in brackets, we get:

[4a, 5a]
$$EIv_1(0) = D_1 = 0 \tag{6a}$$

[4b, 5b]
$$EIv_2(L) = C_2L + D_2 = 0 \tag{6b}$$

[3a, 3b, 5c]
$$\frac{3w_0L}{16}\left(\frac{L}{2}\right)^2 - \frac{w_0}{6}\left(\frac{L}{2}\right)^3 + C_1 = -\frac{w_0L}{16}\left(\frac{L}{2}\right)^2 + C_2 \tag{6c}$$

[4a, 4b, 5d, 6a]
$$\frac{w_0L}{16}\left(\frac{L}{2}\right)^3 - \frac{w_0}{24}\left(\frac{L}{2}\right)^4 + C_1\left(\frac{L}{2}\right)$$

$$= \frac{w_0L}{48}\left(\frac{L}{2}\right)^3 + C_2\left(\frac{L}{2}\right) + D_2 \tag{6d}$$

Combining Eqs. (6b), (6c), and (6d), we get

$$C_1 = -\frac{9}{384}(w_0 L^3) \tag{7a}$$

$$C_2 = \frac{7}{384}(w_0 L^3) \tag{7b}$$

$$D_2 = -\frac{7}{384}(w_0 L^4) \tag{7c}$$

Finally, inserting the constants of integration into Eqs. (3) and (4), we get the following expressions for the slope and deflection in load intervals 1 and 2:

$$v_1' = \frac{w_0 L^3}{384EI}\left[-9 + 72\left(\frac{x}{L}\right)^2 - 64\left(\frac{x}{L}\right)^3\right] \tag{8a}$$

$$v_2' = \frac{w_0 L^3}{384EI}\left[-17 + 48\left(\frac{x}{L}\right) - 24\left(\frac{x}{L}\right)^2\right] \tag{8b}$$

$$v_1 = \frac{w_0 L^4}{384EI}\left[-9\left(\frac{x}{L}\right) + 24\left(\frac{x}{L}\right)^3 - 16\left(\frac{x}{L}\right)^4\right] \tag{8c}$$

$$v_2 = \frac{w_0 L^4}{384EI}\left[1 - 17\left(\frac{x}{L}\right) + 24\left(\frac{x}{L}\right)^2 - 8\left(\frac{x}{L}\right)^3\right] \tag{8d}$$

Ans. (a)

(b) *Determine the maximum deflection of the beam.* As observed in Fig. 2, the maximum deflection occurs in the loaded interval, *AB*. Therefore, we let $v_1'(x_m) = 0$ to determine the location where the slope is zero. This gives the cubic equation

$$-64\left(\frac{x_m}{L}\right)^3 + 72\left(\frac{x_m}{L}\right)^2 - 9 = 0$$

The value of x_m can be found by trial and error or by some other root-finding method. It is

$$x_m = 0.460L$$

Then, the maximum deflection is

$$\delta_{max} = |v_1(0.460L)| = 6.56(10^{-3})\frac{w_0 L^4}{EI} \qquad \textbf{Ans. (b)}$$

Review the Solution The answers have been written in terms of a dimensionless length (x/L) so that all dimensional quantities can be collected into one coefficient term. The coefficient terms:

$$v' \sim \left(\frac{w_0 L^3}{EI}\right) \text{ and } v \sim \left(\frac{w_0 L^4}{EI}\right)$$

are dimensionally correct. As with many beam-deflection problems, the answers to this problem can be found in Appendix E. If they were not, we could check the value of $\delta_{\max}$ by comparing it with the midpoint deflection of a uniformly loaded beam. From Table E.2, that value is

$$|v(L/2)| = \frac{5}{384}\left(\frac{w_0 L^4}{EI}\right)$$

The value of $\delta_{\max}$ should be slightly greater than half of this, which it is.

EXAMPLE 7.4

Use the fourth-order integration method to determine expressions for slope and deflection in intervals AB (interval 1) and BC (interval 2) of the uniform cantilever beam in Fig. 1. Also, determine specific expressions for the slope and deflection at $x = L$.

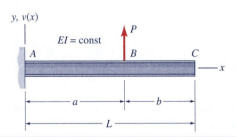

Fig. 1 A concentrated load on a cantilever beam.

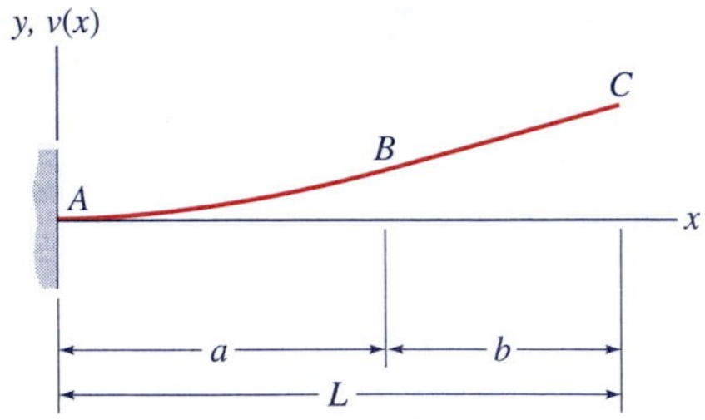

Solution

Sketch the Deflection Curve: As sketched in Fig. 2, the deflection will be upward along the entire length of the beam. Since there is no load to the right of point B, the beam will remain straight from B to C.

Determine $p_1(x)$ and $p_2(x)$: There are no distributed loads on this beam, so for interval 1 and interval 2, respectively,

$$p_1(x) = p_2(x) = 0 \tag{1a,b}$$

Write the Load-Deflection Equations: Combine Eqs. (1) with Eq. 7.12 to give

$$(EIv_1'')'' = 0 \tag{2a}$$

$$(EIv_2'')'' = 0 \tag{2b}$$

Integrate the Differential Equations: For interval 1, the four integrals of Eq. (2a) are:

$$V_1(x) = (EIv_1'')' = C_1 \tag{3a}$$

Fig. 2 The expected deflection curve.

$$M_1(x) = EIv_1'' = C_1x + D_1 \tag{3b}$$

$$EIv_1' = C_1\left(\frac{x^2}{2}\right) + D_1x + E_1 \tag{3c}$$

$$EIv_1 = C_1\left(\frac{x^3}{6}\right) + D_1\left(\frac{x^2}{2}\right) + E_1x + F_1 \tag{3d}$$

For interval 2, the integrals of Eq. (2b) are:

$$V_2(x) = (EIv_2'')' = C_2 \tag{4a}$$

$$M_2(x) = EIv_2'' = C_2x + D_2 \tag{4b}$$

$$EIv_2' = C_2\left(\frac{x^2}{2}\right) + D_2x + E_2 \tag{4c}$$

$$EIv_2 = C_2\left(\frac{x^3}{6}\right) + D_2\left(\frac{x^2}{2}\right) + E_2x + F_2 \tag{4d}$$

Identify the Boundary Conditions and Continuity Conditions: Since there are eight constants of integration to be evaluated, we need a total of eight boundary and/or continuity conditions. The deflection and slope vanish at $x = 0$ and the moment and shear vanish at $x = L$. Therefore,

$$v_1(0) = v_1'(0) = 0 \tag{5a,b}$$

$$V_2(L) = M_2(L) = 0 \tag{6a,b}$$

There is slope continuity and displacement continuity at $x = a$, so

$$v_1'(a) = v_2'(a) \tag{7a}$$

$$v_1(a) = v_2(a) \tag{7b}$$

The final two conditions are the shear and moment conditions that result from equilibrium of the "joint" at B. (They are also listed in Table 7.2.) From the free-body diagram in Fig. 3,

$$V_2(a) - V_1(a) = P \tag{8a}$$

$$M_2(a) = M_1(a) \tag{8b}$$

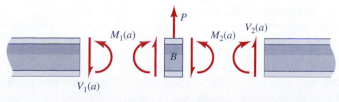

Fig. 3   Equilibrium of the node at B.

481

Evaluate the Constants of Integration: Combining Eqs. (3) through (8) we get the following constants of integration:

$$C_1 = -P, \quad D_1 = Pa, \quad E_1 = F_1 = 0 \qquad (9\text{a--d})$$

$$C_2 = D_2 = 0, \quad E_2 = \frac{Pa^2}{2}, \quad F_2 = -\frac{Pa^3}{6} \qquad (10\text{a--d})$$

Finally, the above constants can be inserted into Eqs. (3c,d) and (4c,d) to give the following slope and deflection equations:

$$v_1'(x) = \frac{Pa^2}{2EI}\left[-\left(\frac{x}{a}\right)^2 + 2\left(\frac{x}{a}\right)\right] \qquad (10\text{a})$$

$$v_1(x) = \frac{Pa^3}{6EI}\left[-\left(\frac{x}{a}\right)^3 + 3\left(\frac{x}{a}\right)^2\right] \qquad (10\text{b})$$

$$v_2'(x) = \frac{Pa^2}{2EI} \qquad (10\text{c})$$

$$v_2(x) = \frac{Pa^3}{6EI}\left[3\left(\frac{x}{a}\right) - 1\right] \qquad (10\text{d})$$

Ans.

Determine the Slope and Deflection at the Tip of the Beam: Evaluating Eqs. (10c) and (10d) at $x = L$, we get

$$v_2'(L) = \frac{Pa^2}{2EI} \qquad (11\text{a})$$

$$v_2(L) = \frac{Pa^2}{6EI}\left(\frac{3L}{a} - 1\right) \qquad (11\text{b})$$

Ans.

Review the Solution This example problem could have been solved easily by the second-order integration method that was illustrated in Example Problems 7.1 and 7.3. Note that **no overall free-body diagrams are needed for a fourth-order solution.** The boundary conditions and continuity conditions for fourth-order solutions may be obtained from local "joint" free-body diagrams, as in Fig. 3. Table 7.1 lists various types of boundary conditions; continuity conditions are listed in Table 7.2.

The above results in Eqs. (11a) and (11b) can be checked by consulting #4 in Table E.1 of Appendix E. These equations will be utilized in Section 7.6 to obtain expressions for the deflection of beams with general distributed loading.

Next, we will use the second-order method and the fourth-order method to solve slope and deflection problems for statically indeterminate beams.

7.4 SLOPE AND DEFLECTION BY INTEGRATION—STATICALLY INDETERMINATE BEAMS

In Section 7.3 we solved for the slope and deflection of statically determinate beams. In this section we will extend the solution procedures of Section 7.3 to solve for the slope and deflection of **statically indeterminate beams.**

For example, compare the simply supported beam in Example Problem 7.3 with the beam in Fig. 7.10. Since the simply supported beam in Example Problem 7.3 is statically determinate, it is possible to determine all reactions using equilibrium equations alone. In Fig. 7.10 there are four reactions, but only three independent equations of equilibrium can be written, so this beam is **statically indeterminate.** The addition of the **rotation constraint** at A gives rise to a **redundant moment reaction,** M_A. Alternatively, we could consider the beam in Fig. 7.10 to be a statically determinate cantilever beam to which an additional support has been added at end B, making the beam statically indeterminate and leading to the name *propped cantilever beam.*

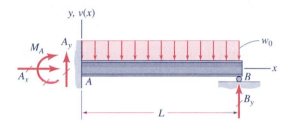

FIGURE 7.10 A statically indeterminate beam.

Each additional constraint, beyond those needed to prevent collapse of a beam, gives rise in an additional redundant reaction, whose value can only be determined by considering the deflection of the entire beam. For example, the redundant reaction M_A in Fig. 7.10 results from the addition of the constraint equation $v'(0) = 0$. That is, in order to maintain a slope of zero at end A of this specific beam with this specific loading, a specific moment M_A is required.

In the example problems that follow, it will be demonstrated that **each redundant constraint, which is expressed by an auxiliary boundary condition or continuity condition, leads to a corresponding redundant reaction.** Thus, for statically indeterminate beams there is a slight modification of the procedure given at the beginning of Section 7.3:

- In Steps 2 and 3, redundant reactions must be included on the free-body diagram(s) and in the equation(s) of equilibrium.
- In Step 5, the redundant constraint conditions must be included along with the other boundary conditions and continuity conditions.
- In Step 6, along with the constants of integration, redundant reactions are evaluated by enforcing boundary conditions and continuity conditions, including the redundant constraint conditions.

EXAMPLE 7.5

A couple, M_B, is applied to a uniform, propped cantilever beam, as shown in Fig. 1. Use the second-order integration method: (a) to determine

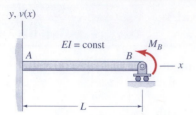

Fig. 1 A uniform, propped-cantilever beam with specified end couple.

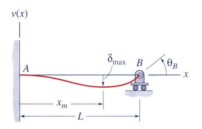

Fig. 2 The expected deflection curve.

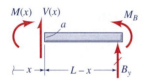

Fig. 3 A free-body diagram.

expressions for the slope and deflection of the beam; and (b) to determine the reactions at A and B. (c) Sketch the shear-force and bending-moment diagrams for the beam.

Solution (a) *Determine expressions for the slope and deflection of the beam.*

Sketch the Deflection Curve: The couple M_B will tend to rotate end B counterclockwise. Therefore, the beam should take the shape indicated in Fig. 2.

Equilibrium—Determine an Expression for $M(x)$: If we take the free-body diagram shown in Fig. 3, we will get an expression for $M(x)$ that includes the given end couple (moment) and the unknown reaction at B. We can let this reaction at B, which arises due to the added constraint at B, be the unknown *redundant*.

$$\left(\sum M \right)_a = 0: \qquad M(x) = M_B + B_y(L - x) \tag{1}$$

Write the Moment-Curvature Equation: From Eq. 7.8 and Eq. (1),

$$EIv'' = M(x) = M_B + B_y(L - x) \tag{2}$$

Integrate the Differential Equation: The two integrations of Eq. (2) give

$$EIv' = M_Bx + B_yLx - B_y\left(\frac{x^2}{2}\right) + C_1 \tag{3a}$$

$$EIv = M_B\left(\frac{x^2}{2}\right) + B_yL\left(\frac{x^2}{2}\right) - B_y\left(\frac{x^3}{6}\right) + C_1x + C_2 \tag{3b}$$

Identify the Boundary Conditions: Since we have three unknowns in Eqs. (3), two constants of integration plus the unknown redundant reaction force B_y, we need a total of three boundary conditions. They are (Table 7.1):

$$v'(0) = v(0) = v(L) = 0 \tag{4a,b,c}$$

Evaluate the Constants: The source equations are identified in brackets.

[4a, 3a] $$EIv'(0) = C_1 = 0 \tag{5a}$$

[4b, 3b, 5a] $$EIv(0) = C_2 = 0 \tag{5b}$$

[4c, 3b, 5a, 5b] $$EIv(L) = M_B\left(\frac{L^2}{2}\right) + B_y\left(\frac{L^3}{3}\right) = 0$$

or $$B_y = -\frac{3}{2}\frac{M_B}{L} \tag{5c}$$

Combining Eqs. (5) with Eqs. (3), we get the following expressions for the slope and deflection of this propped-cantilever beam.

$$v'(x) = \frac{M_B L}{4EI}\left[3\left(\frac{x}{L}\right)^2 - 2\left(\frac{x}{L}\right) \right]$$ (6a)

Ans (a)

$$v(x) = \frac{M_B L^2}{4EI}\left[\left(\frac{x}{L}\right)^3 - \left(\frac{x}{L}\right)^2 \right]$$ (6b)

(b) *Determine the reactions at A and B.* We found the redundant reaction in Eq. (5c). To determine the reactions at $x = 0$, we can use a free-body diagram of the entire beam (Fig. 4).

Equilibrium:

$$\sum F_y = 0: \qquad A_y = \frac{3}{2}\frac{M_B}{L}$$ (7a)

$$\left(\sum M\right)_A = 0: \qquad M_A = M_B - \frac{3}{2}\left(\frac{M_B}{L}\right)L = -\frac{M_B}{2}$$ (7b)

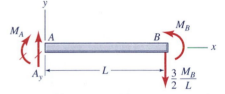

Fig. 4 A free-body diagram.

In summary,

$$A_y = \frac{3}{2}\frac{M_B}{L}, \quad M_A = -\frac{M_B}{2}$$ **Ans. (b)**

(c) *Sketch V and M diagrams.* Since $p(x) = 0$, $V(x)$ is constant (Eq. 5.2); and since $V(x)$ is constant, the slope of the moment diagram is constant (Eq. 5.3). Note that the sign of the moment in Fig. 5 is in agreement with the curvature of the beam as sketched in Fig. 2, that is, the moment-curvature equation, $M = EIv''$, is satisfied.

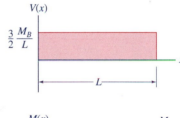

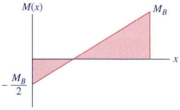

Review the Solution   The deflection and slope expressions in Eqs. (6) have the proper dimensions. From Eq. (6a), the slope at $x = L$ is positive, which is correct. The couple M_B would lift end B up, were it not for the constraint. Therefore, B_y has the correct sign. Finally, M_A has the correct sign according to our preliminary sketch in Fig. 2.

Fig. 5 Shear-force and bending-moment diagrams.

The above solution of Example Problem 7.5 differs little from the second-order solution of statically determinate problems in Section 7.3. The presence of one constraint (boundary condition) beyond the minimum required to prevent collapse of the beam led to an additional unknown constant to be evaluated. **To every additional degree of redundancy there is a corresponding additional boundary condition to use in evaluating the unknowns**.

The next example problem illustrates how to use the fourth-order integration method to solve a statically indeterminate problem of a beam with a distributed load. The beam is clamped at both ends, leading to a **fixed-end solution.** Fixed-end solutions like this enter into the superposition solutions discussed in Section 7.7; several are tabulated in Table E.3 in Appendix E.

EXAMPLE 7.6

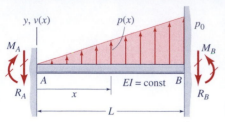

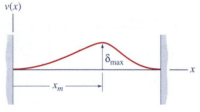

The uniform, linearly elastic beam in Fig. 1 supports a triangularly distributed load. Use the fourth-order integration method: (a) to determine expressions for the shear, the bending moment, the slope, and the deflection of the beam; and (b) to determine the reactions at A and B. (c) Sketch shear-force and bending-moment diagrams.

Fig. 1 A fixed-fixed beam with triangular load.

Fig. 2 The expected deflection curve.

Solution (a) *Determine expressions for the shear, the bending moment, the slope, and the deflection of the beam.*

Sketch the Deflection Curve: The beam will deflect upward, as sketched in Fig. 2.

Determine an Expression for $p(x)$: From similar triangles in Fig. 1, we get

$$\frac{p(x)}{x} = \frac{p_0}{L}$$

or

$$p(x) = p_0\left(\frac{x}{L}\right) \tag{1}$$

Write the Load-Deflection Equation: From Eq. 7.12 and Eq. (1), we get

$$(EIv'')'' = p(x) = p_0\left(\frac{x}{L}\right) \tag{2}$$

Integrate the Differential Equation: We need to integrate Eq. (2) four times, incorporating Eqs. 7.11 and 7.8.

$$(EIv'')' = V(x) = p_0\left(\frac{x^2}{2L}\right) + C_1 \tag{3a}$$

$$EIv'' = M(x) = p_0\left(\frac{x^3}{6L}\right) + C_1x + C_2 \tag{3b}$$

$$EIv' = p_0\left(\frac{x^4}{24L}\right) + C_1\left(\frac{x^2}{2}\right) + C_2x + C_3 \tag{3c}$$

$$EIv = p_0\left(\frac{x^5}{120L}\right) + C_1\left(\frac{x^3}{6}\right) + C_2\left(\frac{x^2}{2}\right) + C_3x + C_4 \tag{3d}$$

Identify the Boundary Conditions: The fixed-fixed boundary conditions are (Table 7.1):

$$v'(0) = v(0) = v'(L) = v(L) = 0 \tag{4a–d}$$

Evaluate the Constants: The source equations are in brackets.

[4a, 3c] $$C_3 = 0 \qquad (5a)$$

[4b, 3d, 5a] $$C_4 = 0 \qquad (5b)$$

[4c, 3c, 5a] $$12C_1L + 24C_2 = -p_0L^2$$

[4d, 3d, 5a, 5b] $$20C_1L + 60C_2 = -p_0L^2$$

From these last two equations we get

$$C_1 = -\frac{3}{20}p_0L \qquad (5c)$$

$$C_2 = \frac{1}{30}p_0L^2 \qquad (5d)$$

We can combine the constants in Eqs. (5) with the expressions in Eqs. (3) to obtain the answers to Part (a).

$$V(x) = \frac{p_0L}{20}\left[10\left(\frac{x}{L}\right)^2 - 3\right] \qquad (6a)$$

$$M(x) = \frac{p_0L^2}{60}\left[10\left(\frac{x}{L}\right)^3 - 9\left(\frac{x}{L}\right) + 2\right] \qquad (6b)$$

Ans. (a)

$$v'(x) = \frac{p_0L^3}{120EI}\left[5\left(\frac{x}{L}\right)^4 - 9\left(\frac{x}{L}\right)^2 + 4\left(\frac{x}{L}\right)\right] \qquad (7a)$$

$$v(x) = \frac{p_0L^4}{120EI}\left[\left(\frac{x}{L}\right)^5 - 3\left(\frac{x}{L}\right)^3 + 2\left(\frac{x}{L}\right)^2\right] \qquad (7b)$$

(b) *Determine the reactions at A and B.* Taking note of the relationship between $V(x)$ and $M(x)$ and the symbols used in Fig. 1, and using Eqs. (6a) and (6b), we get the following reactions:

$$R_A \equiv -V(0) = \frac{3}{20}p_0L \qquad (8a)$$

$$M_A \equiv M(0) = \frac{1}{30}p_0L^2 \qquad (8b)$$

Ans. (b)

$$R_B \equiv V(L) = \frac{7}{20}p_0L \qquad (8c)$$

$$M_B \equiv M(L) = \frac{1}{20}p_0L^2 \qquad (8d)$$

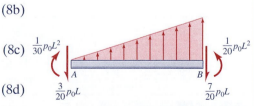

Fig. 3   The reactions.

By showing these reactions on the sketch in Fig. 3 we can see if they look reasonable. The reactions satisfy overall equilibrium, with the larger moment and shear at the more heavily loaded end, *B*. Therefore, our solution seems to be correct.

487

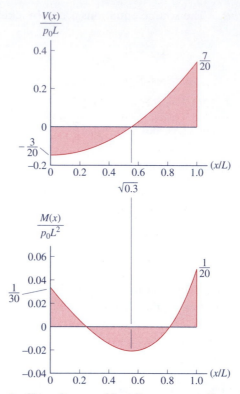

Fig. 4 Shear-force and bending-moment diagrams.

(c) *Sketch V and M diagrams.* Equations (6a) and (6b) can be used to plot the shear-force and bending-moment diagrams, respectively. Note that the sign of the moment $M(x)$ in Fig. 4 is in agreement with the curvature of the beam as sketched in Fig. 2, that is, the moment-curvature equation, $M = EIv''$, is satisfied. From Eq. (6a), the shear force is zero when $(x^*/L) = \sqrt{0.3} = 0.5477$. That is where the moment has its most negative value, $M(x^*/L) = -0.0214\, p_0 L^2$.

Review the Solution The reactions obtained in Part (b) were shown to be reasonable, and the moment diagram in Part (c) is consistent with the assumed deflection shape in Fig. 2. Therefore, the solution seems to be correct.

(See Homework Problems 7.6-37 and 7.6-38 for superposition solutions of this problem.)

*7.5 USE OF DISCONTINUITY FUNCTIONS TO DETERMINE BEAM DEFLECTIONS

In Section 7.2 the *moment-curvature differential equation,* Eq. 7.8, and the *load-deflection differential equation,* Eq. 7.12, were derived, and in Sections 7.3 and 7.4 they were integrated to determine the slope and deflection of beams. In this section, beam-deflection solutions will be based on the load-deflection equation

$$(EIv'')'' = p(x)$$

Load-Deflection (7.12)
Equation repeated

In Example Problems 7.3 and 7.4 you discovered that it is a very lengthy process to solve deflection problems when there are discontinuities in loading and/or support along the length of the beam. Such cases required the writing of an expression for the bending moment for each interval of the beam and the use of continuity conditions at points of discontinuity due to loading or supports. *Discontinuity functions* were introduced in Section 5.5 and were used there to obtain concise expressions for load, shear, and moment for beams. In this section we will illustrate how the use of discontinuity functions greatly simplifies the solution of multi-interval beam-deflection problems.[4] The procedure outlined below applies both to statically determinate beams and to statically indeterminate beams.

One question that arises is, Are constants of integration needed when expressions involving discontinuity functions are integrated? The answer is, Yes, constants of integration are needed to represent the shear at $x = 0$, $V(0)$; the moment at $x = 0$, $M(0)$; the slope of the beam at $x = 0$, $v'(0)$; and the displacement at $x = 0$, $v(0)$. When these constants of integration are introduced, they are introduced with the unit step function in the form $C\langle x \rangle^0$, as indicated in Eqs. 7.17.[5] In subsequent integrals, the Macaulay bracket is then integrated in the usual manner, that is, by using Eqs. 5.11.

$$V(x) = EI(v'')' = \int_{0^+}^{x} p(\xi)d\xi + V(0)\langle x \rangle^0$$

$$M(x) = EIv'' = \int_{0^+}^{x} V(\xi)d\xi + M(0)\langle x \rangle^0$$

$$v'(x) = \int_{0^+}^{x} \left(\frac{M(\xi)}{EI(\xi)} \right) d\xi + v'(0)\langle x \rangle^0$$

$$v(x) = \int_{0^+}^{x} v'(\xi)d\xi + v(0)\langle x \rangle^0$$

(7.17)

The shear-deflection equation, Eq. 7.11, has been incorporated in the first of Eqs. 7.17; the moment-curvature equation, Eq. 7.8, in the second equation. Note that the lower limit of the integrals in Eqs. 7.17 is $x = 0^+$, meaning that integration starts to the right of any load or reaction at the left end of the beam. The constants of integration in Eqs. 7.17 take care of loads and reactions at $x = 0$.

The following straightforward procedure may be used to solve slope/deflection problems for either statically determinate beams or statically indeterminate beams. Therefore, it is a very good method to use either for pencil-and-paper solutions or for computer solutions. However, the displacement-method procedure described in Section 7.7 leads to the more powerful and more generally applicable finite element method.

---

[4]This method of solving beam deflection problems is called *Clebsch's Method* after its developer, the German mathematician A. Clebsch (1833–1872) who proposed the method. An excellent review of the method and its applications may be found in "Clebsch's Method for Beam Deflection," by W. D. Pilkey, [Ref. 7-1].

[5]It is not essential to include the unit step $\langle x \rangle^0$ when introducing the constants of integration, but it is done here in order to maintain uniformity of notation among the various discontinuity terms being integrated.

1. Sketch the expected deflection curve. Draw a free-body diagram of the entire beam, labeling all reactions appropriately.
2. Using information from the *Load* column of Table 5.2, write a discontinuity-function expression for the load $p(x)$. Introduce <u>all</u> reactions as unknowns.
3. Integrate the load expression four times, using information from Table 5.2 or using Eqs. 5.11. Introduce a constant of integration in the form $C\langle x \rangle^0$ with each integral, as indicated in Eq. 7.17.

4. Identify the force-type (shear and moment) and displacement-type (displacement and slope) boundary conditions. Use these, together with the *shear and moment closure equations*

$$V(L^+) = 0, \quad M(L^+) = 0 \tag{7.18}$$

to evaluate all constants of integration.

As the following examples will illustrate, **the shear and moment closure equations automatically enforce overall equilibrium of the beam**. Nevertheless, a free-body diagram of the entire beam should be drawn so that unknown reactions can be labeled clearly.

EXAMPLE 7.7

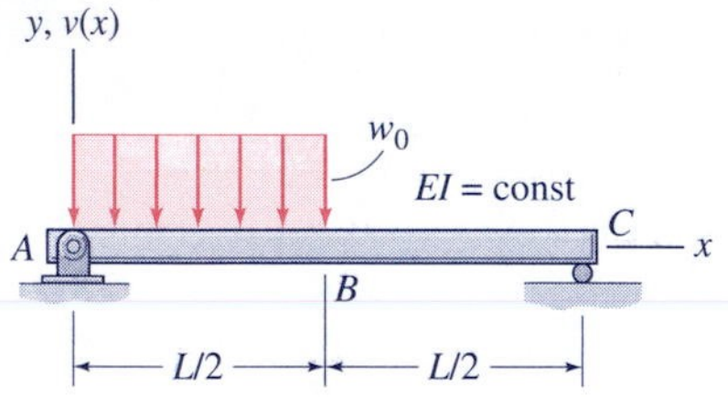

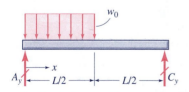

y, v(x)

Fig. 1

Fig. 2 A free-body diagram.

Use discontinuity functions to solve for the deflection of the beam in Example Problem 7.3 (see Fig. 1).

Solution We will follow the procedure outlined above. The sketch of the expected deflection curve may be found as Fig. 2 of Example Problem 7.3.

Free-body Diagram and Load Equation: This is a statically determinate problem, so we could immediately use equilibrium to solve for the reactions. However, these will be provided "automatically" by the shear and moment closure equations as we carry out the steps outlined above in the procedure. Using the discontinuity functions listed under Cases 2 and 3 in Table 5.2 together with the free-body diagram in Fig. 2, we can write the following discontinuity-function expression for $p(x)$.

$$p(x) = A_y\langle x \rangle^{-1} - w_0[\langle x \rangle^0 - \langle x - L/2 \rangle^0] + C_y\langle x - L \rangle^{-1} \tag{1}$$

Integration of the Discontinuity Equations: From the load-deflection equation, Eq. 7.12,

$$(EIv'')'' = p(x) \tag{2}$$

Therefore, we can combine Eqs. (1) and (2) and integrate once to get

$$V(x) = (EIv'')' = A_y\langle x \rangle^0 - w_0[\langle x \rangle^1 - \langle x - L/2 \rangle^1]$$
$$+ C_y\langle x - L \rangle^0 \tag{3a}$$

In this case there is an unknown reaction at $x = 0$, giving a term $A_y\langle x \rangle^0$, so we do not need an additional constant of integration $V(0)\langle x \rangle^0$. Continuing the integrations,

$$M(x) = EIv'' = A_y\langle x\rangle^1 - \frac{w_0}{2}[\langle x\rangle^2 - \langle x - L/2\rangle^2]$$
$$+ C_y\langle x - L\rangle^1 + M(0)\langle x\rangle^0 \tag{3b}$$

Since $M(0) = 0$, we can immediately eliminate the last term in Eq. (3b) and proceed to integrate two more times.

$$EIv' = \frac{A_y}{2}\langle x\rangle^2 - \frac{w_0}{6}[\langle x\rangle^3 - \langle x - L/2\rangle^3]$$
$$+ \frac{C_y}{2}\langle x - L\rangle^2 + EIv'(0)\langle x\rangle^0 \tag{3c}$$

$$EIv = \frac{A_y}{6}\langle x\rangle^3 - \frac{w_0}{24}[\langle x\rangle^4 - \langle x - L/2\rangle^4]$$
$$+ \frac{C_y}{6}\langle x - L\rangle^3 + EIv'(0)\langle x\rangle^1 + EIv(0)\langle x\rangle^0 \tag{3d}$$

Since $v(0) = 0$, we can immediately eliminate the last term in Eq. (3d).

Boundary Conditions and Closure Conditions: We have one remaining force-type boundary condition and one remaining displacement-type boundary condition. They are:

$$M(L) = 0, \qquad v(L) = 0 \tag{4a,b}$$

In addition, we have the shear closure equation

$$V(L^+) = 0 \tag{4c}$$

(We do not need to use $M(L^+) = 0$ since Eq. (4a) takes care of moment closure.)

Let us first apply Eq. (4c), the shear closure equation. Combining Eqs. (3a) and (4c), we get

$$V(L^+) = A_y - w_0(L/2) + C_y = 0 \tag{5a}$$

Note that this is just the equilibrium equation $\Sigma F_y = 0$. Next, let us satisfy Eq. (4a) using $M(x)$ from Eq. (3b). We just get the equilibrium equation $(\Sigma M)_C = 0$. That is,

$$M(L) = A_yL - \frac{w_0}{2}\left[L^2 - \left(\frac{L}{2}\right)^2\right] = 0$$

from which

$$A_y = \frac{3w_0L}{8} \tag{5b}$$

Combining Eqs. (5a) and (5b), we get

$$C_y = \frac{w_0 L}{8} \qquad (5c)$$

(Note that only force-type boundary conditions plus the shear and moment closure conditions are required in order for us to solve for all reactions, even though this is a two-interval beam.)

Equation (4b) can now be evaluated to obtain the constant $v'(0)$. Thus,

$$EIv(L) = \frac{w_0 L}{16} L^3 - \frac{w_0}{24}\left[L^4 - \left(\frac{L}{2}\right)^4\right] + EIv'(0)L = 0$$

so,

$$EIv'(0) = -\frac{9}{384} w_0 L^3 \qquad (5d)$$

Inserting Eqs. (5b), (5c), and (5d) into Eq. (3d), we get

$$EIv(x) = \frac{w_0 L}{16} \langle x \rangle^3 - \frac{w_0}{24}\left(\langle x \rangle^4 - \langle x - L/2 \rangle^4\right)$$
$$+ \frac{w_0 L}{48} \langle x - L \rangle^3 - \frac{3w_0 L^3}{128} \langle x \rangle^1 \qquad \textbf{Ans.} \quad (6)$$

The $\langle x - L \rangle$ term is, of course, zero throughout the length of the beam, so it could be dropped.

Review the Solution By evaluating the Macaulay brackets in Eq. (6) for interval 1 ($0 < x < L/2$) and for interval 2 ($L/2 < x < L$) we see that the resulting expressions are the same as the expressions obtained in Example Problem 7.3, namely

$$v_1 = \frac{w_0 L^4}{384 EI}\left[-9\left(\frac{x}{L}\right) + 24\left(\frac{x}{L}\right)^3 - 16\left(\frac{x}{L}\right)^4\right], \quad (0 \le x \le L/2)$$

$$v_2 = \frac{w_0 L^4}{384 EI}\left[1 - 17\left(\frac{x}{L}\right) + 24\left(\frac{x}{L}\right)^2 - 8\left(\frac{x}{L}\right)^3\right], \quad (L/2 \le x \le L)$$

MDS7.1 & 7.2   **Beam Deflections: Discontinuity-Function Method**

Let us now use discontinuity functions to solve a statically indeterminate, multi-interval deflection problem. Note how the discontinuity-function approach simplifies the solution.

EXAMPLE 7.8

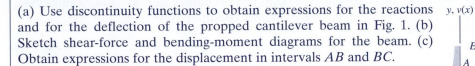

(a) Use discontinuity functions to obtain expressions for the reactions and for the deflection of the propped cantilever beam in Fig. 1. (b) Sketch shear-force and bending-moment diagrams for the beam. (c) Obtain expressions for the displacement in intervals AB and BC.

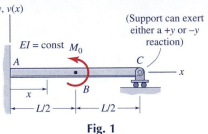

Fig. 1

Solution   We will follow the procedure outlined on page 490 for solving this type of problem.

(a) *Determine the reactions at A and C.*

Sketch the Deflection Curve: If end B were not restrained against vertical displacement, the moment M_0 would cause the beam to deflect upward all along its entire length. Therefore, we can expect a deflected shape *similar* to the curve sketched in Fig. 2. We need a moment diagram to tell just how the curvature changes in going from A to C, but the deflection curve in Fig. 2 should be *approximately* correct.

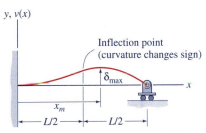

Fig. 2   The expected deflection curve.

Free-body Diagram and Load Equation: Utilizing Cases 1 and 2 of Table 5.2, and referring to the free-body diagram in Fig. 3, we can write the following discontinuity-function expression for the load $p(x)$, including the unknown reactions.

$$p(x) = M_A\langle x\rangle^{-2} + A_y\langle x\rangle^{-1} - M_0\langle x - L/2\rangle^{-2} + C_y\langle x - L\rangle^{-1} \quad (1)$$

Integration of Discontinuity Equations: Integrating Eq. (1), with the aid of Eq. 7.17a, we get

$$V(x) = (EIv'')' = M_A\langle x\rangle^{-1} + A_y\langle x\rangle^0 - \underline{M_0\langle x - L/2\rangle^{-1}}$$
$$+ C_y\langle x - L\rangle^0 \quad (2a)$$

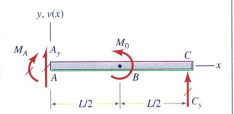

Fig. 3   A free-body diagram.

where the A_y term represents the total shear reaction at A and serves as the constant of integration. The moment terms in Eq. (2a) are underlined because they are singularity terms, not true shear forces. They are nonzero only at the locations where the couples M_A and M_0 are applied.

Continuing with the integrations (as in Eqs. 7.17),

$$M(x) = EIv'' = M_A\langle x\rangle^0 + A_y\langle x\rangle^1 - M_0\langle x - L/2\rangle^0$$
$$+ C_y\langle x - L\rangle^1 \quad (2b)$$

$$EIv' = M_A\langle x\rangle^1 + \frac{A_y}{2}\langle x\rangle^2 - M_0\langle x - L/2\rangle^1$$
$$+ \frac{C_y}{2}\langle x - L\rangle^2 + EIv'(0)\langle x\rangle^0 \quad (2c)$$

$$EIv = \frac{M_A}{2}\langle x\rangle^2 + \frac{A_y}{6}\langle x\rangle^3 - \frac{M_0}{2}\langle x - L/2\rangle^2$$
$$+ \frac{C_y}{6}\langle x - L\rangle^3 + EIv'(0)\langle x\rangle^1 + EIv(0)\langle x\rangle^0 \quad (2d)$$

Boundary Conditions and Closure Conditions: We have five unknowns—A_y, C_y, M_A, $v'(0)$, and $v(0)$—and we have five equations—three displacement-type boundary conditions and two closure equations. The displacement-type boundary conditions are:

$$v'(0) = 0, \quad v(0) = 0, \quad v(L) = 0 \qquad \text{(3a,b,c)}$$

and the closure equations are:

$$V(L^+) = 0, \qquad M(L^+) = 0 \qquad \text{(3d,e)}$$

The $v'(0)$ and $v(0)$ terms in Eqs. (2) are zero because of the fixed boundary at A (Eqs. (3a,b)). Since $\langle x - a \rangle^{-1} = 0$ except at $x = a$, Eq. (2a) can be combined with Eq. (3d) to give

$$V(L^+) = +A_y + C_y = 0 \qquad \text{(4a)}$$

which is just the force-equilibrium equation, $\sum F_y = 0$. From Eq. (2b) and the moment closure equation, Eq. (3e),

$$M(L^+) = M_A + A_y L - M_0 = 0 \qquad \text{(4b)}$$

which is just the moment-equilibrium equation, $(\sum M)_C = 0$. From Eqs. (2d) and (3a) through (3c),

$$EIv(L) = \frac{M_A}{2}L^2 + \frac{A_y}{6}L^3 - \frac{M_0}{2}(L/2)^2 = 0 \qquad \text{(4c)}$$

Finally, combining Eqs. (4a) through (4c) we get

$$A_y = \frac{9}{8}\frac{M_0}{L}, \quad C_y = -\frac{9}{8}\frac{M_0}{L}, \quad M_A = -\frac{1}{8}M_0 \quad \textbf{Ans. (a)} \quad \text{(5)}$$

In summary, the reactions are shown in Fig. 4.

(b) *Sketch V and M diagrams.*

Figures 5a,b show sketches of the shear diagram and the moment diagram, respectively.

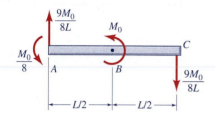

Fig. 4 The reactions to a couple M_0 at $x = L/2$.

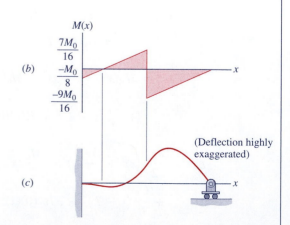

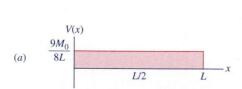

Fig. 5 (*a*) Shear diagram, (*b*) moment diagram, and (*c*) revised deflection curve.

494

(c) *Obtain expressions for $v_1(x)$ and $v_2(x)$.* The deflection equation in discontinuity-function form is obtained by substituting Eqs. (5) into Eq. (2d). Thus,

$$v = \frac{M_0}{EI}\left[-\frac{1}{16}\langle x\rangle^2 + \frac{3}{16L}\langle x\rangle^3 - \frac{1}{2}\langle x - L/2\rangle^2 \right.$$

$$\left. - \frac{3}{16L}\langle x - L\rangle^3\right] \tag{6}$$

Evaluating Eq. (6) for intervals AB, $(0 \le x \le L/2)$, and for BC, $(L/2 \le x \le L)$, we get

$$v_1(x) = \frac{M_0 L^2}{16EI}\left[3\left(\frac{x}{L}\right)^3 - \left(\frac{x}{L}\right)^2\right] \tag{7a}$$

$$v_2(x) = \frac{M_0 L^2}{16EI}\left[3\left(\frac{x}{L}\right)^3 - 9\left(\frac{x}{L}\right)^2 + 8\left(\frac{x}{L}\right) - 2\right] \tag{7b}$$

Ans. (c)

Review the Solution Obviously, the curvature of the beam near $x = 0$ in our "expected deflection" sketch in Fig. 2 does not agree with the negative value that we got for M_A. Either the sketch is not entirely correct, or our solution is not correct, or both! The revised deflection curve in Fig. 5c, which has been drawn with the aid of the moment diagram in Fig. 5b, $\left(M = \dfrac{EI}{\rho}\right)$, is certainly feasible, and is *similar* to our original estimate in Fig. 2, so our solution is probably correct.

7.6   SLOPE AND DEFLECTION OF BEAMS: SUPERPOSITION METHOD

Introduction.   In Sections 7.3 through 7.5 we solved for the slope and deflection of beams by integration, starting with the second-order moment-curvature equation (Eq. 7.8) or with the fourth-order load-deflection equation (Eq. 7.12). These equations are *linear differential equations,* since the deflection function $v(x)$ and its derivatives appear linearly, that is, only to the first power. Therefore, **the slope and deflection of a beam that simultaneously supports several different loads can be obtained by linear superposition, that is, by addition of the effects of the loads acting separately.** For example, the simply supported beam in Fig. 7.11a may be analyzed by adding the solutions for the two separate loads shown in Figs. 7.11b and 7.11c. Single-load solutions like these are available in Table E.2. Superposition holds for *all* quantities: reactions, internal shear and bending moment, slope, deflection, etc. Example Problems 7.9 and 7.10 illustrate this type of superposition applied to statically determinate beams.

Superposition can also be applied in another useful way, as illustrated by Fig. 7.12. The effect of the distributed load can be obtained by summing up the effects of differential loads dP. This may be referred to as **differential-load superposition.**

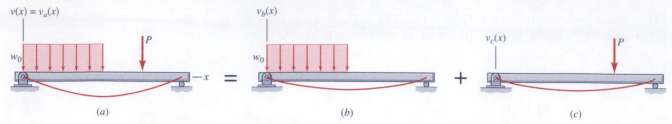

FIGURE 7.11 Superposition of two load cases: $v(x) \equiv v_a(x) = v_b(x) + v_c(x)$.

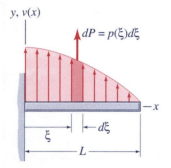

FIGURE 7.12 Superposition of differential loads.

In Fig. 7.12 we can say that a differential load $dP(\xi)$ at location ξ produces a corresponding differential displacement $dv(x, \xi)$ at every location x along the beam, and we can determine the total displacement $v(x)$ by summing up the effect of the entire distributed load by integrating (with respect to ξ) over the loaded portion of the beam. In Example Problem 7.4 we obtained expressions for the slope and deflection of a cantilever beam with a single concentrated load applied at an arbitrary location along the beam. The differential displacement $dv(x, \xi)$ depends on the magnitude and point of application of the differential load, $dP(\xi)$, and also on the point x where dv is being evaluated. Example Problem 7.11 illustrates this superposition procedure.

Finally, the *Method of Superposition* can be applied to statically indeterminate problems as well as to statically determinate problems. In this case there will be constraints on slope and/or displacement (boundary conditions and/or continuity conditions) that must be satisfied. These compatibility equations are used to solve for the unknown redundant reactions.[6] Example Problems 7.12 through 7.15 illustrate the use of the Method of Superposition to solve statically indeterminate beam problems.

Statically Determinate Beams. Example Problems 7.9 and 7.10 illustrate the linear superposition of known solutions to solve statically determinate beam–deflection problems. The following solution procedure is suggested.

SUPERPOSITION PROCEDURE—STATICALLY DETERMINATE BEAMS

1. Carefully study the boundary conditions and the loading given in the problem statement, and sketch the expected deflection curve of the beam.

2. Break the given problem down into *statically determinate subproblems*. The solution of each subproblem must be given in Table E.1 (Cantilever Beams) or in Table E.2 (Simply Supported Beams), or must be obtained directly by integration. Sketch the deflection curve of each of these subproblems.

3. Write *superposition equations* for any quantities that are required by the problem statement, for example, slope, deflection, etc., using subproblem information referred to in Step 2 to express slope and displacement in terms of force-type quantities.

4. Complete the solution (e.g., if requested, determine the maximum deflection).

---

[6]We can refer to the superposition procedure described here as a *Force-Method* procedure because the expressions that are used in the superposition process are displacements and slopes expressed in terms of force-type quantities (distributed forces, concentrated forces, and couples). A *Displacement-Method* solution procedure is presented in Section 7.7.

EXAMPLE 7.9

Determine the maximum deflection of the uniform, linearly elastic cantilever beam in Fig. 1.

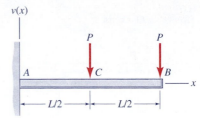

Fig. 1   A cantilever beam with two concentrated loads.

Solution

Sketch the Deflection Curve: Since both loads push the beam downward, the maximum deflection occurs at $x = L$, as indicated in the deflection diagram, Fig. 2.

Select the Subproblems: We only need to examine the cantilever beam candidate solutions in Table E.1, specifically E.1(3) and E.1(4).

Write the Superposition Equation: We only need an equation for the total deflection at B. Thus, from Fig. 3,

$$\delta_{\max} \equiv |v(L)| = \delta_b + \delta_c = |v_b(L)| + |v_c(L)| \qquad (1)$$

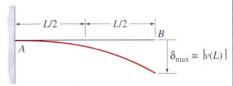

Fig. 2   The expected deflection curve.

where, from Table E.1(4)

$$\delta_b = \frac{5PL^3}{48EI} \qquad (2a)$$

and, from Table E.1(3)

$$\delta_c = \frac{PL^3}{3EI} \qquad (2b)$$

Complete the Solution: From Eqs. (1) and (2),

$$\delta_{\max} = \frac{21PL^3}{48EI} = 0.438\frac{PL^3}{EI} \qquad \textbf{Ans.} \quad (3)$$

Review the Solution   The deflection at the tip of a cantilever beam due to a concentrated tip load is $\delta = \dfrac{PL^3}{3EI}$, as given in Eq. (2b). This is a frequently used expression, and it is useful in estimating the accuracy of answers of many problems, including this one. For example, the deflection of the beam in Fig. 1 should be greater than the deflection of the same beam without the load at $x = L/2$, but less than the deflection would be with both loads at $x = L$. Equation (3) satisfies this inequality:

$$\frac{PL^3}{3EI} < \frac{21PL^3}{48EI} < \frac{2PL^3}{3EI}$$

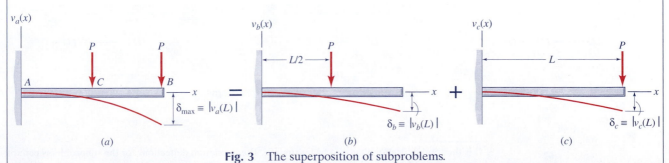

Fig. 3   The superposition of subproblems.

497

EXAMPLE 7.10

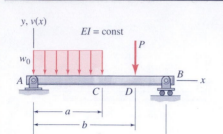

Fig. 1 A simply supported beam with two applied loads.

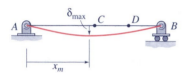

Fig. 2 The expected deflection curve.

A uniform simply supported beam similar to the one in Fig. 7.11 is subjected to a uniform distributed load and a concentrated load as shown in Fig. 1. Consider the particular case: $a = L/2$, $b = 3L/4$, $P = w_0L/2$. Use superposition of solutions from Table E.2 to solve for the maximum deflection of this beam.

Solution

Sketch the Deflection Curve: The beam will deflect downward throughout its entire length, as shown in Fig. 2. The loads in Fig. 1 divide the beam into three intervals: AC, CD, and DB. Certainly, the maximum deflection will not occur in interval DB. If P were zero, x_m would definitely fall in interval AC, but, if there were no distributed load on the beam, x_m would fall between C and D. Therefore, as a part of our solution, we will have to examine an expression for the slope at C to determine whether the beam slopes upward at C, as shown in Fig. 2, or whether it actually slopes downward at C.

Select the Subproblems: Obviously, we only need to examine solutions for simply supported beams. We will use the letter subscripts, like a and b, to denote the constituent subproblems, and we will use number subscripts, progressing from left to right, to denote the interval of validity. For example, $v_{b1}(x)$ denotes the deflection in interval 1 of the beam loaded as shown in Fig. 3b, and so on.

Write the Superposition Equations: Since we are to solve for the maximum deflection, we must first locate the point $x = x_m$ where the <u>total</u> slope v'_a vanishes, that is where $v'_a(x_m) = 0$.[7] From Table E.2(6) and E.2(4), respectively, we can write down the following superposition equations, taking proper account of the signs.

$$v'_{a1}(x) = v'_{b1}(x) + v'_{c1}(x)$$

$$= \frac{-w_0 L^3}{384EI}\left[64\left(\frac{x}{L}\right)^3 - 72\left(\frac{x}{L}\right)^2 + 9\right] \quad (1)$$

$$- \frac{PL^2}{384EI}\left[-48\left(\frac{x}{L}\right)^2 + 15\right]$$

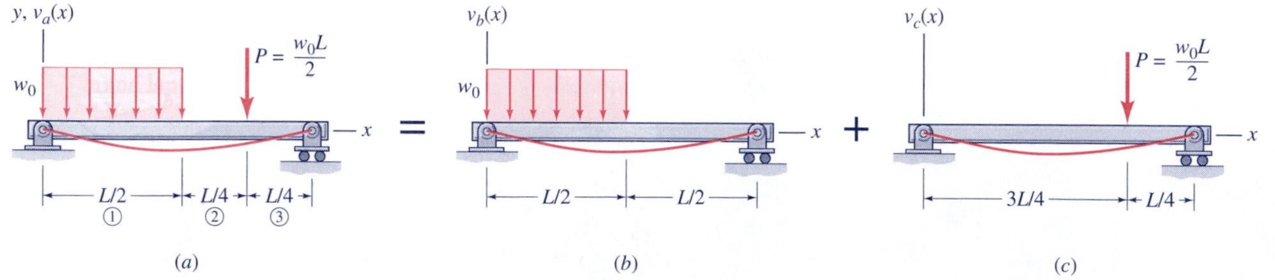

Fig. 3 Superposition of deflection subproblems.

---

[7]It is the total deflection that is to be maximized. The maximum deflections for the subproblems cannot be superimposed because, in general, they do not occur at the same beam location.

Setting $P = \dfrac{w_0 L}{2}$, we get

$$v'_{a1}(x) = \frac{-w_0 L^3}{768EI}\left[128\left(\frac{x}{L}\right)^3 - 192\left(\frac{x}{L}\right)^2 + 33\right] \qquad (2)$$

If $v'_{a1}(L/2) > 0$, we know that the point of zero slope lies in interval 1. Therefore, before writing other superposition equations, let us check the value of $v'_{a1}\left(\dfrac{L}{2}\right)$. Evaluating Eq. (2) at $x = \dfrac{L}{2}$, we obtain

$$v'_{a1}\left(\frac{L}{2}\right) = \frac{-w_0 L^3}{768EI}\left[128\left(\frac{1}{2}\right)^3 - 192\left(\frac{1}{2}\right)^2 + 33\right]$$

$$\qquad (3)$$

$$v'_{a1}\left(\frac{L}{2}\right) = \frac{-w_0 L^3}{768EI}$$

Since $v'_{a1}\left(\dfrac{L}{2}\right) < 0$, the beam must become horizontal in interval 2. So, we use information from Table E.2(4) and E.2(6) to write expressions for the slope and deflection in interval 2. The slope is given by

$$v'_{a2}(x) = v'_{b2}(x) + v'_{c2}(x) \qquad (4a)$$

$$= \frac{-w_0 L^3}{384EI}\left[24\left(\frac{x}{L}\right)^2 - 48\left(\frac{x}{L}\right)^2 + 17\right]$$

$$- \frac{PL^2}{384EI}\left[-48\left(\frac{x}{L}\right)^2 - 15\right]$$

The deflection in interval 2 is given by

$$v_{a2}(x) = v_{b2}(x) + v_{c2}(x) \qquad (4b)$$

$$= \frac{-w_0 L^4}{384EI}\left[8\left(\frac{x}{L}\right)^3 - 24\left(\frac{x}{L}\right)^2 + 17\left(\frac{x}{L}\right) - 1\right]$$

$$+ \frac{PL^3}{384EI}\left[16\left(\frac{x}{L}\right)^3 - 15\left(\frac{x}{L}\right)\right]$$

Setting $P = \dfrac{w_0 L}{2}$ in Eqs. (4), we get

$$v'_{a2}(x) = \frac{w_0 L^3}{768EI}\left[96\left(\frac{x}{L}\right) - 49\right] \qquad (5a)$$

$$v_{a2}(x) = \frac{w_0 L^4}{768EI}\left[48\left(\frac{x}{L}\right)^2 - 49\left(\frac{x}{L}\right) + 2\right] \qquad (5b)$$

Complete the Solution: We can now evaluate Eq. (5a) to determine the value of x_m at which $v'_{a2}(x_m) = 0$. The maximum deflection occurs at this point.

$$v'_{a2}(x_m) = 0 \rightarrow x_m = \frac{49}{96}L \qquad (6a)$$

Then, since x_m falls in interval 2,

$$\delta_{max} \equiv |v_{a2}(x_m)| = \frac{w_0 L^4}{768EI} \left| 48\left(\frac{49}{96}\right)^2 - 49\left(\frac{49}{96}\right) + 2 \right|$$

or

$$\delta_{max} = \frac{w_0 L^4}{768EI}(10.51) = 1.368(10^{-2})\frac{w_0 L^4}{EI} \qquad \textbf{Ans.} \quad (6b)$$

Review the Solution   Since we have a total load of $w_0 L$ on the beam, half of which is distributed, a good check on our answer in Eq. (6b) would be to determine the midspan deflection of a simply supported beam with a uniform distributed load over its total length. From Table E.2(5) we get

$$\delta_{max} = \frac{5w_0 L^4}{384EI} = 1.302(10^{-2})\frac{w_0 L^4}{EI}$$

Thus, our answer in Eq. (6b) appears to be correct, and we will, therefore, assume that we have made no "error" [except that we originally mis-guessed the location of the point of maximum deflection (Fig. 2)].

MDS7.3   **Beam Deflections: Superposition Method**

Next, let us illustrate the *differential-load superposition* approach introduced earlier in Fig. 7.12.

Differential-Load Superposition. As indicated earlier in the discussion of Fig. 7.12, we can superpose differential-load solutions if we have a solution for the desired quantity due to a concentrated load at an arbitrary position on the beam. The steps that may be used are as follows.

PROCEDURE FOR DIFFERENTIAL-LOAD SUPERPOSITION

1. Sketch the load diagram and the deflection curve.
2. In the tables of slopes and deflections of uniform beams, Table E.1 or Table E.2, identify the solution that will provide the desired quantity due to a concentrated force at an arbitrary location on the beam.
3. Form an expression for the distributed load as a function of position along the beam, that is, form an expression for $p(x)$.
4. Use an integral to sum up the effect of the differential load dP, as indicated in Fig. 7.12.

Example Problem 7.11 illustrates the above procedure.

EXAMPLE 7.11

Use the differential-load superposition approach to determine expressions for the deflection and slope at the tip of a uniform, linearly elastic cantilever beam with triangular load, as illustrated in Fig. 1.

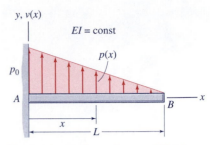

Fig. 1 A cantilever beam with triangular loading.

Solution

Sketch the Deflection Curve: The deflection curve is shown in Fig. 2.

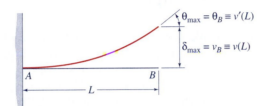

Fig. 2 The expected deflection curve.

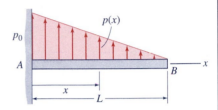

Fig. 3 Load diagram.

Form the Load Expression: Since we only need $v'(L)$ and $v(L)$ and not functions of position like $v'(x)$ and $v(x)$, we do not need to use a dummy variable, like the variable ξ (Greek xi) shown in Fig. 7.12. From similar triangles in Fig. 3, we get

$$\frac{p(x)}{L - x} = \frac{p_0}{L}$$

or

$$p(x) = p_0\left(1 - \frac{x}{L}\right) \tag{1}$$

Form the Deflection Expression and Integrate: We can use the results in Eqs. (10c) and (10d) of Example Problem 7.4. These are also given in Table E.1(4), and illustrated in Fig. 4.

$$dv'(L, x) = \frac{dP x^2}{2EI} \tag{2a}$$

$$dv(L, x) = \frac{dP}{6EI}(3Lx^2 - x^3) \tag{2b}$$

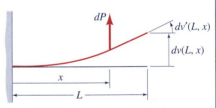

Fig. 4 Deflection due to differential load dP.

In Fig. 4 the increment of load is $dP = p(x)dx$, and since the load extends from $x = 0$ to $x = L$, we have the integrals

$$v'(L) = \frac{1}{2EI}\int_0^L x^2[p(x)dx] \tag{3a}$$

$$v(L) = \frac{1}{6EI}\int_0^L (3Lx^2 - x^3)[p(x)dx] \tag{3b}$$

501

or, combining Eqs. (1) and (3),

$$v'(L) = \frac{P_0}{2EIL} \int_0^L (x^2 L - x^3) dx \tag{4a}$$

$$v(L) = \frac{P_0}{6EIL} \int_0^L (3L^2 x^2 - 4L x^3 + x^4) dx \tag{4b}$$

Finally,

$$v'(L) = \frac{P_0 L^3}{24EI}, \qquad v(L) = \frac{P_0 L^4}{30EI} \qquad \textbf{Ans.} \quad \text{(5a, b)}$$

Review the Solution We can obviously check the above answer by referring to Table E.1(8). But, if the table did not have this particular triangular load, it would undoubtedly have the solution for a uniformly distributed load on a cantilever beam [Table E.1(5)]. We could expect the answer in Eq. (5b) to be somewhat less than half the deflection at the tip of a uniformly loaded cantilever beam, namely $\delta = \dfrac{p_0 L^4}{8EI}$. In fact, our answer is about one-fourth of this value, which seems reasonable.

Statically Indeterminate Beams. As noted earlier in this section, if a beam, or a system of beams, is statically indeterminate, there are more boundary conditions or other constraints than the minimum that is required to prevent collapse of the beam or beam system. For each additional *constraint,* there will be a *redundant force or moment* that can be determined by satisfying the constraint equation(s). The procedure for analyzing statically indeterminate beams by superposition is very similar to the procedure used in Section 3.5 to solve statically indeterminate axial–deformation problems and in Section 4.6 to solve statically indeterminate torsion problems. Consider the propped cantilever beam in Fig. 7.13a.

- The first step is to determine the **degree of statical indeterminacy** of the beam and select the redundant(s) to be used. That is, determine how many constraints must be released to make the beam statically determinate, and decide which specific constraints to release. In Fig. 7.13b the support at B has been removed to form the **released structure,** and in Fig. 7.13c the corresponding **redundant force** is the reaction that is labeled R_B.

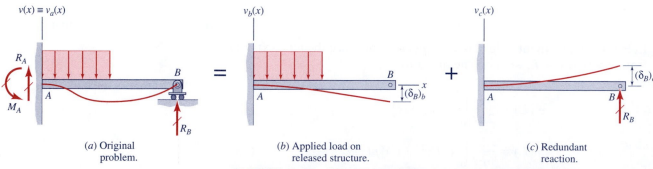

(a) Original problem.   (b) Applied load on released structure.   (c) Redundant reaction.

FIGURE 7.13   The superposition of cantilever-beam subproblems.

- The next step is to obtain the appropriate **force-deformation equations** for each statically determinate subproblem. This can be accomplished either by using one of the integration methods of Section 7.3 or 7.5, or by consulting Table E.1 (cantilever beam solutions) or Table E.2 (simply supported beam solutions). Each subproblem solution satisfies **equilibrium** and is based on **linearly elastic material behavior.** Therefore, any superposition of these subproblem solutions will also satisfy equilibrium and will incorporate linearly elastic behavior.

- Finally, the constraints that were released must be reinstated. This involves writing **compatibility equations in terms of the unknown redundants** and solving them for the redundants.

- Once the redundants have been determined, free-body diagrams and equilibrium equations can be used to determine other reactions. Other quantities that are required can then be obtained by superposition of the subproblem solutions.

The above steps are summarized in the following superposition procedure.[8]

SUPERPOSITION PROCEDURE—STATICALLY INDETERMINATE BEAMS

1. Carefully study the boundary conditions and the loading given in the problem statement, and sketch the expected deflection curve of the beam.

2. Determine the degree of statical indeterminacy, N_R, of the beam (or system), and select and label N_R *redundant forces and/or moments.*

3. Break the given problem down into *statically determinate subproblems,* one for each load on the beam and one for each of the selected redundants. The solution of each selected subproblem must be given in Table E.1 (Cantilever Beams) or in Table E.2 (Simply Supported Beams), or must be obtained directly by integration. Sketch the deflection curve of each of these subproblems.

4. Write *compatibility equations*, one equation for the deflection (slope) corresponding to each redundant force (moment). These express the boundary conditions and/or continuity conditions that are not automatically satisfied by the constituent subproblems.

5. Write *force-deformation equations* that relate the deflection (slope) at each redundant to each load and to each redundant force (moment).

6. Substitute these force-deformation equations into the compatibility equations, and solve for the unknown redundants.

7. Write *superposition equations* for any additional quantities that are required by the problem statement, for example, slope, deflection, etc., using information from the tables to express slope and displacement in terms of force-type quantities.

8. Complete the solution (e.g., if requested, determine the maximum deflection).

In many cases there are alternative ways to construct a superposition solution for a statically indeterminate problem. That is, there may be alternative ways to select the redundant(s) to be used, so long as the required solutions are available (e.g., in Appendix E). The next two example problems illustrate alternative ways to select the redundant and solve a statically indeterminate propped-cantilever-beam problem. In planning a superposition solution, you should mentally explore alternatives and try to solve the problem in the most efficient manner (i.e., Plan the Solution).

---

[8]This is a *Force-Method* procedure, because the primary unknowns are forces and/or moments.

EXAMPLE 7.12

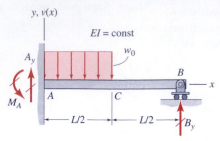

Fig. 1 A propped-cantilever beam.

Solve for the reactions on the beam shown in Fig. 1. Use information on cantilever beams from Table E.1.

Solution

Sketch the Deflection Curve: The expected deflection curve is shown in Fig. 2.

Select the Subproblems: We are to select subproblems from the cantilever-beam table, Table E.1. This means that B_y is to be considered the *redundant reaction*. The constituent subproblems are shown in Fig. 3.

Write the Superposition Equation: The reaction B_y is obtained by first writing a *compatibility equation* for deflection at support B.

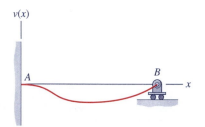

Fig. 2 The expected deflection curve.

$$v(L) = -(\delta_B)_b + (\delta_B)_c = 0 \qquad \textbf{Compatibility} \qquad (1)$$

Write the Subproblem Force-Deformation Equations: From entries E.1(6) and E.1(3) in Appendix E, we get the following equations for the deflection at B due to the distributed load, and the deflection at B due to the redundant reaction, respectively:

$$(\delta_B)_b = \frac{7w_0 L^4}{384EI}$$

$$(\delta_B)_c = \frac{B_y L^3}{3EI}$$

Subproblem Force-Deformation (2a, b)

Write the Compatibility Equation in Terms of Forces: We can eliminate the δ's by substituting Eqs. (2a,b) into Eq. (1), which gives the following **compatibility equation written in terms of forces:**

$$-\frac{7w_0 L^4}{384EI} + \frac{B_y L^3}{3EI} = 0$$

Compatibility in Terms of Forces (1′)

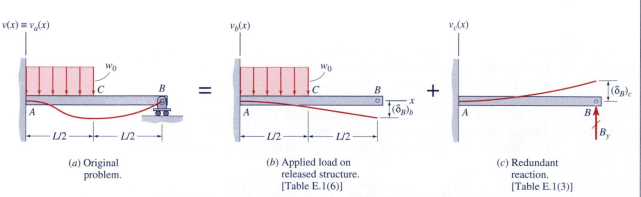

(*a*) Original problem.

(*b*) Applied load on released structure. [Table E.1(6)]

(*c*) Redundant reaction. [Table E.1(3)]

Fig. 3 The superposition of cantilever-beam subproblems.

Finally, we can solve Eq.(1′) for the redundant reaction B_y.

$$B_y = \left(\frac{7}{128}\right) w_0 L \qquad \textbf{Ans.} \quad (3)$$

Complete the Problem: To solve for the reactions at A we need the free-body diagram (Fig. 4) and equilibrium equations.

$$\sum F_y = 0: \qquad A_y = \frac{w_0 L}{2} - \frac{7 w_0 L}{128} = \left(\frac{57}{128}\right) w_0 L$$

$$\left(\sum M\right)_A = 0: \qquad M_A = -\frac{7 w_0 L^2}{128} + \frac{w_0 L^2}{8} = \left(\frac{9}{128}\right) w_0 L^2$$

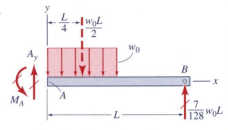

Fig. 4 A free-body diagram.

Therefore, the shear-force and bending-moment reactions at A are

$$A_y = \left(\frac{57}{128}\right) w_0 L, \qquad M_A = \left(\frac{9}{128}\right) w_0 L^2 \qquad \textbf{Ans.} \quad (4)$$

The reactions are shown in Fig. 5.

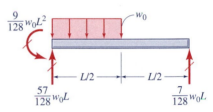

Review the Solution If the support at B were to be removed, the reactions at A would be $A_y = w_0 L/2$, $M_A = w_0 L^2/8$. The presence of a "prop" at B should reduce these values. The amount of reduction exhibited by our answers in Eqs. (4) seems reasonable.

Fig. 5 Summary of reactions on the propped-cantilever beam.

EXAMPLE 7.13

Solve for the moment at A on the propped-cantilever beam in Example Problem 7.12. This time, use information for simply supported beams from Table E.2.

Solution Figures 1 and 2 from Example Problem 7.12 will not be repeated here.

Select the Subproblems: If we select subproblems from the table of simply-supported-beam displacement functions, the moment M_A becomes the redundant reaction. The constituent subproblems are shown in Fig. 1.

Write the Superposition Equation: The reaction M_A is obtained by first writing a *compatibility equation* for the slope (rotation) at A.

$$v'(0) = -(\theta_A)_b + (\theta_A)_c = 0 \qquad \textbf{Compatibility} \quad (1)$$

Write the Subproblem Force-Deformation Equations: From entries E.2(6) and E.2(1) in Appendix E, we get the following equations for the slope at A due to the distributed load acting on the released beam, and

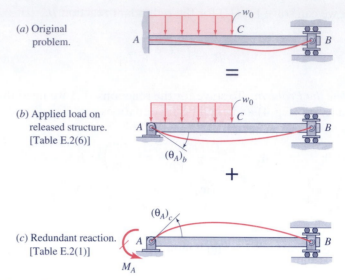

(a) Original problem.

=

(b) Applied load on released structure. [Table E.2(6)]

$(\theta_A)_b$

+

(c) Redundant reaction. [Table E.2(1)]

M_A

Fig. 1 The superposition of simply-supported-beam subproblems.

the slope at A due to the redundant moment reaction, respectively:

$$(\theta_A)_b = \frac{3w_0 L^3}{128EI}$$

$$(\theta_A)_c = \frac{M_A L}{3EI}$$

Subproblem Force- Deformation Behavior (2a,b)

(The word "force" means a force-type quantity, including a concentrated force, a distributed force, or a moment.)

Write the Compatibility Equation in Terms of Forces: We can eliminate the θ's by substituting Eqs. (2a,b) into Eq. (1), which gives the following **compatibility equation written in terms of forces:**

$$-\frac{3w_0 L^3}{128EI} + \frac{M_A L}{3EI} = 0$$

Compatibility in Terms of Forces (1′)

Finally, we can solve Eq. (1′) for the redundant moment reaction M_A.

$$M_A = \frac{9w_0 L^2}{128}$$

Ans.

Review the Solution This is the same expression for M_A that we got in Example Problem 7.12. Other reactions could be obtained from a free-body diagram like Fig. 4 in Example Problem 7.12.

The next example problem illustrates how to solve statically indeterminate beam problems that have more than one redundant.

EXAMPLE 7.14

A concentrated load P is applied to a uniform fixed-fixed beam, as illustrated in Fig. 1. Determine the reactions R_B and M_B.

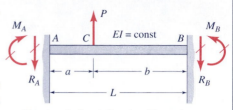

Fig. 1 A fixed-fixed uniform beam.

Solution

Sketch the Deflection Curve: The deflection will be upward, and the slope and deflection are zero at A and B, as indicated in the deflection diagram (Fig. 2).

Select the Subproblems: The results of Example Problem 7.4 [see also Table E.1(4)] can be used, along with the deflection and slope information of Example Problem 7.1 [see also Table E.1(3) and E.1(1)]. These statically determinate cantilever-beam subproblems are shown in Fig. 3.

Write the Superposition Equations: The displacement constraint and the slope constraint must be enforced at end B. The two relevant *compatibility equations* are

$$v_B \equiv v(L) = (\delta_B)_b - (\delta_B)_c + (\delta_B)_d = 0 \tag{1a}$$

$$v'_B \equiv v'(L) = (\theta_B)_b - (\theta_B)_c + (\theta_B)_d = 0 \tag{1b}$$

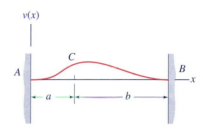

Fig. 2 The expected deflection curve (for $a < b$).

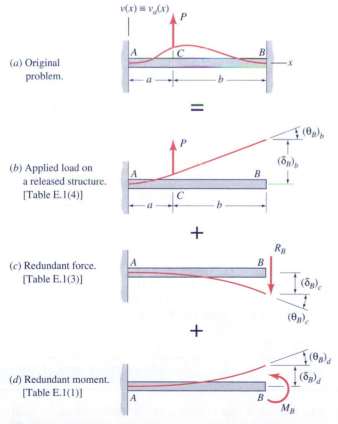

(a) Original problem.

=

(b) Applied load on a released structure. [Table E.1(4)]

+

(c) Redundant force. [Table E.1(3)]

+

(d) Redundant moment. [Table E.1(1)]

Fig. 3 Statically indeterminate superposition involving two redundant reactions.

With *subproblem force-deformation* information from Table E.1, these two compatibility equations become

$$\frac{Pa^3}{6EI}\left(\frac{3L}{a} - 1\right) - \frac{R_B L^3}{3EI} + \frac{M_B L^2}{2EI} = 0 \tag{2a}$$

$$\frac{Pa^2}{2EI} - \frac{R_B L^2}{2EI} + \frac{M_B L}{EI} = 0 \tag{2b}$$

Solving Eqs. (2) simultaneously, we get the following expressions for the two *redundant reactions:*

$$\left. \begin{array}{l} R_B = 3P\left(\dfrac{a}{L}\right)^2 - 2P\left(\dfrac{a}{L}\right)^3 \qquad (3a) \\[4mm] M_B = P\left(\dfrac{a}{L}\right)^2 (L - a) \qquad (3b) \end{array} \right\} \quad \textbf{Ans.}$$

Review the Solution   The above expressions for R_B and M_B have the proper dimensions, F and $F \cdot L$, respectively. If Eqs. (3) are evaluated for $a = 0$, we get $R_B = M_B = 0$, which makes sense. If we evaluate Eqs. (3) at $a = L$, we get $R_B = P$, $M_B = 0$, which also makes sense. Finally, if we evaluate Eq. (3a) at $a = L/2$, we get $R_B = P/2$, which is the correct answer. It appears that our solution is correct.

A similar problem, but for a nonuniform beam, is solved by the *Displacement Method* in Example Problem 7.16.

The statically indeterminate problems in Example Problems 7.12 through 7.14 had displacement and slope constraints of the type $v(L) = 0$, $v'(0) = 0$, and so on. That is, the displacement and slope were zero at certain points along the beam. Figure 7.14 illustrates two situations where the constraint is provided by another flexible structure. The next problem illustrates the solution of statically indeterminate problems of this type.

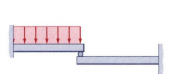

(*a*) The redundant support supplied by a second beam.

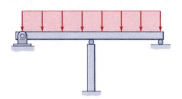

(*b*) The redundant support supplied by an axial-deformation member.

FIGURE 7.14   Beams with flexible supports.

> **EXAMPLE 7.15**

As shown in Fig. 1*a*, a steel beam, *AB*, is designed to be cantilevered from a rigid wall at *A* and supported by a steel hanger rod, *BC*, that is pinned to a rigid support at *C*. The beam is a **W**10 × 12, and the rod diameter is $d = \frac{5}{8}$ in. Use $E_{steel} = 30 \times 10^6$ psi. (a) If the rod is manufactured $\frac{1}{16}$ in. too short, how much stress will be induced in the rod by stretching it, inserting the pin at B, and then releasing the external forces required to mate the parts? (b) How much additional stress is induced in the rod by a uniformly distributed load of 100 lb/ft subsequently applied to the beam, as shown in Fig. 1b? What is the final displacement of B?

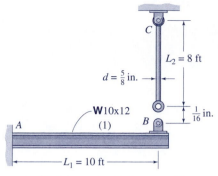

$d = \frac{5}{8}$ in.

W10x12
(1)

$L_2 = 8$ ft

$\frac{1}{16}$ in.

(a) The beam and rod before assembly.

Plan the Solution The rod BC is an axial-deflection member that can be treated as the members in Section 3.7 were treated. In Part (a) the beam will be loaded only by the force of the rod at B. This is a "misfit" problem similar to the ones in Section 3.7. For Part (b) we will have to add the distributed load by superposing another subproblem for the beam. We can follow the same steps that were used in Example 7.12, except for adding the rod at B. We should find that the rod is in tension in Part (a), and that the tension becomes greater in Part (b).

Solution (a) *Determine the initial stress induced in the rod when the rod-beam system is assembled.* Let us follow the same steps used previously in solving statically indeterminate problems by superposition (Force Method).

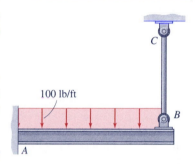

100 lb/ft

(b) The beam-rod system with uniform load.

Fig. 1 A beam-rod system.

Sketch the Deflection Curve: As illustrated in Fig. 2, the rod will pull the beam upward, and the beam will stretch the rod. Note: The short distance between B and the end of the beam is neglected, and the dimension v_B is treated as the deflection at the end of the beam.

Select the Subproblems: Here we can choose a cantilever beam subproblem from Table E.1(3), and we also have an axial-deformation (rod) subproblem. Let T_0 be the "initial force" in the rod, that is, the force in the rod in Fig. 3. Subscript 1 refers to the beam AB, and subscript 2 refers to the rod BC.

Write the Superposition Equation: For the rod and beam to be connected together by a pin at B, we have the following *deformation-compatibility equation*. (The terminology is defined in Figs. 2 and 3.)

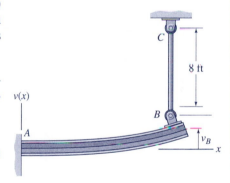

$v(x)$

8 ft

v_B

x

Fig. 2 The deflection curve for Part (a).

$$v_B \equiv v(L_1) = (\delta_B)_1 = \frac{1}{16} \text{ in.} - (\delta_B)_2 \qquad (1)$$

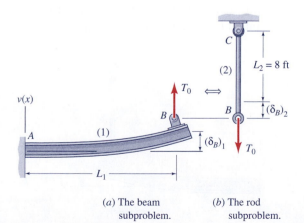

(a) The beam subproblem.

(b) The rod subproblem.

Fig. 3 The subproblems for Part (a).

509

From Table E.1(3) we get the following *force-deformation equation* for the released cantilever-beam subproblem:

$$(\delta_B)_1 = \left(\frac{L^3}{3EI}\right)_1 T_0 \tag{2a}$$

and, from Eq. 3.13, the force-deformation equation for the rod is

$$(\delta_B)_2 \equiv e_2 = \left(\frac{L}{AE}\right)_2 T_0 \tag{2b}$$

Complete Part (a): Combining Eqs. (1) and (2), we get the following *compatibility equation in terms of forces:*

$$\frac{T_0 L_1^3}{3E_1 I_1} + \frac{T_0 L_2}{A_2 E_2} = \frac{1}{16} \text{ in.} \tag{3a}$$

or

$$T_0 \left[\frac{(120 \text{ in.})^3}{3(30 \times 10^6 \text{ psi})(53.8 \text{ in}^4)} + \frac{96 \text{ in.}}{\pi(\frac{5}{16} \text{ in.})^2(30 \times 10^6 \text{ psi})}\right] = \frac{1}{16} \text{ in.} \tag{3b}$$

Therefore, the "initial force" in the rod is

$$T_0 = 170.2 \text{ lb} \tag{4}$$

so the "initial stress" in the rod will be

$$\sigma_0 = \frac{T_0}{A_2} = \frac{170.2 \text{ lb}}{\pi(\frac{5}{16} \text{ in.})^2} = 555 \text{ psi} \qquad \textbf{Ans. (a)} \tag{5}$$

(b) *Determine the effects of adding the distributed load.* We can follow the same steps used in Part (a), adding a distributed-load subproblem for the beam.

Sketch the Deflection Curve: At this point, we do not know whether the beam will still be deflected upward, as it was in Part (a), or whether the distributed load will cause the final deflection at B to be downward. In Fig. 4 we have assumed that the beam deflects upward.

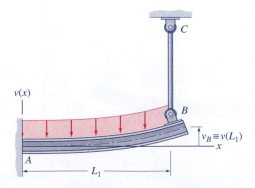

Fig. 4   The deflection curve for Part (b).

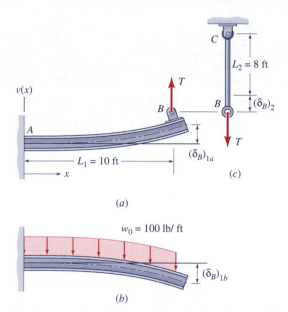

(a)

(b)

(c)

Fig. 5 The beam subproblems (a, b) and the rod subproblem (c).

Select the Subproblems: To the two subproblems in Fig. 3 we need to add a beam with distributed load. Let T (without subscript) be the tension in the rod in Part (b). The subproblems are shown in Fig. 5.

Write the Superposition Equations: Again, for the rod and the beam to be connected by a pin at B, we must satisfy the *compatibility equation*

$$v_B \equiv v(L_1) = (\delta_B)_{1a} - (\delta_B)_{1b} = \frac{1}{16} \text{ in.} - (\delta_B)_2 \tag{6}$$

From Table E.1(3) and E.1(5) we get

$$(\delta_B)_{1a} = \frac{TL_1^3}{3E_1I_1} \tag{7a}$$

$$(\delta_B)_{1b} = \frac{w_0L_1^4}{8E_1I_1} \tag{7b}$$

and from Eq. 3.13 we get

$$(\delta_B)_2 \equiv e_2 = \frac{TL_2}{A_2E_2} \tag{7c}$$

Complete Part (b): Combining Eqs. (6) and (7) we get the following *compatibility equation in terms of forces:*

$$\frac{TL_1^3}{3E_1I_1} - \frac{w_0L_1^4}{8E_1I_1} = \frac{1}{16} \text{ in.} - \frac{TL_2}{A_2E_2} \tag{8a}$$

511

or

$$T\left[\frac{(120 \text{ in.})^3}{3(30 \times 10^6 \text{ psi})(53.8 \text{ in}^4)} + \frac{96 \text{ in.}}{\pi(\frac{5}{16} \text{ in.})^2(30 \times 10^6 \text{ psi})}\right]$$

$$= \frac{1}{16} \text{ in.} + \left[\frac{(100 \text{ lb/bf})(10 \text{ ft})(120 \text{ in.})^3}{8(30 \times 10^6 \text{ psi})(53.8 \text{ in}^4)}\right] \qquad (8b)$$

Therefore,

$$T = 535 \text{ lb} \qquad (9)$$

so the stress in rod BC is

$$\sigma = \frac{T}{A_2} = \frac{535 \text{ lb}}{\pi(\frac{5}{16} \text{ in.})^2} = 1742 \text{ psi}$$

$$\sigma = 1742 \text{ psi} \qquad \textbf{Ans. (b)} \quad (10)$$

We can use Eq. (6) to evaluate the tip deflection.

$$v_B = \frac{1}{16} \text{in.} - (\delta_B)_2 = \frac{1}{16} \text{in.} - \frac{TL_2}{A_2 E_2} \qquad (11)$$

Therefore,

$$v_B = \frac{1}{16} \text{ in.} - \frac{(535 \text{ lb})(96 \text{ in.})}{\pi(\frac{5}{16} \text{ in.})^2(30 \times 10^6 \text{ psi})} = 0.0569 \text{ in.} \qquad \textbf{Ans.(b)}$$

Review the Solution If all of the distributed load (i.e., 1000 lb) were to be applied directly to the rod, it would elongate

$$e = \left(\frac{PL}{AE}\right)_2 = \frac{(1000 \text{ lb})(96 \text{ in.})}{\pi(\frac{5}{16} \text{ in.})^2(30 \times 10^6 \text{ psi})} = 0.01043 \text{ in.}$$

which is much less than 1/16-in. (= 0.0625 in.). Therefore, the rod is so stiff that it acts almost like a rigid support. By comparison, if the full 1000 lb were to be hung from the cantilever beam at B, the deflection of the tip of the beam would be

$$\delta = \frac{PL^3}{3EI} = \frac{(1000 \text{ lb})(120 \text{ in.})^3}{3(30 \times 10^6 \text{ psi})(53.8 \text{ in}^4)} = 0.357 \text{ in.}$$

Thus, we can see that this beam is very flexible in comparison with the rod. As we can see by examining the terms on the left-hand side of Eq. (3b), most of the 1/16-in. gap is closed by deflection of the beam, not by stretching of the rod.

*7.7 SLOPE AND DEFLECTION OF BEAMS: DISPLACEMENT METHOD

As beam-deflection problems get more difficult, solution by the *Method of Superposition,* as described in Section 7.6, may get quite lengthy and tedious, and a solution by the **Displacement Method** is preferable. Stress-analysis and structural-analysis computer programs are generally based on the Displacement Method.

In the Force-Method solutions of the previous section, as with Force-Method solutions in Chapters 3 and 4, *deflections (and slopes) were expressed in terms of forces,* and the deflection (and slope) expressions were superposed. In the case of statically indeterminate structures, displacement (and slope) constraint equations were solved to obtain the redundants. In this section you will discover that, when the Displacement Method is applied to beams, **nodal equilibrium equations are expressed in terms of displacements**, just as was done in Chapters 3 and 4 for axial-deformation and torsion problems, respectively.

Force-Deformation Equations for a Uniform Bernoulli-Euler Beam Element with Linearly Varying Distributed Load.
The first step in formulating a Displacement-Method solution procedure for transversely loaded beams is to establish the *force-deformation relations* for a single beam element (Fig. 7.15).[9]

Unlike the force-deformation equation for a uniform axial-deformation element, with its single axial force F (Eqs. 3.14 and 3.15), or the torque-twist equation for a uniform torsion element with its single torque T (Eqs. 4.20 and 4.21), the shear forces V_i and V_j acting at the ends of the beam element in Fig. 7.15 are not necessarily equal, and neither are the moments M_i and M_j. Therefore, it is first necessary to give distinct "names" to the two ends. We will use the names "i" and "j" to distinguish, respectively, the left end of the beam element and its right end, as indicated in Fig. 7.15a. The *sign convention* adopted for V_i, M_i, V_j, and M_j is that

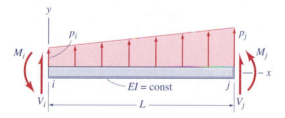

(a) The distributed load and the stress resultants V and M.

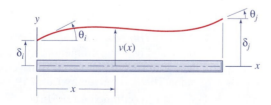

(b) The deflection curve $v(x)$.

FIGURE 7.15 A uniform beam element with transverse loading.

---

[9]This parallels Section 3.4 for the uniform axial-deformation element and Section 4.5 for a uniform element subjected to pure torsion.

positive stress resultants act in the same directions as the corresponding displacements δ_i, θ_i, δ_j, and θ_j,[10] Figures 7.15a and b illustrate these sign conventions.

Combining the force-deformation results in Eqs. 7.16 and the fixed-end reactions given in Eqs. (8) of Example 7.6, we can write the following **force-deformation relations** for the uniform Bernoulli-Euler beam element in Fig. 7.15:

$$V_i = \frac{EI}{L^3}(12\delta_i + 6L\theta_i - 12\delta_j + 6L\theta_j) - \frac{7p_iL}{20} - \frac{3p_jL}{20}$$

$$M_i = \frac{EI}{L^2}(6\delta_i + 4L\theta_i - 6\delta_j + 2L\theta_j) - \frac{p_iL^2}{20} - \frac{p_jL^2}{30}$$

$$V_j = \frac{EI}{L^3}(-12\delta_i - 6L\theta_i + 12\delta_j - 6L\theta_j) - \frac{3p_iL}{20} - \frac{7p_jL}{20}$$

$$M_j = \frac{EI}{L^2}(6\delta_i + 2L\theta_i - 6\delta_j + 4L\theta_j) + \frac{p_iL^2}{30} + \frac{p_jL^2}{20}$$

Force-Deformation Equation with Fixed-End Forces (7.19)

The δ and θ terms in Eqs. 7.19 form the *force-deformation relations* of a uniform beam with only nodal loads, that is, with no distributed loading.[11] The p_i and p_j terms in Eqs. 7.19 are called *fixed-end forces*. As in Example 7.6, these are the shear and moment reactions of a uniform beam whose ends are fixed (i.e., $v(0) = v'(0) = v(L) = v'(L) = 0$).

The deflection curve $v(x)$ along the beam can be expressed in terms of the end displacements (δ_i and δ_j), the end slopes (θ_i and θ_j), and the distributed-load amplitudes (p_i and p_j) by the following equation:[12]

$$v(x) = \delta_i\left[1 - 3\left(\frac{x}{L}\right)^2 + 2\left(\frac{x}{L}\right)^3\right] + L\theta_i\left[\left(\frac{x}{L}\right)^2 - 2\left(\frac{x}{L}\right)^2 + \left(\frac{x}{L}\right)^3\right]$$

$$+ \delta_j\left[3\left(\frac{x}{L}\right)^2 - 2\left(\frac{x}{L}\right)^3\right] + L\theta_j\left[-\left(\frac{x}{L}\right)^2 + \left(\frac{x}{L}\right)^3\right]$$

$$+ \frac{p_iL^4}{120EI}\left[2\left(1 - \frac{x}{L}\right)^2 - 3\left(1 - \frac{x}{L}\right)^3 + \left(1 - \frac{x}{L}\right)^5\right]$$

$$+ \frac{p_jL^4}{120EI}\left[2\left(\frac{x}{L}\right)^2 - 3\left(\frac{x}{L}\right)^3 + \left(\frac{x}{L}\right)^5\right]$$

(7.20)

SUPERPOSITION PROCEDURE—DISPLACEMENT METHOD

1. Sketch the expected deflection curve.

2. Consider the beam to consist of uniform *elements* connected together by *nodes*, or joints. Sketch each element and each node, and label the forces and couples that act on each.

3. *Element Force-Deformation Relations:* Express the shear force and bending moment at the element-to-node interface in terms of the nodal transverse displacement and slope. These are given by Eqs. 7.19.

(continued on p. 515)

---

[10]The sign conventions for V_j and M_i are opposite to the shear and moment sign conventions in Chapter 6. That is, $V(L) = -V_j$, and $M(0) = -M_i$. For small displacements it is permissible to equate the slope $v'(x)$ with the rotation angle θ. Thus, $\theta_i = v'(0)$ and $\theta_j = v'(L)$.

[11]See Homework Problems 7.7-16 and 7.7-17.

[12]See Homework Problem 7.7-17 for a derivation of the δ_i term in Eq. 7.20.

4. *Fixed-End Forces:* If there are distributed loads on an element, include the appropriate *fixed-end force* terms from Eqs. 7.19. (Additional fixed-end forces are given in Table E.3.)

5. *Nodal Equilibrium:* Write force and moment equilibrium equations for each node. Include the external concentrated force and/or couple (if either acts on the node) and the shear and moment from the adjoining beam elements (from Steps 3 and 4).

6. *Geometry of Deformation; Compatibility:* Where an element attaches to a node, express all element end-displacement and slope quantities in terms of the transverse displacement and slope of the node itself.

7. Solve the equilibrium equations for the unknown nodal displacements and slopes.

8. Solve for any other required quantities (reactions, deflection-curve formula, etc).

EXAMPLE 7.16

A fixed-fixed beam has a stepped cross section, as indicated in Fig. 1. A concentrated load P is applied at the point where the cross-sectional properties change from $(EI)_1$ to $(EI)_2$. Using the *Displacement Method,* (a) determine the displacement and slope at node C, and (b) determine the reactions R_B and M_B.

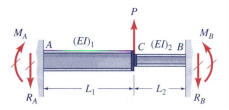

Fig. 1 A stepped, fixed-fixed beam.

Solution (a) *Determine the displacement and slope at node C.*

Sketch the Deflection Curve: The beam will obviously deflect upward from end A to end B. The point of maximum deflection may be to the left of C or to the right of C, depending on the EI values and the L values. In Fig. 2 we assume that the maximum deflection occurs to the right of C, so $\theta_C > 0$.

Separate the Nodes and Elements: Since the key step in a Displacement-Method solution is writing nodal equilibrium equations in terms of the displacement and slope at each node and the external loads applied to the node, let us treat the beam as two elements, (1) and (2), connected together by a node at C, as indicated in Fig. 3.

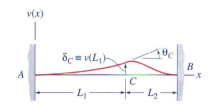

Fig. 2 The expected deflection curve.

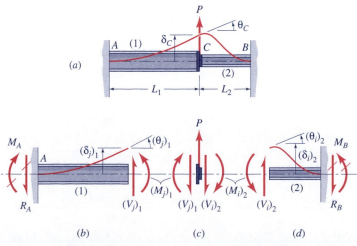

Fig. 3 The elements and connecting node.

Next we directly apply the three fundamentals of deformable-body mechanics—*equilibrium, force-deformation behavior, and geometry of deformation.*

Nodal Equilibrium: Write force and moment equilibrium equations for the node at C, where the transverse displacement and slope are unknown. The free-body diagram of node C is shown in Fig. 3c.

$$\sum F_y = 0: \qquad (V_j)_1 + (V_i)_2 = P \qquad \qquad (1a)$$

$$\left(\sum M\right)_C = 0: \qquad (M_j)_1 + (M_i)_2 = 0 \qquad (1b)$$

Equilibrium

Beam Element Force-Deformation Behavior: Here we want the force-type quantities, that is V and M, to be expressed in terms of displacement-type quantities, that is δ and θ. For this, we can use Eq. 7.19 (or see Table E.3). From Eqs. 7.19 we get

$$(V_j)_1 = \left(\frac{EI}{L^3}\right)_1 [12(\delta_j)_1 - 6L_1(\theta_j)_1] \qquad (2a)$$

$$(M_j)_1 = \left(\frac{EI}{L^2}\right)_1 [-6(\delta_j)_1 + 4L_1(\theta_j)_1] \qquad (2b)$$

Force-
Deformation
Behavior

$$(V_i)_2 = \left(\frac{EI}{L^3}\right)_2 [12(\delta_i)_2 - 6L_2(\theta_i)_2] \qquad (2c)$$

$$(M_i)_2 = \left(\frac{EI}{L^2}\right)_2 [6(\delta_i)_2 + 4L_2(\theta_i)_2] \qquad (2d)$$

Geometry of Deformation: The δ and θ quantities in Eqs. (2) and in Figs. 3b and 3d can be related to the displacement and slope at node C, shown in Fig. 3a. Thus,

$$(\theta_j)_1 = (\theta_i)_2 = \theta_C \qquad (3a)$$

$$(\delta_j)_1 = (\delta_i)_2 = \delta_C \qquad (3b)$$

Geometry of
Deformation

Combining Eqs. (1) through (3) in the order $(3) \to (2) \to (1)$, we get the *equilibrium equations expressed in terms of displacement and slope at node C.*

$$12 \left[\left(\frac{EI}{L^3}\right)_1 + \left(\frac{EI}{L^3}\right)_2\right] \delta_C - 6 \left[\left(\frac{EI}{L^2}\right)_1 - \left(\frac{EI}{L^2}\right)_2\right] \theta_C = P \qquad (4a)$$

$$-6 \left[\left(\frac{EI}{L^2}\right)_1 - \left(\frac{EI}{L^2}\right)_2\right] \delta_C + 4 \left[\left(\frac{EI}{L}\right)_1 + \left(\frac{EI}{L}\right)_2\right] \theta_C = 0 \qquad (4b)$$

Complete Part (a): Equations (4) are two simultaneous, algebraic equations for the unknown joint displacement δ_C and slope θ_C. This is called a *Displacement-Method solution* because the unknowns in Eqs. (4) are displacements. Given numerical values of P, $(EI)_i$, and L_i, we could easily solve for numerical values of these joint displacements.

(b) *Solve for the reactions R_B and M_B.* We can use Eqs. 7.19(c) and 7.19(d) to solve for R_B and M_B using the nodal displacements obtained from the solution of Eqs. (4). Thus,

$$R_B \equiv -(V_j)_2 = -\left(\frac{EI}{L^3}\right)_2 (-12\delta_C - 6L_2\theta_C) \qquad (5a)$$

$$M_B \equiv (M_j)_2 = \left(\frac{EI}{L^2}\right)_2 (6\delta_C + 2L_2\theta_C) \qquad (5b)$$

Review the Solution   As one check to verify that the above solutions will lead to correct answers for δ_C, θ_C, R_B, and M_B, let us take the special case of the symmetrical problem, with $EI = $ const, and $L_1 = L_2 = L/2$. This should lead to the result $\theta_C = 0$. Substituting the above values into Eq. (4b) we get

$$16\left(\frac{EI}{L}\right)\theta_C = 0 \rightarrow \theta_C = 0$$

which is the result that we would expect.

If all of the loads on a beam appear as concentrated loads (forces and/or couples) at nodes (joints), the problem is called a *nodal-load problem*. The preceding example problem is a typical nodal-load beam problem solved by the Displacement Method.

The next example illustrates how distributed loads are treated in a Displacement-Method solution. A key ingredient of a distributed-load solution is the inclusion of appropriate *fixed-end forces* for each individual element (Step 4 in the procedure on p. 515).

EXAMPLE 7.17

The two-span continuous beam in Fig. 1 has a uniform distributed load over span AB. Use the Dsplacement Method: (a) to determine the unknown slopes θ_B and θ_C, and (b) to determine the reactions R_A and M_A at the fixed end A.

Plan the Solution   We need to write moment equilibrium equations for the joints at B and C, where the beam is free to rotate. Equations 7.19 can be used to relate element end moments to the corresponding element end rotations and to the fixed-end moment due to the distributed load on element (1).

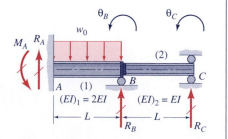

Fig. 1   A two-span continuous beam with uniform distributed loading over span AB.

Solution   (a) *Determine the nodal slopes (rotations) θ_B and θ_C.*

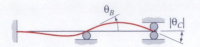

Fig. 2   The expected deflection curve.

Sketch the Deflection Curve: The angles θ_B and θ_C are taken positive counterclockwise, as illustrated in Fig. 1. Therefore, the magnitude of the expected clockwise angle at node C is labeled $|\theta_C|$ in Fig. 2.

Show the Elements and Nodes Separately: All of the reactions, shear forces, and bending moments are shown on the nodes and elements in Fig. 3. As in Example Problem 7.16, we will apply equilibrium, force-deformation behavior, and geometry of deformation to complete our solution.

Nodal Equilibrium: Referring to the free-body diagrams of nodes B and C, we can write the following two moment-equilibrium equations; these correspond to the unknown rotations at B and C.

$$\left(\sum M \right)_B = 0: \qquad (M_j)_1 + (M_i)_2 = 0 \qquad\qquad (1a)$$

Equilibrium

$$\left(\sum M \right)_C = 0: \qquad (M_j)_2 = 0 \qquad\qquad (1b)$$

Element Force-Deformation Equations with Fixed-End Forces: Equation 7.19(d) is used to express $(M_j)_1$ in terms of the unknown slope $(\theta_j)_1$ and the fixed-end moment due to the uniform distributed load $(p_i)_1 = (p_j)_1 = -w_0$. Then Eqs. 7.19(b) and (d) are used to write expressions for the moments $(M_i)_2$ and $(M_j)_2$, respectively. Observe that $(EI)_1 = 2EI$ and that $L_1 = L_2 = L$.

$$(M_j)_1 = \frac{8EI}{L}(\theta_j)_1 - \frac{w_0 L^2}{12} \qquad\qquad (2a)$$

Forces-Deformation Equation with Fixed-End Forces

$$(M_i)_2 = \frac{2EI}{L}[2(\theta_i)_2 + (\theta_j)_2] \qquad\qquad (2b)$$

$$(M_j)_2 = \frac{2EI}{L}[(\theta_i)_2 + 2(\theta_j)_2] \qquad\qquad (2c)$$

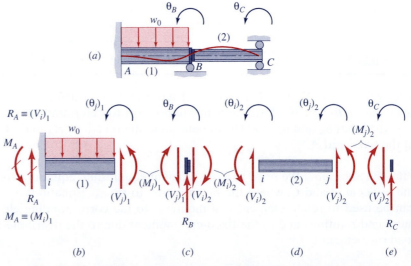

Fig. 3   The continuous beam separated into elements and nodes.

Geometry of Deformation; Compatibility: In Fig. 1 the rotation angles θ_B and θ_C are taken positive counterclockwise in order to be consistent with the sign convention for θ_i and θ_j in Fig. 7.15*b*. Therefore,

$$(\theta_j)_1 = (\theta_i)_2 = \theta_B \qquad \textbf{Geometry of} \qquad (3a)$$
$$(\theta_j)_2 = \theta_C \qquad \textbf{Deformation} \qquad (3b)$$

Equation (3a) is a *compatibility equation* that expresses the continuity of slope at *B*.

Complete Part (a): Finally, we substitute Eqs. (3) → (2) → (1), in Displacement-Method fashion, and get the *equilibrium equations in terms of unknown displacements:*

$$\left[\frac{8EI}{L} + \frac{4EI}{L}\right]\theta_B + \frac{2EI}{L}\theta_C = \frac{w_0L^2}{12} \qquad (4a)$$

$$\frac{2EI}{L}\theta_B + \frac{4EI}{L}\theta_C = 0 \qquad (4b)$$

Finally, solving Eqs. (4) simultaneously, we get

$$\theta_B = \frac{w_0L^3}{132EI}, \qquad \theta_C = -\frac{w_0L^3}{264EI} \qquad \textbf{Ans. (a)} \quad (5)$$

(b) *Determine the reactions R_A and M_A.* As noted in Fig. 3, $R_A \equiv (V_i)_1$ and $M_A \equiv (M_i)_1$. Also, $(p_i)_1 = (p_j)_1 = -w_0$. Using Eqs. 7.19(a) and 7.19(b), together with Eqs. (3) above, we get

$$R_A \equiv (V_i)_1 = \frac{2EI}{L^2}(6\theta_B) + \frac{1}{2}w_0L \qquad (6a)$$

$$M_A \equiv (M_i)_1 = \frac{2EI}{L}(2\theta_B) + \frac{1}{12}w_0L^2 \qquad (6b)$$

Substitution of Eq. (5a) into Eqs. (6) gives

$$R_A = \frac{1}{11}w_0L + \frac{1}{2}w_0L \qquad (7a)$$

$$M_A = \frac{1}{33}w_0L^2 + \frac{1}{12}w_0L^2 \qquad (7b)$$

or

$$R_A = \frac{13}{22}w_0L, \qquad M_A = \frac{15}{132}w_0L^2 \qquad \textbf{Ans. (b)} \quad (8)$$

Review the Solution   If we consider element (1), we can see that the right end of element (1) is somewhere between fully clamped (no rotation) and propped (no moment). The solutions to the reactions for these two statically indeterminate problems are shown in Fig. 4. It can be seen that the reactions R_A and M_A obtained in Eq. (8) do, indeed, lie between the values of the reactions for these two beams.

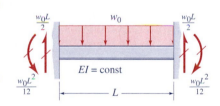

(*a*) A uniform fixed-fixed beam.

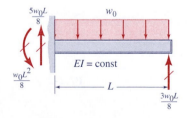

(*b*) A uniform propped-cantilever beam.

Fig. 4   Two beams related to element (1) of this problem.

For all problems in this section, the flexural rigidity of the beam, EI, is constant.

Prob. 7.2-1. For the simply supported beam shown in Fig. P7.2-1, the equation of the deflection curve is

$$v(x) = \frac{M_0 L^2}{3EI}\left[-2\left(\frac{x}{L}\right)^3 + 3\left(\frac{x}{L}\right)^2 - \left(\frac{x}{L}\right)\right]$$

(a) Determine an expression for the maximum deflection, $\delta_{max} \equiv \max |v(x)|$; and (b) use Eqs. 7.8 and 7.13 to determine expressions for the applied moment M_A and for the reactions R_A and R_B.

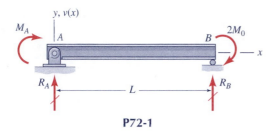

P72-1

Prob. 7.2-2. For the simply supported beam shown in Fig. P7.2-2, the equation of the deflection curve is

$$v(x) = \frac{p_0 L^4}{840EI}\left[\left(\frac{x}{L}\right)^7 - 7\left(\frac{x}{L}\right)^3 + 6\left(\frac{x}{L}\right)\right]$$

(a) Determine an expression for the distributed load, $p(x)$; and (b) use Eq. 7.13 to determine expressions for the reactions R_A and R_B.

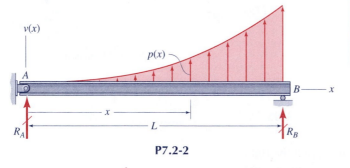

P7.2-2

Prob. 7.2-3. For the cantilever beam shown in Fig. P7.2-3, the equation of the deflection curve is

$$v(x) = \frac{w_0 L^4}{48EI}\left[-2\left(\frac{x}{L}\right)^4 + \left(\frac{x}{L}\right)^3 + 5\left(\frac{x}{L}\right) - 4\right]$$

(a) Determine an expression for the concentrated load, P_A; and (b) determine expressions for the reactions R_B and M_B.

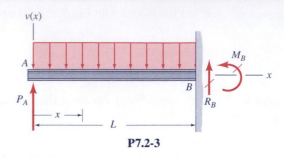

P7.2-3

Prob. 7.2-4. The cantilever beam shown in Fig. P7.2-4 has a downward distributed load that varies linearly from w_1 at $x = 0$ to w_2 at $x = L$. The equation of this beam's deflection curve is

$$v(x) = \frac{w_0 L^4}{120EI}\left[-\left(\frac{x}{L}\right)^5 - 5\left(\frac{x}{L}\right)^4 + 30\left(\frac{x}{L}\right)^3 - 50\left(\frac{x}{L}\right)^2\right]$$

(a) Determine expressions for w_1 and w_2; and (b) determine expressions for the reactions R_A and M_A.

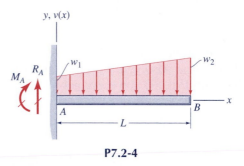

P7.2-4

▼ SECOND-ORDER INTEGRATION METHOD

Problems 7.3-1 through 7.3-17. *These problems are to be solved by* **integrating the second-order differential equation, Eq. 7.8.** *Except for Prob. 7.3-17, the flexural rigidity, EI, is constant for each beam.*

Prob. 7.3-1. For the uniform simply supported beam shown in Fig. P7.3-1. (a) determine the equation of the deflection curve, $v(x)$; (b) determine the slope, θ_A, at the left end; and (c) determine the maximum deflection, $\delta_{max} \equiv \max |v(x)|$. Solve by integrating the second-order differential equation (the moment-curvature equation).

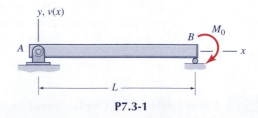

P7.3-1

Prob. 7.3-2. For the uniform simply supported beam shown in Fig. P7.3-2, (a) determine the equation of the deflection curve, $v(x)$; (b) determine the slope, θ_B, at the right end; and (c) determine the maximum deflection, $\delta_{\max} \equiv \max |v(x)|$. Solve by integrating the second-order differential equation (the moment-curvature equation).

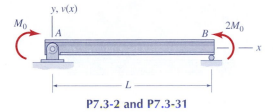

P7.3-2 and P7.3-31

Prob. 7.3-3. A **W**8 × 40 wide-flange beam made of A-36 structural steel ($E = 29 \times 10^3$ ksi) is simply supported, as shown in Fig. P7.3-3. The beam is subjected to a couple M_A that produces a maximum deflection of 0.10 in. (a) What is the value of M_A? (b) What is the maximum flexural stress in the beam under this loading? Solve by integrating the second-order differential equation for deflection, Eq. 7.8, and see Table D.1 for the cross-sectional properties of the beam.

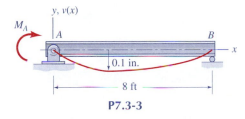

P7.3-3

Prob. 7.3-4. For the uniform simply supported beam in Fig. P7.3-4, (a) determine the equation of the deflection curve, $v(x)$; (b) determine the slope, θ_A, at the left end; and (c) determine the maximum deflection, $\delta_{\max} \equiv \max |v(x)|$. Solve by integrating the second-order differential equation for deflection. (Note: You may have to use a trial and error solution to determine where the maximum deflection occurs.)

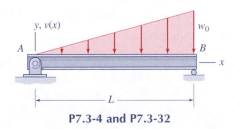

P7.3-4 and P7.3-32

Prob. 7.3-5. The cantilever beam AB in Fig. P7.3-5 supports a triangularly distributed load of maximum intensity w_0. (a) Determine an expression for the deflection curve, $v(x)$, for this beam; and (b) determine expressions for the slope, θ_B, and the deflection, δ_B, at end B. Solve by integrating the second-order differential equation for deflection.

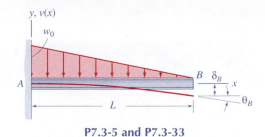

P7.3-5 and P7.3-33

Prob. 7.3-6. For the uniform beam with overhang of length L shown in Fig. P7.3.6, (a) determine the equation, $v(x)$, of the deflection curve for the portion of the beam between the supports A and B; and (b) determine the maximum (upward) deflection between supports A and B, $\delta_{\max} \equiv \max |v(x)|$. Solve by integrating the second-order differential equation for deflection (the moment-curvature equation).

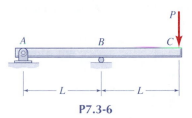

P7.3-6

Prob. 7.3-7. A person standing at end C of a diving board exerts a downward force P, as shown is Fig. P7.3-7. (a) Determine expressions for the deflection curve $v_1(x)$ in segment AB and the deflection curve $v_2(x)$ in the section BC. Solve by integrating the second-order differential equations for deflection. (b) Determine the deflection δ_C where the diver is standing, and (c) determine the maximum (upward) deflection, $\delta_{\max} \equiv \max [v_1(x)]$, in segment AB.

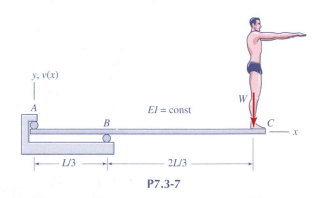

P7.3-7

Prob. 7.3-8. (a) For the (two-interval) simply supported beam in Fig. P7.3-8, determine the equations of the deflection curve, $v_1(x)$ for $0 \le x \le \dfrac{L}{3}$ and $v_2(x)$ for $\dfrac{L}{3} \le x \le L$. Solve by integrating the second-order differential equations for deflection. (b) Determine the location, $\dfrac{x_m}{L}$, of the maximum deflection of the beam, and determine an expression for this maximum deflection, $\delta_{\max}$.

521

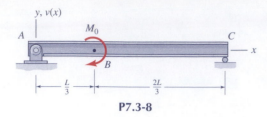

P7.3-8

expressions for the deflection curve $v_1(x)$ in segment AB and the deflection curve $v_2(x)$ in segment BC. Solve by integrating the second-order differential equations for deflection. (b) Determine the deflection δ_C at the right end of the beam. (c) Determine the maximum (upward) deflection, $\delta_{max} \equiv$ max $[v_1(x)]$, in segment AB.

Prob. 7.3-9. A uniform shaft is supported by bearings at A and C and is subjected to a downward load P through the pulley at B. Neglect the width of the pulley and bearings, and assume that the bearings provide only vertical support to the shaft. (a) Determine expressions for the deflection curve $v_1(x)$ in section AB and the deflection curve $v_2(x)$ in section BC. Solve by integrating the second-order differential equations for deflection. (b) The maximum deflection of the shaft between the bearings A and C will occur between B and C, as shown in Fig. P7.3-9. Determine an expression for $\delta_{max} \equiv$ max $|v_2(x)|$.

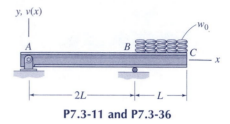

P7.3-11 and P7.3-36

Prob. 7.3-12. For the (two-interval) simply supported beam in Fig. P7.3-12, (a) determine the equations of the deflection curve, $v_1(x)$ for $0 \le x \le \dfrac{L}{2}$ and $v_2(x)$ for $\dfrac{L}{2} \le x \le L$; and (b) determine the vertical displacement at the center of the beam, $\delta_B \equiv |v(L/2)|$. Solve by integrating the second-order differential equation for deflection.

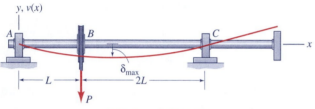

P7.3-9 and P7.3-35

Prob. 7.3-10. An exercise bar with weights rests on a stand as shown in Fig. P7.3-10. Assume that the bar is simply supported on a span of $L = 40$ in. and that each pair of weights exerts a concentrated force $W = 50$ lb at a distance $a = 8$ in. beyond the support points, but neglect the weight of the bar itself. The solid steel bar ($E = 30 \times 10^6$ psi) has a 1 in. diameter. (a) Making use of symmetry of loading and support, and starting with the second-order differential equation for deflection, derive an expression for the deflection curve, $v(x)$, for the portion AB of the bar. Express your answer in terms of W, EI, L, a, and the coordinate x. (b) Determine an expression for the maximum (upward) deflection, $\delta_{max} \equiv v(0)$, and evaluate the expression for the given physical parameters.

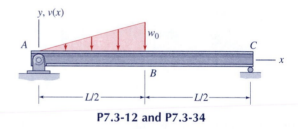

P7.3-12 and P7.3-34

Prob. 7.3-13. The uniform simply supported beam in Fig. P7.3-13 has a symmetric linearly varying distributed load with maximum value w_0 at the center of the beam. From symmetry it can be observed that the slope of the beam must be zero at the center of the beam, that is, $v'(L/2) = 0$. Using this fact, together with the boundary condition $v(0) = 0$, (a) determine an equation for the deflection curve, $v(x)$ for $0 \le x \le \dfrac{L}{2}$. Solve by integrating the second-order differential equation for deflection. (b) Determine an expression for the maximum deflection, $\delta_{max} \equiv |v(L/2)|$.

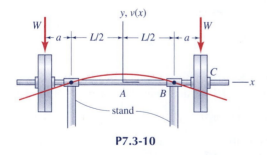

P7.3-10

Prob. 7.3-11. Bags of cement have been stacked on the overhanging segment, BC, of the beam in Fig. P7.3-11 producing a uniform load of intensity w_0 per unit length. (a) Determine

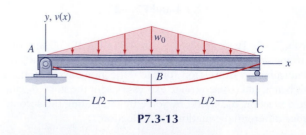

P7.3-13

522

*Prob. 7.3-14. A 10-ft-long **W**8 × 15 wide-flange steel beam *AB* is supported by a roller support at end *A* and by a 6-ft-long rod *BC* of cross-sectional area $A = 1.0 \text{ in}^2$. The beam supports a uniformly distributed load $w_0 = 1$ kip/ft, as shown in Fig. P7.3-14. (a) Neglecting the weight of the beam and the weight of the rod *BC*, determine the elongation of the rod that supports the beam at end *B*. Use $E = 30 \times 10^3$ ksi for the rod and the beam. (b) Determine the location x_m (ft) of the point of maximum deflection of the beam.

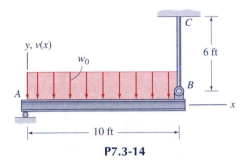

P7.3-14

Prob. 7.3-15. For the (two-interval) cantilever beam in Fig. P7.3-15, (a) determine the equations of the deflection curve, $v_1(x)$ for $0 \le x \le \dfrac{L}{2}$ and $v_2(x)$ for $\dfrac{L}{2} \le x \le L$; and (b) determine the vertical displacement at the right end of the beam, $\delta_C \equiv v(L)$. Solve by integrating the second-order differential equations for deflection.

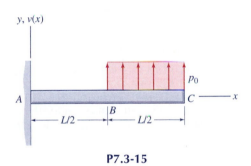

P7.3-15

Prob. 7.3-16. Loads P_A and P_B are applied to a uniform cantilever beam *AC*, as shown in Fig. P7.3-16, causing deflections δ_A and δ_B at *A* and *B*, respectively. Using the *second-order integration method,* determine expressions for P_A and P_B in terms of E, I, L, δ_A, and δ_B.

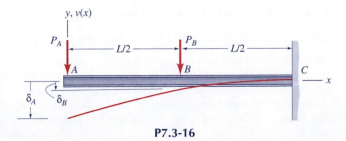

P7.3-16

Prob. 7.3-17. For the (two-interval) cantilever beam in Fig. P7.3-17, (a) determine the equations of the deflection curve, $v_1(x)$ for $0 \le x \le \dfrac{L}{2}$ and $v_2(x)$ for $\dfrac{L}{2} \le x \le L$; and (b) determine the vertical displacement at the left end of the beam, $\delta_A \equiv v(0)$. Note the change in flexural rigidity $EI(x)$ at $x = \dfrac{L}{2}$. Solve by integrating the second-order differential equations for deflection.

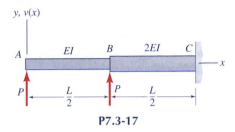

P7.3-17

▼ **BEAM-DEFLECTION APPLICATIONS**

> **Problems 7.3-18 through 7.3-26.** *In solving these problems, you may use deflection formulas given in Tables E.1 and E.2 of Appendix E. The flexural rigidity, EI, is constant for each beam in this group of problems.*

Prob. 7.3-18. The simply supported beam *AB* in Fig. P7.3-18 carries a uniformly distributed load of intensity w_0 on a span of $L = 8$ ft. Determine the maximum deflection, δ_{max}, if the depth of the beam is $h = 6$ in., the maximum flexural stress in the beam is $\sigma_{max} = 12$ ksi, and the beam is made of aluminum with $E = 10 \times 10^3$ ksi.

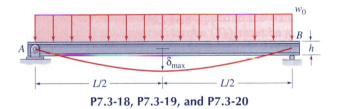

P7.3-18, P7.3-19, and P7.3-20

ᴰProb. 7.3-19. What is the depth h of a uniformly loaded, simply supported beam (see Fig. P7.3-19) if the maximum bending stress is $\sigma_{max} = 8$ ksi, the maximum deflection is $\delta_{max} = 0.20$ in., the span is $L = 10$ ft, and the modulus of elasticity is $E = 30 \times 10^3$ ksi?

Prob. 7.3-20. A **W**360 × 39 wide-flange steel beam is simply supported and carries a uniformly distributed load of $w_0 = 35$ kN/m on a span of $L = 5$ m (see Fig. P7.3-20). For $E = 200$ GPa, (a) determine the maximum deflection of the beam, and (b) determine the maximum flexural stress in the beam. (See Table D.2 of Appendix D for the cross-sectional properties of the beam.)

ᴰProb. 7.3-21. A simply supported beam *AC* carries a concentrated load P at its midspan point *B* (see Fig. P7.3-21). Determine the depth of the beam h in mm if the maximum bending stress is $\sigma_{max} = 50$ MPa, the maximum deflection is $\delta_{max} = 10$ mm, the span is $L = 4$ m, and the modulus of elasticity is $E = 200$ GPa.

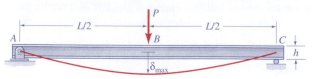

P7.3-21 and P7.3-22

Prob. 7.3-22. A simply supported beam AC carries a concentrated load P at its midspan, B (Fig. P7.3-22). Determine the maximum deflection, δ_{max}, if the span is $L = 8$ ft, the depth of the beam is $h = 6$ in., the maximum flexural stress is $\sigma_{max} = 12$ ksi, and the beam is made of aluminum with $E = 10 \times 10^3$ ksi.

***Prob. 7.3-23.** A $\mathbf{W}8 \times 40$ wide-flange steel beam AD is simply supported and carries two equal concentrated loads of $P = 6$ kips each, as shown in Fig. P7.3-23. The total span of the beam is $L = 12$ ft, and its modulus of elasticity is $E = 30 \times 10^3$ ksi. (a) Determine the maximum deflection of the beam, and (b) determine the maximum flexural stress in the beam. (See Table D.1 of Appendix D for the cross-sectional properties of the beam, and see Table E.2 for appropriate deflection formulas.)

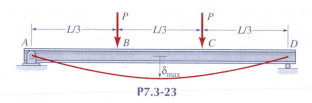

P7.3-23

Prob. 7.3-24. A cantilever beam AB carries a uniformly distributed load of intensity w_0 on a span of $L = 8$ ft (Fig. P7.3-24). Determine the maximum deflection, δ_{max}, if the depth of the beam is $h = 6$ in., the maximum flexural stress in the beam is $\sigma_{max} = 12$ ksi, and the beam is made of steel with $E = 30 \times 10^3$ ksi.

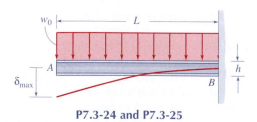

P7.3-24 and P7.3-25

Prob. 7.3-25. What is the maximum bending stress in the uniformly loaded cantilever beam in Fig. P7.3-25 if the depth of the beam is $h = 250$ mm, the maximum deflection is $\delta_{max} = 5$ mm, the span is $L = 3$ m, and the modulus of elasticity is $E = 70$ GPa?

524

***Prob. 7.3-26.** A 10-ft-long $\mathbf{W}6 \times 15$ wide-flange beam AB is supported by two 6-ft-long rods AC and BD of cross-sectional area $A_r = 0.5$ in.^{2}. The beam supports a uniformly distributed load $w_0 = 1$ kip/ft, as shown in Fig. P7.3-26. Determine the maximum deflection of the beam, taking into account the elongation of the two rods but neglecting the weight of the beam and the rods. Use $E = 30 \times 10^3$ ksi for the rods and the beam. (Hint: Solve first for the elongation of the rods, and then use these elongations as the boundary conditions $v(0)$ and $v(L)$ for the beam.)

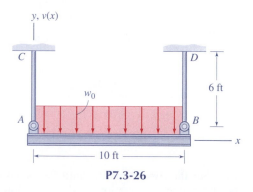

P7.3-26

▼ **FOURTH-ORDER INTEGRATION METHOD**

Problems 7.3-27 through 7.3-37. *These problems are to be solved by* **integrating the fourth-order differential equation,** *Eq. 7.12.* **The flexural rigidity, EI, is constant for each beam.**

Prob. 7.3-27. The simply supported beam in Fig. P7.3-27 is subjected to a distributed load of intensity $p(x) = -w_0 \sin\left(\dfrac{\pi x}{L}\right)$. [Note: This load acts downward, so there is a minus sign in the expression for $p(x)$.] (a) Use the fourth-order method to determine an expression for the deflection curve of this beam, and (b) determine the slope, θ_B, of the beam at end B.

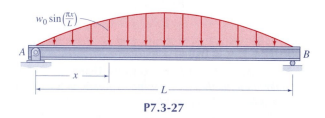

P7.3-27

Prob. 7.3-28. The simply supported beam in Fig. P7.3-28 is subjected to a distributed load of intensity $p(x) = -w_0\left(\dfrac{x}{L}\right)^3$.

[Note: This load acts downward, so there is a minus sign in the expression for $p(x)$.] (a) Use the fourth-order method to determine the equation of the deflection curve, $v(x)$; and (b) determine the maximum deflection, $\delta_{max} \equiv \max |v(x)|$, of this beam. (c) Determine expressions for the reactions A_y and B_y.

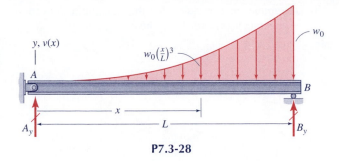

P7.3-28

Prob. 7.3-29. The cantilever beam in Fig. P7.3-29 is subjected to a distributed load of intensity $p(x) = p_0\left[1 - \left(\dfrac{x}{L}\right)^2\right]$. (a) Use the fourth-order method to determine an expression for the deflection curve, $v(x)$; and (b) determine the slope, θ_B, of the righthand end of the beam. (c) Determine expressions for the reactions R_A and M_A.

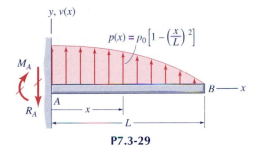

P7.3-29

Prob. 7.3-30. The cantilever beam in Fig. P7.3-30 is subjected to a distributed load of intensity $p(x) = -w_0\left(\dfrac{x}{L}\right)^2$. [Note: This load acts downward, so there is a minus sign in the expression for $p(x)$.] (a) Use the fourth-order method to determine the equation of the deflection curve, $v(x)$; and (b) determine the slope, θ_B, of the righthand end of the beam. (c) Determine expressions for the reactions R_A and M_A.

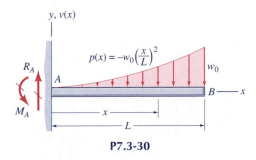

P7.3-30

Prob. 7.3-31. Use the fourth-order method to solve the problem as stated in Prob. 7.3-2.

Prob. 7.3-32. Use the fourth-order method to solve the problem as stated in Prob. 7.3-4.

Prob. 7.3-33. Use the fourth-order method to solve the problem as stated in Prob. 7.3-5.

Prob. 7.3-34. Use the fourth-order method to solve the problem as stated in Prob. 7.3-12.

Prob. 7.3-35. Use the fourth-order method to solve the problem as stated in Prob. 7.3-9.

Prob. 7.3-36. Use the fourth-order method to solve the problem as stated in Prob. 7.3-11.

DProb. 7.3-37. For the cantilever beam in Fig. P7.3-37, (a) use the fourth-order integration method to determine an expression for the tip deflection, δ_B; and then (b) determine an expression for the length L that would make the maximum tensile stress in the beam equal to σ_{allow}, and the corresponding maximum deflection equal to δ_{allow}. Express your answer in terms of $E, \delta_{\text{allow}}, \sigma_{\text{allow}}$, and the depth of the beam, h.

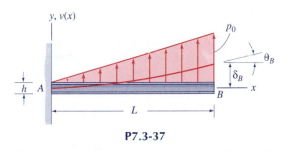

P7.3-37

▼ SECOND-ORDER INTEGRATION METHOD

> **Problems 7.4-1 through 7.4-11.** *These problems are to be solved by* integrating the second-order differential equation, **Eq. 7.8.** *The flexural rigidity, EI, is constant for the beams in Probs. 7.4-1 through 7.4-10.*

Prob. 7.4-1. For the uniformly loaded propped cantilever beam in Fig. P7.4-1, use the second-order integration method (a) to solve for the reactions at A and B; and (b) to determine an expression for the deflection curve $v(x)$.

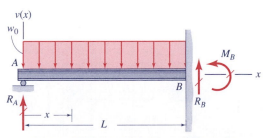

P7.4-1, P7.4-12, P7.6-31, and P11.3-40

Prob. 7.4-2. The propped cantilever beam in Fig. P7.4-2 has a **W8 × 40** cross section and supports a uniformly distributed load of $w_0 = 4$ kips/ft on a span of $L = 10$ ft. (a) Use the second-order integration method to solve for the reactions at A and B. (Solve the problem using symbols, w_0, EI, etc.,

and then substitute numerical values for these quantities.)
(b) Sketch the shear and moment diagrams for this beam.

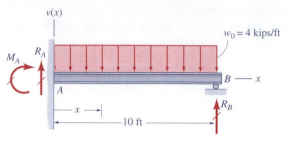

P7.4-2, P7.6-32, P11.5-18, and P11.5-42

Prob. 7.4-3. The propped cantilever beam AB in Fig. P7.4-3 supports a linearly varying load of maximum intensity w_0. (a) Use the second-order integration method to solve for the reactions at A and B and for the deflection curve $v(x)$. (b) Determine an expression for the maximum deflection, $\delta_{max} \equiv \max |v(x)|$. (c) Sketch the shear diagram, $V(x)$.

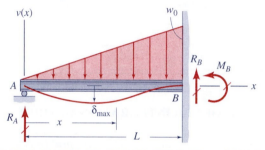

P7.4-3, P7.4-13, P7.6-33, P7.6-34, P11.5-17, and P11.5-41

Prob. 7.4-4. The uniformly loaded beam in Fig. P7.4-4 is completely fixed at ends A and B. (a) Use the second-order integration method to determine the reactions R_A and M_A and to determine an expression for the deflection curve $v(x)$. (b) Sketch the shear diagram, $V(x)$, and the moment diagram, $M(x)$.

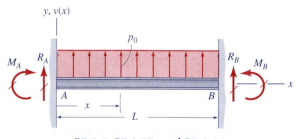

P7.4-4, P7.6-35, and P7.6-36

Prob. 7.4-5. The beam AB in Fig. P7.4-5 supports a linearly varying load of maximum intensity w_0. (a) Use the second-order integration method to determine the reactions at A and B and to determine an expression for the deflection curve $v(x)$.

(b) Determine an expression for the maximum deflection, $\delta_{max} \equiv \max |v(x)|$. (c) Sketch the shear diagram, $V(x)$.

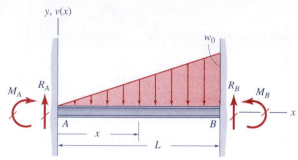

P7.4-5, P7.6-37, P7.6-38, P11.3-41, and P11.5-46

***Prob. 7.4-6.** At end B, the cantilever beam in Fig. P7.4-6 is pinned to a uniform rod whose cross-sectional areas is A_2, whose length is L_2, and whose modulus of elasticity is E_2. The beam supports a uniformly distributed load of intensity w_0; its flexural rigidity is $E_1 I_1$ and its length is L_1. (a) Use the second-order integration method to determine the reactions R_A and M_A at A, and the tension, F_2, in the rod. (b) Determine an expression for the deflection curve, $v(x)$, of the beam.

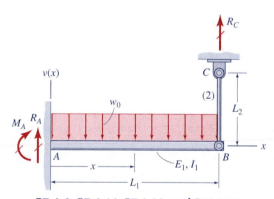

P7.4-6, P7.4-14, P7.6-39, and P11.5-47

Prob. 7.4-7. The propped cantilever beam in Fig. P7.4-7 is subjected to a concentrated load P at distance $L/3$ from end A. (a) Use the second-order integration method to determine the reactions at A and C and the deflection curves $v_1(x)$ and $v_2(x)$ for the segments of the beam to the left of load P and to the right of load P, respectively. (b) Sketch the shear diagram, $V(x)$, and the moment diagram, $M(x)$.

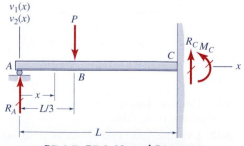

P7.4-7, P7.6-40, and P11.5-19

Prob. 7.4-8. The fixed-fixed beam in Fig. P7.4-8 is subjected to a concentrated couple M_0 at location B, which is at distance a from end A. (a) Use the second-order integration method to determine the reactions at A and C and the deflection curves $v_1(x)$ and $v_2(x)$ for the segments of the beam to the left of B and to the right of B, respectively. (b) Letting $a = L/3$, sketch the shear diagram, $V(x)$, and the moment diagram, $M(x)$, for this beam.

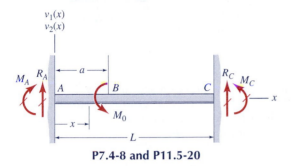

P7.4-8 and P11.5-20

*__Prob. 7.4-9.__ The fixed-fixed beam in Fig. P7.4-9 is subjected to a uniformly distributed load of intensity w_0 over the interval AB (i.e., $0 \le x \le a$). (a) Use the second-order integration method to determine the reactions at A and C and the deflection-curve expressions $v_1(x)$ (for $0 \le x \le a$) and $v_2(x)$ (for $a \le x \le L$). (b) Letting $a = L/2$, sketch the complete shear diagram, $V(x)$, for this beam.

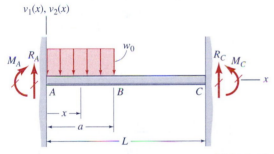

P7.4-9, P7.6-41, P7.6-42, P11.5-23, and P11.5-44

Prob. 7.4-10. The continuous beam in Fig. P7.4-10 is simply supported at ends A and C and is propped at its midpoint B. It supports a uniformly distributed load of intensity w_0 on the span AB. (a) Use the second-order integration method

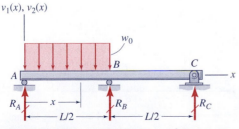

P7.4-10, P7.6-43, P11.5-21, and P11.5-43

to determine the reactions at A, B, and C. (b) Determine expressions for the deflection curves $v_1(x)$ in span AB and $v_2(x)$ in span BC. (c) Sketch the shear diagram, $V(x)$, and the moment diagram, $M(x)$.

*__Prob. 7.4-11.__ The beam AC in Fig. P7.4-11 is fixed to a rigid wall at A and is supported by props at B and C. In span AB the flexural rigidity is EI, but in span BC the flexural rigidity is $2EI$. The beam supports a uniformly distributed load over span BC. (a) Use the second-order integration method to determine the reactions at A, B, and C. (b) Determine expressions for the deflection curves $v_1(x)$ in span AB and $v_2(x)$ in span BC. (c) Sketch the shear diagram, $V(x)$, and the moment diagram, $M(x)$.

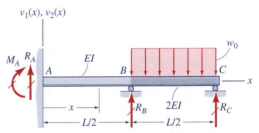

P7.4-11, P7.6-44, and P11.5-22

▼ **FOURTH-ORDER INTEGRATION METHOD**

> **Problems 7.4-12 through 7.4-18.** *These problems are to be solved by* <u>integrating the fourth-order differential equation, Eq. 7.12.</u> *The flexural rigidity, EI, is constant for all beams in this group of problems.*

Prob. 7.4-12. Solve Prob. 7.4-1 using the fourth-order integration method.

Prob. 7.4-13. Solve Prob. 7.4-3 using the fourth-order integration method.

Prob. 7.4-14. Solve Prob. 7.4-6 using the fourth-order integration method.

> **Problems 7.4-15 through 7.4-18.** *For each of these problems,* **use the fourth-order integration method** *to determine the reactions at ends A and B and to determine an expression for the deflection curve v(x).*

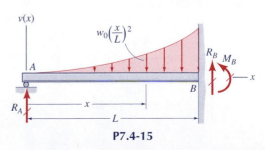

P7.4-15

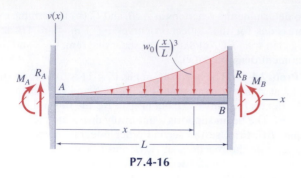

P7.4-16

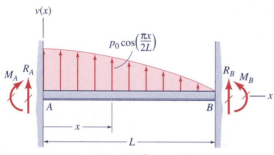

P7.4-17 and P7.6-45

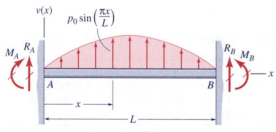

P7.4-18 and P7.6-46

▼ **BEAM DEFLECTIONS:
DISCONTINUITY-FUNCTION METHOD**

MDS 7.1 & 7.2

Problems 7.5-1 through 7.5-20. *Use the discontinuity-function method of Section 7.5 to solve each of the following problems. Let EI = const. for each beam.*

Prob. 7.5-1. For the beam shown in Fig. P7.5-1, (a) determine discontinuity-function expressions for the slope $\theta(x)$ and the deflection $v(x)$ for $0 \le x \le L$; and (b) evaluate the displacement at points B and C.

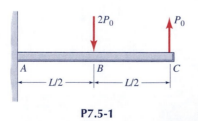

P7.5-1

Prob. 7.5-2. For the beam shown in Fig. P7.5-2, (a) determine discontinuity-function expressions for the slope $\theta(x)$ and the deflection $v(x)$ for $0 \le x \le L$; and (b) evaluate the displacement at points B and C.

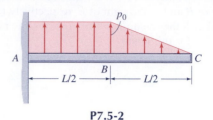

P7.5-2

Prob. 7.5-3. For the beam shown in Fig. P7.5-3, (a) determine discontinuity-function expressions for the slope $\theta(x)$ and the deflection $v(x)$ for $0 \le x \le L$; and (b) evaluate the displacement at points B and C.

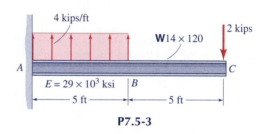

P7.5-3

Prob. 7.5-4. For the beam shown in Fig. P7.5-4, (a) determine discontinuity-function expressions for the slope $\theta(x)$ and the deflection $v(x)$ for $0 \le x \le L$; and (b) evaluate the displacement at points B and C.

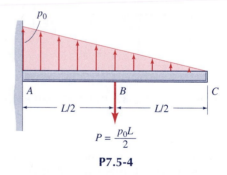

P7.5-4

Prob. 7.5-5. For the beam shown in Fig. P7.5-5, (a) determine discontinuity-function expressions for the slope $\theta(x)$

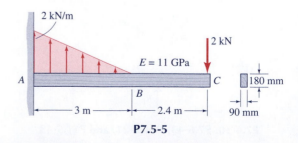

P7.5-5

and the deflection $v(x)$ for $0 \le x \le L$; and (b) evaluate the displacement at points B and C.

Prob. 7.5-6. For the overhanging beam shown in Fig. P7.5-6, (a) determine discontinuity-function expressions for the slope $\theta(x)$ and the deflection $v(x)$ for $0 \le x \le 3L$; (b) determine expressions for the slope and deflection at $x = 0$, that is, for $\theta(0)$ and $v(0)$; and (c) determine an expression for the maximum deflection between the supports B and C.

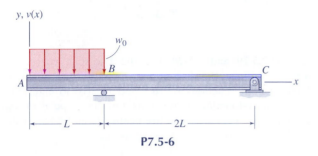

P7.5-6

Prob. 7.5-7. For the simply supported beam shown in Fig. P7.5-7, (a) determine expressions for the end slopes $\theta_A \equiv \theta(0)$ and $\theta_D = \theta(L)$; (b) determine discontinuity-function expressions for $\theta(x)$ and $v(x)$ for $0 \le x \le L$; and (c) determine an expression for the midspan deflection $v(\frac{L}{2})$.

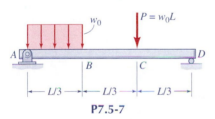

P7.5-7

Prob.7.5-8. For the three-segment overhanging beam shown in Fig. P7.5-8, (a) determine discontinuity-function expressions for the slope $\theta(x)$ and the deflection $v(x)$ for $0 \le x \le \frac{3L}{2}$; and (b) determine the maximum upward deflection of the beam.

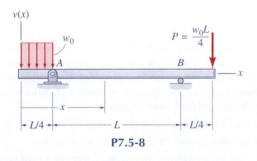

P7.5-8

*Prob. 7.5-9. For the beam shown in Fig. P7.5-9, use the discontinuity-function method to determine an expression for the end couple M_0, in terms of P and L, such that the deflection at end D is zero.

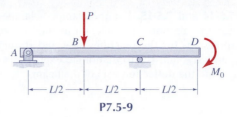

P7.5-9

Probs. 7.5-10 through 7.5-12. For the propped cantilever beams shown in Figs. P7.5-10 through P7.5-12, use the discontinuity-function method (a) to determine expressions for the reactions R_A, R_B, M_B; and (b) to determine expressions for the slope $\theta(x)$ and the deflection $v(x)$ for $0 \le x \le L$.

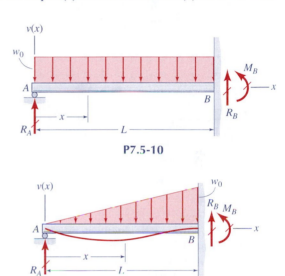

P7.5-10

P7.5-11

P7.5-12

Prob. 7.5-13. For the propped cantilever beam shown, use the discontinuity-function method (a) to determine expressions for the reactions R_A, R_B, and M_B; and (b) to determine expressions for the slope $\theta(x)$ and the deflection $v(x)$ for $0 \le x \le L$.

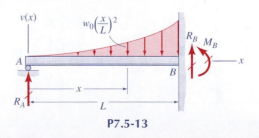

P7.5-13

Probs. 7.5-14 and 7.5-15. For the clamped-clamped beam shown, use the discontinuity-function method (a) to determine expressions for the reactions at A and B, that is, R_A, M_A, R_B, and M_B, and (b) to determine expressions for the slope $\theta(x)$ and the deflection $v(x)$ of the beam.

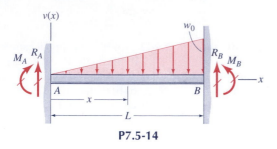

P7.5-14

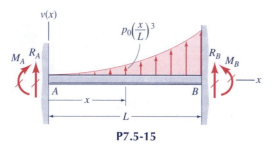

P7.5-15

Probs. 7.5-16 through 7.5-18. For the clamped-clamped beam shown, use the discontinuity-function method (a) to determine expressions for the reactions at A and C, that is, R_A, M_A, R_C, and M_C; (b) to determine discontinuity-function expressions for the slope $\theta(x)$ and the deflection $v(x)$ for $0 \le x \le L$; and (c) to determine an expression for the deflection at point B.

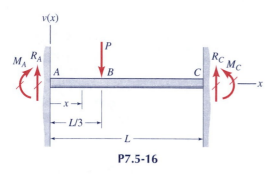

P7.5-16

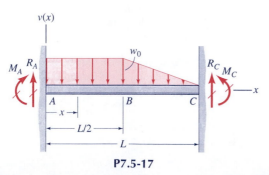

P7.5-17

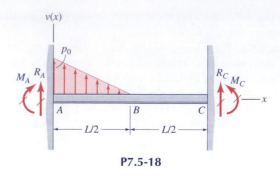

P7.5-18

*Probs. 7.5-19 and 7.5-20. For the two-span continuous beam shown, use the discontinuity-function method (a) to determine the reactions R_A, R_B, and R_C; (b) to determine discontinuity-function expressions for the slope $\theta(x)$ and the deflection $v(x)$ for the entire beam AC; and (c) to evaluate the slope expression $\theta(x)$ at support A.

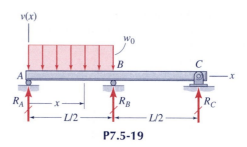

P7.5-19

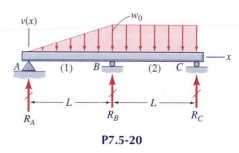

P7.5-20

Problems 7.5-21 through 7.5-24. *A beam and its loading are shown in the referenced figure. For each of these problems, you are to write a computer program that carries out the following steps. Use the programming language or math application software of your own choice, unless your instructor indicates otherwise.*

(a) Input the given problem data: L, E, etc.

(b) Use equilibrium to determine the reactions.

(c) Using discontinuity functions from the Load column of Table 5.2, form an expression for the intensity p(x) of the equivalent distributed load. Include the calculated reactions in your expression for the equivalent load.

(d) Perform a term-by-term integration of the load expression obtained in Part (c) to form a discontinuity-function expression for the shear force $V(x)$. Then, perform a term-by-term integration of the shear expression to obtain a discontinuity-function expression for the bending moment $M(x)$. Plot the moment diagram.

(e) Perform a term-by-term integration of the bending moment expression obtained in Part (d) to form a discontinuity-function expression for the slope $\theta(x)$. Plot the slope diagram for the beam.

(f) Perform a term-by-term integration of the slope expression to obtain a discontinuity-function expression for the deflection $v(x)$. Plot the deflection diagram for the beam.

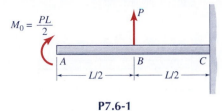

P7.6-1

Probs. 7.6-2 and 7.6-3. Determine expressions for the slope v'_C and the deflection v_C at end C.

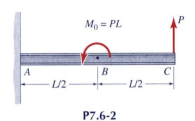

P7.6-2

^{C}*Prob. 7.5-21. (a) Write your computer program for the simply supported beam in Fig. P7.5-21 (see Fig. P5.5-19). (b) Illustrate the use of your computer program for the following data: The beam is a **W**200×59 (see Table D.2 of Appendix D) with $E = 200$ GPa, $L = 10$ m, and $P_B = 25$ kN.

^{C}*Prob. 7.5-22. (a) Write your computer program for the simply supported beam in Fig. P7.5-22 (see Fig. P5.5-20). (b) Illustrate the use of your computer program for the following data: The beam is a **W**250×89 (see Table D.2 of Appendix D) with $E = 200$ GPa, $L = 10$ m, $x_B = 2$ m, $x_C = 4$ m, and $p_0 = 12$ kN/m.

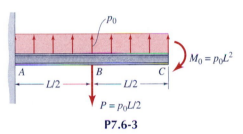

P7.6-3

^{C}*Prob. 7.5-23. (a) Write your computer program for the simply supported beam in Fig. P7.5-23 (see Fig. P5.5-21). (b) Illustrate the use of your computer program for the following data: The beam is a **W**10×60 (see Table D.1 of Appendix D) with $E = 29 (10^3)$ ksi, $L = 12$ ft, $x_B = 2$ ft, $x_C = 5$ ft, and $p_0 = 2$ kips/ft.

Prob. 7.6-4. Determine expressions for the slope v'_A and the deflection v_A at end A.

^{C}*Prob. 7.5-24. (a) Write your computer program for the cantilever beam in Fig. P7.5-24 (see Fig. P5.5-22). (b) Illustrate the use of your computer program for the following data: The beam is a **W**12×50 (see Table D.1 of Appendix D) with $E = 29(10^3)$ ksi, $L = 15$ ft, $x_B = 3$ ft, $x_C = 9$ ft, $P_A = -4$ kips (i.e., 4 kips downward), and $p_0 = 2$ kips/ft.

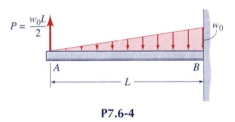

P7.6-4

Problems 7.6-1 through 7.6-57. *Using Tables E.1 and E2, Deflections and Slopes of Beams, apply the Force Method of Superposition to solve each of the problems of this section. Let the flexural rigidity, EI, be constant for each beam, unless stated otherwise.*

Probs. 7.6-5 and 7.6-6. Determine expressions for the slope v'_C and the deflection v_C at end C. (Hint: You can subtract the affect of loading indicated by the dashed lines.)

▼ **SUPERPOSITION METHOD, STATICALLY DETERMINATE BEAMS**

| MDS 7.3 |

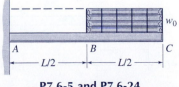

P7.6-5 and P7.6-24

Prob. 7.6-1. Determine expressions for the slope v'_A and the deflection v_A at end A.

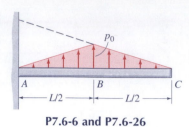

P7.6-6 and P7.6-26

Probs. 7.6-7 and 7.6-8. Determine expressions for the slope v'_A end A and the deflection v_C at section C.

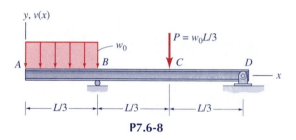

P7.6-7

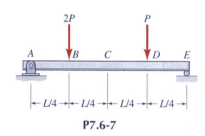

P7.6-8

Probs. 7.6-9 and 7.6-10. (a) Determine expressions for the slope $v'(x)$ and the deflection $v(x)$ in the segment AB; and (b) determine the maximum upward deflection of the beam.

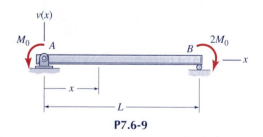

P7.6-9

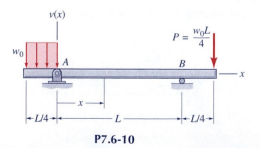

P7.6-10

Prob. 7.6-11. For the antisymmetric loading in Fig. P7.6-11, (a) determine expressions for the slope $v'(x)$ and the deflection $v(x)$ in the segment AB; and (b) determine the maximum upward deflection of the beam, which occurs in segment AB.

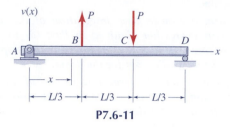

P7.6-11

Prob. 7.6-12. Determine expressions for the slope v'_C and the deflection v_C at end C.

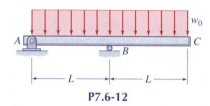

P7.6-12

***Prob. 7.6-13.** Beam AB (beam 1) is cantilevered from a rigid wall at A. Through a roller at B, beam 1 supports one end of simply supported beam BC (beam 2). Both beams have the same flexural rigidity, EI, and both have the same length, L. When there is no load on either beam, the beams are both horizontal. For the two-beam system with a uniform load on beam 2, as shown in Fig. P7.6-13, determine expressions for the following: (a) the common deflection at B, $(v_1)_B = (v_2)_B$; (b) the slope of beam 1 at end B, $(v'_1)_B$; and (c) the slope of beam 2 and end C, $(v'_2)_C$.

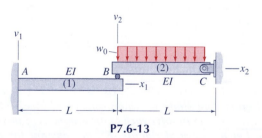

P7.6-13

▼ **SUPERPOSITION APPLICATIONS—STATICALLY DETERMINATE BEAMS**

Prob. 7.6-14. An 8-in. (nominal) standard steel pipe acts as a cantilever beam that supports two 5-kip loads from hanger rods as shown in Fig. P7.6-14. (See Table D.7 for the cross-sectional properties of the pipe.) Determine the deflections v_B at section B and v_C at the right end. Let $E_{\text{steel}} = 29 \times 10^3$ ksi.

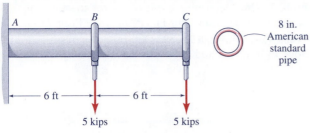

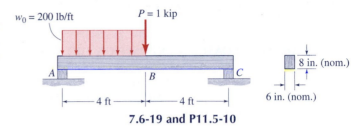

P7.6-14 and P11.5-9

Prob. 7.6-15. Determine the slope v'_C and deflection v_C at end C of the **W**14×120 wide-flange beam in Fig. P7.6-15. (See Table D.1 for the cross-sectional properties of the beam.) Let $E_{steel} = 29 \times 10^3$ ksi.

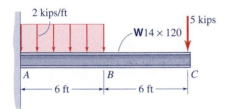

P7.6-15, P11.5-7, and P11.5-32

Prob. 7.6-16. Determine the slope v'_C and deflection v_C at end C of the timber beam in Fig. P7.6-16. Let $E_w = 1.60 \times 10^3$ ksi. (See Table D.8 for the cross-sectional properties of the beam.)

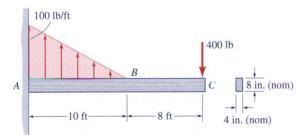

P7.6-16, P11.5-8, and P11.5-33

Prob. 7.6-17. Determine the slope v'_A at end A and the deflection v_B at the midspan section B of the **W**310×143 steel beam in Fig. P7.6-17. (See Table D.2 for the cross-sectional properties of the beam.) Let $E_{steel} = 200$ GPa.

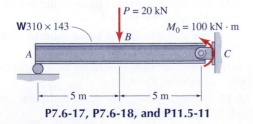

P7.6-17, P7.6-18, and P11.5-11

Prob. 7.6-18. Determine the maximum deflection of the wide-flange beam AC in Fig. P7.6-18.

Prob. 7.6-19. A 6 in. $\times$ 8 in. (nominal) simply supported timber beam supports a uniformly distributed load of $w_0 = 200$ lb/ft from A to B and a concentrated midspan load of 1 kip. (See Table D.8 for the cross-sectional properties of this beam.) Calculate the slope v'_A at end A and the deflection v_B at midspan. Let $E_w = 1,600$ ksi.

7.6-19 and P11.5-10

Prob. 7.6-20. An aluminum-alloy beam has a moment of inertia $I = 50 \times 10^6$ mm^4 and is supported by a pin at A and a 3-m-long aluminum rod that is attached to the beam at C. The cross-sectional area of the rod CD is 200 mm^2. Let $E_a = 70$ GPa. (a) Determine the displacement of the beam at C. (b) By superposing the deflection of a simply supported flexible beam and the deflection that a "rigid" beam would experience due to the stretching of rod CD, determine the total deflection at the load point, B.

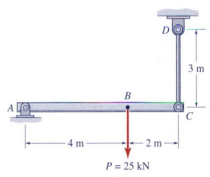

P7.6-20 and P11.5-35

*Prob. 7.6-21. A wide-flange beam supports a uniform load of 2 kips/ft on a 14-ft span. The beam has a moment of inertia of $I = 300$ in^4, and at end B it is supported by a steel rod with a cross-sectional area of 0.50 in^2. Let $E_{steel} = 29 \times 10^3$ ksi. (a) Determine the deflection of the beam at B. (b) Determine expressions for the slope $v'(x)$ and deflection $v(x)$ of the beam by superposing the deflection of a simply supported flexible beam and the deflection that a "rigid" beam would experience due to the stretching of rod BC. (c) Determine the maximum deflection of the beam.

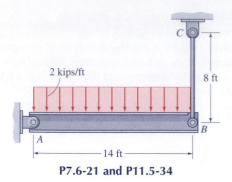

2 kips/ft

8 ft

14 ft

P7.6-21 and P11.5-34

▼ DIFFERENTIAL-LOAD SUPERPOSITION METHOD

> **Problems 7.6-22 through 7.6-30.** *Use the Differential-Load Superposition Method of Example Problem 7.11 to solve problems 7.6-22 through 7.6-30.*

Probs. 7.6-22 and 7.6-23. Determine expressions for the slope v'_B and deflection v_B at end B.

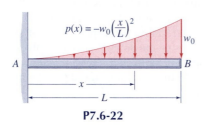

$$p(x) = -w_0\left(\frac{x}{L}\right)^2$$

w_0

x

L

P7.6-22

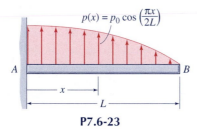

$$p(x) = p_0 \cos\left(\frac{\pi x}{2L}\right)$$

x

L

P7.6-23

Prob. 7.6-24. Determine expressions for the slope v'_C and deflection v_C for the cantilever beam in Fig. P7.6-24 (see Prob. 7.6-5).

Prob. 7.6-25. Determine expressions for the slope v'_D and deflection v_D at end D.

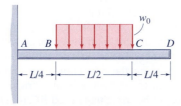

w_0

A B C D

$L/4$ $L/2$ $L/4$

P7.6-25

Prob. 7.6-26. Determine expressions for the slope v'_C and deflection v_C at end C. For Fig. P7.6-26 see Prob. 7.6-6.

Probs. 7.6-27 through 7.6-30. Determine expressions for the slope v'_A at end A and deflection v_B at section B.

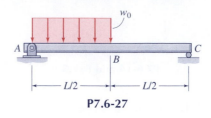

w_0

A B C

$L/2$ $L/2$

P7.6-27

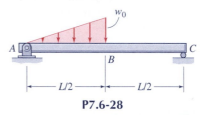

w_0

A B C

$L/2$ $L/2$

P7.6-28

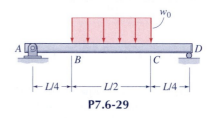

w_0

A B C D

$L/4$ $L/2$ $L/4$

P7.6-29

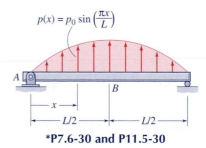

$$p(x) = p_0 \sin\left(\frac{\pi x}{L}\right)$$

A B

x

$L/2$ $L/2$

*P7.6-30 and P11.5-30

▼ SUPERPOSITION METHOD — STATICALLY INDETERMINATE BEAMS

Prob. 7.6-31. Use superposition of beam-deflection solutions from Table E.1 to solve Prob. 7.4-1.

Prob. 7.6-32. Use superposition of beam-deflection solutions from Table E.2 to solve Prob. 7.4-2.

Prob. 7.6-33. Use superposition of beam-deflection solutions from Table E.1 to solve Prob. 7.4-3.

Prob. 7.6-34. Use superposition of beam-deflection solutions from Table E.2 to solve Prob. 7.4-3.

Prob. 7.6-35. Use superposition of beam-deflection solutions from Table E.1 to solve Prob. 7.4-4.

Prob. 7.6-36. Use superposition of beam-deflection solutions from Table E.2 to solve Prob. 7.4-4.

Prob. 7.6-37. Use superposition of beam-deflection solutions from Table E.1 to solve Prob. 7.4-5.

Prob. 7.6-38. Use superposition of beam-deflection solutions from Table E.2 to solve Prob. 7.4-5.

*Prob. 7.6-39. Use superposition of beam-deflection solutions from Table E.1 to solve Prob. 7.4-6.

Prob. 7.6-40. Use superposition of beam-deflection solutions from Table E.2 to solve Prob. 7.4-7.

Prob. 7.6-41. Use superposition of beam-deflection solutions from Table E.1 to solve Prob. 7.4-9.

*Prob. 7.6-42. Use superposition of beam-deflection solutions from Table E.1 to solve Prob. 7.4-9.

Prob. 7.6-43. Use superposition of beam-deflection solutions from Table E.2 to solve Prob. 7.4-10.

Prob. 7.6-44. Use superposition of beam-deflection solutions from Table E.2 to solve Prob. 7.4-11.

Prob. 7.6-45. Use superposition of beam-deflection solutions from Table E.1 to solve Prob. 7.4-17.

Prob. 7.6-46. Use superposition of beam-deflection solutions from Table E.2 to solve Prob. 7.4-18.

Prob. 7.6-47. The cantilever beam AC in Fig. P7.6-47 has additional support from cable CD. The beam supports a uniform load of intensity w_0 over the left half of its length, as shown. Before the load is applied, the beam is straight and the cable is taut, but force-free. Determine an expression that relates the deflection of the beam at end C, v_C, to the load intensity, w_0; the flexural rigidity of the beam, EI; the axial rigidity of the cable, EA; and the lengths of the beam and the cable.

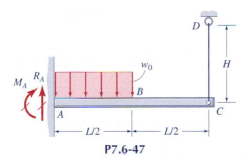

P7.6-47

Prob. 7.6-48. The cantilever beam AC in Fig. P7.6-48 has a moment of inertia $I = 50 \times 10^6$ mm^4 and is supported by rod CD, whose cross-sectional area is $A = 200$ mm^2. Let $E_{beam} = E_{rod}$. A concentrated load $P = 25$ kN is applied to the beam at B. The rod CD is force-free prior to application of the load P. Determine the tension induced in rod CD.

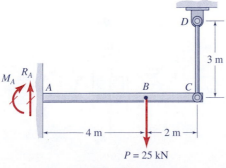

$P = 25$ kN

P7.6-48

Prob. 7.6-49. The simply supported beam AC in Fig. P7.6-49 has additional support from a cable BD that is attached to the beam at its midspan, B. The beam supports a uniform load of intensity w_0 over its entire length, as shown. Before the load is applied, the cable is taut but force-free. Determine an expression that relates the cable tension to the load intensity, w_0; the flexural rigidity of the beam, EI; the axial rigidity of the cable, EA; and the lengths of the beam and the cable.

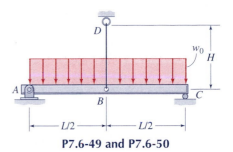

P7.6-49 and P7.6-50

*Prob. 7.6-50. (a) Let T be the tension in cable BD, and draw shear and moment diagrams of the beam AC in Fig. P7.6-50. (b) Determine the relationship between T and the total load w_0L that minimizes the maximum magnitude of the moment in the beam AC. (c) If all values except H are specified (i.e., EI, AE, and L are given), what value of H would give the condition described in Part (b)?

*Prob. 7.6-51. For the two-span uniform beam with distributed loading as shown, use the superposition method of Example Problem 7.14 to solve for the three redundant reactions R_B, R_C, and M_C. Note that you can use subproblem solutions for <u>uniform cantilever beams</u> from Table E.1. You will have to solve three simultaneous equations for the three unknown reactions.

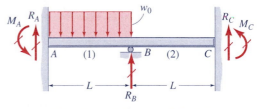

*P7.6-51, *P7.6-52, and P7.7-5

*Prob. 7.6-52. For the two-span uniform beam with distributed loading as shown, use the superposition method of Example Problem 7.14 to solve for the three redundant reactions M_A, R_B, and M_C. Note that you can use subproblem solutions for <u>simply supported uniform beams</u> from Table E.2. You will have to solve three simultaneous equations for the three unknown reactions.

*Probs. 7.6-53 and 7.6-54. Each of these two-span continuous-beam problems involves a nonuniform beam with one degree of static indeterminacy. To solve these by Force-Method Superposition you can treat each of these problems as two simply supported beams that have **compatible slope** where they are joined together at B. One beam has flexural rigidity EI, and the other has flexural rigidity $2EI$. You can find the necessary subproblem solutions in Table E.2. You will have to consider carefully the internal shear force and bending moment acting on each beam at joint B. The free-body diagrams below give you appropriate names for these internal shear forces and bending moments.

For these nonuniform beams, (a) determine the internal moments at B, $(M_1)_B$ and $(M_2)_B$; and (b) determine the reactions R_A, and R_B.

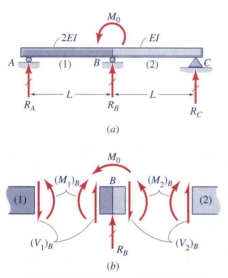

*P7.6-53, P7.7-12, and P11.5-49

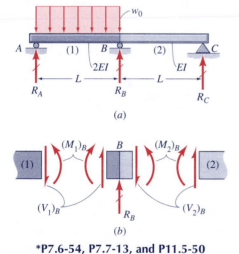

*P7.6-54, P7.7-13, and P11.5-50

*Prob. 7.6-55. This problem involves a nonuniform statically indeterminate beam that has two degrees of static indeterminacy. To solve this problem by Force-Method Superposition you can treat the nonuniform beam as two cantilever beams that have **compatible displacement** and **compatible slope** where they are joined together at B. One

beam has flexural rigidity EI, and the other has flexural rigidity $2EI$. You can find the necessary subproblem solutions in Table E.1. You will have to consider carefully the internal shear force and bending moment acting on each beam at joint B. The free-body diagram of joint B gives you appropriate names for these internal shear forces and bending moments.

For this nonuniform beam, (a) determine the internal moments at B, $(M_1)_B = (M_2)_B$, and the internal shear forces at B, $(V_1)_B$ and $(V_2)_B$; and (b) determine the reactions R_A and M_A.

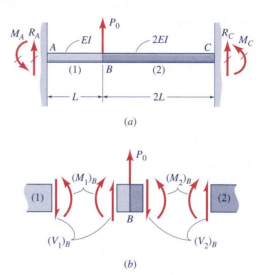

*P7.6-55, P7.7-3, P11.5-25, and P11.5-45

*Probs. 7.6-56 and 7.6-57. Each of these problems involves a two-span nonuniform beam with three degrees of static indeterminacy. To solve these by Force-Method Superposition you can treat each of these problems as two uniform cantilever beams that have **compatible (zero) displacement** and **compatible slope** where they are joined together at B. One beam has flexural rigidity EI, and the other has flexural rigidity $2EI$. You can find the necessary subproblem solutions

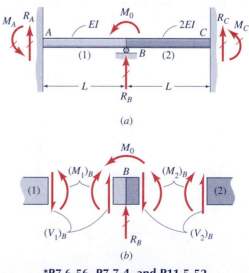

*P7.6-56, P7.7-4, and P11.5-52

in Table E.1. You will have to consider carefully the internal shear force and bending moment acting on each beam at joint B. The free-body diagrams below give you appropriate names for these internal shear forces and bending moments. (These problems are more easily solved by Displacement-Method Superposition, and are among the homework problems for Section 7.7.)

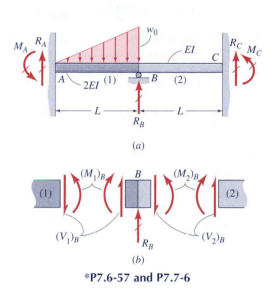

(a)

(b)

*P7.6-57 and P7.7-6

▼ **BEAM DEFLECTIONS: DISPLACEMENT METHOD**   MDS 7.4

Problems 7.7-1 through 7.7-15. *The problems in this section are to be solved by using* **Displacement-Method Superposition.** *The key force-deformation equations are Eqs. 7.19 and 7.20 (p. 514). For distributed-load problems, Table E.3 provides the basic fixed-end forces. The bending stiffness, EI, is constant unless otherwise noted.*

Prob. 7.7-1. A couple M_0 is applied at the propped end of the uniform propped-cantilever beam in Fig. P7.7-1. Determine the reaction R_A.

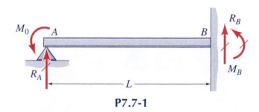

P7.7-1

Prob. 7.7-2 For the fixed-fixed beam shown, (a) determine the transverse displacement δ_B and the slope θ_B at node B, and (b) determine the reactions R_A and M_A at end A. (Note, the lengths are $L_1 = L$ and $L_2 = 2L$.)

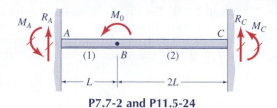

P7.7-2 and P11.5-24

Prob. 7.7-3. For the fixed-fixed beam shown in Fig. P7.7-3, (a) determine the transverse displacement δ_B and the slope θ_B at node B, and (b) determine the reactions R_A and M_A at end A. (Note, the lengths are $L_1 = L$ and $L_2 = 2L$.) For Fig. P7.7-3, see Fig. P7.6-55.

Prob. 7.7-4. An external couple M_0 is applied at node B of the two-span beam shown in Fig. P7.7-4. (a) Determine θ_B, the angle of rotation at B, and (b) determine the reactions R_A, M_A, and R_B. For Fig. P7.7-4, see Fig P7.6-56.

Probs. 7.7-5 through 7.7-7. For the two-span beams with distributed loading as shown, (a) determine θ_B, the angle of rotation at B, and (b) determine the reactions R_A, M_A, and R_B. For Fig. P7.7-5, see Fig. P7.6-51. For Fig. P7.7-6, see Fig. P7.7-6, see Fig. P7.6-57.

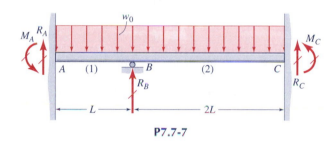

P7.7-7

Probs. 7.7-8 and 7.7-9. For the uniform fixed-fixed beams shown, (a) determine the transverse displacement δ_B and the slope θ_B at node B, and (b) determine the reactions R_A and M_A at end A.

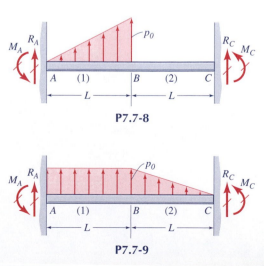

P7.7-8

P7.7-9

Probs. 7.7-10 and 7.7-11. For the two-span beams with distributed loading as shown, (a) determine θ_A and θ_B, the angles of rotation at nodes A and B, respectively, and (b) determine the reactions R_A and R_B at the respective nodes.

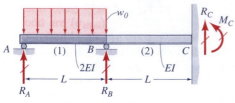

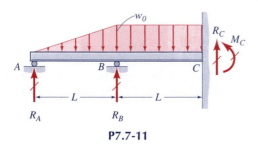

P7.7-10 and P11.5-48

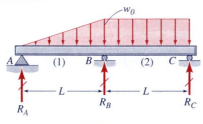

P7.7-11

Probs. 7.7-12 through 7.7-15. For the two-span continuous beams shown, (a) determine the rotation angles at the three supports: θ_A, θ_B, and θ_C; and (b) determine the reaction at A, R_A. For Fig. P7.7-12, see Fig. P7.6-53. For Fig. P7.7-13, see Fig. P7.6-54.

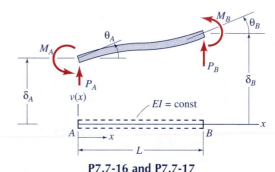

P7.7-14

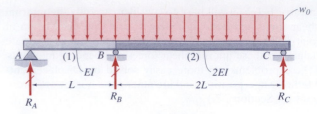

P7.7-15 and P11.5-51

Prob. 7.7-16. Starting with Eqs. 7.16 and using equilibrium and superposition, show that, for a uniform beam with *nodal* (end) *displacements* δ_A, θ_A, δ_B, and θ_B, the corresponding *nodal forces* P_A, M_A, P_B, and M_B, are given by the equations:

$$P_A = \left(\frac{EI}{L^3}\right)(12\delta_A + 6L\theta_A - 12\delta_B + 6L\theta_B)$$

$$M_A = \left(\frac{EI}{L^3}\right)(6L\delta_A + 4L^2\theta_A - 6L\delta_B + 2L^2\theta_B)$$

$$P_B = \left(\frac{EI}{L^3}\right)(-12\delta_A - 6L\theta_A + 12\delta_B - 6L\theta_B)$$

$$M_B = \left(\frac{EI}{L^3}\right)(6L\delta_A + 2L^2\theta_A - 6L\delta_B + 4L^2\theta_B)$$

P7.7-16 and P7.7-17

Prob. 7.7-17. Starting with the second-order deflection equation $EIv''(x) = M(x)$, derive the four equations given in Prob. 7.7-16 for the special case when only δ_A is nonzero, that is, when $\theta_A = \delta_B = \theta_B = 0$, but $\delta_A \neq 0$. (Hint: The boundary conditions are: $v(0) = \delta_A$, $v'(0) = v'(L) = v(L) = 0$.) What is the expression for the deflection curve $v(x)$ due to δ_A?

CHAPTER 7 REVIEW — DEFLECTION OF BEAMS

Section		Suggested Review Problems
7.1	Transverse loads, either concentrated loads or distributed loads, and bending couples cause **beams** to deflect, that is, the beam bends to form a curve that is called the **deflection curve.** The *deflection curve, $v(x)$,* is identified in Fig. 7.1a. Deflection of a Beam (Fig. 7.1a)	
7.2	Analysis of beams involves four distinct quantities: • Two *deflection-type quantities*: **deflection v** and **slope** $$\theta \equiv v' \equiv \frac{dv}{dx};\text{ and}$$ • Two *force-type quantities*: **transverse shear V** and **bending moment M.** Starting with Eq. 7.1, you should go through the steps of the derivation of the **moment-curvature equation,** Eq. 7.8. This is the key equation that is used to relate the deflection of a linearly elastic beam to the loads acting on the beam. Another equation that is very useful for calculating the deflection of a linearly elastic beam is the **load-deflection equation,** Eq. 7.12.	 Notation and Sign Convention for Beam Deflection (Fig. 7.4) $$EIv'' = M(x) \quad \textbf{Moment-curvature equation} \quad (7.8)$$ $$(EIv'')'' = p(x) \quad \textbf{Load-deflection equation} \quad (7.12)$$ The sign conventions for transverse distributed load, $p(x)$; transverse shear, $V(x)$; and bending moment, $M(x)$, are shown in Fig. 7.7. Derive Eq. 7.8.
	You should review Table 7.1 **Boundary Conditions** and Table 7.2 **Continuity Conditions,** which are used in the integration of the second-order Moment-Curvature Equation and the fourth-order Load-Deflection Equation.	The **Moment-Curvature Equation,** Eq. 7.8 above, and the **Shear-Deflection Equation,** Eq. 7.11, are useful in formulating the Boundary Condition shears and moments of Table 7.1, and the shears and moments in the Continuity Conditions of Table 7.2. $$(EIv'')' = V(x) \quad \textbf{Shear-deflection equation} \quad (7.11)$$ 7.2-3

Section			Suggested Review Problems
7.3	Section 7.3 treats the determination of the deflection of **statically determinate beams** by integration of the second-order *Moment-Curvature Equation*, Eq. 7.8, or by integration of the fourth-order *Load-Deflection Equation*, Eq. 7.12.	You should quickly review the **Procedure** near the beginning of Section 7.3. Examples 7.1 and 7.2 treat *single-span beams*; Examples 7.3 and 7.4 treat two-span beams, where it is necessary to use the *continuity conditions* of Table 7.2 in addition to the *boundary conditions* of Table 7.1.	7.3-1 7.3-5 7.3-15 7.3-23 7.3-29
7.4	Section 7.4 treats the analysis of **statically indeterminate beams** by integration of the second-order *Moment-Curvature Equation*, Eq. 7.8, or by integration of the fourth-order *Load-Deflection Equation*, Eq. 7.12.	As is illustrated in Fig. 7.10, statically indeterminate beams have more boundary conditions than are necessary to support the beam. Excess boundary conditions are called **redundants**. A Statically Indeterminate Beam. (Fig. 7.10)	7.4-1 7.4-5 7.4-11 7.4-13
7.6	Section 7.6 treats the analysis of both **statically determinate beams** and **statically indeterminate beams** by the **Superposition Method.**	You should quickly review the procedure titled **Superposition Procedure − Statically Determinate Beams** and Examples 7.9 and 7.10.	7.6-1 7.6-5 7.6-17
	Section 7.6 also treats the analysis of beam deflection by the **Differential-Load Superposition Method.**	Also, quickly review the procedure titled **Procedure for Differential-Load Superposition** and Example 7.11.	7.6-23 7.6-27
	Finally, Section 7.6 discusses the analysis of **statically indeterminate beams** by the **Superposition Method,** which is directly related to the *Basic Force Method* discussed in Sections 3.5 and 4.6.	You should carefully review the procedure titled **Superposition Procedure − Statically Indeterminate Beams** and Examples 7.12 through 7.15.	7.6-33 7.6-34 7.6-35

Sections 7.5 (Use of Discontinuity Functions to Determine Beam Deflections), and 7.7 (Slope and Deflection of Beams: Displacement Method) are "optional" sections.

TRANSFORMATION OF STRESS AND STRAIN; MOHR'S CIRCLE

8

8.1 INTRODUCTION

In earlier chapters you were introduced to *stress analysis*; in particular, you learned about the stress distribution on cross sections of members loaded axially $\left(\sigma = \dfrac{F}{A}\right)$, in torsion $\left(\tau = \dfrac{T\rho}{I_p}\right)$, or in bending $\left(\sigma = \dfrac{-My}{I}, \tau = \dfrac{VQ}{It}\right)$. In Section 2.9 you learned that, even though only axial loading is applied to a member, the stress distribution on an inclined sectioning-plane consists of shear stress as well as normal stress. You also learned that the normal stress and shear stress on an oblique plane are directly related to the axial stress through stress-transformation equations. Similarly, in Section 4.4 an analysis was performed for circular rods in torsion, which explained the very different failures experienced by a ductile torsion rod and by a brittle torsion rod (Fig. 4.17).

Figure 8.1 is a photo of reinforced concrete beams tested to failure. You can see that near the center of the beams the cracking occurs basically on cross sections of the beams, but away from the center of the beams the cracks are quite slanted. In Chapter 9 we will consider the stress distribution in beams that leads to such patterns of cracking. We will also examine other examples of *combined loading* of slender members, that is, loads that produce various combinations of axial force, torque, bending moment, and transverse shear force. But first, in order to completely characterize the **state of stress** produced by a single type of load or by a combination of loads, we must develop **stress-transformation equations.**

We begin, in Section 8.2, by introducing a two-dimensional state of stress called *plane stress*. The stress-transformation equations that are derived in Section 8.3 are a generalization of the stress-transformation equations developed in Sections 2.9 and 4.4. In Section 8.4 we locate maxima and minima of normal stress and shear stress, and in Section 8.5 a very important graphical representation of this stress transformation, called *Mohr's circle,* is introduced.

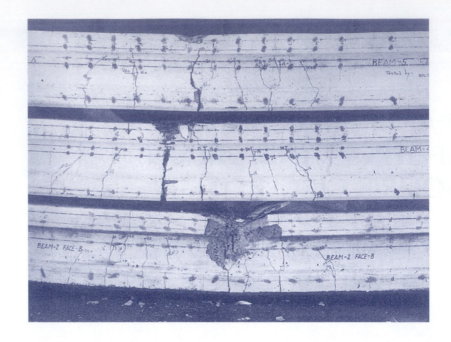

FIGURE 8.1 Typical crack patterns in reinforced concrete beams. (From Prestressed Concrete- A Fundamental Approach, 5th Ed. Update, 2010, Prentice Hall/Pearson Education, Courtesy Dr. Edward G. Nawy, Rutgers University, New Brunswick, New Jersey)

As you will discover in Section 8.3, the *stress-transformation equations* are based solely on *equilibrium;* they do not depend on material properties (linearly elastic behavior, etc.) or on the geometry of deformation. However, there are similar *strain transformation equations;* these are based solely on the *geometry of deformation.* The transformation of strain is discussed in Sections 8.8 through 8.11.

8.2 PLANE STRESS

In general, if we were to pass three mutually orthogonal planes through any point in a deformable body under load (Fig. 8.2*a*), we would find normal and/or shear stresses on all three planes (Fig. 8.2*b*). This is called a *three-dimensional state of stress.* Fortunately, there are many important instances where some of these stresses vanish, so that there is then a *two-dimensional state of stress.* For example, consider the stresses acting on a small element taken from the web of a plate girder of a

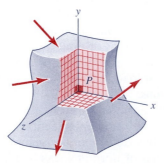

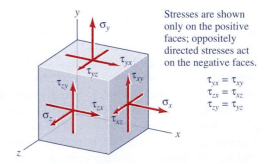

Stresses are shown only on the positive faces; oppositely directed stresses act on the negative faces.

$$\tau_{yx} = \tau_{xy}$$
$$\tau_{zx} = \tau_{xz}$$
$$\tau_{zy} = \tau_{yz}$$

FIGURE 8.2 A general three-dimensional state of stress.

(*a*) A set of three mutually orthogonal planes through an arbitrary point *P*.

(*b*) A three-dimensional state of stress referred to rectangular cartesian axes. (Stresses are shown only on visible faces.)

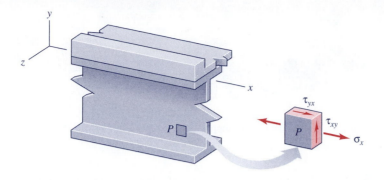

(a) A portion of a bridge girder.

(b) A web element, with stresses shown.

FIGURE 8.3 An example of plane stress.

bridge (Fig. 8.3). Because the web is thin, and because there are no loads applied directly to the surface of the web, the three stresses associated with the z axis (σ_z, $\tau_{zx} = \tau_{xz}$, and $\tau_{zy} = \tau_{yz}$) are all zero. The nonzero stresses are said to lie in the xy plane. Strictly speaking, the nonzero stresses act on planes whose normal vectors lie in the xy plane. Hence, the arrows representing these stresses all lie in the xy plane, as in Fig. 8.3b.

If the stresses σ_z, τ_{xz}, and τ_{yz} vanish everywhere, that is, if

$$\sigma_z = \tau_{xz} = \tau_{yz} = 0 \tag{8.1}$$

the state of stress in a body is said to be **plane stress.** The nonzero stresses implied by Eqs. 8.1 (σ_x, σ_y, $\tau_{xy} = \tau_{yx}$) lie in the xy plane.[1] Figure 8.4 shows three-dimensional and two-dimensional views of an element in plane stress. From moment equilibrium it was found (Eqs. 2.37) that

$$\tau_{yx} = \tau_{xy} \tag{8.2}$$

Since a two-dimensional view can depict the relevant (i.e., nonzero) stress information, it is not really necessary to use a three-dimensional view when depicting plane stress. Special cases of plane stress are *uniaxial stress* (Fig. 8.5a), *pure shear* (Fig. 8.5b), and *biaxial stress* (Fig. 8.5c).

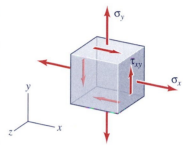

(a) Three-dimensional view.

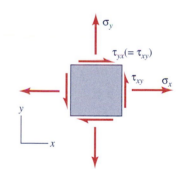

(b) Two-dimensional view.

FIGURE 8.4 A state of plane stress depicted in 3-D and in 2-D.

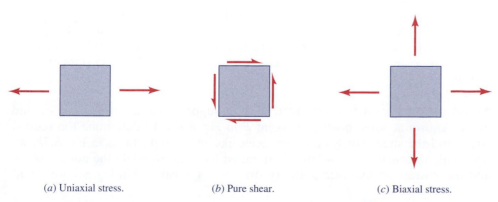

(a) Uniaxial stress.

(b) Pure shear.

(c) Biaxial stress.

FIGURE 8.5 Special cases of plane stress.

---

[1]A state of plane stress would also exist if the stresses were in the xz plane only or the yz plane only.

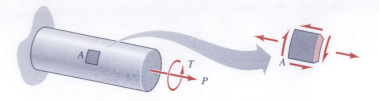

FIGURE 8.6 An example
of two-dimensional stress at the
surface of a deformable body.

(*a*) A rod with axial load and torque.   (*b*) A two-dimensional stress state.

Deformable bodies that are in a state of plane stress are usually thin, plate-like members, like the web of the plate girder in Fig. 8.3.[2] The stress-transformation equations that are derived in Section 8.3 are applicable to such plane-stress problems, but they are also applicable to a much wider class of two-dimensional stress problems, where Eqs. 8.1 are valid locally, but not everywhere in the body. For example, some portions of the surface of a deformable body are subjected to distributed loading or concentrated loading, but at those parts of the surface where no loading is directly applied, the stress state may be classified as *locally two-dimensional,* or *locally plane stress.* A circular rod subjected to an axial load and a torque, as illustrated in Fig. 8.6, is an example of this type of local plane stress. Since there are no stresses on the surface of the body at the point in question, Eq. 8.1 is satisfied locally, and we can treat the stress at this point of the surface as plane stress.

8.3   STRESS TRANSFORMATION FOR PLANE STRESS

State of Stress at a Point.   If the state of plane stress at a point is known with reference to particular coordinates, say the stresses (σ_x, σ_y, and τ_{xy}) in the xy reference coordinates, what are the values of the stresses at the same point in the body if we rotate the frame of reference? That is, what will be the values of ($\sigma_{x'}$, $\sigma_{y'}$, and $\tau_{x'y'}$) on planes oriented along $x'y'$ axes? In Chapter 6 you learned how to use the flexure formula and the shear formula to determine the stresses on the x and y faces of an element located at an arbitrary point (x, y) in a beam, as illustrated in Fig. 8.7a. However, since we have no assurance that the stresses on those particular faces are the critical (i.e., maximum) ones, it is very important for us to be able to determine the stresses on inclined planes, like the n face and the t face in Fig. 8.7b. There is only one unique **state of stress** at a point, say point P in Fig. 8.7, but the state of stress has different representations, depending on the orientation of the axes used. For example, although the stresses (σ_x, σ_y, τ_{xy}) in Fig. 8.7a have different values than the stresses (σ_n, σ_t, τ_{nt}) in Fig. 8.7b, both are representations of the same state of stress at point P.

Notation and Sign Convention.   Let us suppose that the stresses σ_x, σ_y, and τ_{xy} are known at some point (e.g., point P in Fig. 8.7a). To determine the normal stress and the shear stress on other faces, like the n and t faces in Fig. 8.7b, we need only to consider one arbitrarily oriented face and to relate the normal stress and shear stress on that face to the xy stresses. As illustrated in Fig. 8.8. we orient

---

[2]Chapter 2 of *Theory of Elasticity,* by S. P. Timoshenko and J. N. Goodier, [Ref. 8-1], discusses plane stress and plane strain problems.

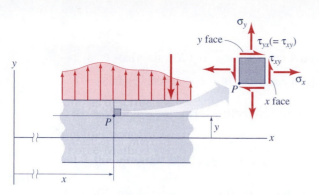

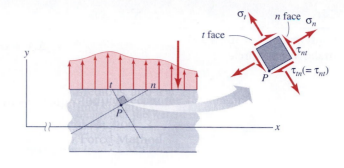

(a) The stress state at point P, referred to x and y axes.   (b) The stress state at point P, referred to n and t axes.

FIGURE 8.7 The state of stress at a point, as represented in two reference frames.

the arbitrary face, which we call the n face, by rotating its outward normal n through the counterclockwise angle $\theta \equiv \theta_{xn}$, starting with the positive x axis.

The t axis is oriented so that the ntz axes, like the xyz axes, form a right-handed coordinate frame. As always, the normal stress σ_n is *positive in tension,* and a *positive shear stress on the $+n$ face, τ_{nt} acts in the $+t$ direction.*

Stress Transformation for Plane Stress.

Let us assume that the stresses σ_x, σ_y, and τ_{xy} are known, and that the stresses σ_n and τ_{nt} are to be determined. As was pointed out earlier, the **stress transformation equations are based solely on equilibrium.** To set up equations of equilibrium, we need a *free-body diagram.* The triangular element PQR in Fig. 8.8, which is repeated as Fig. 8.9a, shows the stresses acting on the three faces of the element. A figure like Fig. 8.9a, which depicts stresses acting on various faces of an element, is called a *stress element.* The stresses act on faces that appear in edge view on the element. The areas of the various faces are given in Fig. 8.9b and, by multiplying each stress by the area of the face on which that stress acts, we obtain the corresponding free-body diagram in Fig. 8.9c.

By writing the equilibrium equations for the free body in Fig. 8.9c and substituting the areas from Fig. 8.9b, and by using the fact that $\tau_{yx} = \tau_{xy}$ (Eq. 8.2), we obtain

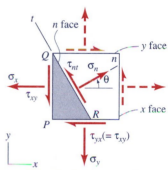

FIGURE 8.8 The relationship of an arbitrarily oriented n face to the reference (xy) axes.

$$^+\!\!\nearrow \sum F_n = 0:$$

$$\sigma_n \Delta A - \sigma_x(\Delta A \cos \theta)\cos \theta - \sigma_y(\Delta A \sin \theta)\sin \theta$$
$$-\tau_{xy}(\Delta A \cos \theta)\sin \theta - \tau_{xy}(\Delta A \sin \theta)\cos \theta = 0$$

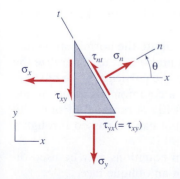

(a) A triangular stress element.

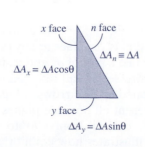

(b) The areas of the respective faces.

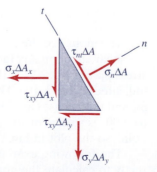

(c) The free-body diagram.

FIGURE 8.9 The free-body diagram based on a plane-stress element.

545

$+\searrow \sum F_t = 0:$

$$\tau_{nt}\Delta A + \sigma_x(\Delta A \cos\theta)\sin\theta - \sigma_y(\Delta A \sin\theta)\cos\theta$$

$$-\tau_{xy}(\Delta A \cos\theta)\cos\theta + \tau_{xy}(\Delta A \sin\theta)\sin\theta = 0$$

Dividing these equations by ΔA and collecting terms, we get

$$\sigma_n = \sigma_x \cos^2\theta + \sigma_y \sin^2\theta + \tau_{xy}(2\sin\theta\cos\theta)$$

$$\tau_{nt} = -(\sigma_x - \sigma_y)\cos\theta\sin\theta + \tau_{xy}(\cos^2\theta - \sin^2\theta) \tag{8.3}$$

Remember that the angle θ in these equations is measured <u>counterclockwise from the x face to the n face</u> (or, equivalently, from the $+x$ axis to the $+n$ axis), as indicated in Fig. 8.9a.

Equations 8.3 can be expressed in a more convenient form by incorporating the following trigonometric identities:

$$2\sin\theta\cos\theta = \sin 2\theta$$

$$\sin^2\theta = \frac{1}{2}(1 - \cos 2\theta) \tag{8.4}$$

$$\cos^2\theta = \frac{1}{2}(1 + \cos 2\theta)$$

Then, from Eqs. 8.3, the **stress-transformation equations for plane stress** become

$$\sigma_n = \left(\frac{\sigma_x + \sigma_y}{2}\right) + \left(\frac{\sigma_x - \sigma_y}{2}\right)\cos 2\theta + \tau_{xy}\sin 2\theta$$

$$\tau_{nt} = -\left(\frac{\sigma_x - \sigma_y}{2}\right)\sin 2\theta + \tau_{xy}\cos 2\theta$$

Stress-Trans-formation Equations (8.5)

It should be emphasized again that, to obtain this stress transformation, we had to multiply stresses times areas to get the *forces* acting on the free body in Fig. 8.9c. It is <u>not</u> correct to just sum the stresses on a stress element, like the one in Fig. 8.9a!

To emphasize the fact that Eqs. 8.5 enable us to compute the normal stress and the shear stress on any face, that is, on a face at any orientation, let us generalize the stress plot of Fig. 2.33 and obtain a plot of σ_n and τ_{nt} for the particular state of stress indicated in Fig. 8.10a. The stresses are all referred to a common stress magnitude σ_0, and their senses are indicated by the arrows in Fig. 8.10a. A range of θ from $-90°$ to $+90°$ is sufficient to represent all possible planes that can be passed through a point, so the plot in Fig. 8.10b extends from $-90°$ to $+90°$.

The following example illustrates how equilibrium equations may be used directly to calculate the normal stress and shear stress on an oblique face.

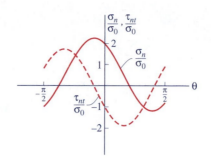

$\sigma_x = 2\sigma_0$
$\sigma_y = -\sigma_0$
$\tau_{xy} = -\sigma_0$

(a) A sample state of plane stress.

(b) Normal stress and shear stress as functions of the orientation of the stress face.

FIGURE 8.10 The normal stress and shear stress corresponding to the sample state of plane stress.

EXAMPLE 8.1

The state of plane stress at a point is indicated in Fig. 1a. Determine the normal stress $\sigma_{x'}$ and the shear stress $\tau_{x'y'}$ on the x' face, which is rotated $30°$ counterclockwise from the x face, as illustrated in Fig. 1b.

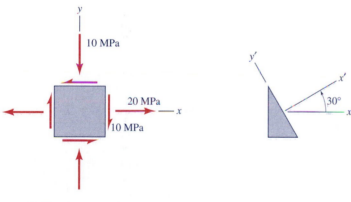

(a) The given state of plane stress at a point.

(b) The orientation of the x' face.

Fig. 1 Stress-transformation data.

Plan the Solution   We can follow the procedure used to derive Eqs. 8.3, using the specific angle of $30°$ rather than the general angle θ.

Solution

Equilibrium: The free-body diagram is shown in Fig. 2.

$+\nearrow \sum F_{x'} = 0:$

$$\sigma_{x'}\Delta A - 20\Delta A_x \cos(30°) + 10\Delta A_x \sin(30°)$$
$$+ 10\Delta A_y \sin(30°) + 10\Delta A_y \cos(30°) = 0$$

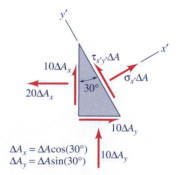

$\Delta A_x = \Delta A \cos(30°)$
$\Delta A_y = \Delta A \sin(30°)$

Fig. 2 The free-body diagram.

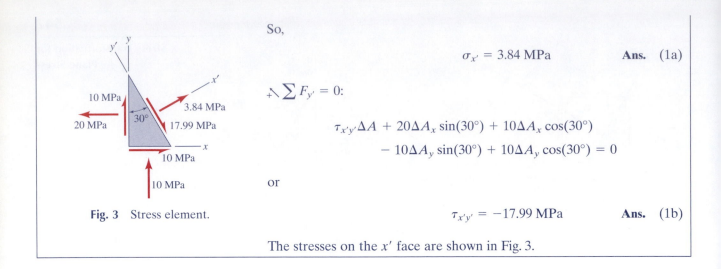

So,

$$\sigma_{x'} = 3.84 \text{ MPa} \qquad \textbf{Ans.} \quad (1a)$$

$$+\nwarrow \sum F_{y'} = 0:$$

$$\tau_{x'y'}\Delta A + 20\Delta A_x \sin(30°) + 10\Delta A_x \cos(30°)$$
$$- 10\Delta A_y \sin(30°) + 10\Delta A_y \cos(30°) = 0$$

or

$$\tau_{x'y'} = -17.99 \text{ MPa} \qquad \textbf{Ans.} \quad (1b)$$

Fig. 3 Stress element.

The stresses on the x' face are shown in Fig. 3.

In Example 8.1, the values of the normal stress and the shear stress on a particular face at a specified angle were obtained by the direct use of a free-body diagram and equilibrium equations. The *stress-transformation equations,* Eqs. 8.5, could also have been used to solve this problem. Starting with Eqs. 8.5 we will: (1) derive expressions for the *maximum normal stress* and the *maximum shear stress*

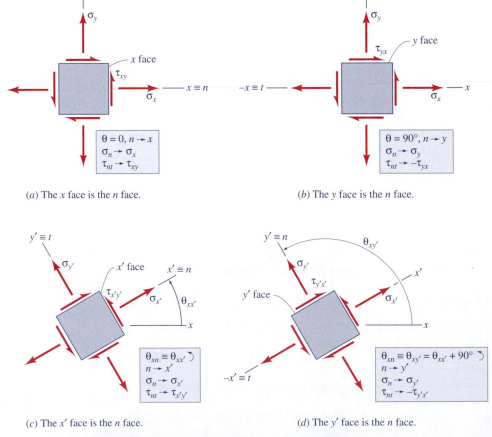

(a) The x face is the n face.

(b) The y face is the n face.

(c) The x' face is the n face.

(d) The y' face is the n face.

FIGURE 8.11 The stresses on orthogonal faces.

at a point (Section 8.4), and (2) develop the *Mohr's circle* graphical representation of the stress-transformation equations (Section 8.5). First, however, formulas for the normal stress and shear stress on two orthogonal faces of a rotated element, like the x' face and the y' face of the $x'y'$ element in Fig. 8.11c, will be derived.

Stresses on Orthogonal Faces. The two stress-transformation equations for plane stress, Eqs. 8.5, relate the normal stress σ_n and the shear stress τ_{nt} on an arbitrary face to a given set of xy stresses. However, we frequently need to make reference to the stresses on two orthogonal faces—for example, the x face and the y face. Figure 8.11a illustrates the case of $\theta = 0$. Then, as indicated on Fig. 8.11a, $n \to x$, $t \to y$, $\sigma_n \to \sigma_x$, and $\tau_{nt} \to \tau_{xy}$. When we consider the y face (Fig. 8.11b), we must remember that the ntz axes form a right-handed coordinate system. Therefore, when $\theta = 90°$, $n \to y$, $t \to -x$, $\sigma_n \to \sigma_y$, and $\tau_{nt} \to -\tau_{yx}(= -\tau_{xy})$. The last of these equivalencies, $\tau_{nt} \to -\tau_{xy}$, results from the fact that the t axis corresponds to the $-x$ axis, not the $+x$ axis.

Now let us consider a pair of arbitrarily oriented orthogonal faces, for example, the x' face and y' face in Fig. 8.11c and Fig. 8.11d. Let the orientation of the faces x' and y' be specified by the one angle $\theta_{xx'}$. Then, $\theta_{xy'} = \theta_{xx'} + 90°$. Figure 8.11c relates σ_n and τ_{nt} to the stresses on the x' face, and Fig. 8.11d relates σ_n and τ_{nt} to the stresses on the y' face. We can calculate these stresses using Eqs. 8.5, noting that $2\theta_{xy'} = 2(\theta_{xx'} + 90°) = 2\theta_{xx'} + 180°$.

$$\sigma_{x'} = \sigma_n(\theta_{xx'}) = \left(\frac{\sigma_x + \sigma_y}{2}\right) + \left(\frac{\sigma_x - \sigma_y}{2}\right)\cos 2\theta_{xx'} + \tau_{xy}\sin 2\theta_{xx'}$$

$$\tau_{x'y'} = \tau_{nt}(\theta_{xx'}) = -\left(\frac{\sigma_x - \sigma_y}{2}\right)\sin 2\theta_{xx'} + \tau_{xy}\cos 2\theta_{xx'}$$

$$\sigma_{y'} = \sigma_n(\theta_{xx'} + 90°) = \left(\frac{\sigma_x + \sigma_y}{2}\right) - \left(\frac{\sigma_x - \sigma_y}{2}\right)\cos 2\theta_{xx'} - \tau_{xy}\sin 2\theta_{xx'}$$

$$\tau_{y'x'} = -\tau_{nt}(\theta_{xx'} + 90°) = -\left(\frac{\sigma_x - \sigma_y}{2}\right)\sin 2\theta_{xx'} + \tau_{xy}\cos 2\theta_{xx'} = \tau_{x'y'}$$

(8.6)

Note that, as expected, $\tau_{y'x'} = \tau_{x'y'}$.

> **EXAMPLE 8.2**

The state of plane stress at a point is indicated in Fig. 1a. Use Eqs. 8.5 to determine the stresses on faces that are rotated 30° counterclockwise from the orientation of the element in Fig. 1a, as illustrated in Fig. 1b. Show the stresses on a rotated stress element.

Plan the Solution For the x' face $\theta = 30°$, while for the y' face $\theta = 30° + 90° = 120°$ (or we could use $\theta = -60°$). This is a straightforward "plug-in" type problem employing Eqs. 8.5 (or 8.6). From Fig. 1a we have:

$$\sigma_x = 20 \text{ MPa}, \quad \sigma_y = -10 \text{ MPa}, \quad \tau_{xy} = -10 \text{ MPa}$$

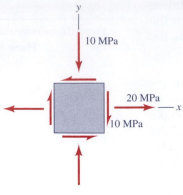

10 MPa

20 MPa
x
10 MPa

(a) The given state of plane stress at a point.

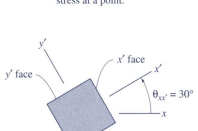

y'

x' face
x'
y' face
$\theta_{xx'} = 30°$
x

(b) An element rotated 30° counterclockwise.

Fig. 1 Stress-transformation data.

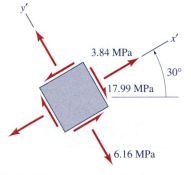

y'
x'
3.84 MPa
30°
17.99 MPa
6.16 MPa

Fig. 2 Stresses on an $x'y'$ element at $\theta = 30°$.

Solution For the x' face, $\theta = 30°$, so $2\theta = 60°$, and, referring to Fig. 8.11c, $n \to x', t \to y'$. Then, Eq. 8.5a gives

$$\sigma_{x'} = \left(\frac{\sigma_x + \sigma_y}{2}\right) + \left(\frac{\sigma_x - \sigma_y}{2}\right)\cos 2\theta + \tau_{xy}\sin 2\theta$$

$$\sigma_{x'} = \left(\frac{20\text{ MPa} - 10\text{ MPa}}{2}\right) + \left(\frac{20\text{ MPa} + 10\text{ MPa}}{2}\right)\cos(60°)$$

$$+ (-10\text{ MPa})\cos(60°)$$

or

$$\sigma_{x'} = 3.84\text{ MPa} \qquad \textbf{Ans.} \quad (1a)$$

Similarly, Eq. 8.5b gives

$$\tau_{x'y'} = -\left(\frac{\sigma_x - \sigma_y}{2}\right)\sin 2\theta + \tau_{xy}\cos 2\theta$$

$$\tau_{x'y'} = -\left(\frac{20\text{ MPa} + 10\text{ MPa}}{2}\right)\sin(60°) + (-10\text{ MPa})\cos(60°)$$

or

$$\tau_{x'y'} = -17.99\text{ MPa} \qquad \textbf{Ans.} \quad (1b)$$

For the y' face, $\theta = 120°$, so $2\theta = 240°$, and, referring to Fig. 8.11d, $n \to y', t \to -x'$. Then, from Eqs. 8.5,

$$\sigma_{y'} \equiv \sigma_n(120°), \qquad \sigma_{y'} = 6.16\text{ MPa}$$
$$\tau_{y'x'} \equiv -\tau_{nt}(120°), \qquad \tau_{y'x'} = -17.99\text{ MPa} \qquad \textbf{Ans.} \quad (2)$$

Equations (1) determine the stresses on the x' face, and Eqs. (2) determine the stresses on the y' face. These results are illustrated in Fig. 2.

Review the Solution This is a very short problem whose results should be verified by first seeing if the magnitude of each answer is reasonable. Then the calculations should be spot-checked for accuracy. In the present case, the stresses given on Fig. 1a are in the same proportions as the stresses in Fig. 8.10a. Hence, we can look at the values of σ_n/σ_0 and τ_{nt}/σ_0 in Fig. 8.10b in order to confirm our answers for this problem.

It is not a coincidence that in Example Problem 8.2, $(\sigma_x + \sigma_y) = (\sigma_{x'} + \sigma_{y'})$. By adding Eqs. 8.6a and 8.6c, we get that, independent of the angle θ,

$$\boxed{\sigma_{x'} + \sigma_{y'} = \sigma_x + \sigma_y} \qquad (8.7)$$

Therefore, the sum of the normal stresses on orthogonal faces is a constant, which is called a *stress invariant*. That is, **the sum of normal stresses on orthogonal faces does not vary with the angle θ.**

It is apparent from Fig. 8.10b that the maxima and minima of normal stress and shear stress at a point do not necessarily act on the x and y faces, that is, on the faces whose stresses are "given." In the next section we will derive equations that enable us to locate the planes on which these maxima (and minima) act.

8.4 PRINCIPAL STRESSES AND MAXIMUM SHEAR STRESS

Since a structural member or machine component may fail because of excessive normal stress or excessive shear stress, it is important to be able to determine the maximum normal stress and the maximum shear stress at a point. Figure 8.10b illustrates the fact that, for a plane-stress situation, there are certain planes on which the maximum and minimum normal stresses act and planes on which the maximum and minimum shear stresses act. In this section we will determine how to locate these planes and how to calculate these special stresses for the plane-stress case. Three-dimensional stress states are discussed in Section 8.6.

Principal Stresses. The maximum and minimum normal stresses are called **principal stresses.**[3] As is evident from Fig. 8.10b, the principal stresses occur on planes that satisfy the equation

$$\frac{d\sigma_n}{d\theta} = 0 \qquad (8.8)$$

where $\sigma_n(\theta)$ is given by Eq. 8.5a. Then,

$$\frac{d\sigma_n}{d\theta} = -(\sigma_x - \sigma_y)\sin 2\theta_p + 2\tau_{xy}\cos 2\theta_p = 0$$

The angles θ_p determine the orientation of the **principal planes,** the planes on which the principal stresses act. They are obtained by solving for the two values of θ_p that satisfy the equation

$$\tan 2\theta_p = \frac{\tau_{xy}}{\left(\dfrac{\sigma_x - \sigma_y}{2}\right)} \qquad (8.9)$$

Figure 8.12a illustrates how to use the tangent value given by Eq. 8.9 to determine the angles θ_p. There are two angles between $0°$ and $360°$ that satisfy Eq. 8.9. As

---

[3]The word principal is often confused with the word principle. Principal stresses are the al type, meaning *most important stresses.*

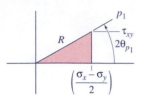

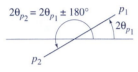

(a) Determination
of angle $2\theta_p$.

$2\theta_{p2} = 2\theta_{p1} \pm 180°$

(b) Two angles satisfy Eq. 8.9.

FIGURE 8.12 Determination
of the angles that locate principal planes.

illustrated by Fig. 8.12b, these two values of $2\theta_p$, labeled $2\theta_{p1}$ and $2\theta_{p2}$, differ by 180°, so the principal planes are oriented at 90° to each other. That is,

$$\theta_{p2} = \theta_{p1} \pm 90° \tag{8.10}$$

To determine the actual values of the principal stresses, that is, the normal stresses on the principal planes, we must substitute θ_{p1} and θ_{p2} into Eq. 8.5a. From Fig. 8.12a we can see that

$$R = \sqrt{\left(\frac{\sigma_x - \sigma_y}{2}\right)^2 + \tau_{xy}^2} \tag{8.11}$$

The quantity R is positive, and it has the units of stress. Also, from Fig. 8.12, the **two principal directions** are given by

$$\sin 2\theta_{p1} = \frac{\tau_{xy}}{R}, \qquad \cos 2\theta_{p1} = \frac{\left(\dfrac{\sigma_x - \sigma_y}{2}\right)}{R} \tag{8.12a}$$

and

$$\sin 2\theta_{p2} = \frac{-\tau_{xy}}{R}, \qquad \cos 2\theta_{p2} = \frac{-\left(\dfrac{\sigma_x - \sigma_y}{2}\right)}{R} \tag{8.12b}$$

Combining Eq. 8.5a with Eqs. 8.11 and 8.12, we get the two expressions for the **two principal stresses:**

$$\begin{aligned} \sigma_1 &\equiv \sigma_n(\theta_{p1}) = \sigma_{\text{avg}} + R \\ \sigma_2 &\equiv \sigma_n(\theta_{p2}) = \sigma_{\text{avg}} - R \end{aligned} \qquad \begin{aligned} &\textbf{Principal} \\ &\textbf{Stresses} \end{aligned} \tag{8.13}$$

where

$$\sigma_{\text{avg}} = \frac{\sigma_x + \sigma_y}{2} \tag{8.14}$$

and where R is given by Eq. 8.11. The designation $\sigma_1 \equiv \sigma_n(\theta_{p1})$ denotes the *maximum normal stress* at the point, while $\sigma_2 \equiv \sigma_n(\theta_{p2})$ denotes the *minimum normal stress* at the point. It is easy to see that σ_1 and σ_2 satisfy Eq. 8.7.

With angles θ_{p1} and θ_{p2} determined by Eqs. 8.12, and σ_1 and σ_2 determined by Eqs. 8.13 (with 8.11 and 8.14), we can show the two principal stresses on a properly oriented element. First, however, let us determine the values of the shear stress on the principal planes by substituting Eqs. 8.12 into Eq. 8.5b. We get

$$\tau_{nt}(\theta_{p1}) = 0, \qquad \tau_{nt}(\theta_{p2}) = 0 \tag{8.15}$$

This illustrates the important fact that **there is no shear stress on principal planes.** Therefore, Fig. 8.13 represents the stresses on a properly oriented **principal-stress element.**

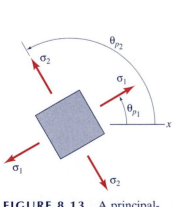

FIGURE 8.13 A principal-
stress element.

In Section 8.5 you will learn a convenient graphical procedure, called *Mohr's circle,* that is very useful for locating principal planes and calculating principal stresses. Both the equations of this section and the Mohr's circle method introduced in Section 8.5 determine *in-plane stresses,* that is, stresses in what has been designated as the *xy* plane. Section 8.6 extends the notion of principal stresses to three dimensions, and it also treats maximum shear stress in three dimensions.

Maximum In-Plane Shear Stress. We can locate the planes of maximum in-plane shear stress and calculate the value of the maximum in-plane shear stress by following the same procedure that we used to determine the principal planes and the principal stresses. By differentiating Eq. 8.5b we obtain

$$\frac{d\tau_{nt}}{d\theta} = -(\sigma_x - \sigma_y)\cos 2\theta_s - 2\tau_{xy}\sin 2\theta_s = 0$$

where θ_s designates a plane on which the shear is a maximum.[4] Therefore,

$$\tan 2\theta_s = -\frac{\left(\dfrac{\sigma_x - \sigma_y}{2}\right)}{\tau_{xy}} \tag{8.16}$$

Figure 8.14 illustrates the fact that two angles, $2\theta_s$, satisfy Eq. 8.16, and that these two angles are $\pm 90°$ from the angles $2\theta_p$ that locate the principal planes. Therefore, **the planes of maximum shear-stress magnitude lie at $\pm 45°$ from principal planes and are oriented at 90° to each other.** From Fig. 8.14 we get

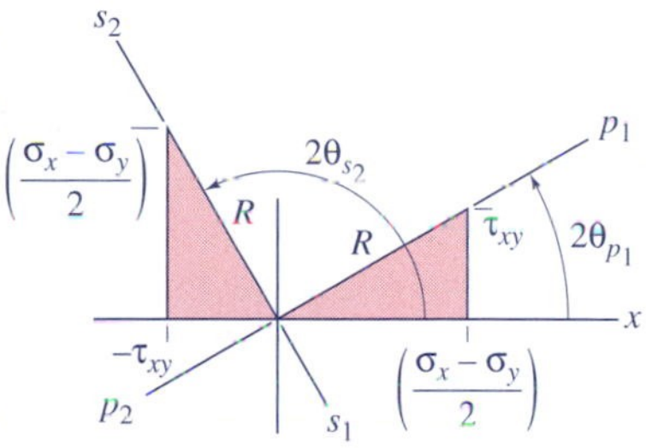

$$\sin 2\theta_{s1} = \frac{-\left(\dfrac{\sigma_x - \sigma_y}{2}\right)}{R}, \qquad \cos 2\theta_{s1} = \frac{\tau_{xy}}{R} \tag{8.17a}$$

and

$$\sin 2\theta_{s2} = \frac{\left(\dfrac{\sigma_x - \sigma_y}{2}\right)}{R}, \qquad \cos 2\theta_{s2} = \frac{-\tau_{xy}}{R} \tag{8.17b}$$

FIGURE 8.14 Determination of the angles that locate planes of maximum shear stress.

Substituting these expressions into Eq. 8.5b, we get the following expressions for **maximum** (and minimum) **in-plane shear stress** τ_{max}:

$$\boxed{\begin{aligned}\tau_{s1} &\equiv \tau_{nt}(\theta_{s1}) = R = \tau_{max} \\ \tau_{s2} &\equiv \tau_{nt}(\theta_{s2}) = -R = -\tau_{max}\end{aligned}} \qquad \begin{aligned}&\textbf{Maximum} \\ &\textbf{In-plane} \\ &\textbf{Shear Stress}\end{aligned} \tag{8.18}$$

Thus, we find that the planes of maximum in-plane shear stress magnitude satisfy the equation

$$\theta_s = \theta_p \pm 45° \tag{8.19}$$

---

[4]Since the maximum (i.e., algebraically largest) shear stress and the minimum (i.e., algebraically least) shear stress have the same magnitude, we will just refer to the "maximum" (see Eq. 8.18).

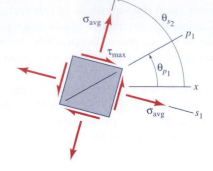

(a) Principal-stress element.

(b) Maximum-shear-stress element.

FIGURE 8.15 The relationship of the planes of maximum shear stress to the principal planes.

and that the shear stresses on these planes are equal in magnitude. Unlike principal stresses σ_1 and σ_2, which may have different magnitudes and different signs, only the magnitude of the maximum shear stress is important. It is labeled τ_{max}.

Before sketching a maximum-shear-stress element, we need to determine the values of normal stress on each of the planes of maximum shear stress. Substituting Eqs. 8.17 into Eq. 8.5a, we get

$$\sigma_{s1} \equiv \sigma_n(\theta_{s1}) = \frac{\sigma_x + \sigma_y}{2} = \sigma_{avg}$$

$$(8.20)$$

$$\sigma_{s2} \equiv \sigma_n(\theta_{s2}) = \frac{\sigma_x + \sigma_y}{2} = \sigma_{avg}$$

Thus, **the planes of maximum shear are not free of normal stress** (unless $\sigma_x = -\sigma_y$). On the contrary, they both have the same normal stress, σ_{avg}. Figure 8.15b depicts the **maximum-shear-stress element** that corresponds to the principal stress element of Fig. 8.13 (repeated as Fig. 8.15a). Note how the **planes of maximum shear are oriented at 45° to the principal planes,** and note that the *maximum-shear diagonal* (in Fig. 8.15b the shear arrows all point to this diagonal) lies along the p_1 principal direction. Mohr's circle, which is discussed in the next section, will provide further insight into the relationships illustrated in Fig. 8.15.

EXAMPLE 8.3

For the plane-stress state depicted in Fig. 1 (same as Fig. 1a of Example 8.1), (a) determine the orientation of the principal planes; determine the principal stresses; and illustrate the principal stresses on a properly oriented principal-stress element. (b) Determine the orientation of the planes of maximum in-plane shear stress; determine the value of the maximum shear stress; and illustrate the maximum in-plane shear stress on a properly oriented element.

Plan the Solution From Fig. 1, we have

$$\sigma_x = 20 \text{ MPa}, \quad \sigma_y = -10 \text{ MPa}, \quad \tau_{xy} = -10\text{MPa}$$

For Part (a), we could determine the two values of $2\theta_p$ from Eq. 8.9, but it will be more instructive to use Eqs. 8.12 to construct a sketch like Fig. 8.12. The values of σ_1 and σ_2 are easily calculated by using Eqs. 8.13.

For Part (b), we could determine the two values of $2\theta_s$ from Eq. 8.16, but, again, it will be more instructive if we use Eqs. 8.17 to construct a sketch like Fig. 8.14. The stresses on the planes of maximum shear are given by Eqs. 8.18.

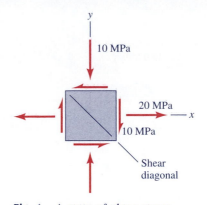

Fig. 1 A state of plane stress.

Solution

(a) Principal Stresses and Principal Planes: To determine the orientation of the principal planes, we will first sketch a figure like Fig. 8.12. To do this, we use Eqs. 8.12a and 8.12b to sketch Fig. 2.

$$\frac{\sigma_x - \sigma_y}{2} = \frac{20\text{ MPa} + 10\text{ MPa}}{2} = 15\text{ MPa}$$

and

$$\tau_{xy} = -10\text{ MPa}$$

The value of R can be computed by referring to Fig. 2 or by using Eq. 8.11. This gives

$$R = \sqrt{\left(\frac{\sigma_x - \sigma_y}{2}\right)^2 + (\tau_{xy})^2} = \sqrt{(15\text{ MPa})^2 + (-10\text{ MPa})^2} \quad (1a)$$

$$R = \sqrt{325}\text{ MPa} = 18.03\text{ MPa} \quad (1b)$$

From Eq. 8.12a, as illustrated in Fig. 2,

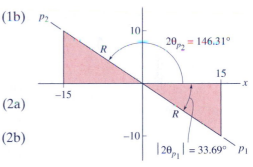

Fig. 2 A sketch representing principal directions.

$$2\theta_{p1} = \left[\sin^{-1}\left(\frac{-10}{\sqrt{325}}\right); \cos^{-1}\left(\frac{15}{\sqrt{325}}\right)\right] = -33.69° \quad (2a)$$

$$\theta_{p1} = -16.8° \qquad\qquad\qquad \textbf{Ans. (a)} \quad (2b)$$

Equation (2a) is to be read, "$2\theta_{p1}$ is the angle whose sine is s and whose cosine is c, as given in the square bracket $[\sin^{-1}(s); \cos^{-1}(c)]$." likewise, from Eq. 8.12b, as illustrated in Fig. 2,

$$2\theta_{p2} = \left[\sin^{-1}\left(\frac{10}{\sqrt{325}}\right); \cos^{-1}\left(\frac{-15}{\sqrt{325}}\right)\right] = 146.31° \quad (3a)$$

$$\theta_{p2} = 73.2° \qquad\qquad\qquad \textbf{Ans. (a)} \quad (3b)$$

From Eq. 8.14,

$$\sigma_{\text{avg}} = \frac{\sigma_x + \sigma_y}{2} = \frac{(20\text{ MPa} - 10\text{ MPa})}{2} = 5\text{ MPa} \quad (4)$$

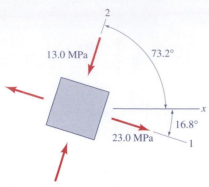

Fig. 3   The principal-stress element.

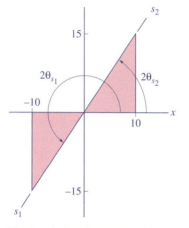

Fig. 4   A sketch representing maximum-shear-stress directions.

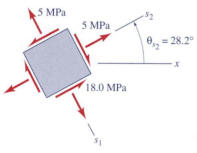

Fig. 5   The maximum in-plane shear-stress element.

and, from Eqs. 8.13,

$$\sigma_1 = \sigma_{\text{avg}} + R = 23.0 \text{ MPa} \qquad \textbf{Ans. (a)} \quad (5a)$$

$$\sigma_2 = \sigma_{\text{avg}} - R = -13.0 \text{ MPa} \qquad \textbf{Ans. (a)} \quad (5b)$$

These stresses are shown on the principal-stress element in Fig. 3.

(b) Maximum In-Plane Shear Stress: To determine the orientation of the planes of maximum in-plane shear stress, we will sketch a figure like Fig. 8.14. (We could also use the fact that the maximum-shear-stress element is rotated 45° from the principal-stress element.) Equations 8.17 enable us to determine the cosine and sine legs of angles $2\theta_{s1}$ and $2\theta_{s2}$ and, from them, to sketch Fig. 4.

From Eq. 8.17a, as illustrated in Fig. 4,

$$2\theta_{s1} = \left[\sin^{-1}\left(\frac{-15}{\sqrt{325}} \right); \cos^{-1}\left(\frac{-10}{\sqrt{325}} \right) \right] = -123.69° \qquad (6a)$$

$$\theta_{s1} = -61.8° \qquad \textbf{Ans. (b)} \quad (6b)$$

Similarly, from Eq. 8.17b, as illustrated in Fig. 4,

$$2\theta_{s2} = \left[\sin^{-1}\left(\frac{15}{\sqrt{325}} \right); \cos^{-1}\left(\frac{10}{\sqrt{325}} \right) \right] = 56.31° \qquad (7a)$$

$$\theta_{s2} = 28.2° \qquad \textbf{Ans. (b)} \quad (7b)$$

From Eqs. 8.18,

$$\tau_{s1} \equiv \tau_{nt}(\theta_{s1}) = R = 18.0 \text{ MPa} \qquad \textbf{Ans. (b)} \quad (8a)$$

$$\tau_{s2} \equiv \tau_{nt}(\theta_{s2}) = -R = -18.0 \text{ MPa} \qquad \textbf{Ans. (b)} \quad (8b)$$

The maximum in-plane shear-stress element is sketched in Fig. 5 using the data in Eq. (4) and in Eqs. (6) through (8).

Review the Solution   The principal stress σ_1 should be $\geq$ the greater normal stress on Fig. 1, which it is; and principal stress σ_2 should be $\leq$ the lesser normal stress on Fig. 1, which it is. The p_1 principal direction should lie between the axis of the greater normal stress, here the x axis, and the shear diagonal. The p_1 direction of $-16.8°$ satisfies this requirement. Therefore, the principal stresses and principal directions are probably correct.

The maximum in-plane shear stress must be greater than the given shear on Fig. 1, and the directions of maximum shear must bisect the principal directions. These requirements are satisfied by Fig. 5, so our maximum shear stresses are also probably correct.

Problems like this one are generally solved by using Mohr's Circle (Section 8.5), which provides a helpful visualization of the entire problem.

Mohr's circle is an ingenious graphical representation of the *plane stress transformation equations*, Eqs. 8.5.[5] It permits an easy visualization of the normal stress and shear stress on arbitrary planes, and it greatly facilitates the solution of plane-stress problems like Example Problems 8.1 through 8.3.

Derivation of Mohr's Circle. Let us begin our derivation of Mohr's circle for plane stress by rewriting Eqs. 8.5 in the form

$$\sigma_n - \sigma_{\text{avg}} = \left(\frac{\sigma_x - \sigma_y}{2}\right)\cos 2\theta + \tau_{xy}\sin 2\theta$$

$$\tau_{nt} = -\left(\frac{\sigma_x - \sigma_y}{2}\right)\sin 2\theta + \tau_{xy}\cos 2\theta \tag{8.21}$$

Squaring both sides of each of these equations, and adding the resulting squares, we get

$$(\sigma_n - \sigma_{\text{avg}})^2 + \tau_{nt}^2 = \left(\frac{\sigma_x - \sigma_y}{2}\right)^2 + \tau_{xy}^2$$

or

$$(\sigma_n - \sigma_{\text{avg}})^2 + \tau_{nt}^2 = R^2 \tag{8.22}$$

This is the equation of a circle in (σ, τ) coordinates, with center at $(\sigma_{\text{avg}}, 0)$ and radius R. The plane-stress transformation equations 8.5 are just parametric equations of a circle, with the parameter being θ, and with the coordinates of point N on the circle representing the normal stress σ_n and shear stress τ_{nt} on the n plane at orientation $\theta \equiv \theta_{xn}$.

To determine more of the properties of Mohr's circle, consider now the circle shown in Fig. 8.16. After discussing many properties of Mohr's circle, using Fig. 8.16, we will suggest a procedure for you to use in constructing a Mohr's circle from given stress data, and we will illustrate how to solve problems like Example Problems 8.1 through 8.3 using Mohr's circle.

Mohr's circle is drawn on a set of rectilinear axes with the horizontal axis (axis of abscissas) representing the normal stress σ, and with the vertical axis (axis of ordinates) representing the shear stress τ. Note that **the positive τ axis is downward.** The sign convention for θ is the same one that was introduced in Section 8.3. The angle $\theta \equiv \theta_{xn}$ is, as previously, measured **counterclockwise** from the x axis to the n axis **on the body** undergoing plane stress. Correspondingly, an angle of $2\theta \equiv 2\theta_{xn}$ is measured **counterclockwise on Mohr's circle.**[6] *Every point on Mohr's*

---

[5]Otto Mohr (1835-1918), a German structural engineer and professor of engineering mechanics, introduced the graphical representation of stress at a point, Mohr's circle; developed a failure criterion (see Sect. 12.3); and made many other important contributions to mechanics of structures and materials [Ref. 8-2]. Mohr's circle is also useful in the computation of strain transformations (Section 8.9) and the computation of moments and products of inertia (Appendix C.3).

[6]The sign convention for Mohr's circle is sometimes a cause of confusion, leading to many different treatments in texts on the mechanics of deformable bodies. Just remember, in this text, τ **Is positive downward**, and **angles are turned in the same direction on both the body (angle θ) and on Mohr's circle (angle 2θ).**

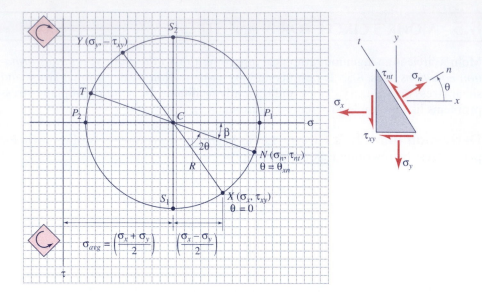

FIGURE 8.16 Properties of Mohr's circle for plane stress.

circle corresponds to the stresses σ and τ on a particular face; for the generic point N, the stresses are (σ_n, τ_m). To emphasize this, we will **label points on the circle with the same label as the face represented by that point**, except that a capital letter will designate the point on Mohr's circle. The x face is represented by point X on the circle; the n face is represented by point N on the circle, and so on.

To reinforce the sign convention for plotting shear stresses on Mohr's circle, the small shear stress icons in Fig. 8.16 indicate that **the shear stress on a face plots as positive shear (i.e., plots downward) if the shear stress on the face would tend to rotate a stress element counterclockwise.** Conversely, the shear stress on a face plots as negative shear (i.e., plots upward) if the shear stress on that face would tend to rotate a stress element clockwise. Note that this sign convention for plotting τ causes the y face to be represented by the point Y at $(\sigma_y, -\tau_{xy})$, just as was explained in Fig. 8.11b.

Before we look at examples of how Mohr's circle is used, let us establish the fact that Mohr's circle does, indeed, provide a graphical representation of the *stress transformation equations*, Eqs. 8.5. Equations 8.23 and 8.24 come directly from the trigonometry of the circle in Fig. 8.16.

$$\sigma_n = \sigma_{\text{avg}} + R \cos \beta, \qquad \tau_{nt} = R \sin \beta \tag{8.23}$$

and

$$\frac{\sigma_x - \sigma_y}{2} = R \cos(2\theta + \beta), \qquad \tau_{xy} = R \sin(2\theta + \beta) \tag{8.24}$$

The angle β on Fig. 8.16 is introduced just to facilitate the derivation that follows. The trigonometric identifies for the cosine and sine of the sum of two angles are

$$\cos(\alpha + \beta) = \cos \alpha \cos \beta - \sin \alpha \sin \beta$$
$$\sin(\alpha + \beta) = \sin \alpha \cos \beta + \cos \alpha \sin \beta \tag{8.25}$$

Letting $\alpha = 2\theta$, we can convert Eqs. 8.24 to the form

$$\frac{\sigma_x - \sigma_y}{2} = R(\cos 2\theta \cos \beta - \sin 2\theta \sin \beta) \qquad (8.26a)$$

$$\tau_{xy} = R(\sin 2\theta \cos \beta + \cos 2\theta \sin \beta) \qquad (8.26b)$$

If we multiply Eq. 8.26a by $\cos 2\theta$ and Eq. 8.26b by $\sin 2\theta$ and add the resulting equations, we get

$$\left(\frac{\sigma_x - \sigma_y}{2}\right)\cos 2\theta + \tau_{xy}\sin 2\theta = R \cos \beta \qquad (8.27)$$

but, combining this with Eq. 8.23a leads to

$$\sigma_n = \sigma_{\text{avg}} + \left(\frac{\sigma_x - \sigma_y}{2}\right)\cos 2\theta + \tau_{xy}\sin 2\theta \qquad \begin{array}{l}(8.5a)\\ (\text{repeated})\end{array}$$

which, of course, is just Eq. 8.5a. Similarly, multiplying Eq. 8.26a by $\sin 2\theta$ and Eq. 8.26b by $\cos 2\theta$ and subtracting the latter from the former we get

$$-\left(\frac{\sigma_x - \sigma_y}{2}\right)\sin 2\theta + \tau_{xy}\cos 2\theta = R \sin \beta \qquad (8.28)$$

but, Eq. 8.23b reveals that this can be written

$$\tau_{nt} = -\left(\frac{\sigma_x - \sigma_y}{2}\right)\sin 2\theta + \tau_{xy}\cos 2\theta \qquad \begin{array}{l}(8.5b)\\ (\text{repeated})\end{array}$$

which is just Eq. 8.5b. Again, we have shown that Eqs. 8.5 are just the parametric equations of a circle in (σ, τ) coordinates.

Having established that **Mohr's circle of stress is a graphical representation of the transformation equations for plane stress,** let us examine other properties that can be easily deduced from Mohr's circle.

Properties of Mohr's Circle. Referring to Fig. 8.16, we can conclude the following:

- The center of Mohr's circle lies on the σ axis at $(\sigma_{\text{avg}}, 0)$.
- Points on the circle that lie above the σ axis (i.e., τ negative) correspond to faces that have a clockwise-acting shear; points that lie below the σ axis (i.e., τ positive) correspond to faces that have a counterclockwise-acting shear, as illustrated by Fig. 8.17.
- The radius of the circle is determined by applying the Pythagorean theorem to the triangle with sides τ_{xy} and $\left(\dfrac{\sigma_x - \sigma_y}{2}\right)$, giving

$$R = \sqrt{\left(\frac{\sigma_x - \sigma_y}{2}\right)^2 + \tau_{xy}^2} \qquad \begin{array}{l}(8.11)\\ \text{repeated}\end{array}$$

(Continued on p. 561)

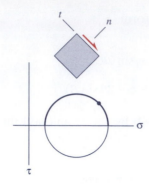

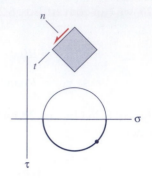

FIGURE 8.17 The Mohr's
circle shear-stress sign
convention.

(*a*) Clockwise shear stress.   (*b*) Counterclockwise shear stress.

PROCEDURE FOR CONSTRUCTING AND USING MOHR'S CIRCLE OF STRESS

To solve plane-stress problems, such as determining the stresses on a particular face (e.g., Example Problems 8.1 and 8.2) or determining principal stresses and maximum in-plane shear stresses (Example Problem 8.3), the following procedure is suggested:

Draw Mohr's Circle[7]

1. Establish a set of (σ, τ) axes, with the same scale on both axes. Remember, the $+\tau$ axis points downward. It is good idea to use paper that has a grid, like graph paper or "engineering paper." Use a scale that will result in a circle of reasonable size.

2. Assuming that σ_x, σ_y, and τ_{xy} are given (or can be determined from a given stress element), locate point X at (σ_x, τ_{xy}) and point Y at $(\sigma_y, -\tau_{xy})$.

3. Connect points X and Y with a straight line, and locate the center of the circle where this line crosses the σ axis at $(\sigma_{\text{avg}}, 0)$.

4. Draw a circle with center at $(\sigma_{\text{avg}}, 0)$ and passing through points X and Y. It is best to use a compass to draw the circle.

Compute the Required Information

5. Form the triangle with sides τ_{xy} and $\left(\dfrac{\sigma_x - \sigma_y}{2}\right)$, and compute

$$R = \sqrt{\left(\frac{\sigma_x - \sigma_y}{2}\right)^2 + \tau_{xy}^2} \qquad (8.11)$$

6. If the stresses on a particular face, call it face n, are required, locate point N on the circle by turning an angle 2θ counterclockwise (or clockwise) on the circle, corresponding to rotating an angle θ counterclockwise (clockwise) from some reference face on the stress element. Using trigonometry, calculate σ_n and τ_{nt}.[8]

7. If the principal stresses and the orientation of the principal planes are required, use

$$\sigma_1 = \sigma_{\text{avg}} + R, \qquad \sigma_2 = \sigma_{\text{avg}} - R \qquad (8.13)$$

to calculate the principal stresses, and use trigonometry to determine some angle, such as $2\theta_{xp1}$ that can be used to locate a principal plane, say p_1, with respect to some known face, say the x face.

8. Use a procedure similar to Step 7 if the maximum in-plane shear stress and the planes of maximum shear stress are required.

---

[7]After you become proficient with Mohr's circle, a simple sketch will suffice. However, at first it is helpful if you draw stress magnitudes (at least roughly) to scale and draw angles the correct size.

[8]Although it was suggested that the circle be accurately drawn to some scale, you should use trigonometry to compute the required answers and only scale magnitudes and angles off of the Mohr's circle as a check of your calculations.

- Two planes that are 90° apart on the physical body are represented by the two points at the extremities of a diameter, such as points X and Y or P_1 and P_2 in Fig. 8.16.

- If we rotate counterclockwise by an angle θ_{ab} to go from face a to face b of the physical body, we must rotate in that same direction through the angle $2\theta_{ab}$ to get from point A on Mohr's circle to point B. Figure 8.18 illustrates this property of the Mohr's circle sign convention. In equations, a positive angle is always counterclockwise.

- The *principal planes* are represented by points P_1 and P_2 at the intersections of Mohr's circle with the σ axis (Fig. 8.16). The corresponding *principal stresses* are $\sigma_1 = \sigma_{\text{avg}} + R$ and $\sigma_2 = \sigma_{\text{avg}} - R$.

- The planes of maximum shear stress are represented by points S_1 and S_2 that lie directly below and above the center of the Mohr's circle (Fig. 8.16). The corresponding stresses are: (σ_{avg}, R) on face s_1, and $(\sigma_{\text{avg}}, -R)$ on face s_2.

- Since the stresses on orthogonal planes n and t are represented by the points at each end of a diameter of Mohr's circle,

$$\sigma_n + \sigma_t = \sigma_x + \sigma_y \qquad \text{(8.7)} \quad \text{repeated}$$

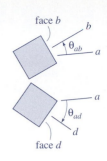

face b

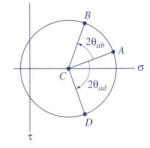

face d

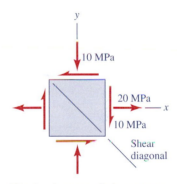

FIGURE 8.18 Consistent angles.

EXAMPLE 8.4

For the plane-stress state in Example Problems 8.1 and 8.2 (Fig. 1), do the following: (a) Draw Mohr's circle. (b) Determine the stresses on all faces of an element that is rotated 30° counterclockwise from the orientation of the stress element in Fig. 1. (c) Determine the orientation of the principal planes; determine the principal stresses. (d) Determine the orientation of the planes of maximum shear stress; determine the value of the maximum shear stress.

Solution   We can just follow the procedure outlined on page 560.

(a) Mohr's Circle: On grid paper (Fig. 2), plot point X at (20 MPa, −10 MPa), and plot point Y at (−10 MPa, 10 MPa). The center of the circle is obtained by connecting X and Y. The diameter crosses the σ axis at C: $(\sigma_{\text{avg}}, 0)$ where

$$\sigma_{\text{avg}} = \frac{20 \text{ MPa} - 10 \text{ MPa}}{2} = 5 \text{ MPa} \qquad (1)$$

The circle is drawn with center at C and passing through points X and Y. The radius R is calculated from the shaded triangle XCB in Fig. 2.

$$R = \sqrt{(15 \text{ MPa})^2 + (10 \text{ MPa})^2} = \sqrt{325} \text{ MPa} = 18.03 \text{ MPa} \qquad (2)$$

(b) Stresses on x′ and y′ Faces: Locate the points on Mohr's circle that correspond to rotating the stress element by 30°. This means rotating 60° counterclockwise from the XY diameter on Mohr's circle. We label these two points X' and Y'.

Fig. 1   A state of plane stress.

y

10 MPa

20 MPa

x

10 MPa

Shear diagonal

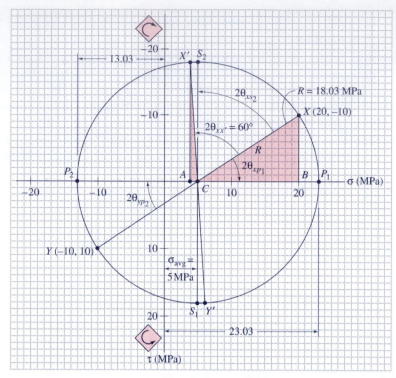

Fig. 2 Mohr's circle.

To determine the stresses at points X' and Y', we need to establish the geometry and trigonometry of the triangle $X'CA$. To determine $\angle X'CA$ we need to first determine the (clockwise) angle $2\theta_{xp1}$, in Fig. 2. Using the triangle, XCB, we get

$$2\theta_{xp1} = \tan^{-1}\left(\frac{10}{15}\right) = 33.69° \tag{3a}$$

$$\theta_{xp1} = 16.8° \circlearrowright \tag{3b}$$

Therefore,

$$\angle X'CA = 180° - 60° - 2\theta_{xp1} = 86.31° \tag{4}$$

From the triangle $X'CA$,

$$\overline{AC} = R\cos(\angle X'CA) = 18.03\cos(86.31°) \tag{5a}$$

or

$$\overline{AC} = 1.16 \text{ MPa} \tag{5b}$$

Therefore,

$$\left.\begin{array}{l} \sigma_{x'} = \sigma_{\text{avg}} - \overline{AC} = 3.84 \text{ MPa} \\ \sigma_{y'} = \sigma_{\text{avg}} + \overline{AC} = 6.16 \text{ MPa} \end{array}\right\} \qquad \textbf{Ans. (b)} \quad (6)$$

Also, from triangle $X'CA$ we get

$$\tau_{x'y'} = -R \sin(\angle X'CA) = -18.03 \sin(86.31°)$$

or

$$\tau_{x'y'} = -18.0 \text{ MPa} \qquad \textbf{Ans. (b)} \quad (7)$$

Equations (6) and (7) are the same answers that we obtained in Example Problem 8.1 by using formulas directly.

(c) Principal Planes and Principal Stresses: We have already calculated θ_{xp1} in Eq. (3). From Fig. 2,

$$2\theta_{yp2} = 2\theta_{xp1} = 33.69° \qquad (8a)$$

$$\theta_{yp2} = 16.8° \, \text{◝} \qquad (8b)$$

Also, from Fig. 2,

$$\sigma_1 = \sigma_{avg} + R = 5 \text{ MPa} + 18.03 \text{ MPa} = 23.0 \text{ MPa}$$

$$\sigma_2 = \sigma_{avg} - R = 5 \text{ MPa} - 18.03 \text{ MPa} = -13.0 \text{ MPa}$$

or

$$\sigma_1 = 23.0 \text{ MPa}, \quad \sigma_2 = -13.0 \text{ MPa} \qquad \textbf{Ans. (c)} \quad (9)$$

(d) Maximum In-Plane Shear Stress: The planes of maximum in-plane shear stress are represented by the points S_1 and S_2 on Mohr's circle. From Fig. 2,

$$2\theta_{xs1} = 90° + 2\theta_{xp1} = 90° + 33.69° = 123.69°$$

so

$$\theta_{xs1} = 61.8° \, \text{◝}$$

Also, by referring to Fig. 2, we see that

$$2\theta_{xs2} = 90° - 2\theta_{xp1} = 90° - 33.69° = 56.31°$$

$$\theta_{xs2} = 28.2° \, \text{◝}$$

The maximum in-plane shear stress occurs on plane s_1 and on plane s_2 and is given by

$$\tau_{s1s2} = R = 18.0 \text{ MPa} \qquad \textbf{Ans. (d)}$$

On the planes of maximum shear stress, the normal stress is

$$\sigma_{s1} = \sigma_{s2} = \sigma_{avg} = 5 \text{ MPa} \qquad \textbf{Ans. (d)}$$

Mohr's Circle—Stress Transformations

8.6 TRIAXIAL STRESS; ABSOLUTE MAXIMUM SHEAR STRESS

Figure 8.2 depicts a general *three-dimensional state of stress,* referred to cartesian (x, y, z) axes, but in Sections 8.2 through 8.5 we dealt only with plane stress—formulating stress transformation equations, determining expressions for principal stresses and maximum in-plane shear stresses, and establishing a graphical representation of the plane-stress transformation equations, called Mohr's circle. We now need to look further at three-dimensional stress states. In particular, we will briefly consider *principal stresses* for a general state of stress, and will then examine *absolute maximum shear stresses* in greater detail.

Principal Stresses and Principal Directions. For a general three-dimensional state of stress at a point, it can be shown that: **there are three principal stresses,** and **the corresponding principal planes are mutually perpendicular.**[9] There is no shear stress on the principal planes. The three **principal stresses** are labeled in the order—maximum, intermediate, and minimum:

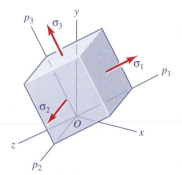

FIGURE 8.19 Principal stresses acting on a three-dimensional element.

$$\sigma_1 \equiv \sigma_{max}, \quad \sigma_2 \equiv \sigma_{int}, \quad \sigma_3 \equiv \sigma_{min} \tag{8.29}$$

that is, $\sigma_1 \geq \sigma_2 \geq \sigma_3$. To each principal stress σ_i there is a unit normal vector that defines the corresponding **principal direction,** that is, the normal to the plane on which that principal stress acts. If we draw the stress element whose faces are all principal planes, we get Fig. 8.19. The principal directions are labeled p_1, p_2, and p_3, with $\sigma_1 \geq \sigma_2 \geq \sigma_3$. Since all faces of this element are free of shear stress, this element is said to be in a state of **triaxial stress.**

Absolute Maximum Shear Stress—General Stress State. In Section 8.4 we examined *maximum in-plane shear stress* for the case of plane stress. For a general state of stress at a point, including the case of plane stress, we need to determine the **absolute maximum shear stress,** that is, the largest-magnitude shear stress acting in any direction on any plane passing through the point.[10] To do so, it is convenient to assume that we already know the principal directions and the principal stresses

---

[9]For a detailed discussion of procedures for determining principal stresses and principal directions, see Sections 75–78 of *Theory of Elasticity,* Third Edition, by S. P. Timoshenko and J. N. Goodier, McGraw-Hill Book Company, New York, 1970, [Ref. 8-1].

[10]For a detailed derivation, see Section 79 of *Theory of Elasticity,* Third Edition, by S. P. Timoshenko and J. N. Goodier, McGraw-Hill Book Company, New York, 1970, [Ref. 8-1].

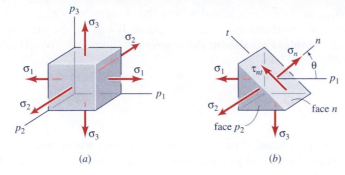

FIGURE 8.20 (*a*) An element in triaxial stress, and (*b*) a wedge for stress transformation based on triaxial stresses.

at the point. Figure 8.20*a* represents the element on which the principal stresses at the point act.

Let us begin our search for the absolute maximum shear stress at a point by examining the stresses on a plane whose normal *n* is perpendicular to the p_2 direction, that is, a plane like the oblique plane shown in Fig. 8.20*b*. Since there is no shear stress at all on the p_2 face, the shear stress on the *n*-face has no p_2-component, as is indicated in Fig. 8.20*b*. Figure 8.21*a* is a two-dimensional view of the wedge in Fig. 8.20*b* showing the forces obtained by multiplying each stress by the area of the face on which it acts. By summing forces on the free-body diagram of Fig. 8.23*a* we obtain a special case of Eqs. 8.5, namely,

$$\sigma_n = \frac{\sigma_1 + \sigma_3}{2} + \frac{\sigma_1 - \sigma_3}{2}\cos 2\theta = \sigma_{\text{avg}} + R\cos 2\theta$$

$$-\tau_{nt} = \frac{\sigma_1 - \sigma_3}{2}\sin 2\theta = R\sin 2\theta$$

(8.30)

It is clear that Eqs. 8.30 locate the point *N* on the Mohr's circle in Fig. 8.21*b*.

The maximum shear stress in the p_1p_3 plane is given by the radius of the circle in Fig. 8.21*b*, that is

$$(\tau_{\text{max}})_{p1p3} \equiv \tau_3 = \frac{\sigma_1 - \sigma_3}{2}$$

(8.31)

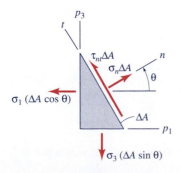

(*a*) A free-body diagram for determining
stresses in the p_1p_3 plane. (σ_2 is omitted
because it is normal to the p_1p_3 plane.)

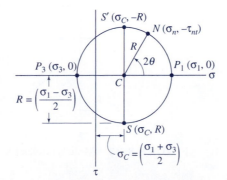

(*b*) Mohr's circle for stresses in the
p_1p_3 plane.

FIGURE 8.21 Transformation of stresses in the p_1p_3 plane.

where $\sigma_1 \equiv \sigma_{\max}$ and $\sigma_3 \equiv \sigma_{\min}$, and where the designation $(\tau_{\max})_{p1p3}$ refers to the maximum in-plane shear stress in the p_1p_3 plane. It can be shown[11] that this shear stress is also the **absolute maximum shear stress** at the point. Therefore,

$$\tau_{\max}^{\text{abs}} = \frac{\sigma_{\max} - \sigma_{\min}}{2} \tag{8.32}$$

and this **absolute maximum shear stress acts on planes whose normal s bisects the 90° angle between the p_1 and p_3 directions,** as illustrated in Fig. 8.22b. The normal stress acting on the planes of maximum shear stress is

$$\sigma_s = \frac{\sigma_{\max} + \sigma_{\min}}{2} \tag{8.33}$$

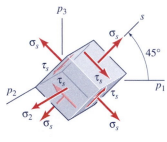

(*a*) Stress element with principal planes as faces.

Note that the τ_s arrows and the σ_s arrows in Fig. 8.22b both have components that point in the p_1 direction.

Stress transformations, like Eqs. 8.30, and Mohr's circles, like Fig. 8.21b, can be developed for faces whose normal n lies in the p_1p_2 plane or the p_2p_3 plane. As shown in Fig. 8.23, the use of Mohr's circle can be extended to three-dimensional stress states by drawing a separate circle through each pair of principal stresses. The absolute maximum shear stress, given by Eq. 8.32, is the radius of the circle that passes through the $\sigma_1 (\equiv \sigma_{\max})$ point and the $\sigma_3 (\equiv \sigma_{\min})$ point.

$$\sigma_s = \left(\frac{\sigma_1 + \sigma_3}{2}\right), \quad \tau_s = \left(\frac{\sigma_1 - \sigma_3}{2}\right)$$

(*b*) The element on which the absolute maximum shear stress acts.

FIGURE 8.22 Planes of absolute maximum shear stress.

Absolute Maximum Shear Stress—Plane-Stress State.

In Section 8.4 we examined the principal stresses and the maximum in-plane shear stresses for the plane-stress states, where the conditions for the plane stress are $\sigma_z = \tau_{xz} = \tau_{yz} = 0$. Since there is no shear stress on the z faces, the z axis is one of the three principal directions at the point. In determining the absolute maximum shear stress, the question, then, is whether the stress $\sigma_z = 0$ is the maximum principal stress (σ_1), the intermediate principal stress (σ_2), or the minimum principal stress (σ_3). Figure 8.24 illustrates these three options. In each case a solid-line Mohr's circle is drawn for the in-plane (i.e., xy plane) stress transformation discussed in Sections 8.4 and 8.5. The dashed-line circles are for the p_1p_3 stress transformation, which

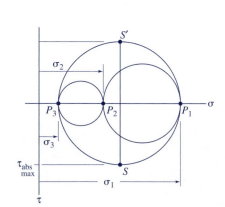

FIGURE 8.23 Mohr's circles for three-dimensional stress at a point.

---

[11]See Section 79 of *Theory of Elasticity,* by S. P. Timoshenko and J. N. Goodier, [Ref. 8-1].

(a) Case I,
$(\sigma_3 \leq \sigma_2 \leq 0)$

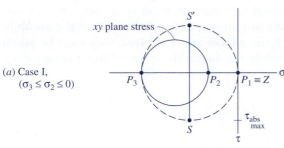

(a1) Mohr's circle.

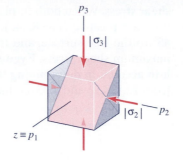

(a2) A maximum shear plane.

(b) Case II,
$(\sigma_3 \leq 0 \leq \sigma_1)$

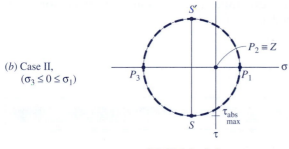

(b1) Mohr's circle.

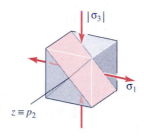

(b2) A maximum shear plane.

(c) Case III,
$(0 \leq \sigma_2 \leq \sigma_1)$

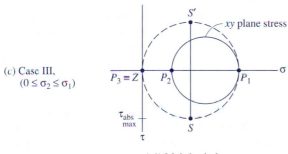

(c1) Mohr's circle.

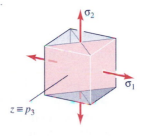

(c2) A maximum shear plane.

FIGURE 8.24 The absolute maximum shear stress for various plane-stress states; planes of absolute maximum shear stress.

leads to the absolute maximum shear stress (see Fig. 8.23). The value of the absolute maximum shear stress is always given by Eq. 8.32, and the corresponding normal stress is always given by Eq. 8.33.

$$
\begin{aligned}
&\text{Case I } (\sigma_3 \leq \sigma_2 \leq 0): && \tau_{\max}^{\text{abs}} = \frac{-\sigma_3}{2}; \sigma_s = \frac{\sigma_3}{2} \\
&\text{Case II } (\sigma_3 \leq 0 \leq \sigma_1): && \tau_{\max}^{\text{abs}} = \frac{\sigma_1 - \sigma_3}{2}; \sigma_s = \frac{\sigma_1 + \sigma_3}{2} \qquad (8.34) \\
&\text{Case III } (0 \leq \sigma_2 \leq \sigma_1): && \tau_{\max}^{\text{abs}} = \frac{\sigma_1}{2}; \sigma_s = \frac{\sigma_1}{2}
\end{aligned}
$$

It should be clear from Fig. 8.24 that **only when the in-plane principal stresses have opposite signs is the maximum in-plane shear stress also the absolute maximum** shear stress. When both in-plane principal stresses are negative (Fig. 8.24a) or when both are positive (Fig. 8.24c), the absolute maximum shear stress acts on planes at 45° to the free surface, and the maximum in-plane shear stress is not the absolute maximum shear stress. **Even though the z faces are stress free, they must be taken into account in determining the absolute maximum shear stress!** But, in every case, from Eqs. 8.32 and 8.33 we have

$$\tau_{\substack{abs \\ max}} = \frac{\sigma_{max} - \sigma_{min}}{2}$$

$$\sigma_s = \frac{\sigma_{max} + \sigma_{min}}{2}$$

(8.35)

The stresses σ_{max} and σ_{min} are signed quantities (i.e., tension positive, compression negative); they are not just magnitudes.

EXAMPLE 8.5

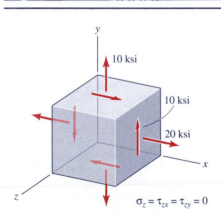

$\sigma_z = \tau_{zx} = \tau_{zy} = 0$

Fig. 1 An element in plane stress.

An element in plane stress has the stresses shown in Fig. 1. (a) Determine the three principal stresses. Use a Mohr's circle to determine in-plane stresses. (b) Determine the maximum in-plane shear stress. (c) Determine the orientation of the principal planes, and sketch the principal-stress element. (d) Determine the absolute maximum shear stress. Show an element oriented so that the absolute maximum shear stress acts on the element.

Plan the Solution We need to determine the principal directions and in-plane principal stresses for the xy plane using the Mohr's circle technique of Section 8.5. From Mohr's circle we can also get the maximum in-plane shear stress. In Part (c) we will have to order the principal stresses $\sigma_1 \geq \sigma_2 \geq \sigma_3$, and compare the three principal stresses in this problem (the two in-plane principal stresses plus $\sigma_z = 0$) with the three cases depicted in Fig. 8.24. The maximum absolute shear stress is calculated using Eq. 8.32.

Solution

(a) Principal Stresses: One of the principal stresses is $\sigma_z = 0$, since $\tau_{zx} = \tau_{zy} = 0$. The other two principal stresses are obtained from the Mohr's circle in Fig. 2.

From triangle XCA we get

$$R = \sqrt{(\overline{CA})^2 + (\overline{XA})^2} = \sqrt{(5 \text{ ksi})^2 + (10 \text{ ksi})^2}$$

So,

$$R = \sqrt{125} \text{ ksi} = 11.18 \text{ ksi} \tag{1}$$

Since all points on the Mohr's circle in Fig. 2 have $\sigma > 0$, $\sigma_z = 0$ is the minimum principal stress. Therefore, the intersections of Mohr's

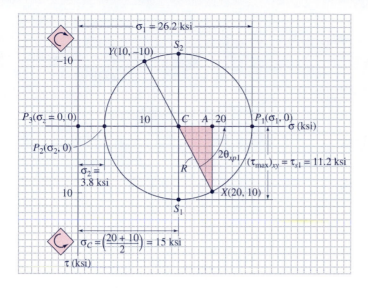

Fig. 2 Mohr's circle for the xy plane, with P_3 shown for reference.

circle with the σ axis are labeled p_1 and p_2. From the circle in Fig. 2,

$$\sigma_1 = \sigma_C + R = 15 \text{ ksi} + 11.2 \text{ ksi} = 26.2 \text{ ksi}$$

$$\sigma_2 = \sigma_C - R = 15 \text{ ksi} - 11.2 \text{ ksi} = 3.8 \text{ ksi}$$ (2)

Therefore, the three principal stresses are

$$\sigma_1 = 26.2 \text{ ksi}, \quad \sigma_2 = 3.8 \text{ ksi}, \quad \sigma_3 = 0 \quad \textbf{Ans.(a)} \quad (3)$$

(b) *Maximum In-Plane Shear Stress:* The maximum shear stress in the xy plane is the shear stress at point S_1 in Fig. 2, or

$$(\tau_{\max})_{xy} = R = 11.2 \text{ ksi} \qquad \textbf{Ans.(b)} \quad (4)$$

(c) *Principal-Stress Element:* To orient the principal-stress element ($p_1 p_2 p_3$ axes) relative to the xyz axes we only need to relate p_1 and p_2 to x and y, since we already know that $p_3 \equiv z$ (since $\sigma_z < \sigma_2 < \sigma_1$). From Fig. 2 we can determine the angle $2\theta_{xp1}$. From triangle XCA we get

$$2\theta_{xp1} = \tan^{-1}\left(\frac{10}{5}\right) = 63.43° \qquad (5a)$$

$$\theta_{xp1} = 31.7° \qquad \textbf{Ans.(c)} \quad (5b)$$

A properly oriented principal-stress element is shown in Fig. 3.

(d) *Absolute Maximum Shear Stress:* The plane-stress Mohr's circle in Fig. 2 corresponds to Case III (Fig. 8.24c). Therefore, we need to construct a $p_1 p_3$ Mohr's circle. For clarity, we will draw another figure, Fig. 4,

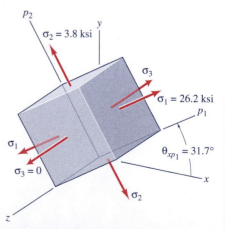

Fig. 3 The principal-stress element.

569

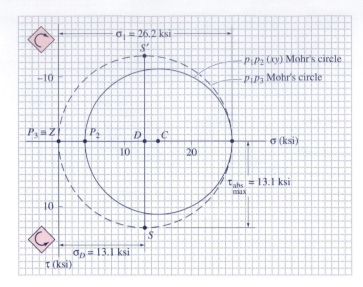

Fig. 4 Mohr's circles for determining τ^{abs}_{max}.

repeating part of Fig. 2. From the dashed-line $p_1 p_3$ Mohr's circle in Fig. 4 we get

$$\tau^{abs}_{max} = \frac{\sigma_1}{2} = 13.1 \text{ ksi} \qquad \textbf{Ans.(d)} \quad (6)$$

Figures 5a through 5c depict the planes of absolute maximum shear stress. First, in Figs. 5a and 5b the orientations of the planes of maximum shear stress at 45° to the p_1 and p_3 axes (faces) are illustrated. Finally, in Fig. 5c a two-dimensional view of the $p_1 p_3$ plane is shown, looking "down" the p_2 axis.

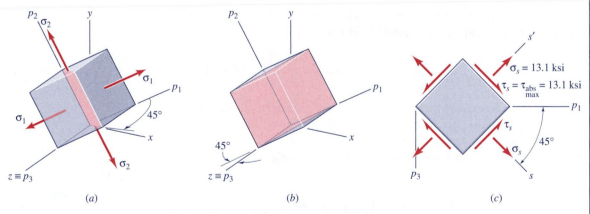

Fig. 5 Planes of absolute maximum shear stress.

Review the Solution We should first check to make sure the points X and Y in Fig. 2 correctly represent the stresses on the x and y faces in Fig. 1, especially making sure that the sign of the shear stress is correct at X and Y. Since the answers in Eqs. (3), (4), (5), and (6) came directly from the Mohr's circles in Figs. 2 and 3, we can visually check to see if they are reasonable.

8.7 PLANE STRAIN

Definitions of extensional strain ϵ and shear strain γ were given in Section 2.12. These strains vary with position in a body and with the orientation of the reference directions. For example, at point P in (or on the surface of) a deformable body, the extensional strains ϵ_x, ϵ_y, and ϵ_z are determined by examining the change in length of short, mutually orthogonal line segments Δx, Δy, and Δz; and the shear strains γ_{xy}, γ_{xz}, and γ_{yz} are determined by the changes in the right angles that originally exist between these lines. Figure 8.25 illustrates these incremental line segments. There are situations when it is necessary to determine the extensional strain ϵ_n or the shear strain γ_{nt}, given the xyz-referenced strains and the n and t directions. In Sections 8.8 through 8.10 we consider only two-dimensional strain analysis, but in Section 8.11 we will return to the topic of three-dimensional strain analysis.

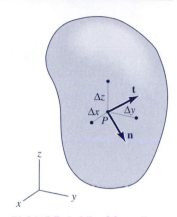

FIGURE 8.25 Mutually orthogonal line segments used in defining extensional strains and shear strains.

One example of two-dimensional strain is called plane strain. If the strains satisfy the equations

$$\epsilon_z = \gamma_{xz} = \gamma_{yz} = 0 \tag{8.36}$$

everywhere in a deformable body, the body is said to be in a **state of plane strain.** The vanishing of extensional strain in the z direction requires that the body somehow be restrained from expanding or contracting in the z direction, and the shear-strain equations further require that all planes that are originally parallel to the xy plane remain plane. One example of plane strain is a very long, cylindrical object, like the dam illustrated in Fig. 8.26. Gravitational loads and upstream water-pressure loads on the dam are all parallel to the xy plane, and because of the (assumed) rigid abutments at its ends, the dam is not free to expand or contract in the z direction. Therefore, it can be assumed that $\epsilon_z = \gamma_{xz} = \gamma_{yz} = 0$ everywhere in the dam.

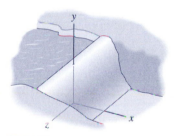

FIGURE 8.26 An example of plane strain.

Plane stress and plane strain are very different and should not be confused. As discussed in Section 8.2, *plane stress* is defined by the equations

$$\sigma_z = \tau_{xz} = \tau_{yz} = 0 \tag{8.1}$$
repeated

and it most frequently occurs in thin, plate-like members, like the web of the plate girder in Fig. 8.3. If we let σ_z (and ΔT) be zero in the last of Eqs. 2.38, we get that, for linearly elastic, isotropic materials,

$$\sigma_z = 0 \rightarrow \epsilon_z = \frac{-\nu}{E}(\sigma_x + \sigma_y) \tag{8.37}$$

On the other hand, if we let ϵ_z (and ΔT) be zero in the last of Eqs. 2.40, we get, for linearly elastic, isotropic materials,

$$\epsilon_z = 0 \rightarrow \sigma_z = \frac{E\nu}{(1+\nu)(1-2\nu)}(\epsilon_x + \epsilon_y) \tag{8.38}$$

Consequently, plane stress generally does not lead to $\epsilon_z = 0$, and plane strain normally requires a nonzero value of σ_z.

571

One very important example of two-dimensional strain analysis is the experimental determination of strains on the surface of a deformable body. Since it is the *strains at a point,* like point A on the surface of the rod in Fig. 8.6a, that are of interest, and since the surface is *locally plane,* we can concentrate on the analysis of strains in a plane.

8.8   TRANSFORMATION OF STRAINS IN A PLANE

Before we derive the strain-transformation equations, let us examine a specific example of how strains vary with the orientation of the reference axes. Figure 8.27 shows the "before-deformation" and "after-deformation" pictures of a plane membrane. Before deformation, a grid of horizontal and vertical lines is marked on the sheet, with a uniform grid spacing of length a. Points P_1, P_2, and P_3 are the origins of axis systems x_1y_1, x_2y_2 and x_3y_3 in the orientations shown. The sheet is deformed by uniformly stretching it to twice its original length in the horizontal direction. At the same time, the sheet is prevented from expanding or contracting in the vertical direction. Hence, each $a \times a$ square before deformation becomes a $2a \times a$ rectangle after deformation. (This would be considered large deformation.) Clearly, the deformation is the same at P_1, P_2, and P_3, but we will get different values of ϵ_n and γ_{nt} depending on the orientation of the reference axes. For example, the shear strain $\gamma_{x1y1} = 0$, but γ_{x2y2} and γ_{x3y3} are clearly not zero, since there are changes in the right angles at P_2 and P_3. Likewise, $\epsilon_{y1} = 0$, but ϵ_{y2} and ϵ_{y3} are not zero, since the lengths of P_2R_2 and P_3R_3 are changed by the deformation.

Whereas the stress-transformation equations, Eqs. 8.3 (or Eqs. 8.5) are based on equilibrium only, the **strain-transformation equations are based solely on the geometry of deformation** (including some small-angle approximations). Figure 8.28a, 8.28b, and 8.28c depict the separate effects of ϵ_x, ϵ_y, and γ_{xy}, respectively. These effects can be added together to get the total expressions for ϵ_n and γ_{nt} as functions of the angle $\theta \equiv \theta_{xn} = \theta_{yt}$ and the three strains ϵ_x, ϵ_y, and γ_{xy}.

Contribution of ϵ_x, ϵ_y, and γ_{xy} to ϵ_n and γ_{nt}. Using Fig. 8.28a, we will derive expressions that relate ϵ_n and γ_{nt} to ϵ_x. The contributions of ϵ_y to ϵ_n and γ_{nt} and of γ_{xy} to ϵ_n and γ_{nt} can be determined in an analogous manner, so these derivations are left as homework problems. (See Homework Problems 8.8-1 and 8.8-2.) Assuming that all strains are small, we can write the following superposition expressions for ϵ_n and γ_{nt}:

$$\epsilon_n = \epsilon_n' + \epsilon_n'' + \epsilon_n''' \tag{8.39a}$$

$$\gamma_{nt} = \gamma_{nt}' + \gamma_{nt}'' + \gamma_{nt}''' \tag{8.39b}$$

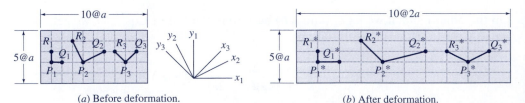

(a) Before deformation.   (b) After deformation.

FIGURE 8.27   An example of how strain quantities depend on the orientation of the reference axes.

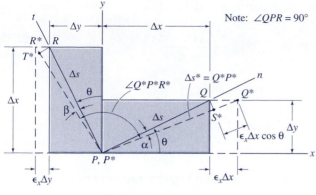
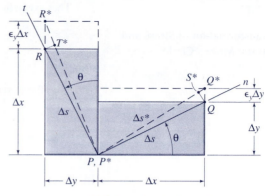

(a) Deformation due to ϵ_x only.

(b) Deformation due to ϵ_y only.

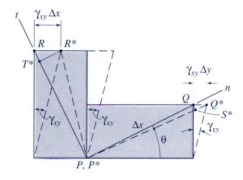

(c) Deformation due to γ_{xy} only.

FIGURE 8.28 The geometry of deformation used to derive strain-transformation equations.

where ϵ_n and γ_{nt} depend on ϵ_x, ϵ_y, and γ_{xy}, and also on the angle θ. That is,

$$\epsilon_n = \epsilon_n(\epsilon_x, \epsilon_y, \gamma_{xy}; \theta)$$

$$\epsilon'_n \equiv \epsilon_n(\epsilon_x; \theta)$$

$$\epsilon''_n \equiv \epsilon_n(\epsilon_y; \theta)$$

$$\epsilon'''_n \equiv \epsilon_n(\gamma_{xy}; \theta)$$

and analogously for γ_{nt}.

To construct Fig. 8.28a, we first construct a rectangle whose diagonal, PQ, is oriented at the angle θ counterclockwise from the x axis and whose length is Δs. The sides of this rectangle have lengths

$$\Delta x = \Delta s \cos \theta, \qquad \Delta y = \Delta s \sin \theta$$

Next, in order to make the angle $\angle QPR$ a right angle, we form a rectangle whose diagonal, PR, is oriented at the angle θ counterclockwise from the y axis and whose length is also Δs. Having constructed the shaded figure that is composed of two rectangles and that represents part of the *undeformed body*, we hold point P fixed and draw dashed lines to show the shape of that part of the deformed body when ϵ_x is positive and $\epsilon_y = \gamma_{xy} = 0$.

573

The extensional strain in the direction n is defined by

$$\epsilon_n(P) = \lim_{\substack{Q \to P \text{ along } n}} \left(\frac{\Delta s^* - \Delta s}{\Delta s} \right) \qquad (2.35) \text{ repeated}$$

and, the shear strain with respect to axes n and t by

$$\gamma_{nt}(P) = \lim_{\substack{Q \to P \text{ along } n \\ R \to P \text{ along } t}} \left(\frac{\pi}{2} - \angle Q^* P^* R^* \right) \qquad (2.36) \text{ repeated}$$

In the present case we can dispense with the limit operation and just write

$$\epsilon_n(P) = \left(\frac{\Delta s^* - \Delta s}{\Delta s} \right) \qquad (8.40)$$

$$\gamma_{nt} = \left(\frac{\pi}{2} - \angle Q^* P^* R^* \right) \qquad (8.41)$$

At the same time, we will make several small-angle approximations.

Contribution of ϵ_x to ϵ_n: We will use the geometry of Fig. 8.28a to develop expressions for $\epsilon'_n(\epsilon_x; \theta)$ and $\gamma'_{nt}(\epsilon_x; \theta)$. From Fig. 8.28a and Eq. 8.40,

$$\epsilon'_n = \frac{\Delta s^* - \Delta s}{\Delta s} = \frac{\overline{Q^* S^*}}{\overline{QP}} \qquad (8.42)$$

To determine the elongation $\overline{Q^* S^*}$, we drop a perpendicular from point Q to the line $P^* Q^*$. In the process, we make the small-angle approximation that $\overline{P^* S^*} \approx \overline{PQ}$, that is, that the perpendicular $\overline{QS^*}$ is (approximately) the arc of a circle of radius Δs with center at P. Since $\overline{QQ^*} = \epsilon_x \Delta x$, the elongation of the rectangle due to the strain ϵ_x, and since angle $\angle QQ^* S^*$ is (approximately) equal to θ, we have

$$\overline{Q^* S^*} = \epsilon_x \Delta x \cos \theta = \epsilon_x (\Delta s \cos \theta) \cos \theta$$

Therefore,

$$\epsilon'_n = \frac{\epsilon_x \Delta s \cos^2 \theta}{\Delta s} = \epsilon_x \cos^2 \theta \qquad (8.43)$$

The contributions ϵ''_n and ϵ'''_n due to ϵ_y and ϵ_{xy}, respectively, can be derived using Figs. 8.28b and 8.28c, respectively. Summing the three contributions (Eq. 8.39a), we get the following **extensional-strain-transformation formula:**

$$\epsilon_n = \epsilon_x \cos^2 \theta + \epsilon_y \sin^2 \theta + \gamma_{xy} \sin \theta \cos \theta \qquad (8.44)$$

Contribution of ϵ_x to γ_{nt}: From Fig. 8.28a and Eq. 8.41 we have

$$\gamma'_{nt} = \frac{\pi}{2} - (\angle Q^* P^* R^*)' = -\alpha - \beta \qquad (8.45)$$

where α and β are angles that are defined in Fig. 8.28a. Again using a small angle approximation, we have

$$\overline{QS^*} \approx \alpha \Delta s, \qquad \overline{RT^*} \approx \beta \Delta s$$

so

$$\gamma'_{nt} \approx -\left(\frac{\overline{QS^*}}{\Delta s}\right) - \left(\frac{\overline{RT^*}}{\Delta s}\right)$$

But,

$$\overline{QS^*} \approx \epsilon_x \Delta x \sin \theta = \epsilon_x (\Delta s \cos \theta) \sin \theta$$

$$\overline{RT^*} \approx \epsilon_x \Delta y \cos \theta = \epsilon_x (\Delta s \sin \theta) \cos \theta$$

Finally, combining these equations, we get

$$\gamma'_{nt} = -2\epsilon_x \sin \theta \cos \theta$$

Combining this expression with the contributions of ϵ_y and γ_{xy} to γ_{nt}, as in Eq. 8.39b, we get the following **shear-strain-transformation formula:**

$$\gamma_{nt} = -2(\epsilon_x - \epsilon_y) \sin \theta \cos \theta + \gamma_{xy}(\cos^2 \theta - \sin^2 \theta) \qquad (8.46)$$

Equations 8.44 and 8.46 are the strain analogs of Eqs. 8.3a and 8.3b, with the one exception that *there is a difference of a factor of two in the shear-strain terms* in Eqs. 8.44 and 8.46 compared with corresponding terms in the stress-transformation equations 8.3a and 8.3b. For example, Eq. 8.46 has the same form as Eq. 8.3b if we divide Eq. 8.46 by two and write it in the form

$$\frac{\gamma_{nt}}{2} = -(\epsilon_x - \epsilon_y) \sin \theta \cos \theta + \frac{\gamma_{xy}}{2} (\cos^2 \theta - \sin^2 \theta)$$

Like the stress-transformation equations, the **strain-transformation equations** can be simplified by incorporating the double-angle trigonometric identities, giving

$$\epsilon_n = \left(\frac{\epsilon_x + \epsilon_y}{2}\right) + \left(\frac{\epsilon_x - \epsilon_y}{2}\right) \cos 2\theta + \left(\frac{\gamma_{xy}}{2}\right) \sin 2\theta \qquad \text{Strain-} \qquad (8.47a)$$

**Trans-
formation
Equations**

$$\frac{\gamma_{nt}}{2} = -\left(\frac{\epsilon_x - \epsilon_y}{2}\right) \sin 2\theta + \left(\frac{\gamma_{xy}}{2}\right) \cos 2\theta \qquad (8.47b)$$

Note that, unlike the shear-stress terms in Eqs. 8.5, **all shear strains in Eqs. 8.47 are divided by two.**

Since the strain-transformation equations, Eqs. 8.47, are completely analogous to the transformation equations for plane stress, Eqs. 8.5, formulas for determining principal directions and principal stresses, and other formulas that are based on Eqs. 8.5, can be converted to formulas for comparable strain-related quantities. In the next section we will use Mohr's circle to solve strain-transformation problems.

The strain-transformation equations, Eqs. 8.47, are completely analogous to the transformation equations for plane stress, on which the derivation of Mohr's circle of stress in Section 8.5 was based, namely, Eqs. 8.5. By following the same procedure that was used in Section 8.5, we obtain the following equations that characterize **Mohr's circle of strain**:

$$(\epsilon_n - \epsilon_{avg})^2 + \left(\frac{\gamma_{nt}}{2}\right)^2 = R^2 \tag{8.48a}$$

where

$$R = \sqrt{\left(\frac{\epsilon_x - \epsilon_y}{2}\right)^2 + \left(\frac{\gamma_{xy}}{2}\right)^2} \tag{8.48b}$$

$$\epsilon_{avg} = \frac{\epsilon_x + \epsilon_y}{2} \tag{8.48c}$$

Equation 8.48a represents the equation of a circle in the $\left(\epsilon, \frac{\gamma}{2}\right)$ plane with center at $(\epsilon_{avg}, 0)$ and radius R (Fig. 8.29). (All strain quantities are dimensionless.) Thus, Eqs. 8.47 are just the parametric equations of this Mohr's circle of strain, with the angle 2θ being the parameter.

To clarify the sign convention of Mohr's circle of strain, particularly the sign convention for shear strain γ, let us recall the sign convention for Mohr's circle of stress. There are two equivalent ways to establish the sign of the τ-coordinate of a point in the (σ, τ) plane:

Method 1: If the ntz axes form a right-handed coordinate system, point N has the coordinates $(\sigma_n, +\tau_{nt})$, while point T has the coordinates $(\sigma_t, -\tau_{nt})$.

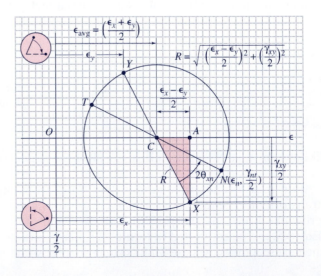

FIGURE 8.29 Mohr's circle of strain.

Method 2: If the point N is plotted with positive shear stress (i.e., downward), then the shear stress τ on the N face would tend to rotate the corresponding nt stress element counterclockwise. (See Fig. 8.17 and also the icons on the τ axis of Fig. 8.16.)

There are two analogous methods for Mohr's circle of strain:

Method 1: If the ntz axes form a right-handed coordinate system, point N has the coordinates $\left(\epsilon_n, +\dfrac{\gamma_{nt}}{2} \right)$, while point T has the coordinates $\left(\epsilon_t, -\dfrac{\gamma_{nt}}{2} \right)$.

Method 2: If the point N is plotted with positive shear strain (i.e., downward), then a line element in the n direction would tend to rotate counterclockwise. (See the icons on the $\dfrac{\gamma}{2}$ axis in Fig. 8.29).

Figure 8.30 illustrates the fact that a positive shear strain at point P represents a reduction in the angle between the n and t axes by an amount γ_{nt}, and, in the process, the incremental line element in the n direction rotates counterclockwise. The icons at the ends of the $\dfrac{\gamma}{2}$ axis in Fig. 8.29 indicate the direction of rotation of a generic line element n, depending on the sign associated with the shear-strain coordinate of point N.

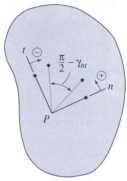

FIGURE 8.30 An explanation of the shear-strain sign convention for Mohr's circle of strain.

PROCEDURE FOR CONSTRUCTING AND USING MOHR'S CIRCLE OF STRAIN

The procedure for constructing Mohr's circle of strain is virtually identical to the procedure for constructing Mohr's circle of stress.

Draw Mohr's Circle: Figure 8.29 illustrates Steps 1 through 6.

1. Establish a set of $\left(\epsilon, \dfrac{\gamma}{2} \right)$ axes, with the same scale on both axes. Use paper that has a grid, and use a scale that results in a circle of reasonable size. The **positive $\dfrac{\gamma}{2}$ axis points downward.**

2. Assuming that ϵ_x, ϵ_y, and γ_{xy} are given, locate the point X at $\left(\epsilon_x, \dfrac{\gamma_{xy}}{2} \right)$ and the point Y at $\left(\epsilon_y, -\dfrac{\gamma_{xy}}{2} \right)$.

3. Connect points X and Y with a straight line, and locate the center of the circle, C, where the line crosses the ϵ axis at $(\epsilon_{\text{avg}}, 0)$.

4. Draw a circle that has its center at $(\epsilon_{\text{avg}}, 0)$ and that passes through points X and Y. (It is a good idea to use a compass in drawing the circle.)

Compute the Required Information:

5. Form the shaded triangle XCA with sides, $\left(\dfrac{\gamma_{xy}}{2} \right)$ and $\left(\dfrac{\epsilon_x - \epsilon_y}{2} \right)$, and compute the radius of the circle.

$$R = \sqrt{\left(\frac{\epsilon_x - \epsilon_y}{2} \right)^2 + \left(\frac{\gamma_{xy}}{2} \right)^2} \qquad \begin{array}{r} (8.48b) \\ \text{repeated} \end{array}$$

6. If the extensional strain in a particular direction, say ϵ_n, is required, locate point N on Mohr's circle by turning an angle 2θ counterclockwise (or clockwise) on the circle, corresponding to rotating an angle θ counterclockwise (or clockwise) from some reference direction such as the x direction. Construct a diameter through point N and the center of the circle, and use trigonometry to calculate the value of ϵ_n (and ϵ_t and γ_{nt} if required).

7. If the *principal strains* and the orientation of the *principal directions of strain* are required, use

$$\epsilon_1 = \epsilon_{\text{avg}} + R$$
$$\epsilon_2 = \epsilon_{\text{avg}} - R \qquad (8.49)$$

to calculate the principal strains, where $\epsilon_{\text{avg}} = \epsilon_C$ is given by Eq. 8.48c. Use trigonometry to determine some angle, like $2\theta_{xp1}$, that can be used to locate a principal direction of strain, say p_1, with respect to some known reference direction. There is no shear strain between the principal directions of strain.

8. Use a procedure similar to Step 7 if the maximum in-plane shear strain and maximum shear directions are required. The directions of maximum shear strain are at $\pm 45°$ to the principal directions of strain.

Figure 8.31 illustrates Steps 7 and 8.

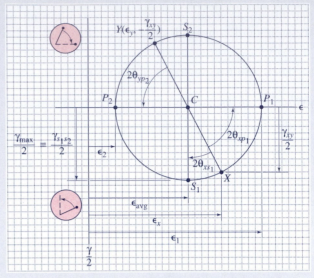

FIGURE 8.31   A Mohr's circle of strain with the principal strains and maximum in-plane shear strains identified.

EXAMPLE 8.6

Fig. 1   The orientation of the n axis.

At a certain point P on a deformable body (Fig. 1), the in-plane strains referred to a set of xy axes are:

$$\epsilon_x = 120\mu, \quad \epsilon_y = -40\mu, \quad \text{and} \quad \gamma_{xy} = -120\mu$$

where μ is the unit of *microstrain* (i.e., 10^{-6} in./in., or microinches per inch). (a) Using a square of unit length in the x and y directions to represent the undeformed body at this point, draw a sketch of this elemental square before and after deformation. (Exaggerate the deformation.) (b) Sketch a Mohr's circle of strain for these in-plane strains. (c) Determine the extensional strain in the direction n that is 30° *clockwise* from the x axis. (d) Determine the principal strains and the principal directions of strain, and sketch a deformed element that is oriented in the principal directions. (e) Finally, compute the maximum in-plane shear strain and the associated extensional strains in the directions of the axes of maximum in-plane shear strain. Sketch a deformed element that is oriented in the directions of maximum in-plane shear strain.

Solution   This example problem can be solved using the eight steps suggested under *Procedure for Constructing and Using Mohr's Circle of Strain.*

(a) State of Strain: The shaded square in Fig. 2 is the undeformed element. The dashed lines indicate the shape of the deformed element. Since the shear strain, γ_{xy}, is negative, the right angle at P increases by

578

a total angle of 120μ. This shear strain angle is apportioned as $\left|\dfrac{\gamma_{xy}}{2}\right|$ clockwise to the x edge PQ and $\left|\dfrac{\gamma_{xy}}{2}\right|$ counterclockwise to the y edge PR.

(b) Mohr's Circle of Strain: On grid paper (Fig. 3), the point X is plotted at $\left(\epsilon_x, +\dfrac{\gamma_{xy}}{2}\right)$, which is the point $(120\mu, -60\mu)$. Point Y is plotted at $\left(\epsilon_y, -\dfrac{\gamma_{xy}}{2}\right)$. Note that the x edge PQ rotates clockwise in Fig. 2, and that this agrees with the icons on the negative (upper) end of the $\dfrac{\gamma}{2}$ axis.

The center of the circle, C, lies at $(\epsilon_{\text{avg}}, 0)$, where

$$\epsilon_{\text{avg}} = \frac{\epsilon_x + \epsilon_y}{2} = \frac{(120\mu - 40\mu)}{2} = 40\mu \tag{1}$$

The radius of the circle is length $\overline{XC}$, which is given by

$$R = \sqrt{\overline{CA}^2 + \overline{XA}^2} = \sqrt{(80\mu)^2 + (60\mu)^2} = 100\mu \tag{2}$$

(c) Extensional Strain ϵ_n: To determine the extensional strain in the n direction shown on Fig. 1, we locate the radial line CN on Mohr's circle in Fig. 3 at an angle $2\theta_{xn} = 60°$ *clockwise* from the radial line CX. From Fig. 3, we get

$$\epsilon_n = \overline{OC} + \overline{CB} = \epsilon_{\text{avg}} + R\cos 2\theta_{p_1 n} \tag{3}$$

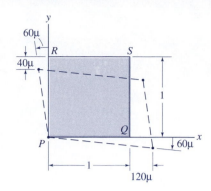

Fig. 2 An example of in-plane strains.

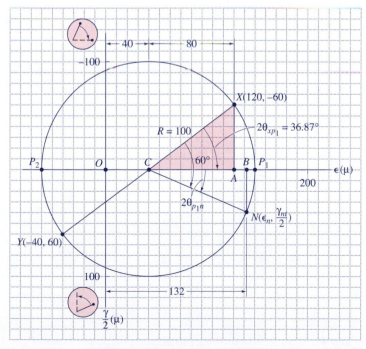

Fig. 3 Mohr's circle of strain.

To determine the angle $2\theta_{p_1n}$, we can use triangle XCA to first determine the angle $2\theta_{xp_1}$.

$$2\theta_{xp_1} = \tan^{-1}\left(\frac{\overline{XA}}{\overline{CA}}\right) = \tan^{-1}\left(\frac{60\mu}{80\mu}\right) = 36.87° \qquad (4)$$

Then,

$$2\theta_{p_1n} = 60° - 2\theta_{xp_1} = 23.13° \qquad (5)$$

So, combining Eqs. (1) through (5), we get

$$\epsilon_n = \overline{OC} + \overline{CB} = 40\mu + 100\mu \cos(23.13°)$$

or

$$\epsilon_n = 132\mu \qquad \text{Ans. (c)} \qquad (6)$$

(d) *Principal Strains and Principal Directions:* To avoid cluttering up Fig. 3, we will repeat the basic Mohr's circle as Fig. 4. The principal directions are represented on Fig. 4 by points P_1 and P_2 at $(\epsilon_1, 0)$ and $(\epsilon_2, 0)$, respectively. From Fig. 4, the principal strains ϵ_1 and ϵ_2 are

$$\epsilon_1 = \epsilon_{\text{avg}} + R = 40\mu + 100\mu = 140\mu$$
$$\epsilon_2 = \epsilon_{\text{avg}} - R = 40\mu - 100\mu = -60\mu \qquad (7)$$

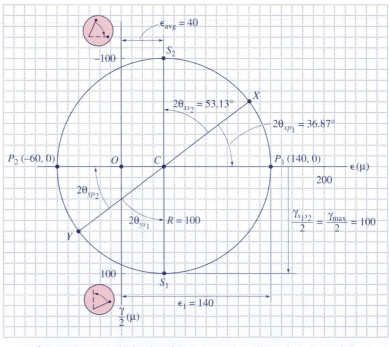

Fig. 4 Mohr's circle showing principal-strain points P_1 and P_2 and maximum-shear-strain points S_1 and S_2.

Points P_1 and P_2 can be located relative to points X and Y, respectively, by *clockwise* angles

$$2\theta_{xp_1} = 2\theta_{yp_2} = 36.87° \,\circlearrowright \qquad (8)$$

as calculated in Eq. (4). Therefore, the *principal strains* and corresponding *principal directions* are:

$$\left. \begin{aligned} \epsilon_1 &= 140\mu, & \theta_{xp_1} &= 18.4° \\ \epsilon_2 &= -60\mu, & \theta_{yp_2} &= 18.4° \end{aligned} \right\} \qquad \textbf{Ans. (d)} \quad (10)$$

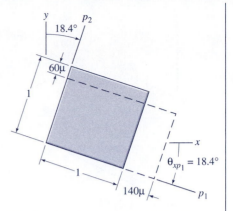

Fig. 5 An element illustrating the principal strains.

To sketch the undeformed and deformed principal-strain elements, let us begin with a unit square oriented at

$$\theta_{xp_1} = \theta_{yp_2} = 18.4° \,\circlearrowright$$

with respect to the xy axes (Fig. 5). The strains ϵ_1 and ϵ_2 are then used in sketching the deformed principal element (dashed lines).

(e) Maximum In-Plane Shear Strain: The points corresponding to maximum in-plane shear strain, points S_1 and S_2, are located at the lowest and highest points on the Mohr's circle in Fig. 4. From Fig. 4,

$$\frac{\gamma_{s_1 s_2}}{2} = R = 100\mu \qquad (11)$$

The directions s_1 and s_2 are determined by the angles $2\theta_{xs_2} = 2\theta_{ys_1}$ on Fig. 4. Thus,

$$2\theta_{xs_2} = 2\theta_{ys_1} = 90 - 2\theta_{xp_1} = 90 - 36.87° = 53.13° \qquad (12)$$

so,

$$\theta_{xs_2} = \theta_{ys_1} = 26.6° \,\circlearrowright \qquad (13)$$

Therefore, the *maximum in-plane shear strain* is given by

$$\gamma_{s_1 s_2} = 200\mu, \quad \theta_{xs_2} = \theta_{ys_1} = 26.6° \qquad \textbf{Ans. (e)} \quad (14)$$

The extensional strain in the s_1 and s_2 directions is

$$\epsilon_{s_1} = \epsilon_{s_2} = \epsilon_{avg} = 40\mu \qquad \textbf{Ans. (e)} \quad (15)$$

To sketch the maximum-shear-strain element, let us begin with a unit square oriented at the angle given in Eq. (14). We draw the element so that $s_1 s_2 z$ forms a right-handed coordinate system. (Note: The $+y$ direction and the $-y$ direction are 180° apart on the deformable body,

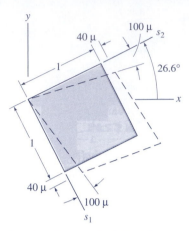

Fig. 6 An element depicting the maximum in-plane shear strain.

so they are 360° "apart" on Mohr's circle, that is, the point Y represents both the $+y$ axis and the $-y$ axis.) To get a right-handed coordinate system $s_1 s_2 z$ in Fig. 6 we should actually rewrite Eq. (14b) as

$$\theta_{xs_2} = \theta_{-ys_1} = 26.6° \, \text{)} \tag{16}$$

Review the Solution The key to correct solution of in-plane strain problems using Mohr's circle is accurate drawing of the circle from information about the strain state, that is, correctly plotting points X and Y. We have to remember to divide all γ values by two before plotting! The Mohr's circle in Fig. 3 is accurately drawn.

It is also useful to sketch the deformed element using information about the state of strain. This has been done in Fig. 2. By comparing Fig. 5 and Fig. 6 with Fig. 2, we can see that all three figures are in agreement; that is, all exhibit stretching in the direction of the QR diagonal in Fig. 2. Therefore, our solutions are probably correct.

 **Mohr's Circle—Strain Transformations**

8.10 MEASUREMENT OF STRAIN; STRAIN ROSETTES

There are several experimental techniques that may be used to measure strain. These are described in texts on *experimental mechanics* or *experimental stress analysis*.[12] The most straightforward technique employs wire or metal foil **electrical-resistance strain gages,** like the ones pictured in Fig. 8.32. Figure 8.32a shows a *metal foil strain gage* and Fig. 8.32b depicts two *strain rosettes*. The gage consists of a wire or foil "grid" on a thin paper or plastic backing that can be bonded (e.g., glued) directly to the surface whose strain is to be measured. Extensional strain along the axis of the grid stretches (or contracts) the metal grid element, causing a change in electrical resistance of the grid. This change in resistance can be converted directly to the extensional strain ϵ_n along the direction n of the axis of the gage.

Since electrical-resistance strain gages can directly measure only extensional strain, it is not possible with a single gage to measure shear strain or, for example, to determine the directions and magnitudes of the principal strains at a point. Therefore, it is common practice to employ a **strain rosette** consisting of three electrical-resistance strain gages mounted on a common backing sheet, like the rosettes depicted in Fig. 8.32b. Let the three gages of a rosette be oriented along axes that are labeled a, b, and c, as shown in Fig. 8.33. The extensional strain along each of these axes can be related to the three strain quantities (ϵ_x, ϵ_y, γ_{xy}), where the xy reference frame may be established in any convenient orientation. For

[12]*Handbook on Experimental Mechanics,* by A. S. Kobayashi, [Ref. 8-3]; *Experimental Solid Mechanics,* by A. Shukla and J. W. Dally, [Ref. 8-4].

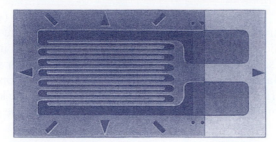

(a) A metal foil strain gage.

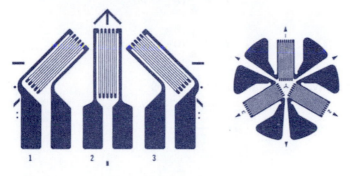

1   2   3

(b) A 45° strain rosette and an equiangular rosette.

FIGURE 8.32 Electrical-resistance strain gages. (Micro-Measurements Division of Measurements Group, Inc.)

example, the axis of the *a* gage may be taken as the *x* axis by setting $\theta_a \equiv 0$. From Eq. 8.44 we get, for arbitrary angles,

$$\boxed{\begin{aligned}\epsilon_a &= \epsilon_x \cos^2 \theta_a + \epsilon_y \sin^2 \theta_a + \gamma_{xy} \sin \theta_a \cos \theta_a \\ \epsilon_b &= \epsilon_x \cos^2 \theta_b + \epsilon_y \sin^2 \theta_b + \gamma_{xy} \sin \theta_b \cos \theta_b \\ \epsilon_c &= \epsilon_x \cos^2 \theta_c + \epsilon_y \sin^2 \theta_c + \gamma_{xy} \sin \theta_c \cos \theta_c\end{aligned}} \tag{8.50}$$

The two common rosette arrangements are the 45° *rectangular rosette* and the 60° *equiangular rosette* shown in Fig. 8.32b. Substituting $\theta_a = 0°$, $\theta_b = 45°$, and $\theta_c = 90°$ into Eqs. 8.50, we get the following equations for the 45° rosette:

$$\begin{aligned}\epsilon_x &= \epsilon_a \\ \epsilon_y &= \epsilon_c \\ \gamma_{xy} &= 2\epsilon_b - \epsilon_a - \epsilon_c\end{aligned} \tag{8.51}$$

For the equiangular rosette, we can let $\theta_a = 0°$, $\theta_b = 60°$, and $\theta_c = 120°$. Then, Eqs. 8.50 give the following:

$$\begin{aligned}\epsilon_x &= \epsilon_a \\ \epsilon_y &= \frac{1}{3}(2\epsilon_b + 2\epsilon_c - \epsilon_a) \\ \gamma_{xy} &= \frac{2}{\sqrt{3}}(\epsilon_b - \epsilon_c)\end{aligned} \tag{8.52}$$

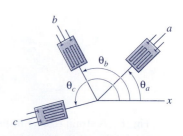

FIGURE 8.33 Notation for a strain rosette with arbitrary angles.

Once strains ϵ_x, ϵ_y, and γ_{xy} have been calculated using Eq. 8.50 (or 8.51 or 8.52), a Mohr's circle of strain can be drawn for the surface strains at the rosette location.

In most instances where strain rosettes are used, there is a need to determine the principal stresses and, perhaps, the absolute maximum shear stress. The rosette is affixed to a surface, which is usually stress-free. In that case, one of the principal stresses is $\sigma_z = 0$. Therefore, if the material constants E and ν are known, we can use Eqs. 2.38 to determine the in-plane principal stresses and also the extensional strain ϵ_z. Then,

$$\epsilon_1 = \frac{1}{E}(\sigma_1 - \nu\sigma_2)$$

$$\epsilon_2 = \frac{1}{E}(\sigma_2 - \nu\sigma_1)$$

which can be solved for σ_1 and σ_2 to give

$$\sigma_1 = \frac{E}{1 - \nu^2}(\epsilon_1 + \nu\epsilon_2)$$

$$\sigma_2 = \frac{E}{1 - \nu^2}(\epsilon_2 + \nu\epsilon_1)$$

(8.53)

In terms of stresses, ϵ_z is given by Eq. 8.37,

$$\epsilon_z = \frac{-\nu}{E}(\sigma_x + \sigma_y) = \frac{-\nu}{E}(\sigma_1 + \sigma_2)$$

And, in terms of the in-plane strains, ϵ_z is given by

$$\epsilon_z = \frac{-\nu(\epsilon_x + \epsilon_y)}{1 - \nu} = \frac{-\nu(\epsilon_1 + \epsilon_2)}{1 - \nu}$$

(8.54)

EXAMPLE 8.7

Fig. 1   A strain rosette installation.

The landing-gear strut of an airplane is instrumented with a 45° strain rosette to measure the strains in the strut during landing. The a-gage is oriented along the axial direction of the strut, as shown in Fig. 1. At one instant during a landing, the measured strains are:

$$\epsilon_a = -700\mu, \qquad \epsilon_b = 0, \qquad \epsilon_c = -100\mu$$

(a) Letting the x axis be oriented along the a-gage and the y axis be oriented along the c-gage, determine the strain values ϵ_x, ϵ_y, and γ_{xy}. (b) Sketch a Mohr's circle of strain. (c) Determine the principal (surface) strains at the rosette location. (d) Determine the principal stresses at the rosette location. Let $E = 15 \times 10^3$ ksi and $\nu = 0.3$.

Plan the Solution   Since this is a standard 45° rosette, we can use Eqs. 8.51 to compute the required xy strains. Then we can use the points

$X\left(\epsilon_x, +\dfrac{\gamma_{xy}}{2}\right)$ and $Y\left(\epsilon_y, -\dfrac{\gamma_{xy}}{2}\right)$ to plot Mohr's circle. The state of stress at the surface is plane stress, so the in-plane principal stresses are given by Eqs. 8.53.

Solution

(a) Rosette Equations: For the 45° rosette, Eqs. 8.51 give

$$\epsilon_x = \epsilon_a = -700\mu$$

$$\epsilon_y = \epsilon_c = -100\mu$$

$$\gamma_{xy} = 2\epsilon_b - \epsilon_a - \epsilon_c = 0 + 700\mu + 100\mu = 800\mu$$

$$\epsilon_x = -700\mu, \quad \epsilon_y = -100\mu, \quad \gamma_{xy} = 800\mu \qquad \textbf{Ans. (a)} \quad (1)$$

(b) Mohr's Circle of Strain: We can use the strain values in Eqs. (1) to plot points $X(\equiv A)$ and $Y(\equiv C)$, and then draw the Mohr's circle of strain, as shown in Fig. 2.

(c) Principal Strains: The values of principal strains in the xy plane are given by

$$(\epsilon_1)_{xy} = \epsilon_{\text{avg}} + R, \qquad (\epsilon_2)_{xy} = \epsilon_{\text{avg}} - R \qquad (2)$$

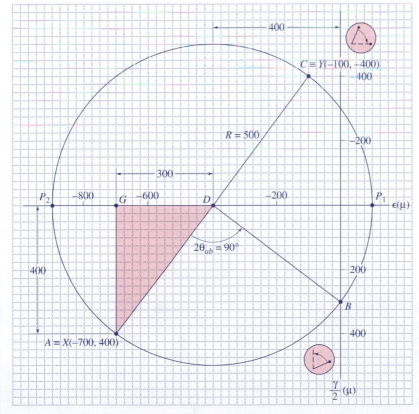

Fig. 2   Mohr's circle of strain.

where

$$\epsilon_{\text{avg}} = \frac{\epsilon_x + \epsilon_y}{2} - \frac{-700\mu - 100\mu}{2} = -400\mu \tag{3}$$

and

$$R = \sqrt{(\overline{GD})^2 + (\overline{GA})^2} = \sqrt{(300\mu)^2 + (400\mu)^2} = 500\mu \tag{4}$$

Therefore, combining Eqs. (2) through (4), we get the following principal strains in the xy plane:

$$(\epsilon_1)_{xy} = 100\mu, \qquad (\epsilon_2)_{xy} = -900\mu \qquad \textbf{Ans. (c)} \tag{5}$$

(d) Principal Stresses: The principal in-plane strains in Eqs. (5) can be substituted into Eqs. 8.53 to give the in-plane (i.e., in the xy plane) principal stresses.

$$(\sigma_1)_{xy} = \frac{E}{1 - \nu^2}(\epsilon_1 + \nu\epsilon_2)_{xy}$$

$$(\sigma_2)_{xy} = \frac{E}{1 - \nu^2}(\epsilon_2 + \nu\epsilon_1)_{xy} \tag{6}$$

Then,

$$(\sigma_1)_{xy} = \frac{(15 \times 10^3 \text{ ksi})}{1 - (0.3)^2}[100 + (0.3)(-900)](10^{-6}) \tag{7a}$$

$$= -2.80 \text{ ksi}$$

$$(\sigma_2)_{xy} = \frac{(15 \times 10^3 \text{ ksi})}{1 - (0.3)^2}[-900 + (0.3)(100)](10^{-6}) \tag{7b}$$

$$= -14.3 \text{ ksi}$$

If we order the principal stresses, including $\sigma_z = 0$, in the order $\sigma_3 \leq \sigma_2 \leq \sigma_1$, we get

$$\sigma_1 = 0 \text{ ksi}, \quad \sigma_2 = -2.8 \text{ ksi}, \quad \sigma_3 = -14.3 \text{ ksi} \qquad \textbf{Ans. (d)}$$

Review the Solution   We have a good check on Parts (a) and (b) by observing that point B, which was not used directly in plotting Mohr's circle, has the correct extensional strain, $\epsilon_b = 0$. We can scale off points P_1 and P_2 on Fig. 2 and see that the values of ϵ_1 and ϵ_2 in Eqs. (5) are correct. It seems strange that $(\sigma_1)_{xy}$ in Eq. (7a) is negative, while ϵ_1 in Eq. (5a) is positive. But, by double-checking Eqs. 8 we see that this result is correct. Hence, we observe that **principal stresses need not always have the same sign as the corresponding principal strains.**

*8.11   ANALYSIS OF THREE-DIMENSIONAL STRAIN

An analysis of the three-dimensional geometry of deformation of a body establishes the fact that, at any point in the body, there are three directions, called **principal-strain directions,** that are mutually perpendicular both before and after deformation.[13] That is, there is no shear strain between the principal strain directions, as is illustrated in Fig. 8.34. (For convenience, the undeformed element in Fig. 8.34 is taken to be a unit cube.) The principal strains are labeled in the order $\epsilon_1 \geq \epsilon_2 \geq \epsilon_3$, that is,

$$\epsilon_1 \equiv \epsilon_{max}, \quad \epsilon_2 \equiv \epsilon_{int}, \quad \epsilon_3 \equiv \epsilon_{min} \qquad (8.55)$$

Their values can be determined by using geometrical equations that are analogous to the equilibrium equations used to determine principal stresses.

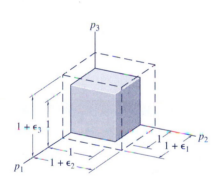

FIGURE 8.34   The principal strains at a point.

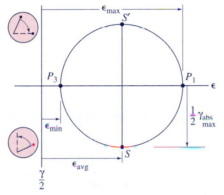

FIGURE 8.35   Mohr's circle for the plane of principal strains ϵ_1 and ϵ_3.

The **absolute maximum shear strain** occurs between two perpendicular axes that lie in the plane of p_1 and p_3, just as in the case of absolute maximum shear stress. Hence, a Mohr's circle of strain can be constructed for the $p_1 p_3$ strain plane, and from this Mohr's circle (Fig. 8.35), the absolute maximum shear strain is found to be

$$\gamma_{\substack{abs \\ max}} = \epsilon_{max} - \epsilon_{min} \qquad (8.56)$$

Plane Strain.   For the special case of *plane strain,*

$$\epsilon_z = \gamma_{xz} = \gamma_{yz} = 0 \qquad \begin{matrix}(8.36) \\ \text{repeated}\end{matrix}$$

---

[13]*Theory of Elasticity,* by S. P. Timoshenko and J. N. Goodier, [Ref. 8-1], Sections 81 and 82.

The determination of the absolute maximum shear strain for this case is completely analogous to the analysis of absolute maximum shear stress for the case of plane stress. For example, if Example Problem 8.6 is a plane-strain problem, then $\epsilon_1 = 140\mu$, $\epsilon_2 = \epsilon_z = 0$, $\epsilon_3 = -60\mu$, and the xy plane corresponds to the p_1p_3 plane when the three-dimensional strain is considered.

Plane Stress. Where a state of plane stress exists in a body, for example at a free surface of the body, the stresses satisfy

$$\sigma_z = \tau_{xz} = \tau_{yz} = 0$$

(8.1)
repeated

If the body is linearly elastic and isotropic, its material behavior satisfies Eqs. 2.38 and 2.40. Therefore, we get

$$\epsilon_z = \frac{-\nu}{E}(\sigma_x + \sigma_y)$$

(8.37)
repeated

and, from Eqs. 2.39 and 8.1,

$$\gamma_{xz} = \gamma_{yz} = 0$$

(8.57)

Therefore, the z direction is a *principal-strain direction*, but the principal strain ϵ_z is not zero. Therefore, it is necessary to determine the two principal strains in the xy plane (e.g., using a Mohr's circle as in Section 8.9) and then to order all three principal strains as in Eq. 8.55 before using Eq. 8.56 to compute the absolute maximum shear strain.

8.12 PROBLEMS

▼ **STRESS-TRANSFORMATION EQUATIONS**

> **Problems 8.3-1 through 8.3-4.** *For each of these problems you are to sketch a free-body diagram like the one in Fig. 2 of Example 8.1, and use equilibrium equations to solve for the normal stress and shear stress on the indicated inclined plane NN.*

Prob. 8.3-1. The state of plane stress at a point is given by the stresses $\sigma_x = 0$ ksi, $\sigma_y = 44.0$ ksi, and $\tau_{xy} = 10.0$ ksi, as indicated on the figure below.

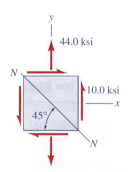

P8.3-1, P8.5-1, and P8.5-35

Prob. 8.3-2. The state of plane stress at a point is given by the stresses $\sigma_x = 48$ MPa, $\sigma_y = -32$ MPa, and $\tau_{xy} = 16$ MPa, as illustrated on the figure below.

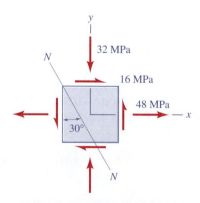

P8.3-2, P8.5-2, and P8.5-35

Prob. 8.3-3. The state of plane stress at a point is given by the stresses $\sigma_x = 3000$ psi, $\sigma_y = -3200$ psi, and $\tau_{xy} = -3000$ psi, as illustrated on the figure below.

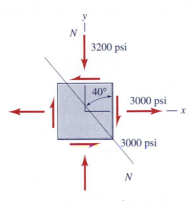

P8.3-3, P8.5-3, and P8.5-35

Prob. 8.3-4. The state of plane stress at a point is given by the stresses $\sigma_x = -48$ MPa, $\sigma_y = -24$ MPa, and $\tau_{xy} = -60$ MPa, as illustrated on the figure below.

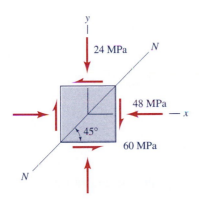

P8.3-4, P8.5-4, and P8.5-35

Problems 8.3-5 through 8.3-12. *For each of these problems, an element in plane stress is subjected to the stresses σ_x, σ_y, and τ_{xy} as indicated. Use the plane-stress-transformation equations, Eqs. 8.6, to determine the stresses $\sigma_{x'}$, $\sigma_{y'}$, and $\sigma_{x'y'}$ on an element that is rotated by the given angle $\theta \equiv \theta_{xx'}$. Show the calculated stresses on an element oriented at this angle.*

Prob. 8.3-5. At the point labeled A in the bracket in Fig. P8.3-5, the given stresses are: $\sigma_x = 4.8$ ksi, $\sigma_y = -1.2$ ksi, and $\tau_{xy} = -4.0$ ksi. The angle is $\theta = 20°$.

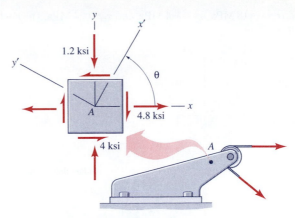

P.8.3-5, P8.3-6, P8.5-5, P8.5-6, and P8.5-23

Prob. 8.3-6. Solve Prob. 8.3-5 for $\theta = -60°$.

Prob. 8.3-7. The given stresses are $\sigma_x = 48$ MPa, $\sigma_y = -32$ MPa, and $\tau_{xy} = 20$ MPa. The angle is $\theta = 30°$.

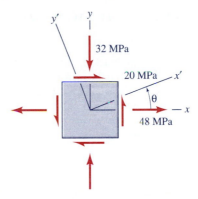

P8.3-7, P8.3-8, P8.5-7, and P8.5-8

Prob. 8.3-8. Solve Prob. 8.3-7 for $\theta = -30°$.

Prob. 8.3-9. The given stresses are $\sigma_x = 3000$ psi, $\sigma_y = -5000$ psi, and $\tau_{xy} = -5000$ psi. The angle is $\theta = 20°$.

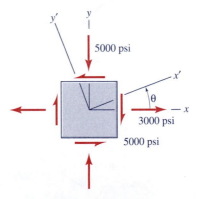

P8.3-9 and P8.3-10

Prob. 8.3-10. Solve Prob. 8.3-9 for $\theta = -20°$.

Prob. 8.3-11. Steel plate girders are often used as the beams that support long multispan bridges. Figure P8.3-11 depicts the region of a plate girder where it transfers its vertical load through a roller support to a concrrte pier. The given stresses are: $\sigma_x = -18$ MPa, $\sigma_y = -4$ MPa, and $\tau_{xy} = 24$ MPa, The angle is $\theta = 30°$.

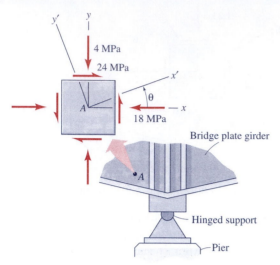

P8.3-11, P8.3-12, P8.5-9, P8.5-10, P8.5-22, and P8.6-11

Prob. 8.3-12. Solve Prob. 8.3-11 for $\theta = -30°$.

Prob. 8.3-13. On a thin bracket, the state of plane stress referred to the $x'y'$ axes has been found to be $\sigma_{x'}$, $\sigma_{y'}$, and $\tau_{x'y'}$, as shown on the figure below. Use the plane-stress-transformation equations to determine the stresses σ_x, σ_y, and τ_{xy}, the stresses referred to the xy axes. Show these stresses on a properly oriented stress element.

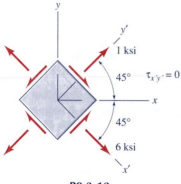

P8.3-13

Prob. 8.3-14. Solve Prob. 8.3-13 for the state of plane stress shown on Fig. P8.3-14.

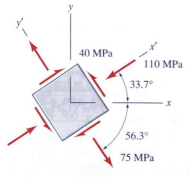

P8.3-14

Prob. 8.3-15. Point A on the surface of a machine component is subjected to plane stress. The stresses σ_x, σ_y, τ_{xy}, $\sigma_{x'}$, and $\tau_{x'y'}$ are known, but the angle $\theta \equiv \theta_{xx'}$ and the normal stress $\sigma_{y'}$ are unknown. Use the information shown on the two stress elements (both are at point A, but are rotated relative to each other) to determine the angle θ and the value of $\sigma_{y'}$.

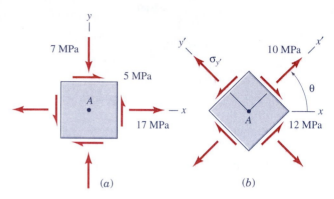

P8.3-15 and P8.5-25

Prob. 8.3-16. Solve Prob. 8.3-15 using the data in Fig. P8.3-16.

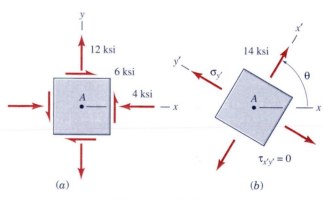

P8.3-16 and P8.5-26

^DProb. 8.3-17. A block of wood is subjected to a vertical compressive stress of magnitude $3\sigma_0$ and, simultaneously, to a horizontal compressive stress of magnitude σ_0, as indicated in Fig. P8.3-17. The wood block will fail if either (a) the

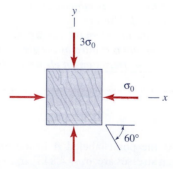

P8.3-17 and P8.5-27

590

compressive stress perpendicular to the grain exceeds 260 psi (C) or (b) the shear stress parallel to the grain exceeds 130 psi. Determine the maximum allowable value of σ_0.

Prob. 8.3-18. A 4-in. by 4-in. by 2-in. (thick) concrete block is subjected to biaxial compression by forces P_x and P_y acting through loading pads, as shown in Fig. P8.3-18a. Determine the compressive axial load P_x (in kips) if the vertical force is $P_y = 8$ kips and compressive normal stresses on the x' and y' planes have the values shown in Fig. P8.3-18b.

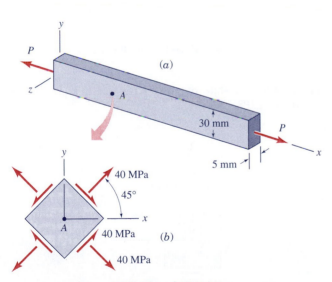

P8.3-18 and P8.5-28

Prob. 8.3-19. The state of plane stress at point A on the surface of an axially loaded bar is shown in Fig. P8.3-19b. Determine the axial load P (in Newtons) if the cross-sectional dimensions of the bar are as indicated in Fig. P8.3-19a.

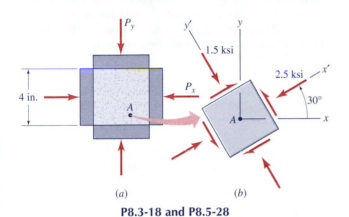

P8.3-19 and P8.5-29

Prob. 8.3-20. The state of plane stress at point A on the surface of an axially loaded bar shown in Fig. P8.3-20. Determine the compressive axial load P (in kips) if the cross-sectional dimensions of the bar are as indicated in Fig. P8.3-20.

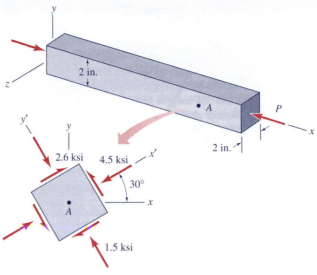

P8.3-20 and P8.5-30

Prob. 8.3-21. The state of plane stress at point A on the surface of a rectangular bar subjected to pure bending in the xy plane is shown in Fig. P8.3-21. Determine the bending moment M (in kip · in.) if the location of A and the cross-sectional dimensions of the bar are as indicated in Fig. P8.3-21. (Hint: Consider Eq. 8.7.)

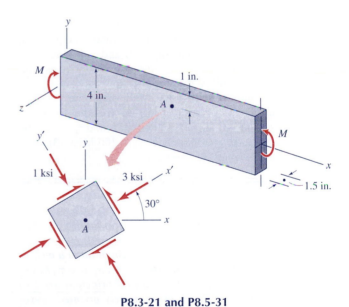

P8.3-21 and P8.5-31

Prob. 8.3-22. At point A on the surface of one side of the backhoe bucket shown in Fig. P8.3-22a, the complete state of stress relative to the xy axes (Fig. P8.3-22b) is to be determined. At point A, the normal stresses on the x' and y' planes are known to be 2.8 ksi in tension and 2.0 ksi in compression, respectively, as indicated in Fig. P8.3-22c. Determine σ_y and τ_{xy}, if, as indicated in Fig. P8.3-22b, $\sigma_x = 3.6$ ksi in compression.

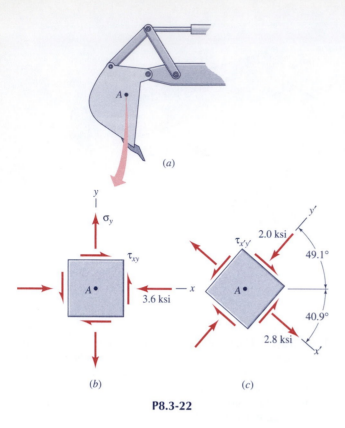

(a)

(b) P8.3-22 *(c)*

Prob. 8.4-5. $\sigma_x = -4$ ksi, $\sigma_y = 12$ ksi, and $\tau_{xy} = 6$ ksi.

Prob. 8.4-6. $\sigma_x = 2{,}400$ psi, $\sigma_y = 14{,}400$ psi, and $\tau_{xy} = 14{,}400$ psi.

Problems 8.4-7 through 8.4-12. *At a certain point in a member subjected to plane stress, the stresses σ_x, σ_y, and τ_{xy} have the values shown on Figs. P8.4-7 through P8.4-12. (a) Determine the <u>principal stresses</u> and show them on a sketch of a properly oriented stress element, and (b) determine the shear stress and the normal stress on the planes of <u>maximum shear stress,</u> and show these on a sketch of a properly oriented stress element.*

Prob. 8.4-7. The state of plane stress for this problem is the stress at point A on the front suspension of an automobile, as shown in Fig. P8.4-7b.

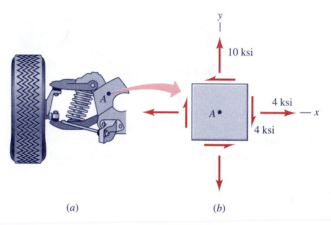

(a) *(b)*

P8.4-7, P8.5-16, and 8.6-12

Prob. 8.4-8. The state of plane stress for this problem is shown in Fig. P8.4-8.

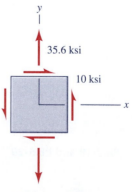

P8.4-8 and P8.5-17

Prob. 8.4-9. The state of plane stress for this problem is shown in Fig. P8.4-9.

---

^cProb. 8.3-23. (a) Write a computer program that: (i) inputs σ_x, σ_y, and τ_{xy}; (ii) uses the stress-transformation equations, Eqs. 8.5, to form expressions for $\sigma_n(\theta)$ and $\tau_{nt}(\theta)$; and (iii) plots $\sigma_n(\theta)$ and $\tau_{nt}(\theta)$ for $-\pi/2 \leq \theta \leq \pi/2$. You may use the programming language or math application software of your own choice, unless your instructor indicates otherwise. (b) Demonstrate your computer program for the following stress states: (1) $\sigma_x = 100$ MPa, $\sigma_y = 0$ MPa, and $\tau_{xy} = 0$ MPa; (2) $\sigma_x = 0$ MPa, $\sigma_y = 0$ MPa, and $\tau_{xy} = 100$ MPa; (3) $\sigma_x = 2$ ksi, $\sigma_y = -1$ ksi, and $\tau_{xy} = -1$ ksi.

▼ PRINCIPAL STRESSES; MAXIMUM SHEAR STRESS

Problems 8.4-1 through 8.4-6. *At a certain point in a member subjected to plane stress, the stresses σ_x, σ_y, and τ_{xy} have the values listed below. (a) Determine the <u>principal stresses</u> and show them on a sketch of a properly oriented stress element, and (b) determine the shear stress and normal stress on the planes of <u>maximum shear stress,</u> and show these on a sketch of a properly oriented stress element.*

Prob. 8.4.1. $\sigma_x = 3000$ psi, $\sigma_y = -5000$ psi, and $\tau_{xy} = -3000$ psi.

Prob. 8.4-2. $\sigma_x = 48$ MPa, $\sigma_y = -32$ MPa, and $\tau_{xy} = 16$ MPa.

Prob. 8.4-3. $\sigma_x = 12$ MPa, $\sigma_y = -8$ MPa, and $\tau_{xy} = -4$ MPa.

Prob. 8.4-4. $\sigma_x = 4.8$ ksi, $\sigma_y = -1.2$ ksi, $\tau_{xy} = -4$ ksi.

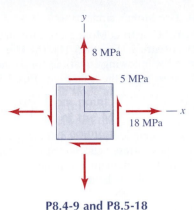

P8.4-9 and P8.5-18

Prob. 8.4-10. The state of plane stress for this problem is shown in Fig. P8.4-10.

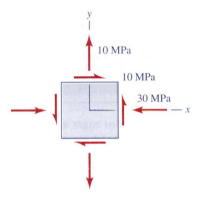

P8.4-10 and P8.5-19

Prob. 8.4-11. The state of plane stress for this problem is shown in Fig. P8.4-11.

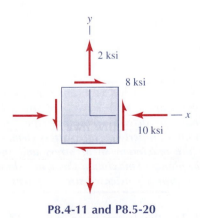

P8.4-11 and P8.5-20

Prob. 8.4-12. The state of plane stress for this problem is shown in Fig. P8.4-12.

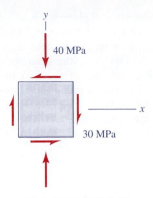

P8.4-12 and P8.5-21

Prob. 8.4-13. At a certain point on the surface of a machine part, the normal stresses on two mutually perpendicular faces (the x and y faces) are 7 MPa (C) and 3 MPa (T), as shown in Fig. P8.4-13. The maximum in-plane shear stress at this point is $\tau_{max} = 10$ MPa. Determine the magnitude, τ, of the shear stress that acts in the direction shown on the x and y faces, and determine the in-plane principal stresses at this point.

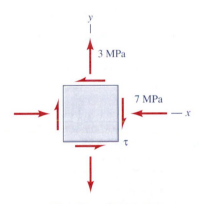

P8.4-13 and P8.5-32

Prob. 8.4-14. At a certain point on the surface of an airplane wing, the state of plane stress can be described by Fig. P8.4-14. The maximum in-plane shear stress at this point is $\tau_{max} = 13$ ksi. (a) Determine the magnitude of the shear stress, τ, on the x and y faces. (b) Determine the two in-plane principal stresses at this point and show them on a properly oriented principal-stress element.

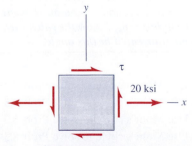

P8.4-14 and P8.5-33

*Prob. 8.4-15. The state of plane stress at a point can be described by a known tensile stress $\sigma_x = 70$ MPa, an unknown tensile stress σ, and an unknown shear stress τ, as indicated in Fig. P8.4-15. At this point the maximum in-plane shear stress is 78 MPa, and one of the two in-plane principal stresses is 22 MPa (T). Determine the values of the two unknown stresses, labeled σ and τ on the figure, and determine the second in-plane principal stress. The stresses act in the directions shown on Fig. P8.4-15, that is, $\sigma_y = \sigma$ and $\tau_{xy} = -\tau$.

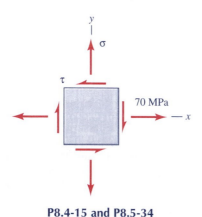

P8.4-15 and P8.5-34

▼ MOHR'S CIRCLE: STRESS TRANSFORMATIONS

> **MDS 8.1–8.3**

> **All problems is Section 8.5 are to be solved by using _Mohr's Circle for Plane Stress_. Consider only the in-plane stresses.**
> **Additional problems on the topic of Mohr's Circle may be found in Chapter 9.**

> **Problems 8.5-1 through 8.5-4. Use _Mohr's circle_ to solve for the normal stress and the shear stress on the indicated inclined plane NN.**

Prob. 8.5-1. Use Mohr's circle to solve Prob. 8.3-1.

Prob. 8.5-2. Use Mohr's circle to solve Prob. 8.3-2.

Prob. 8.5-3. Use Mohr's circle to solve Prob. 8.3-3.

Prob. 8.5-4. Use Mohr's circle to solve Prob. 8.3-4.

> **Problems 8.5-5 through 8.5-10. Use _Mohr's circle_ to solve for the stresses $\sigma_{x'}$, $\sigma_{y'}$, and $\tau_{x'y'}$ on an element that is rotated by the angle $\theta \equiv \theta_{xx'}$. Show the calculated stresses on a stress element oriented at this angle.**

Prob. 8.5-5. Use Mohr's circle to solve Prob. 8.3-5.

Prob. 8.5-6. Use Mohr's circle to solve Prob. 8.3-6.

Prob. 8.5-7. Use Mohr's circle to solve Prob. 8.3-7.

Prob. 8.5-8. Use Mohr's circle to solve Prob. 8.3-8.

Prob. 8.5-9. Use Mohr's circle to solve Prob. 8.3-11.

Prob. 8.5-10. Use Mohr's circle to solve Prob. 8.3-12.

Prob. 8.5-11. (a) Construct Mohr's circle for an element in a uniaxial state of stress (Fig. P8.5-11). (b) Use this Mohr's circle to derive the following equations for the normal stress σ_n and shear stress τ_{nt} on the n-face (see Eqs. 2.30).

$$\sigma_n = \frac{\sigma_x}{2}(1 + \cos 2\theta), \qquad \tau_{nt} = -\left(\frac{\sigma_x}{2}\right)\sin 2\theta$$

(c) Use the Mohr's circle to determine the planes on which the maximum shear stress acts. Sketch a properly oriented maximum-shear-stress element and indicate the normal and shear stresses acting on its faces.

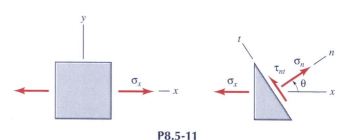

P8.5-11

Prob. 8.5-12. (a) Construct Mohr's circle for an element undergoing pure shear (Fig. P8.5-12). (b) Use this Mohr's circle to derive the following equations for the normal stress σ_n and shear stress τ_{nt} on the n-face.

$$\sigma_n = \tau_{xy} \sin 2\theta, \qquad \tau_{nt} = \tau_{xy} \cos 2\theta$$

(c) Use the Mohr's circle to determine the principal planes. Sketch a properly oriented principal-stress element and show the principal stresses acting on its faces.

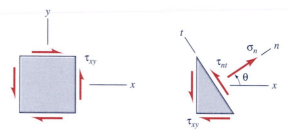

P8.5-12

> **Problems 8.5-13 through 8.5-24. For each of these problems, (a) construct a Mohr's circle of stress. Using this Mohr's circle, (b) determine the principal stresses and show them on a properly oriented stress element, and (c) determine the maximum shear stress and the normal stress on the planes of maximum shear, and show these on a sketch of a properly oriented stress element.**

Prob. 8.5-13. Use the stresses given in Prob. 8.4-1.

Prob. 8.5-14. Use the stresses given in Prob. 8.4-2.

Prob. 8.5-15. Use the stresses given in Prob. 8.4-3.

Prob. 8.5-16. Use the stresses given in Prob. 8.4-7.

Prob. 8.5-17. Use the stresses given in Prob. 8.4-8.

Prob. 8.5-18. Use the stresses given in Prob. 8.4-9.

Prob. 8.5-19. Use the stresses given in Prob. 8.4-10.

Prob. 8.5-20. Use the stresses given in Prob. 8.4-11.

Prob. 8.5-21. Use the stresses given in Prob. 8.4-12.

Prob. 8.5-22. Use the stresses given for point A on the steel plate girder in Prob. 8.3-11.

Prob. 8.5-23. Use the stresses given for point A on the bracket in Prob. 8.3-5.

Prob. 8.5-24. Use the stresses given for point A on the shop crane in Fig. P8.5-24.

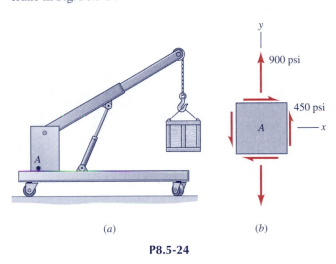

(a)   (b)

P8.5-24

> **Problems 8.5-25 through 8.5-34.** *Use Mohr's circle to solve the named problems from Sections 8.3 and 8.4.*

Prob. 8.5-25. Use Mohr's Circle to solve Prob. 8.3-15.

Prob. 8.5-26. Use Mohr's Circle to solve Prob. 8.3-16.

^D**Prob. 8.5-27.** Use Mohr's Circle to solve Prob. 8.3-17.

*****Prob. 8.5-28.** Use Mohr's Circle to solve Prob. 8.3-18.

Prob. 8.5-29. Use Mohr's Circle to solve Prob. 8.3-19.

Prob. 8.5-30. Use Mohr's Circle to solve Prob. 8.3-20.

Prob. 8.5-31. Use Mohr's Circle to solve Prob. 8.3-21.

Prob. 8.5-32. Use Mohr's Circle to solve Prob. 8.4-13.

Prob. 8.5-33. Use Mohr's Circle to solve Prob. 8.4-14.

Prob. 8.5-34. Use Mohr's Circle to solve Prob. 8.4-15.

> **Problems 8.5-35 and 8.5-36** *are to be solved by use of the Mohr's Circle module of MDSolids.*

^C**Prob. 8.5-35.** Use Mohr's Circle to solve Probs. 8.3-1 through 8.3-4.

^C**Prob. 8.5-36.** Use Mohr's Circle to solve Probs. 8.4-1 through 8.4-4.

▼ **MOHR'S CIRCLE: ABSOLUTE MAXIMUM STRESS** | MDS 8.4–8.5 |

> **Problems 8.6-1 through 8.6-12.** *For the given stress states, (a) Sketch Mohr's circles for three-dimensional stresses, like the circles in Fig. 8.23, and (b) determine the absolute maximum shear stress,* $\tau_{\substack{abs \\ max}}$.
>
> *Additional problems on the topic of absolute maximum shear stress may be found in Chapter 9.*

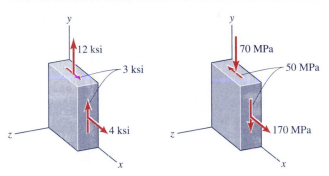

P8.6-1 and P8.6-13   P8.6-2 and P8.6-13

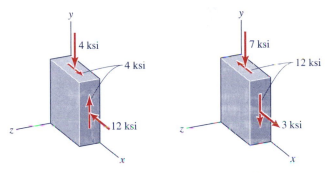

P8.6-3 and P8.6-13   P8.6-4 and P8.6-13

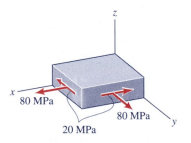

P8.6-5

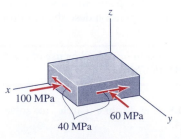

P8.6-6

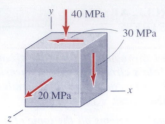

P8.6-7

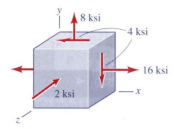

P8.6-8

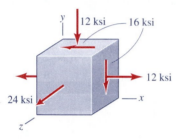

P8.6-9

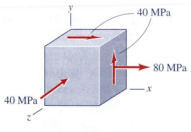

P8.6-10

Prob. 8.6-11 Use the plane stress state at point A on the steel plate girder described in Prob. 8.3-11.

Prob. 8.6-12. Use the plane stress state at point A on the automobile front suspension described in Prob. 8.4-7.

Problem 8.6-13 *is to be solved by use of the* <u>*Mohr's Circle module of MDSolids.*</u>

---

^C**Prob. 8.6-13.** Use Mohr's Circle to solve Probs. 8.6-1 through 8.6-4.

▼ **TRANSFORMATION OF STRAIN—THEORY**

Prob. 8.8-1. Using Fig. 8.28b, derive expressions that relate ϵ_y to ϵ_n and γ_{nt}. (These expressions appear in Eqs. 8.44 and 8.46, respectively.)

Prob. 8.8-2. Using Fig. 8.28c, derive expressions that relate γ_{xy} to ϵ_n and γ_{nt}. (These expressions appear in Eqs. 8.44 and 8.46, respectively.)

*****Prob. 8.8-3.** At point P on the surface of a flat plate, the strains are given by ϵ_x, ϵ_y, and γ_{xy}. Thus, the small rectangle in Fig. P.8.8-3a, whose diagonal PQ defines the direction n, is deformed into the parallelogram shape in Fig. P.8.8-3b. Use trigonometry to derive Eq. 8.44, which relates ϵ_n to the given strains. (Hint: Use the law of cosines.)

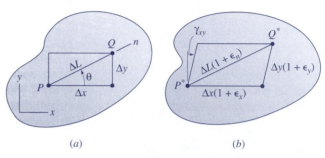

(a)   *(b)*

P8.8-3

Prob. 8.8-4. Following the procedure employed in Section 8.4 to derive expressions for directions of principal stress and the corresponding principal stresses, and starting with Eq. 8.47a, (a) derive expressions for the *directions of principal strain*, that is, the directions θ_{p_1} and θ_{p_2} along which ϵ_n assumes its maximum and minimum values, respectively. (b) Obtain expressions for the *principal strains*, ϵ_1 and ϵ_2, in terms of ϵ_x, ϵ_y, and γ_{xy}.

Prob. 8.8-5. Following the procedure employed in Section 8.4 to derive expressions for the maximum in-plane shear stress, and starting with Eq. 8.47b, obtain an expression for the maximum in-plane shear strain $\gamma_{max} \equiv \gamma_{s_1 s_2}$ in terms of ϵ_x, ϵ_y, and γ_{xy}.

▼ **TRANSFORMATION OF STRAIN—APPLICATIONS**

Prob. 8.8-6. A thin rectangular bar is subjected to an axial load P, as shown in Fig. P8.8-6. Near the center of the bar, line segments AB and AC at 30° and 60°, respectively, to the axis of the bar define the directions n and t. (a) Determine the extensional strains ϵ_x, and ϵ_y. (Note: $\gamma_{xy} = 0$ for axial loading.) (b) Determine the extensional strains ϵ_n and ϵ_t. (c) Determine the shear strain γ_{nt}.

$$E = 30(10^3)\text{ksi}, \quad \nu = 0.3, \quad t \equiv \text{thickness} = 0.50 \text{ in}$$

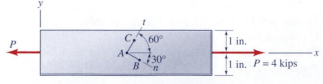

P8.8-6 and P8.9-7

Prob. 8.8-7. Repeat Prob. 8.8-6 for the rectangular bar in Fig. P8.8-7.

$$E = 200\text{GPa}, \quad \nu = 0.3, \quad t \equiv \text{thickness} = 10 \text{ mm}$$

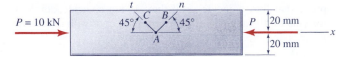

P8.8-7 and P8.9-8

Prob. 8.8-8. A rectangular bar whose cross-sectional dimensions are 5 in. × 1 in. is subjected to an axial load $P = 48$ kips that produces extensional strains $\epsilon_n = 640\mu$, $\epsilon_t = 0$ along the n and t directions indicated in Fig. P8.8-8. Determine the values of Young's modulus, E, and Poisson's ratio, ν, for this bar. Assume linearly elastic, isotropic material behavior.

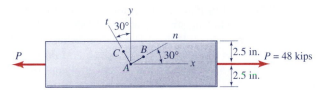

P8.8-8 and P8.9-9

Prob. 8.8-9. The solid, circular aluminum rod in Fig. P8.8-9 has a diameter $d = 40$ mm and can be subjected simultaneously to a torque T in the direction shown and to a tensile axial load P. The modulus of elasticity is $E = 73$ GPa, and the shear modulus is $G = 28$ GPa. (a) Determine the shear strain γ_{xy} that results if $T = 1.0$ kN · m and $P = 0$. (b) For this torsion-only loading, determine the extensional strains ϵ_n and ϵ_t along the respective n and t directions shown in Fig. P8.8-9.

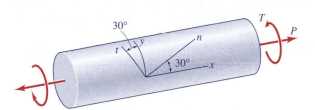

P8.8-9, P8.8-10, P8.9-10, and P8.9-11

*Prob. 8.8-10. For the aluminum rod described in Prob. 8.8-9, (a) Determine the extensional strain ϵ_x and the shear strain γ_{xy} that result if $P = 50$ kN and $T = 500$ N · m. (b) For this loading, determine the extensional strains ϵ_n and ϵ_t along the respective n and t directions shown in Fig. P8.8-10. (Note: There are additional combined-load problems in Chapter 9.)

*Prob. 8.8-11. On the outer surface of a pressure vessel, strains $\epsilon_n = 200\mu$ and $\epsilon_t = 400\mu$ are measured along the n and t directions indicated in Fig. P8.8-11. (a) Determine the extensional strains ϵ_x and ϵ_y. (Note: $\gamma_{xy} = 0$.) (b) Determine

the shear strain γ_{nt}. (c) For pressure vessels, $\sigma_y = 2\sigma_x$ (see Section 9.2). Therefore, what is the value of Poisson's ratio, ν, for this tank? Assume linearly elastic, isotropic material behavior.

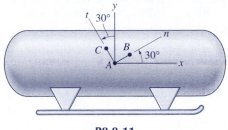

P8.8-11

Prob. 8.8-12. At a certain point on the surface of a pressure vessel (like the one shown in Fig. P8.8-11) the strains are given by: $\epsilon_x = 100\mu$, $\epsilon_y = 500\mu$, and $\gamma_{xy} = 0$. (a) Determine the state of strain for nt axes rotated counterclockwise by $30°$ from the xy axes, as shown in Fig. P8.8-12; that is, determine ϵ_n, ϵ_t, and γ_{nt}. (b) Sketch the shape of the elemental square $ABDC$ after deformation. Let the original lengths of the sides be $\overline{AB} = \overline{AC} = 1$, but exaggerate the deformation.

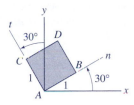

P8.8-12

▼ **MOHR'S CIRCLE: STRAIN TRANSFORMATIONS**　　　　　| MDS 8.6 |

> **Problems 8.9-1 through 8.9-6.** *Use the generic figures below in solving Probs. 8.9-1 through 8.9-6.*

Prob. 8.9-1. An element of material is subjected to the following state of plane strain: $\epsilon_x = 200\mu$, $\epsilon_y = -100\mu$, and $\gamma_{xy} = 0$. (a) Use Mohr's circle to calculate the strains ϵ_n, ϵ_t, and γ_{nt} for an element rotated (counterclockwise) by an angle $\theta = 20°$. (b) Use the given strains to produce a sketch of the deformed xy element, similar to the sketch in Fig. P8.9-1a. (c) Use the calculated strains to produce a sketch of the

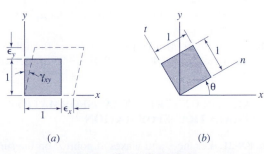

(a)　　　　　　*(b)*

P8.9-1 through P8.9-6

deformed *nt* element, starting with the undeformed unit element in Fig. P8.9-1*b*.

Prob. 8.9-2. Solve Prob. 8.9-1 for the following strains and angle: $\epsilon_x = 200\mu$, $\epsilon_y = -100\mu$, $\gamma_{xy} = -200\mu$, $\theta = 20°$.

Prob. 8.9-3. Solve Prob. 8.9-1 for the following strains and angle: $\epsilon_x = 300\mu$, $\epsilon_y = 750\mu$, $\gamma_{xy} = 450\mu$, $\theta = 30°$.

Prob. 8.9-4. Solve Prob. 8.9-1 for the following strains and angle: $\epsilon_x = 0$, $\epsilon_y = 400\mu$, $\gamma_{xy} = -300\mu$, $\theta = -30°$.

Prob. 8.9-5. Solve Prob. 8.9-1 for the following strains and angle: $\epsilon_x = 50\mu$, $\epsilon_y = 0$, $\gamma_{xy} = 120\mu$, $\theta = 45°$.

Prob. 8.9-6. Solve Prob. 8.9-1 for the following strains and angle: $\epsilon_x = 150\mu$, $\epsilon_y = 300\mu$, $\gamma_{xy} = 200\mu$, $\theta = -22.5°$.

Prob. 8.9-7. Use Mohr's strain circle to solve Prob. 8.8-6.

Prob. 8.9-8. Use Mohr's strain circle to solve Prob. 8.8-7.

Prob. 8.9-9. Use Mohr's strain circle to solve Prob. 8.8-8.

Prob. 8.9-10. Use Mohr's strain circle to solve Prob. 8.8-9.

*Prob. 8.9-11. Use Mohr's strain circle to solve Prob. 8.8-10.

▼ PRINCIPAL STRAINS; MAXIMUM SHEAR STRAIN

> **Problems 8.9-12 through 8.9-18.** *For each of the listed states of plane strain, (a) use <u>Mohr's circle for strain</u> to determine the in-plane principal strains and principal-strain directions, and show how a unit square oriented in the principal-strain directions deforms; and (b) determine the maximum in-plane shear strain.*

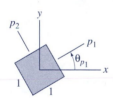

P8.9-12 through P8.9-18

Prob. 8.9-12. $\epsilon_x = 100\mu$, $\epsilon_y = -300\mu$, $\gamma_{xy} = -300\mu$.

Prob. 8.9-13. $\epsilon_x = 400\mu$, $\epsilon_y = -200\mu$, $\gamma_{xy} = -400\mu$.

Prob. 8.9-14. $\epsilon_x = 200\mu$, $\epsilon_y = 500\mu$, $\gamma_{xy} = -300\mu$.

Prob. 8.9-15. $\epsilon_x = 0$, $\epsilon_y = -400\mu$, $\gamma_{xy} = 300\mu$.

Prob. 8.9-16. $\epsilon_x = 100\mu$, $\epsilon_y = 0$, $\gamma_{xy} = 240\mu$.

Prob. 8.9-17. $\epsilon_x = -20\mu$, $\epsilon_y = 220\mu$, $\gamma_{xy} = 100\mu$.

Prob. 8.9-18. $\epsilon_x = 150\mu$, $\epsilon_y = -300\mu$, $\gamma_{xy} = 200\ \mu$.

*Prob. 8.9-19. Draw *Mohr's strain circle* for the following plane-strain state: $\epsilon_x = 90\mu$, $\epsilon_y = 270\mu$, $\epsilon_1, = 330\mu$. Determine the following: γ_{xy}, ϵ_2, θ_{p_1}.

*Prob. 8.9-20. Draw *Mohr's strain circle* for the following: plane strain state: $\epsilon_x = -240\mu$, $\epsilon_y = 0$, $\epsilon_2 = -250\mu$. Determine the following: γ_{xy}, ϵ_1, θ_{p_2}.

▼ COMPARISON OF STRESS TRANSFORMATION AND STRAIN TRANSFORMATION

*Prob. 8.9-21. Let the *x* and *y* axes at point *P* on the surface of a flat plate undergoing plane stress (see Section 8.2) be oriented in the directions of the *principal stresses*, σ_1 and σ_2, respectively, and let the *x'* and *y'* axes be oriented in the directions of the *principal strains*, ϵ_1 and ϵ_2. Assume that the plate is linearly elastic. Do the *axes of principal strain* (the *x'y'* axes) coincide with the *axes of principal stress* (the *xy* axes)? Either prove that they coincide, or, if they do not coincide, derive a formula for the angle θ_{ϵ_1} that orients the ϵ_1 axis with respect to the σ_1 axis.

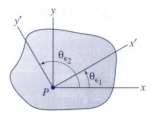

P8.9-21

Prob. 8.9-22. A thin plate is subjected to a state of plane stress with the following stresses: $\sigma_z = 8.4$ ksi, $\sigma_y = 2.4$ ksi, $\tau_{xy} = 0$. ($\sigma_z = \tau_{xz} = \tau_{yz} = 0$ also.) The plate satisfies Hooke's Law, with $E = 10(10^3)$ ksi and $\nu = \frac{1}{3}$. Sketch a *Mohr's stress circle* and also a *Mohr's strain circle*, and comment on the similarities and differences between these two circles.

Prob. 8.9-23. Solve Prob. 8.9-22 for the following state of plane stress: $\sigma_x = 450$ MPa, $\sigma_y = 150$ MPa, $\tau_{xy} = 0$. Let $E = 100$ GPa and $\nu = \frac{1}{3}$.

Prob. 8.9-24. Solve Prob. 8.9-22 for the following state of plane stress: $\sigma_x = 18$ ksi, $\sigma_y = 0$ ksi, $\tau_{xy} = 6$ ksi. Let $E = 10(10^3)$ ksi, $\nu = 0.30$.

▼ MOHR'S CIRCLE: STRAIN ROSETTES

> **MDS 8.7**

Prob. 8.10-1. Starting with Eqs. 8.50, derive Eqs. 8.51 for a 45° strain-gage rosette and Eqs. 8.52 for an equiangular strain-gage rosette.

Prob. 8.10-2. Use Mohr's circle for strain in Fig. P8.10.2 to derive the third equation of Eqs. 8.51, that is,

$$\gamma_{xy} = 2\epsilon_b - \epsilon_a - \epsilon_c$$

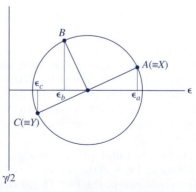

P8.10-2

Prob. 8.10-3. The 45° rectangular rosette in Fig. P8.10-3 was used to obtain the following strains—$\epsilon_a = 175\mu$, $\epsilon_b = 140\mu$, and $\epsilon_c = -105\mu$—at a point on the free surface of a machine component being tested. (a) Determine the in-plane strains ϵ_x, ϵ_y, and γ_{xy}. (b) Use a Mohr's strain circle to determine the in-plane principal strains ϵ_1 and ϵ_2 at the rosette's location. (c) Determine the orientations of the principal-strain axes relative to the orientation of the "a" gage.

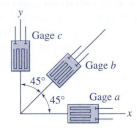

P8.10-3 through P8.10-5

Prob. 8.10-4. Solve Prob. 8.10-3 for the following measured strains: $\epsilon_a = 440\mu$, $\epsilon_b = 510\mu$, and $\epsilon_c = 340\mu$.

Prob. 8.10-5. During stress testing of a new design for a titanium-alloy helicopter transmission case, the following strains were measured on the outer surface of the transmission case by use of a 45° rectangular rosette (see Prob. 8.10-3 for the rosette configuration): $\epsilon_a = -270\mu$, $\epsilon_b = 371\mu$, and $\epsilon_c = 670\mu$. (a) Determine the strain components ϵ_x, ϵ_y, and γ_{xy}. (b) Sketch a Mohr's circle for this strain state, and determine the principal strains and the maximum shear strain at the rosette's location. (c) Letting $E = 100$ GPa and $\nu = 0.33$, determine the principal stresses and the maximum in-plane (i.e., in the plane of the gage) shear stress. Sketch a stress element oriented in the principal directions.

Prob. 8.10-6. The 60° equiangular rosette shown in Fig. P8.10-6 was used to obtain the following strains—$\epsilon_a = 160\mu$, $\epsilon_b = 520\mu$, and $\epsilon_c = 160\mu$—at a point on the free surface of a machine component being tested. (a) Determine the in-plane strains ϵ_x, ϵ_y, and γ_{xy}. (b) Use a Mohr's circle to determine the in-plane principal strains ϵ_1 and ϵ_2 at the rosette's location. (c) Determine the orientations of the principal-strain axes relative to the orientation of the "a" gage.

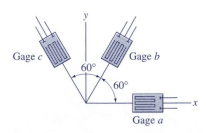

P8.10-6 through P8.10-8

Prob. 8.10-7. Solve Prob. 8.10-6 for the following measured strains: $\epsilon_a = -200\mu$, $\epsilon_b = 450\mu$, and $\epsilon_c = 125\mu$.

Prob. 8.10-8. At a point on the outer surface of a gas-turbine engine, the strains $\epsilon_a = 935\mu$, $\epsilon_b = 167\mu$, and $\epsilon_c = 668\mu$ were measured using a 60° equiangular rosette (see figure with

Prob. 8.10-6). (a) Determine the strain components ϵ_x, ϵ_y, and γ_{xy}. (b) Sketch a Mohr's circle for this state of strain, and determine the principal strains and the maximum shear strain at the rosette's location. (c) Letting $E = 30 \times 10^3$ ksi and $\nu = 0.33$, determine the principal stresses and the maximum in-plane (i.e., in the plane of the gage) shear stress. Sketch a stress element oriented in the principal directions.

Prob. 8.10-9. Starting with Eqs. 8.50, derive expressions for ϵ_x, ϵ_y, and γ_{xy} in terms of measured extensional strains ϵ_a, ϵ_b, and ϵ_c for each of the two rosette orientations shown in Fig. P8.10-9a,b.

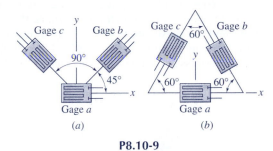

(a)   (b)

P8.10-9

Prob. 8.10-10. Repeat Prob. 8.10.9 for the two rosette configurations shown in Fig. P8.10-10a,b. How do your results compare with Eqs. 8.51 and 8.52, respectively?

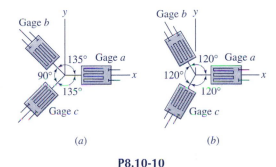

(a)   (b)

P8.10-10

*Prob. 8.10-11. At a point on the surface of a steel machine component, the strain rosette shown in Fig. P8.10-11 measured the following extensional strains: $\epsilon_a = 700\mu$, $\epsilon_b = 560\mu$, and $\epsilon_c = -280\mu$. (a) Determine the strain components, ϵ_x, ϵ_y, and γ_{xy} at the rosette location. (b) Letting $E = 30 \times 10^3$ ksi and $\nu = 0.30$, determine the stress components, σ_x, σ_y, and τ_{xy} at the rosette location. (c) Using a Mohr's circle for strain, determine the principal strains and the maximum shear strain at the point. (d) Determine the principal stresses and the absolute maximum shear stress at the point. (Recall Section 8.6.)

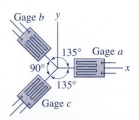

P8.10-11 and P8.10-12

599

Prob. 8.10-12. Use Mohr's circle to determine the extensional strains ϵ_a, ϵ_b, and ϵ_c that would be indicated by the rosette shown in Fig. P8.10-12 (see Prob. 8.10-11) if the two in-plane principal strains at the rosette location are $\epsilon_1 = 736\mu$ and $\epsilon_2 = -184\mu$. The direction of principal strain ϵ_1 is 30° clockwise from the x axis.

*Prob. 8.10-13. At a point on an aluminum-alloy bracket, the strain rosette shown in Fig. P8.10-13 measured the following extensional strains: $\epsilon_a = 592\mu$, $\epsilon_b = -444\mu$, and $\epsilon_c = 740\mu$. (a) Determine the strain components ϵ_x, ϵ_y, and γ_{xy} at the rosette location. (b) Letting $E = 70$ GPa and $\nu = 0.33$, determine the stress components σ_x, σ_y, and τ_{xy} at the rosette location. (c) Using a Mohr's circle for strain, determine the principal strains and the maximum in-plane shear strain at the point. (d) Determine the principal stresses and the absolute maximum shear stress at the point. (Recall Section 8.6.)

Prob. 8.10-14. Use Mohr's circle to determine the extensional strains ϵ_a, ϵ_b, and ϵ_c that would be indicated by the rosette shown in Fig. P8.10-14 (see Prob. 8.10-13) if the two in-plane principal strains at the rosette location are $\epsilon_1 = 580\mu$ and $\epsilon_2 = -260\mu$. The direction of the principal strain ϵ_1 is oriented 15° clockwise from the x axis (i.e., from the orientation of the "a" gage).

Prob. 8.10-15. A *torsion load cell* (to measure torque T) is constructed by mounting two strain gages on a tubular shaft, with the gages oriented at $\pm45°$ to the axis of the tube, as indicated in Fig. P8.10-15. The gages are wired so that the measurement circuit gives an output $\epsilon_t \equiv \epsilon_b - \epsilon_a$. Determine the relationship between the applied torque T and the measured strain difference, ϵ_t, if the tube has the following properties: r_0 = outer radius, r_i = inner radius, E = modulus of elasticity, and ν = Poisson's ratio.

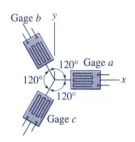

P8.10-13 and P8.10-14

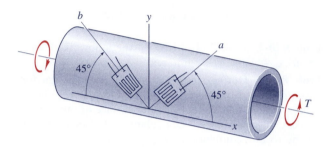

P8.10-15

Section		Suggested Review Problems
8.1	In previous chapters you learned how to analyze deformable bodies loaded axially $\left(\sigma = \dfrac{F}{A}\right)$, in torsion $\left(\tau = \dfrac{T\rho}{I_p}\right)$, or in bending $\left(\sigma = -\dfrac{My}{I}, \tau = \dfrac{VQ}{It}\right)$. In this chapter the **stress transformation equations** are derived so you can combine various types of stress.	
8.2	Section 8.2 defines **plane stress** Figure 8.4 depicts a **state of plane stress** in (a) 3-D, and in (b) 2-D. (a) Three-dimensional view. (b) Two-dimensional view. A state of plane stress (Fig 8.4)	
8.3	In Section 8.3 the **stress transformation equations for plane stress** are derived and applied to various states of plane stress. The stress transformation equations for plane stress are derived by using <u>only</u> the **equations of equilibrium.** (a) A triangular stress element. (b) The areas of the respective faces. (c) The free-body diagram. The free-body diagram for a plane stress element (Fig. 8.9)	Starting with Fig. 8.9c, derive Eqs. 8.5.
	The normal stress σ_n and the shear stress τ_{nt} on the "n-face", whose normal n is at angle θ counterclockwise from the x-axis, are given by Eqs. 8.5. These are called the **stress transformation equations for plane stress.**	$$\left. \begin{array}{l} \sigma_n = \left(\dfrac{\sigma_x + \sigma_y}{2}\right) + \left(\dfrac{\sigma_x - \sigma_y}{2}\right)\cos 2\theta \\[2mm] \qquad + \tau_{xy}\sin 2\theta \\[4mm] \tau_{nt} = -\left(\dfrac{\sigma_x - \sigma_y}{2}\right)\sin 2\theta + \tau_{xy}\cos 2\theta \end{array} \right\} \quad (8.5)$$ 8.3-1 8.3-5 8.3-15

Section			Suggested Review Problems
8.4	In Section 8.4, the equations are derived for the **principal stresses** and corresponding **principal directions.** Principal stresses are the maximum and minimum normal stresses at a point.	The **principal stresses** are: $$\sigma_1 = \sigma_{avg} + R, \quad \sigma_2 = \sigma_{avg} - R \qquad (8.13)$$ where $$\sigma_{avg} = \frac{\sigma_x + \sigma_y}{2} \qquad (8.14)$$ and $$R = \sqrt{\left(\frac{\sigma_x - \sigma_y}{2}\right)^2 + \tau_{xy}^2} \qquad (8.11)$$	8.4-3 8.4-7 8.4-13
	Also derived in Section 8.4 are the equations for the **maximum shear stress** at a point, and the orientation of the **planes of maximum shear stress.**	The **maximum in-plane shear stress** is given by the equation: $$\tau_{max} = R \qquad (8.18)$$	
8.5	**Mohr's Circle,** whose geometry and properties are described in Section 8.5, is a super graphical tool for describing the state of stress at a point. You should carefully review Fig. 8.16 and the table that describes the **Procedure for Constructing and Using Mohr's Circle of Stress.** Note that the positive τ axis points <u>downward.</u> The icons at the two ends of the τ axis indicate how to interpret the sign of shear stress on a face. Angles θ on the stress element become angles 2θ, turned in the same sense on Mohr's circle as on the stress element.	Properties of Mohr's circle for plane stress (Fig. 8.16)	8.5-1 8.5-16 8.5-27
		Points P_1 and P_2 denote the **principal stresses;** points S_1 and S_2 denote the **maximum (and minimum) in-plane shear stresses.**	
8.6	Section 8.6 extends the discussion of state of stress at a point to cover triaxial states of stress.		8.6-9
8.7	Section 8.7 defines **plane strain,** which is quite different than plane stress.		

Section		Suggested Review Problems
8.8	In Section 8.8 the equations for **transformation of strains in a plane,** Eqs. 8.47, are derived. Whereas stress transformation depends solely on equilibrium, **strain-transformation equations are based solely on the geometry of deformation.** $$\epsilon_n = \left(\frac{\epsilon_x + \epsilon_y}{2}\right) + \left(\frac{\epsilon_x - \epsilon_y}{2}\right)\cos 2\theta$$ $$+ \left(\frac{\gamma_{xy}}{2}\right)\sin 2\theta$$ $$\frac{\gamma_{nt}}{2} = -\left(\frac{\epsilon_x - \epsilon_y}{2}\right)\sin 2\theta + \left(\frac{\gamma_{xy}}{2}\right)\cos 2\theta$$ (8.47)	8.8-3 8.8-9
8.9	Section 8.9 shows how a **Mohr's circle of strain** can be used to solve the strain-transformation equations. You should review Fig. 8.29 and the **Procedure for Constructing and Using Mohr's Circle of Strain.** Properties of Mohr's circle for strain (Fig. 8.29)	8.9-1 8.9-3 8.9-15
8.10	Section 8.10 discusses how to measure strain, and introduces **electrical-resistance strain gages,** which are used to measure extensional strain. **Strain rosettes** are also introduced.	8.10-1 8.10-9
Section 8.11 is an "optional" section.		

PRESSURE VESSELS; STRESSES DUE TO COMBINED LOADING

9.1 INTRODUCTION

In previous chapters, specifically Chapters 2, 3, 4, and 6, formulas were derived that relate the normal stress and the shear stress on any cross section of a slender member to the stress resultants on the cross section. Some key formulas that were derived are listed in Table 9.1. In many practical situations, two or more stress resultants occur on a cross section, so we need to determine the **state of stress at a point due to various combined loads.**

The combined effect of normal and shear stresses can be conveniently analyzed (for plane stress) by the use of *Mohr's Circle for stress* (Section 8.5). In this chapter we will investigate the stress distribution in slender members under several combinations of loading. In addition, we will examine the state of stress in thin-wall pressure vessels. The *biaxial state of stress* in thin-wall pressure vessels is also conveniently analyzed by using Mohr's circle.

TABLE 9.1 Formulas for Stresses

Stress Resultant	Symbol	Formula	References
Normal force	F	$\sigma = \dfrac{F}{A}$	Sections 2.2, 3.2
Torsional moment	T	$\tau = \dfrac{T\rho}{I_p}$	Section 4.3
Bending moment	M	$\sigma = \dfrac{-My}{I}$	Section 6.3
Transverse shear force	V	$\tau = \dfrac{VQ}{It}$	Section 6.8

One form of "combined loading" occurs in thin-wall pressure vessels like the cylindrical and spherical tanks shown in Fig. 9.1. Thin-wall pressure vessels vary in size, for example, from 2-in.-diameter hair-spray, shaving-cream, or spray-paint cans to gas-storage tanks that are sixty feet or more in diameter. In this section we will examine cylindrical and spherical tanks under internal (gas) pressure. These are special cases of thin-wall structures called *thin shells*.[1]

Although a metallic or plastic thin-wall pressure vessel does not expand as much as a balloon does when it is pressurized, in both cases the pressurization stretches the "skin" or walls of the pressure vessel, enlarging the vessel. In general, the walls of a vessel are considered to be thin if the radius-to-wall-thickness ratio is ten or more (i.e., $\frac{r}{t} \geq 10$). In that case, the tensile stress in the vessel wall varies insignificantly (<5%) from the inside of the vessel wall to the outside.

Cylindrical Pressure Vessels. The diver's air tank in Fig. 9.1a is an example of a thin-wall cylindrical pressure vessel. Figure 9.2a shows a circular-cylinder pressure vessel with end closures. In the cylindrical section the normal (tensile) stresses in the longitudinal, or axial, direction and in the circumferential, or hoop, direction are called the *axial stress* σ_a and the *hoop stress* σ_h, respectively. We will assume that the vessel contains a pressurized gas whose weight can be neglected, and we will use

(a) A scuba diver with cylindrical air tank.

(b) Spherical oxygen and propellant tanks on a spacecraft.

FIGURE 9.1 Examples of thin-wall pressure vessels.

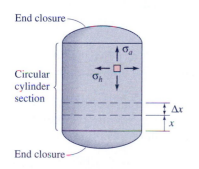

(a) A circular cylinder with end closures.

(b) Shell dimensions.

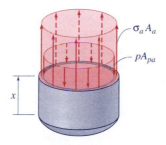

(c) A free-body diagram for determining the axial stress σ_a.

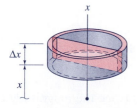

(d) The diametral cutting plane for a hoop-stress free-body diagram.

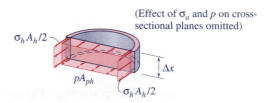

(Effect of σ_a and p on cross-sectional planes omitted)

(e) A free-body diagram for determining the hoop stress σ_h.

FIGURE 9.2 Figures used in the analysis of stresses in circular cylinders under internal pressure.

---

[1]*Thin Elastic Shells* by Harry Kraus, [Ref. 9-1].

appropriate free-body diagrams in relating σ_a and σ_h to the internal pressure p.[2] It is also assumed that σ_a and σ_h are constant through the thickness of the wall of the pressure vessel.

Axial (Longitudinal) Stress: To determine the axial stress, we can employ the free-body diagram in Fig. 9.2c, where a "cut" has been made through the shell wall and the gas at an arbitrary cross section x at some distance from the end closure.[3] Summing the forces in the x direction, we get

$$\sum F_x = 0: \qquad\qquad \sigma_a A_a - p A_{pa} = 0 \qquad\qquad (9.1)$$

The longitudinal stress acts on the cut section of the vessel wall. The gas pressure, however, acts on the gas still occupying the vessel below the cutting plane at x. Thus, Eq. 9.1, together with areas A_a and A_{pa} based on Fig. 9.2b, becomes

$$\sigma_a[\pi(r + t)^2 - \pi r^2] - p\pi r^2 = 0$$

But, since $t \ll r$, this can be approximated by

$$\sigma_a(2\pi rt) - p\pi r^2 = 0$$

so, the **axial stress** is given by

$$\boxed{\sigma_a = \frac{pr}{2t}} \qquad \text{**Axial Stress-Cylinder**} \qquad (9.2)$$

Hoop Stress: The free-body diagram in Fig. 9.2e, which is based on the cross-sectional and longitudinal cutting planes illustrated in Fig. 9.2d, may be used in determining the hoop stress, σ_h. By taking an arbitrary longitudinal cutting plane that cuts the vessel in half (i.e., it contains the longitudinal axis), we obtain two surfaces on which the stress is σ_h. Also, the two σ_h-forces in Fig. 9.2e are parallel. Thus, summing forces in the hoop (i.e., tangential to the circumference) direction, we get

$$\sum F_h = 0: \qquad\qquad \sigma_h A_h - p A_{ph} = 0 \qquad\qquad (9.3)$$

The areas in Eq. 9.3 are $A_h = 2t\Delta x$ and $A_{ph} = 2r\Delta x$, so Eq. 9.3 gives the following expression for **hoop stress:**

$$\boxed{\sigma_h = \frac{pr}{t}} \qquad \text{**Hoop Stress-Cylinder**} \qquad (9.4)$$

The fact that the hoop stress in a cylindrical shell has twice the magnitude of the axial stress accounts for the typical failure mode of cylindrical shells that is depicted in Figs. 9.3. The photo on the left shows the failure of a high-pressure compressed-air aluminum-alloy tank used by divers; on the right is a steel liquid

---

[2]This is the *gage pressure*, that is, the internal pressure minus the external (or atmospheric) pressure. If the pressure is larger on the outside than on the inside, a thin-wall shell may be subject to collapse by buckling. We will consider only internal pressurization.
[3]The effect of the end closure is discussed later in this section.

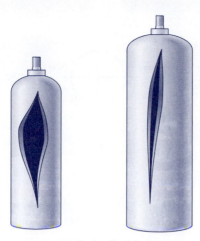

FIGURE 9.3 Typical appearance of failed cylindrical pressure vessels. (Photos courtesy ASM International.)

propane gas (LPG) tank.[4] Because of the catastrophic effects that may accompany pressure-vessel failures, design and construction of pressure vessels is closely regulated by codes (e.g., the ASME Boiler and Pressure Vessel Code [Ref. 9-3]).

Spherical Pressure Vessels. Consider now a thin-wall spherical pressure vessel of inner radius r and wall thickness $t(r/t \geq 10)$. Because of the spherical symmetry, the normal stress will be the same in any direction in the shell, as indicated in Fig. 9.4a. This stress, which is labeled σ_s to identify it with a spherical pressure vessel, can be determined by taking a free-body diagram that consists of half of the spherical vessel plus the pressurized gas occupying the half-sphere. By summing the forces on the free-body diagram in Fig. 9.4b, we get

$$\sigma_s A_s - p A_{ps} = 0 \qquad (9.5)$$

The approximations that led from Eq. 9.1 to Eq. 9.2 hold here as well, so Eq. 9.5 gives the following equation for **normal stress in a spherical pressure vessel:**

$$\boxed{\sigma_s = \frac{pr}{2t}} \quad \begin{array}{l}\textbf{Normal Stress}\\\textbf{in Sphere}\end{array} \qquad (9.6)$$

(a) A spherical pressure vessel with diametral cutting plane.

The stress in a spherical pressure vessel is equal to the longitudinal stress in a cylindrical pressure vessel having the same r/t ratio; and it is just half the value of the hoop stress in the cylinder. Thus, for a given pressure, a spherical pressure vessel can have a thinner wall than a cylindrical vessel can.

State of Stress in Pressure-Vessel Walls. The previous analyses give us the "in-plane" stresses in the walls of cylindrical and spherical pressure vessels. Since there is no shear stress on the cutting planes in Fig. 9.2c, Fig. 9.2e, or Fig. 9.4b, the stresses σ_a, σ_h, and σ_s are all **in-plane principal stresses.** The radial normal stress is

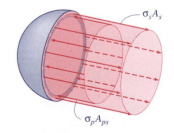

(b) A free-body diagram for determining the normal stress in a spherical shell.

FIGURE 9.4 A spherical shell under internal pressure.

---

[4]These and many other interesting examples of failure are discussed in Ref. 9-2, *Analyzing Failures: The Problems and the Solutions*, V. S. Goel, ed., (1986), ASM International®, Materials Park, OH 44073-0002 (formerly American Society for Metals, Metals Park, OH 44073).

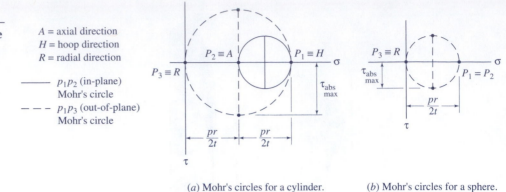

A = axial direction
H = hoop direction
R = radial direction

—————— $p_1 p_2$ (in-plane) Mohr's circle

— — — $p_1 p_3$ (out-of-plane) Mohr's circle

(a) Mohr's circles for a cylinder. (b) Mohr's circles for a sphere.

FIGURE 9.5 Mohr's circles for in-plane and out-of-plane stresses.

the third principal stress in both cases. At the inner surface of the pressure vessel wall, the radial normal stress is $\sigma_r = -p$, since the pressure pushes on the inside surface. On the outer surface the (gage) pressure is zero, so $\sigma_r = 0$. Since r/t is assumed to be equal to, or greater than, ten, the radial stress is much smaller than the in-plane stresses, so it is usually ignored.[5] To determine the state of stress in the walls of cylindrical and spherical pressure vessels, we can use Eqs. 9.2 and 9.4 for cylinders and Eq. 9.6 for spheres to plot a Mohr's circle of stress for each. Since σ_s is the normal stress in any in-plane direction in a spherical vessel, the Mohr's circle for in-plane stresses in Fig. 9.5b degenerates to a single point on the σ axis. From the analysis in Section 8.6 we know that

$$\tau_{\substack{abs \\ max}} = \frac{\sigma_{max} - \sigma_{min}}{2} \qquad \text{(8.32)} \\ \text{repeated}$$

From Figs. 9.5a and 9.5b this gives

$$\left(\tau_{\substack{abs \\ max}}\right)_{cyl} = \frac{pr}{2t}$$

$$\left(\tau_{\substack{abs \\ max}}\right)_{sph} = \frac{pr}{4t} \qquad \text{(9.7)}$$

Effects of End Closures and Other Discontinuities. Of course, to fill a pressure vessel with gas or liquid, or take the gas or liquid out, there must be a hole in the pressure-vessel wall and some sort of "connector," as illustrated in Figs. 9.3 and in Fig. 9.6a. The stress formulas developed above do not apply to stresses in the immediate vicinity of such discontinuities in the pressure-vessel wall.

The vicinity of the joint between the cylindrical section and the end closure of a pressure vessel (Fig. 9.2a) is also a location where the previous analysis does not apply. This is illustrated in Figs. 9.6b and 9.6d. If the end closure and the cylindrical section of the pressure vessel were permitted to expand freely under the effect of internal pressure, the cylinder would expand radially more than the end closure would, as illustrated in Fig. 9.6b. However, since the closure must be welded, or otherwise attached, to the cylinder, both the cylinder and the end closure undergo significant local deformation that involves localized bending stresses. Analysis of the

---

[5]The radial stress can easily be incorporated using the three-dimensional stress analysis approach of Section 8.6.

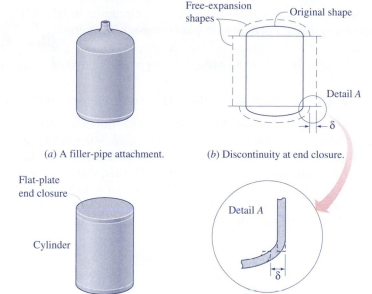

Free-expansion shapes ⟍   Original shape

Detail A

←δ→

(a) A filler-pipe attachment.

(b) Discontinuity at end closure.

Flat-plate end closure ⟍

Cylinder

Detail A

←δ→

(c) A cylinder with flat-plate end closure.

(d) Deformation near cylinder-closure junction.

FIGURE 9.6 Two types of discontinuities in thin-wall pressure vessels.

stress distribution at such discontinuities is beyond the scope of this text.[6] Flat-plate end closures, like the ones illustrated in Fig. 9.6c, are especially undesirable and should be avoided if possible.

EXAMPLE 9.1

A cylindrical pressure vessel 2.50 m in diameter is fabricated by shaping two 10-mm-thick steel plates and butt-welding the plates along helical arcs, as shown in Fig. 1. The maximum internal pressure in the pressure vessel is 1200 kPa. For this pressure level, calculate the following quantities: (a) the axial stress and the hoop stress; (b) the absolute maximum shear stress; and (c) the normal stress, σ_n, perpendicular to the weld line, and the shear stress, τ_{nt}, tangent to the weld line.

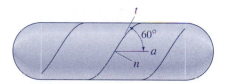

Fig. 1

Solution

(a) In-Plane Stresses: The axial stress and the hoop stress in the cylindrical section of the pressure vessel are given by Eqs. 9.2 and 9.4, respectively. For the given pressure vessel and internal-pressure loading, the axial stress and hoop stress are:

$$\sigma_a = \frac{pr}{2t} = \frac{(1200 \text{ kPa})(1.25 \text{ m})}{2(10 \text{ mm})} = 75 \text{ MPa}$$

$$\sigma_h = \frac{pr}{t} = \frac{(1200 \text{ kPa})(1.25 \text{ m})}{(10 \text{ mm})} = 150 \text{ MPa}$$

---

[6]This topic is treated in textbooks on thin shells; for example, see Section 116 in *Theory of Plates and Shells* by S. Timoshenko and S. Woinowsky-Krieger, [Ref. 9-4].

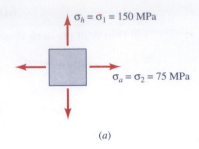

$$\sigma_h = \sigma_1 = 150 \text{ MPa}$$

$$\sigma_a = \sigma_2 = 75 \text{ MPa}$$

(a)

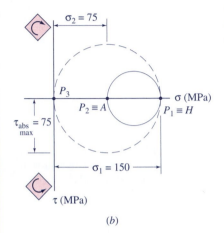

$\sigma_2 = 75$

P_3

$P_2 \equiv A$

σ (MPa)

$P_1 \equiv H$

$\tau_{\text{abs}}_{\text{max}} = 75$

$\sigma_1 = 150$

τ (MPa)

(b)

Fig. 2

Since there is no shear on longitudinal or circumferential cutting surfaces, these stresses are principal in-plane stresses. Therefore,

$$\sigma_1 = \sigma_h = 150 \text{ MPa}, \qquad \sigma_2 = \sigma_a = 75 \text{ MPa} \qquad \textbf{Ans. (a)} \quad (1)$$

These stresses are shown on the biaxial stress element in Fig. 2a.

(b) Absolute Maximum Shear Stress: The in-plane Mohr's circle for stress is shown as the solid-line circle in Fig. 2b. Since both the axial stress and the hoop stress are tension, the absolute maximum shear stress is not the maximum in-plane shear stress. Rather, this is a stress state with $\sigma_1 \geq \sigma_2 \geq 0$, as described in Fig. 8.24c. Since $r/t = 125$, we can neglect the internal pressure, p, in comparison with the in-plane stresses and draw a Mohr's circle passing through P_1 and the origin, P_3. Thus,

$$\tau_{\text{abs}}_{\text{max}} = \frac{\sigma_1}{2} = \frac{150 \text{ MPa}}{2}$$

or

$$\tau_{\text{abs}}_{\text{max}} = 75 \text{ MPa} \qquad \textbf{Ans. (b)} \quad (2)$$

(c) Weld-line Stresses: The in-plane Mohr's circle is redrawn in Fig. 3a. The point N represents the stresses on a face parallel to the weld, since the direction n in Fig. 1 is perpendicular to the weld line. From the Mohr's circle in Fig. 3, we get

$$\sigma_n = \overline{OB} = 112.5 \text{ MPa} - (37.5 \text{ MPa}) \cos 60° = 93.75 \text{ MPa}$$

$$\tau_{nt} = -\overline{NB} = -(37.5 \text{ MPa}) \sin 60° = -32.48 \text{ MPa}$$

or

$$\sigma_n = 93.8 \text{ MPa}, \qquad \tau_{nt} = -32.5 \text{ MPa} \qquad \textbf{Ans. (c)} \quad (3)$$

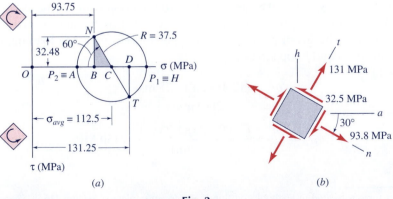

(a)   (b)

Fig. 3

The normal stress σ_t is given by

$$\sigma_t = \overline{OD} = 112.5 \text{ MPa} + (37.5 \text{ MPa}) \cos 60° = 131.25 \text{ MPa}$$

These stresses are shown on the rotated element in Fig. 3b.

   The normal stress and shear stress on the weld can be converted to normal-force-per-unit-length and shear-force-per-unit-length (shear flow) as was done in Section 6.8.

| MDS9.1 | **Stresses in a Cylindrical Pressure Vessel** |

9.3   STRESS DISTRIBUTION IN BEAMS

Recall that in Fig. 8.1 crack patterns were used to illustrate the effect of combined bending moment M and transverse shear V in reinforced-concrete beams under load. We will now examine in greater detail the stress distribution in uniform beams under combined flexure and shear. To illustrate the distribution of stress in a transversely loaded beam, let us consider the simply supported rectangular beam in Fig. 9.7. There are localized stress concentrations at the loading point and at the two support points, so the stresses in the immediate vicinity of these points cannot be computed by using the flexure formula ($\sigma = -My/I$) and the shear formula ($\tau = VQ/It$) developed in Chapter 6. However, according to St. Venant's Principle, which was introduced in Section 2.10, the state of stress away from these stress concentrations can be based on the formulas of *elementary beam theory*. The normal stress and the shear stress may be combined by using the stress-transformation equations of Section 8.3, Eqs. 8.3 or 8.5, or by using Mohr's circle for stress to represent the state of (plane) stress at any point in the beam that is sufficiently distant from the load and support points.

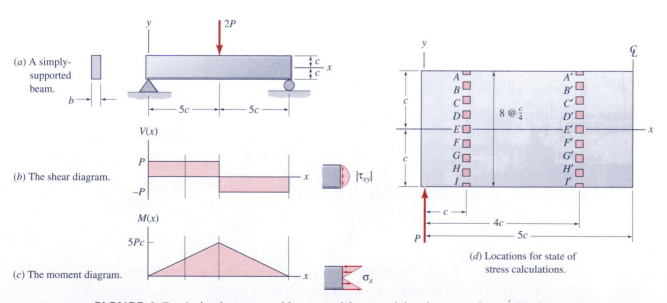

(a) A simply-supported beam.

(b) The shear diagram.

(c) The moment diagram.

(d) Locations for state of stress calculations.

FIGURE 9.7   A simply supported beam used for examining the state of stress in a beam.

For the simply supported beam in Fig. 9.7, we will determine the state of stress at nine equally spaced points at $x = c$, where the bending moment is small, and at nine other equally spaced points at $x = 4c$, where the moment is near its maximum value (Fig. 9.7d). This will permit us to see how the principal stresses σ_1 and σ_2 vary from point to point in this particular beam. We can determine whether, for example, the maximum normal stress is always equal to the maximum flexural stress, or whether some combination of flexural stress and transverse shear stress may lead to a larger normal stress at some point in the beam other than the point(s) where the maximum flexural stress occurs.

For the half-span $0 < x < 5c$ the flexural stress and transverse shear stress are given by

$$\sigma_x = \frac{-Pxy}{I} = -3\sigma_0\left(\frac{xy}{c^2}\right) \tag{9.8}$$

$$\tau_{xy} = -\frac{3\sigma_0}{2}\left(1 - \frac{y^2}{c^2}\right) \tag{9.9}$$

where $\sigma_0 = P/A$. These are illustrated in Figs. 9.7c and 9.7b, respectively. The maximum normal stress occurs at midspan and is $(\sigma_x)_{max} = 15\sigma_0$. Sample Mohr's-circle calculations for point C are illustrated in Fig. 9.8.

The principal stresses at the two cross sections at $x = c$ and $x = 4c$ are tabulated graphically in Fig. 9.9. Note how the orientation of the principal stresses varies from top to bottom of the beam at each cross section, with the principal stresses being parallel to the x and y axes at the top and bottom, and at $\pm 45°$ to the x axis at the neutral axis ($y = 0$). Also note that, even at $x = c$ where the moment is relatively small, the maximum tension (compression) occurs at the bottom (top) of the beam and not at some intermediate height where there is a nonzero transverse shear.

This example is typical of rectangular-beam problems in that the maximum tensile and compressive stresses are simply flexural stresses $\left(\sigma = \frac{-My}{I}\right)$ at sections

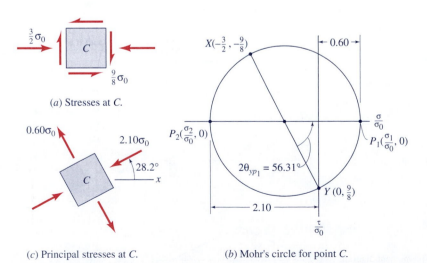

(a) Stresses at C.

(c) Principal stresses at C.   (b) Mohr's circle for point C.

FIGURE 9.8   Determination of the principal stresses at point C in Fig. 9.7d.

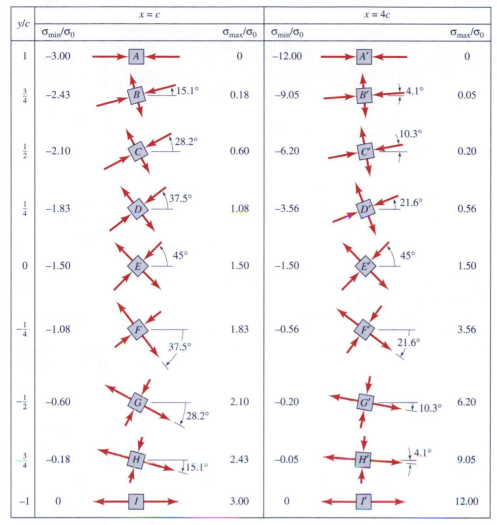

FIGURE 9.9 The principal stresses at two cross sections of the simply supported beam in Fig. 9.7, expressed in terms of $\sigma_0 \equiv P/A$.

where $|M(x)|$ has its maximum value. Therefore, the transverse shear has no effect in determining the maximum normal stresses in the beam. However, in wide-flange beams, where the shear stress τ_{xy} may be large near the outer fibers where σ_x is also large, a complete investigation (i.e., using Mohr's circle) should be conducted to determine the principal stresses at cross sections where V and M are both significant. Such an analysis is carried out in Example Problem 9.2.

Figure 9.9 gives the entire principal-stress picture at a total of 18 points in the beam in Fig. 9.7. It is not possible to provide complete principal-stress information like this (i.e., values of σ_1 and σ_2 and orientation of principal directions) at every point in a beam. However, it is possible to provide two diagrams that permit a visualization of the principal stresses at every point in a beam. One diagram shows the orientation of the two principal directions at every point. Curves, called *stress trajectories,* are drawn so that they are tangent to the principal directions at every point. Since the two principal directions at any point are orthogonal, two stress trajectories pass through every point, and they are perpendicular to each other. Figure 9.10 shows the stress trajectories for a simply supported beam with rectangular

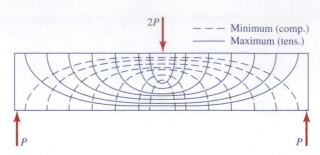

FIGURE 9.10   Stress trajectories for a simply supported beam with a single midspan load.

cross section. (Stress concentrations are not accounted for in this figure.) A typical use of stress trajectories is to determine the direction of the principal tensile stress in beams made of brittle material (e.g., concrete) so that reinforcement can be provided to carry the tensile stresses (recall Fig. 8.1). The stress trajectories show the directions of principal stress, but they provide no information about the magnitudes of the principal stresses. A plot of *stress contours* contains curves that connect points of equal principal stress.

Using the *finite element method* to perform the stress calculations, and using colors to represent different stress magnitudes, it is possible to produce color images of the stress distribution not only in beams, but also in more complex members. The color insert near the beginning of this textbook illustrates stress plots that were produced from finite element solutions.

In Section 6.4, *Design of Beams for Strength,* the sizing of the cross section of a beam to carry specified loads and to have specified supports was based on the maximum flexural stress in the beam. Hence, for example, the maximum moment for the beam in Example Problem 6.5 occurs at $x = 5$ ft, and this moment, together with a given value of allowable stress, was used to determine the most efficient (least weight) cross section of beam to be used for this application. However, since a wide-flange beam was selected for this application, it is possible that the maximum principal stress at a section where both M and V are large may exceed the maximum flexural stress on which the design in Example Problem 6.5 was based. The following example problem explores this possibility.

EXAMPLE 9.2

Just to the left of point B in Fig. 1*a* (same as Fig. 1 of Example Problem 6.5) the transverse shear force is $V = -28$ kips, and the bending moment is $M = -48$ kip · ft. Ignoring any stress concentration due to the support at B, determine the principal stresses at point D in Fig. 1*b*. Compare the maximum tensile stress at D with the maximum flexural stress in the beam, which occurs at $x = 5$ ft, where $M(5\ \text{ft}) = 50$ kip · ft and $V(5\ \text{ft}) = 0$. The beam is a **W**14 × 26. (See Example Problems 6.5 and 6.15.)

Plan the Solution   We can use the flexure formula to determine the normal stress on the cross section at B, and the shear stress distribution on this cross section was determined in Example Problem 6.15. We can use Mohr's circle to combine these and to determine the principal stress magnitudes and directions at point D.

Solution The shear force and bending moment just to the left of section B are shown in Fig. 2a, and the essential cross-sectional dimensions of the **W**14×26 beam are shown in Fig. 2b.

The normal stress at D is

$$(\sigma_x)_D = -\frac{My}{I} = -\frac{(-48 \text{ kip} \cdot \text{ft})(12 \text{ in./ft})(6.535 \text{ in.})}{245 \text{ in}^4}$$

or

$$(\sigma_x)_D = 15.364 \text{ ksi } (T)$$

The shear stress at point D has a magnitude

$$\tau_D \equiv |\tau_{xy}|_D = \frac{|V|Q}{It} = \frac{28 \text{ kips}(5.025 \text{ in.})(0.420 \text{ in.})(6.745 \text{ in.})}{(245 \text{ in}^4)(0.255 \text{ in.})}$$

or

$$\tau_D = 6.380 \text{ ksi } \uparrow$$

We can construct a Mohr's circle for the stresses in the plane of the web at point D. From the Mohr's circle in Fig. 3,

$$R = \sqrt{(7.682)^2 + (6.380)^2} = 9.986 \text{ ksi}$$

$$\left. \begin{array}{l} \sigma_{1_D} = 7.682 \text{ ksi} + 9.986 \text{ ksi} = 17.67 \text{ ksi} \\ \sigma_{2_D} = 7.68 \text{ ksi} - 9.98 \text{ ksi} = -2.30 \text{ ksi} \end{array} \right\} \quad \textbf{Ans.}$$

$$2\theta_{xp_1} = \tan^{-1}\left(\frac{6.380}{7.682}\right) = 39.71°$$

$$\theta_{xp_1} = 19.9°$$

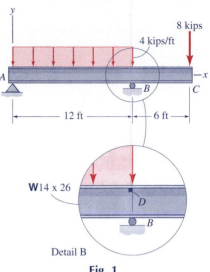

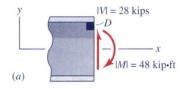

Detail B

Fig. 1

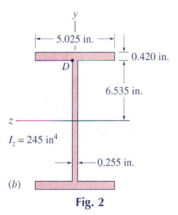

Fig. 2

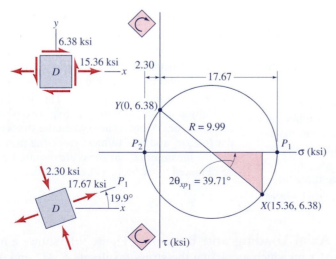

Fig. 3 Mohr's circle for point D.

At $x = 5$ ft, the maximum tensile stress is just the flexural stress at the bottom of the beam (since $V = 0$). Therefore,

$$\sigma_x(5 \text{ ft}, -6.955 \text{ in.}) = \frac{-My}{I} = \frac{-(50 \text{ kip} \cdot \text{ft})(12 \text{ in./ft})(-6.955 \text{ in.})}{245 \text{ in}^4}$$

or

$$\sigma_x(5 \text{ ft}, -6.955 \text{ in.}) = 17.03 \text{ ksi}$$

Therefore, the tensile principal stress at D, 17.67 ksi, is slightly larger than the maximum flexural stress in the beam, 17.03 ksi. However, since an allowable stress of 19 ksi was used in Example Problem 6.5 in selecting the **W**14 × 26 beam cross section, the beam has enough stress margin to be safe, even though the principal stress at D is slightly larger than the maximum flexural stress on which the beam design was originally based.

Review the Solution The calculations in this problem are quite simple, so they can just be rechecked for accuracy. Since there is a significant value of transverse shear force at a cross section where the moment is nearly its maximum value, and since the cross section has heavy flanges and a thin web, we should not be surprised to find principal stresses, like the σ_1 stress at point D, that exceed the maximum flexural stress in the beam.

MDS9.2 & 9.3 **Combined Bending and Shear in Beams**

9.4 STRESSES DUE TO COMBINED LOADS

The analysis Procedure outlined below will now be applied to solve several stress analysis problems that involve various combinations of load types—axial, torsional, and bending.

STRESS-ANALYSIS PROCEDURE FOR COMBINED LOADING

The following three-step procedure will be useful in solving for stresses due to combined loading.

1. *Determine the internal resultants:* This, of course, involves drawing free-body diagrams and writing equilibrium equations. For statically indeterminate problems, material behavior and geometry of deformation must also be considered.

2. *Calculate the individual stresses:* Formulas like those listed in Table 9.1 are used to compute the stress distributions

that result from the various stress resultants. Section 9.2 gives formulas for stresses in thin-wall pressure vessels.

3. *Combine the individual stresses:* This step involves algebraically summing like stresses (e.g., two σ's on the same face), or using Mohr's circle when the stresses are dissimilar (e.g., σ_x and σ_y). In most cases, the principal stresses and the maximum shear stress are required, and these can be obtained from Mohr's circle for stress.

Combined Axial Loading and Bending. Figure 9.11 shows a member with cross section at x on which are acting the stress resultants F, M_y, and M_z. The axial force resultant acts at the centroid of the cross section, and the sign conventions

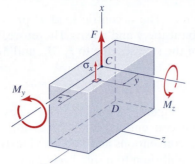

FIGURE 9.11   A member subjected to axial loading and biaxial flexure.

for the axial force and the bending moment components are consistent with those in previous chapters. On any cross section where the y and z axes are *principal axes that pass through the centroid of the cross section*, the normal stress σ_x at point (y, z) in the cross section is given by a combination of Eqs. 2.4 and 6.30, that is, by the equation

$$\sigma_x = \frac{F}{A} + \frac{M_y z}{I_y} - \frac{M_z y}{I_z} \qquad (9.10)$$

EXAMPLE 9.3

An axial compressive load of 800 kips is applied eccentrically to a short rectangular compression member, as shown in Fig. 1. (The effect of eccentric compressive loading on longer members is treated in Section 10.4.) Determine the distribution of normal stress on a cross section, say *ABCD*, that is far enough from the point of load application that stress concentration effects may be neglected. Sketch the stress distribution and identify the location of the neutral axis in cross section *ABCD*.

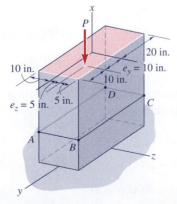

Fig. 1   An eccentrically loaded short compression member.

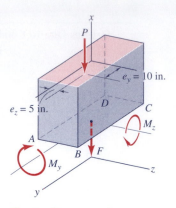

Fig. 2 Stress resultants.

Plan the Solution The eccentric load P produces axial deformation plus bending about the y and z axes. Therefore, this problem involves a superposition of the stresses due to F, M_y, and M_z (Eq. 9.10).

Solution

Stress Resultants: Figure 2 shows the stress resultants on cross section *ABCD*. The sign conventions for stresses and stress resultants is the same as the sign conventions adopted previously in Chapters 2 and 6.

Applying equilibrium to a free-body diagram of the member above section *ABCD* we get the following expressions for the stress resultants:

$$
\begin{aligned}
F &= -P \\
M_y &= -Pe_z \\
M_z &= Pe_y
\end{aligned} \tag{1}
$$

Individual Normal Stresses: Combining Eq. 2.4, for the normal stress due to the axial force F, with Eq. 6.30, for the normal stress due to the bending-moment components, we have Eq. 9.10:

$$
\sigma_x = \frac{F}{A} + \frac{M_y z}{I_y} - \frac{M_z y}{I_z} \tag{2}
$$

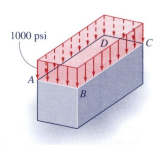

(a) Stress distribution due to F.

Taking each stress contribution separately, and combining Eqs. (1) and (2), we obtain the following:

$$
(\sigma_x)_F = \frac{-P}{A} = \frac{-800 \text{ kips}}{(40 \text{ in.})(20 \text{ in.})} = -1000 \text{ psi}
$$

$$
(\sigma_x)_{M_y} = \frac{(-Pe_z)z}{I_y} = \frac{(-800 \text{ kips})(5 \text{ in.})z}{[\frac{1}{12}(40 \text{ in.})(20 \text{ in.})^3]} = -150z \text{ psi} \tag{3}
$$

$$
(\sigma_x)_{M_z} = \frac{-(Pe_y)y}{I_z} = \frac{-(800 \text{ kips})(10 \text{ in.})y}{[\frac{1}{12}(20 \text{ in.})(40 \text{ in.})^3]} = -75y \text{ psi}
$$

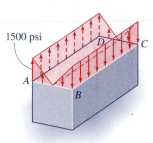

(b) Stress distribution due to M_y.

These stress contributions are sketched in Fig. 3. The maximum values of the bending stresses are

$$
\text{Max}|(\sigma_x)M_y| = 150z|_{z=10 \text{ in}} = 150(10 \text{ in.}) = 1500 \text{ psi}
$$

$$
\text{Max}|(\sigma_x)M_z| = 75y|_{y=20 \text{ m}} = 75(20 \text{ in.}) = 1500 \text{ psi}
$$

Superposition of Stresses: Using Figs. 3a through 3c, we can combine, algebraically, the individual stress contributions at four corners to get

$$
\begin{aligned}
(\sigma_x)_A &= -1000 + 1500 - 1500 = -1000 \text{ psi} \\
(\sigma_x)_B &= -1000 - 1500 - 1500 = -4000 \text{ psi} \\
(\sigma_x)_C &= -1000 - 1500 + 1500 = -1000 \text{ psi} \\
(\sigma_x)_D &= -1000 + 1500 + 1500 = 2000 \text{ psi}
\end{aligned} \tag{4}
$$

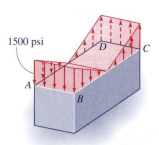

(c) Stress distribution due to M_z.

Fig. 3 Individual stress contributions.

With the aid of these corner stresses, we can sketch the combined stress distribution. Since Eq. (2) is linear in y and z, the *stress surface* will be a plane (which has been "folded" to show tension and compression as they are shown in Fig. 3). In Fig. 4 the member has been rotated about the x axis to provide a better perspective for viewing the stress surface and the *neutral axis,* which is the intersection, RS, of the stress surface with the $ABCD$ plane.

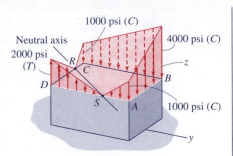

Fig. 4   The combined stresses on section $ABCD$.

The equation of the neutral axis line is given by setting $\sigma_x(y^*, z^*) = 0$, where (y^*, z^*) are coordinates of points on the neutral axis. Combining Eqs. (2) and (3) we get

$$\sigma_x(y^*, z^*) = -1000 - 75y^* - 150z^* = 0 \qquad (5)$$

The intersection, R, of the neutral axis with edge CD is given by setting $y^* = -20$ in. This gives $(y^*, z^*)_R = (-20 \text{ in.}, 3.33 \text{ in.})$. Similarly, $(y^*, z^*)_S = (6.67 \text{ in.}, -10 \text{ in.})$.

Review the Solution   It is obvious from the location of the compressive force P in Fig. 1 that corner B will have the highest compressive stress of any point in the cross section. This is confirmed by Eq. (4b) and by Fig. 4. The load P is far enough from the axis of the member that it actually causes tension at D. This means that between B and D the stress changes from compression to tension. Hence, there is a neutral axis ($\sigma_x = 0$) that passes between A and D. Therefore, Fig. 4 appears to represent accurately the stress distribution due to the eccentric load P.

If a compression member, like the one in Example Problem 9.3, is made of a brittle material like concrete, it is not desirable to permit tensile stresses to occur at any point in the cross section. This requires that the compressive load P be located within a small region near the axis of the member. This region is called the *kern* of the cross section, or *core* of the cross section. For a rectangular cross section, like the one in Fig. 9.11, the kern is bounded by four straight lines. The equations of the four lines that bound the kern can be derived by combining Eqs. (1) and (2) of Example Problem 9.3 to give the normal stress on a cross section like section $ABCD$ in Fig. 9.12. Thus,

$$\sigma_x(y, z) = \frac{-P}{A} + \frac{-Pe_z z}{I_y} + \frac{-Pe_y y}{I_z} \qquad (9.11)$$

This normal stress will be zero at corner D $(-c_y, -c_z)$ if P is applied anywhere along the line "d," whose equation is

$$\left(\frac{c_y}{I_z}\right)e_{yd} + \left(\frac{c_z}{I_y}\right)e_{zd} = \frac{1}{A} \qquad (9.12)$$

Similar equations can be derived for the kern boundary lines a, b, and c. (See Homework Problem 9.4-1.)

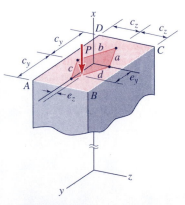

FIGURE 9.12   The kern of a rectangular cross section.

Combined Axial Loading and Torsion.

The next example problem illustrates the superposition of stresses due to combined axial and torsional loading.

EXAMPLE 9.4

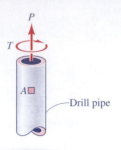

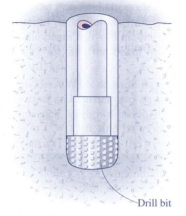

Fig. 1 Portions of an oilwell drill string.

During the drilling of an oil well, the section of the drill pipe at A (above ground level) is under combined loading due to a tensile force $P = 70$ kips and a torque $T = 6$ kip · ft, as illustrated in Fig. 1. The drill pipe has an outside diameter of 4.0 in. and an inside diameter of 3.640 in. Determine the maximum shear stress at point A on the outer surface of the drill pipe. The radial stress at this point is zero. The yield strength in tension of this drill pipe is 95 ksi.

Plan the Solution We can use Mohr's circle to combine the normal stress σ due to force P and the shear stress τ due to T, but we may also need to consider the three-dimensional aspect of the stress at A to determine the absolute maximum shear stress.

Solution

Stress Resultants: The stress resultants are given in the problem statement:

$$F = P = 70 \text{ kips}, \qquad T = 6 \text{ kip} \cdot \text{ft} \tag{1}$$

Individual Stresses: From Eq. 2.4, we get the normal stress

$$\sigma = \frac{F}{A} = \frac{70 \text{ kips}}{\pi[(2 \text{ in.})^2 - (1.820 \text{ in.})^2]} = 32.41 \text{ ksi} \tag{2}$$

From Eq. 4.13, we get the torsional shear stress

$$\tau = \frac{T r_o}{I_p} = \frac{(6 \text{ kip} \cdot \text{ft})(12 \text{ in./ft})(2 \text{ in.})}{\frac{\pi}{2}[(2 \text{ in.})^4 - (1.820 \text{ in.})^4]} = 18.23 \text{ ksi} \tag{3}$$

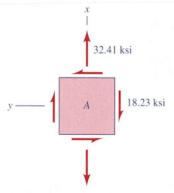

Fig. 2 In-plane stresses at point A.

Figure 2 summarizes the "in-plane" stresses on the surface of the drill pipe at point A. The radial stress, normal to the surface, is zero.

Superposition of Stresses: Mohr's circle may be used to combine the in-plane stresses in Fig. 2. From the Mohr's circle in Fig. 3,

$$R = \sqrt{(16.20 \text{ ksi})^2 + (18.23 \text{ ksi})^2} = 24.39 \text{ ksi} \tag{4}$$

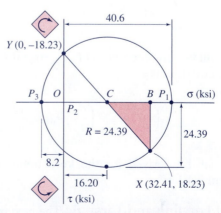

Fig. 3 Mohr's circle.

Then,

$$\sigma_1 = \sigma_{avg} + R = 16.20 + 24.39 = 40.6 \text{ ksi}$$
$$\sigma_3 = \sigma_{avg} - R = 16.20 - 24.39 = -8.2 \text{ ksi} \qquad (5)$$

The in-plane principal stresses are labeled σ_1 and σ_3, since the out-of-plane principal stress, $\sigma_r = 0$, is the intermediate principal stress. Then, from Eq. 8.32,

$$\tau_{abs \atop max} = \frac{\sigma_{max} - \sigma_{min}}{2} = \frac{40.6 \text{ ksi} - (-8.2 \text{ ksi})}{2} \qquad (6a)$$

or

$$\tau_{abs \atop max} = 24.4 \text{ ksi} \qquad \textbf{Ans.} \quad (6b)$$

Review the Solution The calculations in Eqs. (2) and (3) should be rechecked. Points X and Y are plotted correctly, so σ_1 and σ_3 appear to be correct. Finally, since the working stresses in this example should not produce yielding of the drill pipe, the absolute maximum shear stress should be much less than half the tensile yield strength. Therefore, the answer in Eq. (6b) seems reasonable.

MDS9.4 **Shaft Subjected to Combined Axial Loading and Torsion**

Many interesting applications of deformable-body mechanics in the field of oilwell drilling engineering are presented in *Oilwell Drilling Engineering—Principles and Practice,* by H. Rabia, [Ref. 9-5]. See also Ref. 9-6.[7]

General Combined Loading. In the final example problem on stresses due to combined loading, we consider a problem that involves all types of stress resultants: F, T, M, and V.

EXAMPLE 9.5

Wind blowing on a sign produces a pressure whose resultant, P, acts in the $-y$ direction at point C, as shown in Fig. 1. The weight of the sign, W_s, acts vertically through point C, and the thin-wall pipe that supports the sign has a weight W_p.

Following the procedure outlined at the beginning of Section 9.4, determine the principal stresses at points A and B, where the pipe column is attached to its base. Use the following numerical data.

Pipe $OD = 3.50$ in., $A = 2.23$ in^2, $I_y = I_z = 3.02$ in^4, $I_p = 6.03$ in^4,

$W_s = 125$ lb, $W_p = 160$ lb, $P = 75$ lb, $b = 40$ in., $L = 220$ in.

---

[7]The *American Petroleum Institute* (API) maintains standards covering all segments of the oil and gas industry and distributes publications, technical standards, and electronic and online products. http://www.api.org

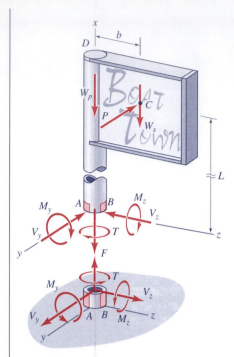

Fig. 1 A cantilevered sign.

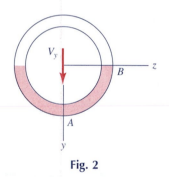

Fig. 2

Plan the Solution It will be a good idea to tabulate the stress resultants, stress formulas, and so forth, so that no stress contribution will be missed. The weight W_s contributes to the axial force, and it also produces a moment about the y axis. The wind force P produces a transverse shear force in the y direction, and it also causes a torque about the x axis and a moment about the z axis. A correct free-body diagram is essential.

Solution

Stress Resultants: All six stress resultants on the cross section at the base of the pipe are shown in Fig. 1. The upper portion of Fig. 1 can serve as a free-body diagram for determining these six stress resultants. The sign convention is the one introduced in Fig. 2.40. Let us tabulate the equilibrium equations and indicate what stress is produced by each stress resultant and label each individual stress.

Individual Stresses: Using the formulas from Table 9.1, we can compute the numerical value of each of the nonzero stresses listed in Table 1.

$$\sigma_{A1} = \sigma_{B1} = \frac{F}{A} = \frac{-(125 \text{ lb}) - (160 \text{ lb})}{2.23 \text{ in}^2} = -128 \text{ psi} \tag{7}$$

The shear stress τ_{B2} is due to the transverse shear force V_y. The basic shear stress formula is

$$\tau_{B2} = \frac{V_y Q}{I_z t} \tag{8}$$

where Q has to be calculated for the shaded area in Fig. 2. In Example Problem 6.16, it was shown that the shear stress in this case (stress at the neutral axis of a thin-wall pipe) is given by

$$\tau = \frac{2V}{A} \tag{9a}$$

TABLE 1   A Table of Stress Resultants and the Stresses Produced

Eq. No.	Equilibrium Equation		Stress at A	Stress at B
(1)	$\sum F_x = 0$	$F = -W_s - W_p$	σ_{A1}	σ_{B1}
(2)	$\sum F_y = 0$	$V_y = -P$	—	τ_{B2}
(3)	$\sum F_z = 0$	$V_z = 0$	—	—
(4)	$\sum M_x = 0$	$T = Pb$	τ_{A4}	τ_{B4}
(5)	$\sum M_y = 0$	$M_y = -W_s b$	—	σ_{B5}
(6)	$\sum M_z = 0$	$M_z = -PL$	σ_{A6}	—

Therefore,

$$\tau_{B2} = \frac{2(75\ \text{lb})}{2.23\ \text{in}^2} = 67\ \text{psi} \tag{9b}$$

$$\tau_{A4} = \tau_{B4} = \frac{Tr_o}{I_p} = \frac{(Pb)r_o}{I_p} \tag{10a}$$

so

$$\tau_{A4} = \tau_{B4} = \frac{(75\ \text{lb})(40\ \text{in.})(1.75\ \text{in.})}{6.03\ \text{in}^4} = 871\ \text{psi} \tag{10b}$$

The flexural stresses due to M_y and M_z are given by Eq. 6.30.

$$\sigma_{B5} = \frac{M_y r_o}{I_y} = \frac{(-W_s b)r_o}{I_y} \tag{11a}$$

$$\sigma_{B5} = \frac{-(125\ \text{lb})(40\ \text{in.})(1.75\ \text{in.})}{3.02\ \text{in}^4} = -2897\ \text{psi} \tag{11b}$$

$$\sigma_{A6} = \frac{-M_z r_o}{I_z} = \frac{-(-PL)r_o}{I_z} \tag{12a}$$

$$\sigma_{A6} = \frac{(75\ \text{lb})(220\ \text{in.})(1.75\ \text{in.})}{3.02\ \text{in}^4} = 9561\ \text{psi} \tag{12b}$$

Superposition of Stresses: Using the above values, and taking proper note of the physical significance of the sign of each term by referring to Fig. 1, we get the stresses shown in Fig. 3.

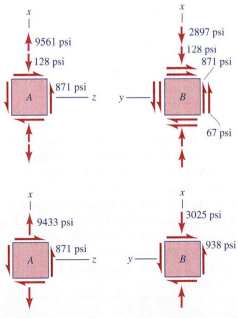

Fig. 3 The states of stress at points A and B.

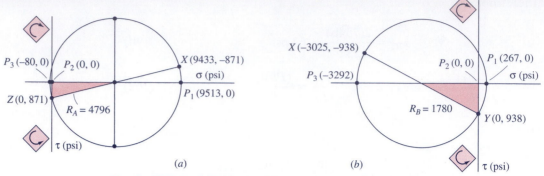

Fig. 4 Mohr's circles for in-plane stresses at points A and B.

Using the stresses shown in Fig. 3, we can construct a Mohr's circle for the states of plane stress at points A and B on the pipe surface. The radial normal stress is $\sigma_r = 0$ at both places. From Fig. 4a.

$$R_A = \sqrt{(9433/2)^2 + (871)^2} = 4796 \text{ psi} \tag{13}$$

$$(\sigma_1)_A = (9433/2) + 4796 = 9513 \text{ psi}$$
$$(\sigma_3)_A = (9433/2) - 4796 = -80 \text{ psi} \tag{14}$$

and, from Fig. 4b,

$$R_B = \sqrt{(-3025/2)^2 + (938)^2} = 1780 \text{ psi} \tag{15}$$

$$(\sigma_1)_B = (-3025/2) + 1780 = 267 \text{ psi}$$
$$(\sigma_3)_B = (-3025/2) - 1780 = -3292 \text{ psi} \tag{16}$$

In summary, the principal stresses at points A and B, rounded to three significant figures, are:

$$(\sigma_1)_A = 9510 \text{ psi}, \quad (\sigma_2)_A = 0, \quad (\sigma_3)_A = -80 \text{ psi}$$
$$(\sigma_1)_B = 267 \text{ psi}, \quad (\sigma_2)_B = 0, \quad (\sigma_3)_B = -3290 \text{ psi}$$

Ans.

Review the Solution By showing all six possible internal resultants at the cross section where stresses are to be calculated, by writing down and solving all six possible equilibrium equations, and by carefully considering what stress(es) is (are) produced by each stress resultant, we have accounted for the effects of all loads on the structure. As noted earlier, we have been careful to make sure that each stress component acts in the direction that "makes sense." For example, the force P bends the pipe in the direction that produces tension at point A, and so forth.

The maximum flexural stress at the base occurs at neither A nor B. Equation 6.30 could be used to combine the flexural stresses due to M_y and M_z, and we would also have to consider the effect of shear stress. (See Homework Problem 9.4-26.)

MDS9.5   **Member Subjected to Combined Axial, Shear, and Bending Stresses**

▼ PRESSURE VESSELS | MDS 9.1 |

> For all pressure-vessel problems for Section 9.2, *the pressure p is the gage pressure, that is, the absolute internal pressure minus the absolute external pressure. All cylinders are right circular cylinders.*

Prob. 9.2-1. A steel oxygen cylinder used by a welder has an inner radius of $r_i = 4$ in. and a wall thickness of 0.5 in. The cylinder is pressurized to $p = 2000$ psi. (a) Determine the axial stress σ_a and the hoop stress σ_h in the cylindrical body of the tank. (b) Determine the tensile force per inch length of the weld between the hemispherical head and the cylindrical body of the tank.

P9.2-1

Prob. 9.2-2. A steel propane tank for a barbecue grill has a 12-in. inside diameter and a wall thickness of $\frac{1}{8}$ in. The tank is pressurized to 200 psi. (a) Determine the axial stress σ_a and the hoop stress σ_h in the cylindrical body of the tank. (b) Determine the tensile force per inch length of the weld between the upper and lower sections of the tank. (c) Determine the absolute maximum shear stress in the cylindrical portion of the tank.

P9.2-2 and P9.2-3

Prob. 9.2-3. Solve Prob. 9.2-2 for a tank with 300-mm inside diameter and a wall thickness of 4 mm, if the tank is pressurized to 1.5 MPa.

Prob. 9.2-4. A scuba diver's aluminum air tank has an outer diameter of $d_o = 7.0$ in. and a wall thickness of $t = 0.5$ in.,

P9.2-4 and P9.2-5

and it is pressurized to a service pressure of $p = 3000$ psi. (a) Determine the principal stresses and the maximum in-plane shear stress in the cylindrical portion of the tank. (b) Determine the absolute maximum shear stress, τ_{abs} .
$_{\text{max}}$

Prob. 9.2-5. Repeat Prob. 9.2-4 for an air tank with an outer diameter of $d_o = 180$ mm and a wall thickness of $t = 13$ mm that it is pressurized to a service pressure of $p = 20$ MPa.

Prob. 9.2-6. The cylindrical portion of a compressed-air tank is fabricated of steel plate that is welded along a helix that makes an angle of $\alpha = 70°$ with respect to the longitudinal axis of the tank. The inside diameter of the cylinder is 48 in., the wall thickness is 0.5 in., and the internal pressure is 240 psi. Determine the following quantities for the cylindrical portion of the tank: (a) the axial stress σ_a and the hoop stress σ_h, (b) the normal stress and shear stress on planes parallel and perpendicular to the weld, (c) the maximum in-plane shear stress, and (d) the absolute maximum shear stress.

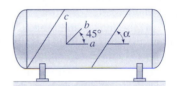

P9.2-6 through P9.2-8

Prob. 9.2-7. Solve Prob. 9.2-6 for a tank with $\alpha = 65°$, $d_i = 1$ m, $t = 20$ mm, and $p = 2$ MPa.

Prob. 9.2-8. A 45° strain gage rosette is placed on the compressed air tank with gage "a" oriented parallel to the axis of the tank. (See Fig. P9.2-8.) The tank is made of steel with Young's modulus $E = 29(10^6)$ psi and Poisson's ratio $\nu = 0.3$. If the inside diameter of the tank is 40 in., the wall thickness is $\frac{3}{8}$ in., and the internal pressure is 180 psi, what would be the readings for ϵ_a, ϵ_b, and ϵ_c?

> **Problems 9.2-9 and 9.2-10** *treat cylindrical tanks used as vertical fluid-storage reservoirs, or standpipes. Problem 9.2-11 treats a water-filled vertical pipe. The hoop stress, σ_h, is given by Eq. 9.4, just as in the case of uniform internal pressure in a circular cylinder. According to Pascal's Law, the fluid develops* **hydrostatic pressure** $p = \gamma h$, *where γ is the* **specific weight** *of the fluid and h is the depth below the fluid surface of the point where the pressure is being calculated. For water, $\gamma = 62.4$ lb/ft^3 = 9.81 kN/m^3.*

^{D}**Prob. 9.2-9.** (See previous Note.) A vertical standpipe has an inside diameter of $d_i = 3$ m and is filled with water to a depth of $h = 5$ m. If the allowable hoop stress is 80 MPa, what is the minimum wall thickness of the tank to the nearest millimeter? (Neglect the restraint that the base exerts on the cylindrical tank.)

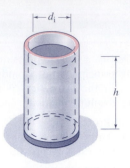

P9.2-9 and P9.2-10

^D**Prob. 9.2-10.** Solve Prob. 9.2-9 for a standpipe with inside diameter of $d_i = 10$ ft that is filled to a depth of $h = 40$ ft. The allowable hoop stress is 12 ksi. (Neglect the restraint that the base exerts on the cylindrical tank).

Prob. 9.2-11. A pump at the base of a vertical steel pipe creates a pressure at the base of the pipe that is sufficient to lift the water in the pipe to a height of 24 ft, where the water is discharged at atmospheric pressure into a cooling tower. If the pipe has an inside diameter of 2 ft and a wall thickness of $\frac{1}{4}$ in., determine (a) the maximum hoop stress in the pipe and (b) the absolute maximum shear stress in the pipe. For Part (b), assume that only the lower 18 ft of the steel pipe is supported by the base, with the upper 6 ft of pipe supported by the elbow attached to the cooling tower. The specific weight of the steel is $\gamma = 490$ lb/ft^3.

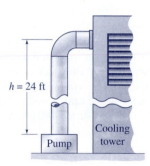

P9.2-11

^D**Prob. 9.2-12.** Hemispherical end caps are welded to a cylindrical main body to form a propane storage tank. The tank has an inside diameter of $d_i = 40$ in. and is to be subjected to a

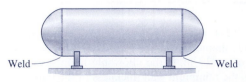

P9.2-12 through P9.2-14

maximum internal pressure of $p = 120$ psi. The allowable tensile stress in the wall of the tank is 12 ksi, and the allowable tensile stress in the weld is 8 ksi. (a) Determine the minimum wall thickness t_c of the cylindrical part of the tank. (b) Determine the minimum wall thickness t_s of the hemispherical end caps. (c) What is the minimum thickness of the weld?

^D**Prob. 9.2-13.** Solve Prob. 9.2-12 for a tank with an inside diameter of $d_i = 750$ mm that is subjected to a maximum internal pressure of $p = 750$ kPa. The allowable tensile stress in the wall of the tank is 80 MPa, and the allowable tensile stress in the weld is 50 MPa.

Prob. 9.2-14. A cylindrical tank with closed ends contains compressed nitrogen gas at a gage pressure of 250 psi. The inside diameter of the tank is 5 ft, and the wall thickness is $\frac{3}{4}$ in. (a) Determine the axial stress σ_a and the hoop stress σ_h. (b) Draw a Mohr's circle for the in-plane stresses in the cylinder wall, and determine the maximum in-plane shear stress. Show a properly oriented maximum-shear-stress element. (c) Calculate the absolute maximum shear stress, $\tau_{\text{abs max}}$, in the wall of the cylindrical tank.

^D**Prob. 9.2-15.** Hydraulic pressure p acts on a piston at A, which in turn exerts a force P on the object at B. The allowable tensile stress in the wall of the hydraulic cylinder is 100 MPa. If the inside diameter of the cylinder is $d_i = 125$ mm, the wall thickness of the cylinder is $t = 6$ mm, and the diameter of the piston rod is $d_r = 20$ mm, determine the maximum force p that can be exerted by the piston rod. (Assume that the stress in the cylinder wall is the only factor that limits the value of p.)

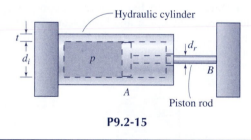

P9.2-15

▼ **COMBINED BENDING AND SHEAR IN BEAMS** | MDS 9.2 & 9.3 |

Prob. 9.3-1. A uniform cantilever beam with rectangular cross section is subjected to a uniform distributed load of intensity w_0, as shown in Fig. P9.3-1. Determine expressions for the two principal stresses and the maximum in-plane shear stress at each of the three indicated points at the root of the beam: $y_A = h/2$, $y_B = h/4$, and $y_C = 0$. Construct a Mohr's circle to obtain your answers for each point. On sketches similar to those in Fig. 9.9, indicate the orientation of the in-plane principal element at each of the three designated points.

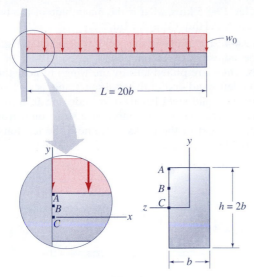

P9.3-1

Prob. 9.3-2. A **W**8 × 40 simply supported steel beam is subjected to a concentrated midspan load of 50 kips, as shown in Fig. P9.3-2. Determine expressions for the two principal stresses and the maximum in-plane shear stress at the three indicated points in the cross section at $x = 2$ ft: $y_A = 4.125$ in., $y_B = 3.565$ in., and $y_C = 2.0$ in. (Note: Point B is in the web, just below the flange.) Construct a Mohr's circle to obtain your answers for each point. On sketches similar to those in Fig. 9.9, indicate the orientation of the in-plane principal element at each of the three designated points.

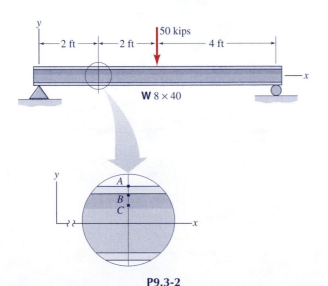

P9.3-2

^{C}Prob. 9.3-3. Use the **MDSolids** modules *Section Properties, Determinate Beams,* and *Mohr's Circle Analysis* to solve Prob. 9.3-2.

^{C}Prob. 9.3-4. Use the **MDSolids** modules *Section Properties, Determinate Beams,* and *Mohr's Circle Analysis* to solve the following problem.

A **W**24 × 94 wide-flange steel cantilever beam supports a distributed load of 1.5 kips/ft and concentrated load of 20 kips, as shown in Fig. P9.3-4. (See Prob. 6.3-25 for this figure.) Let the y axis be upward with its origin at the neutral axis at A. (a) Plot the shear-force and bending-moment diagrams for this beam. (b) Determine the maximum flexural stress in the beam. (c) Using Mohr's circle, determine the principal stresses and the maximum in-plane shear stress at $(x = 8^-$ ft. $y = 11$ in.), that is, the point that is 11 in. above the neutral axis and on the cross section just to the left of B. Note that this point is in the web and is just below the top flange.

^{C}Prob. 9.3-5. Use the **MDSolids** modules *Section Properties, Determinate Beams,* and *Mohr's Circle Analysis* to solve the following problem.

A **W**150 × 24 wide-flange steel beam supports a distributed load of 1 kN/m and concentrated couple of 4 kN · m, as shown in Fig. P9.3-5. (See Prob. 6.3-20 for this figure.) With the origin at the neutral axis at A, let the y axis be upward. (a) Plot the shear-force and bending-moment diagrams for this beam. (b) Determine the maximum flexural stress in the beam. (c) Using Mohr's circle, determine the principal stresses and the maximum in-plane shear stress at $(x = 2^-$ m, $y = 69$ mm), that is, the point that is 69 mm above the neutral axis and on the cross section just to the left of B. Note that this point is in the web and is just below the top flange.

▼ **COMBINED AXIAL LOADING AND BENDING**

Prob. 9.4-1. Show that the boundary line a of the kern of the rectangular cross section in Fig. 9.12 is given by the expression

$$\left(\frac{-c_y}{I_z}\right)e_{ya} + \left(\frac{c_z}{I_y}\right)e_{za} = \frac{1}{A}$$

Note: The formula for the boundary line d is given in Eq. 9.12.

Prob. 9.4-2. The rectangular bar in Fig. P9.4-2 is subjected to bending in the xy plane and, simultaneously, to an axial tensile force P. The state of plane stress at point A at the top edge of the bar is shown in Fig. P9.4-2. If the bending moment is $M = 2$ kN · m, what is the magnitude of the axial force P? (Hint: Consider Eq. 8.7.)

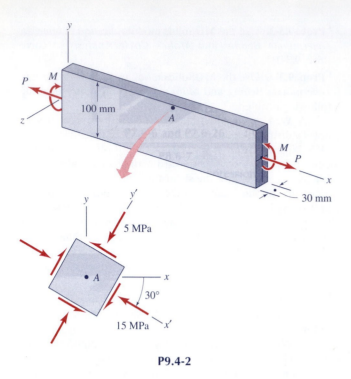

P9.4-2

the value $P = 5$ kips, what is the magnitude of the bending
moment M? (Hint: Consider Eq. 8.7.)

Prob. 9.4-4. The frame of the hacksaw depicted in Fig. P9.4-4
can be adjusted to accommodate either 10-in. or 12-in.
blades, which are pulled taut by the wing nut near the handle. The left-hand end of the hacksaw frame is an L-shaped
rectangular solid steel bar 0.60 in. wide and 0.20 in. thick.
Determine the maximum compressive stress on a cross section in this part of the hacksaw frame if the tension in the
blade is 30 lb.

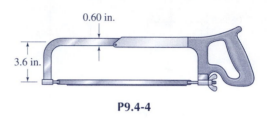

P9.4-4

Prob. 9.4-3. The rectangular bar in Fig. P9.4-3 is subjected to
bending in the xy plane and, simultaneously, to an axial tensile force P. The state of plane stress at point A at the top
edge of the bar is shown in Fig. P9.4-3. If the axial force has

Prob. 9.4-5. One part of the mechanism that controls the
operation of the backhoe bucket in Fig. P9.4-5a is the
(slightly C-shaped) two-force link AB, whose dimensions
are shown in Fig. P9.4-5b. (a) Determine the maximum
tensile stress on the cross section at the center of link AB if
the force exerted on the link by the pins at A and B is $P = 6$ kN.
(b) How much would the maximum tensile stress be if the
link were perfectly straight, with the same cross-sectional
dimensions?

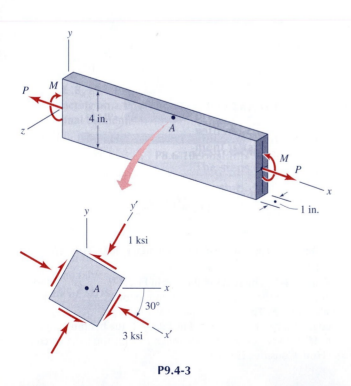

P9.4-3

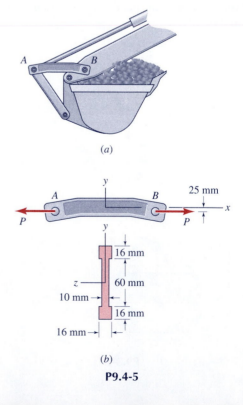

P9.4-5

Prob. 9.4-6. A floor crane, like the one shown in Fig. P9.4-6a, is used to pick up loads and allow them to be moved easily to another location. The vertical column AB is steel rectangular tubing whose dimensions are shown in Fig. P9.4-6b. Determine the maximum tensile stress and the maximum compressive stress on the base cross section at A when the boom BC is in the position shown.

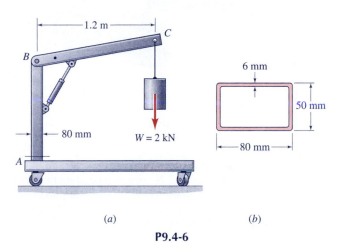

(a)   (b)

P9.4-6

Prob. 9.4-7. Three signal-light clusters are suspended from a tapered horizontal arm that is cantilevered from a tubular steel column as shown in Fig. P9.4-7. The weights of the three signal-light clusters and the horizontal arm are indicated in the figure. The outer diameter of the column is $d_o = 14.00$ in., its inner diameter is $d_i = 13.00$ in., it weighs 75 lb/ft, and it is 20 ft tall. Determine the maximum compressive stress at the base cross section.

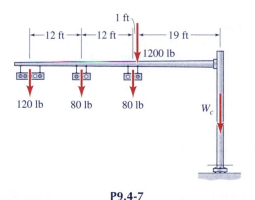

P9.4-7

▼ **COMBINED AXIAL LOADING AND TORSION** | MDS 9.4 |

Prob. 9.4-8. A solid shaft of diameter $d = 2$ in. is subjected simultaneously to an axial load $P = 10$ kips and to a torque $T = 2.5$ kip · in., acting as shown in Fig. P9.4-8. Use Mohr's circle to determine the two principal stresses and the maximum in-plane shear stress at any point on the outer surface of the shaft.

P9.4-8 and P9.4-27

^{D}**Prob. 9.4-9.** A tubular shaft of outer diameter $d_o = 50$ mm and inner diameter $d_i = 40$ mm is subjected simultaneously to a specified torque $T = 250$ N · m and to an axial load P, acting as shown in Fig. P9.4-9. The maximum tensile stress allowed is $\sigma_{allow} = 40$ MPa. Using Mohr's circle, determine the maximum axial load P that can be applied without exceeding this allowable tensile stress.

P9.4-9 and P9.4-28

Prob. 9.4-10. A post-hole digger is mounted on a tractor (not shown). The power unit of the machine applies a torque of 800 lb · in. to the auger, and it also exerts a downward force of 1500 lb on the auger. If the shaft of the auger is a solid circular rod with a diameter of 2.0 in., determine the principal stresses and the maximum shear stress at a typical point A on the surface of the shaft of the auger near the power unit.

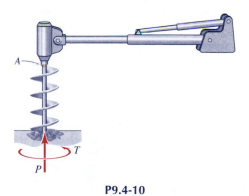

P9.4-10

*****Prob. 9.4-11.** The large earth drill shown in Fig. P9.4-11 is used to make 30-in.-diameter holes for the reinforced-concrete footings that will support the columns of a building. A solid 4 in. × 4 in. steel shaft exerts a downward force $P = 6.4$ kips on the drill bit while, at the same time, it supplies a torque of 15.0 kip · in. to rotate the bit. Use Mohr's circle to determine the two principal stresses and the maximum in-plane shear stress in the square shaft. (Note: Recall that Section 4.10 discusses torsion of noncircular prismatic bars.)

629

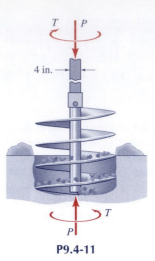

P9.4-11

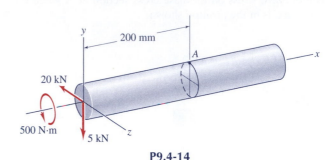

Prob. 9.4-14. A solid circular shaft whose diameter is 50 mm is subjected to a torque of 500 N · m and two components of transverse shear force as shown in Fig. P9.4-14. Use a Mohr's circle to determine the principal stresses and the maximum shear stress on the surface of the shaft at point A.

P9.4-14

Prob. 9.4-12. The shaft of the wind-driven electric power generator shown in Fig. P9.4-12 (see Prob. 4.8-11) has an outer diameter of $d_o = 50$ mm and an inner diameter $d_i = 40$ mm. The generator produces 10 kW of power while the shaft is turning at a speed of 20 RPM. At the same time, the propeller exerts a compressive load of 2 kN on the shaft. Use Mohr's circle to determine the two principal stresses and the maximum in-plane shear stress at any point on the outer surface of the shaft.

ᴰProb. 9.4-15. A solid circular shaft of radius $r = 25$ mm is subjected to a torque $T = 98.2$ N · m and a bending moment M as shown. At point A in the cross section, the maximum compressive stress is -8 MPa. Using a Mohr's circle, determine the value of the bending moment M.

▼ **GENERAL COMBINED LOADING** | MDS 9.5

Prob. 9.4-13. A force of 150 N acts at point B on an L-shaped lug wrench, as shown in Fig. P9.4-13a. The force acts vertically downward, perpendicular to the plane of the wrench. The handle of the lug wrench is a steel rod with a diameter of 12.5 mm, and its planform is shown in Fig. P9.4-13b. Determine the principal stresses and the maximum shear stress at point A, which is on the top of the wrench handle.

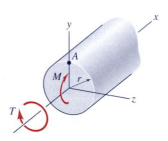

P9.4-15 and P9.4-16

ᴰProb. 9.4-16. A solid circular shaft of radius $r = 1$ in. is subjected to a bending moment $M = 4.71$ kip · in. and a torque T as shown. At point A in the cross section, the maximum shear stress is $\tau_{max} = 6$ ksi. Using a Mohr's circle, determine the value of the torque T.

Prob. 9.4-17. The boom of a crane has a rectangular box section with the dimensions shown. Determine the principal stresses in the boom at points A and B. Neglect the weight of the boom, pulley, and cable.

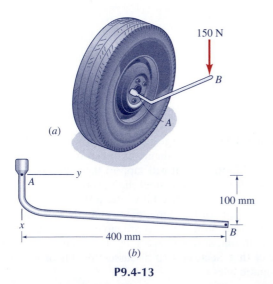

P9.4-13

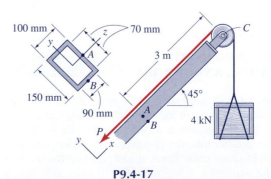

P9.4-17

*Prob. 9.4-18. A vertical force of 40 lb is applied to a pipe wrench, whose handle is parallel to the z axis, as shown in Fig. P9.4-18. Using Mohr's circles, determine the principal stresses at points A and B in the cross section where the pipe threads begin. The pipe has a nominal diameter of 1 in. See Table D.7 of Appendix D for cross-sectional dimensions of the pipe.

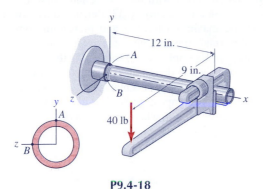

P9.4-18

Prob. 9.4-19. At a particular cross section of the drive axle of a race car, the stress resultants are as shown in Fig. P9.4-19. Use Mohr's circle to determine the principal stresses and the maximum shear stress at the following points in this cross section: (a) Point A, the top point in the cross section, and (b) Point B, the point ($y = 0$, $z = 0.5$ in.) in the cross section.

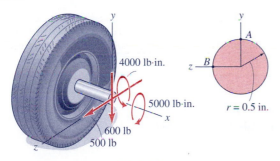

P9.4-19

Prob. 9.4-20. A wide-flange beam is subjected to axial and transverse loads as shown. Determine the principal stresses

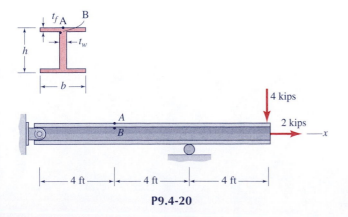

P9.4-20

in the beam at points A and B. The dimensions of the cross section are: $h = 4$ in., $b = 3$ in., $t_f = 0.23$ in., and $t_w = 0.15$ in.

*Prob. 9.4-21. A chair on a ski lift is supported by a steel pipe whose outer diameter is $d_o = 60$ mm and whose inner diameter is $d_i = 52$ mm. The weight of the pipe may be neglected in comparison with the weight of the chair and its occupants, which is $W = 2$ kN. (a) Determine the stresses σ_x, σ_y, and τ_{xy} at point C, which is on the front of the pipe at the indicated cross section. The x axis is parallel to the 45° section of pipe, AB. (b) Using a Mohr's circle, determine the principal stresses and the maximum in-plane shear stress at point C. (c) Determine the maximum tensile stress in the straight section of the pipe, DE.

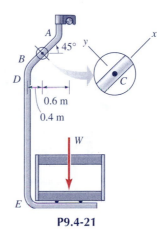

P9.4-21

*Prob. 9.4-22. A section of oilfield drill pipe is being lifted by a crane, as shown in Fig. P9.4-22. The pipe has an outside diameter of $d_o = 4.50$ in., an inside diameter of $d_1 = 3.64$ in., and a total length of $L = 30$ ft. It weighs $w = 20$ lb/ft. Use Mohr's circle to determine the principal stresses at point B when the inclination of the pipe is $\theta = 30°$.

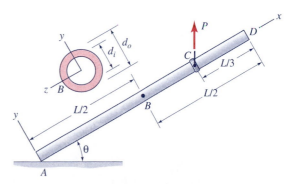

P9.4-22 and P9.4-23

*Prob. 9.4-23. Solve Prob. 9.4-22 for $L = 10$ m, $d_o = 114$ mm, $d_i = 92$ mm, $w = 300$ N/m, and $\theta = 45°$.

Prob. 9.4-24. The frame ABC in Fig. P9.4-24 (see Prob. 1.4-20) consists of **W**150 × 22 wide-flange steel members

631

AB and BC welded together at B. Determine the maximum compressive stress on the cross section at $x_1 = 2$ m, where x_1 is measured along member AB, as shown in Fig. P9.4-24. (See Table D.2 of Appendix D for the cross-sectional properties of the wide-flange shape.)

*Prob. 9.4-25. The L-shaped frame ABC in Fig. P9.4-25a consists of square steel tubing members AB and BC welded together at B. The frame is supported by a fixed pin at A and by a roller at C. Determine the maximum principal stresses and the maximum in-plane shear stress at points D and E (Fig. P9.4-25b) in the cross section at $x = 5$ ft, where x is measured along member AB, as shown in Fig. P9.4-25a.

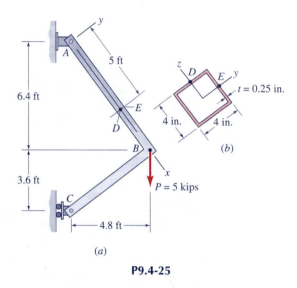

(a)

P9.4-25

following:

$$F = -285 \text{ lb}, V_y = -75 \text{ lb}, V_z = 0,$$

$$T = 3000 \text{ lb} \cdot \text{in.}, M_y = -5000 \text{ lb} \cdot \text{in.}, M_z = -16,500 \text{ lb} \cdot \text{in.}$$

(a) Determine the angle θ that the resultant moment vector M in Fig. P9.4-26 makes with the z axis, and locate the neutral axis of the base cross section. (b) Referring to Section 6.6, determine the maximum tension and maximum compression values of normal stress, σ_x, on the cross section at the base of the signpost in Example 9.5. (c) Neglecting the shear stress on the base cross section due to the transverse shear force V_y, use Mohr's circle to determine the maximum tensile normal stress at the base of the signpost. Indicate the location of the point in the base cross section where this stress occurs.

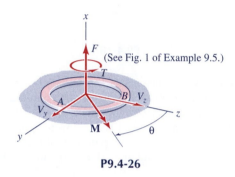

P9.4-26

Problems 9.4-27 and 9.4-28 *are to be solved by use of the Torsion module and the Mohr's Circle module of MDSolids.*

*Prob. 9.4-26. At the base of the cantilevered sign in Fig. 1 of Example 9.5, the stress resultants were found to be the

Prob. 9.4-27. Use **MDSolids** to solve Prob. 9.4-8.
Prob. 9.4-28. Use **MDSolids** to solve Prob. 9.4-9.

Section		Suggested Review Problems
9.1	Chapter 9 illustrates the use of **Mohr's circle** to handle combined stresses.	
9.2	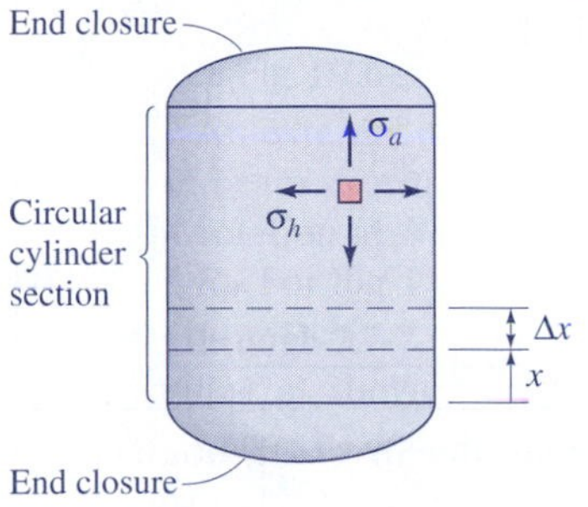 The subject of **Thin-wall Pressure Vessels** is treated in this chapter because shells are subjected to stresses that act in multiple directions. For example, for a **thin cylindrical shell with closed ends,** the stresses are referred to as the *axial stress,* σ_a, and the *hoop stress,* σ_h, as illustrated in Fig. 9.2a. Circular cylinder with end closures. (Fig. 9.2a) The **axial stress** is given by $$\sigma_a = \frac{pr}{2t} \qquad (9.2)$$ The **hoop stress** is given by $$\sigma_h = \frac{pr}{t} \qquad (9.4)$$	9.2-4 9.2-7 9.2-13
	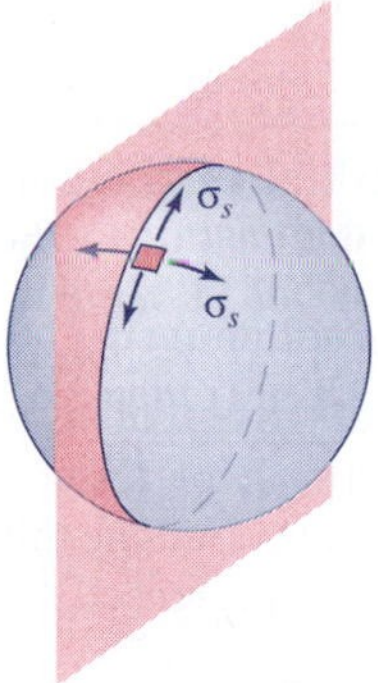 Section 9.2 also treats **thin spherical shells,** which have the same normal stress σ_s that acts in every direction in the shell wall, as illustrated in Fig. 9.4a. A thin spherical shell. (Fig. 9.4a) The **spherical-shell stress** is given by $$\sigma_s = \frac{pr}{2t} \qquad (9.6)$$	Derive Eq. 9.6.

Section		Suggested Review Problems
	Mohr's circles can be employed to determine in-plane shear stress values and absolute maximum shear stress values for cylindrical shells and spherical shells. Mohr's circles are also useful for determining normal stresses and shear stresses along lines that make and angle with the axis of a cylindrical shell.	
	  Mohr's circles for in-plane and out-of-plane stresses. (Fig. 9.5) $$\left(\tau_{\max}^{abs}\right)_{cyl} = \frac{pr}{2t}$$ $$\left(\tau_{\max}^{abs}\right)_{sph} = \frac{pr}{4t} \qquad (9.7)$$	Derive Eqs. 9.7.
9.3	Section 9.3 treats the **Stress Distribution in Beams.** The *flexural stress*, $\sigma = -\dfrac{My}{I}$, and the *transverse shear stress*, $\tau = \dfrac{VQ}{It}$, vary from point to point in a beam. **Mohr's circles** can be employed to determine the magnitude and orientation of such quantities as the *maxim compressive stress,* the *maximum tensile stress,* and the *maximum shear stress* at each point in the beam.	9.3-1
	  Stress trajectories for a simply supported beam. (Fig. 9.10) Figure 9.10 illustrates the trajectories of maximum tension and maximum compression in a particular beam.	
9.4	Section 9.4 treats several special cases of **Stresses Due to Combined Loads.** this includes *Combined Axial Loading and Bending, Combined Axial Loading and Torsion,* and several cases of *General Combined Loading.* In all of these cases **Mohr's Circles** are used to combine the various stresses that are experienced by the structure.	9.4-3 9.4-9

BUCKLING OF COLUMNS

10.1 INTRODUCTION

What constitutes *failure* of a structure or a machine part? As an engineer you must consider several possible *modes of failure* when designing a structure or a machine component. For example, the stress must be kept small enough so that the component will not fail by yielding or by tensile fracture. It might also be important to limit the *deflection* of the component. Furthermore, if the part will be subjected to repeated cycles of loading, the stress must be limited in order to prevent failure by progressive fracture (called *fatigue failure*). To prevent the above types of failures, design criteria based on *strength* (stress) and *stiffness* (deflection) must be taken into consideration. Therefore, the preceding nine chapters were devoted to methods for calculating the stress distribution in, and the deflection of, members subjected to various types of loading. In this chapter we consider another important mode of failure, *buckling*.[1] Figure 10.1 shows the impressive shape into which a thin cylindrical shell deforms when subjected to its axial buckling load. Beams, plates, shells and other structural members may buckle under a variety of loading conditions.[2] The discussion in this chapter, however, is limited to compressive axial loading of slender members.

You can easily perform a buckling "experiment" by applying an axial load to a thin ruler or yard stick (meter stick). Figure 10.2a shows buckling of a thin rod (column), and Fig. 10.2b shows buckling of a truss member under compression. Weights are added until the *critical load* in the compression member, P_{cr}, is reached, and the member suddenly deflects laterally under axial compression. In the analysis of axial deformation in Chapters 2, 3, and 9, we implicitly assumed that, even under compressive loading, the member undergoing axial deformation remained straight, and that the only deformation was a shortening or lengthening of the member. However, as the ruler demonstration shows, at some value of compressive axial load the ruler no longer remains straight, but suddenly deflects laterally, bending

FIGURE 10.1 Buckling failure of an axially compressed thin cylindrical shell. (Photo by W. H. Horton; from Computerized Buckling Analysis of Shells, by D. Bushnell, 1985, With kind permission of Springer Science and Business Media)

---

[1]The term *buckling* will be defined later in this section; it is illustrated in Figs. 10.1 and 10.2.
[2]*Theory of Elastic Stability,* by S. P. Timoshenko and J. M. Gere, [Ref. 10-1]; and *Buckling of Bars, Plates, and Shells,* by D. O. Brush and B. O. Almroth. [Ref. 10-2], treat the buckling of several types of members under various types of loading. *Structural Stability—Theory and Implementation,* by W. F. Chen and E. M. Lui, [Ref. 10-3], presents both theory and design curves.

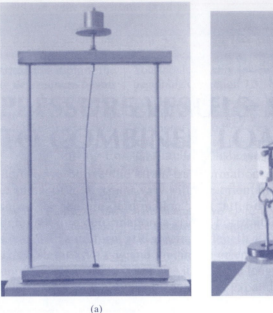
(a)

(b)

FIGURE 10.2 Buckling demonstrations. (Courtesy Roy Craig)

like a beam. This *lateral deflection* caused by *axial compression* is called **buckling.** Buckling failures are often sudden and catastrophic, which makes it all the more important for you to know how they can be prevented.

Stability of Equilibrium. In this section we will examine the basic phenomenon of buckling; then we will examine the buckling behavior of slender columns. In Chapter 1 we noted that static deformable-body-mechanics problems always involve equilibrium; now, however, we must look more closely at the topic of equilibrium and consider the **stability of equilibrium.** This concept can be demonstrated very easily by considering the equilibrium of a ball on three different surfaces, as illustrated in Fig. 10.3. In all three situations the solid-colored ball is in an equilibrium position, that is, it satisfies $\sum F_x = 0$, $\sum F_y = 0$, and $\sum M = 0$. In Fig. 10.3a the ball is said to be in *stable equilibrium* because, if it is slightly displaced to one side and then released, it will move back toward the equilibrium position at the bottom of the "valley." The ball on the top of the "hill" in Fig. 10.3c is also in an equilibrium configuration, because it also satisfies $\sum F_x = 0$, $\sum F_y = 0$, and $\sum M = 0$. In this case, however, the ball is in an *unstable-equilibrium* configuration; if slightly displaced to either side, the ball will tend to move farther from the equilibrium position

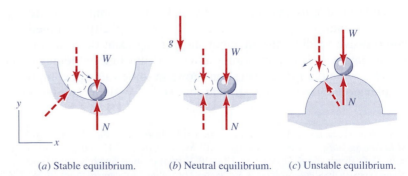

(*a*) Stable equilibrium.   (*b*) Neutral equilibrium.   (*c*) Unstable equilibrium.

FIGURE 10.3   Stability of equilibrium.

at the top of the "hill." Finally, if the ball is on a perfectly flat, level surface, as in Fig. 10.3*b*, it is said to be in a *neutral-equilibrium* configuration. If slightly displaced to either side, it has no tendency to move either away from or toward the original position, since it is in equilibrium in the displaced position as well as the original position.

Buckling. Now, let us see how stability of equilibrium applies to compression members where some form of elastic deformation is possible. In Fig. 10.4 the member *AB* is assumed to be perfectly straight and perfectly rigid, and the supporting pin at *A* is assumed to be frictionless. A force *P* is applied vertically downward at *B*, which is on the axis of the member. The torsional spring at *A* has a spring constant k_θ, so it produces a *restoring moment,* M_{Ar}, at *A* that is directly proportional to the angle of deflection of member *AB* from the vertical. That is,

$$M_{Ar} = k_\theta \theta \qquad (10.1)$$

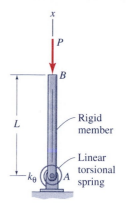

The vertical configuration of member *AB*, shown in Fig. 10.4*a*, is certainly an *equilibrium configuration*. The question is: For what values of *P* is the vertical equilibrium configuration stable? neutral? unstable? To answer this question, we explore what happens when the member *AB* is rotated to one side through a small angle θ, as shown in Fig. 10.4*b*. Let M_{Ad} be the *disturbing moment,* that is, the sum of all moments that tend to make the angle θ increase; and let M_{Ar} be the *restoring moment,* that is, the sum of all moments that tend to make θ decrease. The restoring moment, M_{Ar}, in Fig. 10.4*c* is given by Eq. 10.1; the disturbing moment is given by

(*a*) The vertical equilibrium configuration.

$$M_{Ad} = PL \sin \theta \qquad (10.2)$$

In analogy with the balls in Fig. 10.3, the system has the following requirements for stable equilibrium, neutral equilibrium, and unstable equilibrium, respectively:

Stable Equilibrium: $\qquad M_{Ad} < M_{Ar} \rightarrow PL \sin \theta < k_\theta \theta$

Neutral Equilibrium: $\qquad M_{Ad} = M_{Ar} \rightarrow PL \sin \theta = k_\theta \theta \qquad (10.3)$

Unstable Equilibrium: $\qquad M_{Ad} > M_{Ar} \rightarrow PL \sin \theta > k_\theta \theta$

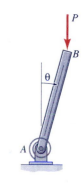

(*b*) A displaced configuration.

Since we are interested in behavior of the system at, and very near, the vertical configuration, we let $\theta \rightarrow 0$ in Eqs. 10.3 and use the approximation $\sin \theta \approx \theta$. Then, Eqs. 10.3 give

Stable Equilibrium: $\qquad P < P_{cr}$

Neutral Equilibrium: $\qquad P = P_{cr} \qquad (10.4)$

Unstable Equilibrium: $\qquad P > P_{cr}$

where

$$\boxed{P_{cr} = \frac{k_\theta}{L}} \qquad (10.5)$$

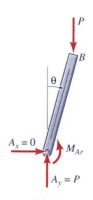

(*c*) A free-body diagram.

The value of the load at which the transition from stable equilibrium to unstable equilibrium occurs is called the **critical load**, P_{cr}. This loss of stability of equilibrium is called **buckling,** so we also call P_{cr} the *buckling load.*

FIGURE 10.4 A simplified model of column buckling.

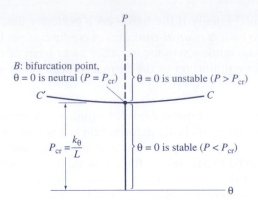

FIGURE 10.5 An equilibrium diagram for an idealized compression member.

Consider now how the buckling load of the compression member AB in Fig. 10.4 could be obtained by performing an experiment. We could grasp member AB (with compressive force P applied), rotate it to the left (or to the right) through a small angle, and then release it. If $P < P_{cr}$, member AB would return to the vertical ($\theta = 0$) equilibrium position. If, on the other hand, $P > P_{cr}$, the column would tend to move farther from the $\theta = 0$ position once it is slightly disturbed. The critical load would be the value of P for which there was no visible tendency for the member either to return to the $\theta = 0$ position or to move farther from it.

A useful way to illustrate the relationship between applied load and stability is the *equilibrium diagram* shown in Fig. 10.5. It is a plot of load P versus deflection angle θ. The point labeled B in Fig. 10.5, where the equilibrium diagram branches, is called the *bifurcation point*.[3] Above the bifurcation point the vertical (i.e., $\theta = 0$) configuration (shown dashed) is an unstable-equilibrium configuration; but there are alternative stable-equilibrium configurations along curves BC and BC', with $\theta \neq 0$. Right at point B, where $P = P_{cr}$, the equilibrium is neutral.

In the next section we will determine the critical load for a straight, uniform, axially loaded, linearly elastic, pin-ended column.

10.2 THE IDEAL PIN-ENDED COLUMN; EULER BUCKLING LOAD

To investigate the stability of real columns with distributed flexibility, as contrasted with the rigid member with torsional spring that was used in the previous section to model column behavior, we begin by considering the *ideal pin-ended column*, as illustrated in Fig. 10.6a.[4] We make the following simplifying assumptions:

- The column is initially perfectly straight, and it is made of linearly elastic material.

- The column is free to rotate, at its ends, about frictionless pins; that is, it is restrained like a simply supported beam. Each pin passes through the centroid of the cross section.

---

[3]The word *bifurcate* means to divide into two branches, or to fork.

[4]It is customary to refer to the pin-ended compression member being analyzed here as a column, and to illustrate it by a vertical member subjected to a downward load. This reflects the very important application to columns in buildings and other structures. However, the analysis applies equally to any pin-ended member, such as a member in a truss (Fig. 10.2b), so long as the member is in compression.

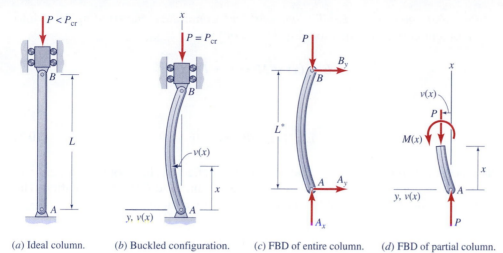

(a) Ideal column. (b) Buckled configuration. (c) FBD of entire column. (d) FBD of partial column.

FIGURE 10.6 Buckling of a pin-ended column.

- The column is symmetric about the xy plane, and any lateral deflection of the column takes place in the xy plane.
- The column is loaded by an axial compressive force P applied by the pins.

Buckled Configuration: If the axial load P is less than the critical load, P_{cr}, the column will remain straight, and it will shorten under a uniform (compressive) axial stress $\sigma = P/A$, as illustrated in Fig. 10.6a.[5] In the straight configuration with $P < P_{cr}$, the column is in *stable equilibrium*. However, if a load $P = P_{cr}$ is applied to the column, the straight configuration becomes a *neutral-equilibrium* configuration, and neighboring configurations, like the buckled shape in Fig. 10.6b, also satisfy equilibrium requirements. Therefore, to determine the value of the critical load, P_{cr}, and the shape of the buckled column, we will determine the value of the load P such that the (slightly) bent shape of the column in Fig. 10.6b is an equilibrium configuration.[6]

Equilibrium of the Buckled Column: First, using the free-body diagram of Fig. 10.6c, we get $A_x = P$ (from $\Sigma F_x = 0$). $A_y = 0$ [from $(\Sigma M)_B = 0$], and $B_y = 0$. Therefore, on the free-body diagram in Fig. 10.6d, we show only a vertical force P acting on the pin at A, and we show no horizontal force $V(x)$ at section x. The sign convention adopted for the moment $M(x)$ in Fig. 10.6d is the sign convention that was used in Chapter 6, namely, that a positive bending moment produces compression in the $+y$ fibers (which are on the left in Fig. 10.6d as a result of our choice of xy axes). From Fig. 10.6d we get

$$\left(\sum M\right)_A = 0: \qquad M(x) = -Pv(x) \qquad (10.6)$$

---

[5]Since column buckling always occurs under compressive normal stress, in the remainder of this chapter we will let σ denote compressive normal stress, digressing from the normal-stress sign convention that has been followed in previous chapters.
[6]This approach is called the *equilibrium method* or *bifurcation method*. There are several alternative ways to determine the critical load. See, for example, *Principles of Structural Stability,* by H. Ziegler, [Ref. 10-4], where four different methods are reviewed.

Differential Equation of Equilibrium, and End Conditions: Substituting $M(x)$ into the moment-curvature equation, Eq. 7.8, we obtain

$$EIv''(x) = M(x) = -Pv(x)$$

or

$$EIv''(x) + Pv(x) = 0 \tag{10.7}$$

This is the *differential equation* that governs the deflected shape of a pin-ended column. It is a homogenous, linear, second-order, ordinary differential equation. The *boundary conditions* (see Table 7.1) for the pin-ended member are

$$v(0) = 0, \quad v(L) = 0 \tag{10.8}$$

Solution of the Differential Equation: The presence of the $v(x)$ term in Eq. 10.7 means that we cannot simply integrate twice to get the solution, as was done in Chapter 7. In fact, only when $EI = $ const is there a simple solution to Eq. 10.7. Therefore, for the remainder of this chapter we will consider only uniform columns.[7] Then Eq. 10.7 is an ordinary differential equation with constant coefficients. Let

$$\lambda^2 = \frac{P}{EI} \tag{10.9}$$

For the uniform column, therefore, Eq. 10.7 becomes

$$v'' + \lambda^2 v = 0 \tag{10.10}$$

The *general solution* to this homogeneous equation is

$$v(x) = C_1 \sin \lambda x + C_2 \cos \lambda x \tag{10.11}$$

We seek a value of λ and constants of integration C_1 and C_2 such that the two boundary conditions of Eqs. 10.8 are satisfied. Thus,

$$v(0) = 0 \rightarrow C_2 = 0$$
$$v(L) = 0 \rightarrow C_1 \sin(\lambda L) = 0 \tag{10.12}$$

Obviously, if we make both C_1 and C_2 equal to zero, the deflection $v(x)$ is zero everywhere (Eq. 10.11), and we just have the original straight configuration. If we want an alternative equilibrium configuration, like the one in Fig. 10.6b, we must pick a value of λ that satisfies Eq. 10.12b with $C_1 \neq 0$, that is, λ must satisfy the *characteristic equation*

$$\sin(\lambda_n L) = 0 \rightarrow \lambda_n = \left(\frac{n\pi}{L}\right), \qquad n = 1, 2, \ldots \tag{10.13}$$

---

[7]Tapered columns are treated in Chapter 9 of *Guide to Stability Design Criteria for Metal Structures*, ed. by T. V. Galambos, [Ref. 10-5].

Combining Eqs. 10.9 and 10.13 gives the following formula for the possible buckling loads:

$$P_n = \frac{n^2 \pi^2 EI}{L^2} \qquad (10.14)$$

The deflection curve that corresponds to each load P_n is obtained by combining Eqs. 10.11 through 10.13 to get

$$v(x) = C \sin\left(\frac{n\pi x}{L}\right) \qquad (10.15)$$

The function that represents the shape of the deflected column is called a *mode shape*, or *buckling mode*. The constant C, which determines the direction (sign) and amplitude of the deflection, is arbitrary, but it must be small. The existence of neighboring equilibrium configurations is analogous to the fact that the ball in Fig. 10.3b can be placed at neighboring locations on the flat, horizontal surface and still be in equilibrium.

The value of P at which buckling will actually occur is obviously the smallest value given by Eq. 10.14 (i.e., $n = 1$). Thus, the *critical load* is

$$\boxed{P_{cr} = \frac{\pi^2 EI}{L^2}} \qquad \begin{array}{l}\textbf{Euler}\\\textbf{Buckling Load}\end{array} \qquad (10.16)$$

and the corresponding **buckling mode** is

$$\boxed{v(x) = C \sin\left(\frac{\pi x}{L}\right)} \qquad \begin{array}{l}\textbf{Buckling}\\\textbf{Mode}\end{array} \qquad (10.17)$$

as illustrated in Fig. 10.7b. The critical load for an ideal column is known as the **Euler buckling load,** after the famous Swiss mathematician Leonhard Euler (1707–1783), who was the first to establish a theory of buckling of columns.[8]

The buckling mode of Eq. 10.17 is sometimes called the *fundamental* (or first) *buckling mode*. Although the column could theoretically buckle in the *second buckling mode*, illustrated in Fig. 10.7c, if a load $P_{cr2} = 4\pi^2 EI/L^2$ were to be applied, this could only happen if there were some lateral bracing at $x = L/2$ to prevent the column from buckling in the first mode (Fig. 10.7b) at the much smaller Euler load of $P_{cr} = \pi^2 EI/L^2$.

Let us examine some important implications of the Euler buckling-load formula, Eq. 10.16. We can express this in terms of **critical** (buckling) **stress.**

$$\sigma_{cr} \equiv \frac{P_{cr}}{A} = \frac{\pi^2 E(Ar^2)}{AL^2}$$

or

$$\boxed{\sigma_{cr} = \frac{\pi^2 E}{(L/r)^2}} \qquad \begin{array}{l}\textbf{Euler}\\\textbf{Buckling Stress}\end{array} \qquad (10.18)$$

(a) Undeflected column.

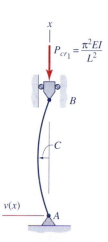

(b) First buckling mode ($n = 1$).

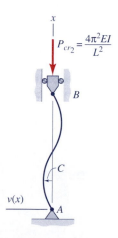

(c) Second buckling mode ($n = 2$).

FIGURE 10.7   Two examples of buckling modes.

---

[8]See Footnote 2 of Chapter 6.

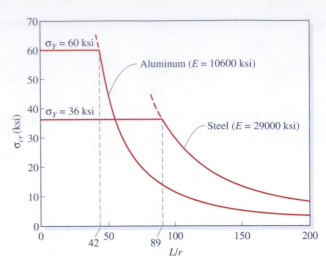

FIGURE 10.8 Graphs of Euler's formula for structural steel and for an aluminum alloy.

where

σ_{cr} = the critical (elastic buckling) stress.

E = the modulus of elasticity.

$r = \sqrt{I/A}$ = the radius of gyration.

L = the length of the member between supports.

The quantity L/r is called the **slenderness ratio** of the column. Curves of σ_{cr} versus L/r for structural steel and for an aluminum alloy are plotted in Fig. 10.8.

From Eq. 10.18 and from Fig. 10.8 we observe the following characteristics of *elastic buckling* of ideal columns:

(a)

(b)

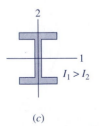

(c)

FIGURE 10.9 Column cross sections.

- The only material property that enters directly into either the elastic buckling load or the buckling stress is the modulus of elasticity, E, which represents the <u>stiffness</u> of the material. Therefore, one way to increase the elastic buckling load of a member would be to use a member that is made of material with a higher E value.

- The elastic buckling load is inversely proportional to the square of the length of the column. Figure 10.8 illustrates this length effect.

- Euler's formula is valid only for "long" columns, that is, columns whose L/r ratio leads to a critical stress below the compressive proportional limit, σ_{PL}. (Since σ_{PL} is generally not available, the compressive yield stress σ_Y is usually substituted for it.) The values of L/r marking the limit of validity of Euler's formula for steel and for an aluminum alloy are illustrated in Fig. 10.8.

- The buckling load can be increased by increasing the value of the cross-sectional moment of inertia, I. This can be done, without increasing the cross-sectional area, by using thin-wall tubular members (e.g., Figs. 10.9a, b). However, if the column wall is too thin, *local buckling* can occur. (Figure 10.1 illustrates buckling of a short, thin-wall member under compression.)

- If the principal moments of inertia of the column cross section are unequal, as illustrated in Fig. 10.9c, the column will buckle about the axis of the cross section that has the least moment of inertia, unless boundary restraints or intermediate bracing force it to do otherwise (see Section 10.3).

- If the slenderness ratio is very large, say $L/r > 200$, the stress at buckling will be very small. Therefore, the strength of the material is underutilized. The design should be modified by, for example, adding lateral bracing or changing the boundary conditions (see Section 10.3).

In the sections that follow, we will explore the following exceptions to the buckling behavior treated in this section: other boundary conditions (Section 10.3), eccentric loading (Section 10.4), imperfections (Section 10.5), and inelastic buckling (Section 10.6).

EXAMPLE 10.1

What is the maximum compressive load that can be applied to an aluminum-alloy compression member (Fig. 1) of length $L = 4$ m if the member is loaded in a manner that permits free rotation at its ends and if a factor of safety of 1.5 against buckling failure is to be applied? Recall, from Eq. 2.26, that

$E = 70$ GPa
$\sigma_Y = 270$ MPa
$r_o = 45$ mm
$r_i = 40$ mm

Fig. 1 Cross section.

$$FS = \frac{\text{failure load}}{\text{allowable load}}$$

Plan the Solution   The allowable load is the critical load divided by the factor of safety. Since the ends of the member are free to rotate (pin-ended), the critical load is determined by Euler's formula. Eq. 10.16, provided that the corresponding stress is less than the yield stress of the material.

Solution   From Eq. 10.16,

$$P_{cr} = \frac{\pi^2 EI}{L^2}$$

$$= \frac{\pi^2 (70 \times 10^9 \text{ N/m}^2)(\frac{\pi}{4})[(0.045)^4 - (0.040)^4] \text{ m}^4}{(4 \text{ m})^2} \tag{1}$$

$$P_{cr} = 52.2 \text{ kN} \qquad\qquad \textbf{Ans.}$$

The corresponding average compressive stress in the member is

$$\sigma_{cr} = \frac{P_{cr}}{A} = \frac{52.2 \text{ kN}}{\pi[(0.045)^2 - (0.040)^2] \text{ m}^2}$$

$$\sigma_{cr} = 39.1 \text{ MPa} < \sigma_Y$$

so this member will undergo elastic buckling. (For this column, $L/r = 133$.)
Finally, the allowable load is

$$P_{allow} = \frac{P_{cr}}{FS} = \frac{52.2 \text{ kN}}{1.5} = 34.8 \text{ kN} \qquad\qquad \textbf{Ans.}$$

10.3 THE EFFECT OF END CONDITIONS ON COLUMN BUCKLING

Rarely, if ever, is a compression load actually applied to a member through frictionless pins. For example, the column might be bolted to a heavy base at the bottom and framed into other members at the top, as illustrated in Fig. 10.10*a*. However, an understanding of the effect of *idealized support conditions,* like those illustrated in Figs. 10.10*b* through 10.10*d*, enables an engineer to estimate the effect that actual end conditions, like those in Fig. 10.10*a*, would have on the buckling load of a real column.

We will begin by deriving an expression for the elastic buckling load of the fixed-pinned column in Fig. 10.10*d*. Then, we will indicate how the concept of the *effective length* of a column can be used to obtain the buckling load of columns with various end conditions.

Buckling Load of an Ideal Fixed-Pinned Column.
We will assume the same ideal conditions that were listed at the beginning of Section 10.2, except that the pin at end *A* of the column is replaced by complete restraint of that end.

Buckled Configuration: To assist us in drawing free-body diagrams of the column in a buckled equilibrium configuration near the straight equilibrium configuration (Fig. 10.11*a*), we sketch a feasible shape of the column that satisfies the prescribed fixed-pinned end conditions (Fig. 10.11*b*).

Equilibrium of the Buckled Column: By examining Fig. 10.11*b*, we can see that the curvature at *A* corresponds to a moment M_A in the sense shown in Fig. 10.11*c*. From $\Sigma F_x = 0$, we get $A_x = P$, and from $(\Sigma M)_A = 0$, we see that the pin at *B* must exert a horizontal force H_B, as indicated on Fig. 10.11*c*. Now we can construct a free-body diagram of the column below section *x* or above section *x*. We have done the latter in Fig. 10.11*d*, from which we get

$$\left(\sum M\right)_O = 0: \qquad M(x) = (L - x)H_B - Pv(x) \qquad (10.19)$$

| (*a*) Actual column. | (*b*) Fixed-free ("flag pole") column. | (*c*) Fixed-fixed column. | (*d*) Fixed-pinned column. |

FIGURE 10.10 Various column end conditions.

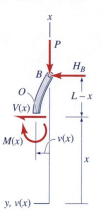

(a) Unbuckled
column.

(b) Buckled shape
of column.

(c) FBD of entire
column.

(d) FBD of partial
column.

FIGURE 10.11   A fixed-
pinned column.

Differential Equation and End Conditions: Substituting $M(x)$ from Eq. 10.19 into the moment-curvature equation, Eq. 7.8, we get

$$EIv''(x) + Pv(x) = H_B L - H_B x \qquad (10.20)$$

As in Section 10.2, let us consider only uniform columns, and let us employ the definition of λ in Eq. 10.9. Then, Eq. 10.20 can be written in the form

$$v'' + \lambda^2 v = \frac{H_B L}{EI} - \frac{H_B x}{EI} \qquad (10.21)$$

Instead of the homogeneous differential equation that we got for the pinned-pinned column (Eq. 10.10), we have obtained a nonhomogeneous, linear, second-order, ordinary differential equation with constant coefficients.

The boundary conditions of the fixed-pinned column are

$$v(0) = 0, \quad v'(0) = 0, \quad v(L) = 0 \qquad (10.22)$$

Solution of the Differential Equation: The solution of Eq. 10.21 under the end conditions of Eq. 10.22 consists of a *complementary solution* plus a *particular solution*. To get the complementary solution, we set the right-hand side of Eq. 10.21 to zero. But this just gives us Eq. 10.10, so we can use Eq. 10.11 as the complementary solution. Since the right-hand side of Eq. 10.21 consists of a constant term and a term that is linear in x, let us try the following particular solution:

$$v_p(x) = C_3 + C_4 x \qquad (10.23)$$

Substituting this into Eq. 10.21, noting that $v_p''(x) = 0$, and recalling that $\lambda^2 = P/EI$, we get

$$\frac{P}{EI}(C_3 + C_4 x) = \frac{H_B L}{EI} - \frac{H_B x}{EI} \qquad (10.24)$$

Therefore, the particular solution is

$$v_p(x) = \frac{H_B L}{P} - \frac{H_B x}{P} \tag{10.25}$$

Finally, the complete *general solution* of Eq. 10.21 is

$$v(x) = \frac{H_B L}{P} - \frac{H_B x}{P} + C_1 \sin \lambda x + C_2 \cos \lambda x \tag{10.26}$$

We have three end conditions, Eqs. 10.22, to be used to determine the four constants — λ, H_B, C_1, and C_2.

$$v(0) = 0 \rightarrow C_2 = -\frac{H_B L}{P}$$

$$v'(0) = 0 \rightarrow C_1 = \frac{H_B}{\lambda P} \tag{10.27}$$

$$v(L) = 0 \rightarrow C_1 \sin \lambda L + C_2 \cos \lambda L = 0$$

Equations 10.27 can be combined to give

$$C_1 [\sin \lambda L - \lambda L \cos \lambda L] = 0 \tag{10.28}$$

This replaces the much simpler condition, Eq. 10.12b, that we got for the pinned-pinned column. Again, there are two solutions, but the solution $C_1 = 0$ leads to $H_B = C_2 = 0$, so we get the "trivial" solution of the straight equilibrium configuration, $v(x) \equiv 0$.

Alternate equilibrium configurations are possible, however, if λ satisfies the following equation:

$$\sin(\lambda_n L) - (\lambda_n L) \cos(\lambda_n L) = 0$$

or

$$\tan(\lambda_n L) = \lambda_n L, \qquad n = 1, 2, \ldots \tag{10.29}$$

This equation is called the *characteristic equation*. It has an infinite number of solutions, but, as in the case of the pinned-pinned column, we are only interested in the smallest value of λL that satisfies Eq. 10.29. One way to solve Eq. 10.29 is to plot $f(\lambda L) \equiv \tan(\lambda L)$ versus λL, and $g(\lambda L) \equiv \lambda L$ versus λL. The smallest value of λL where the curves $f(\lambda L)$ and $g(\lambda L)$ intersect is

$$\lambda_1 L = 4.4934 \tag{10.30}$$

Combining this with Eq. 10.9, we get

$$P_{\text{cr}} = (20.19) \frac{EI}{L^2} \tag{10.31}$$

Thus, replacing the pin at one end of a column by complete restraint raises the buckling load significantly—from the Euler load of Eq. 10.16 to the value given in Eq. 10.31. That is, the buckling load is increased by

$$\left(\frac{20.19 - \pi^2}{\pi^2}\right)(100\%) = 105\%$$

By comparing Eqs. 10.16 and 10.31, we can see that the elastic buckling load of any column can be expressed as some constant times the factor (EI/L^2). Thus, all of the comments regarding the effects of the parameters E, I, and L on the buckling of pinned-pinned columns also hold true for columns with other end conditions.

To get the fundamental buckling-mode shape of the fixed-pinned column, we can combine Eqs. 10.26, 10.27, and 10.30 to get

$$v(x) = C\left\{\sin\left(\frac{4.493x}{L}\right) + 4.493\left[1 - \left(\frac{x}{L}\right) - \cos\left(\frac{4.493x}{L}\right)\right]\right\} \quad (10.32)$$

This mode-shape expression, which is plotted in Fig. 10.12, is not nearly as simple as the $\sin(\pi x/L)$ buckling mode of the pinned-pinned column (Eq. 10.17). However, we again find the *shape* is specified, but the amplitude coefficient is arbitrary.

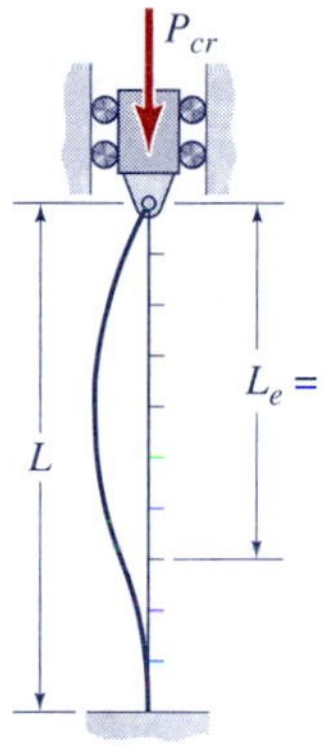

Effective Length of Columns. The Euler buckling load, given by Eq. 10.16, was developed for the ideal pinned-pinned column. We have just found, however, that a change of end conditions leads to an expression for the buckling load that differs from that in the Euler formula only in the value of a multiplicative constant. Therefore, the Euler formula can be extended to give the **elastic buckling load of columns with arbitrary end conditions** if we write it as

$$P_{cr} = \frac{\pi^2 EI}{L_e^2} \qquad \text{Elastic Buckling Load} \qquad (10.33)$$

FIGURE 10.12 The fundamental buckling mode of a fixed-pinned column.

where L_e is the **effective length** of the column. That is, L_e is the length of a pin-ended column having the same buckling load as the actual column. For example, for the fixed-pinned column, we equate the expression for P_{cr} in Eq. 10.31, to the expression in Eq. 10.33 and get

$$\frac{\pi^2}{L_e^2} = \frac{20.19}{L^2}$$

or

$$L_e = 0.70L$$

This effective length of the fixed-pinned column is indicated in Fig. 10.12.

Physically, the effective length of a column is the distance between points of zero moment when the column is deflected in its fundamental elastic buckling mode. Figure 10.13 illustrates the effective lengths of columns with several types of end conditions. (Homework Problem 10.3-13 is a derivation based on Fig. 10.13*e*;

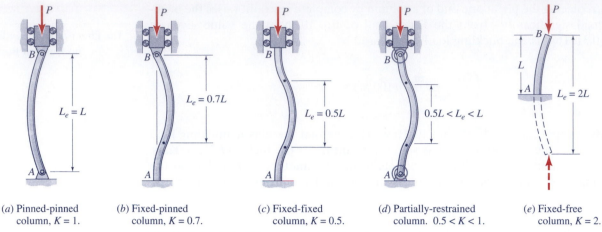

| (a) Pinned-pinned column, K = 1. | (b) Fixed-pinned column, K = 0.7. | (c) Fixed-fixed column, K = 0.5. | (d) Partially-restrained column. 0.5 < K < 1. | (e) Fixed-free column, K = 2. |

FIGURE 10.13 The effective length of various columns.

Problem 10.3-14 is a similar derivation.) Some design codes employ a dimensionless coefficient K, called the *effective-length factor,* where

$$L_e \equiv KL \tag{10.34}$$

Then the *elastic buckling load* is given by

$$P_{cr} = \frac{\pi^2 EI}{(KL)^2} \tag{10.35}$$

The appropriate value of K is listed for each column shown in Fig. 10.13. From Eq. 10.35 we obtain the following expression for the *elastic buckling stress:*

$$\sigma_{cr} = \frac{\pi^2 E}{\left(\dfrac{KL}{r}\right)^2} \qquad \text{**Elastic Buckling Stress**} \tag{10.36}$$

where (KL/r) is the **effective slenderness ratio** of the compression member.

➤ **EXAMPLE 10.2** ◄

A stiff beam BC (assumed to be rigid) is supported by two identical columns whose flexural rigidity is EI (for bending in the xz plane). Assuming that the columns are prevented from rotating at either end by this arrangement and that sidesway is permitted, as illustrated in Fig. 1b, *estimate* the elastic buckling load, P_{cr}, by estimating the effective length of the columns and using Eq. 10.33.

Plan The Solution The problem statement asks us to estimate the effective length and then use Eq. 10.33. We can compare the columns in this problem to the effective-length column samples in Fig. 10.13.

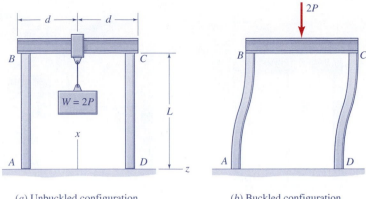

(a) Unbuckled configuration.

(b) Buckled configuration.

Fig. 1

Solution In terms of effective length, the elastic buckling load is given by Eq. 10.33:

$$P_{cr} = \frac{\pi^2 EI}{L_e^2} \tag{1}$$

Figure 1b bears some similarity to Fig. 10.13c, the fixed-fixed column. If we reflect the columns about their tops, as shown in Fig. 2, we have the equivalent of the fixed-fixed column situation of Fig. 10.13c. Therefore,

$$L_e = 0.5(2L) = L \tag{2}$$

so, the critical load for each column is just the Euler buckling load,

$$P_{cr} = \frac{\pi^2 EI}{L^2} \qquad \textbf{Ans.} \quad (3)$$

and the effective-length factor is $K = 1$.

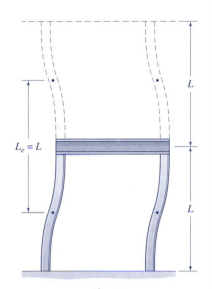

Fig. 2

━━━━━━━━━━━━➤ **EXAMPLE 10.3** ◀━━━━━━━━━

In Example 10.2, buckling of columns AB and CD in the xz plane, as illustrated in Fig. 1a, was considered. However, suppose that there is nothing to prevent the columns from buckling in the y direction, as illustrated in Fig. 1b. For the frame in Fig. 1, determine whether the **W**6 × 20 columns AB and CD will buckle in the xz plane (y axis buckling), or whether they will buckle in the y direction (z axis buckling), and determine the buckling load. Assume that the joints at B and C are rigidly welded joints, that the beam BC is rigid, and that it applies a vertical load P at the centroid of the top of each column.

Let $E = 29(10^3)$ksi, $\sigma_Y = 36$ ksi, $I_y = 13.3$ in^4, $I_z = 41.4$ in^4, $A = 5.87$ in^2, and $L = 16$ ft.

649

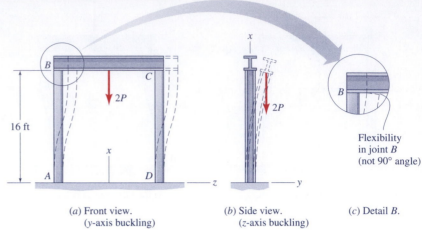

(a) Front view.　　　　　(b) Side view.　　　　　(c) Detail B.
(y-axis buckling)　　　　　(z-axis buckling)

Fig. 1　Two possible buckling modes.

Solution　From Eq. 10.35,

$$P_{cr} = \frac{\pi^2 EI}{(KL)^2} \tag{1}$$

Let P_{cr_y} be the critical load for buckling in the xz plane as illustrated in Fig. 1a, and let P_{cr_z} be the critical load for the buckling mode illustrated in Fig. 1b. Then,

$$P_{cr_y} = \frac{\pi^2 EI_y}{(K_y L)^2}, \qquad P_{cr_z} = \frac{\pi^2 EI_z}{(K_z L)^2} \tag{2}$$

where, from Example 10.2, $K_y = 1$, and from Fig. 10.13e, $K_z = 2$. Then,

$$P_{cr_y} = \frac{\pi^2 [29(10^3)\ \text{ksi}] (13.3\ \text{in}^4)}{[1(16\ \text{ft})(12\ \text{in./ft})]^2} = 103.3\ \text{kips} \tag{3a}$$

$$P_{cr_z} = \frac{\pi^2 [29(10^3)\ \text{ksi}] (41.4\ \text{in}^4)}{[2(16\ \text{ft})(12\ \text{in./ft})]^2} = 80.4\ \text{kips} \tag{3b}$$

Since $P_{cr_z} < P_{cr_y}$, the frame will buckle in the out-of-plane mode indicated in Fig. 1b at a buckling load of

$$P_{cr} = (P_{cr_z})_{cant} = 80.4\ \text{kips} \qquad \textbf{Ans.} \quad (4)$$

Review the Solution　In the problem statement you were told to assume that beam BC is rigid and that the joints at B and C are rigidly welded (i.e., that the angles between the columns and the beam BC remain 90°). Before accepting the value of P_{cr} in Eq. (4) as the true buckling load, let us reconsider these assumptions. Since a real beam BC would not be perfectly rigid, and since the joints at B and C may actually be flexible enough to allow the respective columns to rotate some at B and C relative to the beam BC, as indicated in Fig. 1c, we should take these possibilities into account. A "worst case" assumption would be that the columns are pinned, not fixed, to the beam BC. Since the beam BC is free to translate horizontally, the value of K_y for a cantilever (fixed-free) column would apply; that is, $K_y = 2$. Then, Eq. (3a) would become

$$(P_{cr_y})_{\text{cant}} = \frac{\pi^2 [29(10^3)\text{ ksi}](13.3\text{ in}^4)}{[2(16\text{ ft})(12\text{ in./ft})]^2} = 25.8\text{ kips} \tag{5}$$

By comparing Eqs. (4) and (5) we see the importance of correctly characterizing the end conditions of a column and applying an appropriate factor of safety to account for uncertainty in the end conditions. The most conservative value of buckling load for columns AB and CD is given by Eq. (5), that is

$$P_{cr} = (P_{cr_y})_{\text{cant}} = 25.8\text{ kips}$$

The average compressive stress at this load is

$$\sigma_{cr} = \frac{P_{cr}}{A} = \frac{25.8\text{ kips}}{5.87\text{ in}^2} = 4.40\text{ ksi} < 36\text{ ksi}$$

so the assumption of elastic buckling is valid.

MDS10.3 **Effective Length of Columns**

*10.4 ECCENTRIC LOADING; THE SECANT FORMULA

In Sections 10.2 and 10.3 we considered ideal columns, that is, columns that are initially perfectly straight and whose compressive load is applied through the centroid of the cross section of the member. Such ideal conditions never exist in reality, since perfectly straight structural members cannot be fabricated, and since the point of application of the load seldom, if ever, lies exactly at the centroid of the cross section.

Beam-Column Behavior.[9] Figure 10.14a shows a column with eccentric load applied through a bracket. We will analyze the "pinned-pinned" column with eccentric loading shown in Fig. 10.14b.[10] When the eccentricity, e, is zero, we get the Euler column. When $e \neq 0$, we use the free-body diagram in Fig. 10.14c to get

$$\left(\sum M \right)_A = 0: \qquad M(x) = -P[e + v(x)] \tag{10.37}$$

---

[9]A straight member that is simultaneously subjected to axial compression and lateral bending is called a *beam-column*.
[10]The pin supports at A and B are omitted to simplify this figure.

which is substituted into the moment-curvature equation, Eq. 7.8, to give

$$EIv''(x) + Pv(x) = -Pe \qquad (10.38)$$

As in Eq. 10.9, we consider a uniform member and let $\lambda^2 = P/EI$. Then, Eq. 10.38 may be written as

$$v'' + \lambda^2 v = -\lambda^2 e \qquad (10.39)$$

The particular solution of this differential equation is $v_p(x) = -e = \text{const}$, so the general solution is

$$v(x) = C_1 \sin \lambda x + C_2 \cos \lambda x - e \qquad (10.40)$$

The boundary conditions $v(0) = v(L) = 0$ are used to evaluate the constants in Eq. 10.40.

$$v(0) = 0 \rightarrow C_2 = e$$

$$v(L) = 0 \rightarrow C_1 \sin(\lambda L) + e[\cos(\lambda L) - 1] = 0$$

or, since $(1 - \cos \phi) = 2 \sin^2(\phi/2)$ and $\sin \phi = 2 \sin (\phi/2) \cos (\phi/2)$,

$$C_1 = e \tan\left(\frac{\lambda L}{2}\right)$$

Then,

$$v(x) = e\left[\tan\left(\frac{\lambda L}{2}\right)\sin(\lambda x) + \cos(\lambda x) - 1\right] \qquad (10.41)$$

As indicated in Fig. 10.14*b*, the maximum deflection of the beam-column occurs at $x = L/2$. Its value is

$$v_{\max} = v(L/2) = e[\sec(\lambda L/2) - 1] \qquad (10.42)$$

Unlike an Euler column, which deflects laterally only if P equals or exceeds the Euler buckling load, P_{cr}, lateral deflection of an eccentrically loaded member occurs for any value of load P. It is convenient to illustrate this beam-column deflection by plotting Eq. 10.42, written as

$$v_{\max} = e\left[\sec\left(\frac{\pi}{2}\sqrt{\frac{P}{P_{\mathrm{cr}}}}\right) - 1\right] \qquad (10.43)$$

for several values of e (Fig. 10.15). As P approaches the Euler load, P_{cr}, the lateral deflection of the beam-column increases without bound. In the limit as $e \rightarrow 0$, the curve becomes two straight lines that represent the straight configuration ($P < P_{\mathrm{cr}}$) and the buckled configuration ($P = P_{\mathrm{cr}}$).

The above beam-column analysis is valid only as long as the (compressive) stress does not exceed the compressive proportional limit. It is important to note

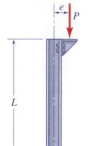

(*a*) Cantilever column.

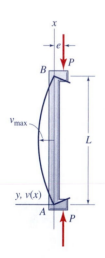

(*b*) Pinned-pinned column.

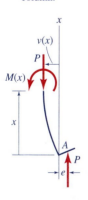

(*c*) Free-body diagram.

FIGURE 10.14 Eccentrically loaded columns.

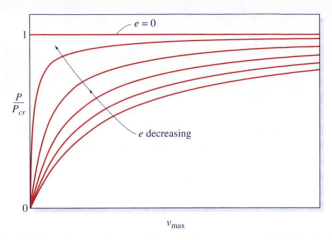

FIGURE 10.15 Load-deflection diagram for an eccentrically loaded column.

that the transverse deflection v is a *nonlinear function* of the load P, even though it depends linearly on the eccentricity e. Thus,

$$v_{\max}(P_1 + P_2) \neq v_{\max}(P_1) + v_{\max}(P_2)$$

as can be seen by examining Eq. 10.43 or Fig. 10.15. This nonlinear load-deflection relationship results from the fact that the bending moment $M(x)$ of the beam-column depends on the deflection $v(x)$, as given by Eq. 10.37. By contrast, the beam deflection produced by lateral loads or by applied couples varies linearly with the value of each load (Chapter 7), since there is no v-term in the bending-moment expression for those types of loading.

A similar analysis of the beam-column behavior of an eccentrically loaded cantilever column, like the one in Fig. 10.14*a*, leads to expressions for $v(x)$ and $v_{\max}$ that are just Eqs. 10.41 and 10.42, respectively, with the length L in these two equations replaced by the effective length of a cantilever column, $L_e = 2L$. (See Homework Prob. 10.4-1.) The behavior of eccentrically loaded columns with other boundary conditions (e.g., those in Figs. 10.13*b*, 10.13*c*, and 10.13*d*) cannot be obtained by simply substituting the effective length L_e for L in these equations.

Secant Formula A member under beam-column action is subjected to a combination of compressive axial force P and bending moment $M(x)$, as indicated by the free-body diagram of Fig. 10.14*c*. The maximum (magnitude) moment occurs at $x = L/2$ and is obtained by combining Eqs. 10.37 and 10.42 in the manner indicated in Fig. 10.16. Thus,

$$M_{\max} = |M(L/2)| = Pe \sec\left(\frac{\lambda L}{2}\right) \tag{10.44}$$

The maximum compressive stress is therefore

$$\sigma_{\max} = \frac{P}{A} + \frac{M_{\max}c}{I} \tag{10.45}$$

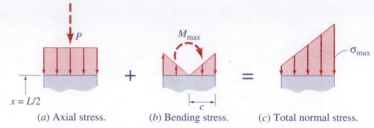

$x = L/2$		c	
(*a*) Axial stress.	(*b*) Bending stress.		(*c*) Total normal stress.

FIGURE 10.16   Superposition of beam-column stresses.

Combining Eqs. 10.44 and 10.45, and recalling the definition of λ in Eq. 10.9, we get

$$\sigma_{\max} = \frac{P}{A}\left[1 + \left(\frac{ec}{r^2}\right)\sec\left(\frac{L}{2r}\sqrt{\frac{P}{AE}}\right)\right] \qquad \textbf{Secant}\ \textbf{Formula} \qquad (10.46)$$

where

$\sigma_{\max}$ = the maximum compressive beam-column stress.

P = the (eccentric) axial compressive load.

A = the cross-sectional area of the compression member.

e = the eccentricity of the load.

c = the distance from the centroid to the extreme fiber where $\sigma_{\max}$ acts (Fig. 10.16*b*).

E = the modulus of elasticity.

I = the area moment of inertia about the centroidal bending axis.

$r = \sqrt{I/A}$ = the radius of gyration.

L = the length of the member.

Equation 10.46 is called the **secant formula.** Although it was derived for simply supported columns, it also holds for cantilever columns, like the one in Fig. 10.14*a*, if the length L in Eq. 10.46 is replaced by the effective length of the cantilever column, $L_e = 2L$. (See Homework Prob. 10.4-1.)

To determine the maximum compressive load that can be applied at a given eccentricity to a column of given length and given material, without causing yielding of the material, we can set $\sigma_{\max} = \sigma_Y$, the yield point in compression, and solve Eq. 10.46 numerically for P/A, the average stress.[11] Figure 10.17 is a plot of the average stress P/A versus the *slenderness ratio* L/r for structural steel for several values of the *eccentricity ratio* ec/r^2. For all curves on Fig. 10.17 except the curve labeled *Euler's formula,* $\sigma_{\max} = \sigma_Y$, and therefore $P = P_Y$, the load at which yielding first occurs.

If $e = 0$ and $\sigma_{\max} = \sigma_Y$, Eq. 10.46 simply gives $\dfrac{P}{A} = \sigma_Y$, and this plots as a horizontal line at stress σ_Y. However, long columns with zero eccentricity buckle

---

[11]We assume that the material remains linearly elastic up to the stress σ_Y, that is, we assume that $\sigma_{PL} = \sigma_Y$.

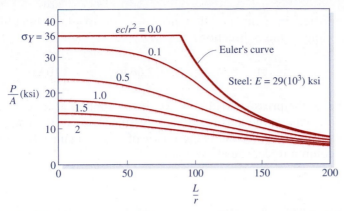

FIGURE 10.17 The average stress, P/A, corresponding to $\sigma_{max} = \sigma_Y$, based on the secant formula (Eq. 10.46).

elastically at the Euler critical load, and their maximum stress is less than σ_Y. Thus, the Euler load is an upper bound on the value of P for long columns, and therefore the Euler curve is shown in Fig. 10.17.[12]

Clearly, the secant formula expresses a nonlinear relationship between the applied compressive load P and the maximum compressive stress σ_{max}. Therefore, the principal of superposition does not apply to the calculation of the stress in beam-columns.[13]

PROCEDURE FOR DETERMINING THE ALLOWABLE LOAD FOR AN ECCENTRICALLY LOADED COLUMN

The *allowable load* for a given eccentrically loaded column may be determined by the following procedure:

1. Obtain, or estimate, the value of the eccentricity e.
2. Substitute the value of e into the secant formula, along with the geometric parameters r, c, A, and L, and the material properties E and σ_Y (i.e., let $\sigma_{max} = \sigma_Y$), and solve for the load P_Y.

3. Divide the load P_Y by the appropriate factor of safety to determine the allowable load. (Note that, because of the nonlinear relationship between load and stress, the factor of safety must be applied directly to the load. It is not applied to the stress to determine an allowable stress from which an allowable load is then calculated!)

The following example problem illustrates the use of the secant formula.

EXAMPLE 10.4

A W6 × 20 structural steel ($E = 29(10^3)$ ksi, $\sigma_Y = 36$ ksi) column is eccentrically loaded as shown in Fig. 1. Assume that the load is applied directly at the top cross section even though it is applied through a bracket that is attached to a flange. Also assume that the column is

---

[12]The behavior of short and medium-length ideal columns is examined further in Section 10.6, *Inelastic Buckling*.

[13]The secant term in Eq. 10.46 approaches unity as $L/r \to 0$. Therefore, for very short members, superposition can be applied, as in Example 9.3.

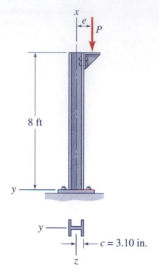

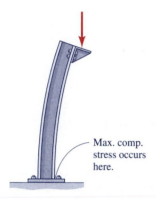

y ⊢H⊣

←|⊦—$c = 3.10$ in.

z

Fig. 1 Eccentrically loaded cantilever column.

Max. comp. stress occurs here.

Fig. 2

supported in a manner that prevents out-of-plane buckling, that is, buckling in the z direction.

$$A = 5.87 \text{ in}^2, \quad r = 2.66 \text{ in.}, \quad c = 3.10 \text{ in.}$$

(a) If a compressive load $P = 20$ kips is applied at an eccentricity $e = 4.0$ in., what is the maximum compressive stress in the column? Where does it occur? (b) What is the factor of safety against initial yielding of the column under the above loading?

Solution

(a) *Maximum Compressive Stress:* As noted earlier the maximum compressive stress in a cantilever column (see Fig. 2) can be calculated directly by using the secant formula, Eq. 10.46, with $L_e = 2L$ replacing the length L. Thus,

$$\sigma_{\max} = \frac{P}{A}\left[1 + \left(\frac{ec}{r^2}\right)\sec\left(\frac{L_e}{2r}\sqrt{\frac{P}{AE}}\right)\right] \tag{1}$$

$$\sigma_{\max} = \frac{20}{5.87}\left[1 + \left(\frac{4.0(3.10)}{(2.66)^2}\right)\sec\left(\frac{2(96)}{2(2.66)}\sqrt{\frac{20}{5.87(29,000)}}\right)\right]$$

$$= 9.866 \text{ ksi}$$

$$\sigma_{\max} = 9.87 \text{ ksi} \qquad\qquad \textbf{Ans.}$$

(b) *Factor of Safety:* Since the secant formula is nonlinear, we must determine the value of load P_Y that makes $\sigma_{\max} = \sigma_Y$, the compressive yield stress. We may be able to get an estimate from Fig. 10.17. For this column

$$\frac{L_e}{r} = \frac{2(96 \text{ in.})}{(2.66 \text{ in.})^2} = 72.2$$

$$\frac{ec}{r^2} = \frac{(4.0 \text{ in.})(3.10 \text{ in.})}{(2.66 \text{ in.})^2} = 1.753$$

By interpolating between the curves for $\left(\frac{ec}{r^2}\right) = 1.0$ and $\left(\frac{ec}{r^2}\right) = 2.0$ at $\frac{L_e}{r} \approx 75$ on Fig. 10.17, we get $\frac{P_Y}{A} \approx 12$ ksi. Therefore, since $A = 5.87 \text{ in}^2$, we can expect a solution of $P_Y \approx 70$ kips.

We need to solve for the value of P_Y that satisfies Eq. (1) with $\sigma_{\max} = \sigma_Y = 36$ ksi, that is, the equation

$$36 \text{ ksi} = \frac{P_Y}{(5.87 \text{ in}^2)}\left[1 + 1.753 \sec\left(36.1\sqrt{\frac{P_Y}{(5.87 \text{ in}^2)(29,000 \text{ ksi})}}\right)\right]$$

or

$$211 = P_Y\left[1 + 1.753 \sec\left(0.0875\sqrt{P_Y}\right)\right] \tag{2}$$

Therefore, the calculated load that will cause first yielding is

$$P_Y = 64.2 \text{ kips}$$

Since the actual load on the column is 20 kips, the factor of safety with respect to yielding is

$$FS_Y = \frac{P_Y}{P} = \frac{64.2 \text{ kips}}{20 \text{ kips}} = 3.21$$

$$FS_Y = 3.2 \qquad\qquad \textbf{Ans.}$$

Note that this factor of safety, which is based on <u>loads</u> (Eq. 2.26), is not equal to the ratio of stresses σ_Y/σ_{max}.

---

*10.5   IMPERFECTIONS IN COLUMNS

The previous section indicated how the behavior of a straight column is affected if the load is applied eccentrically rather than at the centroid of the cross section of the member. The behavior of a column is also affected by any initial lack of straightness of the axis of the column, or *initial imperfection* of the column. The ideal pin-ended column of Fig. 10.6a is replaced, in Fig. 10.18a, by the column that has an initial crookedness $v_0(x)$.

Although $v_0(x)$ is usually small, its exact functional form differs from column to column and is unknown. However, we can get an idea of the effect of initial crookedness by assuming that $v_0(x)$ can be represented by

$$v_0(x) = \delta_0 \sin\left(\frac{\pi x}{L}\right) \qquad\qquad (10.47)$$

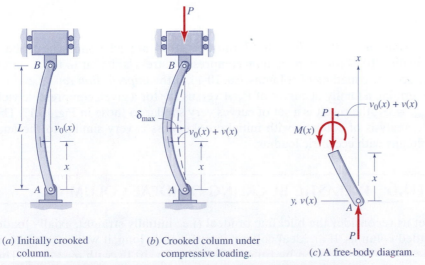

(a) Initially crooked column.

(b) Crooked column under compressive loading.

(c) A free-body diagram.

FIGURE 10.18   A column with initial crookedness $v_0(x)$.

which has the same shape as the fundamental buckling mode of an ideal column (Eq. 10.17). From the free-body diagram of Fig. 10.18c we get

$$M(x) = -P[v(x) + v_0(x)] \qquad (10.48)$$

where $v(x)$ is the deflection caused by the load P. Combining Eqs. 10.47 and 10.48 with the moment-curvature equation, Eq. 7.8, gives the differential equation

$$EIv''(x) + Pv(x) = -P\delta_0 \sin\left(\frac{\pi x}{L}\right) \qquad (10.49)$$

The solution to Eq. 10.49 with the boundary conditions $v(0) = v(L) = 0$ is

$$v(x) = \left(\frac{\alpha\delta_0}{1 - \alpha}\right) \sin\left(\frac{\pi x}{L}\right) \qquad (10.50)$$

where

$$\alpha \equiv \frac{P}{P_{cr}} = \frac{PL^2}{\pi^2 EI} \qquad (10.51)$$

From Eq. 10.50 we can determine the maximum deflection, δ_{max}, and the maximum bending moment, M_{max}, as follows:

$$\delta_{max} = \delta_0 + v(L/2) = \frac{\delta_0}{1 - \alpha} \qquad (10.52)$$

$$M_{max} = P\delta_{max} = \frac{P\delta_0}{1 - \alpha} \qquad (10.53)$$

Then,

$$\sigma_{max} = \frac{P}{A} + \frac{M_{max}c}{I} = \frac{P}{A}\left[1 + \frac{\delta_0 c}{r^2(1 - \alpha)}\right] \qquad (10.54)$$

where $r^2 = 1/A$.

Since $\alpha = P/P_{cr}$, Eqs. 10.52 through 10.54 are all nonlinear in the load P. Equation 10.54 for the maximum compressive stress is similar to the secant formula for eccentric loading of columns, Eq. 10.46. If the *imperfection ratio* $\delta_0 c/r^2$ is used to determine a family of curves of P_Y/A versus L/r for a given compressive yield stress $\sigma_{max} = \sigma_Y$, the result is a set of curves very similar to those in Fig. 10.17. Therefore, the analysis of columns with initial crookedness is very similar to the analysis of columns with eccentric loading.

*10.6 INELASTIC BUCKLING OF IDEAL COLUMNS

Let us reconsider the buckling of ideal (i.e., initially straight, axially loaded) pin-ended columns. If an ideal column is sufficiently long, it will buckle elastically at the critical stress given by Euler's formula, Eq. 10.18, with $\sigma_{cr} \leq \sigma_{PL}$. Therefore, we can determine the value of the slenderness ratio, L/r, above which elastic

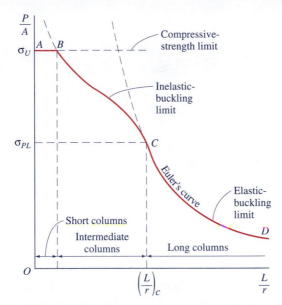

FIGURE 10.19 Buckling behavior of an ideal column.

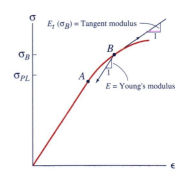

FIGURE 10.20 Compressive stress-strain diagram.

buckling occurs, by setting $\sigma_{cr} = \sigma_{PL}$ in Eq. 10.18. This gives the **critical slenderness ratio**

$$\left(\frac{L}{r}\right)_c = \sqrt{\frac{\pi^2 E}{\sigma_{PL}}} \tag{10.55}$$

Columns for which $(L/r) > (L/r)_c$ are called **long columns.** Long columns fail at their *elastic buckling limit,* namely, the Euler buckling stress.

If an ideal column is very short, it will not buckle at all, but it will simply fail by crushing of the material at the ultimate compressive stress σ_U. Columns that fail in this manner are called **short columns,** and they fail at their *compressive-strength limit,* as indicated on Fig. 10.19. Between short columns and long columns lies a range of columns, called **intermediate columns,** whose failure mode is referred to as *inelastic buckling.* They fail when the average stress P/A reaches the *inelastic-buckling limit,* indicated by the curve BC in Fig. 10.19.

Let us now examine the phenomenon of inelastic buckling to determine the inelastic buckling limit of an ideal column whose compressive stress-strain diagram has the form shown in Fig. 10.20.[14] Assume that the column fails in the range of intermediate columns, so buckling does not occur at a stress below the proportional limit σ_{PL}. And assume that when its average compressive stress P/A reaches the value σ_B (point B on Fig. 10.20), the column is just on the verge of inelastic buckling. There are two principal theories that predict the value of σ_B at which a column will buckle inelastically. The simplest theory is the **tangent-modulus theory,** developed in 1889 by F. Engesser, a German engineer. The tangent-modulus formula is

---

[14]Because actual columns are not perfectly straight and are not loaded exactly along the centroidal axis, and because stresses (called *residual stresses*) are induced into columns when they are manufactured, it is not possible to determine an analytical expression for the inelastic buckling limit of real columns. However, use of the analytical expressions presented here, together with the results of many column buckling experiments, makes it possible to obtain empirical column-design formulas, several of which are discussed in Section 10.7.

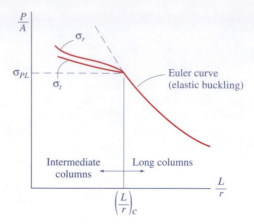

FIGURE 10.21 Critical stress versus slenderness ratio.

obtained by replacing Young's modulus, E, in Euler's buckling formula by the *tangent modules E_t*, which is the slope of the compressive stress-strain curve. At stress σ_B the tangent modulus is given by

$$E_t(\sigma_B) = \frac{d\sigma}{d\epsilon}\bigg|_{\sigma=\sigma_B} \tag{10.56}$$

The meaning of E_t is illustrated on Fig. 10.20. Thus, the *tangent-modulus buckling stress* is given by

$$\sigma_t = \frac{\pi^2 E_t}{(L/r)^2} \qquad \text{\textbf{Tangent-Modulus}} \atop \textbf{Buckling Formula} \tag{10.57}$$

The tangent-modulus curve for inelastic buckling is labeled σ_t in Fig. 10.21.

A slightly higher inelastic buckling stress σ_r is predicted by the *reduced-modulus theory*. In the reduced-modulus formula

$$\sigma_r = \frac{\pi^2 E_r}{(L/r)^2} \tag{10.58}$$

the *reduced modulus E_r ($E_t < E_r < E$)* depends on E, E_t, and also the shape of the cross section. Since actual buckling tests of intermediate-length columns result in buckling loads that tend to be close to the tangent-modulus load, and also since the tangent-modulus load is far easier to calculate than the reduced-modulus load, the tangent-modulus theory is the generally preferred theory for inelastic buckling of columns [Ref. 10-6].

The following example illustrates the use of the tangent-modulus theory to determine the inelastic buckling stress.

EXAMPLE 10.5

A 4 × 4 × 1/2 tubular column, whose cross section is shown in Fig. 1, has an effective length of 100 in. ($A = 6.36$ in^2, $r = 1.39$ in.). The column is made of a material whose compressive stress-strain curve is given by the curve *ABC* in Fig. 2. The corresponding curve of tangent modulus E_t versus compressive stress σ is the curve *DEFG* in Fig. 2, with values of the tangent modulus on the upper scale. Compute the buckling load for this column.

Solution Let us first determine whether elastic or inelastic buckling governs, by using Eq. 10.55 to determine L_c/r. From the tangent-modulus curve of Fig. 2 we can estimate that $E = 30(10^3)$ ksi, and $\sigma_{PL} \approx 29$ ksi. Then,

$$\frac{L_c}{r} = \sqrt{\frac{\pi^2 E}{\sigma_{PL}}} = \sqrt{\frac{\pi^2(30 \times 10^3 \text{ ksi})}{29 \text{ ksi}}} = 101.0$$

The actual value of L/r is

$$\frac{L}{r} = \frac{100 \text{ in.}}{1.39 \text{ in.}} = 71.9$$

Since $L/r < L_c/r$, inelastic buckling occurs.

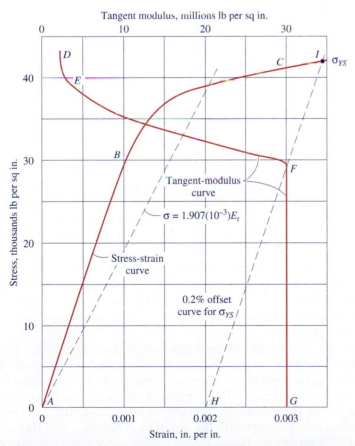

Fig. 2 The compressive stress-strain curve and the tangent-modulus curve for a structural material.

From the tangent-modulus formula, Eq. 10.57, we get

$$\sigma_t = \frac{\pi^2 E_t}{(L/r)^2} = \frac{\pi^2 E_t}{(71.9)^2} = 0.001907 \, E_t \qquad (1)$$

This straight line is plotted in Fig. 2. Its intersection with the tangent-modulus curve gives the solution for the tangent-modulus stress. First, try $\sigma_1 = 33$ ksi. From Fig. 2, we estimate that the corresponding value of the tangent modulus is $E_1 \equiv E_t(\sigma_1) \approx 17(10^3)$ ksi. From Eq. (1),

$$\sigma_{t1} = 0.001907\left[17(10^3) \text{ ksi}\right] = 32.4 \text{ ksi}$$

which is lower than our first-guess value of 33 ksi. We need a slightly higher value of E_t, so try $E_t(\sigma_2) \approx 17.2(10^3)$ ksi. Then Eq. (1) gives

$$\sigma_{t1} = 32.8 \text{ ksi}$$

which agrees with the value taken from the tangent-modulus curve. Then,

$$P_t = \sigma_t A = (32.8 \text{ ksi})(6.36 \text{ in}^2) = 209 \text{ kips}$$

or, rounded to two significant figures,

$$P_t = 210 \text{ kips} \qquad \textbf{Ans.}$$

10.7 DESIGN OF CENTRALLY LOADED COLUMNS

In Sections 10.2 through 10.6 we examined the behavior of columns of known geometry (perfectly straight or with a specified form of imperfection), known material behavior (free of any residual stress and having a known compressive stress-strain diagram), known end conditions (pinned, fixed, or free), and known line of action of load. For real columns, all of these factors, and more, are subject to variability (e.g., see Ref. 10-5), and this variability must be properly accounted for when a column is designed. Therefore, **design codes** typically specify **column-design formulas** that are obtained by curve-fitting data from laboratory tests of many real columns and that incorporate appropriate factors of safety, effective-length factors, and other modifying factors.

In this section we will consider several *representative* column-design formulas for centrally loaded columns of steel, aluminum alloy, and wood. Each design formula applies only to columns made of a specific material and having an effective slenderness ratio in a specific range. The purpose of this section is just to illustrate the basic aspects of the **column-design process.** In some cases the code notation has been changed to be consistent with the notation in Sections 10.2 through 10.6 (e.g., to the notation σ for stress). Because column-design formulas and procedures are updated periodically by the various code-writing agencies, the reader should consult the current edition of the appropriate design code or specification before designing a column for actual fabrication and use. The names, mailing addresses, or Web addresses of several major code-writing societies and agencies are listed in the References for Chapter 10.

The **allowable stress** σ_{allow} is related to the **critical stress** σ_{cr}, or buckling stress, by the equation

$$\sigma_{\text{allow}} = \frac{\sigma_{\text{cr}}}{FS} \qquad (10.59)$$

where FS is the **factor of safety.** The **allowable load** is then given by the allowable stress times the area of the cross section of the column.

$$P_{\text{allow}} = \sigma_{\text{allow}} A \qquad (10.60)$$

The column-design formulas presented in Eqs. 10.65, 10.66, and 10.67 can be used in a very straightforward manner to determine the **allowable load** for a column of a given material with a given effective length and given cross section. On the other hand, **column design**—that is, determination of a suitable cross section for a column of given material and given effective length to support a given load—is an iterative procedure, which is described below.

COLUMN DESIGN PROCEDURE

The basic steps of this design procedure are:

Step 1: Select a trial cross section.

Step 2: Determine the value of the effective slenderness ratio by comparing the values $(KL/r)_1$ and $(KL/r)_2$ for buckling in the directions of the principal axes of the cross section. (For a wood column with rectangular cross section, use KL/d instead of KL/r.)

Step 3: Using the column-design formula that is appropriate to the selected material and to the effective slenderness ratio calculated in Step 2, determine the allow-

able stress σ_{allow}. Multiply σ_{allow} by the cross-sectional area of the trial cross section to determine the allowable load.

Step 4: If the allowable load calculated in Step 3 is equal to or slightly greater than the given design load, accept the trial section. If the allowable load is less than the given design load, repeat Steps 1 through 3 for another trial section. If the allowable load is much greater than required, try a smaller cross section.

Structural Steel Columns. The design of columns of structural steel is based on formulas proposed by the Structural Stability Research Council (SSRC), formerly called the Column Research Council (CRC). The American Institute of Steel Construction (AISC), an organization that publishes specifications for use in the design of steel structures, has adopted the SSRC formulas and prescribed the safety factors to be used in the design of steel columns [Ref. 10-7].

For long columns, the Euler formula, Eq. 10.36, is used. This formula is theoretically valid as long as σ_{cr} does not exceed the proportional limit of the material in compression. (The yield stress σ_Y may be substituted for σ_{PL}.) However, it has been observed that the process of rolling that is used to produce structural-steel shapes such as wide-flange sections produces large residual compressive stresses so that yielding may occur at a compressive stress as low as half the nominal compressive yield stress. Therefore, the AISC limits the range of validity of the Euler formula to those values of KL/r for which $\sigma_{cr} \leq 0.5\ \sigma_Y$, that is, for values of $KL/r \geq (KL/r)_c$, with

$$\left(\frac{KL}{r}\right)_c = \left(\frac{\pi^2 E}{0.5\sigma_Y}\right)^{\frac{1}{2}} \tag{10.61}$$

The value of E given by AISC is $29(10^3)$ ksi.[15]

Combining Eqs. 10.36 and 10.61, we can write Euler's formula in the form

$$\frac{\sigma_{cr}}{\sigma_Y} = \frac{\pi^2 E}{\sigma_Y \left(\dfrac{KL}{r}\right)^2} = \frac{\left(\dfrac{KL}{r}\right)_c^2}{2\left(\dfrac{KL}{r}\right)^2}, \qquad \left(\frac{KL}{r}\right) \geq \left(\frac{KL}{r}\right)_c \tag{10.62}$$

This equation is plotted in Fig. 10.22, where its range of validity is indicated by the solid-line portion of the curve labeled *Euler's formula*.

For shorter columns, that is, for values of KL/r that do not exceed $(KL/r)_c$, the Column Research Council proposed use of the parabolic curve[16]

$$\frac{\sigma_{cr}}{\sigma_Y} = 1 - \frac{\left(\dfrac{KL}{r}\right)^2}{2\left(\dfrac{KL}{r}\right)_c^2}, \qquad \left(\frac{KL}{r}\right) \leq \left(\frac{KL}{r}\right)_c \tag{10.63}$$

This curve has a horizontal tangent at $KL/r = 0$, and it merges smoothly with the Euler curve at $(KL/r)_c$, as shown in Fig. 10.22.

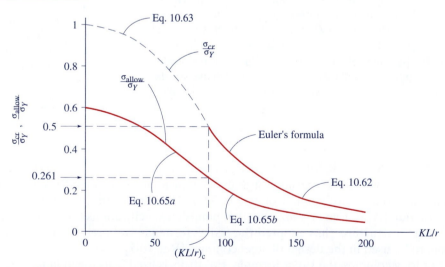

FIGURE 10.22 Design curves for buckling of structural steel columns.

---

[15]This value is for room-temperature conditions (70°F); the assumed value of E decreases linearly to a value of $25(10^3)$ ksi at 900°F.

[16]This is sometimes called the *Johnson column formula*.

$$FS_1 = \frac{5}{3} + \frac{3\left(\dfrac{KL}{r}\right)}{8\left(\dfrac{KL}{r}\right)_c} - \frac{\left(\dfrac{KL}{r}\right)^3}{8\left(\dfrac{KL}{r}\right)_c^3}, \qquad \left(\frac{KL}{r}\right) \leq \left(\frac{KL}{r}\right)_c \qquad (10.64a)$$

and, corresponding to Eq. 10.62 the factor of safety is the constant

$$FS_2 = \frac{23}{12}, \qquad \left(\frac{KL}{r}\right) \geq \left(\frac{KL}{r}\right)_c \qquad (10.64b)$$

The AISC formulas for σ_{allow} are obtained by combining Eqs. 10.60 and 10.62 through 10.64 to give

$$\frac{\sigma_{\text{allow}}}{\sigma_Y} = \frac{1}{FS_1}\left[1 - \frac{\left(\dfrac{KL}{r}\right)^2}{2\left(\dfrac{KL}{r}\right)_c^2}\right], \qquad \left(\frac{KL}{r}\right) \leq \left(\frac{KL}{r}\right)_c \qquad (10.65a)$$

$$\frac{\sigma_{\text{allow}}}{\sigma_Y} = \frac{1}{FS_2}\left[\frac{\left(\dfrac{KL}{r}\right)_c^2}{2\left(\dfrac{KL}{r}\right)^2}\right], \qquad \left(\frac{KL}{r}\right)_c \leq \left(\frac{KL}{r}\right) \leq 200 \qquad (10.65b)$$

These formulas are also plotted in Fig. 10.22.

Finally, it is noted in Eq. 10.65b and on Fig. 10.22 that the AISC restricts columns to values of $KL/r \leq 200$.

Aluminum-Alloy Columns. The Aluminum Association provides specifications for the design of aluminum-alloy structures. Euler's formula is used for long columns, and straight lines are prescribed for short and intermediate columns. The general form of these curves is indicated in Fig. 10.23. Specific design formulas, which depend on alloy, temper, and usage, are provided by the Aluminum Association [Ref. 10-8]. For the alloy 2014-T6, an alloy used in building structures, the column design formulas are:

$$\sigma_{\text{allow}} = 28 \text{ ksi}, \qquad 0 \leq \frac{KL}{r} \leq 12 \qquad (10.66a)$$

$$\sigma_{\text{allow}} = \left[30.7 - 0.23\left(\frac{KL}{r}\right)\right] \text{ ksi}, \qquad 12 < \frac{KL}{r} \leq 55 \qquad (10.66b)$$

$$\sigma_{\text{allow}} = \frac{54{,}000 \text{ ksi}}{\left(\dfrac{KL}{r}\right)^2}, \qquad 55 < \frac{KL}{r} \qquad (10.66c)$$

These formulas incorporate a factor of safety.

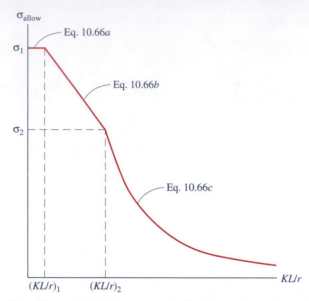

FIGURE 10.23 The typical design curve for buckling of aluminum-alloy columns.

Wood Columns. The design of wood structural members is governed by the *National Design Specification for Wood Construction* published by the American Forest & Paper Association [Ref. 10-9]. The *National Design Specification* (currently *NDS-2005*) provides a single formula for the column stability factor covering the effective slenderness ratio range of $0 \leq KL/d \leq 50$. The allowable compressive stress for a wood column is given by

$$\sigma_{\text{allow}} = F_c \left[\frac{1 + (F_{cE}/F_c)}{2c} - \sqrt{\left[\frac{1 + (F_{cE}/F_c)}{2c} \right]^2 - \frac{F_{cE}/F_c}{c}} \right] \qquad (10.67)$$

In the above formula,

F_c = the value of allowable stress for compression parallel to the grain of the wood.

$F_{cE} = \dfrac{K_{cE}E}{(KL/d)^2}$ = the reduced Euler buckling stress.

E = the modulus of elasticity.

K_{cE} = 0.30 for visually graded lumber.

c = 0.8 for sawn lumber.

The values of F_c and E vary greatly with type of wood, moisture content, duration of load, and other factors, with typical values of F_c being in the range of 400 psi to 2000 psi, and typical values of E ranging from 800 ksi to about 1800 ksi. The effective slenderness ratio, KL/d, is taken as the larger of the two values $(KL/d)_1$ and $(KL/d)_2$ for buckling in the direction of the principal axes of the cross section, with d_1 and d_2 being the finished dimensions of the rectangular cross section.

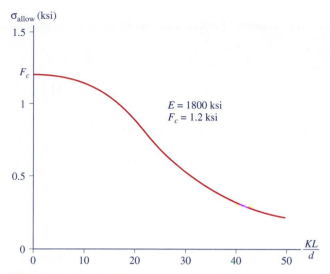

FIGURE 10.24   A design curve for buckling of wood columns.

The wood column design curve, based on Eq. 10.67 with $E = 1800$ ksi and $F_c = 1.2$ ksi, is plotted in Fig. 10.24.

The **column-design procedure** outlined earlier will now be illustrated.

EXAMPLE 10.6

Determine the size of a square S4S (surfaced on four sides) structural-lumber column of 10 ft unbraced length to carry an axial load of 10 kips. Assume that the square ends of the column give an effective length factor $K = 1$, that $E = 1800$ ksi, and that $F_c = 1.2$ ksi.

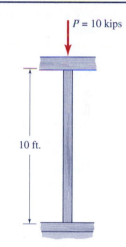

Fig. 1

Solution

Step 1: Try a nominal 6×6 in. section (5.5 in. $\times$ 5.5 in. finished dimensions).

Step 2: Since this is a square cross section

$$\left(\frac{KL}{d}\right) = \frac{1(10 \text{ ft})(12 \text{ in./ft})}{5.5 \text{ in.}} = 21.82$$

for both principal directions. Then,

$$F_{cE} = \frac{K_{cE}E}{(KL/d)^2} = \frac{0.3(1800)}{(21.82)^2} = 1.134 \text{ ksi}$$

Step 3: From the wood-column design formula, Eq. 10.67,

$$\sigma_{\text{allow}} = F_c \left[\frac{1 + (F_{cE}/F_c)}{2c} - \sqrt{\left[\frac{1 + (F_{cE}/F_c)}{2c} \right]^2 - \frac{F_{cE}/F_c}{c}} \right] = 0.8054 \text{ ksi}$$

$$P_{\text{allow}} = \sigma_{\text{allow}} A = (0.8054 \text{ ksi})(5.5 \text{ in.})(5.5 \text{ in.}) = 24.4 \text{ kips} > 10 \text{ kips}$$

Step 4: The 6×6 in. section is obviously adequate, but let us see if a 4×4 in. section would be sufficient.

Step 1: Try a nominal 4×4 in. section (3.5 in. $\times$ 3.5 in. finished dimensions).

Step 2:

$$\left(\frac{KL}{d}\right) = \frac{1(10 \text{ ft})(12 \text{ in./ft})}{3.5 \text{ in.}} = 34.3$$

Step 3: From Eq. 10.67, $\sigma_{\text{allow}} = 0.4154$ ksi. Then,

$$P_{\text{allow}} = \sigma_{\text{allow}} A = (0.4154 \text{ ksi})(3.5 \text{ in.})(3.5 \text{ in.}) = 5.09 \text{ kips} < 10 \text{ kips}$$

Step 4: Since we were asked to select S4S structural lumber with square cross section, we must use the 6×6 in. section. **Ans.**

The design formulas presented in this section are only applicable to centrally loaded columns. The design of eccentrically loaded columns is beyond the scope of this book.

> MDS10.4 – 10.9 | **Design of Centrally Loaded Columns**

10.8 PROBLEMS

Problems 10.1-1 through 10.1-4. *In solving the problems for Section 10.1, assume that the bars are rigid, that the springs are linearly elastic, and that the displacements and angles of rotation are small. Also, neglect gravity, and assume that buckling takes place in the plane of the figure.*

Prob. 10.1-1. Determine the critical load, P_{cr}, for the bar-spring system shown in Fig. P10.1-1. The load P remains vertical as the bar rotates about the pin at A. The force in spring BD is proportional to the elongation of the spring, with spring constant k. That is, $F_s = ke$, where e is the elongation of the spring. The spring is unstretched when $\theta = 0$.

Prob. 10.1-2. Determine the critical load, P_{cr}, for the bar-spring system shown in Fig. P10.1-2. The load P remains horizontal. End A is free to move horizontally, and point B moves in a circular arc about C. A frictionless pin connects bars AB and BC at point B.

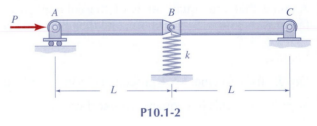

P10.1-2

Prob. 10.1-3. Determine the critical load, P_{cr}, for the bar-spring system shown in Fig. P10.1-3. The load P remains vertical.

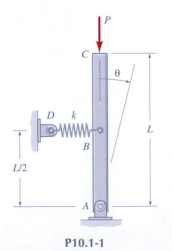

P10.1-1

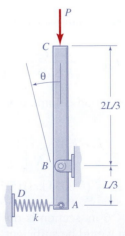

P10.1-3

Prob. 10.1-4. Determine the critical load, P_{cr}, for the bar-spring system shown in Fig. P10.1-4. The load P remains horizontal. End A is free to move horizontally, and point B moves in a circular arc about C. A frictionless pin connects bars AB and BC at point B. Like the rotational spring in Fig. 10.4, the rotational spring at C exerts a restoring moment $(M_C)_r = k_\theta \theta$ when bar BC rotates away from the horizontal position.

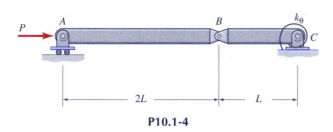

P10.1-4

Prob. 10.1-5. Modify the two-bar system in Fig. P10.1-4 by moving the rotational spring, with spring constant k_θ, from end C to the pin joint at B, so that there will be a restoring moment at B that is proportional to the angle between bar AB and bar BC. Neglect gravity, and assume that the entire two-bar system ABC is horizontal when the spring is attached across the joint at B and the horizontal load P is applied at the moveable end A. Determine an expression for the critical load P_{cr} for this modified two-bar system. Include a sketch of the modified two-bar system, and express your answer in terms of the spring stiffness k_θ and the length L.

▼ **EULER BUCKLING LOAD** | MDS 10.1 & 10.2

Problems 10.2-1 through 10.2-19. *In solving the problems for Section 10.2, assume that the compression member in question is an ideal slender, prismatic, pin-ended, elastic column.*

Prob. 10.2-1. Determine the critical load, P_{cr}, of a square tubular steel column that is 20 ft long and that has the cross section shown in Fig. P10.2-1. Let $E_{st} = 29(10^3)$ ksi and $\sigma_Y = 46$ ksi.

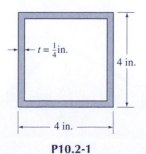

P10.2-1

Prob. 10.2-2. Determine the critical load, P_{cr}, of a steel pipe column that is 20 ft long; that has a $\frac{1}{4}$-in. wall thickness, as shown in Fig. P10.2-2; and that weighs the same amount per foot as the square tubular column in Fig. P10.2-1. Let $E_{st} = 29(10^3)$ ksi and $\sigma_Y = 46$ ksi.

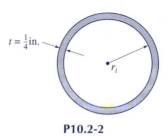

P10.2-2

D Prob. 10.2-3. A builder plans to nail together two boards to form a column. Which of the configurations in Fig. P10.2-3 would give the greater buckling load, configuration A, or configuration B? Assume that the ends of the columns are "pinned" in a manner that permits buckling in any direction. Show calculations to support your answer.

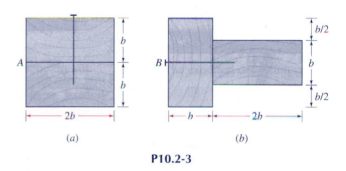

P10.2-3

D Prob. 10.2-4. (a) Calculate the critical load, P_{cr}, for a **W**8 × 35 steel column having a length of 24 ft. Let $E_{st} = 29(10^3)$ ksi and $\sigma_Y = 36$ ksi, and assume that the column is free to buckle in any direction. See Table D.1 of Appendix D for wide-flange section properties. (b) Would a **W**12 × 35 column be better (i.e., have a higher critical load) than the 8-in. section? (c) If $\sigma_Y = 50$ ksi rather than 36 ksi, would there be any effect on the critical load?

D Prob. 10.2-5. A pin-ended column that is 16 ft long must support a compressive axial load of $P = 300$ kips with a factor of safety with respect to buckling of 2.0. Assume that the column has "pinned" ends that allow buckling in any direction. From Table D.1 of Appendix D, select the lightest **W** shape that will support the load. Let $E_{st} = 29(10^3)$ ksi and $\sigma_Y = 36$ ksi.

D Prob. 10.2-6. A solid-waste compactor, which is illustrated in Fig. P10.2-6, can be purchased in a "standard" configuration or a "heavy-duty" configuration. One difference between the designs is the size of the push rod that is used. The push rod may be considered to be a pin-ended column of length 1.5 m. The push rod on the standard model is a rectangular bar with dimensions $b = 15$ mm, $h = 30$ mm. It is made of steel with $E_{st} = 200$ GPa, $\sigma_Y = 250$ MPa. (a) Determine the compaction-force capacity of the standard model if the push rod is designed with a factor of safety of 3.0 with respect to elastic buckling. (b) If the push rod of the heavy-duty model is made of the same steel, and if it also has a rectangular cross section with $h = 2b$, what dimension b is required (to the nearest mm) if the heavy-duty model is to have a capacity of 50 kN?

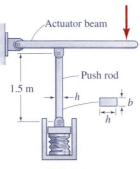

P10.2-6

Prob. 10.2-7. Three pin-ended columns have the same length and the same cross-sectional area. They are made of the same material, and they are free to buckle in any direction. The cross sections of the columns are: (1) a circle, (2) a square, and (3) an equilateral triangle. Determine expressions for the elastic buckling loads—P_1, P_2, and P_3—of the respective columns.

*DProb. 10.2-8. An equal-leg angle section of length $L = 12$ ft is to be used to support a compressive load of $P = 20$ kips with a factor of safety of 2.5 against elastic buckling.[17] Assume that the ends of the column are pinned in a manner that permits it to buckle in any direction. From Table D.5 of Appendix D, select the lightest steel section that meets the design requirements. Let $E_{st} = 29(10^3)$ ksi and $\sigma_Y = 36$ ksi.

*DProb. 10.2-9. Solve Prob. 10.2-8 if the load is $P = 10$ kips and the length is $L = 10$ ft.

*DProb. 10.2-10. From Table D.7 of Appendix D, select a standard steel pipe that would meet the design requirements of Prob. 10.2-8.

Prob. 10.2-11. It is suspected that a small control rod on the spacecraft depicted in Fig. P10.2-11 "failed" as a result of elastic buckling when the rod became exposed to direct sun-

light that increased its temperature uniformly by an amount ΔT. Assume that the ends of the pushrod are pinned to structure that is "rigid," and assume that the control rod is stress-free when $\Delta T = 0$. Determine an expression for the value of ΔT that would be required to cause the buckling failure. Express your answer in terms of cross-sectional and material properties of the bar, that is, in terms of I, A, α, etc., and the length L.

*Prob. 10.2-12. A "rigid" horizontal beam AD is supported by two pin-ended columns as shown in Fig. P10-2.12. The columns have the same cross-sectional dimensions and are made of the same material (i.e., same I and same E). (a) Determine an expression for the load W_a that will cause elastic buckling of one of the columns, and indicate which column will buckle first. (b) Determine an expression for load W_b that would cause the second column to buckle also. Assume that the force in the column that buckles first remains constant after that column buckles.

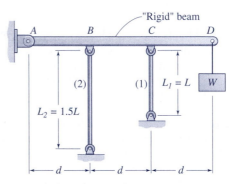

P10.2-12 and P10.2-13

*Prob. 10.2-13. Solve Prob. 10.2-12 with $L_1 = L_2 = L$.

DProb. 10.2-14. An aluminum-alloy column is 15 ft long and is pinned at both ends. What is the allowable compressive load P_{allow} if the factor of safety against elastic buckling is 2.75 and the column is a 6.00×4.00 standard I-beam section weighing 4.692 lb/ft? (See Table D.9 of Appendix D.) Let $E_{al} = 10.0(10^3)$ ksi and $\sigma_{PL} = 38$ ksi.

DProb. 10.2-15. The truss shown in Fig. P10.2-15 is used as part of a rig to pull pilings out of the ground. If the two truss members are steel pipes with outer diameter of $d_o = 2.875$ in. and wall thickness of $t = 0.275$ in., determine the largest tension that can be exerted on the piling without causing elastic buckling of a truss member. The cable that exerts tension T on the piling passes over a pulley that is free to

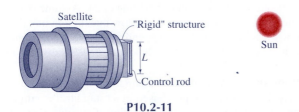

P10.2-11

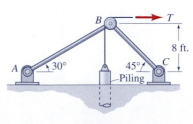

P10.2-15

[17]Because of the difficulty of applying axial loads exactly passing through the centroid of an angle section, such members are not recommended for use as columns.

rotate about the same pin that connects the two truss members together at B. Let $E_{st} = 29(10^3)$ ksi and $\sigma_Y = 36$ ksi.

ᴰProb. 10.2-16. The truss shown in Fig. P10.2-16 is made of steel rods, each of which has a solid, circular cross section with a diameter of 1 in. The ends of the members are pinned. Determine the maximum load W that can be supported at D without causing elastic buckling of any member of the truss. Let $E_{st} = 29(10^3)$ ksi and $\sigma_Y = 36$ ksi.

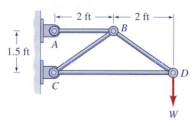

P10.2-16

ᴰProb. 10.2-17. The compression member BC of the truss ABC in Fig. P10.2-17 is a square solid steel bar with 1.75 in. × 1.75 in. cross section. The tension member AB is a solid steel rod that has a circular cross section with diameter d. Let $E_{st} = 29(10^3)$ ksi and $\sigma_Y = 36$ ksi. (a) Determine the maximum load W that can be supported at B without causing elastic buckling of member BC. (b) Determine the diameter of member AB if it is to yield in tension just when the load W reaches the value determined in Part (a). The yield stress in tension is $\sigma_Y = 36$ ksi.

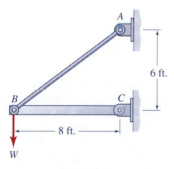

P10.2-17

ᴰProb. 10.2-18. Member AB of the truss in Fig. P10.2-18 has a fixed length L_1, is pinned to a fixed support at A, and remains horizontal. Determine an expression for the load W

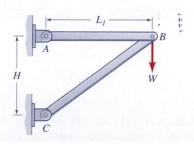

P10.2-18

that would produce elastic buckling of member BC. Express your answer in terms of the distance H, which determines the (variable) location of support point C and also determines the length of member BC.

ᴰProb. 10.2-19. The steel compression strut BC of the frame ABC in Fig. P10.2-19 is a steel tube with an outer diameter of $d_o = 48$ mm and a wall thickness of $t = 5$ mm. Determine the factor of safety against elastic buckling if a distributed load of 10 kN/m is applied to the horizontal frame member AB as shown. Let $E_{st} = 210$ GPa and $\sigma_Y = 340$ MPa.

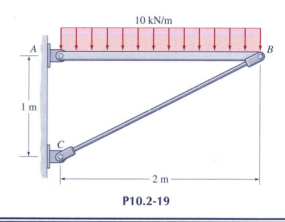

P10.2-19

▼ **EFFECTIVE LENGTH OF COLUMNS**   **MDS 10.3**

> **Problems 10.3-1 through 10.3-14.** *In solving the problems of Section 10.3, assume that the compression member in question is an ideal slender, prismatic, elastic column with the stated end conditions. Also, assume that deflections and slopes are small.*

Prob. 10.3-1. A **W**14 × 26 wide-flange steel column has a length of $L = 18$ ft and is supported at its ends so that it is free to buckle in any direction. Using the effective-length method, calculate the elastic buckling load, P_{cr}, for columns with the following end conditions: (a) pinned-pinned. (b) fixed-pinned, and (c) fixed-fixed. See Figs. 10.13a, b, and c, respectively, for columns with these end conditions, and see Table D.1 of Appendix D for the properties of the **W**14 × 26 section. Let $E_{st} = 29(10^3)$ ksi and $\sigma_Y = 50$ ksi.

Prob. 10.3-2. Solve Prob. 10.3-1 for a standard steel pipe column with nominal diameter of $d = 3$ in. and a length of $L = 20$ ft. See Table D.7 of Appendix D for the cross-sectional properties of the pipe column. Let $E_{st} = 29(10^3)$ ksi and $\sigma_Y = 50$ ksi.

Prob. 10.3-3. Solve Prob. 10.3-1 for a steel pipe column with an outer diameter of $d_o = 60$ mm, a wall thickness of $t = 4$ mm, and a length of $L = 4$ m. Let $E_{st} = 210$ GPa and $\sigma_Y = 340$ MPa.

ᴰProb. 10.3-4. Two 4 × 3 × $\frac{1}{2}$ steel angles are attached together as shown in Fig. P10.3-4 to form a column with a tee cross section. (a) Determine the radii of gyration of the cross section about the y axis and about the z axis using the cross sectional properties given in Table D.6 of Appendix D. (b) If the column has fixed-free end conditions, has a length of $L = 8$ ft, and is free to buckle in any direction, what is the factor of safety of the column against elastic buckling if the compressive axial load on the column is $P = 20$ kips? Let $E_{st} = 29(10^3)$ ksi and $\sigma_Y = 36$ ksi.

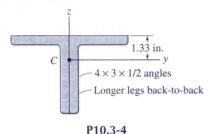

P10.3-4

DProb. 10.3-5. A square, tubular steel column has the dimensions shown in Fig. P10.3.5, has a length of $L = 3$ m, and has an axial compression load of $P = 50$ kN. If the column has fixed-free end conditions and is free to buckle in any direction, what is the factor of safety with respect to elastic buckling? Let $E_{st} = 210$ GPa and $\sigma_Y = 250$ MPa.

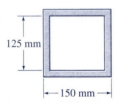

P10.3-5

Prob. 10.3-6. The $6 \times 6 \times \frac{1}{2}$ equal-leg angle section in Fig. P10.3-6 has a cross-sectional area $A = 5.75$ in^2 and radii of gyration $r_y = r_z = 1.86$ in. The minimum radius of gyration for this section is $r_m = 1.18$ in. If this section is to be used as a fixed-fixed column of length $L = 22$ ft, what is the largest compressive load that can be applied through the centroid of the cross section without causing elastic buckling?[18] Let $E_{st} = 29(10^3)$ ksi and $\sigma_Y = 36$ ksi.

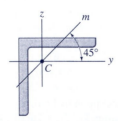

P10.3-6 and P10.3-7

Prob. 10.3-7. Solve Prob. 10.3-6 for an equal-leg steel column with cross-sectional area $A = 9700$ mm^2, and with radii of

---

[18]Because of the difficulty of applying axial loads exactly passing through the centroid of an angle section, such members are not recommended for use as columns.

gyration $r_y = r_z = 62$ mm and $r_m = 40$ mm. The section is to be used as a fixed-free column of length $L = 4$ m. Let $E_{st} = 210$ GPa and $\sigma_Y = 340$ MPa.

Prob. 10.3-8. Determine the elastic buckling load of a 12-ft-long 6×6 timber column that can be considered to be fixed at its base and free at the top. Let the modulus of elasticity (in compression parallel to the grain) be $E_w = 1.6(10^3)$ ksi, and let $\sigma_Y = 6$ ksi. (Recall that the finished dimensions of the wood column are smaller than the nominal dimensions stated above. See Table D.8 of Appendix D.)

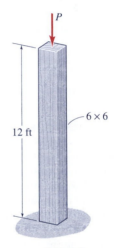

P10.3-8

Prob. 10.3-9. Rigid walls at each end of a solid slender rod AB provide fixed-fixed boundary conditions for the member in Fig. P10.3-9. At the reference temperature, T_0, the rod is perfectly stress-free. (a) Derive a formula that expresses the uniform increase in temperature, ΔT_{cr}, required to cause elastic buckling of the compression member. Express your answer in terms of the following material and geometric parameters: the coefficient of thermal expansion, α, and the slenderness ratio, L/r. (b) Determine the value of ΔT_{cr} in °F required to cause elastic buckling of a stainless steel rod with a diameter of $d = 1$ in. and a length of $L = 4$ ft. The coefficient of thermal expansion for stainless steel is $\alpha = 9.6 \times 10^{-6}$/°F.

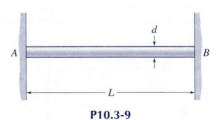

P10.3-9

Prob. 10.3-10. A straight, slender rod is fixed to a rigid support at end A and pinned to a rigid support at end B, as shown in Fig. P10.3-10. At the reference temperature, T_0, the rod is perfectly stress-free. (a) Derive a formula that expresses the uniform increase in temperature, ΔT_{cr}, required to cause elastic buckling of the compression member. Express your answer in terms of the following material and geometric parameters: the coefficient of thermal expansion, α, and the slenderness ratio, r/L. (b) Determine the value of ΔT_{cr} in °C required to cause elastic buckling of an aluminum rod with a diameter of $d = 20$ mm and a length of $L = 1$ m. The coefficient of thermal expansion for aluminum is $\alpha = 23 \times 10^{-6}/°C$.

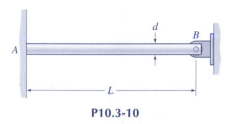

P10.3-10

Prob. 10.3-11. A rigid beam AB that supports a uniformly distributed load of intensity w (force per unit length) is pinned to a rigid support at end A and is supported by a square timber column at end B. Neglecting the weight of the beam AB, derive a formula for the value of the distributed load, w_{cr}, that would cause elastic buckling of the support column BC. Since the exact end conditions of the column are not known, express your answer in terms of the following material and geometric parameters: the modulus of elasticity of the wood column, E_w; the (finished) cross-sectional dimension of the square column, b; the length a of the horizontal beam AB; and the effective length of the column, KL.

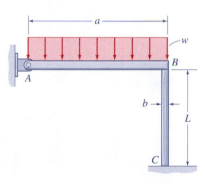

P10.3-11

*Prob. 10.3-12. A beam AB that supports a uniformly distributed load of intensity w (force per unit length) is simply supported at both ends, and it rests on a pipe column at its center C. Neglecting the weight of the beam AB, derive a formula for the value of the distributed load, w_{cr}, that would cause elastic buckling of the support column CD. Assume that the column is fixed to its base at D and pinned to the beam at C. Express your answer in terms of the following material and geometric parameters: the modulus of elasticity, E (same for the beam AB and the column CD); the

length of the beam, a, and the length of the column, L; the moment of inertia of the beam, I_b, and of the column, I_c; and the cross-sectional area of the column, A_c. (Note: This must be solved as a statically indeterminate problem. Use the fact that the deflection of the beam at C and the shortening of the column just prior to buckling are equal.)

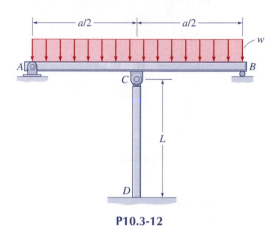

P10.3-12

*Prob. 10.3-13. Following a procedure similar to the one used in Fig. 10.11 and in Eqs. 10.20 through 10.32 to determine an expression for the elastic buckling load of a fixed-pinned column, derive an expression for the buckling load of a fixed-free column, as shown in Fig. 10.13e and in Fig. P10.3-13. (Hint: Take a free-body diagram of length $(L - x)$ from the top. The maximum deflection, δ, can be eliminated by setting $v(L) \equiv \delta$.)

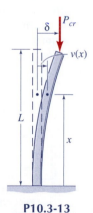

P10.3-13

*Prob. 10.3-14. Following a procedure similar to the one used in Fig. 10.11 and in Eqs. 10.20 through 10.32 to determine an expression for the elastic buckling load of a fixed-pinned column, derive an expression for the buckling load of the fixed-free column shown in Fig. P10.3-14. The portion of the column from A to B is flexible, with modulus of elasticity E and moment of inertia I, but the portion of the column from B to the point of application of the load P can be considered to be perfectly rigid. (Hint: Take a free-body diagram from section x to the top of the column, where the

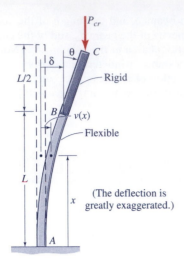

P10.3-14

force P is applied. Use the end conditions that $v(L) \equiv \delta$ and $v'(L) \equiv \theta$.)

> **Problems 10.4-1 through 10.4-12.** *In solving these problems, assume that the compression member in question is an ideal slender, prismatic, elastic column; and assume that bending occurs in the xy plane.*

Prob. 10.4-1. The cantilever column in Fig. P10.4-1 has a vertical load P that is applied at an eccentricity e with respect to the axis of the column. Starting with a free-body diagram, formulate the differential equation that governs the beam-column action of this member. Then, solving the differential equation, determine expressions for (a) the maximum transverse deflection, v_{max}, and (b) the maximum bending moment, M_{max}.

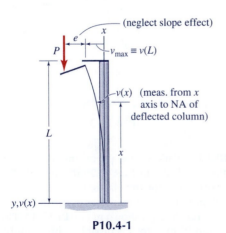

P10.4-1

Prob. 10.4-2. A compressive load $P = 2$ kips is applied parallel to the axis of the column in Fig. P10.4-2 at an eccentricity $e = 0.2$ in. from the axis. The steel column ($E = 30 \times 10^3$

ksi) has a circular cross section with diameter $d = 1$ in.; its length is $L = 5$ ft. (a) Determine the maximum deflection, v_{max}, of this eccentrically loaded column, and (b) determine the maximum bending moment, M_{max}.

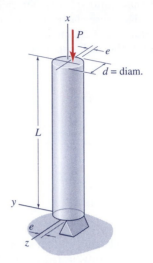

P10.4-2, P10.4-7, P10.4-9, and P10.4-15

Prob. 10.4-3. A compressive load $P = 100$ kN is applied parallel to the axis of the column in Fig. P10.4-3 at an eccentricity $e = 30$ mm from the axis. The aluminum-alloy column ($E = 70$ GPa) has a thin-wall, square, box cross section with outer dimension $b = 130$ mm and wall thickness $t = 10$ mm. Its length is $L = 4$ m. (a) Determine the maximum deflection, v_{max}, of this eccentrically loaded column, and (b) determine the maximum bending moment, M_{max}.

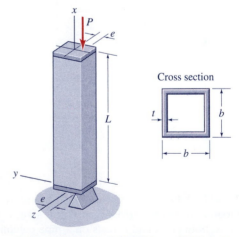

P10.4-3, P10.4-8, P10.4-10, and P10.4-16

[D]Prob. 10.4-4. A load $P = 120$ kips is applied at an eccentricity $e = 8$ in. in the plane of the web of the **W**12 × 50 wide-flange steel column shown in Fig. P10.4-4. Let $E = 30 \times 10^3$ ksi, and see Table D.1 for the cross-sectional properties of this member. What is the maximum length, L_{allow}, that this column can have if the transverse deflection at the top of the column must not exceed $v_{max} = 1$ in.? (Note: Since this is a cantilever column, you will need to consider its "effective length.")

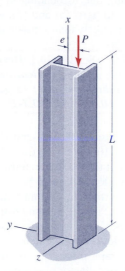

P10.4-4 and P10.4-5

^{D}Prob. 10.4-5. Repeat Prob. 10.4-4 for a **W360 $\times$ 79** wide-flange steel column as follows: $E = 70$ GPa, $P = 300$ kN, $e = 240$ mm, and $v_{max} = 25$ mm. (See Table D.2 for the cross-sectional properties of this column.)

Prob. 10.4-6. A compressive load P is applied in the mid-plane of a square pinned-end column at an eccentricity e from the axis of the column, as shown in Fig. P10.4-6. (a) If $P = P_{cr}/4$ and $e = b/4$, determine the maximum deflection of the column, v_{max}. (P_{cr} is the Euler buckling load.) (b) For the same conditions as in (a), determine an expression for the maximum bending moment M_{max}.

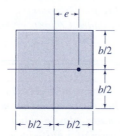

P10.4-6

^{D}Prob. 10.4-7. Consider the steel column in Fig. P10.4-7 (see Prob. 10.4-2). Let $d = 1$ in., $e = 0.2$ in., $L = 5$ ft, and $E = 30 \times 10^3$ ksi. (a) If the applied compressive load is $P = 2.5$ kips, what is the magnitude of the maximum normal stress, σ_{max}? (b) What is the allowable compressive load, P_{allow}, if the steel has a yield strength of $\sigma_Y = 36$ ksi and the factor of safety with respect to yielding of the material is $FS = 3$?

^{D}Prob. 10.4-8. Consider the aluminum-alloy column in Fig. P10.4-8 (see Prob. 10.4-3). Let $b = 130$ mm, $t = 10$ mm,

$e = 30$ mm, $L = 4$ m, and $E = 70$ GPa. (a) If the applied compressive load is $P = 200$ kN, what is the magnitude of the maximum normal stress, σ_{max}? (b) What is the allowable compressive load, P_{allow}, if the steel has a yield strength of $\sigma_Y = 250$ MPa and the factor of safety with respect to yielding of the material is $FS = 2.5$?

^{*D}Prob. 10.4-9. Consider the steel column in Fig. P10.4-9 (see Prob. 10.4-2). Let $E = 30 \times 10^3$ ksi, $L = 8$ ft, $\sigma_Y = 36$ ksi, and the factor of safety with respect to yielding be $FS = 2.5$. (a) If the maximum compressive load to be applied to the column is $P = 2.0$ kips, and it is to be applied at an eccentricity $e = 0.25$ in., what is the required (i.e., minimum) column diameter, d, to the nearest 0.10 in.? (b) If the diameter of the column is $d = 2.0$ in., and the maximum compressive load to be applied to the column is $P = 5.0$ kips, what (to the nearest 0.10 in.) is the maximum eccentricity, e_{max}, at which this load may be applied?

^{*D}Prob. 10.4-10. Consider the aluminum-alloy column in Fig. P10.4-10 (see Prob. 10.4-3). Let $E = 70$ GPa, $L = 6$ m, $t = 15$ mm, $\sigma_Y = 250$ MPa, and the factor of safety with respect to yielding be $FS = 2$. (a) If the maximum compressive load to be applied to the column is $P = 250$ kN, and it is to be applied at an eccentricity $e = 50$ mm, what is the required (i.e., minimum) column outer cross-sectional dimension, b, to the nearest mm? (b) If the outer cross-sectional dimension of the column is $b = 180$ mm, and the maximum compressive load to be applied to the column is $P = 250$ kN, what, to the nearest 1 mm, is the maximum eccentricity, e_{max}, at which this load may be applied?

^{*D}Prob. 10.4-11. A compressive load P is applied parallel to the axis of the steel pin-ended pipe column (d_o = outer diameter; t = wall thickness) in Fig. P10.4-11 at an eccentricity e from the axis. Bending is restricted to the plane containing the axis of the column and the line of action of the loading. Let $E = 30 \times 10^3$ ksi, $L = 10$ ft, $t = 0.25$ in., and $e = 6$ in. (a) If the outer diameter of the column is $d_o = 6$ in., and the compressive load is $P = 25$ kips, what is the

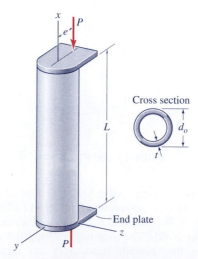

P10.4-11 and P10.4-12

maximum normal stress in the column? (b) If the outer diameter of the column is $d_o = 6$ in., the yield strength of the material is $\sigma_Y = 50$ ksi, and the factor of safety with respect to yielding is $FS = 2.5$, what is the allowable load, P_{allow}? (c) If the allowable load is $P_{allow} = 10$ kips, the yield strength of the steel is $\sigma_Y = 50$ ksi, and the factor of safety with respect to yielding is $FS = 2.5$, what is the required (minimum) outer diameter, d_o, to the nearest 0.1 in.?

*D**Prob. 10.4-12.** For the steel pin-ended pipe column in Fig. P10.4-12, let $E = 210$ GPa, $L = 4$ m, $t = 10$ mm, and $e = 50$ mm. (a) If the outer diameter of the column is $d_o = 220$ mm, and the compressive load is $P = 500$ kN, what is the maximum normal stress in the column? (b) If the outer diameter of the column is $d_o = 220$ mm, the yield strength of the material is $\sigma_Y = 250$ MPa, and the factor of safety with respect to yielding is $FS = 2.5$, what is the allowable load P_{allow}? (c) If the allowable load is $P_{allow} = 200$ kN, the yield strength of the steel is $\sigma_Y = 250$ MPa, and the factor of safety with respect to yielding is $FS = 2.5$, what is the required (minimum) outer diameter, d_o, to the nearest mm?

*Prob. 10.4-13.** A compressive load P is applied parallel to the x axis of the slender, rectangular aluminum-alloy member in Fig. P10.4-13. The load acts in the xy plane at an eccentricity e from the z axis. The member is supported at its ends in a manner that permits bending in any direction. Let $E = 70$ GPa, $L = 1$ m, $b = 10$ mm, $h = 40$ mm, and $e = 10$ mm. (a) Determine the load $(P_{cr})_y$ for elastic buckling of the member for bending in the xz plane. (b) If the yield strength of the material is $\sigma_Y = 270$ MPa, what load $(P_Y)_z$ would cause yielding due to bending in the xy plane? (c) From your answers to Parts (a) and (b), what can you conclude about the most likely failure mode of this member?

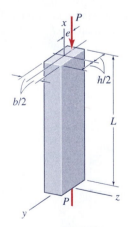

P10.4-13 and P10.4-14

*Prob. 10.4-14.** A compressive load P is applied parallel to the x axis of the slender, rectangular titanium-alloy member in Fig. P10.4-14. The load acts in the xy plane at an eccentricity e from the z axis. The member is supported at its ends in a manner that permits bending in any direction. Let

$E = 16 \times 10^3$ ksi, $L = 20$ in., $b = 0.25$ in., $h = 0.50$ in., and $e = 0.25$ in. (a) Determine the load $(P_{cr})_y$ for elastic buckling of the member for bending in the xz plane, (b) If the yield strength of the material is $\sigma_Y = 120$ ksi, what load $(P_Y)_z$ would cause yielding due to bending in the xy plane? (c) From your answers to parts (a) and (b), what can you conclude about the most likely failure mode of this member?

*C**Prob. 10.4-15.** Consider the pin-ended column in Prob. 10.4-2, with $E = 30 \times 10^3$ ksi, $L = 8$ ft, $d = 2$ in., and $e = 0.5$ in. (a) Generate a plot of the maximum normal stress, σ_{max}, versus the compressive load, P, for $0 \le P \le 15$ kips. Assume that the maximum normal stress remains less than σ_Y for this range of loading, (b) Plot the maximum transverse displacement, v_{max}, for values of the compressive load $0 \le P \le 15$ kips. (c) Based on the plots you have generated in (a) and (b), write n short paragraph discussing the difference between using Eq. 2.26 to define "factor of safety" and using Eq. 2.27 to define "factor of safety." (Note that the *Procedure for Determining the Allowable Load for an Eccentrically Loaded Column* specifies the use of Eq. 2.26.)

*C**Prob. 10.4-16.** Consider the pin-ended column in Prob. 10.4-3, with $E = 70$ GPa, $L = 4$ m, $b = 130$ mm, $t = 10$ mm, and $e = 30$ mm. (a) Generate a plot of the maximum normal stress, σ_{max}, versus the compressive load, P, for the $0 \le P \le 350$ kN. Assume that the maximum normal stress remains less than σ_Y for this range of loading. (b) Plot the maximum transverse displacement, v_{max}, for values of the compressive load $0 \le P \le 350$ kN. (c) Based on the plots you have generated in (a) and (b), write a short paragraph discussing the difference between using Eq. 2.26 to define "factor of safety" and using Eq. 2.27 to define "factor of safety." (Note that the *Procedure for Determining the Allowable Load for an Eccentrically Loaded Column*, given in Section 10.4, specifies the use of Eq. 2.26.)

Problems 10.5-1 through 10.5-5. *In solving these problems, assume that the compression member in question is a slender, prismatic, elastic column with an initial deflection in the form of a half-sine curve (Eq. 10.47) with amplitude δ_0. Also assume that the compressive load P acts at the centroid of the cross section.*

Prob. 10.5-1. A compressive load $P = 2$ kips is applied parallel to the axis of an imperfect pin-ended column, like the one shown in Fig. 10.18. The amplitude of the initial imperfection is $\delta_0 = v_0(L/2) = 0.25$ in. The steel column ($E = 30 \times 10^3$ ksi) has a circular cross section with diameter $d = 1.25$ in., and its length is $L = 6$ ft. (a) Determine the maximum total deflection, δ_{max}, of this eccentrically loaded column, and (b) determine the maximum normal stress, σ_{max}.

Prob. 10.5-2. A compressive load $P = 200$ kN is applied parallel to the axis of an imperfect pin-ended column. like the one shown in Fig. 10.18. The amplitude of the initial imperfection is $\delta_0 \equiv v_0(L/2) = 30$ mm. The aluminum-alloy column ($E = 70$ GPa) has a thin-wall, square, box cross section with outer dimensions $b = 200$ mm and wall thickness $t = 10$ mm, Its length is $L = 5$ m. (a) Determine the maximum total deflection, δ_{max}, of this eccentrically loaded column, and (b) determine the maximum normal stress, σ_{max}.

Prob. 10.5-3. A compressive load P is applied parallel to the axis of a pin-ended, imperfect elastic (modulus E) column, like the one shown in Fig. 10.18. The cross section of the column is a square with edge dimension b, and its length is L. (a) If $P = P_{cr}/4$ and $\delta_0 = b/4$. determine an expression for the maximum total deflection of the column, δ_{max} (P_{cr} is the Euler buckling load.) Assume that $\sigma_{max} < \sigma_Y$. (b) For the same conditions as in (a), determine an expression for the maximum normal stress σ_{max}.

D*Prob. 10.5-4. A compressive load P is applied to an imperfect. steel, pin-ended pipe column, like the one shown in Fig. 10.18 (d_o = outer diameter, t = wall thickness). Let $E = 30 \times 10^3$ ksi, $L = 10$ ft, $t = 0.25$ in., and $\delta_0 = 0.5$ in. (a) If the outer diameter of the column is $d_o = 6$ in., and the compressive load is $P = 25$ kips, what is the maximum normal stress in the column? (b) If the outer diameter of the column is $d_o = 6$ in., the yield strength of the material is $\sigma_Y = 50$ ksi, and the factor of safety with respect to yielding is $FS = 2.5$, what is the allowable load P_{allow}? (c) If the allowable load is $P_{allow} = 80$ kips, the yield strength of the steel is $\sigma_Y = 50$ ksi, and the factor of safety with respect to yielding is $FS = 2.5$, what is the required (minimum) outer diameter, d_o, to the nearest 0.1 in.?

D*Prob. 10.5-5. A compressive load P is applied to an imperfect, steel, pin-ended pipe column, like the one shown in Fig. 10.18. Let $E = 210$ GPa, $L = 3$ m, $t = 20$ mm, and $\delta_0 = 10$ mm. (a) If the outer diameter of the column is $d_o = 100$ mm, and the compressive load is $P = 300$ kN, what is the maximum normal stress in the column? (b) If the outer diameter of the column is $d_o = 100$ mm, the yield strength of the material is $\sigma_Y = 270$ MPa, and the factor of safety with respect to yielding is $FS = 3.0$, what is the allowable load P_{allow}? (c) If the allowable load is $P_{allow} = 300$ kN, the yield strength of the steel is $\sigma_Y = 270$ MPa, and the factor of safety with respect to yielding is $FS = 3$, what is the required (minimum) outer diameter, d_o, to the nearest mm?

Computer Exercises—Section 10.5. *Develop a computer program (e.g., using a mathematical programming language or a spreadsheet program) to generate the plots required in Probs. 10.5-6 and 10.5-7. In solving these problems, assume that the compression member in question is a slender, prismatic, elastic column with an initial deflection in the form of a half-sine curve (Eq. 10.47) with amplitude δ_0. Also assume that the compressive load P acts at the centroid of the cross section.*

***C*Prob. 10.5-6.** A compressive load P is applied to an imperfect pin-ended column, like the one shown in Fig. 10.18. The steel column ($E = 30 \times 10^3$ ksi, $\sigma_Y = 36$ ksi) has a circular cross section with diameter $d = 2$ in. (a) Let the compressive load be $P = P_Y$, the load that makes $\sigma_{max} = \sigma_Y$. For an initial imperfection $\delta_0 = 0.25$ in., generate a plot of the average normal stress, P/A, versus the normalized length, L/r, for $10 \leq L/r \leq 200$. (b) Change δ_0 from 0.25 in. to 0.5 in., and generate a second plot. (c) Write a short paragraph discussing the effect of initial imperfection amplitude and column length on the ability of columns to carry compressive loads.

***C*Prob. 10.5-7.** A compressive load P is applied to an imperfect pin-ended column, like the one shown in Fig. 10.18. The steel column ($E = 200$ GPa, $\sigma_Y = 340$ MPa) has a square, box cross section with cross-sectional dimensions $b = 100$ mm and $t = 12.5$ mm. (a) Let the compressive load be $P = P_Y$, the load that makes $\sigma_{max} = \sigma_Y$. For an initial imperfection $\delta_0 = 10$ mm, generate a plot of the average normal stress P/A, versus the normalized length, L/r, for $10 \leq L/r \leq 200$. (b) Change δ_0 from 10 mm to 20 mm, and generate a second plot. (c) Write a short paragraph discussing the effect of initial imperfection amplitude and column length on the ability of columns to carry compressive loads.

Problems 10.6-1 through 10.6-3. *In solving these problems, assume that the compression member in question is a slender, prismatic ideal column made of the material whose compressive stress-strain properties are plotted.*

Prob. 10.6-1. A 6-in. (nominal diameter) standard-weight steel pipe column has an effective length of 120 in. (See Table D.7 for the cross-sectional properties of this pipe column.) The steel of which the column is made has a compressive stress-strain curve that is given by the curve ABC in Fig. 2 of Example 10.5. The compressive tangent modulus curve for this material is the curve $DEFG$ in this same figure. (a) How long would the column have to be for it to buckle elastically (i.e., according to the Euler formula for elastic buckling)? (b) Calculate the buckling load for this column. (Assume that $\sigma_{EL} = \sigma_{PL}$, and estimate σ_{PL} from the σ vs E_t curve.)

Prob. 10.6-2. A 4 in. $\times$ 4 in. $\times \frac{3}{8}$ in. (actual dimensions; see Fig. P10.6-2a) aluminum-alloy box column has an effective length of 72 in. The material of which the column is made has a compressive stress-strain curve that is given by the curve ABC in Fig. P10.6-2b. The compressive tangent modulus curve (i.e., the σ vs E_t curve) for this material is the curve $DEFG$ in this same figure. (a) How long would the column have to be for it to buckle elastically (i.e., according to the Euler formula for elastic buckling)? (b) Calculate the buckling load for this column. (Assume that $\sigma_{EL} = \sigma_{PL}$, and estimate σ_{PL} from the σ vs E_t curve.)

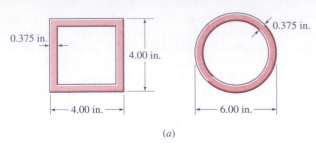

(a)

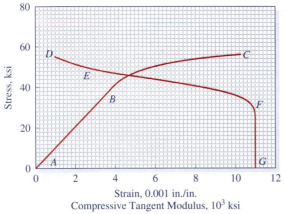

(b)

P10.6-2 and P10.6-3

Prob. 10.6-3. An aluminum-alloy pipe column has the material properties shown in Fig. 10.6-3b. The (actual) dimensions of the pipe column, shown in Fig. 10.6-3a, are: outer diameter d_o = 6.000 in., wall thickness t = 0.375 in., and length L = 8 ft. (a) Calculate the buckling load for this column. (Assume that $\sigma_{EL} = \sigma_{PL}$, and estimate σ_{PL} from the σ vs E_t curve.) (b) How long would the column have to be for it to buckle elastically (i.e., according to the Euler formula for elastic buckling)?

▼ DESIGN OF COLUMNS | **MDS 10.4–10.9**

> **Problems for Section 10.7.** *In solving these problems, assume that the compressive axial load is centrally applied at the ends of the column and, unless indicated otherwise, that the column is supported in a manner that permits buckling in any direction.*

ᴰProb. 10.7-1. Determine the allowable compressive axial loads P (= P_{allow}) for pin-supported **W**8 × 40 wide-flange steel columns having the following lengths: L = 12 ft, L = 16 ft, and L = 24 ft. Let E = 29 × 10³ ksi and σ_Y = 36 ksi.

ᴰProb. 10.7-2. Determine the allowable compressive axial loads P (= P_{allow}) for pin-supported **W**12 × 65 wide-flange steel columns having the following lengths: L = 10 ft, L = 20 ft, and L = 30 ft. Let E = 29 × 10³ ksi and σ_Y = 50 ksi.

ᴰProb. 10.7-3. Determine the allowable compressive axial loads P (= P_{allow}) for pin-supported **W**14 × 53 wide-flange steel columns having the following lengths: L = 18 ft. L = 22 ft. and L = 26 ft. Let E = 29 × 10³ ksi and σ_Y = 36 ksi.

ᴰProb. 10.7-4. Determine the allowable compressive axial loads P (= P_{allow}) for pin-supported **W**200 × 59 wide-flange steel columns having the following lengths: L = 4 m, L = 6 m, and L = 8 m. Let E = 200 GPa and σ_Y = 250 MPa.

ᴰProb. 10.7-5. Determine the allowable compressive axial loads P (= P_{allow}) for pin-supported **W**250 × 45 wide-flange steel columns having the following lengths: L = 4 m, L = 5 m, and L = 6 m. Let E = 200 GPa and σ_Y = 250 MPa.

ᴰProb. 10.7-6. Determine the allowable compressive axial loads P (= P_{allow}) for pin-supported **W**310 × 97 wide-flange steel columns having the following lengths: L = 6 m, L = 8 m, and L = 10 m. Let E = 200 GPa and σ_Y = 340 MPa.

ᴰProb. 10.7-7. Determine the maximum allowable length for a pin-supported **W**10 × 60 wide-flange steel column if the compressive axial load is (a) 150 kips; (b) 300 kips. Let E = 29 × 10³ ksi and σ_Y = 36 ksi.

ᴰProb. 10.7-8. Determine the maximum allowable length for a pin-supported **W**12 × 50 wide-flange steel column if the compressive axial load is (a) 100 kips; (b) 200 kips. Let E = 29 × 10³ ksi and σ_Y = 36 ksi.

ᴰProb. 10.7-9. Determine the maximum allowable length for a pin-supported **W**16 × 100 wide-flange steel column if the compressive axial load is (a) 300 kips; (b) 450 kips. Let E = 29 × 10³ ksi and σ_Y = 50 ksi.

ᴰProb. 10.7-10. The 27-ft-long **W**12 × 50 wide-flange steel column shown in Fig. P10.7-10 has a "fixed" base and receives equal downward (compressive) loading from the beams that frame into it at the top. Consider only buckling in the xy plane. Because of the flexibility of the adjoining beams, the end conditions of the column fall somewhere between "fixed-fixed" and "pinned-pinned." It is estimated

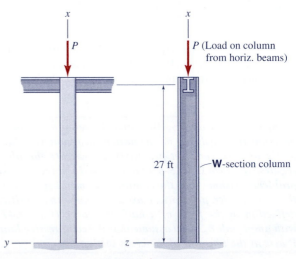

P10.7-10, P10.7-11, and P10.7-12

that the effective length factor, K, for the beam falls in the range $0.6 \leq K \leq 0.8$. (Recall that $K = 0.5$ for a fixed-ended column and $K = 1.0$ for a pin-ended column.) Using the AISC formulas, determine the allowable loads for the following effective-length factors: $K = 0.6, K = 0.7,$ and $K = 0.8$. Let $E = 29 \times 10^3$ ksi and $\sigma_Y = 36$ ksi.

DProb. 10.7-11. Repeat Prob. 10.7-10 for a 24-ft-long **W**16 × 40 wide-flange steel column.

DProb. 10.7-12. Repeat Prob. 10.7-10 for a 28-ft-long **W**8 × 40 wide-flange steel column.

DProb. 10.7-13. Using the Aluminum Association column design formulas, determine the allowable compressive axial loads $P\,(= P_{\text{allow}})$ for 2.5 in. × 2.5 in. (i.e., $b = 2.5$ in. in Fig. P10.7-13) pin-supported 2014-T6 aluminum-alloy columns having the following lengths: $L = 3$ ft, $L = 4$ ft, and $L = 5$ ft.

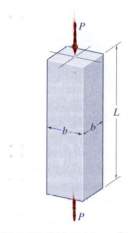

P10.7-13, P10.7-14, and P10.7-15

DProb. 10.7-14. Using the Aluminum Association column design formulas, determine the maximum allowable length for the 3 in. × 3 in. pin-supported 2014-T6 aluminum-alloy column in Fig. P10.7-14 if the compressive axial load is (a) 150 kips; (b) 175 kips.

DProb. 10.7-15. To the nearest $\frac{1}{16}$ in., determine the minimum cross-sectional dimension b for a square cross-section, pin-supported 4-ft-long 2014-T6 aluminum-alloy column (Fig. P10.7-15) if the compressive axial load is (a) 150 kips; (b) 175 kips. Use the Aluminum Association column design formulas.

DProb. 10.7-16. Determine the maximum allowable length for the 2 in. × 4 in. pin-supported 2014-T6 aluminum-alloy column in Fig. P10.7-16 if the compressive axial load is (a) 100 kips; (b) 150 kips. Use the Aluminum Association column design formulas, and assume that the column is free to buckle in any direction.

DProb. 10.7-17. To the nearest $\frac{1}{16}$ in., determine the minimum cross-sectional dimension b for a $b \times 2b$ rectangular-cross-section pin-supported 2-ft-long 2014-T6 aluminum-alloy column (Fig. P10.7-17) if the compressive axial load is 100 kips. Use the Aluminum Association column design formulas, and assume that the column is free to buckle in any direction.

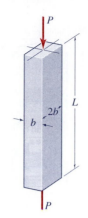

P10.7-16 and P10.7-17

DProb. 10.7-18. Determine the maximum allowable length for the pin-supported 2014-T6 aluminum-alloy pipe column in Fig. P10.7-18 if $d_o = 5$ in., $t = 0.5$ in., and if the compressive axial load is (a) 75 kips; (b) 150 kips. Use the Aluminum Association column design formulas.

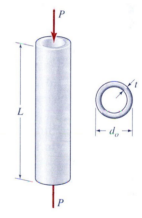

P10.7-18 and P10.7-19

DProb. 10.7-19. The pin-supported 2014-T6 aluminum-alloy pipe column in Fig. P10.7-19 is $L = 10$ ft long, (a) If the outer diameter is $d_o = 6$ in. and the wall thickness is $t = 0.5$ in., what is the allowable compressive axial load? (b) To the nearest $\frac{1}{16}$ in., determine the minimum wall thickness t if the outer diameter of the column is $d_o = 8$ in. and the compressive axial load is 250 kips. Use the Aluminum Association column design formulas.

DProb. 10.7-20. Determine the allowable compressive axial loads $P\,(= P_{\text{allow}})$ for 4 × 4 (nominal dimensions; see Table D.8 for actual finish dimensions) S4S timber columns (Fig. P10.7-20) having the following lengths: $L = 6$ ft and $L = 8$ ft. Use the NDS design formula with $E = 1800$ ksi and $F_c = 1.2$ ksi, and assume "pin supports" (i.e., $K = 1$).

DProb. 10.7-21. Determine the maximum allowable length for the 4 × 4 (nominal dimensions; see Table D.8) S4S timber columns in Fig. P10.7-21 if the compressive axial load is (a) 6 kips; (b) 9 kips; and (c) 12 kips. Use the NDS design formula with $E = 1800$ ksi and $F_c = 1.2$ ksi, and assume "pin supports" (i.e., $K = 1$).

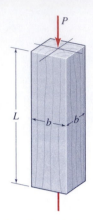

P10.7-20 and P10.7-21

ᴰ**Prob. 10.7-22.** The timber column shown in Fig. P10.7-22 frames into heavy beams at its top and at its base, so that partial fixity with $K = 0.8$ can safely be assumed. The column is 12 ft long and must support an axial compressive load of $P = 30$ kips. From Table D.8, select the minimum size of square-cross-section $S4S$ timber that can be used for this column. Use the NDS design formula with $E = 1800$ ksi and $F_c = 1.2$ ksi.

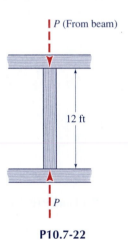

P10.7-22

ᴰ**Prob. 10.7-23.** Determine the allowable compressive axial loads $P (= P_{allow})$ for 4×6 (nominal dimensions; see Table D.8 for actual finish dimensions) $S4S$ timber columns (Fig. P10.7-23) having the following lengths: $L = 6$ ft, $L = 8$ ft, and $L = 10$ ft. Use the NDS design formula with $E = 1800$ ksi and $F_c = 1.2$ ksi, and assume "pin supports" (i.e., $K = 1$).

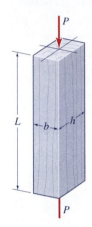

P10.7-23 and P10.7-24

ᴰ**Prob. 10.7-24.** From Table D.8, select the minimum-size $6 \times h_{nom}$ (nominal dimensions) $S4S$ timber column 8-ft long (Fig. P10.7-24) that will support an axial compressive load of up to $P = 40$ kips, Use the NDS design formula with $E = 1800$ ksi and $F_c = 1.2$ ksi, and assume "pin supports" (i.e., $K = 1$).

Section		Suggested Review Problems
10.1	In this chapter you learned that there are systems for which it is necessary to examine the **stability of equilibrium–unstable, neutrally stable,** or **stable.** To assess the stability of a structure, it is **slightly displaced** from its original equilibrium configuration to discover the load required to produce an **alternate equilibrium configuration.** An important example of this is the **buckling of columns,** that is, behavior of slender members in compression. The phenomenon of buckling can be illustrated by the simplified model shown in Fig. 10.4.	10.1-3
	  (a) The vertical equilibrium configuration. (b) A displaced configuration. (c) A free-body diagram. A simplified model of column buckling (Fig. 10.4)	
10.2	The study of the stability of real columns with distributed flexibility begins with the study of the **ideal pin-ended column,** for which the buckling load is called the **Euler buckling load** P_{cr} (Eq. 10.16). The alternate equilibrium configuration is called the **buckling mode** (Eq. 10.17). P_{cr} the lowest load at which buckling can occur, and $v_1(x)$ is called the **fundamental mode shape.** Note that only the shape is determined; not the amplitude.	Derive Eq. 10.16 and Eq. 10.17. 10.2-3 10.2-7 10.2-15
	  (a) Ideal column. (b) Buckled configuration. (c) FBD of entire column. (d) FBD of partial column. Buckling of an ideal pin-ended column (Fig. 10.6) $$P_{cr} = \frac{\pi^2 EI}{L^2} \qquad (10.16)$$ $$v_1(x) = C\sin\left(\frac{\pi x}{L}\right) \qquad (10.17)$$	

Section		Suggested Review Problems
	The *Euler buckling load* can be divided by the cross-sectional area A of the column to give the **Euler buckling stress** (Eq. 10.18). $$\sigma_{cr} = \frac{\pi^2 E}{(L/r)^2} \quad (10.18)$$ where $r = \sqrt{I/A}$ is called the *radius of gyration* of the column. The ratio L/r is the key buckling parameter, as is seen in the plots of the **Euler formula** in Fig. 10.8. Notice how buckling stress (and load) decrease with increasing L/r ratio.	Graphs of the Euler formula for structural steel and an aluminum alloy. (Fig. 10.8)	
10.3	Section 10.3 discusses the effect that various end conditions (e.g., fixed end, free end, etc.) have on the buckling load of a column. Then, $$\sigma_{cr} = \frac{\pi^2 E}{(KL/r)^2} \quad (10.36)$$ where KL is called the **effective length** of the column. Figure 10.13 depicts the effective length of columns with various end conditions.	The effective length of various columns. (Fig. 10.13)	10.3-1 10.3-13
10.7	Section 10.7 discusses the **design of centrally loaded columns.** You should review the **column design procedure,** which is an iterative procedure.	Column design is governed by formulas provided by the various engineering societies. You should review the examples of column design for **structural steel columns, aluminum-alloy columns, and wood columns.**	10.7-3 10.7-13 10.7-21

Sections 10.4 (**Eccentric Loading**), 10.5 (**Imperfections**), and 10.6 (**Inelastic Buckling**) are "optional" sections.

ENERGY METHODS

11

11.1 INTRODUCTION

In Chapters 1 through 10 we employed the three fundamental concepts of deformable-body mechanics (*equilibrium, geometry of deformation,* and *constitutive behavior of materials*) to examine the response of several types of structural members to applied loads and/or temperature changes. We determined the distribution of normal stress and shear stress in members and the deformation of the members. We also examined the stability of members undergoing axial compression. We turn now to the important topic of **energy methods** in deformable-body mechanics. Before the advent of the digital computer, energy methods were the most powerful tools available for solving deflection problems, especially statically indeterminate problems. And now, energy methods form a basis of the finite-element method, the most popular and most powerful current method for analyzing deformable bodies (machines, structures, etc.). You will see that the energy methods presented in this chapter again incorporate the three essentials of deformable-body mechanics— equilibrium, geometry of deformation, and constitutive behavior of materials.

In mechanics the term *work* refers to a quantity that is basically (*force × distance*). When work is done on a deformable body, some or all of the work done on the body goes into *strain energy* stored in the body. For example, when you stretch a rubber band by pulling on it, the work that you do on the rubber band is stored as strain energy. When you release the applied force, the rubber band releases this energy as it returns to its undeformed shape.

In this chapter we will define a number of work-energy terms: work of external forces, complementary work, strain energy, complementary strain energy, virtual displacements, virtual forces, and virtual work. The *Work-Energy Principle* will be employed to calculate the deflection of members subjected to single static loads. Several more powerful energy methods—*Castigliano's Second Theorem,* the *Principal of Virtual Displacements,* and the *Principle of Virtual Forces*—will also be introduced and will be used to solve more complex problems, such as statically indeterminate problems for structures with several loads acting simultaneously. The relationship of the energy principles to the *Displacement Method* and the *Force Method,* introduced in Chapter 3 and used in subsequent chapters, will be pointed out. Finally, energy methods will be used to solve several simple *impact-loading* problems.

This section is devoted to providing the definitions of important work-energy terms that will be used in later sections of this chapter. There, several work-energy principles are introduced, and deformable-body mechanics problems are solved by use of these work-energy principles.

Figure 11.1 shows members with three types of applied loads—axial (Fig. 11.1a). torsional (Fig. 11.1b), and bending (Fig. 11.1c). Below each of these figures is a plot of the load-deformation curve that might typically be obtained in each case by slowly increasing the magnitude of the applied load. In the definitions that follow we will refer primarily to the axial-loading case, but the definitions can easily be extended to the other two cases in Fig. 11.1 as well as to more general loading by forces and couples.

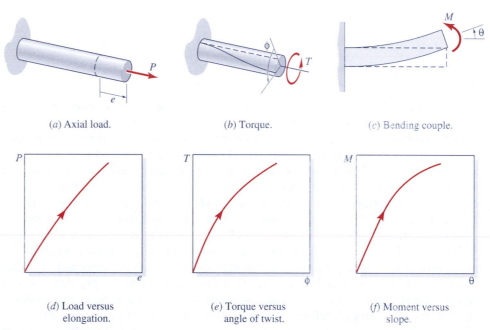

(a) Axial load.

(b) Torque.

(c) Bending couple.

(d) Load versus elongation.

(e) Torque versus angle of twist.

(f) Moment versus slope.

FIGURE 11.1 Several load-deformation cases. (To conserve space, only the part of the load-deformation curves corresponding to positive loads are shown. Negative loading is, however, equally permissible.)

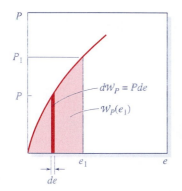

FIGURE 11.2 Load versus elongation curve for axial deformation; the area representing the work $\mathcal{W}$.

Work. Consider the case of axial loading of a slender rod by a single load P, as indicated in Fig. 11.1a. The **work** done <u>by</u> the force P to elongate the rod by an amount e_1 will be designated by $\mathcal{W}_P(e_1)$, and it is given by the integral of force times distance

$$\mathcal{W}_P(e_1) = \int_0^{e_1} P(e) \, de \qquad \textbf{Work} \qquad (11.1)$$

On Fig. 11.2 this appears as the area (shaded red) <u>below</u> the $P - e$ curve up to $e = e_1$. Since the force varies with elongation (i.e., with "distance"), the integral expression in Eq. 11.1 must be used to calculate the work done as the force is increased up to the value $P_1 \equiv P(e_1)$.

In a similar manner, the work done by a torsional couple T and by a bending couple M, respectively, are

$$W_T(\phi_1) = \int_0^{\phi_1} T(\phi)\,d\phi, \quad W_M(\theta_1) = \int_0^{\theta_1} M(\theta)\,d\theta \qquad (11.2)$$

If the load-displacement relationship is linear, the curves in Fig. 11.2 become straight lines, and the area under the curve is triangular. Therefore, for the linear case.

$$W_P = \frac{1}{2}Pe, \quad W_T = \frac{1}{2}T\phi, \quad W_M = \frac{1}{2}M\theta \qquad (11.3)$$

EXAMPLE 11.1

The axial deformation member in Figs. 1a and 1b has the linear load-elongation curve shown in Fig. 1c. Using the load-elongation curve, explain why $W(\Delta_1 + \Delta_2) \neq W(\Delta_1) + W(\Delta_2)$.

Plan the Solution The work done by an axial load acting on a linearly elastic member is given by Eq. 11.3a and is represented by the triangular area under the load-elongation curve.

Solution From Eq. 11.3a and Fig. 1c, $W(\Delta_1)$ is the lower-left triangular area

$$W(\Delta_1) = \frac{1}{2}P_1\Delta_1 \qquad (1)$$

The work done by force P, over an elongation Δ_2 only, $W(\Delta_2)$, is the upper-right triangular area in Fig. 1c. given by

$$W(\Delta_2) = \frac{1}{2}P_2\Delta_2 \qquad (2)$$

and $W(\Delta_1 + \Delta_2)$ is the entire area under the load-elongation curve up to $\Delta = \Delta_1 + \Delta_2$.

$$W(\Delta_1 + \Delta_2) = \frac{1}{2}(P_1 + P_2)(\Delta_1 + \Delta_2)$$

$$= \frac{1}{2}P_1\Delta_1 + \left[\frac{1}{2}P_1\Delta_2 + \frac{1}{2}P_2\Delta_1\right] + \frac{1}{2}P_2\Delta_2 \qquad (3)$$

$$= \frac{1}{2}P_1\Delta_1 + [P_1\Delta_2] + \frac{1}{2}P_2\Delta_2$$

where the equation $P = ke$ has been used to obtain the $P_1\Delta_2$ term from the bracketed term on the preceding line. Therefore, due to the presence of the terms in square brackets in Eq. (3),

$$W(\Delta_1 + \Delta_2) \neq W(\Delta_1) + W(\Delta_2) \qquad (4)$$

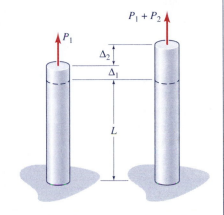

(a) Load P_1 applied. (b) Load P_2 added.

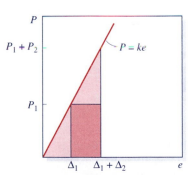

(c) Load versus elongation curve.

Fig. 1 An illustration of the work due to multiple loads.

The term in square brackets in Eq. (3) corresponds to the darker-shaded rectangular area in Fig. 1c. It is the additional work done by the first load P_1 when the second load P_2 is applied, stretching the rod an additional amount Δ_2.

Review the Solution   The graphical interpretation of the equations agrees with the terms of the equations.

The lesson that we learn from this example is that work is not a linear function of load (or deformation), so we must exercise care in the calculation of work, especially when multiple loads are involved.

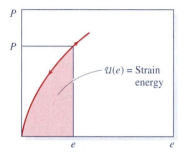

FIGURE 11.3   An illustration of strain energy.

Strain Energy.   As indicated in Section 2.5, a material is said to be *elastic* (or, to behave elastically) if the load-deflection path traced on unloading is the same as the load-deflection path traced on loading. The arrows in Fig. 11.3 denote the fact that the same load-elongation curve is traced when the load is decreasing as when it is increasing. The work done on an elastic member during loading is stored in the member as strain energy, and that strain energy is recovered upon unloading. For an elastic body, the work, W, done <u>on</u> the body <u>by</u> the applied force is stored as **elastic strain energy,** $\mathcal{U}$. That is,

$$\mathcal{U}(e) = \mathcal{W}(e)|_{\text{done on elastic body}} \tag{11.4}$$

Since the strain energy is equal to the work done on the elastic body, it is represented by the area that is shaded red in Fig. 11.3.

Work and energy are both expressed in **units** of *force × length*. In the USCS system of units, work and energy are expressed in pound-inches (lb · in.), pound-feet (lb · ft), kip-inches (kip · in.), or kip-feet (kip · ft).[1] In the SI system, the unit of work and energy is the joule (J), which is equal to one newton-meter (1 J = 1 N · m).

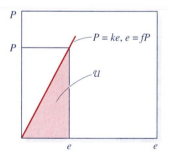

FIGURE 11.4   The load versus elongation diagram of a linearly elastic member.

Linearly Elastic Behavior.   When the material is linearly elastic, the load versus deflection curve has the form illustrated in Fig. 11.4. In this case it is convenient to use Eqs. 3.14 and 3.15 and write

$$P = ke, \qquad e = fP \tag{11.5}$$

where k and f are the **stiffness coefficient** and **flexibility coefficient,** respectively, ($f = 1/k$). In this case the triangular area that represents the strain energy $\mathcal{U}$ can be expressed in any of the following ways:

$$\mathcal{U} = \frac{1}{2}Pe = \frac{1}{2}ke^2 = \frac{1}{2}fP^2 \tag{11.6}$$

It is usually convenient to think of $\mathcal{U}$ as being a function of e, that is, $\mathcal{U} = \frac{1}{2}ke^2$.

However, for linearly elastic members the alternative forms listed in Eqs. 11.6 may sometimes be useful (see Section 11.5), and it is sometimes convenient to think of the strain energy stored in a linearly elastic member as $\left(\frac{1}{2}P\right)e$, that is, the average force times the total elongation.

---

[1]Note that, although these are the same as the units used to express the moment of a force (e.g., bending moment or torque), the meaning is quite different. For this reason, some authors express the USCS units of work and energy as inch-pounds (in.-lb), foot-pounds (ft-lb), inch-kips (in.-k), or foot-kips (ft-k).

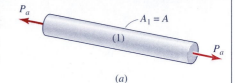
EXAMPLE 11.2

The two linearly elastic rods in Fig. 1 have the same modulus E and total length, $2L$, and are to be compared on the basis of the elastic strain energy that is stored in each: (a) when the maximum stress in each of the two rods is the same: and (b) when the total elongation of each of the two rods is the same.

Plan the Solution Since $\sigma_i = F_i/A_i$, the maximum axial stress in rod (b) occurs in section (3), and since $A_1 = A_3 = A$, the loads P_a and P_b will be equal in Part (a). In Part (b) we want $e_b = e_a$. Let us begin by plotting P–e curves, like Fig. 11.4, for both rods. Then we can easily compare the values of $\mathcal{U}_a$ and $\mathcal{U}_b$ for the two cases. We can make use of Eqs. 11.5 (or 3.14 and 3.15) to relate P to e for each rod.

Solution Let subscripts 1 through 3 refer to the elements labeled in Fig. 1, and let subscripts a and b refer to the rods in Figs. 1a and 1b, respectively.

Equilibrium: The free-body diagrams for the three rods elements are shown in Fig. 2.

$$\left(\sum F\right)_1 = 0: \qquad F_1 = P_a$$

$$\left(\sum F\right)_2 = 0: \qquad F_2 = P_b \qquad (1)$$

$$\left(\sum F\right)_3 = 0: \qquad F_3 = P_b$$

Force-Elongation Relations: From Eqs. 11.5, the equations

$$e_i = f_i F_i = \left(\frac{L}{AE}\right)_i F_i, \qquad F_i = k_i e_i = \left(\frac{AE}{L}\right)_i e_i \qquad (2)$$

give the force-elongation relations of the individual elements. For *compatibility of deformation* we get

$$e_a = e_1, \qquad e_b = e_2 + e_3 \qquad (3)$$

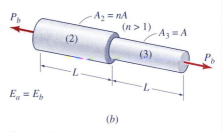

$E_a = E_b$

(b)

Fig. 1 Two rods undergoing axial deformation.

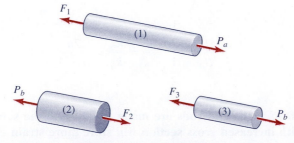

Fig. 2 Free-body diagrams.

Therefore,

$$e_a = \left(\frac{2L}{AE}\right)P_a \equiv f_a P_a$$

$$e_b = \left(\frac{L}{nAE}\right)P_b + \left(\frac{L}{AE}\right)P_b = \left(\frac{n+1}{n}\right)\left(\frac{L}{AE}\right)P_b \equiv f_b P_b \tag{4}$$

and, therefore,

$$P_a = \frac{1}{2}\left(\frac{AE}{L}\right)e_a \equiv k_a e_a$$

$$P_b = \left(\frac{n}{n+1}\right)\left(\frac{AE}{L}\right)e_b \equiv k_b e_b \tag{5}$$

Let us plot Eqs. (5) on the same $P - e$ graph (Fig. 3). Since $n > 1$, $k_b > k_a$, so the slope of the $P_b - e_b$ line is greater than the slope of the $P_a - e_a$ line in Fig. 3. That is, due to its larger cross section, rod (b) is stiffer.

Strain Energy: From Eqs. 11.6 and Fig. 11.4, strain energy is the triangular area under the load-deflection curve. The area is given by

$$\mathcal{U} = \frac{1}{2}ke^2 = \frac{1}{2}fP^2 \tag{6}$$

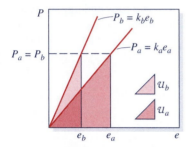

(a) Strain energy relationship when $P_b = P_a$.

Complete Part (a), the equal-stress case: The maximum stresses in the two rods are equal if $P_a = P_b$. From Fig. 3a it is clear that $\mathcal{U}_a > \mathcal{U}_b$. That is, **the rod with enlarged cross section stores <u>less</u> strain energy than the uniform rod does when the maximum stress in each of the two rods is the same.** In equation form,

$$\frac{\mathcal{U}_b}{\mathcal{U}_a} = \frac{\frac{1}{2}f_b P_b^2}{\frac{1}{2}f_a P_a^2} = \frac{f_b}{f_a} = \frac{n+1}{2n} < 1 \qquad \textbf{Ans.(a)} \tag{7}$$

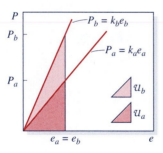

(b) Strain energy relationship when $e_b = e_a$.

Fig. 3 Strain energy comparisons.

Complete Part (b), the equal-elongation case: If the elongation of the two rods is the same, there must be a greater force acting on the stiffer rod. That is, $P_b > P_a$, as seen in Fig. 3b. In this case. $\mathcal{U}_b > \mathcal{U}_a$.

$$\frac{\mathcal{U}_b}{\mathcal{U}_a} = \frac{\frac{1}{2}k_b e_b^2}{\frac{1}{2}k_a e_a^2} = \frac{k_b}{k_a} = \frac{2n}{n+1} > 1 \qquad \textbf{Ans.(b)} \tag{8}$$

Therefore, **if the two rods are made to elongate the same amount, the rod with increased cross section will store <u>more</u> strain energy than the uniform rod does.**

Strain Energy Density for Linearly Elastic Bodies. To introduce the concept of strain energy, the simple case of axial deformation of an elastic member by
a single load P was considered above. However, to treat more complex situations we
need to consider how the stored strain energy is distributed throughout the deformed body. This leads us to the topic of strain energy per unit volume, or, simply,
strain energy density. Consider a small volume element dV in a linearly elastic rod
undergoing axial deformation, as illustrated in Fig. 11.5.

Let $d\mathcal{U}$ be the strain energy stored in an elemental volume of a linearly elastic
body, like dV in Fig. 11.5. Furthermore, let the strain energy stored in the elemental
volume be expressed in the form

$$d\mathcal{U} = \bar{u}\, dV$$

where $\bar{u}$ is the **strain energy density,** that is, the strain energy per unit volume at
the location of the differential volume dV. Then, the **total strain energy** is obtained
by summing the $d\mathcal{U}$s over the volume of the entire body, giving the integral
expression

$$\mathcal{U} = \int_V \bar{u}\, dV \qquad \text{\textbf{Total Strain Energy}} \qquad (11.7)$$

For the uniaxial stress state depicted in Fig. 11.5 we have

$$d\mathcal{U} = \left(\tfrac{1}{2}\sigma_x dy\, dz\right)\left(\epsilon_x dx\right) \equiv \bar{u}_{\sigma_x} dV$$

$$\text{avg. force} \times \text{displ.}$$

where

$$\bar{u}_{\sigma_x} = \frac{1}{2}\sigma_x \epsilon_x \qquad (11.8)$$

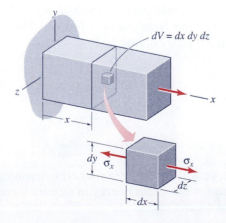

FIGURE 11.5   An element
of volume subjected to uniaxial stress σ_x.

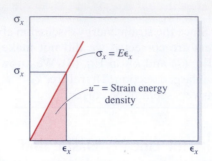

FIGURE 11.6   Strain energy density for a uniaxial stress state.

is the strain energy density for linearly elastic deformation due to a uniaxial stress state σ_x. Since $\sigma_x = E\epsilon_x$, Eq. 11.8 can also be written in the following alternative uniaxial stress-state forms:

$$\bar{u}_{\sigma_x} = \frac{1}{2}\frac{\sigma_x^2}{E} = \frac{1}{2}E\epsilon_x^2 \tag{11.9}$$

This strain energy density is depicted in the stress-strain diagram in Fig. 11.6. The strain energy density is the <u>area below the stress-strain curve</u>.

Consider now an elemental volume dV subjected only to shear stress $\tau_{xy} = \tau_{yx}$, as shown in Fig. 11.7. For this case we have

$$d\mathcal{U} = (\tfrac{1}{2}\tau_{xy}dx\,dz)\,(\gamma_{xy}dy) \equiv \bar{u}_{\tau_{xy}}dV$$

$$\text{avg. force} \times \text{displ.}$$

Therefore, the strain energy density related to τ_{xy} is

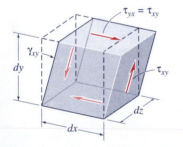

FIGURE 11.7   An elemental volume undergoing deformation due to shear stress $\tau_{xy} = \tau_{yx}$.

$$\bar{u}_{\tau_{xy}} = \frac{1}{2}\tau_{xy}\gamma_{xy} \tag{11.10}$$

Since $\tau_{xy} = G\gamma_{xy}$, alternative forms of Eq. 11.10 are

$$\bar{u}_{\tau_{xy}} = \frac{1}{2}\frac{\tau_{xy}^2}{G} = \frac{1}{2}G\gamma_{xy}^2 \tag{11.11}$$

In a similar manner, we could derive expressions for the strain energy density associated with the remaining components of stress. The general expression for **strain energy density** in a linearly elastic body is

$$\bar{u} = \frac{1}{2}(\sigma_x\epsilon_x + \sigma_y\epsilon_y$$
$$+ \sigma_z\epsilon_z + \tau_{xy}\gamma_{xy}$$
$$+ \tau_{xz}\gamma_{xz} + \tau_{yz}\gamma_{yz})$$

Strain Energy Density $\tag{11.12}$

In the next section we will use this strain-energy-density expression as the basis for deriving expressions for the strain energy in slender members undergoing axial deformation, torsion, and bending.

11.3 ELASTIC STRAIN ENERGY FOR VARIOUS TYPES OF LOADING

In order to apply work-energy methods to solve problems, we must first obtain expressions for the elastic strain energy stored when various types of loads are applied—axial loads, torsional loads, and bending loads. We will restrict our attention in this section to linearly elastic behavior.

Axial Deformation. In Section 3.2 the basic equations for normal stress and extensional strain were developed. Recall that, for axial deformation of a member such as the one illustrated in Fig. 11.8, and for the special case of $E \equiv E(x)$ (i.e., the material is homogeneous at any cross section, but its properties may vary with x),

$$\epsilon_x(x) = \frac{du(x)}{dx} \tag{11.13}$$

and

$$\sigma_x = \frac{F(x)}{A(x)} \tag{11.14}$$

where $u(x)$ is the displacement of the cross section at x and $F(x)$ is the (internal) force on the cross section at x. Therefore, since $dV = A(x)dx$ for the axial-deformation member, we can combine Eqs. 11.7, 11.9, 11.13, and 11.14 to get the following expressions for **strain energy due to axial deformation:**

$$\mathcal{U} = \frac{1}{2}\int_0^L \frac{F^2 dx}{EA} \tag{11.15}$$

and

$$\mathcal{U} = \frac{1}{2}\int_0^L EA(u')^2 dx \tag{11.16}$$

where $()' \equiv \dfrac{d()}{dx}$; and F, E, A, and u may all be functions of x. We will have occasion to use both of the above expressions for the strain energy of linearly elastic bars undergoing axial deformation.

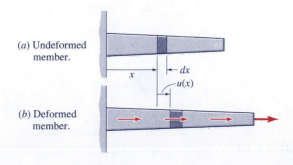

(a) Undeformed member.

x — dx

$u(x)$

(b) Deformed member.

FIGURE 11.8 A member undergoing axial deformation.

FIGURE 11.9 An end-loaded uniform rod.

For a uniform linearly elastic rod subjected to axial end loads, as illustrated in Fig. 11.9, Eq. 11.15 gives

$$U = \frac{P^2 L}{2EA} \qquad (11.17)$$

Torsion of Circular Members. Although it is possible to obtain expressions for strain energy stored in torsional members with various cross sections, here we will only consider the simplest case of linearly elastic circular rods that are homogeneous at each cross section [i.e., $G \equiv G(x)$], as shown in Fig. 11.10. Recall from Eqs. 4.1 and 4.12 that, for this case, the shear strain γ is given by

$$\gamma(x, \rho) = \rho \frac{d\phi(x)}{dx} \qquad (11.18)$$

and the shear stress τ is related to the internal torque at a cross section by the equation

$$\tau(x, \rho) = \frac{T(x)\rho}{I_P(x)} \qquad (11.19)$$

Combining Eqs. 11.7, 11.11, 11.18 and 11.19, we obtain the following expression for the **strain energy due to torsion,** expressed in terms of the internal torque T:

$$U = \frac{1}{2} \int_0^L \int_A \frac{1}{G(x)} \left[\frac{T(x)\rho}{I_P(x)} \right]^2 dA\, dx$$

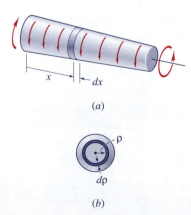

FIGURE 11.10 A circular rod undergoing torsional deformation.

FIGURE 11.11   A uniform, end-loaded torsion member.

or, since $I_p = \int_A \rho^2 dA$,

$$U = \frac{1}{2} \int_0^L \frac{T^2 dx}{GI_p}$$ (11.20)

An alternative expression for U in terms of the twist rate ϕ' is

$$U = \frac{1}{2} \int_0^L \int_A G(x) \left[\rho \frac{d\phi(x)}{dx} \right]^2 dA dx$$

or

$$U = \frac{1}{2} \int_0^L GI_p(\phi')^2 dx$$ (11.21)

For a uniform circular linearly elastic rod subjected to end torques T, as shown in Fig. 11.11, Eq. 11.20 gives

$$U = \frac{T^2 L}{2GI_p}$$ (11.22)

Bending of Beams—Flexural Strain Energy. A beam, like the one in Fig. 11.12, has flexural stress σ_x and shear stress τ_{xy} that vary with position in the beam. From Eq. 11.12 it can be seen that the contributions of normal stress and shear stress to the strain energy can be treated separately. We begin by considering the contribution of the flexural stress σ_x (and strain ϵ_x), and we employ the Bernoulli-Euler beam theory of Sections 6.2 and 6.3. Recall from Eq. 6.3 that the

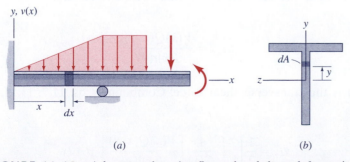

(a)   (b)

FIGURE 11.12   A beam undergoing flexural and shear deformation.

extensional strain ϵ_x is related to the local radius of curvature, $\rho(x)$, of the deflection curve and to the deflection $v(x)$ by

$$\epsilon_x(x, y) = \frac{-y}{\rho(x)} \approx -y\frac{d^2v(x)}{dx^2} \tag{11.23}$$

where y is the distance from the neutral axis of the cross section. When Young's modulus is independent of position in the cross section, that is, $E \equiv E(x)$, the flexural stress σ_x is related to the internal bending moment, $M(x)$, by the flexure formula, Eq. 6.13, that is, by

$$\sigma_x(x, y) = \frac{-M(x)y}{I(x)} \tag{11.24}$$

From Eqs. 11.7, 11.9, 11.23, and 11.24 we get the following equations for the strain energy due to flexure:

$$\mathcal{U}_\sigma = \frac{1}{2}\int_0^L \int_A \frac{1}{E(x)}\left[\frac{-M(x)y}{I(x)}\right]^2 dA\,dx$$

or

$$\boxed{\mathcal{U}_\sigma = \frac{1}{2}\int_0^L \frac{M^2 dx}{EI}} \tag{11.25}$$

where the subscript σ emphasizes the fact that this is the **strain energy due to flexural stress.** In terms of the curvature $v''(x)$, the strain energy due to flexure is given by

$$\mathcal{U} = \frac{1}{2}\int_0^L \int_A E(x)\left[\frac{-y}{\rho(x)}\right]^2 dA\,dx = \frac{1}{2}\int_0^L \int_A E(x)\left(\frac{d^2v(x)}{dx^2}\right)^2 y^2 dA\,dx$$

or

$$\boxed{\mathcal{U}_\sigma = \frac{1}{2}\int_0^L EI(v'')^2\, dx} \tag{11.26}$$

Bending of Beams—Shear-Strain Energy. The shear stress in a beam also contributes to the strain energy that is stored in the beam. In Section 6.8 an equilibrium analysis led to the shear-stress formula of Eq. 6.67, namely

$$\tau_{xy} = \frac{V(x)Q(x, y)}{I(x)t(y)} \tag{11.27}$$

where $V(x)$ is the transverse shear force. Combining Eqs. 11.7, 11.11, and 11.21, we get

$$\mathcal{U}_\tau = \frac{1}{2}\int_0^L \int_A \frac{1}{G(x)}\left[\frac{V(x)Q(x, y)}{I(x)t(y)}\right]^2 dA\,dx$$

or

$$\mathcal{U}_\tau = \frac{1}{2}\int_0^L \left[\frac{V^2(x)}{G(x)I^2(x)} \int_A \frac{Q^2(x, y)}{t^2(y)}\, dA \right] dx \qquad (11.28)$$

To simplify this expression for $\mathcal{U}_\tau$, let us define a new cross-sectional property f_s, called the *form factor for shear*. Let

$$f_s(x) \equiv \frac{A(x)}{I^2(x)}\int_A \frac{Q^2(x, y)}{t^2(y)}\, dA \qquad (11.29)$$

(The form factor is a dimensionless number that depends only on the shape of the cross section, so it rarely actually varies with x.) Combining Eqs. 11.28 and 11.29 we get the following expression for the **strain energy due to shear in bending:**

$$\mathcal{U}_\tau = \frac{1}{2}\int_0^L \frac{f_s V^2 dx}{GA} \qquad (11.30)$$

The form factor for shear must be evaluated for each shape of cross section. For example, for a rectangular cross section of width b and height h, the expression

$$Q = \frac{b}{2}\left(\frac{h^2}{4} - y^2\right)$$

was obtained in Example Problem 6.14. Therefore, from Eq. 11.29 we get

$$f_s = \frac{bh}{(\frac{1}{12}bh^3)^2}\int_{-h/2}^{h/2} \frac{1}{b^2}\left[\frac{b}{2}\left(\frac{h^2}{4} - y^2\right)\right]^2 b\, dy = \frac{6}{5} \qquad (11.31)$$

The form factor for other cross-sectional shapes is determined in a similar manner. Several of these are listed in Table 11.1. The approximation for an I-section or box section is based on assuming that the shear force is uniformly distributed over the depth of the web(s).

TABLE 11.1   Form Factor f_s for Shear

Section	f_s
Rectangle	$\dfrac{6}{5}$
Circle	$\dfrac{10}{9}$
Thin tube	2
I-section or box section	$\approx \dfrac{A}{A_{\text{web}}}$

EXAMPLE 11.3

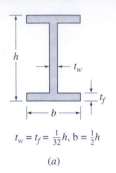

$t_w = t_f = \frac{1}{32}h$, $b = \frac{1}{2}h$

(a)

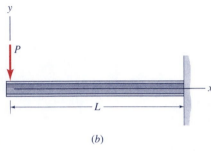

(b)

Fig. 1 A wide-flange beam.

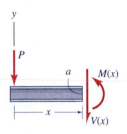

Fig. 2 Free-body diagram.

Using the form factor approximation for I-sections in Table 11.1, determine the ratio of the shear strain energy $\mathcal{U}_\tau$ to the flexural strain energy $\mathcal{U}_\sigma$ for the wide-flange beam shown in Fig. 1.[2] Assume that $E = 2.6G$, and express your answer in terms of the depth-to-length ratio h/L.

Plan the Solution   We can use equilibrium to determine expressions for $M(x)$ and $V(x)$ that can be substituted into Eqs. 11.25 and 11.30, respectively, to calculate $\mathcal{U}_\sigma$ and $\mathcal{U}_\tau$.

Solution

Equilibrium: From the free-body diagram in Fig. 2.

$$\Sigma F_y = 0: \qquad\qquad V(x) = -P$$

$$\left(\Sigma M\right)_a = 0: \qquad\qquad M(x) = -Px \qquad\qquad (1)$$

Strain Energy: From Eq. 11.25, the strain energy due to flexure is given by

$$\mathcal{U}_\sigma = \frac{1}{2}\int_0^L \frac{M^2 dx}{EI} \qquad\qquad (2a)$$

and from Eq. 11.30, the strain energy due to transverse shear is

$$\mathcal{U}_\tau = \frac{1}{2}\int_0^L \frac{f_s V^2 dx}{GA} \qquad\qquad (2b)$$

From Table 11.1, f_s can be approximated by

$$f_s = \frac{A}{A_{web}} \qquad\qquad (3)$$

For the cross section shown in Fig. 1a,

$$A = 2bt_f + (h - 2t_f)t_w = \frac{62}{1024}h^2 = 0.060547h^2$$

$$I = \frac{1}{12}[bh^3 - (b - t_w)(h - 2t_f)^3] = 0.009480h^4 \qquad (4)$$

$$f_s = \frac{A}{A_w} = \frac{62}{30} = 2.07$$

---

[2]The shape of this cross section is approximately that of a **W**8 × 13 beam.

Combining Eqs. (1) and (2), we get

$$\mathcal{U}_\tau = \frac{1}{2}\frac{f_s}{GA}\int_0^L (-P)^2 dx = \frac{f_s P^2 L}{2GA}$$

$$\mathcal{U}_\sigma = \frac{1}{2}\left(\frac{1}{EI}\right)\int_0^L (-Px)^2 dx = \frac{P^2 L^3}{6EI}$$

(5)

Therefore,

$$\frac{\mathcal{U}_\tau}{\mathcal{U}_\sigma} = \frac{\dfrac{f_s P^2 L}{2GA}}{\dfrac{P^2 L^3}{6EI}}$$

(6)

Using the given ratio of $E/G = 2.6$, together with other values from Eqs. (4), we get

$$\frac{\mathcal{U}_\tau}{\mathcal{U}_\sigma} = \frac{3EIf_s}{GAL^2} \approx 2.5\left(\frac{h}{L}\right)^2 \qquad \textbf{Ans.} \quad (7)$$

A value of $\mathcal{U}_\tau/\mathcal{U}_\sigma = 0.05$ (i.e., 5%) corresponds to an L/h ratio of approximately 7, which is a relatively short beam. Therefore, **shear strain energy and, correspondingly, shear deformation, may be neglected except in the case of short, stubby beams.**

Review the Solution The answer in Eq. (7) is dimensionless, as it should be. Furthermore, the dependence on h/L seems reasonable. That is, shear becomes important when the length of the beam is too short to have large values of bending moment. Finally, we should check the answer by rechecking each formula and calculation.

11.4 WORK-ENERGY PRINCIPLE FOR CALCULATING DEFLECTIONS

In the remaining sections of this chapter, we will examine several energy principles, that is, several ways in which energy methods can be used to solve deformable-body problems. The simplest (but most limited) of these is the *principle of work and energy*. We will consider only the case of slowly applied loading, so that the kinetic energy (e.g., 1/2 mass × velocity-squared) can be ignored. We will not consider other forms of energy such as thermal energy, chemical energy, and electromagnetic energy. Therefore, *if the stresses in a body do not exceed the elastic limit, all of the work done on a body by external forces is stored in the body as elastic strain energy.* The formula

$$\boxed{\mathcal{W}_{\text{ext}} = \mathcal{U}} \qquad \begin{array}{l}\textbf{Work-Energy}\\ \textbf{Principle}\end{array} \qquad (11.32)$$

is a statement of the **Work-Energy Principle.** This principle is not restricted to linearly elastic behavior. The next two example problems illustrate the use of the above Work-Energy Principle to **calculate the displacement due to a single load applied to a deformable body.**

EXAMPLE 11.4

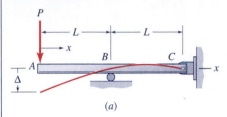

(a)

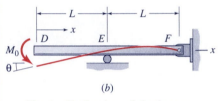

(b)

Fig. 1 Deflection of the beams.

Use the Work-Energy Principle to calculate the transverse deflection of the beam in Fig. 1a and the end slope of the beam in Fig. 1b. Assume that both beams remain linearly elastic under the given loads, and let EI = constant. Neglect shear strain energy of the beams.

Plan the Solution We can use Eqs. 11.3a and 11.3c to give the work done by the external loads P and M_0, respectively. The strain energy in each case can be determined by the use of Eq. 11.25. Finally, the deflection Δ and slope θ will result from applying the Work-Energy Principle, Eq. 11.32.

Solution In the following solution, let subscript a apply to the beam in Fig. 1a and subscript b to the beam in Fig. 1b, and let subscript 1 apply to the span $0 < x < L$ and the subscript 2 apply to the span $L < x < 2L$.

Equilibrium: To obtain expressions for $M(x)$ for each beam, we first draw the necessary free-body diagrams.
From Fig. 2a,

$$\left(\sum M\right)_C = 0: \qquad\qquad B_y = 2P \qquad\qquad (1a)$$

From Fig. 2b,

$$\left(\sum M\right)_{C_1} = 0: \qquad\qquad M_{a1} = -Px \qquad\qquad (1b)$$

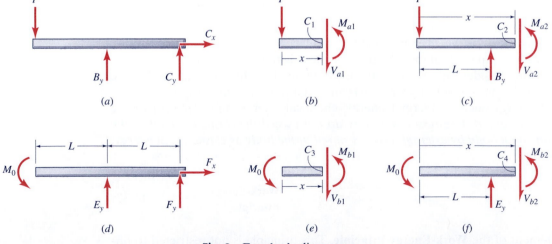

(a)   (b)   (c)

(d)   (e)   (f)

Fig. 2 Free-body diagrams.

From Fig. 2c,

$$\left(\sum M\right)_{C_2} = 0: \quad M_{a2} = -Px + 2P(x - L) = P(x - 2L) \tag{1c}$$

From Fig. 2d,

$$\left(\sum M\right)_F = 0: \quad E_y = \frac{M_0}{L} \tag{1d}$$

From Fig. 2e,

$$\left(\sum M\right)_{C_3} = 0: \quad M_{b1} = -M_0 \tag{1e}$$

From Fig. 2f,

$$\left(\sum M\right)_{C_4} = 0: \quad M_{b2} = -M_0 + \frac{M_0}{L}(x - L) = \frac{M_0}{L}(x - 2L) \tag{1f}$$

Strain Energy: The strain energy stored in a beam due to flexure is given by Eq. 11.25 (modified for length $2L$),

$$\mathcal{U} = \frac{1}{2}\int_0^{2L} \frac{M^2 dx}{EI} = \frac{1}{2}\int_0^{L} \frac{M_1^2 dx}{EI} + \frac{1}{2}\int_L^{2L} \frac{M_2^2 dx}{EI} \tag{2}$$

Combining Eqs. (1) and (2) we get

$$\mathcal{U}_a = \frac{1}{2}\int_0^{L} \frac{(-Px)^2 dx}{EI} + \frac{1}{2}\int_L^{2L} \frac{[P(x - 2L)]^2 dx}{EI} = \frac{P^2 L^3}{3EI} \tag{3a}$$

$$\mathcal{U}_b = \frac{1}{2}\int_0^{L} \frac{(-M_0)^2 dx}{EI} + \frac{1}{2}\int_L^{2L} \frac{\left[\frac{M_0}{L}(x - 2L)\right]^2 dx}{EI} = \frac{2M_0^2 L}{3EI} \tag{3b}$$

Work of External Loads: Equations 11.3a and 11.3c give

$$W_a = \frac{1}{2}P\Delta, \qquad W_b = \frac{1}{2}M_0\theta \tag{4a-b}$$

Work-Energy Principle: Applying the Work-Energy Principle, Eq. 11.32, we get

$$W_a = \mathcal{U}_a: \qquad \Delta = \frac{2}{3}\frac{PL^3}{EI} \qquad \textbf{Ans.} \tag{5a}$$

$$W_b = \mathcal{U}_b: \qquad \theta = \frac{4}{3}\frac{M_0 L}{EI} \qquad \textbf{Ans.} \tag{5b}$$

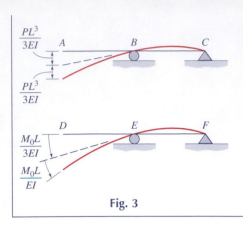

Review the Solution The preceding answers can be checked by applying the method of superposition of deflections discussed in Section 7.6 and using data from Tables E.1 and E.2 of Appendix E. The results are shown in Fig. 3.

A "ballpark" estimate of each answer can be obtained by just ignoring the slope at B and E. From Table E.1 this would give estimates of

$$\Delta > \frac{PL^3}{3EI}, \quad \theta > \frac{M_0 L}{EI}$$

Calculating the displacement of various points in a "complex" structure, like a planar truss, is not a simple matter. However, if the "right question" is asked, the Work-Energy Principle provides a simple, straightforward way to calculate displacements, as the next example problem illustrates.

EXAMPLE 11.5

For the two-bar truss in Fig. 1, use the Work-Energy Principle to compute the displacement Δ_B of node B in the direction of the load P. Let $A_1 = A_2 = 1.2 \text{ in}^2$, and $E_1 = E_2 = 10 \times 10^3$ ksi. The members remain linearly elastic under the given loading.

Plan the Solution For each member in the truss, we can use Eq. 11.17 to determine the strain energy, using the member axial forces obtained from equilibrium of the joint at B. Since strain energy is a scalar quantity, the total strain energy is the sum of the strain energies in the two members. Since the structure remains linearly elastic, and since Δ_B is in the same direction as P, the work of the external load is given by Eq. 11.3a, as in the previous example. Finally, the Work-Energy Principle, Eq. 11.32, will lead to an equation for the unknown displacement Δ_B.

Fig. 1 A two-member planar truss.

Solution

Equilibrium: Since the truss in Fig. 1 is statically determinate, we can compute the member forces by using only equilibrium equations and the free-body diagram shown in Fig. 2.

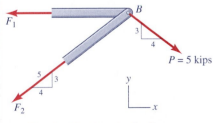

Fig. 2 The free-body diagram.

$$\sum F_y = 0: \qquad -\frac{3}{5}F_2 - \frac{3}{5}P = 0 \rightarrow F_2 = -P \tag{1a}$$

$$\sum F_x = 0: \qquad -F_1 - \frac{4}{5}(-P) + \frac{4}{5}P = 0 \rightarrow F_1 = \frac{8}{5}P \tag{1b}$$

Strain Energy: Since strain energy is a scalar quantity,

$$\mathcal{U} = \mathcal{U}_1 + \mathcal{U}_2 \tag{2}$$

For each member, the strain energy is given by Eq. 11.17. Therefore.

$$U = \frac{F_1^2 L_1}{2A_1 E_1} + \frac{F_2^2 L_2}{2A_2 E_2}$$

$$U = \frac{(8 \text{ kips})^2 (40 \text{ in.}) + (-5 \text{ kips})^2 (50 \text{ in.})}{2(1.2 \text{ in}^2)(10 \times 10^3 \text{ ksi})} \qquad (3)$$

Therefore,

$$U = 0.15875 \text{ kip} \cdot \text{in.} \qquad (4)$$

Work of External Load: The work of the external load P is

$$\mathcal{W}_{\text{ext}} = \frac{1}{2} P\Delta_B = \frac{1}{2}(5 \text{ kips})(\Delta_B \text{ in.}) = 2.5\Delta_B \text{ kip} \cdot \text{in.} \qquad (5)$$

since, as noted above, the structure is linear, and Δ_B is in the same direction as P.

Work-Energy Principle: From Eq. 11.32,

$$\mathcal{W}_{\text{ext}} = U \qquad (6)$$

Therefore,

$$2.5\Delta_B = 0.15875$$

and, finally,

$$\Delta_B = 0.0635 \text{ in.} \qquad \textbf{Ans.} \quad (7)$$

Review the Solution We could solve this problem by the method discussed in Section 3.10. However, we just need to see if our answer is reasonable. We know that the elongation of a single axial member is given by $e_i = F_i L_i / A_i E_i$ (Eq. 3.13). Therefore,

$$e_1 = \frac{(8 \text{ kips})(40 \text{ in.})}{(1.2 \text{ in}^2)(10 \times 10^3 \text{ ksi})} = 0.0267 \text{ in.}$$

$$e_2 = \frac{(-5 \text{ kips})(50 \text{ in.})}{(1.2 \text{ in}^2)(10 \times 10^3 \text{ ksi})} = -0.0208 \text{ in.}$$

Without carrying out the geometrical calculations necessary to relate e_1 and e_2 to Δ_B, we can see that our answer in Eq. (7) is reasonable.

The Work-Energy Principle provides a relatively simple, straightforward procedure for calculating the displacement of a body at the point of application of a single load and in the direction of the load, as illustrated in Example Problems 11.4 and 11.5. However, suppose the two loads, P and M_0, are applied slowly and

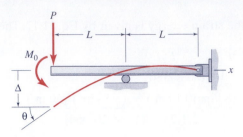

FIGURE 11.13   A beam with two loads.

simultaneously to a beam, as illustrated in Fig. 11.13. Following the procedure of Example Problem 11.4, we have

$$M_1(x) = -Px - M_0$$

$$M_2(x) = \left(P + \frac{M_0}{L}\right)(x - 2L)$$

(11.33)

Then, substitution of these moments into Eq. 11.25 gives

$$\mathcal{U} = \frac{P^2 L^3}{3EI} + \frac{5PM_0 L^2}{6EI} + \frac{2M_0^2 L}{3EI}$$

(11.34)

The work done by P and M_0 when they are applied simultaneously to the beam is

$$\mathcal{W}_{\text{ext}} = \frac{1}{2}P\Delta + \frac{1}{2}M_0\theta$$

(11.35)

It is clear that we cannot use the Work-Energy Principle, $\mathcal{W}_{\text{ext}} = \mathcal{U}$, to determine either Δ or θ since there are two unknowns and only one equation.

In conclusion, **the Work-Energy Principle is useful only for determining the displacement at the point of application of a single load and in the direction of that load.** And, of course, it applies only to elastic deformation. Therefore, we now turn to more powerful energy methods.

11.5   CASTIGLIANO'S SECOND THEOREM; THE UNIT-LOAD METHOD

In this section we examine two closely related energy methods that may be used for calculating displacements of linearly elastic deformable bodies—*Castigliano's Second Theorem*[3] and the *Unit-Load Method*. Prior to the advent of the *finite element method*, these two energy methods were the most powerful techniques available for calculating deflections of moderately complex statically indeterminate structures.

Castigliano's Second Theorem. We restrict our attention to deformable bodies that behave linearly under the action of independent loads, $P_1, P_2, \ldots, P_N$, plus, possibly, distributed loads and/or additional concentrated loads. Furthermore, let us designate the displacements that correspond to the above N loads as $\Delta_1, \Delta_2, \ldots, \Delta_N$.

---

[3]The First Theorem and Second Theorem of Italian engineer Alberto Castigliano (1847–1884) appeared in his 1873 dissertation for the engineer's degree at Turin. Italy. [Ref. 11-1] Castigliano's First Theorem is discussed in Section 11.7.

That is , let Δ_i be the displacement at the point of application of load P_i in the direction of load P_i.

Castigliano's Second Theorem can be stated as follows:[4]

Among all possible equilibrium configurations of a linearly elastic deformable body or system, the actual. configuration is the one for which

$$\boxed{\begin{array}{l} \Delta_i = \dfrac{\partial \mathcal{U}(P_1, P_2, \ldots, P_N)}{\partial P_i} \\[2mm] i = 1, 2, \ldots, N \end{array}}$$

Castigliano's
Second   (11.36)
Theorem

where Δ_i is the displacement corresponding to force P_i, and $\mathcal{U}$ is the strain energy expressed as a function of the loads.

In words, Eq. 11.36 says that *the partial derivative of the strain energy of a structure with respect to any load is the displacement that corresponds to that load.* As noted in Eq. 11.36, the strain energy must be expressed as a function of the loads. The terms "load" and "displacement" are to be understood in the generalized sense (e.g., a force and the corresponding translational displacement, or a torque and the corresponding angle of twist where the torque is applied).

Typically, the strain energy is related to the loads through expressions like Eqs. 11.15, 11.20, 11.25, and 11.30, that is, through expressions like

$$\mathcal{U} = \int_0^L \frac{F^2 dx}{2AE} + \int_0^L \frac{T^2 dx}{2GI_p} + \int_0^L \frac{M^2 dx}{2EI} + \int_0^L \frac{f_s V^2 dx}{2GA} \tag{11.37}$$

where $F(x)$, $T(x)$, $M(x)$, and $V(x)$ are related to the P's by equilibrium equations, thus guaranteeing that <u>only equilibrium states are considered.</u> It is convenient to combine Eqs. 11.36 and 11.37 initially, which gives

$$\Delta_i = \int_0^L \frac{F(x)\dfrac{\partial F(x)}{\partial P_i} dx}{AE} + \int_0^L \frac{T(x)\dfrac{\partial T(x)}{\partial P_i} dx}{GI_p}$$
$$+ \int_0^L \frac{M(x)\dfrac{\partial M(x)}{\partial P_i} dx}{EI} + \int_0^L \frac{f_s V(x)\dfrac{\partial V(x)}{\partial P_i} dx}{GA} \tag{11.38}$$

Of course, if a system consists of several members, Eqs. 11.37 and 11.38 will have to include summations over all members in the system. This will be illustrated in example problems involving planar trusses.

Castigliano's Second Theorem Applied to Statically Determinate Problems.
Figure 11.13 was used to point out a severe limitation of the Work-Energy Principle, namely that it can be used to calculate deflections only when just a single load is applied. We will use the beam in Fig. 11.13 to illustrate the case with which displacements can be calculated using Castigliano's Second Theorem, even when multiple loads are applied. As noted in the statement of the theorem, we must express the strain energy $\mathcal{U}$ as a function of the loads, incorporating equilibrium in the process.

---

[4]The theorem is stated here without derivation. It is a special case of the *Cross-Engesser Theorem,* which is derived and discussed in Section 11.8.

EXAMPLE 11.6

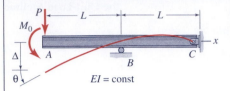

Fig. 1

EI = const

Use Castigliano's Second Theorem to solve for the slope θ and displacement Δ at end A of the uniform, linearly elastic beam shown in Fig. 1. Neglect shear deformation of the beam.

Plan the Solution   We can use Castigliano's Second Theorem, Eq. 11.36, to determine θ and Δ directly. The strain energy for this situation was given in Eq. 11.34 (Section 11.4) in conjunction with Fig. 11.13.

Solution   In Eq. 11.34 the strain energy for the beam in Fig. 1 was given as a function of the loads P and M_0 as

$$ \mathcal{U} = \frac{P^2 L^3}{3EI} + \frac{5PM_0 L^2}{6EI} + \frac{2M_0^2 L}{3EI} \tag{1} $$

Castigliano's Second Theorem. Eq. 11.36, gives the following equations for determining Δ and θ:

$$ \Delta = \frac{\partial \mathcal{U}}{\partial P}, \qquad \theta = \frac{\partial \mathcal{U}}{\partial M_0} \tag{2} $$

Therefore, combining Eqs. (1) and (2) we get

$$ \left. \begin{aligned} \Delta &= \frac{2PL^3}{3EI} + \frac{5M_0 L^2}{6EI} \\ \theta &= \frac{5PL^2}{6EI} + \frac{4M_0 L}{3EI} \end{aligned} \right\} \qquad \textbf{Ans.} \quad (3) $$

Review the Solution   The first check we should make is to check the dimensions of each term to be sure that the answers are dimensionally correct. We see that they are. See Example Problem 11.4 for additional magnitude checks that could be performed, for example, using the following data from Table E.1 of Appendix E for the two separate load cases in Fig. 2.

$$ \Delta_P = \frac{PL^3}{3EI} \qquad\qquad \Delta_M = \frac{M_0 L^2}{2EI} $$

$$ \theta_P = \frac{PL^2}{2EI} \qquad\qquad \theta_M = \frac{M_0 L}{EI} $$

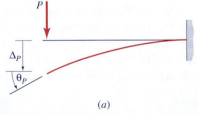

(a)

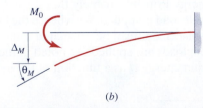

(b)

Fig. 2

Castigliano's Second Theorem Applied to Statically Indeterminate Problems.

When a structure is statically indeterminate and has independent external loads, $P_1, P_2, \ldots, P_N$, and we write the strain energy in terms of forces using equilibrium relations to give us $F(x)$, $T(x)$, $M(x)$, and so forth, the strain energy will also involve unknown redundant forces $R_1, R_2, \ldots, R_{N_R}$. That is, we will get $U = U(P_1, P_2, \ldots, P_N; R_1, R_2, \ldots, R_{N_R})$. The displacements that correspond to the N_R redundant forces are all zero (i.e., we imagine that the system is cut so that the redundants that we have chosen appear as known forces, but the system is not really cut). Therefore, we have N_R equations that enable us to solve for the N_R redundants. We also get N equations to solve for the displacements under the N external loads. Thus, **Castigliano's Second Theorem** *applied to statically indeterminate bodies* is given by the following two equations:

> *Among all possible equilibrium configurations of a statically indeterminate, linearly elastic body, the actual equilibrium configuration is the one that satisfies the equations*

$$0 = \frac{\partial U}{\partial R_i}, \qquad i = 1, 2, \ldots, N_R$$

$$\Delta_i = \frac{\partial U}{\partial P_i}, \qquad i = 1, 2, \ldots, N$$

Castigliano's Second Theorem   (11.39)

where U *is the strain energy of the body expressed as a function of the* N *applied loads* P_i *and the* N_R *redundant forces* R_i, *and where* Δ_i *is the displacement component of the point of application of load* P_i *in the direction of that load.*[5]

To illustrate the use of these equations, we will solve a problem similar to Example Problem 3.20 of Section 3.10. This example problem illustrates the use of a **dummy load** to obtain the displacement when there is no real load that corresponds to the required displacement.

⊳⊳⊳ **EXAMPLE 11.7** ⊲⊲⊲

A three-element truss has the configuration shown in Fig. 1. Each member has a cross-sectional area A and is made of material whose modulus is E. A horizontal load, P_{xB}, is applied to the truss at node B.

Using Castigliano's Second Theorem in the form of Eqs. 11.39, with F_2 as the redundant force, solve for (a) the member forces F_1, F_2, and F_3; (b) the horizontal displacement u_B at B; and (c) the vertical displacement v_B at B.

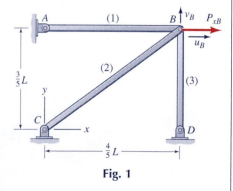

Fig. 1

Plan the Solution   To determine F_2 we will need to use the first of Eqs. 11.39, letting the single redundant (R_1) be F_2. We can use the second of Eqs. 11.39 to solve directly for the displacement u_B that corresponds to the horizontal load P_{xB}. Since there is no load at B corresponding to the vertical displacement v_B that is required in Part (c), we will need to introduce a dummy load P_{yB}. After carrying out the differentiation with respect to P_{yB} to get v_B, we can then set $P_{yB} = 0$.

---

[5] Equation 11.39a is sometimes referred to as the *Principle of Least Work*.

Equation 11.17 gives the strain energy stored in a two-force, linearly elastic rod.

Solution   (a) *Use Castigliano's Second Theorem to determine the member forces.*

Equilibrium: The strain energy expression that we utilize in Eqs. 11.39 must be expressed in terms of forces in a manner such that equilibrium is automatically satisfied. Therefore, we begin by writing equilibrium equations for the free body in Fig. 2. To determine v_B [in Part (c)] we will need to have a force acting in the y direction at B. Therefore, on the free-body diagram in Fig. 2, a "dummy force" P_{yB} is included. We let F_2 be the redundant force, and use equilibrium equations to write F_1 and F_3 in terms of F_2.

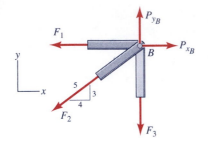

Fig. 2   Free-body diagram.

$$\sum F_x = 0: \qquad\qquad F_1 = P_{xB} - \frac{4}{5}F_2$$

$$\tag{1}$$

$$\sum F_y = 0: \qquad\qquad F_3 = P_{yB} - \frac{3}{5}F_2$$

Strain Energy: Equation 11.6c can be used to express the strain energy in member i in terms of the axial force F_i. Thus,

$$u_i = \frac{1}{2}f_iF_i^2 \tag{2}$$

where $f_i = (L/AE)_i$, the flexibility coefficient for member i. Combining Eqs. (1) and (2) and summing for the three-element system, we get

$$u = \frac{1}{2}f_1\left(P_{xB} - \frac{4}{5}F_2\right)^2 + \frac{1}{2}f_2F_2^2 + \frac{1}{2}f_3\left(P_{yB} - \frac{3}{5}F_2\right)^2 \tag{3}$$

Castigliano's Second Theorem: To determine the redundant force F_2, we use the first of Eqs. 11.39, that is,

$$\frac{\partial u}{\partial F_2} = 0 \tag{4}$$

Differentiating the strain energy expression in Eq. (3) with respect to F_2, with $P_{yB} = 0$, we get

$$-\frac{4}{5}f_1\left(P_{xB} - \frac{4}{5}F_2\right) + f_2F_2 - \frac{3}{5}f_3\left(-\frac{3}{5}F_2\right) = 0$$

Solving for F_2, we get the expression

$$F_2 = \frac{20f_1P_{xB}}{16f_1 + 25f_2 + 9f_3} \tag{5}$$

The flexibility coefficients are

$$f_1 = \frac{L_1}{A_1 E_1} = \frac{4}{5}\left(\frac{L}{AE}\right) = \frac{4}{5}f, \quad f_2 = f, \quad f_3 = \frac{3}{5}f$$

Finally, the redundant force F_2 is

$$F_2 = \frac{20\left(\frac{4}{5}f\right)P_{xB}}{16\left(\frac{4}{5}f\right) + 25f + 9\left(\frac{3}{5}f\right)} = \frac{10}{27}P_{xB}$$

This answer can be combined with Eqs. (1) to give

$$F_1 = \frac{19}{27}P_{xB}, \quad F_2 = \frac{10}{27}P_{xB}, \quad F_3 = \frac{-2}{9}P_{xB} \quad \textbf{Ans. (a)} \quad (6)$$

(b) *Determine the horizontal displacement u_B.* We use the second of Eqs. 11.39 in the form

$$u_B = \frac{\partial \mathcal{U}}{\partial P_{xB}} \tag{7}$$

Combining Eqs. (3) and (7) (with $P_{yB} = 0$) gives

$$u_B = f_1\left(P_{xB} - \frac{4}{5}F_2\right) = \left(\frac{4}{5}f\right)\left(\frac{19}{27}P_{xB}\right)$$

or

$$u_B = \frac{76}{135}\frac{P_{xB}L}{AE} \quad \textbf{Ans. (b)} \quad (8)$$

(c) *Determine the vertical displacement v_B.* Corresponding to the dummy load P_{yB}, we write the second of Eqs. 11.39 in the form

$$v_B = \left.\frac{\partial \mathcal{U}}{\partial P_{yB}}\right|_{P_{yB}=0} \tag{9}$$

Thus,

$$v_B = f_3\left(-\frac{3}{5}F_2\right) = \left(-\frac{3}{5}\right)\left(\frac{3}{5}f\right)\left(\frac{10}{27}P_{xB}\right)$$

or

$$v_B = -\frac{2}{15}\frac{P_{xB}L}{AE} \quad \textbf{Ans. (c)} \quad (10)$$

The element forces in Eqs. (6) all have the dimensions of force, and the displacements in Eqs. (8) and (10) also have the correct dimensions. The signs of F_1 and F_2 indicate tension, which is correct for P_{xB} acting to the right: and with F_2 pulling down on node B, F_3 must be pushing up, as it is. To check further, we could re-solve the problem using the Displacement Method, as in Example 3.20, or we could use the Force Method, which should give us the same expression for F_2 that we got in Eq. (5).

The Unit-Load Method.

Suppose that we need to calculate the value of the displacement Δ at point C in the beam shown in Fig. 11.14a. In Fig. 11.14a are shown all of the real loads applied to the beam. (There could be a real load corresponding to the desired displacement Δ, but the load P in Fig. 11.14a could also be a dummy load that is actually zero.) To calculate the displacement Δ we could use Eq. 11.38, but instead we will derive and use a variation of Eq. 11.38 called the *Unit-Load Method*.

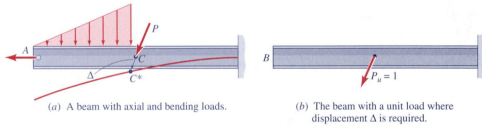

(a) A beam with axial and bending loads.

(b) The beam with a unit load where displacement Δ is required.

FIGURE 11.14 Illustration of the unit-load method.

Since this is a linear problem, the internal forces $F(x)$, $T(x)$, and so forth are linear functions of each of the loads, including P. Therefore, if we evaluate $\dfrac{\partial F(x)}{\partial P}$, the answer that we get is the same as if we were to evaluate $F(x)$ for single load of $P_u = 1$. This leads to a method for calculating displacements called the **Unit-Load Method**. Figure 11.14b shows the beam with a unit load for the value of P where displacement Δ is to be obtained. Let $F(x)$, $T(x)$, etc, be the *internal resultants due to the real loads* (Fig. 11.14a); and let $F_u(x)$, $T_u(x)$, etc, be the *internal resultants due to the unit load*, $P_u = 1$ (Fig. 11.14b). Then Eq. 11.38 can be written,

$$
\Delta = \int_0^L \frac{F F_u dx}{AE} + \int_0^L \frac{T T_u dx}{GI_p}
$$
$$
+ \int_0^L \frac{M M_u dx}{EI} + \int_0^L \frac{f_s V V_u dx}{GA}
$$

Unit-Load Method (11.40)

◀────── **EXAMPLE 11.8** ──────▶

Use the Unit-Load Method to determine the slope, θ, at the tip of the uniform, linearly elastic cantilever beam with linearly varying distributed load, as shown in Fig. 1. Neglect shear deformation.

Plan the Solution We can use the bending moment term in Eq. 11.40. We will need to put a unit couple at A corresponding to the desired angle θ.

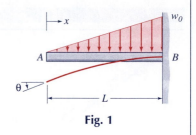

Fig. 1

Solution

Unit Load: To calculate the angle θ, we need to place a unit couple at A, as shown in Fig. 2.

Fig. 2 Unit-load case.

Equilibrium: We need $M(x)$ based on the real loads in Fig. 1 and $M_u(x)$ based on the unit couple in Fig. 2. Figure 3 shows the appropriate free-body diagrams.

$$\left(\sum M\right)_{C_1} = 0: \qquad M(x) = -\left(\frac{w_0 x}{L}\right)\left(\frac{x}{2}\right)\left(\frac{x}{3}\right) = -\frac{w_0 x^3}{6L} \qquad (1a)$$

$$\left(\sum M\right)_{C_2} = 0: \qquad M_u(x) = -1 \qquad (1b)$$

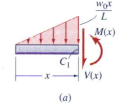

(a)

Unit-Load Calculation: From Eq. 11.40,

$$\theta = \int_0^L \frac{M M_u \, dx}{EI} = \int_0^L \frac{1}{EI}\left(-\frac{w_0 x^3}{6L}\right)(-1)\, dx \qquad (2)$$

$$\theta = \frac{w_0 L^3}{24EI} \qquad \textbf{Ans.} \quad (3)$$

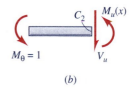

(b)

Fig. 3 Free-body diagrams.

Review the Solution The solution in Eq. (3) is dimensionally correct. For this simple problem we can check the answer in Table E.1 of Appendix E.

As a second example of the Unit-Load Method, and also as an example of a statically indeterminate problem, consider a statically indeterminate truss similar to the one in Example Problem 11.7. (Note that we must employ a unit load corresponding to each redundant force that we choose, and then set the resulting displacements of the redundants to zero.)

For trusses, the unit-load equation can be written as the following sum of member contributions:

$$\Delta = \sum_{i=1}^N f_i F_i F_{ui} = \sum_{i=1}^N \left(\frac{F F_u L}{AE}\right)_i \qquad \begin{array}{l}\textbf{Unit-Load}\\ \textbf{Method}\\ \textbf{(Trusses)}\end{array} \qquad (11.41)$$

EXAMPLE 11.9

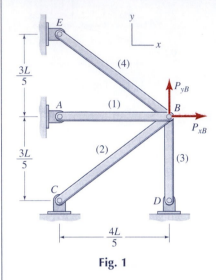

Fig. 1

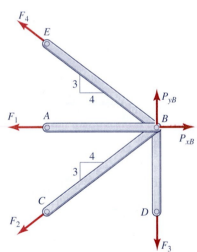

Fig. 2 Free-body diagram with real forces.

A four-element truss has the configuration shown in Fig. 1. Each member has a cross-sectional area A and modulus of elasticity E. Horizontal and vertical loads are applied at node B. Use the Unit-Load method to determine the member forces F_1 through F_4. Let F_2 and F_4 be the two redundant forces.

Plan the Solution We will need to write equilibrium equations for the real forces. We can use a free-body diagram for this. Since there are four unknown forces and only two equations of equilibrium, there are two redundant forces, which are to be F_2 and F_4. We can employ the Unit-Load Method of Eq. 11.41. However, since there is no displacement at C corresponding to F_2 or at E corresponding to F_4, we need to apply a unit load at each of these places and set the corresponding displacement in the unit-load equation to zero.

Solution

Equilibrium: The equilibrium equations relating the real forces can be obtained by use of Fig. 2. Thus,

$$\sum F_x = 0: \qquad F_1 = P_{xB} - \frac{4}{5}F_2 - \frac{4}{5}F_4$$

$$\sum F_y = 0: \qquad F_3 = P_{yB} - \frac{3}{5}F_2 + \frac{3}{5}F_4 \qquad (1)$$

Since we have two redundants, we need two unit-load cases. The free-body diagrams for these are shown in Fig. 3. We will designate the two unit-load cases by $(\,\cdot\,)'$ and $(\,\cdot\,)''$ superscripts, rather than by subscript u. From Fig. 3a,

$$F_2' = 1, \qquad\qquad F_4' = 0$$

$$\sum F_x = 0: \qquad F_1' = -\frac{4}{5} \qquad (2)$$

$$\sum F_y = 0: \qquad F_3' = -\frac{3}{5}$$

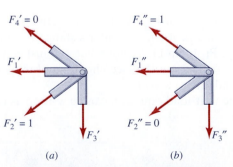

(a)   (b)

Fig. 3 Unit-load free-body diagrams.

From Fig. 3b,

$$F_2'' = 0, \qquad F_4'' = 1$$

$$\sum F_x = 0: \qquad\qquad F_1'' = -\frac{4}{5} \qquad\qquad (3)$$

$$\sum F_y = 0: \qquad\qquad F_3'' = \frac{3}{5}$$

Unit-Load Calculation: Since the members of this truss are uniform, we can use Eq. 11.41, that is,

$$\Delta = \sum_{i=1}^{4} f_i F_i F_{ui} \qquad\qquad (4)$$

But, we have two unit-load cases corresponding to redundants F_2 and F_3. Therefore, from Eq. (4) we have

$$0 = \sum_{i=1}^{4} f_i F_i F_i', \qquad 0 = \sum_{i=1}^{4} f_i F_i F_i'' \qquad\qquad (5)$$

The flexibility coefficients are given by

$$f_i = \left(\frac{L}{AE}\right)_i$$

so

$$f_1 = \frac{4}{5}\left(\frac{L}{AE}\right) \equiv \frac{4}{5}f, \quad f_2 = f_4 = f, \quad f_3 = \frac{3}{5}f \qquad\qquad (6)$$

Substituting Eqs. (1), (2), (3), and (6) into Eqs. (5) we get

$$0 = \left(\frac{4}{5}f\right)\left(P_{xB} - \frac{4}{5}F_2 - \frac{4}{5}F_4\right)\left(-\frac{4}{5}\right) + fF_2(1)$$

$$+ \left(\frac{3}{5}f\right)\left(P_{yB} - \frac{3}{5}F_2 + \frac{3}{5}F_4\right)\left(-\frac{3}{5}\right) + 0$$

$$\qquad\qquad (7)$$

$$0 = \left(\frac{4}{5}f\right)\left(P_{xB} - \frac{4}{5}F_2 - \frac{4}{5}F_4\right)\left(-\frac{4}{5}\right) + 0$$

$$+ \left(\frac{3}{5}f\right)\left(P_{yB} - \frac{3}{5}F_2 + \frac{3}{5}F_4\right)\left(\frac{3}{5}\right) + fF_4(1)$$

This gives us two simultaneous algebraic equations to solve for the redundants F_2 and F_4. Once F_2 and F_4 are determined, the remaining member forces F_1 and F_3 can be obtained from Eqs. (1), the equilibrium equations for the real forces. Equations (7) simplify to

$$216F_2 + 37F_4 = 80P_{xB} + 45P_{yB}$$
$$37F_2 + 216F_4 = 80P_{xB} - 45P_{yB} \tag{8}$$

Solving for F_2 and F_4 we get

$$\left.\begin{array}{l} F_2 = \dfrac{80}{253}P_{xB} + \dfrac{45}{179}P_{yB} \\[2ex] F_4 = \dfrac{80}{253}P_{xB} - \dfrac{45}{179}P_{yB} \end{array}\right\} \qquad \textbf{Ans.} \quad (9)$$

and, combining Eqs. (1) and (9) we get

$$\left.\begin{array}{l} F_1 = \dfrac{125}{253}P_{xB} \\[2ex] F_3 = \dfrac{125}{179}P_{yB} \end{array}\right\} \qquad \textbf{Ans.} \quad (10)$$

Review the Solution All answers in Eqs. (9) and (10) obviously have the correct dimension of force. A positive force P_{xB} should put members 1, 2, and 4 in tension; and since the vertical components of the resulting values of F_2 and F_4 balance out, F_3 should not depend on P_{xB}. Equations (9) and (10) exhibit these effects of P_{xB}. A similar analysis shows that Eqs. (9) and (10) properly reflect the effect of P_{xB} also.

To calculate displacements for a truss with many members, it is convenient to tabulate the contributions of the various members to the sum in Eq. 11.41, as illustrated by the following example problem.

> **EXAMPLE 11.10**

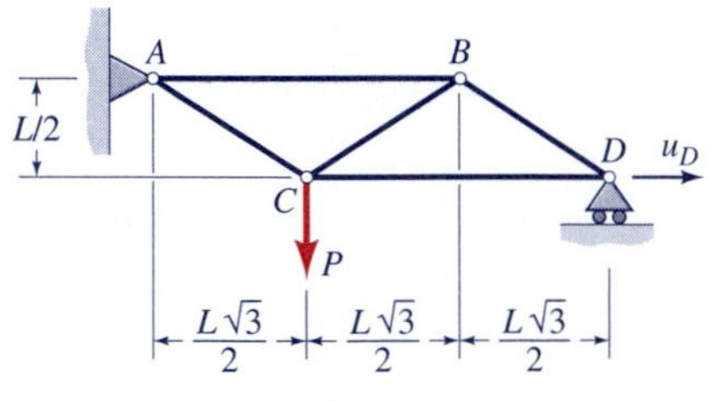

Fig. 1

Determine the horizontal displacement u_D at node D that results when load P is applied downward at node C, as shown in Fig. 1. All members have the same value of AE.

Plan the Solution We will need to use equilibrium of the truss joints to determine the F_i's due to the applied load P, and we will also use equilibrium to compute the F_{ui}'s due to a unit horizontal force at D. Then, u_D will be given by Eq. 11.41.

Solution

Equilibrium of Real Forces: Either the *Method of Joints* or the *Method of Sections* can be used to determine the F_i's due to P. The relevant free-body diagrams are not shown, but the results are shown in Fig. 2.

Equilibrium of Virtual Forces Due to Unit Load: For a unit load applied at D in the direction of u_D, equilibrium solutions give the F_{ui}'s shown in Fig. 3.

Unit-Load Deflection Calculation: Table 1 gives the values of the terms in the unit-load equation (Eq. 11.41).

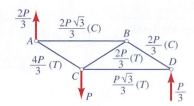

Fig. 2 Real forces F_i.

$$u_D = \sum_{i=1}^{5} \frac{F_i F_{ui} L_i}{A_i E_i} \tag{1}$$

$$u_D = -\frac{2}{3} \frac{PL}{AE} \qquad \textbf{Ans.} \tag{2}$$

Review the Solution The answer is dimensionally correct. It also makes sense that pulling down on node C would tend to pull node D to the left, so the sign of the answer is reasonable.

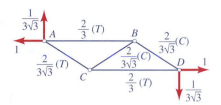

Fig. 3 Forces F_{ui} due to a unit load.

TABLE 1 Unit-Load Deflection Calculations

(1) Member	(2) L_i	(3) F_i	(4) F_{ui}	(5) $F_i F_{ui} L_i$
AB	$\sqrt{3}L$	$\dfrac{-2P\sqrt{3}}{3}$	$\dfrac{2}{3}$	$-\dfrac{4}{3}PL$
AC	L	$\dfrac{4P}{3}$	$\dfrac{2\sqrt{3}}{9}$	$\dfrac{8\sqrt{3}}{27}PL$
BC	L	$\dfrac{2P}{3}$	$\dfrac{-2\sqrt{3}}{9}$	$\dfrac{-4\sqrt{3}}{27}PL$
BD	L	$\dfrac{-2P}{3}$	$\dfrac{2\sqrt{3}}{9}$	$-\dfrac{4\sqrt{3}}{27}PL$
CD	$\sqrt{3}L$	$\dfrac{P\sqrt{3}}{3}$	$\dfrac{2}{3}$	$\dfrac{2}{3}PL$
				$\displaystyle\sum_{i=1}^{5} = -\dfrac{2}{3}PL$

*11.6 VIRTUAL WORK

Of the two energy methods discussed so far, *Castigliano's Second Theorem* (Section 11.5) was shown to have much broader applicability than the *Work-Energy Principle* (Section 11.4). For example, it was shown that Castigliano's Second Theorem can be used to calculate deflections of statically indeterminate structures having several loads. In Sections 11.6 and 11.7 we now introduce very powerful strain-energy methods that are based on the concepts of *virtual displacements* and *virtual work*. Then, in Section 11.8 complementary-energy methods that are based on *virtual forces* and *complementary virtual work* are presented.[6]

---

[6]In the present solid-mechanics context, the relevant definition of the word *virtual* is "being in essence or effect, but not in fact." [Ref.: *Webster's New Collegiate Dictionary*.]

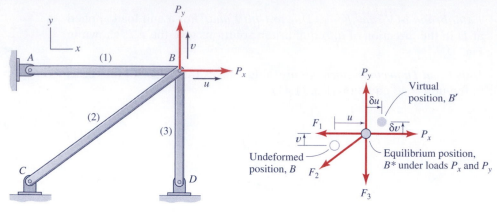

(a) A three-bar planar truss. (b) Real displacements and virtual displacements.

FIGURE 11.15 An example of real displacements and virtual displacements.

Virtual Displacements and Virtual Work. Consider the pin that connects the three two-force members of the planar truss in Fig. 11.15a to be a "particle" that is acted on by the forces shown in Fig. 11.15b. (A similar truss was analyzed in Example Problem 3.20.) When there are no loads on the truss, the truss is unde-formed, and the pin is at its *initial position, B*. When loads P_x and P_y are applied, the pin moves a distance u in the x direction and a distance v in the y direction to its (true) equilibrium position, B^*. That is, u and v are the components of the *real dis-placement* of pin B. Now, imagine that the pin moves infinitesimal distances δu in the x direction and δv in the y direction. These infinitesimal, imaginary displace-ments are called **virtual displacements.**[7]

> *A* **virtual displacement** *is an infinitesimal, imaginary displacement, denoted by the symbol $\delta(\,\cdot\,)$, that is consistent with all kinematic constraints. That is, virtual displacements must satisfy the same boundary conditions and compatibility condi-tions that the real displacements must satisfy. Real forces are not altered by virtual displacements.*

(Note carefully the distinction between the true displacements u and v caused by the loads, and the infinitesimal virtual displacements that "exist" only in our imagi-nation.)

> **Virtual work,** *designated by δW, is the work done by real forces when virtual displace-ments occur. The real forces are assumed to remain constant.*

To be precise, virtual work is the work that would be done if the points of applica-tion of the forces were to actually move through the (imaginary) virtual displace-ments. Hence, for the example in Fig. 11.15b,

$$\delta W = \left(\sum F_x \right) \delta u + \left(\sum F_y \right) \delta v \tag{11.42}$$

---

[7]Although the Greek letter δ has previously been used to denote displacements themselves, in the remainder of this chapter the symbol δ will be reserved for use as the *virtual operator,* much the same as the letter d is the *differential operator* in dx, dV, and so forth. Capital delta (Δ) will be used for displacement.

Equilibrium, and the Principle of Virtual Displacements. At position B^* the pin in Fig. 11.15b is in static equilibrium. From Newton's Second Law,

$$\sum F_x = 0, \quad \sum F_y = 0 \tag{11.43}$$

at the equilibrium position. Therefore, the following statement holds:

> *If a particle is in equilibrium under a system of forces, then for any virtual displacement, the virtual work δW is zero.*

The converse, which is called the **Principle of Virtual Displacements,** is also true.[8]

> *If the virtual work done on a particle is zero for any arbitrary kinematically admissible virtual displacements, then the particle is in equilibrium.*

Thus, if, for <u>arbitrary</u> virtual displacement from a given configuration.

$$\boxed{\delta W = 0} \quad \begin{array}{l}\textbf{Principle of Virtual} \\ \textbf{Displacements}\end{array} \tag{11.44}$$

then the given configuration is an equilibrium configuration. Referring back to Eq. 11.42, if $\delta W = 0$ regardless of the (nonzero) values that we might pick for δu and δv, then it must be true that $\sum F_x = \sum F_y = 0$.

Virtual Work for Deformable Bodies. The Principle of Virtual Displacements can be extended to any collection of particles—we just get more virtual displacements and more equilibrium equations, but only one virtual work. For example, if there were two particles, we would get

$$\delta W = \left[\left(\sum F_x\right)_1 \delta u_1 + \left(\sum F_y\right)_1 \delta v_1\right] + \left[\left(\sum F_x\right)_2 \delta u_2 + \left(\sum F_y\right)_2 \delta v_2\right]$$

But, we are interested in applying virtual work methods to deformable bodies. Let the (deformable) spring AB and (rigid) block C in Fig. 11.16 comprise a system that is in equilibrium under the action of a static load P. Since every particle in the spring AB and block C is in equilibrium, we can say that $\delta W = 0$ applies to the entire system. But the force P is the only external force that can do work, because it is the only external force whose point of application can move. Therefore, the virtual work done by the external force is

$$\delta W_{\text{ext}} = P\delta u$$

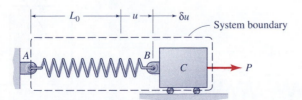

FIGURE 11.16   A system in equilibrium.

---

[8]This is also sometimes called the *Principle of Virtual Work.*

In addition, if block C were to undergo a virtual displacement δu, there would be some virtual work done by the internal forces in the deformable spring. Thus, we can state the following **Principle of Virtual Displacements for a deformable system:**

> *If the virtual work done on a deformable system during any kinematically admissible displacements (changes from a given configuration) of the system is zero, then the system is in equilibrium in the given configuration.*

$$\delta W = \delta W_{ext} + \delta W_{int} = 0 \tag{11.45}$$

In Eq. 11.45, δW_{int} is the virtual work done by internal forces in the system under consideration.

What is δW_{int} and how is it calculated? Let us isolate just the spring in Fig. 11.16 and call it our "deformable system," as depicted in Fig. 11.17a. And let us suppose that this is a nonlinear elastic spring whose force versus elongation curve is shown in Fig. 11.17b. The system is in equilibrium in Fig. 11.17a because the applied force $F_s(u_1)$ corresponds to the displacement u_1. Therefore, we can say that

$$\delta W_{ext} + \delta W_{int} = 0$$

We know that $\delta W_{ext}|_{u_1} = F_s(u_1)\delta u$, so

$$\delta W_{int}|_{u_1} = -F_x(u_1)\delta u$$

That is, the virtual work done by the internal forces in an elastic deformable body when the body undergoes virtual displacements is equal to the negative of the virtual work done by the external forces acting on the deformable body alone. Therefore, referring to Fig. 11.17b, we see that the virtual work of internal forces and the strain energy are related by

$$\delta W_{int} = -\delta \mathcal{U} \tag{11.46}$$

where $\delta \mathcal{U}$ is the infinitesimal, imaginary increment in the strain energy that would result from the body undergoing a virtual displacement δu, as indicated in Fig. 11.17b.

We will need to obtain $\delta \mathcal{U}$ when the strain energy is expressed as a function of several displacements, say $\mathcal{U} = \mathcal{U}(q_1, q_2, \ldots, q_n)$, where the q's are independent

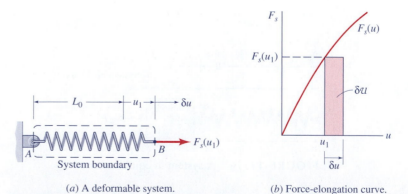

(a) A deformable system.   (b) Force-elongation curve.

FIGURE 11.17   A nonlinear, elastic spring treated as a deformable system.

displacements. Then $\delta \mathcal{U}$ can be determined in the same manner as the differential $d\mathcal{U}$, namely by

$$\delta\mathcal{U} = \frac{\delta\mathcal{U}}{\partial q_1}\,\delta q_1 + \frac{\partial\mathcal{U}}{\partial q_2}\,\delta q_2 + \cdots + \frac{\partial\mathcal{U}}{\partial q_n}\,\delta q_n \tag{11.47}$$

Equation 11.47 gives the **virtual strain-energy change** due to virtual displacements $\delta q_1, \delta q_2,$ etc. Its use is illustrated in the following section.

*11.7 STRAIN-ENERGY METHODS

Principle of Virtual Displacements. By combining Eqs. 11.45 and 11.46 we arrive at the **Principle of Virtual Displacements** *applied to elastic bodies,* which can be stated as follows:

> *Among all kinematically admissible configurations of an elastic deformable body, the actual* **equilibrium configuration** *is the one satisfies Eq. 11.48 when the body undergoes arbitrary virtual displacements from that configuration.*

$$\boxed{\delta\mathcal{W}_{\text{ext}} = \delta\mathcal{U}} \qquad \begin{array}{l}\text{\textbf{Principle of Virtual}}\\ \text{\textbf{Displacements}}\end{array} \tag{11.48}$$

A **kinematically admissible configuration** is a configuration that satisfies all relevant kinematic boundary conditions and all relevant equations of deformation compatibility.

In the next example problem we apply the Principle of Virtual Displacements to a problem that could also be solved by the Work-Energy Principle of Section 11.4. Then, in Example Problem 11.12, Eq. 11.48 will be used to solve a deflection problem that cannot be solved by the method of Section 11.4.

> ► **EXAMPLE 11.11** ◄

For the two-segment rod in Fig. 1, use the Principle of Virtual Displacements for deformable bodies to solve for the following: (a) the displacement u_B of node B; and (b) the internal forces F_1 and F_2 in the two elements of the statically indeterminate rod system.

The rod properties are $A_1, E_1, L_1,$ and $A_2, E_2, L_2,$ respectively.

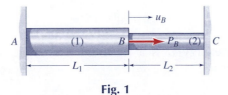

Fig. 1

Plan the Solution Using the compatibility of displacements at node B and using a strain energy formula from Eq. 11.6, we can determine an expression for $\mathcal{U}(u_B)$, the strain energy of the two-bar system in terms of the displacement of node B. The load P_B does virtual work when node B moves (actually, when it is imagined to move) through a virtual displacement δu_B. Thus, we can apply Eq. 11.48 to get an equation of equilibrium for node B.

Solution   (a) *Determine u_B, the displacement of node B.*

Strain Energy: It will simplify notation if we use the element stiffness coefficients

$$k_1 = \frac{A_1 E_1}{L_1}, \qquad k_2 = \frac{A_2 E_2}{L_2} \tag{1}$$

Then, using Eq. 11.6b we can write the total strain energy as

$$\mathcal{U} = \mathcal{U}_1 + \mathcal{U}_2 = \frac{1}{2} k_1 e_1^2 + \frac{1}{2} k_2 e_2^2 \tag{2}$$

where e_i is the elongation of rod element i.

Geometry of Deformation: According to the Principle of Virtual Displacements, we are to consider only "kinematically admissible deformations." Therefore, the kinematics of deformation, including the compatibility of the displacements of the two rods at node B, must be incorporated into the strain energy so that all kinematic requirements will be automatically satisfied.

$$e_1 = u_B, \qquad e_2 = -u_B \tag{3}$$

Then, combining Eqs. (2) and (3), we get the strain energy in a form that automatically guarantees that all kinematic requirements are satisfied:

$$\mathcal{U} = \frac{1}{2} (k_1 + k_2) u_B^2 \tag{4}$$

Virtual Work of the External Force: The point of application (node B) of external force P_B is imagined to move through a virtual displacement δu_B (in the same direction as u_B), doing virtual work

$$\delta W_{\text{ext}} = P_B \delta u_B \tag{5}$$

Principle of Virtual Displacements: From Eq. 11.48.

$$\delta W_{\text{ext}} = \delta \mathcal{U} \tag{6}$$

Since the strain energy $\mathcal{U}$ is a function of a single displacement u_B, Eq. 11.47 reduces to

$$\delta \mathcal{U} = \frac{d\mathcal{U}}{du_B} \delta u_B = (k_1 + k_2) u_B \delta u_B \tag{7}$$

Substituting Eqs. (5) and (7) into Eq. (6) we get

$$P_B \delta u_B = (k_1 + k_2) u_B \delta u_B$$

or

$$[(k_1 + k_2)u_B - P_B]\delta u_B = 0 \tag{8}$$

The Principle of Virtual Displacements states that the system is in equilibrium if Eq. (8) is satisfied for *any arbitrary virtual displacement,* in this case for arbitrary δu_B. Therefore, the expression in square brackets in Eq. (8) must vanish, giving

$$u_B = \frac{P_B}{k_1 + k_2} \qquad \textbf{Ans.} \quad (9)$$

where k_1 and k_2 are given in Eqs. 1.
(b) *Determine the element forces, F_1 and F_2.* From Eq. 3.15.

$$F_i = k_i e_i \tag{10}$$

Therefore, combining Eqs. (3), (9), and (10), we get

$$F_1 = \frac{k_1 P_B}{k_1 + k_2}, \qquad F_2 = \frac{-k_2 P_B}{k_1 + k_2} \qquad \textbf{Ans.} \quad (11)$$

Review the Solution We could check this solution by using the methods of Chapter 3. Since k_i has the dimensions of F/L, the dimensions in Eq. (9) are correct. The force P_B has to stretch rod AB and also compress rod BC. Therefore, the denominator in Eq. (9) correctly reflects the addition of k_1 and k_2 in resisting P_B.

Castigliano's First Theorem. The Principle of Virtual Displacements. Eq. 11.48, can be used to formulate the equations of equilibrium of any elastic body. When the deformation of the body can be characterized by a finite number of independent displacements, Eq. 11.47 provides a convenient procedure for expressing δU in terms of the virtual displacements δq_i. The virtual work of external forces can then be expressed in the form

$$\delta W_{\text{ext}} = \sum_{i=1}^{N} P_i \delta q_i \tag{11.49}$$

where N is the number of independent displacements, q_i. The P_i's are called **generalized forces.** The correct values or expressions for the P_i's are determined by formulating δW_{ext}.

Combining Eqs. 11.47 and 11.49, we get the following form of the Principle of Virtual Displacements:

$$\sum_{i=1}^{N} \left(P_i - \frac{\partial U}{\partial q_i} \right) \delta q_i = 0$$

But, since the above sum is required to be zero for any arbitrary δq_i's and since the δq_i's are independent, we get

$$\boxed{P_i = \frac{\partial U}{\partial q_i}, \quad i = 1, 2, \ldots, N} \qquad \begin{array}{l}\textbf{Castigliano's} \\ \textbf{First Theorem}\end{array} \qquad (11.50)$$

This equation is called **Castigliano's First Theorem.** It can be viewed as an alternative form of the Principle of Virtual Displacements when the deformation of the elastic system can be characterized by N discrete displacement coordinates, q_i. Equation 11.50 produces *equilibrium equations written in terms of displacements*.

The following example problem could be solved by applying the Principle of Virtual Displacements. However, we will use the slightly shorter route of using Castigliano's First Theorem. This example problem demonstrates the power of Castigliano's First Theorem to provide solutions to problems that cannot be solved by the Work-Energy Principle of Section 11.4.

EXAMPLE 11.12

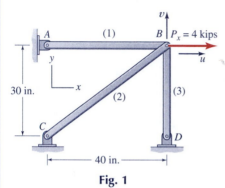

Fig. 1

A three-element truss has the configuration shown in Fig. 1. A horizontal load of 4 kips is applied to the truss through a pin at node B. Each member has a cross-sectional area of 1.2 in², and all of them are made of aluminum ($E = 10 \times 10^3$ ksi). Use Castigliano's First Theorem to determine u and v, the horizontal and vertical displacements, respectively, at node B.

Plan the Solution The strain energy of an element can be written in terms of its elongation by using Eq. 11.6b. The elongation of each element can be related to the nodal displacements u and v through Eq. 3.26. Since there are two nodal displacements, we will have two corresponding virtual displacements δu and δv. Using Castigliano's First Theorem, Eq. 11.50, we will get *two equilibrium equations to solve simultaneously* for u and v.

Solution

Strain Energy: We can borrow part of this solution from Example Problem 3.20 (Section 3.10).

$$k_1 = \left(\frac{AE}{L}\right)_1 = 300 \text{ kips/in.}, \quad \cos\theta_1 = 1.0, \quad \sin\theta_1 = 0.0$$

$$k_2 = \left(\frac{AE}{L}\right)_2 = 240 \text{ kips/in.}, \quad \cos\theta_2 = 0.8, \quad \sin\theta_2 = 0.6 \quad (1)$$

$$k_3 = \left(\frac{AE}{L}\right)_3 = 400 \text{ kips/in.}, \quad \cos\theta_3 = 0.0, \quad \sin\theta_3 = 1.0$$

where θ_i is defined in Fig. 2.

Using Eq. 11.6b, we can write the total strain energy in the three-bar truss as

$$\mathcal{U} = \mathcal{U}_1 + \mathcal{U}_2 + \mathcal{U}_3 = \frac{1}{2}k_1 e_1^2 + \frac{1}{2}k_2 e_2^2 + \frac{1}{2}k_3 e_3^2 \quad (2)$$

From Eq. 3.26, the elongation of each element can be related to the nodal displacements u and v by

$$e_i = u\cos\theta_i + v\sin\theta_i \quad (3)$$

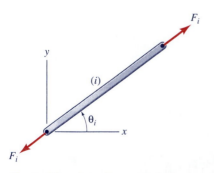

Fig. 2 Notations for truss element.

Then,

$$e_1 = u, \quad e_2 = 0.8u + 0.6v, \quad e_3 = v \tag{4}$$

Combining Eqs. (2) and (4), together with k's from Eq. (1), we get

$$\mathcal{U} = [150u^2 + 120(0.8u + 0.6v)^2 + 200v^2] \tag{5}$$

Virtual Work of the External Force: The horizontal force P_x acting through a horizontal virtual displacement of δu inches does virtual work given by

$$\delta \mathcal{W}_{ext} = 4\delta u \text{ kip} \cdot \text{in}$$

For this case, Eq. 11.49 takes the form

$$\delta \mathcal{W}_{ext} = P_x \delta u + P_y \delta v \tag{6}$$

where

$$P_x = 4 \text{ kips}, \qquad P_y = 0 \text{ kips} \tag{7}$$

Castigliano's First Theorem: From Eq. 11.50, the two equations of equilibrium for this truss are given by

$$\frac{\partial \mathcal{U}}{\partial u} = P_x, \qquad \frac{\partial \mathcal{U}}{\partial v} = P_y \tag{8}$$

so, differentiating Eq. (5) with respect to u and v and combining the results with Eqs. (7) and (8), we get

$$\boxed{\begin{array}{l} 453.6u + 115.2v = 4 \text{ kips} \\ 115.2u + 486.4v = 0 \text{ kips} \end{array}} \tag{9}$$

The left-hand sides of these two equations are exactly the same as those of the *nodal equilibrium equations in terms of nodal displacements* that we got in Eq. (4) of Example Problem 3.20 by employing the *Displacement Method.* The right-hand sides are just the nodal force components P_x and P_y. The solution of Eqs. (9) is

$$u = 9.38(10^{-3}) \text{ in.}$$
$$v = -2.22(10^{-3}) \text{ in.} \qquad \textbf{Ans.} \quad (10)$$

Review the Solution   We got essentially the same equations of equilibrium by energy methods as we got by using the Displacement Method in Example Problem 3.20.

Although only linearly elastic systems have been considered in the example problems used to illustrate the Principle of Virtual Displacements and Castigliano's First Theorem, these methods only require that the system be elastic, not necessarily linear.[9]

---

[9]See Ref. 11-2, Section 10.6, for an example that treats a truss with nonlinearly elastic members.

Displacement Method and Finite-Element Analysis. Note that the primary solution in Example Problem 11.11 was a solution for the nodal displacement u_B, and recall from earlier chapters (e.g., Chapters 3, 4, and 7) that this was a characteristic of *Displacement-Method* solutions. Also, the use of Castigliano's First Theorem in Example Problem 11.12 produced essentially the same two nodal equilibrium equations that were obtained by use of the Displacement Method in Example Problem 3.21. We can conclude that the **Principle of Virtual Displacements and Castigliano's First Theorem provide alternative ways to formulate a Displacement-Method solution for deformable-body mechanics problems.** The Principle of Virtual Displacements is the basis upon which most **Finite Element Analysis** computer codes are developed.

*11.8 COMPLEMENTARY-ENERGY METHODS

In the previous section, two *Strain-Energy Methods*—the *Principle of Virtual Displacements* and a related form called *Castigliano's First Theorem*—were introduced and were shown to be very useful energy methods. They were also shown to be directly related to the *Displacement Method* and, therefore, to *Finite Element Analysis*. In the present section *Complementary-Energy Methods* are introduced and are shown to be closely related to the *Force Method,* which was introduced in Section 3.9.

Complementary Work and Complementary Strain Energy. The **complementary work** $\mathcal{W}^c$ done by an axial force P is defined by the following integral:

$$\mathcal{W}^c(P_1) = \int_0^{P_1} e(P)\,dP \tag{11.51}$$

As illustrated in Fig. 11.18a, this represents the area <u>above</u> the load-elongation curve. Note that complementary work is expressed as a function of force, not as a function of displacement, as work is. As in Eqs. 11.2, similar complementary work integrals can be written for torsional and bending loads.

For an elastic body, *the complementary work $\mathcal{W}^c$ done on the body is stored as complementary strain energy $\mathcal{U}^c$*, which is, therefore, defined as

$$\mathcal{U}^c(P) = \mathcal{W}^c(P)|_{\text{done on elastic body}} \tag{11.52}$$

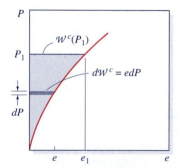

(a) The area representing complementary work.

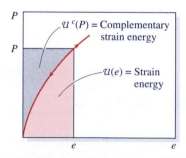

(b) The areas representing strain energy and complementary strain energy.

FIGURE 11.18 Complementary work and complementary strain energy.

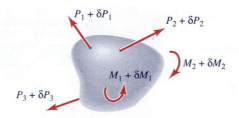

FIGURE 11.19 An illustration of virtual forces.

As indicated in Fig. 11.18b complementary strain energy is the area above the load-elongation curve up to the value of the applied load. For linearly elastic axial deformation it is convenient to express the **complementary strain energy** of a member in the form

$$\mathcal{U}^c(F_i) = \frac{1}{2} f_i F_i^2$$

Complementary Strain Energy (11.53)

Virtual Forces and Complementary Virtual Work.

Virtual displacements, δu, δq_i, and so forth, were defined in Section 11.6 and used in the formulation of analytical methods based on virtual work. Recall that a virtual displacement was defined as an imaginary, infinitesimal displacement that does not affect the forces in a body and that is consistent with the geometric compatibility conditions of the body. In contrast, a **virtual force** is defined as follows:

> A **virtual force,** δF, is an infinitesimal, imaginary force that, together with other virtual forces applied to a body in equilibrium, does not affect the equilibrium state of the body.

Figure 11.19 illustrates real forces P_i and virtual forces δP_i (including real couples M_i and virtual couples δM_i) applied to a deformable body. The body in Fig. 11.19 is in equilibrium under the action of the (real) applied loads $P_1, P_2 \ldots, M_1$, and so forth. The virtual forces and couples are arbitrary, but they must also satisfy the same equilibrium equations as the real forces. Therefore, when virtual forces are applied as external loads, **virtual stresses** are induced within the body so that equilibrium of every part of the body is maintained. Figure 11.20a shows an axial load P_1 applied to a rod, and Fig. 11.20b depicts the resulting virtual normal stress $\delta\sigma_x$ required at cross-section x to maintain equilibrium of the free body in Fig. 11.20b when P_1 is increased to $P_1 + \delta P$.

Complementary virtual work is defined as follows:

> **Complementary virtual work,** designated by δW^c, is the work done by virtual forces without altering the configuration of the body on which the virtual forces act.

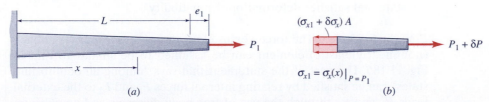

FIGURE 11.20 Virtual force and virtual stress.

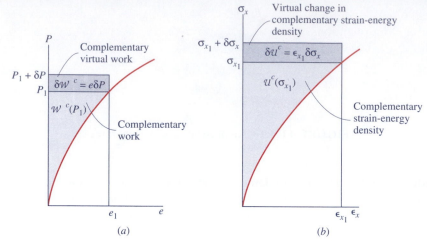

FIGURE 11.21 (a) Complementary work and complementary virtual work; (b) complementary strain-energy density and virtual change in complementary strain-energy density.

Complementary virtual work is illustrated in Fig. 11.21a. As indicated in Fig. 11.21b, **complementary strain-energy density** is the area <u>above</u> the stress-strain curve.

Principle of Virtual Forces.
The **Principle of Virtual Forces,** or the *Principle of Complementary Virtual Work,* states that:

> *Among all possible equilibrium states, the actual state of a body is the one that satisfies deformation compatibility.*

In equation form, the *Principle of Virtual Forces* is given by

$$\delta \mathcal{W}^c_{\text{ext}} = \delta \mathcal{U}^c \qquad \text{**Principle of Virtual Forces**} \qquad (11.54)$$

In Example Problem 11.13 this principle is applied to a statically determinate problem, illustrating how the Principle of Virtual Forces leads to deformation-compatibility equations. In Example Problem 11.14 it is applied to a statically indeterminate problem.

EXAMPLE 11.13

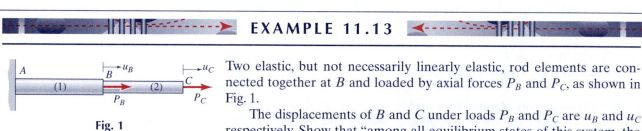

Fig. 1

Two elastic, but not necessarily linearly elastic, rod elements are connected together at B and loaded by axial forces P_B and P_C, as shown in Fig. 1.

The displacements of B and C under loads P_B and P_C are u_B and u_C respectively. Show that "among all equilibrium states of this system, the state that satisfies Eq. 11.54 is the kinematically admissible state, that is, the state that satisfies deformation compatibility."

Plan the Solution The force versus elongation (F_i vs e_i) curve for each (nonlinearly elastic) element can be assumed to be similar to that in Fig. 11.18b. The part of the statement that says "among all equilibrium states" can be satisfied by relating internal forces F_1 and F_2 to the external loads P_B and P_C through the use of free-body diagrams and equilibrium equations. The "deformation compatibility" in this problem consists of relating the elongations e_1 and e_2 to the displacements u_B and u_C (e.g., see Chapter 3). The complementary virtual work terms will be $u_B\delta P_B, e_1\,\delta F_1$, and so forth.

Solution

Equilibrium: Both real forces and virtual forces must satisfy equilibrium. Since this is a statically determinate rod system, we can directly relate the internal forces F_1 and F_2 to the external forces P_B and P_C through equilibrium equations based on the free-body diagrams in Fig. 2. The same is true for virtual forces. Equilibrium must hold for the real forces and also when arbitrary virtual forces are added. Therefore, we must have

$$\left(\sum F\right)_1 = 0:\qquad F_1 = P_B + P_C \qquad \text{and} \qquad \delta F_1 = \delta P_B + \delta P_C$$

$$\tag{1}$$

$$\left(\sum F\right)_2 = 0:\qquad F_2 = P_C \qquad \text{and} \qquad \delta F_2 = \delta P_C$$

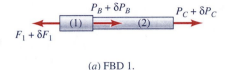

$F_1 + \delta F_1$

(a) FBD 1.

(b) FBD 2.

Fig. 2 Free-body diagrams.

Principle of Virtual Forces: For each element the internal forces are assumed to be related to the elongations by a curve having the form indicated in Fig. 3. Therefore, for element i the virtual force δF_i does an amount of complementary virtual work $e_i\delta F_i$, as indicated in Fig. 3. This is the change in complementary energy in element i.

The Principle of Virtual Forces, Eq. 11.54, states that

$$\delta \mathcal{W}^c = \delta \mathcal{U}^c \tag{2}$$

Writing out the contributions to each side of this equation we get

$$u_B\delta P_B + u_C\delta P_C = e_1\delta F_1 + e_2\delta F_2 \tag{3}$$

The virtual forces must satisfy the virtual-force equilibrium equations in Eqs. (1). Substituting these into Eq. (3) gives

$$u_B\delta P_B + u_C\delta P_C = e_1(\delta P_B + \delta P_C) + e_2(\delta P_C) \tag{4}$$

or

$$(e_1 - u_B)\delta P_B + (e_1 + e_2 - u_C)\delta P_C = 0 \tag{5}$$

Then, since δP_B and δP_C are arbitrary [as long as equilibrium is satisfied; which it is by the step from Eq. (3) to Eq. (4)], Eq. (5) requires that

$$u_B = e_1, \qquad u_C = e_1 + e_2 \qquad \textbf{Ans.} \quad (6)$$

These are the equations of deformation compatibility for this problem.

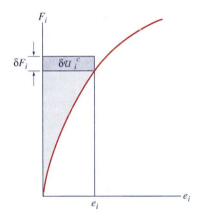

Fig. 3 Force-elongation curve for an element.

Review the Solution We started by guaranteeing that the virtual forces would always satisfy equilibrium, applied the Principle of Virtual Forces, and arrived at sensible expressions for the kinematic relationships between nodal displacements and elongations. Therefore, we illustrated the statement of the Principle of Virtual Forces, as requested. Although we allowed for possible nonlinear material behavior, this application did not involve material properties directly, since we were not asked to solve for u_B and u_C in terms of the loads.

Next we consider a statically indeterminate problem. This will enable us to illustrate the relationship between the Principle of Virtual Force and the Force Method, which was discussed in Chapters 3, 4, and 7.

EXAMPLE 11.14

The external force P is applied to the rigid block in Fig. 1 in such a manner that the two rods that support the block undergo equal elongations. Let the rod elements be linearly elastic with flexibility coefficients f_i (see Eq. 3.14). Use the Principle of Virtual Forces, Eq. 11.54: (a) to obtain the compatibility equations for this system, and (b) to obtain the internal forces in the two rods.

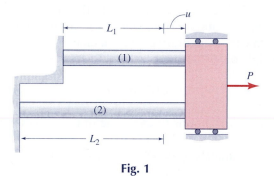

Fig. 1

Plant the Solution   Part (a) should be a repeat, more or less, of Example Problem 11.13. In Part (b) we can employ the linearly elastic force-elongation relationships, $e_i = f_i F_i$, and see if this gets the desired expressions for F_1 and F_2. This problem is statically indeterminate, however, so we can expect to have to identify one of the F_i's as a redundant force.

Solution

(a) Obtain the compatibility equations.

Equilibrium: The block in Fig. 2 must be in equilibrium under the action of the real forces, and also when virtual forces are added to the real forces. Therefore,

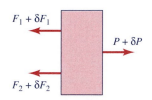

Fig. 2   Free-body diagram.

$$\sum F = 0: \quad F_1 + F_2 = P \quad \text{and} \quad \delta F_1 + \delta F_2 = \delta P \tag{1}$$

Since there is only one equilibrium equation for real forces (and, correspondingly only one equilibrium equation for virtual forces), this is a statically indeterminate problem. Let us select F_1 as the redundant force (and δF_1 as the redundant virtual force). Then, Eqs. (1) can be expressed in the form

$$F_2 = P - F_1 \qquad \text{and} \qquad \delta F_2 = \delta P - \delta F_1 \qquad (2)$$

Principle of Virtual Forces: The Principle of Virtual Forces, Eq. 11.54, states that

$$\delta W_{\text{ext}}^c = \delta \mathcal{U}^c \qquad (3)$$

Complete Part (a) by determining the compatibility equations: If we write $\delta \mathcal{U}^c$ in terms of e_1 and e_2 directly, as we did in Example Problem 11.13, we will get the requested compatibility equations.

$$u \delta P = e_1 \delta F_1 + e_2 \delta F_2 \qquad (4)$$

Substituting Eq. (2b) into Eq. (4), we get

$$u \delta P = e_1 \delta F_1 + e_2 (\delta P - \delta F_1) \qquad (5)$$

or

$$(u - e_2)\delta P - (e_1 - e_2)\delta F_1 = 0 \qquad (6)$$

But, since equilibrium is satisfied for any δP and any δF_1 because of the step from Eq. (4) to Eq. (5), Eq. (6) requires that

$$e_2 = u, \qquad e_1 = e_2 \qquad \textbf{Ans. (a)} \quad (7)$$

Therefore, the Principle of Virtual Forces has produced a complete set of compatibility equations for this problem.

(b) Determine the axial forces F_1 and F_2. If we insert the force-elongation relations for the two rod elements, we will be able to solve for the forces in the rods. In the present problem, the rods are linearly elastic. We need the equations in the form $e = e(F)$. Therefore, let us use Eq. 3.14 written as

$$e_1 = f_1 F_1, \qquad e_2 = f_2 F_2 \qquad (8)$$

Substituting Eqs. (2a) and (8) into Eq. (7b) we get

$$f_1 F_1 = f_2 (P - F_1)$$

and, solving for the redundant force F_1, we get

$$F_1 = \frac{f_2 P}{f_1 + f_2} \qquad \textbf{Ans. (b)} \quad (9a)$$

This redundant force can be substituted back into Eq. (2a) to give the other internal force

$$F_2 = \frac{f_1 P}{f_1 + f_2} \qquad \textbf{Ans.} \quad (9b)$$

Note that the use of the *Principle of Virtual Forces* leads us naturally to a *Force-Method* solution of this statically indeterminate problem.

Review the Solution Using the Principle of Virtual Forces, we arrived at the compatibility equations that we could have written down "by inspection." However, we also arrived at a solution for internal forces that we can verify by noting that the solution for the forces in Eqs. (9) satisfies equilibrium, $F_1 + F_2 = P$, and compatibility, $f_1 F_1 = f_2 F_2$.

The Crotti-Engesser Theorem. The Principle of Virtual Displacements, Eq. 11.48, was rephrased as Castigliano's First Theorem by incorporating the expression

$$\delta W_{\text{ext}} = \sum_{i=1}^{N} P_i \delta q_i$$

for the virtual work of external forces. In an analogous manner, when bodies are loaded by discrete forces the complementary virtual work can be written in the form

$$\delta W^c = \sum_{i=1}^{N} u_i \delta P_i \qquad (11.55)$$

where the P_i's can include couples, and where u_i is the displacement (or rotation) that corresponds to the force (or couple) P_i. Since the complementary strain energy is fundamentally a function of the applied loads, an expression analogous to Eq. 11.47 can be written for $\delta \mathcal{U}^c$, as follows:

$$\delta \mathcal{U}^c = \sum_{i=1}^{N} \frac{\partial \mathcal{U}^c}{\partial P_i} \delta P_i \qquad (11.56)$$

Inserting Eqs. 11.55 and 11.56 into Eq. 11.54, the Principle of Virtual Forces, and noting that the virtual forces are independent, we get

$$u_i = \frac{\partial \mathcal{U}^c}{\partial P_i}, \qquad i = 1, 2, \ldots, N \qquad \begin{array}{l}\textbf{Crotti-Engesser}\\ \textbf{Theorem}\end{array} \quad (11.57)$$

This is called the **Crotti-Engesser Theorem.**[10] It applies to nonlinearly elastic bodies as well as linearly elastic bodies, as is illustrated by the next example problem.

---

[10]The Crotti-Engesser Theorem is named after the Italian engineer Francesco Crotti (1839–1896) who developed it in 1878, and the German engineer Friedrich Engesser (1848–1931), who independently derived the theorem in 1889. Engesser introduced the notion of *complementary energy*. [Ref. 11-1]

EXAMPLE 11.15

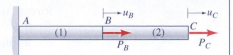

Axial loads P_B and P_C are applied, as shown in Fig. 1a, to a two-element rod system whose elements both behave according to the nonlinear force-elongation curve in Fig. 1b. Using the Crotti-Engesser Theorem, determine expressions for u_B and u_C in terms of the loads P_B and P_C and the material constant D.

(a) A 2-element rod system.

Plant the Solution The complementary strain energy of an element is the shaded area above the curve in Fig. 1b. We can obtain $\mathcal{U}_i^c(F_i)$ by integration. To get $\mathcal{U}^c(P_B, P_C)$ we must relate the F_i's to the P's by equilibrium equations. Then we can directly apply Eq. 11.57, the Crotti-Engesser Theorem.

Solution

Equilibrium: This is the same configuration as the one in Example Problem 11.13, where we got

$$F_1 = P_B + P_C \quad \text{and} \quad \delta F_1 = \delta P_B + \delta P_C$$
$$F_2 = P_C \quad \text{and} \quad \delta F_2 = \delta P_C \tag{1}$$

Complementary Strain Energy: From Eqs. 11.51 and 11.52,

$$\mathcal{U}_i^c(F_i) = \int_0^{F_i} e\,dF = \int_0^{F_i} DF^3\,dF = \frac{1}{4}DF_i^4 \tag{2}$$

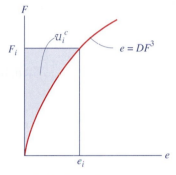

(b) Nonlinear element force-elongation curve.

Fig. 1

For the two-element system, then,

$$\mathcal{U}^c = \frac{1}{4}DF_1^4 + \frac{1}{4}DF_2^4 \tag{3}$$

which, when combined with Eqs. (1), gives

$$\mathcal{U}^c = \frac{D}{4}\left[(P_B + P_C)^4 + P_C^4\right] \tag{4}$$

Crotti-Engesser Theorem: Since Eq. (4) incorporates both equilibrium and force-elongation information, we apply the Crotti-Engesser Theorem, which takes care of the geometric compatibility aspect of the solution. Then,

$$\left.\begin{aligned} u_B &= \frac{\partial \mathcal{U}^c}{\partial P_B} = D(P_B + P_C)^3 \\ u_C &= \frac{\partial \mathcal{U}^c}{\partial P_C} = D\left[(P_B + P_C)^3 + P_C^3\right] \end{aligned}\right\} \quad \textbf{Ans. (5)}$$

Review the Solution For this simple example, we can see "by inspection" that $u_B = e_1$, and $u_C = e_1 + e_2$. Then, using $e_i = DF_i^3$ from Fig. 1b, and substituting the equilibrium expressions for the F_i's, we get the answers in Eq. (5).

The Crotti-Engesser Theorem Applied to Statically Indeterminate Systems. Example Problem 11.14 treats a statically indeterminate problem from the standpoint of the Principle of Virtual Forces. Let us select a set of N_R redundant forces to use in solving a statically indeterminate problem, and let these be called $R_1, R_2, \ldots, R_{N_R}$. Then the complementary strain energy can be expressed in terms of the applied loads and these redundants. The displacements that correspond to the redundant forces are all zero (i.e., the system is not really cut, but we imagine that it is cut so that the redundants that we have chosen appear as known forces). Therefore, the **Crotti-Engesser Theorem applied to statically indeterminate systems** is given by the following two equations. First, the N_R redundant forces are determined by

$$\frac{\partial \mathcal{U}^c(P_1, P_2, \ldots, P_N; R_1, \ldots, R_{N_R})}{\partial R_i} = 0, \quad i = 1, 2, \ldots, N_R \qquad (11.58)$$

Then, the displacements that correspond to the loads P_i can be obtained from

$$u_i = \frac{\partial \mathcal{U}^c(P_1, P_2, \ldots, P_N; R_1, \ldots, R_{N_R})}{\partial P_i}, \quad i = 1, 2, \ldots, N \qquad (11.59)$$

We will now use Eq. 11.58 to re-solve Example Problem 11.14.

EXAMPLE 11.16

For the system in Example Problem 11.14, which consists of two linearly elastic rods connected to a rigid block, use Eq. 11.58 to solve for the redundant force F_1.

Plant the Solution   For linearly elastic rods, Eq. 11.53 gives an expression for $\mathcal{U}^c$. If we incorporate the equilibrium equation (see Example Problem 11.14) into this we will get $\mathcal{U}^c$ in terms of P and F_1 ($R_1 \equiv F_1 \equiv$ redundant force). Then we can apply Eq. 11.58 to get an equation for F_1.

Solution

Equilibrium: From Eq. (2) of Example Problem 11.14,

$$F_2 = P - F_1 \qquad (1)$$

Complementary Strain Energy: From Eq. 11.53,

$$\mathcal{U}^c = \frac{1}{2} f_1 F_1^2 + \frac{1}{2} f_2 F_2^2 \qquad (2)$$

Selecting F_1 as the redundant and combining Eqs. (1) and (2), we get

$$\mathcal{U}^c = \frac{1}{2} f_1 F_1^2 + \frac{1}{2} f_2 (P - F_1)^2 \qquad (3)$$

Crotti-Engesser Theorem: With F_1 taken as the redundant, Eq. 11.58 becomes

$$\frac{\partial \mathcal{U}^c}{\partial F_1} = 0 \qquad (4)$$

Then, the result of combining Eqs. (3) and (4) is

$$f_1 F_1 - f_2(P - F_1) = 0$$

or

$$F_1 = \frac{f_2 P}{f_1 + f_2} \qquad \textbf{Ans.} \quad (5)$$

which is the same solution (a Force-Method solution) that we got in Example Problem 11.14.

Review the Solution No further checking is necessary, since we got the same answer that we got in Example Problem 11.14. If we take $\partial \mathcal{U}^c / \partial P$ using Eq. (3), we directly get the equation for the displacement, that is,

$$u = f_2(P - F_1)$$

into which we can substitute F_1 from Eq. (5).

By comparing Eqs. 11.36 and 11.57 you will note that Castigliano's Second theorem is a special case of the Crotti-Engesser Theorem for the case of linearly elastic behavior, when the complementary strain energy $\mathcal{U}^c$ is equal to the strain energy $\mathcal{U}$, as illustrated in Fig. 11.22. For use with Castigliano's theorem (Eq. 11.36) the strain energy is written in terms of forces instead of displacements. We will therefore refer to Castigliano's Second Theorem and the Unit-Load Method as *Special Linear Methods*. Table 11.2 summarizes the features of the *Strain-Energy Methods* of Section 11.7, the *Complementary-Energy Methods* of Section 11.8, and the *Special Linear Methods* of Section 11.5.

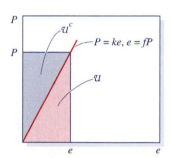

FIGURE 11.22 The load versus elongation diagram of a linearly elastic member.

TABLE 11.2 **Comparison of Energy Methods**

Strain-Energy Methods	Complementary-Energy Methods	Special Linear Methods
Functional dependencies: $\delta W(\text{displ.}), \delta \mathcal{U}(\text{displ.})$	Functional dependencies: $\delta W^c(\text{forces}), \delta \mathcal{U}^c(\text{forces})$	Functional dependencies: $\delta W(\text{forces}), \delta \mathcal{U}(\text{forces})$
Compatibility implicitly required.	Equilibrium implicitly required.	Equilibrium implicitly required.
Uses virtual displacements.	Uses virtual forces.	Uses virtual forces.
Produces equilibrium equations in terms of displacements.	Produces compatibility equations in terms of forces.	Produces compatibility equations in terms of forces.
Material may be nonlinear.	Material may be nonlinear.	Material must be linear.

So far we have assumed that all loads are applied slowly until each reaches its maximum value and then remains at this *static-load* value. When forces are applied more rapidly, like wave loading of an offshore oil platform or impact loading of an automobile during a collision, it is necessary to turn to the topic of structural dynamics to determine the time-dependent behavior of the dynamically loaded deformable body.[11] Here we make simplifying assumptions that enable us to see that there is definitely a difference between the response of a deformable body to static loading and its response to dynamic loading.

We begin by employing a "massless" linear spring as the deformable member. Later we will generalize to bars, beams, and other deformable bodies. In the first case we consider a weight W that falls from height h onto the massless, linear spring with spring constant k (i.e., $P = k\Delta$). In the second case we consider a mass that is moving with speed v when it strikes the spring.

A. Gravitational Potential Energy Converted to Strain Energy of a "Massless" Linear Spring.

Let weight W be dropped from height h onto a massless, linear spring, and assume that no energy is lost during the initial contact of the weight with the spring. Three positions are identified in Fig. 11.23a: (1) the position where the weight is released from rest, (2) the position where the weight initially contacts the spring, and (3) the lowest position reached by the weight, where the spring has its maximum compression, Δ_{max}, and where the weight is (momentarily) stopped.

If no energy is lost during the impact, we can equate the gravitational potential energy lost by the weight in falling through the distance $(h + \Delta_{max})$ to the increase in strain energy in the spring when it is compressed by an amount Δ_{max}. Then,

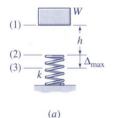

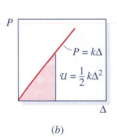

(a)

$$W(h + \Delta_{max}) = \frac{1}{2} k \Delta_{max}^2 \tag{11.60}$$

Solving for the positive root of this quadratic equation in Δ_{max}, we get

$$\Delta_{max} = \left(\frac{W}{k}\right) + \left[\left(\frac{W}{k}\right)^2 + 2h\left(\frac{W}{k}\right)\right]^{1/2} \tag{11.61}$$

But W/k is the static compression, Δ_{st}, that would occur if the weight W were slowly lowered onto the spring. Therefore, Eq. 11.61 can be recast in the convenient form

$$\Delta_{max} = \Delta_{st}\left[1 + \left(1 + \frac{2h}{\Delta_{st}}\right)^{1/2}\right] \tag{11.62}$$

This equation says that $\Delta_{max} \geq 2\Delta_{st}$, with

$$\Delta_{max} = 2\Delta_{st} \tag{11.63}$$

FIGURE 11.23 Impact loading—falling-weight case.

---

[11]For example, see *Fundamentals of Structural Dynamics*, 2nd Edition, by Roy R. Craig, Jr. and Andrew J. Kurdila, [Ref. 11-3].

if the weight is suddenly released when it is just touching the spring, that is, at position (2) in Fig. 11.23a.

At position (3) in Fig. 11.23a, where the spring is at its maximum compression, Δ_{max}, the force exerted on the spring by the weight (and vice versa) has a magnitude

$$P_{max} = k\Delta_{max} \tag{11.64}$$

where Δ_{max} is given by Eq. 11.61 or 11.62.

B. Kinetic Energy Converted to Strain Energy of a "Massless" Linear Spring.

Figure 11.24 shows a mass that is moving with speed v at the instant when it makes contact with a massless, linear spring. If we assume that no energy is lost in the impact process, then energy is conserved, and the kinetic energy of the mass at position (1) in Fig. 11.24 is converted to strain energy stored in the spring in position (2). Then,

$$\frac{1}{2}Mv^2 = \frac{1}{2}k\Delta_{max}^2 \tag{11.65}$$

or

$$\boxed{\Delta_{max} = \sqrt{\frac{Mv^2}{k}}} \tag{11.66}$$

As in Case A, the maximum force exerted on the spring by the mass is given by Eq. 11.64, with Δ_{max}, in the present case, given by Eq. 11.66.

C. Impact on Deformable Bodies.

Figure 11.25 depicts as sliding collar that drops from height h, makes contact with a flange on the end of a linearly elastic rod, and stretches the rod by an amount Δ_{max}. The analysis of Case A can be applied to this system and other deformable bodies if we make the following assumptions:

- The mass of the impacted deformable body is negligible in comparison with the impacting mass.
- The impacting mass is rigid.
- No energy is lost in the impact.

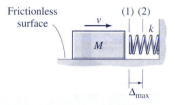

FIGURE 11.24 Impact loading—moving-mass case.

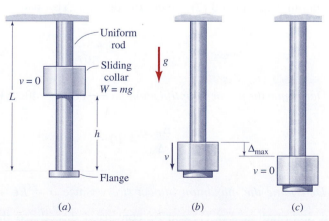

FIGURE 11.25 Impact loading of a uniform rod.

These assumptions lead to conservative answers. That is, the deformation and stresses calculated are greater than the actual values would be if energy losses and other factors hinted at above were completely accounted for.

With the preceding assumptions, we do not have to consider conservation of momentum upon impact or consider stress waves in the impacted body; we can just apply the results obtained previously for a "massless" spring. Equations 11.62 and 11.66 can be used to determine Δ_{max} for the respective two types of impact, with an equivalent stiffness. k, determined for the particular elastic body impacted. However, to determine the maximum stress caused by the impact, we must apply the dynamic load P_{max} of Eq. 11.64 to the particular body impacted (e.g., rod, beam, etc.). The next two example problems illustrate the effect of impact loading on deformable bodies.

EXAMPLE 11.17

For the rod and sliding collar of Fig. 11.25, (a) determine an expression for Δ_{max} as a function of W, A, E, h, and L. (b) If the weight is dropped from a height of $h = 40\Delta_{st}$, determine the value of the *impact amplification factor* Δ_{max}/Δ_{st}, (c) Determine the maximum impact stress σ_{max} in terms of the static stress σ_{st}, the drop height h, and the rod parameters.

Solution

(a) Determine Δ_{max}, the maximum displacement. We could use either Eq. 11.61 or 11.62, taking Δ to be positive for elongation, rather than for compression as in the original derivation. Let us use Eq. 11.62.

$$\Delta_{max} = \Delta_{st}\left[1 + \left(1 + \frac{2h}{\Delta_{st}}\right)^{1/2}\right] \qquad (1)$$

If the weight were to be lowered slowly onto the flange of the rod, we would get the static elongation

$$\Delta_{st} = \frac{W}{k} = \frac{WL}{AE} \qquad (2)$$

Combining Eqs. (1) and (2), we get the desired expression

$$\Delta_{max} = \frac{WL}{AE}\left[1 + \left(1 + \frac{2AEh}{WL}\right)^{1/2}\right] \qquad \textbf{Ans. (a)} \quad (3)$$

(b) Determine the impact amplification factor. For $h = 40\Delta_{st}$, Eq. (1) gives

$$\frac{\Delta_{max}}{\Delta_{st}} = 10 \qquad \textbf{Ans. (b)} \quad (4)$$

(c) Determine the maximum impact stress. Since $\sigma = E\epsilon = E\left(\dfrac{\Delta}{L}\right)$ for axial deformation, Eq. (1) can be converted directly to the following

equation for σ_{max}:

$$\sigma_{max} = \sigma_{st}\left[1 + \left(1 + \frac{2Eh}{L\sigma_{st}}\right)^{1/2}\right] \qquad \text{Ans. (c)} \quad (5)$$

It is left as an exercise for the reader (Probs. 11.9-8, 11.9-9) to show that impact loading of a rod with enlarged cross section over a portion of its length produces a higher maximum stress than the stress given by Eq. (5) for a uniform rod. Thus, from the standpoint of elastic energy absorption, a rod with uniform cross section is preferable to a rod with nonuniform cross section.

EXAMPLE 11.18

A diver is springing on the diving board shown in Fig. 1. On a particular bounce, the diver reaches a height h above the end of the board. Treat the diver as a rigid mass that falls from a height h, and assume that the diving board is much lighter than the diver (e.g., a fiberglass board might be very light) and is straight when the diver strikes it at the very end. (a) Determine the maximum deflection of the tip of the diving board. Express your answer in terms of W, h, and the parameters of the diving board—E, I, and L. (b) Determine the maximum flexural stress caused by the diver's "impact." Express your answer in terms of the maximum static flexural stress and the parameters of the diving board.

Neglect shear deformation and neglect any energy loss during impact. Also neglect the mass of the diving board, and assume that $EI = $ const.

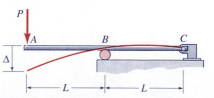

Fig. 1

Plan the Solution   We can use the results of Example Problem 11.4 to determine an expression for k to use in Eq. 11.62 for Δ_{max}. In Part (b) we can use Eq. 11.64 to determine the maximum force that the diver exerts on the beam. Then we can use the flexure formula, Eq. 6.13, with the maximum bending moment produced by P_{max}.

Solution

(a) *Determine the maximum deflection at the tip of the diving board.* This deflection can be determined from Eq. 11.62, which we can write in the form

$$\Delta_{max} = \frac{W}{k}\left[1 + \left(1 + \frac{2kh}{W}\right)^{1/2}\right] \qquad (1)$$

From Example Problem 11.4, the deflection Δ at the tip of the beam (diving board) in Fig. 2 is

$$\Delta = \frac{2}{3}\frac{PL^3}{EI} \qquad (2)$$

Fig. 2   The deflection caused by a tip load.

Since k is defined by $P = k\Delta$, for this beam configuration and loading

$$k = \frac{3EI}{2L^3} \tag{3}$$

Therefore, Eqs. (1) and (3) may be combined, giving the desired answer

$$\Delta_{\max} = \frac{2WL^3}{3EI}\left[1 + \left(1 + \frac{3EIh}{WL^3}\right)^{1/2}\right] \qquad \textbf{Ans. (a)} \quad (4)$$

(b) Determine the maximum flexural stress. For the free body in Fig. 3,

$$\left(\sum M\right)_B = 0: \qquad\qquad C_y = -P$$

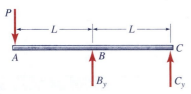

Fig. 3   Free-body diagram.

Therefore, the beam loading is symmetric about B, and the maximum bending moment along the beam occurs at B, where $M_B = -PL$. We obtain the maximum impact force at the tip, A, by combining Eq. 11.64 and Eq. (3), giving

$$P_{\max} = \left(\frac{3EI}{2L^3}\right)\Delta_{\max} \tag{5}$$

so the maximum bending moment is

$$(M_B)_{\max} = -\left(\frac{3EI}{2L^2}\right)\Delta_{\max} \tag{6}$$

Let c be the distance from the neutral axis to the top fibers of the beam. Then, from the flexure formula, Eq. 6.13,

$$\sigma_{\max} = \frac{-(M_B)_{\max}c}{I} = \left(\frac{3Ec}{2L^2}\right)\Delta_{\max} \tag{7}$$

Let σ_{st} be the maximum flexural stress in the beam under static loading (e.g., with the diver standing on the end of the diving board). Then, $(M_B)_{\text{st}} = -WL$, so

$$\sigma_{\text{st}} = \frac{WLc}{I} \tag{8}$$

Finally, we can combine Eqs. (4), (7), and (8) to get the following expression relating $\sigma_{\max}$ and σ_{st}:

$$\sigma_{\max} = \sigma_{\text{st}}\left[1 + \left(1 + \frac{3Ehc}{\sigma_{\text{st}}L^2}\right)^{1/2}\right] \qquad \textbf{Ans. (b)} \quad (9)$$

(Note that this expression differs from the answer for axial impact on a rod, whereas the same expression, Eq. 11.62, relates $\Delta_{\max}$ to Δ_{st} for both problems.)

Prob. 11.3-1. A uniform aluminum-alloy rod with cross-sectional area $A = 1$ in^2 and length $L = 50$ in. is subjected to an axial load P, as shown in Fig. P11.3-1. Let $E_{al} = 10(10^3)$ ksi. (a) Sketch a load versus elongation diagram (i.e., load P versus elongation e) for elongations from 0 in. to 0.250 in. (Hint. Recall Eq. 3.15.) (b) Calculate the strain energy U stored in the rod when $P = 40$ kips, and indicate on your $P - e$ diagram of Part (a) the area that represents this strain energy.

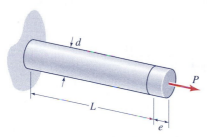

P11.3-1 and P11.3-2

Prob. 11.3-2. A uniform steel rod with a diameter of $d = 30$ mm and a length $L = 1.5$ m is subjected to an axial load P as shown in Fig. P11.3-2. Let $E_{st} = 200$ GPa and $\sigma_Y = 250$ MPa. (a) Determine the minimum value of P at which yielding would occur. (b) Calculate the strain energy U stored in the rod when P reaches the yield load P_Y determined in Part (a).

DProb. 11.3-3. The stepped rod shown in Fig. P11.3-3 is subjected to equal axial loads of magnitude P at section B and at section C. (a) If $A_1 = 2A, A_2 = A, L_1 = L_2 = L$, and $E_1 = E_2 = E$, determine: (1) the strain energy U_1 in the rod AB, (2) the strain energy U_2 in rod BC, and (3) the total strain energy U. (b) Would <u>less</u> strain energy be stored in rod AC if the rod were uniform, that is, if $A_1 = A_2 = A$? Justify your yes or no answer.

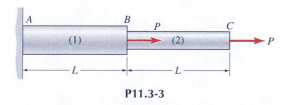

P11.3-3

Prob. 11.3-4. Determine the strain energy U stored in the uniform member shown in Fig. P11.3-4 if the modulus of elasticity is E and the cross-sectional area is A.

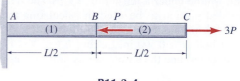

P11.3-4

DProb. 11.3-5. (a) Determine the strain energy U stored in the uniform aluminum rod shown in Fig. P11.3-5a. Let $E_{al} = 10(10^3)$ ksi and $P_B = 6$ kips. (b) Determine the strain energy stored in the rod if $P_E = 6$ kips and the diameter over half of the length of the rod is increased to $d_2 = 1.25$ in., as illustrated in Fig. P11.3-5b.

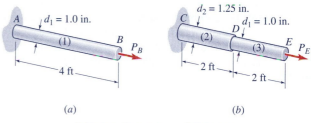

(a)   (b)

P11.3-5, P11.3-6, and P11.4-1

Prob. 11.3-6. (a) Determine the value of the force P_E that would cause the stepped rod in Fig. P11.3-6b to have the same elongation as the rod in Fig. P11.3-6a when $P_B = 6$ kips. Let $E_{al} = 10(10^3)$ ksi. (b) Determine the strain energy stored in each of the two rods under the loads specified in Part (a).

DProb. 11.3-7. Let U_a be the strain energy stored in a uniform rod of cross-sectional area A and length L whose modulus of elasticity is E, as illustrated in Fig. P11.3-7a. If the cross-sectional area of the rod is changed to αA ($\alpha < 1$ means that the area is decreased; $\alpha > 1$ means that the area is increased) over a portion of the length equal to λL ($0 \le \lambda \le 1$) determine an expression for the ratio of strain energy in the stepped bar to the strain energy in the uniform bar, that is, determine an expression for U_b/U_a in terms of the parameters α and λ.

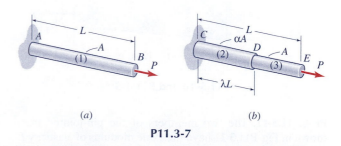

(a)   (b)

P11.3-7

*Prob. 11.3-8. Two types of steel bolts are being considered for a particular application for which energy-storage capacity is important. Bolt A has threads along its entire length, but bolt B has a larger-diameter shank over a 3-in. portion of its original length. When the nut is tightened, a 3.5-in. section of each bolt will be under a tension of 5 kips. The diameter of the threaded portion to be used in calculating the tensile stress is 0.432 in., and the diameter of the unthreaded shank portion of bolt B is 0.500 in. Letting $E_{st} = 30(10^3)$ ksi for both bolts, determine the strain energy stored in each bolt.

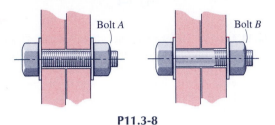

P11.3-8

Prob. 11.3-9. Determine an expression for the strain energy U stored in a solid conical frustum (Fig. P11.3-9) having a maximum diameter of d_A, a minimum diameter of d_B, and a length of L. Let the modulus of elasticity be E and the axial load be P.

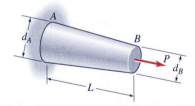

P11.3-9 and P11.4-2

Prob. 11.3-10. The width of the 0.5-in.-thick steel plate shown in Fig. P11.3-10 ($E = 29(10^3)$ ksi) varies linearly from $h_1 = 4$ in. to $h_2 = 2$ in. over a length $L = 12$ in. (a) Determine the strain energy U_a stored in this plate when it is subjected to a load $P = 2$ kips. (b) Determine the strain energy U_b stored in the plate if the load P is increased to the value that causes initial yielding in the plate. Let $\sigma_Y = 36$ ksi.

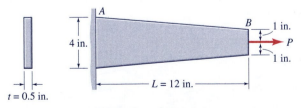

P11.3-10 and P11.4-3

Prob. 11.3-11. The two members of the pin-jointed truss shown in Fig. P11.3-11 have the same modulus of elasticity E and the same cross-sectional area A. Determine the total strain energy U stored in the truss in terms of the load P, the length L, E, and A.

738

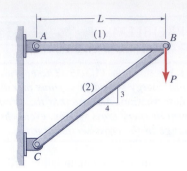

P11.3-11 and P11.4-4

Prob. 11.3-12. The cross-sectional areas of the four members of the pin-jointed truss shown in Fig. P11.3-12 are $A_1 = A_2 = 1$ in^2, $A_3 = A_4 = 1.5$ in^2, Determine the total strain energy U stored in the truss when a vertical load $P = 2$ kips is applied at D. All members are made of steel with $E_{st} = 30(10^3)$ ksi.

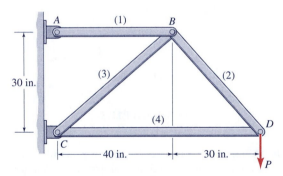

P11.3-12 and P11.4-5

Prob. 11.3-13. The cross-sectional areas of the three members of the pin-jointed truss shown in Fig. P11.3-13 are $A_1 = A_2 = 500$ mm^2, and $A_3 = 800$ mm^2. Determine the total strain energy U stored in the truss when a load $P = 5.2$ kN is applied at B, as shown. All members are made of aluminum alloy with $E_{al} = 70$ GPa.

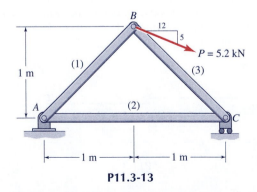

P11.3-13

Prob. 11.3-14. A uniform rod AB of length L and cross-sectional area A hangs under its own weight from a rigid

support at A as illustrated in Fig. P11.3-14. If the rod is made of linearly elastic material with modulus of elasticity E and specific weight γ (i.e., weight per unit volume), determine the strain energy $\mathcal{U}$ stored in the rod in terms of γ, E, A, and L.

P11.3-14

*Prob. 11.3-15.** Determine an expression for the strain energy $\mathcal{U}$ stored in a vertically hanging solid conical frustum having a maximum diameter d_1, a minimum diameter d_2, and a length L (Fig. P11.3-15). Let the modulus of elasticity be E and the specific weight (i.e., the weight per unit volume) be γ.

P11.3-15

> **Problems 11.3-16 through 11.3-25.** *These problems deal with the strain energy stored in various members undergoing torsional deformation. Assume that the stress distribution obeys Eq. 4.11 at every cross section, even where there is a sudden change in the diameter of the cross section.*

Prob. 11.3-16. A uniform aluminum-alloy rod with circular cross section of diameter $d = 1$ in. and length $L = 50$ in. is subjected to an end torque T as shown in Fig. P11.3-16. Let $G = 4(10^3)$ ksi. (a) Sketch a torque versus total angle-of-

P11.3-16

twist diagram for torques ranging from 0 kip · in. to 10 kip · in. (Hint: Recall Eq. 4.21.) (b) Calculate the strain energy $\mathcal{U}$ stored in the rod when $T = 4$ kip · in., and indicate on your $T - \phi$ diagram of Part (a) the area that represents this strain energy.

Prob. 11.3-17. The stepped rod in Fig. P11.3-17 is subjected to equal external torques T_0 at section B and at end C. The shear modulus of the rod is G, and the dimensions of the rod are $d_1 = 1.5d$, $d_2 = d$, and $L_1 = L_2 = L$, (a) Determine (1) the strain energy $\mathcal{U}_1$ stored in rod AB. (2) the strain energy $\mathcal{U}_2$ stored in rod BC, and (3) the total strain energy $\mathcal{U}$. (b) Would <u>less</u> strain energy be stored in rod AC if the rod were uniform, that is, if $d_1 = d_2 = d$? Justify your answer.

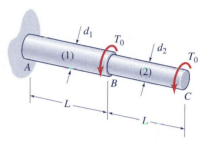

P11.3-17

Prob. 11.3-18. A uniform torsion rod is subjected to an end torque T, as shown in Fig. P11.3-18. The rod has a diameter $d = 30$ mm and a length $L = 1.5$ m. If the maximum shear stress in the rod is 60 MPa, how much strain energy is stored in the rod? Let $G = 70$ GPa.

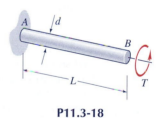

P11.3-18

Prob. 11.3-19. A uniform tubular aluminum-alloy shaft $(G = 3.8(10^3)$ ksi$)$ with outside diameter $d_o = 1.25$ in, and length $L = 2$ ft is subjected to a torque $T = 3$ kip · in. (Fig. P11.3-19). (a) If the maximum shear stress in the shaft is 10 ksi, determine the value of the inside diameter, d_i, of the shaft, and (b) determine the strain energy stored in the shaft.

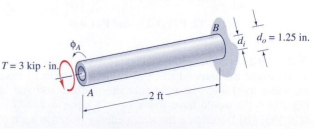

P11.3-19 and P11.4-6

Prob. 11.3-20. Determine the strain energy stored in the tapered bar in Fig. P11.3-20. Express your answer in terms of T_0, L, d_A, d_B, and the shear modulus G.

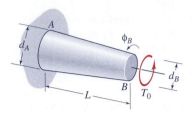

P11.3-20 and P11.4-7

Prob. 11.3-21. The uniform torsion rod shown in Fig. 11.3-21 is subjected to a distributed torque of constant magnitude t_0 (torque per unit length). Derive an expression for the total strain energy stored in the rod. Express your answer in terms of t_0, d, L, and the shear modulus G.

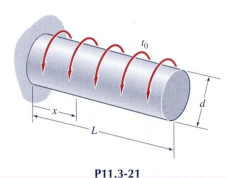

P11.3-21

Prob. 11.3-22. (a) Determine the strain energy U stored in the uniform aluminum rod AB shown in Fig. P11.3-22a. Let $G = 4(10^3)$ ksi and $T_B = 5$ kip $\cdot$ in. (b) Determine the strain energy stored in the rod CE if $T_E = 5$ kip $\cdot$ in, and the diameter over half the length of the rod is $d_2 = 1.25$ in., as illustrated in Fig. P11.3-22b.

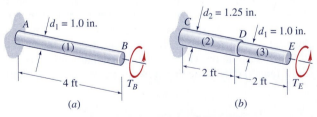

P11.3-22, P11.3-23, and P11.4-8

Prob. 11.3-23. (a) Determine the value of the torque T_E that would cause the stepped rod CE in Fig. P11.3-23b to have the same total angle of twist that the uniform rod AB in Fig. P11.3-23a would have when $T_B = 5$ kip $\cdot$ in. Let $G = 4(10^3)$ ksi. (b) Determine the strain energy stored in each of the two rods under the torques specified in Part (a).

Prob. 11.3-24. A torque T_B is applied to the uniform rod AC as shown in Fig. P11.3-24. The rod is attached to rigid supports at ends A and C. Determine the total strain energy $U(\phi_B)$ stored in the rod. Express your answer in terms of the length L, the polar moment of inertia I_p, the shear modulus G, and the angle of twist ϕ_B. The torque T_B will not appear explicitly in your answer.

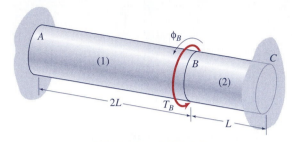

P11.3-24 and P11.4-9

Prob. 11.3-25. Torques T_B and T_C are applied to the uniform rod AD, as shown in Fig. P11.3-25. The rod is attached to rigid supports at A and D. Determine an expression for the total strain energy in the rod, $U(\phi_B, \phi_C)$ where ϕ_B and ϕ_C are the angles of twist at sections B and C, respectively. Express your answer in terms of ϕ_B, ϕ_C, the length L, the polar moment of inertia I_p, and the shear modulus G. The torques T_B and T_C will not appear explicitly in your answer.

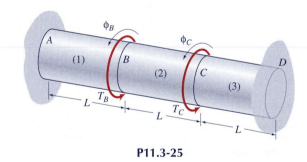

P11.3-25

DProb. 11.3-26. Four gears are attached to a shaft that transmits torques as shown in Fig. P11.3-26. (a) Determine the

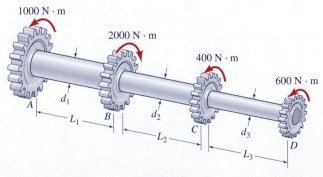

P11.3-26

required diameters d_1 through d_3 if the allowable shear stress for each segment of the shaft is 80 MPa. (b) Determine the strain energy stored in the shaft if the three segments have the diameters determined in Part (a), the shear modulus of the shaft is $G = 80$ GPa, and the shaft lengths are $L_1 = L_2 = L_3 = 1$ m.

ᴰProb. 11.3-27. A stepped steel shaft is subjected to the three torques shown in Fig. P11.3-27. (a) Determine the required diameters d_1 through d_3 if the allowable shear stress for each segment of the shaft is 12 ksi. (b) Determine the strain energy stored in the shaft if the three segments have the diameters determined in Part (a) and the shear modulus of the shaft is $G = 11.5(10^3)$ ksi.

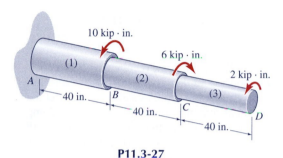

P11.3-27

Prob. 11.3-28. A tubular shaft has the dimensions shown in Fig. P11.3-28 and is made of brass with a shear modulus $G = 40$ GPa. A strain gage mounted at 45° to the axis of the shaft gives a strain reading of $\epsilon = 2.00(10^{-3})$. (a) Determine the value of the applied torque T that produces the given extensional-strain reading. (Recall Section 4.4.) (b) Determine the strain energy stored in the shaft under the stated loading.

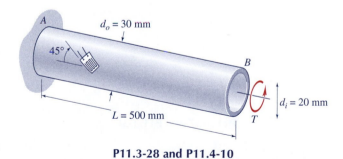

P11.3-28 and P11.4-10

***Prob. 11.3-29.** The two-segment solid stepped shaft in Fig. P11.3-29 is stress-free when it is welded to rigid structures at ends A and C. The diameters of the segments are $d_1 = 1.250$ in. and $d_2 = 0.875$ in., and the shaft is made of steel with a shear modulus $G = 11(10^3)$ ksi. With torque T_B applied at node B, as shown, a strain gage mounted on segment 1 at 45° to the axis of the shaft gives a strain reading of $\epsilon = 1.20(10^{-3})$. (a) Determine the value of the applied torque T_B that produces the given extensional-strain reading.

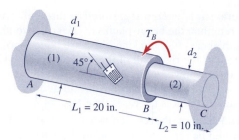

P11.3-29 and P11.4-11

(Recall Section 4.4.) (b) Determine the strain energy stored in the shaft under the given loading.

Prob. 11.3-30. A stepped, solid shaft has the dimensions shown in Fig. P11.3-30 and is made of material with shear modulus G. The shaft is stress-free when its ends are fixed to rigid structures at ends A and D. A single torque T_C is applied to the shaft at node C. Determine an expression for the total energy in the rod. $\mathcal{U}(\phi_B, \phi_C)$, where ϕ_B and ϕ_C are the angles of rotation at sections B and C, respectively. Express your answer in terms of ϕ_B, ϕ_C, the length L, the polar moment of inertia I_p, and the shear modulus G. The torque T_C will not appear explicitly in your answer.

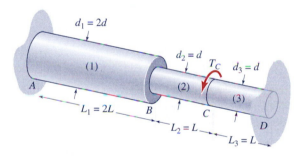

P11.3-30 and P11.4-12

> **Problems 11.3-31 through 11.3-42** *deal with the strain energy stored in various linearly elastic beams. Except for Probs. 11.3-31 and 11.3-36, all beams are uniform, that is, EI = const.*

ᴰProb. 11.3-31. A wood cantilever beam with rectangular cross section has a concentrated load P applied at end A as shown in Fig. P11.3-31a. (a) Determine an expression for

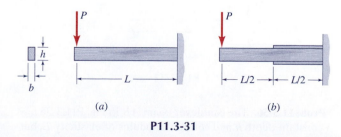

(a)   (b)

P11.3-31

the flexural strain energy, $\mathcal{U}_{\sigma a}$, stored in this uniform beam. (b) To strengthen the beam, planks of the same wood and of width b, thickness $h/4$, and length $L/2$ are bonded to the top and bottom of the right half of the original beam, giving the stepped beam shown in Fig. P11.3-31b. Determine the flexural strain energy, $\mathcal{U}_{\sigma b}$, for this beam, and discuss why $\mathcal{U}_{\sigma b}$ is less (or more, if that is the case) than $\mathcal{U}_{\sigma a}$.

Prob. 11.3-32. Let the concentrated load P in Example Problem 11.3 be replaced by a uniformly distributed load of intensity w_0 over the entire length L. (a) Determine an expression for the strain-energy ratio, $\mathcal{U}_\tau/\mathcal{U}_\sigma$ similar to the first part of Eq. (7) in this example. (b) Using the dimensions of the wide-flange beam in Example Problem 11.3, simplify your answer in Part (a) to the form given in the second part of Eq. (7) of the example.

> **For Problems 11.3-33 through 11.3-35.** (a) determine the flexural strain energy, $\mathcal{U}_\sigma$, and (b) determine the bending shear strain energy, $\mathcal{U}_\tau$.

Prob. 11.3-33. For the beam in Fig. P11.3-33, $E = 200$ GPa and $G = 80$ GPa.

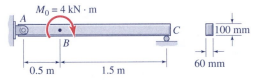

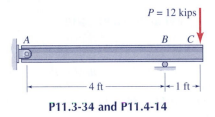

P11.3-33 and P11.4-13

Prob. 11.3-34. For the **W**8 × 21 beam in Fig. P11.3-34, $E = 30(10^3)$ ksi and $G = 11.5(10^3)$ ksi.

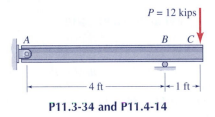

P11.3-34 and P11.4-14

Prob. 11.3-35. For the beam in Fig. P11.3-35. $E = 70$ GPa and $G = 26$ GPa.

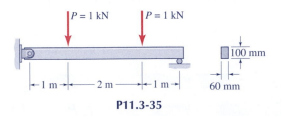

P11.3-35

***Prob. 11.3-36.** The cantilever beam AB in Fig. P11.3-36 has a constant depth h and constant modulus of elasticity E, but

its width varies linearly from b_0 at $x = 0$ to $3b_0$ at $x = L$. Determine an expression for the flexural strain energy, $\mathcal{U}_\sigma$, stored in this beam when a concentrated transverse load P is applied at $x = 0$.

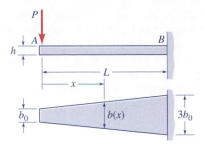

P11.3-36 and P11.4-17

^{D}Prob. 11.3-37. A uniform, rectangular beam AC is to be simply supported and is to support a concentrated transverse load P at its midspan, as shown in Fig. P11.3-37a. The cross-sectional dimensions are b and $2b$. (a) Compare the flexural strain energy $\mathcal{U}_{\sigma b}$ that would be stored in the beam if it is placed in orientation "b" (i.e., with the $2b$ dimension vertical) with the flexural strain energy $\mathcal{U}_{\sigma c}$ that would be stored in the beam if it is placed in orientation "c," with the $2b$ dimension horizontal. (b) Compare the maximum flexural stress values for the two configurations.

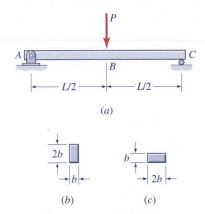

P11.3-37 and P11.3-38

^{D}Prob. 11.3-38. (a) Repeat Prob. 11.3-37, but this time compare the shear strain energy $\mathcal{U}_{\tau b}$ that would be stored in the beam if it is placed in orientation "b" (i.e., with the $2b$ dimension vertical) with the shear strain energy $\mathcal{U}_{\tau c}$ that would be stored in the beam if it is placed in orientation "c," with the $2b$ dimension horizontal. (b) Compare the maximum shear stress values for the two configurations.

Prob. 11.3-39. Determine an expression for the flexural strain energy, $\mathcal{U}_\sigma$, stored in the uniform, simply supported beam shown in Fig. P11.3-39. The flexural rigidity. EI, is constant.

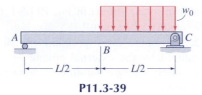

P11.3-39

Prob. 11.3-40. Determine an expression for the flexural strain energy, $\mathcal{U}_\sigma$, stored in the uniform, propped-cantilever beam shown in Fig. P11.3-40. (see Prob. 7.4-1 for this figure.) Express your answer as a function of w_0, the intensity of the applied uniformly distributed load, and R_A, the redundant reaction at A. The flexural rigidity, EI, is constant.

Prob. 11.3-41. Determine an expression for the flexural strain energy, $\mathcal{U}_\sigma$, stored in the fixed-fixed beam AB shown in Fig. P11.3-41. (See Prob. 7.4-5 for this figure.) Express your answer as a function of w_0, the maximum intensity of the linearly varying distributed load, and R_A and M_A, the redundant reactions at $x = 0$. The flexural rigidity, EI, is constant.

*__Prob. 11.3-42.__ A uniform cantilever beam AB has an inclined concentrated load P applied at the centroid of the cross section at end A as shown in Fig. P11.3-42. Determine an expression for the total strain energy, $\mathcal{U}_\sigma$, associated with normal stress σ. Express your answer as a function of the load P, the inclination angle α, the modulus of elasticity E, and the dimensions of the beam—b, h, and L. Start your solution with Eqs. 11.7 and 11.9a.

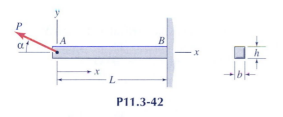

P11.3-42

In Problems 11.4-1 through 11.4-20 *use the* **Work-Energy Principle** *to calculate the required linear or angular displacement.*

Prob. 11.4-1. Using the data and figures of Prob. 11.3-5, determine the displacement Δ_B due to the load $P_B = 6$ kips acting on the uniform rod in Fig. P11.3-5a and the displacement Δ_E due to the load $P_E = 6$ kips acting on the stepped rod in Fig. P11.3-5b.

Prob. 11.4-2. Derive an expression for the displacement Δ_B due to the axial load P acting at end B of the tapered bar in Fig. P11.3-9. Let $d_A = 2d$ and $d_B = d$.

Prob. 11.4-3. Using the data and figures of Prob. 11.3-10, determine the displacement Δ_B due to a load $P = 2$ kips acting on the tapered bar in Fig. P11.3-10.

Prob. 11.4-4. Determine the vertical displacement Δ_B in the direction of the load P acting on the two-member truss in Fig. P11.3-11.

Prob. 11.4-5. Determine the vertical displacement Δ_D in the direction of the load P acting on the four-member truss in Fig. P11.3-12.

Prob. 11.4-6. Using the data and figure of Prob. 11.3-19, determine the angle of twist, ϕ_A, at end A of the tubular shaft in Fig. P11.3-19.

Prob. 11.4-7. If the end diameters of the tapered shaft Fig. P11.3-20 are $d_A = 2d$ and $d_B = d$, determine the angle of twist of the shaft, ϕ_B.

Prob. 11.4-8. Use the *Work-Energy Principle* to solve Prob. 11.3-23a.

Prob. 11.4-9. For the torsion rod in Fig. P11.3-24, determine an expression relating the twist angle ϕ_B and the applied torque T_B.

Prob. 11.4-10. Using the data and figure of Prob. 11.3-28, determine the angle of rotation of end B, ϕ_B, when the strain gage reads $\epsilon = 2.0(10^{-3})$.

Prob. 11.4-11. Using the data and figure of Prob. 11.3-29, determine the angle of rotation of joint B, ϕ_B, when the strain gage reads $\epsilon = 1.20(10^{-3})$.

*__Prob. 11.4-12.__ The stepped, solid shaft in Fig. P11.3-30 is made of a material with shear modulus G and is subjected to a single applied torque T_C. Determine an expression for the resulting angle of rotation, ϕ_C, at joint C.

Prob. 11.4-13. Considering only flexural strain energy and using the data and figure for the uniform rectangular beam in Prob. 11.3-33, determine the angle of rotation θ_B in the direction of the moment $M_0 = 4$ kN · m applied at point B.

Prob. 11.4-14. Considering only flexural strain energy, and using the data and figure for the $\mathbf{W}8 \times 21$ wide-flange beam in Fig. P11.3-34, determine the vertical displacement Δ_C in the direction of the load $P = 12$ kips acting at end C.

Prob. 11.4-15. The uniform, cantilever beam with rectangular cross section, shown in Fig. P11.4-15, is made of linearly elastic material with modulus of elasticity E and Poisson's ratio ν. Determine expressions for $\Delta_{B\sigma}$ and $\Delta_{B\tau}$, the contributions to the displacement at B that are related, respectively, to the flexural strain energy $\mathcal{U}_\sigma$ and the shear strain energy $\mathcal{U}_\tau$.

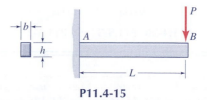

P11.4-15

Prob. 11.4-16. Use the *Work-Energy Principle* to obtain an expression for the deflection Δ_B (flexure only) under the load P for the simply supported beam in Fig. P11.4-16, $EI =$ const.

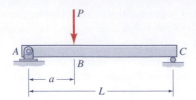

P11.4-16, P11.5-1, and P11.5-26

*Prob. 11.4-17. Determine an expression for the tip deflection Δ_A, for the tapered beam in Prob. 11.3-36. Consider flexure only.

Prob. 11.4-18. Use the *Work-Energy Principle* to determine the slope θ_A under the couple M_0 applied at end A of the uniform simply supported beam in Fig. P11.4-18. Let EI = const, and consider flexure only.

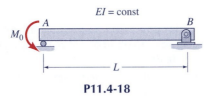

P11.4-18

Prob. 11.4-19. Use the *Work-Energy Principle* to determine the deflection Δ_B under the load P on the **W**150 × 24 wide-flange, simply supported beam in Fig. P11.4-19. Let E = 200 GPa, and consider flexure only.

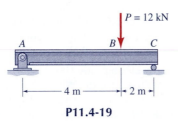

P11.4-19

Prob. 11.4-20. Use the *Work-Energy Principle* to determine the slope θ_A under the couple M_0 = 2100 kip · in. applied at end A of the **W**16 × 100 wide-flange beam in Fig. P11.4-20. Let E = 30 × 10³ ksi, and consider flexure only.

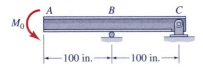

P11.4-20, P11.5-3, and P11.5-27

In Problems 11.5-1 through 11.5-16 *use* Castigliano's Second Theorem, *Eq. 11.36 (or Eq. 11.38), to solve each of these statically determinate problems.*

Prob. 11.5-1. Determine an expression for the vertical displacement. Δ_B, under the load at point B on the simply

supported uniform beam shown in Fig. P11.5-1. (See Prob. 11.4-16 for this figure.) EI = const.

Prob. 11.5-2. Determine an expression for the rotation angle θ_B produced by the external couple M_0 applied at point B of the simply supported uniform beam shown in Fig. P11.5-2. EI = const.

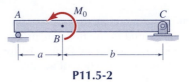

P11.5-2

Prob. 11.5-3. Determine the rotation angle θ_A produced by the external couple M_0 applied at end A of the beam AC shown in Fig. P11.5-3. (See Prob. 11.4-20 for this figure.) EI = const.

Prob. 11.5-4. Determine the vertical deflection Δ_C under the load P applied at end C of the beam in Fig. P11.5-4. Let E_{wood} = 1,600 ksi, and see Table D.8 for the cross-sectional dimensions of the 6 × 8 wood beam.

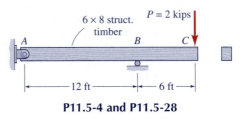

P11.5-4 and P11.5-28

Prob. 11.5-5. Determine the vertical deflection Δ_B under the load P applied at point B to the **W**310 × 143 structural steel beam in Fig. P11.5-5. Let E_{steel} = 200 GPa.

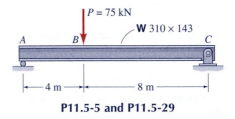

P11.5-5 and P11.5-29

Prob. 11.5-6. Determine an expression for the vertical deflection Δ_A under the load P acting at end A of the uniform beam in Fig. P11.5-6. Let EI = const.

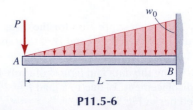

P11.5-6

Prob. 11.5-7. Use *Castigliano's Second Theorem* to solve Prob. 7.6-15.

Prob. 11.5-8. Use *Castigliano's Second Theorem* to solve Prob. 7.6-16.

Prob. 11.5-9. Use *Castigliano's Second Theorem* to solve Prob. 7.6-14.

Prob. 11.5-10. Use *Castigliano's Second Theorem* to solve Prob. 7.6-19.

Prob. 11.5-11. Use *Castigliano's Second Theorem* to solve Prob. 7.6-17.

***Prob. 11.5-12.** Two segments, AB and BC, of $4 \times 2 \times 0.1875$ structural steel tubing are welded together at B to form the L-shaped frame ABC shown in Fig. P11.5-12. Use *Castigliano's Second Theorem* to determine the vertical displacement Δ_C under the load $P = 500$ lbs at C. Let $E = 29(10^3)$ ksi. (Be sure to include the effects of both bending and stretching of segment AB on the displacement at C.)

$$A = 2.02 \text{ in}^2, \qquad I = 1.29 \text{ in}^4$$

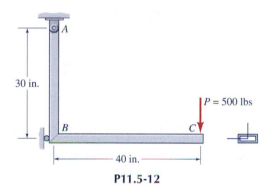

P11.5-12

Prob. 11.5-13. Determine the horizontal component, u_B, of the displacement of point B due to flexure of the curved beam shown in Fig. P11.5-13. Assume that the radius of curvature, R, of the centerline of the bar is large relative to the radial depth of the cross section so that the straight-beam elastic flexure formula holds. Express your answer in terms of P, R, E, and I.

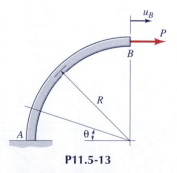

P11.5-13

Prob. 11.5-14. Determine an expression for the vertical displacement Δ_C under the vertical load P that acts at end C of the equal-leg, right-angle-joined pipe ABC shown in

Fig. P11.5-14. The points A, B, and C lie in a horizontal plane. Let the modulus of elasticity, E, the shear modulus, G, and the moment of inertia, I, be constant over the lengths AB and BC, and ignore the joint dimensions at B. Neglect shear deformation due to bending of segments AB and BC. (Note: Segment AB undergoes both bending and torsion. The polar moment of inertia is $I_p = 2I$.)

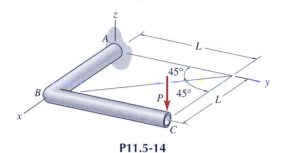

P11.5-14

Prob. 11.5-15. For the truss in Fig. P11.5-15, determine the horizontal displacement u_B due to the horizontal load P_x applied at joint B. Let $E = 30(10^3)$ ksi, $P_x = 1.5$ kips, $A_1 = 1$ in^2, $A_2 = 2$ in^2, $a = 24$ in., and $\beta = 30°$.

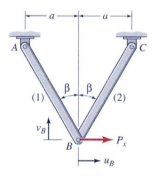

P11.5-15 and P11.5-36

Prob. 11.5-16. For the truss in Fig. P11.5-16, determine the downward vertical displacement v_B due to the weight W that is suspended from the pin at B. Let $E = 200$ GPa, $A_1 = 800$ mm^2, $A_2 = 1600$ mm^2, and $W = 10$ kN.

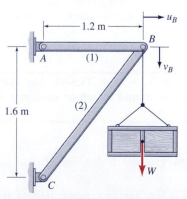

P11.5-16 and P11.5-37

Prob. 11.5-17. The uniform (EI = const) propped-cantilever beam in Fig. P7.4-3 supports a linearly varying load of maximum intensity w_0 (force per unit length). Use *Castigliano's Second Theorem*: (a) to determine the (redundant) reaction R_A, and (b) to determine the slope, θ_A, of the beam at end A.

Prob. 11.5-18. For the uniformly loaded propped-cantilever beam in Fig. P7.4-2, determine expressions for: (a) the (redundant) reaction R_B, and (b) the slope of the beam at end B.

****Prob. 11.5-19.** The uniform (EI = const) propped-cantilever beam in Fig. P7.4-7 supports a concentrated load P at point B. Use *Castigliano's Second Theorem*: (a) to determine the reaction R_A at end A, and (b) to determine the vertical displacement, Δ_B, of the beam at the point of application of load P.

****Prob. 11.5-20.** The uniform fixed-fixed beam in Fig. P7.4-8 is subjected to a concentrated load P at distance a from end A. Use *Castigliano's Second Theorem*: (a) to determine the reaction R_A and M_A at end A, and (b) to determine the vertical displacement, Δ_B, of the beam at the point of application of load P.

Prob. 11.5-21. The uniform continuous beam in Fig. P7.4-10 supports a uniformly distributed load of intensity w_0 on the span AB. Use *Castigliano's Second Theorem*: (a) to determine the reaction R_B at the central support, and (b) to determine the slope θ_C of the beam at end C.

****Prob. 11.5-22.** For the non-uniform beam AC in Fig. P7.4-11, use *Castigliano's Second Theorem*: (a) to determine the reactions R_A and M_A at end A, and (b) to determine the slope θ_B of the beam at the central support, B.

Prob. 11.5-23. For the uniform fixed-fixed beam AC in Fig. P7.4-9, use *Castigliano's Second Theorem* to determine the fixed-end reactions R_A and M_A.

Prob. 11.5-24. For the uniform fixed-fixed beam AC in Fig. P7.7-2, use *Castigliano's Second Theorem*: (a) to solve for the fixed-end reaction, R_A and M_A, and (b) to solve for the transverse displacement Δ_B at node B.

Prob. 11.5-25. For the nonuniform fixed-fixed beam shown in Fig. P7.6-55a, use *Castigliano's Second Theorem*: (a) to solve for the fixed-end reactions R_A and M_A, and (b) to solve for the transverse displacement Δ_B at node B.

Prob. 11.5-26. Use the *Unit-Load Method* to solve Prob. 11.5-1.

Prob. 11.5-27. Use the *Unit-Load Method* to solve Prob. 11.5-3.

Prob. 11.5-28. Use the *Unit-Load Method* to solve Prob. 11.5-4.

Prob. 11.5-29. Use the *Unit-Load Method* to solve Prob. 11.5-5.

****Prob. 11.5-30.** For the uniform simply supported beam in Fig. P11.5-30, determine the vertical displacement Δ_B at midspan and the slope θ_C at end C. (See Prob. 7.6-30 for Fig. P11.5-30.)

Prob. 11.5-31. For the uniform cantilever beam in Fig P11.5-31, use the *Unit-Load Method* to determine the vertical displacement Δ_A and the slope θ_A at the free end A. The maximum load intensity (at B) is w_0 (force per unit length).

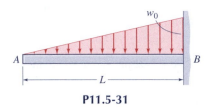

P11.5-31

Prob. 11.5-32. Use the *Unit-Load Method* to solve Prob. 7.6-15.

Prob. 11.5-33. Use the *Unit-Load Method* to solve Prob. 7.6-16.

****Prob. 11.5-34.** For the beam-rod system described in Prob. 7.6-21 and shown in Fig. P7.6-21, use the *Unit-Load Method* to determine the vertical displacement Δ_B and slope θ_B of end B when beam AB is subjected to a uniformly distributed load as shown.

Prob. 11.5-35. For the beam described in Prob. 7.6-20 and shown in Fig. P7.6-20, use the *Unit-Load Method* to determine the vertical displacements Δ_B and Δ_C under the load P and at end C, respectively. Consider both bending of the beam AC and stretching of the rod CD.

Prob. 11.5-36. Use the *Unit-Load Method* to determine the <u>vertical</u> displacement v_B of joint B of the two-member truss described in Prob. 11.5-15 and shown in Fig. P11.5-15.

Prob. 11.5-37. Use the *Unit-Load Method* to determine the <u>horizontal</u> displacement u_B of joint B of the two-member truss described in Prob. 11.5-16 and shown in Fig. P11.5-16.

Prob. 11.5-38. The symmetric five-member pin-jointed truss shown in Fig, P11.5-38 has a height H = 2 m and a span

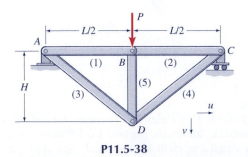

P11.5-38

$L = 6$ m. A load $P = 60$ kN acts vertically through the joint at B. The cross-sectional area of each tension member is 1200 mm^2 and of each compression member is 3200 mm^2. The truss members are made of steel with $E_{\text{steel}} = 200$ GPa. Use the *Unit-Load Method:* (a) to calculate the vertical displacement, v_B, of joint B, and (b) to calculate the horizontal displacement, u_C, of joint C.

Prob. 11.5-39. Each of the six members of the planar truss shown in Fig. P11.5-39 has an axial rigidity AE. The truss is loaded by equal vertical forces P at joints D and E. Use the *Unit-Load Method* to determine the horizontal and vertical components, u_E and v_E, respectively, of the displacement of joint E.

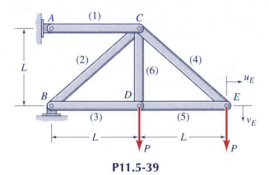

P11.5-39

Problems 11.5-40 through 11.5-55. *Use the Unit-Load Method to solve the following statically indeterminate problems.*

Prob. 11.5-40. By combining Eqs. 11.39 and 11.40, show that the *Unit-Load Method* for statically indeterminate structures can be expressed by the following equations:

$$0 = \Delta_{ri} = \int_0^L \frac{F F_{ri}\, dx}{AE} + \int_0^L \frac{T T_{ri}\, dx}{G I_p}$$

$$+ \int_0^L \frac{M M_{ri}\, dx}{EI} + \int_0^L \frac{f_s V V_{ri}\, dx}{GA}, i = 1, 2, \ldots, N_R$$

$$\Delta_{ui} = \int_0^L \frac{F F_{ui}\, dx}{AE} + \int_0^L \frac{T T_{ui}\, dx}{G I_p}$$

$$+ \int_0^L \frac{M M_{ui}\, dx}{EI} + \int_0^L \frac{f_s V V_{ui}\, dx}{GA}, i = 1, 2, \ldots, N$$

where F_{ri}, T_{ri}, and so forth, are the distributions of force, torque, and so forth, due to a unit value of the redundant R_i; and F_{ui}, T_{ui}, and so forth, are the distributions of force, torque, and so forth, due to a unit value of force (or couple) P_i where the linear displacement (or angular displacement) is to be determined. (The latter unit "forces" may be "dummy forces" if no actual load is applied where displacement is to be determined.)

Prob. 11.5-41. Use the *Unit-Load Method* to solve Prob. 11.5-17. Let R_A be the redundant force.

Prob. 11.5-42. Use the *Unit-Load Method* to solve Prob. 11.5-18. Let R_B be the redundant force.

Prob. 11.5-43. Use the *Unit-Load Method* to solve Prob. 11.5-21, Let R_B be the redundant force.

Prob. 11.5-44. Use the *Unit-Load Method* to solve Prob. 11.5-23. Let R_A and M_A be the redundant reactions.

Prob. 11.5-45. Use the *Unit-Load Method* to solve Prob. 11.5-25. Let R_A and M_A be the redundant reactions.

Prob. 11.5-46. Use the *Unit-Load Method* to determine the (redundant) reactions R_A and M_A for the uniform fixed-fixed beam with linearly varying load, as shown in Fig. P7.4-5. Let $EI = $ const.

Prob. 11.5-47. For the uniformly loaded beam in Fig. P7.4-6, use the *Unit-Load Method:* (a) to determine the (redundant) force F_2 in the hanger rod, and (b) to determine the vertical displacement, v_B, of the rod-supported end B.

Prob. 11.5-48. For the two-span beam AC in Fig. P7.7-10, use the *Unit-Load Method:* (a) to determine the two (redundant) reactions R_A and R_B, and (b) to determine the slope, θ_A, of the beam at end A.

Prob. 11.5-49. For the nonuniform two-span continuous beam in Fig. P7.6-53a, use the *Unit-Load Method:* (a) to determine the (redundant) reaction R_A, and (b) to determine the slope angle, θ_A, at end A.

Prob. 11.5-50. For the nonuniform two-span continuous beam in Fig. P7.6-54a, use the *Unit-Load Method:* (a) to determine the (redundant) reaction R_A, and (b) to determine the slope angle, θ_A, at end A.

Prob. 11.5-51. For the nonuniform two-span continuous beam in Fig. P7.7-15, use the *Unit-Load Method:* (a) to determine the (redundant) reaction R_A, and (b) to determine the slope angle, θ_A, at end A.

*Prob. 11.5-52. For the two-span nonuniform beam AC in Fig. P7.6-56a, use the *Unit-Load Method*: (a) to determine three redundant reactions R_A, M_A, and R_B; and (b) to determine the slope angle, θ_B, at the central support B.

Prob. 11.5-53. The truss members in Fig. P3.10-13 have the following axial rigidities: $(AE)_2 = (AE)_3 = AE$, and $(AE)_1 = 2AE$. A horizontal load P is applied to the truss at joint A. (a) Use the *Unit-Load Method* to determine the (redundant) force F_1 in member (1): then, determine the other two member forces. (b) Use the *Unit-Load Method* to determine u_A and v_A, the horizontal displacement and the vertical displacement, respectively, of joint A.

Prob. 11.5-54. Each of the three truss members in Fig. P3.10-16 has a length L, and modulus of elasticity E. The cross-sectional areas of the members are $A_1 = A_2 = A$, and $A_3 = 2A$. A horizontal load P acts on the truss at joint A. (a) Use the *Unit-Load Method* to determine the (redundant) force F_3 in member (3); then, determine the other two member forces. (b) Use the *Unit-Load Method* to determine u_A and v_A, the horizontal displacement and the vertical displacement, respectively, of joint A.

Prob. 11.5-55. For the truss in Fig. P3.10-19: $A_1 = A_2 = A_3 = 1.0$ in^2, $E_1 = E_2 = E_3 = 30(10^3)$ ksi, and $P = 15$ kips. (a) Use the *Unit-Load Method* to determine the (redundant) force F_1 in member (1): then, determine the other two member forces. (b) Use the *Unit-Load Method* to determine u_A and

v_A, the horizontal displacement and the vertical displacement, respectively, of joint A.

Problems 11.7-1 through 11.7-10. *Use the* **Principle of Virtual Displacements,** *Eq. 11.48, to solve each of these problems.*

Prob. 11.7-1. (a) For the two-segment rod in Fig. P11.7-1, use the *Principle of Virtual Displacements* to determine an expression for the axial displacement u_B at the point of application of the axial force P. (b) Determine the axial forces F_1 and F_2 in segments (1) and (2) respectively.

$$A_1 = 2A, \quad A_2 = A, \quad E = \text{const}$$

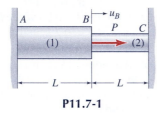

P11.7-1

Prob. 11.7-2. (a) For the two-segment rod in Fig. P11.7-2, use the *Principle of Virtual Displacements* to determine an expression for the axial displacement u_B at the point of application of the axial force P. (b) Determine the axial forces F_1 and F_2 in segments (1) and (2) respectively.

$$A_1 = 1.0 \text{ in}^2, \quad A_2 = 1.2 \text{ in}^2$$

$$E = 10(10^3) \text{ ksi}, \quad P = 20 \text{ kips}$$

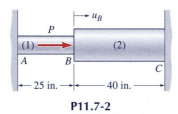

P11.7-2

Prob. 11.7-3. A rigid beam is supported by three columns, as shown in Fig. P11.7-3. (a) Use the *Principle of Virtual Displacements* to determine the vertical displacement v (taken positive downward) of the beam when a vertical load $P = 540$ kN is applied. (b) Determine the axial forces $F_1 = F_3$ and F_2 carried by the respective columns.

$$A_1 = A_3 = 3500 \text{ mm}^2$$

$$A_2 = 2000 \text{ mm}^2, \quad E = 200 \text{ GPa}$$

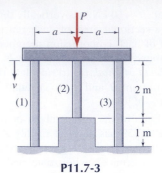

P11.7-3

Prob. 11.7-4. A single axial force P is applied at node C of the three-segment rod shown in Fig. P11.7-4. (a) Using the *Principle of Virtual Displacements,* determine expressions for the axial displacements u_B and u_C of nodes B and C respectively. (b) Determine the forces F_1, F_2, and F_3 in the three segments of the rod.

$$A_1 = A_3 = 2A, \quad A_2 = A, \quad E = \text{const}$$

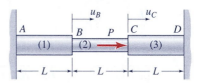

P11.7-4 and P11.7-12

Prob. 11.7-5. Use the *Principle of Virtual Displacements* to solve Prob. 3.4-1.

Problems 11.7-6 through 11.7-10. *For each of the pin-jointed trusses shown, (a) use the* **Principle of Virtual Displacements** *to determine the horizontal and vertical displacements of joint D, u_D and v_D, respectively, and (b) determine the axial force in member (1), that is, the force in member AD.*

Prob. 11.7-6. For the truss in Fig. P11.7-6, $A_1 = A_2 = A_3 = 1.0 \text{ in}^2$, $E = 30(10^3)$ ksi, and $P_x = 10$ kips.

Prob. 11.7-7. For the truss in Fig. P11.7-7, $A_1 = A_2 = 1000 \text{ mm}^2$, $A_3 = 2000 \text{ mm}^2$, $E = 70$ GPa, and $P_x = 100$ kN.

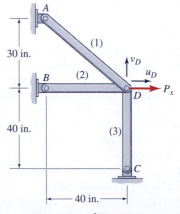

P11.7-6 and P11.7-17

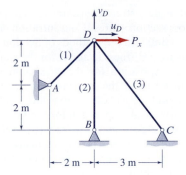

P11.7-7 and P11.7-18

Prob. 11.7-8. For the truss in Fig. P11.7-8, $A_1 = A_2 = A_3 = A$, and $E = $ const.

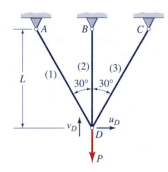

P11.7-8, P11.7-9, and P11.7-19

Prob. 11.7-9. For the truss in Fig. P11.7-9, $A_1 = A_2 = A$, $A_3 = 2A$, and $E = $ const.

Prob. 11.7-10. For the truss in Fig. P11.7-10, $A_1 = A_2 = A_3 = A$, and $E = $ const.

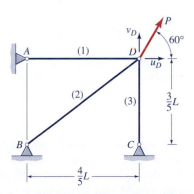

P11.7-10 and P11.7-20

Prob. 11.7-11. Use *Castigliano's First Theorem* to solve Prob. 3.8-4. Use u_B and u_C as the two displacement unknowns.

Prob. 11.7-12. Use *Castigliano's First Theorem* to solve Prob. 11.7-4. Use u_B and u_C as the two displacement unknowns.

Prob. 11.7-13. Use *Castigliano's First Theorem* to solve Prob. 3.8-14. Let θ be the displacement variable, and express *Castigliano's First Theorem* in the form:

$$\frac{d\mathcal{U}}{d\theta} = M_0 = P(a + b)$$

Prob. 11.7-14. Use *Castigliano's First Theorem* to solve Prob. 3.8-15. Let θ be the displacement variable, and express *Castigliano's First Theorem* in the form:

$$\frac{d\mathcal{U}}{d\theta} = M_0 = P(a + b)$$

Prob. 11.7-15. The rigid beam AC in Fig. P11.7-15 (see Prob. 3.6-17 for this figure) is supported by three vertical rods that are attached to the beam at points A, B, and C. When the rods are initially attached to the beam, the three rods are stress-free. (a) Use *Castigliano's First Theorem* to solve for the vertical displacements u_A and u_C of points A and C when downward loads $P_A = 8$ kips and $P_C = 2$ kips are applied to the beam at ends A and C, respectively. (b) Calculate the axial stresses, σ_1, σ_2, and σ_3 resulting from the given loading,

$$A = 1.0 \text{ in}^2, \quad L = 30 \text{ in.}, \quad E = 30(10^3) \text{ ksi}$$

Prob. 11.7-16. Solve Prob. 11.7-15 with $P_A = 16$ kN and $P_C = 4$ kN and with the following properties of the structure:

$$A = 500 \text{ mm}^2, \quad L = 1 \text{ m}, \quad E = 100 \text{ GPa}$$

Prob. 11.7-17. Use *Castigliano's First Theorem* to solve Prob. 11.7-6.

Prob. 11.7-18. Use *Castigliano's First Theorem* to solve Prob. 11.7-7.

Prob. 11.7-19. Use *Castigliano's First Theorem* to solve 11.7-9.

Prob. 11.7-20. Use *Castigliano's First Theorem* to solve Prob. 11.7-10.

Prob. 11.9-1. If, instead of being released from rest at position (1) in Fig. 11.23a, the weight W has a downward speed v at position (1), determine an expression (similar to Eq. 11.62) for the maximum deflection of the spring.

DProb. 11.9-2. A uniform rod like the one shown in Fig. 11.25 has a diameter d and length L, and it is to be used to absorb the impact loading from a falling mass. Determine the total amount of elastic strain energy that can be absorbed by the rod if it is made of the following metals: (a) 2014-T6 aluminum alloy, (b) 6061-T6 aluminum alloy, (c) ASTM-A36 structural steel, and (d) Ti-6AL-4V titanium alloy.

Let $d = 1$ in, and $L = 40$ in., and express your answers in units of kip · in.

DProb. 11.9-3. Repeat Prob. 11.9-2 for a rod whose dimensions are $d = 30$ mm and $L = 1$ m. Express your answers in units of kN · m.

> **In Problems 11.9-4 through 11.9-11** *assume that the material remains linearly elastic and that the three assumptions regarding Impact on Deformable Bodies are satisfied.*

Prob. 11.9-4. As shown in Fig. P.11.9-4, a weight $W = 200$ lb is dropped from a height $h = 20$ in. and lands on the top of a 4-in.-square aluminum alloy post ($E = 10 \times 10^3$ ksi and $\sigma_Y = 60$ ksi) whose length is $L = 2$ ft. The aluminum post rests on a rigid base. Determine the maximum compressive stress in the post, the maximum shortening of the post, and the impact magnification factor, Δ_{max}/Δ_{st}.

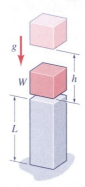

P11.9-4 and P11.9-5

Prob. 11.9-5. From what height h_1 must the weight $W = 200$ lb in Prob. 11.9-4 be dropped if the impact causes a maximum stress in the post of $\sigma_Y/4$? From what height h_2 must it be dropped if the impact causes a maximum stress in the post of $\sigma_Y/2$?

Prob. 11.9-6. The collar mass in Fig. 11.25 slides down a uniform 6061-T6 aluminum-alloy rod of diameter d, impacting against the flange at the bottom of the rod. Determine the maximum stress in the rod if: (a) the collar is released from rest at distance $h = h_a$ above the collar: (b) the collar is released from rest at $h = 0$, that is, when it is just in contact with the collar; and (c) the collar is lowered slowly onto the flange. Let $m = 50$ kg, $d = 25$ mm, $L = 1.25$ m, and $h_a = 0.5$ m.

Prob. 11.9-7. Repeat Prob. 11.9-6 Parts (a) through (c) using the following data for the rod and the mass: $W = 500$ lb, $d = 2$ in., $L = 50$ in., and $h = 20$ in. (d) From what height would the 500-lb weight have to be dropped to cause the rod to yield? (e) What weight, if dropped from a height of $h = 20$ in., would cause the rod to yield?

*Prob. 11.9-8. The stepped rod in Fig. P11.9-8 has cross-sectional areas $2A$ and A. (a) Determine an expression for the maximum elongation of the rod as a function of W, E, A, L, and h. Neglect the energy stored in the short transition

segment, and neglect stress-concentration effects due to the change of cross section. (Stress Concentration is discussed in Section 12.2.) Hint: Use the methods of Section 3.3 to determine an equivalent stiffness factor, k, for the stepped bar. (b) Determine an expression for the maximum impact axial stress, σ_{max}, in terms of the maximum static stress, σ_{st}, the drop height h, and the rod dimensions and modulus of elasticity. (c) Determine an expression for the ratio of the maximum impact stress determined in Part (b) to the maximum stress that a uniform bar of cross-sectional area A and length $2L$ would experience under impact loading by a weight W dropped from height h.

Prob. 11.9-9. The stepped ASTM-A36 structural steel rod in Fig. P11.9-9 has equal-length segments of length L, with cross-sectional areas A (diameter $= d$) and $2A$. Determine the maximum axial impact stress in the rod if a collar of weight W is dropped from height h above the flange at the bottom of the rod. Neglect the energy stored in the short transition segment, and neglect stress-concentration effects due to the change of cross section. (The topic of Stress Concentration is discussed in Section 12.2.) Hint: Use the methods of Section 3.3 to determine an equivalent stiffness factor, k, for the stepped rod. Let $d = 0.5$ in., $L = 20$ in., $W = 20$ lb, and $h = 2$ in.

*Prob. 11.9-10. A drop test is used to test the impact performance of automobile bumpers. Assume that the bumper is a uniform simply supported beam and that mass m is dropped from height h, impacting the beam at distance aL from support A ($0 < a \le 1/2$), as shown in Fig. P11.9-10. (a) Derive an expression that relates the maximum flexural stress due to impact, σ_{max}, to the drop height and other parameters: W, E, I, c, a, and L. (b) Specialize your answer to Part (a) for the case $a = 1/2$.

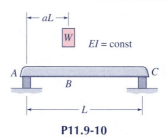

P11.9-10

*Prob. 11.9-11. A mass m is dropped from height h, impacting the uniform cantilever beam AC at point B, at a distance aL from the cantilevered end as shown in Fig. P11.9-11, A. Determine an expression that relates the maximum tip deflection. $(\Delta_C)_{max}$, to the drop height h and location ($0 < a \le 1$) and to other parameters: m, E, I, c, and L.

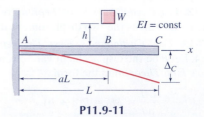

P11.9-11

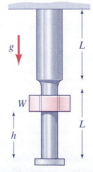

P11.9-8 and P11.9-9

Section			Suggested Review Problems
11.2	Chapter 11 discusses **energy methods** as applied to *axial deformation, torsion, and bending*. Most derivations and examples are applied to axial deformation, as will be done in this Chapter Review. **Work** and **energy** are two related mechanical quantities; both have the dimension of *force × displacement*.	**Work–W** $$W_p(e_1) = \int_0^{e_1} P(e)\,de \qquad (11.1)$$ Axial deformation. Load versus elongation. (Fig. 11.1a) (Fig. 11.2)	
	The **total strain energy** that is stored in an elastic member is equal to the area under the load-displacement curve ($P - e$ curve for axial deformation).	**Strain Energy–U** $$U(e) = W(e)_{\text{done on elastic body}} \qquad (11.4)$$ Strain energy for axial deformation of an elastic member. (Fig. 11.3)	
	Energy methods are most frequently applied to linearly elastic bodies. In this case, the **strain energy density** $\bar{u}$ is useful. For a linearly elastic isotropic material, the strain energy density is given by Eq. 11.12 and the total strain energy by Eq. 11.13.	**Strain Energy Density $\bar{u}$** Strain energy density for uniaxial stress. (Fig. 11.6) The **strain energy density** is given by $$\bar{u} = \frac{1}{2}(\sigma_x \epsilon_x + \sigma_y \epsilon_y + \sigma_z \epsilon_z$$ $$+ \tau_{xy}\gamma_{xy} + \tau_{xz}\gamma_{xz} + \tau_{yz}\gamma_{yz}) \qquad (11.12)$$ and, the **total strain energy** is given by $$U = \int_V \bar{u}\,dV \qquad (11.7)$$	

Section			Suggested Review Problems
11.3	Section 11.3 illustrates the calculation of the strain energy stored in members undergoing *axial deformation, torsion,* and *bending*. You should review this section and note the formulas presented in it (e.g., Eqs. 11.15, 11.16, 11.20, 11.21, 11.25, and 11.26).		11.3-3 11.3-9 11.3-13 11.3-23 11.3-33
11.4	Section 11.4 presents the **work-energy principle.** This principle, which is not restricted to linearly elastic behavior, is useful only for determining the displacement at the point of application of a <u>single</u> load and in the direction of that load.	**Work-Energy Principle** If the stresses in a body do not exceed the elastic limit, all of the work done <u>on</u> the body by external forces is stored in the body as strain energy. $$\mathcal{W}_{\text{ext}} = \mathcal{U} \qquad (11.32)$$	11.4-5 11.4-9
11.5	Section 11.5-presents two very important energy principles: • **Castigliano's Second theorem** and the • **Unit-Load Method**. In Section 11.5, Castigliano's Second theorem is applied to *statically determinate systems* and to *statically indeterminate systems*.	**Castigliano's Second Theorem** Among all possible configurations of a linearly elastic, deformable body, or system, the actual configuration is the one for which $$\Delta_i = \frac{\partial \mathcal{U}(P_1, P_2, \ldots, P_N)}{\partial P_i} \qquad (11.36)$$ where Δ_i is the displacement corresponding to the force P_i, and $\mathcal{U}$ is the strain energy expressed as a function of the loads.	11.5-1 11.5-6 11.5-14

Sections 11.6 through 11.8 are *optional sections* that present other important energy methods and suggest how these methods are related to the **displacement method,** the **force method,** and the **finite element method.** Section 11.9 is an *optional section* that discusses energy solutions for problems that involve **impact loading.**

SPECIAL TOPICS RELATED TO DESIGN

12

12.1 INTRODUCTION

In previous chapters many **design problems** have been presented. Typically you have been asked to determine the size of the cross section of a member or connector (bolt, pin, etc.) given the load and given the allowable stress or deflection. In this chapter you will be introduced to several additional important design-related topics that you will encounter again in future courses on design of structures and/or machines. Section 12.2 on **Stress Concentrations** discusses the effect of holes, notches, or changes of cross section on the stresses due to axial loading, torsion, or bending. Section 12.3 on **Failure Theories** indicates how mechanical properties obtained from simple uniaxial tension or compression testing may be used to predict yielding or brittle fracture of members subjected to more complex loading conditions. Finally, in Section 12.4 on **Fatigue and Fracture** you will learn how repeated loading and unloading of a member causes small cracks to grow in length, eventually leading to a fatigue failure.

12.2 STRESS CONCENTRATIONS

The key formulas

$$\sigma = \frac{F}{A}, \qquad \tau = \frac{T\rho}{I_p}, \qquad \sigma = -\frac{My}{I}$$

enable one to calculate the normal stress due to axial loading, the shear stress due to torsion, and the flexural stress due to bending, respectively. These formulas may be used only so long as the cross section of the member is relatively uniform, that is, there are no abrupt changes in cross section. In Section 2.10 it was noted that these formulas do not hold in the immediate vicinity of points of application of load, but that away from that vicinity, *St. Venant's Principle* ensures the validity of the above formulas for bodies of uniform cross section. The increase of stress due

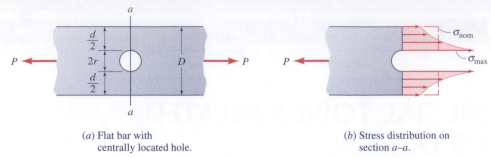

(a) Flat bar with
centrally located hole.

(b) Stress distribution on
section a–a.

FIGURE 12.1 Stress concentration in an axially loaded flat bar with circular hole.

to localized application of a load or due to nonuniformity of the cross section is called a **stress concentration.** In the design of load-bearing members it is important to avoid stress concentrations whenever possible and to properly account for them when they do exist, particularly if the member is made of brittle material or is subjected to a fluctuating load (see Section 12.4). We will briefly examine the effect of stress concentrations in members subjected to axial loading, torsional loading, and bending.

Stress Concentration—Axial Loading.

The last picture in the color insert depicts the stress-concentration effects due to a centrally located circular hole in an axially loaded flat bar, as computed by use of the finite element method. Holes, notches, or abrupt changes in cross section produce stress concentrations in axially-loaded members.

Figure 12.1b shows the stress concentration due to a *centrally located circular hole* in a flat bar in tension. The maximum normal stress occurs at the edge of the hole on the cross section $a - a$, which passes through the center of the hole. In Figure 12.2 a *shoulder fillet* of radius r is used to smooth the transition between the wider portion of the bar and the narrower portion. The maximum normal stress for this case occurs on section $b - b$, where the fillet joins the narrower part of the bar.

The stress distribution due to a stress concentration like the ones in Figs. 12.1 and 12.2 may be determined analytically through the use of the theory of elasticity [Ref. 12-1], or numerically by using finite element analysis [Refs. 12-2, 12-3] (see the last color-insert photo), or experimentally through the use of photoelasticity [Ref. 12-4]. The exact distribution of stress is not of great importance, but the maximum value of stress is very important. This maximum stress, σ_{max}, may be

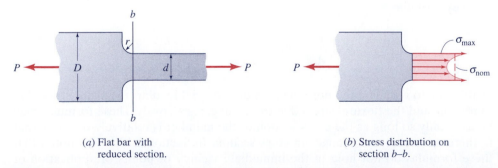

(a) Flat bar with
reduced section.

(b) Stress distribution on
section b–b.

FIGURE 12.2 Stress-concentration in an axially loaded flat bar with abrupt change in cross section. (Copyright © Boeing)

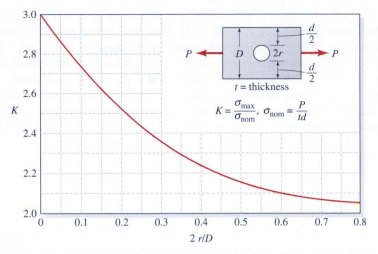

FIGURE 12.3 Stress-concentration factor K for a flat bar with centrally located circular hole (tension). (Adapted from Ref. 12-6. Reprinted by permission of J. Wiley & Sons. Inc.)

related to the average stress on the *net* cross section, the *nominal stress* σ_{nom}, by defining the **stress-concentration factor** as

$$K = \frac{\sigma_{max}}{\sigma_{nom}} \tag{12.1}$$

For linearly elastic behavior, the stress-concentration factor is a function of the geometry of the member and the type of load applied (i.e., axial, torsion, bending). Formulas for the stress-concentration factor for many types of loading and geometry are given in Ref. 12-5; and many graphs may be found in Ref. 12-6, from which the graphs in this section were obtained. Values of the stress-concentration factor given in these references and in the figures presented in this section are based on linearly elastic behavior and are valid only as long as the computed value σ_{max} does not exceed the proportional limit of the material.

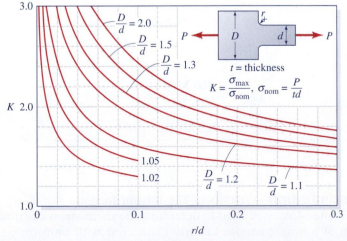

FIGURE 12.4 Stress concentration factor K for a flat bar with shoulder fillets (tension). (Adapted from Ref. 12-6. Reprinted by permission of J. Wiley & Sons, Inc.)

Figure 12.3 shows the effect of hole size on the stress-concentration factor for a flat bar with centrally located circular hole. Note that as the radius of the hole approaches zero, the stress-concentration factor approaches the value three. So, a very small hole can have a very damaging effect on a member. Fatigue cracks, which are discussed in Section 12.4, frequently are initiated at just such small holes.

From Fig. 12.4 it may be observed that as $r \to 0$ the stress concentration factor increases rapidly. With no fillet (i.e., for $r = 0$) the theoretical stress concentration factor would be infinitely large because of the sharp 90° reentrant corner. Therefore, good design practice requires that such sharp corners be avoided and that generous fillets be provided.

EXAMPLE 12.1

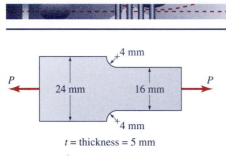

t = thickness = 5 mm

Fig. 1

An aluminum bar has the dimensions shown in Fig. 1. If the allowable stress is $\sigma_{allow} = 200$ MPa, determine the maximum axial force P_{allow} that can be carried by the bar.

Solution Figure 12.4 may be used to estimate the stress concentration factor for this bar with shoulder fillets. For the bar in Fig. 1,

$$\frac{D}{d} = \frac{24 \text{ mm}}{16 \text{ mm}} = 1.50$$

$$\frac{r}{d} = \frac{4 \text{ mm}}{16 \text{ mm}} = 0.25$$

Since there is a curve for $D/d = 1.50$, we can directly estimate that $K = 1.74$.

The allowable load is based on the average stress in the smaller cross section, so

$$P_{allow} = \sigma_{nom} A_{min} = \sigma_{nom}(80 \text{ mm}^2)$$

and when the stress reaches σ_{allow} at the points of stress concentration,

$$K\sigma_{nom} = \sigma_{allow} = 200 \text{ MPa}$$

Therefore,

$$P_{allow} = \left(\frac{200 \text{ MPa}}{1.74}\right)(80 \text{ mm}^2) = 9.20 \text{ kN}$$

$$P_{allow} = 9.20 \text{ kN} \qquad \textbf{Ans.}$$

Stress Concentration—Torsion. The torsion formula $\tau = T\rho/I_p$, which was derived in Chapter 4, may be used to determine the shear-stress distribution on the cross section of a homogeneous linearly elastic rod with uniform circular cross section. In order for this formula to be valid, the cross section in question must not be near a point where torque is applied to the rod by a gear, pulley, or flange, or to a point where there is a sudden change in the diameter of the cross section. Figure 12.5 shows three such situations where stress concentration occurs in circular rods loaded in torsion.

Applications like the two shown in Fig. 12.5a and 12.5b are considered in courses on machine design. Here we will only consider the stress concentration in the vicinity of a change in cross section, as illustrated in Fig. 12.5c and Fig. 12.6. Based on St. Venant's Principle, the shear-stress distributions at sections $a - a$ and $c - c$, which are more than 1 diameter away from the region of diameter change, may be determined by the torsion formula

$$\tau = \frac{T\rho}{I_p} \tag{12.2}$$

These linear shear-stress distributions are illustrated in Figs. 12.6b and 12.6d, respectively. But, as indicated in Fig. 12.6c, the maximum shear stress in a stepped torsion rod with shoulder fillet occurs at section $b - b$, where the fillet joins the smaller-diameter portion of the rod. The **stress-concentration factor for torsion** is defined as

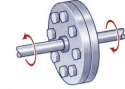

(a) Flanged connection.

(b) Gear on shaft.

$$K = \frac{\tau_{\max}}{\tau_{\text{nom}}} = \tau_{\max}\left(\frac{\pi d^3}{16T}\right) \tag{12.3}$$

(c) Stepped shaft.

since the nominal maximum shear stress would occur at the outer fibers of the smaller cross section, as indicated in Fig. 12.6d.

Figure 12.7 gives the value of the stress-concentration factor K for various geometries of stepped shafts. As was discussed above for axial loading of a stepped bar, the stress-concentration factor for torsion of a stepped shaft approaches infinity as the fillet radius r approaches zero. Therefore, in order to minimize the effect of the torsional shear-stress concentration characterized by Fig. 12.7, one should provide a fillet with the largest radius that is practical for the particular design. (The application of the stress-concentration factor to torsion problems is virtually identical to the application to axial deformation problems, as illustrated in Example Problem 12.1. Therefore, no torsion example is provided, although there are homework exercises on this topic.)

FIGURE 12.5 Three situations that produce stress concentration in circular shafts subjected to torsion.

Stress Concentration—Bending. In Chapter 6 the flexure formula, namely $\sigma = -My/I$, was derived for bending of a uniform beam having a plane of symmetry. This formula is valid only so long as there is no abrupt change in cross section and no concentrated load or reaction at or very near the cross section of interest. Here we will consider only two cases where stress concentrations occur in beams—the

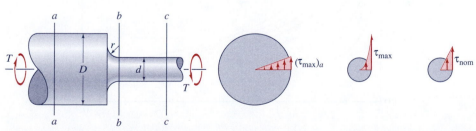

(a) Stepped shaft subjected to torsion.   (b) Section $a - a$.   (c) Section $b - b$.   (d) Section $c - c$.

FIGURE 12.6 A stepped torsion rod.

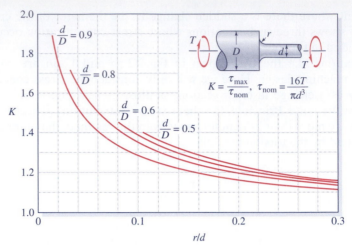

FIGURE 12.7 Torsional stress-concentration factor K for a stepped shaft with shoulder fillet. (Adapted from Ref. 12-6. Reprinted by permission of J. Wiley & Sons, Inc.)

case of a rectangular beam with abrupt change of cross section (Fig. 12.8), and that of a rectangular beam with symmetric, U-shaped notches (Fig. 12.9). Other cases are considered in Refs. [12-5, 12-6].

The **stress-concentration factor for bending** is defined as

$$K = \frac{\sigma_{max}}{\sigma_{nom}} = \sigma_{max}\left(\frac{td^2}{6M}\right) \tag{12.4}$$

with σ_{nom} being the nominal maximum stress at the reduced cross section, based on the flexure formula. Figures 12.8 and 12.9 present values of the stress-concentration factor K based on the data from Ref. [12-6]. These figures clearly indicate the desirability of avoiding fillets and notches with small radius r.

The stress-concentration factors presented in Figs. 12.8 and 12.9 are based on linearly elastic behavior, so values of σ_{max} obtained by using Eq. 12.4 are valid only

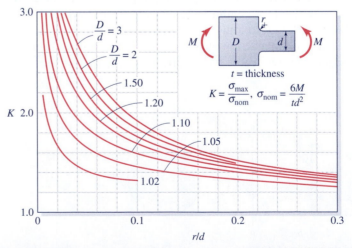

FIGURE 12.8 Stress concentration factors for pure bending of flat bars with fillets (Adapted from Ref. 12-6. Reprinted by permission of J. Wiley & Sons, Inc.)

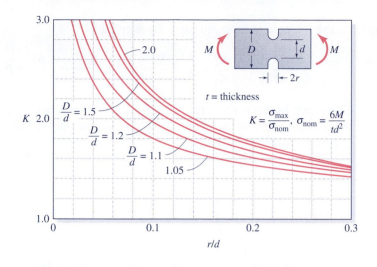

$$K = \frac{\sigma_{max}}{\sigma_{nom}}, \quad \sigma_{nom} = \frac{6M}{td^2}$$

$t = \text{thickness}$

FIGURE 12.9 Stress-concentration factors for pure bending of flat bars with U-shaped notches. (Adapted from Ref. 12-6. Reprinted by permission of J. Wiley & Sons, Inc.)

if the proportional limit of the material is not exceeded. Otherwise, if the material is ductile, plastic deformation will occur in the vicinity of the stress concentration, resulting in stresses that are smaller than the value of σ_{max} given by Eq. 12.4.

EXAMPLE 12.2

The flat bars in Fig. 1 are subjected to pure bending. By what percentage can the maximum moment in the bar be increased by removing material in order to convert the deep grooves of Fig. 1a to the semicircular grooves as illustrated in Fig. 1b? Assume that both bars have the same thickness t and are made of the same material (i.e., they have the same allowable stress).

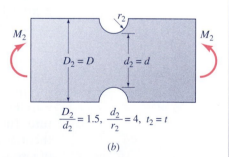

$$\frac{D_1}{d_1} = 1.5, \quad \frac{d_1}{r_1} = 8, \quad t_1 = t$$

(a)

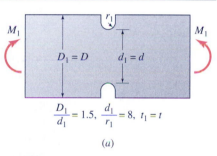

$$\frac{D_2}{d_2} = 1.5, \quad \frac{d_2}{r_2} = 4, \quad t_2 = t$$

(b)

Fig. 1

Solution The stress concentration factors for the two beams in Fig. 1 can be estimated from the curves in Fig. 12.9. The required parameters are

$$\frac{D_1}{d_1} = \frac{D_2}{d_2} = 1.5, \quad \frac{r_1}{d_1} = \frac{1}{8}, \quad \frac{r_2}{d_2} = \frac{1}{4}$$

From Fig. 12.9 we estimate that

$$K_1 = 2.06, \quad K_2 = 1.61$$

From Eq. 12.4,

$$\sigma_{max} = K\sigma_{nom} = K\left(\frac{6M}{td^2}\right) \tag{1}$$

But, since $\sigma_{max} = \sigma_{allow}$ is the same for both bars, and since they have the same dimensions t and d, Eq. (1) gives

$$\frac{M_2}{M_1} = \frac{K_1}{K_2}$$

so, the increase in moment capacity is given by

$$\frac{M_2 - M_1}{M_1} = \frac{K_1}{K_2} - 1 = \frac{K_1 - K_2}{K_2} = \frac{2.06 - 1.61}{1.61} = 0.28 \qquad \textbf{Ans.}$$

Therefore, by decreasing the sharpness of the notches we can increase the applied moment by about 28% without increasing the value of the maximum stress in the beam.

*12.3 FAILURE THEORIES

The design engineer is faced with two distinct tasks. The first task is to analyze the behavior of proposed designs subjected to specified loadings. For simple structural members and machine components the formulas in this book may be used to calculate stress and deformation. If there are stress concentrations, the procedures of Section 12.2 must be applied. For more complex members the *finite element method* [Refs. 12-2, 12-3] is generally used to obtain the stress distribution and deformation. In some cases, solutions may be obtained by using the theory of elasticity or the theory of plates and shells. The other important task of the design engineer is to *determine what values of stress and/or deformation would constitute* **failure** *of the object being designed.*[1] That is the subject we address in this section.

If a tension test is performed on a specimen of *ductile material,* the specimen may be said to fail when the axial stress reaches the yield stress σ_Y, that is, the **criterion of failure** is *yielding*. If the specimen is made of *brittle material,* the usual failure criterion is *brittle fracture* at the ultimate tensile stress σ_U. But a machine component or structural element is invariably subjected to a *multiaxial state of stress,* for which it is more difficult to designate what value of stress produces failure.

A tension test is relatively easy to perform using the procedures described in Section 2.4, and test results are published for many materials. But in order to apply the results of a tension test (or a compression test, or a torsion test) to a member that is subjected to multiaxial loading, it is necessary to consider the actual mechanism of failure. That is, was failure caused by the maximum normal stress reaching a critical value? Or was it due to maximum shear stress that reached a critical value, or to strain energy or some other quantity having reached its critical value? In the tension test, the criterion for failure can be easily stated in terms of the principal (tensile) stress σ_1, but for multiaxial stress we must consider the actual cause of the failure and say what combinations of stress would constitute failure.

In this section we consider four theories of failure.[2] Two of these apply to materials that behave in a ductile manner, that is, to materials that yield before they fracture. The other two theories apply to brittle materials. For plane stress, the failure theories are expressed in terms of the principal stresses σ_1 and σ_2. For triaxial states of stress, $\sigma_1, \sigma_2,$ and σ_3 are used.

Ductile Materials.
Two theories of failure for ductile materials will be discussed, the maximum-shear-stress theory and the maximum-distortion-energy theory.

---

[1]In the remainder of this section we will only consider failures due to excessive static stress, not failures due to excessive deflection or due to buckling.
[2]The theories presented here apply only to homogeneous, isotropic materials.

Maximum-Shear-Stress Theory:[3] When a flat bar of ductile material, like mild steel, is tested in tension, it is observed that the mechanism that is actually responsible for yielding is *slip,* that is, shearing along planes of maximum shear stress at 45° to the axis of the member. Initial yielding is associated with the appearance of the first slip line on the surface of the specimen, and as the strain increases, more slip lines appear until the entire specimen has yielded. If this slip is assumed to be the actual mechanism of failure, then the stress that best characterizes this failure is the shear stress on the slip planes. Figure 12.10 shows a Mohr's circle of stress for this uniaxial stress state, indicating that the shear stress on the slip planes has a magnitude of $\sigma_Y/2$. Therefore, if it is postulated that in a ductile material under any state of stress (uniaxial, biaxial, or triaxial), failure occurs when the shear stress on any plane reaches the value $\sigma_Y/2$, then the **failure criterion for the maximum-shear-stress theory** may be stated as

$$\tau_{\substack{abs \\ max}} = \frac{\sigma_Y}{2} \qquad (12.5)$$

where σ_Y is the yield stress determined by a simple tension test. Using Eq. 8.32 we can express Eq. 12.5 in terms of principal stresses as

$$\sigma_{max} - \sigma_{min} = \sigma_Y \qquad (12.6)$$

where σ_{max} is the maximum principal stress and σ_{min} is the minimum principal stress.[4]

For the case of plane stress, the *maximum-shear-stress failure criterion* may be stated in terms of the in-plane principal stresses σ_1 and σ_2 as follows:[5]

$$\left.\begin{array}{l} |\sigma_1| = \sigma_Y \text{ if } |\sigma_1| \geq |\sigma_2| \\ |\sigma_2| = \sigma_Y \text{ if } |\sigma_2| \geq |\sigma_1| \end{array}\right\} \quad \text{and if } \sigma_1 \text{ and } \sigma_2 \text{ have same sign}$$

$$|\sigma_1 - \sigma_2| = \sigma_Y \qquad \text{if } \sigma_1 \text{ and } \sigma_2 \text{ have opposite signs} \qquad (12.7)$$

Equations 12.7 may be represented in the convenient graphical forms shown in Fig. 12.11. For a member undergoing plane stress, the state of stress at every point in the body can be represented by a *stress point* (σ_1, σ_2) in the $\sigma_1 - \sigma_2$ plane, as indicated in Fig. 12.11.[6] If the state of stress for any point in the body corresponds to a stress point that lies outside the hexagon of Fig. 12.11 or on its boundary, failure is said to have occurred according to the maximum-shear-stress theory.

Maximum-Distortion-Energy Theory:[7] Although the maximum-shear-stress theory provides a reasonable hypothesis for yielding in ductile materials, the maximum-distortion-energy theory correlates better with test data and is therefore generally preferred. In this theory, yielding is assumed to occur when the energy associated

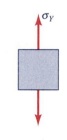

(a) Mohr's circle for $\sigma_1 = \sigma_Y$.

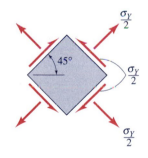

(b) Principal-stress element.

(c) Maximum-shear-stress element.

FIGURE 12.10 Principal stresses and maximum shear stresses for a uniaxial stress test.

---

[3]The names of C. A. Coulomb, H. Tresca, and J. J. Guest are associated with this theory of failure [Ref. 12-7].
[4]Note that this theory of failure ignores the normal stresses acting on the planes of maximum shear stress.
[5]For the case of plane stress, the out-of-plane principal stress is called σ_3 ($\sigma_3 \equiv 0$), even though it is not necessarily the minimum principal stress.
[6]Although σ_1 and σ_2 are principal stresses, Eqs. 12.7 and Fig. 12.11 do not require that the principal axes be labeled such that $\sigma_1 \geq \sigma_2$, as was done in Chapter 8.
[7]This theory is also called the *maximum-octahedral-shear-stress theory.* Credit for this theory is generally given to M. T. Huber, R. von Mises, and H. Hencky, although it was earlier conjectured by J. Clerk Maxwell [Ref. 12-7].

with change of shape of a body undergoing multiaxial loading is equal to the energy of distortion in a tensile specimen when yielding occurs at the uniaxial yield stress σ_Y.

Consider the strain energy stored in an element of volume, like the one shown in Fig. 12.12a. The strain energy density due to multiaxial loading is given by Eq. 11.12. which can be written, using the three principal axes, in the form

$$\bar{u} = \frac{1}{2}(\sigma_1\epsilon_1 + \sigma_2\epsilon_2 + \sigma_3\epsilon_3) \tag{12.8}$$

Combining Eq. 12.8 with Hooke's Law (Eq. 2.38 with $\Delta T = 0$) we get

$$\bar{u} = \frac{1}{2E}\left[\sigma_1^2 + \sigma_2^2 + \sigma_3^2 - 2\nu(\sigma_1\sigma_2 + \sigma_2\sigma_3 + \sigma_1\sigma_3)\right] \tag{12.9}$$

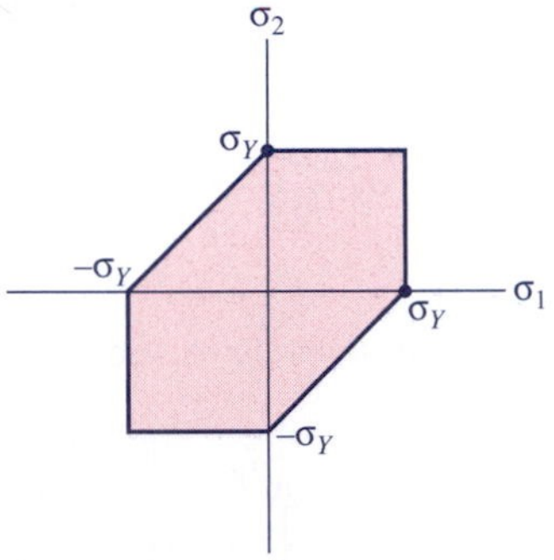

A portion of this strain energy can be associated with the *change of volume* of the element, and the remainder of the strain energy is associated with *change of shape*, that is with *distortion*. The change of volume is produced by the average stress $\sigma_{\text{avg}} = \frac{1}{3}(\sigma_1 + \sigma_2 + \sigma_3)$, as illustrated in Fig. 12.12b. The net stresses shown in Fig. 12.12c produce distortion without any change of volume.

Experiments have shown that materials do not yield when they are exposed to *hydrostatic stresses*[8] of extremely large magnitude. Therefore, it has been postulated that the stresses that actually cause yielding are the stresses that produce distortion. This hypothesis constitutes the **maximum-distortion-energy yield (failure) criterion,** which states:

Yielding of a ductile material occurs when the distortion energy per unit volume equals or exceeds the distortion energy per unit volume when the same material yields in a simple tension test.

When the distortion-producing stresses of Fig. 12.12c are substituted into Eq. 12.9 we get the following expression for the *distortion-energy density,*

$$\bar{u}_d = \frac{1}{12G}\left[(\sigma_1 - \sigma_2)^2 + (\sigma_2 - \sigma_3)^2 + (\sigma_1 - \sigma_3)^2\right] \tag{12.10}$$

•Experimental data from tension test.

FIGURE 12.11 Failure hexagon for the maximum-shear-stress theory (plane stress).

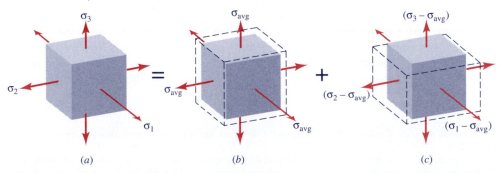

(a)   (b)   (c)

FIGURE 12.12 (a) Triaxial stress state. (b) Stresses producing volume change. (c) Stresses producing distortion.

---

[8]Figure 12.12b represents a hydrostatic state of stress, that is, equal stresses in all three principal directions.

The distortion energy density in a tensile test specimen at the yield stress σ_Y is

$$(\bar{u}_d)_Y = \frac{1}{6G}\sigma_Y^2 \qquad (12.11)$$

since $\sigma_1 = \sigma_Y$ and $\sigma_2 = \sigma_3 = 0$. Therefore, yielding occurs when the distortion energy for general loading, given by Eq. 12.10, equals or exceeds the value of $(\bar{u}_d)_Y$ in Eq. 12.11. Therefore, the *maximum-distortion-energy failure criterion* can be stated in terms of the three principal stresses as

$$\frac{1}{2}\left[(\sigma_1 - \sigma_2)^2 + (\sigma_2 - \sigma_3)^2 + (\sigma_1 - \sigma_3)^2\right] = \sigma_Y^2 \qquad (12.12a)$$

In terms of the normal stresses and shear stresses on three arbitrary mutually orthogonal planes, the **maximum-distortion-energy failure criterion** can be shown to have the form[9]

$$\frac{1}{2}\left[(\sigma_x - \sigma_y)^2 + (\sigma_y - \sigma_z)^2 + (\sigma_x - \sigma_z)^2 + 6(\tau_{xy}^2 + \tau_{yz}^2 + \tau_{xz}^2)\right] = \sigma_Y^2 \qquad (12.12b)$$

For the case of plane stress, the corresponding expressions for the *maximum-distortion-energy yield criterion* can easily be obtained from Eqs. 12.12 by setting $\sigma_3 = \sigma_z = \tau_{xz} = \tau_{yz} = 0$. In terms of the principal stresses, then,

$$\sigma_1^2 - \sigma_1\sigma_2 + \sigma_2^2 = \sigma_Y^2 \qquad (12.13)$$

This is the equation of an ellipse in the $\sigma_1 - \sigma_2$ plane, as depicted in Fig. 12.13. For comparison purposes, the failure hexagon for the maximum-shear-stress yield theory is also shown in dashed lines in Fig. 12.13. At the six vertices of the hexagon the two failure theories coincide; that is, both theories predict that yielding will occur if the state of (plane) stress at a point corresponds to any one of these six stress states. Otherwise, the maximum-shear-stress theory gives the more conservative (i.e., smaller-valued) estimate of the stresses required to produce yielding, since the hexagon falls either on or inside the ellipse.

A convenient way to apply the maximum-distortion-energy theory is to take the square root of the left-hand side of Eq. 12.12a (or Eq. 12.12b) to form an equivalent stress quantity that is called the **Mises equivalent stress.** Either of the following two equations can be used to compute the Mises equivalent stress, σ_M:

$$\sigma_M = \frac{\sqrt{2}}{2}\left[(\sigma_1 - \sigma_2)^2 + (\sigma_2 - \sigma_3)^2 + (\sigma_1 - \sigma_3)^2\right]^{1/2} \qquad (12.14a)$$

or

$$\sigma_M = \frac{\sqrt{2}}{2}\left[(\sigma_x - \sigma_y)^2 + (\sigma_y - \sigma_z)^2 + (\sigma_x - \sigma_z)^2 + 6(\tau_{xy}^2 + \tau_{yz}^2 + \tau_{xz}^2)\right]^{1/2} \qquad (12.14b)$$

For the case of plane stress, the corresponding expressions for the Mises equivalent stress can easily be obtained from Eqs. 12.14 by setting $\sigma_3 = \sigma_z = \tau_{xz} = \tau_{yz} = 0$.

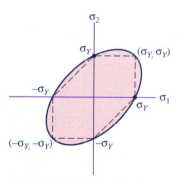

• Experimental data from tension test.
−− Maximum-shear-stress criterion.

FIGURE 12.13 Failure ellipse for the maximum-distortion-energy theory (plane stress).

---

[9]See Sections 78 and 90 of Ref. 12-1.

By comparing the value of the Mises equivalent stress at any point with the value of the tensile yield stress, σ_Y, it can be determined whether yielding is predicted to occur according to the maximum-distortion-energy theory of failure. Therefore, the Mises equivalent stress is widely used when calculated stresses are presented in tabular form or in the form of *color stress plots,* as has been done for the finite element analysis results shown in the last picture in the color-photo insert.

EXAMPLE 12.3

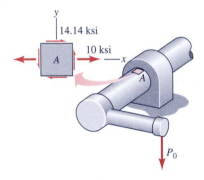

Fig. 1

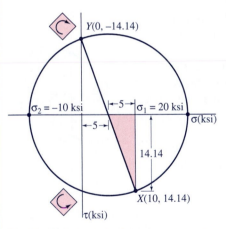

Fig. 2 Mohr's circle for stress state at point A.

A force P_0 kips applied by a lever arm to the shaft in Fig. 1 produces stresses at the critical point A having the values shown on the element in Fig. 1. Determine the load $P_S \equiv c_S P_0$ that would cause the shaft to fail according to the maximum-shear-stress theory, and determine the load $P_D \equiv c_D P_0$ that would cause failure according to the maximum-distortion-energy theory. The shaft is made of steel with $\sigma_Y = 36$ ksi.

Solution   It will be helpful it we construct a Mohr's circle for the plane stress state in Fig. 1 and also sketch the failure envelopes for the two failure theories. Figure 2 is a Mohr's circle for the given stresses at A.

Since σ_1 is positive and σ_2 is negative, we only need to sketch the fourth quadrant of the failure envelope. This is shown in Fig. 3, Since the stresses at point A are proportional to load, the stresses due to any load cP_0 will lie along the radial line identified in Fig. 3 as the *load line.* This line passes through the origin of the $\sigma_1 - \sigma_2$ plane and through the stress point ($\sigma_{1P} = 20$ ksi, $\sigma_{2P} = -10$ ksi) that corresponds to the principal stresses due to load P_0. Failure according to the maximum-shear-stress theory occurs at the stress state marked S in Fig. 3, and failure according to the maximum-distortion-energy theory occurs at point D.

Maximum-Shear-Stress Theory: Point S is the intersection of the load line given by

$$\sigma_1 = -2\sigma_2 \tag{1}$$

and the maximum-shear-stress boundary line

$$\sigma_1 - \sigma_2 = \sigma_Y = 36 \text{ ksi} \tag{2}$$

Solving Eqs. (1) and (2) for σ_1 and σ_2 we get

$$\sigma_{1s} = 24 \text{ ksi}, \qquad \sigma_{2s} = -12 \text{ ksi}$$

Combining these values with σ_1 and σ_2 of Fig. 2 we get

$$c_S = \frac{P_S}{P_0} = \frac{\sigma_{1S}}{c_S P_0} = \frac{24 \text{ ksi}}{20 \text{ ksi}} = 1.2$$

$$c_S = 1.2 \qquad\qquad \textbf{Ans.}$$

Maximum-Distortion-Energy Theory: Point D in Fig. 3 is the intersection of the load line, given by Eq. (1), and the ellipse given by Eq. 12.13. Thus,

$$\sigma_{1D}^2 - \sigma_{1D}\sigma_{2D} + \sigma_{2D}^2 = \sigma_Y^2 = (36 \text{ ksi})^2 \qquad (3)$$

Combining Eqs. (1) and (3) gives

$$7\sigma_{2D}^2 = (36 \text{ ksi})^2$$

Then, since $\sigma_1 > 0$ and $\sigma_2 < 0$, we get

$$\sigma_{1D} = 27.21 \text{ ksi}, \qquad \sigma_{2D} = -13.61 \text{ ksi}$$

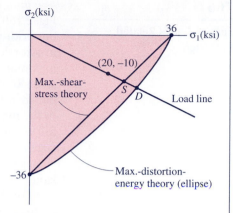

Fig. 3 Fourth quadrant of failure envelope.

Comparing these stresses with the principal stresses produced by P_0 gives

$$c_D = \frac{P_D}{P_0} = \frac{\sigma_{1D}}{\sigma_{1P}} = \frac{27.21 \text{ ksi}}{20 \text{ ksi}} = 1.36$$

$$c_D = 1.36 \qquad\qquad\qquad \textbf{Ans.}$$

We could get this same result by noting that the factor of safety against yielding, according to the maximum-distortion-energy yield criterion, is given by

$$FS_d = \sigma_Y/\sigma_M$$

where σ_M is the Mises equivalent stress for the case of plane stress. For the original load P_0,

$$\sigma_M = [\sigma_1^2 - \sigma_1\sigma_2 + \sigma_2^2]^{1/2} = [(20)^2 - (20)(-10) + (-10)^2]^{1/2} = 26.46 \text{ ksi}$$

Then,

$$FS_d = \frac{\sigma_Y}{\sigma_M} = \frac{36 \text{ ksi}}{26.46 \text{ ksi}} = 1.36$$

In summary, a 20% increase in the load would cause failure according to the maximum-shear-stress theory, but a 36% increase would be required to cause failure according to the maximum-distortion-energy theory. That is, under load P_0 the member would have a factor of safety $FS_s = 1.2$ with respect to failure according to the maximum-shear-stress theory and a factor of safety of $FS_d = 1.36$ with respect to failure according to the maximum-distortion-energy of failure.

MDS12.1	**Failure Theories**

Brittle Materials. Two theories of failure for brittle materials are presented, the maximum-normal-stress theory and Mohr's failure theory.

Maximum-Normal-Stress Theory:[10] It was stated in Section 2.4 and illustrated in Fig. 2.16*b*, that in a tension test, a brittle material fails suddenly by *fracture*, without prior yielding. And it was stated in Section 4.4 and illustrated in Fig. 4.17*b*, that in a torsion test, a bar made of brittle material also fails by fracture on planes of maximum tensile stress. Experiments have shown that the value of the normal stress on the failure plane for this biaxial state of stress is not significantly different than the fracture stress σ_U in a uniaxial tensile test. Therefore, the hypothesis of the **maximum-normal-stress theory** is that an object made of brittle material will fail when the maximum principal stress in the material reaches the ultimate normal stress that the material can sustain in a uniaxial tension test. This theory also assumes that compression failures occur at the same ultimate stress value as do tension failures.

For the case of *plane stress*, the **maximum-normal-stress failure criterion** is given by the equations

$$|\sigma_1| = \sigma_U \quad \text{or} \quad |\sigma_2| = \sigma_U \tag{12.15}$$

These equations may be plotted on the $\sigma_1 - \sigma_2$ plane, as shown in Fig. 12.14.[11]

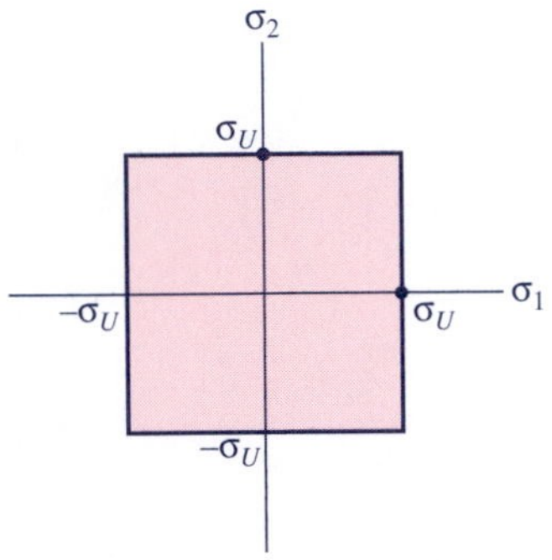

Mohr's Failure Criterion:[12] If the ultimate compressive strength of a brittle material is not equal to its ultimate strength in tension, the maximum-normal-stress theory should not be used. An alternative failure theory was proposed by Otto Mohr and is called **Mohr's failure criterion.** Figure 12.15*a* shows Mohr's circles for a uniaxial tensile test and for a uniaxial compression test for a brittle material having a tensile ultimate strength σ_{TU} and an ultimate strength in compression of magnitude σ_{CU}. By Mohr's theory, when σ_1 and σ_2 have the same sign, failure occurs if either of the following stress limits is reached:

$$\sigma_{\max} = \sigma_{TU} \quad \text{or} \quad \sigma_{\min} = -\sigma_{CU} \tag{12.16}$$

•Experimental data from tension test.

FIGURE 12.14 Failure diagram for the maximum-normal-stress theory (plane stress).

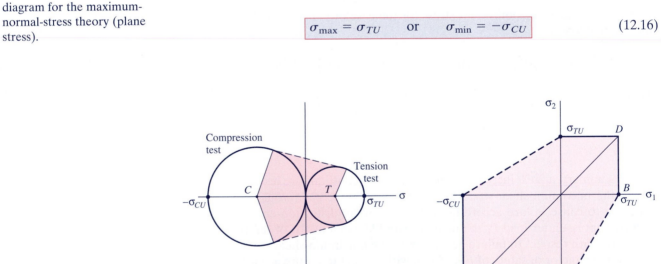

(a)   (b)

FIGURE 12.15 Mohr's failure criterion (plane stress).

[10] Also called *Rankine's Theory* after W. J. M. Rankine (1820–1872), an eminent professor of engineering at Glasgow University in Scotland.

[11] As in Figs. 12.11 and 12.13 σ_1 is not necessarily the greater principal stress.

[12] This theory is named for the German engineer Otto Mohr (1835–1918), the developer of Mohr's circle.

These equations are plotted as solid-line boundaries BD and CA, respectively, in Fig. 12.15b. (Note that above the diagonal line CD, $\sigma_2 > \sigma_1$, which would violate the naming convention of principal stresses that $\sigma_1 \geq \sigma_2$. Thus, only stress states that fall on or below this diagonal line CD need be considered.) For cases where σ_1 and σ_2 have opposite signs, Mohr proposed that the failure boundary be determined by drawing tangents to the tension and compression circles, as illustrated by the dashed lines in Fig. 12.15a. It can be shown (see Example 12.4) that the principal stresses for all circles that have centers on the σ-axis and are tangent to the dashed tangent lines in Fig. 12.15a plot as points on the dashed-line boundaries in Fig. 12.15b. Stress states with $\sigma_1 > 0$ and $\sigma_2 < 0$ fall in the fourth quadrant in Fig. 12.15b. For this case, Mohr's failure criterion states that failure occurs if the stress state falls on the dashed line AB in Fig. 12.15b, that is, if

$$\boxed{\frac{\sigma_1}{\sigma_{TU}} = \frac{\sigma_2}{\sigma_{CU}} + 1} \qquad (12.17)$$

The combination of Eqs. 12.16 and 12.17 constitute **Mohr's Failure Criteria.**

If torsion-test data or other plane-stress failure data are available, the failure boundary AB in the fourth quadrant of Fig. 12.15b can be modified to incorporate these experimental data (e.g., see *Modified Mohr Theory* in Ref. [12-8]).

EXAMPLE 12.4

Show that, for $\sigma_1 > 0$ and $\sigma_2 < 0$, the stress points that lie on the dashed line AB in Fig. 12.15b correspond to the principal stresses on Mohr's circles that have centers lying between points C and T in Fig. 12.15a and are tangent to the dashed lines in this figure.

Solution Let us redraw Fig. 12.15a and add an intermediate circle as specified in the problem statement (Fig. 1). The equation of the dashed-line AB in Fig. 12.15b is

$$\sigma_2 = \sigma_{CU}\left(\frac{\sigma_1}{\sigma_{TU}} - 1\right) \qquad (1)$$

We are to prove that this equation is the equation that relates the principal stresses σ_1 and σ_2 for the circle with center at N in Fig. 1. From Fig. 1,

$$\sigma_1 = \sigma_n + \overline{NN'}, \qquad \sigma_2 = \sigma_n - \overline{NN'}, \qquad (2)$$

where σ_n is the normal stress corresponding to point N. Also, from Fig. 1, the centers of the tension-test Mohr's circle and the compression-test Mohr's circle are

$$\sigma_t = \frac{\sigma_{TU}}{2}, \qquad \sigma_c = -\frac{\sigma_{CU}}{2} \qquad (3)$$

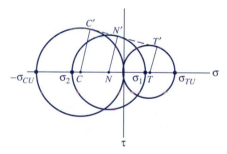

Fig. 1

and their radii are

$$\overline{TT'} = \frac{\sigma_{TU}}{2}, \qquad \overline{CC'} = \frac{\sigma_{CU}}{2} \qquad (4)$$

Let

$$\sigma_n = \sigma_c + k(\sigma_t - \sigma_c)$$

That is,

$$\sigma_n = -\frac{\sigma_{CU}}{2} + k\left(\frac{\sigma_{TU}}{2} + \frac{\sigma_{CU}}{2}\right) \qquad (5)$$

Then, the radius $\overline{NN'}$ will be linearly related to the radii $\overline{CC'}$ and $\overline{TT'}$ by the equation

$$\overline{NN'} = \overline{CC'} - k(\overline{CC'} - \overline{TT'})$$

or

$$\overline{NN'} = \frac{\sigma_{CU}}{2} - k\left(\frac{\sigma_{CU}}{2} - \frac{\sigma_{TU}}{2}\right) \qquad (6)$$

Combining Eqs. (2), (5), and (6), we get

$$\sigma_1 = k\sigma_{TU}, \qquad \sigma_2 = \sigma_{CU}(k - 1) \qquad (7)$$

The elimination of k from Eqs. (7) produces Eq. 1, the desired equation of the dashed line AB in Fig. 12.15b. QED.

*12.4   FATIGUE AND FRACTURE

At some time or other you have probably held a paper clip in your hands and bent the paper-clip wire back and forth several times until the wire finally broke in two. The failure did not occur when the paper clip was first bent, even though the wire experienced very large plastic deformation. Instead, failure occurred after a few reversals of flexural stress in the wire. This type of failure is called **fatigue failure.**[13] If the failure occurs after a few cycles of loading, perhaps up to a thousand cycles, it is called *low-cycle fatigue*. However, many metal components experience fatigue failure only after millions of stress cycles. In any case, when failure occurs at a stress level that is less than the level that would produce fracture under a single static application of load, the failure is called a fatigue failure.

---

[13]Fatigue failure of axles of railway cars was a source of great concern and study in the 1800s. The term *fatigue* was used by Poncelet to denote this type of failure due to cyclic stresses. A Wöler (1819–1914), a German railway engineer, is credited with developing the first fatigue testing machine. [Ref. 12-7]

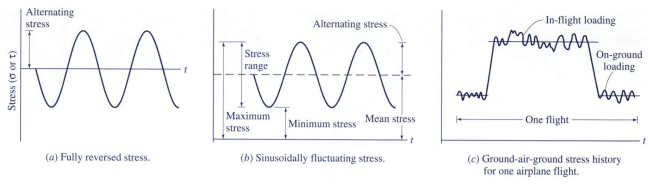

(a) Fully reversed stress.

(b) Sinusoidally fluctuating stress.

(c) Ground-air-ground stress history for one airplane flight.

FIGURE 12.16 Typical cyclic stresses that can produce fatigue failure.

The types of fluctuating stresses that lead to fatigue failure in metals are illustrated in Fig. 12.16. Figure 12.16a is the type of fully reversed stress that would be experienced by a railway-car axle as the train moves at constant speed along the track. Figure 12.16b shows a superposition of a (constant) mean stress and a sinusoidally varying stress; Fig. 12.16c illustrates the more complicated type of fluctuating stress experienced, for example, by an airplane wing component during a single flight. The fracture that occurs as a result of such fluctuating stresses usually begins at a point of stress concentration (Section 12.2) like the edge of the fastener hole in the splice plate shown in Fig. 12.17a and the airplane wing leading-edge nose cap in Fig. 12.17b. The crack is initiated in a region of high stress intensity, usually at some microscopic flaw or imperfection. Stress cycles cause the fatigue crack to grow slowly in size until the crack reaches a **critical crack length,** at which point the crack propagates at an explosive rate leaving the component unable to sustain load. Figure 12.17c shows the typical "beach-sand markings" of a fatigue fracture surface. Along the bottom edge of this photo, two points of initiation of the fatigue

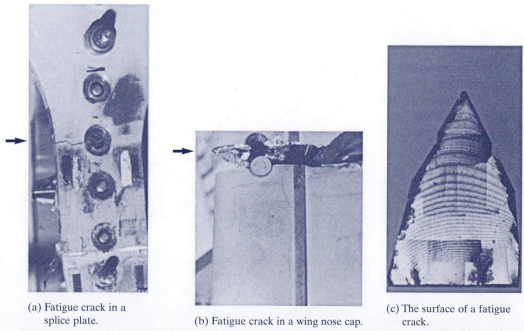

(a) Fatigue crack in a splice plate.

(b) Fatigue crack in a wing nose cap.

(c) The surface of a fatigue crack.

FIGURE 12.17 Examples of fatigue cracks. (Used with permission, Lockheed Martin, 1993.)

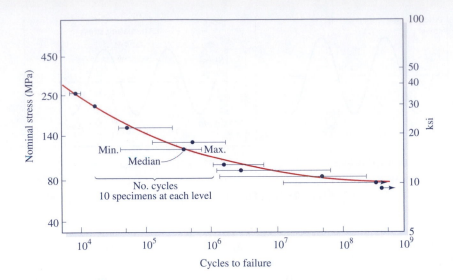

FIGURE 12.18 An S-N diagram for 7075-T6 aluminum-alloy notched tensile specimens. (From Ref. 12-10).

crack can be identified; the "ridges" on the fracture surface indicate the progress of the crack as it propagated with successive cycles of stress.

To characterize the behavior of a material under repeated cycles of loading, **fatigue tests** at various levels of stress are performed, and the results are plotted as an **S-N diagram,** or *endurance curve.*[14] Figure 12.18 is an S-N diagram based on a series of tests on nominally identical test specimens. (Note that the diagram is a log-log plot.) Several tests (e.g., ten tests at each stress level for Fig. 12.18) are conducted with cyclic stress whose maximum amplitude is slightly less than the ultimate static strength of the material. The number of cycles of stress required to cause fracture at that stress level is recorded for each test. Similar test series are conducted at progressively lower levels of stress, and the results plotted as the number of cycles to cause fatigue failure in each test. The scatter band for the tests at various stress levels are indicated on Fig. 12.18.

The endurance curves for some materials have the form illustrated in Fig. 12.19, where there is a stress level, called the *fatigue level,* or **endurance limit,** of the

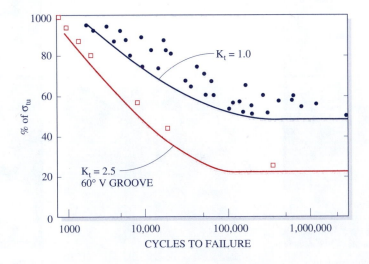

FIGURE 12.19 An S-N diagram for steel round-bar rotating-beam specimens. (From Ref. 12-11, maintained and published by CINDAS/ Purdue University under Cooperative Research and Development Agreement with the U.S. Air Force. Used with permission.)

[14]Test procedures to produce S-N diagrams are described in ASTM Standards E466-E468. [Ref. 12-9]

material, below which a virtually infinite number of stress cycles can be sustained without resulting in a fatigue failure. Many steel alloys exhibit this type of behavior. Fatigue limits for steel alloys, based on 10^7 stress cycles, are typically 35%-60% of the ultimate tensile strength of the steel.[15] As indicated in Fig. 12.19, notches or other imperfections obviously degrade fatigue strength. Some materials, including many aluminum alloys, do not exhibit a clearly defined endurance limit. In such cases, if the material continues to undergo cyclic stressing, it will eventually fail, regardless of how small the stress is. Tabulated "endurance limits" for aluminum alloys indicate that they can sustain up to 5×10^8 stress cycles if the stress does not-exceed about 25% of their ultimate tensile strength.[16]

From Figs. 12.18 and 12.19 and the preceding discussion, it is clear that **fatigue strength** must be taken into consideration in the design of a component that is to be subjected to cycles of stress during its service lifetime. One approach to fatigue design is called the **safe-life design philosophy.** Using endurance limit data and data from fatigue tests of the actual structural component under realistic stress cycles (see Fig. 12.16*c*), the design engineer establishes a safe lifetime for the component, that is, the number of stress cycles that the component will be allowed to experience before it must be retired from service.

As the desire has arisen to extend the useful lifetime of machines and structures (e.g., commercial and military aircraft), and as the discipline of *fracture mechanics* has matured, a new design philosophy has gained in importance—the **fail-safe design philosophy,** or *damage-tolerant design philosophy*. This philosophy involves the use of extensive fatigue testing, including careful observation of the initiation and propagation of cracks, to determine the complex relationship between cyclic loading and crack propagation. Then, with proper inspection, fatigue cracks can be prevented from reaching critical length and damaged parts can be replaced before leading to catastrophic failure of the entire structure or machine.[17] Figure 12.20

FIGURE 12.20 An airplane horizontal tail section undergoing full-scale fatigue testing. (Copyright © Boeing, 1995.)

---

[15]*Structural Alloys Handbook.* [Ref. 12-11]

[16]*Aluminum Standards and Data.* [Ref. 12-12] (See the source of these data for restrictions on their use in design.)

[17]Reference 12-13 discusses fracture mechanics and describes fatigue-fracture control strategies. Reference 12-14 discusses fatigue design methodologies in general, and Reference 12-15 applies the fail-safe and safe-life fatigue-design methods to aircraft structural design.

shows an airplane horizontal tail section undergoing full-scale fatigue testing to identify the fatigue-critical parts of the structure and to determine how many stress cycles could be applied before some component of the tail would fail by fatigue. This information enables the airplane's designers to ensure that no fatigue failure will occur while the aircraft is in service.

In your future courses on machine design or design of structures, you will have the opportunity to study and apply *safe-life design* and *fail-safe design* in much greater depth.

12.5 PROBLEMS

Problems 12.2-1 through 12.2-8. *In solving these stress-concentration problems, assume that the axial load is centrally applied at the ends of the flat bar and that the material remains linearly elastic.*

ᴰProb. 12.2-1. For the flat tension bar with central hole shown in Fig. P12.2-1, let $D = 3$ in., $t = 0.5$ in., and $P = 8$ kips. Determine the maximum normal stress in the bar for the following hole diameters: 0.5 in., 1.0 in., and 2.0 in.

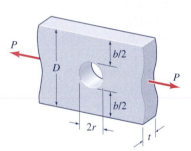

P12.2-1 through P12.2-4

ᴰProb. 12.2-2. For the flat tension bar with central hole shown in Fig. P12.2-2, let $r = 20$ mm, $t = 12$ mm, and $P = 40$ kN. Determine the maximum normal stress in the bar for the following bar widths: $D = 100$ mm, $D = 125$ mm, and $D = 150$ mm.

ᴰProb. 12.2-3. For the flat tension bar with central hole shown in Fig. P12.2-3, let $r = 15$ mm, $t = 20$ mm, and $\sigma_{max} = 75$ MPa. Determine the maximum axial load that can be applied to the bar for the following bar widths: $D = 130$ mm, $D = 150$ mm, and $D = 170$ mm.

ᴰProb. 12.2-4. For the flat tension bar with central hole shown in Fig. P12.2-4, let $D = 2$ in., $t = 0.5$ in., and $\sigma_{max} = 12$ ksi. Determine the maximum axial load that can be applied to the bar for the following hole diameters: 0.75 in., 1.0 in., and 1.50 in.

ᴰProb. 12.2-5. For the flat tension bar with stepped cross section shown in Fig. P12.2-5, the fillets are quarter circles having the largest radius that is consistent with the two widths D and d (i.e., $r = (D-d)/2$). Let $D = 4.0$ in. and $t = 1.0$ in.

(a) If the smaller width is $d = 3$ in. and the axial load is $P = 15$ kips, what is the maximum normal stress in the bar? (b) If the bar must be able to carry an axial load $P = 15$ kips without the maximum normal stress exceeding $\sigma_{max} = 8$ ksi, what is the minimum width, d, that can be used for the reduced-width cross section?

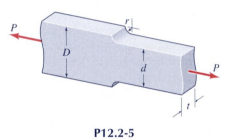

P12.2-5

ᴰProb. 12.2-6. For the flat tension bar with stepped cross section shown in Fig. P12.2-6, the fillets are quarter circles of radius $r = 10$ mm. Let $D = 80$ mm and $t = 20$ mm for this bar. (a) If the smaller width is $d = 50$ mm and the axial load is $P = 25$ kN, what is the maximum normal stress in the bar? (b) If the bar must be able to carry an axial load $P = 35$ kN without the maximum normal stress exceeding $\sigma_{max} = 60$ MPa, what is the minimum width, d, of the reduced-width cross section?

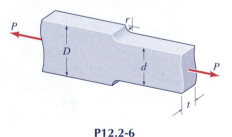

P12.2-6

ᴰProb. 12.2-7. For the flat tension bar with stepped cross section shown in Fig. P12.2-7, the hole is centered in the wide portion of the bar and the fillets are quarter circles. Let $D = 4$ in., $d = 2.5$ in., and $t = 0.5$ in. for this bar, and let the maximum axial load be $P = 10$ kips. What is the maximum hole

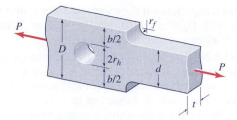

P12.2-7 and P12.2-8

radius and what is the minimum fillet radius that may be accommodated if the maximum normal stress in the bar is not to exceed $\sigma_{max} = 18$ ksi? Select radii that are multiples of 0.05 in.

DProb. 12.2-8. For the flat tension bar with stepped cross section shown in Fig. P12.2-8, the hole is centered in the wide portion of the bar and the fillets are quarter circles. Let $D = 100$ mm, $t = 20$ mm, and $d = 65$ mm for this bar, and let the maximum axial load be $P = 70$ kN. What is the maximum hole radius and what is the minimum fillet radius that may be accommodated if the maximum normal stress in the bar is not to exceed $\sigma_{max} = 125$ MPa? Select radii that are multiples of 1.0 mm.

Problems 12.2-9 through 12.2-14. *In solving these torsional stress-concentration problems, consider stepped, circular shafts as shown here, and assume that the material remains linearly elastic.*

DProb. 12.2-9. A stepped torsion bar with major diameter $D = 65$ mm and minor diameter $d = 40$ mm is subjected to a torque $T = 100$ N · m. (a) Determine the maximum shear stress in the shaft for the following size fillets: $r_1 = 5$ mm and $r_2 = 10$ mm. (b) Compare the shear stress results you obtained in Part (a) with the maximum shear stress the shaft would experience if it were of uniform diameter $D = d = 40$ mm.

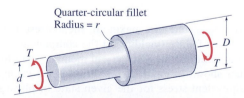

Quarter-circular fillet
Radius = r

P12.2-9 through P12.2-14

DProb. 12.2-10. A stepped torsion bar has a major diameter $D = 1.5$ in. and a full quarter-circular fillet (i.e., $r = (D\text{-}d)/2$). It is subjected to a torque $T = 1200$ lb · in. (a) Determine the maximum shear stress in the shaft for the following minor diameters: $d_1 = 1$ in., and $d_2 = 1.25$ in. (b) Compare the shear stress results you obtained in Part (a) with the maximum shear stresses the shafts would experience if they

were of uniform diameters $D_1 = d_1 = 1$ in. and $D_2 = d_2 = 1.25$ in., respectively.

DProb. 12.2-11. A stepped torsion bar with major diameter $D = 2$ in. and minor diameter $d = 1.25$ in. is subjected to a torque $T = 3750$ lb · in. If the allowable maximum shear stress in the shaft is $\tau_{max} = 12$ ksi, what is the minimum radius that may be used for the quarter-circular fillet at the junction of the two segments of the shaft? The fillet radius must be chosen as some multiple of 0.05 in. Show your calculations for at least three different radii.

DProb. 12.2-12. A stepped torsion bar with major diameter $D = 50$ mm and (quarter-circular) fillet radius 7 mm is subjected to a torque $T = 400$ N · m. If the allowable maximum shear stress in the shaft is $\tau_{max} = 70$ MPa, what is the minimum diameter that may be used for the smaller-diameter segment of the shaft? The diameter must be chosen as some multiple of 1 mm. Show your calculations for at least three different diameters.

DProb. 12.2-13. A stepped torsion bar with major diameter $D = 2$ in. and minor diameter $d = 1.5$ in. has a full quarter-circular fillet at the junction of the two segments (i.e., $r = (D - d)/2$). If the allowable maximum shear stress in the shaft is $\tau_{max} = 16$ ksi, and the shaft rotates at a constant angular speed of 500 rpm, what is the maximum power (hp) that may be delivered by the shaft?

DProb. 12.2-14. In Example Prob. 4.18 it was determined that a $\frac{5}{8}$-in.-diameter shaft would be required to transmit 10 hp at 875 rpm if the allowable maximum shear stress in the shaft is 20 ksi. Can a stepped shaft with major diameter $D = 1$ in., and minor diameter $d = 0.625$ in. (i.e., $\frac{5}{8}$ in.) be used instead of a uniform $\frac{5}{8}$-in.-diameter shaft? If so, what is the minimum radius that may be used for the quarter-circular fillet at the junction of the two segments of the shaft? The fillet radius must be chosen as some multiple of $\frac{1}{16}$ in. Show your calculations for at least three different fillet radii.

Problems 12.2-15 through 12.2-20. *In solving these stress-concentration problems, assume that the material remains linearly elastic.*

DProb. 12.2-15. The stepped-cross-section beam in Fig. P12.2-15 has a constant thickness $t = 15$ mm, a major depth $D = 60$ mm, and a minor depth $d = 40$ mm; and it is subjected to a bending moment $M = 100$ N · m. (a) Determine the maximum bending (normal) stress in the

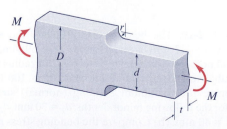

P12.2-15, P12.2-16, and P12.2-17

beam for the following sizes of quarter-circular fillets; $r_1 = 5$ mm, and $r_2 = 10$ mm. (b) Compare the bending stress results you obtained in Part (a) with the maximum flexural stress the beam would experience if it were of uniform depth $D = d = 40$ mm.

ᴰProb. 12.2-16. The stepped-cross-section beam in Fig. P12.2-16 has a major depth $D = 1.5$ in., and it has full quarter-circular fillets joining the major depth and the minor depth (i.e., $d = D - 2r$). The beam has a constant thickness $t = 0.5$ in., and is subjected to a bending moment $M = 1200$ lb · in. (a) Determine the maximum bending (normal) stress in the beam for the following minor depths: $d_1 = 1$ in., and $d_2 = 1.25$ in. (b) Compare the bending stress results you obtained in Part (a) with the maximum flexural stresses the beams would experience if they were of uniform depths $D_1 = d_1 = 1$ in. and $D_2 = d_2 = 1.25$ in., respectively.

ᴰProb. 12.2-17. The stepped-cross-section beam in Fig. P12.2-17 has a major depth $D = 2.5$ in., and it has full quarter-circular fillets joining the major depth and the minor depth (i.e., $d = D - 2r$). The beam has a constant thickness $t = 1.0$ in., and is subjected to a bending moment $M = 5$ kip · in. If the allowable maximum normal stress is $\sigma_{max} = 12$ ksi, determine the smallest minor depth d that may be used. The depth d must be chosen as some multiple of 0.10 in. Show your calculations for at least three different values of d.

ᴰProb. 12.2-18. The beam in Fig. P12.2-18 has a constant thickness $t = 0.75$ in., a minor depth $d = 2$ in., and symmetrical notches with semicircular roots of radius $r = 0.125$ in. The beam is subjected to a bending moment $M = 2500$ lb · in. (a) Determine the maximum bending (normal) stress in the beam for the following major depths: $D_1 = 2.25$ in., $D_2 = 2.5$ in., and $D_3 = 2.75$ in. (b) Compare the bending stress results you obtained in Part (a) with the maximum flexural stress that a uniform beam of depth $D = d = 2$ in. (without notches) would experience if subjected to the same bending moment, $M = 2500$ lb · in.

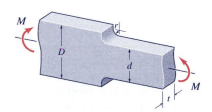

P12.2-18, P12.2-19, and P12.2-20

ᴰProb. 12.2-19. The beam in Fig. P12.2-19 has a constant thickness $t = 12$ mm, a major depth $D = 50$ mm, and symmetrical notches with semicircular roots of radius $r = 5$ mm. It is subjected to a bending moment $M = 100$ N · m. (a) Determine the maximum bending (normal) stress in the beam for the following minor depths: $d_1 = 30$ mm, $d_2 = 35$ mm, and $d_3 = 40$ mm. (b) Compare the bending stress results you obtained in Part (a) with the maximum flexural stress the

beam would experience if they were uniform beams (i.e., no notches), with $D_1 = d_1 = 30$ mm, $D_2 = d_2 = 35$ mm, and $D_3 = d_3 = 40$ mm, respectively.

ᴰProb. 12.2-20. The beam in Fig. P12.2-20 has a constant thickness $t = 0.5$ in., a major depth $D = 2$ in., a minor depth $d = 1.5$ in., and symmetrical notches with semicircular roots of radius r. The beam is subjected to a bending moment $M = 1000$ lb · in. If the allowable maximum normal stress is $\sigma_{max} = 12$ ksi, determine the smallest notch radius r that may be used. The radius r must be chosen as some multiple of $\frac{1}{32}$ in. Show your calculations for at least three different values of r.

▼ **FAILURE THEORIES** | MDS 12.1 |

> **Problems 12.3-1 through 12.3-13.** *In solving these problems, assume that the members are made of materials that behave in a* **ductile** *manner.*

Prob. 12.3-1. At a point in a thin plate (Fig. P12.3-1) the stresses σ_x, σ_y, and τ_{xy} are known, and the corresponding principal stresses, σ_1 and σ_2, are of opposite sign. Express the failure criterion of the maximum-shear-stress theory in terms of σ_x, σ_y, and τ_{xy}.

Prob. 12.3-2. At a point in a thin plate (Fig. P12.3-2) the stresses σ_x, σ_y, and τ_{xy} are known, and the corresponding principal stresses, σ_1 and σ_2, are the opposite sign. Express the failure criterion of the maximum-distortion-energy theory in terms of σ_x, σ_y, and τ_{xy}.

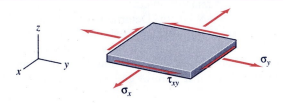

P12.3-1 and P12.3-2

Prob. 12.3-3. Would the maximum-shear-stress theory of failure predict failure (yielding) if the components of plane stress at a point in a steel structural member were to reach the values shown in Fig. P12.3-3? What is the value of the Mises equivalent stress for the given state of plane stress? Would failure be predicted by the maximum-distortion-energy theory?

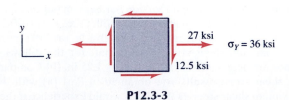

P12.3-3

Prob. 12.3-4. Solve Prob. 12.3-3 for the stresses indicated in Fig. P12.3-4.

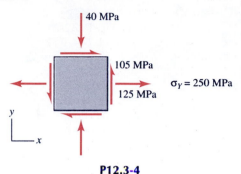

P12.3-4

Prob. 12.3-5. Solve the problem stated in Example Problem 12.3 if the force P_0 produces the stresses depicted in Fig. P12.3-5.

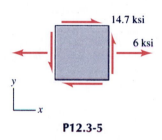

P12.3-5

Prob. 12.3-6. Would the maximum-shear-stress theory of failure predict failure (yielding) if the components of plane stress at a point in a member with yield stress σ_Y were to reach the values shown in Fig. P12.3-6? Would failure be predicted by the maximum-distortion-energy theory?

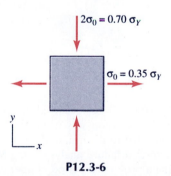

P12.3-6

^DProb. 12.3-7. The components of plane stress at a point on the surface of a member made of soft bronze ($\sigma_Y = 175$ MPa) are shown in Fig. P12.3-7. (a) For this state of stress, what is the factor of safety, FS_s, as predicted by the failure criterion of the maximum-shear-stress theory of failure? (b) What is the value of the Mises equivalent stress for the given state of plane stress, and what factor of safety, FS_d, is predicted by the failure criterion of the maximum-distortion-energy theory of failure?

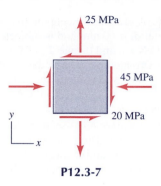

P12.3-7

^DProb. 12.3-8. The components of stress at a highly stressed point on the brass propeller shaft of a power boat ($\sigma_Y = 60$ ksi) are shown in Fig. P12.3-8. (a) For this state of stress, what factor of safety, FS_s, is predicted by the failure criterion of the maximum-shear-stress theory of failure? (b) What is the value of the Mises equivalent stress for the given state of plane stress, and what is the factor of safety, FS_d, predicted by the failure criterion of the maximum-distortion-energy theory of failure?

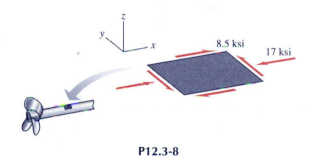

P12.3-8

^DProb. 12.3-9. A section of double-extra-strong pipe made of steel with yield strength $\sigma_Y = 50$ ksi is subjected to a bending moment $M = 35$ kip · in. and torque $T = 175$ kip · in., as shown in Fig. P12.3-9. The outer diameter of the pipe is $d_o = 3.5$ in., and the inner diameter is $d_i = 2.3$ in. (a) For this loading of the pipe, what factor of safety, FS_s, is predicted by the failure criterion of the maximum-shear-stress theory of failure? (b) What is the value of the Mises equivalent stress for the given state of plane stress, and what is the factor of safety, FS_d, predicted by the failure criterion of the maximum-distortion-energy theory of failure?

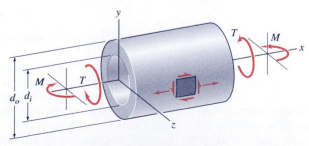

P12.3-9 and P12.3-10

ᴰ*Prob. 12.3-10. A section of steel pipe ($\sigma_Y = 340$ MPa) has an inner diameter $d_i = 60$ mm and is subjected to a bending moment $M = 7$ kN · m and torque $T = 7.8$ kN · m, as shown in Fig. P12.3-10. This member is to be designed in accordance with the maximum-distortion-energy criterion of failure, with a factor of safety $FS_d = 2.0$. To the nearest millimeter, what outer diameter, d_o, is required?

ᴰ*Prob. 12.3-11. A square-cross-section aluminum-alloy bar ($\sigma_Y = 40$ ksi) is subjected to a compressive axial force of magnitude $P = 48$ kips and a torque $T = 13$ kip · in., as shown in Fig. P12.3-11. This member is to be designed in accordance with the maximum-shear-stress criterion of failure, with a factor of safety $FS_s = 2.0$. To the nearest 0.1 in., what is the minimum allowable cross-sectional dimension, b? (Remember to use the theory of torsion of noncircular prismatic bars, which is discussed in Section 4.10.)

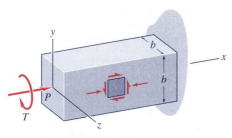

P12.3-11

Prob. 12.3-12. A rod with circular cross section and yield stress σ_Y is subjected to a bending moment M and a torque T, as shown in Fig. P12.3-12. Bending occurs in the xz plane. Express the maximum-shear-stress criterion of failure for this bar in terms of M, T, the rod diameter d, and the yield stress σ_Y.

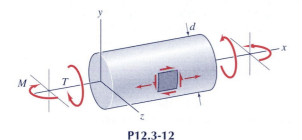

P12.3-12

ᴰ*Prob. 12.3-13. The handlebar of a bench-press machine has the "longhorn" shape shown in Fig. P12.3-13. You are to analyze the configuration depicted in Fig. P12.3-13b, where the (curved) axis of the handlebar lies in the xy plane and the force P exerted by the athlete's hand acts in the z direction. The handlebar is formed from a solid circular-cross-section steel bar having a diameter $d = 1$ in. and yield strength $\sigma_Y = 50$ ksi. If the handlebar is to be designed according to the maximum-shear-stress failure criterion with a factor of safety

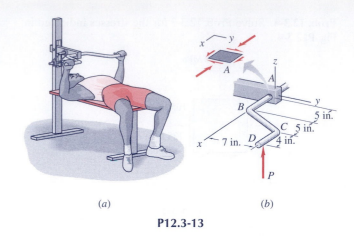

(a)   (b)

P12.3-13

of $FS_s = 2.5$, what is the allowable total force (i.e., $2P$) that should be advertised in the specifications for this bench-press machine?

Problems 12.3-14 and 12.3-15. *In solving these problems, assume that the members are made of materials that behave in a brittle manner.*

Prob. 12.3-14. A rod with circular cross section and ultimate tensile strength σ_{TU} is subjected to a tensile force P and a torque T, as shown in Fig. P12.3-14. Express the maximum-normal-stress criterion of failure for this bar in terms of P, T, the rod diameter d, and the ultimate tensile strength, σ_{TU}.

P12.3-14

*Prob. 12.3-15. After failures occurred in several cast-iron bearing housings, a decision was made to use strain-gage rosettes (Section 8.10) to determine the operating stresses, and then to perform a failure analysis using Mohr's failure criterion. During a prolonged period of operation, the most critical combination of stresses was determined to be ($\sigma_x = 0$, $\sigma_y = 115$ MPa, $\tau_{xy} = 75$ MPa); and the tensile and compressive ultimate strengths of the cast iron were determined to be $\sigma_{TU} = 170$ MPa and $\sigma_{CU} = 655$ MPa, respectively. (a) Determine the principal stresses σ_1 and σ_2 corresponding to the given stress state. (b) Construct a Mohr failure diagram, like the one in Fig. 12.15b, for the cast iron. (c) Using the results you obtained in Parts (a) and (b), can you explain why failures have been occurring in the cast-iron bearing housings? Show your calculations.

Section		Suggested Review Problems
12.2	The topic of Section 12.2 is **Stress Concentrations.** This is applied to stress concentrations in: • **Axial deformation,** • **Torsion,** and • **Bending.** The increase in stress due to a localized application of a load or due to a nonuniformity of the cross section is called a **stress concentration.** Figure 12.1*b* illustrates how a cross-sectional discontinuity, like a hole, causes an increase in stress near the hole, and Fig. 12.3 shows how the amount of stress increase depends on the size of the hole.	 Stress concentration in an axially loaded flat bar with central hole. (Fig. 12.1*b*) Stress concentration factor K for an axially loaded flat bar with central hole. (Fig. 12.3)	12.2-3
	Additional plots in Section 12.2 provide values of the stress-concentration factor K for the following cases:	• Flat bar in tension, with shoulder fillets (Fig. 12.4), • Stepped shaft in torsion, with shoulder fillets (Fig. 12.7) • Flat bar in pure bending, with shoulder fillets (Fig. 12.8), and • Flat bar in pure bending, with symmetric U-shaped notches (Fig. 12.9) Unlike the stress-concentration-factor plot in Fig. 12.3. which is a single curve, the stress-concentration-factor plots listed above all have multiple curves, which are related to some geometric parameter of the member in question.	12.2-7 12.2-11 12.2-15 12.2-19

Section 12.3 **Failure Theories** and Section 12.4 **Fatigue and Fracture** are *optional sections* that present additional important design-related topics.

NUMERICAL ACCURACY; APPROXIMATIONS

A

A.1 NUMERICAL ACCURACY; SIGNIFICANT DIGITS

The engineering quantities that enter into deformable-body mechanics problems (e.g., force, length, strain) can usually be measured to an *accuracy* of about 1 part in 100 (1%) or, in some cases, perhaps to 1 part in 1000 (0.1%). In engineering calculations the accuracy of a number is indicated by the number of significant digits used in stating the number. A *significant digit* is any digit from 1 to 9, or any zero that is not used to show the position of the decimal point. For example, the numbers 27, 4.5, 0.30, and 0.0091 each has two significant digits. These could also be written in powers-of-ten form as $27, 45 \times 10^{-1}, 30 \times 10^{-2}$, and 91×10^{-4}, again indicating that each number has two significant digits. Zeros immediately to the left of the decimal point can lead to some ambiguity regarding the number of digits that are significant. For example, to indicate that the number 30,000 has two significant digits, rather than just one, it would be preferable to write it as 30×10^3.

It is tempting, when one uses a calculator or computer to make engineering computations, to record all of the digits that are displayed in the computed result, but this could give an unwarranted impression of the true accuracy of the number. For example, it would be reasonable to see the value of a force stated as $F = 426$ lb, but it would be quite unreasonable to see the same force recorded as $F = 426.379$ lb.

In engineering problems five types of numbers are encountered:

- Exact numbers (e.g., the 32 in the formula $I_p = \pi d^4/32$).
- "Formally exact" numbers (e.g., the value of π, or the value of $\sin \theta$).[1]
- Given data.
- Intermediate results of calculations.
- Final results of calculations.

In this book the final results of calculations are reported according to the following rules that are standard practice in engineering:

---

[1]The values of π and of trigonometric functions are calculated to many significant digits (ten or more) within the calculator or computer.

- Numbers that begin with the digits 2 through 9 are recorded to three significant digits.
- Numbers that begin with the digit 1 are recorded to four significant digits.

In some instances (e.g., when data is read from a graph) fewer significant digits may be recorded. Given data are assumed to be accurate to the number of significant digits indicated above, even though fewer digits may actually be stated. Intermediate results, if recorded for use in further calculations, are recorded to several additional digits in order to preserve numerical accuracy.

A.2 APPROXIMATIONS

Power-series approximation formulas of two types are useful (e.g., in the discussion of extensional strain in Section 2.3) — a *trigonometric approximation* and a *binomial-expansion approximation*. Series expansions of the trigonometric functions $\sin \theta$, $\cos \theta$, and $\tan \theta$ are:

$$\sin \theta = \theta - \frac{\theta^3}{3!} + \frac{\theta^3}{5!} - \cdots$$

$$\cos \theta = 1 - \frac{\theta^2}{2!} + \frac{\theta^4}{4!} - \cdots$$

$$\tan \theta = \theta + \frac{\theta^3}{3} + \frac{2\theta^3}{15} + \cdots$$

Therefore, if $\theta \ll 1$, we can replace the trigonometric functions by the approximations:

$$\sin \theta \approx \theta, \quad \cos \theta \approx 1, \quad \tan \theta \approx \theta \tag{A-1}$$

A small quantity β may appear in an expression of the form $(1 + \beta)^n$. The series expansion for this expression is

$$(1 + \beta)^n = 1 + n\beta + \frac{n(n-1)}{2!}\beta^2 + \cdots$$

Therefore, if $\beta \ll 1$, the approximation

$$(1 + \beta)^n \approx 1 + n\beta \tag{A-2}$$

may be used. Equations (A-1) and (A-2) will be useful in many strain-displacement analyses to reduce complex, nonlinear strain-displacement equations to simpler linear, small-displacement forms (e.g., see Example Prob. 2.5 and the discussion of *Small-Displacement/Small-Strain Behavior* that follows Example Prob. 2.5).

SYSTEMS OF UNITS

B.1 INTRODUCTION

The physical quantities encountered in science and engineering (e.g., force, mass, length, time, and temperature) must be expressed in some system of units. The *International System of Units*[2] (SI) was established by international agreement to provide a uniform system of units for measurement throughout the world. In the United States, the *U.S. Customary System of Units* (USCS) is the most commonly used system. Other important systems of units are non-SI forms of the metric system and the British Imperial System of Units. For the foreseeable future, engineers in the United States should be able to use either USCS or SI units, so both are used throughout this book.

There are three classes of units: base units, derived units, and supplementary units. The *base units* are defined in terms of specific physical *standards*. For example, the standard unit of *mass,* the kilogram (kg), is defined by a bar of platinum-iridium alloy that is kept at the International Bureau of Weights and Measures in Sèvres, France. The *second*(s) is defined as the duration of 9 192 631 770 periods of the radiation corresponding to the transition between the two hyperfine levels of the ground state of the cesium 133 atom. The *meter* (m) is defined to be the distance traveled by light in a vacuum during a time interval 1/299 792 458 of a second. Plane angles and solid angles comprise a second group of units, called *supplementary units.* Most units are *derived units,* which are related by an algebraic formula to base units and, in some instances, to supplementary units.

B.2 SI UNITS

The International System of Units has four base units that are of importance in mechanics: The unit of mass is the *kilogram* (kg), the unit of length is the *meter* (m), the unit of time is the *second* (s), and the unit of temperature is the *kelvin*

---

[2]*Guide for the Use of the International System of Units (SI)*, NIST Special Publication 811, National Institute of Standards and Technology, Gaithersburg, MD, 20899-0001.

TABLE B-1. SI Units

Class	Quantity	Name of Unit	SI Symbol	Unit Formula
Base	Length	meter	m	—
	Mass	kilogram	kg	—
	Time	second	s	—
	Temperature	kelvin	K	—
Derived	Area	square meter	m^2	—
	Volume	cubic meter	m^3	—
	Force	newton	N	$kg \cdot m/s^2$
	Stress, pressure	pascal	Pa	N/m^2
	Work, energy	joule	J	$N \cdot m$
	Power	watt	W	$N \cdot m/s$
Supplementary	Plane angle	radian	rad	—

(K).[3] The unit of force is the newton (N). This derived unit is based on Newton's Second Law, $F = Ma$. Thus, a *newton* is defined as the force required to give one kilogram of mass an acceleration of one meter per second squared:

$$1 \text{ N} = 1 \text{ kg} \cdot \text{m/s}^2$$

The SI units that are pertinent to the mechanics of deformable bodies are listed, according to class, in Table B-1.

In the SI system (and in other metric systems), prefixes denoting powers of ten are used to modify basic units (e.g., millimeter, kilonewton). Table B-2 lists the prefixes that are preferred in SI usage. Prefixes that do not signify powers of 10^3 or 10^{-3} (e.g., centi $= 10^{-2}$) should be avoided. In units represented by fractions, use a prefix in the numerator, rather than in the denominator (except that the symbol for kilogram (kg) may appear in the denominator). For example, if the calculation of a certain normal stress were to give the result $\sigma = 12.6$ N/mm^2, this result should be recorded as 12.6 MPa ($\equiv$ 12.6 MN/m^2).

TABLE B-2. SI Prefixes

Prefix	Symbol	Multiplication Factor	Exponential Form
giga	G	1 000 000 000	10^9
mega	M	1 000 000	10^6
kilo	k	1 000	10^3
milli	m	0.001	10^{-3}
micro	μ	0.000 001	10^{-6}
nano	n	0.000 000 001	10^{-9}

---

[3]Neither the word *degrees* nor the symbol ° is part of the name of the SI unit of temperature, as they are in other systems of units (e.g., degrees Fahrenheit, °F).

In solving problems that involve stress ($= F/A$), it is convenient to use newtons (N) or kilonewtons (kN) for the SI units of *force* and mm² for *area*, noting that

- $1 \text{ N/mm}^2 \equiv 1 \text{ MPa}$
- $1 \text{ kN/mm}^2 \equiv 1 \text{ GPa}$

In SI usage, as shown in the third column of Table B-2, a space, rather than a comma, is used to separate the digits of a numeric value into groups of three, both before and after the decimal point. When there are four digits, the space is optional (e.g., 5943 m, or 5 943 m).

B.3 U.S. CUSTOMARY UNITS; CONVERSION OF UNITS

In the U.S. Customary System of Units (sometimes called the British gravitational system) the base units are the following: the foot (ft) for length, the pound (lb) for force, the second (s) for time, and the degree Fahrenheit (°F) for temperature. The foot is defined by U.S. statute as exactly 0.3048 m. The pound is defined as the weight at sea level and at a latitude of 45° of a platinum standard that is kept at the Bureau of Standards in Washington, DC. This platinum standard has a mass of 0.453 592 43 kg.

In the U.S. Customary System of Units, the unit of mass is a derived unit, called the *slug*. One slug is the mass that would be accelerated at 1 ft/sec squared by a force of 1 lb:

$$1 \text{ slug} = 1 \text{ lb} \cdot \text{s}^2/\text{ft}$$

Conversion factors between SI units and USCS units are given in the table inside the front cover of this book.

GEOMETRIC PROPERTIES OF PLANE AREAS

C

C.1   FIRST MOMENTS OF AREA; CENTROID

Definitions. The solutions of most problems in this book involve one or more geometric properties of plane areas[4]—area, centroid, second moment, etc. The total *area* of a plane surface enclosed by bounding curve B is defined by the integral

$$A = \int_A dA \tag{C-1}$$

which is understood to mean a summation of differential areas dA over two spatial variables, such as y and z Fig. C-1.

The *first moments* of the area A about the y and z axes, respectively, are defined as

$$Q_y = \int_A z\,dA, \quad Q_z = \int_A y\,dA \tag{C-2}$$

Q_y and Q_z are called first moments because the distances z and y appear to the first power in the defining integrals.

The *centroid* of an area is its "geometric center." The coordinates $(\bar{y}, \bar{z})$ of the centroid C (Fig. C-2) are defined by the first-moment equations

$$\bar{y}A = \int_A y\,dA, \quad \bar{z}A = \int_A z\,dA \tag{C-3}$$

For simple geometric shapes (e.g., rectangles, triangles, circles) there are closed-form formulas for the geometric properties of plane areas. A number of these are

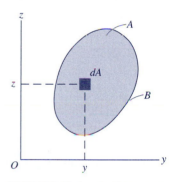

FIGURE C-1   A plane area.

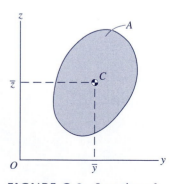

FIGURE C-2   Location of the centroid of an area.

---

[4]The word *area* is used in two senses: In one sense, the word refers to the portion of a plane surface that lies within a prescribed bounding curve, like the area bounded by the closed curve B in Fig. C-1; in the second sense, the word refers to the quantity of surface within the bounding curve [Eq. (C-1)].

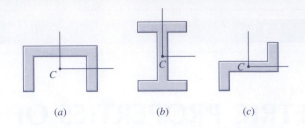

FIGURE C-3 Three types of area symmetry.

(a) (b) (c)

given in a table inside the back cover of this book.[5] An area may possess one of the three symmetry properties illustrated in Fig. C-3. If an area has *one axis of symmetry*, like the vertical axis of the *C*-section in Fig. C-3*a*, the centroid of the area lies on that axis. If the area has *two axes of symmetry*, like the wide-flange shape in Fig. C-3*b*, then the centroid lies at the intersection of those axes. Finally, if the area is *symmetric about a point*, like the *Z*-section in Fig. C-3*c*, the center of symmetry is the centroid of the area.

Composite Areas. Many structural shapes are composed of several parts, each of which is a simple geometric shape. For example, each of the areas in Fig, C-3 can be treated as a composite area made up of three rectangular areas. Since the integrals in Eqs. (C-1) through (C-3) represent summations over the total area A, they can be evaluated by summing the contributions of the constituent areas A_i, giving

$$A = \sum_i A_i, \quad Q_z = \bar{y}A = \sum_i \bar{y}_i A_i, \quad Q_y = \bar{z}A = \sum_i \bar{z}_i A_i \qquad \text{(C-4)}$$

Note that $\bar{y}$ in Fig. C-4 can be determined directly from the symmetry of the figure about the z' axis.

COMPOSITE-AREA PROCEDURE FOR LOCATING THE CENTROID

1. Divide the composite area into simpler areas for which there exist formulas for area and for the coordinates of the centroid. (See the table inside the back cover.)
2. Establish a convenient set of reference axes (y, z).
3. Determine the area, A, using Eq. (C-4a).
4. Calculate the coordinates of the composite centroid, $(\bar{y}, \bar{z})$, using Eqs. (C-4b, c).

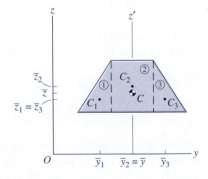

FIGURE C-4 A composite area.

---

[5]The reader may consult textbooks on integral calculus or statics for exercises in evaluating the integrals in Eqs. C-1 through C-3 for specific shapes.

EXAMPLE C-1

Locate the centroid of the L-shaped area in Fig. 1.

Solution A—Addition Method   Following the procedure outlined above, we divide the L-shaped area into two rectangles, as shown in Fig. 2. The y and z axes are located along the outer edges of the area, with the origin at the lower-left corner. Since the composite area consists of only two areas, the composite centroid, C, lies between C_1 and C_2 on the line joining the two centroids, as illustrated in Fig. 2.

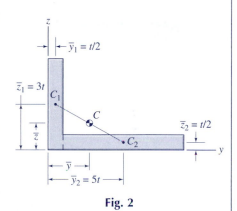

Fig. 1

Area: From Eq. (C-4a),

$$A = A_1 + A_2 = (6t)(t) + (8t)(t) = 14t^2 \qquad (1)$$

Centroid: From Eqs. (C-4b) and (C-4c),

$$\bar{y}A = \bar{y}_1 A_1 + \bar{y}_2 A_2 = (t/2)(6t^2) + 5t(8t^2) = 43t^2$$

$$\bar{y} = \frac{43t^3}{14t^2} = \frac{43}{14}t = 3.07t \qquad \textbf{Ans.} \quad (2)$$

$$\bar{z}A = \bar{z}_1 A_1 + \bar{z}_2 A_2 = (3t)(6t^2) + (t/2)(8t^2) = 22t^3$$

$$\bar{z} = \frac{22t^3}{14t^2} = \frac{22}{14}t = 1.57t \qquad \textbf{Ans.} \quad (3)$$

Fig. 2

Solution B—Subtraction Method   Sometimes (although not in this particular example) it is easier to solve composite-area problems by treating the area as the net area obtained by subtracting one or more areas from a larger area. Then, in Eqs. (C-4), the A_i's of the removed areas are simply taken as negative areas. This method will now be applied to the L-shaped area in Fig. 1 by treating it as a larger rectangle from which a smaller rectangle is to be subtracted (Fig. 3). Area A_1 is the large rectangle $PQRS$; area A_2 is the smaller unshaded rectangle. The composite centroid, C, lies along the line joining the two centroids, C_1 and C_2, but it does not fall between them.

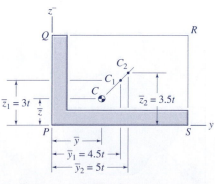

Fig. 3

Area: From Eq. (C-4a),

$$A = A_1 + A_2 = (9t)(6t) + [-(8t)(5t)] = 14t^2 \qquad (4)$$

Centroid: From Eqs. (C-4b) and (C-4c),

$$\bar{y}A = \bar{y}_1A_1 + \bar{y}_2A_2 = (4.5t)(54t^2) + [(5t)(-40t^2)] = 43t^3$$

$$\bar{y} = 43t^3/14t^2 = 3.07t \qquad \textbf{Ans.} \quad (5)$$

$$\bar{z}A = \bar{z}_1A_1 + \bar{z}_2A_2 = (3t)(54t^2) + [(3.5t)(-40t^2)] = 22t^3$$

$$\bar{z} = 22t^3/14t^2 = 1.57t \qquad \textbf{Ans.} \quad (6)$$

C.2 MOMENTS OF INERTIA OF AN AREA

Definitions of Moments of Inertia. The *moments of inertia* of a plane area (Fig. C-5) about axes y and z in the plane are defined by the integrals

$$I_y = \int_A z^2\,dA, \quad I_z = \int_A y^2\,dA \qquad (C-5)$$

These are called the *moment of inertia with respect to the y axis* and the *moment of inertia with respect to the z axis,* respectively. Since each integral involves the square of the distance of the elemental area dA from the axis involved, these quantities are called *second moments of area.* These moments of inertia appear primarily in formulas for bending of beams (see Chapter 6).

The moments of inertia defined in Eqs. (C-5) are with respect to axes that lie in the plane of the area under consideration. The second moment of area about the x axis, that is, with respect to the origin O, is called the *polar moment of inertia* of the area. It is defined by

$$I_\rho = \int_A \rho^2\,dA \qquad (C-6)$$

Since, by the Pythagorean theorem, $\rho^2 = y^2 + z^2$, I_ρ is related to I_y and I_z by

$$I_\rho = I_y + I_z \qquad (C-7)$$

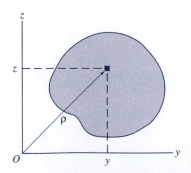

FIGURE C-5 A plane area.

Since Eqs. (C-5) and (C-6) involve squares of distances, I_y, I_z, and I_p are always positive. All have the dimension of $(\text{length})^4$ — in^4, mm^4, etc.

A table listing formulas for coordinates of the centroid and for moments of inertia of a variety of shapes may be found inside the back cover of this book. The most useful formulas for moments of inertia and for polar moment of inertia are derived here.

Moments of Inertia of a Rectangle: For the rectangle in Fig. C-6a, Eq. (C-5a) gives

$$I_y = \int_A z^2 dA = \int_{-h/2}^{h/2} z^2(b\,dz) = b\frac{z^3}{3}\Big|_{-h/2}^{h/2} = \frac{bh^3}{12}$$

where the y axis passes through the centroid and is parallel to the two sides of length b, I_z may be derived in an analogous manner, so the moments of inertia of a rectangle for the two centroidal axes parallel to the sides of the rectangle are:

$$\boxed{I_y = \frac{bh^3}{12}, \quad I_z = \frac{hb^3}{12}} \tag{C-8}$$

Polar Moment of Inertia of a Circle about its Center: Letting $dA = 2\pi\rho\,d\rho$, the area of the dark-shaded ring in Fig. C-6b, and using Eq. (C-6), we can determine the polar moment of inertia of a circle about its center:

$$I_\rho = \int_A \rho^2 dA = \int_0^r \rho^2(2\pi\rho\,d\rho) = \frac{\pi r^4}{2}$$

$$\boxed{I_\rho = \frac{\pi r^4}{2} = \frac{\pi d^4}{32}} \tag{C-9}$$

Radii of Gyration.

A length called the *radius of gyration* is defined for each moment of inertia by the formulas

$$\boxed{r_y = \sqrt{\frac{I_y}{A}}, \quad r_z = \sqrt{\frac{I_z}{A}}} \tag{C-10}$$

These lengths are used to simplify several formulas in Chapters 6 and 10. If these formulas are written in the form

$$I = Ar^2$$

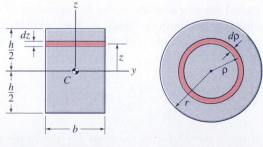

(a) A rectangular area.   (b) A circular area.

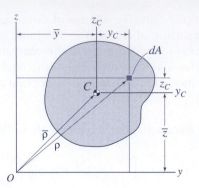

FIGURE C-7   Notation for deriving parallel-axis theorems.

then it is clear that the radius of gyration is the distance at which the entire area could be concentrated and still give the same value I, of moment of inertia about a given axis.

Parallel-Axis Theorems for Moments of Inertia.

Let the (y, z) pair of axes be parallel to centroidal axes (y_C, z_C), as shown in Fig. C-7. The centroid, C, is located with respect to the (y, z) axes by the centroidal coordinates $(\bar{y}, \bar{z})$. Since, from Fig. C-7, $z = \bar{z} + z_C$, the moment of inertia I_y is given by

$$I_y = \int_A z^2 dA = \int_A (\bar{z} + z_C)^2 dA$$

$$= \bar{z}^2 A + 2\bar{z} \int_A z_C dA + \int_A z_C^2 dA$$

$$\boxed{I_y = \bar{z}^2 A + I_{y_C}} \tag{C-11a}$$

The term $\int_A z_C dA$ vanishes since the y_C axis passes through the centroid; the term I_{y_C} is the *centroidal moment of inertia* about the y_C axis. The $\bar{z}^2 A$ term is the moment of inertia that area A would have about the y axis if all of the area were to be concentrated at the centroid. Since this term is always zero or positive, the centroidal moment of inertia is the minimum moment of inertia with respect to all parallel axes.

By the same procedure that was used to obtain Eq. (C-11a), we get

$$\boxed{I_z = \bar{y}^2 A + I_{z_C}} \tag{C-11b}$$

Equations C-11 are called *parallel-axis theorems for moments of inertia*.

As a simple example of calculations based on the parallel-axis theorem, let us determine the moment of inertia of the rectangle in Fig. C-8 about the y' axis along an edge of length b. From Eq. (C-11a),

$$I_{y'} = (\bar{z}')^2 A + I_{y_C} = \left(\frac{h}{2}\right)^2 (bh) + \frac{bh^3}{12} = \frac{bh^3}{3} \tag{C-12}$$

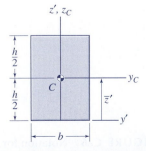

FIGURE C-8   A rectangular area with two sets of axes.

In a similar manner, a *parallel-axis theorem for the polar moment of inertia* may be derived. From Eq. (C-6) and Fig. C-7, the polar moment of inertia about point O is

$$I_{\rho O} = \int_A \rho^2 dA = \int_A (y^2 + z^2) dA$$

$$= \int_A [(\bar{y} + y_C)^2 + (\bar{z} + z_C)^2] dA$$

$$= \int_A (\bar{y}^2 + 2\bar{y}y_C + y_C^2 + \bar{z}^2 + 2\bar{z}z_C + z_C^2) dA$$

$$\boxed{I_{\rho O} = \bar{\rho}^2 A + I_{\rho_C}} \qquad (C\text{-}13)$$

since $\bar{y}^2 + \bar{z}^2 = \bar{\rho}^2$, $\int_A y_C dA = \int_A z_C dA = 0$, and $\int_A (y_C^2 + z_C^2) dA = I_{\rho_C}$. Note that Eq. (C-13) follows easily from Eq. (C-7) and Eqs. (C-11).

Moments of Inertia of Composite Areas.

The moments of inertia of a composite area, like the one in Fig. C-4, may be computed by summing the contributions of the individual areas:

$$\boxed{I_y = \sum_i (A_y)_i, \quad I_z = \sum_i (I_z)_i} \qquad (C\text{-}14)$$

As an efficient procedure for calculating moments of inertia of composite areas, the following is suggested.

COMPOSITE-AREA PROCEDURE FOR CALCULATING SECOND MOMENTS

1. Divide the composite area into simpler areas for which there exist formulas for centroidal coordinates and moments of inertia. (See the table inside the back cover.)
2. Locate the centroid of each constituent area and establish centroidal reference axes (y_C, z_C) parallel to the given (y, z) axes.

3. Employ Eqs. (C-11) to compute the moments of inertia of the constituent areas with respect to the (y, z) axes and Eq. (C-14) to sum them.

The next example problem illustrates this procedure.

EXAMPLE C-2

Determine the centroidal moment of inertia I_y for the L-shaped section in Example C-1. (Here, in Fig. 1, the origin of the (y, z) reference frame is at the centroid of the composite area. The centroidal reference axes for the rectangular "legs" of the L-shaped area are (y_1, z_1) and (y_2, z_2), respectively.)

Solution   We can combine Eqs. (C-14) with the parallel axis theorems, Eqs. (C-11), to compute the required moments of inertia.

$$I_y = (I_y)_1 + (I_y)_2 = [(I_{y_C})_1 + A_1 \bar{z}_1^2] + [(I_{y_C})_2 + A_2 \bar{z}_2^2] \qquad (1)$$

where $(I_{y_C})_i$ is the moment of inertia of area A_i about its own centroidal y axis, and $\bar{z}_i$ is the z-coordinate of the centroid C_i measured in the (y, z)

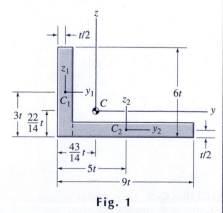

Fig. 1

reference frame with origin at the composite centroid, C. Referring to Fig. 1, we get

$$I_y = (I_y)_1 + (I_y)_2$$

$$= \left[\frac{1}{12}(t)(6t)^3 + (t)(6t)\left(3t - \frac{22}{14}t\right)^2 \right]$$

$$+ \left[\frac{1}{12}(8t)(t)^3 + (t)(8t)\left(\frac{1}{2}t - \frac{22}{14}t\right)^2 \right]$$

$$= 18t^4 + \frac{600}{49}t^4 + \frac{2}{3}t^4 + \frac{450}{49}t^4$$

$$= \frac{842}{21}t^4 = 40.1t^4 \qquad\qquad \textbf{Ans} \quad (2)$$

C.3 PRODUCT OF INERTIA OF AN AREA

Definition of Product of Inertia. Another geometric property of plane areas is called the *product of inertia,* which is defined by (refer to Fig. C-1)

$$I_{yz} = \int_A yz\,dA \qquad\qquad (C\text{-}15)$$

The product of inertia is required in the study of bending of unsymmetric beams (Section 6.6).

As an example, let us determine the product of inertia of a rectangular area with respect to two sets of axes (Fig. C-9).

From Eq. (C-15) and Fig. C-9(a),

$$I_{yz} = \int_A yz\,dA = \int_0^h \int_0^b yz\,dy\,dz = \frac{b^2 h^2}{4} \qquad\qquad (C\text{-}16)$$

Now consider the product of inertia with respect to the (y', z') axes in Fig. C-9b. The y' axis is an axis of symmetry, and it passes through the centroid C.

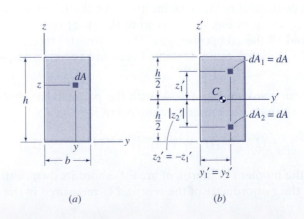

FIGURE C-9 A rectangular area with two sets of axes.

(a)   (b)

C-8

As is clear from Fig. C-9b, when either reference axis is an axis of symmetry of the area, like the y' axis in this figure, the product of inertia is zero, since

$$I_{y'z'} = \int_A y'z'dA$$

and, because of symmetry (since $y_2' = y_1'$, but $z_2' = -z_1'$), the contributions of dA_1 and dA_2 to the integral cancel each other. Therefore,

$$I_{yz} = 0 \tag{C-17}$$

if either they y axis or the z axis is an axis of symmetry of the area.

Parallel-Axis Theorem for Product of Inertia of an Area.

The procedure used to derive parallel-axis theorems for moments of inertia, leading to Eqs. (C-11) and (C-13), may be applied to derive a parallel-axis theorem for products of inertia. From Eq. (C-15) and Fig. C-7,

$$I_{yz} = \int_A yz \, dA = \int_A (\bar{y} + y_C)(\bar{z} + z_C)dA$$

$$= \bar{y}\bar{z}A + \bar{y}\int_A z_C dA + \bar{z}\int_A y_C dA + I_{y_C z_C}$$

Therefore, since y_C and z_C are coordinates in a centroidal reference frame, the *parallel-axis theorem for products of inertia of an area* is

$$\boxed{I_{yz} = \bar{y}\bar{z}A + I_{y_C z_C}} \tag{C-18}$$

Just as for the moments of inertia, I_{yz} has one term that represents the product of inertia of an area A concentrated at the centroid, plus a *centroidal product of inertia* $I_{y_C z_C}$.

Product of Inertia for Composite Areas.

The summations for moments of inertia in Eqs. (C-14) are readily extended to the product of inertia of an area composed of several constituent areas:

$$\boxed{I_{yz} = \sum_i (I_{yz})_i} \tag{C-19}$$

▶ EXAMPLE C-3 ◀

For the L-shaped area in Example C-2, use the composite-area procedure to determine the centroidal product of inertia, I_{yz}. (Note: Here the (y, z) reference frame is a centroidal reference frame for the whole area; (y_1, z_1) and (y_2, z_2) are centroidal reference frames for the constituent areas A_1 and A_2, respectively.) The centroidal product of inertia relative to the (y, z) axes is given by

$$I_{yz} = (I_{y_1 z_1} + A_1 \bar{y}_1 \bar{z}_1) + (I_{y_2 z_2} + A_2 \bar{y}_2 \bar{z}_2)$$

It is very important to note that $\bar{y}_1, \bar{z}_1$, etc., are signed values, that is, some of them could be negative. By Eq. (C-17), $I_{y_1z_1} = I_{y_2z_2} = 0$. Therefore,

$$I_{yz} = A_1\bar{y}_1\bar{z}_1 + A_2\bar{y}_2\bar{z}_2$$

$$= (6t^2)\left(-\frac{43}{14}t + \frac{1}{2}t\right)\left(3t - \frac{22}{14}t\right)$$

$$+ (8t^2)\left(5t - \frac{43}{14}t\right)\left(-\frac{22}{14}t + \frac{1}{2}t\right)$$

or

$$I_{yz} = -\frac{270}{7}t^4 \qquad\qquad \textbf{Ans.}$$

Note that, since the centroid C_1 lies in the second quadrant and C_2 lies in the fourth quadrant, both A_1 and A_2 make negative contributions to I_{yz}.

C.4 AREA MOMENTS OF INERTIA ABOUT INCLINED AXES; PRINCIPAL MOMENTS OF INERTIA

In some applications, especially in unsymmetric bending of beams (Section 6.6), it is necessary to determine the moments and products of inertia relative to inclined axes (y', z') when I_y, I_z, and I_{yz} are known. The *coordinate transformation* relating coordinates (y', z') to coordinates (y, z) can be deduced from Fig. C-10.

The angle θ is measured **positive counterclockwise from y to y'** (and z to z').

$$y' = y \cos \theta + z \sin \theta$$
$$z' = -y \sin \theta + z \cos \theta$$

(C-20)

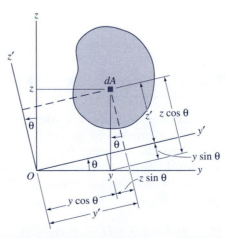

FIGURE C-10   Transformation of coordinates in a plane.

From Eqs. (C-5), (C-15), and (C-20),

$$I_{y'} = \int_A (z')^2 dA = \int_A (-y \sin \theta + z \cos \theta)^2 dA$$

$$\text{(C-21)}$$

$$I_{y'z'} = \int_A y'z' \, da = \int_c (y \cos \theta + z \sin \theta)(-y \sin \theta + z \cos \theta) dA$$

Expanding each of the above integrands and recognizing that $\int_A y^2 \, dA = I_z$, and so forth, we get

$$I_{y'} = I_y \cos^2 \theta + I_z \sin^2 \theta - 2I_{yz} \sin \theta \cos \theta$$

$$I_{y'z'} = (I_y - I_z) \sin \theta \cos \theta + I_{yz}(\cos^2 \theta - \sin^2 \theta)$$

These equations may be simplified by using the trigonometric identities $\sin 2\theta = 2 \sin \theta \cos \theta$ and $\cos 2\theta = \cos^2 \theta - \sin^2 \theta$. Thus,

$$I_{y'} = \frac{I_y + I_z}{2} + \frac{I_y - I_z}{2} \cos 2\theta - I_{yz} \sin 2\theta$$

$$I_{y'z'} = \frac{I_y - I_z}{2} \sin 2\theta + I_{yz} \cos 2\theta$$

$$\text{(C-22)}$$

Note the similarity between these equations and the stress-transformation equations, Eqs. 8-5.[6]

Principal Moments of Inertia. From Eqs. (C-22) it may be seen that $I_{y'}$ and $I_{y'z'}$ depend on the angle θ. We will now determine the orientations of the y' axis for which $I_{y'}$ takes on its maximum and minimum values. The axes having these orientations are called the *principal axes of inertia* of the area, and the corresponding moments of inertia are called the *principal moments of inertia*. To each point O in an area, there is a specific set of principal axes passing through that point. The principal axes that pass through the centroid of the area, called the *centroidal principal axes*, are the most important. The orientations of the centroidal principal axes for several unequal-leg angles are given in Appendix D.6.

The moment of inertia $I_{y'}$ will have a maximum, or minimum, value if the y' axis is oriented at an angle $\theta = \theta_p$ that satisfies the equation

$$\frac{dI_{y'}}{d\theta} = -2\left(\frac{I_y - I_z}{2}\right) \sin 2\theta - 2I_{yz} \cos 2\theta = 0$$

Therefore,

$$\tan 2\theta_p = \frac{-I_{yz}}{\left(\dfrac{I_y - I_z}{2}\right)}$$

$$\text{(C-23)}$$

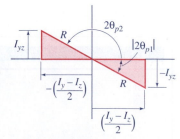

Figure C-11 illustrates how to use the tangent value given by Eq. (C-23) to determine the angles θ_p. There are two distinct angles that satisfy Eq. (C-23). As

FIGURE C-11 Orientation of the principal axes of inertia.

---

[6] There is a sign difference between the τ_{nt}-type terms and the I_{yz}-type terms, however.

illustrated by Fig. C-11, these two values of $2\theta_p$, labeled $2\theta_{p_1}$ and $2\theta_{p_2}$, differ by 180°, so the principal axes are oriented at 90° to each other (as they must be).

From Fig. C-11, the hypotenuse of either of the shaded triangles is given by

$$R = \sqrt{\left(\frac{I_y - I_z}{2}\right)^2 + I_{yz}^2} \tag{C-24}$$

Also, from Fig. C-11, the angles $2\theta_{p_1}$ and $2\theta_{p_2}$ satisfy

$$\sin 2\theta_{p_1} = \frac{-I_{yz}}{R}, \qquad \cos 2\theta_{p_1} = \frac{\left(\dfrac{I_y - I_z}{2}\right)}{R} \tag{C-25a}$$

$$\sin 2\theta_{p_2} = \frac{I_{yz}}{R}, \qquad \cos 2\theta_{p_2} = \frac{-\left(\dfrac{I_y - I_{yz}}{2}\right)}{R} \tag{C-25b}$$

Substituting these sines and cosines into the equation for $I_{y'}$, Eq. (C-22a), we get the following expressions for the two principal moments of inertia:

$$\boxed{\begin{aligned} I_{max} \equiv I_{p_1} &= \frac{I_y + I_z}{2} + \sqrt{\left(\frac{I_y - I_z}{2}\right)^2 + I_{yz}^2} \\ I_{min} \equiv I_{p_2} &= \frac{I_y + I_z}{2} - \sqrt{\left(\frac{I_y - I_z}{2}\right)^2 + I_{yz}^2} \end{aligned}} \tag{C-26}$$

If Eqs. (C-25a) or Eqs. (C-25b) are substituted into Eq. (C-22b), it is found that

$$I_{p_1 p_2} = 0 \tag{C-27}$$

That is, the product of inertia with respect to the principal axes of inertia is equal to zero.

By adding Eqs. (C-26a) and (C-26b) we get

$$I_{p_1} + I_{p_2} = I_y + I_z \tag{C-28}$$

Thus, the sum of the moments of inertia about any pair of mutually perpendicular axes passing through a given point in a given plane is constant.

EXAMPLE C-4

For the L-shaped area in Fig. 1 of Example C-2, (a) determine the orientation of the centroidal principal axes and show the orientation on a sketch. (b) Determine the principal moments of inertia.

$$I_y = \frac{5894}{147} t^4 = 40.10t^4, \qquad I_z = \frac{33,103}{294} t^4 = 112.60t^4$$

$$I_{yz} = \frac{-270}{7} t^4 = -38.57t^4$$

Solution   (a) From Eq. (C-23),

$$\tan 2\theta_p = \frac{-I_{yz}}{\left(\dfrac{I_y - I_z}{2}\right)} = \frac{-\left(\dfrac{-270}{7}\right)}{\dfrac{11,788 - 33,103}{2(294)}} = -1.064$$

$$2\theta_{p_1} = 133.22°, \quad 2\theta_{p_2} = -46.78°$$

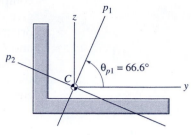

Fig. 1

Then, as illustrated in Fig. 1,

$$\theta_{p_1} = 66.6°, \quad \theta_{p_2} = -23.4° \qquad \text{**Ans.**}$$

(b) From Eq. (C-26a),

$$I_{p_1} = \frac{I_y + I_z}{2} + \sqrt{\left(\frac{I_y - I_z}{2}\right)^2 + I_{yz}^2}$$

$$= \frac{40.10t^4 + 112.60t^4}{2} + \sqrt{\left(\frac{40.10t^4 - 112.60t^4}{2}\right)^2 + (-38.57t^4)^2}$$

$$= 129.28t^4$$

or

$$I_{p_1} = 129.3t^4 \qquad \text{**Ans.**}$$

Similarly, from Eq. (C-26b),

$$I_{p_2} = 23.4t^4 \qquad \text{**Ans.**}$$

Mohr's Circle for Moments and Products of Inertia. Equations (C-22) have the same basic form as Eqs. 8.5, which were used to develop Mohr's circle for stress.[7] Therefore, by a procedure that is virtually identical to that in Section 8.5, it can be shown that a Mohr's circle plotted as in Fig. C-12 can be used to

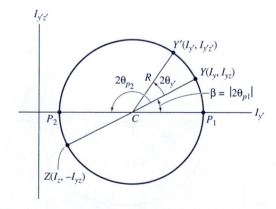

FIGURE C-12   Mohr's circle for moments and products of inertia.

---

[7]There is a difference between the signs preceding the τ_{xy}-type terms in Eqs. 8.5 and the signs preceding the corresponding I_{yz}-type terms in Eqs. (C-22). Thus, for Mohr's circle for moments and products of inertia, the $I_{y'z'}$ axis is *positive upward*, not positive downward, as it was for Mohr's circle for stress.

compute $I_{y'}$ and $I_{y'z'}$ for any (y', z') axes located at angle θ counterclockwise from the given (y, z) axes. And the Mohr's circle provides a convenient way to calculate the orientation of the principal axes of inertia and the principal moments of inertia, I_{p_1} and I_{p_2}, given moments of inertia I_y and I_z and the corresponding product of inertia I_{yz}. To an angle θ measured counterclockwise (or clockwise) on the planar area A, there corresponds an angle 2θ measured counterclockwise (or clockwise) on Mohr's circle.

The following procedure will facilitate your calculation of moments and products of inertia with respect to rotated axes.

MOHR'S-CIRCLE PROCEDURE FOR MOMENTS AND PRODUCTS OF INERTIA

1. Establish a set of Mohr's-circle axes $(I_{y'}, I_{y'z'})$, as shown in Fig. C-12. (Note that the positive $I_{y'z'}$ axis is counterclockwise 90° from the $I_{y'}$ axis, unlike the τ_{nt} axis for Mohr's circle of stress in Chapter 8.)
2. Plot points $Y:(I_y, +I_{yz})$ and $Z:(I_z, -I_{yz})$, respectively.
3. Draw a straight line joining points Y and Z. The intersection of the YZ line with $I_{y'}$ axis is the center of the Mohr's circle passing through points Y and Z.

4. Point Y', located at angle $2\theta_{y'}$ counterclockwise from the line CY, as shown in Fig. C-12, locates the point whose coordinates are $(I_{y'}, +I_{y'z'})$.
5. Points P_1 and P_2 locate the two principal axes at $2\theta_{p_1}$ and $2\theta_{p_2}$, respectively, as shown in Fig, C-12. The principal moments of inertia are I_{p_1} and I_{p_2}, which are also given by Eqs. (C-26).

EXAMPLE C-5

(a) Draw the Mohr's circle for the centroidal moments and products of inertia for the L-shaped area in Fig. 1 of Example C-2, given that:

$$I_y = \frac{5894}{147} t^4 = 40.10t^4, \quad I_z = \frac{33{,}103}{294} t^4 = 112.60t^4$$

$$I_{yz} = \frac{-270}{7} t^4 = -38.57t^4$$

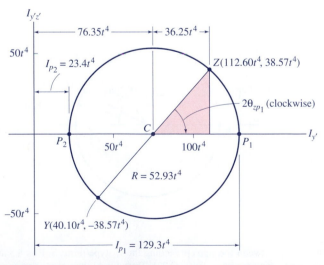

Fig. 1   Mohr's circle for centroidal inertias of an L-shaped area.

(b) Use the Mohr's circle constructed in Part (a) to compute the principal moments of inertia I_{p_1} and I_{p_2} and to locate the principal axes. Show the orientation of the principal axes on a sketch.

Solution (a) Sketch Mohr's circle and calculate the principal moments of inertia. Points Y and Z are plotted and Mohr's circle is then drawn (Fig. 1). From the circle,

$$I_{avg.} = \frac{40.10t^4 + 112.60t^4}{2} = 76.35t^4$$

$$R = \sqrt{\left(\frac{112.60t^4 - 40.10t^4}{2}\right)^2 + (38.57t^4)^2} = 52.93t^4$$

$$I_{p_1} = I_{avg.} + R = 129.3t^4$$

$$I_{p_2} = I_{avg.} - R = 23.4t^4 \qquad\qquad \textbf{Ans. (a)}$$

(b) Determine the orientation of the principal axes and show them on a sketch.

$$\tan|2\theta_{zp_1}| = \frac{38.57t^4}{\left(\dfrac{112.60t^4 - 40.10t^4}{2}\right)} = 1.064$$

Therefore, $2\theta_{zp_1} = 46.78°$ (clockwise), so

$$\theta_{zp_1} = \theta_{yp_2} = 23.4° \text{ clockwise} \qquad\qquad \textbf{Ans. (b)}$$

Note that the orientations of the principal axes in Fig. 2 are such that the contributions to $I_{p_1p_2}$ of the areas in the four quadrants cancel out, giving $I_{p_1p_2} = 0$.

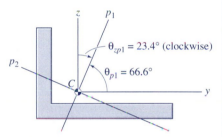

Fig. 2   Principal axes of inertia.

The results obtained from Mohr's circle are the same as those obtained by the use of formulas in Example C-4. However, mistakes are less likely to be made if Mohr's circle is carefully drawn and it is recalled that an angle 2θ on Mohr's circle corresponds to an angle θ on the planar area A, and that angles are taken in the same sense on Mohr's circles as on the planar area.

MDS 6.1   **Beam Cross-Sectional Properties**—*Section Properties* is an MDS computer program module for calculating section properties of plane areas: area, location of centroid, moments of inertia, product of inertia, orientation of principal axes, etc., properties that are defined and illustrated in **Appendix C.** The *Section Properties* module is closely linked with the *Flexure* module.

SECTION PROPERTIES OF SELECTED STRUCTURAL SHAPES

D

Tables D.1 through D.10 give the cross-sectional properties of structural shapes made of aluminum, steel and wood. The tables for steel shapes and for aluminum shapes were compiled from more extensive tables given in the following references; they are used with permission:

- Table D.1 and Tables D.3–D.7: *Manual of Steel Construction — Load and Resistance Factor Design,* American Institute of Steel Construction, Inc., 400 N. Michigan Ave., Chicago, IL, 60611-4185.
- Table D.2: *Metric Properties of Structural Shapes,* American Institute of Steel Construction, Inc., 400 N. Michigan Ave., Chicago, IL, 60611-4185.
- Tables D.9 and D.10: *The Aluminum Design Manual,* The Aluminum Association Inc., 900 19th Street, NW, Washington, DC, 20006.

The cross-sectional axes in the following tables (e.g., $X - X$) are those of the original sources of the tabular data.

Nomenclature

I = moment of inertia

J = polar moment of inertia (pipe sections)

S = elastic section modulus

Z = plastic section modulus

$r = \sqrt{I/A}$ = radius of gyration

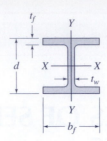

D.1. Properties of Steel Wide-Flange (W) Shapes (U.S. Customary Units)

Designation*	Area A in^2	Depth d in.	Flange Width b_f in.	Flange Thickness t_f in.	Web Thickness t_w in.	Axis $X-X$ I_x in^4	Axis $X-X$ S_x in^3	Axis $X-X$ r_x in.	Axis $Y-Y$ I_y in^4	Axis $Y-Y$ S_y in^3	Axis $Y-Y$ r_y in.	Plastic Modulus Z_x in^3
W36×230	67.6	35.90	16.470	1.260	0.760	15000	837	14.9	940	114	3.73	943
×150	44.2	35.85	11.975	0.940	0.625	9040	504	14.3	270	45.1	2.47	581
W33×201	59.1	33.68	15.745	1.150	0.715	11500	684	14.0	749	95.2	3.56	772
×130	38.3	33.09	11.510	0.855	0.580	6710	406	13.2	218	37.9	2.39	467
W30×173	50.8	30.44	14.985	1.065	0.655	8200	539	12.7	598	79.8	3.43	605
× 90	26.4	29.53	10.400	0.610	0.470	3620	245	11.7	115	22.1	2.09	283
W27×146	42.9	27.38	13.965	0.975	0.605	5630	411	11.4	443	63.5	3.21	461
× 84	24.8	26.71	9.960	0.640	0.460	2850	213	10.7	106	21.2	2.07	244
W24× 94	27.7	24.31	9.065	0.875	0.515	2700	222	9.87	109	24.0	1.98	254
× 62	18.2	23.74	7.040	0.590	0.430	1550	131	9.23	34.5	9.80	1.38	153
W21×101	29.8	21.36	12.290	0.800	0.500	2420	227	9.02	248	40.3	2.89	253
× 73	21.5	21.24	8.295	0.740	0.455	1600	151	8.64	70.6	17.0	1.81	172
× 50	14.7	20.83	6.530	0.535	0.380	984	94.5	8.18	24.9	7.64	1.30	110
W18×130	38.2	19.25	11.160	1.200	0.670	2460	256	8.03	278	49.9	2.70	291
× 76	22.3	18.21	11.035	0.680	0.425	1330	146	7.73	152	27.6	2.61	163
W18× 60	17.6	18.24	7.555	0.695	0.415	984	108	7.47	50.1	13.3	1.69	123
× 50	14.7	17.99	7.495	0.570	0.355	800	88.9	7.38	40.1	10.7	1.65	101
W16×100	29.4	16.97	10.425	0.985	0.585	1490	175	7.10	186	35.7	2.51	198
× 67	19.7	16.33	10.235	0.665	0.395	954	117	6.96	119	23.2	2.46	130
× 50	14.7	16.26	7.070	0.630	0.380	659	81.0	6.68	37.2	10.5	1.59	92.0
× 40	11.8	16.01	6.995	0.505	0.305	518	64.7	6.63	28.9	8.25	1.57	72.9
W14×176	51.8	15.22	15.650	1.310	0.830	2140	281	6.43	838	107	4.02	320
×120	35.3	14.48	14.670	0.940	0.590	1380	190	6.24	495	65.7	3.74	212
× 82	24.1	14.31	10.130	0.855	0.510	882	123	6.05	148	29.3	2.48	139
× 53	15.6	13.92	8.060	0.660	0.370	541	77.8	5.89	57.7	14.3	1.92	87.1
× 26	7.69	13.91	5.025	0.420	0.255	245	35.3	5.65	8.91	3.54	1.08	40.2
W12×152	44.7	13.71	12.480	1.400	0.870	1430	209	5.66	454	72.8	3.19	243
× 96	28.2	12.71	12.160	0.900	0.550	833	131	5.44	270	44.4	3.09	147
× 65	19.1	12.12	12.000	0.605	0.390	533	87.9	5.28	174	29.1	3.02	96.8
× 50	14.7	12.19	8.080	0.640	0.370	394	64.7	5.18	56.3	13.9	1.96	72.4
× 35	10.3	12.50	6.560	0.520	0.300	285	45.6	5.25	24.5	7.47	1.54	51.2
× 22	6.48	12.31	4.030	0.425	0.260	156	25.4	4.91	4.66	2.31	0.847	29.3
W10× 60	17.6	10.22	10.080	0.680	0.420	341	66.7	4.39	116	23.0	2.57	74.6
× 45	13.3	10.10	8.020	0.620	0.350	248	49.1	4.32	53.4	13.3	2.01	54.9
× 30	8.84	10.47	5.810	0.510	0.300	170	32.4	4.38	16.7	5.75	1.37	36.6
× 12	3.54	9.87	3.960	0.210	0.190	53.8	10.9	3.90	2.18	1.10	0.785	12.6
W 8× 48	14.1	8.50	8.110	0.685	0.400	184	43.3	3.61	60.9	15.0	2.08	49.0
× 40	11.7	8.25	8.070	0.560	0.360	146	35.5	3.53	49.1	12.2	2.04	39.8
× 35	10.3	8.12	8.020	0.495	0.310	127	31.2	3.51	42.6	10.6	2.03	34.7
× 21	6.16	8.28	5.270	0.400	0.250	75.3	18.2	3.49	9.77	3.71	1.26	20.4
× 15	4.44	8.11	4.015	0.315	0.245	48.0	11.8	3.29	3.41	1.70	0.876	13.6
W 6× 25	7.34	6.38	6.080	0.455	0.320	53.4	16.7	2.70	17.1	5.61	1.52	18.9
× 20	5.87	6.20	6.020	0.365	0.260	41.1	13.4	2.66	13.3	4.41	1.50	14.9
× 16	4.74	6.28	4.030	0.405	0.260	32.1	10.2	2.60	4.43	2.20	0.966	11.7
× 15	4.43	5.99	5.990	0.260	0.230	29.1	9.72	2.56	9.32	3.11	1.46	10.8

*W(nominal depth in inches) × (weight in pounds per foot)

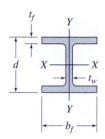

D.2. Properties of Steel Wide-Flange (W) Shapes (SI Units)

Designation*	Area A mm^2	Depth d mm	Flange Width b_f mm	Flange Thickness t_f mm	Web Thickness t_w mm	Axis $X-X$ I_x 10^6mm^4	S_x 10^3mm^3	r_x mm	Axis $Y-Y$ I_y 10^6mm^4	S_y 10^3mm^3	r_y mm	Plastic Modulus Z_x 10^3mm^3
W920×342	43 600	912	418.0	32.0	19.30	6250	13 700	379	390	1870	94.6	15 400
×223	28 500	911	304.0	23.9	15.90	3770	8280	364	112	737	62.7	9540
W840×299	38 100	855	400.0	29.2	18.20	4790	11 200	355	312	1560	90.5	12 600
×193	24 700	840	292.0	21.7	14.70	2780	6620	335	90.3	618	60.5	7620
W760×257	32 800	773	381.0	27.1	16.60	3420	8850	323	250	1310	87.3	9930
×134	17 000	750	264.0	15.5	11.90	1500	4000	297	47.7	361	53.0	4630
W690×217	27 700	695	355.0	24.8	15.40	2340	6730	291	185	1040	81.7	7570
×125	16 000	678	253.0	16.3	11.70	1190	3510	273	44.1	349	52.5	4010
W610×140	17 900	617	230.0	22.2	13.10	1120	3630	250	45.1	392	50.2	4150
× 92	11 800	603	179.0	15.0	10.90	646	2140	234	14.4	161	34.9	2510
W530×150	19 200	543	312.0	20.3	12.70	1010	3720	229	103	660	73.2	4150
×109	13 900	539	211.0	18.8	11.60	667	2470	219	29.5	280	46.1	2830
× 74	9490	529	166.0	13.6	9.65	410	1550	208	10.4	125	33.1	1810
W460×193	24 700	489	283.0	30.5	17.00	1020	4170	203	115	813	68.2	4760
×113	14 400	463	280.0	17.3	10.80	556	2400	196	63.3	452	66.3	2670
× 74	9460	457	190.0	14.5	9.02	333	1460	188	16.6	175	41.9	1650
W410×149	19 000	431	265.0	25.0	14.90	619	2870	180	77.7	586	63.9	3250
×100	12 700	415	260.0	16.9	10.00	398	1920	177	49.5	381	62.4	2130
× 74	9510	413	180.0	16.0	9.65	275	1330	170	15.6	173	40.5	1510
× 60	7600	407	178.0	12.8	7.75	216	1060	169	12.0	135	39.7	1200
W360×262	33 400	387	398.0	33.3	21.10	894	4620	164	350	1760	102	5260
×179	22 800	368	373.0	23.9	15.00	575	3130	159	207	1110	95.3	3480
×122	15 500	363	257.0	21.7	13.00	365	2010	153	61.5	479	63.0	2270
× 79	10 100	354	205.0	16.8	9.40	227	1280	150	24.2	236	48.9	1430
× 39	4960	353	128.0	10.7	6.48	102	578	143	3.75	58.6	27.5	661
W310×226	28 900	348	317.0	35.6	22.10	596	3430	144	189	1190	80.9	3980
×143	18 200	323	309.0	22.9	14.00	348	2150	138	113	731	78.8	2420
× 97	12 300	308	305.0	15.4	9.91	222	1440	134	72.9	478	77.0	1590
× 74	9480	310	205.0	16.3	9.40	165	1060	132	23.4	228	49.7	1190
× 52	6670	318	167.0	13.2	7.62	119	748	134	10.3	123	39.3	841
× 33	4180	313	102.0	10.8	6.60	65.0	415	125	1.92	37.6	21.4	480
W250× 89	11 400	260	256.0	17.3	10.70	143	1100	112	48.4	378	65.2	1230
× 67	8560	257	204.0	15.7	8.89	104	809	110	22.2	218	50.9	901
× 45	5700	266	148.0	13.0	7.62	71.1	535	112	7.03	95	35.1	602
× 18	2280	251	101.0	5.3	4.83	22.5	179	99.3	0.919	18.2	20.1	208
W200× 71	9100	216	206.0	17.4	10.20	76.6	709	91.7	25.4	247	52.8	803
× 59	7580	210	205.0	14.2	9.14	61.2	583	89.9	20.4	199	51.9	653
× 52	6640	206	204.0	12.6	7.87	52.7	512	89.1	17.8	175	51.8	569
× 31	3980	210	134.0	10.2	6.35	31.3	298	88.7	4.10	61.2	32.1	335
× 22	2860	206	102.0	8.0	6.22	20.0	194	83.6	1.42	27.8	22.3	222
W150× 37	4730	162	154.0	11.6	8.13	22.2	274	68.5	7.07	91.8	38.7	310
× 24	3060	160	102.0	10.3	6.60	13.4	168	66.2	1.83	35.9	24.5	192
× 22	2860	152	152.0	6.6	5.84	12.1	159	65.0	3.87	50.9	36.8	176

*W(nominal depth in mm) × (mass in kg/m)

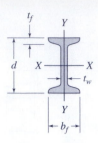

D.3. Properties of American Standard (S) Beams (U.S. Customary Units)

Designation*	Area A in²	Depth d in.	Flange Width b_f in.	Flange Thickness t_f in.	Web Thickness t_w in.	Axis $X-X$ I_x in⁴	Axis $X-X$ S_x in³	Axis $X-X$ r_x in.	Axis $Y-Y$ I_y in⁴	Axis $Y-Y$ S_y in³	Axis $Y-Y$ r_y in.	Plastic Modulus Z_x in³
S24 × 100	29.3	24.00	7.245	0.870	0.745	2390	199	9.02	47.7	13.2	1.27	240
× 90	26.5	24.00	7.125	0.870	0.625	2250	187	9.21	44.9	12.6	1.30	222
× 80	23.5	24.00	7.000	0.870	0.500	2100	175	9.47	44.2	12.1	1.34	204
S20 × 96	28.2	20.30	7.200	0.920	0.800	1670	165	7.71	50.2	13.9	1.33	198
× 75	22.0	20.00	6.385	0.795	0.635	1280	128	7.62	29.8	9.32	1.16	153
S18 × 70	20.6	18.00	6.251	0.691	0.711	926	103	6.71	24.1	7.72	1.08	125
× 54.7	16.1	18.00	6.001	0.691	0.461	804	89.4	7.07	20.8	6.94	1.14	105
S15 × 50	14.7	15.00	5.640	0.622	0.550	486	64.8	5.75	15.7	5.57	1.03	77.1
× 42.9	12.6	15.00	5.501	0.622	0.411	447	59.6	5.95	14.4	5.23	1.07	69.3
S12 × 50	14.7	12.00	5.477	0.659	0.687	305	50.8	4.55	15.7	5.74	1.03	61.2
× 35	10.3	12.00	5.078	0.544	0.428	229	38.2	4.72	9.87	3.89	0.980	44.8
S10 × 35	10.3	10.00	4.944	0.491	0.594	147	29.4	3.78	8.36	3.38	0.901	35.4
× 25.4	7.46	10.00	4.661	0.491	0.311	124	24.7	4.07	6.79	2.91	0.954	28.4
S 8 × 23	6.77	8.00	4.171	0.426	0.441	64.9	16.2	3.10	4.31	2.07	0.798	19.3
× 18.4	5.41	8.00	4.001	0.426	0.271	57.6	14.4	3.26	3.73	1.86	0.831	16.5
S 6 × 17.25	5.07	6.00	3.565	0.359	0.465	26.3	8.77	2.28	2.31	1.30	0.675	10.6
× 12.5	3.67	6.00	3.332	0.359	0.232	22.1	7.37	2.45	1.82	1.09	0.705	8.47
S 4 × 9.5	2.79	4.00	2.796	0.293	0.326	6.79	3.39	1.56	0.903	0.646	0.569	4.04
× 7.7	2.26	4.00	2.663	0.293	0.193	6.08	3.04	1.64	0.764	0.574	0.581	3.51

*S(nominal depth in inches) × (weight in pounds per foot)

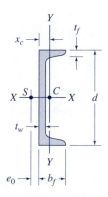

D.4. Properties of American Standard (C) Channels (U.S. Customary Units)

Designation*	Area A in²	Depth d in.	Flange Width b_f in.	Flange Thickness t_f in.	Web Thickness t_w in.	Centroid Loc'n. x_c in.	Shear Center Loc'n. e_o in.	Axis X–X I_x in⁴	Axis X–X S_x in³	Axis X–X r_x in.	Axis Y–Y I_y in⁴	Axis Y–Y S_y in³	Axis Y–Y r_y in.
C15×50	14.7	15.00	3.716	0.650	0.716	0.798	0.583	404	53.8	5.24	11.0	3.78	0.867
×40	11.8	15.00	3.520	0.650	0.520	0.777	0.767	349	46.5	5.44	9.23	3.37	0.886
×33.9	9.96	15.00	3.400	0.650	0.400	0.787	0.896	315	42.0	5.62	8.13	3.11	0.904
C12×30	8.82	12.00	3.170	0.501	0.510	0.674	0.618	162	27.0	4.29	5.14	2.06	0.763
×25	7.35	12.00	3.047	0.501	0.387	0.674	0.746	144	24.1	4.43	4.47	1.88	0.780
×20.7	6.09	12.00	2.942	0.501	0.282	0.698	0.870	129	21.5	4.61	3.88	1.73	0.799
C10×30	8.82	10.00	3.033	0.436	0.673	0.649	0.369	103	20.7	3.42	3.94	1.65	0.669
×25	7.35	10.00	2.886	0.436	0.526	0.617	0.494	91.2	18.2	3.52	3.36	1.48	0.676
×20	5.88	10.00	2.739	0.436	0.379	0.606	0.637	78.9	15.8	3.66	2.81	1.32	0.692
×15.3	4.49	10.00	2.600	0.436	0.240	0.634	0.796	67.4	13.5	3.87	2.28	1.16	0.713
C 9×20	5.88	9.00	2.648	0.413	0.448	0.583	0.515	60.9	13.5	3.22	2.42	1.17	0.642
×15	4.41	9.00	2.485	0.413	0.285	0.586	0.682	51.0	11.3	3.40	1.93	1.01	0.661
×13.4	3.94	9.00	2.433	0.413	0.233	0.601	0.743	47.9	10.6	3.48	1.76	0.962	0.669
C 8×18.75	5.51	8.00	2.527	0.390	0.487	0.565	0.431	44.0	11.0	2.82	1.98	1.01	0.599
×13.75	4.04	8.00	2.343	0.390	0.303	0.553	0.604	36.1	9.03	2.99	1.53	0.854	0.615
×11.5	3.38	8.00	2.260	0.390	0.220	0.571	0.697	32.6	8.14	3.11	1.32	0.781	0.625
C 7×14.75	4.33	7.00	2.299	0.366	0.419	0.532	0.441	27.2	7.78	2.51	1.38	0.779	0.564
×12.25	3.60	7.00	2.194	0.366	0.314	0.525	0.538	24.2	6.93	2.60	1.17	0.703	0.571
× 9.8	2.87	7.00	2.090	0.366	0.210	0.540	0.647	21.3	6.08	2.72	0.968	0.625	0.581
C 6×13	3.83	6.00	2.157	0.343	0.437	0.514	0.380	17.4	5.80	2.13	1.05	0.642	0.525
×10.5	3.09	6.00	2.034	0.343	0.314	0.499	0.486	15.2	5.06	2.22	0.866	0.564	0.529
× 8.2	2.40	6.00	1.920	0.343	0.200	0.511	0.599	13.1	4.38	2.34	0.693	0.492	0.537
C 5× 9	2.64	5.00	1.885	0.320	0.325	0.478	0.427	8.96	3.56	1.83	0.632	0.450	0.489
× 6.7	1.97	5.00	1.750	0.320	0.190	0.484	0.552	7.49	3.00	1.95	0.479	0.378	0.493
C 4× 7.25	2.13	4.00	1.721	0.296	0.321	0.459	0.386	4.59	2.29	1.47	0.433	0.343	0.450
× 5.4	1.59	4.00	1.584	0.296	0.184	0.457	0.502	3.85	1.93	1.56	0.319	0.283	0.449
C 4× 6	1.76	3.00	1.596	0.273	0.356	0.455	0.322	2.07	1.38	1.08	0.305	0.268	0.416
× 5	1.47	3.00	1.498	0.273	0.258	0.438	0.392	1.85	1.24	1.12	0.247	0.233	0.410
× 4.1	1.21	3.00	1.410	0.273	0.170	0.436	0.461	1.66	1.10	1.17	0.197	0.202	0.404

*C(nominal depth in inches) × (weight in pounds per foot)

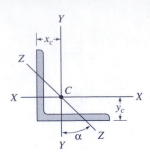

D.5. Properties of Steel Angle Sections–Equal Legs (U.S. Customary Units)

Size and Thickness in.	Weight per ft lb	Area A in²	Axis $X-X$				Axis $Y-Y$				Axis $Z-Z$	
			I_x in⁴	S_x in³	r_x in.	y_c in.	I_y in⁴	S_y in³	r_y in.	x_c in.	r_z in.	tan α
L8×8×1	51.0	15.0	89.0	15.8	2.44	2.37	89.0	15.8	2.44	2.37	1.56	1.000
$\frac{3}{4}$	38.9	11.4	69.7	12.2	2.47	2.28	69.7	12.2	2.47	2.28	1.58	1.000
$\frac{1}{2}$	26.4	7.75	48.6	8.36	2.50	2.19	48.6	8.36	2.50	2.19	1.59	1.000
L6×6×1	37.4	11.0	35.5	8.57	1.80	1.86	35.5	8.57	1.80	1.86	1.17	1.000
$\frac{3}{4}$	28.7	8.44	28.2	6.66	1.83	1.78	28.2	6.66	1.83	1.78	1.17	1.000
$\frac{1}{2}$	19.6	5.75	19.9	4.61	1.86	1.68	19.9	4.61	1.86	1.68	1.18	1.000
$\frac{3}{8}$	14.9	4.36	15.4	3.53	1.88	1.64	15.4	3.53	1.88	1.64	1.19	1.000
L5×5×$\frac{7}{8}$	27.2	7.98	17.8	5.17	1.49	1.57	17.8	5.17	1.49	1.57	0.973	1.000
$\frac{3}{4}$	23.6	6.94	15.7	4.53	1.51	1.52	15.7	4.53	1.51	1.52	0.975	1.000
$\frac{1}{2}$	16.2	4.75	11.3	3.16	1.54	1.43	11.3	3.16	1.54	1.43	0.983	1.000
$\frac{3}{8}$	12.3	3.61	8.74	2.42	1.56	1.39	8.74	2.42	1.56	1.39	0.990	1.000
L4×4×$\frac{3}{4}$	18.5	5.44	7.67	2.81	1.19	1.27	7.67	2.81	1.19	1.27	0.778	1.000
$\frac{1}{2}$	12.8	3.75	5.56	1.97	1.22	1.18	5.56	1.97	1.22	1.18	0.782	1.000
$\frac{3}{8}$	9.8	2.86	4.36	1.52	1.23	1.14	4.36	1.52	1.23	1.14	0.788	1.000
$\frac{1}{4}$	6.6	1.94	3.04	1.05	1.25	1.09	3.04	1.05	1.25	1.09	0.795	1.000
L3$\frac{1}{2}$×3$\frac{1}{2}$×$\frac{1}{2}$	11.1	3.25	3.64	1.49	1.06	1.06	3.64	1.49	1.06	1.06	0.683	1.000
$\frac{3}{8}$	8.5	2.48	2.87	1.15	1.07	1.01	2.87	1.15	1.07	1.01	0.687	1.000
$\frac{1}{4}$	5.8	1.69	2.01	0.794	1.09	0.968	2.01	0.794	1.09	0.968	0.694	1.000
L3×3×$\frac{1}{2}$	9.4	2.75	2.22	1.07	0.898	0.932	2.22	1.07	0.898	0.932	0.584	1.000
$\frac{3}{8}$	7.2	2.11	1.76	0.833	0.913	0.888	1.76	0.833	0.913	0.888	0.587	1.000
$\frac{1}{4}$	4.9	1.44	1.24	0.577	0.930	0.842	1.24	0.577	0.930	0.842	0.592	1.000
L2$\frac{1}{2}$×2$\frac{1}{2}$×$\frac{1}{2}$	7.7	2.25	1.23	0.724	0.739	0.806	1.23	0.724	0.739	0.806	0.487	1.000
$\frac{3}{8}$	5.9	1.73	0.984	0.566	0.753	0.762	0.984	0.566	0.753	0.762	0.487	1.000
$\frac{1}{4}$	4.1	1.19	0.703	0.394	0.769	0.717	0.703	0.394	0.769	0.717	0.491	1.000
L2×2×$\frac{3}{8}$	4.7	1.36	0.479	0.351	0.594	0.636	0.479	0.351	0.594	0.636	0.389	1.000
$\frac{1}{4}$	3.19	0.938	0.348	0.247	0.609	0.592	0.348	0.247	0.609	0.592	0.391	1.000
$\frac{1}{8}$	1.65	0.484	0.190	0.131	0.626	0.546	0.190	0.131	0.626	0.546	0.398	1.000

1. The $Z - Z$ axis is the axis of minimum moment of inertia, that is, $I_2 \equiv I_z = Ar_z^2$.
2. The **product of inertia,** I_{xy}, for these angle cross sections may be calculated to within the accuracy permitted by the tabular data by use of the formula

$$I_{xy} = Ar_z^2 - I_x$$

The product of inertia is negative for all table entries because of the orientation of the angle cross section relative to the xy axes.

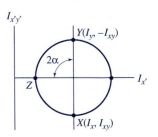

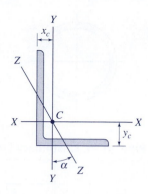

D.6. Properties of Steel Angle Sections–Unequal Legs (U.S. Customary Units)

Size and Thickness in.	Weight per ft lb	Area A in^2	Axis X – X				Axis Y – Y				Axis Z – Z	
			I_x in^4	S_x in^3	r_x in.	y_c in.	I_y in^4	S_y in^3	r_y in.	x_c in.	r_z in.	tan α
L8×6×1	44.2	13.0	80.8	15.1	2.49	2.65	38.8	8.92	1.73	1.65	1.28	0.543
$\frac{3}{4}$	33.8	9.94	63.4	11.7	2.53	2.56	30.7	6.92	1.76	1.56	1.29	0.551
$\frac{1}{2}$	23.0	6.75	44.3	8.02	2.56	2.47	21.7	4.79	1.79	1.47	1.30	0.558
L8×4×1	37.4	11.0	69.6	14.1	2.52	3.05	11.6	3.94	1.03	1.05	0.846	0.247
$\frac{3}{4}$	28.7	8.44	54.9	10.9	2.55	2.95	9.36	3.07	1.05	0.953	0.852	0.258
$\frac{1}{2}$	19.6	5.75	38.5	7.49	2.59	2.86	6.74	2.15	1.08	0.859	0.865	0.267
L6×4×$\frac{3}{4}$	23.6	6.94	24.5	6.25	1.88	2.08	8.68	2.97	1.12	1.08	0.860	0.428
$\frac{1}{2}$	16.2	4.75	17.4	4.33	1.91	1.99	6.27	2.08	1.15	0.987	0.870	0.440
$\frac{3}{8}$	12.3	3.61	13.5	3.32	1.93	1.94	4.90	1.60	1.17	0.941	0.877	0.446
L5×3×$\frac{1}{2}$	12.8	3.75	9.45	2.91	1.59	1.75	2.58	1.15	0.829	0.750	0.648	0.357
$\frac{3}{8}$	9.8	2.86	7.37	2.24	1.61	1.70	2.04	0.888	0.845	0.704	0.654	0.364
$\frac{1}{4}$	6.6	1.94	5.11	1.53	1.62	1.66	1.44	0.614	0.861	0.657	0.663	0.371
L4×3×$\frac{1}{2}$	11.1	3.25	5.05	1.89	1.25	1.33	2.42	1.12	0.864	0.827	0.639	0.543
$\frac{3}{8}$	8.5	2.48	3.96	1.46	1.26	1.28	1.92	0.866	0.879	0.782	0.644	0.551
$\frac{1}{4}$	5.8	1.69	2.77	1.00	1.28	1.24	1.36	0.599	0.896	0.736	0.651	0.558
L3$\frac{1}{2}$×2$\frac{1}{2}$×$\frac{1}{2}$	9.4	2.75	3.24	1.41	1.09	1.20	1.36	0.760	0.704	0.705	0.534	0.486
$\frac{3}{8}$	7.2	2.11	2.56	1.09	1.10	1.16	1.09	0.592	0.719	0.660	0.537	0.496
$\frac{1}{4}$	4.9	1.44	1.80	0.755	1.12	1.11	0.777	0.412	0.735	0.614	0.544	0.506
L3×2×$\frac{1}{2}$	7.7	2.25	1.92	1.00	0.924	1.08	0.672	0.474	0.546	0.583	0.428	0.414
$\frac{3}{8}$	5.9	1.73	1.53	0.781	0.940	1.04	0.543	0.371	0.559	0.539	0.430	0.428
$\frac{1}{4}$	4.1	1.19	1.09	0.542	0.957	0.993	0.392	0.260	0.574	0.493	0.435	0.440

1. The $Z - Z$ axis is the axis of minimum moment of inertia, that is, $I_2 \equiv I_z = Ar_z^2$.

2. The **product of inertia,** I_{xy}, for these angle cross sections may be calculated to within the accuracy permitted by the tabular data by use of the formula

$$I_{xy} = \frac{(I_y - I_x)}{2}\tan(2\alpha)$$

The product of inertia is negative for all table entries because of the orientation of the angle cross section relative to the xy axes.

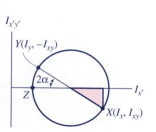

D.7. Properties of Standard-Weight Steel Pipe* (U.S. Customary Units)

Dimensions					Properties					
Nominal Diameter in.	Outside Diameter in.	Inside Diameter in.	Wall Thickness in.	Weight per ft (plain ends) lbs	Area in^2	I in^4	S in^3	r in.	J in^4	Z in^3
$\frac{1}{2}$	0.840	0.622	0.109	0.85	0.250	0.017	0.041	0.261	0.034	0.059
$\frac{3}{4}$	1.050	0.824	0.113	1.13	0.333	0.037	0.071	0.334	0.074	0.100
1	1.315	1.049	0.133	1.68	0.494	0.087	0.133	0.421	0.175	0.187
$1\frac{1}{4}$	1.660	1.380	0.140	2.27	0.669	0.195	0.235	0.540	0.389	0.324
$1\frac{1}{2}$	1.900	1.610	0.145	2.72	0.799	0.310	0.326	0.623	0.620	0.448
2	2.375	2.067	0.154	3.65	1.07	0.666	1.561	0.787	1.33	0.761
$2\frac{1}{2}$	2.875	2.469	0.203	5.79	1.70	1.530	1.06	0.947	3.06	1.45
3	3.500	3.068	0.216	7.58	2.23	3.02	1.72	1.16	6.03	2.33
$3\frac{1}{2}$	4.000	3.548	0.226	9.11	2.68	4.79	2.39	1.34	9.58	3.22
4	4.500	4.026	0.237	10.79	3.17	7.23	3.21	1.51	14.5	4.31
5	5.563	5.047	0.258	14.62	4.30	15.2	5.45	1.88	30.3	7.27
6	6.625	6.065	0.280	18.97	5.58	28.1	8.50	2.25	56.3	11.2
8	8.625	7.981	0.322	28.55	8.40	72.5	16.8	2.94	145.	22.2
10	10.750	10.020	0.365	40.48	11.9	161.	29.9	3.67	321.	39.4
12	12.750	12.000	0.375	49.56	14.6	279.	43.8	4.38	559.	57.4

*Steel pipe is also available in extra-strong and double-extra-strong weights. Based on a table compiled by the American Institute of Steel Construction and published in the *Manual of Steel Construction.*

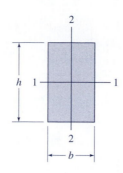

D.8. Properties of Structural Lumber* (U.S. Customary Units)

Sectional properties of American Standard Lumber (S4S)*

Nominal Dimensions $b \times h$	Net Dimensions $b \times h$	Area $A = blt$	Axis 1 – 1		Axis 2 – 2		Weight per Lineal Foot
			Moment of Inertia $I_1 = \frac{bh^3}{12}$	Section Modulus $S_1 = \frac{bh^2}{6}$	Moment of Inertia $I_2 = \frac{hb^3}{12}$	Section Modulus $S_2 = \frac{hb^2}{6}$	
in.	in.	in^2	in^4	in^3	in^4	in^3	lb
2×4	1.5×3.5	5.25	5.36	3.06	0.98	1.31	1.3
2×6	1.5×5.5	8.25	20.80	7.56	1.55	2.06	2.0
2×8	1.5×7.25	10.88	47.63	13.14	2.04	2.72	2.6
2×10	1.5×9.25	13.88	98.93	21.39	2.60	3.47	3.4
2×12	1.5×11.25	16.88	177.98	31.64	3.16	4.22	4.1
3×4	2.5×3.5	8.75	8.93	5.10	4.56	3.65	2.1
3×6	2.5×5.5	13.75	34.66	12.60	7.16	5.73	3.3
3×8	2.5×7.25	18.13	79.39	21.90	9.44	7.55	4.4
3×10	2.5×9.25	23.13	164.89	35.65	12.04	9.64	5.6
3×12	2.5×11.25	28.13	296.63	52.73	14.65	11.72	6.8
4×4	3.5×3.5	12.25	12.51	7.15	12.51	7.15	3.0
4×6	3.5×5.5	19.25	48.53	17.65	19.65	11.23	4.7
4×8	3.5×7.25	25.38	111.15	30.66	25.90	14.80	6.2
4×10	3.5×9.25	32.38	230.84	49.91	33.05	18.89	7.9
4×12	3.5×11.25	39.38	415.28	73.83	40.20	22.97	9.6
6×6	5.5×5.5	30.25	76.3	27.7	76.3	27.7	7.4
6×8	5.5×7.5	41.25	193.4	51.6	104.0	37.8	10.0
6×10	5.5×9.5	52.25	393.0	82.7	131.7	47.9	12.7
6×12	5.5×11.5	63.25	697.1	121.2	159.4	58.0	15.4
8×8	7.5×7.5	56.25	263.7	70.3	263.7	70.3	13.7
8×10	7.5×9.5	71.25	535.9	112.8	334.0	89.1	17.3
8×12	7.5×11.5	86.25	950.5	165.3	404.3	107.8	21.0

*S4S = surfaced four sides.

All properties and weights are for dressed lumber sizes.

Specific weight = 35 lb/ft^3.

Axes 1 − 1 and 2 − 2 are principal centroidal axes.

Based on a table compiled by the National Forest Products Association.

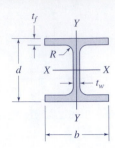

D.9. Properties of Aluminum Association Standard I-Beams (U.S. Customary Units)

| Size | | Area* | Weight† | Flange Thickness | Web Thickness | Fillet Radius | Section Properties | | | | | |
| | | | | | | | Axis $X-X$ | | | Axis $Y-Y$ | | |
Depth d in.	Width b in.	in^2	lb/ft	t_f in.	t_w in.	R in.	I in^4	S in^3	r in.	I in^4	S in^3	r in.
3.00	2.50	1.392	1.637	0.20	0.13	0.25	2.24	1.49	1.27	0.52	0.42	0.61
3.00	2.50	1.726	2.030	0.26	0.15	0.25	2.71	1.81	1.25	0.68	0.54	0.63
4.00	3.00	1.965	2.311	0.23	0.15	0.25	5.62	2.81	1.69	1.04	0.69	0.73
4.00	3.00	2.375	2.793	0.29	0.17	0.25	6.71	3.36	1.68	1.31	0.87	0.74
5.00	3.50	3.146	3.700	0.32	0.19	0.30	13.94	5.58	2.11	2.29	1.31	0.85
6.00	4.00	3.427	4.030	0.29	0.19	0.30	21.99	7.33	2.53	3.10	1.55	0.95
6.00	4.00	3.990	4.692	0.35	0.21	0.30	25.50	8.50	2.53	3.74	1.87	0.97
7.00	4.50	4.932	5.800	0.38	0.23	0.30	42.89	12.25	2.95	5.78	2.57	1.08
8.00	5.00	5.256	6.181	0.35	0.23	0.30	59.69	14.92	3.37	7.30	2.92	1.18
8.00	5.00	5.972	7.023	0.41	0.25	0.30	67.78	16.94	3.37	8.55	3.42	1.20
9.00	5.50	7.110	8.361	0.44	0.27	0.30	102.02	22.67	3.79	12.22	4.44	1.31
10.00	6.00	7.352	8.646	0.41	0.25	0.40	132.09	26.42	4.24	14.78	4.93	1.42
10.00	6.00	8.747	10.286	0.50	0.29	0.40	155.79	31.16	4.22	18.03	6.01	1.44
12.00	7.00	9.925	11.672	0.47	0.29	0.40	255.57	42.60	5.07	26.90	7.69	1.65
12.00	7.00	12.153	14.292	0.62	0.31	0.40	317.33	52.89	5.11	35.48	10.14	1.71

*Areas and section properties listed are based on nominal dimensions.

†Weight per foot is based on nominal dimensions and a density of 0.098 pound per cubic inch, which is the density of alloy 6061.

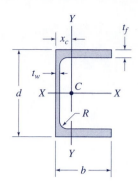

D.10. Properties of Aluminum Association Standard Channels (U.S. Customary Units)

Size		Area*	Weight†	Flange Thickness	Web Thickness	Fillet Radius	Section Properties						
							Axis X–X			Axis Y–Y			
Depth d in.	Width b in.	in^2	lb/ft	t_f in.	t_w in.	R in.	I in^4	S in^3	r in.	I in^4	S in^3	r in.	x_c in.
2.00	1.00	0.491	0.577	0.13	0.13	0.10	0.288	0.288	0.766	0.045	0.064	0.303	0.298
2.00	1.25	0.911	1.071	0.26	0.17	0.15	0.546	0.546	0.774	0.139	0.178	0.397	0.471
3.00	1.50	0.965	1.135	0.20	0.13	0.25	1.41	0.94	1.21	0.22	0.22	0.47	0.49
3.00	1.75	1.358	1.597	0.26	0.17	0.25	1.97	1.31	1.20	0.42	0.37	0.55	0.62
4.00	2.00	1.478	1.738	0.23	0.15	0.25	3.91	1.95	1.63	0.60	0.45	0.64	0.65
4.00	2.25	1.982	2.331	0.29	0.19	0.25	5.21	2.60	1.62	1.02	0.69	0.72	0.78
5.00	2.25	1.881	2.212	0.26	0.15	0.30	7.88	3.15	2.05	0.98	0.64	0.72	0.73
5.00	2.75	2.627	3.089	0.32	0.19	0.30	11.14	4.45	2.06	2.05	1.14	0.88	0.95
6.00	2.50	2.410	2.834	0.29	0.17	0.30	14.35	4.78	2.44	1.53	0.90	0.80	0.79
6.00	3.25	3.427	4.030	0.35	0.21	0.30	21.04	7.01	2.48	3.76	1.76	1.05	1.12
7.00	2.75	2.725	3.205	0.29	0.17	0.30	22.09	6.31	2.85	2.10	1.10	0.88	0.84
7.00	3.50	4.009	4.715	0.38	0.21	0.30	33.79	9.65	2.90	5.13	2.23	1.13	1.20
8.00	3.00	3.526	4.147	0.35	0.19	0.30	37.40	9.35	3.26	3.25	1.57	0.96	0.93
8.00	3.75	4.923	5.789	0.41	0.25	0.35	52.69	13.17	3.27	7.13	2.82	1.20	1.22
9.00	3.25	4.237	4.983	0.35	0.23	0.35	54.41	12.09	3.58	4.40	1.89	1.02	0.93
9.00	4.00	5.927	6.970	0.44	0.29	0.35	78.31	17.40	3.63	9.61	3.49	1.27	1.25
10.00	3.50	5.218	6.136	0.41	0.25	0.35	83.22	16.64	3.99	6.33	2.56	1.10	1.02
10.00	4.25	7.109	8.360	0.50	0.31	0.40	116.15	23.23	4.04	13.02	4.47	1.35	1.34
12.00	4.00	7.036	8.274	0.47	0.29	0.40	159.76	26.63	4.77	11.03	3.86	1.25	1.14
12.00	5.00	10.053	11.822	0.62	0.35	0.45	239.69	39.95	4.88	25.74	7.60	1.60	1.61

*Areas and section properties listed are based on nominal dimensions.

†Weight per foot is based on nominal dimensions and a density of 0.098 pound per cubic inch, which is the density of alloy 6061.

DEFLECTIONS AND SLOPES OF BEAMS; FIXED-END ACTIONS

E.1. Deflections and Slopes of Cantilever Uniform Beams*

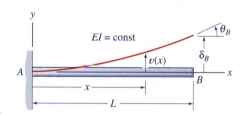

Notation

$v(x)$ = deflection in the y direction

$v'(x)$ = slope of the deflection curve

$\delta_B \equiv v(L)$ = deflection at end B

$\theta_B \equiv v'(L)$ = slope at end B

1

$$v = \frac{M_0 x^2}{2EI} \qquad v' = \frac{M_0 x}{EI}$$

$$\delta_B = \frac{M_0 L^2}{2EI} \qquad \theta_B = \frac{M_0 L}{EI}$$

2

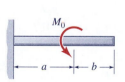

$$v = \frac{M_0 x^2}{2EI} \qquad v' = \frac{M_0 x}{EI} \qquad 0 \le x \le a$$

$$v = \frac{M_0 a}{2EI}(2x - a) \qquad v' = \frac{M_0 a}{EI} \qquad a \le x \le L$$

$$\delta_B = \frac{M_0 a}{2EI}(2L - a) \qquad \theta_B = \frac{M_0 a}{EI}$$

3

$$v = \frac{Px^2}{6EI}(3L - x) \qquad v' = \frac{Px}{2EI}(2L - x)$$

$$\delta_B = \frac{PL^3}{3EI} \qquad \theta_B = \frac{PL^2}{2EI}$$

4

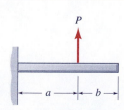

$$v = \frac{Px^2}{6EI}(3a - x) \qquad v' = \frac{Px}{2EI}(2a - x) \qquad 0 \le x \le a$$

$$v = \frac{Pa^2}{6EI}(3x - a) \qquad v' = \frac{Pa^2}{2EI} \qquad a \le x \le L$$

$$\delta_B = \frac{Pa^2}{6EI}(3L - a) \qquad \theta_B = \frac{Pa^2}{2EI}$$

(continued)

5

$$v = \frac{p_0 x^2}{24EI}(6L^2 - 4Lx + x^2)$$

$$v' = \frac{p_0 x}{6EI}(3L^2 - 3Lx + x^2)$$

$$\delta_B = \frac{p_0 L^4}{8EI} \qquad \theta_B = \frac{p_0 L^3}{6EI}$$

6

$$v = \frac{p_0 x^2}{24EI}(6a^2 - 4ax + x^2) \qquad 0 \le x \le a$$

$$v' = \frac{p_0 x}{6EI}(3a^2 - 3ax + x^2) \qquad 0 \le x \le a$$

$$v = \frac{p_0 a^3}{24EI}(4x - a) \qquad v' = \frac{p_0 a^3}{6EI} \qquad a \le x \le L$$

$$\delta_B = \frac{p_0 a^3}{24EI}(4L - a) \qquad \theta_B = \frac{p_0 a^3}{6EI}$$

7

$$v = \frac{p_0 x^3}{120LEI}(20L^3 - 10L^2 x + x^3)$$

$$v' = \frac{p_0 x}{24LEI}(8L^3 - 6L^2 x + x^3)$$

$$\delta_B = \frac{11p_0 L^4}{120EI} \qquad \theta_B = \frac{p_0 L^3}{8EI}$$

8

$$v = \frac{p_0 x^2}{120LEI}(10L^3 - 10L^2 x + 5Lx^2 - x^3)$$

$$v' = \frac{p_0 x}{24LEI}(4L^3 - 6L^2 x + 4Lx^2 - x^3)$$

$$\delta_B = \frac{p_0 L^4}{30EI} \qquad \theta_B = \frac{p_0 L^3}{24EI}$$

9

$$p(x) = p_0 \cos\left(\frac{\pi x}{2L}\right)$$

$$v = \frac{p_0 L}{3\pi^4 EI}\left(48L^3\cos\frac{\pi x}{2L} - 48L^3 + 3\pi^3 Lx^2 - \pi^3 x^3\right)$$

$$v' = \frac{p_0 L}{\pi^3 EI}\left(2\pi^2 Lx - \pi^2 x^2 - 8L^2\sin\frac{\pi x}{2L}\right)$$

$$\delta_B = \frac{2p_0 L^4}{3\pi^4 EI}(\pi^3 - 24) \qquad \theta_B = \frac{p_0 L^3}{\pi^3 EI}(\pi^2 - 8)$$

*Beam-deflection theory is covered in Chapter 7. The sign convention used here is the same as in Chapter 7.

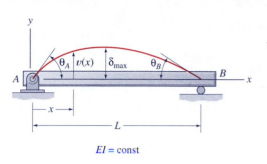

Notation

$v(x)$ = deflection in the y direction

$v'(x)$ = slope of the deflection curve

$\theta_A \equiv v'(0)$ = slope (angle) at end A

$\theta_B \equiv -v'(L)$ = angle of rotation at end B

x_m = distance from end A to the point of maximum deflection

$\delta_C \equiv |v(L/2)|$ = deflection at the center of the beam

$\delta_{max} \equiv \max|v(x)|$ = maximum deflection

$EI = $ const

1

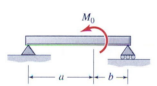

$$v = \frac{M_0 x}{6LEI}(2L^2 - 3Lx + x^2)$$

$$v' = \frac{M_0}{6LEI}(2L^2 - 6Lx + 3x^2)$$

$$\theta_A = \frac{M_0 L}{3EI} \qquad \theta_B = \frac{M_0 L}{6EI}$$

$$x_m = L\left(1 - \frac{\sqrt{3}}{3}\right) \text{ and } \delta_{max} = \frac{M_0 L^2}{9\sqrt{3}EI}$$

2

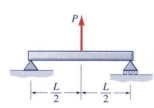

$$v = \frac{-M_0 x}{6LEI}(6aL - 3a^2 - 2L^2 - x^2) \qquad 0 \le x \le a$$

$$v' = \frac{-M_0}{6LEI}(6aL - 3a^2 - 2L^2 - 3x^2) \qquad 0 \le x \le a$$

$$\theta_A = \frac{-M_0}{6LEI}(6aL - 3a^2 - 2L^2) \qquad \theta_B = \frac{-M_0}{6LEI}(3a^2 - L^2)$$

3

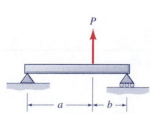

$$v = \frac{Px}{48EI}(3L^2 - 4x^2) \qquad 0 \le x \le \frac{L}{2}$$

$$v' = \frac{P}{16EI}(L^2 - 4x^2) \qquad 0 \le x \le \frac{L}{2}$$

$$\delta_C = \delta_{max} = \frac{PL^3}{48EI} \qquad \theta_A = \theta_B = \frac{PL^2}{16EI}$$

4

$$v = \frac{Pbx}{6LEI}(L^2 - b^2 - x^2) \qquad 0 \le x \le a$$

$$v' = \frac{Pb}{6LEI}(L^2 - b^2 - 3x^2) \qquad 0 \le x \le a$$

$$\theta_A = \frac{Pab(L + b)}{6LEI}$$

$$\theta_B = \frac{Pab(L + a)}{6LEI}$$

$$\text{If } a \ge b, x_m = \sqrt{\frac{L^2 - b^2}{3}} \text{ and } \delta_{max} = \frac{Pb(L^2 - b^2)^{3/2}}{9\sqrt{3}LEI}$$

(continued)

5

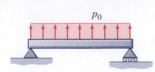

$$v = \frac{p_0 x}{24EI}(L^3 - 2Lx^2 + x^3)$$

$$v' = \frac{p_0}{24EI}(L^3 - 6Lx^2 + 4x^3)$$

$$\delta_C = \delta_{max} = \frac{5p_0 L^4}{384EI} \qquad \theta_A = \theta_B = \frac{p_0 L^3}{24EI}$$

6

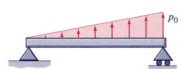

$$v = \frac{p_0 x}{24LEI}(a^4 - 4a^3 L + 4a^2 L^2 + 2a^2 x^2 - 4aLx^2 + Lx^3) \qquad 0 \le x \le a$$

$$v' = \frac{p_0}{24LEI}(a^4 - 4a^3 L + 4a^2 L^2 + 6a^2 x^2 - 12aLx^2 + 4Lx^3) \qquad 0 \le x \le a$$

$$v = \frac{p_0 a^2}{24LEI}(-a^2 L + 4L^2 x + a^2 x - 6Lx^2 + 2x^3) \qquad a \le x \le L$$

$$v' = \frac{p_0 a^2}{24LEI}(4L^2 + a^2 - 12Lx + 6x^2) \qquad a \le x \le L$$

$$\theta_A = \frac{p_0 a^2}{24LEI}(2L - a)^2 \qquad \theta_B = \frac{p_0 a^2}{24LEI}(2L^2 - a^2)$$

7

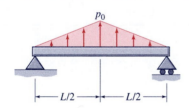

$$v = \frac{p_0 x}{360LEI}(7L^4 - 10L^2 x^2 + 3x^4)$$

$$v' = \frac{p_0}{360LEI}(7L^4 - 30L^2 x^2 + 15x^4)$$

$$\theta_A = \frac{7p_0 L^3}{360EI} \qquad \theta_B = \frac{p_0 L^3}{45EI}$$

$$x_m = 0.5193 \, L \qquad \delta_{max} = 0.00652\frac{p_0 L^4}{EI}$$

8

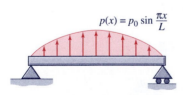

$$v = \frac{p_0 x}{960LEI}(5L^2 - 4x^2)^2 \qquad 0 \le x \le \frac{L}{2}$$

$$v' = \frac{p_0}{192LEI}(5L^2 - 4x^2)(L^2 - 4x^2) \qquad 0 \le x \le \frac{L}{2}$$

$$\delta_C = \delta_{max} = \frac{p_0 L^4}{120LEI} \qquad \theta_A = \theta_B = \frac{5p_0 L^3}{192EI}$$

9

$p(x) = p_0 \sin\frac{\pi x}{L}$

$$v = \frac{p_0 L^4}{\pi^4 EI} \sin\left(\frac{\pi x}{L}\right)$$

$$v' = \frac{p_0 L^3}{\pi^3 EI} \cos\left(\frac{\pi x}{L}\right)$$

$$\delta_C = \delta_{max} = \frac{p_0 L^4}{\pi^4 EI} \qquad \theta_A = \theta_B = \frac{p_0 L^3}{\pi^3 EI}$$

*Beam-deflection theory is covered in Chapter 7. The sign convention used here is the same as in Chapter 7.

E.3. Fixed-End Actions for Uniform Beams*

Loading	Shear and Moment Reactions[†]	
	End A	**End B**

	Loading	End A	End B
1		$R_A = \dfrac{6M_0a}{L^2}\left(1 - \dfrac{a}{L}\right)$ $M_A = M_0\left(-1 + 4\dfrac{a}{L} - 3\dfrac{a^2}{L^2}\right)$	$R_B = -\dfrac{6M_0a}{L^2}\left(1 - \dfrac{a}{L}\right)$ $M_B = \dfrac{M_0a}{L}\left(2 - 3\dfrac{a}{L}\right)$
2		$R_A = -\dfrac{Pb^2}{L^3}(3a + b)$ $M_A = -\dfrac{Pab^2}{L^2}$	$R_B = -\dfrac{Pa^2}{L^3}(a + 3b)$ $M_B = \dfrac{Pba^2}{L^2}$
3		$R_A = -\dfrac{p_0L}{2}$ $M_A = -\dfrac{p_0L^2}{12}$	$R_B = -\dfrac{p_0L}{2}$ $M_B = \dfrac{p_0L^2}{12}$
4		$R_A = -\dfrac{3p_0L}{20}$ $M_A = -\dfrac{p_0L^2}{30}$	$R_B = -\dfrac{7p_0L}{20}$ $M_B = \dfrac{p_0L^2}{20}$
5		$R_A = \dfrac{6EI}{L^2}\theta_B$ $M_A = \dfrac{2EI}{L}\theta_B$	$R_B = -\dfrac{6EI}{L^2}\theta_B$ $M_B = \dfrac{4EI}{L}\theta_B$
6		$R_A = -\dfrac{12EI}{L^3}\delta_B$ $M_A = -\dfrac{6EI}{L^2}\delta_B$	$R_B = \dfrac{12EI}{L^3}\delta_B$ $M_B = -\dfrac{6EI}{L^2}\delta_B$

*This table provides fixed-end actions for use with Section 7.7.
[†]For all fixed-end actions tabulated, the positive sense for shear and moment reactions are the same as those shown in the first table entry.

MECHANICAL PROPERTIES OF SELECTED ENGINEERING MATERIALS

F

Mechanical properties of engineering materials vary significantly as a result of heat treatment, mechanical working, moisture content, and various other factors. The properties listed in Tables F.1 through F.3 are representative values that are intended for educational purposes only, not for commercial design of members. Additional information is available from suppliers of engineering materials, from various websites, and from several other sources, including the following: *Annual Book of ASTM,* American Society for Testing Materials, Philadelphia, PA; *Metals Handbook* ASM International, Materials Park, OH; *Aluminum and Aluminum Alloys* and *Aluminum Standards and Data,* The Aluminum Association, Washington, DC; *Wood Handbook,* U.S. Department of Agriculture, Washington, DC; *Marks' Standard Handbook for Mechanical Engineers,* McGraw-Hill, Inc., New York, NY; and *CRC Materials Science and Engineering Handbook,* CRC Press, Inc., Boca Raton, FL.

TABLE F.1. Specific Weight and Mass Density

Material	Specific Weight (γ)			Mass Density (ρ)	
	lb/in^3	lb/ft^3	kN/m^3	slugs/ft^3	kg/m^3
Aluminum Alloys	0.096–0.103	165–180	26–28	5.2–5.5	2600–2800
Alloy 2014-T6	0.101	175	27	5.4	2800
Alloy 6061-T6	0.098	170	27	5.3	2700
Brass	0.301–0.313	520–540	82–85	16–17	8300–8600
Red Brass (85% Cu, 35% Zn)	0.313	540	85	17	8600
Cast Iron	0.252–0.266	435–460	68–72	13–14	7000–7400
Gray, ASTM-A48	0.260	450	71	14	7200
Malleable, ASTM-47	0.264	456	72	14	7300
Steel	0.284	490	77	15.2	7850
Titanium	0.162	280	44	8.7	4500
Concrete					
Plain	0.081–0.087	140–150	22–44	4.4–4.7	2200–2400
Lightweight	0.052–0.067	90–115	14–18	2.8–3.6	1400–1800
Reinforced	0.087	150	24	4.7	2400
Glass	0.087–0.104	150–180	24–28	4.7–5.6	2400–2900
Plastics					
Nylon, type 6/6 (molding cpd.)	0.041	70	11	2.2	1100
Polycarbonate	0.043	75	12	2.3	1200
Vinyl, rigid PVC	0.048–0.052	82–90	12.9–14.1	2.6–2.8	1320–1440
Wood	0.014–0.026	25–45	3.9–7.1	0.78–1.4	400–720
Douglas Fir	0.019	32	5.0	1.0	510
Southern Pine	0.022	38	6.0	1.2	610

TABLE F.2.   Modulus of Elasticity, Shear Modulus of Elasticity, and Poisson's Ratio

Material	Modulus of Elasticity (E)		Shear Modulus of Elasticity (G)		Poisson's Ratio (ν)
	10^3 ksi	GPa	10^3 ksi	GPa	
Aluminum Alloys	10.0–11.4	70–79	3.8–4.3	26–30	0.33
Alloy 2014-T6	10.6	73	4.0	27	0.33
Alloy 6061-T6	10.0	70	3.8	26	0.33
Brass					
Red Brass (85% Cu, 15% Zn)					
Cold-rolled	15	100	5.6	37	0.34
Annealed	15	100	5.6	37	0.34
Cast Iron					
Gray, ASTM-A48	10	70	4.1	29	0.22
Malleable, ASTM-47	24	165	9.4	65	0.27
Steel					
Structural, ASTM-A36	29	200	11.2	78	0.29
Stainless, AISI 302					
Cold-rolled	28	195	10.8	75	0.30
Annealed	28	195	10.8	75	0.30
High-strength, low alloy, ASTM-A242	29	200	11.2	78	0.29
Quenched & tempered, ASTM-A514	29	200	11.2	78	0.29
Titanium					
Alloy (6% Al, 4% V)	16.5	115	6.2	43	0.33
Concrete[1]					0.1–0.2
Medium Strength	3.6	25	—	—	—
High Strength	4.5	31	—	—	—
Glass	8.7	60	—	—	0.2–0.3
Plastics					
Nylon, type 6/6 (molding cpd.)	0.4	2.8	—	—	0.4
Polycarbonate	0.35	2.4	—	—	—
Vinyl, rigid PVC	0.4	2.8	—	—	—
Wood[2]					
Douglas Fir	1.75	12	—	—	—
Southern Pine	1.75	12	—	—	—

[1]Concrete properties are for compression.
[2]Timber properties are for loading parallel to the grain.

TABLE F.3. Yield Strength, Ultimate Strength, Percent Elongation in 2 Inches, and Coefficient of Thermal Expansion

Material	Yield Strength $(\sigma_Y)^{1,2}$		Ultimate Strength $(\sigma_U)^1$		Percent Elongation over 2 in. Gage Length	Coefficient of Thermal Expansion (α)	
	ksi	MPa	ksi	MPa		$10^{-6}/°F$	$10^{-6}/°C$
Aluminum Alloys						11.7–13.3	21–24
Alloy 2014-T6	60	410	70	480	13	12.8	23.0
Alloy 6061-T6	40	275	45	310	17	13.1	23.6
Brass							
Red Brass (85% Cu, 15% Zn)							
Cold-rolled	60	410	75	520	4	10.4	19
Annealed	15	100	40	275	50	10.4	19
Cast Iron							
Gray, ASTM-A48	—	—	25	170	0.5	6.7	12
Malleable, ASTM-47	33	230	50	345	10	6.7	12
Steel							
Structural, ASTM-A36	36	250	58	400	20	6.5	12
Stainless, AISI 302							
Cold-rolled	75	520	125	860	12	9.6	17
Annealed	38	260	95	655	50	9.6	17
High-strength, low alloy, ASTM-A242	50	345	70	480	22	6.5	12
Quenched & tempered, ASTM-A514	100	690	110	760	18	6.5	12
Titanium							
Alloy (6% Al, 4% V)	120	830	130	900	10	5.3	9.5
Concrete[3]							
Medium Strength	—	—	4	28	—	5.5	10
High Strength	—	—	6	40	—	5.5	10
Glass[4]	—	—	(4)	(4)	—	3–6	5–11
Plastics[5]							
Nylon, type 6/6 (molding cpd.)	8.0	55	11	75	50	17	30
Polycarbonate	8.5	60	9.5	65	110	3.8	6.8
Vinyl (rigid PVC)	6	40	7	50	1–10	28–33	50–60
Wood[5]							
Douglas Fir	—	—	7.5	50	—	—	—
Southern Pine	—	—	8.5	60	—	—	—

[1]For ductile metals, the strength in compression is generally assumed to be equal to the tensile strength.
[2]For most metals, this is the 0.2% offset value.
[3]Concrete properties are for loading in compression.
[4]Glass properties vary widely. For example, glass fibers may have tensile strengths to 1,000 ksi (7000 MPa) or more.
[5]Timber properties are for loading in compression parallel to the grain.

COMPUTATIONAL MECHANICS

G

There are two primary ways in which the computer can be of valuable assistance to you as you study the topics in this *Mechanics of Materials* text. First, you can use math application software or a spreadsheet program or a programming language of your own choice to write a computer program to solve a mechanics of materials problem. There are twenty such exercises in the book. For example, there are three computer-based design problems (Probs. 2.8-16 through 2.8-18) that ask you to use the computer to obtain a plot that will enable you to select the best design for a simple truss structure. By writing your own computer programs for several of these exercises, you will not only gain experience in programming, but you should also gain valuable experience in organizing efficient, systematic solutions of mechanics of materials problems.

Use of the award-winning **MDSolids** educational software is another way in which the computer can be of valuable assistance to you. MDSolids, which is available from the website www.wiley.com/college/craig, is described in Appendix G.1.[1] Since MDSolids is written in Visual Basic, it is only available for use on computers running a Windows operating system (Windows 95, Windows 98, Windows NT, Windows XP, Windows Vista, or Windows 7).[2]

G.1 MDSolids

The **MDSolids** software package and ninety special MDSolids-based example problems that are provided with *Mechanics of Materials, 3rd edition* will enable you to use the computer for solving problems in axial deformation, torsion, bending, combined-loading, and buckling of the type treated in Chapters 2 through 12 of the book. MDSolids provides both systematic problem-solving procedures and a user-friendly graphical interface, and through its use you can gain valuable insight into the behavior of structural members and systems under various loading conditions.

---

[1]The MDSolids educational software package is copyrighted by its author, Dr. Timothy A. Philpot. It was a winner of the 1998 Premier Award for Excellence in Engineering Education Courseware. The MDSolids website is (www.mdsolids.com).
[2]Windows is a registered trademark of the Microsoft Corporation.

MDSolids Modules. MDSolids consists of *modules*, which are similar to book chapters in that each module focuses on specific mechanics of materials concepts and problem-solving methods. There are currently <u>twelve</u> MDSolids modules:

- Basic Stress and Strain Problems
- Beam and Strut Axial Structures
- Truss Analysis and Stresses
- Statically Indeterminate Axial Structures
- Torsion Members
- Determinate Beams
- Flexure
- Section Properties
- Column Buckling
- Mohr's Circle Analysis
- General Analysis of Axial, Torsion, and Beam Structures
- Pressure Vessels

The modules can be accessed in any sequence. MDSolids is powerful enough so that many different structural configurations and loadings can be analyzed with each separate module, but the modules are also coordinated so that results from one module are available for use in related modules.

How to Acquire MDSolids. MDSolids is an educational software package developed specifically for the introductory mechanics of materials course. The version of MDSolids that is provided with this *Mechanics of Materials* textbook consists of two closely integrated parts: (1) the basic MDSolids educational software, and (2) a special supplement of ninety example problems.

MDSolids is available from either the Instructor Companion Website or the Student Companion Website at www.wiley.com/college/craig. Students who wish to download the software must first register on the Wiley Student Companion Website. For details, please see the registration card that is provided in this book.

Features of the MDSolids Software: Some of the key features of the basic MDSolids software are:

- *Versatility:* As indicated above, MDSolids has computational modules pertaining to all of the topics taught in a typical mechanics of materials course. The scope of MDSolids offers routines to help students at all levels of understanding.
- *Ease of Input:* Graphic cues are provided to guide users in entering data, so that the student is able to define a problem intuitively and directly without the need for a user's manual.
- *Visual Communication:* Each MDSolids routine features a picture, sketch, or plot that graphically depicts important aspects of the problem. For a number of topics, including stresses in beams, deflection of beams, Mohr's Circle for stress and strain, and others, plots that show the results are generated.
- *Correct Solution and Intermediate Results:* MDSolids is an "electronic solutions manual," giving not only the correct solution for each problem but also providing intermediate steps that can be used to confirm the problem-solving approach.

- *Text-based Explanations:* Many of the MDSolids modules provide extra explanations to describe in words how the calculations are performed.
- *Help Files:* The MDSolids Help Files contain instructions for using the software, but, more importantly, they contain theoretical background and practical suggestions for solving various types of problems.

Features of the Special MDSolids-based Example Problems: To accompany this *Mechanics of Materials* textbook, the author of MDSolids has created ninety (90) additional special example problems. The key features of these special example problems are:

- *Close Ties with the Textbook:* At the point in the book where one of these example problems can serve to supplement the text discussion and the text example(s), there is an MDS icon that "points" to one or more of these special MDS example problems. In the homework problem sections, an MDS icon precedes the group of homework problems of the type that is covered by the identified special MDS example problem(s).
- *Complete, Textbook-style Format:* The special example problems are written in the same style and with the same nomenclature, sign conventions, etc. as the example problems in the book. They are not just computer-generated numerical solutions.
- *Broad Range of Topics and Level of Difficulty:* There are special MDSolids examples for Chapters 2 through 12. Many of these are illustration-of-concept type examples that will be of special help to those students who need a "tutor" to supplement the book and the instructor's lectures. There are also "end-to-end" type examples that illustrate the power of the linking of MDSolids modules (e.g., section properties, shear-force and bending-moment in beams, flexural and shear stresses in beams, and Mohr's circle).
- *Close Ties with the MDSolids Modules:* The special example problems are treated as help files of the basic MDSolids software. For example, the user can close an example problem and immediately be in the appropriate MDSolids module(s) with the data of the example problem already entered into the correct MDSolids input boxes. See "Suggestions on How to Use MDSolids" below.

Suggestions on How to Use MDSolids. Ideally, the MDSolids software modules and the special example problems will initially be used in the following manner. After reading the text material on a topic and studying the example problem(s) in the book, look at the problem statement of a relevant special MDS example; next, attempt to work the example by hand; then, consult the complete, textbook-style solution; and, finally, move directly to the relevant MDSolids module(s) and enter different input to see how the different input affects the solution. At each step you will get accurate, immediate, and often pictorial feedback. Then, you can move on to the assigned homework or to related homework-type problems, working them by hand and using MDSolids as an "electronic solutions manual" to provide feedback. Once you are familiar with how MDSolids works, you will be able to go directly to the MDSolids modules and obtain solutions to a broad range of mechanics of materials problems, including complex design-type problems.

ANSWERS TO SELECTED ODD-NUMBERED PROBLEMS

A file entitled "ANSWERS TO SELECTED ODD-NUMBERED PROBLEMS" can be found on the Student Companion Website at www.wiley.com/college/craig.

REFERENCES

1-1 Craig, R. R. Jr., and Kurdila, A. J., *Fundamentals of Structural Dynamics*, 2nd edition, John Wiley & Sons, Inc., New York, NY, 2006.

1-2 Hughes, T. J. R., *The Finite Element Method — Linear Static and Dynamic Finite Element Analysis*, Dover Publications, Inc., Mineola, NY, 2000.

1-3 Szabo, B., and Babuska, I., *Introduction to Finite Element Analysis: Formulation, Verification and Validation*, John Wiley & Sons, Inc., New York, NY, 2011.

1-4 Meriam, J. L., and Kraige, L. G., *Statics*, John Wiley & Sons, Inc., New York, NY, 1997.

1-5 Crandall, S. H., et. al., *An Introduction to the Mechanics of Solids*, McGraw-Hill, Inc., New York, 1978.

2-1 Timoshenko, S. P., *History of Strength of Materials*, McGraw-Hill, Inc., New York, 1953.

2-2 Drucker, D. C., *Introduction to Mechanics of Deformable Solids*, McGraw-Hill, Inc., New York, 1967.

2-3 Ramaley, D., and McHenry, D., *Stress-Strain Curves for Concrete Strained Beyond Ultimate Load*, Lab. Rept. No. Sp-12, U.S. Bureau of Reclamation, Denver, CO, 1947. (Reprinted in: Ferguson, P. M., Breen, J. E., and Jirsa, J. O., *Reinforced Concrete Fundamentals*, 5th edition, John Wiley & Sons, Inc., New York, 1988.)

2-4 Lee, S. M., ed., *International Encyclopedia of Composites*, VCH Publishers, Inc., New York, 1989.

2-5 *Engineered Materials Handbook, Vol. 1 — Composites*, ASM International, Materials Park, OH, 44073-0002, 1987.

2-6 Carswell, T. S., and Nason, H. K., "Effect of Environmental Conditions on the Mechanical Properties of Organic Plastics," *Symposium on Plastics*, American Society for Testing Materials, Philadelphia, 1944. (Reprinted in: Callister, W. D. Jr., *Materials Science and Engineering*, 3rd edition, John Wiley & Sons, Inc., New York, 1994.)

2-7 Gordon, J. E., *Structures, or Why Things Don't Fall Down*, Da Capo Press, New York, 1978.

2-8 *Manual of Steel Construction — Load & Resistance Factor Design*, American Institute of Steel Construction, One East Wacker Drive, Suite 3100, Chicago, 60601-2001.

2-9 Shinozuka, M., and Yao, J. T. P., eds., *Probabilistic Methods in Structural Engineering*, American Society of Civil Engineers, New York, 1981.

2-10 Madsen, H. O., Krenk, S., and Lind, N. C., *Methods of Structural Safety*, Prentice-Hall, Inc., Englewood Cliffs, NJ, 1986.

2-11 Lewis, E. E., *Introduction to Reliability Engineering*, John Wiley & Sons, New York, 1987.

2-12 Timoshenko, S. P., and Goodier, J. N., *Theory of Elasticity*, 3rd edition, McGraw-Hill, Inc., New York, 1970.

3-1 Bathe, Klaus-Jürgen, *Finite Element Procedures*, Prentice-Hall, Inc., Englewood Cliffs, NJ, 1996.

3-2 Cook, R. D., *Finite Element Modeling for Stress Analysis*, John Wiley & Sons, Inc., New York, NY, 1995.

4-1 Oden, J. T., and Ripperger, E. A., *Mechanics of Elastic Structures*, 2nd edition, McGraw-Hill, Inc., New York, 1981.

4-2 Timoshenko, S. P., and Goodier, J. N., *Theory of Elasticity*, 3rd edition, McGraw-Hill, Inc., New York, 1970.

6-1 Timoshenko, S. P., *History of Strength of Materials*, McGraw-Hill, Inc., New York, 1953.

6-2 *Manual of Steel Construction — Allowable Stress Design*, and *Manual of Steel Construction — Load & Resistance Factor Design*, American Institute of Steel Construction, One East Wacker Drive, Suite 3100, Chicago, IL, 60601-2001.

6-3 *The Aluminum Design Manual*, The Aluminum Association, 900 19th Street, NW, Suite 300, Washington, DC, 20006.

6-4 Timoshenko, S. P., and Goodier, J. N., *Theory of Elasticity*, 3rd edition, McGraw-Hill, Inc., New York, 1970.

6-5 Popov, E. P., *Engineering Mechanics of Solids*, Prentice Hall, Englewood Cliffs, NJ, 1990.

7-1 Pilkey, W. D., "Clebsch's Method for Beam Deflection," *J. of Engineering Education*, Vol. 54, No. 5, January 1964, pp.170–174.

8-1 Timoshenko, S. P., and Goodier, J. N., *Theory of Elasticity*, 3rd edition, McGraw-Hill, Inc., New York, 1970.

8-2 Timoshenko, S. P., *History of Strength of Materials*, McGraw-Hill, Inc., New York, 1953.

8-3 Kobayashi, A. S., *Handbook on Experimental Mechanics*, Society for Experimental Mechanics, Inc., Bethel, CT, 1993.

8-4 Shukla, A. and Dally, J. W., *Experimental Solid Mechanics*, College House Enterprises, Knoxville, TN, www.collegehouse-books.com, 2010.

9-1 Kraus, H., *Thin Elastic Shells*, John Wiley & Sons, Inc., New York, 1967.

9-2 *Analyzing Failures: The Problems and Solutions*, V. S. Goel, ed., ASM International, Materials Park, OH, 1986.

9-3 *ASME Boiler and Pressure Vessel Code, An Internationally Recognized Code*, The American Society of Mechanical Engineers, New York, NY.

9-4 Timoshenko, S. P., and Woinowsky-Krieger, S., *Theory of Plates and Shells*, McGraw-Hill, Inc., New York, 1959.

9-5 Rabia, H., *Oilwell Drilling Engineering — Principles and Practice*, Graham & Trotman, Ltd., London, 1985.

10-1 Timoshenko, S. P., and Gere, J. M., *Theory of Elastic Stability*, 2nd edition, McGraw-Hill, Inc., New York, 1961.

10-2 Brush, D. O., and Almroth, B. O., *Buckling of Bars, Plates, and Shells*, McGraw-Hill, Inc., New York, 1975.

10-3 Chen, W. F., and Lui, E. M., *Structural Stability — Theory and Implementation*, Elsevier, New York, 1987.

10-4 Ziegler, H., *Principles of Structural Stability*, Blaisdell Publishing Co., Waltham, MA, 1968.

10-5 Galambos, T. V., ed., *Guide to Stability Design Criteria for Metal Structures*, 4th edition, John Wiley & Sons, Inc., New York, 1988.

10-6 Duberg, J. E., "Inelastic Buckling," Chapter 52 in *Handbook of Engineering Mechanics*, edited by W. Flügge, McGraw-Hill, Inc., New York, 1962.

10-7 *Steel Construction Manual,* 13th edition (combines ASD and LRFD design methods), American Institute of Steel Construction, www.aisc.org, 2006.

10-8 *Aluminum Design Manual (2010)*, The Aluminum Association, www.aluminum.org, 2010.

10-9 *National Design Specification for Wood Construction (2005)*, American Wood Council, www.awc.org, 2005.

11-1 Timoshenko, S. P., *History of Strength of Materials*, McGraw-Hill, Inc., New York, 1953.

11-2 Gere, J. M., and Timoshenko, S. P., *Mechanics of Materials*, 3rd edition, PWS-KENT Publishing Company, Boston, 1984.

11-3 Craig, R. R. Jr., *Structural Dynamics—An Introduction to Computer Methods*, John Wiley & Sons, Inc., New York, 1981.

12-1 Timoshenko, S. P., and Goodier, J. N., *Theory of Elasticity*, 3rd edition, McGraw-Hill, Inc., New York, 1970.

12-2 Bathe, Klaus-Jürgen, *Finite Element Procedures*, Prentice-Hall, Inc., Englewood Cliffs, NJ, 1996.

12-3 Cook, R. D., *Finite Element Modeling for Stress Analysis*, John Wiley & Sons, Inc., New York, 1995.

12-4 Kobayashi, A. S., *Handbook on Experimental Mechanics*, Society for Experimental Mechanics, Inc., Bethel, CT, 1993.

12-5 Roark, R. J., and Young, W. C., *Roark's Formulas for Stress and Strain*, 6th edition, McGraw-Hill, Inc., New York, 1988.

12-6 Peterson, R. E., *Stress Concentration Factors*, John Wiley & Sons, Inc., New York, 1974.

12-7 Timoshenko, S. P., *History of Strength of Materials*, McGraw-Hill, Inc., New York, 1953.

12-8 Juvinall, R. C., *Fundamentals of Machine Component Design*, John Wiley & Sons, Inc., New York, 1983.

12-9 *ASTM Standards*, American Society for Testing and Materials, 1916 Race Street, Philadelphia, PA 19103-1187.

12-10 Hardrath, H. F., Utley, E. C., and Guthrie, D. E., *Rotating-Beam Fatigue Tests of Notched and Unnotched 7075-T6 Aluminum-Alloy Specimens Under Stress of Constant and Varying Amplitudes*, NASA TN D-210, December 1959. (Reprinted in: Hertzberg, R. W., *Deformation and Fracture Mechanics of Engineering Materials*, 3rd edition, John Wiley & Sons, Inc., New York, 1989.)

12-11 *Structural Alloys Handbook*, CINDAS/USAF CRDA, Purdue University, 1293 Potter Engineering Center, West Lafayette, IN 47907-1293.

12-12 *Aluminum Standards and Data*, The Aluminum Association, 900 19 Street, NW, Washington, DC, 1993.

12-13 Rolfe, S. T., and Barsom, J. M., *Fracture and Fatigue Control in Structures, Applications of Fracture Mechanics*, Prentice-Hall, Inc., Englewood Cliffs, NJ, 1977.

12-14 Osgood, C. C., *Fatigue Design*, Pergamon Press, Inc., Elmsford, NY, 1982.

12-15 Niu, M. C. Y., *Airframe Structural Design*, Technical Book Company, Los Angeles, 1988.

INDEX